BIOCHEMISTRY

Second Edition

BIOCHEMISTRY

Lubert Stryer

STANFORD UNIVERSITY

W. H. FREEMAN AND COMPANY
San Francisco

DESIGNER: *Robert Ishi*
ILLUSTRATOR: *Donna Salmon*
ILLUSTRATION COORDINATOR: *Audre W. Loverde*
PRODUCTION COORDINATOR: *William Murdock*
COMPOSITOR: *York Graphic Services*
PRINTER AND BINDER: *Arcata Book Group*

Library of Congress Cataloging in Publication Data

Stryer, Lubert.
    Biochemistry.

    Includes bibliographies and index.
    1. Biological chemistry. I. Title. [DNLM:
1. Biochemistry. QU4 *S*928b]
QP514.2.S66   1981   574.19′2   80-24699
ISBN 0-7167-1226-1

9 8 7 6 5 4 3

To my teachers

Paul F. Brandwein
Daniel L. Harris
Douglas E. Smith
Elkan R. Blout
Edward M. Purcell

CONTENTS

LIST OF TOPICS

Part II GENERATION AND STORAGE OF METABOLIC ENERGY   233

Part III BIOSYNTHESIS OF MACROMOLECULAR PRECURSORS 455

CHAPTER 20. Biosynthesis of Membrane Lipids and Steroid Hormones 457

CHAPTER 21. Biosynthesis of Amino Acids and Heme 485

PREFACE
TO THE SECOND EDITION

The pace of discovery in biochemistry has been exceptionally rapid during the past several years. This progress has greatly enriched our understanding of the molecular basis of life and has opened many new areas of inquiry. The sequencing of DNA, the construction and cloning of new combinations of genes, the elucidation of metabolic control mechanisms, and the unraveling of membrane transport and transduction processes are some of the highlights of recent research. One of my aims in this edition has been to weave new knowledge into the fabric of the text. I have sought to enhance the book's teaching effectiveness by centering the exposition of new material on common themes wherever feasible and by citing recurring motifs. I have also tried to convey a sense of the intellectual power and beauty of the discipline of biochemistry.

I am indebted to Thomas Emery, Henry Epstein, Alexander Glazer, Roger Kornberg, Robert Martin, and Jeffrey Sklar for their counsel, criticism, and encouragement in the preparation of this edition. Robert Baldwin, Charles Cantor, Richard Caprioli, David Eisenberg, Alan Fersht, Robert Fletterick, Herbert Friedmann, Horace Jackson, Richard Keynes, Sung-Hou Kim, Aaron Klug, Arthur Kornberg, Daniel Koshland, Jr., Samuel Latt, Vincent Marchesi, David Nelson, Garth Nicolson, Vernon Oi, Robert Renthal, Carl Rhodes, Frederic Richards, James Rothman, Peter Sargent, Howard Schachman, Joachim Seelig, Eric Shooter, Elizabeth Simons, James Spudich, Theodore Steck, Thomas Steitz, Judit C.-P. Stenn, Robert Trelstad, Christopher Walsh, Simon Whitney, and Bernhard Witkop also gave valuable advice.

Patricia Mittelstadt edited both editions of this text. I deeply appreciate her critical and sustained contributions. I am indebted to Donna Salmon for her outstanding drawings. David Clayton,

David Dressler, John Heuser, Lynne Mercer, Kenneth Miller, George Palade, Nigel Unwin, and Robley Williams generously provided many fine electron micrographs. Betty Hogan typed the manuscript and played an indispensable role in its preparation. Cary Leiden and Karen Marzotto carefully read the proofs. I also wish to thank Michael Graves for his excellent photographic work.

My wife, Andrea, and my sons, Michael and Daniel, have cheerfully allowed this text to become a member of the family. I am deeply grateful to them for their patience and buoyancy. Andrea provided much advice on style and design, as she did for the first edition.

I have been heartened by the many letters that I have received from readers of the first edition. Their comments and criticisms have enlighted, stimulated, and encouraged me. I look forward to a continuing dialogue with readers in the years ahead.

August 1980 *Lubert Stryer*

This book is an outgrowth of my teaching of biochemistry to under-graduates, graduate students, and medical students at Yale and Stanford. My aim is to provide an introduction to the principles of biochemistry that gives the reader a command of its concepts and language. I also seek to give an appreciation of the process of discovery in biochemistry. My exposition of the principles of biochemistry is organized around several major themes:

1. Conformation—exemplified by the relationship between the three-dimensional structure of proteins and their biological activity

2. Generation and storage of metabolic energy

3. Biosynthesis of macromolecular precursors

4. Information—storage, transmission, and expression of genetic information

5. Molecular physiology—interaction of information, conformation, and metabolism in physiological processes

The elucidation of the three-dimensional structure of proteins, nucleic acids, and other biomolecules has contributed much in recent years to our understanding of the molecular basis of life. I have emphasized this aspect of biochemistry by making extensive use of molecular models to give a vivid picture of architecture and dynamics at the molecular level. Another stimulating and heartening aspect of contemporary biochemistry is its increasing interaction with medicine. I have presented many examples of this interplay. Discussions of molecular diseases such as sickle-cell anemia and of the mechanism of action of drugs such as penicillin enrich the teaching of biochemistry. Finally, I have tried to define several challenging areas of inquiry in biochemistry today, such as the molecular basis of excitability.

In writing this book, I have benefitted greatly from the advice, criticism, and encouragement of many colleagues and students. Leroy Hood, Arthur Kornberg, Jeffrey Sklar, and William Wood gave me invaluable counsel on its overall structure. Richard Caprioli, David Cole, Alexander Glazer, Robert Lehman, and Peter Lengyel read much of the manuscript and made many very helpful suggestions. I am indebted to Frederic Richards for sharing his thoughts on macromolecular conformation and for extensive advice on how to depict three-dimensional structures. Deric Bownds, Thomas Broker, Jack Griffith, Hugh Huxley, and George Palade made available to me many striking electron micrographs. I am also very thankful for the advice and criticism that were given at various times in the preparation of this book by Richard Dickerson, David Eisenberg, Moises Eisenberg, Henry Epstein, Joseph Fruton, Michel Goldberg, James Grisolia, Richard Henderson, Harvey Himel, David Hogness, Dale Kaiser, Samuel Latt, Susan Lowey, Vincent Marchesi, Peter Moore, Allan Oseroff, Jordan Pober, Russell Ross, Edward Reich, Mark Smith, James Spudich, Joan Steitz, Thomas Steitz, and Alan Waggoner.

I am grateful to the Commonwealth Fund for a grant that enabled me to initiate the writing of this book. The interest and support of Robert Glaser, Terrance Keenan, and Quigg Newton came at a critical time. One of my aims in writing this book has been to achieve a close integration of word and picture and to illustrate chemical transformations and three-dimensional structures vividly. I am especially grateful to Donna Salmon, John Foster, and Jean Foster for their work on the drawings, diagrams, and graphs. Many individuals at Yale helped to bring this project to fruition. I particularly wish to thank Margaret Banton and Sharen Westin for typing the manuscript, William Pollard for photographing space-filling models, and Martha Scarf for generating the computer drawings of molecular structures on which many of the illustrations in this book are based. John Harrison and his staff at the Kline Science Library helped in many ways.

Much of this book was written in Aspen. I wish to thank the Aspen Center of Physics and the Given Institute of Pathobiology for their kind hospitality during several summers. I have warm memories of many stimulating discussions about biochemistry and molecular aspects of medicine that took place in the lovely garden of the Given Institute and while hiking in the surrounding wilderness areas. The concerts in Aspen were another source of delight, especially after an intensive day of writing.

I am deeply grateful to my wife, Andrea, and to my children, Michael and Daniel, for their encouragement, patience, and good spirit during the writing of this book. They have truly shared in its gestation, which was much longer than expected. Andrea offered advice on style and design and also called my attention to the remark of the thirteenth-century Chinese scholar Tai T'ung (*The Six Scripts: Principles of Chinese Writing*): "Were I to await perfection, my book would never be finished."

I welcome comments and criticisms from readers.

October 1974 *Lubert Stryer*

BIOCHEMISTRY

MOLECULES AND LIFE

Biochemistry is the study of the molecular basis of life. There is much excitement and activity in biochemistry today for several reasons. *First, the chemical bases of some central processes are now understood.* The discovery of the double-helical structure of deoxyribonucleic acid (DNA), the elucidation of the genetic code, the determination of the three-dimensional structure of some protein molecules, and the unraveling of the central metabolic pathways are some of the outstanding achievements of biochemistry. *Second, there are common molecular patterns and principles that underlie the diverse expressions of life.* Organisms as different as the bacterium *Escherichia coli* and humans have many common features at the molecular level. They use the same building blocks to construct macromolecules. The flow of genetic information from DNA to ribonucleic acid (RNA) to protein is essentially the same in both species. Both use adenosine triphosphate (ATP) as the currency of energy. *Third, biochemistry is having an increasing effect on medicine.* For example, assays for enzyme activity now play an important role in clinical diagnosis. The level of certain enzymes in the serum is a valuable criterion of whether a patient has recently had a myocardial infarction. Furthermore, biochemistry is beginning to provide a basis for the rational design of drugs. Most significant, the molecular mechanisms of diseases are being elucidated, as exemplified by sickle-cell anemia and the large number of inborn errors of metabolism that have been characterized. *Fourth, the rapid development of biochemistry in recent years has enabled investigators to tackle some of the most challenging and fundamental problems in biology and medicine.* How does a single cell

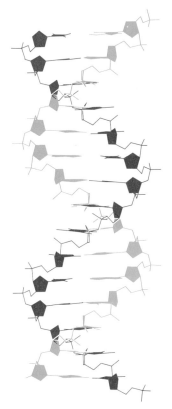

Figure 1-1
Model of the DNA double helix. The diameter of the helix is about 20 Å.

light microscope is about 0.2 μm (2000 Å), which corresponds to the size of many subcellular organelles. Mitochondria, the major generators of ATP in aerobic cells, can just be resolved by the light microscope. Most of our knowledge of biological structure in the 1 Å (0.1 nm) to 10^4 Å (1 μm) range has come from electron microscopy and x-ray diffraction.

The molecules of life are constantly in flux. Chemical reactions in biological systems are catalyzed by enzymes, which typically convert substrate into product in milliseconds (msec, 10^{-3} sec). Some enzymes act even more rapidly, in times as short as a few microseconds (μsec, 10^{-6} sec). Many conformational changes in biological macromolecules also are rapid. For example, the unwinding of the DNA double helix, which is essential for its replication and expression, is a microsecond event. The rotation of one domain of a protein with respect to another can take place in nanoseconds (nsec, 10^{-9} sec). Many noncovalent interactions between groups in macromolecules are formed and broken in nanoseconds. Even more rapid processes can be probed with very short light pulses from lasers. It is remarkable that the primary event in vision—a change in structure of the light-absorbing group—occurs within a few picoseconds (psec, 10^{-12} sec) after the absorption of a photon. Clearly, the dynamics of biological systems span a broad range in time (Figure 1-6). An appreciation of the span of evolutionary time is equally

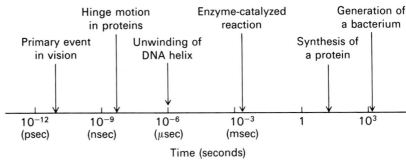

Figure 1-6
Typical rates of some processes in biological systems.

important. Life on earth arose some 3.5×10^9 years ago, which corresponds to 1.1×10^{17} sec.

We will also be concerned with energy changes in molecular events (Figure 1-7). The ultimate source of energy for life is the sun. The energy of visible light, say green photons, is 57 kilocalories per mole (kcal/mol). ATP, the universal currency of energy, has a usable energy content of about 12 kcal/mol. In contrast, the average energy of each vibrational degree of freedom in a molecule is much smaller, 0.6 kcal/mol at 25°C. This amount of energy is much less than that needed to dissociate covalent bonds (e.g., 83 kcal/mol for a C—C bond). Hence, the covalent framework of biomolecules is

MOLECULES AND LIFE

Biochemistry is the study of the molecular basis of life. There is much excitement and activity in biochemistry today for several reasons. *First, the chemical bases of some central processes are now understood.* The discovery of the double-helical structure of deoxyribonucleic acid (DNA), the elucidation of the genetic code, the determination of the three-dimensional structure of some protein molecules, and the unraveling of the central metabolic pathways are some of the outstanding achievements of biochemistry. *Second, there are common molecular patterns and principles that underlie the diverse expressions of life.* Organisms as different as the bacterium *Escherichia coli* and humans have many common features at the molecular level. They use the same building blocks to construct macromolecules. The flow of genetic information from DNA to ribonucleic acid (RNA) to protein is essentially the same in both species. Both use adenosine triphosphate (ATP) as the currency of energy. *Third, biochemistry is having an increasing effect on medicine.* For example, assays for enzyme activity now play an important role in clinical diagnosis. The level of certain enzymes in the serum is a valuable criterion of whether a patient has recently had a myocardial infarction. Furthermore, biochemistry is beginning to provide a basis for the rational design of drugs. Most significant, the molecular mechanisms of diseases are being elucidated, as exemplified by sickle-cell anemia and the large number of inborn errors of metabolism that have been characterized. *Fourth, the rapid development of biochemistry in recent years has enabled investigators to tackle some of the most challenging and fundamental problems in biology and medicine.* How does a single cell

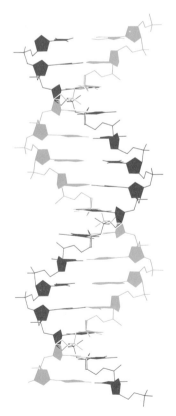

Figure 1-1
Model of the DNA double helix. The diameter of the helix is about 20 Å.

give rise to cells as different as those in muscle, the brain, and the liver? How do cells find each other in forming a complex organ? How is the growth of cells controlled? What is the mechanism of memory? How does light elicit a nerve impulse in the retina? What are the causes of cancer? What are the causes of schizophrenia?

MOLECULAR MODELS

The interplay between the three-dimensional structure of biomolecules and their biological function is the unifying motif of this book. Three types of atomic models will be used to depict molecular architecture: space-filling, ball-and-stick, and skeletal. The *space-filling* models are the most realistic. The size and configuration of an atom in a space-filling model are determined by its bonding properties and van der Waals radius (Figure 1-2). The colors of the atoms are set by convention.

Hydrogen: white	Oxygen: red
Carbon: black	Phosphorus: yellow
Nitrogen: blue	Sulfur: yellow

Space-filling models of several simple molecules are illustrated in Figure 1-3.

Figure 1-2
Space-filling models of hydrogen, carbon, nitrogen, oxygen, phosphorus, and sulfur atoms.

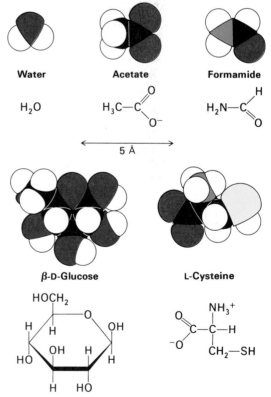

Figure 1-3
Space-filling models of water, acetate, formamide, glucose, and cysteine.

Ball-and-stick models are not as realistic as space-filling models because the atoms are depicted as spheres of smaller radius than the van der Waals radius. However, the bonding arrangement is easier to see than in space-filling models because the bonds are explicitly represented by sticks. The taper of a stick tells whether the direction of the bond is in front of the plane of the page or behind it. More of a complex structure can be seen in ball-and-stick models than in space-filling models. An even simpler image is achieved with *skeletal models,* which show only the molecular framework. In these models, atoms are not explicitly shown. Rather, their positions are implied by the junctions of bonds and their termini. Skeletal models are frequently used to depict large biological macromolecules, such as protein molecules having several thousand atoms. Space-filling, ball-and-stick, and skeletal models of ATP are compared in Figure 1-4.

SPACE, TIME, AND ENERGY

In considering molecular structure, it is important to have a sense of scale (Figure 1-5). The angstrom (Å) unit, which is equal to 10^{-10} meter (m) or 0.1 nanometer (nm), is customarily used as the measure of length at the atomic level. The length of a C—C bond, for example, is 1.54 Å. Small biomolecules, such as sugars and amino acids, are typically several angstroms long. Biological macromolecules, such as proteins, are at least tenfold larger. For example, hemoglobin, the oxygen-carrying protein in red blood cells, has a diameter of 65 Å. Another tenfold increase in size brings us to assemblies of macromolecules. Ribosomes, the protein-synthesizing machinery of the cell, have diameters of about 300 Å. The span from 100 Å (10 nm) to 1000 Å (100 nm) also encompasses most viruses. Cells are typically a hundred times as large, in the range of micrometers (μm). For example, a red blood cell is 7 μm (7 $\times$ 10^4 Å) long. It is important to note that the limit of resolution of the

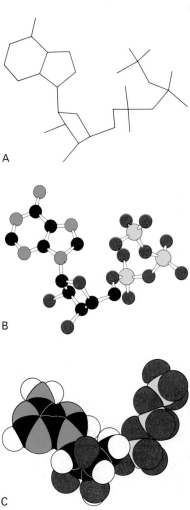

A

B

C

10 Å

Figure 1-4
Comparison of (A) skeletal, (B) ball-and-stick, and (C) space-filling models of ATP. Hydrogen atoms are not shown in models A and B.

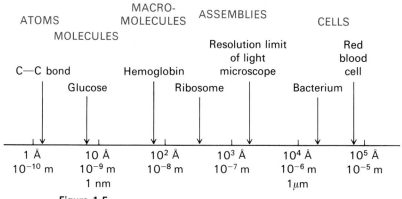

Figure 1-5
Dimensions of some biomolecules, assemblies, and cells.

light microscope is about 0.2 μm (2000 Å), which corresponds to the size of many subcellular organelles. Mitochondria, the major generators of ATP in aerobic cells, can just be resolved by the light microscope. Most of our knowledge of biological structure in the 1 Å (0.1 nm) to 10^4 Å (1 μm) range has come from electron microscopy and x-ray diffraction.

The molecules of life are constantly in flux. Chemical reactions in biological systems are catalyzed by enzymes, which typically convert substrate into product in milliseconds (msec, 10^{-3} sec). Some enzymes act even more rapidly, in times as short as a few microseconds (μsec, 10^{-6} sec). Many conformational changes in biological macromolecules also are rapid. For example, the unwinding of the DNA double helix, which is essential for its replication and expression, is a microsecond event. The rotation of one domain of a protein with respect to another can take place in nanoseconds (nsec, 10^{-9} sec). Many noncovalent interactions between groups in macromolecules are formed and broken in nanoseconds. Even more rapid processes can be probed with very short light pulses from lasers. It is remarkable that the primary event in vision—a change in structure of the light-absorbing group—occurs within a few picoseconds (psec, 10^{-12} sec) after the absorption of a photon. Clearly, the dynamics of biological systems span a broad range in time (Figure 1-6). An appreciation of the span of evolutionary time is equally

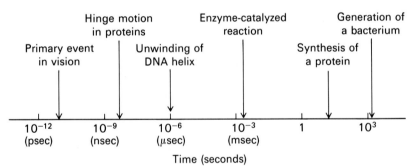

Figure 1-6
Typical rates of some processes in biological systems.

important. Life on earth arose some 3.5×10^9 years ago, which corresponds to 1.1×10^{17} sec.

We will also be concerned with energy changes in molecular events (Figure 1-7). The ultimate source of energy for life is the sun. The energy of visible light, say green photons, is 57 kilocalories per mole (kcal/mol). ATP, the universal currency of energy, has a usable energy content of about 12 kcal/mol. In contrast, the average energy of each vibrational degree of freedom in a molecule is much smaller, 0.6 kcal/mol at 25°C. This amount of energy is much less than that needed to dissociate covalent bonds (e.g., 83 kcal/mol for a C—C bond). Hence, the covalent framework of biomolecules is

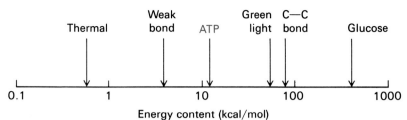

Figure 1-7
Some biologically important energies.

stable in the absence of enzymes and inputs of energy. On the other hand, thermal energy is enough to make and break noncovalent bonds in biological systems, which typically have an energy of only a few kilocalories per mole. Energy generation and storage will be considered in detail in Part II of this book.

DESIGN OF THIS BOOK

This book has five parts, each centered on a major theme:

> I: Conformation and Dynamics
> II: Generation and Storage of Metabolic Energy
> III: Biosynthesis of Macromolecular Precursors
> IV: Information
> V: Molecular Physiology

The interplay of three-dimensional structure and biological activity as exemplified in proteins is the major theme of Part I. The structure and function of myoglobin and hemoglobin, the oxygen-carrying proteins in vertebrates, are presented in detail because these proteins illustrate some general principles. Hemoglobin is especially interesting because its binding of oxygen is regulated by specific molecules in its environment. The molecular pathology of hemoglobin, particularly sickle-cell anemia, is also presented. We then turn to enzymes and consider how they recognize substrates and enhance reaction rates by factors of a million or more. The enzymes lysozyme, carboxypeptidase A, and chymotrypsin are examined in detail because many general principles of catalysis have been elucidated from their study. Rather different facets of the theme of conformation emerge in the chapter on collagen and elastin, two connective-tissue proteins. The final chapter in Part I is an introduction to biological membranes, which are organized assemblies of lipids and proteins. A major function of membranes in biological systems is to create compartments.

Part II deals with the generation and storage of metabolic energy. First, the overall strategy of metabolism is presented. Cells convert energy from fuel molecules into ATP. In turn, ATP drives

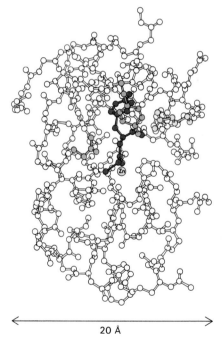

20 Å

Figure 1-8
Structure of an enzyme-substrate complex. Glycyltyrosine (shown in red) is bound to carboxypeptidase A, a digestive enzyme. Only a quarter of the enzyme is shown. [After W. N. Lipscomb. *Proc. Robert A. Welch Found. Conf. Chem. Res.* 15(1971):141.]

most energy-requiring processes in cells. In addition, reducing power in the form of nicotinamide adenine dinucleotide phosphate (NADPH) is generated for use in biosyntheses. The metabolic pathways that carry out these reactions are then presented in detail. For example, the generation of ATP from glucose requires a sequence of three series of reactions—glycolysis, the citric acid cycle, and oxidative phosphorylation. The last two are also common to the generation of ATP from the oxidation of fats and some amino acids, the other major fuels. We see here an illustration of molecular economy. Two storage forms of fuel molecules, glycogen and triacylglycerols (neutral fats), are also discussed in Part II. The concluding topic of this part of the book is photosynthesis, in which the primary event is the light-activated transfer of an electron from one substance to another against a chemical potential gradient.

Part III deals with the biosynthesis of macromolecular precursors, starting with the synthesis of membrane lipids and steroids. The pathway for the synthesis of cholesterol, a 27-carbon steroid, is of particular interest because all of its carbon atoms come from a 2-carbon precursor. The reactions leading to the synthesis of selected amino acids and the heme group are discussed in the next chapter. The control mechanisms in these pathways are of general significance. The biosynthesis of nucleotides, the activated precursors of DNA and RNA, is then considered. The final chapter in this part deals with the integration of metabolism. How are energy-yielding and energy-consuming reactions coordinated to meet the needs of an organism?

The storage, transmission, and expression of genetic information constitute the central theme of Part IV. The experiments that revealed that DNA is the genetic material and the discovery of the DNA double helix are discussed first, followed by a presentation of the enzymatic mechanism of DNA replication. We then turn to the expression of genetic information in DNA, starting with the evidence that messenger RNA is the information-carrying intermediate in protein synthesis. The process of transcription, which is the synthesis of RNA according to instructions given by a DNA template, is then considered. This brings us to the genetic code, which is the relationship between the sequence of bases in DNA (or its messenger RNA transcript) and the sequence of amino acids in a protein. The code, which is the same in all organisms, is beautiful in its simplicity. Three bases constitute a codon, the unit that specifies one amino acid. Codons on messenger RNA are read sequentially by transfer RNA molecules, which are the adaptors in protein synthesis. We then examine the mechanism of protein synthesis, a process called translation because it converts the four base-pairing letters of nucleic acids into the twenty-letter alphabet of proteins. Translation is carried out by the coordinated interplay of more than a hundred kinds of macromolecules, centered on ribosomes. The next chapter deals with the control of gene expression in bacteria.

10 Å

Figure 1-9
Model of CDP-diacylglycerol, an activated intermediate in the synthesis of some membrane lipids.

The focus here is on the lactose and tryptophan operons of *E. coli,* which are now understood in detail. This is followed by a discussion of the recent research of gene expression in higher organisms. They contain far more DNA than do bacteria and have a distinct nucleus, enabling their cells to become differentiated. Viral multiplication and assembly are considered next. Viral assembly exemplifies some general principles of how biological macromolecules form highly ordered structures. Some viruses that cause cancers in experimental animals are considered here as well. Part IV concludes with a chapter on the formation of new genes by recombination. The significance of DNA cloning is discussed in this chapter.

Part V, titled Molecular Physiology, is a transition from biochemistry to physiology. Many of the concepts that were developed in the preceding parts of the book are used here, because physiology involves the interplay of information, conformation, and metabolism. We start with the assembly of cell membranes and bacterial-cell envelopes. How do cells specify the positions of proteins synthesized by them? Then we turn to the molecular basis of the immune response. How does an organism recognize a foreign substance? The next chapter deals with the problem of how chemical-bond energy is transformed into coordinated motion. Recent studies have shown that myosin and actin, the major proteins in muscle, have a contractile role in most cells of higher organisms. The molecular basis of hormone action is then considered, with an emphasis on some common themes. The transport of ions, such as Na^+, K^+, and Ca^{2+}, and molecules is presented next. Molecular pumps in membranes transport these ions to generate gradients that are at the heart of excitability. The final chapter deals with sensory processes and considers such questions as: How are action potentials propagated by nerve cells and transmitted across synapses? How is a retinal rod cell triggered by a single photon? How do bacteria detect and move toward nutrients in their environment?

One of the most satisfying features of biochemistry is that it continually enriches our understanding of biological processes at all levels of organization.

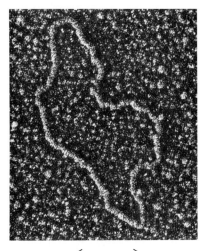

←———— 2000 Å ————→
(0.2 μm)

Figure 1-10
Electron micrograph of a DNA molecule. [Courtesy of Dr. Thomas Broker.]

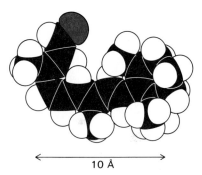

←———— 10 Å ————→

Figure 1-11
Model of 11-*cis*-retinal, the light-absorbing group in rhodopsin. The isomerization of this chromophore by light is the first event in visual excitation.

PART I

CONFORMATION AND DYNAMICS

exemplified by the relation between the three-dimensional structure of proteins and their biological activities

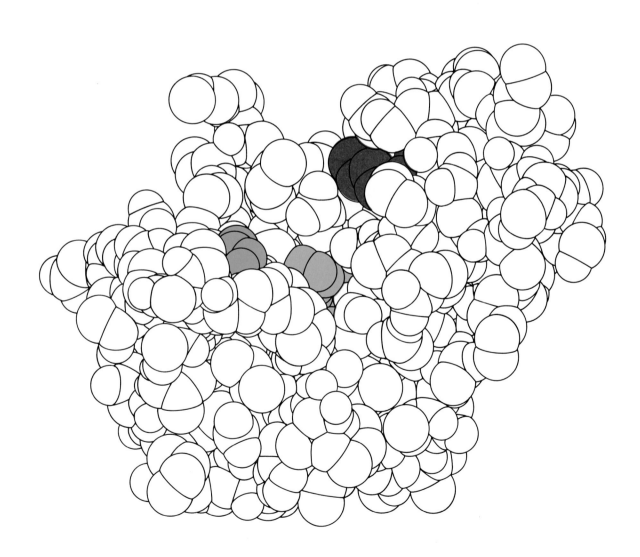

CHAPTER 2

INTRODUCTION TO
PROTEIN STRUCTURE AND FUNCTION

Proteins play crucial roles in virtually all biological processes. The significance and remarkable scope of their functions are exemplified in:

1. *Enzymatic catalysis.* Nearly all chemical reactions in biological systems are catalyzed by specific macromolecules called enzymes. Some of these reactions, such as the hydration of carbon dioxide, are quite simple. Others, such as the replication of an entire chromosome, are highly intricate. Nearly all enzymes exhibit enormous catalytic power. They usually enhance reaction rates by at least a millionfold. Indeed, chemical transformations rarely occur at perceptible rates in vivo in the absence of enzymes. Several thousand enzymes have been characterized, and many of them have been crystallized. The striking fact is that all known enzymes are proteins. Thus, proteins play the unique role of determining the pattern of chemical transformations in biological systems.

2. *Transport and storage.* Many small molecules and ions are transported by specific proteins. For example, hemoglobin transports oxygen in erythrocytes, whereas myoglobin, a related protein, transports oxygen in muscle. Iron is carried in the plasma of blood

Figure 2-1
Photomicrograph of a crystal of hexokinase, a key enzyme in the utilization of glucose. [Courtesy of Dr. Thomas Steitz and Dr. Mark Yeager.]

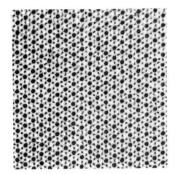

Figure 2-2
Electron micrograph of a cross section of insect flight muscle showing a hexagonal array of two kinds of protein filaments. [Courtesy of Dr. Michael Reedy.]

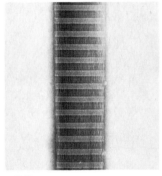

Figure 2-3
Electron micrograph of a fiber of collagen. [Courtesy of Dr. Jerome Gross and Dr. Romaine Bruns.]

by transferrin and is stored in the liver as a complex with ferritin, a different protein.

3. *Coordinated motion.* Proteins are the major component of muscle. Muscle contraction is accomplished by the sliding motion of two kinds of protein filaments. On the microscopic scale, such coordinated motions as the movement of chromosomes in mitosis and the propulsion of sperm by their flagella also are produced by contractile assemblies consisting of proteins.

4. *Mechanical support.* The high tensile strength of skin and bone is due to the presence of collagen, a fibrous protein.

5. *Immune protection.* Antibodies are highly specific proteins that recognize and combine with such foreign substances as viruses, bacteria, and cells from other organisms. Proteins thus play a vital role in distinguishing between self and nonself.

6. *Generation and transmission of nerve impulses.* The response of nerve cells to specific stimuli is mediated by receptor proteins. For example, rhodopsin is the photoreceptor protein in retinal rod cells. Receptor molecules that can be triggered by specific small molecules, such as acetylcholine, are responsible for transmitting nerve impulses at synapses—that is, at junctions between nerve cells.

7. *Control of growth and differentiation.* Controlled sequential expression of genetic information is essential for the orderly growth and differentiation of cells. Only a small fraction of the genome of a cell is expressed at any one time. In bacteria, repressor proteins are important control elements that silence specific segments of the DNA of a cell. A quite different way in which proteins act in differentiation is exemplified by nerve growth factor, a protein complex that guides the formation of neural networks in higher organisms.

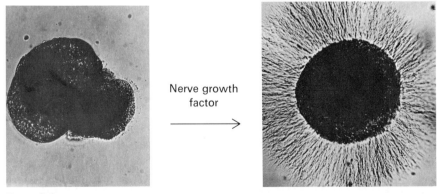

Nerve growth factor

Figure 2-4
Photomicrograph of a ganglion showing the proliferation of nerves after addition of nerve growth factor, a complex of proteins. [Courtesy of Dr. Eric Shooter.]

Amino acids are the basic structural units of proteins. An amino acid consists of an amino group, a carboxyl group, a hydrogen atom, and a distinctive R group bonded to a carbon atom, which is called the α-carbon (Figure 2-5). An R group is referred to as a *side*

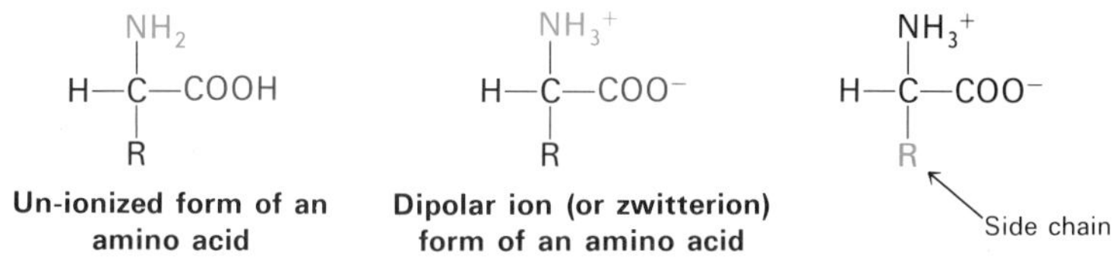

Un-ionized form of an amino acid   Dipolar ion (or zwitterion) form of an amino acid   Side chain

Figure 2-5
Structure of the un-ionized and zwitterion forms of an amino acid.

chain for reasons that will be evident shortly. Amino acids in solution at neutral pH are predominantly *dipolar ions* (or *zwitterions*) rather than un-ionized molecules. In the dipolar form of an amino acid, the amino group is protonated ($-NH_3^+$) and the carboxyl group is dissociated ($-COO^-$). The ionization state of an amino acid varies with pH (Figure 2-6). In acid solution (e.g., pH 1), the

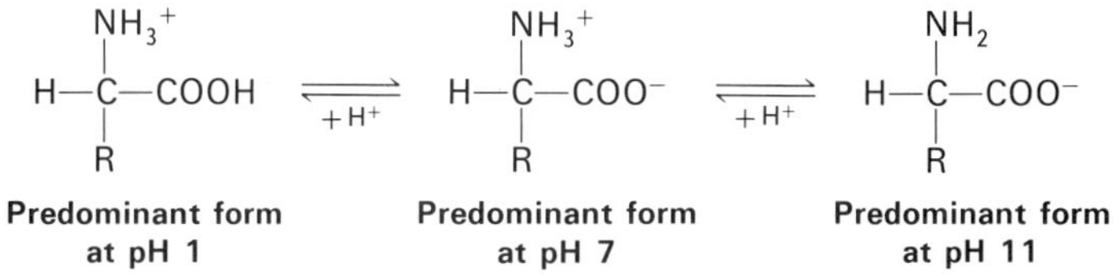

Predominant form at pH 1   Predominant form at pH 7   Predominant form at pH 11

Figure 2-6
Ionization states of an amino acid as a function of pH.

carboxyl group is un-ionized ($-COOH$) and the amino group is ionized ($-NH_3^+$). In alkaline solution (e.g., pH 11), the carboxyl group is ionized ($-COO^-$) and the amino group is un-ionized ($-NH_2$). The concept of pH and the acid-base properties of amino acids are discussed further in the Appendix to this chapter.

The tetrahedral array of four different groups about the α-carbon atom confers optical activity on amino acids. The two mirror-image forms are called the L-isomer and the D-isomer (Figure 2-7). *Only* L-*amino acids are constituents of proteins.* Hence, the designation of the optical isomer will be omitted and the L-isomer implied in discussions of proteins herein, unless otherwise noted.

Twenty kinds of side chains varying in *size, shape, charge, hydrogen-bonding capacity, and chemical reactivity* are commonly found in proteins. Indeed, all proteins in all species, from bacteria to humans, are constructed from the same set of twenty amino acids. This fun-

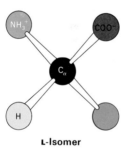

L-Isomer

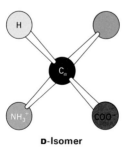

D-Isomer

Figure 2-7
Absolute configurations of the L- and D-isomers of amino acids.

damental alphabet of proteins is at least two billion years old. The remarkable range of functions mediated by proteins results from the diversity and versatility of these twenty kinds of building blocks. In subsequent chapters, we will explore ways in which this alphabet is used to create the intricate three-dimensional structures that enable proteins to participate in so many biological processes.

Let us look at this repertoire of amino acids. The simplest one is glycine, which contains a hydrogen atom as its side chain (Figure 2-8). Alanine has a methyl group as its side chain. The other amino

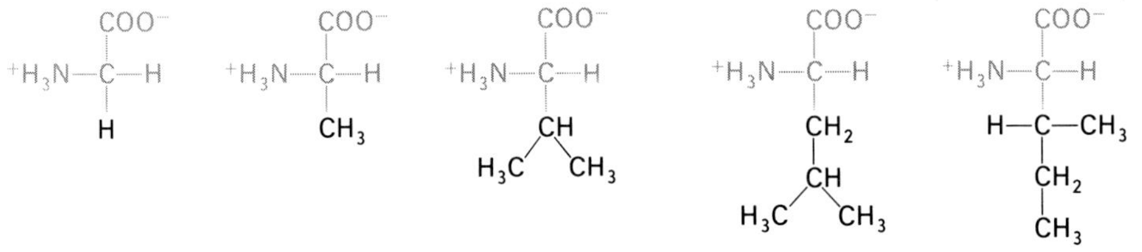

Figure 2-8
Amino acids having aliphatic side chains.

Glycine (Gly), Alanine (Ala), Valine (Val), Leucine (Leu), Isoleucine (Ile)

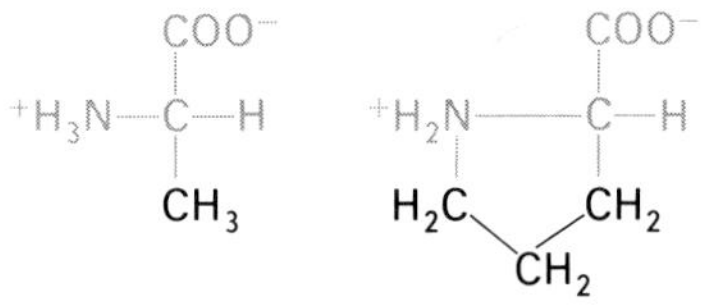

Alanine (Ala), Proline (Pro)

Figure 2-9
Proline differs from the other common amino acids in that it has a secondary amino group.

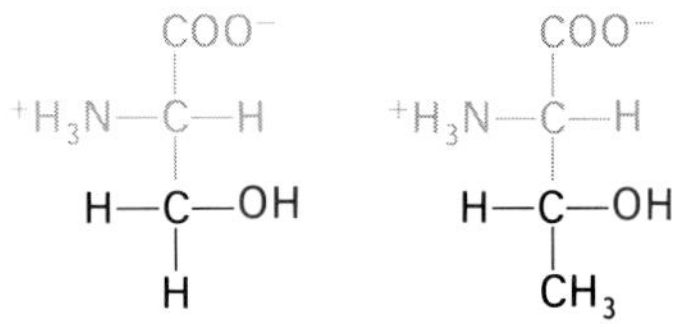

acids that have hydrocarbon side chains are valine, leucine, isoleucine, and proline. However, proline differs from the other amino acids in the basic set of twenty in that it contains a secondary rather than a primary amino group (Figure 2-9). Strictly speaking, proline is an imino acid rather than an amino acid. The side chain of proline is bonded to both the amino group and the α-carbon, which results in a cyclic structure.

Two amino acids, serine and threonine, contain aliphatic hydroxyl groups (Figure 2-10).

There are three common aromatic amino acids: phenylalanine, tyrosine, and tryptophan (Figure 2-11).

Serine (Ser), Threonine (Thr)

Figure 2-10
Serine and threonine have aliphatic hydroxyl side chains.

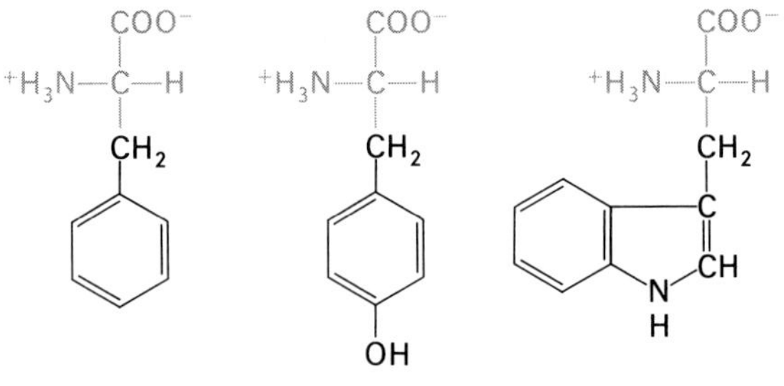

Phenylalanine (Phe), Tyrosine (Tyr), Tryptophan (Trp)

Figure 2-11
Phenylalanine, tyrosine, and tryptophan have aromatic side·chains.

The side chains of the amino acids mentioned so far are uncharged at physiological pH. We turn now to some charged side chains. Lysine and arginine are positively charged at neutral pH, whereas whether histidine is positively charged or neutral depends on its local environment. These basic amino acids are shown in Figure 2-12. The negatively charged side chains are those of glu-

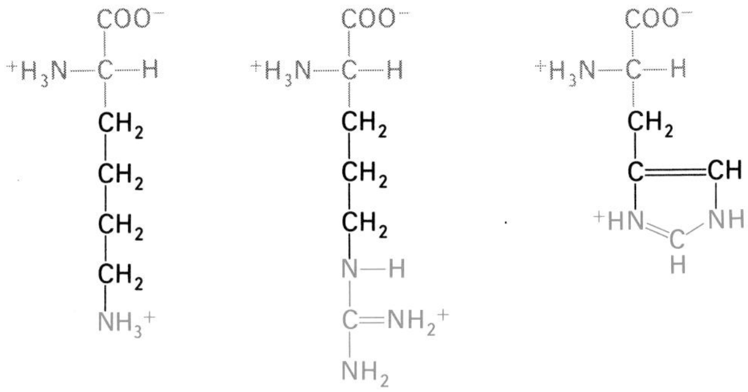

Lysine
(Lys)

Arginine
(Arg)

Histidine
(His)

Figure 2-12
Lysine, arginine, and histidine have basic side chains.

tamic acid and aspartic acid (Figure 2-13). These amino acids will be called glutamate and aspartate to emphasize the fact that they are negatively charged at physiological pH. The uncharged derivatives of glutamate and aspartate are glutamine and asparagine (Figure 2-14), each of which contains a terminal amide group rather than a carboxylate. Finally, there are two amino acids whose side chains contain a sulfur atom: methionine and cysteine (Figure 2-15). As will be discussed shortly, cysteine plays a special role in some proteins by forming disulfide cross-links.

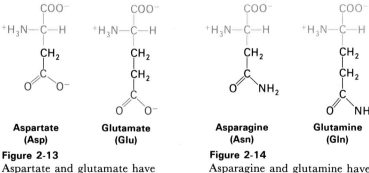

Aspartate
(Asp)

Glutamate
(Glu)

Figure 2-13
Aspartate and glutamate have acidic side chains.

Asparagine
(Asn)

Glutamine
(Gln)

Figure 2-14
Asparagine and glutamine have amide side chains.

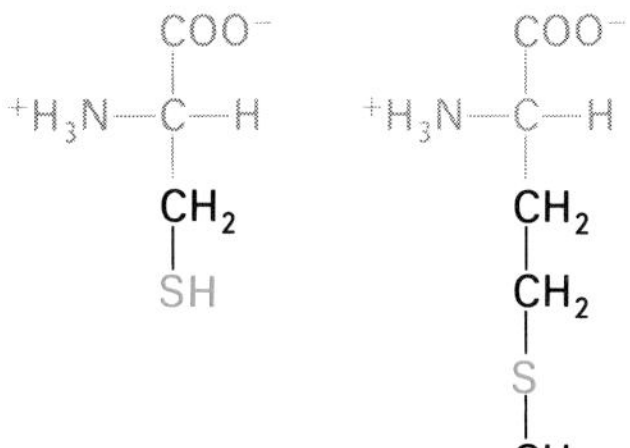

Cysteine
(Cys)

Methionine
(Met)

Figure 2-15
Cysteine and methionine have sulfur-containing side chains.

Table 2-1
Abbreviations for amino acids

Amino acid	Three-letter abbreviation	One-letter symbol
Alanine	Ala	A
Arginine	Arg	R
Asparagine	Asn	N
Aspartic acid	Asp	D
Asparagine or aspartic acid	Asx	B
Cysteine	Cys	C
Glutamine	Gln	Q
Glutamic acid	Glu	E
Glutamine or glutamic acid	Glx	Z
Glycine	Gly	G
Histidine	His	H
Isoleucine	Ile	I
Leucine	Leu	L
Lysine	Lys	K
Methionine	Met	M
Phenylalanine	Phe	F
Proline	Pro	P
Serine	Ser	S
Threonine	Thr	T
Tryptophan	Trp	W
Tyrosine	Tyr	Y
Valine	Val	V

SPECIAL AMINO ACIDS SUPPLEMENT THE BASIC SET OF TWENTY

Some proteins contain special amino acids that are formed by modification of a common amino acid following its incorporation into the polypeptide chain. For example, collagen contains hydroxyproline, a hydroxylated derivative of proline (Figure 2-16). The added

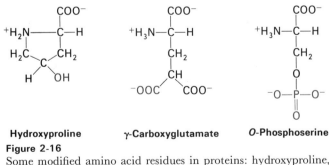

Hydroxyproline   **γ-Carboxyglutamate**   *O*-**Phosphoserine**
Figure 2-16
Some modified amino acid residues in proteins: hydroxyproline, γ-carboxyglutamate, and phosphoserine. Groups added after the polypeptide chain is synthesized are shown in red.

hydroxyl group stabilizes the collagen fiber, as will be discussed later (p. 192). The biological importance of this modification is evident in scurvy, which results from insufficient hydroxylation of collagen. Another special amino acid is γ-carboxyglutamate. Defective carboxylation of glutamate in prothrombin, a clotting protein, can lead to hemorrhage (p. 176). The most ubiquitous modified amino acid in proteins is phosphoserine. The action of some hormones is mediated by the phosphorylation and dephosphorylation of specific serine residues in a variety of proteins (p. 368).

AMINO ACIDS ARE LINKED BY PEPTIDE BONDS TO FORM POLYPEPTIDE CHAINS

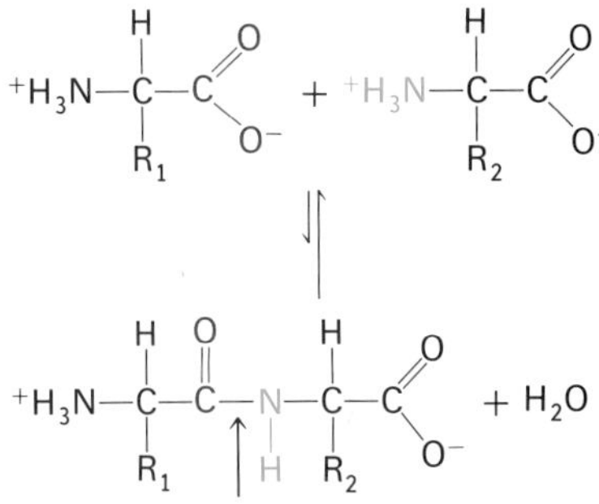

In proteins, the α-carboxyl group of one amino acid is joined to the α-amino group of another amino acid by a *peptide bond* (also called an amide bond). The formation of a dipeptide from two amino acids by loss of a water molecule is shown in Figure 2-17. The equilibrium of this reaction lies far on the side of hydrolysis rather than synthesis. Hence, the biosynthesis of peptide bonds requires an input of free energy, whereas their hydrolysis is thermodynamically downhill.

Many amino acids, usually more than a hundred, are joined by peptide bonds to form a *polypeptide chain,* which is an unbranched structure (Figure 2-18). An amino acid unit in a polypeptide is

Figure 2-17
Formation of a peptide bond.

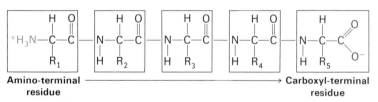

Figure 2-18
A pentapeptide. The constituent amino acid residues are outlined. The chain starts at the amino end.

called a *residue.* A polypeptide chain has direction because its building blocks have different ends—namely, the α-amino and the α-carboxyl groups. By convention, *the amino end is taken to be the beginning of a polypeptide chain.* The sequence of amino acids in a polypeptide chain is written starting with the amino-terminal residue. Thus, in the tripeptide alanine-glycine-tryptophan, alanine is the amino-terminal residue and tryptophan is the carboxyl-terminal residue. Note that tryptophan-glycine-alanine is a different tripeptide.

A polypeptide chain consists of a regularly repeating part, called the *main chain,* and a variable part, comprising the distinctive *side chains* (Figure 2-19). The main chain is sometimes termed the backbone.

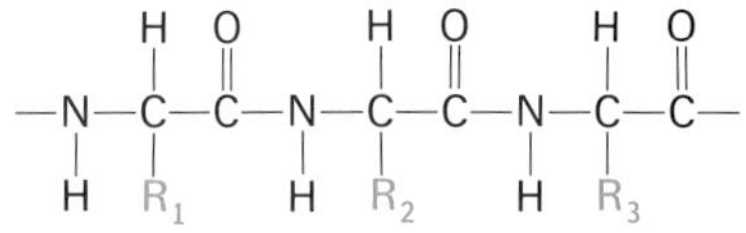

Figure 2-19
A polypeptide chain is made up of a regularly repeating *backbone* and distinctive *side chains* (R_1, R_2, R_3, shown in green).

18

In some proteins, a few side chains are cross-linked by *disulfide bonds*. These cross-links are formed by the oxidation of cysteine residues. The resulting disulfide is called cystine (Figure 2-20). No other covalent cross-links are generally found in proteins.

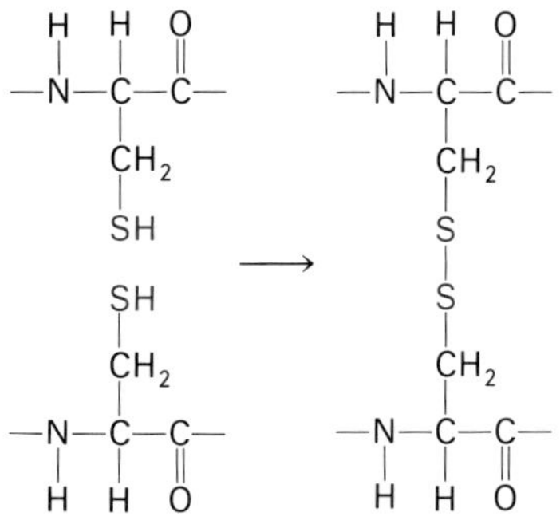

PROTEINS CONSIST OF ONE OR MORE POLYPEPTIDE CHAINS

Many proteins, such as myoglobin, consist of a single polypeptide chain. Others contain two or more chains, which may be either identical or different. For example, hemoglobin is made up of two chains of one kind and two of another kind. These four chains are held together by noncovalent forces. Alternatively, the polypeptide chains of some multichain proteins are linked by disulfide bonds. The two chains of insulin, for example, are joined by two disulfide bonds.

PROTEINS CAN BE PURIFIED BY A VARIETY OF TECHNIQUES

The purification of a protein is an indispensable step toward the elucidation of its mechanism of action. Several thousand proteins have been isolated in pure form. Proteins can be separated from each other and from other kinds of molecules on the basis of such characteristics as *size, solubility, charge, and specific binding affinity*. In purifying a protein, various separation methods are tried and their efficiency is evaluated by assaying for a distinctive property of the protein of interest. The assay for an enzyme, for example, is usually based on its specific catalytic activity. The total amount of protein is also measured so that the degree of purification obtained in a particular step can be determined.

Proteins can be separated from small molecules by dialysis through a semipermeable membrane (Figure 2-21). Molecules

Figure 2-20
A disulfide bridge (—S—S—) is formed from the sulfhydryl groups (—SH) of two cysteine residues. The product is a *cystine* residue.

Cysteine   Cystine

Large molecules cannot traverse the membrane.

Small molecules can traverse the membrane.

Semipermeable membrane

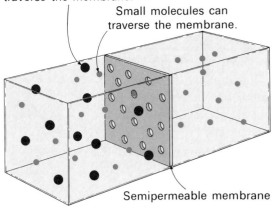

Figure 2-21
Separation of molecules on the basis of size by dialysis.

whose mass is greater than about 15 kilodaltons (kdal) are retained inside a typical dialysis bag, whereas smaller molecules and ions traverse the pores of such a dialysis membrane and emerge in the dialysate outside the bag. Separations on the basis of size can also be achieved by the technique of *gel-filtration chromatography* (Figure 2-22). The sample is applied to the top of a column consisting of an

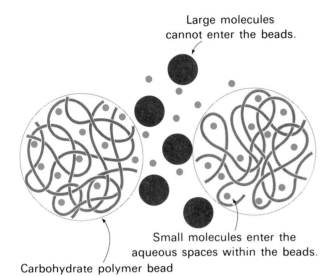

Large molecules
cannot enter the beads.

Small molecules enter the
aqueous spaces within the beads.

Carbohydrate polymer bead

Figure 2-22
Separation of molecules on the basis
of size by gel-filtration chromatography.

insoluble but highly hydrated carbohydrate polymer in the form of beads, which are typically 0.1 mm in diameter. Sephadex is a commonly used commercial preparation. Small molecules can enter these beads, but large ones cannot. The result is that small molecules are distributed both in the aqueous solution inside the beads and between them, whereas large molecules are located only in the solution between the beads. Large molecules flow more rapidly through this column and emerge first because a smaller volume is accessible to them.

Proteins can also be separated on the basis of their net charge by *ion-exchange chromatography*. If a protein has a net positive charge at pH 7, it will usually bind to an ion-exchange column containing carboxylate groups, whereas a negatively charged protein will not. Such a positively charged protein can be released from the column by adding sodium chloride or another salt to the eluting buffer. Sodium ions compete with positively charged groups on the protein for binding to the column. Proteins that have a low density of net positive charge will tend to emerge first, followed by those having a higher charge density. Factors other than net charge also influence the behavior of proteins on ion-exchange columns. The net charge of a protein also influences its rate of migration in an electric field. This principle is exploited in *electrophoresis,* a technique that will be discussed in more detail in a subsequent chapter (p. 90). The very high resolving power of electrophoresis is worth noting here. It is

Dalton—
A unit of mass very nearly equal to that of a hydrogen atom (precisely equal to 1.0000 on the atomic mass scale).
The terms "dalton" and "molecular weight" are used interchangeably; for example, a 20,000-dalton protein has a molecular weight of 20,000.
Named after John Dalton (1766–1844), who developed the atomic theory of matter.
Kilodalton (kdal)—
A unit of mass equal to 1000 daltons. Most proteins have a mass of between 10 and 100 kdal.

remarkable that more than a thousand different proteins in the simple bacterium *E. coli* can be resolved in a single experiment by two-dimensional electrophoresis (Figure 2-23).

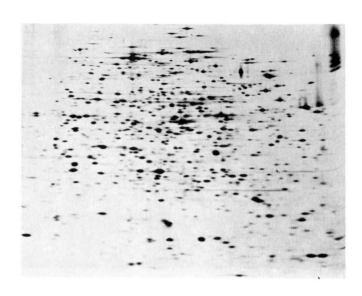

Figure 2-23
Two-dimensional electrophoresis of the proteins from *E. coli*. More than a thousand different proteins from this bacterium have been resolved. These proteins were separated according to their isoelectric pH in the horizontal direction and their molecular weight in the vertical direction. [Courtesy of Dr. Patrick H. O'Farrell.]

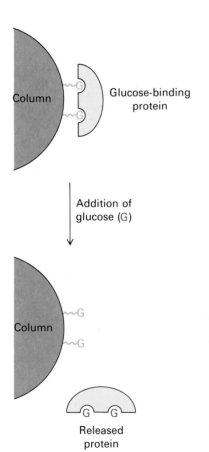

Figure 2-24
Affinity chromatography of concanavalin A (shown in yellow) on a column containing covalently attached glucose residues (G).

Affinity chromatography is another powerful and generally applicable means of purifying proteins. This technique takes advantage of the high affinity of many proteins for specific chemical groups. For example, concanavalin A, a plant protein, can be purified by passing a crude extract through a column that contains covalently attached glucose residues. Concanavalin A binds to such a column because it has affinity for glucose, whereas most other proteins are not adsorbed. The bound concanavalin A can then be released from this column by adding a concentrated solution of glucose. The glucose in solution displaces the column-attached glucose residues from binding sites on concanavalin A (Figure 2-24). In general, affinity chromatography can be effectively used to isolate a protein that recognizes group X by (1) convalently attaching X or a derivative of it to a column, (2) adding a mixture of proteins to this column, which is then washed with buffer to remove unbound proteins, and (3) eluting the desired protein by adding a high concentration of a soluble form of X.

PROTEINS HAVE UNIQUE AMINO ACID SEQUENCES THAT ARE SPECIFIED BY GENES

In 1953, Frederick Sanger determined the amino acid sequence of insulin, a protein hormone (Figure 2-25). *This work is a landmark in biochemistry because it showed for the first time that a protein has a precisely defined amino acid sequence.* This accomplishment also stimulated other scientists to carry out sequence studies of a wide variety of proteins. Indeed, the complete amino acid sequences of several

Figure 2-25
Amino acid sequence of bovine insulin.

hundred proteins are now known. The striking fact is that each protein has a unique, precisely defined amino acid sequence. A series of incisive studies in the late 1950s and early 1960s revealed that the amino acid sequences of proteins are genetically determined. The sequence of nucleotides in DNA, the molecule of heredity, specifies a complementary sequence of nucleotides in RNA, which in turn specifies the amino acid sequence of a protein (p. 597). Furthermore, proteins are synthesized from their constituent amino acids by a common mechanism.

The importance of determining amino acid sequences of proteins is fourfold. First, the determination of the sequence of a protein is a significant step toward the elucidation of the molecular basis of its biological activity. A sequence is particularly informative if it is considered together with other chemical and physical data. Second, the sequences and detailed three-dimensional structures of numerous proteins need to be known so that the rules governing the folding of polypeptide chains into highly specific three-dimensional forms can be deduced. The amino acid sequence is the link between the genetic message in DNA and the three-dimensional structure that is the basis of a protein's biological function. Third, alterations in amino acid sequence can produce abnormal function and disease. Fatal disease, such as sickle-cell anemia, can result from a change in a single amino acid in a single protein. Sequence determination is thus part of molecular pathology, an emerging area of medicine. Fourth, the amino acid sequence of a protein reveals much about its evolutionary history. Amino acid sequences of unrelated proteins are very different. Proteins resemble one another in their amino acid sequences only if they have a common ancestor. Consequently, molecular events in evolution can be traced from amino acid sequences.

Protein—
A word coined by Jöns J. Berzelius in 1838 to emphasize the importance of this class of molecules. Derived from the Greek word *proteios,* which means "of the first rank."

EXPERIMENTAL METHODS FOR THE DETERMINATION OF AMINO ACID SEQUENCE

Let us first consider how the sequence of a short peptide could be determined. Suppose that the peptide has six amino acid residues in the following sequence:

Ala-Gly-Asp-Phe-Arg-Gly

Ninhydrin

Fluorescamine

The abbreviations used are the standard ones given in Table 2-1 (p. 16). First, the *amino acid composition* of the peptide is determined. The peptide is hydrolyzed into its constituent amino acids by heating it in 6 N HCl at 110°C for 24 hours. The amino acids in the hydrolysate are separated by ion-exchange chromatography on a column of sulfonated polystyrene. The separated amino acids are detected by the color produced on heating with *ninhydrin:* α-amino acids give an intense blue color, whereas imino acids, such as proline, give a yellow color. The technique is very sensitive; it can detect even a microgram of an amino acid, which is about the amount present in a thumbprint. The quantity of amino acid is proportional to the optical absorbance of the solution after heating with ninhydrin. As little as a few nanograms of an amino acid can be detected using *fluorescamine,* which reacts with its α-amino group to form a highly fluorescent product. The identity of the amino acid is revealed by its elution volume, which is the volume of buffer used to remove the amino acid from the column (Figure 2-26). A comparison of the chromatographic patterns of the hydrolysate with that of a standard mixture of amino acids shows that the amino acid composition of the peptide is

$$(Ala, Arg, Asp, Gly_2, Phe)$$

The parentheses denote that this is the amino acid composition of the peptide, not its sequence.

The amino-terminal residue of a protein or peptide can be identified by labeling it with a compound with which it forms a stable

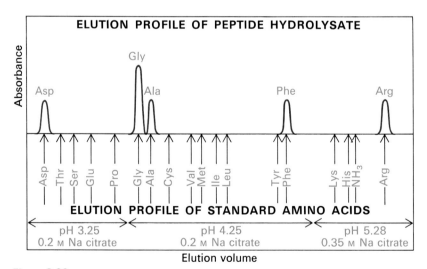

Figure 2-26
Different amino acids in a peptide hydrolysate can be separated by ion-exchange chromatography on a sulfonated polystyrene resin (such as Dowex-50). Buffers of increasing pH are used to elute the amino acids from the column. Aspartate, which has an acidic side chain, is first to emerge, whereas arginine, which has a basic side chain, is the last.

covalent link (Figure 2-27). *Fluorodinitrobenzene* (FDNB), first used for this purpose by Sanger, reacts with the uncharged α-NH$_2$ group to form a yellow dinitrophenyl (DNP) derivative of the peptide. The bond between the DNP and the terminal amino group is stable under conditions that hydrolyze peptide bonds. Hydrolysis of the DNP-peptide in 6 N HCl yields a DNP-amino acid, which is identified as DNP-alanine by its chromatographic properties.

Dansyl chloride is now often used to identify amino-terminal residues. It reacts with amino groups to form highly fluorescent and stable sulfonamide derivatives. A few nanograms of an amino-terminal residue can be identified after acid hydrolysis of peptide bonds.

Although the DNP and dansyl methods for the determination of the amino-terminal residue are powerful, they cannot be used repetitively on the same peptide because the peptide is totally de-

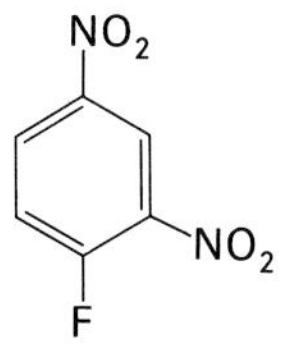

Fluorodinitrobenzene

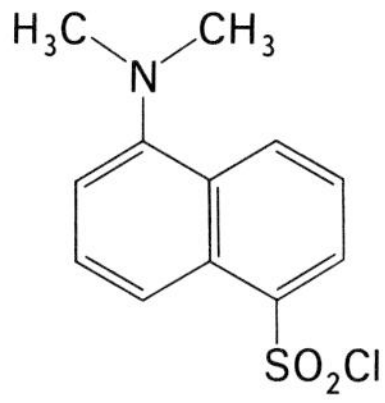

Dansyl
chloride

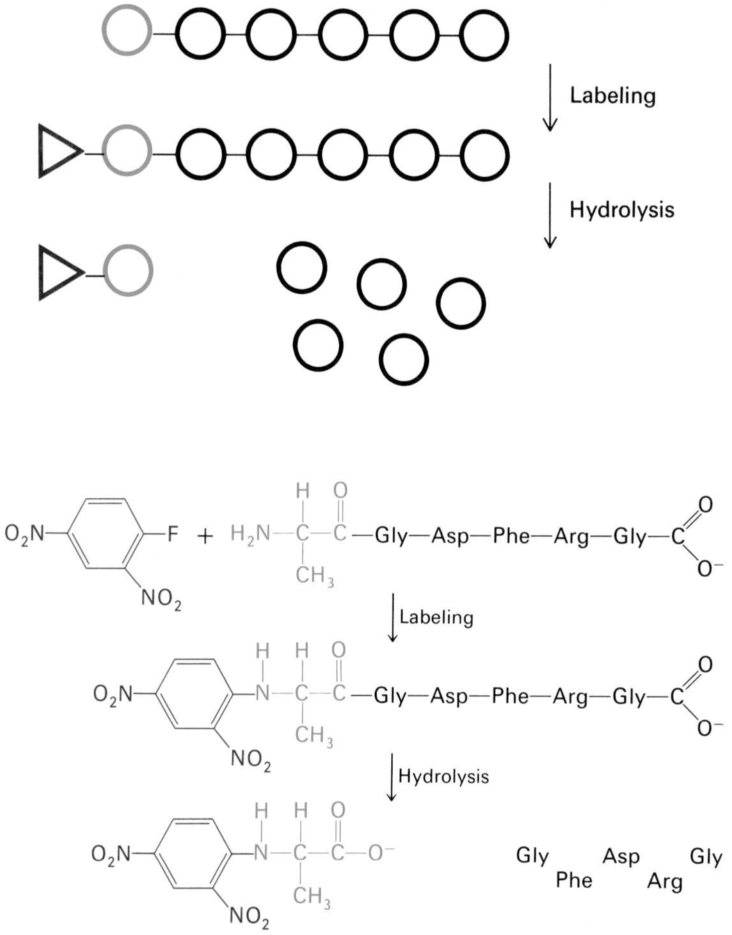

Figure 2-27
Determination of the amino-terminal residue of a peptide. Fluorodinitrobenzene (Sanger's reagent) is used to label the peptide, which is then hydrolyzed. The DNP-amino acid (DNP-alanine in this example) is identified by its chromatographic characteristics.

EDMAN DEGRADATION

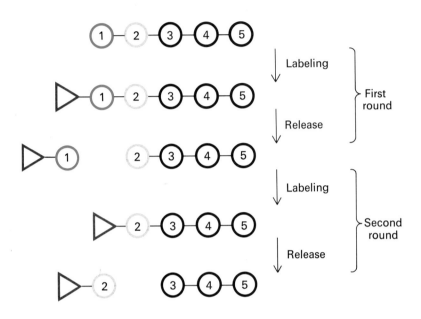

Figure 2-28
The Edman degradation. The labeled amino-terminal residue (PTH-alanine in the first round) can be released without hydrolyzing the rest of the peptide. Hence, the amino-terminal residue of the shortened peptide (Gly-Asp-Phe-Arg-Gly) can be determined in the second round. Three more rounds of the Edman degradation reveal the complete sequence of the original peptide.

graded in the acid-hydrolysis step. Pehr Edman devised a method for labeling the amino-terminal residue and cleaving it from the peptide without disrupting the peptide bonds between the other amino acid residues. The *Edman degradation* sequentially removes one residue at a time from the amino end of a peptide (Figure 2-28). *Phenyl isothiocyanate* reacts with the uncharged terminal amino group of the peptide to form a phenylthiocarbamoyl derivative. Under mildly acidic conditions, a cyclic derivative of the terminal amino acid is liberated, which leaves an intact peptide shortened by one amino acid. The cyclic compound is a phenylthiohydantoin (PTH) amino acid. The PTH-amino acid can be identified by chromatographic procedures. Furthermore, the amino acid composition of the shortened peptide:

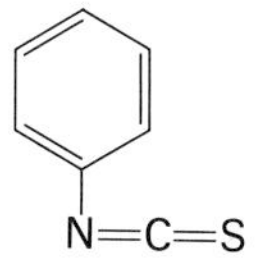

$$(Arg, Asp, Gly_2, Phe)$$

can be compared with that of the original peptide:

$$(Ala, Arg, Asp, Gly_2, Phe)$$

The difference between these analyses is one alanine residue, which shows that alanine is the amino-terminal residue of the original peptide. The Edman procedure can then be repeated on the shortened peptide. The amino acid analysis after the second round of degradation is

$$(Arg, Asp, Gly, Phe)$$

showing that the second residue from the amino end is glycine. This conclusion can be confirmed by chromatographic identification of PTH-glycine obtained in the second round of the Edman degradation. Three more rounds of the Edman degradation will reveal the complete sequence of the original peptide.

The experimental strategy for determining the amino acid sequence of proteins is to divide and conquer. A protein is *specifically cleaved into, smaller peptides* that can be readily sequenced by the Edman method. Specific cleavage can be achieved by chemical or enzymatic methods. For example, Bernhard Witkop and Erhard Gross discovered that cyanogen bromide (CNBr) splits the polypeptide chain only on the carboxyl side of methionine residues (Figure 2-29). A protein that has ten methionines will usually yield eleven

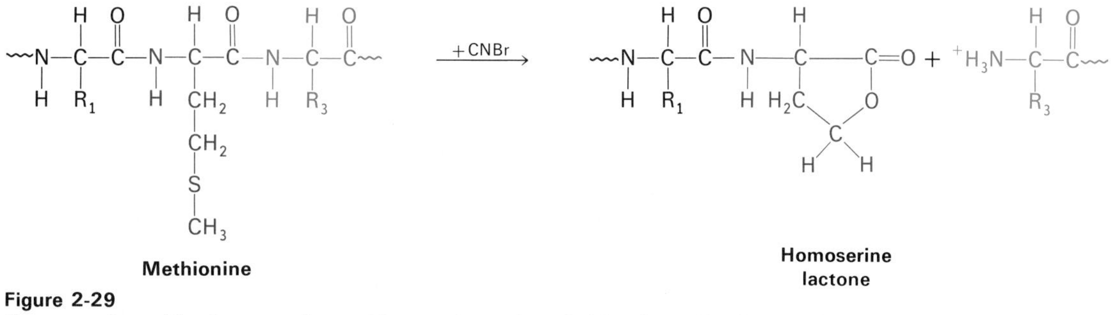

Phenyl isothiocyanate

Figure 2-29
Cyanogen bromide cleaves polypeptides on the carboxyl side of methionine residues.

peptides on cleavage with CNBr. Highly specific cleavage is also obtained with trypsin, a proteolytic enzyme from intestinal juice. Trypsin cleaves polypeptide chains on the carboxyl side of arginine and lysine residues (Figure 2-30). A protein that contains nine

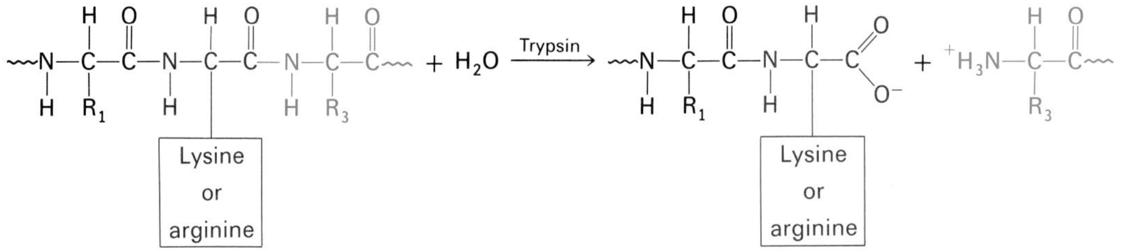

Figure 2-30
Trypsin hydrolyzes polypeptides on the carboxyl side of arginine and lysine residues.

lysines and seven arginines will usually yield seventeen peptides on digestion with trypsin. Each of these tryptic peptides, except for the carboxyl-terminal peptide of the protein, will end with either arginine or lysine. Several other ways of specifically cleaving polypeptide chains are given in Table 2-2.

Table 2-2
Specific cleavage of polypeptides

Reagent	Cleavage site
Chemical cleavage	
Cyanogen bromide	Carboxyl side of methionine residues
Hydroxylamine	Asparagine–glycine bonds
2-Nitro-5-thiocyanobenzoate	Amino side of cysteine residues
Enzymatic cleavage	
Trypsin	Carboxyl side of lysine and arginine residues
Clostripain	Carboxyl side of arginine residues
Staphylococcal protease	Carboxyl side of aspartate and glutamate residues (glutamate only under certain conditions)

The peptides obtained by specific chemical or enzymatic cleavage are separated by chromatographic methods. The sequence of each purified peptide is then determined by the Edman method. At this point, the amino acid sequences of segments of the protein are known, but the order of these segments is not yet defined. The necessary additional information is obtained from what are called *overlap peptides* (Figure 2-31). An enzyme different from trypsin is

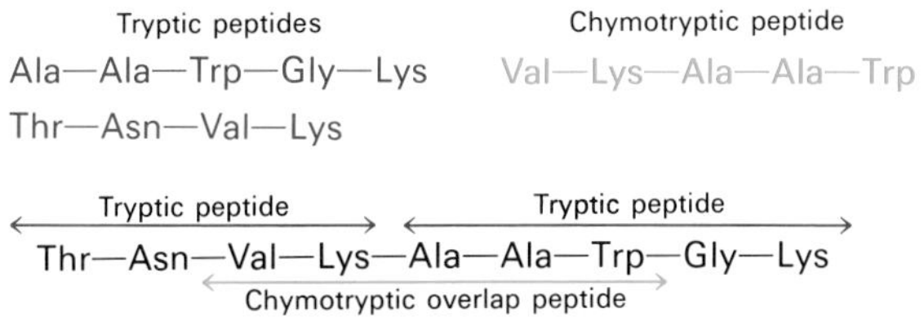

Tryptic peptides   Chymotryptic peptide

Ala—Ala—Trp—Gly—Lys   Val—Lys—Ala—Ala—Trp

Thr—Asn—Val—Lys

← Tryptic peptide → ← Tryptic peptide →

Thr—Asn—Val—Lys—Ala—Ala—Trp—Gly—Lys

← Chymotryptic overlap peptide →

Figure 2-31
The peptide obtained by chymotryptic digestion overlaps two tryptic peptides, which thus establishes their order.

used to split the polypeptide chain at different linkages. For example, chymotrypsin cleaves preferentially on the carboxyl side of aromatic and other bulky nonpolar residues. Because these chymotryptic peptides overlap two or more tryptic peptides, they can be used to establish their order. The entire amino acid sequence of the protein is then determined.

These methods apply to a protein that consists of a single polypeptide chain without any disulfide bonds. Additional steps in the elucidation of sequence are necessary if a protein has disulfide bonds or more than one chain. For a protein made up of two or more polypeptide chains held together by noncovalent bonds, denaturing agents, such as urea or guanidine hydrochloride, are used to dissociate the chains. The dissociated chains must be separated before sequence determination can begin. For polypeptide chains that are covalently linked by disulfide bonds, as in insulin, oxidation with performic acid is used to cleave the disulfide bonds, yielding cysteic acid residues (Figure 2-32).

Analyses of protein structures have been markedly accelerated by the development of the *sequenator,* an instrument for the automatic determination of amino acid sequence. A thin film of protein in a spinning cylindrical cup is subjected to the Edman degradation. The reagents and extracting solvents are passed over the immobilized film of protein, and the released PTH-amino acid is identified by high-pressure liquid chromatography. One cycle of the Edman degradation is carried out in less than two hours. The sequenator can determine the amino acid sequence of a polypeptide or a protein containing as many as a hundred residues.

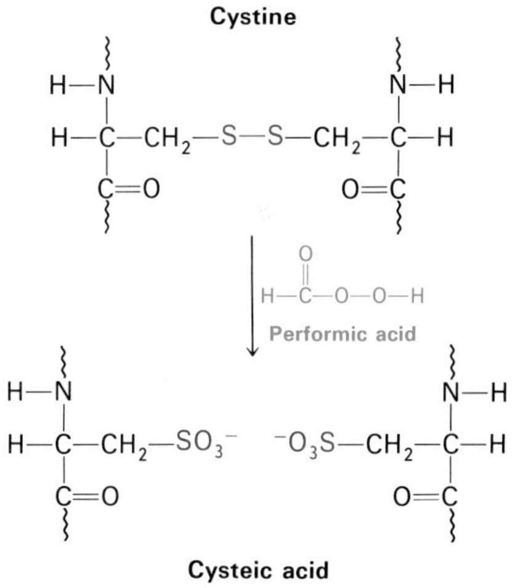

Figure 2-32
Performic acid cleaves disulfides.

CONFORMATION OF POLYPEPTIDE CHAINS

A striking characteristic of proteins is that they have well-defined three-dimensional structures. A stretched-out or randomly arranged polypeptide chain is devoid of biological activity, as will be discussed shortly. Function arises from *conformation,* which is the three-dimensional arrangement of atoms in a structure. Amino acid sequences are important because they specify the conformation of proteins.

In the late 1930s, Linus Pauling and Robert Corey initiated x-ray crystallographic studies of the precise structure of amino acids and peptides. Their aim was to obtain a set of standard bond distances

Angstrom (Å)—
A unit of length equal to 10^{-10} meter.
$1\ \text{Å} = 10^{-10}\ \text{m} = 10^{-8}\ \text{cm}$
$= 10^{-4}\ \mu\text{m} = 10^{-1}\ \text{nm}$
Named after Anders J. Ångström (1814–1874), a spectroscopist.

and bond angles for these building blocks and then use this information to predict the conformation of proteins. One of their important findings was that *the peptide unit is rigid and planar*. The hydrogen of the substituted amino group is nearly always *trans* to the oxygen of the carbonyl group (Figure 2-33). There is no freedom of

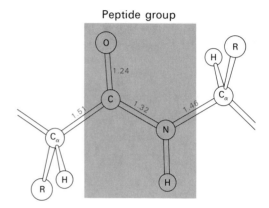

Peptide group

Figure 2-33
The peptide group is a rigid planar unit. Standard bond distances (in Å) are shown.

rotation about the bond between the carbonyl carbon atom and the nitrogen atom of the peptide unit because this link has partial double-bond character (Figure 2-34). The length of this bond is 1.32 Å, which is between that of a C—N single bond (1.49 Å) and a C=N double bond (1.27 Å). In contrast, the link between the α-carbon atom and a carbonyl carbon atom is a pure single bond. The bond between an α-carbon atom and a nitrogen atom also is a pure single bond. Consequently, *there is a large degree of rotational freedom about these bonds on either side of the rigid peptide unit* (Figure 2-35). Rotations about these bonds are designated by the angles ψ and ϕ (Figure 2-36). The conformation of the main chain of a polypeptide is fully defined when ψ and ϕ for each amino acid residue are known.

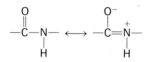

Figure 2-34
The peptide group is planar because the carbon–nitrogen bond has partial double-bond character.

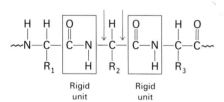

Rigid unit   Rigid unit

Figure 2-35
There is considerable freedom of rotation about the bonds joining the peptide groups to the α-carbon atoms.

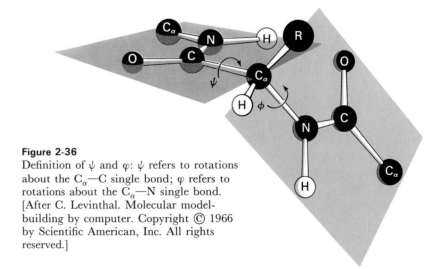

Figure 2-36
Definition of ψ and φ: ψ refers to rotations about the C_α—C single bond; φ refers to rotations about the C_α—N single bond. [After C. Levinthal. Molecular model-building by computer. Copyright © 1966 by Scientific American, Inc. All rights reserved.]

PERIODIC STRUCTURES: THE ALPHA HELIX, BETA PLEATED SHEET, AND COLLAGEN HELIX

Can a polypeptide chain fold into a regularly repeating structure? To answer this question, Pauling and Corey evaluated a variety of potential polypeptide conformations by building precise molecular models of them. They adhered closely to the experimentally observed bond angles and distances for amino acids and small peptides. In 1951, they proposed two periodic polypeptide structures, called the α helix and β pleated sheet.

The α *helix* is a rodlike structure. The tightly coiled polypeptide main chain forms the inner part of the rod, and the side chains extend outward in a helical array (Figures 2-37 and 2-38). The α

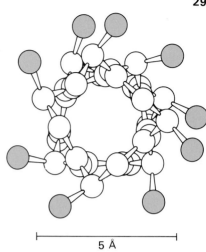

5 Å

Figure 2-38
Cross-sectional view of an α helix. Note that the side chains (shown in green) are on the outside of the helix. The van der Waals radii of the atoms are larger than shown here; hence there is almost no free space inside the helix.

Figure 2-37
Models of a right-handed α helix: (A) only the α-carbon atoms are shown on a helical thread; (B) only the backbone nitrogen (N), α-carbon (C_α), and carbonyl carbon (C) atoms are shown; (C) entire helix. Hydrogen bonds (denoted in part C by · · · in red) between NH and CO groups stabilize the helix.

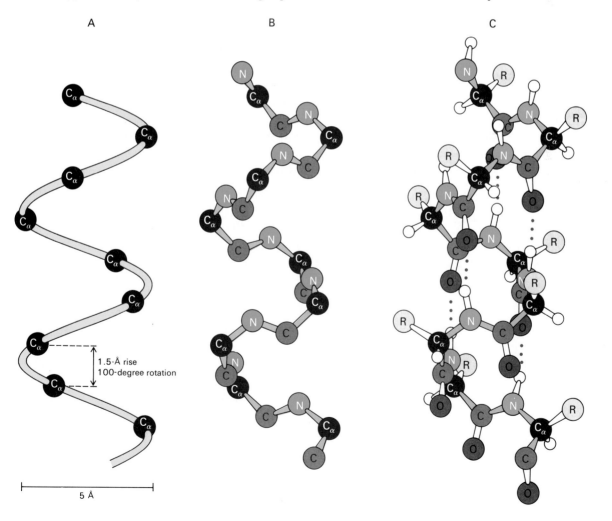

A

B

C

1.5-Å rise
100-degree rotation

5 Å

helix is stabilized by hydrogen bonds between the NH and CO groups of the main chain. The CO group of each amino acid is hydrogen bonded to the NH group of the amino acid that is situated four residues ahead in the linear sequence (Figure 2-39). Thus,

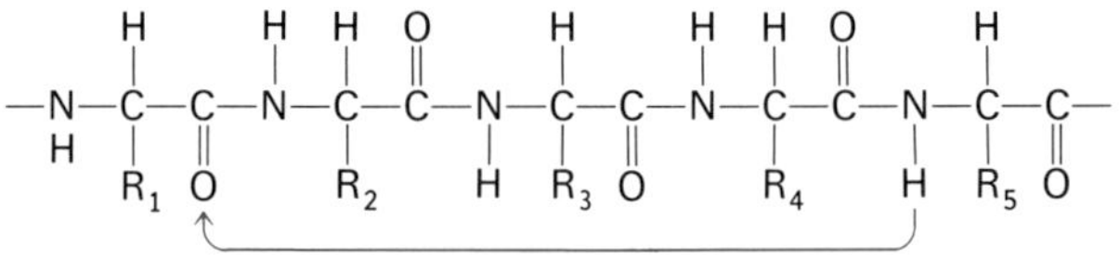

Figure 2-39
In the α helix, the NH group of residue n is hydrogen bonded to the CO group of residue $(n - 4)$.

"When we consider that the fibrous proteins of the epidermis, the keratinous tissues, the chief muscle protein, myosin, and now the fibrinogen of the blood all spring from the same peculiar shape of molecule, and are therefore probably all adaptations of a single root idea, we seem to glimpse one of the great coordinating facts in the lineage of biological molecules."

K. BAILEY, W. T. ASTBURY,
AND K. M. RUDALL
Nature, 1943

all the main-chain CO and NH groups are hydrogen bonded. Each residue is related to the next one by a translation of 1.5 Å along the helix axis and a rotation of 100°, which gives 3.6 amino acid residues per turn of helix. Thus, amino acids spaced three and four apart in the linear sequence are spatially quite close to one another in an α helix. In contrast, amino acids two apart in the linear sequence are situated on opposite sides of the helix and so are unlikely to make contact. The pitch of the α helix is 5.4 Å, the product of the translation (1.5 Å) and the number of residues per turn (3.6). The screw-sense of a helix can be right-handed (clockwise) or left-handed (counterclockwise); the α helices found in proteins are right-handed.

The α-helix content of proteins of known three-dimensional structure is highly variable. In some, such as myoglobin and hemoglobin, the α helix is the major structural motif. Other proteins, such as the digestive enzyme chymotrypsin, are virtually devoid of α helix. The single-stranded α helix discussed above is usually a rather short rod, typically less than 40 Å in length. A variation of the α-helical theme is used to construct much longer rods, extending to 1000 Å or more. Two or more α helices can entwine around each other to form a cable. Such α-*helical coiled coils* are found in several proteins: keratin in hair, myosin and tropomyosin in muscle, epidermin in skin, and fibrin in blood clots. The helical cables in these proteins serve a mechanical role in forming stiff bundles of fibers.

The structure of the α helix was deduced by Pauling and Corey six years before it was actually to be seen in the x-ray reconstruction of the structure of myoglobin. *The elucidation of the structure of the α helix is a landmark in molecular biology because it demonstrated that the conformation of a polypeptide chain can be predicted if the properties of its components are rigorously and precisely known.*

In the same year, Pauling and Corey discovered another periodic structural motif, which they named the β pleated sheet (β because it was the second structure they elucidated, the α helix having been the first). The β *pleated sheet* differs markedly from the α helix in that it is a sheet rather than a rod. The polypeptide chain in the β pleated sheet is almost fully extended (Figure 2-40) rather than

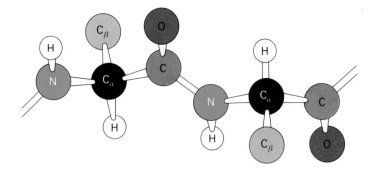

Figure 2-40
Conformation of a dipeptide unit in a β pleated sheet. The polypeptide chain is almost fully stretched out.

being tightly coiled as in the α helix. The axial distance between adjacent amino acids is 3.5 Å, in contrast with 1.5 Å for the α helix. Another difference is that the β pleated sheet is stabilized by hydrogen bonds between NH and CO groups in *different* polypeptide strands, whereas in the α helix the hydrogen bonds are between NH and CO groups in the *same* polypeptide chain. Adjacent strands in a β pleated sheet can run in the same direction (*parallel β sheet*) or in opposite directions (*antiparallel β sheet*). For example, silk fibroin consists almost entirely of stacks of antiparallel β sheets (Figure 2-41). Such β-sheet regions are a recurring structural motif in many proteins. Structural units comprising from two to five parallel or antiparallel β strands are especially common.

The *collagen helix,* a third periodic structure, will be discussed in detail in Chapter 9. This specialized structure is responsible for the high tensile strength of collagen, the major component of skin, bone, and tendon.

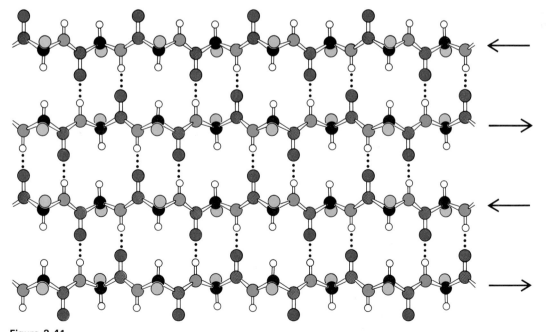

Figure 2-41
Antiparallel β pleated sheet. Adjacent strands run in opposite directions. Hydrogen bonds between NH and CO groups of adjacent strands stabilize the structure. The side chains (shown in green) are above and below the plane of the sheet.

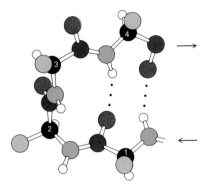

Figure 2-42
Structure of a β-turn. The CO group of residue 1 of the tetrapeptide shown here is hydrogen bonded to the NH group of residue 4, which results in a hairpin turn.

Performic acid

POLYPEPTIDE CHAINS CAN REVERSE DIRECTION BY MAKING β-TURNS

Most proteins have compact, globular shapes due to frequent reversals of the direction of their polypeptide chains. Analyses of the three-dimensional structures of numerous proteins have revealed that many of these chain reversals are accomplished by a common structural element called the *β-turn*. The essence of this hairpin turn is that the CO group of residue n of a polypeptide is hydrogen bonded to the NH group of residue $(n + 3)$ (Figure 2-42). Thus, a polypeptide chain can abruptly reverse its direction.

LEVELS OF STRUCTURE IN PROTEIN ARCHITECTURE

In discussing the architecture of proteins, it is convenient to refer to four levels of structure. *Primary structure* is simply the sequence of amino acids and location of disulfide bridges, if there are any. The primary structure is thus a complete description of the covalent connections of a protein. *Secondary structure* refers to the steric relationship of amino acid residues that are close to one another in the linear sequence. Some of these steric relationships are of a regular kind, giving rise to a periodic structure. The α helix, the β pleated sheet, and the collagen helix are examples of secondary structure. *Tertiary structure* refers to the steric relationship of amino acid residues that are far apart in the linear sequence. It should be noted that the dividing line between secondary and tertiary structure is arbitrary. Proteins that contain more than one polypeptide chain display an additional level of structural organization, namely *quaternary structure,* which refers to the way in which the chains are packed together. Each polypeptide chain in such a protein is called a *subunit.* Another useful term is *domain,* which refers to a compact, globular unit of protein structure. Many proteins fold into domains having masses that range from 10 to 20 kdal. The domains of large proteins are usually connected by relatively flexible regions of polypeptide chain.

AMINO ACID SEQUENCE SPECIFIES THREE-DIMENSIONAL STRUCTURE

Insight into the relation between the amino acid sequence of a protein and its conformation came from the work of Christian Anfinsen on ribonuclease, an enzyme that hydrolyzes RNA. Ribonuclease is a single polypeptide chain consisting of 124 amino acid residues (Figure 2-43). It contains four disulfide bonds, which can be irreversibly oxidized by *performic acid* to give cysteic acid residues

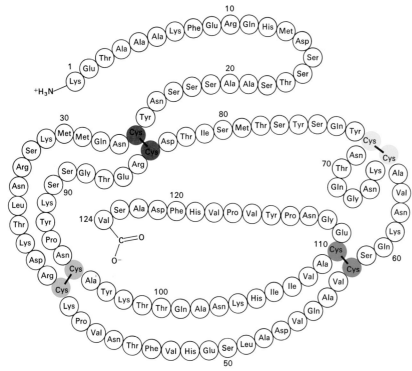

Figure 2-43
Amino acid sequence of bovine ribonuclease. The four disulfide bonds are shown in color. [After C. H. W. Hirs, S. Moore, and W. H. Stein. *J. Biol. Chem.* 235(1960):633.]

(Figure 2-32). Alternatively, these disulfide bonds can be cleaved reversibly by reducing them with a reagent such as *β-mercaptoethanol*, which forms mixed disulfides with cysteine side chains (Figure 2-44). In the presence of a large excess of *β*-mercap-

$$HO-CH_2-CH_2-SH$$
β-Mercaptoethanol

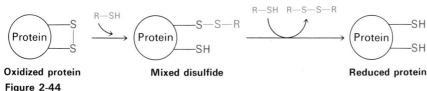

Oxidized protein   **Mixed disulfide**   **Reduced protein**

Figure 2-44
Reduction of the disulfide bonds in a protein by an excess of a sulfhydryl reagent such as *β*-mercaptoethanol.

toethanol, the mixed disulfides also are reduced, so that the final product is a protein in which the disulfides (cystines) are fully converted into sulfhydryls (cysteines). However, it was found that ribonuclease at 37°C and pH 7 cannot be readily reduced by *β*-mercaptoethanol unless the protein is partially unfolded by denaturing agents such as *urea* or *guanidine hydrochloride*. Although the mecha-

$$H_2N-\overset{\overset{\displaystyle O}{\|}}{C}-NH_2$$
Urea

$$H_2N-\overset{\overset{\displaystyle NH_2^+\ Cl^-}{\|}}{C}-NH_2$$
Guanidine hydrochloride

nism of action of these denaturing agents is not fully understood, it is evident that they disrupt noncovalent interactions. Polypeptide chains devoid of cross-links usually assume a *random-coil* conformation in 8 M urea or 6 M guanidine HCl, as evidenced by physical properties such as viscosity and optical rotary spectra. When ribonuclease was treated with β-mercaptoethanol in 8 M urea, the product was a fully reduced, randomly coiled polypeptide chain *devoid of enzymatic activity*. In other words, ribonuclease was *denatured* by this treatment (Figure 2-45).

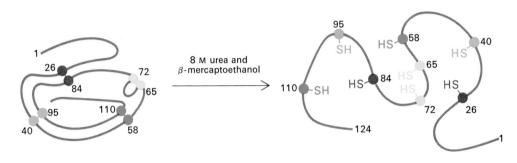

Native ribonuclease　　　　　　**Denatured reduced ribonuclease**

Figure 2-45
Reduction and denaturation of ribonuclease.

Anfinsen then made the critical observation that the denatured ribonuclease, freed of urea and β-mercaptoethanol by dialysis, slowly regained enzymatic activity. He immediately perceived the significance of this chance finding: the sulfhydryls of the denatured enzyme became oxidized by air and the enzyme spontaneously refolded into a catalytically active form. Detailed studies then showed that nearly all of the original enzymatic activity was regained if the sulfhydryls were oxidized under suitable conditions (Figure 2-46). All of the measured physical and chemical properties of the refolded enzyme were virtually identical with those of the native enzyme. These experiments showed that *the information needed to specify the complex three-dimensional structure of ribonuclease is contained in its amino acid sequence*. Subsequent studies of other proteins have

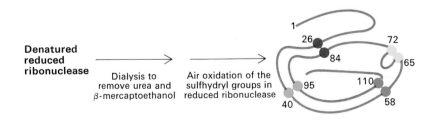

Native ribonuclease

Figure 2-46
Renaturation of ribonuclease.

established the generality of this principle, which is a central one in molecular biology: *sequence specifies conformation.*

A quite different result was obtained when reduced ribonuclease was reoxidized while it was still in 8 M urea. This preparation was then dialyzed to remove the urea. Ribonuclease reoxidized in this way had only 1% of the enzymatic activity of the native protein. Why was the outcome of this experiment different from the one in which reduced ribonuclease was reoxidized in a solution free of urea? The reason is that wrong disulfide pairings were formed when the random-coil form of the reduced molecule was reoxidized. There are 105 different ways of pairing eight cysteines to form four disulfides; only one of these combinations is enzymatically active. The 104 wrong pairings have been picturesquely termed "scrambled" ribonuclease. Anfinsen then found that "scrambled" ribonuclease spontaneously converted into fully active, native ribonuclease when trace amounts of β-mercaptoethanol were added to the aqueous solution of the reoxidized protein (Figure 2-47). The added

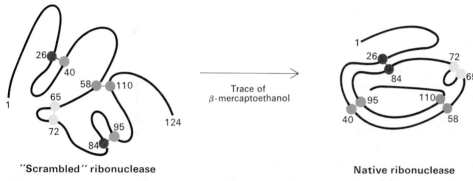

"Scrambled" ribonuclease　　　　　**Native ribonuclease**

Figure 2-47
Formation of native ribonuclease from "scrambled" ribonuclease in the presence of a trace of β-mercaptoethanol.

β-mercaptoethanol catalyzed the rearrangement of disulfide pairings until the native structure was regained, which took about ten hours. This process was driven entirely by the decrease in free energy as the "scrambled" conformations were converted into the stable, native conformation of the enzyme. *Thus, the native form of ribonuclease appears to be the thermodynamically most stable structure.*

Anfinsen (1964) wrote:

> It struck me recently that one should really consider the sequence of a protein molecule, about to fold into a precise geometric form, as a line of melody written in canon form and so designed by Nature to fold back upon itself, creating harmonic chords of interaction consistent with biological function. One might carry the analogy further by suggesting that the kinds of chords formed in a protein with scrambled disulfide bridges, such as I mentioned earlier, are dissonant, but that, by giving an opportunity for rearrangement by the addition of mercaptoethanol, they modulate to give the pleasing harmonics of the

native molecule. Whether or not some conclusion can be drawn about the greater thermodynamic stability of Mozart's over Schoenberg's music is something I will leave to the philosophers of the audience.

PROTEINS FOLD BY THE ASSOCIATION OF α-HELICAL AND β-STRAND SEGMENTS

How are the harmonic chords of interaction created in the conversion of an unfolded polypeptide chain into a folded protein? One possibility a priori is that all possible conformations are searched to find the energetically most favorable form. How long would such a random search take? Consider a small protein with 100 residues. If each residue can assume three different conformations, the total number of structures is 3^{100}, which is equal to 5×10^{47}. If it takes 10^{-13} seconds to convert one structure into another, the total search time would be $5 \times 10^{47} \times 10^{-13}$ seconds, which is equal to 5×10^{34} seconds, or 1.6×10^{27} years! Note that this length of time is a minimal estimate because the actual number of possible conformations per residue is greater than three and the time that it takes to change from one conformation into another is probably considerably longer than 10^{-13} seconds. Clearly, it would take much too long for even a small protein to fold by randomly trying out all possible conformations to determine which one is energetically best.

How, then, do proteins fold in a few seconds or minutes? The answer is not yet known, but a plausible hypothesis is that *small stretches of secondary structure serve as intermediates in the folding process.* According to this model, short segments (~15 residues) of an unfolded polypeptide chain flicker in and out of their native α-helical or β-sheet form. These transient structures find each other by diffusion and stabilize each other by forming a complex (Figure 2-48). For example, two α helices, two β strands, or an α helix and a β strand may come together. These αα, ββ, and αβ complexes, which are called *folding units,* then act as nuclei to stabilize other flickering elements of secondary structure. This model is supported by several lines of experimental evidence. The first is that the tendency of a polypeptide to adopt a regular secondary structure depends to a large degree on its amino acid composition. The formation of an α helix is favored by glutamate, methionine, alanine, and leucine residues, whereas β-sheet formation is enhanced by valine, isoleucine, and tyrosine residues. Second, the transition from a random coil to an α helix can occur in less than a microsecond. Thus, short segments of secondary structure can be formed very rapidly. Third, the postulated folding units (αα, ββ, and αβ complexes) are in fact major elements in protein structure. The challenge now is to directly detect and identify the evanescent intermediates in folding and recreate the pathway that gives form and function to the polypeptide chain.

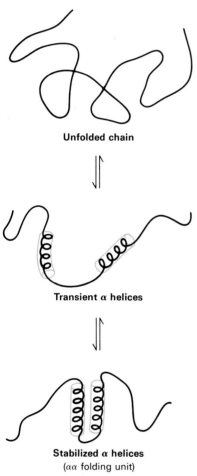

Unfolded chain

Transient α helices

Stabilized α helices
(αα folding unit)

Figure 2-48
Postulated step in protein folding. Two segments of an unfolded polypeptide chain transiently become α helical. These helices are then stabilized by the formation of a complex between the two segments.

Proteins play key roles in nearly all biological processes. All enzymes, the catalysts of chemical reactions in biological systems, are proteins. Hence, proteins determine the pattern of chemical transformations in cells. Proteins mediate a wide range of other functions, such as transport and storage, coordinated motions, mechanical support, immune protection, excitability, and the control of growth and differentiation.

The basic structural units of proteins are amino acids. All proteins in all species from bacteria to humans are constructed from the same set of twenty amino acids. The side chains of these buildings blocks differ in size, shape, charge, hydrogen-bonding capacity, and chemical reactivity. They can be grouped as follows: (a) aliphatic side chains—glycine, alanine, valine, leucine, isoleucine, and proline; (b) hydroxyl aliphatic side chains—serine and threonine; (c) aromatic side chains—phenylalanine, tyrosine, and tryptophan; (d) basic side chains—lysine, arginine, and histidine; (e) acidic side chains—aspartic acid and glutamic acid; (f) amide side chains—asparagine and glutamine; and (g) sulfur side chains—cysteine and methionine.

Many amino acids, usually more than a hundred, are joined by peptide bonds to form a polypeptide chain. A peptide bond links the α-carboxyl group of one amino acid and the α-amino group of the next one. In some proteins, a few side chains are cross-linked by disulfide bonds, which result from the oxidation of cysteine residues. A protein consists of one or more polypeptide chains. Each kind of protein has a unique amino acid sequence that is genetically determined. The amino acid sequence of a protein is elucidated in the following way. First, its amino acid composition is determined by ion-exchange chromatography of an acid hydrolysate of the protein. The amino-terminal residue is identified by using an end-group reagent such as dansyl chloride. Second, the protein is specifically cleaved into small peptides. For example, trypsin hydrolyzes proteins on the carboxyl side of lysine and arginine residues. Third, the amino acid sequence of these peptides is then determined by the Edman technique, which successively removes the amino-terminal residue. Finally, the order of the peptides is established from the amino acid sequences of overlap peptides.

The critical determinant of the biological function of a protein is its conformation, which is defined as the three-dimensional arrangement of the atoms of a molecule. Three regularly repeating conformations of polypeptide chains are known: the α helix, the β pleated sheet, and the collagen helix. Short segments of the α helix and the β pleated sheet are found in many proteins. An important principle is that the amino acid sequence of a protein specifies its three-dimensional structure, as was first shown for ribonuclease. Reduced, unfolded ribonuclease spontaneously forms the correct disulfide pairings and regains full enzymatic activity when oxidized by air following removal of mercaptoethanol and urea. Proteins fold by the association of short polypeptide segments that transiently adopt α-helical or β-sheet forms.

SELECTED READINGS

WHERE TO START

Dickerson, R. E. and Geis, I., forthcoming. *Proteins: Structure, Function, and Evolution.* Benjamin. [A lucid and well-illustrated introduction to proteins.]

Moore, S., and Stein, W. H., 1973. Chemical structures of pancreatic ribonuclease and deoxyribonuclease. *Science* 180:458–464.

Anfinsen, C. B., 1973. Principles that govern the folding of protein chains. *Science* 181:223–230.

BOOKS ON PROTEIN CHEMISTRY

Cantor, C. R., and Schimmel, P. R., 1980. *Biophysical Chemistry.* Freeman. [Chapters 2 and 5 in Part I and Chapters 20 and 21 in Part III give an excellent account of the principles of protein conformation.]

Cooper, T. G., 1977. *The Tools of Biochemistry.* Wiley. [A valuable guide to experimental methods in protein chemistry and other facets of biochemistry. Principles of procedures are clearly presented.]

Neurath, H., and Hill, R. L., (eds.), 1976. *The Proteins* (3rd ed.). Academic Press. [A multivolume treatise that contains many fine articles.]

Haschemeyer, R. H., and Haschemeyer, A. E. V., 1973. *Proteins: A Guide to Study by Physical and Chemical Methods.* Wiley-Interscience.

Hirs, C. H. W., and Timasheff, S. N., (eds.), 1977. *Enzyme Structure,* Part E. Methods in Enzymology, vol. 47. Academic Press. [An excellent collection of authoritative articles on amino acid analysis, end-group methods, chemical and enzymatic cleavage, peptide separation, sequence analysis, chemical modification, and peptide synthesis.]

Schultz, G. E., and Schirmer, R. H., 1979. *Principles of Protein Structure.* Springer-Verlag.

AMINO ACID SEQUENCE DETERMINATION

Konigsberg, W. H., and Steinman, H. M., 1977. Strategy and methods of sequence analysis. *In* Neurath, H., and Hill, R. L., (eds.), *The Proteins* (3rd ed.), vol. 3, pp. 1–178. Academic Press.

Niall, H., 1977. Automated methods for sequence analysis. *In* Neurath, H., and Hill, R. L., (eds.), *The Proteins* (3rd ed.), vol. 3, pp. 179–238. Academic Press.

Needleman, S. B., (ed.), 1977. *Advanced Methods in Protein Sequence Determination.* Springer-Verlag.

Dayhoff, M. O., (ed.), 1972. *Atlas of Protein Sequence and Structure.* National Biomedical Research Foundation, Silver Springs, Maryland. [A valuable compilation of all known amino acid sequences. This atlas is revised every few years.]

COVALENT MODIFICATION OF PROTEINS

Uy, R., and Wold, F., 1977. Post-translational covalent modification of proteins. *Science* 198:890–896.

Hershko, A., and Fry, M., 1975. Post-translational cleavage of polypeptide chains: role in assembly. *Ann. Rev. Biochem.* 44:775–797.

Glazer, A. N., Delange, R. J., and Sigman, D. S., 1975. *Chemical Modification of Proteins.* North-Holland.

FOLDING OF PROTEINS

Creighton, T. E., 1978. Experimental studies of protein folding and unfolding. *Prog. Biophys. Mol. Biol.* 33:231–297.

Baldwin, R. L., 1975. Intermediates in protein folding reactions and the mechanism of protein folding. *Ann. Rev. Biochem.* 44:453–475.

Chou, P. Y., and Fasman, C. D., 1978. Empirical predictions of protein conformation. *Ann. Rev. Biochem.* 47:251–276.

Sternburg, M. J. E., and Thornton, J. M., 1978. Prediction of protein structure from amino acid sequence, *Nature* 271:15–20.

Karplus, M., and Weaver, D. L., 1976. Protein-folding dynamics. *Nature* 260:404–406.

Levitt, M., and Chothia, C., 1976. Structural patterns in globular proteins. *Nature* 261:552–558.

Chothia, C., 1975. Structural invariants in protein folding. *Nature* 254:304–308.

Anfinsen, C. B., 1964. On the possibility of predicting tertiary structure from primary sequence. *In* Sela, M., (ed.), *New Perspectives in Biology,* pp. 42–50. American Elsevier.

REVIEW ARTICLES

Advances in Protein Chemistry.
Annual Review of Biochemistry.
Annual Review of Biophysics and Bioengineering.

APPENDIX
Acid-Base Concepts

Ionization of Water

Water dissociates into hydronium (H_3O^+) and hydroxyl (OH^-) ions. For simplicity, we refer to the hydronium ion as a hydrogen ion (H^+) and write the equilibrium as

$$H_2O \rightleftharpoons H^+ + OH^-$$

The equilibrium constant K_{eq} of this dissociation is given by

$$K_{eq} = \frac{[H^+][OH^-]}{[H_2O]} \tag{1}$$

in which the terms in brackets denote molar concentrations. Because the concentration of water (55.5 M) is changed little by ionization, expression 1 can be simplified to give

$$K_w = [H^+][OH^-] \tag{2}$$

in which K_w is the ion product of water. At $25\,°C$, K_w is 1.0×10^{-14}.

Note that the concentrations of H^+ and OH^- are reciprocally related. If the concentration of H^+ is high, then the concentration of OH^- must be low, and vice versa. For example, if $[H^+] = 10^{-2}$ M, then $[OH^-] = 10^{-12}$ M.

Definition of Acid and Base

An acid is a proton donor. A base is a proton acceptor.

$$Acid \rightleftharpoons H^+ + Base$$
$$CH_3{-}COOH \rightleftharpoons H^+ + CH_3{-}COO^-$$

Acetic acid   Acetate

$$NH_4^+ \rightleftharpoons H^+ + NH_3$$

Ammonium ion   Ammonia

The species formed by the ionization of an acid is its conjugate base. Conversely, protonation of a base yields its conjugate acid. Acetic acid and acetate ion are a conjugate acid-base pair.

Definition of pH and pK

The pH of a solution is a measure of its concentration of H^+. The pH is defined as

$$pH = \log_{10}\frac{1}{[H^+]} = -\log_{10}[H^+] \tag{3}$$

The ionization equilibrium of a weak acid is given by

$$HA \rightleftharpoons H^+ + A^-$$

The apparent equilibrium constant K for this ionization is

$$K = \frac{[H^+][A^-]}{[HA]} \tag{4}$$

The pK of an acid is defined as

$$pK = -\log K = \log\frac{1}{K} \tag{5}$$

Inspection of equation 4 shows that *the pK of an acid is the pH at which it is half-dissociated.*

Henderson-Hasselbalch Equation

What is the relationship between pH and the ratio of acid to base? A useful expression can be derived starting with equation 4. Rearrangement of that equation gives

$$\frac{1}{[H^+]} = \frac{1}{K}\frac{[A^-]}{[HA]} \tag{6}$$

Taking the logarithm of both sides of equation 6 gives

$$\log\frac{1}{[H^+]} = \log\frac{1}{K} + \log\frac{[A^-]}{[HA]} \tag{7}$$

Substituting pH for $\log 1/[H^+]$ and pK for $\log 1/K$ in equation 7 yields

$$pH = pK + \log\frac{[A^-]}{[HA]} \tag{8}$$

which is commonly known as the Henderson-Hasselbalch equation.

The pH of a solution can be calculated from equation 8 if the molar proportion of A⁻ to HA and the pK of HA are known. Consider a solution of 0.1 M acetic acid and 0.2 M acetate ion. The pK of acetic acid is 4.8. Hence, the pH of the solution is given by

$$pH = 4.8 + \log \frac{0.2}{0.1} = 4.8 + \log 2$$

$$= 4.8 + 0.3 = 5.1$$

Conversely, the pK of an acid can be calculated if the molar proportion of A⁻ to HA and the pH of the solution are known.

Buffering Power

An acid-base conjugate pair (such as acetic acid and acetate ion) has an important property: it resists changes in the pH of a solution. In other words, it acts as a *buffer*. Consider the addition of OH⁻ to a solution of acetic acid (HA):

$$HA + OH^- \rightleftharpoons A^- + H_2O$$

A plot of the dependence of the pH of this solution on the amount of OH⁻ added is called a *titration curve*. Note that there is an inflection point in the curve at pH 4.8, which is the pK of acetic acid.

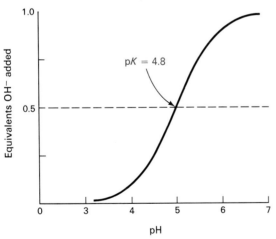

Figure 2-49
Titration curve of acetic acid.

In the vicinity of this pH, a relatively large amount of OH⁻ produces little change in pH. In general, a weak acid is most effective in buffering against pH changes in the vicinity of its pK value.

pK Values of Amino Acids

An amino acid such as glycine contains two ionizable groups: an α-carboxyl group and a protonated α-amino group. As base is added, these two groups are titrated (Figure 2-50). The pK of the

Figure 2-50
Titration of the ionizable groups of an amino acid.

α-COOH group is 2.3, whereas that of the α-NH₃⁺ group is 9.6. The pK values of these groups in other amino acids are similar. Some amino acids, such as aspartic acid, also contain an ionizable side chain. The pK values of ionizable side chains in amino acids range from 3.9 (aspartic acid) to 12.5 (arginine).

Table 2-3
pK values of some amino acids

Amino acid	pK values (25°C)		
	α-COOH group	α-NH₃⁺ group	Side chain
Alanine	2.3	9.9	
Glycine	2.4	9.8	
Phenylalanine	1.8	9.1	
Serine	2.1	9.2	
Valine	2.3	9.6	
Aspartic acid	2.0	10.0	3.9
Glutamic acid	2.2	9.7	4.3
Histidine	1.8	9.2	6.0
Cysteine	1.8	10.8	8.3
Tyrosine	2.2	9.1	10.9
Lysine	2.2	9.2	10.8
Arginine	1.8	9.0	12.5

After J. T. Edsall and J. Wyman. *Biophysical Chemistry* (Academic Press, 1958), ch. 8.

PROBLEMS

1. The following reagents are often used in protein chemistry:

CNBr   Dansyl chloride
Urea   6 N HCl
β-Mercaptoethanol   Ninhydrin
Trypsin   Phenyl isothiocyanate
Performic acid   Chymotrypsin

Which one is the best suited for accomplishing each of the following tasks?

(a) Determination of the amino acid sequence of a small peptide.

(b) Identification of the amino-terminal residue of a peptide (of which you have less than 10^{-7} g).

(c) Reversible denaturation of a protein devoid of disulfide bonds. Which additional reagent would you need if disulfide bonds were present?

(d) Hydrolysis of peptide bonds on the carboxyl side of aromatic residues.

(e) Cleavage of peptide bonds on the carboxyl side of methionines.

(f) Hydrolysis of peptide bonds on the carboxyl side of lysine and arginine residues.

2. What is the pH of each of the following solutions?

(a) 10^{-3} N HCl.

(b) 10^{-2} N NaOH.

(c) A mixture of equal volumes of 0.1 M acetic acid and 0.03 M sodium acetate.

(d) A mixture of equal volumes of 0.1 M glycine and 0.05 M NaOH.

(e) A mixture of equal volumes of 0.1 M glycine and 0.05 M HCl.

3. What is the ratio of base to acid at pH 4, 5, 6, 7, and 8 for an acid with a pK of 6?

4. Tropomyosin, a muscle protein, is a two-stranded α-helical coiled coil. The protein's mass is 70 kdal. The average residue is about 110 dal. What is the length of the molecule?

5. Anhydrous hydrazine has been used to cleave peptide bonds in proteins. What are the reaction products? How might this technique be used to identify the carboxyl-terminal amino acid?

6. The amino acid sequence of human adrenocorticotropin, a polypeptide hormone, is

Ser-Tyr-Ser-Met-Glu-His-Phe-Arg-Trp-Gly-Lys-Pro-Val-Gly-Lys-Lys-Arg-Arg-Pro-Val-Lys-Val-Tyr-Pro-Asp-Ala-Gly-Glu-Asp-Gln-Ser-Ala-Glu-Ala-Phe-Pro-Leu-Glu-Phe

(a) What is the approximate net charge of this molecule at pH 7? Assume that its side chains have the pK values given in Table 2-3 (p. 40) and that the pKs of the terminal —NH$_3^+$ and —COOH groups are 7.8 and 3.6, respectively.

(b) How many peptides result from the treatment of the hormone with cyanogen bromide?

7. Ethyleneimine reacts with cysteine side chains in proteins to form S-aminoethyl derivatives. The peptide bonds on the carboxyl side of these modified cysteine residues are susceptible to hydrolysis by trypsin. Why?

8. An enzyme that catalyzes disulfide-sulfhydryl exchange reactions has been isolated. Inactive "scrambled" ribonuclease is rapidly converted into enzymatically active ribonuclease by this enzyme. In contrast, insulin is rapidly inactivated by this enzyme. What does this important observation imply about the relation between the amino acid sequence and the three-dimensional structure of insulin?

For additional problems, see W. B. Wood, J. H. Wilson, R. M. Benbow, and L. E. Hood, *Biochemistry: A Problems Approach* (Benjamin, 1974), ch. 2; and R. Montgomery and C. A. Swenson, *Quantitative Problems in the Biochemical Sciences,* 2nd ed. (Freeman, 1976), chs. 6, 7, and 8.

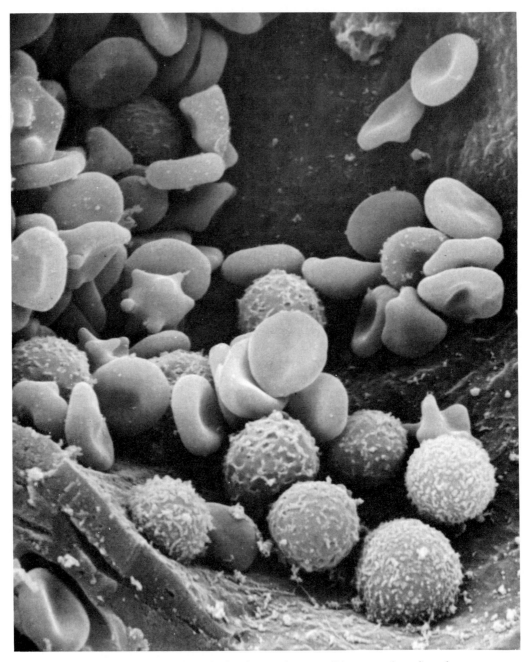

Scanning electron micrograph showing erythrocytes (biconcave-shaped) and leucocytes (rounded) in a small blood vessel. [From *Tissues and Organs* by R. G. Kessel and R. H. Kardon. W. H. Freeman and Company. Copyright © 1979.]

CHAPTER 3

OXYGEN TRANSPORTERS: MYOGLOBIN AND HEMOGLOBIN

The transition from anaerobic to aerobic life was a major step in evolution because it uncovered a rich reservoir of energy. Eighteen times as much energy is extracted from glucose in the presence of oxygen than in its absence. Vertebrates have evolved two principal mechanisms for supplying their cells with a continuous and adequate flow of oxygen. The first is a circulatory system that actively delivers oxygen to the cells. In the absence of a circulatory system, the size of an aerobic organism would be limited to about a millimeter because the diffusion of oxygen over larger dimensions would be too slow to meet cellular needs. The second major adaptation is the acquisition of *oxygen-carrying molecules* to overcome the limitation imposed by the low solubility of oxygen in water. The oxygen carriers in vertebrates are the proteins *hemoglobin* and *myoglobin*. Hemoglobin, which is contained in red blood cells, serves as the oxygen carrier in blood. The presence of hemoglobin increases the oxygen transporting capacity of a liter of blood from 5 to 250 ml of O_2. Hemoglobin also plays a vital role in the transport of carbon dioxide and hydrogen ion. Myoglobin, which is located in muscle, serves as a reserve supply of oxygen and facilitates the movement of oxygen within muscle.

Figure 3-1
Scanning electron micrograph of an erythrocyte. [Courtesy of Dr. Mark Goldman and Dr. Robert Leif.]

OXYGEN BINDS TO A HEME PROSTHETIC GROUP

The capacity of myoglobin or hemoglobin to bind oxygen depends on the presence of a nonpolypeptide unit, namely a *heme group*. The heme also gives myoglobin and hemoglobin their distinctive color. Indeed, many proteins require tightly bound, specific nonpolypeptide units for their biological activities. Such a unit is called a *prosthetic group*. A protein without its characteristic prosthetic group is termed an *apoprotein*.

The heme consists of an organic part and an iron atom. The organic part, *protoporphyrin*, is made up of four *pyrrole* groups. The four pyrroles are linked by methene bridges to form a tetrapyrrole ring. Four methyl, two vinyl, and two propionate side chains are attached to the tetrapyrrole ring. These substituents can be arranged in fifteen different ways. Only one of these isomers, called protoporphyrin IX, is present in biological systems.

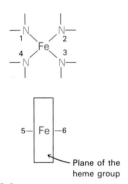

Figure 3-2
The iron atom in heme can form six bonds. Four of these bonds are in the plane of the heme. The fifth bond is on one side of this plane, and the sixth bond on the other side. The locations of the bonded atoms are sometimes called coordination positions.

Protoporphyrin IX

Heme
(Fe-protoporphyrin IX)

The iron atom in heme binds to the four nitrogens in the center of the protoporphyrin ring (Figures 3-2 and 3-3). The iron can form

Figure 3-3
The heme group in myoglobin (yellow for Fe, blue for N, red for O, and black for C).

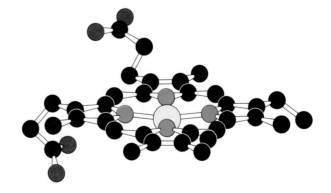

two additional bonds, one on either side of the heme plane. These bonding sites are termed the fifth and sixth coordination positions. The iron atom in the heme can be in the ferrous ($+2$) or the ferric ($+3$) oxidation state. The corresponding forms of hemoglobin are called *ferrohemoglobin* and *ferrihemoglobin,* respectively. Ferrihemoglobin is also called methemoglobin. Only ferrohemoglobin, the $+2$ oxidation state, can bind oxygen. The same nomenclature applies to myoglobin.

X-RAY CRYSTALLOGRAPHY REVEALS THREE-DIMENSIONAL STRUCTURE IN ATOMIC DETAIL

The elucidation of the three-dimensional structure of myoglobin by John Kendrew and of hemoglobin by Max Perutz are landmarks in molecular biology. These studies came to fruition in the late 1950s and showed that x-ray crystallography can reveal the structures of molecules as large as proteins. The largest structure solved before 1957 was Vitamin B_{12}, which is an order of magnitude smaller than myoglobin (17.8 kdal) and hemoglobin (66 kdal). The determination of the three-dimensional structures of these proteins was a great stimulus to the field of protein crystallography. The detailed three-dimensional structures of more than fifty proteins are now known, and elucidation of the structures of many others is in progress. *X-ray crystallography is making a profound contribution to our understanding of protein structure and function because it is the only method that can reveal the three-dimensional position of most of the atoms in a protein.* Electron microscopy can be highly informative about the structures of biological macromolecules, but the technique is not yet effective in delineating molecular architecture in atomic detail.

Before turning to the three-dimensional structures of myoglobin and hemoglobin, let us consider some basic aspects of the x-ray crystallographic method. First, crystals of the protein are needed. Myoglobin, for example, is crystallized by adding ammonium sulfate to a concentrated solution of the protein (Figure 3-4). Ammonium sulfate at a concentration of 3 M markedly reduces the solubility of myoglobin and leads to its crystallization. The solubility of

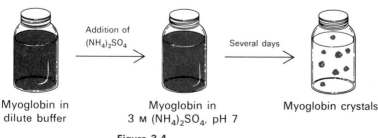

Myoglobin in
dilute buffer

Addition of
$(NH_4)_2SO_4$

Myoglobin in
3 M $(NH_4)_2SO_4$, pH 7

Several days

Myoglobin crystals

Figure 3-4
Crystallization of myoglobin.

most proteins is reduced on addition of high concentrations of any salt. This effect is called *salting-out*. Crystals of myoglobin (Figure 3-5) produced in this way may be several millimeters long.

The three components in an x-ray crystallographic experiment are a *source of x-rays,* a *protein crystal,* and a *detector* (Figure 3-6). A

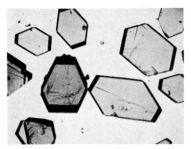

Figure 3-5
Photograph of myoglobin crystals.
[Courtesy of Dr. John Kendrew.]

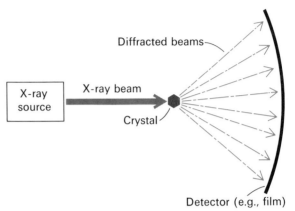

Figure 3-6
Essence of an x-ray crystallographic experiment: an x-ray beam, a crystal, and a detector.

beam of x-rays of wavelength 1.54 Å is produced by accelerating electrons against a copper target. A narrow beam of x-rays strikes the protein crystal. Part of the beam goes through the crystal without any change in direction. The rest is *scattered* in a variety of directions. The scattered (or *diffracted*) beams can be detected by photographic film, the blackening of the emulsion being proportional to the intensity of the scattered x-ray beam. The basic physical principles underlying the technique are:

1. Electrons scatter x-rays. The amplitude of the wave scattered by an atom is proportional to its number of electrons. Thus, a carbon atom scatters six times as strongly as a hydrogen atom.

2. The scattered waves recombine. Each atom contributes to each scattered beam. The scattered waves reinforce one another if they are in phase and they cancel one another if they are out of phase.

3. The way in which the scattered waves recombine depends only on the atomic arrangement.

STEPS IN THE X-RAY STUDY OF MYOGLOBIN

Kendrew chose myoglobin for x-ray analysis for several reasons: it is a rather small protein (17.8 kdal), it is easily prepared in quantity, and it is readily crystallized. Furthermore, myoglobin had the addi-

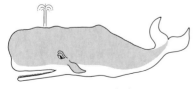

A sperm whale.

tional advantage of being closely related to hemoglobin, which was already being studied by his colleague Perutz. Myoglobin from the skeletal muscle of the sperm whale was chosen because it is stable and forms excellent crystals. The skeletal muscle of diving mammals, such as whale, seal, and porpoise, is particularly rich in myoglobin, which serves as a store of oxygen during a dive.

The myoglobin crystal is mounted in a capillary and positioned in a precise orientation with respect to the x-ray beam and the film. Precessional motion of the crystal yields an x-ray photograph in which there is a regular array of spots. The x-ray photograph shown in Figure 3-7 is a two-dimensional section through a three-dimensional array of spots. For myoglobin, there are 25,000 of these spots.

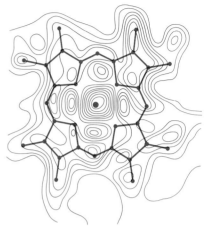

Figure 3-8
Section from the electron-density map of myoglobin showing the heme group. The peak of the center of this section corresponds to the position of the iron atom. [From J. C. Kendrew. The three-dimensional structure of a protein molecule. Copyright © 1961 by Scientific American, Inc. All rights reserved.]

Figure 3-7
X-ray precessional photograph of a myoglobin crystal.

The intensity of each spot is measured. These intensities are the basic experimental data of an x-ray crystallographic analysis.

The next step is to reconstruct an image of myoglobin from the observed intensities. This is accomplished by a mathematical relation called a Fourier series, which is a sum of sine and cosine terms. However, the scattered intensities of myoglobin crystals provide only part of the information needed to calculate the Fourier series. The essential missing data—namely, the phases of the scattered beams—are obtained from the complete diffraction patterns of myoglobin crystals that contain heavy atoms such as uranium and lead at one or two sites in the molecule. The stage is now set for the calculation of an electron-density map using high-speed computers. Fortunately, the development of fast computers with large memories kept pace with the needs of the myoglobin project. The final Fourier synthesis for myoglobin contained about a billion terms.

The Fourier synthesis gives the density of electrons at a large number of regularly spaced points in the crystal. This three-dimensional electron-density distribution is represented by a series of parallel sections stacked on top of each other. Each section is a transparent plastic sheet on which the electron-density distribution is represented by contour lines (Figure 3-8). This mode of representing electron density is analogous to the use of contour lines in geological survey maps to depict altitude (Figure 3-9). The next step is

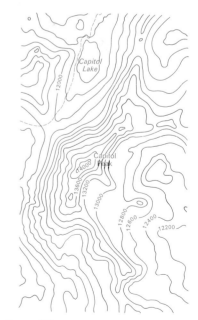

Figure 3-9
Section from a U.S. Geological Survey Map of the Capitol Peak Quadrangle, Colorado.

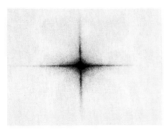

Figure 3-10
Effect of resolution on the quality of a reconstructed image, as shown by an optical analog of x-ray diffraction: (A) the Parthenon; (B) a diffraction pattern of the Parthenon; (C and D) images reconstructed from the pattern in part B. More data were used to obtain D than C, which accounts for the higher quality of the image in D. [Courtesy of Dr. Thomas Steitz (part A) and Dr. David De Rosier (part B).

to interpret the electron-density map. A critical factor is the *resolution* of the x-ray analysis, which is determined by the number of scattered intensities used in the Fourier synthesis. The fidelity of the image depends on the resolution of the Fourier synthesis, as shown by the optical analogy in Figure 3-10. The analysis of myoglobin was carried out in three stages. In the first, completed in 1957, only the innermost 400 spots of the diffraction pattern were used, corresponding to a resolution of 6 Å. As will be seen shortly, this low-resolution electron-density map revealed the course of the polypeptide chain but few other structural details. The reason is that polypeptide chains pack together so that their centers are between 5 and 10 Å apart. Maps at higher resolution are needed to delineate groups of atoms, which lie from 2.8 to 4.0 Å apart, and individual atoms, which are between 1.0 and 1.5 Å apart. Such maps of myoglobin were obtained in 1959 at a resolution of 2 Å (10,000 intensities) and in 1962 at a resolution of 1.4 Å (25,000 intensities). The ultimate resolution of an x-ray analysis is determined by the degree of perfection of the crystal. For proteins, this limiting resolution is usually not much better than 2 Å.

In 1957, Kendrew and his colleagues saw what no one had ever seen before: a three-dimensional picture of a protein molecule in all its complexity. The model derived from their 6-Å Fourier synthesis contained a set of high-density rods of just the dimensions expected for a polypeptide chain (Figure 3-11). The molecule appeared very compact. Closer examination showed that the molecule consisted of a complicated and intertwining set of these rods, going straight for a distance, then turning a corner and going off in a new direction. The location of the iron atom of the heme was also evident because it contains many more electrons than does any other atom in the structure. The most striking features of the molecule were its irregularity and its total lack of symmetry.

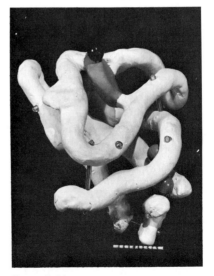

Figure 3-11
Model of myoglobin at low resolution. [Courtesy of Dr. John Kendrew.]

The high-resolution electron-density map of myoglobin, obtained two years later, contained a wealth of structural detail. The positions of 1200 of the 1260 nonhydrogen atoms were clearly defined to a precision of better than 0.3 Å. The course of the main chain and

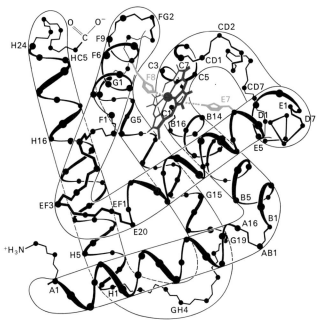

Figure 3-12
Model of myoglobin at high resolution. Only the α-carbon atoms are shown. The heme group is shown in red. [After R. E. Dickerson. In *The Proteins,* H. Neurath, ed., 2nd ed., vol. 2 (Academic Press, 1964), p. 634.]

Val-	Leu-	Ser-	Glu-	Gly-	Glu-	Trp-	Gln-	Leu-	Val -
NA1	NA2	A1	A2	A3	A4	A5	A6	A7	A8
Leu-	His-	Val-	Trp-	Ala-	Lys-	Val-	Glu-	Ala-	Asp-
A9	A10	A11	A12	A13	A14	A15	A16	AB1	B1
Val-	Ala-	Gly-	His-	Gly-	Gln-	Asp-	Ile-	Leu-	Ile -
B2	B3	B4	B5	B6	B7	B8	B9	B10	B11
Arg-	Leu-	Phe-	Lys-	Ser-	His-	Pro-	Glu-	Thr-	Leu-
B12	B13	B14	B15	B16	C1	C2	C3	C4	C5
Glu-	Lys-	Phe-	Asp-	Arg-	Phe-	Lys-	His-	Leu-	Lys-
C6	C7	CD1	CD2	CD3	CD4	CD5	CD6	CD7	CD8
Thr-	Glu-	Ala-	Glu-	Met-	Lys-	Ala-	Ser-	Glu-	Asp-
D1	D2	D3	D4	D5	D6	D7	E1	E2	E3
Leu-	Lys-	Lys-	His-	Gly-	Val-	Thr-	Val-	Leu-	Thr-
E4	E5	E6	E7	E8	E9	E10	E11	E12	E13
Ala-	Leu-	Gly-	Ala-	Ile-	Leu-	Lys-	Lys-	Lys-	Gly-
E14	E15	E16	E17	E18	E19	E20	EF1	EF2	EF3
His-	His-	Glu-	Ala-	Glu-	Leu-	Lys-	Pro-	Leu-	Ala-
EF4	EF5	EF6	EF7	EF8	F1	F2	F3	F4	F5
Gln-	Ser-	His-	Ala-	Thr-	Lys-	His-	Lys-	Ile-	Pro-
F6	F7	F8	F9	FG1	FG2	FG3	FG4	FG5	G1
Ile -	Lys-	Tyr-	Leu-	Glu-	Phe-	Ile-	Ser-	Glu-	Ala-
G2	G3	G4	G5	G6	G7	G8	G9	G10	G11
Ile -	Ile -	His-	Val-	Leu-	His-	Ser-	Arg-	His-	Pro-
G12	G13	G14	G15	G16	G17	G18	G19	GH1	GH2
Gly-	Asn-	Phe-	Gly-	Ala-	Asp-	Ala-	Gln-	Gly-	Ala-
GH3	GH4	GH5	GH6	H1	H2	H3	H4	H5	H6
Met-	Asn-	Lys-	Ala-	Leu-	Glu-	Leu-	Phe-	Arg-	Lys-
H7	H8	H9	H10	H11	H12	H13	H14	H15	H16
Asp-	Ile -	Ala-	Ala-	Lys-	Tyr-	Lys-	Glu-	Leu-	Gly-
H17	H18	H19	H20	H21	H22	H23	H24	HC1	HC2
Tyr-	Gln-	Gly							
HC3	HC4	HC5							

10
20
30
40
50
60
70
80
90
100
110
120
130
140
150
153

Figure 3-13
Amino acid sequence of sperm whale myoglobin. The label below each residue in the sequence refers to its position in an α-helical region or a nonhelical region. For example, B4 is the fourth residue in the B helix; EF7 is the seventh residue in the nonhelical region between the E and F helices. [After A. E. Edmundson. *Nature* 205 (1965):883; H. C. Watson. *Prog. Stereochem.* 4(1969):299–333.]

the position of the heme group are shown in Figure 3-12. Some important features of myoglobin are:

1. Myoglobin is *extremely compact*. The overall dimensions are about 45 × 35 × 25 Å. There is very little empty space inside.

2. About 75% of the main chain is folded in an *α-helical conformation*. All of the α helices are right-handed. There are eight major helical segments, referred to as A, B, C, . . . , H. The first residue in helix A is designated A1, the second A2, and so forth (Figure 3-13). There are five nonhelical segments between helical regions (named CD, e.g., if located between the C and D helices). There are two other nonhelical regions: the two residues at the amino-terminal end (named NA1 and NA2) and the five residues at the carboxyl-terminal end (named HC1 through HC5).

3. Not all the factors responsible for termination of helical segments are known. An important one is the occurrence of a proline

residue. *Proline cannot be accommodated in an α helix* (except at one of its ends). The five-membered pyrrolidone ring simply does not fit within a straight stretch of α helix. There are four prolines in myoglobin, whereas there are eight terminations of helices. Clearly, other factors are involved. For example, the interaction of the OH group of serine or threonine with the main-chain carbonyl group sometimes disrupts an α helix.

4. The peptide groups are *planar*. The carbonyl group is *trans* to the main-chain NH. Also, the bond angles and distances are like those in dipeptides and related compounds.

5. The inside and outside are well defined. *The interior consists almost entirely of nonpolar residues* such as leucine, valine, methionine, and phenylalanine. There are no glutamic, aspartic, glutamine, asparagine, lysine, or arginine side chains in the interior of the molecule. Residues that have both a polar and a nonpolar part, such as threonine, tyrosine, and tryptophan, are oriented so that their nonpolar portions point inward. The only polar residues inside myoglobin are two histidines, which have a critical function at the active site. The outside of the molecule contains both polar and nonpolar residues.

CH_2

H_3C CH OH
Outside
Threonine

CH_2
OH
Outside
Tyrosine

CH_2
N
H
Outside
Tryptophan

THE OXYGEN-BINDING SITE IN MYOGLOBIN

The heme group is located in a crevice in the myoglobin molecule. The highly polar propionate side chains of the heme are on the surface of the molecule. At physiological pH, these carboxylates are ionized. The rest of the heme is inside the molecule, where it is surrounded by nonpolar residues except for two histidines. The iron atom of the heme is directly bonded to one of these histidines, namely residue F8 (Figures 3-14 and 3-15). This histidine, which occupies the fifth coordination position, is called the proximal histidine. The iron atom is about 0.3 Å out of the plane of the porphyrin, on the side of histidine F8. *The oxygen-binding site is on the other side of the heme plane, at the sixth coordination position.* A second histidine residue (E7), termed the distal histidine, is nearby. This E7 distal histidine is not bonded to the heme. A section of an electron-density map that includes the heme is shown in Figure 3-16.

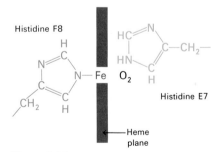

Figure 3-14
Schematic diagram of the oxygen-binding site in myoglobin: the fifth coordination position is occupied by histidine F8 (the proximal histidine); O_2 is bound at the sixth coordination position; and histidine E7 (the distal histidine) is near the sixth coordination position.

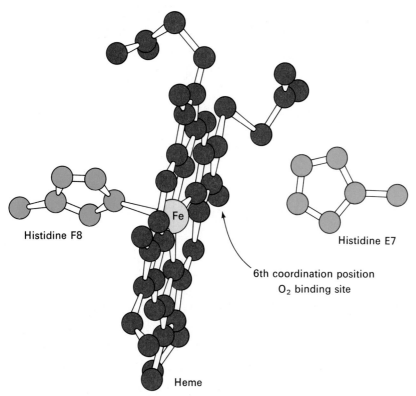

Figure 3-15
Model of the oxygen-binding site in myoglobin showing the heme group, the
proximal histidine (F8), and the distal histidine (E7).

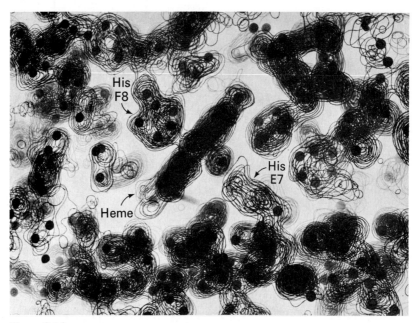

Figure 3-16
Section of an electron-density map of myoglobin near the oxygen-binding site.
Electron density extending across the lower part of the map is the E helix.
[Courtesy of Dr. John Kendrew.]

Three physiologically pertinent forms of myoglobin have been investigated: deoxymyoglobin, oxymyoglobin, and ferrimyoglobin. Their conformations appear to be very similar, except at the sixth coordination position (Table 3-1). In ferrimyoglobin, this site is

Table 3-1
Heme environment

| Form | Oxidation state of Fe | Occupant | |
		5th coordination position	6th coordination position
Deoxymyoglobin	+2	His F8	Empty
Oxymyoglobin	+2	His F8	O_2
Ferrimyoglobin	+3	His F8	H_2O

occupied by water; in deoxymyoglobin, it is empty; in oxymyoglobin, it is occupied by O_2. The axis of the bound O_2 is at an angle to the iron–oxygen bond (Figure 3-17). The iron atom moves toward the heme plane by about 0.2 Å on oxygenation. It is important to note that the heme group is not a rigid structure. Indeed, the motion of the iron atom on oxygenation plays a key role in the physiological action of hemoglobin, as will be discussed in the next chapter (p. 76).

The residues in contact with the heme come from different segments of the linear sequence of amino acids. Contributions are made by residues ranging from C4 to H14, corresponding to amino acids 39 and 138 in the linear sequence. Thus, *the heme-binding site is very much a three-dimensional entity.*

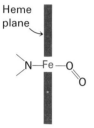

Figure 3-17
Bent, end-on orientation of O_2 in oxymyoglobin. The angle between the O_2 axis and the Fe–O bond is 121 degrees.

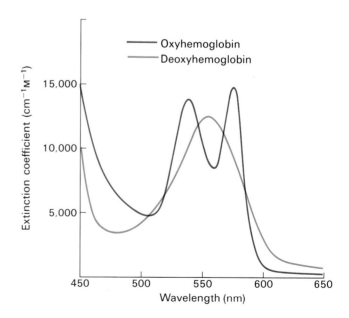

Figure 3-18
Visible absorption spectra of myoglobin and hemoglobin change markedly on binding oxygen. Myoglobin and hemoglobin have very similar visible absorption spectra.

A HINDERED HEME ENVIRONMENT IS ESSENTIAL FOR REVERSIBLE OXYGENATION

The oxygen-binding site comprises only a small fraction of the volume of the myoglobin molecule. Indeed, oxygen is directly bonded only to the iron atom of the heme. Why is the polypeptide portion of myoglobin needed for oxygen transport and storage? The answer comes from a consideration of the oxygen-binding properties of an isolated heme group. In water, an isolated ferrous heme group can bind oxygen, but it does so for only a fleeting moment. The reason is that in water the ferrous heme is very rapidly oxidized to ferric heme, which cannot bind oxygen. A complex of O_2 sandwiched between two hemes is an intermediate in this reaction. *The heme group in myoglobin is much less susceptible to oxidation because two myoglobin molecules cannot readily associate to form a heme-O_2-heme complex.* The formation of this sandwich is sterically blocked by the distal histidine and other residues surrounding the sixth coordination site. The strongest evidence for the importance of steric factors in determining the rate of oxidation of heme comes from studies of synthetic model compounds. James Collman synthesized *picket-fence iron porphyrin complexes* (Figure 3-19) that mimic the oxygen-binding sites of myoglobin and hemoglobin. These compounds have a protective enclosure for binding O_2 on one side of the porphyrin ring, whereas the other side is left unhindered so that it can bind a base. In fact, when the base is a substituted imidazole, the oxygen affinity of the picket-fence compound (Figure 3-20) is like that of myoglobin. Furthermore, the picket fence stabilizes the ferrous form of this iron porphyrin and thus enables it to reversibly bind oxygen for long periods. The critical difference between this model compound and ordinary heme is the presence of the picket fence, which blocks the formation of the sandwich dimer.

Thus, myoglobin has created a special microenvironment for its

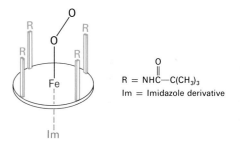

Figure 3-19
Schematic diagram of a picket-fence iron porphyrin with a bound O_2. The picket fence prevents two of these porphyrins from coming together to form the major intermediate in oxidation.

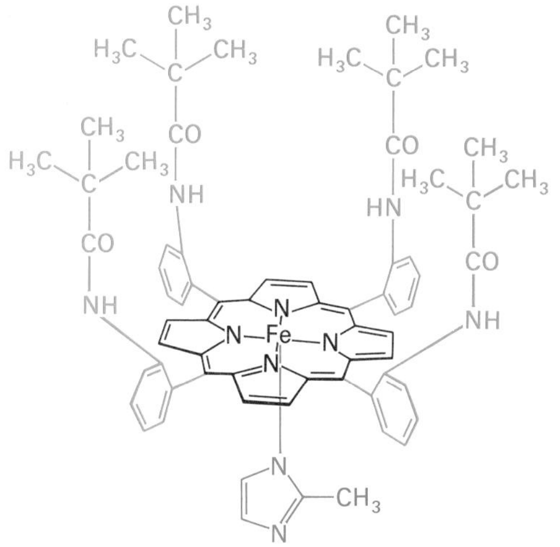

Figure 3-20
Structural formula of a picket-fence iron porphyrin. [After J. Collman, J. I. Brauman, E. Rose, and K. S. Suslick. *Proc. Nat. Acad. Sci.* 75(1978):1053.]

prosthetic group, which confers distinctive properties on it. In general, *the function of a prosthetic group is partly dependent on its polypeptide environment.* For example, the same heme group has quite a different function in cytochrome c, a protein in the terminal oxidation chain in the mitochondria of all aerobic organisms. In cytochrome c, the heme is a reversible carrier of electrons rather than of oxygen. The heme has yet another function in the enzyme catalase, where it catalyzes the conversion of hydrogen peroxide into water and oxygen.

CARBON MONOXIDE BINDING IS DIMINISHED BY THE PRESENCE OF THE DISTAL HISTIDINE

Carbon monoxide is a poison because it combines with ferromyoglobin and ferrohemoglobin and thereby interferes with oxygen transport. The heme group has an intrinsically high affinity for CO. An isolated heme in solution typically binds CO some 25,000 times as strongly as O_2. However, for myoglobin and hemoglobin, the binding affinity for CO is only about 200 times as great as for O_2. How do these proteins suppress the innate preference of heme for carbon monoxide? The answer comes from x-ray crystallographic and infra-red spectroscopic studies of complexes of CO and O_2 with myoglobin and model iron porphyrins. In the very high affinity complexes of CO with iron porphyrins, the Fe, C, and O atoms are in a linear array (Figure 3-21). In contrast, the CO axis is at an angle to the Fe–C bond in carbonmonoxymyoglobin. The

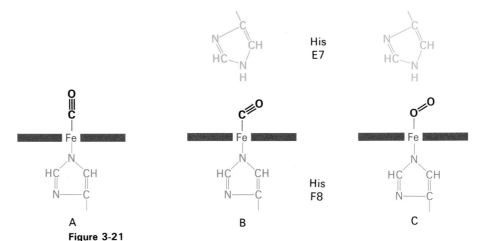

Figure 3-21
Structural basis of the diminished affinity of myoglobin and hemoglobin for carbon monoxide: (A) linear mode of binding of CO to isolated iron porphyrins; (B) bent mode of binding of CO to myoglobin and hemoglobin in which the distal histidine (E7) prevents CO from binding linearly and so the affinity for CO is markedly reduced; (C) bent mode of binding of O_2 in myoglobin and hemoglobin. Isolated iron porphyrins also bind O_2 in a bent mode.

linear binding of CO is prevented mainly by the steric hindrance of the distal histidine. On the other hand, the O_2 axis is at an angle to the Fe–O bond both in oxymyoglobin and in model compounds. *Thus, the protein forces the CO to bind at an angle rather than in line. This bent geometry in the globins weakens the interaction of CO with the heme, whereas it optimizes the binding of O_2.*

The decreased affinity of myoglobin and hemoglobin for CO is biologically important. Carbon monoxide was a potential hazard long before the emergence of industrialized societies. It is produced endogenously in the breakdown of heme (p. 507). Thus, the generation of CO within an organism is inherently linked to its use of heme. The level of endogenously formed carbon monoxide is such that about 1% of the sites in myoglobin and hemoglobin are blocked by CO. This degree of inhibition is certainly tolerable. In contrast, endogenously produced CO would lead to massive poisoning if the relative affinity of these proteins for CO approached the value characteristic of isolated iron porphyrins. This problem was solved by the evolution of heme proteins that discriminate between O_2 and CO by sterically imposing a bent and hence weaker mode of binding for CO.

STRUCTURES OF MYOGLOBIN IN SOLUTION AND IN THE CRYSTALLINE STATE ARE VERY SIMILAR

High-resolution x-ray analysis of myoglobin provided a magnificently detailed, though static, picture of the molecule. It is pertinent to ask whether the structure of myoglobin in the crystalline state is like that in solution. The crystalline state of myoglobin differs from the solution state in its high ionic strength, 3 M $(NH_4)_2SO_4$, and in the existence of interactions between different myoglobin molecules. Do these factors distort the structure of myoglobin so that the picture that emerges from the x-ray analysis is not relevant toward an understanding of its biological function? The answer to this question is a decisive no. There are several lines of evidence that indicate that *the structures in solution and in the crystalline state are very similar.*

1. The crystalline state of myoglobin is functionally active. Crystalline myoglobin can bind oxygen. However, the reaction is not as fast as it is in solution. The crystalline state is generally more sluggish in its reactivity.

2. The absorption spectrum of the heme group of myoglobin is virtually the same in solution and in the crystal. The absorption spectrum is a very sensitive index of the detailed environment of the heme group.

3. The α-helix content of a molecule in solution can be estimated from optical rotatory dispersion and circular dichroism measurements. The α-helix content of myoglobin in solution determined in this way agrees well with the α-helix content derived from the electron-density map of the crystal. It seems improbable that a major change in conformation would be unaccompanied by a change in the amount of helix present.

4. X-ray analysis of seal myoglobin shows that this molecule has a tertiary structure very much like that of sperm whale myoglobin. In contrast, the crystal lattices of these myoglobins are quite different, and so the interactions between myoglobin molecules in the two crystals are dissimilar. Thus, it is unlikely that major distortions are imposed by the crystal lattice.

NONPOLAR INTERACTIONS ARE IMPORTANT IN STABILIZING THE CONFORMATION OF MYOGLOBIN

X-ray analysis reveals the structure of myoglobin but does not directly explain why myoglobin folds in that way. Indeed, one of the challenges in protein chemistry is to discover the rules that determine how the amino acid sequence specifies the three-dimensional structure. One generalization emerging from the structure of myoglobin is that the inside consists of tightly packed nonpolar residues. Residues such as valine, leucine, isoleucine, methionine, and phenylalanine are *hydrophobic*. Given a choice between water and a nonpolar environment, they markedly prefer a nonpolar milieu. Furthermore, there are van der Waals attractive forces between these closely packed side chains. Thus, a significant driving force in the folding of myoglobin is the tendency of these residues to flee from water. These hydrophobic side chains are thermodynamically more stable when clustered in the interior of the molecule than when extended into the aqueous surroundings. This stability arises in the folding process because of the gain in entropy when ordered water molecules around the exposed hydrophobic groups are dispersed (p. 130). The polypeptide chain therefore folds spontaneously in aqueous solution so that its hydrophobic side chains are buried and its polar, charged chains are on the surface.

UNFOLDED MYOGLOBIN CAN SPONTANEOUSLY REFOLD INTO AN ACTIVE MOLECULE

What is the effect of the heme group on the three-dimensional structure of myoglobin? This question was answered by preparing *apomyoglobin*, which is myoglobin devoid of its heme group. Apomyoglobin was obtained by lowering the pH of a myoglobin solution to 3.5. The heme group binds only weakly to the protein at this acidic pH, and so it can be removed by extraction with an organic solvent. The aqueous phase, which contained the apomyoglobin, was then neutralized. Optical rotatory dispersion measurements showed that the α-helix content of apomyoglobin at neutral pH is 60%, which is appreciably lower than that of myoglobin (75%). Hydrodynamic studies indicated that apomyoglobin is less compact than myoglobin. Furthermore, apomyoglobin is much less stable than myoglobin. Thus, it is evident that *the presence of the heme group has an appreciable effect on the structure of myoglobin.*

Is the folding of apomyoglobin and the subsequent insertion of the heme group a spontaneous process? Apomyoglobin at neutral pH was unfolded by the addition of urea or guanidine. The α-helix content under these conditions (e.g., 8 M urea) was nearly zero. The urea was then removed by dialysis. The α-helix content increased to 60%, which showed that apomyoglobin had refolded. The subsequent addition of heme to this solution of apomyoglobin resulted in the formation of a biologically active molecule (Figure 3-22). On reduction to the ferrous state, the refolded myoglobin was fully effective in reversibly binding oxygen. Thus, *the complex three-dimensional structure of myoglobin is inherent in the amino acid sequence of apomyoglobin,* when it is combined with the heme prosthetic group. This finding reinforced the generality of the principle first discovered in studies of the renaturation of ribonuclease: *amino acid sequence specifies conformation.*

HEMOGLOBIN CONSISTS OF FOUR POLYPEPTIDE CHAINS

We turn now to hemoglobin, a related protein. Hemoglobin is made up of four polypeptide chains, in contrast with myoglobin, which is a single-chain protein. The four chains are held together by noncovalent interactions. There are four binding sites for oxygen on the hemoglobin molecule, because each chain contains one heme group. Hemoglobin A, the principal hemoglobin in adults, consists of two chains of one kind, called α chains, and two of another kind, called β chains. Thus, *the subunit structure of hemoglobin A is $\alpha_2\beta_2$.* There is a minor hemoglobin in adults, called hemoglobin A_2, which constitutes about 2% of the total hemoglobin. Hemoglobin A_2 has the subunit structure $\alpha_2\delta_2$. Fetuses have distinctive hemoglobins. An embryonic $\alpha_2\epsilon_2$ hemoglobin has been detected early in fetal life. This is followed by hemoglobin F, which has the subunit structure $\alpha_2\gamma_2$. The biological significance of these different hemoglobins is an intriguing question, which will be considered in the next chapter. The α chain, which is common to these hemoglobins, contains 141 amino acid residues. The β, δ, and γ chains have

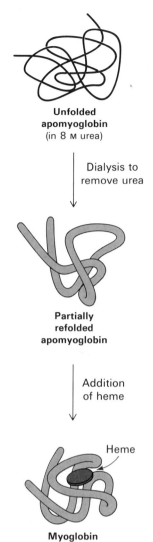

Unfolded apomyoglobin (in 8 M urea)

Dialysis to remove urea

Partially refolded apomyoglobin

Addition of heme

Heme

Myoglobin

Figure 3-22
Formation of myoglobin from unfolded apomyoglobin.

rather similar amino acid sequences (Figure 3-23) and a common length of 146 residues.

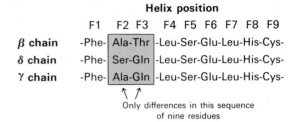

Figure 3-23
The β, γ, and δ chains of human hemoglobins have similar amino acid sequences, as shown here for residues F1 to F9.

X-RAY ANALYSIS OF HEMOGLOBIN

As mentioned earlier, the three-dimensional structure of hemoglobin A was determined by Max Perutz and his colleagues. This monumental accomplishment had its start in 1936, when Perutz left Austria to pursue graduate work in Cambridge, England. Perutz joined the laboratory of John Bernal, who had taken the first x-ray pictures of protein crystals two years earlier. Bernal and Dorothy Crowfoot Hodgkin, a graduate student, had obtained excellent diffraction patterns from the enzyme pepsin, thus showing that proteins in fact possess well-defined structures. They envisioned in 1934 the high promise of x-ray crystallography for "arriving at far more detailed conclusions about protein structure than previous physical or chemical methods have been able to give." However, more than twenty years passed before this promise was realized. When Perutz chose hemoglobin as his thesis subject, the largest structure that had been solved was the dye phthalocyanin, which contains 58 atoms. Thus, Perutz chose to tackle a molecule one hundred times as large. It was little wonder that "my fellow students regarded me with a pitying smile. . . . Fortunately, the examiners of my doctoral thesis did not insist on a determination of the structure, otherwise I should have had to remain a graduate student for twenty-three years." Fortunately, Lawrence Bragg, who with his father in 1912 was the first to use x-ray crystallography to solve structures, became director of the Cavendish Laboratory at this time and supported the project. He wrote: "I was frank about the outlook. It was like multiplying a zero probability that success would be achieved by an infinity of importance if the structure came out; the result of this mathematic operation was anyone's guess." Success came in 1959, when Perutz obtained a low-resolution electron-density image of horse oxyhemoglobin. Since then, high-resolution maps have been obtained for both human and horse oxyhemoglobin and deoxyhemoglobin. Human and horse hemoglobins are very similar.

The hemoglobin molecule is nearly spherical, with a diameter of 55 Å. The four chains are packed together in a tetrahedral array (Figure 3-24). The heme groups are located in crevices near the

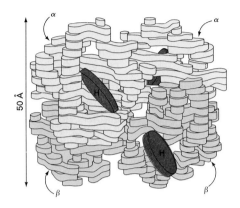

Figure 3-24
Model of hemoglobin at low resolution. The α chains in this model are yellow, the β chains blue, and the heme groups red. [After M. F. Perutz. The hemoglobin molecule. Copyright © 1964 by Scientific American, Inc. All rights reserved.]

exterior of the molecule, one in each subunit. The four oxygen-binding sites are far apart; the distance between the two closest iron atoms is 25 Å. Each α chain is in contact with both β chains. In contrast, there are few interactions between the two α chains or between the two β chains.

THE ALPHA AND BETA CHAINS OF HEMOGLOBIN CLOSELY RESEMBLE MYOGLOBIN

The three-dimensional structures of myoglobin and the α and β chains of human hemoglobin are strikingly similar (Figure 3-25).

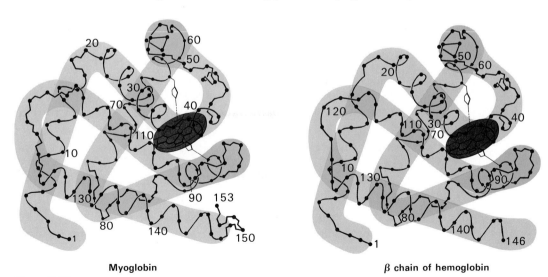

Myoglobin

β chain of hemoglobin

Figure 3-25
Comparison of the conformations of the main chain of myoglobin and the β chain of hemoglobin. The similarity of their conformations is evident. [From M. F. Perutz. The hemoglobin molecule. Copyright © 1964 by Scientific American, Inc. All rights reserved.]

The close resemblance in the folding of their main chains was unexpected because there are many differences in the amino acid sequences of these three polypeptide chains. In fact, the sequence is the same for the three chains at only 24 out of 141 positions, which shows that quite different amino acid sequences can specify very similar three-dimensional structures (Figure 3-26).

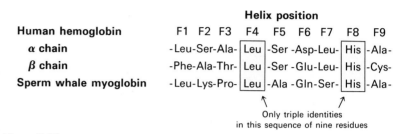

Figure 3-26
Comparison of the amino acid sequences of sperm whale myoglobin and the α and β chains of human hemoglobin, for residues F1 to F9. The amino acid sequences of these three polypeptide chains are much less alike than are their three-dimensional structures.

It is evident that the three-dimensional form of sperm whale myoglobin and of the α and β chains of human hemoglobin has broad biological significance. In fact, this structure seems to be common to all vertebrate myoglobins and hemoglobins. *The intricate folding of the polypeptide chain first evident in myoglobin is nature's fundamental design for an oxygen carrier: it places the heme in an environment that enables it to carry oxygen reversibly.*

CRITICAL RESIDUES IN THE AMINO ACID SEQUENCE

The amino acid sequences of hemoglobins from more than twenty species (ranging from lamprey to humans) are known. A comparison of these sequences shows considerable variability at most positions. However, there are nine positions in the sequence that contain the same amino acid in all or nearly all species studied thus far (Table 3-2). These conserved positions are especially important for the function of the hemoglobin molecule. Several of them are directly involved in the oxygen-binding site. Another conserved residue is tyrosine HC2, which stabilizes the molecule by forming a hydrogen bond between the H and F helices. A glycine (B6) is conserved because of its small size: a side chain larger than a hydrogen atom would not allow the B and E helices to approach each other as closely as they do (Figure 3-27). Proline C2 is important because it terminates the C helix.

The positions of the nonpolar residues in the interior of hemoglobin vary considerably. However, the change is always of one non-

Table 3-2
Conserved amino acid residues in hemoglobin

Position	Amino acid	Role
F8	Histidine	Proximal heme-linked histidine
E7	Histidine	Distal histidine near the heme
CD1	Phenylalanine	Heme contact
F4	Leucine	Heme contact
B6	Glycine	Allows the close approach of the B and E helices
C2	Proline	Helix termination
HC2	Tyrosine	Cross-links the H and F helices
C4	Threonine	Uncertain
H10	Lysine	Uncertain

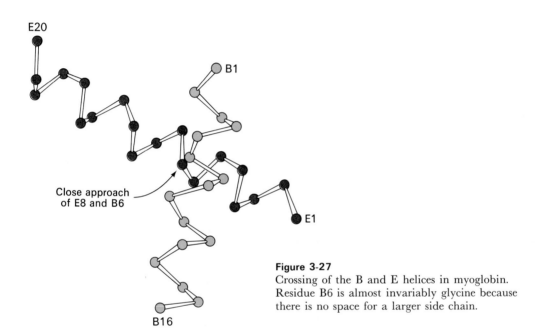

Figure 3-27
Crossing of the B and E helices in myoglobin. Residue B6 is almost invariably glycine because there is no space for a larger side chain.

polar residue for another (as from alanine to isoleucine). Thus, *the striking nonpolar character of the interior of the molecule is conserved.* As discussed previously, the reversible oxygenation of the heme group depends on its location in a nonpolar niche where it is protected from water. Also, the nonpolar core of the hemoglobin molecule is important in stabilizing its three-dimensional structure.

In contrast, the residues on the surface of the molecule are highly variable. Indeed, few are consistently positively or negatively charged. It might be thought that the prolines would be constant because of their role as helix-breakers. However, this is not so. Only one proline is constant, yet the lengths and directions of the helices in all myoglobins are very similar. Obviously, there must be other ways of terminating or bending α helices.

HEMOGLOBIN: A STEP UP IN EVOLUTION

Thus far, the structural similarity of myoglobin and hemoglobin has been emphasized. However, these molecules are functionally quite different. The subunits of hemoglobin have the same structural design as myoglobin. But in coming together to form an $\alpha_2\beta_2$ tetramer, new properties of profound biological importance emerge. They will be discussed in the next chapter.

SUMMARY

Myoglobin and hemoglobin are the oxygen-carrying proteins in vertebrates. Myoglobin facilitates the transport of oxygen in muscle and serves as a reserve store of oxygen in that tissue. Hemoglobin, which is packaged in erythrocytes, is the oxygen carrier in blood. These proteins can bind oxygen because they contain a tightly bound heme prosthetic group. Heme is a substituted porphyrin with a central iron atom. The iron atom in heme can be in the ferrous ($+2$) or ferric ($+3$) state. Only the ferrous form can bind O_2.

Myoglobin, a single polypeptide chain of 153 residues (17.8 kdal), has a compact shape. The inside of myoglobin consists almost exclusively of nonpolar residues, whereas the outside contains both polar and nonpolar residues. About 75% of the polypeptide chain is α helical. There are eight helical segments. The single ferrous heme group is located in a nonpolar niche, which protects it from oxidation to the ferric form. The iron atom of the heme is directly bonded to a nitrogen atom of a histidine side chain. This proximal histidine (F8) occupies the fifth coordination position. The sixth coordination position on the other side of the heme plane is the binding site for O_2. A second histidine, termed the distal histidine (E7), is next to this site. The proximal histidine increases the oxygen affinity of the heme, whereas the distal hisitidine sterically diminishes the binding of carbon monoxide. The distal histidines and other residues near the sixth coordination position also inhibit the oxidation of the heme to the ferric state.

Functionally active myoglobin can be reformed from unfolded apomyoglobin and heme. This renaturation experiment shows that the conformation of myoglobin is inherent in its amino acid sequence, as was first demonstrated for ribonuclease. The sequestration of nonpolar residues in the interior of myoglobin seems to be an important driving force in the folding of this molecule.

Hemoglobin consists of four polypeptide chains, each with a heme group. Hemoglobin A, the predominant hemoglobin in adults, has the subunit structure $\alpha_2\beta_2$. Hemoglobin A_2, a minor hemoglobin in adults, has the subunit structure $\alpha_2\delta_2$, whereas that of hemoglobin F, a fetal hemoglobin, is $\alpha_2\gamma_2$. The three-dimensional structures of the α and β chains of hemoglobin are strikingly similar to that of myoglobin, although their amino acid sequences are quite different. A comparison of the amino acid sequences of hemoglobins from many species shows that nine positions are virtually invariant. This group of conserved amino acid residues includes several near the heme, such as the proximal and distal histidines. The markedly nonpolar character of the interior of each subunit is also conserved.

SELECTED READINGS

WHERE TO START

Kendrew, J. C., 1961. The three-dimensional structure of a protein molecule. *Sci. Amer.* 205(6):96–11. [Available as *Sci. Amer.* Offprint 121.]

Perutz, M. F., 1964. The hemoglobin molecule. *Sci. Amer.* 211(5):64–76. [Offprint 196.]

Kendrew, J. C., 1963. Myoglobin and the structure of proteins. *Science* 139:1259–1266.

X-RAY CRYSTALLOGRAPHY

Matthews, B. W., 1977. X-ray structure of proteins. *In* Neurath, H., and Hill, R. L., (eds.), *The Proteins* (3rd ed.), vol. 3, pp. 404–590. Academic Press.

Holmes, K. C., and Blow, D. M., 1965. *The Use of X-ray Diffraction in the Study of Protein and Nucleic Acid Structure.* Wiley-Interscience. [An excellent introduction to one of the most important techniques in biochemistry.]

Glusker, J. P., and Trueblood, K. N., 1972. *Crystal Structure Analysis: A Primer.* Oxford University Press. [A lucid and concise introduction to x-ray crystallography in general.]

STRUCTURE OF MYOGLOBIN AND HEMOGLOBIN

Watson, H. C., 1969. The stereochemistry of the protein myoglobin. *Prog. Stereochem.* 4:299–333. [A detailed presentation of the three-dimensional structure of sperm whale ferrimyoglobin, including the atomic coordinates of all nonhydrogen atoms in the structure.]

Takano, T., 1977. Structure of myoglobin refined at 2.0 Å resolution. *J. Mol. Biol.* 110:537–584. [Structural analyses of sperm whale ferrimyoglobin and deoxymyoglobin.]

Phillips, S. E. V., 1978. Structure of oxymyoglobin. *Nature* 273:247–248.

Steigemann, W., and Weber, E., 1979. Structure of erythrocruorin in different ligand states refined at 1.4 Å resolution. *J. Mol. Biol.* 127:309–338. [A high-resolution structural analysis of a monomeric insect hemoglobin that resembles myoglobin and hemoglobin.]

Perutz, M. F., 1976. Structure and mechanism of haemoglobin. *Brit. Med. Bull.* 32:195–208.

REFOLDING OF MYOGLOBIN

Harrison, S. C., and Blout, E. R., 1965. Reversible conformational changes of myoglobin and apomyoglobin. *J. Biol. Chem.* 240:299–303.

PHYSIOLOGICAL ASPECTS

Wittenberg, J. B., 1970. Myoglobin-facilitated oxygen diffusion: role of myoglobin in oxygen entry into muscle. *Physiol. Rev.* 50:559–636.

Garby, L., and Meldon, J., 1977. *The Respiratory Functions of Blood.* Plenum.

MOLECULAR EVOLUTION

Zuckerkandl, E., 1965. The evolution of hemoglobin. *Sci. Amer.* 212(5):110–118. [Offprint 1012.]

Dayhoff, M. O., Hunt, L. T., McLaughlin, P. J., and Jones, D. D., 1972. Gene duplications in evolution: the globins. *In* Dayhoff, M. O., (ed.), *Atlas of Protein Sequence and Structure,* vol. 5, pp. 17–30. National Biomedical Research Foundation, Silver Springs, Maryland.

MODEL SYSTEMS

Collman, J. P., 1977. Synthetic models for the oxygen-binding hemoproteins. *Acc. Chem. Res.* 10:265–272. [A highly informative review of picket-fence porphyrins.]

Collman, J. P., Brauman, J. I., Halbert, T. R., and Suslick, K. S., 1976. Nature of O_2 and CO binding to metalloporphyrins and heme proteins. *Proc. Nat. Acad. Sci.* 73:3333–3337. [Presents the structural basis for the decreased binding of carbon monoxide by myoglobin and hemoglobin.]

Traylor, T. G., 1978. Hemoprotein oxygen transport: models and mechanisms. *In* Van Tamelen, E. E., (ed.), *Bioorganic Chemistry,* vol. 4, pp. 437–468. Academic Press.

PROBLEMS

1. The average volume of a red blood cell is 87 cubic microns. The mean concentration of hemoglobin in red cells is 34 g/100 ml.
   (a) What is the weight of the hemoglobin contained in a red cell?
   (b) How many hemoglobin molecules are there in a red cell?
   (c) Could the hemoglobin concentration in red cells be much higher than the observed value? (Hint: Suppose that a red cell contained a crystalline array of hemoglobin molecules 65 Å apart in a cubic lattice).

2. How much iron is there in the hemoglobin of a 70-kg adult? Assume that the blood volume is 70 ml/kg of body weight and that the hemoglobin content of blood is 16 g/100 ml.

3. The myoglobin content in some human muscles is about 8 g/kg of muscle. In sperm whale, the myoglobin content is about 80 g/kg of muscle.
   (a) How much O_2 is bound to myoglobin in human muscle and in that of sperm whale? Assume that the myoglobin is saturated with O_2.
   (b) The amount of oxygen dissolved in tissue water (in equilibrium with venous blood at 37°C) is about 3.5×10^{-5} M. What is the ratio of oxygen bound to myoglobin to that directly dissolved in the water of sperm whale muscle?

4. The amino acid composition of sperm whale myoglobin is

Ala 5	Gln 5	Leu 18	Ser 6
Arg 4	Glu 14	Lys 19	Thr 5
Asn 2	Gly 11	Met 2	Trp 2
Asp 6	His 12	Phe 6	Tyr 3
Cys 0	Ile 9	Pro 4	Val 8

   (a) Estimate the net charge of ferromyoglobin at pH 2, 7, and 9.
   (b) Estimate the isoelectric point of myoglobin. The isoelectric point is defined as the pH at which there is no net charge.

5. A solution of myoglobin is treated with cyanogen bromide.
   (a) What are the cleavage products? Refer to the amino acid sequence of myoglobin.
   (b) In aqueous solution, the cleavage products have a low content of α helix. What does this finding reveal about the stability of α helices in aqueous solution?

6. The binding of O_2 to myoglobin can be described by the simple equilibrium

   $$Mb + O_2 \rightleftharpoons MbO_2$$

   Let [Mb] denote the concentration of deoxymyoglobin; $[MbO_2]$, the concentration of oxymyoglobin; pO_2, the concentration of O_2 (expressed as the partial pressure of O_2); and P_{50}, the pO_2 at which $[Mb] = [MbO_2]$. Derive an equation analogous to the Henderson-Hasselbalch equation that gives the ratio of oxymyoglobin to deoxymyoglobin as a function of pO_2 and P_{50}.

7. The equilibrium constant K for the binding of oxygen to myoglobin is 10^{-6} M, where K is defined as

   $$K = [Mb][O_2]/[MbO_2]$$

   The rate constant for the combination of O_2 with myoglobin is 2×10^7 M^{-1} sec^{-1}.
   (a) What is the rate constant for the dissociation of O_2 from oxymyoglobin?
   (b) What is the mean duration of the oxymyoglobin complex?

8. The absorbance A of a solution is defined as

   $$A = \log_{10}(I_0/I)$$

   in which I_0 is the incident light intensity and I is the transmitted light intensity. The absorbance is related to the molar absorption coefficient (extinction coefficient) ϵ (in cm^{-1} M^{-1}), concentration c (in M), and path length l (in cm) by

   $$A = \epsilon l c$$

   The extinction coefficient of myoglobin at 580 nm is 15,000 cm^{-1} M^{-1}. What is the absorbance of a 1 mg/ml solution across a 1-cm path? What percentage of the incident light is transmitted by this solution?

CHAPTER 4

HEMOGLOBIN:
AN ALLOSTERIC PROTEIN

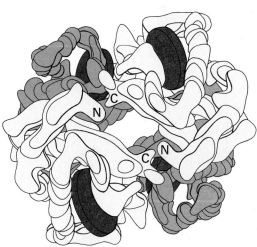

Figure 4-1
Model of hemoglobin at low resolution. The α chains in this model are yellow, the β chains are blue, and the heme groups red. [After M. F. Perutz. The hemoglobin molecule. Copyright © 1964 by Scientific American, Inc. All rights reserved.]

New properties emerged in tetrameric hemoglobin in its evolution from monomeric myoglobin. Hemoglobin is a much more intricate molecule than myoglobin. First, hemoglobin transports H^+ and CO_2 in addition to O_2. Second, the binding of oxygen by hemoglobin is regulated by specific molecules in its environment, namely H^+, CO_2, and organic phosphate compounds. These regulatory molecules greatly affect the oxygen-binding properties of hemoglobin even though they are bound to sites on the protein that are far from the hemes. Indeed, interactions between spatially distinct sites, termed *allosteric interactions,* occur in many proteins. Allosteric effects play a critical role in the control and integration of molecular events in biological systems. Hemoglobin is the best understood of the allosteric proteins, and so it is rewarding to examine its structure and function in some detail.

FUNCTIONAL DIFFERENCES BETWEEN MYOGLOBIN AND HEMOGLOBIN

Hemoglobin is an allosteric protein, whereas myoglobin is not. This difference is expressed in three ways:

1. The binding of O_2 to hemoglobin enhances the binding of additional O_2 to the same hemoglobin molecule. In other words, O_2 binds cooperatively to hemoglobin. In contrast, the binding of O_2 to myoglobin is not cooperative.

2. The affinity of hemoglobin for oxygen depends on pH, whereas that of myoglobin is independent of pH. The CO_2 molecule also affects the oxygen-binding characteristics of hemoglobin.

3. The oxygen affinity of hemoglobin is further regulated by organic phosphates such as diphosphoglycerate. The result is that hemoglobin has a lower affinity for oxygen than does myoglobin.

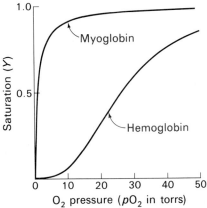

Figure 4-2
Oxygen dissociation curves of myoglobin and hemoglobin. Saturation of the oxygen-binding sites is plotted as a function of the partial pressure of oxygen surrounding the solution.

Torr—
  A unit of pressure equal to that exerted by a column of mercury 1 mm high at 0°C and standard gravity (1 mm Hg).
  Named after Evangelista Torricelli (1608–1647), the inventor of the mercury barometer.

THE BINDING OF OXYGEN TO HEMOGLOBIN IS COOPERATIVE

The saturation Y is defined as the fractional occupancy of the oxygen-binding sites. The value of Y can range from 0 (all sites empty) to 1 (all sites filled). A plot of Y versus pO_2, the partial pressure of oxygen, is called an *oxygen dissociation curve*. The oxygen dissociation curves of myoglobin and hemoglobin differ in two ways (Figures 4-2 and 4-3). For any given pO_2, Y is higher for myoglobin than for hemoglobin. In other words, *myoglobin has a higher affinity for oxygen than does hemoglobin.* The oxygen affinity can be characterized by a quantity called P_{50}, which is the partial pressure of oxygen at which 50% of sites are filled (i.e., at which $Y = 0.5$). For myoglobin, P_{50} is typically 1 torr, whereas for hemoglobin P_{50} is 26 torrs.

The second difference is that *the oxygen dissociation curve of myoglobin has a hyperbolic shape, whereas that of hemoglobin is sigmoidal.* As will be discussed shortly, the sigmoidal shape of the curve for hemoglobin is ideally suited to its physiological role as an oxygen carrier in blood. In molecular terms, the sigmoidal shape means that the binding of oxygen to hemoglobin is cooperative—that is, the binding of oxygen at one heme facilitates the binding of oxygen to the other hemes.

Let us consider the oxygen dissociation curves of these molecules in quantitative terms, starting with myoglobin because it is simpler. The binding of oxygen to myoglobin (Mb) can be described in terms of a simple equilibrium:

$$MbO_2 \rightleftharpoons Mb + O_2 \qquad (1)$$

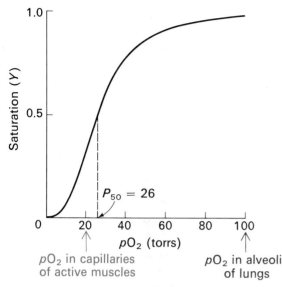

Figure 4-3
Oxygen dissociation curve of hemoglobin.
Typical values for pO_2 in the capillaries
of active muscle and in the alveoli of the
lung are marked on the x-axis. Note that
P_{50} for hemoglobin under physiological
conditions lies between these values.

The equilibrium constant K for the dissociation of oxymyoglobin is

$$K = \frac{[\text{Mb}][\text{O}_2]}{[\text{MbO}_2]} \qquad (2)$$

in which $[\text{MbO}_2]$ is the concentration of oxymyoglobin, $[\text{Mb}]$ is the concentration of deoxymyoglobin, and $[\text{O}_2]$ is the concentration of uncombined oxygen, all in moles per liter. The fractional saturation Y is defined as

$$Y = \frac{[\text{MbO}_2]}{[\text{MbO}_2] + [\text{Mb}]} \qquad (3)$$

Substitution of equation 2 into equation 3 yields

$$Y = \frac{[\text{O}_2]}{[\text{O}_2] + K} \qquad (4)$$

Because oxygen is a gas, it is convenient to express its concentration in terms of pO_2, the partial pressure of oxygen (in torrs) in the atmosphere surrounding the solution. Equation 4 then becomes

$$Y = \frac{pO_2}{pO_2 + P_{50}} \qquad (5)$$

Equation 5 plots as a hyperbola. In fact, the oxygen dissociation curve calculated from equation 5, taking P_{50} to be 1 torr, closely matches the experimentally observed curve for myoglobin.

In contrast, because the curve for hemoglobin is sigmoidal, none of the curves described by equation 5 can match it. In molecular terms, this means that the binding of O_2 is cooperative. Consider an extreme model in which the only species are deoxyhemoglobin and hemoglobin (Hb) containing four bound molecules of O_2:

$$Hb(O_2)_4 \rightleftharpoons Hb + 4\,O_2 \qquad (6)$$

The equilibrium constant for this hypothetical reaction is

$$K = \frac{[Hb][O_2]^4}{[Hb(O_2)_4]} \qquad (7)$$

and so

$$Y = \frac{(pO_2)^4}{(pO_2)^4 + (P_{50})^4} \qquad (8)$$

A plot of equation 8 yields a sigmoidal curve (Figure 4-4). However, note that this calculated curve is steeper than the observed curve for hemoglobin. In other words, the model expressed by equation 6 is too extreme.

How, then, can we characterize a binding process having an intermediate degree of cooperativity? In 1913, Archibald Hill showed that the curve obtained from the oxygen-binding data for hemoglobin agrees with the equation derived for the hypothetical equilibrium:

$$Hb(O_2)_n \rightleftharpoons Hb + n\,O_2 \qquad (9)$$

This expression yields

$$Y = \frac{(pO_2)^n}{(pO_2)^n + (P_{50})^n} \qquad (10)$$

which can be rearranged to give

$$\frac{Y}{1-Y} = \left(\frac{pO_2}{P_{50}}\right)^n \qquad (11)$$

This equation states that the ratio of oxyheme (Y) to deoxyheme ($1 - Y$) is equal to the nth power of the ratio of pO_2 to P_{50}. Taking the logarithm of both sides of equation 11 gives

$$\log\left(\frac{Y}{1-Y}\right) = n \log pO_2 - n \log P_{50} \qquad (12)$$

Note that a plot of log $[Y/(1 - Y)]$ versus log pO_2 yields a straight line with a slope of n. This representation is called a *Hill plot* and its slope n at the midpoint of the binding ($Y = 0.5$) is called the *Hill coefficient*.

Myoglobin gives a linear Hill plot with $n = 1.0$, whereas $n = 2.8$ for hemoglobin (Figure 4-5). The slope of 1.0 for myoglobin means that O_2 molecules bind independently of each other, as indicated in equation 1. In contrast, the Hill coefficient of 2.8 for hemoglobin means that *the binding of oxygen in hemoglobin is cooperative*. Binding at one heme facilitates the binding of oxygen at the other hemes on

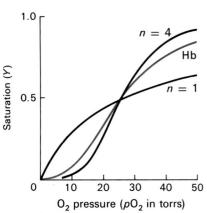

Figure 4-4
The oxygen saturation curve for hemoglobin lies between the curves calculated for $n = 1$ (noncooperative binding) and $n = 4$ (totally cooperative binding).

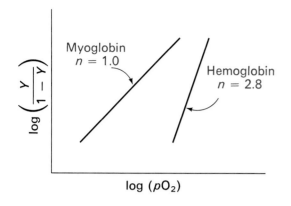

Figure 4-5
Hill plot for the binding of O_2 to myoglobin and hemoglobin. The slope of 2.8 for hemoglobin indicates that it binds oxygen cooperatively, in contrast with myoglobin, which has a slope of 1.0.

the same tetramer. Conversely, the unloading of oxygen at one heme facilitates the unloading of oxygen at the others. In other words, there is communication among the heme groups of a hemoglobin molecule. The cooperative binding of oxygen by hemoglobin is sometimes called *heme-heme interaction*. The mechanism of this interaction will be discussed shortly.

THE COOPERATIVE BINDING OF OXYGEN BY HEMOGLOBIN ENHANCES OXYGEN TRANSPORT

The cooperative binding of oxygen by hemoglobin makes hemoglobin a more efficient oxygen transporter. The oxygen saturation of hemoglobin changes more rapidly with changes in the partial pressure of O_2 than it would if the oxygen-binding sites were independent of each other. Let us consider a specific example. Assume that the alveolar pO_2 is 100 torrs, and that the pO_2 in the capillary of an active muscle is 20 torrs. Let $P_{50} = 30$ torrs, and take $n = 2.8$. Then Y in the alveolar capillaries will be 0.97, and Y in the muscle capillaries will be 0.25. The oxygen delivered will be proportional to the difference in Y, which is 0.72. Let us now make the same calculation for a hypothetical oxygen carrier for which P_{50} is also 30 torrs, but in which the binding of oxygen is not cooperative ($n = 1$). Then $Y_{alveoli} = 0.77$, and $Y_{muscle} = 0.41$, and so $\Delta Y = 0.36$. Thus, *the cooperative binding of oxygen by hemoglobin enables it to deliver twice as much oxygen as it would if the sites were independent.*

H+ AND CO_2 PROMOTE THE RELEASE OF O_2 (THE BOHR EFFECT)

Myoglobin shows no change in oxygen binding over a broad range of pH, nor does CO_2 have an appreciable effect. In hemoglobin, acidity enhances the release of oxygen. In the physiological range, a lowering of pH shifts the oxygen dissociation curve to the right, so

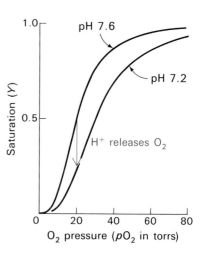

Figure 4-6
Effect of pH on the oxygen affinity of hemoglobin. Lowering the pH from 7.6 to 7.2 results in the release of O_2 from oxyhemoglobin.

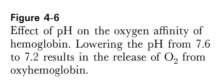

Figure 4-7
A summary of the Bohr effect. The actual mechanism and stoichiometry are more complex than indicated in this diagram.

2,3-Diphosphoglycerate
(DPG)

that the oxygen affinity is decreased (Figure 4-6). Increasing the concentration of CO_2 (at constant pH) also lowers the oxygen affinity. In rapidly metabolizing tissue, such as contracting muscle, much CO_2 and acid are produced. *The presence of higher levels of CO_2 and H^+ in the capillaries of such metabolically active tissue promotes the release of O_2 from oxyhemoglobin.* This important mechanism for meeting the higher oxygen needs of metabolically active tissues was discovered by Christian Bohr in 1904.

The reciprocal effect, discovered ten years later by John B. S. Haldane, occurs in the alveolar capillaries of the lungs. The high concentration of O_2 there unloads H^+ and CO_2 from hemoglobin, just as the high concentration of H^+ and CO_2 in active tissues drives off O_2. These linkages between the binding of O_2, H^+, and CO_2 are known as the *Bohr effect* (Figure 4-7).

DPG LOWERS THE OXYGEN AFFINITY

The oxygen affinity of hemoglobin within red cells is lower than that of hemoglobin in free solution. As early as 1921, Joseph Barcroft wondered, "Is there some third substance present . . . which forms an integral part of the oxygen-hemoglobin complex?" Indeed there is. Reinhold Benesch and Ruth Benesch showed in 1967 that 2,3-diphosphoglycerate (DPG) binds to hemoglobin and has a large effect on its oxygen affinity. DPG is present in human red cells at about the same molar concentration as hemoglobin. In the absence of DPG, the P_{50} of hemoglobin is 1 torr, like that of myoglobin. In its presence, P_{50} becomes 26 torrs (Figure 4-8). Thus, *DPG reduces the oxygen affinity of hemoglobin by a factor of 26.* DPG is physiologically important. In its absence, hemoglobin would unload little oxygen in passing through tissue capillaries, because the pO_2 there is about 26 torrs.

DPG exerts its effect on the oxygen affinity of hemoglobin by binding to deoxyhemoglobin but not to oxyhemoglobin. The binding of oxygen and DPG to hemoglobin is mutually exclusive. To a good approximation, the oxygenation of hemoglobin in the presence of DPG can be represented by the equation

$$\text{Hb-DPG} + 4\,O_2 \rightleftharpoons \text{Hb}(O_2)_4 + \text{DPG}$$

CLINICAL SIGNIFICANCE OF DPG

The discovery of the role of DPG in oxygen transport has led to new insights in several areas of clinical medicine. For example, it has been known for some time that blood stored in *acid-citrate-dextrose, a conventional medium, has an increased oxygen affinity: P_{50} is 16 instead of 26 torrs. The reason is now apparent. The oxygen affinity increases during storage because there is a concomitant decrease in the level of DPG, from 4.5 mM to less than 0.5 mM in ten days. The oxygen affinity of stored blood may be of critical clinical importance. If a patient is given a large volume of blood with high oxygen affinity, the amount of oxygen unloaded in his tissues may be compromised. Transfused red cells that are totally depleted of DPG can regain half the normal level within 24 hours, but this restitution may not be rapid enough to be effective in a severely ill patient. Thus, under certain conditions, it may be advantageous to give transfusions of red cells that have a normal oxygen affinity. The DPG level of deficient cells cannot be restored by adding DPG to them because red-cell membranes are not permeable to this highly charged molecule. However, the decrease in DPG level of stored cells can be prevented if the medium in which they are suspended contains *inosine*. This uncharged nucleoside molecule can penetrate the red-cell membrane. Inside the red cell, inosine is converted into DPG by a complex series of reactions (see p. 335). Inosine is now frequently used in this way to preserve the functional integrity of stored blood.

Adaptive mechanisms in disorders of oxygen delivery to the tissues (termed *hypoxia*) have also been illuminated by the work on DPG. Let us consider a patient who has severe obstructive pulmonary emphysema. In this disease, airflow in the bronchioles is blocked, and consequently the arterial blood is not fully saturated with oxygen. Indeed, the pO_2 of the arterial blood of such a patient may be only 50 torrs, half the normal value. There is a compensatory shift in his oxygen dissociation curve, which is due to an increase in the level of DPG from 4.5 to 8.0 mM. At this higher DPG level, P_{50} is 31 torrs, rather than 26 torrs. With a P_{50} of 31 torrs, the arterial saturation (Y_A) is 0.82, whereas the venous saturation (Y_V) is 0.49; hence, the change in oxygen saturation (ΔY) is 0.33. For a normal P_{50} of 26 torrs, $Y_A = 0.86$, $Y_V = 0.60$, and $\Delta Y = 0.26$. The

Figure 4-8
Diphosphoglycerate decreases the oxygen affinity of hemoglobin.

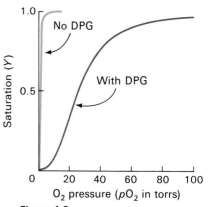

Inosine

shift of the oxygen dissociation curve is advantageous to this patient because it increases ΔY from 0.26 to 0.33. Thus, the increased DPG level leads to a 27% increase in the amount of oxygen delivered to the tissues.

In some *metabolic disorders of the red cell,* the level of DPG is abnormal and there is a concomitant change in the oxygen affinity of hemoglobin. These disorders are discussed in a later chapter (p. 277).

The mechanism of *high-altitude adaptation* is a challenging problem in respiratory physiology. Changes in the DPG level may play a role in this adaptive process. When a person goes from sea level to 4500 meters, the concentration of DPG in red cells increases from 4.5 to 7.0 mM within two days and there is a concomitant decrease in the oxygen affinity. The arterial oxygen saturation is diminished by the lowered P_{50}, but more oxygen is transported (ΔY is larger) because of a greater degree of unloading in the capillary circulation. These changes in DPG level and P_{50} are reversed on returning to sea level.

FETAL HEMOGLOBIN HAS A HIGH OXYGEN AFFINITY

A fetus has its own kind of hemoglobin, called hemoglobin F ($\alpha_2\gamma_2$), which differs from adult hemoglobin A ($\alpha_2\beta_2$), as mentioned previously. An important property of hemoglobin F is that it has a higher oxygen affinity under physiological conditions than does hemoglobin A (Figure 4-9). The higher oxygen affinity of hemoglo-

Figure 4-9
Fetal red blood cells have a higher oxygen affinity than do maternal red blood cells. In the presence of DPG, the oxygen affinity of fetal hemoglobin is higher than that of maternal hemoglobin.

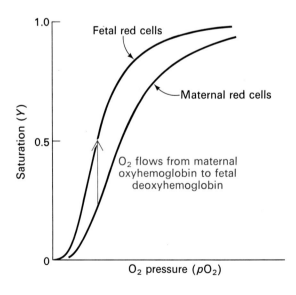

bin F optimizes the transfer of oxygen from the maternal to the fetal circulation. Hemoglobin F is oxygenated at the expense of hemoglobin A on the other side of the placental circulation.

The higher oxygen affinity of fetal blood has been known for many years, but its basis has only recently been elucidated. *Hemoglobin F binds DPG less strongly than does hemoglobin A and consequently has a higher oxygen affinity*. Indeed, in the absence of DPG, the oxygen affinities of the two hemoglobins are in reverse order. This observation was puzzling until it was appreciated that DPG must be present when the oxygen dissociation curves are measured because DPG is present in both fetal and adult red cells.

SUBUNIT INTERACTION IS REQUIRED FOR ALLOSTERIC EFFECTS

Let us now consider the structural basis of the allosteric effects. Hemoglobin can be dissociated into its constituent chains. The properties of the isolated α chain are very much like those of myoglobin. The α chain by itself has a high oxygen affinity, a hyperbolic oxygen dissociation curve, and oxygen-binding characteristics that are insensitive to pH, CO_2 concentration, and DPG level. The isolated β chain readily forms a tetramer, β_4, which is called hemoglobin H. Like the α chain and myoglobin, β_4 entirely lacks the allosteric properties of hemoglobin. In short, *the allosteric properties of hemoglobin arise from subunit interaction. The functional unit of hemoglobin is a tetramer consisting of two kinds of polypeptide chains.*

THE QUATERNARY STRUCTURE OF HEMOGLOBIN CHANGES MARKEDLY ON OXYGENATION

In 1937, Felix Haurowitz observed that crystals of deoxyhemoglobin shatter when they are oxygenated. The crystal shatters because the structure of oxyhemoglobin is appreciably different from that of deoxyhemoglobin. The oxygenated molecules can no longer fit precisely on the crystal lattice of the deoxygenated hemoglobin. In contrast, crystals of myoglobin, like crystals of β_4, are unaffected by oxygenation.

X-ray crystallographic studies have shown that oxy and deoxyhemoglobin differ markedly in quaternary structure (Figure 4-10). The oxygenated molecule is more compact. For instance, the distance between the iron atoms of the β chains decreases from 39.9 to 33.4 Å on oxygenation. The changes in the contacts between the α and β chains are of special interest. There are two kinds of contact regions between

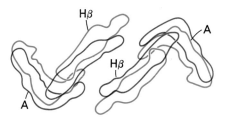

Figure 4-10
Projection of part of the electron-density maps of oxyhemoglobin (shown in red) and deoxyhemoglobin (shown in blue) at a resolution of 5.5 Å. The A and H helices of the two β chains of hemoglobin are shown here. The center of the diagram corresponds to the central cavity of the molecule. One of the conformational changes accompanying oxygenation—a movement of the H helices toward each other—is shown here. [After M. F. Perutz. *Nature* 228(1970):738.]

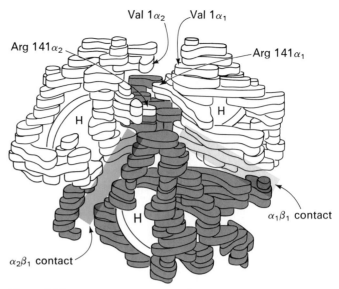

Figure 4-11
Model of oxyhemoglobin at low resolution showing the two kinds of interfaces between α and β chains. The α chains are white, the β chains black. The three hemes that can be seen in this view of the molecule are labeled H. The $\alpha_2\beta_1$ contact region is shown in blue, the $\alpha_1\beta_1$ contact region in yellow. [After M. F. Perutz and L. F. TenEyck. *Cold Spring Harbor Symp. Quant. Biol.* 36(1971):296.]

the α and β chains, called the $\alpha_1\beta_1$ and the $\alpha_1\beta_2$ contacts (Figure 4-11). In the transition from oxy to deoxyhemoglobin, there is little change in the $\alpha_1\beta_1$ contact. In contrast, there are large structural changes at the $\alpha_1\beta_2$ contact. One pair of $\alpha\beta$ subunits rotates relative to the other $\alpha\beta$ pair by 15 degrees on oxygenation (Figure 4-12). Some atoms at this interface shift by as much as 6 Å. In fact, the $\alpha_1\beta_2$ contact region is designed to act as a switch between two alternative structures. The two forms of the dove-tailed $\alpha_1\beta_2$ inter-

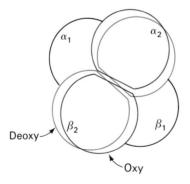

Figure 4-12
Schematic diagram showing the change in quaternary structure on oxygenation. One pair of $\alpha\beta$ subunits shifts with respect to the other by a rotation of 15° and a translation of 0.8 Å. The oxy form of the rotated $\alpha\beta$ subunit is shown in red and the deoxy form in blue. [After J. Baldwin and C. Chothia. *J. Mol. Biol.* 129(1979):192. Copyright by Academic Press Inc. (London) Ltd.]

face are stabilized by different sets of hydrogen bonds (Figure 4-13).

The $\alpha_1\beta_2$ contact is closely connected to the heme groups. Hence, structural changes in this region can be expected to affect the hemes. The importance of the $\alpha_1\beta_2$ contact is reinforced by the finding that most residues in it are the same in all species. Also, almost all mutations in the $\alpha_1\beta_2$ contact diminish heme-heme interaction, whereas $\alpha_1\beta_1$ contact mutations do not.

DEOXYHEMOGLOBIN IS CONSTRAINED BY SALT LINKS BETWEEN DIFFERENT CHAINS

In oxyhemoglobin, the carboxyl-terminal residues of all four chains have almost complete freedom of rotation. In contrast, these terminal groups are anchored in deoxyhemoglobin (Figure 4-14). First,

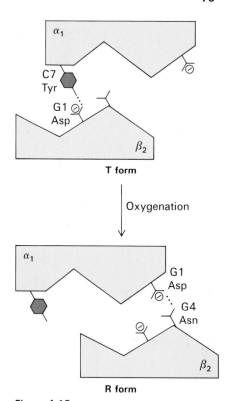

Figure 4-13
The $\alpha_1\beta_2$ interface switches from the T to the R form on oxygenation. The dove-tailed construction of this interface allows the two subunits to readily slide across each other.

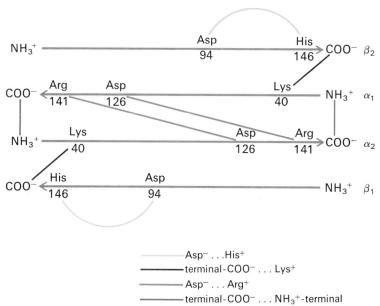

Figure 4-14
Cross-links between different subunits in deoxyhemoglobin. These non-covalent, electrostatic interactions are disrupted on oxygenation.

the carboxyl terminus of the α_1 chain interacts with the amino terminus of the α_2 chain. Second, the arginine side chain of this carboxyl-terminal residue is linked to an aspartate on the α_2 chain. Third, the carboxyl terminus of the β_1 chain is linked to a lysine side chain on the α_2 chain. Finally, the imidazole side chain of the carboxyl-terminal residue of the β_1 chain interacts with an aspartate side chain on the same chain. These pairs of residues are held together by noncovalent, electrostatic interactions between oppo-

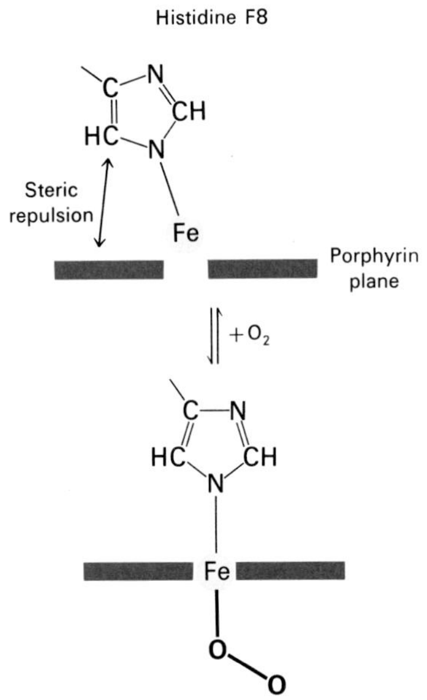

Figure 4-15
The iron atom moves into the plane of
the heme on oxygenation. The
proximal histidine (F8) is pulled along
with the iron atom and becomes less
tilted. The iron atom is out-of-plane
because of steric repulsion between a
carbon atom of the imidazole ring and
a nitrogen atom of the heme.

sitely charged groups, which are often called *salt links*. Deoxyhemoglobin is a tauter, more constrained molecule than oxyhemoglobin
because of these eight salt links.

*The quaternary structure of deoxyhemoglobin is termed the T (tense) form;
that of oxyhemoglobin, the R (relaxed) form.* The designations R and T
are generally used to describe alternative quaternary structures of
allosteric proteins, the T form having a lower affinity for the substrate.

THE IRON ATOM MOVES INTO THE PLANE
OF THE PORPHYRIN ON OXYGENATION

The conformational changes discussed thus far take place at some
distance from the heme. Now let us see what happens at the heme
group itself on oxygenation. *In deoxyhemoglobin, the iron atom is about
0.6 Å out of the heme plane because of steric repulsion between the proximal
histidine and the nitrogen atoms of the porphyrin* (Figure 4-15). *On oxygenation, the iron atom moves into the plane of the porphyrin so that it can form a
strong bond with O_2.* The energy gained in forming this bond more
than pays for the energy lost in bringing the proximal histidine
closer to the heme plane. Structural studies of many synthetic iron
porphyrins have shown that the iron atom is out of plane in five-
coordinated compounds, whereas it is in plane or nearly so in
six-coordinated complexes. The electron spin of the iron atom
also affects its location, with high spin favoring an out-of-plane
geometry.

MOVEMENT OF THE IRON ATOM IS TRANSMITTED
TO OTHER SUBUNITS BY THE PROXIMAL HISTIDINE

How does the movement of the iron atom into the plane of the
heme favor the switch in quaternary structure from T to R? A key
side chain in the transmission of structural change from the heme of
one subunit to the other subunits is the proximal histidine, which is
pulled by the iron atom (Figure 4-16). The heme group and proximal histidine make intimate contact with some fifteen side chains
and so the structures of the F helix, the EF corner, and the FG
corner change on oxygenation. These changes are then transmitted
to the subunit interfaces. The expulsion of tyrosine HC2 from the
pocket between the F and H helices leads to the rupture of interchain salt links. Consequently, the equilibrium between the two
quaternary structures is shifted to the R form on oxygenation.
Thus, *a structural change within a subunit (oxygenation) is translated into
structural changes at the interfaces between subunits. The binding of oxygen at
one heme site is thereby communicated to parts of the molecule that are far
away.*

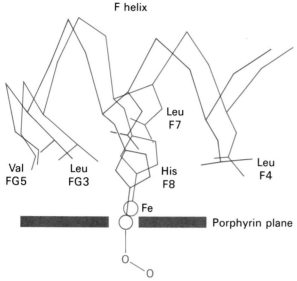

F helix

Leu
F7

Val
FG5

Leu
FG3

His
F8

Leu
F4

Fe

Porphyrin plane

O
O

Figure 4-16
Conformational changes induced by the movement of the
iron atom on oxygenation. The proximal histidine is pulled
by the iron atom and becomes less tilted. The oxygenated
structure is shown in red and the deoxygenated structure in
blue. [After J. Baldwin and C. Chothia. *J. Mol. Biol.*
129(1979):192. Copyright by Academic Press Inc. (London)
Ltd.]

MECHANISM OF THE COOPERATIVE BINDING OF OXYGEN

Why does the fourth molecule of O_2 bind to hemoglobin some 300
times as tightly as the first one? A qualitative answer to this question is depicted in Figure 4-17. Deoxyhemoglobin is a taut molecule, constrained by its eight salt links between the four subunits.
Oxygenation cannot occur unless some of these salt links are broken
so that the iron atom can move into the plane of the heme group.
The number of salt links that need to be broken for the binding of

Figure 4-17
Postage-stamp analogy of the oxygenation of hemoglobin. Two perforated
edges must be torn to remove the first
stamp. Only one perforated edge must
be torn to remove the second stamp,
and one edge again to remove the
third stamp. The fourth stamp is then
free.

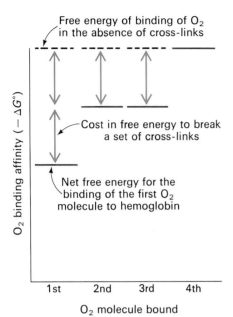

Figure 4-18
Free energy of binding of successive molecules of O_2 for the model depicted in Figure 4-17. The binding affinity $(-\Delta G°)$ increases because fewer cross-links must be broken to bind the fourth O_2 molecule than the first one.

an O_2 molecule depends on whether it is the first, second, third, or fourth to be bound. More salt links must be broken to permit the entry of the first O_2 than of subsequent ones. Because energy is required to break salt links, the binding of the first O_2 is energetically less favorable than that of subsequent oxygen molecules. In this scheme, the binding affinity of the second and third O_2 molecules is intermediate between that of the first and last (Figure 4-18). This sequential increase in oxygen affinity would give the sigmoidal oxygen-binding curve that is experimentally observed.

This allosteric mechanism can be expressed in more definitive structural terms. Nearly all molecules of deoxyhemoglobin are in the T form. The binding of successive molecules of O_2 makes the R form increasingly probable. Thus, the first molecule of O_2 binds to the T form, whereas the fourth molecule of O_2 binds to the R form. *The T form has a 300-fold lower affinity for O_2 than does the R form because the movement of the proximal histidine into the heme plane is more constrained in the T form.* Specifically, the proximal histidine is more tilted in the T form (see Figure 4-16) and so there is greater steric repulsion in moving toward the heme plane. The more symmetric position of the proximal histidine in the R form allows it to approach the heme with less hindrance. In addition, the O_2 binding site in the β subunits is more obstructed by valine E11 in the T form than in the R form.

DPG DECREASES OXYGEN AFFINITY BY CROSS-LINKING DEOXYHEMOGLOBIN

DPG binds specifically to deoxyhemoglobin in the ratio of one DPG per hemoglobin tetramer. This is a very interesting stoichiometry. A protein with a subunit structure of $\alpha_2\beta_2$ is usually expected to have at least two binding sites for any small molecule. The finding of one binding site immediately suggested that DPG binds on the symmetry axis of the hemoglobin molecule in the central cavity, where the four subunits are near each other (Figure 4-19). The binding site for DPG is contributed by the following positively charged residues on *both* β chains: the α-amino group, lysine EF6, and histidine H21. These groups can interact with the strongly negatively charged DPG, which carries nearly four negative charges at physiological pH. DPG is stereochemically complementary to a constellation of six positively charged groups in the β chains that face the central cavity of the hemoglobin molecule (Figure 4-20). DPG binds more weakly to fetal hemoglobin than to hemoglobin A because residue H21 in fetal hemoglobin is serine rather than histidine.

On oxygenation, DPG is extruded because the central cavity becomes too small. Specifically, the gap between the H helices of the β chains becomes narrowed. Also, the distance between the α-amino groups increases from 16 to 20 Å, so that they can no longer

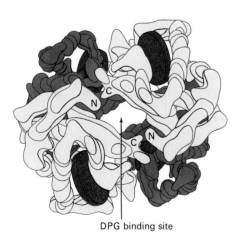

DPG binding site

Figure 4-19
The binding site for DPG is in the central cavity of deoxyhemoglobin. [After M. F. Perutz. The hemoglobin molecule. Copyright © 1964 by Scientific American, Inc. All rights reserved.]

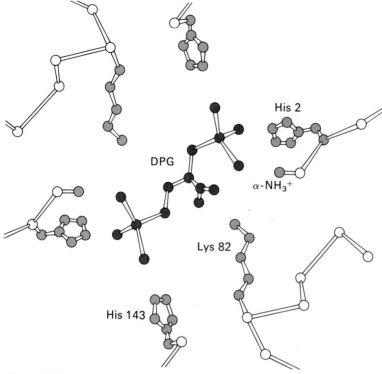

Figure 4-20
Mode of binding of DPG to human deoxyhemoglobin. DPG interacts with three positively charged groups on each β chain. [After A. Arnone. *Nature* 237(1972):148.]

make contact with the phosphates of the DPG molecule.

The basis of the effect of DPG in decreasing oxygen affinity is now evident. *DPG stabilizes the deoxyhemoglobin quaternary structure by cross-linking the β chains.* In other words, DPG shifts the equilibrium toward the T form. As mentioned previously, deoxyhemoglobin has eight salt links that anchor its carboxyl-terminal residues. These must be broken for oxygenation to occur. The binding of DPG contributes additional cross-links that must be broken, and so the oxygen affinity of hemoglobin is reduced by the binding of DPG.

CO$_2$ BINDS TO THE TERMINAL AMINO GROUPS OF HEMOGLOBIN AND LOWERS ITS OXYGEN AFFINITY

In aerobic metabolism, about 0.8 of a molecule of CO_2 is produced per molecule of O_2 consumed. Most of the CO_2 is transported as *bicarbonate*, which is formed within red cells by the action of *carbonic anhydrase:*

$$CO_2 + H_2O \rightleftharpoons HCO_3^- + H^+$$

Much of the H^+ generated by this reaction is taken up by deoxyhemoglobin as part of the Bohr effect. In addition, CO_2 is carried by hemoglobin in the form of *carbamate*. The un-ionized form of the α-amino groups of hemoglobin can react reversibly with CO_2:

$$R—NH_2 + CO_2 \rightleftharpoons R—NHCOO^- + H^+$$

The bound carbamates form salt bridges that stabilize the T form. *Hence, the binding of CO_2 lowers the oxygen affinity of hemoglobin.* Conversely, CO_2 binds more tightly to deoxyhemoglobin than to oxyhemoglobin.

MECHANISM OF THE BOHR EFFECT

How does hemoglobin take up protons in converting from the oxy into the deoxy form? The affinity of some sites for H^+ must be increased as a consequence of this transition. Specifically, the pKs of some groups must be *raised* in the transition from oxy to deoxyhemoglobin; an increase in pK means stronger binding of H^+. About 0.5 H^+ is taken up by hemoglobin for each molecule of O_2 that is released. This uptake of H^+ helps to buffer the pH in active tissues.

Which groups have their pKs raised? Let us consider the possible binding sites for H^+ in hemoglobin (Figure 4-21 and Table 4-1) and then narrow the choice. The side-chain carboxylates of glutamate

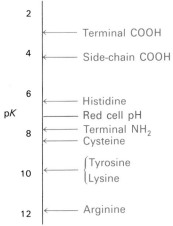

Figure 4-21
Typical pKs of acidic groups in proteins. The environment of a particular group can make its actual pK higher or lower than the value given here.

Table 4-1
pK values of ionizable groups in proteins

Group	Acid $\rightleftharpoons$ Base + H$^+$	Typical pK
Terminal carboxyl	$—COOH \rightleftharpoons —COO^- + H^+$	3.1
Aspartic and glutamic acid	$—COOH \rightleftharpoons —COO^- + H^+$	4.4
Histidine	(imidazole ring) $—NH^+ \rightleftharpoons —N + H^+$	6.5
Terminal amino	$—NH_3^+ \rightleftharpoons —NH_2 + H^+$	8.0
Cysteine	$—SH \rightleftharpoons —S^- + H^+$	8.5
Tyrosine	(phenol) $—OH \rightleftharpoons —O^- + H^+$	10.0
Lysine	$—NH_3^+ \rightleftharpoons —NH_2 + H^+$	10.0
Arginine	$—N—C(NH_2^+)(NH_2) \rightleftharpoons —N—C(NH)(NH_2) + H^+$	12.0

and aspartate normally have p*K*s of about 4. It is unlikely that their p*K*s will be raised to a value between 7 and 8, which would be necessary for a group to participate in the Bohr effect. Similarly, p*K*s of the side chains of tyrosine, lysine, and arginine are usually higher than 10, and so it is unlikely that the p*K* of any of them would be shifted far enough. Hence, the most plausible candidates for the groups participating in the Bohr effect are histidine, cysteine, and the terminal amino group, which normally have p*K* values near 7.

The identification of the specific groups responsible for the Bohr effect was achieved by combining information derived from chemical and x-ray studies. The carboxyl-terminal histidines of the β chains (histidine 146β) were implicated by x-ray studies. This suggestion was tested by preparing a hemoglobin devoid of histidine 146 on its β chains. Carboxypeptidase B, a specific proteolytic enzyme that splits the peptide bond involving carboxyl-terminal basic amino acids in some polypeptide chains, was used for this purpose. The modified hemoglobin produced in this way had only half of the Bohr effect of normal hemoglobin. Hence, *histidine 146 on each β chain appears to make a major contribution to the Bohr effect.*

The role of the terminal amino groups of hemoglobin in the Bohr effect was ascertained by specifically modifying them with cyanate to form a *carbamoylated* derivative, which can no longer bind H^+. The Bohr effect was unchanged when the terminal amino groups of only the β chains were carbamoylated. In contast, the Bohr effect was significantly reduced when the terminal amino groups of the α chains were carbamoylated.

$$\mathrm{R{-}NH_2} \;+\; {}^{-}\mathrm{NCO} \;+\; \mathrm{H^+} \longrightarrow \mathrm{R{-}NH{-}C({=}O){-}NH_2}$$

Terminal amino group | **Cyanate** | **Carbamoylated derivative**

The x-ray results provide a concrete picture of how these groups might function in the Bohr effect. In oxyhemoglobin, histidine 146β rotates freely, whereas, in deoxyhemoglobin, this terminal residue participates in a number of interactions. Of particular significance is the interaction of its imidazole ring with the negatively charged aspartate 94 on the same β chain. The close proximity of this negatively charged group enhances the likelihood that the imidazole group will bind a proton (Figure 4-22). In other words, the proximity of aspartate 94 raises the p*K* of histidine 146. Thus, *in the transition from oxy to deoxyhemoglobin, histidine 146 acquires a greater affinity for H^+ because its local charge environment is altered.*

The environment of the terminal amino groups of the α chains changes in a similar way on deoxygenation. In oxyhemoglobin,

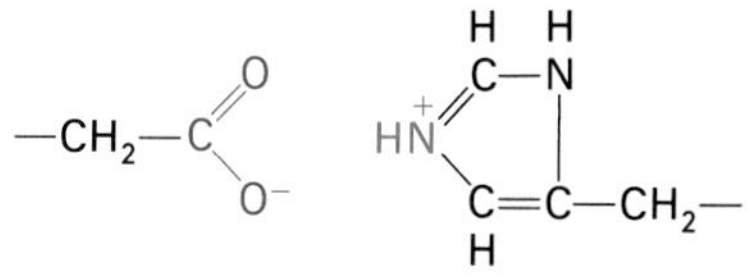

Asp 94 **His 146**

Figure 4-22
Aspartate 94 raises the p*K* of histidine 146 in deoxyhemoglobin but not in oxyhemoglobin. The proximity of the negative charge on aspartate 94 favors protonation of histidine 146 in deoxyhemoglobin.

these groups are free. In deoxyhemoglobin, the terminal amino group of one α chain interacts with the carboxyl-terminal group of the other α chain. *The proximity of a negatively charged carboxylate residue in deoxyhemoglobin increases the affinity of this terminal amino group for H^+.* X-ray results have implicated a third group in the Bohr effect, namely histidine 122 of the α chains.

In summary, three pairs of proton-binding groups (a terminal amino group and two histidines) have different environments in oxy and deoxyhemoglobin. Their immediate environments are more negatively charged in deoxyhemoglobin. Consequently, these groups take up H^+ when oxygen is released.

COMMUNICATION WITHIN A PROTEIN MOLECULE

We have seen that the binding of O_2, H^+, CO_2, and DPG are linked. These molecules are bound to spatially distinct sites that communicate with each other by means of conformational changes within the protein. The binding sites are separate because these molecules are structurally very different. *The interplay between the different sites is mediated by changes in quaternary structure.* In fact, every known allosteric protein consists of two or more polypeptide chains. The contact region between two chains can amplify and transmit conformational changes from one subunit to another. An allosteric protein does not have fixed properties. Rather, its functional characteristics are responsive to specific molecules in its environment. Consequently, allosteric interactions have immense importance in cellular function. *In the step from myoglobin to hemoglobin, a macromolecule capable of perceiving information from its environment has emerged.*

SUMMARY

New properties appear in tetrameric hemoglobin that are not present in monomeric myoglobin. Hemoglobin transports H^+ and CO_2 in addition to O_2. Furthermore, the binding of these molecules is regulated by allosteric interactions, which are defined as interactions between spatially distinct sites that are transmitted by conformational changes in the protein. Indeed, hemoglobin is the best-understood allosteric protein. Hemoglobin exhibits three kinds of allosteric effects. First, the oxygen-binding curve of hemoglobin is sigmoidal, which means that the binding of oxygen is cooperative. The binding of oxygen to one heme facilitates the binding of oxygen to the other hemes on the same molecule. More oxygen is transported because it is bound cooperatively. Second, H^+ and CO_2 promote the release of O_2 from hemoglobin, an effect that is physiologically important in enhancing the release of O_2 in metabolically active tissues such as muscle. Conversely, O_2 promotes the release of H^+ and CO_2 in the alveolar capillaries of the lungs. These allosteric linkages between the binding of H^+, CO_2, and O_2 are known as the Bohr effect. Third, the affinity of hemoglobin for O_2 is further regulated by 2,3-diphosphoglycerate (DPG), a small molecule with a high density of negative charge. DPG can bind to deoxyhemoglobin but not to oxyhemoglobin. Hence, DPG lowers the oxygen affinity of hemoglobin. DPG plays a role in adaptation to high altitude and hypoxia. Fetal hemoglobin has a higher oxygen affinity than does adult hemoglobin because it binds less DPG.

The allosteric properties of hemoglobin arise from interactions between its α and β subunits. The T (tense) quaternary structure is constrained by salt links between different subunits, which gives it a low affinity for O_2. These intersubunit salt links are absent from the R form, which has a high affinity for O_2. On oxygenation, the iron atom moves into the plane of the heme, pulling the proximal histidine with it. This motion cleaves some of the salt links, and the equilibrium is shifted from T to R. The fourth molecule of O_2 binds much more strongly than the first because fewer salt links remain to be broken. DPG binds to positively charged groups on the two β chains surrounding the central cavity of hemoglobin. The T form is stabilized by the binding of DPG and so the oxygen affinity is decreased. Carbon dioxide, another allosteric effector, binds to the terminal amino groups of all four chains by forming a readily reversible carbamate linkage. The hydrogen ions participating in the Bohr effect are bound to three pairs of sites. The environments of a pair of terminal amino groups and two pairs of histidine side chains become more negatively charged in deoxyhemoglobin, which causes them to take up H^+ when O_2 is released. Like DPG, CO_2 and H^+ lower the oxygen affinity by stabilizing the T form.

SELECTED READINGS

WHERE TO START

Perutz, M. F., 1964. The hemoglobin molecule. *Sci. Amer.* 211(11):2–14. [A delightful account of the x-ray crystallography effort on hemoglobin before 1964. Available as *Sci. Amer.* Offprint 196.]

Perutz, M. F., 1978. Hemoglobin structure and respiratory transport. *Sci. Amer.* 239(6):92–125. [A look at hemoglobin fourteen years later. The structural bases of its allosteric properties are discussed in this fine article.]

REVIEWS AND BOOKS

Perutz, M. F., 1979. Regulation of oxygen affinity of hemoglobin: influence of structure of the globin on the heme iron. *Ann. Rev. Biochem.* 48:327–386.

Bauer, C., 1974. On the respiratory function of hemoglobin. *Rev. Physiol. Biochem. Pharmacol.* 70:1–29.

Bunn, H. F., Forget, B. G., and Ranney, H. M., 1977. *Human Hemoglobins.* Saunders.

Garby, L., and Meldon, J., 1977. *The Respiratory Functions of Blood.* Plenum.

Antonini, E., and Brunori, M., 1971. *Hemoglobin and Myoglobin in Their Reactions with Ligands*. North-Holland.

BOHR EFFECT AND ACTION OF DIPHOSPHOGLYCERATE

Kilmartin, J. V., and Rossi-Bernardi, L., 1973. Interaction of hemoglobin with hydrogen ions, carbon dioxide, and organic phosphates. *Physiol. Rev.* 53:836–890.

Kilmartin, J. V., 1976. Interaction of haemoglobin with protons, CO_2, and 2,3-diphosphoglycerate. *Brit. Med. Bull.* 32:209–222.

Benesch, R., and Benesch, R. E., 1969. Intracellular organic phosphates as regulators of oxygen release by haemoglobin. *Nature* 221:618–622.

Lenfant, C., Torrance, J. D., Woodson, R. D., Jacobs, P., and Finch, C. A., 1970. Role of organic phosphates in the adaptation of man to hypoxia. *Fed. Proc.* 29:1115–1117.

Tyuma, I., and Shimizu, K., 1970. Effect of organic phosphates on the difference in oxygen affinity between fetal and adult human hemoglobin. *Fed. Proc.* 29:1112–1114.

Morpurgo, G., Arese, P., Bosia, A., Pescarmona, G. P., Luzzana, M., Modiano, G., and Ranjet, S. K., 1976. Sherpas living permanently at high altitude: a new pattern of adaptation. *Proc. Nat. Acad. Sci.* 73:747–751.

MECHANISM OF ALLOSTERIC INTERACTIONS

Perutz, M. F., 1970. Stereochemistry of cooperative effects of haemoglobin. *Nature* 228:726–739. [Presentation of a stimulating hypothesis for the mechanism of the allosteric properties of hemoglobin.]

Hopfield, J. J., 1973. Relationship between structure, cooperativity, and spectra in a model of hemoglobin action. *J. Mol. Biol.* 77:207–222.

Gelin, B. R., and Karplus, M., 1977. Mechanism of tertiary structural change in hemoglobin. *Proc. Nat. Acad. Sci.* 74:801–805.

Baldwin, J., and Chothia, C., 1979. Hemoglobin: the structural changes related to ligand binding and its allosteric mechanism. *J. Mol. Biol.* 129:175–220.

PROBLEMS

1. What is the effect of each of the following treatments on the oxygen affinity of hemoglobin A in vitro?
   (a) Increase in pH from 7.2 to 7.4.
   (b) Increase in pCO_2 from 10 to 40 torrs.
   (c) Increase in [DPG] from 2×10^{-4} to 8×10^{-4} M.
   (d) Dissociation of $\alpha_2\beta_2$ into monomer subunits.

2. What is the effect of each of the following treatments on the number of H^+ bound to hemoglobin A in vitro?
   (a) Increase in pO_2 from 20 to 100 torrs (at constant pH and pCO_2).
   (b) Reaction of hemoglobin with excess cyanate (at constant pH).

3. The erythrocytes of birds and turtles contain a regulatory molecule different from DPG. This substance is also effective in reducing the oxygen affinity of human hemoglobin stripped of DPG. Which of the following substances would you predict to be most effective in this regard?
   (a) Glucose 6-phosphate.
   (b) Inositol hexaphosphate.
   (c) HPO_4^{2-}.
   (d) Malonate.
   (e) Arginine.
   (f) Lactate.

4. The oxygen dissociation curve of hemoglobin can be described to a good approximation by the empirical equation

$$Y = \frac{(pO_2)^n}{(pO_2)^n + (P_{50})^n}$$

   in which $n = 2.8$. The derivation of this equation was discussed on page 68.

   (a) The amount of oxygen transported is proportional to ΔY, the change in saturation. Calculate ΔY for a hemoglobin molecule that goes from the lung to active muscle. Let $P_{50} = 26$ torrs, $P_{lung} = 100$ torrs, and $P_{muscle} = 20$ torrs.
   (b) Suppose that the binding of oxygen to hemoglobin were noncooperative but that P_{50} remained at 26 torrs. What would ΔY be for such a noncooperative oxygen binder?

5. The pK of an acid depends in part on its environment. Predict the effect of the following environmental changes on the pK of a glutamic acid side chain.
   (a) A lysine side chain is brought into close proximity.
   (b) The terminal carboxyl group of the protein is brought into close proximity.
   (c) The glutamic acid side chain is shifted from the outside of the protein to a nonpolar site inside.

6. The oxygen affinities of picket-fence iron porphyrins with various bases at the fifth coordination position have been measured. With 1-methylimidazole, P_{50} is 0.49 torr, whereas with 1,2-dimethylimidazole, P_{50} is 38 torrs. Why does the extra methyl group in dimethylimidazole produce such a large decrease in oxygen affinity?

7. The concept of linkage is crucial for the understanding of many biochemical processes. Consider a protein molecule P that can bind A or B or both:

$$\begin{array}{ccc}
\text{P} & \xrightleftharpoons{K_A} & \text{PA} \\
{\scriptstyle K_B}\updownarrow & & \updownarrow{\scriptstyle K_{AB}} \\
\text{PB} & \xrightleftharpoons[K_{BA}]{} & \text{PAB}
\end{array}$$

The dissociation constants for these equilibria are defined as

$$K_A = \frac{[P][A]}{[PA]} \qquad K_B = \frac{[P][B]}{[PB]}$$

$$K_{BA} = \frac{[PB][A]}{[PAB]} \qquad K_{AB} = \frac{[PA][B]}{[PAB]}$$

   (a) Suppose $K_A = 5 \times 10^{-4}$ M, $K_B = 10^{-3}$ M, and $K_{BA} = 10^{-5}$ M. Is the value of the fourth dissociation constant K_{AB} defined? If so, what is it?
   (b) What is the effect of A on the binding of B? What is the effect of B on the binding of A?

8. Carbon monoxide combines with hemoglobin to form CO-hemoglobin. Crystals of CO-hemoglobin are isomorphous with those of oxyhemoglobin. Each heme in hemoglobin can bind one carbon monoxide molecule, but O_2 and CO cannot simultaneously bind to the same heme. The binding affinity for CO is about 200 times as great as that for oxygen. Exposure for 1 hour to a CO concentration of 0.1% in inspired air leads to the occupancy by CO of about half of the heme sites in hemoglobin, which is frequently fatal.

An interesting problem was posed (and partially solved) by J. S. Haldane and J. G. Priestley in 1935:

> If the action of CO were simply to diminish the oxygen-carrying power of the hemoglobin, without other modification of its properties, the symptoms of CO poisoning would be very difficult to understand in the light of other knowledge. Thus, a person whose blood is half-saturated with CO is practically helpless, as we have just seen; but a person whose hemoglobin percentage is simply diminished to half by anemia may be going about his work as usual.

What is the key to this seeming paradox?

9. A protein molecule P reversibly binds a small molecule L. The dissociation constant K for the equilibrium

$$P + L \rightleftharpoons PL$$

is defined as

$$K = \frac{[P][L]}{[PL]}$$

The protein transports the small molecule from a region of high concentration $[L_A]$ to one of low concentration $[L_B]$. Assume that the concentrations of unbound small molecules remain constant. The protein goes back and forth between regions A and B.
   (a) Suppose $[L_A] = 10^{-4}$ M and $[L_B] = 10^{-6}$ M. What value of K yields maximal transport? One way of solving this problem is to write an expression for ΔY, the change in saturation of the ligand-binding site in going from region A to B. Then, take the derivative of ΔY with respect to K.
   (b) Treat oxygen transport by hemoglobin in a similar way. What value of P_{50} would give a maximal ΔY? Assume that the P in the lungs is 100 torrs, whereas P in the tissue capillaries is 20 torrs. Compare your calculated value of P_{50} with the physiological value of 26 torrs.

CHAPTER 5

MOLECULAR DISEASE: SICKLE-CELL ANEMIA

In 1904, James Herrick, a Chicago physician, examined a twenty-year-old black college student who had been admitted to the hospital because of a cough and fever. The patient felt weak and dizzy and had a headache. For about a year he had been experiencing palpitation and shortness of breath. Also, for several years he had participated less in physical activities. Three years earlier, there had been a discharge of pus from one ear, lasting six months. Since childhood, he had frequently had sores on his legs that were slow to heal.

On physical examination, the patient appeared rather well developed physically and was intelligent. There was a tinge of yellow in the whites of his eyes, and the visible mucous membranes were pale. His lymph nodes were enlarged. There were about twenty scars on his legs and thighs. His heart was distinctly abnormal, in that it was enlarged and a murmur was detected. Herrick noted that "the heart's action reminded one of a heart under strong stimulation, though no history of ingestion of a stimulant of any kind was obtainable."

The laboratory examination included careful scrutiny of the stools to determine whether any parasites were present, a strong possibility in a patient who had grown up in the tropics. None were found. The sputum showed no tubercle bacilli. The urine con-

Figure 5-1
Sickle cells from the blood of a patient with sickle-cell anemia, as viewed under a light microscope. [Courtesy of Dr. Frank Bunn.]

tained cellular debris indicative of diseased kidneys. The blood was highly abnormal:

	Observed	Normal range
Red cell count	2.6×10^6/ml	4.6–6.2×10^6/ml
Hemoglobin content	8 g/100 ml	14–18 g/100 ml
White cell count	15,250/ml	4,000–10,000/ml

The patient was definitely anemic, his hemoglobin content being half of the normal value. The red cells varied greatly in size, many abnormally small ones being evident. Also, many nucleated red cells were present. They are the precursors of normal red cells, which do not contain nuclei. Herrick described the unusual red cells in these terms: "The shape of the red cells was very irregular, but *what especially attracted attention was the large number of thin, elongated, sickle-shaped and crescent-shaped forms.* These were seen in fresh specimens, no matter in what way the blood was spread on the slide. . . . They were not seen in specimens of blood taken at the same time from other individuals and prepared under exactly similar conditions. They were surely not artifacts, nor were they any form of parasite."

The treatment was supportive, consisting of rest and nourishing food. The patient left the hospital four weeks later, less anemic and feeling much better. However, his blood still exhibited a "tendency to the peculiar crescent-shape in the red corpuscles though this was by no means as noticeable as before."

Herrick was puzzled by the clinical picture and laboratory findings. Indeed, he waited six years before publishing the case history and then candidly asserted that "not even a definite diagnosis can be made." He noted the chronic nature of the disease, and the diversity of abnormal physical and laboratory findings: cardiac enlargement, a generalized swelling of lymph nodes, jaundice, anemia, and evidence of kidney damage. He concluded that the disease could not be explained on the basis of an organic lesion in any one organ. He singled out the abnormal blood picture as the key finding and titled his case report *Peculiar Elongated and Sickle-Shaped Red Blood Corpuscles in a Case of Severe Anemia.* Herrick suggested that "some unrecognized change in the composition of the corpuscle itself may be the determining factor."

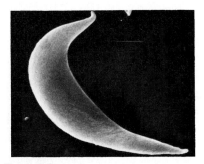

Figure 5-2
Scanning electron micrograph of an erythrocyte from a patient with sickle-cell anemia. [Courtesy of Dr. Jerry Thornwaite and Dr. Robert Leif.]

SICKLE-CELL ANEMIA IS A CHRONIC, HEMOLYTIC DISEASE THAT IS GENETICALLY TRANSMITTED

Other cases of this disease, called sickle-cell anemia, were found soon after the publication of Herrick's description. Indeed, sickle-cell anemia is not a rare disease. It is a significant public health problem whenever there is a substantial black population. The incidence of sickle-cell anemia among blacks is about four per thousand. In the past, sickle-cell anemia has usually been a fatal disease, often before age thirty, as a result of infection, renal failure, cardiac failure, or thrombosis.

The sickled red cells become trapped in the small blood vessels. Circulation is impaired, resulting in the damage of multiple organs, particularly bone and kidney. The sickled cells are more fragile than normal cells. They hemolyze readily and consequently have a shorter life than normal cells. The resulting anemia is usually severe. The chronic course of the disease is punctuated by crises in which the proportion of sickled cells is especially high. During such a crisis, the patient may go into shock.

Sickle-cell anemia is genetically transmitted. Patients with *sickle-cell anemia are homozygous for an abnormal gene* located on one of the autosomal chromosomes. Offspring who receive the abnormal gene from one parent but its normal allele from the other have *sickle-cell trait*. Such *heterozygous* people are usually not symptomatic. Only 1% of the red cells in a heterozygote's venous circulation are sickled, in contrast with about 50% in a homozygote. However, sickle-cell trait, which occurs in about one of ten blacks, is not entirely benign. Vigorous physical activity at high altitude, air travel in unpressurized planes, and anesthesia can be potentially hazardous to a person who has sickle-cell trait. The reason will be evident shortly.

THE SOLUBILITY OF DEOXYGENATED SICKLE HEMOGLOBIN IS ABNORMALLY LOW

Herrick correctly surmised the location of the defect in sickle-cell anemia. Red cells from a patient with this disease will sickle on a microscope slide in vitro if the concentration of oxygen is reduced. In fact, the hemoglobin in these cells is itself defective. Deoxygenated sickle-cell hemoglobin has an abnormally low solubility, which is about one twenty-fifth of the solubility of normal deoxygenated hemoglobin. A fibrous precipitate is formed if a concentrated solution of sickle-cell hemoglobin is deoxygenated (Figure 5-3). This precipitate deforms red cells and gives them their sickle shape. Sickle-cell hemoglobin is commonly referred to as hemoglobin S (Hb S) to distinguish it from hemoglobin A (Hb A), the normal adult hemoglobin.

HEMOGLOBIN S HAS AN ABNORMAL ELECTROPHORETIC MOBILITY

In 1949, Pauling and his associates examined the physical-chemical properties of hemoglobin from normal people and from those with sickle-cell trait or sickle-cell anemia. Their experimental approach

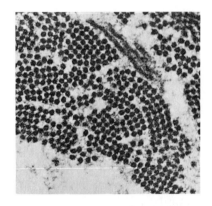

2000 Å

Figure 5-3
Electron micrographs of fibers of deoxygenated sickle-cell hemoglobin. The upper figure is a cross section and the lower a longitudinal section. The fibers are 170 Å thick. [From J. T. Finch, M. F. Perutz, J. F. Bertles, and J. Döbler, *Proc. Nat. Acad. Sci.* 70(1973):719.]

was to search for differences in these hemoglobins by measuring their mobilities in an electric field. This technique is called *electrophoresis*. The velocity of migration (V) of a protein (or of any other molecule) in an electric field depends on the strength of the electric field (E), the net electric charge on the protein (Z), and the frictional resistance (f). The frictional resistance is a function of the size and shape of the protein. The migration velocity is related to these variables by

$$V = \frac{EZ}{f} \tag{1}$$

The *isoelectric point* is the pH at which there is no net electric charge on a protein. At this pH, the electrophoretic mobility is zero because Z in equation 1 is zero. Below the isoelectric pH, the molecule is positively charged; above the isoelectric pH, it is negatively charged. The isoelectric points for normal and sickle-cell hemoglobin are:

	Normal	Sickle-cell anemia	Difference
Oxyhemoglobin	6.87	7.09	0.22
Deoxyhemoglobin	6.68	6.91	0.23

Thus, the electrophoretic difference between normal and sickle-cell hemoglobin is the same in both the oxygenated and deoxygenated forms.

The different electrophoretic velocities of hemoglobins A and S could arise from a change either in the net charge Z or in the frictional coefficient f. A purely frictional effect (produced by a change in shape) would cause one species to move more slowly than the other throughout the entire pH range. This is not the case, because the slope of the plot of the electrophoretic velocity versus pH is the same for hemoglobins A and S. Furthermore, other physical-chemical experiments, namely velocity sedimentation and free diffusion measurements, showed that the frictional coefficient is the same for the oxygenated forms of these hemoglobins.

These observations suggested that *there is a difference in the number or kind of ionizable groups in the two hemoglobins.* How many are in each? An estimate was made from the acid-base titration curve of hemoglobin. In the neighborhood of pH 7, this curve is nearly linear. A change of one pH unit in the hemoglobin solution produces a change of about thirteen charges. The difference in isoelectric pH of 0.23 therefore corresponds to about three charges per hemoglobin molecule. It was concluded that *sickle-cell hemoglobin has between two and four more net positive charges per molecule than normal hemoglobin.*

Is the difference in the polypeptide chain or in the heme group? The porphyrin isolated from sickle-cell hemoglobin was shown to be normal on the bases of its melting point and x-ray crystallo-

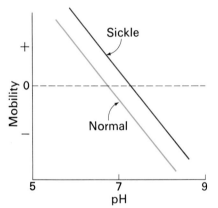

Figure 5-4
Electrophoretic mobility of sickle-cell hemoglobin and of normal hemoglobin as a function of pH.

graphic pattern. Hence, it was deduced that *the difference between the two hemoglobins must be in their polypeptide chains.*

Patients with sickle-cell anemia (who are homozygous for the sickle gene) have hemoglobin S but no hemoglobin A. In contrast, people with sickle-cell trait (who are heterozygous for the sickle gene) have both kinds of hemoglobin in approximately equal amounts (Figure 5-5). Thus, Pauling's study revealed "a clear case of a change produced in a protein molecule by an allelic change in a single gene" involved in its synthesis. This was the first demonstration of a *molecular disease.*

FINGERPRINTING: DETECTION OF THE ALTERED AMINO ACID IN SICKLE HEMOGLOBIN

The electrophoretic analysis showed that hemoglobin S has from two to four more positive charges than does hemoglobin A. A difference of this kind could arise in several ways:

Hb A	Hb S
Neutral amino acid ⟶	Positively charged amino acid
Negatively charged amino acid ⟶	Positively charged amino acid
Negatively charged amino acid ⟶	Neutral amino acid

The breakthrough in the elucidation of the precise change came in 1954, when Vernon Ingram devised a new technique for detecting amino acid substitutions in proteins. The hemoglobin molecule was split into smaller units for analysis because it was anticipated that it would be easier to detect an altered amino acid in a peptide containing about 20 residues than in a protein ten times as large. Trypsin was used to specifically cleave hemoglobin on the carboxyl side of its lysine and arginine residues. Because the $\alpha\beta$ half of hemoglobin contains a total of 27 lysine and arginine residues, 28 different peptides were formed by *tryptic digestion.* The next step was to separate these peptides. This was accomplished by a *two-dimensional procedure* (Figure 5-6). The mixture of peptides was placed in a spot

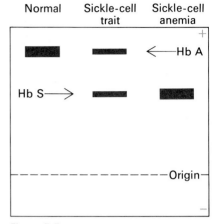

Figure 5-5
Starch-gel electrophoresis pattern at pH 8.6 of hemoglobin isolated from a normal person, from a person with sickle-cell trait, and from a person with sickle-cell anemia.

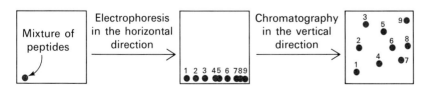

Figure 5-6
A mixture of peptides is resolved by electrophoresis in the horizontal direction, followed by partition chromatography in the vertical direction.

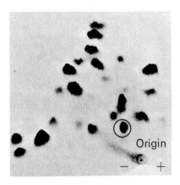

Hemoglobin A

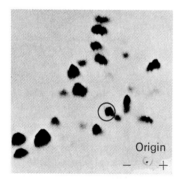

Hemoglobin S

Figure 5-7
Comparison of the ninhydrin-stained fingerprints of hemoglobin A and hemoglobin S. The position of the peptide that is different in these hemoglobins is circled in red. [Courtesy of Dr. Corrado Baglioni.]

at one corner of a large sheet of filter paper. *Electrophoresis* was first carried out in one direction, separating the peptides according to their net charge. However, electrophoresis did not suffice to separate the peptides completely. There was much overlapping. A second separation technique, namely paper chromatography, was carried out perpendicular to the direction of electrophoresis. In this procedure, the end of the filter paper closest to the peptides is immersed in a mixture of organic solvents and water at the bottom of a sealed glass jar. The solvent ascends into the paper. Each peptide now has a choice: it can either migrate with the solvent, which is quite nonpolar, or cling to the hydrated cellulose support, which is highly polar. This separatory technique is called *partition chromatography*. A highly nonpolar peptide will partition with the rising solvent and will consequently move with the solvent front to the top of the paper, whereas a markedly polar peptide will remain near the bottom of the paper. Paper chromatography and electrophoresis are complementary techniques because they separate peptides on the bases of different characteristics—namely, their degree of polarity and net charge, respectively. This sequence of steps—selective cleavage of a protein into small peptides, followed by their separation in two dimensions—is called *fingerprinting*.

The resulting fingerprint was highly revealing. The peptide spots were made visible by staining the fingerprint with ninhydrin. A comparison of the maps for hemoglobins A and S showed that *all but one of the peptide spots matched*. The one spot that was different was eluted from each fingerprint and then shown to be a single peptide consisting of eight amino acids. Amino acid analysis indicated that this peptide in hemoglobin S differed from the one in hemoglobin A by a single amino acid.

A SINGLE AMINO ACID IN THE BETA CHAIN IS ALTERED

The α and β chains were separated by ion-exchange chromatography. Fingerprints of the separate chains showed that the change in hemoglobin S is in its β chains. In fact, the difference is located in the amino-terminal tryptic peptide of the β chain. Ingram determined the sequence of this peptide and showed that *hemoglobin S contains valine instead of glutamate at position 6 of the β chain:*

Hemoglobin A	Val-His-Leu-Thr-Pro-Glu-Glu-Lys-
Hemoglobin S	Val-His-Leu-Thr-Pro-Val-Glu-Lys-
	β1 2 3 4 5 6 7 8

SICKLE HEMOGLOBIN HAS STICKY PATCHES ON ITS SURFACE

The side chain of valine is distinctly nonpolar, whereas that of glutamate is highly polar. The substitution of valine for glutamate at position 6 of the β chains places a nonpolar residue on the outside of hemoglobin S (Figure 5-8). *This alteration markedly reduces the solubility of deoxygenated hemoglobin S but has little effect on the solubility of oxygenated hemoglobin S.* This fact is crucial to an understanding of the clinical picture of sickle-cell anemia and sickle-cell trait.

The molecular basis of sickling can be visualized as follows:

1. The substitution of valine for glutamic acid gives hemoglobin S a sticky patch on the outside of each of its β chains (Figure 5-9). This sticky patch is present on both oxy and deoxyhemoglobin S and is missing from hemoglobin A.

2. There is a site on deoxyhemoglobin S that is complementary to the sticky patch (Figure 5-9). Its location is not yet known. The

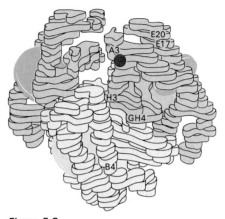

Figure 5-8
Model of deoxyhemoglobin A at low resolution. The α chains are shown in yellow, the β chains in blue. The position of the amino acid change in hemoglobin S is marked in red. Note that this site is at the surface of the molecule. [After J. T. Finch, M. F. Perutz, J. F. Bertles, and J. Döbler. *Proc. Nat. Acad. Sci.* 70(1973):721.]

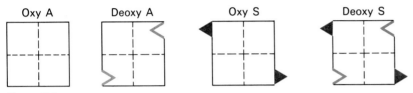

Figure 5-9
The red triangle represents the sticky patch that is present on both oxy and deoxyhemoglobin S but not on either form of hemoglobin A. The complementary site is represented by an indentation that can accommodate the triangle. This complementary site is present in deoxyhemoglobin S and is probably also present in deoxyhemoglobin A.

complementary site on one deoxyhemoglobin S molecule interacts with the sticky patch on another deoxyhemoglobin S molecule, which results in the formation of long aggregates that distort the red cell.

3. In oxyhemoglobin S, the complementary site is masked. The sticky patch is present, but it cannot bind to another hemoglobin S because the complementary site is unavailable.

4. Thus, *sickling occurs when there is a high concentration of the deoxygenated form of hemoglobin S* (Figure 5-10).

These facts account for several clinical characteristics of sickle-cell anemia. A vicious circle is set up when sickling occurs in a small

Figure 5-10
Interaction of a sticky patch on deoxyhemoglobin S with the complementary site on another deoxyhemoglobin S molecule to form long aggregates. One strand of the helical fiber is represented here.

blood vessel. The blockage of the vessel creates a local region of low oxygen concentration. Hence, more hemoglobin goes into the deoxy form and so more sickling occurs. A person with sickle-cell trait is usually asymptomatic because not more than half of his hemoglobin is hemoglobin S. This is too low a concentration for extensive sickling at normal oxygen levels. However, if the oxygen level is unusually low (as at high altitude), sickling can occur in someone who has sickle-cell trait.

DEOXYHEMOGLOBIN S FORMS LONG HELICAL FIBERS

As described previously, deoxyhemoglobin S forms fibrous precipitates that deform red cells and give them their sickle shape (Figure 5-2). Two kinds of fibers have been seen by electron microscopy: a fiber with a diameter of 170 Å (Figure 5-3) and, more frequently, one with a diameter of 215 Å (Figure 5-11). It seems likely that the more prevalent fibrous precipitate is a four-stranded helix built

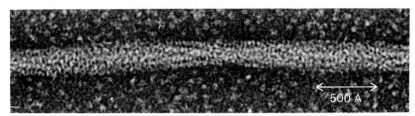

Figure 5-11
Electron micrograph of a negatively stained fiber of deoxyhemoglobin S. [From G. Dykes, R. H. Crepeau, and S. J. Edelstein. *Nature* 272(1978):509.]

from ten external and four internal hemoglobin S molecules (Figure 5-12). A significant feature of this fourteen-stranded helix is that each hemoglobin S molecule makes contact with at least eight others. It is evident that the fiber is stabilized by multiple interactions. The one involving valine 6 of the β chain tips the thermodynamic balance of the deoxy state toward fiber formation, but it is not the only important contact.

THE RATE OF FIBER FORMATION DEPENDS STRONGLY ON THE CONCENTRATION OF DEOXYHEMOGLOBIN S

The kinetics of formation of deoxyhemoglobin S fibers are important in determining whether a red cell becomes sickled during its passage through the capillary circulation, which takes about a second. The most important determinant is the concentration of deoxyhemoglobin S, as shown by in vitro studies that took advantage of the fact that deoxyhemoglobin S is markedly more soluble at low

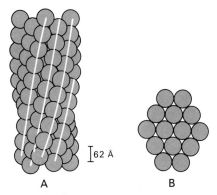

62 Å

A   B

Figure 5-12
A fourteen-stranded helical model of the deoxyhemoglobin S fiber: (A) axial view; (B) cross-sectional view. Each circle represents a hemoglobin S tetramer. [After G. Dykes, R. H. Crepeau, and S. J. Edelstein. *Nature* 272(1978):509.]

temperatures than at high ones. The polymerization of deoxy-hemoglobin S was initiated by rapidly shifting the temperature of a solution from 4° to 37°C. The formation of fibers was then measured by monitoring the change in a physical property, such as the increase in light scattering. At a sufficiently low concentration of deoxyhemoglobin S, the formation of fibers is delayed by many minutes (Figure 5-13). In fact, the delay time τ depends on the concentration c (more precisely, on the thermodynamic activity) of deoxyhemoglobin S according to

$$1/\tau = k(c/C_s)^n$$

in which C_s is the solubility of hemoglobin S at equilibrium, k is a constant that depends on environmental conditions, and n is the power dependence of this polymerization reaction. The striking experimental finding is that n is a large number, about 10. In other words, *the rate of fiber formation is proportional to the tenth power of the activity of deoxyhemoglobin S. Thus, fiber formation is a highly concerted reaction.* These kinetic data indicate that nucleation is the rate-limiting phase in fiber formation. The fiber grows rapidly once a critical cluster of about ten deoxyhemoglobin S molecules has been formed (Figure 5-14). Such a nucleus may correspond to a major part of one turn of the fourteen-stranded helix (Figure 5-12). These findings have important clinical implications. They demonstrate that *kinetic,* as well as thermodynamic, factors are important in sickling. A red cell that is supersaturated with respect to deoxy-hemoglobin S will not sickle if the lag time for fiber formation is longer than the transit time from the peripheral capillaries to the alveoli of the lungs, where reoxygenation occurs. The very strong dependence of the polymerization rate on the concentration of deoxyhemoglobin S accounts for the fact that people with sickle-cell trait are usually asymptomatic. The concentration of deoxy-hemoglobin S in the red cells of these heterozygotes is about a factor of two less than in homozygotes, and so their rate of fiber formation is about a thousandfold slower ($2^{10} = 1024$).

HIGH INCIDENCE OF THE SICKLE GENE IS DUE TO THE PROTECTION CONFERRED AGAINST MALARIA

The frequency of the sickle gene is as high as 40% in certain parts of Africa. Until recently, most homozygotes have died before reaching adulthood, and so there must have been strong selective pressures to maintain the high incidence of the gene. James Neel proposed that the heterozygote enjoys advantages not shared by either the normal homozygote or the sickle-cell homozygote. In fact, Anthony Allison found that people with sickle-cell trait are protected against the

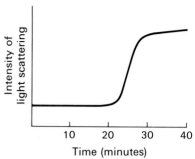

Figure 5-13
Kinetics of formation of fibers of deoxy-hemoglobin S in vitro as monitored by the change in light scattering. The reaction was initiated by rapidly raising the temperature from 4° to 37°C. The long lag corresponds to the nucleation phase of the reaction.

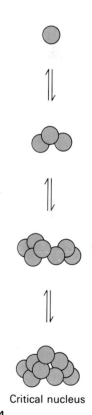

Critical nucleus

Figure 5-14
Nucleation phase in the formation of deoxyhemoglobin S fibers. The assembly of these nuclei is much slower than their subsequent growth. [After J. Hofrichter, P. D. Ross, and W. A. Eaton. *Proc. Nat. Acad. Sci.* 71(1974):4864.]

most lethal form of malaria. The incidence of malaria and the frequency of the sickle gene in Africa are definitely correlated (Figure 5-15). This is a clear-cut example of *balanced polymorphism*—the heterozygote is protected against malaria and does not suffer from sickle-cell disease, whereas the normal homozygote is vulnerable to malaria.

STRATEGIES FOR THE DESIGN OF ANTISICKLING DRUGS

Information about the molecular basis of sickle-cell anemia is now being used to design specific, nontoxic drugs to prevent or lessen its clinical consequences. Three types of therapeutic strategies are being pursued. The first group of prospective drugs inhibits the formation of hemoglobin S fibers. Synthetic oligopeptides with sequences corresponding to the amino-terminal region of the β^S chain and other contact sites are being evaluated as *stereospecific gelation inhibitors*. Compounds that decrease the concentration of the deoxy form of hemoglobin S by increasing its oxygen affinity are a second type of antisickling agent. For example, *cyanate enters red cells and increases oxygen affinity by irreversibly carbamoylating the α-amino groups of hemoglobin*. Hemoglobin reacts rapidly with cyanate because $HN{=}C{=}O$ (isocyanic acid), the un-ionized form, is a reactive analog of $O{=}C{=}O$. Recall that the terminal-amino groups of hemoglobin participate in the reversible binding of CO_2 (p. 79). Cyanate was a promising antisickling drug until extensive clinical trials uncovered its toxic side effects. For example, some patients treated with cyanate developed peripheral nerve damage, probably because proteins other than hemoglobin were carbamoylated. The search is on for less toxic modifiers of hemoglobin. A third potential approach to treating sickle-cell anemia is to *decrease the total concentration of hemoglobin S in the red cell*. For example, this might be achieved by increasing the volume of the red cell. Even a small increase in the red-cell volume would have a large antisickling effect because the rate of fiber formation depends on a high power of the concentration of deoxyhemoglobin S. Thus, ion pumps and channels in the red-cell membrane are potential targets in the therapy of sickle-cell anemia.

MOLECULAR PATHOLOGY OF HEMOGLOBIN

More than a hundred abnormal hemoglobins have been discovered by the examination of patients with clinical symptoms and by electrophoretic surveys of normal populations. In northern Europe, one of three hundred persons is heterozygous for a variant of hemoglobin A. The frequency of any one mutant allele is less than 10^{-4},

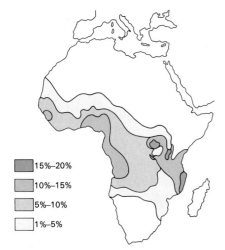

15%–20%
10%–15%
5%–10%
1%–5%

Figure 5-15
Frequency of the sickle-cell gene in Africa. High frequencies are restricted to regions where malaria is a major cause of death. [After A. C. Allison. Sickle cells and evolution. Copyright © 1956 by Scientific American, Inc. All rights reserved.]

which is lower by several orders of magnitude than the frequency of the sickle allele in regions where malaria is endemic. In other words, most abnormal hemoglobins do not confer a selective advantage on the person. They are almost always neutral or harmful.

The abnormal hemoglobins are of several types:

1. *Altered exterior.* Nearly all substitutions on the surface of the hemoglobin molecule are harmless. Hemoglobin S is a striking exception.

2. *Altered active site.* The defective subunit cannot bind oxygen because of a structural change near the heme that directly affects oxygen binding.

3. *Altered tertiary structure.* The polypeptide chain is prevented by the amino acid substitution from folding into its normal conformation. These hemoglobins are usually unstable.

4. *Altered quaternary structure.* Some mutations at subunit interfaces result in the loss of allosteric properties. These hemoglobins usually have an abnormal oxygen affinity.

HEMOGLOBIN M: ACTIVE-SITE MUTATION

Substitution of tyrosine for the proximal or distal heme histidine results in the stabilization of the heme in the ferric form, which can no longer bind oxygen (Figure 5-16). The tyrosine side chain is ionized in this complex with the ferric ion of the heme. This change can occur in either the α or β chain. Indeed, all four kinds of mutants have been observed. Mutant hemoglobins characterized by a permanent ferric state of two of the hemes are called *hemoglobin M*. The letter M signifies that the altered chains are in the *methemoglobin* (ferrihemoglobin) form. The patients are usually cyanotic. The disease has been seen only in heterozygous form, because homozygosity would almost certainly be lethal.

POLAR GROUPS IN THE HEME POCKET
IMPAIR THE BINDING OF HEME

A different kind of active-site mutation occurs when a polar group is substituted for a nonpolar one in the heme crevice. There are sixty interatomic contacts between the polypeptide chain and the heme group. These contacts are nonpolar in nature. The conservation of these interactions in normal hemoglobins of many species suggests that most of them are essential for the function of the hemoglobin molecule. Indeed, mutations at these heme-binding sites almost always have harmful consequences. For example, in

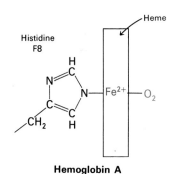

Figure 5-16
Substitution of tyrosine for the proximal histidine (F8) results in the formation of a hemoglobin M. The negatively charged oxygen atom of tyrosine is coordinated to the iron atom, which is in the ferric state. Water, rather than O_2, is bound at the sixth coordination position.

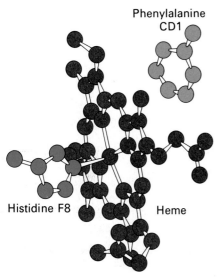

Figure 5-17
Location of phenylalanine CD1, one of the conserved residues of hemoglobin. The aromatic ring of this residue is in contact with the heme.

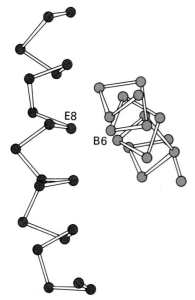

Figure 5-18
A hemoglobin molecule can have a normal structure only if the residue at B6 is glycine. There is no space for a larger side chain because this is where the B and E helices cross.

hemoglobin Hammersmith, the phenylalanine residue at CD1 (Figure 5-17) is replaced by serine. This change impairs the binding of the heme, probably because the polar serine residue enables water to enter the crevice in which the heme is normally bound.

SOME MUTATIONS PRODUCE UNSTABLE HEMOGLOBINS RESULTING FROM DISTORTED TERTIARY STRUCTURE

Amino acid substitutions at sites far from the heme can prevent the molecule from folding into its normal conformation, which severely disrupts its function. An interesting example is hemoglobin Riverdale-Bronx, which has arginine at B6 in place of glycine. The arginine is too large to fit into the narrow region where the B and E helices usually cross (Figure 5-18). In fact, B6 is glycine in all known normal myoglobins and hemoglobins. The effect of this substitution is to make hemoglobin Riverdale unstable.

Hemoglobin Gun Hill is one of the few known abnormal hemoglobins in which a mutation affects more than one amino acid residue in a chain. In this hemoglobin, five amino acids in each of the β chains are deleted. These residues correspond to the last amino acid in the F helix and the next four in the corner between the F and G helices. This deletion mutation removes a section of the polypeptide chain that forms essential contacts with the heme group. As a result there is no heme in the β chains. The clinical picture is a hemolytic anemia.

ALLOSTERIC INTERACTIONS ARE IMPAIRED IN SOME SUBUNIT INTERFACE MUTANTS

Effective oxygen transport requires that the allosteric properties of hemoglobin be fully operative. The integrity of the heme-binding site and the normal folding of a hemoglobin subunit are necessary but not sufficient for optimal function. Residues at contacts between subunits also are important because some of them transmit information between different subunits. Contacts between like chains in hemoglobin are tenuous and polar, whereas those between unlike chains are extensive and predominantly nonpolar. There are two different kinds of interfaces between the α and β chains, named $\alpha_1\beta_1$ and $\alpha_1\beta_2$. As noted previously, the $\alpha_1\beta_2$ *contacts* seem to play a key role in transmitting allosteric interactions because there is considerable movement at these interfaces on oxygenation. This inference is reinforced by the finding that *nearly all hemoglobins with replacements at the $\alpha_1\beta_2$ contact have diminished cooperativity. Furthermore, the oxygen affinities of these hemoglobins are usually abnormal.*

For example, hemoglobin Kempsey has a P_{50} of 15 torrs, in contrast with the normal value of 26 torrs. In this mutant, the G1 aspartate on the β chain is replaced by asparagine. How does this substitution lead to an abnormally high oxygen affinity? Recall that the $\alpha_1\beta_2$ interface in deoxyhemoglobin is normally stabilized by a hydrogen bond between the G1 aspartate and a tyrosine residue on the α chain (p. 75). This hydrogen bond contributes to the stability of the T (tense) state that is characteristic of normal deoxyhemoglobin. *The absence of this interaction in hemoglobin Kempsey lessens the stability of its T state and so it has a higher oxygen affinity. Conversely, mutations that impair the stability of the R (relaxed) state would be expected to reduce the oxygen affinity.* In fact, the P_{50} of hemoglobin Kansas is 70 torrs. Patients with this hemoglobin are markedly cyanotic because their arterial oxygen saturation is only about 0.6 rather than 0.97. The substitution of threonine for G4 asparagine in hemoglobin Kansas prevents this residue from forming a hydrogen bond that normally stabilizes oxyhemoglobin.

IMPACT OF THE DISCOVERY OF MUTANT HEMOGLOBINS

Mutations affecting oxygen transport have had a major impact on molecular biology, medicine, genetics, and anthropology. Their significance is threefold:

1. *They are sources of insight into relationships between the structure and function of normal hemoglobin.* Mutations of a single amino acid residue are highly specific chemical modifications provided by nature. They show which parts of the molecule are critical for function.

2. *The discovery of mutant hemoglobins has revealed that disease can arise from a change of a single amino acid in one kind of polypeptide chain.* The concept of molecular disease, which is now an integral part of medicine, had its origins in the incisive studies of sickle-cell hemoglobin.

3. *The finding of mutant hemoglobins has enhanced our understanding of evolutionary processes.* Mutations are the raw materials of evolution; the studies of sickle-cell anemia have shown that a mutation may be simultaneously beneficial and harmful. The disease of an individual may be a concomitant of the evolutionary process, as has been picturesquely described by Zuckerkandl and Pauling (1962):

> Subjectively, to evolve must most often have amounted to suffering from a disease. And these diseases were of course molecular. The appearance of the concept of good and evil, interpreted by man as his painful expulsion from Paradise, was probably a molecular disease that turned out to be evolution.

SUMMARY

The change of a single amino acid in a single protein caused by gene mutation can produce disease. The best-understood molecular disease is sickle-cell anemia. Hemoglobin S, the abnormal hemoglobin in patients with this disease, consists of two normal α chains and two mutant β chains. The amino acid change in the β chains of hemoglobin S is the replacement of glutamate at position 6 by valine. This substitution of a nonpolar side chain for a polar one drastically reduces the solubility of deoxyhemoglobin S. In contrast, the solubility of oxyhemoglobin S is quite normal. Deoxygenated hemoglobin S forms fibrous precipitates that deform the red cell and give it a sickle shape. These red cells are thereby destroyed and so the clinical picture is a chronic hemolytic anemia. Sickle-cell anemia occurs when a person is homozygous for the mutant sickle gene for the β chain. Heterozygotes synthesize a mixture of hemoglobin A and hemoglobin S. The heterozygous condition, called sickle-cell trait, is relatively asymptomatic. About one of ten blacks is heterozygous for the sickle gene. This high incidence of a gene that is harmful in homozygotes is due to the fact that it is beneficial in heterozygotes, a state known as balanced polymorphism. People with sickle-cell trait are protected against the most lethal form of malaria.

More than a hundred mutant hemoglobins have been discovered by the examination of the hemoglobin of patients with hematologic symptoms and by surveys of normal populations. Hemoglobins with electrophoretic mobilities different from that of hemoglobin A are fingerprinted and the amino acid sequence of the new peptide in the fingerprint is then determined. Several classes of mutant hemoglobins are known. First, nearly all substitutions on the surface of the hemoglobin molecule are harmless. Hemoglobin S is the striking exception. Second, the oxygen-binding site is often impaired by substitutions near the heme. For example, substitution of tyrosine for the proximal or distal histidine results in the formation of a hemoglobin M, which is locked in the ferric (met) state and therefore unable to bind oxygen. Third, alterations in the interior of the molecule often distort the tertiary structure and result in an unstable molecule, as in the hemoglobin that contains arginine in place of glycine at a site where glycine is required because of its small size. Fourth, alterations at subunit interfaces often lead to the loss of allosteric properties. The oxygen affinity is changed by amino acid substitutions that preferentially stabilize the T or R state.

SELECTED READINGS

WHERE TO START

Allison, A. C., 1956. Sickle cells and evolution. *Sci. Amer.* 195(2):87–94. [Available as *Sci. Amer.* Offprint 1065.]

Perutz, M. F., and Lehmann, H., 1968. Molecular pathology of human haemoglobin. *Nature* 219:902–909. [A landmark analysis of more than fifty mutant hemoglobins. Clinical symptoms and the inferred alterations in structure are correlated.]

SICKLE-CELL ANEMIA

Dean, J., and Schechter, A. N., 1978. Sickle-cell anemia: molecular and cellular bases of therapeutic approaches. *New Engl. J. Med.* 299:752–763, 804–811, and 863–870. [An excellent review of structural, genetic, and clinical aspects of sickle-cell anemia.]

Herrick, J. B., 1910. Peculiar elongated and sickle-shaped red blood corpuscles in a case of severe anemia. *Arch. Intern. Med.* 6:517–521.

Neel, J. V., 1949. The inheritance of sickle-cell anemia. *Science* 110:64–66.

Pauling, L., Itano, H. A., Singer, S. J., and Wells, I. C., 1949. Sickle cell anemia: a molecular disease. *Science* 110:543–548.

Ingram, V. M., 1957. Gene mutation in human haemoglobin: the chemical difference between normal and sickle cell haemoglobin. *Nature* 180:326–328.

Pasvol, G., Weatherall, D. J., and Wilson, R. J. M., 1978. Cellular mechanism for the protective effect of haemoglobin S against P. falciparum malaria. *Nature* 274:701–703.

Behe, M. J., and Englander, S. W., 1978. Sickle hemoglobin gelation: reaction order and critical nucleus size. *Biophys. J.* 23:129–145.

Eaton, W. A., Hofrichter, J., and Ross, P. D., 1976. Delay time of gelation: a possible determinant of clinical severity in sickle cell disease. *Blood* 47:621–627.

OTHER MUTANT HEMOGLOBINS

Bunn, H. F., Forget, B. G., and Ranney, H. M., 1977. *Human Hemoglobins.* Saunders.

Caughey, W. S., (ed.), 1978. *Biochemical and Clinical Aspects of Hemoglobin Abnormalities.* Academic Press.

Lehmann, H., and Huntsman, R. G., 1974. *Man's Haemoglobins* (2nd ed.), Lippincott.

Winslow, R. M., and Anderson, W. F., 1978. The hemoglobinopathies. *In* Stanbury, J. B., Wyngaarden, J. B., and Fredrickson, D. S., (eds.), *The Metabolic Basis of Inherited Disease* (4th ed.), pp. 1465–1507. McGraw-Hill.

Bellingham, A. J., 1976. Hemoglobins with altered oxygen affinity. *Brit. Med. Bull.* 32:234–238.

DISEASE AND EVOLUTION

Zuckerkandl, E., and Pauling, L., 1962. Molecular disease, evolution, and genic heterogeneity. *In* Kasha, M., and Pullman, B., (eds.), *Horizons in Biochemistry,* pp. 189–225. Academic Press.

Cavalli-Sforza, L. L., and Bodmer, W. F., 1971. *The Genetics of Human Populations.* Freeman. [An excellent discussion of genetic polymorphism is given in chapters 4 and 5.]

PROBLEMS

1. A hemoglobin with an abnormal electrophoretic mobility is detected in a screening program. Fingerprinting after tryptic digestion reveals that the amino acid substitution is in the β chain. The normal amino-terminal tryptic peptide (Val-His-Leu-Thr-Pro-Glu-Glu-Lys) is missing. A new tryptic peptide consisting of six amino acid residues is found. Valine is the amino terminal residue of this peptide.

   (a) Which amino acid substitutions are consistent with these data?

   (b) What is the electrophoretic mobility of this hemoglobin likely to be in relation to Hb A and Hb S at pH 8?

2. The electrophoretic mobilities at pH 8.6 of a series of mutant Hb A molecules are shown below.

Match positions a, b, c, and d with the following mutant hemoglobins:

Hb D (α68)	Lysine instead of asparagine
Hb J (β69)	Aspartate instead of glycine
Hb N (β95)	Glutamate instead of lysine
Hb C (β6)	Lysine instead of glutamate

3. Some mutations in a hemoglobin gene affect all three of the hemoglobins A, A_2, and F, whereas others affect only one of them. Why?

4. The starch-gel electrophoresis pattern of hemoglobin from a person with sickle-cell trait contains two major bands. One of them is Hb A $(\alpha_2\beta_2)$ and the other is Hb S $(\alpha_2\beta_2^S)$. The absence of a third band with an intermediate electrophoretic mobility initially suggested that the hybrid molecule $\alpha_2\beta\beta^S$ is not present. However, it was subsequently found that this hybrid hemoglobin is in fact present in solution. Why is it absent from the gel pattern? [Hint: Consider the effect of an electric field on the equilibrium $2(\alpha_2\beta\beta^S) \rightleftharpoons \alpha_2\beta_2 + \alpha_2\beta_2^S$.]

5. Hb A inhibits the formation of long fibers of Hb S and the subsequent sickling of the red cell upon deoxygenation. Why does Hb A have this effect?

6. At high ionic strength, hemoglobin splits at its $\alpha_1\beta_2$ interfaces into $\alpha\beta$ dimers. Predict the dissociability of the oxy and deoxy forms of hemoglobin Kempsey relative to that of hemoglobin A.

CHAPTER 6

INTRODUCTION TO ENZYMES

Chemical reactions in biological systems rarely occur in the absence of a catalyst. These catalysts are specific proteins called enzymes. The striking characteristics of all enzymes are their catalytic power and specificity. Furthermore, the activity of many enzymes is regulated. In addition, some enzymes are intimately involved in the transformation of different forms of energy. Let us examine these highly distinctive and biologically crucial properties of enzymes.

ENZYMES HAVE ENORMOUS CATALYTIC POWER

Enzymes accelerate reactions by factors of at least a million. Indeed, most reactions in biological systems do not occur at perceptible rates in the absence of enzymes. Even a reaction as simple as the hydration of carbon dioxide is catalyzed by an enzyme.

$$CO_2 + H_2O \rightleftharpoons H_2CO_3$$

Otherwise, the transfer of CO_2 from the tissues into the blood and then to the alveolar air would be incomplete. Carbonic anhydrase, the enzyme that catalyzes this reaction, is one of the fastest known. Each enzyme molecule can hydrate 10^5 molecules of CO_2 in one

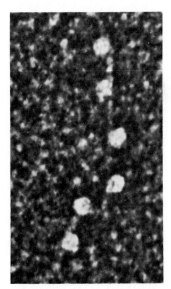

Figure 6-1
Electron micrograph of DNA polymerase I molecules (white spheres) bound to a threadlike synthetic DNA template. [Courtesy of Dr. Jack Griffith.]

second. This catalyzed reaction is 10^7 times faster than the uncatalyzed reaction.

ENZYMES ARE HIGHLY SPECIFIC

Enzymes are highly specific both in the reaction catalyzed and in their choice of reactants, which are called *substrates*. An enzyme usually catalyzes a single chemical reaction or a set of closely related reactions. The degree of specificity for substrate is usually high and sometimes virtually absolute.

Let us consider *proteolytic enzymes* as an example. The reaction catalyzed by these enzymes is the hydrolysis of a peptide bond.

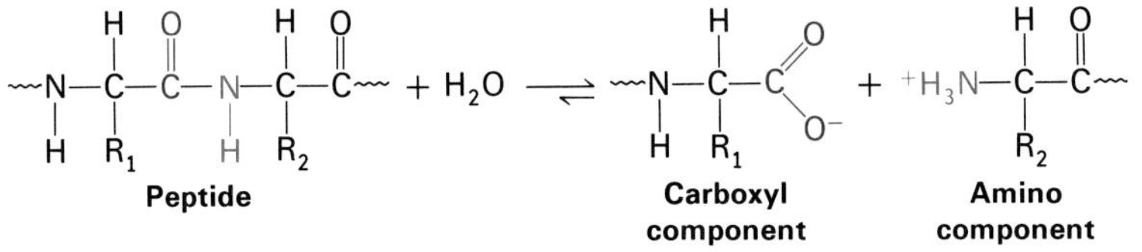

$$\sim\!\!N\!-\!\!\underset{R_1}{\overset{H}{C}}\!-\!\!\overset{O}{C}\!-\!\!N\!-\!\!\underset{R_2}{\overset{H}{C}}\!-\!\!\overset{O}{C}\!\!\sim\;+\;H_2O\;\rightleftharpoons\;\sim\!\!N\!-\!\!\underset{R_1}{\overset{H}{C}}\!-\!\!\overset{O}{C}\!\!\diagdown_{O^-}\;+\;{}^+H_3N\!-\!\!\underset{R_2}{\overset{H}{C}}\!-\!\!\overset{O}{C}\!\!\sim$$

Peptide   Carboxyl component   Amino component

Most proteolytic enzymes also catalyze a different but related reaction, namely the hydrolysis of an ester bond.

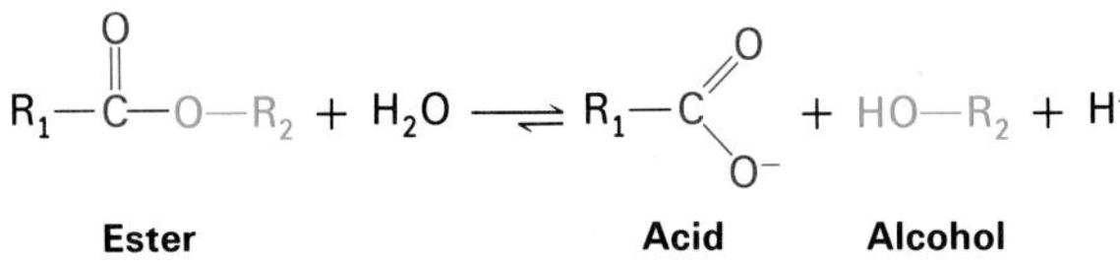

$$R_1\!-\!\!\overset{O}{\overset{\|}{C}}\!-\!O\!-\!R_2\;+\;H_2O\;\rightleftharpoons\;R_1\!-\!\!\overset{O}{C}\!\!\diagdown_{O^-}\;+\;HO\!-\!R_2\;+\;H^+$$

Ester   Acid   Alcohol

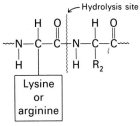

Figure 6-2
Specificity of trypsin.

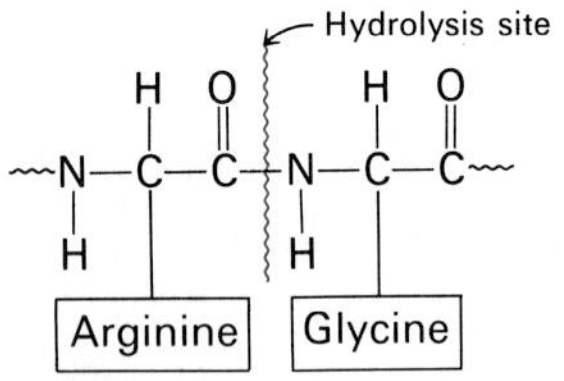

Proteolytic enzymes vary markedly in their degree of substrate specificity. Subtilisin, which comes from certain bacteria, is quite undiscriminating about the nature of the side chains adjacent to the peptide bond to be cleaved. Trypsin, as was mentioned in Chapter 2, is quite specific in that it splits peptide bonds on the carboxyl side of lysine and arginine residues only (Figure 6-2). Thrombin, an enzyme participating in blood clotting, is even more specific than trypsin. The side chain on the carboxyl side of the susceptible peptide bond must be arginine, whereas the one on the amino side must be glycine (Figure 6-3).

Another example of the high degree of specificity of enzymes is provided by DNA polymerase I. This enzyme synthesizes DNA by linking together four kinds of nucleotide building blocks. The sequence of nucleotides in the DNA strand that is being synthesized is determined by the sequence of nucleotides in another DNA strand that serves as a template (see Figure 6-1). DNA polymerase I is remarkably precise in carrying out the instructions given by the template. The wrong nucleotide is inserted into a new DNA strand less than once in a million times.

Figure 6-3
Specificity of thrombin, a clotting factor.

Some enzymes are synthesized in an *inactive precursor form* and are activated at a physiologically appropriate time and place. The digestive enzymes exemplify this kind of control. For example, trypsinogen is synthesized in the pancreas and is activated by peptide-bond cleavage in the small intestine to form the active enzyme trypsin (Figure 6-4). This type of control is also repeatedly used in

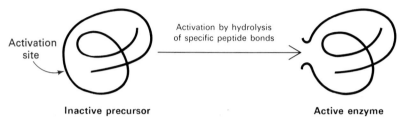

Figure 6-4
Zymogen activation by hydrolysis of specific peptide bonds.

the sequence of enzymatic reactions leading to the clotting of blood. The enzymatically inactive precursors of proteolytic enzymes are called *zymogens*.

Another mechanism that controls activity is the covalent insertion of a small group on an enzyme. This control mechanism is called *covalent modification*. For example, the activities of the enzymes that synthesize and degrade glycogen are regulated by the attachment of a phosphoryl group to a specific serine residue on these enzymes (p. 368). This modification can be reversed by hydrolysis. Specific enzymes catalyze the insertion and removal of phosphoryl and other modifying groups.

A different kind of regulatory mechanism affects many reaction sequences resulting in the synthesis of small molecules such as amino acids. The enzyme that catalyzes the first step in such a biosynthetic pathway is inhibited by the ultimate product (Figure 6-5). The biosynthesis of isoleucine in bacteria illustrates this type

$$-CH_2-O-\overset{\overset{\displaystyle O}{\|}}{\underset{\underset{\displaystyle O^-}{|}}{P}}-O^-$$

**Phosphorylated derivative
of a serine residue**

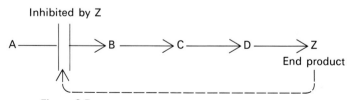

Figure 6-5
Feedback inhibition of the first enzyme in a pathway by reversible binding of the final product.

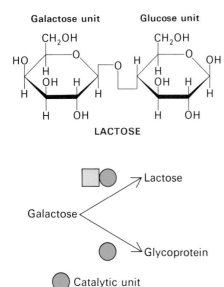

Figure 6-6
Lactose, a sugar consisting of a galactose and a glucose residue, is synthesized by an enzyme that contains a catalytic subunit and a specificity-modifier subunit. A different reaction is catalyzed by the catalytic subunit alone.

of control, which is called *feedback inhibition.* Threonine is converted into isoleucine in five steps, the first of which is catalyzed by threonine deaminase. This enzyme is inhibited when the concentration of isoleucine reaches a sufficiently high level. Isoleucine binds to a regulatory site on the enzyme, which is distinct from its catalytic site. The inhibition of threonine deaminase is mediated by an *allosteric interaction,* which is reversible. When the level of isoleucine drops sufficiently, threonine deaminase becomes active again, and consequently isoleucine is again synthesized.

The specificity of some enzymes is under physiological control. The synthesis of lactose by the mammary gland is a particularly striking example (Figure 6-6). Lactose synthetase, the enzyme that catalyzes the synthesis of lactose, consists of a catalytic subunit and a modifier subunit. The catalytic subunit by itself cannot synthesize lactose. It has a different role, which is to catalyze the attachment of galactose to a protein that contains a covalently linked carbohydrate chain. *The modifier subunit alters the specificity of the catalytic subunit* so that it links galactose to glucose to form lactose. The level of the modifier subunit is under hormonal control. During pregnancy, the catalytic subunit is formed in the mammary gland, but little modifier subunit is formed. At the time of birth, hormonal levels change drastically, and the modifier subunit is synthesized in large amounts. The modifier subunit then binds to the catalytic subunit to form an active lactose synthetase complex that produces large amounts of lactose. This system clearly shows that hormones can exert their physiological effects by altering the specificity of enzymes.

ENZYMES TRANSFORM DIFFERENT KINDS OF ENERGY

In many biochemical reactions, *the energy of the reactants is converted into a different form with high efficiency.* For example, in photosynthesis, light energy is converted into chemical-bond energy. In mitochondria, the free energy contained in small molecules derived from foods is converted into a different currency, that of adenosine triphosphate (ATP). The chemical-bond energy of ATP is then utilized in many different ways. In muscular contraction, the energy of ATP is converted into mechanical energy. Cells and organelles have pumps that utilize ATP to transport molecules and ions against chemical and electrical gradients. These transformations of energy are carried out by enzyme molecules that are integral parts of highly organized assemblies.

ENZYMES DO NOT ALTER REACTION EQUILIBRIA

An enzyme is a catalyst and consequently it cannot alter the equilibrium of a chemical reaction. This means that an enzyme accelerates the forward and reverse reaction by precisely the same factor.

Consider the interconversion of A and B. Suppose that in the absence of enzyme the forward rate (k_F) is 10^{-4} sec^{-1} and the reverse rate (k_R) is 10^{-6} sec^{-1}. The equilibrium constant K is given by the ratio of these rates:

$$A \underset{10^{-6}\ sec^{-1}}{\overset{10^{-4}\ sec^{-1}}{\rightleftharpoons}} B$$

$$K = \frac{[B]}{[A]} = \frac{k_F}{k_R} = \frac{10^{-4}}{10^{-6}} = 100$$

The equilibrium concentration of B is 100 times that of A, whether or not enzyme is present. However, it would take several hours to approach this equilibrium without enzyme, whereas equilibrium would be attained within a second when enzyme is present. Thus, enzymes accelerate the attainment of equilibria but do not shift their positions.

ENZYMES DECREASE THE ACTIVATION ENERGIES OF REACTIONS CATALYZED BY THEM

A chemical reaction, $A \rightleftharpoons B$, goes through a *transition state* that has a higher energy than either A or B. The rate of the forward reaction depends on the temperature and on the difference in free energy between that of A and the transition state, which is called the *Gibbs free energy of activation* and symbolized by $\Delta G^{\ddagger}$ (Figure 6-7A).

$$\Delta G^{\ddagger} = G_{\text{transition state}} - G_{\text{substrate}}$$

The reaction rate is proportional to the fraction of molecules that have a free energy equal to or greater than $\Delta G^{\ddagger}$. The proportion of molecules that have an energy equal to or greater than $\Delta G^{\ddagger}$ increases with temperature.

Enzymes accelerate reactions by decreasing $\Delta G^{\ddagger}$, the activation barrier. The combination of substrate and enzyme creates a new reaction pathway whose transition-state energy is lower than it would be if the reaction were taking place in the absence of enzyme (Figure 6-7B).

FORMATION OF AN ENZYME-SUBSTRATE COMPLEX IS THE FIRST STEP IN ENZYMATIC CATALYSIS

The making and breaking of chemical bonds by an enzyme are preceded by the formation of an *enzyme-substrate* (ES) complex. The substrate is bound to a specific region of the enzyme called the *active site*. Most enzymes are highly selective in their binding of substrates. Indeed, the catalytic specificity of enzymes depends in large part on the specificity of the binding process. Furthermore, the control of enzymatic activity may also take place at this stage.

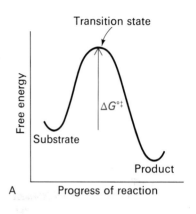

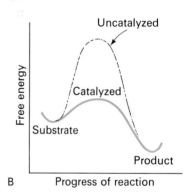

Figure 6-7
A. Definition of $\Delta G^{\ddagger}$, the free energy of activation.
B. Enzymes accelerate catalysis by reducing $\Delta G^{\ddagger}$.

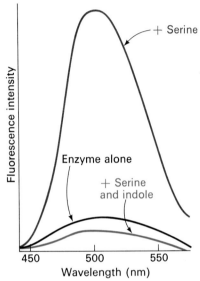

Figure 6-8
Fluorescence intensity of the pyridoxal phosphate group at the active site of tryptophan synthetase changes upon addition of serine and indole, the substrates.

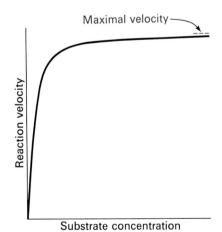

Figure 6-9
Velocity of an enzyme-catalyzed reaction as a function of the substrate concentration.

The existence of ES complexes has been shown in a variety of ways:

1. ES complexes have been *directly visualized by electron microscopy and x-ray crystallography*. Complexes of nucleic acids and their polymerase enzymes are evident in electron micrographs (Figure 6-3). Detailed information concerning the location and interactions of glycyl-L-tyrosine, a substrate of carboxypeptidase A, has been obtained from x-ray studies of that ES complex.

2. The *physical properties* of an enzyme, such as its solubility or heat stability, frequently change upon formation of an ES complex.

3. The *spectroscopic characteristics* of many enzymes and substrates change upon formation of an ES complex just as the absorption spectrum of deoxyhemoglobin changes markedly when it binds oxygen or when it is oxidized to the ferric state, as described previously (Figure 3-18). These changes are particularly striking if the enzyme contains a colored prosthetic group. Tryptophan synthetase, a bacterial enzyme that contains a pyridoxal phosphate prosthetic group, affords a nice illustration. This enzyme catalyzes the synthesis of L-tryptophan from L-serine and indole. The addition of L-serine to the enzyme produces a marked increase in the fluorescence of the pyridoxal phosphate group (Figure 6-8). The subsequent addition of indole, the second substrate, quenches this fluorescence to a level lower than that of the enzyme alone. Thus, fluorescence spectroscopy reveals the existence of an enzyme-serine complex and of an enzyme-serine-indole complex. Other spectroscopic techniques, such as nuclear and electron magnetic resonance, also are highly informative about ES interactions.

4. A high degree of *stereospecificity* is displayed in the formation of ES complexes. For example, D-serine is *not* a substrate of tryptophan synthetase. Indeed, the D-isomer does not even bind to the enzyme. This implies that the substrate-binding site has a very well defined shape.

5. ES complexes can sometimes be *isolated in pure form*. For an enzyme that catalyzes the reaction $A + B \rightleftharpoons C$, it is sometimes possible to isolate an EA complex. This can be done if the enzyme has a sufficiently high affinity for A and if B is absent from the mixture.

6. At a constant concentration of enzyme, the reaction rate increases with increasing substrate concentration until a maximal velocity is reached (Figure 6-9). In contrast, uncatalyzed reactions do not show this saturation effect. In 1913, Leonor Michaelis interpreted the *maximal velocity of an enzyme-catalyzed reaction* in terms of the formation of a discrete ES complex. At a sufficiently high substrate concentration, the catalytic sites are filled and so the reaction rate reaches a maximum. This is the oldest and most general evidence for the existence of ES complexes.

The active site of an enzyme is the region that binds the substrates (and the prosthetic group, if any) and contributes the residues that directly participate in the making and breaking of bonds. These residues are called the *catalytic groups*. Although enzymes differ widely in structure, specificity, and mode of catalysis, a number of generalizations concerning their active sites can be stated:

1. *The active site takes up a relatively small part of the total volume of an enzyme.* Most of the amino acid residues in an enzyme are not in contact with the substrate. This raises the intriguing question of why enzymes are so big. Nearly all enzymes are made up of more than 100 amino acid residues, which gives them a mass greater than 10 kdal and a diameter of more than 25 Å.

2. *The active site is a three-dimensional entity.* The active site of an enzyme is not a point, a line, or even a plane. It is an intricate three-dimensional form made up of groups that come from different parts of the linear amino acid sequence—indeed, residues far apart in the linear sequence may interact more strongly than adjacent residues in the amino acid sequence, as has already been seen for myoglobin and hemoglobin. In lysozyme, an enzyme that will be discussed in more detail in the next chapter, the important groups in the active site are contributed by residues numbered 35, 52, 62, 63, and 101 in the linear sequence of 129 amino acids.

3. *Substrates are bound to enzymes by relatively weak forces.* ES complexes usually have equilibrium constants that range from 10^{-2} to 10^{-8} M, corresponding to free energies of interaction ranging from -3 to -12 kcal/mol. These values should be compared with the strengths of covalent bonds, which are between -50 and -110 kcal/mol.

4. *Active sites are clefts or crevices.* In all enzymes of known structure, substrate molecules are bound to a cleft or crevice from which water is usually excluded unless it is a reactant. The cleft also contains several polar residues that are essential for binding and catalysis. The nonpolar character of the cleft enhances the binding of substrate. In addition, the cleft creates a microenvironment in which certain polar residues acquire special properties essential for their catalytic role.

5. *The specificity of binding depends on the precisely defined arrangement of atoms in an active site.* A substrate must have a matching shape to fit into the site. Emil Fischer's metaphor of the lock and key (Figure 6-10), stated in 1890, has proved to be an essentially correct and highly fruitful way of looking at the stereospecificity of catalysis. However, recent work suggests that the active sites of some enzymes are not rigid. In such an enzyme, the shape of the active site is

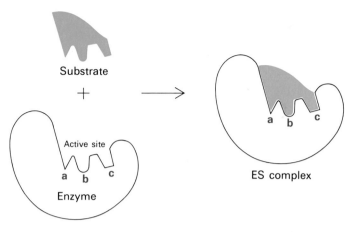

Figure 6-10
Lock-and-key model of the interaction of substrates and enzymes. The active size of the enzyme by itself is complementary in shape to that of the substrate.

modified by the binding of substrate. The active site has a shape complementary to that of the substrate only *after* the substrate is bound. This process of dynamic recognition is called *induced fit* (Figure 6-11). Furthermore, some enzymes preferentially bind a strained form of the substrate corresponding to the transition state.

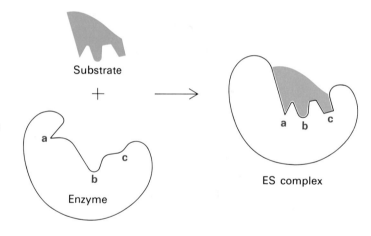

Figure 6-11
Induced-fit model of the interaction of substrates and enzymes. The enzyme changes shape upon binding substrate. The active site has a shape complementary to that of the substrate only *after* the substrate is bound.

THE MICHAELIS-MENTEN MODEL ACCOUNTS FOR THE KINETIC PROPERTIES OF MANY ENZYMES

For many enzymes, the rate of catalysis, V, varies with the substrate concentration, [S], in a manner shown in Figure 6-12. At a fixed concentration of enzyme, V is almost linearly proportional to [S] when [S] is small. At high [S], V is nearly independent of [S]. In 1913, Leonor Michaelis and Maud Menten proposed a simple model to account for these kinetic characteristics. The critical feature in their treatment is that a specific ES complex is a necessary intermediate in catalysis. The model proposed, which is the simplest one that accounts for the kinetic properties of many enzymes, is

$$E + S \underset{k_2}{\overset{k_1}{\rightleftharpoons}} ES \xrightarrow{k_3} E + P \qquad (1)$$

An enzyme, E, combines with S to form an ES complex, with a rate constant k_1. The ES complex has two possible fates. It can dissociate to E and S, with a rate constant k_2, or it can proceed to form product P, with a rate constant k_3. It is assumed that none of the product reverts to the initial substrate, a condition that holds in the initial stage of a reaction before the concentration of product is appreciable.

We want an expression that relates the rate of catalysis to the concentrations of substrate and enzyme and the rates of the individual steps. The starting point is that the catalytic rate is equal to the product of the concentration of the ES complex and k_3:

$$V = k_3[ES] \qquad (2)$$

Now we need to express ES in terms of known quantities. The rates of formation and breakdown of ES are given by

$$\text{Rate of formation of ES} = k_1[E][S] \qquad (3)$$

$$\text{Rate of breakdown of ES} = (k_2 + k_3)[ES] \qquad (4)$$

We are interested in the catalytic rate under steady-state conditions. In a *steady state,* the concentrations of intermediates stay the same while the concentrations of starting materials and products are changing. This occurs when the rates of formation and breakdown of the ES complex are equal. On setting the right-hand sides of equations 3 and 4 equal,

$$k_1[E][S] = (k_2 + k_3)[ES] \qquad (5)$$

By rearranging equation 5,

$$[ES] = \frac{[E][S]}{(k_2 + k_3)/k_1} \qquad (6)$$

Equation 6 can be simplified by defining a new constant, K_M, called the *Michaelis constant:*

$$K_M = \frac{k_2 + k_3}{k_1} \qquad (7)$$

and substituting it into equation 6, which then becomes

$$[ES] = \frac{[E][S]}{K_M} \qquad (8)$$

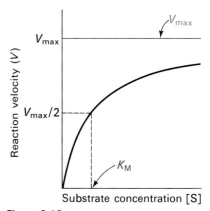

Figure 6-12
A plot of the reaction velocity, V, as a function of the substrate concentration, [S], for an enzyme that obeys Michaelis-Menten kinetics (V_{max} is the maximal velocity and K_M is the Michaelis constant).

Now let us examine the numerator of equation 8. The concentration of uncombined substrate, [S], is very nearly equal to the total substrate concentration, provided that the concentration of enzyme is much lower than that of the substrate. The concentration of uncombined enzyme, [E], is equal to the total enzyme concentration, E_T, minus the concentration of the ES complex.

$$[E] = [E_T] - [ES] \tag{9}$$

On substituting this expression for [E] in equation 8,

$$[ES] = ([E_T] - [ES])[S]/K_M \tag{10}$$

Solving equation 10 for [ES] gives

$$[ES] = [E_T]\frac{[S]/K_M}{1 + [S]/K_M} \tag{11}$$

or

$$[ES] = [E_T]\frac{[S]}{[S] + K_M} \tag{12}$$

By substituting this expression for [ES] into equation 2, we get

$$V = k_3[E_T]\frac{[S]}{[S] + K_M} \tag{13}$$

The maximal rate, V_{max}, is attained when the enzyme sites are saturated with substrate—that is, when [S] is much greater than K_M—so that $[S]/([S] + K_M)$ approaches 1. Thus,

$$V_{max} = k_3[E_T] \tag{14}$$

Substituting equation 14 into equation 13 yields the Michaelis-Menten equation:

$$V = V_{max}\frac{[S]}{[S] + K_M} \tag{15}$$

This equation accounts for the kinetic data given in Figure 6-12. At low substrate concentration, when [S] is much less than K_M, $V = [S]V_{max}/K_M$; that is, the rate is directly proportional to the substrate concentration. At high substrate concentration, when [S] is much greater than K_M, $V = V_{max}$; that is, the rate is maximal, independent of substrate concentration.

The meaning of K_M is evident from equation 15. When [S] = K_M, then $V = V_{max}/2$. Thus, K_M *is equal to the substrate concentration at which the reaction rate is half of its maximal value.*

V_{max} AND K_M CAN BE DETERMINED BY VARYING THE SUBSTRATE CONCENTRATION

The Michaelis constant, K_M, and the maximal rate, V_{max}, can be readily derived from rates of catalysis at different substrate concentrations if an enzyme operates according to the simple scheme given in equation 1. It is convenient to transform the Michaelis-Menten equation into one that gives a straight line plot. This can be done by taking the reciprocal of both sides of equation 15 to give

$$\frac{1}{V} = \frac{1}{V_{max}} + \frac{K_M}{V_{max}} \cdot \frac{1}{[S]} \tag{16}$$

A plot of $1/V$ versus $1/[S]$ yields a straight line with an intercept of $1/V_{max}$ and a slope of K_M/V_{max} (Figure 6-13).

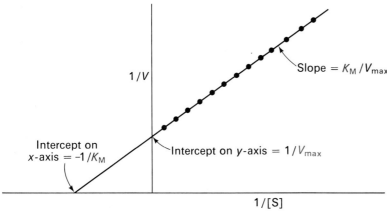

Figure 6-13
A double-reciprocal plot of enzyme kinetics: $1/V$ is plotted as a function of $1/[S]$. The slope is K_M/V_{max}, the intercept on the y-axis is $1/V_{max}$, and the intercept on the x-axis is $-1/K_M$.

SIGNIFICANCE OF K_M AND V_{max} VALUES

The K_M values of enzymes range widely (Table 6-1). For most enzymes, K_M lies between 10^{-1} and 10^{-6} M. The K_M value for an

Table 6-1
K_M values of some enzymes

Enzyme	Substrate	K_M
Chymotrypsin	Acetyl-L-tryptophanamide	5×10^{-3} M
Lysozyme	Hexa-N-acetylglucosamine	6×10^{-6} M
β-Galactosidase	Lactose	4×10^{-3} M
Threonine deaminase	Threonine	5×10^{-3} M
Carbonic anhydrase	CO_2	8×10^{-3} M
Penicillinase	Benzylpenicillin	5×10^{-5} M
Pyruvate carboxylase	Pyruvate	4×10^{-4} M
	HCO_3^-	1×10^{-3} M
	ATP	6×10^{-5} M
Arginine-tRNA synthetase	Arginine	3×10^{-6} M
	tRNA	4×10^{-7} M
	ATP	3×10^{-4} M

enzyme depends on the particular substrate and also on environmental conditions such as the temperature and ionic strength. The Michaelis constant, K_M, has two meanings. First, K_M is the concentration of substrate at which half the active sites are filled. Once the K_M is known, the fraction of sites filled, f_{ES}, at any substrate concentration can be calculated from

$$f_{ES} = \frac{V}{V_{max}} = \frac{[S]}{[S] + K_M} \qquad (17)$$

Second, K_M is related to the rate constants of the individual steps in the catalytic scheme given in equation 1. In equation 7, K_M is defined as $(k_2 + k_3)/k_1$. Consider a limiting case in which k_2 is much greater than k_3. This means that dissociation of the ES complex to E and S is much more rapid than formation of E and product. Under these conditions ($k_2 \gg k_3$),

$$K_M = \frac{k_2}{k_1} \qquad (18)$$

The dissociation constant of the ES complex is given by

$$K_{ES} = \frac{[E][S]}{[ES]} = \frac{k_2}{k_1} \qquad (19)$$

In other words, K_M *is equal to the dissociation constant of the ES complex if* k_3 *is much smaller than* k_2. When this condition is met, K_M is a measure of the strength of the ES complex: a high K_M indicates weak binding; a low K_M indicates strong binding. It must be stressed that K_M indicates the affinity of the ES complex only when k_2 is much greater than k_3. This is in fact the case of many, but not all, enzymes.

The maximal rate, V_{max}, reveals the turnover number of an enzyme if the concentration of active sites $[E_T]$ is known, because

$$V_{max} = k_3[E_T] \qquad (20)$$

For example, a 10^{-6} M solution of carbonic anhydrase catalyzes the formation of 0.6 M H_2CO_3 per second when it is fully saturated with substrate. Hence, k_3 is 6×10^5 sec^{-1}. The kinetic constant k_3 is called the *turnover number*. The turnover number of an enzyme is *the number of substrate molecules converted into product per unit time when the enzyme is fully saturated with substrate*. The turnover number of 600,000 sec^{-1} for carbonic anhydrase is one of the largest known. Each round of catalysis occurs in a time equal to $1/k_3$, which is 1.7 microseconds for carbonic anhydrase. The turnover numbers of most enzymes for their physiological substrates fall in the range of 1 to 10^4 per second (Table 6-2).

Table 6-2
Maximum turnover numbers of some enzymes

Enzyme	Turnover number (per second)
Carbonic anhydrase	600,000
3-Ketosteroid isomerase	280,000
Acetylcholinesterase	25,000
Penicillinase	2,000
Lactate dehydrogenase	1,000
Chymotrypsin	100
DNA polymerase I	15
Tryptophan synthetase	2
Lysozyme	0.5

When the substrate concentration is much greater than K_M, the rate of catalysis is equal to k_3, the turnover number, as described in the preceding section. However, most enzymes are not saturated with substrate under physiological conditions. The $[S]/K_M$ ratio is typically between .01 and 1.0. When $[S] \ll K_M$, the enzymatic rate is much less than k_3 because most of the active sites are unoccupied. Is there a suitable parameter to characterize the kinetics of an enzyme under these conditions? Indeed there is, as can be shown by combining equations 2 and 8 to give

$$V = \frac{k_3}{K_M}[E][S] \qquad (21)$$

When $[S] \ll K_M$, the concentration of free enzyme, $[E]$, is nearly equal to the total concentration of enzyme $[E_T]$, and so

$$V = \frac{k_3}{K_M}[S][E_T] \qquad (22)$$

Thus, when $[S] \ll K_M$, the enzymatic velocity depends on the value of k_3/K_M and on $[S]$.

Are there any physical limits on the value of k_3/K_M? Note that this ratio depends on k_1, k_2, and k_3, as can be shown by substituting for K_M:

$$k_3/K_M = \frac{k_3 k_1}{k_2 + k_3} \qquad (23)$$

The ultimate limit on the value of k_3/K_M is set by k_1, the rate of formation of the ES complex. *This rate cannot be faster than the diffusion-controlled encounter of an enzyme and its substrate.* Diffusion limits the value of k_1 so that it cannot be higher than between 10^8 and 10^9 M^{-1} sec^{-1}. Hence, the upper limit on k_3/K_M is between 10^8 and 10^9 M^{-1} sec^{-1}.

This restriction also pertains to enzymes having more complex reaction pathways than that of equation 1. Their maximal catalytic rate when substrate is saturating, denoted by k_{cat}, depends on several rate constants rather than on k_3 alone. The pertinent parameter for these enzymes is k_{cat}/K_M. In fact, *the k_{cat}/K_M ratios of a number of enzymes, such as acetylcholinesterase, carbonic anhydrase, and triosephosphate isomerase, are between 10^8 and 10^9 M^{-1} sec^{-1}, which shows that they have attained kinetic perfection. Their catalytic velocity is restricted only by the rate at which they encounter substrate in the solution.* Any further gain in catalytic rate can come only by decreasing the time for diffusion.

Indeed, some series of enzymes are associated into organized assemblies (p. 293) so that the product of one enzyme is very rapidly found by the next enzyme. Thus, the limit imposed by the rate of diffusion in solution can be partially overcome by sequestering substrates and products in the confined volume of a multienzyme complex.

ENZYMES CAN BE INHIBITED BY SPECIFIC MOLECULES

The inhibition of enzymatic activity by specific small molecules and ions is important because it serves as a major control mechanism in biological systems. Also, many drugs and toxic agents act by inhibiting enzymes. Furthermore, enzyme inhibition can be a source of insight into the mechanism of enzyme action. Enzyme inhibition can be either reversible or irreversible. In *irreversible inhibition,* the inhibitor is covalently linked to the enzyme or bound so tightly that its dissociation from the enzyme is very slow. The action of nerve gas poisons on acetylcholinesterase, an enzyme that plays an important role in the transmission of nerve impulses, is an example of irreversible inhibition. Diisopropylphosphofluoridate (DIPF), one of these agents, reacts with a critical serine residue at the active site on the enzyme to form an inactive diisopropylphosphoryl enzyme (Figure 6-14). Alkylating reagents, such as

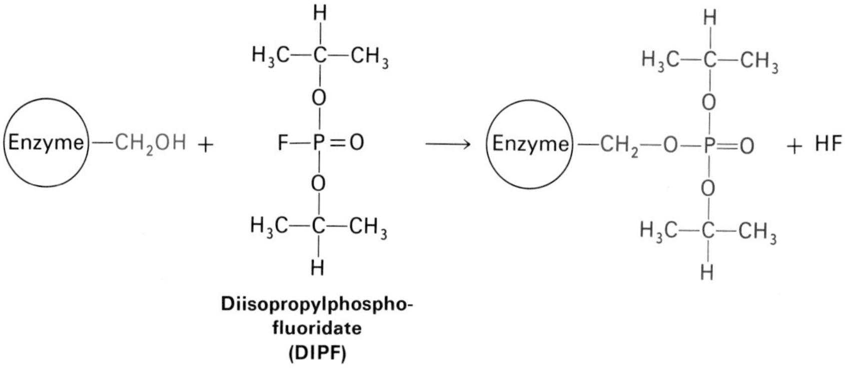

**Diisopropylphospho-
fluoridate
(DIPF)**

Figure 6-14
Inactivation of chymotrypsin and acetylcholinesterase by diisopropylphosphofluoridate (DIPF).

iodoacetamide, may irreversibly inhibit enzymatic activity by modifying cysteine and other side chains (Figure 6-15).

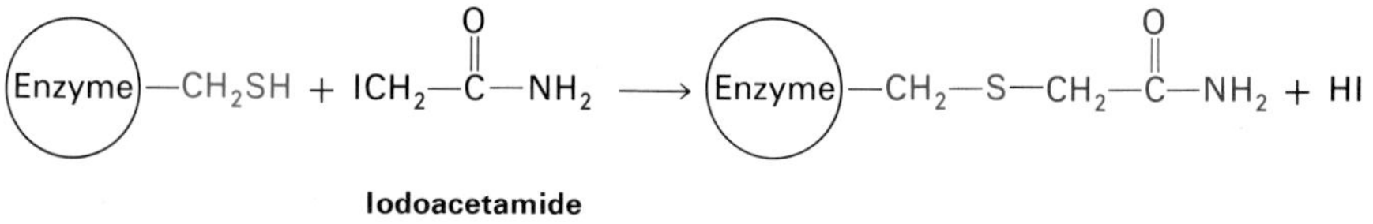

Iodoacetamide

Figure 6-15
Inactivation of an enzyme with a critical cysteine residue by iodoacetamide.

In contrast, *reversible inhibition* is characterized by a rapid equilibrium of the inhibitor and enzyme. The simplest type of reversible inhibition is competitive inhibition. A *competitive inhibitor* resembles the substrate and binds to the active site of the enzyme (Figure 6-16). The substrate is then prevented from binding to the same active site. In other words, the binding of substrate and a competitive inhibitor are mutually exclusive events. *A competitive inhibitor*

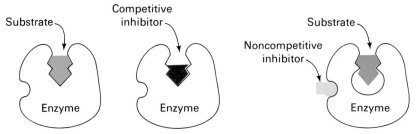

Figure 6-16
Distinction between a *competitive* inhibitor and a *noncompetitive* inhibitor: (left) enzyme-substrate complex; (middle) a competitive inhibitor prevents the substrate from binding; (right) a noncompetitive inhibitor does not prevent the substrate from binding.

diminishes the rate of catalysis by reducing the proportion of enzyme molecules that have a bound substrate. A classic example of competitive inhibition is the action of malonate on succinate dehydrogenase, an enzyme that dehydrogenates succinate. Malonate differs from succinate in having one rather than two methylene groups. A physiologically important example of competitive inhibition is found in the formation of 2,3-diphosphoglycerate from 1,3-diphosphoglycerate. Diphosphoglycerate mutase, the enzyme catalyzing this isomerization, is competitively inhibited by low levels of 2,3-diphosphoglycerate. In fact, it is not uncommon for the product of an enzymatic reaction to be a competitive inhibitor of the substrate because of their structural resemblance.

$$
\begin{array}{cc}
COO^- & COO^- \\
| & | \\
CH_2 & CH_2 \\
| & | \\
CH_2 & COO^- \\
| & \\
COO^- & \\
\textbf{Succinate} & \textbf{Malonate}
\end{array}
$$

$$
\begin{array}{ccc}
\begin{matrix} O \\ \| \\ C-OPO_3{}^{2-} \\ | \\ H-C-OH \\ | \\ H_2C-OPO_3{}^{2-} \end{matrix}
& \rightleftharpoons &
\begin{matrix} O \\ \| \\ C-O^- \\ | \\ H-C-OPO_3{}^{2-} \\ | \\ H_2C-OPO_3{}^{2-} \end{matrix} \\
\textbf{1,3-Diphosphoglycerate} & & \textbf{2,3-Diphosphoglycerate}
\end{array}
$$

In *noncompetitive inhibition,* which is also reversible, the inhibitor and substrate can bind simultaneously to an enzyme molecule. This means that their binding sites do not overlap. A noncompetitive inhibitor acts by decreasing the turnover number of an enzyme rather than by diminishing the proportion of enzyme molecules that have a bound substrate. More complex patterns of inhibition are produced when the inhibitor affects both the binding of substrate and the turnover number of the enzyme.

Enzyme activity may also be inhibited by interactions between sites on different subunits of an oligomeric enzyme. This kind of inhibition, which is called *allosteric inhibition,* is very important physiologically. It will be discussed shortly.

COMPETITIVE AND NONCOMPETITIVE INHIBITION ARE KINETICALLY DISTINGUISHABLE

Measurements of the rates of catalysis at different concentrations of substrate and inhibitor serve to distinguish between competitive and noncompetitive inhibition. In *competitive inhibition,* the intercept of the plot of $1/V$ versus $1/[S]$ is the same in the presence and absence of inhibitor, although the slope is different (Figure 6-17).

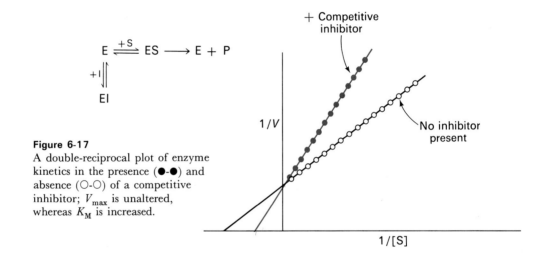

Figure 6-17
A double-reciprocal plot of enzyme kinetics in the presence (●-●) and absence (○-○) of a competitive inhibitor; V_{max} is unaltered, whereas K_M is increased.

This reflects the fact that V_{max} is not altered by a competitive inhibitor. *The hallmark of competitive inhibition is that the inhibition can be overcome at a sufficiently high substrate concentration.* Substrate and inhibitor compete for the same site. At a sufficiently high substrate concentration, virtually all the active sites are filled by substrate, and the enzyme is fully operative. The increase in the slope of the $1/V$ versus $1/[S]$ plot indicates the strength of binding of competitive inhibitor. In the presence of a competitive inhibitor, equation 16 is replaced by

$$\frac{1}{V} = \frac{1}{V_{max}} + \frac{K_M}{V_{max}}\left(1 + \frac{[I]}{K_i}\right)\left(\frac{1}{[S]}\right) \tag{24}$$

in which $[I]$ is the concentration of inhibitor and K_i is the dissociation constant of the enzyme-inhibitor complex:

$$E + I \rightleftharpoons EI$$

$$K_i = \frac{[E][I]}{[EI]} \tag{25}$$

In other words, the slope of the plot is increased by the factor $(1 + [I]/K_i)$ in the presence of a competitive inhibitor. Consider an enzyme with a K_M of 10^{-4} M. In the absence of inhibitor, $V = V_{max}/2$ when $[S] = 10^{-4}$ M. In the presence of 2×10^{-3} M competitive inhibitor that is bound to the enzyme with a K_i of 10^{-3} M, the apparent K_M will be 3×10^{-4} M. Consequently, $V = V_{max}/4$.

In *noncompetitive inhibition* (Figure 6-18), V_{max} is decreased and so the intercept on the *y*-axis is increased. The slope, which is equal to

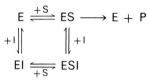

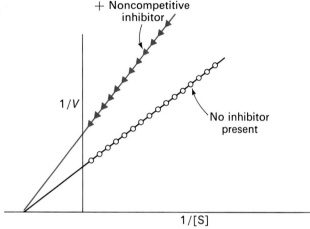

Figure 6-18
A double-reciprocal plot of enzyme kinetics in the presence (▲-▲-▲) and absence (O-O-O) of a noncompetitive inhibitor; K_M is unaltered by the noncompetitive inhibitor, whereas V_{max} is decreased.

K_M/V^I_{max}, is increased by the same factor. In contrast with V_{max}, K_M is not affected by this kind of inhibition. *Noncompetitive inhibition cannot be overcome by increasing the substrate concentration.* The maximal velocity in the presence of a noncompetitive inhibitor, V^I_{max}, is given by

$$V^I_{max} = \frac{V_{max}}{1 + [I]/K_i} \qquad (26)$$

TREATMENT OF ETHYLENE GLYCOL POISONING BY COMPETITIVE INHIBITION

About fifty deaths occur annually from the ingestion of ethylene glycol, a constituent of permanent-type automobile antifreeze. Ethylene glycol itself is not lethally toxic. Rather, the harm is done by oxalic acid, an oxidation product of ethylene glycol. The first step in this conversion is the oxidation of ethylene glycol by alcohol dehydrogenase (Figure 6-19). This reaction can be effectively inhib-

Figure 6-19
Formation of oxalic acid from ethylene glycol is inhibited by ethanol.

ited by the administration of a nearly intoxicating dose of ethanol. The basis for this effect is that *ethanol is a competing substrate and so it blocks the oxidation of ethylene glycol to aldehyde products.* The ethylene glycol is then excreted harmlessly. The same rationale is used in the therapy of methanol poisoning.

ALLOSTERIC ENZYMES DO NOT OBEY MICHAELIS-MENTEN KINETICS

The Michaelis-Menten model has greatly affected the development of enzyme chemistry. Its virtues are its simplicity and broad applicability. However, the kinetic properties of many enzymes cannot be accounted for by the Michaelis-Menten model. An important group consists of the *allosteric enzymes,* which often display sigmoidal plots of the reaction velocity, V, versus substrate concentration, [S], rather than the hyperbolic plots predicted by the Michaelis-Menten equation (eq. 15). Recall that the oxygen-binding curve of myoglobin is hyperbolic, whereas that of hemoglobin is sigmoidal. The situation with enzymes is analogous. In allosteric enzymes, one active site in an enzyme molecule can affect another active site in the same enzyme molecule. A possible outcome of this interaction across subunits is that the binding of substrate becomes cooperative, which would give a sigmoidal plot of V versus [S]. In addition, the activity of allosteric enzymes may be altered by regulatory molecules that are bound to sites other than the catalytic sites, just as oxygen binding in hemoglobin is affected by DPG, H^+, and CO_2.

THE CONCERTED MODEL FOR ALLOSTERIC INTERACTIONS

An elegant and incisive model for allosteric enzymes was proposed in 1965 by Jacques Monod, Jeffries Wyman, and Jean-Pierre Changeux. Let us apply their approach to an allosteric enzyme made up of two identical subunits, each having one active site. Suppose that a subunit can exist in either of two conformations, called R and T. The R (relaxed) state has a high affinity for substrate, whereas the T (tense) state has a low affinity (Figure 6-20). Recall that these designations were used to describe alternative quaternary structures of hemoglobin (p.76). Forms R and T are interconvertible. *An important assumption of this model is that both subunits must be in the same conformational state, so that the symmetry of the dimer is conserved.* Thus, RR and TT are allowed conformations, but RT is not permitted. In the absence of substrate, the two allowed states are symbolized as T_0 and R_0, and L is the ratio of their concentrations.

T form
(Low affinity for substrate)

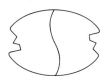

R form
(High affinity for substrate)

Figure 6-20
Schematic representation of the R and T forms of an allosteric enzyme.

$$R_0 \rightleftharpoons T_0 \tag{27}$$

$$L = T_0/R_0 \tag{28}$$

For simplicity, let us assume that the substrate does not bind to the T state. The R state of the dimer can bind one or two substrate molecules; these species are denoted by R_1 and R_2, respectively.

$$R_0 + S \rightleftharpoons R_1 \tag{29}$$

$$R_1 + S \rightleftharpoons R_2 \tag{30}$$

$$K_R = \frac{2[R_0][S]}{[R_1]} = \frac{[R_1][S]}{2[R_2]} \tag{31}$$

In this model, the microscopic dissociation constant, K_R, is the same for the binding of the first and second substrate molecules to the R form of the dimeric enzyme molecule. The factor of two in equation 31 takes into account the fact that a substrate can bind to either of two sites in R_0 to form R_1 and, likewise, the fact that a substrate can be released from either of two sites in R_2 to form R_1.

We want an expression that gives us the *fractional saturation, Y*, which is the fraction of active sites that have a bound substrate, as a function of the substrate concentration.

$$Y = \frac{[\text{occupied sites}]}{[\text{total sites}]} = \frac{[R_1] + 2[R_2]}{2([T_0] + [R_0] + [R_1] + [R_2])} \tag{32}$$

Substituting equations 27 through 31 into equation 32 gives the desired expression for Y:

$$Y = \left(\frac{[S]}{K_R}\right)\frac{1 + [S]/K_R}{L + (1 + [S]/K_R)^2} \tag{33}$$

Let us plot equation 33 with $K_R = 10^{-5}$ M and $L = 10^4$. Such a plot of Y versus $[S]$ is sigmoidal rather than hyperbolic (see Figure 6-23 on the next page). In other words, *the binding of substrate is cooperative.* If the turnover number per active site is the same for ES complexes in R_1 and R_2, then the plot of reaction velocity versus substrate concentration will also be sigmoidal, because

$$V = YV_{\text{max}} \tag{34}$$

Let us look at this binding process (Figure 6-21). In the absence of substrate, nearly all the enzyme molecules are in the T form. Specifically, there is only one molecule in the R form for every 10^4 molecules in the T form, in the above example. The addition of substrate shifts this conformational equilibrium in the direction of the R form, because substrate binds only to the R form. When substrate binds to one site, the other site on the same enzyme molecule must also be in the R form, according to the basic postulate of this model. In other words, the transition from T to R or vice versa

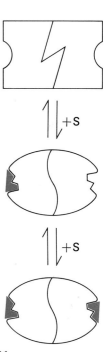

Figure 6-21
Concerted model for the cooperative binding of substrate in an allosteric enzyme. The low affinity form, TT, switches to the high affinity form, RR, upon binding the first substrate molecule.

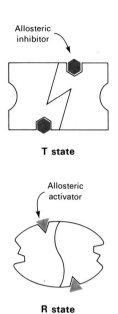

T state

R state

Figure 6-22
In the concerted model, an allosteric inhibitor (represented by a hexagon) stabilizes the T state, whereas an allosteric activator (represented by a triangle) stabilizes the R state.

Figure 6-23
Saturation, *Y*, as a function of substrate concentration, [S], according to the concerted model (eq. 33). The effects of an allosteric activator and inhibitor are also shown.

is *concerted*. Hence, *the proportion of enzyme molecules in the R form increases progressively as more substrate is added, and so the binding of substrate is cooperative.* When the active sites are fully saturated, all of the enzyme molecules are in the R form.

The effects of allosteric activators and inhibitors can readily be accounted for by this concerted model. An allosteric inhibitor binds preferentially to the T form, whereas an allosteric activator binds preferentially to the R form (Figure 6-22). Consequently, *an allosteric inhibitor shifts the $R \rightleftharpoons T$ conformational equilibrium toward T, whereas an allosteric activator shifts it toward R.* These effects can be expressed quantitatively by a change in the allosteric equilibrium constant, *L*, which is a variable in equation 33. An allosteric inhibitor increases *L*, whereas an allosteric activator decreases *L*. These effects are shown in Figure 6-23, in which *Y* is plotted versus [S] for these values of *L*: 10^3 (activator present), 10^4 (no activator, no inhibitor), and 10^5 (inhibitor present). The fractional saturation, *Y*, at all values of [S] is decreased by the presence of the inhibitor and increased by the presence of activator.

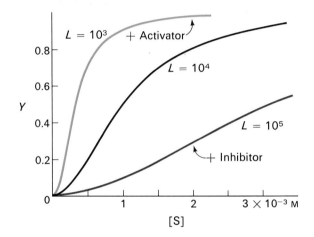

It is useful to define two terms at this point. *Homotropic* effects refer to allosteric interactions between identical ligands (bound molecules or ions), whereas *heterotropic* effects refer to interactions between different ligands. In the preceding example, the cooperative binding of substrate to enzyme is a homotropic effect. In contrast, the effect of an activator or inhibitor on the binding of substrate is heterotropic because interactions take place between different kinds of molecules. In the concerted model for allosteric interactions, homotropic effects are necessarily positive (cooperative), whereas heterotropic effects can be either positive or negative.

THE SEQUENTIAL MODEL FOR ALLOSTERIC INTERACTIONS

Allosteric interactions can also be accounted for by a *sequential model,* which has been developed by Daniel Koshland, Jr. The simplest form of this model makes three assumptions:

1. There are only two conformational states (R and T) accessible to any one subunit.

2. The binding of substrate changes the shape of the subunit to which it is bound. However, the conformations of the other subunits in the enzyme molecule are not appreciably altered.

3. The conformational change elicited by the binding of substrate in one subunit can increase or decrease the substrate-binding affinity of the other subunits in the same enzyme molecule.

The sequential model pictures the binding process in an allosteric enzyme to occur as shown in Figure 6-24. The binding is cooperative if the affinity of RT for substrate is greater than that of TT.

The simple sequential model differs from the concerted model in several ways. First, an equilibrium between the R and T forms in the absence of substrate is not assumed in the former. Rather, the conformational transition from T to R is *induced* by the binding of substrate. Second, the conformational change from T to R in different subunits of an enzyme molecule is sequential, not concerted. The hybrid species RT is prominent in the sequential model but excluded in the concerted model. The concerted model supposes that symmetry is essential for the interaction of subunits in oligomeric proteins and therefore requires that it be conserved in allosteric transitions. In contrast, the sequential model assumes that subunits can interact even if they are in different conformational states. Finally, these models differ in that homotropic interactions are necessarily positive in the concerted model but can be either positive or negative in the sequential model. Whether the second substrate molecule can be bound more or less tightly than the first depends on the nature of the distortion induced by the binding of the first substrate molecule.

Which model is correct? For some allosteric proteins, the concerted model fits well, whereas for others the sequential model appears applicable. However, neither model is satisfactory for yet another group of allosteric proteins. For these proteins, the assumption that there are only two significant conformational states (R and T) appears to be too restrictive. More complex models are needed to account for the allosteric properties of these proteins.

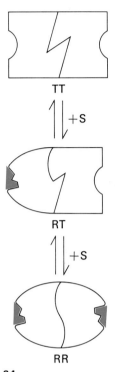

Figure 6-24
Sequential model for the cooperative binding of substrate in an allosteric enzyme. The empty active site in RT has a higher affinity for substrate than do the sites in TT.

ELECTROSTATIC, HYDROGEN, AND VAN DER WAALS BONDS IN ENZYME-SUBSTRATE COMPLEXES

Reversible molecular interactions in biological systems are mediated by three different kinds of forces. The folding of macromolecules, the binding of substrates to enzymes, and the interactions of cells—indeed, all molecular interactions in biological systems—require the interplay of *electrostatic bonds, hydrogen bonds,* and *van der*

Waals bonds. These three fundamental noncovalent bonds differ in their geometrical requirements, strength, and specificity. Furthermore, they are affected in different ways by the presence of water, which has a profound influence. Let us consider the characteristics of each of these three basic bonds.

CHARGED SUBSTRATES CAN BIND TO OPPOSITELY CHARGED GROUPS ON ENZYMES

A charged group on a substrate can interact with an oppositely charged group on an enzyme. The force of such an *electrostatic interaction* is given by Coulomb's law:

$$F = \frac{q_1 q_2}{r^2 D}$$

in which q_1 and q_2 are the charges of the two groups, r is the distance between them, and D is the dielectic constant of the medium. An electrostatic interaction is strongest in a vacuum (where D is 1) and is weakest in a medium such as water (where D is 80).

The binding of glycyl-L-tyrosine to carboxypeptidase A, a proteolytic enzyme that cleaves carboxyl-terminal residues, is an example of an electrostatic interaction. The negatively charged terminal carboxylate group of the dipeptide substrate interacts with the positively charged guanidinium group of an arginine residue of the enzyme. The distance between these oppositely charged groups is 2.8 Å.

Substrate   **Arginine side chain of enzyme**

This kind of interaction is also called an ionic bond, salt linkage, salt bridge, or ion pair. All these terms have the same meaning: an electrostatic interaction between oppositely charged groups. A negatively charged substrate can form an electrostatic bond with the positively charged side chain of a lysine or arginine residue. The imidazole group of a histidine residue and the terminal amino group are also potential binding sites for a negatively charged substrate if their pKs render them positively charged at the pH of the medium. For a positively charged substrate, the potential binding sites on the enzyme are the negatively charged carboxylate groups of aspartate and glutamate and the terminal carboxylate of the polypeptide chain.

SUBSTRATES BIND TO ENZYMES BY PRECISELY DIRECTED HYDROGEN BONDS

Many substrates are uncharged, yet they bind to enzymes with high affinity and specificity. The significant interactions for these substrates and indeed also for most charged substrates are hydrogen bonds. *In a hydrogen bond, a hydrogen atom is shared by two other atoms.* The atom to which the hydrogen is more tightly linked is called the hydrogen donor, whereas the other atom is the hydrogen acceptor. In fact, a hydrogen bond can be considered an intermediate in the transfer of a proton from an acid to a base. The acceptor atom has a partial negative charge that attracts the hydrogen atom. In short, the hydrogen bond is reminiscent of a ménage à trois.

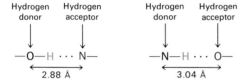

The donor atom in a hydrogen bond in biological systems is an oxygen or nitrogen atom that has a covalently attached hydrogen atom. The acceptor atom is either oxygen or nitrogen. The kinds of hydrogen bonds formed and their bond lengths are given in Table 6-3. The bond energies range from about 3 to 7 kcal/mol. Hydrogen bonds are stronger than van der Waals bonds but much weaker than covalent bonds. The length of a hydrogen bond is intermediate between that of a covalent bond and a van der Waals bond. *An important feature of hydrogen bonds is that they are directional.* The strongest hydrogen bond is one in which the donor, hydrogen, and acceptor atoms are colinear. If the acceptor atom is at an angle to the line joining the donor and hydrogen atoms, the bond becomes weaker with increasing angle.

Hydrogen bonds have already been encountered in the discussion of the structure of myoglobin and hemoglobin. In the α helix, the peptide —NH and —CO groups are hydrogen bonded to each other. The nitrogen atom is the hydrogen donor, whereas the oxygen atom is the hydrogen acceptor. The distance between the nitrogen and oxygen atoms is 2.9 Å. The hydrogen atom is closer to the nitrogen than to the oxygen atom by 0.9 Å.

Another example of a hydrogen bond in myoglobin and hemoglobin is the link between the hydroxyl group of tyrosine HC2 and the peptide carbonyl of FG4. The oxygen atom of the hydroxyl group of tyrosine is the hydrogen donor, whereas the oxygen atom of the peptide carbonyl is the hydrogen acceptor.

Table 6-3
Typical hydrogen-bond distances

Bond	Distance (Å)
O—H···O	2.70
O—H···O⁻	2.63
O—H···N	2.88
N—H···O	3.04
N⁺—H···O	2.93
N—H···N	3.10

O—H ··· O
Strong hydrogen bond

O
⋮
O—H
Weak hydrogen bond

The role of hydrogen bonding in the interaction of substrates with enzymes is nicely illustrated by the binding of the uridine portion of the substrate to pancreatic ribonuclease, an enzyme that cleaves ribonucleic acid (Figure 6-25). Three hydrogen bonds are involved:

Figure 6-25
Hydrogen-bond interactions in the binding of a substrate to ribonuclease. [After F. M. Richards, H. W. Wyckoff, and N. Allewell. In *The Neurosciences: Second Study Program,* F. O. Schmitt, ed. (Rockefeller University Press, 1970), p. 970.]

1. One of the C=O groups of the uridine ring is hydrogen bonded to a peptide N—H.

2. The N—H group of the uridine ring is hydrogen bonded to the —OH of a threonine residue.

3. The other ring C=O is hydrogen bonded to the —OH group of a serine residue.

PROTEINS ARE RICH IN HYDROGEN-BONDING POTENTIALITY

Amino acid side chains and the peptide main chain can form a variety of different kinds of hydrogen bonds. In fact, eleven of the twenty fundamental amino acids can form hydrogen bonds through their side chains. It is convenient to group these residues according to the kinds of hydrogen bonds they can form.

1. The side chains of tryptophan and arginine can serve as *hydrogen-bond donors only.*

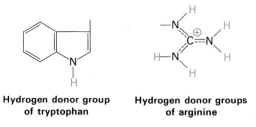

**Hydrogen donor group
of tryptophan** **Hydrogen donor groups
of arginine**

2. The side chains of asparagine, glutamine, serine, and threonine can serve as *hydrogen-bond donors and acceptors,* as can the peptide group.

3. The hydrogen-bonding capabilities of lysine (and the terminal amino group), aspartic and glutamic acid (and the terminal carboxyl group), tyrosine, and histidine vary with pH. These groups can serve as both hydrogen-bond acceptors and donors over a certain range of pH, and as acceptors or donors (but not both) at other pH values, as shown for aspartate and glutamate in Figure 6-26. *The hydrogen-bonding modes of these ionizable residues are pH dependent.*

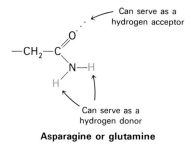

Asparagine or glutamine

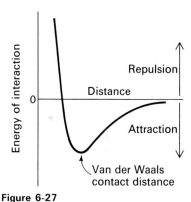

Figure 6-26
The hydrogen-bonding potentialities of aspartate and glutamate.

VAN DER WAALS INTERACTIONS ARE IMPORTANT WHEN THERE IS STERIC COMPLEMENTARITY

There is a nonspecific attractive force between any two atoms when they are from about 3 to 4 Å apart. This interaction, called a van der Waals bond, is weaker and less specific than electrostatic and hydrogen bonds but no less important in biology. The basis of the van der Waals bond is that the distribution of electronic charge around an atom changes with time. At any instant, the charge distribution is not perfectly symmetric. This transient asymmetry in the electronic charge around an atom will alter the electron distribution around its neighboring atoms. The attraction between a pair of atoms increases as they come closer, until they are separated by the van der Waals *contact distance* (Figure 6-27). At a shorter distance, very strong repulsive forces become dominant because the outer electron clouds overlap. The contact distance between an oxygen and carbon atom, for example, is 3.4 Å, which is obtained by adding 1.4 and 2.0 Å, the contact radii of the O and C atoms, respectively.

The van der Waals bond energy for a pair of atoms is about 1 kcal/mol. This is considerably weaker than a hydrogen or electrostatic bond, which is in the 3 to 7 kcal/mol range. Thus, a single van der Waals bond counts for very little. Its strength is only a little more than the average thermal energy of molecules at room temperature (0.6 kcal/mol). Van der Waals forces become significant in binding only when numerous substrate atoms can simultaneously come close to numerous enzyme atoms. The van der Waals force fades rapidly when the distance between a pair of atoms is even 1 Å

Figure 6-27
Energy of a van der Waals interaction as a function of the distance between two atoms.

Table 6-4
Van der Waals contact radii of atoms

Atom	Radius (Å)
H	1.2
C	2.0
N	1.5
O	1.4
S	1.85
P	1.9

greater than their contact distance. Numerous atoms of a substrate can interact with many atoms of an enzyme only if their shapes match. In other words, an effective van der Waals interaction between a substrate and an enzyme can occur only if they are *sterically complementary*. Thus, though there is virtually no specificity in a single van der Waals interaction, *specificity arises when there is an opportunity to make a large number of van der Waals bonds simultaneously*. Repulsions between atoms closer than the van der Waals contact distance are as important as the attractive forces in the generation of specificity.

THE BIOLOGICALLY IMPORTANT PROPERTIES OF WATER ARE ITS POLARITY AND COHESIVENESS

Thus far, no consideration has been given to the effect of water on the three basic kinds of bonds. In fact, water is an active participant in molecular interactions in biology. Two properties of water are especially important in this regard:

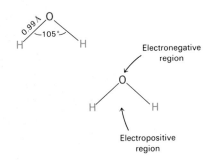

1. *Water is a polar molecule.* The shape of the molecule is triangular, not linear, and so there is an asymmetrical distribution of charge. The oxygen nucleus draws electrons away from the hydrogen nuclei, which leaves the region around them with a net positive charge. If a tetrahedron is described about the oxygen atom, with the hydrogen nuclei at two of the corners, then the other two corners are electronegative. The water molecule is thus an electrically polar structure.

2. *Water molecules have a high affinity for each other.* In a group of water molecules clustered together, a positively charged region in one molecule tends to orient itself toward a negatively charged region in one of its neighbors. Each of its two electronegative regions attracts a proton of a neighbor molecule. Each of its own protons attracts the oxygen end of a neighbor. Thus, each oxygen molecule is the center of a tetrahedron of other oxygens, the O—O distance being 2.76 Å.

The structure of a form of ice is shown in Figure 6-28. When ice melts, its highly regular crystalline structure breaks in many places. However, only about 15% of the bonds are broken. In liquid water, each H_2O is hydrogen bonded to 3.4 neighbors on the average. Thus, liquid water has a partially ordered structure. The hydrogen-bonded aggregates are constantly forming and breaking up. The striking feature of water is that it can *bond with itself*. Indeed, the number of protons that form the positive ends of the hydrogen bonds around any given oxygen atom is equal to the number of unshared electron pairs that form the negative ends. The tetrahedral arrangement of these bonds gives rise to extensive three-dimensional structures. *Water is highly cohesive.*

Figure 6-28
Structure of a form of ice.

The polarity and hydrogen-bonding capabilities of water make it a highly interacting molecule. The consequence is that water weakens electrostatic and hydrogen-bond interactions between other molecules. Water is a very effective competitor in these polar interactions. Consider the action of water on the hydrogen bonding of a carbonyl and amide group (Figure 6-29). The hydrogen atoms of water can replace the NH group as hydrogen-bond donors and the oxygen atom of water can replace the carbonyl oxygen. Thus, a strong hydrogen-bond interaction between a CO and NH group occurs only if water is excluded.

Water diminishes the strength of electrostatic interactions by a factor of 80, the dielectric constant of water, compared with the same interactions in a vacuum. The high dielectric constant of water is an expression of its polarity and its capacity to form an oriented solvent shell around an ion. This attenuates the electrostatic interaction of one ion with another (Figure 6-30). These ori-

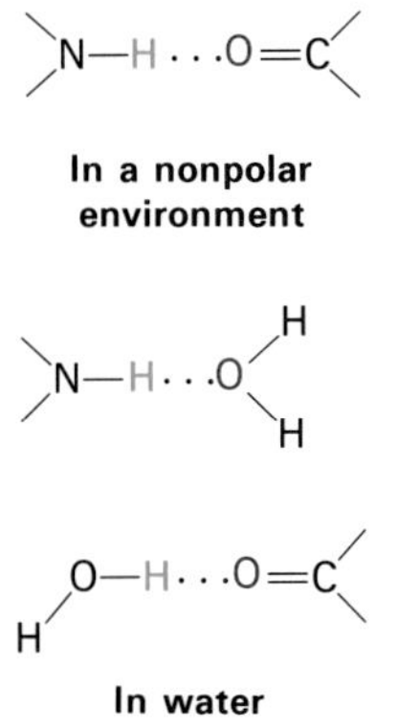

Figure 6-29
Water competes for hydrogen bonds.

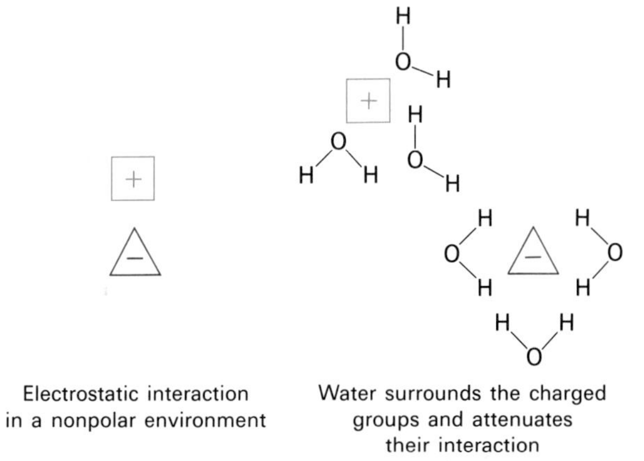

Electrostatic interaction in a nonpolar environment

Water surrounds the charged groups and attenuates their interaction

Figure 6-30
Water attenuates electrostatic interactions between charged groups.

ented solvent shells produce electric fields of their own, which oppose the fields produced by the ions. The consequence is that electrostatic interactions of these ions are markedly weakened. Water has an unusually high dielectric constant.

HYDROPHOBIC INTERACTIONS: NONPOLAR GROUPS TEND TO ASSOCIATE IN WATER

The sight of dispersed oil droplets coming together in water to form a single large oil drop is a familiar one. An analogous process occurs at the atomic level: *nonpolar molecules or groups tend to cluster together in water*. These associations are called *hydrophobic interactions*. In a figurative sense, water tends to squeeze nonpolar molecules together.

Let us examine the basis of hydrophobic interactions, which are a

Table 6-5
Dielectric constants of some solvents

Substance	Dielectric constant $(20°C)$
Hexane	1.9
Benzene	2.3
Diethyl ether	4.3
Chloroform	5.1
Acetone	21.4
Ethanol	24
Methanol	33
Water	80
Hydrogen cyanide	116

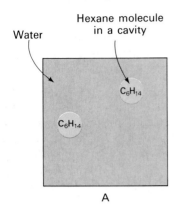

Water

Hexane molecule
in a cavity

C_6H_{14}

C_6H_{14}

A

Two hexane molecules
in a single cavity

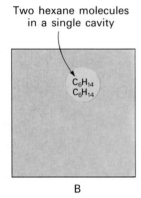

C_6H_{14}
C_6H_{14}

B

Figure 6-31
A schematic representation of two molecules of hexane in a small volume of water: (A) the hexane molecules occupy different cavities in the water structure and (B) they occupy the same cavity, which is energetically more favorable.

major driving force in the folding of macromolecules, the binding of substrates to enzymes, and many other molecular interactions in biological systems. Consider the introduction of a nonpolar molecule, such as hexane, into water. A cavity is created, which transiently disrupts some hydrogen bonds between water molecules. The displaced water molecules then reorient themselves to form a maximum number of new hydrogen bonds. This is accomplished but at a price: the number of ways of forming favorable hydrogen bonds in the cage of water around the hexane molecule are much fewer than in pure water. The water molecules around the hexane molecule are much more ordered than elsewhere in the solution, and so the entropy of the solution decreases. Now let us consider the arrangement of two hexane molecules in water. Do they sit in two small cavities (Figure 6-31A) or in a single larger one (Figure 6-31B)? The experimental fact is that the two hexane molecules come together and occupy a single large cavity. This association is driven by the release of some of the oriented water molecules around the separated hexanes. Thus, the basis of a hydrophobic interaction is the increase in entropy resulting from the enhanced freedom of released water molecules. *Nonpolar solute molecules are driven together in water not primarily because they have a high affinity for each other but because water bonds strongly to itself.*

SUMMARY

The catalysts in biological systems are enzymes, which invariably are proteins. Enzymes are highly specific and have great catalytic power. They usually enhance reaction rates by a factor of at least 10^7. Enzymes do not alter reaction equilibria. Rather, they serve as catalysts by reducing the activation energy of chemical reactions. The Michaelis-Menten model accounts for the kinetic properties of some enzymes. In this model, an enzyme (E) combines with a substrate (S) to form an enzyme-substrate (ES) complex, which can proceed to form a product (P) or to dissociate into E and S.

$$\text{E} + \text{S} \underset{k_2}{\overset{k_1}{\rightleftharpoons}} \text{ES} \overset{k_3}{\longrightarrow} \text{E} + \text{P}$$

The rate (V) of formation of product is given by the Michaelis-Menten equation:

$$V = V_{max} \frac{[\text{S}]}{[\text{S}] + K_M}$$

in which V_{max} is the rate when the enzyme is fully saturated with substrate, and K_M, the Michaelis constant, is the substrate concentration at which the reaction rate is half maximal. The maximal rate, V_{max}, is equal to the product of k_3 and the total concentration of enzyme. The kinetic constant k_3, called the turnover number, is the number of substrate molecules converted into product per unit time at a single catalytic site when the enzyme is fully saturated with substrate. Turnover numbers for most enzymes are between 1 and 10^4 per second.

Enzymes can be inhibited by specific small molecules or ions. In irreversible inhibition, the inhibitor is covalently linked to the enzyme or bound so tightly that its dissociation from the enzyme is very slow. In contrast, reversible inhibition is characterized by a rapid equilibrium between enzyme and inhibitor. A competitive inhibitor prevents the substrate from binding to the active site. It reduces the reaction velocity by diminishing the proportion of enzyme molecules that have a bound substrate. In noncompetitive inhibition, the inhibitor decreases the turnover number. Competitive inhibition can be distinguished from noncompetitive inhibition by determining whether the inhibition can be overcome by raising the substrate concentration.

The catalytic activity of many enzymes is regulated in vivo. Allosteric interactions, which are defined as interactions between spatially distinct sites, are particularly important in this regard. All known allosteric enzymes consist of two or more subunits. Allosteric interactions are mediated by conformational changes that are transmitted from one subunit to another. Allosteric enzymes usually exhibit sigmoidal rather than hyperbolic plots of V versus [S]. Two limiting models—the concerted model and the sequential model—have been postulated to account for some properties of these enzymes.

Reversible molecular interactions in biological systems are due to electrostatic bonding, hydrogen bonding, and van der Waals bonding. These interactions are profoundly affected by the presence of water. Some important characteristics of water are its polarity, capacity to serve as both a hydrogen donor and a hydrogen acceptor, and cohesiveness. Consequently, water diminishes the strength of electrostatic and hydrogen bonding between other molecules and ions. In contrast, water accentuates the interaction of nonpolar molecules. Substrates are bound to enzymes at active-site clefts from which water is largely excluded when the substrate is bound. The exclusion of water strengthens electrostatic and hydrogen bonds between enzyme and substrate. A significant part of the binding energy comes from the association of nonpolar parts of the substrate with those of the active site. The specificity of the enzyme-substrate interaction arises from hydrogen bonding, which is highly directional, and the shape of the active site, which rejects molecules that do not have a complementary shape. The recognition of substrate by many enzymes is a dynamic process accompanied by conformational changes at the active site of the enzyme.

SELECTED READINGS

WHERE TO START

Koshland, D. E., Jr., 1973. Protein shape and biological control. *Sci. Amer.* 229(4):52–64. [An excellent introduction to the importance of conformational flexibility for the specificity and regulation of enzyme action. Available as *Sci. Amer.* Offprint 1280.]

BOOKS ON ENZYMES

Fersht, A., 1977. *Enzyme Structure and Mechanism.* Freeman. [A concise and lucid introduction to enzyme action, with emphasis on physical principles.]

Walsh, C., 1979. *Enzymatic Reaction Mechanisms.* Freeman. [An excellent account of the chemical basis of enzyme action. The author demonstrates that the large number of enzyme-catalyzed reactions in biological systems can be grouped into a small number of types of chemical reactions.]

Boyer, P. D., (ed.), 1970. *The Enzymes* (3rd ed.). Academic Press. [This multivolume treatise on enzymes contains a wealth of information. Volumes 1 and 2 (available in a paperback edition) deal with general aspects of enzyme structure, mechanism, and regulation. Volume 3 and subsequent ones contain detailed and authoritative articles on individual enzymes.]

ENZYME KINETICS AND MECHANISMS

Fersht, A. R., 1974. Catalysis, binding, and enzyme-substrate complementarity. *Proc. Roy. Soc.* B187:397–407.

Jencks, W. P., 1975. Binding energy, specificity, and enzymic catalysis: the Circe effect. *Advan. Enzymol.* 43:219–410.

Knowles, J. R., and Albery, W. J., 1976. Evolution of enzyme function and the development of catalytic efficiency. *Biochemistry* 15:5631–5640.

Warshel, A., 1978. Energetics of enzyme catalysis. *Proc. Nat. Acad. Sci.* 75:5250–5254. [Presentation of a hypothesis that emphasizes the importance of electrostatic interactions.]

ALLOSTERIC INTERACTIONS

Monod, J., Changeux, J.-P., and Jacob, F., 1963. Allosteric proteins and cellular control systems. *J. Mol. Biol.* 6:306–329. [A classic paper that introduced the concept of allosteric interactions.]

Monod, J., Wyman, J., and Changeux, J.-P., 1965. On the nature of allosteric transitions. *J. Mol. Biol.* 12:88–118. [Presentation of the concerted model for allosteric transitions.]

Koshland, D. E., Jr., Nemethy, G., and Filmer, D., 1966. Comparison of experimental binding data and theoretical models in proteins containing subunits. *Biochemistry* 5:365–385. [Presentation of the sequential model for allosteric transitions.]

MOLECULAR INTERACTIONS AND BINDING

Richards, F. M., Wyckoff, H. W., and Allewell, N., 1970. The origin of specificity in binding: a detailed example in a protein-nucleic acid interaction. *In* Schmitt, F. O., (ed.), *The Neurosciences: Second Study Program,* pp. 901–912. Rockefeller University Press.

Davidson, N., 1967. Weak interactions and the structure of biological macromolecules. *In* Quarton, G. C., Melnechuk, T., and Schmitt, F. O., (eds.), *The Neurosciences: A Study Program,* pp. 46–56. Rockefeller University Press.

Tanford, C., 1978. The hydrophobic effect and the organization of living matter. *Science* 200:1012–1018.

PROBLEMS

1. The hydrolysis of pyrophosphate to orthophosphate is important in driving forward biosynthetic reactions such as the synthesis of DNA. This hydrolytic reaction is catalyzed in *E. coli* by a pyrophosphatase that has a mass of 120 kdal and consists of six identical subunits. Purified enzyme has a V_{max} of 2800 units per milligram of enzyme. For this enzyme, a unit of activity is defined as the amount of enzyme that hydrolyzes 10 μmol of pyrophosphate in 15 minutes at 37°C under standard assay conditions.

   (a) How many moles of substrate are hydrolyzed per second per milligram of enzyme when the substrate concentration is much greater than K_M?

   (b) How many moles of active site are there in 1 mg of enzyme? Assume that each subunit has one active site.

(c) What is the turnover number of the enzyme? Compare this value with others mentioned in this chapter.

2. Penicillin is hydrolyzed and thereby rendered inactive by penicillinase, an enzyme present in some resistant bacteria. The mass of this enzyme in *Staphylococcus aureus* is 29.6 kdal. The amount of penicillin hydrolyzed in 1 minute in a 10-ml solution containing 10^{-9} g of purified penicillinase was measured as a function of the concentration of penicillin. Assume that the concentration of penicillin does not change appreciably during the assay.

[Penicillin]	Amount hydrolyzed (moles)
0.1×10^{-5} M	0.11×10^{-9}
0.3×10^{-5} M	0.25×10^{-9}
0.5×10^{-5} M	0.34×10^{-9}
1.0×10^{-5} M	0.45×10^{-9}
3.0×10^{-5} M	0.58×10^{-9}
5.0×10^{-5} M	0.61×10^{-9}

(a) Make a $1/V$ versus $1/[S]$ plot of these data. Does penicillinase appear to obey Michaelis-Menten kinetics? If so, what is the value of K_M?
(b) What is the value of V_{max}?
(c) What is the turnover number of penicillinase under these experimental conditions? Assume one active site per enzyme molecule.

3. The kinetics of an enzyme are measured as a function of substrate concentration in the presence and absence of 2×10^{-3} M inhibitor (I).

[S]	Velocity ($\mu mol/min$)	
	No inhibitor	Inhibitor
0.3×10^{-5} M	10.4	4.1
0.5×10^{-5} M	14.5	6.4
1.0×10^{-5} M	22.5	11.3
3.0×10^{-5} M	33.8	22.6
9.0×10^{-5} M	40.5	33.8

(a) What are the values of V_{max} and K_M in the absence of inhibitor? in its presence?
(b) What type of inhibition is this?

(c) What is the binding constant of this inhibitor?
(d) If $[S] = 1 \times 10^{-5}$ M and $[I] = 2 \times 10^{-3}$ M, what fraction of the enzyme molecules have a bound substrate? a bound inhibitor?
(e) If $[S] = 3 \times 10^{-5}$ M, what fraction of the enzyme molecules have a bound substrate in the presence and absence of 2×10^{-3} M inhibitor? Compare this ratio with the ratio of the reaction velocities under the same conditions.

4. The kinetics of the enzyme discussed in problem 3 are measured in the presence of a different inhibitor. The concentration of this inhibitor is 10^{-4} M.

[S]	Velocity ($\mu mol/min$)	
	No inhibitor	Inhibitor
0.3×10^{-5} M	10.4	2.1
0.5×10^{-5} M	14.5	2.9
1.0×10^{-5} M	22.5	4.5
3.0×10^{-5} M	33.8	6.8
9.0×10^{-5} M	40.5	8.1

(a) What are the values of V_{max} and K_M in the presence of this inhibitor? Compare them with those obtained in problem 3.
(b) What type of inhibition is this?
(c) What is the dissociation constant of this inhibitor?
(d) If $[S] = 3 \times 10^{-5}$ M, what fraction of the enzyme molecules have a bound substrate in the presence and absence of 10^{-4} M inhibitor?

5. The plot of $1/V$ versus $1/[S]$ is sometimes called a Lineweaver-Burk plot. Another way of expressing the kinetic data is to plot V versus $V/[S]$, which is known as an Eadie-Hofstee plot.

(a) Rearrange the Michaelis-Menten equation to give V as a function of $V/[S]$.
(b) What is the significance of the slope, the y-intercept, and the x-intercept in a plot of V versus $V/[S]$?
(c) Make a sketch of a plot of V versus $V/[S]$ in the absence of an inhibitor, in the presence of a competitive inhibitor, and in the presence of a noncompetitive inhibitor.

6. For allosteric enzymes, low concentrations of competitive inhibitors frequently act as activators. Why? (Hint: Consider the analogy of CO-hemoglobin.)

7. The hormone progesterone contains two ketone groups. Little is known about the properties of the receptor protein that recognizes progesterone. At pH 7, which amino acid side chains might form hydrogen bonds with progesterone? (Assume that the side chains in the receptor protein have the same pKs as in the amino acids in aqueous solution).

8. Suppose that two substrates A and B compete for an enzyme. Derive an expression relating the ratio of the rates of utilization of A and B, V_A/V_B, to the concentrations of these substrates and their values of k_3 and K_M. (Hint: express V_A as a function of k_3/K_M for substrate A, and do the same for V_B.) Is specificity determined by K_M alone?

For additional problems, see W. B. Wood, J. H. Wilson, R. M. Benbow, and L. E. Hood, *Biochemistry: A Problems Approach* (Benjamin, 1974), chs. 6 and 7; and R. Montgomery and C. A. Swenson, *Quantitative Problems in the Biochemical Sciences,* 2nd ed. (Freeman, 1976), ch. 11.

CHAPTER 7

MECHANISMS OF ENZYME ACTION: LYSOZYME AND CARBOXYPEPTIDASE

In 1922, Alexander Fleming, a bacteriologist in London, had a cold. He was not one to waste a moment and consequently used his cold as an opportunity to do an experiment. He allowed a few drops of his nasal mucus to fall on a culture plate containing bacteria. He was excited to find some time later that the bacteria near the mucus had been dissolved away and thought that the mucus might contain the universal antibiotic he was seeking. Fleming showed that the antibacterial substance was an enzyme, which he named lysozyme—*lyso* because of its capacity to lyse bacteria and *zyme* because it was an enzyme. He also discovered a small round bacterium that was particularly susceptible to lysozyme, which he named *Micrococcus lysodeikticus* because it was a displayer of lysis ("deiktikos" means able to show). Fleming found that tears are a rich source of lysozyme. Volunteers provided tears after they suffered a few squirts of lemon—an "ordeal by lemon." The *St. Mary's Hospital Gazette* published a cartoon showing children coming for a few pennies to Fleming's laboratory, where one attendant administered the beatings while another collected their tears! Fleming was disappointed to find that lysozyme was not effective against the most harmful bacteria. But seven years later, he did discover a highly effective antibiotic: penicillin—a striking illustration of Pasteur's comment that chance favors the prepared mind.

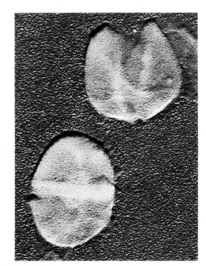

Figure 7-1
Electron micrograph of the isolated cell wall of *Micrococcus lysodeikticus*. [Courtesy of Dr. Nathan Sharon.]

CH₂OH

N-Acetylglucosamine
(NAG)

CH₂OH

N-Acetylmuramic acid
(NAM)

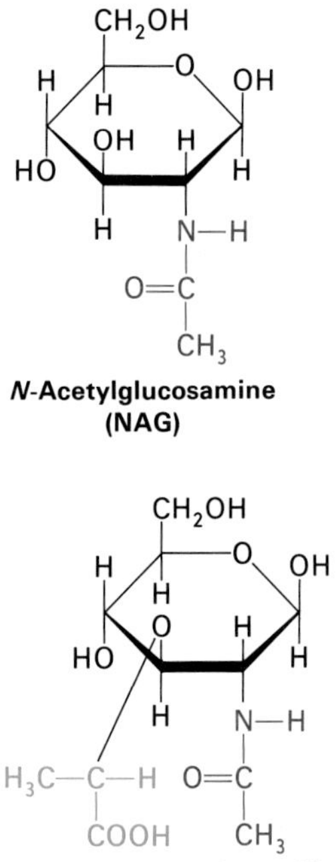

Figure 7-2
Sugar residues in the polysaccharide of bacterial cell walls.

Lysozyme dissolves certain bacteria by cleaving the polysaccharide component of their cell walls. The function of the cell wall in bacteria is to confer mechanical support. A bacterial cell devoid of a cell wall usually bursts because of the high osmotic pressure inside the cell. The detailed structure of bacterial cell walls will be discussed in a later chapter. Let us consider here the structure of the polysaccharide portion.

The cell-wall polysaccharide is made up of two kinds of sugars: N-*acetylglucosamine* (NAG) and N-*acetylmuramic acid* (NAM). NAM and NAG are derivatives of glucosamine in which the amino group is acetylated (Figure 7-2). In NAM, a lactyl side chain is attached to C-3 of the sugar ring by an ether bond. In bacterial cell walls, NAM and NAG are joined by *glycosidic linkages* between C-1 of one sugar and C-4 of the other. The oxygen atom in a glycosidic bond can be located either above or below the plane of the sugar ring. In the α configuration, the oxygen is below the plane of the sugar; in the β configuration, it is above (see Chapter 12 for a more detailed discussion of the properties and nomenclature of sugars). All of the glycosidic bonds of the cell-wall polysaccharide have a β *configuration* (Figure 7-3). NAM and NAG alternate in sequence. Thus, the cell-

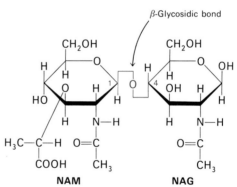

Figure 7-3
NAM is linked to NAG by a $\beta(1\rightarrow4)$ glycosidic bond.

wall polysaccharide is an alternating polymer of NAM and NAG residues joined by $\beta(1 \rightarrow 4)$ glycosidic linkages. Different polysaccharide chains are cross-linked by short peptides that are attached to some of the NAM residues.

Lysozyme hydrolyzes the glycosidic bond between C-1 of NAM and C-4 of NAG (Figure 7-4). The other glycosidic bond, between C-1 of NAG and C-4 of NAM, is not cleaved. *Chitin,* a polysaccharide found in the shell of crustaceans, is also a substrate for lysozyme. Chitin consists only of NAG residues joined by $\beta(1 \rightarrow 4)$ glycosidic links.

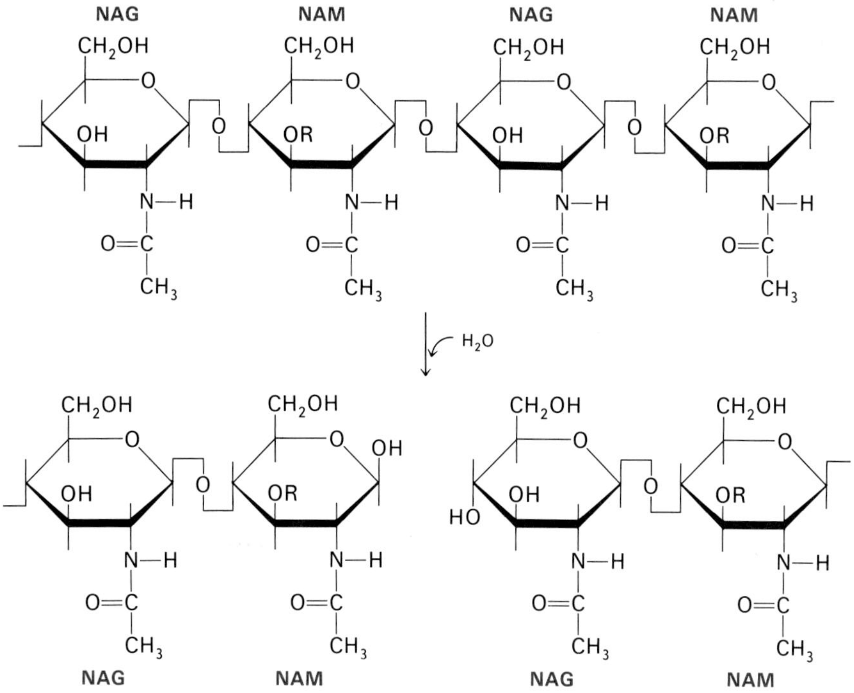

Figure 7-4
Lysozyme hydrolyzes the glycosidic bond between NAM and NAG (R refers to the lactyl group of NAM).

THREE-DIMENSIONAL STRUCTURE OF LYSOZYME

Lysozyme is a relatively small enzyme. The enzyme from hen egg white, a rich source, is made up of a single polypeptide chain of 129 amino acids and has a mass of 14.6 kdal. The enzyme is cross-linked by four disulfide bridges, which contribute to its high stability. The amino acid sequence of lysozyme is shown in Figure 7-5.

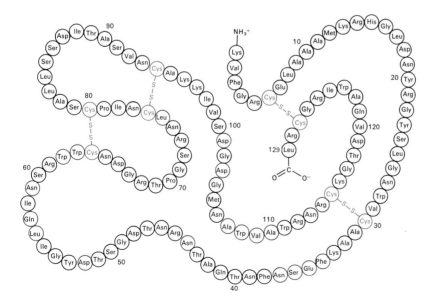

Figure 7-5
Amino acid sequence of hen egg-white lysozyme. Residues that are part of the active site are circled in red. [After R. E. Canfield and A. K. Liu. *J. Biol. Chem.* 240(1965):2000; D. C. Phillips. *Sci. Amer.* (5)215(1966):79.]

In 1965, David Phillips and his colleagues determined the three-dimensional structure of lysozyme. Their high-resolution electron-density map was the first for an enzyme molecule. Lysozyme is a compact molecule, roughly ellipsoidal in shape, with dimensions $45 \times 30 \times 30$ Å. The folding of this molecule is complex (Figure 7-7). There is much less α helix than in myoglobin and hemoglobin.

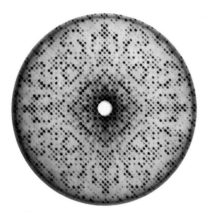

Figure 7-6
X-ray precessional photograph of a lysozyme crystal. [Courtesy of Dr. David Phillips.]

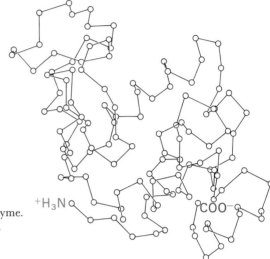

Figure 7-7
Three-dimensional structure of lysozyme. Only the α-carbon atoms are shown. [Courtesy of Dr. David Phillips.]

In a number of regions, the polypeptide chain is in an extended conformation. In one of these sections, the chain doubles back on itself, and the two strands are hydrogen bonded through their peptide groups. This hairpin section is similar to a regularly repeating secondary structure mentioned earlier—namely, the *antiparallel β pleated sheet* seen in silk protein. The interior of lysozyme, like that of myoglobin and hemoglobin, is almost entirely nonpolar. Hydrophobic interactions evidently play an important role in the folding of lysozyme, as they do for most proteins.

FINDING THE ACTIVE SITE IN LYSOZYME

Knowledge of the detailed three-dimensional structure of lysozyme did not immediately reveal the catalytic mechanism. In fact, the location of the active site was not obvious on looking at the electron-density map. Lysozyme does not contain a prosthetic group and thus lacks a built-in marker at its active site, in contrast with such proteins as myoglobin and hemoglobin. The essential information needed to identify the active site, specify the mode of binding of substrate, and elucidate the enzymatic mechanism came from an x-ray crystallographic study of the interaction of lysozyme with inhibitors. When the three-dimensional structure of a protein is known, the mode of binding of small molecules can often be determined quite readily by x-ray crystallographic methods. These experiments are feasible because protein crystals are quite porous. Rather large inhibitor molecules can diffuse in the channels between different protein molecules and find their way to specific binding sites. The electron density corresponding to the additional molecule can be calculated directly from the intensities of the x-ray reflections (using the phases already determined for the native protein crystal) if the crystal structure is not markedly altered. This technique is called the *difference Fourier method*.

Ideally, one would like to use the difference Fourier method to elucidate the structure of an enzyme-substrate (ES) complex undergoing catalysis. However, under ordinary conditions, the conversion of bound substrate into product is much more rapid than the diffusion of new substrate into the crystal. This difficulty can sometimes be circumvented by cooling the crystal to a low temperature (e.g., $-50°C$) to slow the catalytic process. This experimental approach is called *cryoenzymology*. Alternatively, much information can be derived from a study of a complex of an enzyme with an unreactive (or very slowly reactive) analog of its substrate. For lysozyme, this was achieved with the trimer of N-acetylglucosamine (tri-NAG or NAG$_3$), whose structure is shown in Figure 7-8. Oligomers of N-

Tri-*N*-acetylglucosamine
(Tri-NAG)

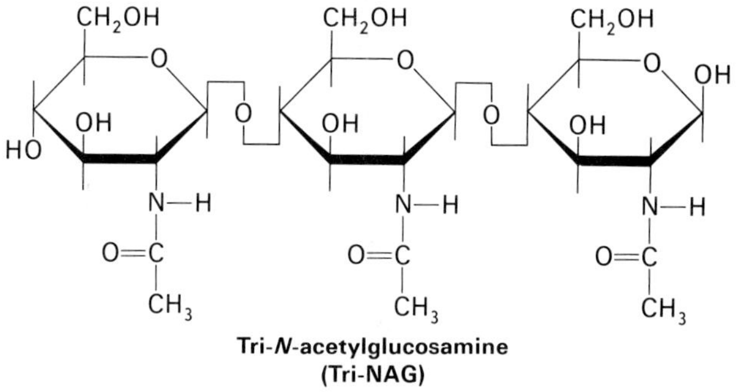

Figure 7-8
Formula of tri-*N*-acetylglucosamine
(tri-NAG or NAG$_3$), a competitive
inhibitor of lysozyme.

acetylglucosamine consisting of fewer than five residues are hydrolyzed very slowly or not at all. However, they do bind to the active site of the enzyme. Indeed, tri-NAG is a potent competitive inhibitor of lysozyme.

MODE OF BINDING OF A COMPETITIVE INHIBITOR

The x-ray study of the tri-NAG lysozyme complex showed the location of the active site, revealed the interactions responsible for the specific binding of substrate, and led to the proposal of a detailed enzymatic mechanism. Tri-NAG binds to lysozyme in a cleft at the surface of the enzyme and occupies about half of the cleft. Tri-NAG is bound to lysozyme by hydrogen-bond and van der Waals interac-

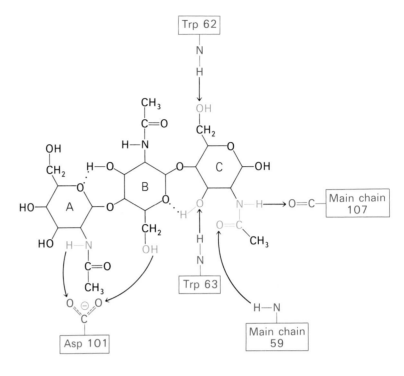

Figure 7-9
Hydrogen bonds between tri-NAG and lysozyme. The groups on tri-NAG involved in hydrogen bonding are shown in blue; those on lysozyme, in red.

tions. Electrostatic interactions cannot occur because tri-NAG lacks ionic groups.

The *hydrogen-bond interactions* between tri-NAG and lysozyme are shown in Figure 7-9. The carboxylate group of aspartate 101 is hydrogen bonded to residues A and B. The most specific and extensive hydrogen bonds are made between the enzyme and sugar residue C. There are four hydrogen bonds. The NH of the indole ring of tryptophan 62 is hydrogen bonded to the oxygen attached to C-6. The adjacent amino acid residue, tryptophan 63, is similarly hydrogen bonded to the oxygen attached to C-3. The ring of tryptophan 62 moves by 0.75 Å when tri-NAG binds to the enzyme. There are very good hydrogen bonds between the acetamido side-chain CO and NH groups of sugar residue C and the main-chain NH and CO groups of amino acid residues 59 and 107, respectively.

There are a large number of *van der Waals contacts* between tri-NAG and the enzyme. Sugar residue B is involved in few polar contacts with the enzyme but is closely associated with the indole ring of tryptophan 62. Residue A has rather tenuous contacts with the enzyme.

FROM STRUCTURE TO ENZYMATIC MECHANISM

1. *How would a substrate bind?* As noted previously, the mode of binding of an effective substrate cannot be established directly by the x-ray crystallographic method. However, the structure of an enzyme–competitive-inhibitor complex may provide key clues to

how a substrate binds to the enzyme. The finding that tri-NAG fills only half of the cleft in lysozyme was a very suggestive starting point. It was assumed that the observed binding of tri-NAG as an inhibitor involves interactions that are also utilized in the binding of substrate. It seemed likely that additional sugar residues, which would fill the rest of the cleft, are required for the formation of a reactive ES complex. There was space for three additional sugar residues. This was encouraging, because it was known that the hexamer of N-acetylglucosamine (hexa-NAG) is rapidly hydrolyzed by the enzyme.

Three additional sugar residues, named D, E, and F, were fitted into the cleft by careful model building (Figure 7-10). Residues E and F went in nicely, making a number of good hydrogen bonds and van der Waals contacts. However, residue D would not fit unless it was distorted. Its C-6 and O-6 atoms came too close to several groups on the enzyme unless the ring was distorted from its normal conformation, which has the appearance of a chair.

2. *Which bond is cleaved?* The rate of hydrolysis of oligomers of N-acetylglucosamine increases strikingly as the number of residues is increased from four to five—that is, from NAG_4 to NAG_5 (Table 7-1). There is a further increase in the cleavage rate when the sixth

Table 7-1
Effectiveness of oligomers of
N-acetylglucosamine as substrates

Substrate	Relative rate of hydrolysis
NAG_2	0
NAG_3	1
NAG_4	8
NAG_5	4,000
NAG_6	30,000
NAG_8	30,000

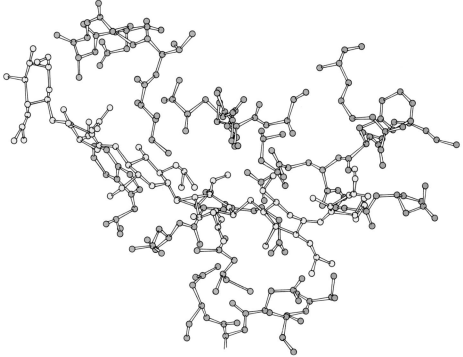

Figure 7-10
Mode of binding of hexa-NAG (shown in yellow) to lysozyme. The locations of sugar residues A, B, and C (left) are those observed in the tri-NAG–lysozyme complex, whereas those of residues D, E, and F (right) are inferred by model building. The two residues shown in green directly participate in catalysis.

residue is added (NAG_6), but there is no change as the number of residues is increased to eight. This finding is consistent with the crystallographic results, which show that the active-site cleft would be fully occupied by six sugar residues.

Which bond of hexa-NAG is cleaved? Because tri-NAG is stable, the A—B bond (i.e., the glycosidic bond between the A and B sugar residues) cannot be the site of cleavage. Similarly, the B—C bond cannot be the site of cleavage. A second crucial piece of evidence is that site C cannot be occupied by NAM. NAG fits nicely into site C, but NAM is too large because of its lactyl side chain. The bond cleaved in bacterial cell walls is the NAM–NAG link. Hence, the C—D bond cannot be the site of cleavage if the bacterial cell-wall polysaccharide binds to the enzyme in the same manner as hexa-NAG. The inability of NAM to fit into site C excludes yet another cleavage site: the E—F bond. The cell-wall polysaccharide is an alternating polymer of NAM and NAG, and so NAM cannot occupy site E if it cannot occupy site C.

These observations eliminated the A—B, B—C, C—D, and E—F bonds as possible sites of enzymatic cleavage of the hexamer substrate. Hence, *the D—E bond was the only remaining candidate for the cleavage site* (Figure 7-11).

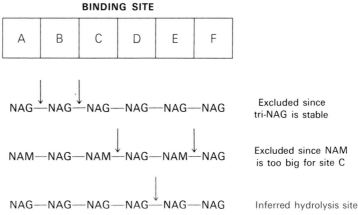

Figure 7-11
Steps in deducing that the glycosidic bond between sugar residues D and E is the one cleaved by lysozyme.

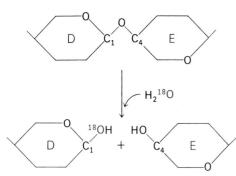

Figure 7-12
Hydrolysis in ^{18}O water showed that lysozyme cleaves the C_1—O bond rather than the O—C_4 bond. (Only the skeletons of the D and E residues are shown here.)

3. *Which groups on the enzyme participate in catalysis?* The inference that it is the D—E bond that is split was an important step toward defining the groups on the enzyme that carry out the hydrolysis reaction. However, it was necessary to localize the site of cleavage even more precisely: on which side of the glycosidic oxygen atom is the bond cleaved? The question was answered by carrying out the enzymatic hydrolysis in the presence of water enriched with ^{18}O, the stable heavy isotope of oxygen (Figure 7-12). The sugars isolated contained ^{18}O attached to C-1 of the D sugar. In contrast, the hydroxyl group attached to C-4 of the E sugar contained the ordi-

nary isotope of oxygen. *Hence, the bond split is the one between C-1 of residue D and the oxygen of the glycosidic linkage to residue E.* This experiment illustrates the use of isotopes in elucidating enzymatic mechanisms. In the absence of an isotopic marker, it would have been difficult, if not impossible, to establish the precise site of cleavage.

A search was then made for possible catalytic groups close to the glycosidic bond that is cleaved. As mentioned in the preceding chapter, a *catalytic group* is one that directly participates in the formation or breakage of covalent bonds. The most plausible candidates are groups that can serve as *hydrogen-bond donors or acceptors.* The donation or abstraction of a hydrogen ion is a critical step in most enzymatic reactions. The only plausible catalytic residues near the glycosidic bond that is cleaved by lysozyme are aspartic 52 and glutamic 35. The aspartic acid residue is on one side of the glycosidic linkage, whereas the glutamic acid residue is on the other. These two acidic side chains have markedly different environments. Aspartic 52 is in a distinctly polar environment, where it serves as a hydrogen-bond acceptor in a complex network of hydrogen bonds. In contrast, glutamic 35 lies in a nonpolar region. Thus, it seemed likely that at pH 5, which is the pH optimum for the hydrolysis of chitin by lysozyme, *aspartic 52 is in the ionized COO⁻ form, whereas glutamic 35 is in the un-ionized COOH form.* The nearest oxygen of each of these acid groups is located about 3 Å away from the glycosidic linkage (Figure 7-13).

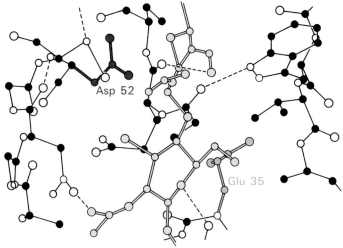

Figure 7-13
Structure of part of the active site of lysozyme. The D and E rings of the hexa-NAG substrate are shown in yellow. The side chains of aspartate 52 (red) and glutamic 35 (green) are in close proximity. [After W. N. Lipscomb. *Proc. Robert A. Welch Found. Conf. Chem. Res.* 15(1971):150.]

A CARBONIUM ION INTERMEDIATE IS CRITICAL FOR CATALYSIS

Phillips and his colleagues proposed a detailed catalytic mechanism for lysozyme based on the preceding structural data. The essential steps are:

1. The —COOH group of glutamic 35 donates an H$^+$ to the bond between C-1 of the D ring and the glycosidic oxygen atom, which thereby cleaves this bond (Figure 7-14).

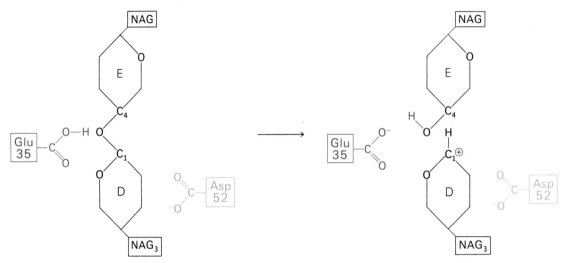

Figure 7-14
The first step in the proposed catalytic mechanism for lysozyme is the transfer of an H$^+$ from Glu 35 to the oxygen atom of the glycosidic bond. The glycosidic bond is thereby cleaved, and a carbonium ion intermediate is formed.

2. This creates a positive charge on C-1 of the D ring. This transient species is called a *carbonium ion* because it contains a positively charged carbon atom.

3. The dimer of NAG consisting of residues E and F diffuses away from the enzyme.

4. The carbonium ion intermediate then reacts with OH$^-$ from the solvent (Figure 7-15). Tetra-NAG, consisting of residues A, B, C, and D, diffuses away from the enzyme.

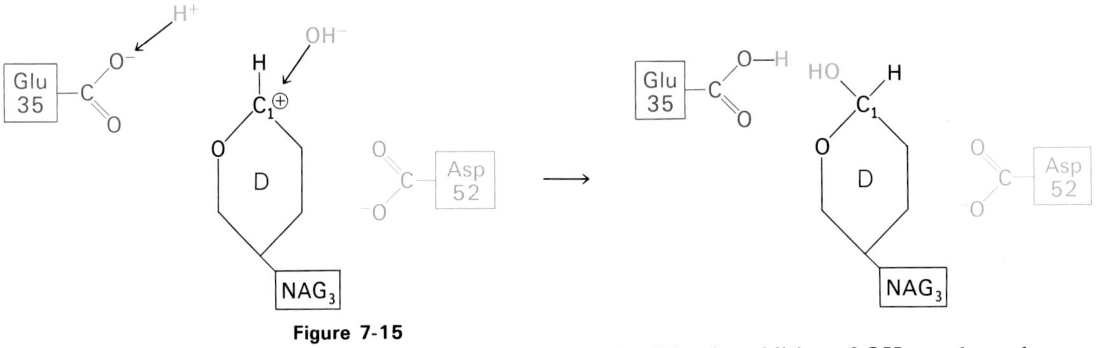

Figure 7-15
The hydrolysis reaction is completed by the addition of OH$^-$ to the carbonium ion intermediate and of H$^+$ to the side chain of Glu 35.

5. Glutamic 35 becomes protonated, and the enzyme is ready for another round of catalysis.

The critical elements of this catalytic scheme are:

1. *General acid catalysis.* A proton is transferred from glutamic 35, which is un-ionized and optimally located 3 Å away from the glycosidic oxygen atom.

2. *Promotion of the formation of the carbonium ion intermediate.* The enzymatic reaction is markedly facilitated by two different factors that stabilize the carbonium ion intermediate:

    a. The electrostatic factor is the presence of a negatively charged group 3 Å away from the carbonium ion intermediate. Aspartic 52, which is in the negatively charged carboxylate form, interacts electrostatically with the positive charge on C-1 of ring D.

    b. The geometrical factor is the distortion of ring D (Figure 7-16). Hexa-NAG fits best into the active-site cleft if sugar residue

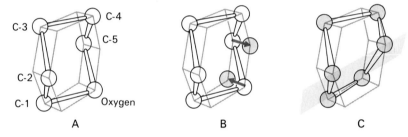

Figure 7-16
Distortion of the D ring of the substrate of lysozyme into a half-chair form:
(A) a sugar residue in the normal chair form; (B) on binding to lysozyme, the ring oxygen atom and C-5 of sugar residue D move so that C-1, C-2, C-5 and O are in the same plane, as shown in part C. [After D. C. Phillips. The three-dimensional structure of an enzyme molecule. Copyright © 1966 by Scientific American, Inc. All rights reserved.]

D is distorted out of its customary chair conformation into a half-chair form. This distortion enhances catalysis because the half-chair geometry markedly promotes carbonium ion formation. The planarity of carbon atoms 1, 2, and 5 and the ring oxygen atom in the half-chair form enables the positive charge to be shared between C-1 and the ring oxygen atom. Thus, *in the process of binding, the enzyme forces the substrate to assume the geometry of the transition state,* which is a carbonium ion.

EXPERIMENTAL SUPPORT FOR THE PROPOSED MECHANISM

The mode of binding of substrate and the mechanism of catalysis proposed on the basis of the crystallographic studies have been

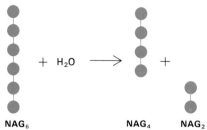

Figure 7-17
Hexa-NAG is hydrolyzed to
tetra-NAG and di-NAG.

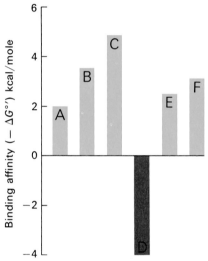

Figure 7-18
Contributions made by each of the six
sugars of hexa-NAG to the standard
free energy of binding of this substrate.
The binding of residue D *costs* free
energy because it must be distorted to
fit into the active site.

tested in a variety of chemical experiments. All of the experimental results thus far support the crystallographic hypothesis. A number of experimental findings are especially pertinent in this regard:

1. *Cleavage pattern.* Hexa-NAG is split into tetra-NAG and di-NAG, which confirms the crystallographic hypothesis that cleavage occurs between the fourth and fifth residues of a hexamer (Figure 7-17).

2. *Binding affinity.* The contributions of each of the six sugar residues to the total free energy of binding to a hexameric substrate have been determined from measurements of binding equilibria (Figure 7-18). The striking finding is that sugar residue D makes a negative contribution to the binding affinity. The cost of binding residue D is about 4 kcal/mol. This result confirms the crystallographic inference that residue D is bound in a distorted form. A price must be paid for the distortion of residue D from its customary chair form into a half-chair form. Also of interest is the finding that residue C makes the largest positive contribution to the binding affinity. The crystallographic model shows that residue C makes a large number of hydrogen bonds and van der Waals interactions.

3. *Transition-state analogs.* The distortion of sugar residue D into a half-chair form is an important aspect of the postulated enzymatic mechanism because the half-chair form is assumed to be the conformation of the transition state. As noted above, this idea is supported by the finding that the binding of residue D costs energy, the price of distortion. Further evidence comes from a study of a transition-state analog of the substrate—that is, a compound that has the geometry of the transition state in catalysis *before* it is bound to the enzyme, as well as when it is bound. The D ring of the lactone analog of tetra-NAG (Figure 7-19) has a half-chair conformation in

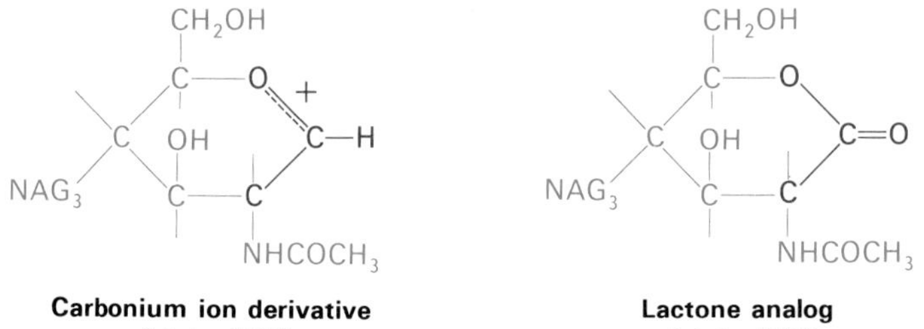

**Carbonium ion derivative
of tetra-NAG**

**Lactone analog
of tetra-NAG**

Figure 7-19
The lactone analog of tetra-NAG resembles the transition-state intermediate in the reaction catalyzed by lysozyme because its D ring has a conformation like that of a half-chair form.

crystals of the tetrasaccharide alone. When bound to lysozyme, the C-1, C-2, C-4, C-6, and O atoms of the D ring of this analog are coplanar. This sofa conformation is similar to the postulated half-chair form for the transition state, and so the lactone analog binds to lysozyme with little distortion, in contrast with tetra-NAG. In fact, the binding of this lactone analog of tetra-NAG to subsites A through D of lysozyme is 3600 times as strong as that of tetra-NAG. This finding suggests that *the distortion of the D ring of a normal substrate could accelerate catalysis by a factor of the order of 3600.*

This factor in catalysis was clearly foreseen by Pauling in a lecture that he gave in 1948:

> I think that enzymes are molecules that are complementary in structure to the activated complexes of the reactions that they catalyze, that is, to the molecular configuration that is intermediate between the reacting substances and the products of reaction for these catalyzed processes. The attraction of the enzyme molecule for the activated complex would thus lead to a decrease in its energy and hence to a decrease in the energy of activation of the reaction and to an increase in the rate of reaction.

4. *pH dependence of the catalytic rate.* The rate of hydrolysis of chitin is most rapid at pH 5 (Figure 7-20). The enzymatic activity drops sharply on either side of this optimal pH. The decrease on the alkaline side is due to the ionization of glutamic 35, whereas the decrease in rate on the acid side reflects the protonation of aspartate 52. Lysozyme is active only when glutamic 35 is un-ionized and aspartate 52 is ionized.

5. *Selective chemical modification.* Lysozyme remains active when all of its carboxyl groups except those of glutamic 35 and aspartate 52 are esterified. Glutamic 35 and aspartate 52 are unmodified if this reaction is taking place in the presence of substrate. When the substrate is removed, aspartate 52 also becomes esterified (whereas glutamic 35 remains unaltered). The modification of aspartate 52 produces totally inactive enzyme, which supports the proposal that the precise position of the carboxylate group of aspartate 52 is important in stabilizing the carbonium ion intermediate.

6. *Transglycosylation.* Hexa-NAG and di-NAG are formed at a slow rate when tetra-NAG is added to lysozyme (Figure 7-21). This

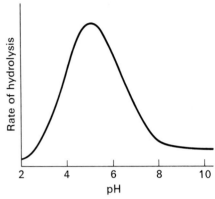

Figure 7-20
The rate of hydrolysis of chitin (poly-NAG) by lysozyme as a function of pH.

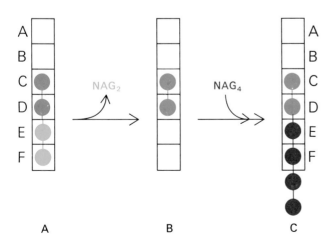

Figure 7-21
The occurrence of a *glycosyl-enzyme intermediate* is supported by the finding that lysozyme catalyzes a slow transglycosylation reaction. NAG_4 (shown in red) adds to the glycosyl-enzyme intermediate (shown in blue in part B) to yield NAG_6.

reaction, called a transglycosylation, supports an important feature of the proposed enzymatic mechanism—namely, the occurrence of a *glycosyl-enzyme intermediate*. In the usual hydrolytic reaction, this glycosyl-enzyme intermediate reacts with OH^-. In transglycosylation, the reaction is with another sugar: ROH. The transglycosylation reaction is specific because the acceptor is bound to sites E and F of the active-site cleft. Moreover, the glycosidic bond formed has the β configuration, the same as the substrate. The proposed geometry of the catalytic intermediate is supported by this finding.

CARBOXYPEPTIDASE A: A ZINC-CONTAINING PROTEOLYTIC ENZYME

Let us now turn to carboxypeptidase A, a digestive enzyme that hydrolyzes the carboxyl-terminal peptide bond in polypeptide chains. Hydrolysis occurs most readily if the carboxyl-terminal residue has an aromatic or a bulky aliphatic side chain (Figure 7-22).

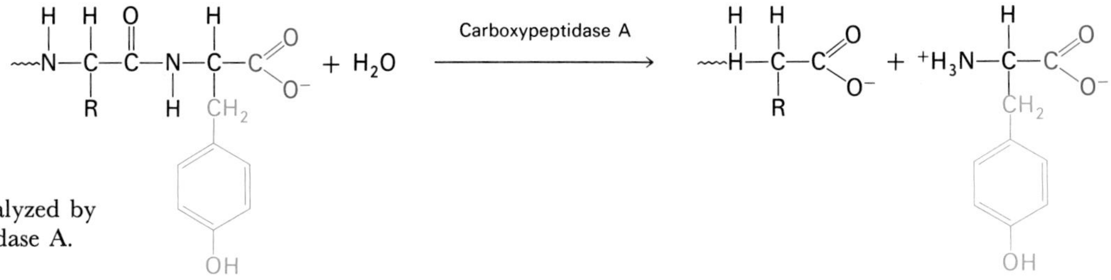

Figure 7-22
Reaction catalyzed by carboxypeptidase A.

The catalytic mechanism of this enzyme is especially interesting because it is very different from that of lysozyme. Two aspects of the catalytic mechanism of carboxypeptidase A, which will be discussed shortly, are particularly noteworthy:

1. *Induced fit.* The binding of substrate is accompanied by quite large alterations in the structure of the enzyme.

2. *Electronic strain.* The enzyme contains a zinc atom and other groups at its active site that induce rearrangements in the electronic distribution of the substrate, thereby rendering it more susceptible to hydrolysis.

The three-dimensional structure of carboxypeptidase A at a resolution of 2 Å was solved by William Lipscomb in 1967 (Figure 7-23). This enzyme is a single polypeptide chain of 307 amino acid residues. Carboxypeptidase A has a compact shape that can be

Figure 7-23
Three-dimensional structure of carboxypeptidase A. Only
the α-carbon atoms and the zinc ion (shaded circle near the
center) are shown. [From W. N. Lipscomb. *Proc. Robert A.
Welch Found. Conf. Chem. Res.* 15(1971):134.]

approximated by an ellipsoid of dimensions 50 × 42 × 38 Å. The
enzyme contains regions of α helix (38%) and of β pleated sheet
(17%). There is a tightly bound zinc ion, which is essential for
enzymatic activity. This zinc ion is located in a groove near the
surface of the molecule, where it is coordinated to a tetrahedral
array of two histidine side chains, a glutamate side chain, and a
water molecule (Figure 7-24). A large pocket near the zinc ion ac-
commodates the side chain of the terminal residue of the peptide
substrate.

BINDING OF SUBSTRATE INDUCES LARGE STRUCTURAL
CHANGES AT THE ACTIVE SITE OF CARBOXYPEPTIDASE A

The mode of binding of substrates to carboxypeptidase A has been
deduced from the structure of a complex of glycyltyrosine and this
enzyme. Glycyltyrosine is a slowly hydrolyzed substrate. Its binding

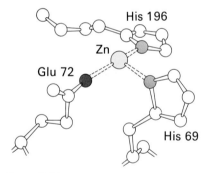

Figure 7-24
A zinc ion is coordinated to two histidine side chains and a glutamate side
chain at the active site of carboxypeptidase A. A water molecule coordinated to the zinc is not shown here.
[After D. M. Blow and T. A. Steitz.
X-ray diffraction studies of enzymes,
Ann. Rev. Biochem. 39:78. Copyright ©
1970 by Annual Reviews, Inc. All
rights reserved.]

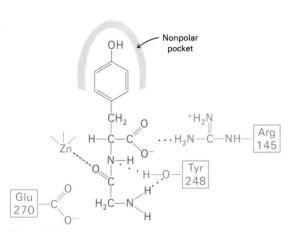

Figure 7-25
Schematic representation of the binding of glycyltyrosine to the active site of carboxypeptidase A. The proposed catalytically active complex is shown here.

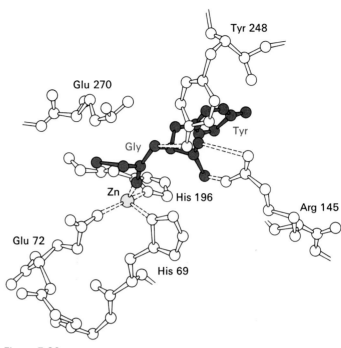

Figure 7-26
Three-dimensional structure of glycyltyrosine at the active site of carboxypeptidase A. Glycyltyrosine, the substrate, is shown in red. [After D. M. Blow and T. A. Steitz. X-ray diffraction studies of enzymes, *Ann. Rev. Biochem.* 39:79. Copyright © 1970 by Annual Reviews, Inc. All rights reserved.]

(Figures 7-25 and 7-26) can be described in terms of five interactions.

1. The negatively charged terminal carboxylate of glycyltyrosine interacts electrostatically with the positively charged side chain of arginine 145.

2. The tyrosine side chain of the substrate binds to a nonpolar pocket in the enzyme.

3. The NH hydrogen of the peptide bond to be cleaved is hydrogen bonded to the OH group of the aromatic side chain of tyrosine 248.

4. The carbonyl oxygen of the peptide bond to be cleaved is coordinated to the zinc ion.

5. The terminal amino group of the substrate is hydrogen bonded through an intervening water molecule to the side chain of glutamate 270. This interaction probably does not occur in productive ES complexes, and in fact it may account for the very slow hydrolysis of glycyltyrosine.

The binding of glycyltyrosine is accompanied by a structural rearrangement of the active site (Figure 7-27). In fact, *the catalytic*

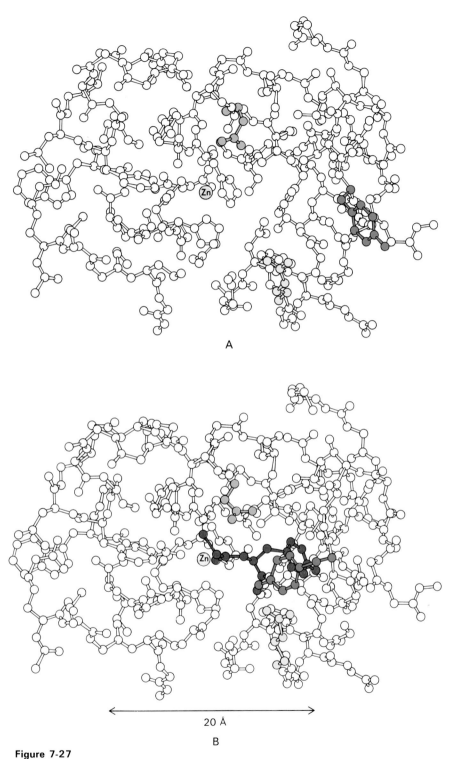

A

B

20 Å

Figure 7-27
The structure of carboxypeptidase A changes upon binding substrate: (A) enzyme alone (Arg 145 is shown in yellow, Glu 270 in green, and Tyr 248 in blue); (B) enzyme-substrate complex (glycyltyrosine, the substrate, is shown in red). Only part of the enzyme is shown here. [After W. N. Lipscomb. *Proc. Robert A. Welch Found. Conf. Chem. Res.* 15(1971):140–141.]

groups of the enzyme are brought into the correct orientation by the binding of substrate, as originally proposed by Koshland in his *induced-fit model* of enzyme action. The guanidinium group of arginine 145 moves 2 Å, as does the carboxylate group of glutamate 270. The binding of the carbonyl group of the substrate to the zinc ion displaces a bound water molecule. At least four other water molecules are displaced from the nonpolar pocket when the tyrosine side chain of the substrate binds them. The largest conformational change is the movement of the phenolic hydroxyl group of tyrosine 248 by 12 Å, a distance equal to about a quarter of the diameter of the protein. This motion is accomplished primarily by a facile rotation about a single carbon–carbon bond. The hydroxyl group of tyrosine 248 moves from the surface of the molecule to the vicinity of the peptide bond of the substrate. An important consequence of this motion is that it closes the active-site cavity and completes its conversion from a water-filled region into a hydrophobic one. These structural changes may be initiated by the binding of arginine 145 to the terminal carboxylate group of the substrate.

ELECTRONIC STRAIN ACCELERATES CATALYSIS BY CARBOXYPEPTIDASE A

Lipscomb has proposed a catalytic mechanism for carboxypeptidase A based on x-ray crystallographic studies. The productive ES complex is postulated to have the structure shown in Figure 7-28. In this proposed mechanism, the hydroxyl group of tyrosine 248 donates a proton to the NH of the peptide bond to be split. The carbonyl carbon atom of this peptide bond is attacked by the carboxylate group of glutamate 270, which acts as a nucleophile. The resulting anhydride of glutamate 270 and the acid component of the substrate is hydrolyzed in a subsequent step.

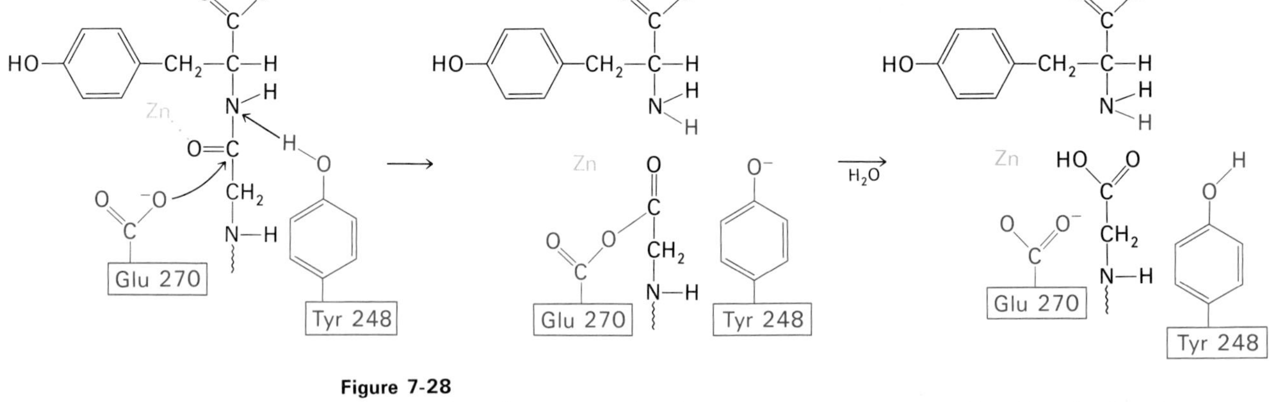

Figure 7-28
A proposed catalytic mechanism for carboxypeptidase A in which Glu 270 directly attacks the carbonyl carbon atom of the susceptible peptide bond and Tyr 248 donates a proton to the NH group of this peptide. The resulting anhydride is then hydrolyzed.

An alternative mechanism that is also compatible with the x-ray data is shown in Figure 7-29. In this scheme, glutamate 270 activates a water molecule. The resulting OH⁻ directly attacks the carbonyl carbon atom of the susceptible peptide bond. At the same time, tyrosine 248 donates a proton to the NH group of the susceptible peptide bond, which results in its hydrolysis. This mechanism differs from the one in Figure 7-28 in that the susceptible peptide bond of the substrate, rather than an anhydride intermediate, is directly hydrolyzed by water. Recent chemical and spectroscopic studies suggest that peptide substrates are hydrolyzed by this direct mechanism, whereas ester substrates form an anhydride with glutamate 270, which is subsequently hydrolyzed.

What is the role of zinc in this catalytic scheme? The carbonyl group of the susceptible peptide bond is pointed toward the zinc ion so that the C=O bond is polarized more than usual, which renders the carbonyl carbon atom more vulnerable to nucleophilic attack. The induction of a dipole is enhanced by the nonpolar environment of the zinc ion, which increases its effective charge. The proximity of the negative charge on glutamate 270 also contributes to the induction of a large dipole in the carbonyl group. Thus, *carboxypeptidase A induces electronic strain in its substrate to accelerate catalysis.*

The need for substrate-induced structural changes in the active site of carboxypeptidase A can now be appreciated. The bound substrate is surrounded on all sides by catalytic groups of the enzyme. This arrangement promotes catalysis for the reasons cited above. *It is evident that a substrate could not enter such an array of catalytic groups (nor could a product leave) unless the enzyme were flexible.* In general, a flexible enzyme has a much larger repertoire of potential conformations that can be exploited for catalysis and selected in the course of evolution than does a rigid enzyme. Furthermore, an induced fit may contribute to the specificity of an enzyme. A substrate for carboxypeptidase A must have a terminal carboxylate group. The enzyme tests for the presence of this group in the following way. If present, the terminal carboxylate forms a salt link with arginine 145, which triggers the movement of tyrosine 248 into a catalytically active position. If there is no terminal carboxylate, this movement fails to occur, and the enzyme is inactive. In other words, *induced fit can serve as a dynamic recognition process.*

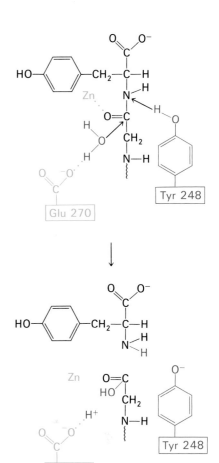

Figure 7-29
An alternative catalytic mechanism for carboxypeptidase A. Tyr 248 has the same role as in the mechanism shown in Figure 7-28. However, Glu 270 activates a water molecule, which attacks the carbonyl carbon atom of the susceptible peptide bond. Hydrolysis is direct; an anhydride is not formed.

SUMMARY

Lysozyme is a small enzyme that cleaves the polysaccharide component of bacterial cell walls. This polysaccharide is an alternating polymer of *N*-acetylglucosamine (NAG) and *N*-acetylmuramic acid (NAM) residues joined by $\beta(1 \rightarrow 4)$ glycosidic linkages. Lysozyme hydrolyzes the glycosidic bond between C-1 of NAM and C-4 of NAG. Oligomers of *N*-acetylglucosamine are also hydrolyzed by lysozyme. Hexa-NAG and higher oligomers are good substrates, whereas tri-NAG and di-NAG are hydrolyzed at very slow rates. Tri-NAG is an effective competitive inhibitor. The three-dimensional structure of lysozyme and of the tri-NAG–lysozyme complex are known at atomic resolution. Tri-NAG occupies half of a cleft that runs across the enzyme, to which it is bound by many hydrogen bonds and van der Waals interactions. The mode of binding of hexa-NAG, a good substrate, was deduced by model building starting with the structure of the tri-NAG complex.

A plausible catalytic mechanism has been proposed for lysozyme. First, the critical groups in catalysis are the un-ionized carboxyl of glutamic 35 and the ionized carboxylate of aspartate 52. Both are about 3 Å from the glycosidic linkage that is hydrolyzed; namely, the one between residues D and E of a hexameric substrate. Second, glutamate 35 donates an H^+ to the bond between C-1 of the D ring and the glycosidic oxygen atom and thereby cleaves this bond. Carbon-1 of the D ring becomes positively charged; this transient species is called a *carbonium* ion. Third, this carbonium ion reacts with OH^- from the solvent, and glutamic 35 becomes protonated again. Lysozyme is ready for another round of catalysis after the products diffuse away. Fourth, catalysis is markedly enhanced by two factors that promote the formation of the carbonium ion intermediate. The electrostatic factor is the proximity of the negatively charged side chain of aspartate 52. The geometrical factor is the distortion of ring D into a half-chair form, which enables the positive charge in the carbonium ion to be shared between C-1 and the ring oxygen atom.

Carboxypeptidase A, a digestive enzyme that hydrolyzes the carboxyl-terminal peptide in polypeptides, exemplifies some different principles of catalysis. The structure of this enzyme and its complex with glycyltyrosine, a substrate analog, are known at atomic resolution. The binding of glycyltyrosine induces large structural changes at the active site, which convert it from a water-filled region into a hydrophobic one. Carboxypeptidase A exemplifies the role of *induced fit* in catalysis. Another noteworthy feature of this enzyme is that it contains a zinc ion at its active site, which is essential for catalysis. The carbonyl atom of the susceptible peptide bond is polarized by the zinc so that it becomes more vulnerable to nucleophilic attack. This is an example of the induction of *electronic strain* in a substrate. In catalysis by carboxypeptidase A, a water molecule activated by glutamate 270 directly attacks the carbonyl group of the susceptible peptide bond. At the same time, tyrosine 248 donates a proton to the NH group of this peptide bond, which results in its hydrolysis.

SELECTED READINGS

WHERE TO START

Phillips, D. C., 1966. The three-dimensional structure of an enzyme molecule. *Sci. Amer.* 215(5):78–90. [A superb article on the three-dimensional structure and catalytic mechanism of lysozyme. Available as *Sci. Amer.* Offprint 1055.]

Lipscomb, W. N., 1971. Structures and mechanisms of enzymes. *Proc. Robert A. Welsh Found. Conf. Chem. Res.* 15:131–156. [A beautifully illustrated and incisive discussion of the structure and mechanism of carboxypeptidase A and several other enzymes.]

STRUCTURE AND ENZYMATIC MECHANISM OF LYSOZYME

Osserman, E. F., Canfield, R. E., and Beychok, S., (eds.), 1974. *Lysozyme.* Academic Press.

Imoto, T., Johnson, L. N., North, A. C. T., Phillips, D. C., and Rupley, J. A., 1972. Vertebrate lysozymes. *In* Boyer, P. D., (ed.), *The Enzymes* (3rd ed.), vol. 7, pp. 666–868. Academic Press.

Dahlquist, F. W., Rand-Meir, T., and Raftery, M. A., 1968. Demonstration of carbonium ion intermediate during lysozyme catalysis. *Proc. Nat. Acad. Sci.* 61:1194–1198.

Rupley, J. A., 1967. The binding and cleavage by lysozyme of *N*-acetylglucosamine oligosaccharides. *Proc. Roy. Soc.* (B)167:416–428. [Presents results on hydrolysis in ^{18}O water, $G°$ for the binding of oligosaccharides, and rates of hydrolysis by lysozyme.]

Ford, L. O., Johnson, L. N., Mackin, P. A., Phillips, D. C., and Tjian, R., 1974. Crystal structure of a lysozyme-tetrasaccharide lactone complex. *J. Mol. Biol.* 88:349–371.

ROLE OF STRAIN IN CATALYSIS

Pauling, L., 1948. Nature of forces between large molecules of biological interest. *Nature* 161:707–709. [Includes a visionary statement of the importance of strain in enzymatic catalysis.]

Wolfenden, R., 1972. Analog approaches to the structure of the transition state in enzyme reactions. *Acc. Chem. Res.* 5:10–18.

Leinhard, G. E., 1973. Enzymatic catalysis and transition-state theory. *Science* 180:149–154.

Secemski, I. I., and Lienhard, G. E., 1971. The role of strain in catalysis by lysozyme. *J. Amer. Chem. Soc.* 93:3549–3550.

Schindler, M., and Sharon, N., 1976. A transition state analog of lysozyme catalysis prepared from the bacterial cell wall tetrasaccharide. *J. Biol. Chem.* 251:4330–4335.

Warshel, A., and Levitt, M., 1976. Theoretical studies of enzymatic reactions: dielectric, electrostatic, and steric stabilization of the carbonium ion in the reaction of lysozyme. *J. Mol. Biol.* 103:227–249. [These calculations suggest that electrostatic factors are more important than steric factors in promoting the formation of the carbonium ion in lysozyme.]

STRUCTURE AND ENZYMATIC MECHANISM OF CARBOXYPEPTIDASE A

Quiocho, F. A., and Lipscomb, W. N., 1971. Carboxypeptidase A: a protein and an enzyme. *Advan. Protein Chem.* 25:1–78. [A review of the crystallographic studies of this enzyme. Includes atomic coordinates.]

Makinen, M. W., Yamamura, K., and Kaiser, E. T., 1976. Mechanism of action of carboxypeptidase in ester hydrolysis. *Proc. Nat. Acad. Sci.* 73:3882–3886. [Presents evidence for a covalent acyl-enzyme intermediate in the hydrolysis of an ester substrate.]

Breslow, R., and Wernick, D. L., 1977. Unified picture of mechanisms of catalysis by carboxypeptidase A. *Proc. Nat. Acad. Sci.* 74:1303–1307. [Presents evidence that hydrolysis of peptide substrates is direct and proposes that esters and peptides are hydrolyzed by different mechanisms.]

PROBLEMS

1. Predict the relative rates of hydrolysis by lysozyme of these oligosaccharides (G stands for an *N*-acetylglucosamine residue, and M for *N*-acetylmuramic acid):
   (a) M-M-M-M-M-M
   (b) G-M-G-M-G-M
   (c) M-G-M-G-M-G

2. Predict on the basis of the data given in Figure 7-18 which of the sugar binding sites A to F on lysozyme will be occupied in the major complex with each of these oligosaccharides (same abbreviations as in Problem 1):
   (a) G-G   (c) G-G-G-G
   (b) G-M

3. Suppose that hexa-NAG is synthesized so that the glycosidic oxygen between its D and E sugar residues is labeled with ^{18}O. Where will this isotope appear in the products formed by hydrolysis with lysozyme?

4. An analog of tetra-NAG containing —H in place of —CH_2OH at C-5 of residue D binds to lysozyme much more strongly than does tetra-NAG. Propose a structural basis for this difference in binding affinity.

5. Compare the coordination of the zinc atom in carboxypeptidase A with that of the iron atom in oxymyoglobin and oxyhemoglobin.
   (a) Which atoms are directly bonded to these metal ions?
   (b) Which side chains contribute these metal-binding groups?
   (c) Which other side chains in proteins are potential metal-binding groups?

6. Experimental evidence for the direct mechanism (Figure 7-29) of peptide hydrolysis by carboxypeptidase A has come from oxygen-18 exchange studies. *N*-Benzoylglycine labeled with ^{18}O in its carboxyl group was incubated with carboxypeptidase A. In the presence of L-phenylalanine, ^{18}O was transferred to H_2O. In contrast, no ^{18}O was transferred when L-β-phenyllactic acid was added to the incubation mixture instead of L-phenylalanine. What is the simplest interpretation of this result?

CHAPTER 8

ZYMOGEN ACTIVATION: DIGESTIVE ENZYMES AND CLOTTING FACTORS

Lysozyme acquires full enzymatic activity as it spontaneously folds into its characteristic three-dimensional form. In contrast, many other proteins are synthesized as inactive precursors that are subsequently activated by cleavage of one or a few specific peptide bonds. If the active protein is an enzyme, the inactive precursor is called a *zymogen* (or a *proenzyme*).

Activation of proteins by specific proteolysis recurs frequently in biological systems. Several examples follow.

1. The *digestive enzymes* that hydrolyze proteins are synthesized as zymogens in the stomach and pancreas (Table 8-1).

Table 8-1
Gastric and pancreatic zymogens

Site of synthesis	Zymogen	Active enzyme
Stomach	Pepsinogen	Pepsin
Pancreas	Chymotrypsinogen	Chymotrypsin
Pancreas	Trypsinogen	Trypsin
Pancreas	Procarboxypeptidase	Carboxypeptidase
Pancreas	Proelastase	Elastase

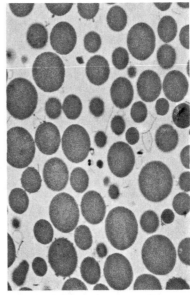

Figure 8-1
Electron micrograph of zymogen granules in an acinar cell of the pancreas. [Courtesy of Dr. George Palade.]

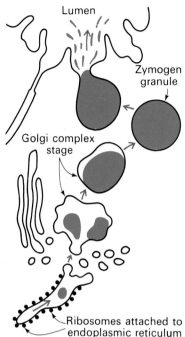

Figure 8-2
Diagrammatic representation of the secretion of zymogens by an acinar cell of the pancreas. [After a drawing kindly provided by Dr. George Palade.]

2. *Blood clotting* is mediated by a cascade of proteolytic activations that assures a rapid and amplified response to trauma.

3. Some protein hormones are synthesized as inactive precursors. For example, *insulin* is derived from *proinsulin* by proteolytic removal of a peptide.

4. The fibrous protein *collagen,* which is present in skin and bone, is derived from *procollagen,* a soluble precursor.

CHYMOTRYPSINOGEN IS ACTIVATED BY SPECIFIC CLEAVAGE OF A SINGLE PEPTIDE BOND

Chymotrypsin is a digestive enzyme that hydrolyzes proteins in the small intestine. Its inactive precursor, *chymotrypsinogen,* is synthesized in the pancreas, as are several other zymogens and digestive enzymes. Indeed, the pancreas is one of the most active organs in synthesizing proteins. The enzymes and zymogens are synthesized in the acinar cells of the pancreas (Figure 8-2). The proteins travel from the endoplasmic reticulum to the Golgi apparatus, where they are covered with a membrane made up of lipid and protein. These *zymogen granules* appear in electron micrographs as very dense bodies because they have a high concentration of protein (see Figure 8-1). The zymogen granules accumulate at the apex of the acinar cell and are secreted into a duct leading into the duodenum when stimulated by a hormonal or nerve impulse signal.

Chymotrypsinogen is a single polypeptide chain consisting of 245 amino acid residues. It is cross-linked by five disulfide bonds. Chymotrypsinogen is virtually devoid of enzymatic activity. It is converted into a fully active enzyme when the peptide bond joining arginine 15 and isoleucine 16 is cleaved by trypsin (Figure 8-3). The

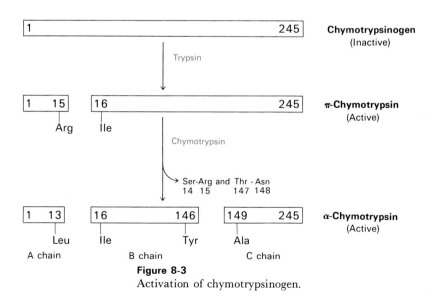

Figure 8-3
Activation of chymotrypsinogen.

resulting active enzyme, called π-chymotrypsin, then acts on other π-chymotrypsin molecules. Two peptides are removed to yield α-chymotrypsin, the stable form of the enzyme. The additional cleavages made in the conversion of π- into α-chymotrypsin are superfluous, because π-chymotrypsin is already fully active. The striking feature of this activation process is that *cleavage of a single specific peptide bond transforms the protein from a catalytically inactive form into one that is fully active.*

THREE-DIMENSIONAL STRUCTURE OF CHYMOTRYPSIN

An understanding of this remarkable activation process depends on a detailed knowledge of the structure and catalytic mechanism of chymotrypsin. Fortunately, much is known about this enzyme from chemical and x-ray crystallographic studies. In fact, chymotrypsin is one of the most thoroughly studied of all enzymes, and so it is rewarding to look at it in some detail.

Chymotrypsin consists of three polypeptide chains connected by two interchain disulfide bonds (Figure 8-4). The mass of the en-

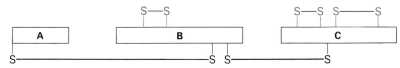

Figure 8-4
α-Chymotrypsin contains two interchain disulfide bonds and three intrachain disulfide bonds.

zyme is about 25 kdal. The three-dimensional structure of the enzyme at 2-Å resolution (Figure 8-5) is known from the x-ray crystallographic studies of David Blow and his colleagues. The molecule is

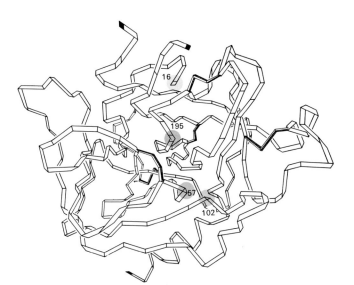

Figure 8-5
Three-dimensional structure of α-chymotrypsin. Only the α-carbon atoms are shown. Catalytically important residues are marked in color. [From D. M. Blow. In *The Enzymes*, P. D. Boyer, ed., 3rd ed., vol. 3 (Academic Press, 1971), p. 194.]

a compact ellipsoid of dimensions $51 \times 40 \times 40$ Å. All charged groups are on the surface of the molecule except for three that play a critical role in catalysis. The folding of the molecule is complex. Chymotrypsin contains very little α helix, in contrast with myoglobin and hemoglobin. The chains tend to be fully extended and often run parallel to each other, separated by about 5 Å. There is extensive hydrogen bonding between the peptide groups of adjacent strands. Parts of the molecule have a secondary structure that resembles an antiparallel pleated sheet, as was also found in lysozyme.

CHYMOTRYPSIN IS SPECIFIC FOR AROMATIC AND BULKY NONPOLAR SIDE CHAINS

The biological role of chymotrypsin is to catalyze the hydrolysis of proteins in the small intestine (Figure 8-6). The equilibrium of this

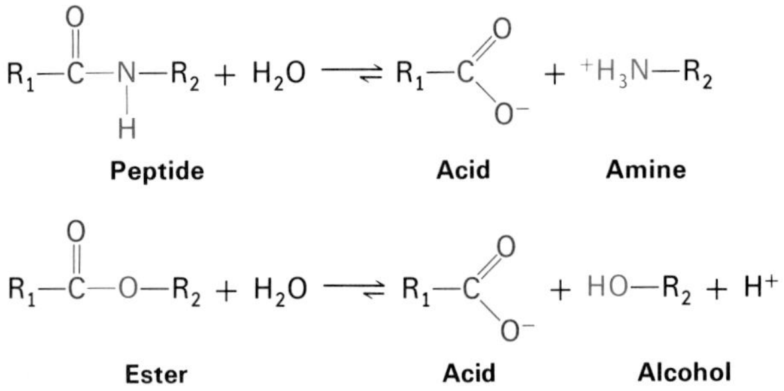

Peptide   Acid   Amine

Ester   Acid   Alcohol

Figure 8-6
Chymotrypsin catalyzes the hydrolysis of peptide and ester bonds.

reaction overwhelmingly favors hydrolysis ($>99\%$). Chymotrypsin does not cleave all peptide bonds at a significant rate. Rather, it is selective for peptide bonds on the *carboxyl side* of the *aromatic side chains* tyrosine, tryptophan, and phenylalanine and of large *hydrophobic residues such as methionine* (Figure 8-7).

Chymotrypsin also hydrolyzes *ester bonds*. Although this reaction is not important physiologically, it is of interest because of its close relationship to peptide-bond hydrolysis (Figure 8-6). Indeed, much of our knowledge of the catalytic mechanism of chymotrypsin comes from studies of the hydrolysis of simple esters.

PART OF THE SUBSTRATE IS COVALENTLY BOUND TO CHYMOTRYPSIN DURING CATALYSIS

Chymotrypsin catalyzes the hydrolysis of peptide or ester bonds in two distinct stages. This was first revealed by studies of the kinetics of hydrolysis of *p*-nitrophenyl acetate. The liberation of *p*-nitrophenol, one of the products, is clearly biphasic when large amounts of

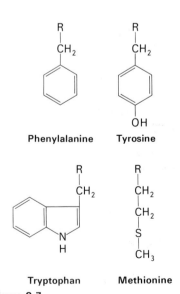

Phenylalanine   Tyrosine

Tryptophan   Methionine

Figure 8-7
Chymotrypsin preferentially hydrolyzes peptide bonds on the carboxyl side of aromatic and bulky nonpolar side chains.

enzyme are used (Figure 8-8). There is an initial *rapid burst* of *p*-nitrophenol product, followed by its formation at a much *slower steady-state rate.*

The first step is the combination of *p*-nitrophenyl acetate with chymotrypsin to form an enzyme-substrate (ES) complex (Figure 8-9). The ester bond of this substrate is cleaved. One of the prod-

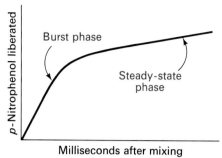

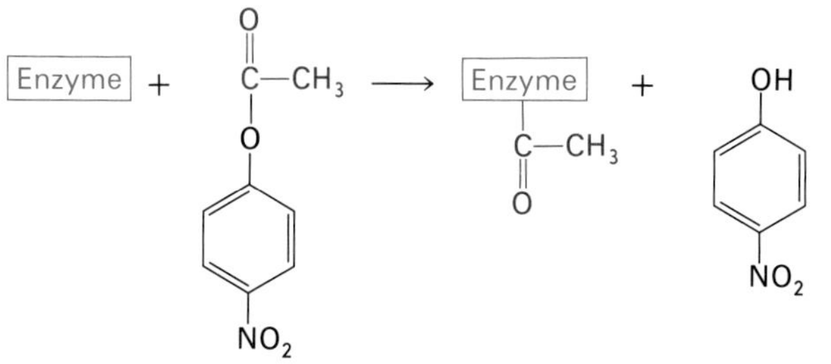

Figure 8-8
Two phases in the formation of *p*-nitrophenol are evident following the mixing of chymotrypsin and *p*-nitrophenyl acetate.

Figure 8-9
Acylation: formation of the acetyl-enzyme intermediate.

p-Nitrophenyl acetate   Acetyl-enzyme intermediate   p-Nitrophenol

ucts, *p*-nitrophenol, is then released from the enzyme, whereas the acetyl group of the substrate becomes covalently attached to the enzyme. Water then attacks the acetyl-enzyme complex to yield acetate ion and regenerate the enzyme (Figure 8-10). The initial

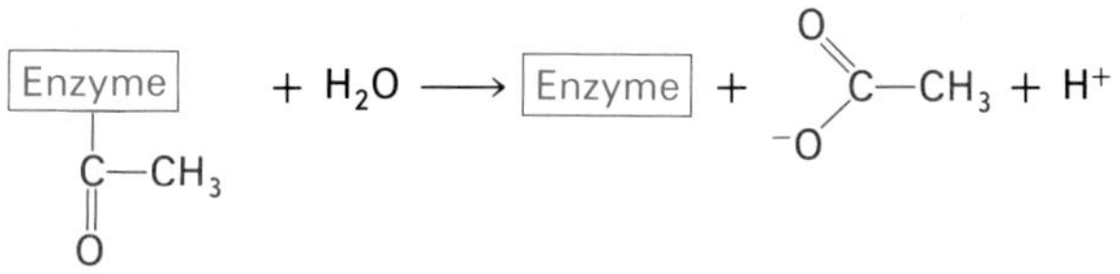

Figure 8-10
Deacylation: hydrolysis of the acetyl-enzyme intermediate.

Acetyl-enzyme intermediate   Acetate

rapid burst of *p*-nitrophenol production corresponds to the formation of the acetyl-enzyme complex. This step is called *acylation.* The slower steady-state production of *p*-nitrophenol corresponds to the hydrolysis of the acetyl-enzyme complex to regenerate the free enzyme. This second step, called *deacylation,* is rate-limiting in the hydrolysis of esters by chymotrypsin. In fact, the acetyl-enzyme complex is sufficiently stable to be isolated under appropriate conditions. The catalytic mechanism of chymotrypsin can thus be represented by the following scheme, in which P_1 is the amine (or alcohol) component of the substrate, $E—P_2$ is the covalent intermediate, and P_2 is the acid component of the substrate.

A distinctive feature of this mechanism is the occurrence of a covalent intermediate. In the reaction discussed above, an acetyl group is covalently bonded to the enzyme. In general, the group attached to chymotrypsin at the $E—P_2$ stage is an acyl group. Thus, $E—P_2$ is an *acyl-enzyme intermediate.*

$$E + S \rightleftharpoons ES \xrightarrow{\quad} E—P_2 \xrightarrow{\quad} E$$
$$\qquad\qquad\;\; P_1 \qquad\quad\; P_2$$

THE ACYL GROUP IS ATTACHED TO AN UNUSUALLY REACTIVE SERINE RESIDUE ON THE ENZYME

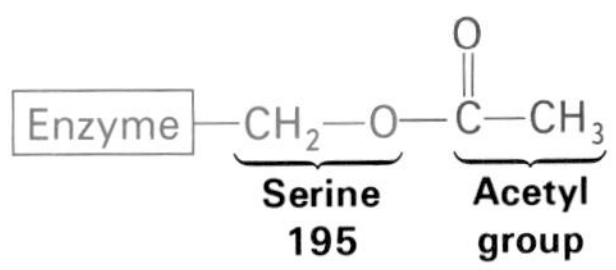

The site of attachment of the acyl group was identified following the isolation of E—P$_2$, which is quite stable at pH 3. The acyl group is linked to the oxygen atom of a specific serine residue, namely serine 195. This serine residue is unusually reactive. It can be specifically labeled with *organic fluorophosphates,* such as diisopropylphosphofluoridate (DIPF). DIPF reacts only with serine 195 to form an inactive *diisopropylphosphoryl-enzyme complex,* which is indefinitely stable (Figure 8-11). The remarkable reactivity of serine 195 is highlighted by the fact that the other 27 serine residues in chymotrypsin are untouched by DIPF.

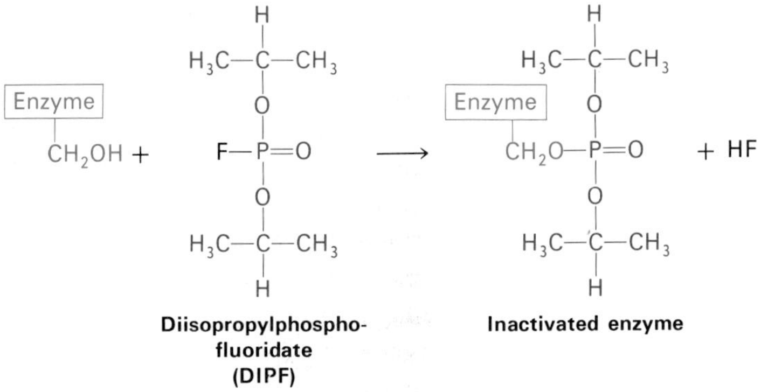

Figure 8-11
Diisopropylphosphofluoridate (DIPF) inactivates chymotrypsin by forming a diisopropylphosphoryl derivative of serine 195.

Chymotrypsin is not the only enzyme to be inactivated by DIPF. Numerous other proteolytic enzymes, such as trypsin, elastase, thrombin, and subtilisin, react specifically with DIPF and are thereby inactivated. The reaction takes place at a unique serine residue, as in chymotrypsin. Hence, these enzymes are called the *serine proteases.* DIPF also reacts with the serine residue of *acetylcholinesterase,* an enzyme crucial for the transmission of nerve impulses at certain synapses. In fact, as mentioned in Chapter 6, the inactivation of acetylcholinesterase by DIPF is the basis for its use in *insecticides* and *nerve gases.*

DEMONSTRATION OF THE CATALYTIC ROLE OF HISTIDINE 57 BY AFFINITY LABELING

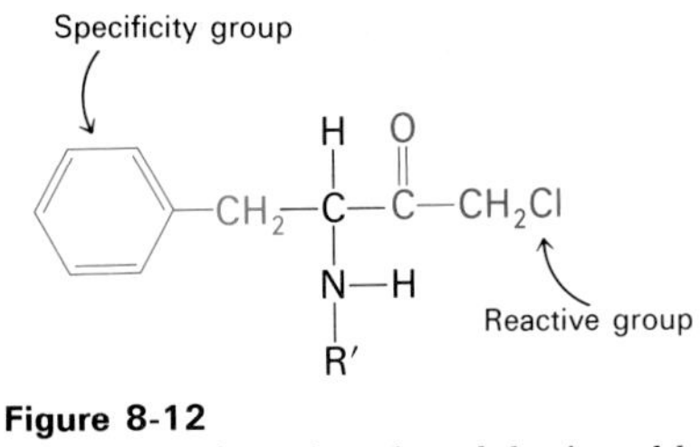

The importance of a second residue in catalysis was shown by *affinity-labeling* studies. The strategy was to react chymotrypsin with a molecule that (1) specifically binds to the active site because it resembles a substrate and then (2) forms a stable covalent bond with a group on the enzyme that is in close proximity. These criteria are met by tosyl-L-phenylalanine chloromethyl ketone (TPCK), whose structure is shown in Figure 8-12. The phenylalanine side

Figure 8-12
Structure of tosyl-L-phenylalanine chloromethyl ketone (TPCK), an affinity-labeling reagent for chymotrypsin (R′ represents a tosyl group).

chain of TPCK enables it to bind specifically to chymotrypsin. The reactive group in TPCK is the chloromethyl ketone function. TPCK attacks chymotrypsin only at histidine 57, which is alkylated at one of its ring nitrogens (Figure 8-13). The TPCK derivative of chymotrypsin is enzymatically inactive. Three lines of evidence indicated that histidine 57 is part of the active site. First, the affinity-labeling reaction was highly stereospecific; the D-isomer of TPCK was totally ineffective. Second, the reaction was inhibited when a competitive inhibitor of chymotrypsin, β-phenylpropionate, was present. Third, the rate of inactivation by TPCK varied with pH in nearly the same way as did the rate of catalysis.

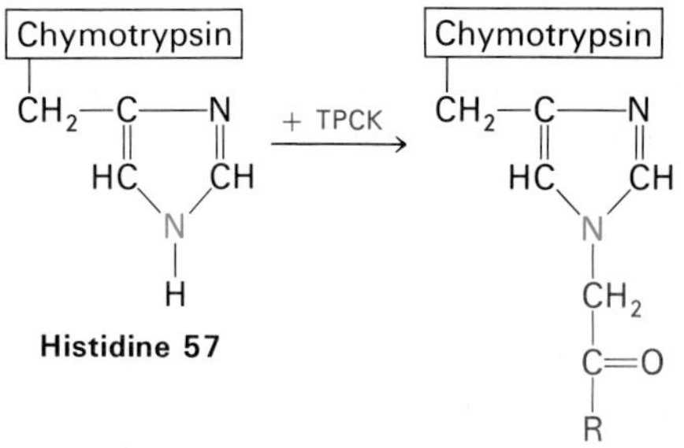

Figure 8-13
Alkylation of histidine 57 in chymotrypsin by TPCK.

A CHARGE RELAY NETWORK SERVES AS A PROTON SHUTTLE DURING CATALYSIS

The catalytic activity of chymotrypsin depends on the unusual reactivity of serine 195. A —CH$_2$OH group is usually quite unreactive under physiological conditions. What makes it so reactive in the active site of chymotrypsin? A plausible explanation has emerged from x-ray studies of the three-dimensional structure of the enzyme. As could be expected from the affinity-labeling work, histidine 57 is adjacent to serine 195. The carboxyl side chain of aspartate 102 is also nearby (Figure 8-14).

In fact, these three residues interact to enhance the catalytic power of chymotrypsin. Aspartate 102 is hydrogen bonded to histidine 57, which is in turn hydrogen bonded to serine 195. These three residues form a *charge relay network*. As described on page 165, this network plays a critical role in catalysis by transiently binding a proton (Figure 8-15). The buried carboxylate of aspartate 102

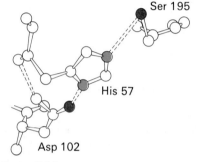

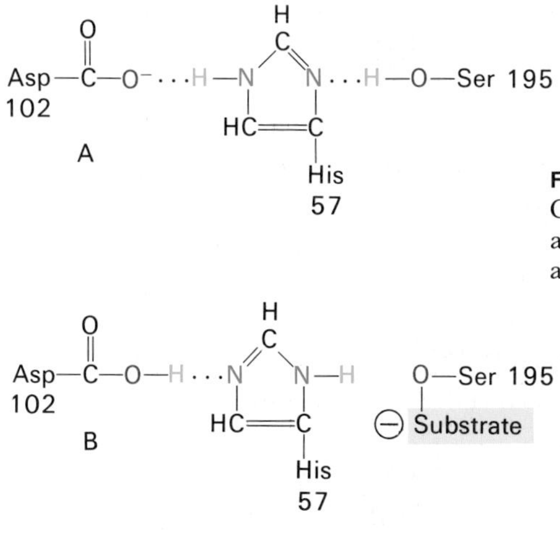

Figure 8-14
Conformation of the charge relay system in chymotrypsin. [After D. M. Blow and T. A. Steitz. X-ray diffraction studies of enzymes, *Ann. Rev. Biochem.* 39(1970):86. Copyright © 1970 by Annual Reviews Inc. All rights reserved.]

Figure 8-15
Charge relay network in chymotrypsin: (A) enzyme alone; (B) on addition of a substrate, aspartate 102 and histidine 57 transiently bind a proton.

polarizes the imidazole group of histidine 57, which enhances its capacity to act as a *proton shuttle*. Aspartate 102 and histidine 57 are poised to accept the proton from the serine 195 hydroxyl group during the nucleophilic attack by this oxygen on the substrate.

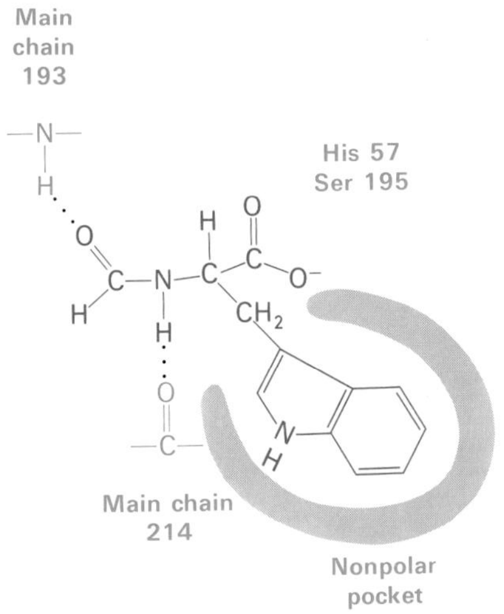

Figure 8-16
Schematic representation of the binding of formyl-L-tryptophan, a substrate analog, to chymotrypsin.

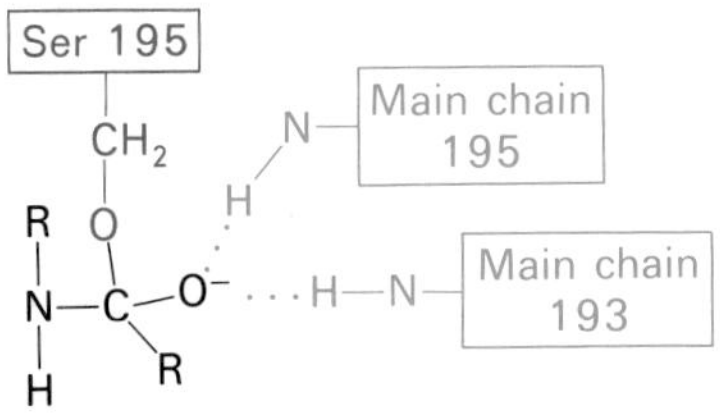

Figure 8-17
The tetrahedral transition-state intermediate in the acylation and deacylation reactions of chymotrypsin. The hydrogen bonds formed by two NH groups from the main chain of the enzyme are critical in stabilizing this intermediate. This site is called the *oxyanion hole.*

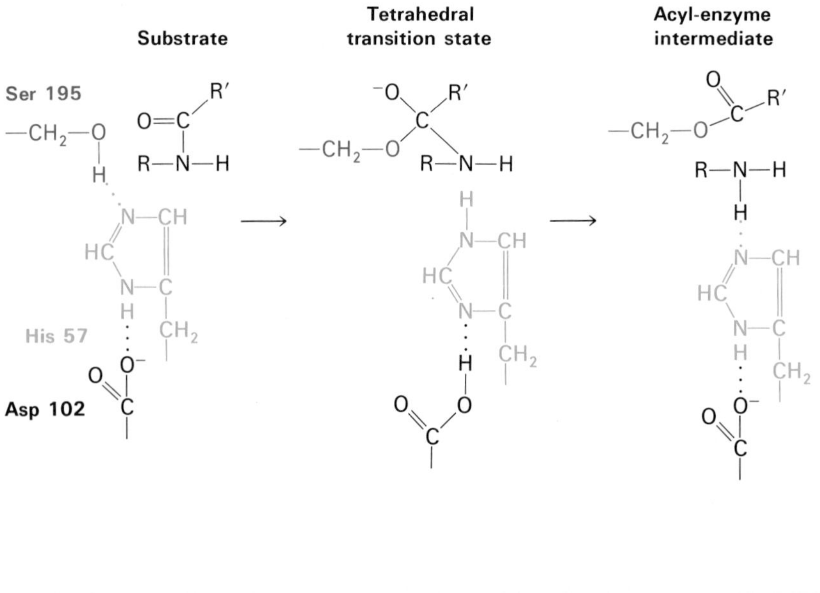

Figure 8-18
First stage in the hydrolysis of a peptide by chymotrypsin: *acylation.* A tetrahedral transition state is formed. The amine component then rapidly diffuses away, which leaves an acyl-enzyme intermediate.

CHYMOTRYPSIN CONTAINS A DEEP POCKET FOR THE BINDING OF AN AROMATIC SIDE CHAIN

Crystallographic studies of complexes of chymotrypsin with substrate analogs have shown the location of the specificity site and the likely orientation of the susceptible peptide bond of a good substrate. Formyl-L-tryptophan binds to chymotrypsin with its indole side chain fitted neatly into a pocket near serine 195 (Figure 8-16). This deep cleft accounts for the specificity of chymotrypsin for aromatic and other bulky hydrophobic side chains. Crystallographic analyses of complexes of chymotrypsin and polypeptide substrate analogs show extensive hydrogen bonding between the main chain of the substrate and that of the enzyme. The pattern of hydrogen bonds is like that of an antiparallel β pleated sheet.

A TRANSIENT TETRAHEDRAL INTERMEDIATE IS FORMED DURING CATALYSIS

A plausible catalytic mechanism for chymotrypsin has been deduced from extensive x-ray crystallographic and chemical data. In this mechanism, *histidine 57 and serine 195 participate directly in the cleavage of the susceptible peptide bond of the substrate.* Peptide-bond hydrolysis starts with an attack by the oxygen atom of the hydroxyl group of serine 195 on the carbonyl carbon atom of the susceptible peptide bond. The carbon–oxygen bond of this carbonyl group becomes a single bond and the oxygen atom acquires a net negative charge. The four atoms bonded to the carbonyl carbon are arranged as in a tetrahedron. The formation of this *transient tetrahedral intermediate* from a planar amide group is made possible by hydrogen bonds between the negatively charged carboxyl oxygen atom (called an *oxyanion*) and two main-chain NH groups (Figure 8-17). The other essential event in the formation of this tetrahedral intermediate is the transfer of a proton from serine 195 to histidine 57 (Figure 8-18). This proton transfer is markedly facilitated by the presence of the charge relay network. Aspartate 102 precisely orients the imidazole ring of histidine 57 and partially neutralizes the charge on this ring that develops during the transition state. The proton stored by the histidine-aspartate couple of this network is then donated to the nitrogen atom of the susceptible peptide bond. As a result, the susceptible peptide bond is cleaved. At this stage, the amine component is hydrogen bonded to histidine 57, whereas the acid component of the substrate is esterified to serine 195. The *acylation stage* of the hydrolytic reaction is now completed.

The next stage is *deacylation* (Figure 8-19). The amine component

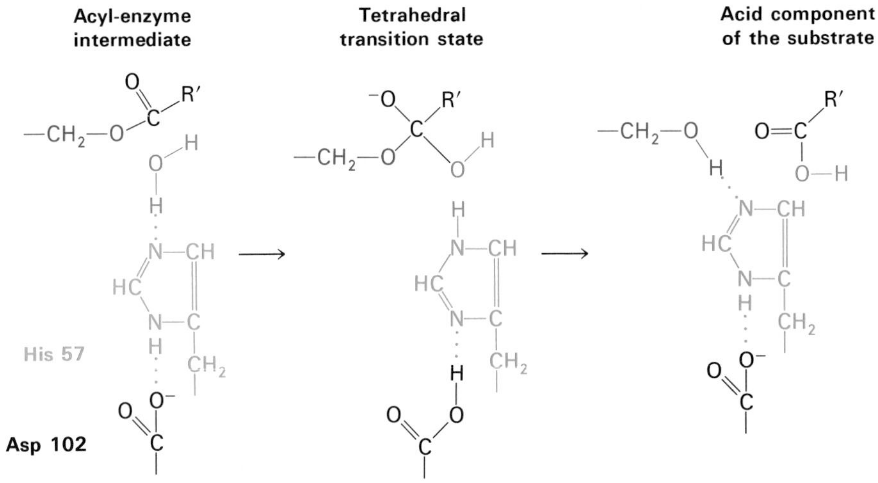

Figure 8-19
Second stage in the hydrolysis of a peptide of chymotrypsin: *deacylation.* The acyl-enzyme intermediate is hydrolyzed by water. Note that deacylation is essentially the reverse of acylation, with water replacing the amine component of the substrate.

of the substrate diffuses away, and a water molecule takes its place at the active site. In essence, *deacylation is the reverse of acylation, with H_2O substituting for the amine component.* First, the charge relay network draws a proton away from water. The resulting OH^- ion simultaneously attacks the carbonyl carbon atom of the acyl group that is attached to serine 195. As in acylation, a transient tetrahedral intermediate is formed. Histidine 57 then donates a proton to the oxygen atom of serine 195, which results in the release of the acid component of the substrate. This acid component diffuses away and the enzyme is ready for another round of catalysis.

MECHANISM OF ZYMOGEN ACTIVATION

We now return to the question of how cleavage of a single peptide bond in chymotrypsinogen converts it into an active enzyme. The three-dimensional structure of chymotrypsinogen has been elucidated by Joseph Kraut and some of the conformational changes in activation have been identified:

1. Hydrolysis of the peptide bond between arginine 15 and isoleucine 16 creates new carboxyl- and amino-terminal groups.

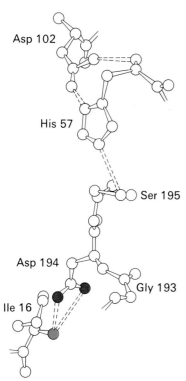

Figure 8-20
Environment of aspartate 194 and iso-leucine 16 in chymotrypsin. The electrostatic interaction between the carboxylate of Asp 194 (red) and the α-NH₂ group of Ile 16 (blue) is essential for the activity of chymotrypsin. These groups are adjacent to the charge relay network. [After D. M. Blow and T. A. Steitz. X-ray diffraction studies of enzymes, *Ann. Rev. Biochem.* 39(1970):86. Copyright © 1970 by Annual Reviews Inc. All rights reserved.]

2. The newly formed *amino-terminal group of isoleucine 16 turns inward and interacts with aspartate 194* in the interior of the chymotrypsin molecule (Figure 8-20). Protonation of this amino group stabilizes the active form of chymotrypsin, as shown by the dependence of enzyme activity on pH.

3. This electrostatic interaction between a positively charged amino group and a negatively charged carboxylate ion in a nonpolar region triggers a number of conformational changes. Methionine 192 moves from a deeply buried position in the zymogen to the surface of the molecule in the enzyme, and residues 187 and 193 become more extended. These changes result in the formation of the *substrate specificity site* for aromatic and bulky nonpolar groups. One side of this site is made up of residues 189 through 192. *This cavity for part of the substrate is not fully formed in the zymogen.*

4. The tetrahedral transition-state in catalysis by chymotrypsin is stabilized by hydrogen bonds between the negatively charged carboxyl oxygen atom and two main-chain NH groups (see Figure 8-17). One of these NH groups is not appropriately located in chymotrypsinogen, and so *the oxyanion hole is incomplete in the zymogen.*

5. The conformational changes elsewhere in the molecule are very small. Thus, *the switching on of enzymatic activity in a protein can be accomplished by discrete, highly localized conformational changes that are triggered by the hydrolysis of a single peptide bond.*

TRYPSIN AND ELASTASE: VARIATIONS ON A THEME

Trypsin and elastase are like chymotrypsin in a number of respects:

1. They are secreted by the pancreas as zymogens and are activated by cleavage of a single peptide bond. The new amino terminus turns inward and interacts electrostatically with the carboxylate of aspartate 194.

2. About 40% of the amino acid sequences of these three enzymes are identical. The degree of identity is even higher for amino acid residues located in the interior of these enzymes.

3. They are inhibited by fluorophosphates such as DIPF. Trypsin and elastase also have an active-site serine residue. In fact, the amino acid sequence around this serine is the same in all three enzymes: Gly-Asp-Ser-Gly-Gly-Pro.

4. X-ray studies have shown that the tertiary structures of these three enzymes are very similar (Figure 8-21). Elastase and trypsin have charge relay networks just like the one in chymotrypsin. Each also contains an oxyanion hole.

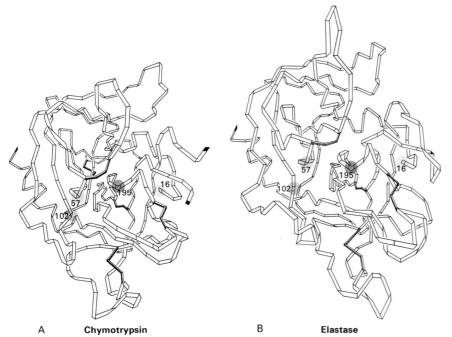

A   **Chymotrypsin**   **B**   **Elastase**

Figure 8-21
Comparison of the conformation of the main chains of (A) chymotrypsin and
(B) elastase. The locations of the charge relay network (residues 102, 57, and
195) and of the α-amino group of residue 16 are shown in color to emphasize
the similarity of these enzymes. [After B. S. Hartley and D. M. Shotton. In
The Enzymes, P. D. Boyer, ed., 3rd ed., vol. 3 (Academic Press, 1971), p. 362.]

5. The catalytic mechanisms of these three enzymes are nearly
identical. They are highly effective catalysts because they bind the
transition state of the reaction. The oxyanion hole and the charge relay network are the essential structural elements of each enzyme in
promoting the formation of the transient tetrahedral intermediate.

Although similar in structure and mechanism, these enzymes
differ strikingly in specificity. Chymotrypsin requires an aromatic
or bulky nonpolar side chain. Trypsin requires a lysine or arginine
residue. Elastase cannot cleave either of these kinds of substrates.
The specificity of elastase is directed toward the smaller, uncharged
side chains. X-ray studies have shown that these different substrate
specificities arise from quite small structural changes in the binding
site (Figure 8-22). In chymotrypsin, a nonpolar pocket serves as a
niche for the aromatic or bulky nonpolar side chain. In trypsin, one
residue in this pocket is changed relative to chymotrypsin: a serine
is replaced by an aspartate. This aspartate in the nonpolar pocket of
trypsin can form a strong electrostatic bond with a positively
charged lysine or arginine side chain of a substrate. In elastase, the
pocket no longer exists because the two glycine residues lining it in
chymotrypsin are replaced by the much bulkier valine and threonine.

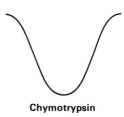

Chymotrypsin

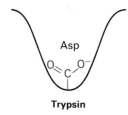

Trypsin

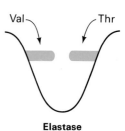

Elastase

Figure 8-22
A highly simplified representation of
part of the substrate-binding sites in
chymotrypsin, trypsin, and elastase.

The structural changes accompanying the activation of trypsinogen are somewhat different from those associated with the activation of chymotrypsinogen. The independent x-ray analyses of Robert Huber and Robert Stroud have shown that the conformation of four stretches of polypeptide, comprising about 15% of the molecule, changes markedly on activation. *These regions, called the activation domain, are very floppy in the zymogen, whereas they have a well-defined conformation in trypsin.* Furthermore, the oxyanion hole in trypsinogen is too far from histidine 57 to promote the formation of the tetrahedral transition state.

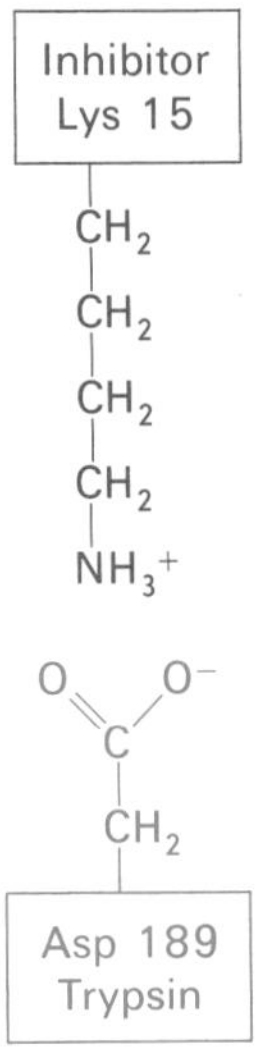

PANCREATIC TRYPSIN INHIBITOR BINDS VERY TIGHTLY TO THE ACTIVE SITE OF TRYPSIN

The activity of the pancreatic proteases is controlled in two distinct ways. The conversion of a zymogen into a protease by cleavage of a single peptide bond is a precise means of switching on enzymatic activity. However, this activation step is irreversible and so a different mechanism is needed to stop proteolysis. This is accomplished by specific protease inhibitors. For example, *pancreatic trypsin inhibitor,* a 6-kdal protein, inhibits trypsin by binding very tightly to its active site (Figure 8-23). The dissociation constant of this complex is 10^{-13} M, which corresponds to a standard free-energy of binding of about -18 kcal/mol. A remarkable feature of this interaction is that the complex is not dissociated by treatment with 8 M urea or 6 M guanidine hydrochloride. These denaturing agents almost always dissociate protein oligomers into their constituent subunits. The reason for the exceptional stability of the complex is that pancreatic trypsin inhibitor is a very effective substrate analog. X-ray analyses show that the side chain of lysine 15 of this inhibitor binds to the aspartate side chain in the specificity pocket of the enzyme. In addition, there are many hydrogen bonds between the main chain of trypsin and that of its inhibitor. The array of hydrogen bonds is like that made by a true substrate. Most important, the carbonyl group of lysine 15 and the surrounding atoms of the inhibitor fit snugly in the oxyanion hole of the enzyme. However, the complex is frozen at this intermediate stage of catalysis, probably because histidine 57 of trypsin cannot move to donate a proton to the amide nitrogen of the inhibitor. The peptide bond between lysine 15 and alanine 16 in pancreatic trypsin inhibitor is cleaved but at a very slow rate. The half life of this trypsin-inhibitor complex is several months. *The inhibitor has very high affinity for trypsin because its structure is almost perfectly complementary to that of the active site of trypsin.* The restricted conformational flexibility in the binding site of the inhibitor contributes to the block in catalysis and accounts for the very slow dissociation of the complex.

Figure 8-23
A prominent feature of the interaction of pancreatic trypsin inhibitor with trypsin is an electrostatic bond between lysine 15 of the inhibitor and aspartate 189 of the enzyme. The —NH$_3$$^+$ group of lysine 15 is also hydrogen bonded to several oxygen atoms in the specificity pocket of trypsin.

DIVERGENT AND CONVERGENT EVOLUTION
OF THE SERINE PROTEASES

Chymotrypsin, elastase, and trypsin have similar overall amino acid sequences because they evolved from a common ancestor. The gene for this common ancestral enzyme probably duplicated several times. Mutations in these genes eventually gave rise to the present-day proteins. The acquisition of different specificities by mutations of genes descended from a common ancestor is called *divergent evolution.*

A different kind of evolutionary process is revealed by comparing chymotrypsin with subtilisin, a bacterial enzyme. Subtilisin is also a serine protease. However, the amino acid sequences of chymotrypsin and subtilisin are very different, indicating that they arose independently in evolution. For example, chymotrypsin contains five disulfide bonds, whereas subtilisin has none. The sequence around the active-site serine is

    -Gly-Thr-Ser*-Met-Ala-Ser   (in subtilisin)
    -Gly-Asp-Ser*-Gly-Gly-Pro   (in chymotrypsin)

The three-dimensional structures of these enzymes are entirely different. It was therefore surprising to find that *subtilisin has a charge relay network* that resembles the one in chymotrypsin in consisting of an aspartic, a histidine, and a serine residue. Furthermore, subtilisin contains an oxyanion hole, which stabilizes the tetrahedral transition state. The similarity of the charge relay network, oxyanion hole, and catalytic mechanism in subtilisin and chymotrypsin is a striking example of *independent convergent evolution* at the level of an enzyme molecule. Why have the vertebrate pancreatic serine proteases and bacterial subtilisin attained the same solution by independent, parallel evolutionary processes? As described on page 170, there appear to be only a few ways of efficiently cleaving peptide bonds. Studies of the several hundred known proteolytic enzymes indicate that there may be only four major types of catalytic mechanisms and that the serine proteases are especially prevalent.

ACTIVATIONS OF THE PANCREATIC ZYMOGENS
ARE COORDINATED

The digestion of proteins in the duodenum requires the concurrent action of several proteolytic enzymes, because each is specific for a limited number of side chains. Thus, the zymogens must be switched on at the same time. Coordinated control is achieved by the action of *trypsin as the common activator of all the pancreatic zymo-*

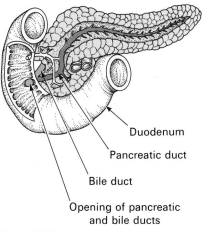

Val—(Asp)₄—Lys—Ile—Val∼∼
Trypsinogen

Enteropeptidase

Val—(Asp)₄—Lys   Ile—Val∼∼
Trypsin

Duodenum

Pancreatic duct

Bile duct

Opening of pancreatic
and bile ducts

Figure 8-24
Diagram of the pancreas and its relationship to the small intestine.

gens—trypsinogen, chymotrypsinogen, proelastase, and procarboxypeptidase. How is enough trypsin produced to initiate these activation processes? The cells that line the duodenum produce an enzyme, *enteropeptidase*, that hydrolyzes a unique lysine–isoleucine peptide bond in trypsinogen as it enters the duodenum. The small amount of trypsin produced in this way activates trypsinogen and the other zymogens. Thus, *the formation of trypsin by enteropeptidase is the master activation step.*

PREMATURE ACTIVATION OF ZYMOGENS CAN BE LETHAL, AS IN PANCREATITIS

The pancreas is protected in several ways from the proteolytic action of the enzymes it synthesizes. All of the pancreatic enzymes that hydrolyze proteins are synthesized as inactive zymogens. Furthermore, they are packaged in granules surrounded by membranes made of protein and lipid. As a further safeguard, the pancreatic secretion contains an inhibitor (p. 168) that renders inactive any small amounts of trypsin that may be present in the pancreas.

The pancreas also secretes lipases. One of them, phospholipase A_2, is synthesized as prophospholipase A_2, which is activated by trypsin. The other fat-splitting enzymes become active in the presence of bile salts, which they encounter in the duodenum. These steroids solubilize lipids, which makes them susceptible to enzymatic cleavage by lipases. Bile is synthesized in the liver, stored in the gall bladder, and released into the duodenum, where it acts in concert with the pancreatic lipases.

Acute pancreatitis is a serious and sometimes lethal disease that is characterized by the *premature activation of the proteolytic and lipolytic enzymes of the pancreas.* In pancreatitis, these enzymes are unleashed while they are still inside the pancreas. The effect is to destroy the pancreas itself and its blood vessels. Trauma to the acinar tissue can cause acute pancreatitis.

SERINE, ZINC, THIOL, AND CARBOXYL PROTEASES ARE THE MAJOR FAMILIES OF PROTEOLYTIC ENZYMES

Thus far, consideration has been given to two classes of proteolytic enzymes: the *zinc proteases*, exemplified by carboxypeptidase A (p. 148), and the *serine proteases*. The *thiol proteases* are another widely distributed group. Papain, derived from papaya, contains an active-site cysteine, which plays a role analogous to that of serine 195 in chymotrypsin. Catalysis proceeds through a thioester intermediate and is facilitated by a nearby histidine side chain. The other major family of proteolytic enzymes comprises the *carboxyl proteases,* which are also called *acid proteases* because most of them are active only in an acid environment. The best-known member of this family is *pepsin,* the principal protease in gastric juice. Pepsin, which has a mass of 34.6 kdal, is formed by the cleavage of a 44-residue peptide from the amino-terminal end of pepsinogen, the zymogen. This activation occurs spontaneously at pH 2 or below, or it can be catalyzed by pepsin. Consequently, pepsinogen is converted into pepsin within a few seconds after it has been secreted into the gastric lumen. The active site of pepsin contains two aspartate residues. One of them must be ionized and the other un-ionized for the enzyme to be active, which results in a pH optimum of between 2 and 3. Carboxyl proteases with similar structures and enzymatic properties have been isolated from lysosomes, which carry out intracellular digestion (p. 463), and from a variety of molds (Figure 8-25). A common feature of the carboxyl proteases is that they are inhibited by very low concentrations (about 10^{-10} M) of *pepstatin,* a hexapeptide transition-state analog.

CLOTTING OCCURS BY A CASCADE OF ZYMOGEN ACTIVATIONS

The activation of a precursor protein by peptide-bond cleavage is a basic means of control that recurs in many different biochemical systems. We now turn to the role of zymogen activations in clot formation, which is one of three mechanisms of hemostasis. The other two are a rapid constriction of the injured vessel and the aggregation of platelets to form a plug on the injured surface of the blood vessel.

A clot is formed by a series of transformations involving more than ten different proteins. A striking feature of the process is the occurrence of a *series of zymogen activations.* In this enzymatic *cascade,* the activated form of one factor catalyzes the activation of the next factor. Very small amounts of the initial factors are needed because of the catalytic nature of the activation process. The numerous steps yield a large amplification, assuring a rapid response to trauma.

CLOT FORMATION REQUIRES THE INTERPLAY OF TWO KINDS OF ENZYMATIC PATHWAYS

In 1863, Joseph Lister showed that blood stayed fluid in the excised jugular vein of an ox, but that it rapidly clotted when it was transferred to a glass vessel. The abnormal surface triggered components already present in the blood. Consequently, this pathway of clotting is called the *intrinsic* one. Clotting can also be triggered by the addition of substances that are normally not present in the blood. Extracts from many tissues, particularly brain, rapidly cause clot

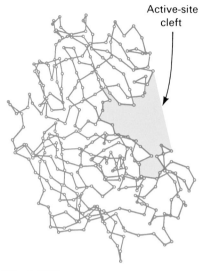

Active-site cleft

Figure 8-25
Structure of the acid protease from *Rhizopus,* a mold. The α-carbon atoms are shown. The active site is located in the groove between the two lobes, which can accommodate about eight amino acid residues of the substrate. Pepsin has a very similar structure. [From E. Subramanian. *Trends Biochem. Sci.* 3(1978):2.]

formation when added to plasma. This pathway of clotting is called the *extrinsic* one.

Clotting involves the interplay of the intrinsic and extrinsic pathways (Figure 8-26). Both are needed for proper clotting, as evidenced by various clotting disorders in which there is a deficiency of

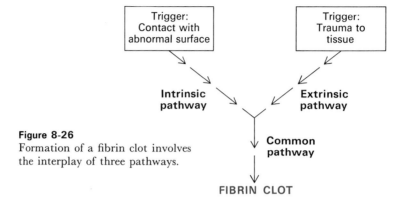

Figure 8-26
Formation of a fibrin clot involves the interplay of three pathways.

a single protein in one of the pathways. Furthermore, the intrinsic and extrinsic pathways converge on a final *common pathway* that results in the production of a *fibrin clot*.

FIBRINOGEN IS CONVERTED BY THROMBIN INTO A FIBRIN CLOT

The best-characterized part of the clotting process is the conversion of fibrinogen into fibrin by thrombin, a proteolytic enzyme. Fibrinogen is a much larger and more elongated protein than the ones discussed thus far (e.g., lysozyme and chymotrypsin). Electron micrographs show that fibrinogen is made up of three nodules connected by two rods (Figure 8-27). Fibrinogen is 460 Å long and has a mass of 340 kdal, which makes it about ten times as large as chymotrypsin. It consists of six polypeptide chains. There are pairs of three kinds of chains: one of them is called Aα, the second Bβ, and the third γ.

Fibrinogen, a highly soluble molecule in the plasma, is converted into insoluble fibrin monomer by the proteolytic action of thrombin. Thrombin cleaves four *arginine–glycine peptide bonds* in fibrinogen. Four peptides are released: an A peptide of 18 residues from each of the two α chains and a B peptide of 20 residues from each of the two β chains. These A and B peptides are called *fibrinopeptides*. A fibrinogen molecule devoid of these fibrinopeptides is called *fibrin monomer*. It has the subunit structure $(\alpha\beta\gamma)_2$ and about 97% of the amino acid residues of fibrinogen.

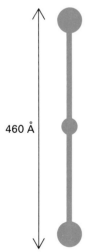

460 Å

Figure 8-27
Diagram of a fibrinogen molecule, based on electron micrographs.

Fibrin monomers have a much lower solubility than the parent fibrinogen molecules. They spontaneously associate to form *fibrin,* which has the form of long, insoluble fibers. Electron micrographs and low-angle x-ray patterns show that fibrin has a periodic structure that repeats every 230 Å (Figure 8-28). Because fibrinogen is about 460 Å long, it seems likely that fibrin monomers come together to form a half-staggered array (Figure 8-29).

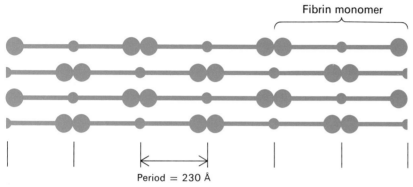

Figure 8-29
Proposed arrangement of fibrin monomers in a fibrin clot. This half-staggered array of fibrin monomers would yield the observed 230-Å period.

Figure 8-28
Electron micrograph of fibrin. The 230-Å period along the fiber axis is half the length of a fibrinogen molecule. [Courtesy of Dr. Henry Slayter.]

Why do fibrin monomers aggregate, whereas their parent fibrinogen molecules stay in solution? A definitive answer to this question should come from detailed structural studies now in progress. The fibrinopeptides of all vertebrate species studied thus far have a *large net negative charge.* Aspartate and glutamate residues are found in abundance. An unusual negatively charged derivative of tyrosine, namely *tyrosine-O-sulfate,* is found in fibrinopeptide B. The presence of these and other negatively charged groups in the fibrinopeptides probably keeps fibrinogen molecules apart. *Their release by thrombin gives fibrin monomers a different surface-charge pattern, leading to their specific aggregation.* Recall that a change of a single charged group, glutamate to valine, causes the aggregation of deoxyhemoglobin molecules in sickle-cell anemia.

Tyrosine-O-sulfate

THE FIBRIN CLOT IS STRENGTHENED
BY COVALENT CROSS-LINKS

The clot produced by the spontaneous aggregation of fibrin monomer is quite fragile. It is subsequently stabilized by the formation of covalent cross-links between the side chains of different molecules in the fibrin fiber. In fact, *peptide bonds are formed between specific glutamine and lysine side chains in a transamidation reaction* (Figure 8-30).

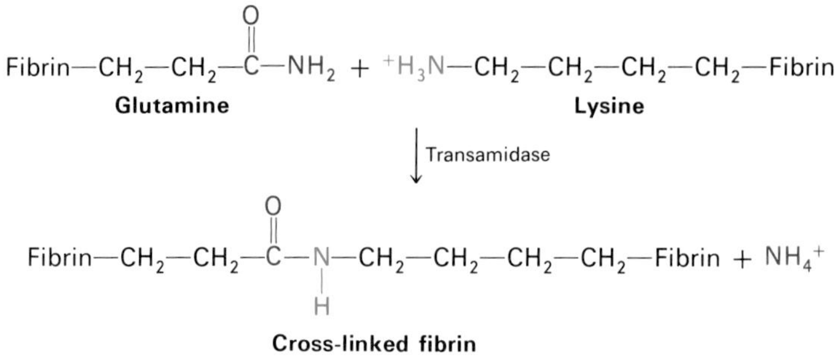

$$\text{Fibrin—CH}_2\text{—CH}_2\text{—C(=O)—NH}_2 \; + \; {}^{+}\text{H}_3\text{N—CH}_2\text{—CH}_2\text{—CH}_2\text{—CH}_2\text{—Fibrin}$$

Glutamine　　　　　　　　　　**Lysine**

$\downarrow$ Transamidase

$$\text{Fibrin—CH}_2\text{—CH}_2\text{—C(=O)—N(H)—CH}_2\text{—CH}_2\text{—CH}_2\text{—CH}_2\text{—Fibrin} \; + \; \text{NH}_4^{+}$$

Cross-linked fibrin

Figure 8-30
Fibrin is cross-linked by transamidation.

This cross-linking reaction is catalyzed by a transamidase enzyme. Cross-links of this type are rarely found in proteins. The importance of this unusual bond in fibrin is highlighted by the finding that patients with a deficiency of this transamidase enzyme have a pronounced tendency to bleed.

THROMBIN IS HOMOLOGOUS TO TRYPSIN

The specificity of thrombin for arginine–glycine bonds suggests that thrombin might resemble trypsin. Indeed it does, as shown by amino acid sequence studies. Thrombin has a mass of 33.7 kdal and consists of two chains. The A chain of 49 residues exhibits no detectable homology to the pancreatic enzymes. The B chain, however, is quite similar in sequence to trypsin, chymotrypsin, and elastase. The sequence around its active-site serine is Gly-Asp-Ser-Gly-Gly-Pro, the same as that in the pancreatic serine proteases. Moreover, thrombin also contains a charge relay network. Its three-dimensional structure is not yet known, but one important feature of its specificity site is already apparent. Thrombin, like trypsin, contains an aspartate residue at the bottom of its substrate-binding cleft. This negatively charged group undoubtedly forms an electrostatic bond with the positively charged arginine side chain. Thrombin is much more specific than trypsin. Thrombin cleaves certain arginine–glycine bonds, whereas trypsin cleaves most peptide bonds following arginine or lysine residues.

Thrombin, like the pancreatic serine proteases, is synthesized as a 66-kdal zymogen called *prothrombin*. Proteolytic cleavage of an arginine–threonine bond releases a 32-kdal fragment from the amino terminus of prothrombin (Figure 8-31). Cleavage of an arginine–leucine bond then yields active thrombin. An ion pair like the one between the positively charged amino group of isoleucine 16 and

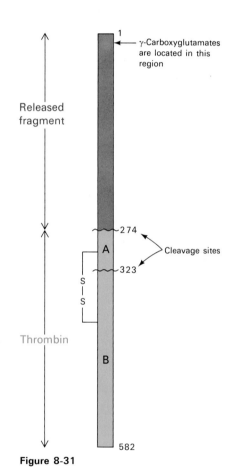

Figure 8-31
Structure of prothrombin. Cleavage of two peptide bonds (Arg 274–Thr 275 and Arg 323–Ile 324) yields thrombin. The released amino-terminal fragment of prothrombin is shown in red. All of the γ-carboxyglutamate residues are in this fragment. The A and B chains of thrombin are joined by a disulfide bond.

the negatively charged aspartate 194 in chymotrypsin is also present in thrombin.

Similarities in amino acid sequence indicate that *thrombin is evolutionarily related to the pancreatic serine proteases.* This is reinforced by the presence of a charge relay system and a similar activation mechanism. It is noteworthy that prothrombin is formed in the liver, which has a common embryological origin with the pancreas.

VITAMIN K IS REQUIRED FOR THE SYNTHESIS OF PROTHROMBIN

Vitamin K has been known for many years to be essential for the synthesis of prothrombin and several other clotting factors. Recent studies of the abnormal prothrombin synthesized in the absence of vitamin K or in the presence of vitamin K antagonists, such as dicoumarol, have revealed the mode of action of this vitamin. *Dicoumarol* is found in spoiled sweet clover and causes a fatal hemorrhagic disease in cattle fed on this hay. This coumarin derivative is used clinically as an *anticoagulant* to prevent thromboses in patients prone to clot formation. Dicoumarol and such related vitamin K antagonists as *warfarin* also serve as effective rat poisons. Cows fed

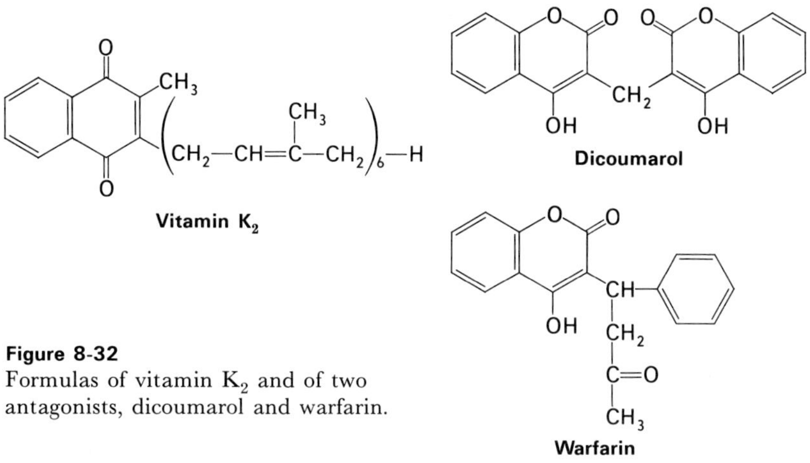

Figure 8-32
Formulas of vitamin K_2 and of two antagonists, dicoumarol and warfarin.

dicoumarol contain an abnormal prothrombin that does not bind Ca^{2+}, in contrast with normal prothrombin. This difference was puzzling for some time because abnormal prothrombin has the same number of amino acid residues and gives the same amino acid analysis after acid hydrolysis as does normal prothrombin. Fragmentation of normal prothrombin showed that its capacity to bind Ca^{2+} resides in its amino-terminal region (Figure 8-33). The electrophoretic mobility of an amino-terminal peptide from abnormal prothrombin was then found to be markedly different from that of the corresponding peptide from the normal molecule. Nuclear

	NH$_2$
1	Ala
	Asn
	Lys
	Gly
	Phe
	Leu
	Gla
	Gla
	Val
10	Arg
	Lys
	Gly
	Asn
	Leu
	Gla
	Arg
	Gla
	Cys
	Leu
20	Gla
	Gla
	Pro
	Cys
	Ser
	Arg
	Gla
	Gla
	Ala
	Phe
30	Gla
	Ala
	Leu
	Gla
	Ser
35	Leu

Figure 8-33
Amino acid sequence of the amino-terminal region of prothrombin [γ-carboxyglutamate (Gla) residues are shown in red].

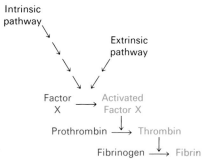

γ-Carboxyglutamate

Intrinsic pathway

Extrinsic pathway

Factor X → Activated Factor X

Prothrombin → Thrombin

Fibrinogen → Fibrin

Figure 8-34
Final steps in the formation of a fibrin clot. These three reactions constitute the common pathway. Factor X is activated by products of the intrinsic and extrinsic pathways.

magnetic resonance studies of these peptides revealed that normal prothrombin contains γ-*carboxyglutamate,* a previously unknown amino acid, whereas abnormal prothrombin lacks this modified amino acid. In fact, *the first ten glutamate residues in the amino-terminal region of prothrombin are carboxylated to γ-carboxyglutamate by a vitamin-K–dependent enzyme system.* This modified amino acid was unrecognized until recently because the γ-carboxy group is lost on acid hydrolysis, yielding glutamate.

PROTHROMBIN ON PHOSPHOLIPID SURFACES IS ACTIVATED BY FACTOR X_a

The cascading nature of the clotting process is shown in Figure 8-34. Prothrombin is activated by a proteolytic enzyme called Factor X_a. Clotting factors are assigned Roman numerals for ease of discussion. The subscript "a" means that the factor is in the active form. Prothrombin is converted into thrombin by the proteolytic action of Factor X_a. This activation is accelerated by Factor V, which is not itself an enzyme. Factor V can be regarded as a *modifier protein.*

The vitamin-K–dependent carboxylation reaction converts glutamate, a weak chelator of Ca^{2+}, into γ-carboxyglutamate, a much stronger chelator. The binding of Ca^{2+} by prothrombin anchors it to phospholipid membranes derived from blood platelets following injury. The functional significance of the binding of prothrombin to phospholipid surfaces is that it brings prothrombin in close proximity to factors X_a and V, which accelerates the activation of prothrombin by more than a factor of 10^4. The amino-terminal fragment of prothrombin, which contains the Ca^{2+}-binding sites, is released in this activation step. Consequently, thrombin is freed from the phospholipid surface and can therefore activate fibrinogen in the plasma.

How is Factor X activated? This activation step takes place where the intrinsic and extrinsic pathways meet. The products of both the intrinsic and the extrinsic pathways are proteolytic enzymes that activate Factor X.

HEMOPHILIA AND OTHER BLEEDING DISORDERS HAVE REVEALED SOME OF THE EARLY STEPS IN CLOTTING

Biochemical studies of the early steps in clotting are more difficult than those of the late ones because the early clotting factors are present only in small amounts. The amount of fibrinogen in 1 ml of blood is 3 mg, but the amount of Factor X is only 0.01 mg. The concentrations of some of the earlier clotting factors are even lower. Furthermore, these proteins are highly labile. Some of the important breakthroughs in the elucidation of the pathways of clotting have therefore come from studies of patients with bleeding disorders.

The best known of these diseases is *hemophilia*. This clotting disorder is genetically transmitted as a sex-linked recessive characteristic. Heterozygous females are asymptomatic carriers. A famous carrier of this disease was Queen Victoria, who transmitted it to the royal families of Prussia, Spain, and Russia (Figure 8-35).

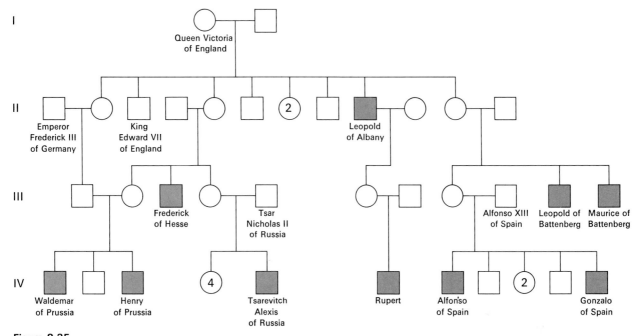

Figure 8-35
Pedigree of hemophilia in the royal families of Europe. All of Queen Victoria's children, but not all individuals in later generations, are included in this diagram. Females are symbolized by circles, normal males by white squares, and hemophilic males by red squares. [After C. Stern. *Principles of Human Genetics,* 3rd ed. (W.H. Freeman and Company). Copyright ©️ 1973.]

In 1904, the tsarevich Alexis was born. He was the first male heir to have been born to a reigning Russian tsar since the seventeenth century, which was taken as an omen of hope. Four healthy daughters had previously been born to Tsar Nicholas II and Empress Alexandra, a granddaughter of Queen Victoria. The mood changed six weeks after the birth of Alexis, as reflected in his father's diary: "A hemorrhage began this morning without the slightest cause from the navel of our small Alexis. It lasted with but a few interruptions until evening." The evidence grew stronger and more ominous as Alexis started to crawl and toddle, which caused large, blue swellings on his legs and arms. The hemorrhages became more serious and there was little that physicians could do to alleviate the pain. The anguished empress then turned to Rasputin, who was reputed to be a miracle man. No man knew less about molecular disease, no man ever profited more from it. Rasputin held a position of great power in the Russian court for many years because the empress placed great faith in his healing abilities.

Why did Nicholas and Alexandra marry when it was already known that her brother, nephews, and uncle had hemophilia? In 1873, Grandidier, a French physician, counseled that "all members of bleeder families should be advised against marriage." Indeed, the hereditary nature of the disease was fully appreciated as early as 1803, by John Otto:

> About seventy or eighty years ago, a woman by the name of Smith settled in the vicinity of Plymouth, New Hampshire, and transmitted the following idiosyncrasy to her decendants. . . . It is a surprising circumstance that the males only are subject to this strange affection, and that all of them are not liable to it. . . . Although the females are exempt, they are still capable of transmitting it to their male children.

J. B. S. Haldane has suggested that "kings are carefully protected against disagreeable realities. . . . The hemophilia of the Tsarevich was a symptom of the divorce between royalty and reality."

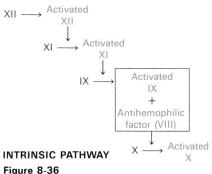

INTRINSIC PATHWAY
Figure 8-36
Intrinsic pathway of clotting. Inactive forms of factors are shown in red, active ones in green. Activated factors catalyze the activation of other factors.

INTRINSIC PATHWAY OF CLOTTING

The defect in hemophilia is in the intrinsic pathway (Figure 8-36). A protein called *antihemophilic factor* is missing or has a markedly reduced activity. Antihemophilic factor (VIII) acts in concert with Factor IX_a, a proteolytic enzyme, to activate Factor X. Antihemophilic factor is not itself an enzyme; rather, it is a modifier protein.

Factor IX was discovered in an interesting way. A boy named Stephen Christmas had a clotting disorder that was clinically indistinguishable from classical hemophilia in its symptoms and mode of inheritance. The clotting time of his blood measured in a glass tube was prolonged, as in hemophilia. His blood was then mixed with the blood from a hemophiliac. The striking result was that the mixed blood had a nearly normal clotting time. Thus, it was evident that Christmas could not have the same molecular defect as the hemophiliac patient, despite the virtually identical clinical picture. The clotting factors missing in these patients must have been different, because their plasmas complemented each other in clotting. In this way, a new factor termed IX, or *Christmas factor,* was discovered. In general, a *complementation test is a powerful means of determining whether two inactive systems lack the same component.* The power of the complementation approach is that the components of a system can be enumerated before they are isolated and purified. Complementation tests are extensively used in bacterial and viral genetics.

Christmas factor is two steps removed from the initial trigger in the intrinsic pathway. This reaction sequence is initiated by the contact of Factor XII with an abnormal surface. Activated Factor XII$_a$, together with kallikrein and kinin, then converts Factor XI into its active form. In turn, Christmas factor (IX) is activated by XI$_a$. In these activation steps, zymogens are converted into active enzymes.

EXTRINSIC PATHWAY OF CLOTTING

As in the intrinsic pathway, the end result of the extrinsic pathway is the activation of Factor X. The extrinsic pathway (Figure 8-37) seems to be relatively simple. Trauma to the blood vessel releases a lipoprotein called *tissue factor*. A complex of tissue factor and Factor VII then catalyzes the activation of X. Tissue factor is a modifier protein, since Factor VII is inactive in the absence of tissue factor. Thus, there are at least three modifier proteins in the clotting process: Factor V, Factor VIII, and tissue factor.

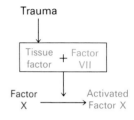

EXTRINSIC PATHWAY
Figure 8-37
Extrinsic pathway of clotting.

Table 8-2
Blood coagulation factors

Factor	Pathway	Function of active form
Hageman factor (XII)	Intrinsic	
Kallikrein	Intrinsic	Activate XI
Kinin	Intrinsic	
Plasma thromboplastin antecedent (PTA) (XI)	Intrinsic	Activates IX
Christmas factor (IX)	Intrinsic	Both required to activate X
Antihemophilic factor (VIII)	Intrinsic	
Tissue factor	Extrinsic	Both required to activate X
Proconvertin (VII)	Extrinsic	
Stuart factor (X)	Common	Activates II
Accelerin (V)	Common	Stimulates activation of II
Prothrombin (II)	Common	Activates fibrinogen (I)
Fibrinogen (I)	Common	FORMS FIBRIN CLOT
Fibrin-stabilizing factor (XIII)	Common	Stabilizes clot by cross-linking fibrin

CONTROL OF CLOTTING: A CHALLENGING PROBLEM

It is evident that the clotting process must be precisely regulated. There is a fine line between hemorrhage and thrombosis. Clotting must occur rapidly yet remain confined to the area of injury. Why are there so many stages? Why are there distinct intrinsic and ex-

Figure 8-38
Electron micrograph of an ordered
array of modified fibrinogen. The axial
repeat distance is 225 Å, as in fibrin.
[Courtesy of Dr. Carolyn Cohen.]

trinsic pathways and how are they related? What are the mechanisms that normally limit clot formation to the site of injury? These intriguing questions are of interest for their own sake and also because of their pertinence to preventing and treating cardiovascular diseases, which constitute a major health problem.

The lability of activated clotting factors contributes significantly to the control of clotting. The activated clotting factors have a short lifetime because they are diluted by blood flow, removed by the liver, degraded by proteases, and inactivated by specific inhibitors. The most important inhibitor is *antithrombin III,* a plasma protein that inactivates thrombin by forming an irreversible complex with it. This complex is probably like the stable acyl-enzyme complex formed between trypsin and its pancreatic inhibitor (p. 168). Antithrombin III also inhibits clotting Factors IX_a, X_a, and XI_a. The inhibitory action of antithrombin III is enhanced by *heparin,* a negatively charged polysaccharide (p. 201) found in mast cells near the walls of blood vessels. Heparin acts as an *anticoagulant* by increasing the rate of formation of the irreversible complex between thrombin and antithrombin III.

SUMMARY

The activation of a protein by proteolytic cleavage of one or a few peptide bonds is a recurring control mechanism in biological systems. If the active protein is an enzyme, the inactive precursor is called a zymogen (or a proenzyme). The best-understood zymogen activation is the conversion of chymotrypsinogen into chymotrypsin. Trypsin hydrolyzes the peptide bond between residues 15 and 16 in chymotrypsinogen and thereby converts it into a fully active chymotrypsin. The newly formed amino-terminal group of isoleucine 16 turns inward and forms an electrostatic bond with aspartate 194 in the interior. This interaction triggers a series of localized conformational changes that result in the creation of a pocket for the binding of the aromatic (or bulky nonpolar) side chain of the substrate. In addition, an oxyanion hole that stabilizes the transition state is formed.

A highly reactive serine residue (serine 195) plays a critical role in the catalytic mechanism of chymotrypsin. The first stage in the hydrolysis of a peptide substrate is the formation of a covalent *acyl-enzyme intermediate,* in which the carboxyl component of the substrate is esterified to the hydroxyl group of serine 195. This acyl-enzyme intermediate is hydrolyzed in the second stage of catalysis. Histidine 57 also plays a key role in both the acylation and deacylation reactions. In fact, serine 195 is made a strong nucleophile by a series of hydrogen bonds. Serine 195 is hydrogen bonded to histidine 57, which is in turn hydrogen bonded to aspartate 102 in the inside of the enzyme. These three residues constitute a *charge relay network,* which accelerates catalysis by a factor of about 10^3. Such a charge relay network is also present in trypsin, elastase, and thrombin. Indeed, these enzymes are like chymotrypsin in amino acid sequence, conformation, and catalytic mechanism, but they differ in specificity. These variations on a theme probably arose by divergent evolution from a common ancestor. The two other major families of proteolytic enzymes are the zinc, thiol, and carboxyl proteases.

Zymogen activations also play a major role in the control of blood clotting. A striking feature of the clotting process is that it occurs by a cascade of zymogen conversions, in which the activated form of one clotting factor catalyzes the activation of the next precursor. Clotting involves the interplay of two reaction sequences, called the intrinsic and extrinsic pathways. Both are needed for normal clotting. They meet at a common pathway that results in the formation of a fibrin clot. Fibrinogen, a highly soluble molecule in the plasma, is converted into fibrin by the hydrolysis of four arginine–glycine bonds. This reaction is catalyzed by thrombin, an enzyme that is similar to trypsin. Two A and two B peptides comprising 3% of the fibrinogen are released. The resulting fibrin monomer then spontaneously forms long, insoluble fibers called fibrin. The fibrin clot is subsequently strengthened by the formation of covalent cross-links as a result of transamidation reactions between specific glutamine and lysine side chains. Vitamin K is required for the carboxylation of glutamate residues in prothrombin and several other clotting factors. The binding of calcium ion by γ-carboxyglutamate anchors prothrombin to platelet membranes, which accelerates its activation by factors X_a and V.

SELECTED READINGS

WHERE TO START

Stroud, R. M., 1974. A family of protein-cutting proteins. *Sci. Amer.* 231(1):24–88. [An excellent introduction to the structure, catalytic properties, and evolution of serine proteases. Available as *Sci. Amer.* Offprint 1301.]

McKusick, V. A., 1965. The royal hemophilia. *Sci. Amer.* 213(2):88–95.

SERINE PROTEASES AND THEIR ZYMOGENS

Blow, D. M., 1976. Structure and mechanism of chymotrypsin. *Acc. Chem. Res.* 9:145–152.

Huber, R., and Bode, W., 1978. Structural basis of the activation and action of trypsin. *Acc. Chem. Res.* 11:114–122.

Kraut, J., 1977. Serine proteases: structure and mechanism of catalysis. *Ann. Rev. Biochem.* 46:331–358.

Stroud, R. M., Kossiakoff, A. A., and Chambers, J. L., 1977. Mechanisms of zymogen activation. *Ann. Rev. Biophys. Bioeng.* 6:177–193.

Boyer, P. D., (ed.), 1971. *The Enzymes* (3rd ed.), vol. 3. Academic Press. [This volume contains many excellent articles on proteolytic enzymes, zymogens, and inhibitors.]

ACID PROTEASES AND THEIR ZYMOGENS

Tang, J., 1976. Pepsin and pepsinogen: models for carboxyl (acid) proteases and their zymogens. *Trends Biochem. Sci.* 1:205–208.

TRYPSIN INHIBITORS

Laskowski, M., Jr., and Kato, I., 1980. Protein inhibitors of proteinases. *Ann. Rev. Biochem.* 49:593–626.

Ruhlmann, A., Kukla, D., Schwager, P., Bartels, K., and Huber, R., 1973. Structure of the complex formed by bovine trypsin and bovine pancreatic trypsin inhibitor. *J. Mol. Biol.* 89:73–101.

Sweet, R. M., Wright, H. T., Janin, J., Chothia, C. H., and Blow, D. M., 1974. Crystal structure of the complex of porcine trypsin with soybean trypsin inhibitor (Kunitz) at 2.6-Å resolution. *Biochemistry* 13:4212–4228.

PROTEASES IN BIOLOGICAL PROCESS

Reich, E., Rifkin, D. B., and Shaw, E., (eds.), 1975. *Proteases and Biological Control.* Cold Spring Harbor Laboratory. [A fine collection of articles on the role of proteases in a wide range of biological processes such as fertilization, cell migration, tumor invasiveness, morphogenesis, and metamorphosis.]

Fong, D., and Bonner, J. T., 1979. Proteases in cellular slime mold development: evidence for their involvement. *Proc. Nat. Acad. Sci.* 76:6481–6485.

BLOOD CLOTTING

Jackson, C. M., and Nemerson, Y., 1980. Blood coagulation. *Ann. Rev. Biochem.* 49:765–811.

Ratnoff, O. D., 1978. Hereditary disorders of hemostasis. *In* Stanbury, J. B., Wyngaarden, J. B., and Fredrickson, D. S., (eds.), *The Metabolic Basis of Inherited Disease* (4th ed.), pp. 1755–1791. McGraw-Hill.

Davie, E. W., and Hanahan, D. J., 1977. Blood coagulation proteins. *In* Putnam, F. W., (ed.), *The Plasma Proteins* (2nd ed.), pp. 422–544. Academic Press.

Ogston, D., and Bennett, B., (eds.), 1977. *Haemostasis: Biochemistry, Physiology, and Pathology.* Wiley.

Stenflo, J., and Suttie, J. W., 1977. Vitamin K-dependent formation of γ-carboxyglutamic acid. *Ann. Rev. Biochem.* 46:157–172.

Link, K. P., 1943. The anticoagulant from spoiled sweet clover hay. *Harvey Lectures* 39:162–216. [A delightful account of the discovery of dicoumarol and its anticoagulant action.]

Doolittle, R. F., 1973. Structural aspects of the fibrinogen to fibrin conversion. *Advan. Protein Chem.* 27:1–109.

Tooney, N. M., and Cohen, C., 1977. Crystalline states of a modified fibrinogen. *J. Mol. Biol.* 110:363–385.

Massie, R. K., 1967. *Nicholas and Alexandra.* Dell. [A fascinating account of the interplay of hemophilia and Russian history.]

PROBLEMS

1. Let us compare *lysozyme, carboxypeptidase A,* and *chymotrypsin.*
   (a) Which of these enzymes requires a metal ion for catalytic activity?
   (b) Which of these enzymes is a single polypeptide chain?
   (c) Which of these enzymes is rapidly inactivated by DIPF?
   (d) Which of these enzymes is formed by proteolytic cleavage of an identified proenzyme?

2. The transfer of a proton from an enzyme to its substrate is often a key step in catalysis.
   (a) Does this occur in the proposed catalytic mechanisms for chymotrypsin, lysozyme, and carboxypeptidase A?
   (b) If so, identify the proton donor in each case.

3. A substrate is frequently attacked by a nucleophilic group of an enzyme. What is the nucleophile in the proposed mechanism for chymotrypsin and for carboxypeptidase A?

4. Cite the factors that contribute to the catalytic power of enzymes.

5. TPCK is an affinity-labeling reagent for chymotrypsin. It inactivates chymotrypsin by alkylating histidine 57.
   (a) Design an affinity-labeling reagent for trypsin that resembles TPCK.
   (b) How would you test its specificity?
   (c) Which other serine protease might be inactivated by this affinity-labeling reagent for trypsin?

6. Elastase is specifically inhibited by an aldehyde derivative of one of its substrates:

$$N\text{-Acetyl}-Pro-Ala-Pro-N-\underset{CH_3}{\overset{H}{\underset{|}{\overset{|}{C}}}}-\overset{O}{\underset{H}{\overset{\diagup}{C}}}$$

   In fact, this aldehyde is an analog of the transition state for catalysis by elastase.
   (a) Which residue at the active site of elastase is most likely to form a covalent bond with this aldehyde?
   (b) What type of covalent link would be formed?

7. Boron acids are another kind of transition-state analog for enzymes that form acyl-enzyme intermediates. Acetylcholinesterase is an enzyme that catalyzes the hydrolysis of the ester bond in acetylcholine:

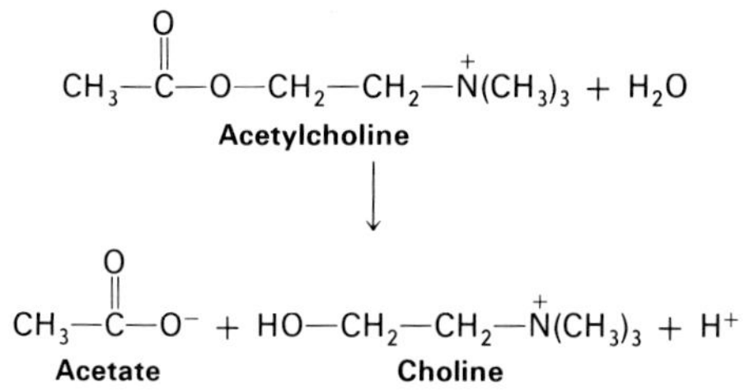

$$CH_3-\overset{\overset{O}{\|}}{C}-O-CH_2-CH_2-\overset{+}{N}(CH_3)_3 + H_2O$$

Acetylcholine

↓

$$CH_3-\overset{\overset{O}{\|}}{C}-O^- + HO-CH_2-CH_2-\overset{+}{N}(CH_3)_3 + H^+$$

Acetate   **Choline**

Acetylcholinesterase is specifically inhibited by this boron analog of acetylcholine:

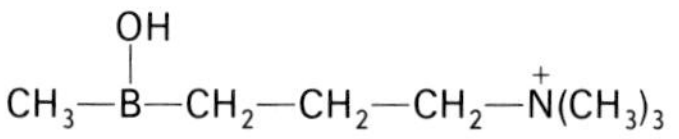

$$CH_3-\overset{\overset{OH}{|}}{B}-CH_2-CH_2-CH_2-\overset{+}{N}(CH_3)_3$$

Boron analog of acetylcholine

How might this boron analog be covalently attached to the active site of acetylcholinesterase?

8. The synthesis of Factor X, like that of prothrombin, requires vitamin K. Factor X also contains γ-carboxyglutamate residues in its amino-terminal region. However, activated Factor X, in contrast with thrombin, retains this region of the molecule. What is a likely functional consequence of this difference between the two activated species?

9. Antithrombin III forms an irreversible complex with thrombin but not with prothrombin. What is the most likely reason for this difference in reactivity?

10. Fibrin, like myosin, keratin, and epidermin, contains rods made up of intertwined α helices. These α-helical coiled coils (p. 30) contain repeating heptapeptide units (*abcdefg*) in which residues *a* and *d* are hydrophobic. Propose a reason for this regularity.

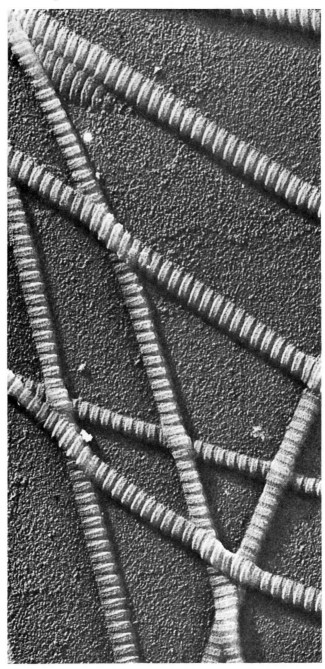

Figure 9-1
Electron micrograph of intact collagen fibrils obtained from skin.
The preparation was shadowed with chromium. The period
along the fiber axis is 640 Å. [Courtesy of Dr. Jerome Gross.]

CHAPTER 9

CONNECTIVE-TISSUE PROTEINS: COLLAGEN, ELASTIN, AND PROTEOGLYCANS

All multicellular organisms contain collagen, a family of fibrous proteins. Indeed, it is the most abundant protein in mammals, constituting a quarter of the total. Collagen is the major fibrous element of skin, bone, tendon, cartilage, blood vessels, and teeth. It is present to some extent in nearly all organs and serves to hold cells together in discrete units. In addition to its structural role in mature tissue, collagen has a directive role in developing tissue. *Collagen is distinctive in forming insoluble fibers that have a high tensile strength* (Figure 9-1). Furthermore, the basic structural motif of collagen is modified to meet the specialized needs of particular tissues. Four variations on a common theme have been characterized (Table 9-1).

Collagen
Derived from the Greek words meaning *to produce glue.*

Table 9-1
Types of collagen

Type	Composition	Distribution
I	$[\alpha 1(\text{I})]_2\alpha 2$	Skin, tendon, bone, cornea
II	$[\alpha 1(\text{II})]_3$	Cartilage, intervertebral disc, vitreous body
III	$[\alpha 1(\text{III})]_3$	Fetal skin, cardiovascular system
IV	$[\alpha 1(\text{IV})]_3$	Basement membrane

TROPOCOLLAGEN IS THE BASIC STRUCTURAL UNIT OF COLLAGEN

The insolubility of collagen fibers was for many years an impasse in the chemical characterization of collagen. The breakthrough came when it was found that collagen from the tissues of young animals can be extracted in soluble form because it is not yet extensively cross-linked. The absence of covalent cross-links in immature collagen makes it feasible to extract the basic structural unit, called *tropocollagen*, in intact form.

Tropocollagen has a mass of about 285 kdal and consists of *three polypeptide chains* of the same size. The chain composition depends on the type of collagen (Table 9-1). Type I collagen, the most prevalent species, consists of two chains of one kind, termed $\alpha 1(I)$, and one of another, termed $\alpha 2$. The other types of collagen have three identical chains. Each of the three strands in collagen consists of about a thousand amino acid residues. Thus, the basic structural unit of collagen is very large—indeed, more than ten times the size of chymotrypsin.

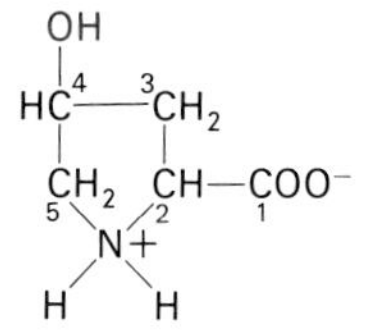

4-Hydroxyproline
(Hyp)

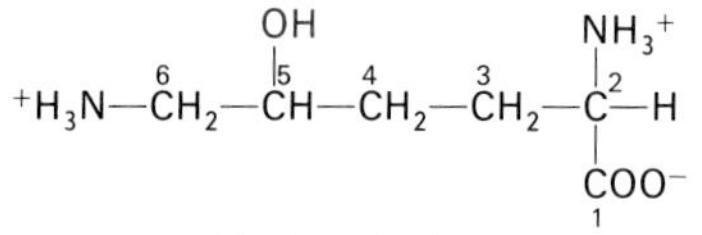

5-Hydroxylysine
(Hyl)

13
-Gly-Pro-Met-Gly-Pro-Ser-Gly-Pro-Arg-
22
-Gly-Leu-Hyp-Gly-Pro-Hyp-Gly-Ala-Hyp-
31
-Gly-Pro-Gln-Gly-Phe-Gln-Gly-Pro-Hyp-
40
-Gly-Glu-Hyp-Gly-Glu-Hyp-Gly-Ala-Ser-
49
-Gly-Pro-Met-Gly-Pro-Arg-Gly-Pro-Hyp-
58
-Gly-Pro-Hyp-Gly-Lys-Asn-Gly-Asp-Asp-

Figure 9-2
Amino acid sequence of part of the $\alpha 1(I)$ chain of collagen. Every third residue is glycine in a region spanning more than a thousand residues.

COLLAGEN HAS AN UNUSUAL AMINO ACID COMPOSITION AND SEQUENCE

The proportion of *glycine* residues in all collagen molecules is nearly one-third, which is unusually high for a protein. In hemoglobin, for example, the glycine content is 5%. Also, *proline* is present to a much greater extent in collagen than in most other proteins. Furthermore, collagen contains two amino acids that are present in very few other proteins, namely *hydroxyproline* and *hydroxylysine*.

The amino acid sequence of collagen is remarkably regular: *nearly every third residue is glycine*. Moreover, the sequence glycine–proline–hydroxyproline recurs frequently (Figure 9-2). In contrast, globular proteins rarely exhibit regularities in their amino acid sequences. Two other proteins with some regularly repeating sequences are silk fibroin and elastin.

SOME PROLINE AND LYSINE RESIDUES BECOME HYDROXYLATED

Hydroxyproline and hydroxylysine are not incorporated in the polypeptide chains of collagen by the usual mechanisms of protein synthesis. If ^{14}C-labeled hydroxyproline is administered to a rat, none of the radioactivity appears in the collagen that is synthesized. In contrast, the hydroxyproline in collagen is radioactive if ^{14}C-proline is given. Thus, proline is a precursor of the hydroxyproline residues in collagen, whereas exogenous hydroxyproline is not.

Certain proline residues in collagen are converted into hydroxy-proline by *prolyl hydroxylase,* an enzyme with a ferrous ion at its active site. The oxygen atom that becomes attached to C-4 of proline comes from O_2. The other oxygen atom of O_2 emerges in succinate, which is formed from α-ketoglutarate, an obligatory substrate in this reaction (Figure 9-3). Hence, prolyl hydroxylase is a *dioxygenase.* A noteworthy feature of this hydroxylation reaction is the need for a *reducing agent* such as *ascorbate* (p. 192) to maintain the iron atom in the ferrous state.

This hydroxylation reaction is highly specific. Proline as a free amino acid is not a substrate. Rather, hydroxylation takes place at specific sites on relatively large polypeptide chains before they become helical. *Proline can be hydroxylated at C-4 only if it is situated on the amino side of a glycine residue.* In addition, a few proline residues are hydroxylated at C-3 by a different enzyme. In contrast, these prolines are located on the carboxyl side of glycine residues.

A small proportion of the lysine residues in collagen become hydroxylated at C-5 by the action of *lysyl hydroxylase.* As in the hydroxylation of proline, molecular oxygen, α-ketoglutarate, and ascorbate are required. Lysines modified in this way are always on the amino side of glycine residues.

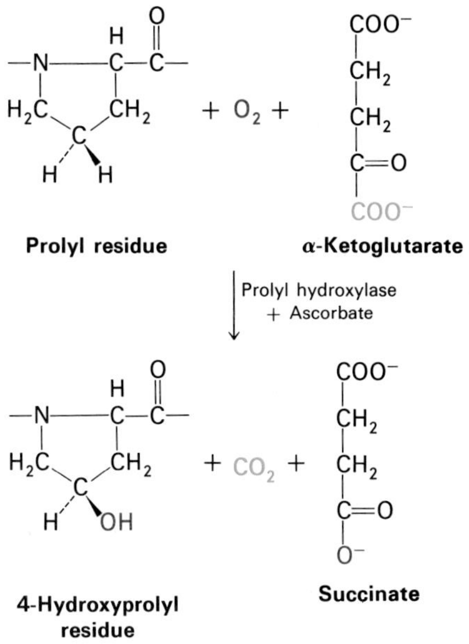

Figure 9-3
Hydroxylation of a proline residue at C-4 by the action of prolyl hydroxylase, an enzyme that activates molecular oxygen.

SUGARS ARE ATTACHED TO HYDROXYLYSINE RESIDUES

Collagen contains carbohydrate units covalently attached to its hydroxylysine residues. A disaccharide of glucose and galactose is commonly found (Figure 9-4). These sugars are inserted by the

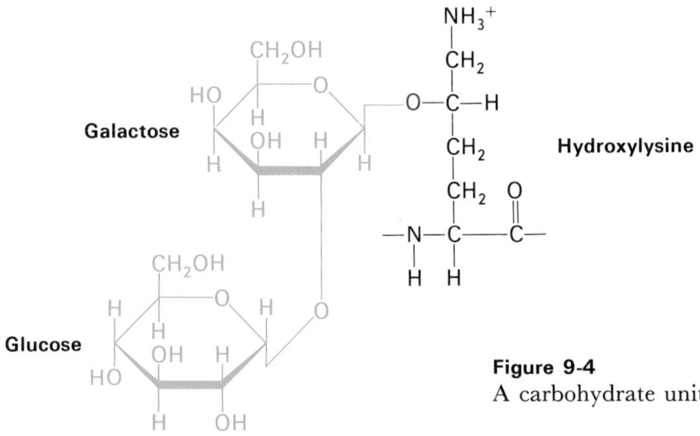

Figure 9-4
A carbohydrate unit in collagen.

sequential action of *galactosyl transferase* and *glucosyl transferase.* These glycosylating enzymes are specific for hydroxylysine residues in nascent collagen, before it has assumed a helical conformation. The number of carbohydrate units per tropocollagen depends on the tissue. For example, tendon collagen (type I) contains 6, whereas the lens capsule (type IV) has 110.

TROPOCOLLAGEN IS A TRIPLE-STRANDED HELICAL ROD

Let us turn to the conformation of the basic structural unit of the type I collagen fiber. Electron microscopic and hydrodynamic studies have shown that *tropocollagen has the shape of a rod 3000 Å long and 15 Å in diameter.* It is one of the longest known proteins. Its elongated character can be appreciated by noting that tropocollagen is sixty times as long as the diameter of chymotrypsin, but its diameter is less than half that of chymotrypsin. Each of the three polypeptide chains is in a helical conformation (Figure 9-5). Furthermore, the three helical strands wind around each other to form a stiff cable. Indeed, the strength of a collagen fiber is remarkable: a load of at least 10 kg is needed to break a fiber 1 mm in diameter.

The helical motif of the individual chains of the triple-stranded collagen cable is nicely illustrated by a model compound, *poly-L-proline.* This synthetic polypeptide has a helical form (Figure 9-6)

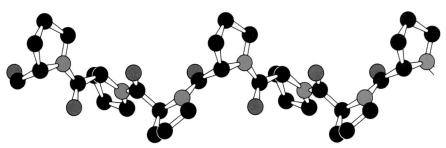

Figure 9-6
Model of a helical form of poly-L-proline (type II *trans* helix). This kind of helix is the basic structural motif of each of the three strands of tropocollagen.

that is entirely different from the α helix. There are no hydrogen bonds in the poly-L-proline helix. Instead, *this helix is stabilized by steric repulsion of the pyrrolidone rings of the proline residues.* The pyrrolidone rings keep out of each other's way when the polypeptide chain assumes this helical form (known as the type II *trans* helix), which is much more open than the tightly coiled α helix. The rise per residue along the helix axis of poly-L-proline is 3.12 Å, whereas it is 1.5 Å in the α helix. There are three residues per turn in the poly-L-proline helix.

Let us now look at the conformation of a single strand of the triple-stranded collagen helix (Figure 9-7). Each strand has a helical conformation like that shown for poly-L-proline. The three strands wind around each other to form a *superhelical cable* (Figure 9-8). The rise per residue in this superhelix is 2.9 Å and the number of residues per turn is nearly 3.3. *The three strands are hydrogen bonded to*

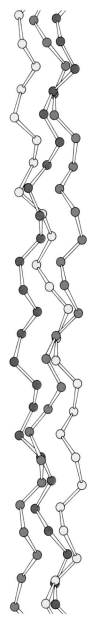

Figure 9-5
Model of the triple-stranded collagen helix. Only the α-carbon atoms are shown.

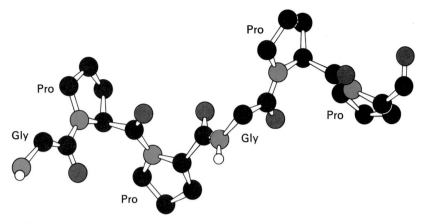

Figure 9-7
Conformation of a single strand of the collagen triple helix. The sequence shown here is -Gly-Pro-Pro-Gly-Pro-Pro.

each other. The hydrogen donors are the peptide NH groups of glycine residues, and the hydrogen acceptors are the peptide CO groups of residues on the other chains. The direction of the hydrogen bond is transverse to the long axis of the tropocollagen rod. The hydroxyl groups of hydroxyproline residues and water molecules also participate in hydrogen bonding, which stabilizes the triple helix. A space-filling model of the triple-stranded collagen helix is shown in Figure 9-9.

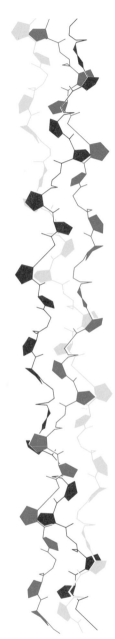

Figure 9-9
A space-filling model of the collagen triple helix. [Courtesy of Dr. Alexander Rich.]

Figure 9-8
Skeletal model of the triple-stranded collagen helix. The repeating sequence shown here is -Gly-Pro-Pro.

GLYCINE IS CRITICAL BECAUSE IT IS SMALL

We are now in a position to see why glycine occupies every third position in the amino acid sequence of tropocollagen. The inside of the triple-stranded helical cable is very crowded (Figure 9-10). In

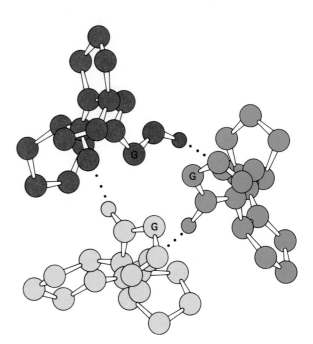

Figure 9-10
Cross section of a model of collagen. Each strand is hydrogen bonded to the other two strands (··· denotes a hydrogen bond). The α-carbon atom of a glycine residue in each strand is labeled G. Every third residue must be glycine because there is no space near the helix axis (center) for a larger amino acid residue. Note that the pyrrolidone rings are on the outside. Each strand of the triple helix is shown in a different color.

fact, *the only residue that can fit in an interior position is glycine.* There are three residues per turn of the helix, and so every third residue on each strand must be glycine. The two amino acid residues on either side of glycine are located on the outside of the cable, where the bulky rings of proline and hydroxyproline residues can be readily accommodated.

We might think of glycine as an insignificant amino acid, the least important of its kind because its side chain consists of only a hydrogen atom. In protein structure, however, simplicity is sometimes advantageous. Glycine is a very important amino acid precisely because it occupies very little space and thereby allows different polypeptide strands to come together. We have already seen this in myoglobin and hemoglobin, where glycine B6 is invariant because it allows the close approach of the B and E helices. In chymotrypsin, the substrate-binding cavity can accommodate a large aromatic group because two of the residues lining the cavity are glycine. In the evolution of cytochrome *c*, an electron carrier, the amino acid residue that has been most conserved is glycine. Here, again, we see in collagen *the critical role of glycine in increasing the range of allowed conformations in the folding of polypeptide chains.*

THE STABILITY OF THE COLLAGEN HELIX DEPENDS ON COOPERATIVE INTERACTIONS

If a solution of tropocollagen is heated, large changes in physical properties occur at a characteristic temperature (Figure 9-11). For example, the viscosity of the solution drops sharply, indicating that the molecules have lost their rodlike shape. The altered optical rotatory properties reveal that the helical structure of the individual strands has been destroyed. Thus, thermal motion can overcome the forces that stabilize the triple-stranded helix, yielding a disrupted structure called gelatin, which is a random coil. This structural transition occurs abruptly at a certain temperature, in a manner analogous to the melting of a crystal. The term melting is carried over into biochemistry to denote the loss of a highly ordered structure that takes place within a narrow temperature range. The tropocollagen helix is highly ordered in the direction of the long axis of the rod. The abruptness of the transition with increasing temperature reveals that the triple-stranded helix is stabilized by *cooperative interactions*. In other words, the helical form is the consequence of *many reinforcing bonds,* each of which is relatively weak. The formation of any one of these stabilizing bonds very much depends on whether adjacent bonds are also made. A zipper is an example of a cooperative structure. In subsequent chapters, a variety of other highly cooperative macromolecular structures such as DNA, viruses, and cell membranes will be discussed.

The temperature at which half of the helical structure is lost is called the *melting temperature* (T_m). The T_m of tropocollagen is a criterion of the stability of its helical structure. For intact collagen fibers, the comparable index is the *shrinkage temperature* (T_s). Collagens from different species have different melting temperatures. In fact, the T_m or T_s of a collagen is related to the body temperature of the source species (Table 9-2). Collagens from icefish have the lowest T_m, whereas those from warm-blooded animals have the highest T_m. This difference in thermal stability is correlated with the content of the imino acids (proline plus hydroxyproline) in the collagen. *The higher the imino acid content, the more stable the helix.* The

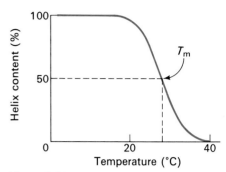

Figure 9-11
Melting curve of a collagen molecule.

Table 9-2
Dependence of thermal stability on imino acid content

Source	Proline plus hydroxyproline (per 1000 residues)	Thermal stability (°C)		Body temperature (°C)
		T_s	T_m	
Calf skin	232	65	39	37
Shark skin	191	53	29	24–28
Cod skin	155	40	16	10–14

imino acid content of collagens increased in the evolution of warm-blooded species from cold-blooded ones.

In short, the stability of the helical form of a single strand of tropocollagen depends on the locking effect of proline and hydroxyproline residues. The triple helix is further stabilized by transverse hydrogen bonding and van der Waals interactions between residues on different strands. The superhelix is sterically allowed because glycine occupies every third position in the amino acid sequence.

Studies of the melting temperatures of synthetic polypeptide models of collagen have been sources of insight into the biological significance of the hydroxylation of proline. The T_m of poly-(Pro-Pro-Gly) is 24°C, whereas that of poly(Pro-Hyp-Gly) is 58°C, which shows that *the triple helix is markedly stabilized by hydroxylation.* The properties of unhydroxylated collagen prepared by incubating tendon cells with α,α'-bipyridyl, an iron chelator that inhibits prolyl hydroxylase, support this conclusion. The unhydroxylated collagen synthesized by these cells does not form a triple helix at 37°C, whereas it rapidly becomes helical when cooled to below 24°C.

DEFECTIVE HYDROXYLATION IS ONE OF THE BIOCHEMICAL LESIONS IN SCURVY

The importance of the hydroxylation of collagen becomes evident in *scurvy.* A vivid description of this disease was given by Jacques Cartier in 1536, when it afflicted his men as they were exploring the Saint Lawrence River:

> Some did lose all their strength, and could not stand on their feet. . . . Others also had all their skins spotted with spots of blood of a purple colour: then did it ascend up to their ankles, knees, thighs, shoulders, arms, and necks. Their mouths became stinking, their gums so rotten, that all the flesh did fall off, even to the roots of the teeth, which did also almost all fall out.

The means of preventing scurvy was succinctly stated by James Lind, an English physician, in 1753:

> Experience indeed sufficiently shows that as greens or fresh vegetables, with ripe fruits, are the best remedies for it, so they prove the most effectual preservatives against it.

Lind urged the inclusion of lemon juice in the diet of sailors. His advice was adopted by the British Navy some forty years later.

Scurvy is caused by a dietary deficiency of ascorbic acid (vitamin C). Primates and guinea pigs have lost the ability to synthesize ascorbic acid and so they must acquire it from their diets. Ascorbic acid, an effective reducing agent (Figure 9-12), maintains prolyl hydroxylase in an active form, probably by keeping its iron atom in the reduced ferrous state. Collagen synthesized in the absence of ascorbic acid is insufficiently hydroxylated and hence has a lower

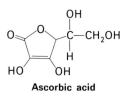

Ascorbic acid

Ascorbate

Dehydroascorbic acid

Figure 9-12
Formulas of ascorbic acid (vitamin C) and ascorbate, its ionized form. The pK_a of the acidic hydroxyl group of ascorbic acid is 4.2. Dehydroascorbic acid is the oxidized form of ascorbate.

melting temperature. This abnormal collagen cannot properly form fibers and thus causes the skin lesions and blood-vessel fragility that are so prominent in scurvy.

PROCOLLAGEN IS THE BIOSYNTHETIC PRECURSOR OF COLLAGEN

The triple-stranded type I collagen helix is rapidly assembled in vivo. In contrast, the process takes days in vitro from a solution of the constituent $\alpha 1(I)$ and $\alpha 2$ chains. Furthermore, the yield is low. Why is there such a disparity between in vitro and in vivo assembly of the tropocollagen helix? The answer is suggested by comparing the refolding of denatured chymotrypsinogen and chymotrypsin. The unfolded zymogen spontaneously refolds into the correct three-dimensional structure, whereas the enzyme does not. The reason is that part of the structure needed to specify the three-dimensional form is missing in chymotrypsin; namely, the two dipeptides that were cleaved in the activation process.

It was surmised that the $\alpha 1(I)$ and $\alpha 2$ chains do not spontaneously form the correct tropocollagen structure because part of the necessary information is missing. This is indeed so. *The constituent chains of collagen are synthesized in the form of larger precursors.* The precursor of the $\alpha 1(I)$ chain, called *pro-$\alpha 1(I)$*, has a mass of 140 kdal, in contrast with 95 kdal for $\alpha 1(I)$. Additional peptides are located at both the amino-terminal and the carboxyl-terminal ends of the pro-$\alpha 1(I)$ chain (Figure 9-13). Similarly, the $\alpha 2$ chain is synthesized

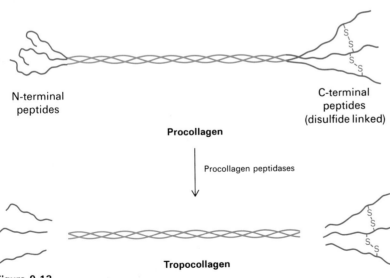

N-terminal
peptides

C-terminal
peptides
(disulfide linked)

Procollagen

Procollagen peptidases

Tropocollagen

Figure 9-13
Schematic diagram of the conversion of procollagen into collagen by the excision of the amino-terminal peptide (about 15 kdal) and the carboxyl-terminal peptide (about 30 kdal) from each of the three chains. The carboxyl-terminal regions of the three strands of procollagen are linked by disulfide bonds.

as a 140-kdal precursor called *pro-α2*. The additional peptide segments in pro-α1(I) and pro-α2 have an amino acid composition that is quite different from that of the rest of the chain. They are not rich in glycine, hydroxyproline, and proline. The amino-terminal peptides of pro-α1(I) and of pro-α2 are distinctive in containing intrachain disulfides. Furthermore, *the carboxyl-terminal regions of these precursor chains are linked by interchain disulfides that are absent from collagen.*

THE ADDITIONAL PEPTIDE REGIONS OF THE PRECURSOR CHAINS ARE ENZYMATICALLY EXCISED

The pro-α1(I) and pro-α2 chains are synthesized by fibroblast cells and secreted into the extracellular spaces of connective tissues. The additional amino-terminal and carboxyl-terminal peptide regions on these precursor chains are then cleaved by specific proteolytic enzymes called *procollagen peptidases.*

Defects in the conversion of precursor chains into α1 and α2 can lead to generalized disorders of connective tissue. For example, patients with the *Ehlers-Danlos syndrome* have stretchable skin, hypermobile joints, and short stature. Patients with one form of this disease have significant amounts of procollagen in extracts of their skin and tendons (Figure 9-14). Furthermore, their fibroblasts have decreased levels of procollagen peptidase. *Dermatosparaxis,* a genetically transmitted (recessive) disease of cattle, is caused by the absence of a procollagen peptidase. The dermis of an afflicted animal is very fragile because it contains disorganized collagen bundles. A significant proportion of its mature type I collagen consists of α1(I) and α2 chains that have retained the amino-terminal regions of procollagen. This disease demonstrates that excision of these amino-terminal peptides is important for the orderly formation of collagen fibers.

THE COLLAGEN FIBER IS A STAGGERED ARRAY OF TROPOCOLLAGEN MOLECULES

Tropocollagen fibers spontaneously associate in a specific way to form collagen fibers in the extracellular space. The overall geometry of the fiber has been deduced from x-ray and electron-microscopic studies. Collagen fibers exhibit cross striations every 680 Å (as shown in Figure 9-1). In contrast, the length of a tropocollagen molecule is 3000 Å. A fiber period several times as small as the length of the constituent molecules indicates that tropocollagen molecules in adjacent rows cannot be in register. Instead, adjacent rows of tropocollagens are displaced by approximately one-fourth of the length of the basic unit. *The fundamental structural design of the*

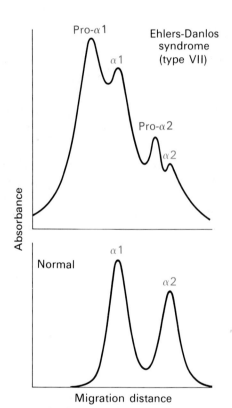

Figure 9-14
Acrylamide gel electrophoresis patterns of skin extracts. Appreciable amounts of procollagen (pro-α1 and pro-α2) are present in acid extracts of skin from patients with type VII Ehlers-Danlos syndrome, whereas none is evident in controls from normal persons. [After J. R. Lichtenstein, G. R. Martin, L. D. Kohn, P. H. Byers, and V. A. McKusick. *Science* 182(1973):299. Copyright 1973 by the American Association for the Advancement of Science.]

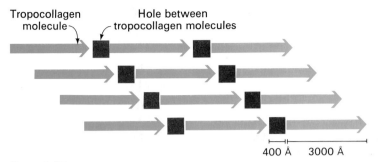

Figure 9-15
A schematic representation of the basic structural design of a collagen fiber. Tropocollagen molecules (shown as blue arrows) form a quarter-staggered array. The gaps between tropocollagen molecules along a row (represented by red squares) may be nucleation sites in bone formation.

Figure 9-16
A passage from the Fugue in D Major, Well-tempered Clavier, by J. S. Bach. [From R. Erickson, *The Structure of Music: A Listener's Guide* (Noonday Press, 1955), p. 130.]

collagen fiber is a quarter-staggered array of tropocollagen molecules (Figure 9-15). This arrangement is reminiscent of a musical fugue (Figure 9-16).

An interesting structural feature is that the tropocollagen molecules along a row are not linked end-to-end. There is a gap of about 400 Å between the end of one tropocollagen molecule and the start of another (Figure 9-15). This gap may play a role in *bone formation*. Bone consists of an organic phase, which is nearly entirely collagen, and an inorganic phase, which is calcium phosphate. Specifically, the structure is like that of hydroxylapatite, which has the composition $Ca_{10}(PO_4)_6(OH)_2$. Collagen is required for the deposition of calcium phosphate crystals to form bone. In fact, the initial crystals are found at intervals of about 680 Å, which is the period of the collagen fibers. It is possible that the gaps between the tropocollagen molecules along a row are the *nucleation sites* for the mineral phase of bone.

PROCOLLAGEN PEPTIDASES CONTROL FIBER FORMATION

Procollagen molecules do not spontaneously associate into fibers as do tropocollagen molecules. Excision of the amino-terminal and carboxyl-terminal peptides is essential for the formation of normal collagen fibers. Thus, *the formation of collagen fibers is analogous to the formation of fibrin fibers.* Procollagen is akin to fibrinogen, tropocollagen to fibrin monomer, and the procollagen peptidases to thrombin. In both systems, specific proteolytic cleavages are required for fiber formation.

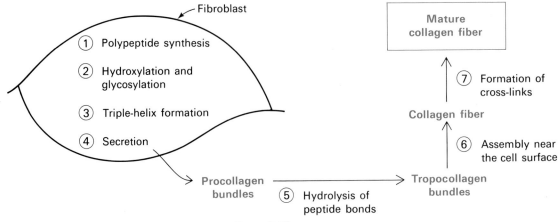

Figure 9-17
Steps in the formation of mature collagen fibers.

Procollagen, rather than tropocollagen, is secreted by fibroblasts (Figure 9-17). The pathway of biosynthesis and secretion is like that of the pancreatic zymogens (p. 158) and other exported proteins (p. 712). Collagen fiber formation takes place in the extracellular fluid near the cell surface rather than inside the fibroblast because of the external location of the procollagen peptidases. The additional peptide regions of the precursor chains prevent the premature formation of fibers. They may also guide the transport of procollagen across the permeability barrier of the fibroblast. An earlier role of the extra peptide regions is to facilitate the alignment of the three chains and promote formation of the triple helix. The interchain disulfide bonds may be particularly important in this regard.

COLLAGEN FIBERS ARE STRENGTHENED BY CROSS-LINKS

Collagen, like fibrin, is stabilized by the formation of covalent cross-links. Two kinds of cross-links are formed in the collagen fiber: intramolecular (within a tropocollagen unit) and intermolecular (between different tropocollagen units). The bonds formed appear to be unique to collagen and elastin, a related protein.

The intramolecular cross-links in collagen are derived from lysine side chains. The first step is the conversion of the ϵ-amino terminus of lysine into an aldehyde, a reaction catalyzed by lysyl oxidase. Two of these aldehydes then undergo an *aldol condensation* (Figure 9-18). In this condensation reaction, an enolate ion derived from one aldehyde adds to the carbonyl group of the other aldehyde. The location of this intramolecular cross-link is interesting. The amino acid sequence of the amino-terminal segment of the $\alpha 1(I)$ chain of collagen shows that every third residue is glycine starting at position

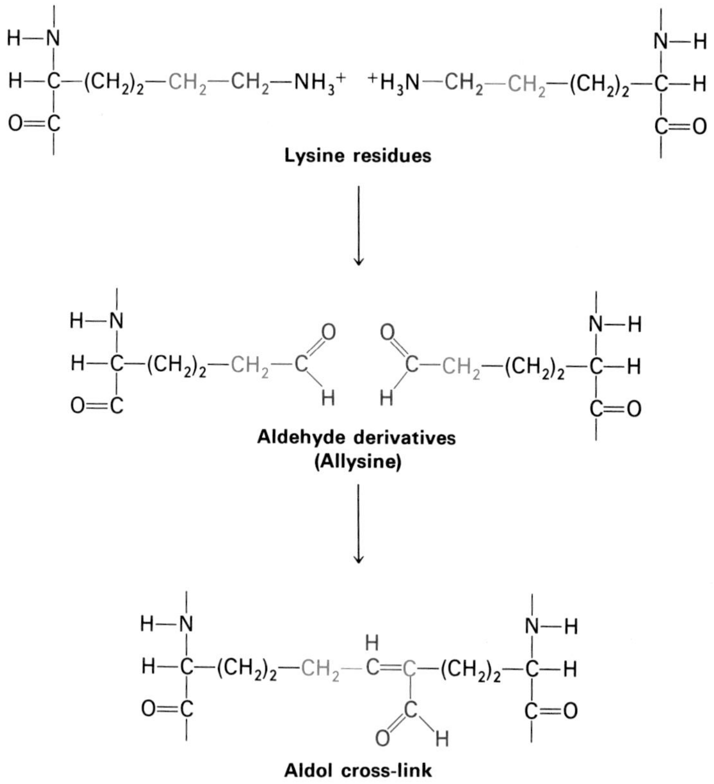

Lysine residues

**Aldehyde derivatives
(Allysine)**

Aldol cross-link

Figure 9-18
Formation of an aldol cross-link from two lysine side chains.

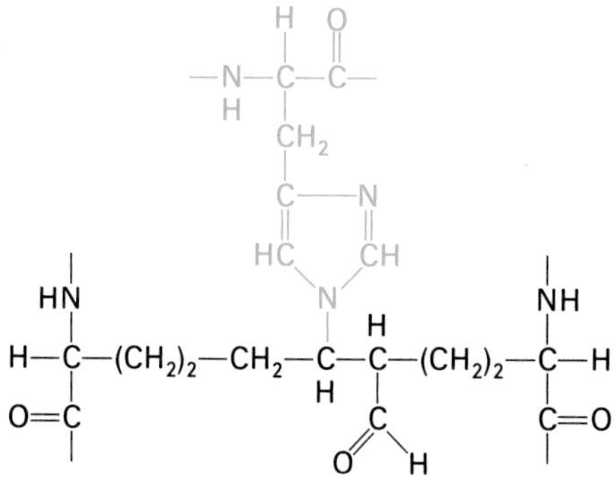

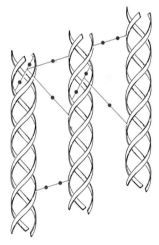

Histidine–aldol cross-link
Figure 9-19
A histidine side chain can insert into the carbon–carbon double bond in an aldol cross-link to yield a histidine–aldol cross-link.

13. Thus, lysine 5 is in a nonhelical region of the polypeptide chain, near the start of the triple-stranded helix. The flexibility of this nonhelical segment may facilitate the formation of this cross-link.

This aldol cross-link can react with a histidine side chain to form a histidine–aldol cross-link (Figure 9-19). The aldehyde group in the aldol–histidine cross-link can form a Schiff base with yet another side chain, such as that of hydroxylysine. In this way, four side chains can be covalently bonded to each other. Some of these cross-links can be between side chains on different tropocollagen molecules (Figure 9-20).

The extent and type of cross-linking varies with the physiological function and age of the tissue. The collagen in the Achilles' tendon of mature rats is highly cross-linked, whereas that of the flexible tail tendon is much less so. The importance of cross-linking in conferring mechanical strength on collagen fibers is evident in *lathyrism*, a disease of animals caused by the ingestion of seeds of *Lathyris odoratus*, the sweet pea. The toxic agent is *β-aminopropionitrile*, which inhibits the transformation of lysyl side chains into aldehydes. Hence, the collagen in these animals is extremely fragile.

Figure 9-20
Intramolecular and intermolecular cross-links in collagen. [After M. L. Tanzer. *Science* 180(1973):562. Copyright 1973 by the American Association for the Advancement of Science.]

$$N{\equiv}C{-}CH_2CH_2{-}NH_3^+$$
β-Aminopropionitrile

COLLAGENASES ARE ENZYMES THAT SPECIFICALLY DEGRADE COLLAGEN

Figure 9-21
Collagen is rapidly removed from the tail fin of a tadpole during metamorphosis. [From C. M. Lapiere and J. Gross. In *Mechanisms of Hard Tissue Destruction*, R. F. Sognnaes, ed., p. 665. Copyright 1963 by the American Association for the Advancement of Science.]

A collagenase is an enzyme that cleaves peptide bonds located in the characteristic helical regions of collagen. Collagen molecules are otherwise very resistant to enzymatic attack. Two types of collagenases are known:

1. One type is formed by certain microorganisms. *Clostridium histolyticum*, a bacterium that causes *gas gangrene*, secretes a collagenase that splits each polypeptide chain of collagen at more than two hundred sites. The bond cleaved by this bacterial enzyme is

$$-X\!\!\downarrow\!\!Gly-Pro-Y-$$

This enzyme contributes to the invasiveness of this highly pathogenic clostridium by *destroying the connective-tissue barriers of the host.* The bacteria are unaffected by the enzyme because they contain no collagen.

2. *Tissue collagenases,* the other type, have been found in amphibian and mammalian tissues undergoing growth or remodeling. This enzyme activity was first demonstrated in the tail fin of tadpoles undergoing metamorphosis (Figure 9-21). Large amounts of collagen are resorbed in that tissue in the course of a few days. Collagenolytic activity was assayed by culturing small pieces of metamorphosing tail fin on a layer of fibrous collagen, which is opaque. The formation of a clear zone within several hours revealed the presence of a high level of collagenase (Figure 9-22). High enzyme activity was then found in mammalian tissue undergoing rapid reorganization, as in the uterus following pregnancy. It is evident that the activity of tissue collagenases must be precisely controlled in timing and magnitude.

The specificity of tadpole collagenase is remarkable. It cleaves tropocollagen across its three chains at a unique site near amino acid residue 750 in the sequence of 1000 residues. The one-quarter and three-quarter length fragments spontaneously unfold at body temperature and are then cleaved by other proteolytic enzymes.

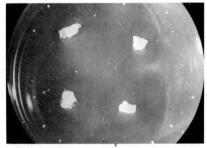

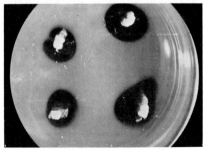

Figure 9-22
Explants from tadpole fin were placed on the surface of a reconstituted collagen gel (top). After 24-hour incubation at 37°C, there was a clear zone around each explant, indicative of collagenase activity (bottom). [From C. M. Lapiere and J. Gross. In *Mechanisms of Hard Tissue Destruction*, R. F. Sognnaes, ed., p. 681. Copyright 1963 by the American Association for the Advancement of Science.]

ELASTIN IS A RUBBERLIKE PROTEIN IN ELASTIC FIBERS

Elastin is found in most connective tissues in conjunction with collagen and polysaccharides. *It is the major component of elastic fibers, which can stretch to several times their length and then rapidly return to their original size and shape when the tension is released.* Large amounts of elastin are found in the walls of blood vessels, particularly in the arch of the aorta near the heart, and in ligaments. The elastic ligaments prominent in the necks of grazing animals are an especially rich source of elastin. There is relatively little elastin in skin, tendon, and loose connective tissue.

The amino acid composition of elastin is highly distinctive. As in collagen, one-third of the residues are glycine. Also, elastin is rich in proline. In contrast with collagen, elastin contains very little hydroxyproline, no hydroxylysine, and few polar amino acids. Elastin is very rich in nonpolar aliphatic residues, such as alanine, valine, leucine, and isoleucine.

Mature elastin contains many cross-links, which render it highly insoluble and therefore difficult to analyze. However, *a soluble precursor of elastin* called *proelastin* has been isolated from *copper-deficient* pigs. Copper deficiency blocks the formation of aldehydes, which are essential for cross-linking, as in collagen. Amino acid sequence studies of proelastin, which has a mass of about 125 kdal, are in progress. Some interesting results have been obtained thus far. There are several regions that are highly rich in alanine and lysine residues. In particular, each of the sequences

-Lys-Ala-Ala-Lys-

and

-Lys-Ala-Ala-Ala-Lys-

occurs several times. Cross-links are formed in these regions. The aldol cross-link found in collagen (Figure 9-18) is also present in elastin. Another type of cross-link derived from lysine residues is *lysinonorleucine* (Figure 9-23), which also occurs in both collagen and elastin. *Desmosine,* found only in elastin, is derived from four lysine

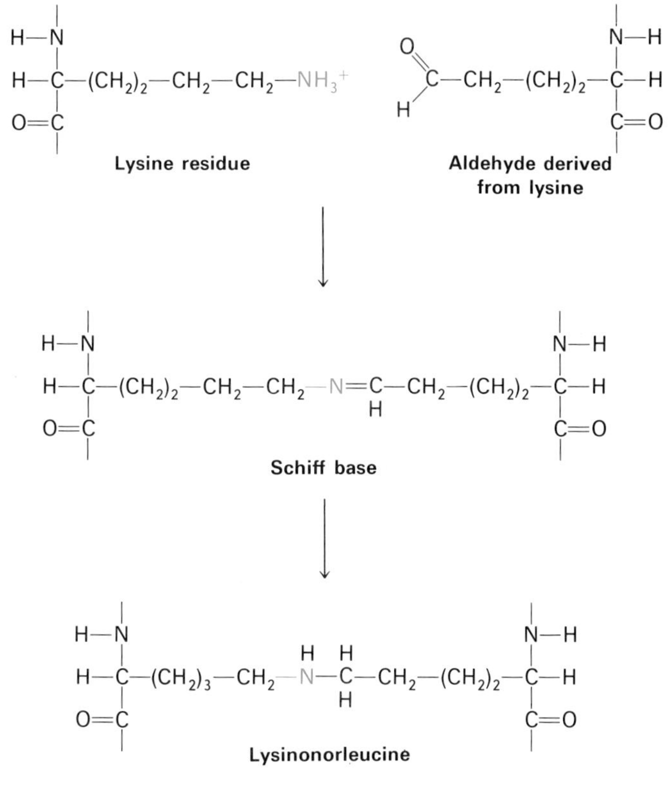

Figure 9-23
Formation of a lysinonorleucine cross-link in collagen or elastin.

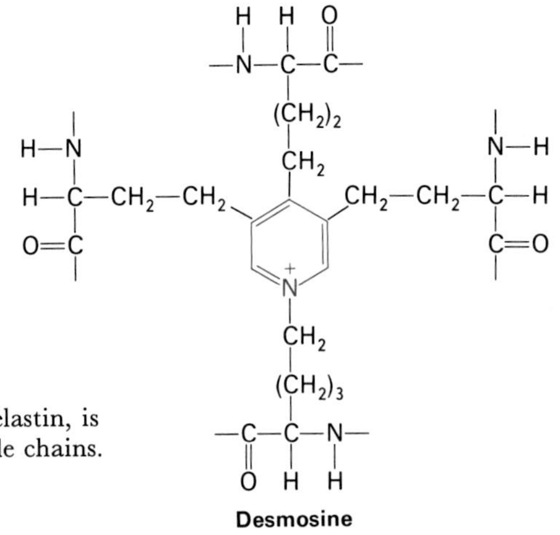

Figure 9-24
Desmosine, a cross-link in elastin, is
derived from four lysine side chains.

-Pro-Gly-Val-Gly-Val-Pro-Gly-Val-Gly-Val-

-Pro-Gly-Val-Gly-Val-Pro-Gly-Val-Ser-Val-

-Pro-Gly-Val-Gly-Val-Pro-Gly-Val-Gly-Val-

Figure 9-25
Part of the amino acid sequence of a
soluble precursor of elastin. The repeating nature of this sequence is evident.

side chains (Figure 9-24). *These cross-links may be important in enabling elastin fibers to return to their original size and shape after stretching.*

Regions of elastin between cross-links are rich in glycine, proline, and valine. The amino acid sequence of some of these regions displays regularities (Figure 9-25).

The elucidation of the conformation of these regular regions and their relationship to the remarkable elasticity of elastin are challenging areas of inquiry. Detailed information concerning the molecular basis of the function of this protein is likely to be important for the elucidation of some cardiovascular diseases in view of the role of elastin in arterial dynamics.

PROTEOGLYCANS FORM THE GROUND SUBSTANCE OF CONNECTIVE TISSUE

Connective tissues are also rich in *proteoglycans,* which consist of polysaccharide (about 95%) and protein (about 5%) units. These very large polyanions bind water and cations and thereby form the ground substance of connective tissue. Proteoglycans are important in determining the viscoelastic properties of joints and of other structures that are subject to mechanical deformation. *Glycosaminoglycans,* the polysaccharide chains of proteoglycans, are made up of *disaccharide repeating units* containing a derivative of an *amino sugar,* either glucosamine or galactosamine. At least one of the sugars in the disaccharide has a *negatively charged* carboxylate or sulfate group. Hyaluronate, chondroitin sulfate, keratan sulfate, heparan sulfate, and heparin are the major glycosaminoglycans (Figure 9-26). Heparan sulfate is like heparan except that it has fewer *N*- and *O*-sulfate groups and more *N*-acetyl groups.

In the proteoglycan from cartilage (Figure 9-27), keratan sulfate and chondroitin sulfate are covalently attached to a polypeptide backbone called the *core protein.* About 140 of these subunits are

Chondroitin 6-sulfate

Keratan sulfate

Heparin

Dermatan sulfate

Hyaluronate

Figure 9-26
Structural formulas of the repeating disaccharide units of some major glycosaminoglycans. The negatively charged groups are shown in red.

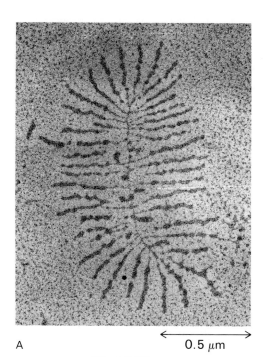

A

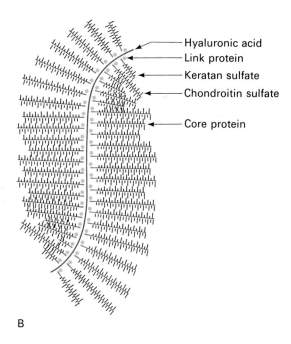

Hyaluronic acid
Link protein
Keratan sulfate
Chondroitin sulfate
Core protein

B

Figure 9-27
A. Electron micrograph of a proteoglycan aggregate from bovine fetal epiphyseal cartilage. Proteoglycan monomers arise laterally at regular intervals from the opposite sides of an elongated central filament of hyaluronate. [Courtesy of Dr. Joseph Buckwalter and Dr. Lawrence Rosenberg.]
B. Schematic diagram. [After L. Rosenberg. In *Dynamics of Connective Tissue Macromolecules,* M. Burleigh and R. Poole, eds. (North-Holland, 1975), p. 105.]

noncovalently bound at intervals of 300 Å to a very long filament of *hyaluronate*. This interaction is promoted by a small *link protein*. The entire complex has a mass of the order of 10^5 kdal and a length of several microns. Proteoglycans from some other tissues have a different structure, suggesting that this class of very large biomolecules has several as yet undiscovered functions.

SUMMARY

Collagen is a family of proteins having a very high tensile strength. It is the major fibrous component of skin, bone, tendon, cartilage, blood vessels, and teeth. Four types of collagens with different tissue distributions have been characterized. The basic structural unit of collagen is tropocollagen, which consists of three strands, each about 1000 residues long. Collagen is unusually rich in glycine and proline. Furthermore, collagen contains hydroxyproline and hydroxylysine, which are present in very few other proteins. The amino acid sequence of collagen is highly distinctive: nearly every third residue is glycine. Tropocollagen is a triple-stranded helical rod, 3000 Å long and 15 Å in diameter. The helical motif of each of its chains is entirely different from that of an α helix. There are hydrogen bonds between NH groups of glycine residues on each strand and CO groups on the other two strands. The stability of the collagen helix also depends on the locking effect of its proline and hydroxyproline residues and on hydrogen bonding by the hydroxyl group of hydroxyproline. The inside of this triple-stranded helix is crowded, and so every third residue in the amino acid sequence must be glycine.

Proteolytic activation plays an important role in the biosynthesis of collagen. The three chains of type I tropocollagen are synthesized as larger precursors called pro-α1(I) and pro-α2. Some proline residues in these procollagen chains are converted into hydroxyproline· by prolyl hydroxylase, an enzyme that requires O_2, Fe^{2+}, and α-ketoglutarate for activity; a reducing agent such as ascorbic acid is also needed. Collagen synthesized in the absence of ascorbic acid is less stable because it is insufficiently hydroxylated, which results in scurvy. Lysine is hydroxylated by a different enzyme. Sugars are then attached to hydroxylysine residues on the precursor chains. These hydroxylation and glycosylation reactions take place inside the fibroblast. Procollagens are secreted into the extracellular medium. Amino-terminal and carboxyl-terminal regions on these precursors are then cleaved by procollagen peptidase to yield tropocollagen, which spontaneously associates into fibers. The basic design of a collagen fiber is a quarter-staggered array of tropocollagen molecules. Finally, collagen fibers are cross-linked to give them additional strength. For example, some lysine side chains in collagen are oxidized to aldehydes, which condense to give aldol cross-links.

Elastin is an insoluble, rubberlike protein in the elastic fibers of connective tissues, which can be reversibly stretched to several times their initial length. Connective tissues such as ligaments and the arch of the aorta contain much elastin. Like collagen, elastin is rich in proline and glycine. Unlike collagen, elastin contains very little hydroxyproline and no hydroxylysine. Its amino acid composition is highly nonpolar. The amino acid sequence of elastin displays certain regularities. For example, the sequence -Pro-Gly-Val-Gly-Val- recurs frequently. Elastin is synthesized as a soluble precursor, which then becomes cross-linked in a variety of ways. Desmosine, one of these cross-links, is derived from four lysine residues. Aldehyde intermediates participate in the formation of cross-links in elastin, as they do in collagen.

Proteoglycans, the other major macromolecular component of connective tissue, consist of polysaccharide and protein units. The polysaccharide chains, called glycosaminoglycans, are made up of disaccharide repeating units that have a high density of negative charge. Proteoglycans form the ground substance and are important in determining the viscoelastic properties of connective tissues.

SELECTED READINGS

WHERE TO START

Eyre, D. R., 1980. Collagen: molecular diversity in the body's protein scaffold. *Science* 207:1315–1322.

Bornstein, P., and Traub, W., 1979. The chemistry and biology of collagen. *In* Neurath, H., and Hill, R. L., (eds.), *The Proteins* (3rd ed.), vol. 4, pp. 411–632. Academic Press.

Gross, J., 1973. Collagen biology: structure, degradation, and disease. *Harvey Lectures* 68:351–432.

Ross, R., and Bornstein, P., 1971. Elastic fibers in the body. *Sci. Amer.* 224(6)44–52. [Available as *Sci. Amer.* Offprint 1225.]

BOOKS

Ramachandran, G. N., and Reddi, A. H., (eds.), 1976. *Biochemistry of Collagen.* Plenum.

Burleigh, P. M. C., and Poole, A. R., (eds.), 1975. *Dynamics of Connective Tissue Macromolecules.* American Elsevier.

BIOSYNTHESIS OF COLLAGEN

Prockop, D. J., Kivirikko, K. I., Tuderman, L., and Guz-man, N. A., 1979. The biosynthesis of collagen and its disorders. *New Engl. J. Med.* 301:13–23 and 77–85.

Fessler, J. H., and Fessler, L. I., 1978. Biosynthesis of collagen. *Ann. Rev. Biochem.* 47:129–162.

Jackson, D. S., 1978. Biosynthesis of collagen. *In* Arnstein, H. R. V., (ed.), *Amino Acid and Protein Biosynthesis II*, International Review of Biochemistry, vol. 18, pp. 233–259. University Park Press.

Fessler, L. I., Morris, N. P., and Fessler, J. H., 1975. Procollagen: biological scission of amino and carboxyl extension peptides. *Proc. Nat. Acad. Sci.* 72:4905–4909.

Bruns, R. R., Hulmes, D. J. S., Therrien, S. F., and Gross, J., 1979. Procollagen segment-long-spacing crystallites: their role in collagen fibrillogenesis. *Proc. Nat. Acad. Sci.* 76:313–317.

Tanzer, M. L., 1973. Cross-linking of collagen. *Science* 180:561–566.

HYDROXYLATION OF COLLAGEN

Cardinale, G. J., and Udenfriend, S., 1974. Prolyl hydroxylase. *Advan. Enzymol.* 41:245–300.

Hayaishi, O., (ed.), 1974. *Molecular Mechanisms of Oxygen Activation.* Academic Press.

204

Berg, R. A., and Prockop, D. J., 1973. The thermal transition of a non-hydroxylated form of collagen: evidence for a role for hydroxyproline in stabilizing the triple-helix of collagen. *Biochem. Biophys. Res. Comm.* 52:115–120.

COLLAGENASES

Gross, J., and Nagai, Y., 1965. Specific degradation of the collagen molecule by tadpole collagenolytic enzyme. *Proc. Nat. Acad. Sci.* 54:1197–1204.

Seifter, S., and Harper, E., 1971. The collagenases. *In* Boyer, P. D., (ed.), *The Enzymes* (3rd ed.), vol. 3, pp. 649–697. Academic Press.

ELASTIN

Sandberg, L. B., 1976. Elastin structure in health and disease. *Int. Rev. Connect. Tissue Res.* 7:159–210.

Urry, D. W., 1974. On the molecular basis for vascular calcification. *Perspect. Biol. Med.* 18:68–84. [Discusses the binding of calcium ion by elastin.]

GLYCOSAMINOGLYCANS

Lindahl, U., and Höök, M., 1978. Glycosaminoglycans and their binding to biological macromolecules. *Ann. Rev. Biochem.* 47:385–417.

CONNECTIVE-TISSUE DISEASES

Pinnell, S. R., 1978. Disorders of collagen. *In* Stanbury, J. B., Wyngaarden, J. B., and Fredrickson, D. S., (eds.), *The Metabolic Basis of Inherited Disease* (4th ed.), pp. 1366–1394. McGraw-Hill.

Trelstad, R. L., Rubin, D., and Gross, J., 1977. Osteogenesis imperfecta congenita: evidence for a generalized molecular disorder of collagen. *Lab. Invest.* 36:501–508.

Krane, S. M., Pinnell, S. R., and Erbe, R. W., 1972. Lysyl-protocollagen hydroxylase deficiency in fibroblasts from siblings with hydroxylysine-deficient collagen. *Proc. Nat. Acad. Sci.* 69:2899–2903.

Lapiere, C. M., Lenaers, A., and Kohn, L. D., 1971. Procollagen peptidase: an enzyme excising the coordination peptides of procollagen. *Proc. Nat. Acad. Sci.* 68:3054–3058. [Demonstration of the molecular defect in dermatosparaxis, an inherited connective-tissue disorder in cattle.]

Major, R. H., (ed.), 1945. *Classic Descriptions of Disease* (3rd ed.). Thomas. [Cartier's description of scurvy is given on page 587 of this volume.]

Nishikimi, M., and Udenfriend, S., 1977. Scurvy as an inborn error of ascorbic acid biosynthesis. *Trends Biochem. Sci.* 2:111–112.

PROBLEMS

1. Poly-L-proline, a synthetic polypeptide, can adopt a helical conformation like that of a single strand in the collagen triple helix.
   (a) Poly-L-proline cannot form a triple helix. Why?
   (b) Poly(Gly-Pro-Pro) can form a triple helix like that of collagen. Predict the thermal stability of the triple helix of poly(Gly-Pro-Gly) relative to that of poly(Gly-Pro-Pro).
   (c) Would you expect poly(Gly-Pro-Gly-Pro) to form a triple helix like that of collagen?

2. Consider the amino acid sequence

   -Gly-Leu-Pro-Gly-Pro-Pro-Gly-Ala-Pro-Gly-

   (a) Which residues are susceptible to hydroxylation at C-4 by prolyl hydroxylase?
   (b) Which peptide bonds are most susceptible to hydrolysis by the collagenase from *Clostridium histolyticum?*

3. Several kinds of covalent cross-links in proteins have been discussed in the preceding chapters. Identify the covalent cross-links, if there are any, in the following proteins.
   (a) Ribonuclease.
   (b) Hemoglobin.
   (c) Fibrin.
   (d) Collagen.
   (e) Elastin.

4. Prolyl hydroxylase can decarboxylate α-ketoglutarate in the absence of a peptidyl prolyl substrate. Ferrous ion, O_2, and ascorbate are needed for this reaction. What does this finding imply about the mechanism of hydroxylation by this enzyme?

CHAPTER **10**

INTRODUCTION TO
BIOLOGICAL MEMBRANES

We now turn to biological membranes, which are organized assemblies consisting mainly of proteins and lipids. The functions carried out by membranes are indispensable for life. Membranes give cells their individuality by separating them from their environment. *Membranes are highly selective permeability barriers* rather than impervious walls because they contain specific molecular *pumps* and *gates*. These transport systems regulate the molecular and ionic composition of the intracellular medium. Eucaryotic cells also contain internal membranes that form the boundaries of organelles such as mitochondria, chloroplasts, and lysosomes. Functional specialization in the course of evolution has been closely linked to the formation of these compartments.

Membranes also control the flow of information between cells and their environment. They contain *specific receptors for external stimuli*. The movement of bacteria toward food, the response of target cells to hormones such as insulin, and the perception of light are examples of processes in which the primary event is the detection of a signal by a specific receptor in a membrane. In turn, *some membranes generate signals,* which can be chemical or electrical. Thus, membranes play a central role in biological communication.

The two most important *energy conversion processes* in biological systems are carried out by membrane systems that contain highly ordered arrays of enzymes and other proteins. *Photosynthesis,* in which light is converted into chemical-bond energy, occurs in the

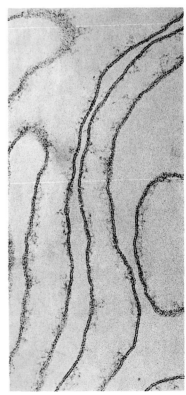

Figure 10-1
Electron micrograph of a preparation of plasma membranes from red blood cells. [Courtesy of Dr. Vincent Marchesi.]

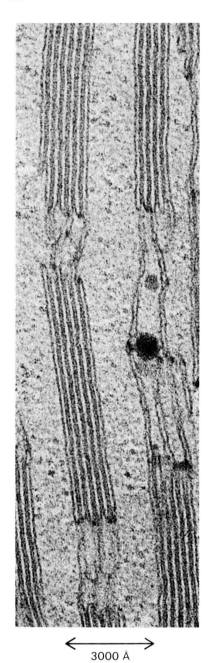

← 3000 Å →

Figure 10-2
Light is converted into chemical-bond energy by photosynthetic assemblies in the thylakoid membranes of chloroplasts. [Courtesy of Dr. Myron Ledbetter. From *Introduction to the Fine Structure of Plant Cells* by M. C. Ledbetter and K. R. Porter (Springer-Verlag, 1970).]

inner membranes of chloroplasts (Figure 10-2), whereas *oxidative phosphorylation*, in which adenosine triphosphate (ATP) is formed by the oxidation of fuel molecules, takes place in the inner membranes of mitochondria. These and other membrane processes will be discussed in detail in later chapters. This chapter deals with some essential features that are common to most biological membranes.

COMMON FEATURES OF BIOLOGICAL MEMBRANES

Membranes are as diverse in structure as they are in function. However, they do have in common a number of important attributes:

1. Membranes are *sheetlike structures,* only a few molecules thick, that form *closed boundaries* between compartments of different composition. The thickness of most membranes is between 60 and 100 Å.

2. Membranes consist mainly of *lipids* and *proteins.* The weight ratio of protein to lipid in most biological membranes ranges from 1:4 to 4:1. Membranes also contain *carbohydrates* that are linked to lipids and proteins.

3. *Membrane lipids are relatively small molecules* that have both a hydrophilic and a hydrophobic moiety. These lipids spontaneously form *closed bimolecular sheets* in aqueous media. These *lipid bilayers* are barriers to the flow of polar molecules.

4. *Specific proteins mediate distinctive functions of membranes.* Proteins serve as pumps, gates, receptors, energy transducers, and enzymes. Membrane proteins are intercalated into lipid bilayers, which create a suitable environment for the action of these proteins.

5. Membranes are *noncovalent assemblies.* The constituent protein and lipid molecules are held together by many noncovalent interactions, which are cooperative in character.

6. Membranes are *asymmetric.* The inside and outside faces of membranes are different.

7. Membranes are *fluid structures.* Lipid molecules diffuse rapidly in the plane of the membrane, as do proteins, unless anchored by specific interactions. Membranes can be regarded as *two-dimensional solutions of oriented proteins and lipids.*

PHOSPHOLIPIDS ARE THE MAJOR CLASS OF MEMBRANE LIPIDS

Lipids are a group of biomolecules that are strikingly different from amino acids and proteins. By definition, lipids are water-insoluble biomolecules that have high solubility in organic solvents such as chloroform. Lipids have a variety of biological roles: they serve as fuel molecules, as highly concentrated energy stores, and as components of membranes. The first two roles of lipids will be discussed in Chapter 18. Here, the concern is with lipids as membrane constituents. The three major kinds of membrane lipids are *phospholipids, glycolipids,* and *cholesterol.*

Let us start with phospholipids, because they are abundant in all biological membranes. Phospholipids are derived from either *glycerol,* a three-carbon alcohol, or *sphingosine,* a more complex alcohol. Phospholipids derived from glycerol are called *phosphoglycerides.* A phosphoglyceride consists of a glycerol backbone, two fatty acid chains, and a phosphorylated alcohol.

The *fatty acid chains* in phospholipids and glycolipids usually contain an even number of carbon atoms, typically between 14 and 24. The 16- and 18-carbon fatty acids are the most common ones. In animals, the hydrocarbon chain in fatty acids is unbranched. Fatty acids may be saturated or unsaturated. The configuration of double bonds in unsaturated fatty acids is nearly always *cis.* As will be discussed shortly, the length and the degree of unsaturation of fatty acid chains in membrane lipids have a profound effect on membrane fluidity. The structures of the ionized form of two common fatty acids—palmitic acid (C_{16}, saturated) and oleic acid (C_{18}, one double bond)—are shown below. They will be referred to as palmitate and oleate to emphasize the fact that they are ionized under physiological conditions. The nomenclature of fatty acids is discussed in Chapter 18.

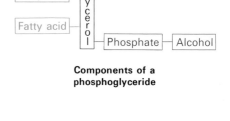

Components of a phosphoglyceride

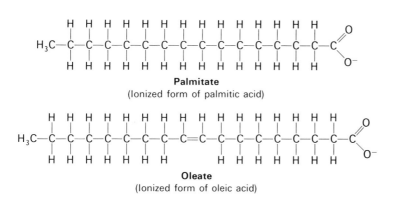

Palmitate
(Ionized form of palmitic acid)

Oleate
(Ionized form of oleic acid)

Figure 10-3
Space-filling models of (A) palmitate (C_{16}, saturated) and (B) oleate (C_{18}, unsaturated). The *cis* double bond in oleate produces a bend in the hydrocarbon chain.

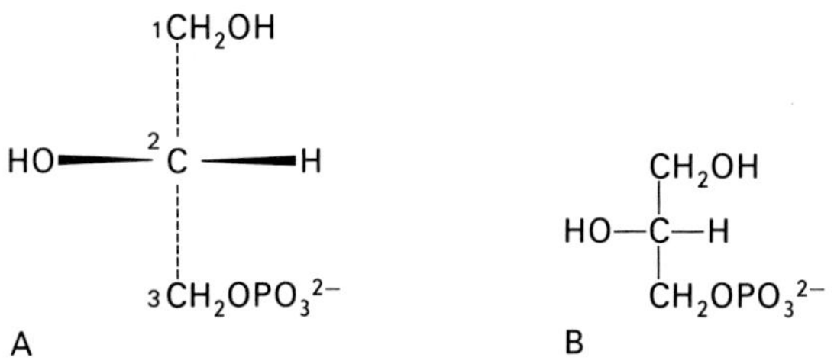

$_1CH_2OH$

$$HO\blacksquare{^2C}\blacksquare H$$

$_3CH_2OPO_3^{2-}$

A

CH_2OH

$HO{-}C{-}H$

$CH_2OPO_3^{2-}$

B

Figure 10-4
Absolute configuration of the glycerol 3-phosphate moiety of membrane lipids:
(A) H and OH, attached to C-2, are in front of the plane of the page,
whereas C-1 and C-3 are behind it; (B) Fischer representation of this structure. In a Fischer projection, horizontal bonds denote bonds in front, whereas
vertical bonds denote bonds behind the plane of the page.

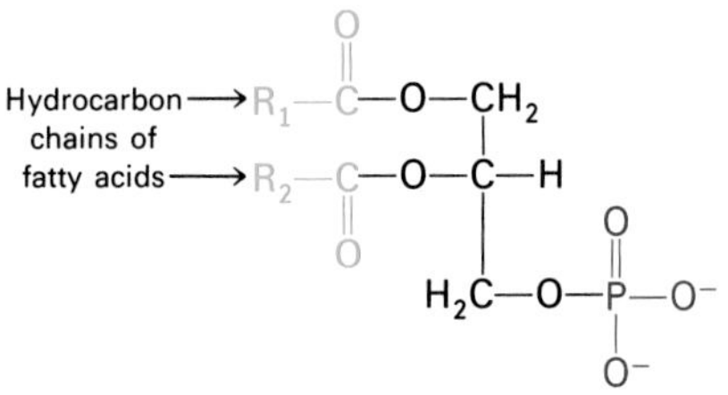

In phosphoglycerides, the hydroxyl groups at C-1 and C-2 of
glycerol are esterified to the carboxyl groups of two fatty acid
chains. The C-3 hydroxyl group of the glycerol backbone is esterified to phosphoric acid. The resulting compound, called *phosphatidate* (or diacylglycerol-3-phosphate), is the simplest phosphoglyceride. Only small amounts of phosphatidate are present in
membranes. However, it is a key intermediate in the biosynthesis of
the other phosphoglycerides.

The major phosphoglycerides are derivatives of phosphatidate.
The phosphate group of phosphatidate becomes esterified to the
hydroxyl group of one of several alcohols. The common alcohol
moieties of phosphoglycerides are serine, ethanolamine, choline,
glycerol, and inositol.

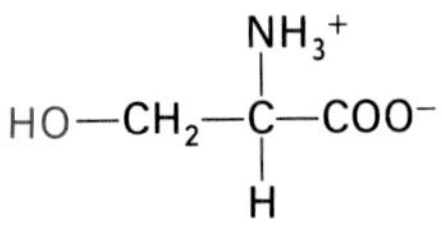

Hydrocarbon chains of fatty acids → R_1, R_2

Phosphatidate
(Diacylglycerol 3-phosphate)

$HO{-}CH_2{-}\overset{NH_3^+}{\underset{H}{C}}{-}COO^-$

Serine

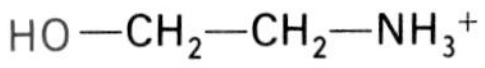

$HO{-}CH_2{-}CH_2{-}NH_3^+$

Ethanolamine

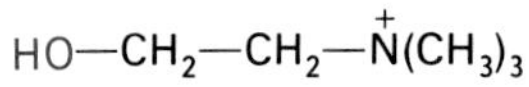

$HO{-}CH_2{-}CH_2{-}\overset{+}{N}(CH_3)_3$

Choline

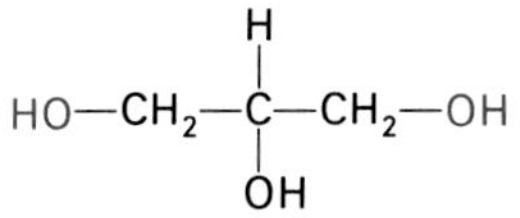

$HO{-}CH_2{-}\overset{H}{\underset{OH}{C}}{-}CH_2{-}OH$

Glycerol

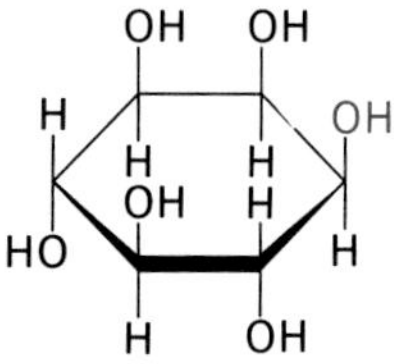

Inositol

Now let us link some of these components to form phosphatidyl
choline, a phosphoglyceride found in most membranes of higher
organisms.

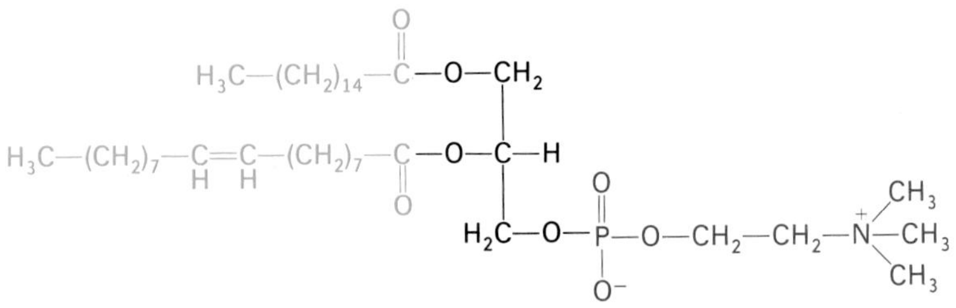

A phosphatidyl choline
(1-Palmitoyl-2-oleoyl-phosphatidyl choline)

The structural formulas of the other principal phosphoglycerides—namely, phosphatidyl ethanolamine, phosphatidyl serine, phosphatidyl inositol, and diphosphatidyl glycerol—are given in Figure 10-5.

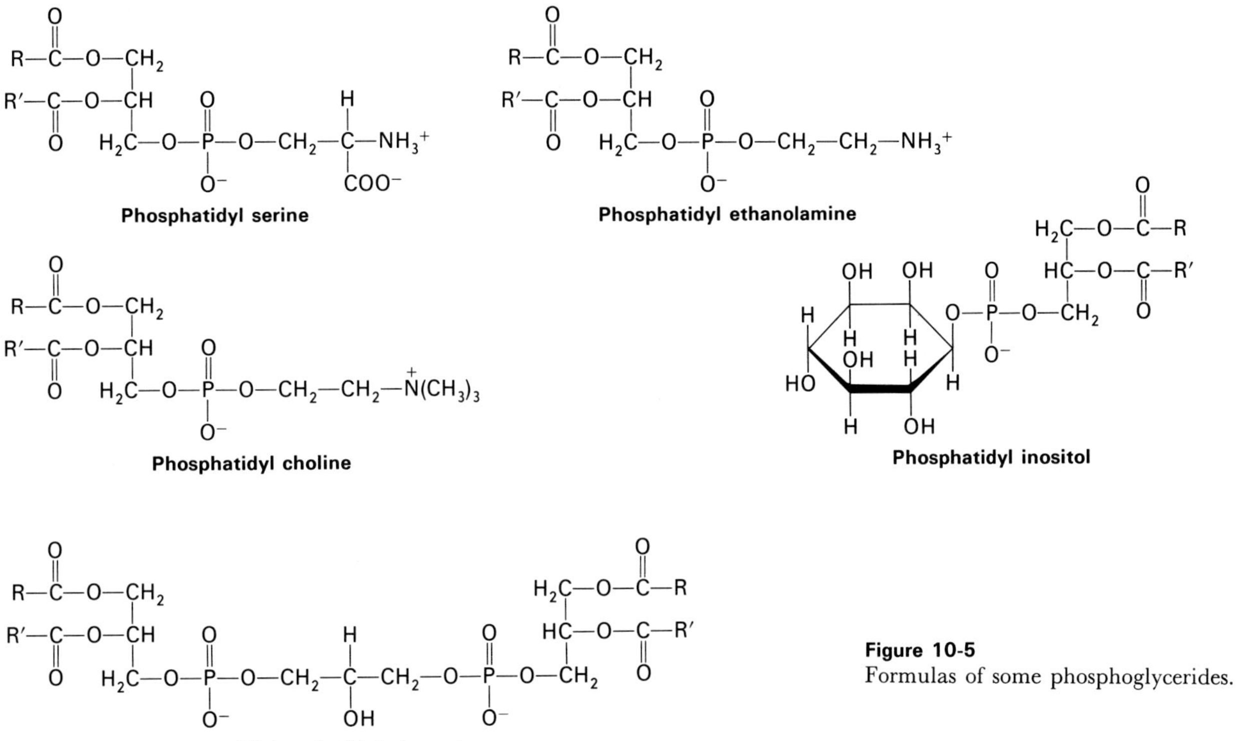

Phosphatidyl serine

Phosphatidyl ethanolamine

Phosphatidyl choline

Phosphatidyl inositol

Diphosphatidyl glycerol

Figure 10-5
Formulas of some phosphoglycerides.

Sphingomyelin is the only phospholipid in membranes that is not derived from glycerol. Instead, the backbone in sphingomyelin is *sphingosine,* an amino alcohol that contains a long, unsaturated hydrocarbon chain. In sphingomyelin, the amino group of the sphingosine backbone is linked to a fatty acid by an amide bond. In addition, the primary hydroxyl group of sphingosine is esterified to phosphoryl choline. As will be shown shortly, the conformation of sphingomyelin resembles that of phosphatidyl choline.

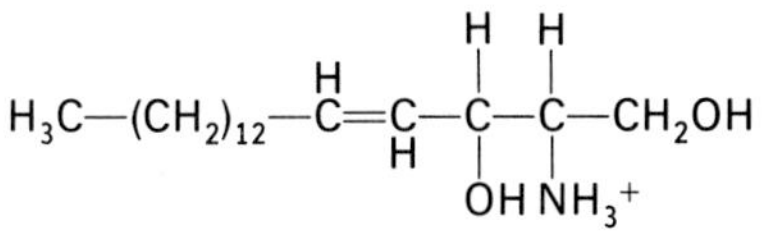

Sphingosine

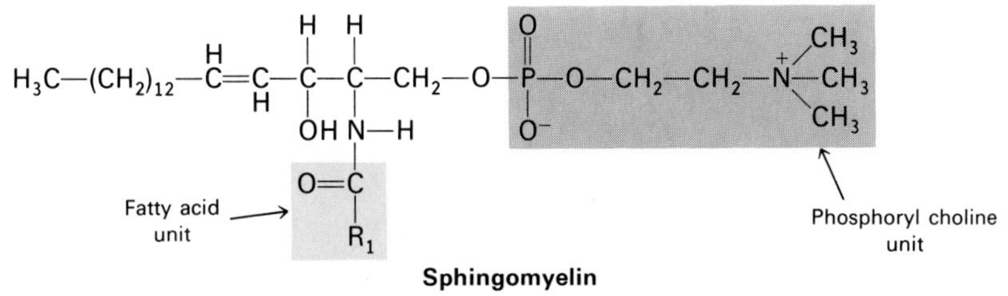

Sphingomyelin

Fatty acid unit

Phosphoryl choline unit

MANY MEMBRANES ALSO CONTAIN GLYCOLIPIDS AND CHOLESTEROL

Glycolipids, as their name implies, are *sugar-containing lipids.* In animal cells, glycolipids, like sphingomyelin, are derived from sphingosine. The amino group of the sphingosine backbone is acylated by a fatty acid, as in sphingomyelin. Glycolipids differ from sphingomyelin in the nature of the unit that is linked to the primary hydroxyl group of the sphingosine backbone. In glycolipids, one or more sugars (rather than phosphoryl choline) are attached to this group. The simplest glycolipid is *cerebroside,* in which there is only one sugar residue, either glucose or galactose. More complex glycolipids, such as *gangliosides,* contain a branched chain of as many as seven sugar residues.

Cerebroside
(A glycolipid)

Another important lipid in some membranes is *cholesterol.* This sterol is present in eucaryotes but not in most procaryotes. The plasma membranes of eucaryotic cells are usually rich in cholesterol, whereas the membranes of their organelles typically have lesser amounts of this neutral lipid.

Cholesterol

PHOSPHOLIPIDS AND GLYCOLIPIDS READILY FORM BILAYERS

The repertoire of membrane lipids is extensive, perhaps even bewildering at first sight. However, they possess a critical common structural theme: *membrane lipids are amphipathic molecules.* They contain both a *hydrophilic* and a *hydrophobic* moiety (Table 10-1).

Table 10-1

211

Chapter 10
INTRODUCTION TO MEMBRANES

Hydrophobic and hydrophilic units of membrane lipids

Membrane lipid	Hydrophobic unit	Hydrophilic unit
Phosphoglycerides	Fatty acid chains	Phosphorylated alcohol
Sphingomyelin	Fatty acid chain and hydrocarbon chain of sphingosine	Phosphoryl choline
Glycolipid	Fatty acid chain and hydrocarbon chain of sphingosine	One or more sugar residues
Cholesterol	Entire molecule except for OH group	OH group at C-3

Let us look at a space-filling model of a phosphoglyceride, such as phosphatidyl choline (Figure 10-6). Its overall shape is roughly rectangular. The two fatty acid chains are approximately parallel to one another, whereas the phosphoryl choline moiety points in the opposite direction. Sphingomyelin has a similar conformation (Figure 10-7). The sugar group of a glycolipid occupies nearly the same

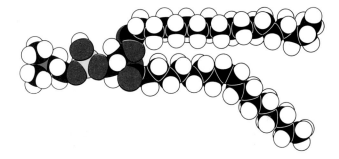

Figure 10-6
Space-filling model
of a phosphatidyl
choline molecule.

Figure 10-7
Space-filling model of a
sphingomyelin molecule.

position as the phosphoryl choline unit of sphingomyelin. Therefore, the following shorthand has been adopted to represent these membrane lipids. The hydrophilic unit, also called the *polar head group*, is represented by a circle, whereas the hydrocarbon tails are depicted by straight or wavy lines (Figure 10-8).

Now let us consider the arrangement of phospholipids and glycolipids in an aqueous medium. It is evident that their polar head groups will have affinity for water, whereas their hydrocarbon tails will avoid water. This can be accomplished by forming a *micelle,* in

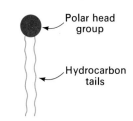

Polar head
group

Hydrocarbon
tails

Figure 10-8
Symbol for a phospholipid or
glycolipid molecule.

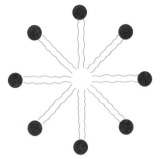

Figure 10-9
Diagram of a section of a micelle formed from phospholipid molecules.

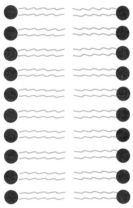

Figure 10-10
Diagram of a section of a bilayer membrane formed from phospholipid molecules.

which the polar head groups are on the surface and the hydrocarbon tails are sequestered inside (Figure 10-9).

Another arrangement that satisfies both the hydrophilic and hydrophobic preferences of these membrane lipids is a *bimolecular sheet,* which is also called a *lipid bilayer* (Figure 10-10). *In fact, the favored structure for most phospholipids and glycolipids in aqueous media is a bimolecular sheet rather than a micelle.* The preference for a bilayer structure is of critical biological importance. A micelle is a limited structure, usually less than 200 Å in diameter. In contrast, a bimolecular sheet can have macroscopic dimensions, such as a millimeter (10^7 Å). Phospholipids and glycolipids are key membrane constituents because they readily form extensive bimolecular sheets. Furthermore, these sheets serve as permeability barriers, yet they are quite fluid.

The formation of lipid bilayers is a *self-assembly process.* In other words, the structure of a bimolecular sheet is inherent in the structure of the constituent lipid molecules, specifically in their amphipathic character. The formation of lipid bilayers from glycolipids and phospholipids is a rapid and spontaneous process in water. *Hydrophobic interactions are the major driving force for the formation of lipid bilayers.* Recall that hydrophobic interactions also play a dominant role in the folding of proteins in aqueous solution. Water molecules are released from the hydrocarbon tails of membrane lipids as these tails become sequestered in the nonpolar interior of the bilayer. This release of water results in a large increase in entropy. Furthermore, there are *van der Waals attractive forces* between the hydrocarbon tails. These van der Waals forces favor close packing of the hydrocarbon tails. Finally, there are favorable *electrostatic* and *hydrogen-bonding* interactions between the polar head groups and water molecules. Thus, lipid bilayers are stabilized by the full array of forces that mediate molecular interactions in biological systems.

LIPID BILAYERS ARE NONCOVALENT, COOPERATIVE STRUCTURES

Another important feature of lipid bilayers is that they are *cooperative structures.* They are held together by many *reinforcing, noncovalent interactions.* Phospholipids and glycolipids cluster together in water to minimize the number of exposed hydrocarbon chains. A pertinent analogy is the clustering together of sheep in the cold to minimize the area of exposed body surface. Clustering is also favored by the van der Waals attractive forces between adjacent hydrocarbon chains. These energetic factors have three significant biological consequences: (1) lipid bilayers have an inherent tendency to be *extensive;* (2) lipid bilayers will tend to *close on themselves* so that there are no ends with exposed hydrocarbon chains, which results in the formation of a compartment; and (3) lipid bilayers are *self-sealing* because a hole in a bilayer is energetically unfavorable.

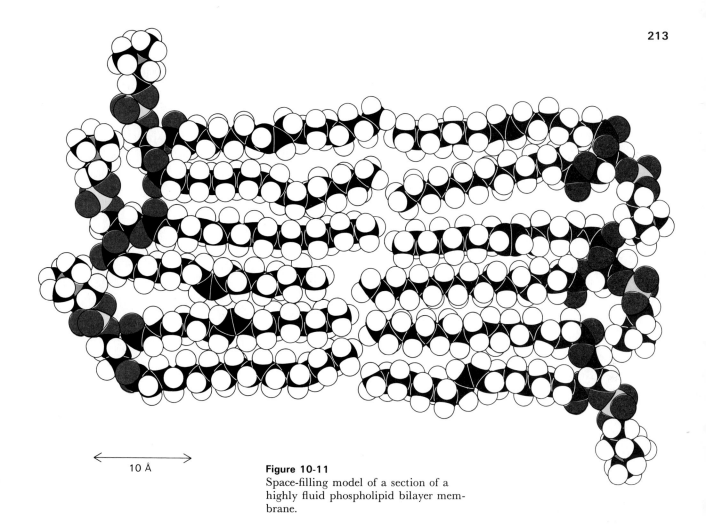

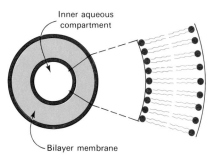

Figure 10-11
Space-filling model of a section of a highly fluid phospholipid bilayer membrane.

10 Å

LIPID BILAYERS ARE HIGHLY IMPERMEABLE TO IONS AND MOST POLAR MOLECULES

The permeabilities of lipid bilayers have been measured in two well-defined *synthetic systems: lipid vesicles* and *planar bilayer membranes*. These model systems have been sources of insight into a major function of biological membranes—namely, their role as permeability barriers. The key finding is that lipid bilayers are inherently impermeable to ions and most polar molecules.

Lipid vesicles (also known as *liposomes*) are aqueous compartments enclosed by a lipid bilayer (Figure 10-12). They can be formed by suspending a suitable lipid, such as phosphatidyl choline, in an aqueous medium. This mixture is then *sonicated* (i.e., agitated by high-frequency sound waves) to give a dispersion of closed vesicles that are quite uniform in size. Alternatively, vesicles can be prepared by rapidly mixing a solution of lipid in ethanol with water.

Inner aqueous compartment

Bilayer membrane

Figure 10-12
Diagram of a lipid vesicle.

This can be accomplished by injecting the lipid through a fine needle. Vesicles formed by these methods are nearly spherical in shape and have a diameter of about 500 Å. Larger vesicles (of the order of 10^4 Å, or 1 μm, in diameter) can be prepared by slowly evaporating the organic solvent from a suspension of phospholipid in a mixed solvent system.

Ions or molecules can be trapped in the aqueous compartment of lipid vesicles by forming them in the presence of these substances (Figure 10-13). For example, if vesicles 500 Å in diameter are

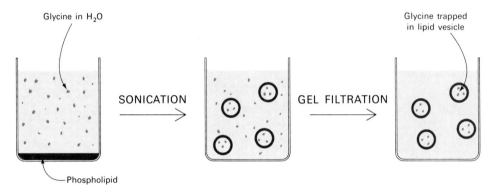

Glycine in H$_2$O

SONICATION

GEL FILTRATION

Glycine trapped
in lipid vesicle

Phospholipid

Figure 10-13
Preparation of a suspension of lipid vesicles containing glycine molecules.

formed in a 0.1 M glycine solution, about two thousand molecules of glycine will be trapped in each inner aqueous compartment. These glycine-containing vesicles can be separated from the surrounding solution of glycine by dialysis or by gel-filtration chromatography. The permeability of the bilayer membrane to glycine can then be determined by measuring the rate of efflux of glycine from the inner compartment of the vesicle to the ambient solution. These lipid vesicles are valuable not only for permeability studies. They fuse with the plasma membrane of many kinds of cells and can thus be used to introduce a wide variety of impermeable substances into cells. The selective fusion of lipid vesicles with particular kinds of cells is a promising means of controlling the delivery of drugs to target cells.

The other well-defined synthetic membrane is a *planar bilayer membrane*. This structure can be formed across a 1-mm hole in a partition between two aqueous compartments. Such a membrane is very well suited for electrical studies because of its large size and simple geometry. Paul Mueller and Donald Rudin showed that a large bilayer membrane can be readily formed in the following way. A fine paint brush is dipped into a membrane-forming solution, such as phosphatidyl choline in decane. The tip of the brush is then stroked across a hole (1 mm in diameter) in a partition between two aqueous media. The lipid film across the hole thins spontaneously; the excess lipid forms a torus at the edge of the hole. A planar bilayer membrane consisting primarily of phosphatidyl choline is formed within a few minutes. The electrical conduction properties of this macroscopic bilayer membrane can be readily studied by inserting electrodes into each aqueous compartment (Figure 10-14). For example, its permeability to ions is determined by measuring the current across the membrane as a function of the applied voltage.

Permeability studies of lipid vesicles and electrical conductance measurements of planar bilayers have shown that *lipid bilayer membranes have a very low permeability for ions and most polar molecules.* Water is a conspicuous exception to this generalization; it readily traverses such membranes. The range of measured permeability coefficients is very wide (Figure 10-15). For example, Na^+ and K^+ traverse these membranes 10^9 times more slowly than does H_2O. Tryptophan, a zwitterion at pH 7, crosses the membrane 10^3 times more slowly than indole, a structurally related molecule that lacks ionic groups. *The permeability coefficients of small molecules are correlated with their solubility in a nonpolar solvent relative to their solubility in water.* This rela-

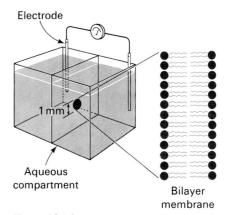

Figure 10-14
Experimental arrangement for the study of planar bilayer membranes. A bilayer membrane is formed across a 1-mm hole in a septum that separates two aqueous compartments.

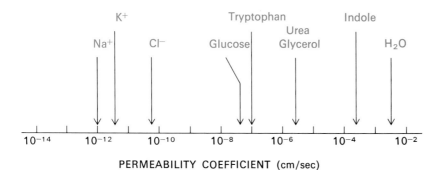

Increasing permeability

Figure 10-15
Permeability coefficients of some ions and molecules in lipid bilayer membranes.

tionship suggests that a small molecule might traverse a lipid bilayer membrane in the following way: first, it sheds its solvation shell of water; then, it becomes dissolved in the hydrocarbon core of the membrane; finally, it diffuses through this core to the other side of the membrane, where it becomes resolvated by water.

PROTEINS CARRY OUT MOST MEMBRANE PROCESSES

We now turn to membrane proteins, which are responsible for most of the dynamic processes carried out by membranes. Membrane lipids form a permeability barrier and thereby establish compartments, whereas *specific proteins mediate distinctive membrane functions,* such as transport, communication, and energy transduction. Membrane lipids create the appropriate environment for the action of such proteins.

Membranes differ in their protein content. Myelin, a membrane that serves as an insulator around certain nerve fibers, has a low

$$H_3C\text{---}(CH_2)_{10}\text{---}CH_2OSO_3^-\ Na^+$$

**Sodium dodecyl sulfate
(SDS)**

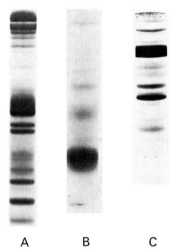

| A | B | C |

Figure 10-16
SDS–polyacrylamide-gel patterns of
(A) the plasma membrane of erythrocytes, (B) the disc membranes of retinal rod cells, and (C) the sarcoplasmic
reticulum membrane of muscle cells.
[Courtesy of Dr. Theodore Steck (part
A) and Dr. David MacLennan
(part C).]

content of protein (18%). Lipid, the major molecular species in myelin, is well suited for insulation. In contrast, the plasma membranes of most other cells are much more active. They contain many pumps, gates, receptors, and enzymes. The protein content of these plasma membranes is typically 50%. Membranes involved in energy transduction, such as the internal membranes of mitochondria and chloroplasts, have the highest content of protein, typically 75%.

The major proteins in a membrane can be readily visualized by *SDS–polyacrylamide-gel electrophoresis*. In this technique, the membrane to be analyzed is first solubilized in a 1% solution of sodium dodecyl sulfate (SDS). This detergent disrupts most protein-protein and protein-lipid interactions. The solution is layered on top of an acrylamide gel containing SDS, which is then subjected to an electric field for a few hours. The electrophoretic mobility of many proteins in this gel depends on their mass rather than on their net charge in the absence of SDS. The negative charge contributed by SDS molecules bound to the protein is much larger than the net charge of the protein itself. A pattern of bands appears when the gel is stained with a dye such as coomassie blue. A few micrograms of a protein can be visualized in this way. The gel electrophoresis patterns of three membranes—the plasma membrane of erythrocytes, the photoreceptor membrane of retinal rod cells, and the sarcoplasmic reticulum membrane of muscle—are shown in Figure 10-16. The gel patterns reveal that these three membranes have very different protein compositions. Furthermore, they differ in the number of kinds of polypeptide chains and in distribution of mass. In short, *membranes that perform different functions have different proteins.*

FUNCTIONAL MEMBRANE SYSTEMS CAN BE RECONSTITUTED FROM PURIFIED COMPONENTS

Numerous membrane proteins have been solubilized and purified (see Chapters 34 to 37). Some of them are active in detergent solution. For example, rhodopsin, a photoreceptor protein, has the same 500-nm absorption band in detergent solution as it does in the retinal disc membrane. Furthermore, the prosthetic group of this protein undergoes the same structural change on illumination in both environments. Calsequestrin, a calcium-binding protein from the sarcoplasmic reticulum of muscle, retains this ion-binding property when it is removed from its membrane environment. The calcium pump (called the Ca^{2+} ATPase) from this membrane system has also been isolated. In fact, functionally active vesicles can be formed from a mixture of phospholipid and the pump protein. These vesicles accumulate Ca^{2+} if ATP, the energy source for the pump, is provided. *The reconstitution of functionally active membrane systems from purified components is a powerful experimental approach in the elucidation of membrane processes.*

SOME MEMBRANE PROTEINS ARE DEEPLY IMBEDDED IN THE LIPID BILAYER

Some membrane proteins can be released by relatively mild means, such as extraction by a high ionic strength solution (e.g., 1 M NaCl). Other membrane proteins are bound much more tenaciously and can be separated only by using a detergent (Figure 10-17) or an

$$H_3C—(CH_2)_{10}—CH_2—\overset{\overset{\displaystyle CH_3}{|}}{\underset{\underset{\displaystyle CH_3}{|}}{N^+}}—O^-$$

**Dodecyldimethylamine oxide
(DDAO, Ammonyx LO)**

$$H_3C—(CH_2)_{10}—CH_2—\overset{\overset{\displaystyle CH_3}{|}}{\underset{\underset{\displaystyle CH_3}{|}}{N^+}}—CH_3 \quad Br^-$$

**Dodecyltrimethylammonium bromide
(DTAB)**

$$H_3C—\overset{\overset{\displaystyle H_3C}{|}}{\underset{\underset{\displaystyle H_3C}{|}}{C}}—CH_2—\overset{\overset{\displaystyle CH_3}{|}}{\underset{\underset{\displaystyle CH_3}{|}}{C}}—\langle \rangle—O—(CH_2—CH_2—O)_n—H$$

Polyoxyethylene p-t-octyl phenol
(Triton X series)

Sodium cholate

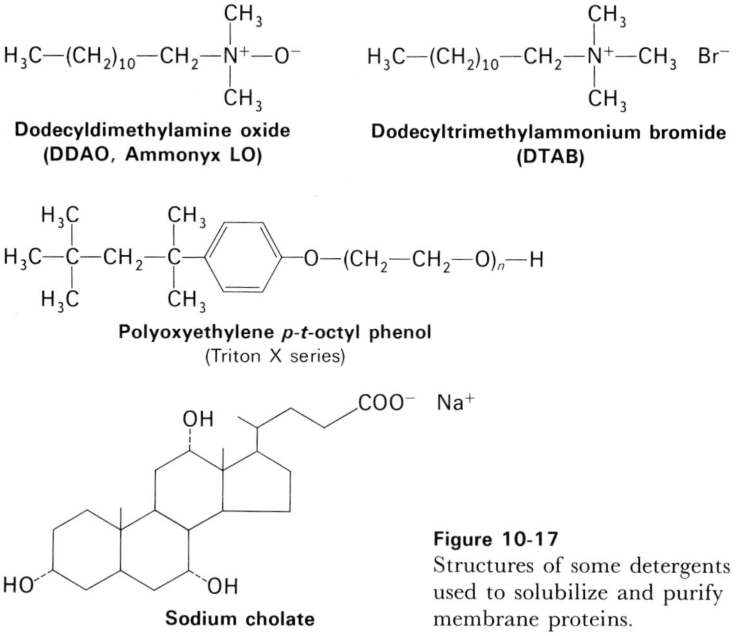

Figure 10-17
Structures of some detergents used to solubilize and purify membrane proteins.

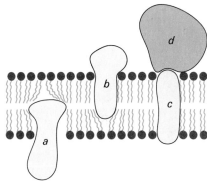

Figure 10-18
Integral membrane proteins (*a, b, c*) interact extensively with the hydrocarbon region of the bilayer. Peripheral membrane proteins (*d*) bind to the surface of integral membrane proteins.

organic solvent. Membrane proteins can be classified as being either *peripheral* or *integral* on the basis of this difference in dissociability (Figure 10-18). Integral proteins interact extensively with the hydrocarbon chains of membrane lipids and so they can be released only by agents that compete for these nonpolar interactions. In contrast, peripheral proteins are bound to membranes by electrostatic and hydrogen-bond interactions and so these polar interactions can be disrupted by adding salts or by changing the pH. It now appears that most peripheral membrane proteins are bound to the surfaces of integral proteins.

Freeze-fracture electron microscopy is a valuable technique for ascertaining whether proteins are located in the interior of biological membranes. Cells or membrane fragments are rapidly frozen at liquid-nitrogen temperatures. The frozen membrane is then fractured by the impact of a microtome knife. Cleavage usually occurs along a plane in the middle of the bilayer (Figure 10-19). Hence, extensive regions *within* the lipid bilayer are exposed. These exposed regions can then be shadowed with carbon and platinum, which yields a replica of the interior of the bilayer. The external surfaces of membranes can also be viewed by combining freeze-fracture and

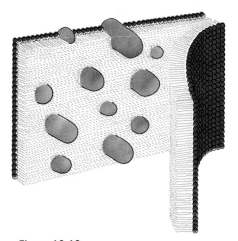

Figure 10-19
Technique of freeze-fracture electron microscopy. The cleavage plane passes through the middle of the bilayer membrane. [After S. J. Singer. *Hosp. Pract.* 8(1973):81.]

deep-etching techniques. First, the interior of a frozen membrane is exposed by fracturing. The ice that covers one of the adjacent membrane surfaces is then sublimed away; this process is termed deep-etching. The combined technique, called *freeze-etching electron microscopy*, provides a view of the interior of a membrane and of both its surfaces. An attractive feature of the freeze-etching technique is that fixatives and dehydrating agents are not required.

Freeze-etching studies have provided direct evidence for the presence of integral proteins in many biological membranes. Erythrocyte membrane, for example, contains a high density of globular particles, approximately 75 Å in diameter, in its interior (Figure 10-20). The inside of the sarcoplasmic reticulum membrane is also

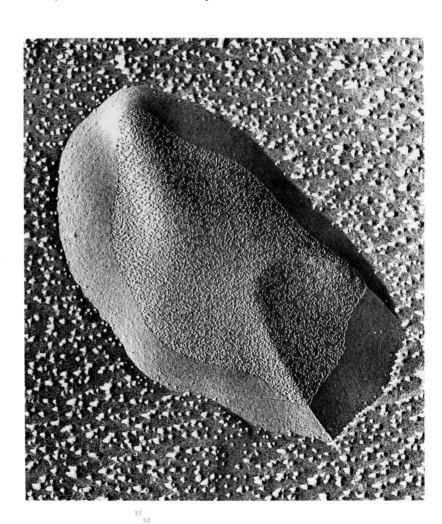

Figure 10-20
Freeze-etch electron micrograph of the plasma membrane of red blood cells. The interior of the membrane, which has been exposed by fracture of the membrane, is rich in globular particles that have a diameter of about 75 Å. These particles are integral membrane proteins. [Courtesy of Dr. Vincent Marchesi.]

rich in globular particles. In contrast, synthetic bilayers formed from phosphatidyl choline yield smooth fracture faces. Also, the fracture faces of large areas of myelin membranes are smooth, as might be expected for a relatively inert membrane that serves primarily as an insulator.

Erythrocytes have been choice objects of inquiry in studies of membranes because of their ready availability and relative simplicity. They lack organelles and thus have only a single membrane, the plasma membrane. Nearly all of the cytoplasmic contents of these cells can be released by osmotic hemolysis to give *ghosts,* which are quite pure plasma membranes. The SDS–polyacrylamide-gel pattern of this membrane preparation is shown in Figure 10-21. More than ten bands are evident when this gel is stained with coomassie blue. The major ones are numbered and referred to as bands 1, 2, 3, 4.1, 4.2, 5, 6, and 7 (Figure 10-21). Staining with the periodic-acid–Schiff (PAS) reagent brings out several proteins that are rich in carbohydrate. These bands are designated as PAS-1, PAS-2, PAS-3, and PAS-4.

What are the locations of these proteins in the red-cell membrane? The proteins corresponding to bands 1, 2, 4.1, 4.2, 5, and 6 can be extracted from the membranes by altering the ionic strength or pH of the medium, and so they are peripheral proteins. Furthermore, they are unaffected when intact red cells or sealed ghosts are incubated with a variety of proteolytic enzymes. In contrast, they are extensively digested when leaky ghosts are treated with proteases. *Hence, it can be concluded that these peripheral proteins are located on the cytoplasmic face of the red-cell membrane.* Band 6 is glyceraldehyde 3-phosphate dehydrogenase, a glycolytic enzyme (p. 263), whereas band 5 is actin, a key protein in muscle contraction and cell motility (p. 816). Bands 1 and 2, called *spectrin,* associate to form an extended filamentous network. Spectrin in concert with other proteins probably *stabilizes and regulates the shape of the red-cell membrane,* which changes markedly in going through small blood vessels (Figure 10-22). Furthermore, red cells are subject to great mechanical stress when pumped by the heart.

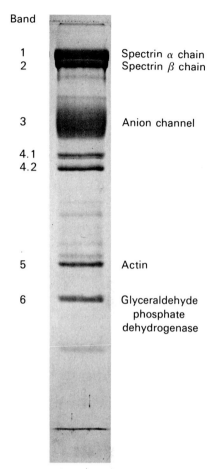

Band

1 — Spectrin α chain
2 — Spectrin β chain

3 — Anion channel

4.1
4.2

5 — Actin

6 — Glyceraldehyde phosphate dehydrogenase

Figure 10-21
SDS-polyacrylamide-gel pattern of the erythrocyte membranes. The gel was stained with coomassie blue. [Courtesy of Dr. Vincent Marchesi.]

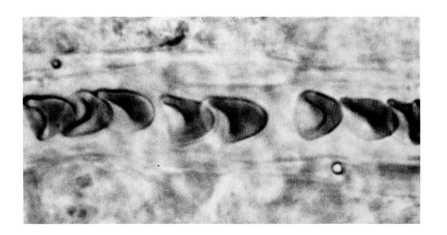

Figure 10-22
Erythrocytes are reversibly deformed as they flow through a small blood vessel. [From P. I. Brånemark. *Intravascular Anatomy of Blood Cells in Man* (Basel: S. Karger AG), 1971.]

In contrast, bands 3 and 7 and all four PAS bands can be dissociated from the red-cell membrane only by detergents or organic solvents. Hence, they are integral membrane proteins. This inference is reinforced by freeze-fracture electron microscopy (Figure 10-20), which shows that some red-cell membrane proteins are deeply embedded in the hydrocarbon region of the membrane.

THE ANION CHANNEL AND GLYCOPHORIN SPAN THE RED-CELL MEMBRANE

The red-cell membrane contains an *anion channel* that makes it permeable to HCO_3^- and Cl^-. The rapid exchange of these ions across the membrane is essential for the transport of CO_2 by red cells. Recent studies have shown that the anion channel is a dimer of the 95-kdal *band-3 protein*, which comprises a quarter of the total protein in a red-cell membrane. The location and orientation of the band-3 protein have been investigated by subjecting intact red cells, leaky ghosts, and inside-out vesicles to proteolysis by chymotrypsin. The cleavage pattern depends on whether chymotrypsin has access to the outer surface, the inner surface, or both surfaces of these membranes. These experiments revealed that *the band-3 protein is located on both sides of the red-cell membrane and that all molecules of this protein point in the same direction* (Figure 10-23). Moreover, its carbohydrate units

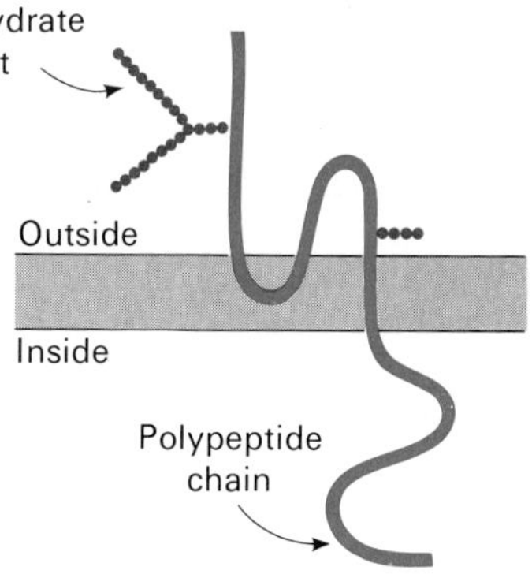

Figure 10-23
Schematic diagram of a model of the band-3 protein (anion channel) in red-cell membranes. [After R. S. Weinstein, J. K. Khodadad, and T. L. Steck. *J. Supramol. Struct.* 8(1978):325.]

are located on the outer surface. The transmembrane nature of the band-3 protein fits nicely with its function as a channel.

The other well-characterized transmembrane protein in red cells is *glycophorin A,* which consists of sixteen oligosaccharide units attached to a single polypeptide chain. Indeed, 60% of the mass of this glycoprotein is carbohydrate, and so its name (derived from the Greek words meaning "to carry sugar") is apt. The abundance of carbohydrate in glycophorin is the reason that it stains intensely with the PAS reagent. In fact, glycophorin corresponds to the *PAS-1 band*. Proteolytic, chemical-modification, and electron-microscopic studies have revealed that glycophorin A consists of three domains: (1) an amino-terminal region containing all of the carbohydrate units, which is located on the outer face of the mem-

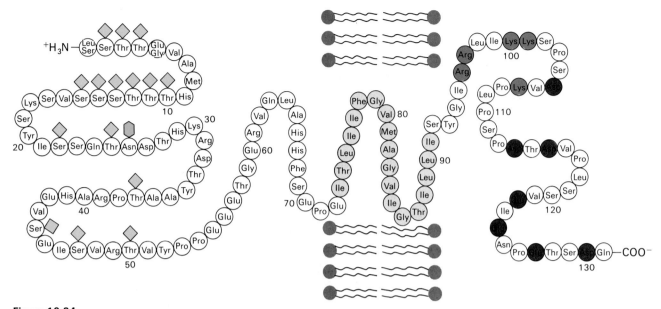

Figure 10-24
Amino acid sequence and transmembrane disposition of glycophorin A from the red-cell membrane. The fifteen carbohydrate units that are attached to serine or threonine residues are shown in light green and the one attached to an asparagine side chain in dark green. The hydrophobic residues buried in the bilayer are shown in yellow. The carboxyl-terminal part of the molecule, located inside the cell, is rich in negatively charged (red) and positively charged (blue) residues. The carbohydrate units in the amino-terminal portion of the molecule outside the cell are rich in negatively charged sialic acid groups. [Courtesy of Dr. Vincent Marchesi.]

brane; (2) a hydrophobic middle region, which is buried within the hydrocarbon core of the membrane; and (3) a carboxyl-terminal region rich in polar and ionized side chains, which is exposed on the inner face of the red-cell membrane (Figure 10-24). Although much is known about the structure of glycophorin, its function is still an enigma.

CARBOHYDRATE UNITS ARE LOCATED ON THE EXTRACELLULAR SIDE OF PLASMA MEMBRANES

Membranes of eucaryotic cells are usually between 2% and 10% carbohydrate, in the form of *glycolipids* and *glycoproteins*. As mentioned earlier (p. 210), the glycolipids of higher organisms are derivatives of sphingosine with one or more sugar residues. In membrane glycoproteins, one or more chains of sugar residues are attached to serine, threonine, or asparagine side chains of the protein, usually through *N*-acetylglucosamine or *N*-acetylgalactosamine. The location of these carbohydrate groups in membranes can be determined by specific labeling techniques. *Lectins,* which are plant proteins with high affinity for specific sugar residues, are valuable probes in this regard. For example, *concanavalin A* binds to internal and nonreducing terminal α-mannosyl residues, whereas *wheat-germ agglutinin*

N-Acetylglucosamine
linked to an
asparagine residue

N-Acetylgalactosamine
linked to a
serine residue

binds to terminal *N*-acetylglucosamine residues. These lectins can be readily seen in electron micrographs if they are conjugated to ferritin, a protein with a very electron dense core of iron hydroxide. The ferritin conjugate of concanavalin A binds specifically to the outer surface of the erythrocyte membrane and not to the inner cytoplasmic surface. The same asymmetry has been observed in the binding of a wide variety of lectins to many kinds of cell membranes. Thus, the localization of the carbohydrate units of the anion channel and glycophorin (Figures 10-23 and 10-24) exemplifies a fundamental theme of membrane structure: *sugar residues are always located on the external surface of plasma membranes. Indeed, the sugar residues in the plasma membranes of all mammalian cells studied thus far are located exclusively on the external surface* (Figure 10-25).

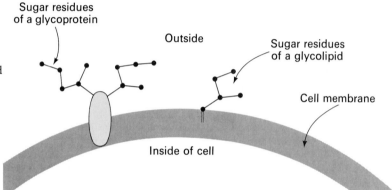

Figure 10-25
Sugar residues of glycoproteins and glycoplipids are usually located on the outside surface of mammalian plasma membranes.

One possible role of carbohydrate groups is to orient glycoproteins in membranes. Because sugars are highly hydrophilic, the sugar residues of a glycoprotein or of a glycolipid will tend to be located at a membrane surface, rather than in the hydrocarbon core. The cost in free energy of inserting an oligosaccharide chain into the hydrocarbon core of a membrane is very high. Consequently, there is a high barrier to the rotation of a glycoprotein from one side of a membrane to the other. The carbohydrate moieties of membrane glycoproteins help to maintain the asymmetric character of biological membranes.

Carbohydrates on cell surfaces may also be important in *intercellular recognition*. The interaction of different cells to form a tissue and the detection of foreign cells by the immune system of a higher organism are examples of processes that depend on the recognition of one cell surface by another. Carbohydrates have the potential for great structural diversity. An enormous number of patterns of surface sugars is possible because (1) monosaccharides can be joined to each other through any of several hydroxyl groups, (2) the C-1 linkage can have either an α or a β configuration, and (3) extensive branching is possible. Indeed, many more different oligosaccharides can be formed from four sugars than oligopeptides from four amino acids.

Biological membranes are not rigid structures. On the contrary, lipids and many membrane proteins are constantly in lateral motion. The rapid movement of membrane proteins has been visualized by means of fluorescence microscopy. Human cells and mouse cells in culture can be induced to fuse with each other. The resulting hybrid cell is called a *heterokaryon*. Part of the plasma membrane of this heterokaryon comes from a mouse cell, the rest from a human cell. Do the membrane proteins derived from the mouse and human cells stay segregated in the heterokaryon or do they intermingle? This question was answered by using fluorescent-labeled antibodies as markers that could be followed by light microscopy. An antibody specific for mouse membrane proteins was labeled to show a green fluorescence, and an antibody specific for human membrane proteins was labeled to show a red fluorescence (Figure 10-26). In a newly formed heterokaryon, half of the surface displayed green fluorescing patches, the other half red. However, in less than an hour (at 37°C), the red and green fluorescing patches became completely intermixed. This experiment revealed that a *membrane protein can diffuse through a distance of several microns in approximately one minute.*

A more general and quantitative method for measuring the lateral mobility of membrane molecules in intact cells is the *fluorescence photobleaching recovery technique* (Figure 10-27). First, a cell-surface component is specifically labeled with a fluorescent chromophore. A small region of the cell surface (∼3 μm²) is viewed through a fluorescent microscope. The fluorescent molecules in this region are then destroyed by a very intense light pulse from a laser. The fluorescence of this region is subsequently monitored as a function of time using a light level sufficiently low to prevent further bleaching. If the labeled component is mobile, bleached molecules leave and unbleached molecules enter the illuminated region, which results

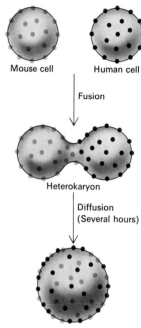

Figure 10-26
Diagram showing the fusion of a mouse cell and a human cell, followed by diffusion of membrane components in the plane of the plasma membrane. The green and red fluorescing markers are completely intermingled after several hours.

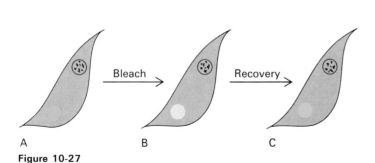 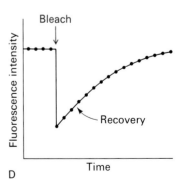

Figure 10-27
Fluorescence photobleaching recovery technique: (A) fluorescence from a labeled cell-surface component in a small illuminated region of a cell; (B) the fluorescent molecules are bleached by an intense light pulse; (C) the fluorescence intensity recovers as bleached molecules diffuse out and unbleached molecules diffuse into the illuminated region; (D) the rate of recovery depends on the diffusion coefficient.

in an increase in the fluorescence intensity. The rate of recovery of the fluorescence level depends on the lateral mobility of the fluorescent-labeled component, which can be expressed in terms of a diffusion coefficient D. The average distance s (in cm) traversed in two dimensions in time t (sec) depends on D (cm^2 sec^{-1}) according to

$$s = (4\,Dt)^{1/2}$$

The diffusion coefficient of lipids in a variety of membranes is about 10^{-8} cm^2 sec^{-1}. Thus, a phospholipid molecule diffuses an average distance of 2×10^{-4} cm, or $2\,\mu m$, in 1 second. This means that a *lipid molecule can travel from one end of a bacterium to the other in a second.* The magnitude of the observed diffusion coefficient indicates that the viscosity of the membrane is about one hundred times that of water, rather like that of olive oil.

In contrast, proteins vary markedly in their lateral mobility. *Some proteins are nearly as mobile as lipids, whereas others are virtually immobile.* For example, the photoreceptor protein rhodopsin, a highly mobile protein, has a diffusion coefficient of 4×10^{-9} cm^2 sec^{-1}. At the other extreme is *fibronectin,* a peripheral glycoprotein that participates in cell-substratum interactions, for which D is less than 10^{-12} cm^2 sec^{-1}. Some of the less-mobile proteins may be anchored to submembranous cytoplasmic filaments.

MEMBRANE PROTEINS DO NOT ROTATE ACROSS BILAYERS

The spontaneous rotation of lipids from one side of a membrane to the other is a very slow process, in contrast with their movement parallel to the plane of the bilayer. The transition of a molecule from one membrane surface to the other is called *transverse diffusion,* or *flip-flop,* whereas diffusion in the plane of a membrane is termed *lateral diffusion.* The flip-flop of phospholipid molecules in phosphatidyl choline vesicles has been directly measured by electron spin resonance techniques, which showed that *a phospholipid molecule flip-flops once in several hours* (see problem 5, p. 231, for the experimental design). Thus, a phospholipid molecule takes about 10^9 times as long to flip-flop across a 50-Å membrane as it takes to diffuse a distance of 50 Å in the lateral direction.

The free-energy barriers to the flip-flopping of protein molecules are even larger than for lipids because proteins have more extensive polar regions. In fact, *the flip-flop of a protein molecule has not been observed.* Hence, *membrane asymmetry can be preserved for long periods.*

FLUID MOSAIC MODEL OF BIOLOGICAL MEMBRANES

In 1972, S. Jonathan Singer and Garth Nicolson proposed a fluid mosaic model for the gross organization of biological membranes. The essence of their model is that membranes are *two-dimensional*

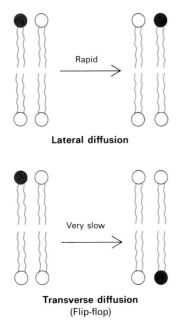

Lateral diffusion

Transverse diffusion
(Flip-flop)

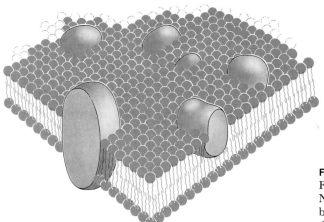

Figure 10-28
Fluid mosaic model. [After S. J. Singer and G. L. Nicolson. *Science* 175(1972):723. Copyright 1972 by the American Association for the Advancement of Science.]

solutions of oriented globular proteins and lipids (Figure 10-28). This proposal is supported by a wide variety of experimental observations. The major features of this model are:

1. Most of the membrane phospholipid and glycolipid molecules are in bilayer form. This lipid bilayer has a dual role: it is both a *solvent* for integral membrane proteins and a *permeability barrier*.

2. A small proportion of membrane lipids interact specifically with particular membrane proteins and may be essential for their function.

3. Membrane proteins are free to diffuse laterally in the lipid matrix unless restricted by special interactions, whereas they are not free to rotate from one side of a membrane to the other.

MEMBRANES ARE ASYMMETRIC

Membranes are structurally and functionally asymmetric, as exemplified by the orientations of glycophorin and the anion channel and, more generally, by the external localization of membrane carbohydrates. The outer and inner surfaces of *all known biological membranes have different components and different enzymatic activities.* A clear-cut example is provided by the pump that regulates the concentration of Na^+ and K^+ ions in cells. This transport system is located in the plasma membrane of nearly all cells in higher organisms. The Na^+-K^+ pump assembly is oriented in the plasma membrane so that it pumps Na^+ out of the cell and K^+ into it (Figure 10-29). Furthermore, ATP must be on the inside of the cell to drive the pump. Ouabain, a specific inhibitor of the pump, is effective only if it is located outside.

As will be discussed in detail in a later chapter (p. 718), membrane proteins have a unique orientation because they are synthesized and inserted in an asymmetric manner. This absolute asym-

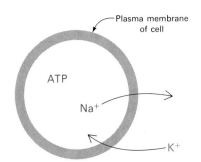

Figure 10-29
Asymmetry of the Na^+-K^+ transport system in plasma membranes.

metry is preserved by the absence of transmembrane movement of these proteins throughout their lifetime in the membrane. Lipids, too, are asymmetrically distributed as a consequence of their mode of biosynthesis, but this asymmetry is usually not absolute, except in the case of glycolipids. For example, in the red-cell membrane, sphingomyelin and phosphatidylcholine are preferentially located in the outer leaflet of the bilayer, whereas phosphatidyl ethanolamine and phosphatidyl serine are mainly in the inner leaflet. Large amounts of cholesterol are present in both leaflets. The functional significance of lipid asymmetry is not yet understood.

MEMBRANE FLUIDITY IS CONTROLLED BY FATTY ACID COMPOSITION AND CHOLESTEROL CONTENT

The fatty acyl chains of lipid molecules in bilayer membranes can exist in an ordered, rigid state or in a relatively disordered, fluid state. In the ordered state, all of the C—C bonds have a *trans* conformation, whereas in the disordered state, some are in the *gauche* conformation (Figure 10-30).

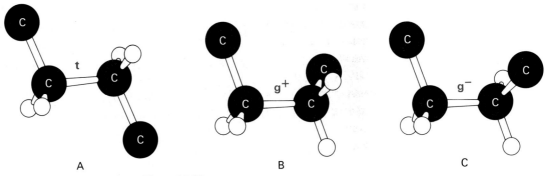

Figure 10-30
Conformation of C—C bonds in fatty acyl chains: (A) *trans* (t) conformation; (B and C) a 120-degree rotation yields a *gauche* (g) conformation, which can be either g^+ (clockwise rotation) or g^- (counterclockwise rotation). A gauche conformation bends a hydrocarbon chain by 120 degrees.

The transition from the rigid (all *trans*) to the fluid (partially *gauche*) state occurs as the temperature is raised above T_m, the melting temperature. *This transition temperature depends on the length of the fatty acyl chains and on their degree of unsaturation.* The rigid state is favored by the presence of saturated fatty acyl residues because their straight hydrocarbon chains interact very favorably with each other (Figure 10-31A). On the other hand, *a cis double bond produces a bend in the hydrocarbon chain. This bend interferes with a highly ordered packing of fatty acyl chains* and so T_m is lowered (Figure 10-31B). The length of the fatty acyl chain also affects the transition temperature. Long hydrocarbon chains interact more strongly than do short ones. Specifically, each additional —CH_2— group makes a favorable contribution of about −0.5 kcal/mol to the free energy of interaction of two adjacent hydrocarbon chains.

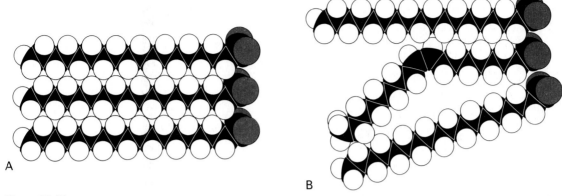

Figure 10-31
The highly ordered packing of fatty acid chains is disrupted by the presence of *cis* double bonds. These space-filling models show the packing of (A) three molecules of stearate (C_{18}, saturated) and (B) a molecule of oleate (C_{18}, unsaturated) between two molecules of stearate.

Procaryotes regulate the fluidity of their membranes by varying the number of double bonds and the length of their fatty acyl chains. For example, the ratio of saturated to unsaturated fatty acyl chains in the *E. coli* membrane decreases from 1.6 to 1.0 as the growth temperature is lowered from 42° to 27°C. This decrease in the proportion of saturated residues prevents the membrane from becoming too rigid at the lower temperature. *In eucaryotes, cholesterol also is a key regulator of membrane fluidity.* Cholesterol prevents the crystallization of fatty acyl chains by fitting between them. In fact, cholesterol abolishes the phase transition. An opposite effect of cholesterol is to sterically block large motions of fatty acyl chains and thereby make the membrane less fluid. Thus, cholesterol moderates the fluidity of membranes.

THREE-DIMENSIONAL IMAGES OF MEMBRANES CAN BE RECONSTRUCTED FROM ELECTRON MICROGRAPHS

The power of x-ray crystallography in elucidating the three-dimensional structure of soluble proteins is clearly established, as exemplified by the oxygen carriers and enzymes that were discussed in previous chapters. What about x-ray crystallographic analyses of membrane proteins? The difficulty is that three-dimensional crystals of integral membrane proteins have not yet been obtained. However, some membrane proteins give rise to ordered lattices in the plane of the membrane—in other words, they form two-dimensional crystals. These crystalline sheets are suitable for structural analyses by electron microscopy, as shown for the *purple membrane* of *Halobacterium halobium,* a salt-loving bacterium. This specialized region of the cell membrane contains *bacteriorhodopsin,* a 25-kdal protein, which converts light into a transmembrane proton gradient that is used to synthesize ATP (p. 451). Crystalline sheets with diameters as large as 1 μm can be obtained. The significance of having about 20,000 bacteriorhodopsin molecules in this crystalline array is that an image can be obtained with a very low dose of electrons and so there is little radiation damage. Furthermore, unstained samples can be analyzed to obtain a high-resolution image. A single electron micrograph of a crystalline sheet of the purple membrane yields a view of the structure projected onto a plane. The next stage is to tilt the sample and recombine the information from some twenty electron micrographs by Fourier techniques.

Using this technique, Richard Henderson and Nigel Unwin reconstructed a three-dimensional image of the purple membrane at 7-Å resolution (Figure 10-32). *The protein in this membrane contains seven closely packed α helices, which extend nearly perpendicular to the plane of the membrane for most of its 45-Å width.* Lipid bilayer regions fill the spaces between the protein molecules. The purple membrane probably exemplifies a structural motif that will be found in other integral membrane proteins. In particular, it seems likely that other membrane pumps and channels will have α-helical segments spanning the bilayer.

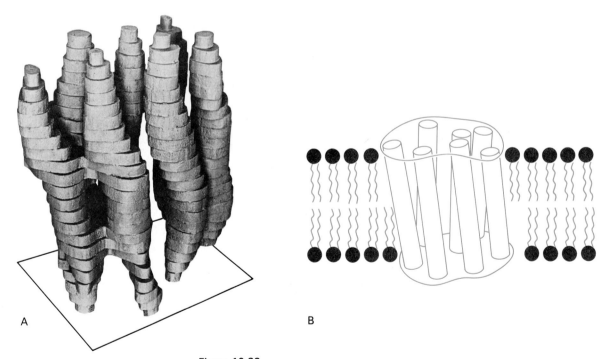

A

B

Figure 10-32
A. Model of bacteriorhodopsin constructed from a 7-Å three-dimensional map. B. Interpretative diagram showing the arrangement of α-helical segments in the lipid bilayer. The connections between these helices are not yet known. [Courtesy of Dr. Richard Henderson and Dr. Nigel Unwin.]

Biological membranes are sheetlike structures, typically 75 Å wide, that are composed of protein and lipid molecules held together by noncovalent interactions. Membranes are highly selective permeability barriers. They create closed compartments, which may be an entire cell or an organelle within a cell. Pumps and gates in membranes regulate the molecular and ionic compositions of these compartments. Membranes also control the flow of information between cells. For example, some membranes contain receptors for hormones such as insulin. Furthermore, membranes are intimately involved in such energy conversion processes as photosynthesis and oxidative phosphorylation.

The major classes of membrane lipids are phospholipids, glycolipids, and cholesterol. Phosphoglycerides, a type of phospholipid, consist of a glycerol backbone, two fatty acid chains, and a phosphorylated alcohol. The fatty acid chains usually contain between 14 and 24 carbon atoms; they may be saturated or unsaturated. Phosphatidyl choline, phosphatidyl serine, and phosphatidyl ethanolamine are major phosphoglycerides. Sphingomyelin, a different type of phospholipid, contains a sphingosine backbone instead of glycerol. Glycolipids are sugar-containing lipids derived from sphingosine. A common feature of these membrane lipids is that they are amphipathic molecules. They spontaneously form extensive bimolecular sheets in aqueous solutions because they contain both a hydrophilic and a hydrophobic moiety. These lipid bilayers are highly impermeable to ions and most polar molecules, yet they are quite fluid, which enables them to act as a solvent for membrane proteins.

Distinctive membrane functions such as transport, communication, and energy transduction are mediated by specific proteins. Some membrane proteins are deeply imbedded in the hydrocarbon region of the lipid bilayer. Transmembrane proteins such as the band-3 protein from erythrocytes can serve as ion channels. Membranes are structurally and functionally asymmetric, as exemplified by the directionality of ion transport systems in them and the localization of sugar residues on the external surface of mammalian plasma membranes. Membranes are dynamic structures in which proteins and lipids diffuse rapidly in the plane of the membrane (lateral diffusion), unless restricted by special interactions. In contrast, the rotation of proteins and lipids from one side of a membrane to the other (transverse diffusion, or flip-flop) is usually very slow. The degree of fluidity of membranes partly depends on the chain length and the extent to which their constituent fatty acids are unsaturated.

SELECTED READINGS

WHERE TO START

Singer, S. J., and Nicolson, G. L., 1972. The fluid mosaic model of the structure of cell membranes. *Science* 175:720–731.

Raff, M. C., 1976. Cell-surface immunology. *Sci. Amer.* 234(5):30–39. [A lucid and well-illustrated article showing how antibodies have contributed to our understanding of cell-surface structure and dynamics. Available as *Sci. Amer.* Offprint 1338.]

Rothman, J. E., and Lenard, J., 1977. Membrane asymmetry. *Science* 195:743–753.

BOOKS AND REVIEWS

Bretscher, M. S., and Raff, M. C., 1975. Mammalian plasma membranes. *Nature* 258:43–49.

Bretscher, M. S., 1973. Membrane structure: some general principles. *Science* 181:622–629.

Ashwell, G., and Morell, A. G., 1977. Membrane glycoproteins and recognition phenomena. *Trends Biochem. Sci.* 2:76–78.

Tyrrell, D. A., Heath, T. D., Colley, C. M., and Ryman, B. E., 1976. New aspects of liposomes. *Biochim. Biophys. Acta* 457:259–302.

Poste, G., and Nicolson, G. L., (eds.), 1977. *Dynamic Aspects of Cell Surface Organization.* North-Holland. [Contains articles on membrane dynamics, cell-surface immunology, and changes in cell-surface proteins during cell growth and malignant transformation.]

Tanford, C., 1980. *The Hydrophobic Effect: Formation of Micelles and Biological Membranes,* 2nd ed. Wiley-Interscience. [A lucid statement of thermodynamic aspects of membrane structure.]

MEMBRANE DYNAMICS

Frye, C. D., and Edidin, M., 1970. The rapid intermixing of cell surface antigens after formation of mouse-human heterokaryons. *J. Cell Sci.* 7:319–335.

Kornberg, R. D., and McConnell, H. M., 1971. Inside-outside transitions of phospholipids in vesicle membranes. *Biochemistry* 10:1111–1120.

Cone, R. A., 1972. Rotational diffusion of rhodopsin in the visual receptor membrane. *Nature New Biol.* 236:39–43.

Poo, M., and Cone, R. A., 1974. Lateral diffusion of rhodopsin in the photoreceptor membrane. *Nature* 247:438–441.

Elson, E. L., and Schlessinger, J., 1979. Long-range motions on cell surfaces. *In* Schmitt, F. O., and Worden, F. G., (eds.), *The Neurosciences: Fourth Study Program,* pp. 691–701. MIT Press. [Discussion of the fluorescence photobleaching recovery technique and its use in measuring the lateral diffusion of cell-surface components.]

Seelig, A., and Seelig, J., 1977. Effect of a single *cis* double bond on the structure of a phospholipid bilayer. *Biochemistry* 16:45–50.

ELECTRON MICROSCOPY OF MEMBRANES

Branton, D., 1966. Fracture faces of frozen membranes. *Proc. Nat. Acad. Sci.* 55:1048–1056.

Henderson, R., and Unwin, P. N. T., 1975. Three-dimensional model of purple membrane obtained by electron microscopy. *Nature* 257:28–32. [The first view of a membrane protein. At 7-Å resolution, transmembrane helices are evident.]

Unwin, P. N. T., and Henderson, R., 1975. Molecular structure determination by electron microscopy of unstained crystalline specimens. *J. Mol. Biol.* 94:425–440. [Presentation of the experimental strategy for determining the structure of membrane proteins that form highly ordered two-dimensional arrays.]

RED-CELL MEMBRANE

Steck, T. L., 1974. The organization of proteins in the human red blood cell membrane. *J. Cell Biol.* 62:1–19.

Marchesi, V. T., Furthmayr, H., and Tomita, M., 1976. The red cell membrane. *Ann. Rev. Biochem.* 45:667–698.

Steck, T. L., 1978. The band 3 protein of the human red cell membrane: a review. *J. Supramol. Struct.* 8:311–324.

SYNTHETIC AND RECONSTITUTED MEMBRANES

Fleischer, S., and Packer, L., (eds.), 1974. *Methods in Enzymology,* vol. 32. Academic Press. [Contains excellent articles on experimental methods in membrane research, including procedures for the preparation of model and reconstituted membranes.]

Darszon, A., Vandenberg, E. A., Ellisman, M. H., and Montal, M., 1979. Incorporation of membrane proteins into large single bilayer vesicles. *J. Cell Biol.* 81:446–452.

Racker, E., 1977. Perspectives and limitations of resolutions-reconstitution experiments. *J. Supramol. Struct.* 6:215–228.

PROBLEMS

1. How many phospholipid molecules are there in a 1-μm^2 region of a phospholipid bilayer membrane? Assume that a phospholipid molecule occupies 70 Å^2 of the surface area.

2. Bacterial phospholipids contain cyclopropane fatty acid residues.

Predict the effect of the cyclopropane ring on the packing of hydrocarbon chains in the interior of a bilayer. Would these cyclopropane fatty acid groups tend to make a membrane more or less fluid?

3. What is the average distance traversed by a membrane lipid in 1 μsec, 1 msec, and 1 sec? Assume a diffusion coefficient of 10^{-8} cm^2/sec.

4. The diffusion coefficient D of a rigid spherical molecule is given by

$$D = kT/(6\pi\eta r)$$

in which η is the viscosity of the solvent, r is the radius of the sphere, k is the Boltzman constant (1.38 $\times$ 10^{-16} erg/deg), and T is the absolute temperature. What is the diffusion coefficient at 37°C of a 100-kdal protein in a membrane that has an effective viscosity of 1 poise (1 poise = 1 erg sec/cm^3)? What is the average distance traversed by this protein in 1 μsec, 1 msec, and 1 sec? Assume that this protein is an unhydrated, ripid sphere of density 1.35 g/cm^3.

5. R. D. Kornberg and H. M. McConnell (1971) investigated the transverse diffusion (flip-flop) of phospholipids in a bilayer membrane by using a paramagnetic analog of phosphatidyl choline, called *spin-labeled phosphatidyl choline*.

The nitroxide (NO) group in spin-labeled phosphatidyl choline gives a distinctive paramagnetic resonance spectrum. This spectrum disappears when nitroxides are converted into amines by reducing agents such as ascorbate.

Lipid vesicles containing phosphatidyl choline (95%) and the spin-labeled analog (5%) were prepared by sonication and purified by gel-filtration chromatography. The outside diameter of these liposomes was about 250 Å. The amplitude of the paramagnetic resonance spectrum decreased to 35% of its initial value within a few minutes of the addition of ascorbate. There was no detectable change in the spectrum within a few minutes after the addition of a second aliquot of ascorbate. However, the amplitude of the residual spectrum decayed exponentially with a half-time of 6.5 hr. How would you interpret these changes in the amplitude of the paramagnetic spectrum?

PART II

GENERATION AND STORAGE
OF METABOLIC ENERGY

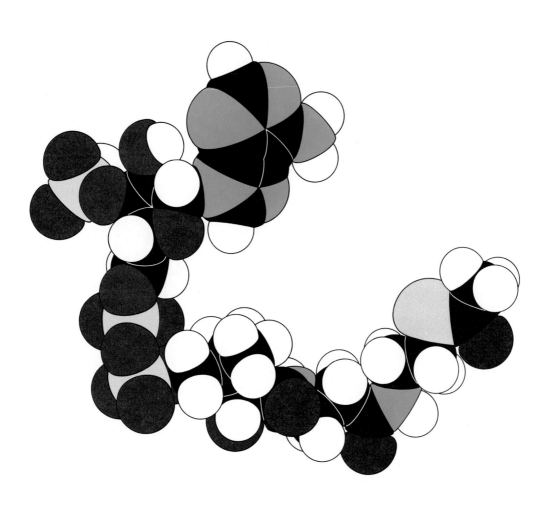

CHAPTER 11

METABOLISM:
BASIC CONCEPTS AND DESIGN

The concepts of conformation and dynamics developed in Part I—especially those aspects dealing with the specificity and catalytic power of enzymes, the regulation of their catalytic activity, and the formation of compartments by membranes—enable us to turn now to two major questions of biochemistry:

1. *How do cells extract energy and reducing power from their environments?*

2. *How do cells synthesize the building blocks of their macromolecules?*

These processes are carried out by a highly integrated network of chemical reactions, which are collectively known as *metabolism.*

There are at least a thousand chemical reactions in even as simple an organism as *E. coli.* The array of reactions may seem overwhelming at first glance. However, closer scrutiny reveals that metabolism has a *coherent design containing many common motifs.* The number of reactions in metabolism is large, but the number of *kinds* of reactions is relatively small. The mechanisms of these reactions are usually quite simple. For example, a double bond is often formed by dehydration. Furthermore, a group of about a hundred molecules plays a central role in all forms of life. Moreover, metabolic pathways are regulated in common ways. The purpose of this chapter is to introduce some general principles and motifs of metabolism.

FREE ENERGY IS THE MOST USEFUL THERMODYNAMIC FUNCTION IN BIOCHEMISTRY

Let us start by reviewing some thermodynamic relationships that are essential for an understanding of metabolism. In thermodynamics, a *system* is the matter within a defined region. The matter in the rest of the universe is called the *surroundings*. *The first law of thermodynamics states that the total energy of a system and its surroundings is a constant.* In other words, energy is conserved. The mathematical expression of the first law is

$$\Delta E = E_B - E_A = Q - W \qquad (1)$$

in which E_A is the energy of a system at the start of a process and E_B at the end of the process, Q is the heat absorbed by the system, and W is the work done by the system. An important feature of equation 1 is that *the change in energy of a system depends only on the initial and final states and not on the path of the transformation.*

The first law of thermodynamics cannot be used to predict whether a reaction can occur spontaneously. Some reactions do occur spontaneously although ΔE is positive. In such cases, the system absorbs heat from its surroundings so that the sum of the energies of the system and its surroundings remains the same. It is evident that a function different from ΔE is required. One such function is the *entropy* (S), which is a measure of the *degree of randomness or disorder of a system.* The entropy of a system increases (ΔS is positive) when it becomes more disordered. *The second law of thermodynamics states that a process can occur spontaneously only if the sum of the entropies of the system and its surroundings increases.*

$$(\Delta S_{\text{system}} + \Delta S_{\text{surroundings}}) > 0 \text{ for a spontaneous process} \quad (2)$$

Note that the entropy of a system can decrease during a spontaneous process, provided that the entropy of the surroundings increases so that their sum is positive. For example, the formation of a highly ordered biological structure is thermodynamically feasible because the decrease in entropy in such a system is more than offset by an increase in entropy in its surroundings.

One difficulty in using entropy as a criterion of whether a biochemical process can occur spontaneously is that the entropy changes of chemical reactions are not readily measured. Furthermore, the criterion of spontaneity given in equation 2 requires that both the entropy change of the surroundings and that of the system of interest be known. These difficulties are obviated by using a different thermodynamic function called the *free energy,* which is denoted by the symbol G (or F, in the older literature). In 1878, Josiah Willard Gibbs created the free-energy function by combining the first and second laws of thermodynamics. The basic equation is

$$\Delta G = \Delta H - T\Delta S \qquad (3)$$

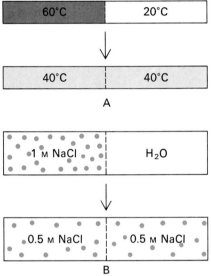

Figure 11-1
Examples of processes that are driven by an increase in the entropy of a system: (A) diffusion of heat; and (B) diffusion of a solute.

in which ΔG is the change in free energy of a system undergoing a transformation at constant pressure (P) and temperature (T), ΔH is the change in enthalpy of this system, and ΔS is the change in entropy of this system. Note that the properties of the surroundings do not enter into this equation. The enthalpy change is given by

$$\Delta H = \Delta E + P\Delta V \tag{4}$$

The volume change, ΔV, is small for nearly all biochemical reactions, and so ΔH is nearly equal to ΔE. Hence,

$$\Delta G \cong \Delta E - T\Delta S \tag{5}$$

Thus, the ΔG of a reaction depends both on the change in internal energy and on the change in entropy of the system.

The change in free energy (ΔG) of a reaction, in contrast with the change in internal energy (ΔE) of a reaction, is a valuable criterion of whether it can occur spontaneously.

1. *A reaction can occur spontaneously only if* ΔG *is negative.*

2. A system is at equilibrium and no net change can take place if ΔG is zero.

3. A reaction cannot occur spontaneously if ΔG is positive. An input of free energy is required to drive such a reaction.

Two additional points need to be emphasized here. First, the ΔG of a reaction depends only on the free energy of the products (the final state) minus that of the reactants (the initial state). *The ΔG of a reaction is independent of the path of the transformation.* The mechanism of a reaction has no effect on ΔG. For example, the ΔG for the oxidation of glucose to CO_2 and H_2O is the same whether it occurs by combustion in vitro or by a series of many enzyme-catalyzed steps in a cell. Second, ΔG *provides no information about the rate of a reaction*. A negative ΔG indicates that a reaction can occur spontaneously, but it does not signify that it will occur at a perceptible rate. As previously discussed (p. 107), the rate of a reaction depends on the *free energy of activation* ($\Delta G^{\ddagger}$), which is unrelated to ΔG.

STANDARD FREE-ENERGY CHANGE OF A REACTION AND ITS RELATION TO THE EQUILIBRIUM CONSTANT

Consider the reaction

$$A + B \Longleftrightarrow C + D$$

The ΔG of this reaction is given by

$$\Delta G = \Delta G^{\circ} + RT \log_e \frac{[C][D]}{[A][B]} \tag{6}$$

in which ΔG° is the *standard free-energy change*, R is the gas constant, T is the absolute temperature, and [A], [B], [C], and [D] are the molar concentrations (more rigorously, the activities) of the reactants. $\Delta G°$ is the free-energy change for this reaction under standard conditions—that is, when each of the reactants A, B, C, and D is present at a concentration of 1.0 M. Thus, the ΔG of a reaction depends on the *nature* of the reactants (expressed in the $\Delta G°$ term of equation 6) and on their *concentrations* (expressed in the logarithmic term in equation 6).

A convention has been adopted to simplify free-energy calculations for biochemical reactions. The standard state is defined as having a pH of 7. Consequently, the activity of H^+ corresponding to a pH of 7 has a value of 1 in equations 6 and 9. Also, the activity of water is taken to be 1 in these equations. The *standard free-energy change at pH 7,* denoted by the symbol $\Delta G°'$, will be used throughout this book. The *kilocalorie* will be used as the unit of energy.

The relationship between the standard free energy and the equilibrium constant of a reaction can be readily derived. At equilibrium, $\Delta G = 0$. Equation 6 then becomes

$$0 = \Delta G°' + RT \log_e \frac{[C][D]}{[A][B]} \tag{7}$$

and so

$$\Delta G°' = -RT \log_e \frac{[C][D]}{[A][B]} \tag{8}$$

The equilibrium constant under standard conditions, K'_{eq}, is defined as

$$K'_{eq} = \frac{[C][D]}{[A][B]} \tag{9}$$

Substituting equation 9 into equation 8 gives

$$\Delta G°' = -RT \log_e K'_{eq} \tag{10}$$
$$\Delta G°' = -2.303\, RT \log_{10} K'_{eq} \tag{11}$$

which can be rearranged to give

$$K'_{eq} = 10^{-\Delta G°'/(2.303\, RT)} \tag{12}$$

Substituting $R = 1.98 \times 10^{-3}\, kcal\, mol^{-1}\, deg^{-1}$ and $T = 298°K$ (corresponding to 25°C) gives

$$K'_{eq} = 10^{-\Delta G°'/1.36} \tag{13}$$

when $\Delta G°'$ is expressed in kcal/mol. Thus, the standard free energy and the equilibrium constant of a reaction are related by a simple expression. For example, an equilibrium constant of 10 corresponds to a standard free-energy change of -1.36 kcal/mol at 25°C (Table 11-1).

Let us calculate $\Delta G°'$ and ΔG for the isomerization of dihydroxyacetone phosphate to glyceraldehyde 3-phosphate as an example.

Table 11-1
Relationship between $\Delta G°'$ and K'_{eq} (at 25°C)

K'_{eq}	$\Delta G°'$ (kcal/mol)
10^{-5}	6.82
10^{-4}	5.46
10^{-3}	4.09
10^{-2}	2.73
10^{-1}	1.36
1	0
10	-1.36
10^2	-2.73
10^3	-4.09
10^4	-5.46
10^5	-6.82

This reaction occurs in glycolysis (p. 255). At equilibrium, the ratio of glyceraldehyde 3-phosphate to dihydroxyacetone phosphate is .0475 at 25°C (298°K) at pH 7. Hence, $K'_{eq} = .0475$. The standard free-energy change for this reaction is then calculated from equation 11:

$$\Delta G^{\circ\prime} = -2.303\, RT \log_{10} K'_{eq}$$
$$= -2.303 \times 1.98 \times 10^{-3} \times 298 \times \log_{10}(.0475)$$
$$= +1.8 \text{ kcal/mol}$$

Now let us calculate ΔG for this reaction when the initial concentration of dihydroxyacetone phosphate is 2×10^{-4} M and the initial concentration of glyceraldehyde 3-phosphate is 3×10^{-6} M. Substituting these values into equation 6 gives

$$\Delta G = 1.8 \text{ kcal/mol} + 2.303\, RT \log_{10} \frac{3 \times 10^{-6}\text{ M}}{2 \times 10^{-4}\text{ M}}$$
$$= 1.8 \text{ kcal/mol} - 2.5 \text{ kcal/mol}$$
$$= -0.7 \text{ kcal/mol}$$

This negative value for the ΔG indicates that the isomerization of dihydroxyacetone phosphate to glyceraldehyde 3-phosphate can occur spontaneously when these species are present at the concentrations stated above. Note that ΔG for this reaction is negative although $\Delta G^{\circ\prime}$ is positive. *It is important to stress that whether the ΔG for a reaction is larger, smaller, or the same as $\Delta G^{\circ\prime}$ depends on the concentrations of the reactants. The criterion of spontaneity for a reaction is ΔG, not $\Delta G^{\circ\prime}$.*

A THERMODYNAMICALLY UNFAVORABLE REACTION CAN BE DRIVEN BY A FAVORABLE ONE

An important thermodynamic fact is that *the overall free-energy change for a series of reactions is equal to the sum of the free-energy changes of the individual steps.* Consider the reactions

A $\rightleftharpoons$ B + C	$\Delta G^{\circ\prime} = +5 \text{ kcal/mol}$
B $\rightleftharpoons$ D	$\Delta G^{\circ\prime} = -8 \text{ kcal/mol}$
A $\rightleftharpoons$ C + D	$\Delta G^{\circ\prime} = -3 \text{ kcal/mol}$

Under standard conditions, A cannot be spontaneously converted into B and C because ΔG is positive. However, the conversion of B into D under standard conditions is thermodynamically feasible. Because free-energy changes are additive, the conversion of A into C and D has a $\Delta G^{\circ\prime}$ of -3 kcal/mol, which means that it can occur spontaneously under standard conditions. Thus, *a thermodynamically unfavorable reaction can be driven by a thermodynamically favorable one.* These reactions are *coupled* by B, the common intermediate. We will encounter many examples of energy coupling in metabolism.

$$\begin{array}{ccc}
\begin{array}{c} CH_2OH \\ | \\ C{=}O \\ | \\ CH_2OPO_3{}^{2-} \end{array} & \rightleftharpoons & \begin{array}{c} O\!\!\diagup\!\!\diagdown H \\ C \\ | \\ H{-}C{-}OH \\ | \\ CH_2OPO_3{}^{2-} \end{array} \\[6pt]
\textbf{Dihydroxyacetone} & & \textbf{Glyceraldehyde} \\
\textbf{phosphate} & & \textbf{3-phosphate}
\end{array}$$

ATP IS THE UNIVERSAL CURRENCY OF FREE ENERGY IN BIOLOGICAL SYSTEMS

Living things require a continual input of free energy for three major purposes: the performance of mechanical work in muscle contraction and other cellular movements, the active transport of molecules and ions, and the synthesis of macromolecules and other biomolecules from simple precursors. The free energy used in these processes, which maintain an organism in a state that is far from equilibrium, is derived from the environment. *Chemotrophs* obtain this energy by the oxidation of foodstuffs, whereas *phototrophs* obtain it by trapping light energy. The free energy derived from the oxidation of foodstuffs and from light is partly transformed into a special form before it is used for motion, active transport, and biosyntheses. This special carrier of free energy is *adenosine triphosphate* (ATP). The central role of ATP in energy exchanges in biological systems was perceived by Fritz Lipmann and by Herman Kalckar in 1941.

ATP is a nucleotide consisting of an adenine, a ribose, and a triphosphate unit (Figure 11-2; see p. 511 for a discussion of nucleotide nomenclature). The active form of ATP is usually a complex of ATP with Mg^{2+} or Mn^{2+}. In considering the role of ATP as an energy carrier, we can focus on its triphosphate moiety. *ATP is an energy-rich molecule because its triphosphate unit contains two phosphoanhydride bonds.* A large amount of free energy is liberated when ATP is hydrolyzed to adenosine diphosphate (ADP) and orthophosphate (P_i) or when ATP is hydrolyzed to adenosine monophosphate (AMP) and pyrophosphate (PP_i). The $\Delta G^{\circ\prime}$ for these reactions depends on the ionic strength of the medium and on the concentra-

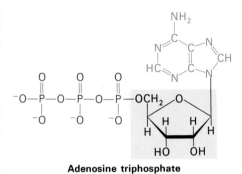

Adenosine triphosphate (ATP)

Figure 11-2
Adenosine triphosphate consists of an adenine (blue), a ribose (yellow), and a triphosphate (red) unit.

Figure 11-3
Structures of ATP, ADP, and AMP. (Adenosine consists of adenine linked to ribose.)

Adenosine triphosphate (ATP)   **Adenosine diphosphate (ADP)**   **Adenosine monophosphate (AMP)**

tions of Mg^{2+} and Ca^{2+}. We will use a value of -7.3 kcal/mol. Under typical cellular conditions, the actual ΔG for these hydrolyses is approximately -12 kcal/mol.

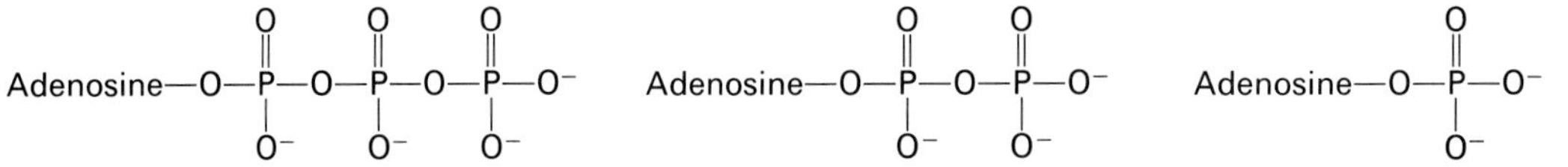

$$ATP + H_2O \rightleftharpoons ADP + P_i + H^+ \qquad \Delta G^{\circ\prime} = -7.3 \text{ kcal/mol}$$
$$ATP + H_2O \rightleftharpoons AMP + PP_i + H^+ \qquad \Delta G^{\circ\prime} = -7.3 \text{ kcal/mol}$$

ATP, AMP, and ADP are interconvertible. The enzyme adenylate kinase (also called myokinase) catalyzes the reaction

$$ATP + AMP \rightleftharpoons ADP + ADP$$

The free energy liberated in the hydrolysis of an anhydride bond of ATP is used to drive reactions that require an input of free energy, such as muscle contraction. In turn, ATP is formed from ADP and P_i when fuel molecules are oxidized in chemotrophs or when light is trapped by phototrophs. *This ATP-ADP cycle is the fundamental mode of energy exchange in biological systems.*

Some biosynthetic reactions are driven by nucleotides that are analogous to ATP, namely guanosine triphosphate (GTP), uridine triphosphate (UTP), and cytidine triphosphate (CTP). The diphosphate forms of these nucleotides are denoted by GDP, UDP, and CDP, respectively. Enzymes catalyze the transfer of the terminal phosphoryl group from one nucleotide to another, as in the reactions

$$ATP + GDP \rightleftharpoons ADP + GTP$$
$$ATP + GMP \rightleftharpoons ADP + GDP$$

ATP IS CONTINUOUSLY FORMED AND CONSUMED

ATP serves as the principal *immediate donor of free energy* in biological systems rather than as a storage form of free energy. In a typical cell, an ATP molecule is consumed within a minute following its formation. *The turnover of ATP is very high.* For example, a resting human consumes about 40 kg of ATP in 24 hours. During strenuous exertion, the rate of utilization of ATP may be as high as 0.5 kg per minute. Motion, active transport, signal amplification, and biosyntheses can occur only if ATP is continuously regenerated from ADP (Figure 11-4). Phototrophs harvest the free energy in light to generate ATP, whereas chemotrophs form ATP by the oxidation of fuel molecules.

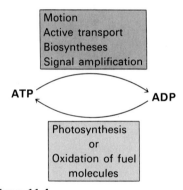

Figure 11-4
The ATP-ADP cycle is the fundamental mode of energy exchange in biological systems.

STRUCTURAL BASIS OF THE HIGH GROUP-TRANSFER POTENTIAL OF ATP

Let us compare the standard free energy of hydrolysis of ATP with that of a phosphate ester, such as glycerol 3-phosphate:

$$ATP + H_2O \rightleftharpoons ADP + P_i + H^+ \qquad \Delta G^{\circ\prime} = -7.3 \text{ kcal/mol}$$
$$\text{Glycerol 3-phosphate} + H_2O \rightleftharpoons \text{glycerol} + P_i$$
$$\Delta G^{\circ\prime} = -2.2 \text{ kcal/mol}$$

The magnitude of $\Delta G^{\circ\prime}$ for the hydrolysis of glycerol 3-phosphate is much smaller than that of ATP. This means that ATP has a stronger tendency to transfer its terminal phosphoryl group to water than does glycerol 3-phosphate. In other words, ATP has a higher *phosphate group transfer potential* than does glycerol 3-phosphate.

Glycerol 3-phosphate

What is the structural basis of the high phosphate group transfer potential of ATP? The structures of ATP and its hydrolysis products, ADP and P_i, must be examined to answer this question because $\Delta G°'$ depends on the *difference* in free energies of the products and reactants. Two factors prove to be important in this regard: *electrostatic repulsion* and *resonance stabilization*. At pH 7, the triphosphate unit of ATP carries about four negative charges. These charges repel each other strongly because they are in close proximity. *The electrostatic repulsion between these negatively charged groups is reduced when ATP is hydrolyzed. The other factor contributing to the high group-transfer potential of ATP is that ADP and P_i enjoy greater resonance stabilization than does ATP.* For example, orthophosphate has a number of resonance forms of similar energy (Figure 11-5). In contrast, the terminal portion of ATP has fewer significant resonance forms per phosphate group. Forms of the type shown in Figure 11-6 are unlikely to occur because the two phosphorus atoms compete for electron pairs on oxygen. Furthermore, there is a positive charge on an oxygen atom adjacent to a positively charged phosphorus atom, which is electrostatically unfavorable.

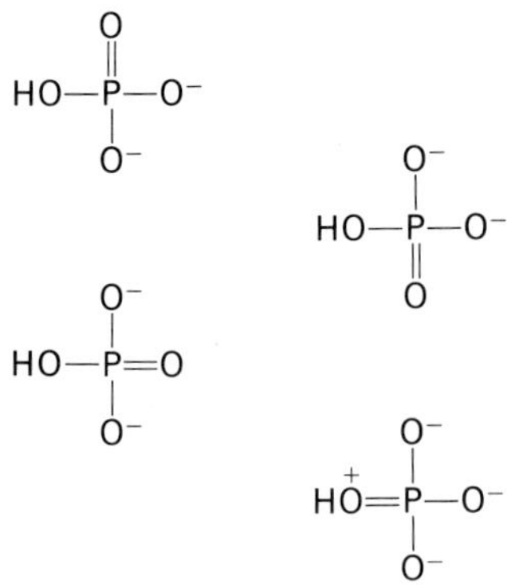

Various other compounds in biological systems have a high phosphate group transfer potential. In fact, some of them, such as phosphoenolpyruvate, acetyl phosphate, and creatine phosphate (Figure 11-7), have a higher group-transfer potential than does ATP. This means that phosphoenolpyruvate can transfer its phosphoryl group to ADP to form ATP. In fact, this is one of the ways in which ATP is generated in the breakdown of sugars. It is significant that ATP has a group-transfer potential that is intermediate among the biologically important phosphorylated molecules (Table 11-2). This intermediate position enables ATP to function efficiently as a carrier of phosphoryl groups.

ATP is often called a high-energy phosphate compound and its phosphoanhydride bonds are referred to as high-energy bonds. It should be noted that there is nothing special about the bonds themselves. *They are high-energy bonds in the sense that much free energy is released when they are hydrolyzed,* for the reasons given above. Lipmann's term "high-energy bond" and his symbol $\sim$P for a compound having a high phosphate group transfer potential are vivid, concise, and useful notations. In fact, Lipmann's squiggle did much to stimulate interest in bioenergetics.

Figure 11-5
Significant resonance forms of orthophosphate.

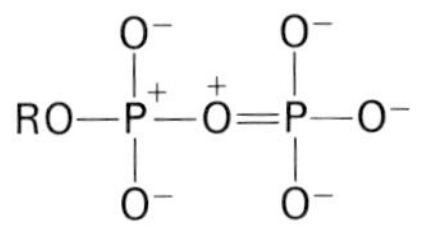

Figure 11-6
An improbable resonance form of the terminal portion of ATP.

Table 11-2
Free energies of hydrolysis of some phosphorylated compounds

Compound	$\Delta G°'$ (kcal/mol)
Phosphoenolpyruvate	−14.8
Carbamoyl phosphate	−12.3
Acetyl phosphate	−10.3
Creatine phosphate	−10.3
Pyrophosphate	−8.0
ATP (to ADP)	−7.3
Glucose 1-phosphate	−5.0
Glucose 6-phosphate	−3.3
Glycerol 3-phosphate	−2.2

Note: Phosphoenolpyruvate has the highest phosphate group transfer potential of the compounds listed.

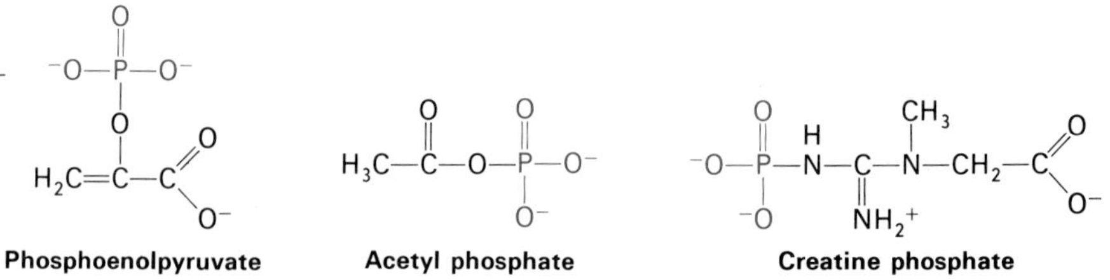

Phosphoenolpyruvate **Acetyl phosphate** **Creatine phosphate**

Figure 11-7
Compounds with a higher phosphate group transfer potential than that of ATP.

An understanding of the role of ATP in energy coupling can be enhanced by considering a chemical reaction that is thermodynamically unfavorable without an input of free energy. Suppose that the standard free energy of the conversion of A into B is $+4 \text{ kcal/mol}$.

$$A \rightleftharpoons B \qquad \Delta G^{\circ\prime} = +4 \text{ kcal/mol}$$

According to equation 13 (p. 238), the equilibrium constant K'_{eq} of this reaction at $25°C$ is given by

$$\frac{[B]_{eq}}{[A]_{eq}} = K'_{eq} = 10^{-\Delta G^{\circ\prime}/1.36} = 1.15 \times 10^{-3}$$

Thus, A cannot be spontaneously converted into B when the molar ratio of B to A is equal to or greater than 1.15×10^{-3}. However, A can be converted into B when the $[B]/[A]$ ratio is higher than 1.15×10^{-3} if the reaction is coupled to the hydrolysis of ATP. The new overall reaction is

$$A + ATP + H_2O \rightleftharpoons B + ADP + P_i + H^+$$
$$\Delta G^{\circ\prime} = -3.3 \text{ kcal/mol}$$

Its standard free-energy change of -3.3 kcal/mol is the sum of $\Delta G^{\circ\prime}$ for the conversion of A into B ($+4 \text{ kcal/mol}$) and for the hydrolysis of ATP (-7.3 kcal/mol). The equilibrium constant of this coupled reaction is

$$K'_{eq} = \frac{[B]_{eq}}{[A]_{eq}} \cdot \frac{[ADP]_{eq}[P_i]_{eq}}{[ATP]_{eq}} = 10^{3.3/1.36}$$
$$= 2.67 \times 10^2$$

At equilibrium, the ratio of $[B]$ to $[A]$ is given by

$$\frac{[B]_{eq}}{[A]_{eq}} = K'_{eq} \frac{[ATP]_{eq}}{[ADP]_{eq}[P_i]_{eq}}$$

The ATP-generating system of cells maintains the $[ATP]/[ADP][P_i]$ ratio at a high level, typically of the order of 500. For this ratio,

$$\frac{[B]_{eq}}{[A]_{eq}} = 2.67 \times 10^2 \times 500 = 1.34 \times 10^5$$

which means that the hydrolysis of ATP enables A to be converted into B until the $[B]/[A]$ ratio reaches a value of 1.34×10^5. This equilibrium ratio is strikingly different from the value of 1.15×10^{-3} for the reaction $A \rightleftharpoons B$ that does not include ATP hydrolysis. In other words, the coupled hydrolysis of ATP has changed the equilibrium ratio of B to A by a factor of about 10^8.

We see here the thermodynamic essence of the role of ATP as an energy-coupling agent. *Cells maintain a high level of ATP by using light or oxidizable substrates as sources of free energy. The hydrolysis of an ATP molecule then changes the equilibrium ratio of products to reactants of a coupled reaction by a very large factor, of the order of 10^8.* More generally, the hydrolysis of n ATP molecules changes the equilibrium ratio of a coupled reaction (or sequence of reactions) by a factor of 10^{8n}. For example, the hydrolysis of three ATP molecules in a coupled reaction changes the equilibrium ratio by a factor of 10^{24}. Thus, *a thermodynamically unfavorable reaction sequence can be converted into a favorable one by coupling it to the hydrolysis of a sufficient number of ATP molecules.* It should also be emphasized that A and B in the preceding coupled equation have a very general significance. For example, A and B may represent *different conformations of a protein,* as in muscle contraction. Alternatively, A and B may refer to the *concentrations of an ion or molecule on the outside and inside of a cell,* as in the active transport of a nutrient. Furthermore, A and B may designate *different chemical species,* as in the biosynthesis of complex molecules from simple precursors. The chapters ahead will deal with many remarkable mechanisms of energy coupling in biological processes.

Figure 11-8
Structure of the oxidized form of nicotinamide adenine dinucleotide (NAD^+) and of nicotinamide adenine dinucleotide phosphate ($NADP^+$). In NAD^+, R = H; in $NADP^+$, R = PO_3^{2-}.

NADH AND FADH₂ ARE THE MAJOR ELECTRON CARRIERS IN THE OXIDATION OF FUEL MOLECULES

Chemotrophs derive free energy from the oxidation of fuel molecules, such as glucose and fatty acids. In aerobic organisms, the ultimate electron acceptor is O_2. However, electrons are not transferred directly from fuel molecules and their breakdown products to O_2. Instead, these substrates transfer electrons to special carriers, which are either *pyridine nucleotides* or *flavins.* The reduced forms of these carriers then transfer their high-potential electrons to O_2 by means of an electron-transport chain located in the inner membrane of mitochondria. ATP is formed from ADP and P_i as a result of this flow of electrons. This process, called *oxidative phosphorylation* (Chapter 14), is the major source of ATP in aerobic organisms. Alternatively, the high-potential electrons derived from the oxidation of fuel molecules can be used in biosyntheses that require *reducing power* in addition to ATP.

Nicotinamide adenine dinucleotide (NAD^+) is a major electron acceptor in the oxidation of fuel molecules (Figure 11-8). The reactive part of NAD^+ is its nicotinamide ring. *In the oxidation of a substrate, the nicotinamide ring of NAD^+ accepts a hydrogen ion and two electrons, which are equivalent to a hydride ion.* The reduced form of this carrier is called NADH.

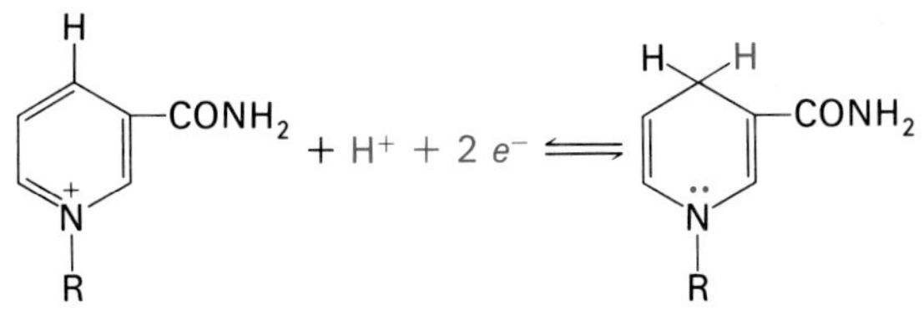

NAD^+ is the electron acceptor in many reactions of the type

In this dehydrogenation, one hydrogen atom of the substrate is directly transferred to NAD^+, whereas the other appears in the solvent. Both electrons lost by the substrate are transferred to the nicotinamide ring.

The other major electron carrier in the oxidation of fuel molecules is *flavin adenine dinucleotide* (Figure 11-9). The abbreviations for the oxidized and reduced forms of this carrier are FAD and $FADH_2$, respectively. FAD is the electron acceptor in reactions of the type

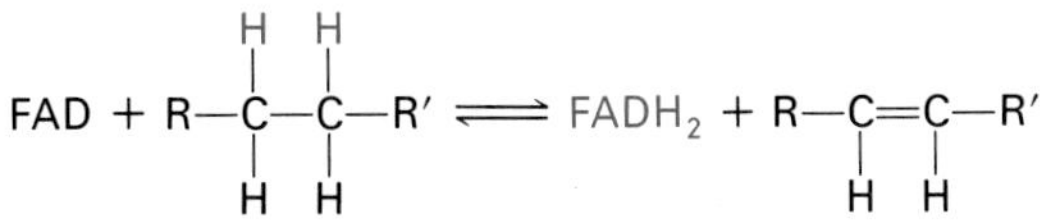

The reactive part of FAD is its isoalloxazine ring (Figure 11-10). FAD, like NAD^+, is a two-electron acceptor. However, unlike NAD^+, FAD accepts both of the hydrogen atoms lost by the substrate. The electron-transfer potentials of NADH and $FADH_2$ and the thermodynamics of oxidation-reduction reactions will be discussed in Chapter 14.

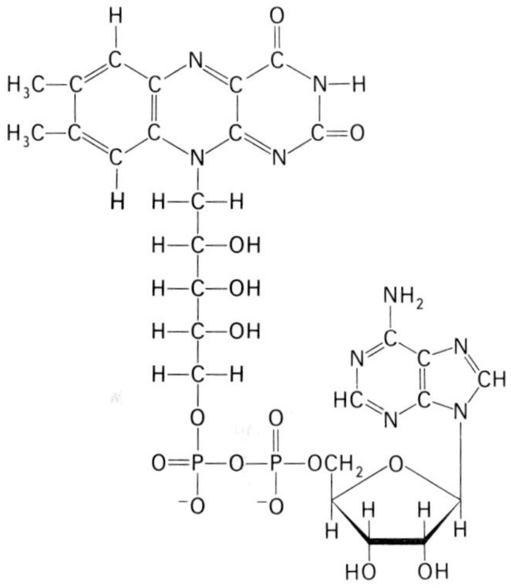

Figure 11-9
Structure of the oxidized form of flavin adenine dinucleotide (FAD).

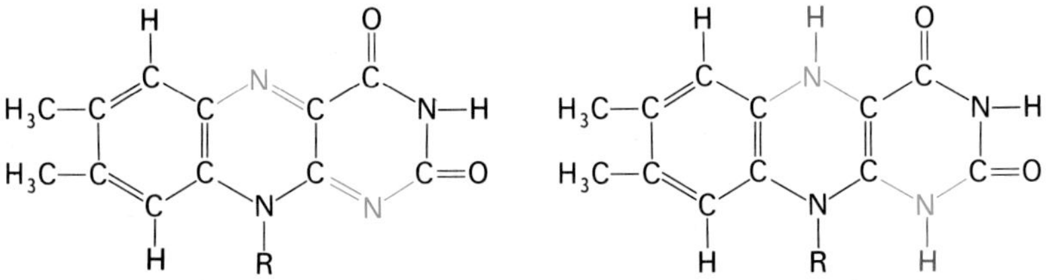

Oxidized form
(FAD)

Reduced form
(FADH$_2$)

Figure 11-10
Structures of the reactive parts of FAD and FADH$_2$.

NADPH IS THE MAJOR ELECTRON DONOR IN REDUCTIVE BIOSYNTHESES

In most biosyntheses, the precursors are more oxidized than the products. Hence, reductive power is needed in addition to ATP. For example, in the biosynthesis of fatty acids, the keto group of an added C_2 unit is reduced to a methylene group in several steps. This sequence of reactions requires an input of four electrons.

$$R-CH_2-\overset{\overset{\displaystyle O}{\|}}{C}-R' + 4\ H^+ + 4\ e^- \longrightarrow R-CH_2-CH_2-R' + H_2O$$

The electron donor in most reductive biosyntheses is the reduced form of nicotinamide adenine dinucleotide phosphate (NADPH) (Figure 11-9). NADPH differs from NADH in that the 2'-hydroxyl group of its adenosine moiety is esterified with phosphate. The oxidized form of NADPH is denoted as $NADP^+$. NADPH carries electrons in the same way as NADH. However, *NADPH is used almost exclusively for the reductive biosyntheses, whereas NADH is used primarily for the generation of ATP.* The extra phosphate group on NADPH is a tag that reveals the designated purpose of the molecule to discerning enzymes. The biological significance of the distinction between NADPH and NADH will be discussed later (p. 333).

It is important to note that NADH, NADPH, and $FADH_2$ react very slowly with O_2 in the absence of catalysts. Likewise, ATP is hydrolyzed at a slow rate in the absence of a catalyst. These molecules are kinetically quite stable, although there is a large thermodynamic driving force for the reaction of these electron carriers with O_2 and of ATP with water. *The stability of these molecules in the absence of specific catalysts is essential for their biological function because it enables enzymes to control the flow of free energy and reductive power.*

COENZYME A IS A UNIVERSAL CARRIER OF ACYL GROUPS

Coenzyme A is another central molecule in metabolism. In 1945, Lipmann found that a heat-stable cofactor was required in many enzyme-catalyzed acetylations. This cofactor was named *coenzyme A* (CoA), the A standing for *acetylation*. It was isolated, and its structure was determined several years later (Figure 11-11). The terminal sulfhydryl group in CoA is the reactive site. Acyl groups are linked to CoA by a thioester bond. The resulting derivative is called an *acyl CoA*. An acyl group often linked to CoA is the acetyl unit; this derivative is called *acetyl CoA*. The $\Delta G^{\circ\prime}$ for the hydrolysis of acetyl CoA has a large negative value:

$$\text{Acetyl CoA} + H_2O \rightleftharpoons \text{acetate} + \text{CoA} + H^+$$
$$\Delta G^{\circ\prime} = -7.5 \text{ kcal/mol}$$

$$R-\overset{\overset{\displaystyle O}{\|}}{C}-S-CoA$$

Acyl CoA

$$H_3C-\overset{\overset{\displaystyle O}{\|}}{C}-S-CoA$$

Acetyl CoA

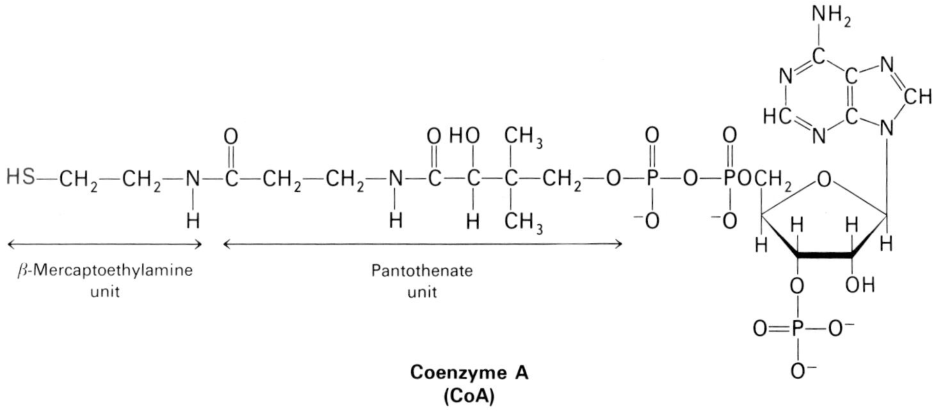

Figure 11-11
Structure of coenzyme A (CoA).

In other words, *acetyl CoA has a high acetyl group transfer potential.* CoA is a carrier of activated acetyl or other acyl groups, just as ATP is a carrier of activated phosphoryl groups.

We will encounter other carriers of activated groups in our consideration of metabolism. Several of them are listed in Table 11-3.

Table 11-3
Some activated carriers in metabolism

Carrier molecule	Group carried in activated form
ATP	Phosphoryl
NADH and NADPH	Electrons
$FADH_2$	Electrons
Coenzyme A	Acyl
Lipoamide	Acyl
Thiamine pyrophosphate	Aldehyde
Biotin	CO_2
Tetrahydrofolate	One-carbon units
S-Adenosylmethionine	Methyl
Uridine diphosphate glucose	Glucose
Cytidine diphosphate diacylglycerol	Phosphatidate

These carriers mediate the interchange of activated groups in a wide variety of biochemical reactions. Indeed, they have very similar roles in all forms of life. Their universal presence is one of the unifying motifs of biochemistry.

MOST WATER-SOLUBLE VITAMINS ARE COMPONENTS OF COENZYMES

Lipmann has commented that "doctors like to prescribe vitamins and millions of people take them, but it requires a good deal of biochemical sophistication to understand why they are needed and how the organism uses them." Vitamins are organic molecules that are needed in small amounts in the diets of higher animals. These molecules serve nearly the same roles in all forms of life, but higher animals have lost the capacity to synthesize them. There are two groups of vitamins: the fat-soluble ones, designated by the letters A, D, E, and K, and the water-soluble ones, which are referred to as the vitamin B complex. Most of the biochemical roles of the water-soluble vitamins are known. In fact, most of them are components of coenzymes (Table 11-4). For example, riboflavin (Vitamin B_2) is a precursor of FAD, and pantothenate is a component of coenzyme A.

Table 11-4
Coenzyme derivatives of some water-soluble vitamins

Vitamin	Coenzyme derivative
Thiamine (Vitamin B_1)	Thiamine pyrophosphate
Riboflavin (Vitamin B_2)	Flavin adenine dinucleotide and flavin mononucleotide
Nicotinate (niacin)	Nicotinamide adenine dinucleotide
Pyridoxine, pyridoxal, and pyridoxamine (Vitamin B_6)	Pyridoxal phosphate
Pantothenate	Coenzyme A
Biotin	Covalently attached to carboxylases
Folate	Tetrahydrofolate
Cobalamin (Vitamin B_{12})	Cobamide coenzymes

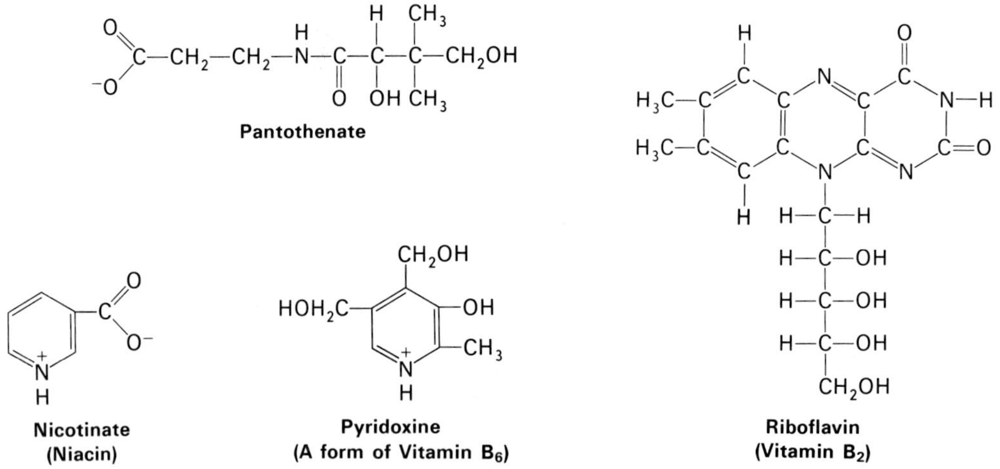

Figure 11-12
Structures of some water-soluble vitamins.

Less is known about the molecular basis of action of the fat-soluble vitamins (Figure 11-13). Vitamin K, which is required for normal clotting, participates in the carboxylation of glutamate residues to γ-carboxyglutamate (p. 174). Vitamin A (retinol) is the precursor of retinal, the light-absorbing group in visual pigments (p. 897). A deficiency of this vitamin results in night blindness. Furthermore, young animals require Vitamin A for growth. The metabolism of calcium and phosphorus is regulated by a hormone that is derived from vitamin D (p. 478). A deficiency in vitamin D impairs bone formation in growing animals. Infertility in rats is a consequence of vitamin E deficiency. This vitamin protects unsaturated membrane lipids from oxidation.

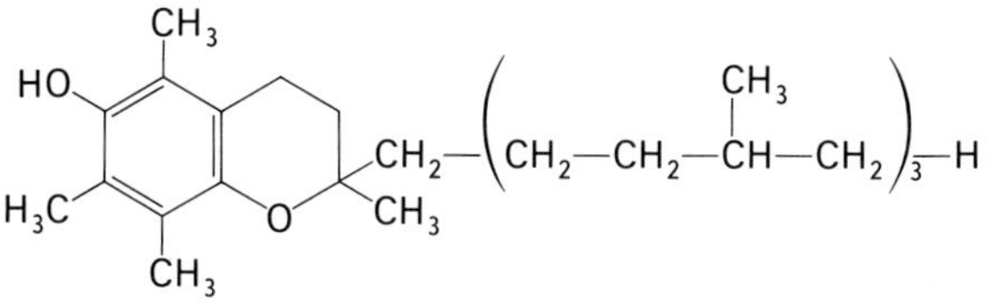

α-Tocopherol
(Vitamin E)

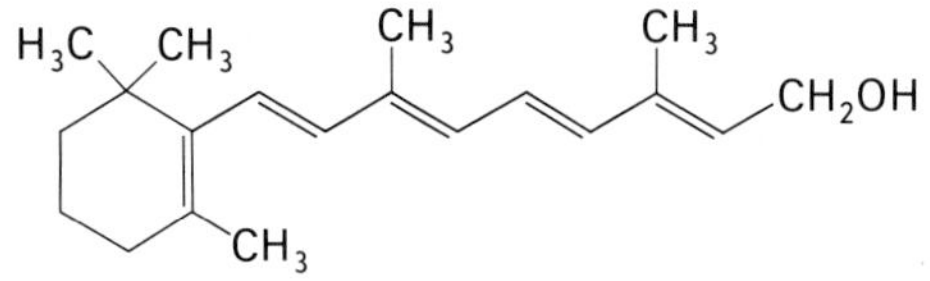

Retinol
(Vitamin A)

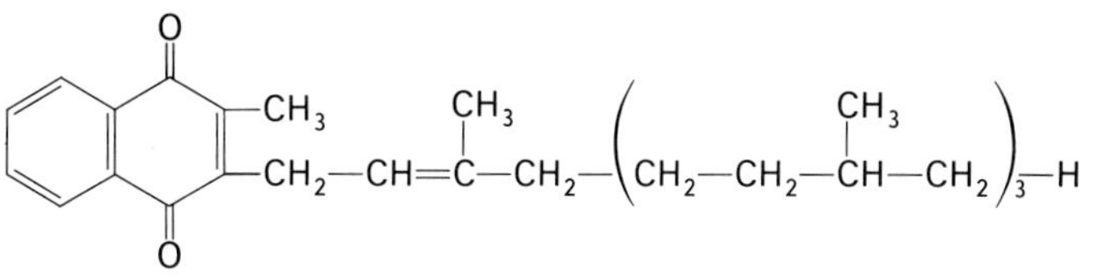

Vitamin K₁

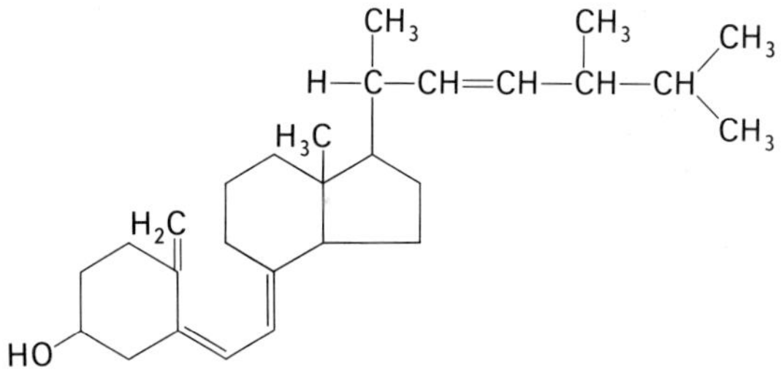

Calciferol
(Vitamin D₂)

Figure 11-13
Structures of some fat-soluble vitamins.

STAGES IN THE EXTRACTION OF ENERGY FROM FOODSTUFFS

Let us take an overview of the process of energy generation in higher organisms before considering these reactions in detail in subsequent chapters. Hans Krebs described three stages in the generation of energy from the oxidation of foodstuffs. *In the first stage, large molecules in food are broken down into smaller units.* Proteins are hydrolyzed to their twenty kinds of constituent amino acids, polysaccharides are hydrolyzed to simple sugars such as glucose, and fats are hydrolyzed to glycerol and fatty acids (Figure 11-14). No useful energy is generated in this phase. *In the second stage, these numerous small molecules are degraded to a few simple units that play a central role in metabolism.* In fact, most of them—sugars, fatty acids, glycerol, and several amino acids—are converted into the acetyl unit of acetyl CoA. Some ATP is generated in this stage, but the amount is small compared with that obtained from the complete oxidation of the acetyl unit of acetyl CoA. *The third stage consists of the citric acid cycle and oxidative phosphorylation,* which are the final common pathways in the oxidation of fuel molecules. Acetyl CoA brings acetyl units into this cycle, where they are completely oxidized to CO_2. Four pairs of electrons are transferred to NAD^+ and FAD for each acetyl group that is oxidized. Then, ATP is generated as electrons flow from the reduced forms of these carriers to O_2, a process called oxidative phosphorylation. Most of the ATP generated by the degradation of foodstuffs is formed in this third stage.

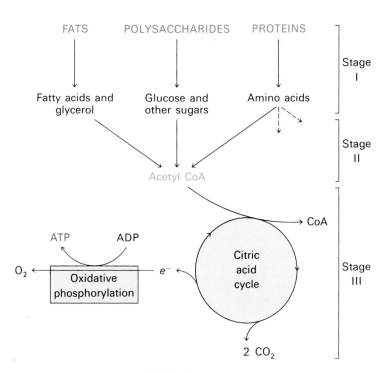

Figure 11-14
Stages in the extraction of energy from foodstuffs.

METABOLIC PROCESSES ARE REGULATED BY A VARIETY OF MECHANISMS

Even the simplest bacterial cell has the capacity for carrying out more than a thousand interdependent reactions. It is evident that this complex network must be rigorously regulated. Furthermore, metabolic control must be flexible, because the external environment of cells is not constant. Studies of a wide range of organisms have shown that there are a number of mechanisms for the control of metabolism. It must be emphasized that the central metabolic pathways have been almost fully elucidated, but knowledge concerning their regulation is still in its infancy. Few problems in biochemistry today are as intellectually challenging and important.

A major mechanism of metabolic regulation is the control of the *amounts* of certain enzymes. This mode of regulation has been extensively studied in bacteria. The control of the *rate of synthesis* of β-galactosidase and other proteins needed for the utilization of lactose is a classic example, which will be discussed in detail in Chapter 28. Recent studies have shown that the *rate of degradation* of some enzymes is controlled as well. Metabolic regulation is also achieved by control of the *catalytic activities* of certain enzymes. One general and important mechanism is *reversible allosteric control*. For example, the first reaction in many biosynthetic pathways is allosterically inhibited by the ultimate product of the pathway, an interaction called *feedback inhibition*. The activity of some enzymes is also modulated by *covalent modifications*, such as the phosphorylation of a specific serine residue.

An important general principle of metabolism is that *biosynthetic and degradative pathways are almost always distinct*. This separation is necessary for energetic reasons, as will be evident in subsequent chapters. It also facilitates the control of metabolism. In eucaryotes, metabolic regulation and flexibility are also enhanced by *compartmentation*. For example, fatty acid oxidation occurs in mitochondria, whereas fatty acid synthesis occurs in the cytosol (the soluble part of the cytoplasm). Compartmentation segregates these opposed reactions.

Many reactions in metabolism are controlled in part by the *energy status* of the cell. One index of the energy status is the *energy charge*, which is proportional to the mole fraction of ATP plus half the mole fraction of ADP, given that ATP contains two anhydride bonds, whereas ADP contains one. Hence, the energy charge is defined as

$$\text{Energy charge} = \frac{[\text{ATP}] + \frac{1}{2}[\text{ADP}]}{[\text{ATP}] + [\text{ADP}] + [\text{AMP}]}$$

The energy charge can have a value ranging from 0 (all AMP) to 1 (all ATP). Daniel Atkinson has shown that *ATP-generating pathways are inhibited by a high energy charge, whereas ATP-utilizing pathways are stimulated by a high energy charge*. In plots of the reaction rates of such pathways versus the energy charge, the curves are steepest near an energy charge of 0.9, where they usually intersect (Fig. 11-15). It is evident that the control of these pathways is designed to maintain the energy charge within rather narrow limits. In other words, *the energy charge, like the pH of a cell, is buffered*. The energy charge of most cells is in the range of 0.80 to 0.95. An alternative index of the

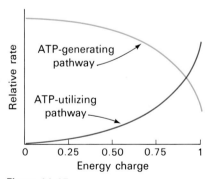

Figure 11-15
Effect of the energy charge on the relative rates of a typical ATP-generating (catabolic) pathway and a typical ATP-utilizing (anabolic) pathway.

energy status is the *phosphorylation potential,* which is defined as

$$\text{Phosphorylation potential} = \frac{[\text{ATP}]}{[\text{ADP}][\text{P}_i]}$$

The phosphorylation potential, in contrast with the energy charge, depends on the concentration of P_i and is directly related to the free energy available from ATP.

SUMMARY

Cells extract energy from their environments and convert foodstuffs into cell components by a highly integrated network of chemical reactions called metabolism. The most valuable thermodynamic concept for understanding the energetics of metabolism is free energy, which is a measure of the capacity of a system to do useful work at constant pressure and temperature. A reaction can occur spontaneously only if the change in free energy (ΔG) is negative. The ΔG for a reaction is independent of path and depends only on the nature of the reactants and their activities (which can sometimes be approximated by their concentrations). The free-energy change of a reaction occurring when reactants and products are at unit activity is called the standard free-energy change ($\Delta G°$). Biochemists usually use $\Delta G°'$, which is the standard free-energy change at pH 7. ATP, the universal currency of energy in biological systems, is an energy-rich molecule because it contains two anhydride bonds. The electrostatic repulsion between these negatively charged groups is reduced when ATP is hydrolyzed. Also, ADP and P_i are stabilized by resonance more than is ATP. The hydrolysis of ATP shifts the equilibrium of a coupled reaction by a factor of about 10^8.

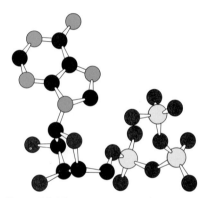

Figure 11-16
Model of adenosine triphosphate (ATP).

The basic strategy of metabolism is to form ATP, NADPH, and macromolecular precursors. ATP is consumed in muscle contraction and other motions of cells, active transport, and biosyntheses. NADPH, which carries two electrons at a high potential, provides reducing power in the biosynthesis of cell components from more oxidized precursors. ATP and NADPH are continuously generated and consumed. There are three stages in the extraction of energy from foodstuffs in aerobic organisms. In the first stage, large molecules are broken down into smaller ones, such as amino acids, sugars, and fatty acids. In the second stage, these small molecules are degraded to a few simple units that have a pervasive role in metabolism. One of them is the acetyl unit of acetyl CoA, a carrier of activated acyl groups. The third stage of metabolism is the citric acid cycle and oxidative phosphorylation, in which fuels are completely oxidized to CO_2 and ATP is generated as electrons flow to O_2, the ultimate electron acceptor.

Metabolism is regulated in a variety of ways. The amounts of some critical enzymes are controlled by regulation of the rate of protein synthesis and degradation. In addition, the catalytic activities of some enzymes are regulated by allosteric interactions (as in feedback inhibition) and by covalent modification. Compartmentation and distinct pathways for biosynthesis and degradation also contribute to metabolic regulation. The energy charge, which depends on the relative amounts of ATP, ADP, and AMP, plays a role in metabolic regulation. A high energy charge inhibits ATP-generating (catabolic) pathways, whereas it stimulates ATP-utilizing (anabolic) pathways.

SELECTED READINGS

OVERVIEWS OF METABOLISM

Krebs, H. A., and Kornberg, H. L., 1957. *Energy Transformations in Living Matter*. Springer-Verlag. [Includes a valuable appendix by K. Burton containing thermodynamic data.]

Wood, W. B., 1974. *The Molecular Basis of Metabolism*. Unit 3 in Biocore. McGraw-Hill.

THERMODYNAMICS

Klotz, I. M., 1967. *Energy Changes in Biochemical Reactions*. Academic Press. [A concise introduction, full of insight.]

Bray, H. G., and White, K., 1966. *Kinetics and Thermodynamics in Biochemistry* (2nd ed.). Academic Press.

Hill, T. L., 1977. *Free Energy Transduction in Biology*. Academic Press.

Ingraham, L. L., and Pardee, A. B., 1967. Free energy and entropy in metabolism. *In* Greenberg, D. M., (ed.), *Metabolic Pathways* (3rd ed.), vol. 1, pp. 1–46. Academic Press.

Alberty, R. A., 1968. Effect of pH and metal ion concentration on the equilibrium hydrolysis of adenosine triphosphate to adenosine diphosphate. *J. Biol. Chem.* 243:1337–1343.

Jencks, W. P., 1970. Free energies of hydrolysis and decarboxylation. *In* Sober, H. A., *Handbook of Biochemistry* (2nd ed.), pp. J181–J186. Chemical Rubber Co.

REGULATION OF METABOLISM

Newsholme, E. A., and Start, C., 1973. *Regulation in Metabolism*. Wiley. [An excellent treatment of metabolic regulation in mammals.]

Atkinson, D. E., 1977. *Cellular Energy Metabolism and Its Regulation*. Academic Press. [Contains a detailed account of the concept of energy charge and of other aspects of metabolic control.]

Erecińska, M., and Wilson, D. F., 1978. Homeostatic regulation of cellular energy metabolism. *Trends Biochem. Sci.* 3:219–223. [Review of the regulatory role of the phosphorylation potential in mitochondria.]

HISTORICAL ASPECTS

Kalckar, H. M., (ed.), 1969. *Biological Phosphorylations*. Prentice-Hall. [A valuable collection of many classical papers on bioenergetics.]

Fruton, J. S., 1972. *Molecules and Life*. Wiley-Interscience. [Perceptive and scholarly essays on the interplay of chemistry and biology since 1800. Metabolism and bioenergetics are among the topics treated in detail.]

Lipmann, F., 1971. *Wanderings of a Biochemist*. Wiley-Interscience. [Contains reprints of some of the author's classic papers and several delightful essays.]

PROBLEMS

1. What is the direction of each of the following reactions when the reactants are initially present in equimolar amounts? Use the data given in Table 11-2.

    (a) ATP + creatine $\rightleftharpoons$
              creatine phosphate + ADP

    (b) ATP + glycerol $\rightleftharpoons$
              glycerol 3-phosphate + ADP

    (c) ATP + pyruvate $\rightleftharpoons$
              phosphoenolpyruvate + ADP

    (d) ATP + glucose $\rightleftharpoons$
              glucose 6-phosphate + ADP

2. What information do the $\Delta G^{\circ\prime}$ data given in Table 11-2 provide about the relative rates of hydrolysis of pyrophosphate and acetyl phosphate?

3. Consider the reaction

    ATP + pyruvate $\rightleftharpoons$
              phosphoenolpyruvate + ADP

    (a) Calculate $\Delta G^{\circ\prime}$ and K'_{eq} at 25°C for this reaction, using the data given in Table 11-2.

    (b) What is the equilibrium ratio of pyruvate to phosphoenolpyruvate if the ratio of ATP to ADP is 10?

4. Calculate $\Delta G^{\circ\prime}$ for the isomerization of glucose 6-phosphate to glucose 1-phosphate. What is the equilibrium ratio of glucose 6-phosphate to glucose 1-phosphate at 25°C?

5. The formation of acetyl CoA from acetate is an ATP-driven reaction:

    Acetate + ATP + CoA $\rightleftharpoons$
              acetyl CoA + AMP + PP_i

    (a) Calculate $\Delta G^{\circ\prime}$ for this reaction, using data given in this chapter.

    (b) The PP_i formed in the above reaction is rapidly hydrolyzed in vivo because of the ubiquity of inorganic pyrophosphatase. The $\Delta G^{\circ\prime}$ for the hydrolysis of PP_i is -8 kcal/mol. Calculate the $\Delta G^{\circ\prime}$ for the overall reaction. What effect does the hydrolysis of PP_i have on the formation of acetyl CoA?

6. The pK of an acid is a measure of its proton group transfer potential.
    (a) Derive a relationship between ΔG° and pK.
    (b) What is the ΔG° for the ionization of acetic acid, which has a pK of 4.8?

7. What is the common structural feature of ATP, FAD, NAD$^+$, and CoA?

8. Fibrinogen contains tyrosine-O-sulfate. Propose an activated form of sulfate that could react in vivo with the aromatic hydroxyl group of a tyrosine residue in a protein to form tyrosine-O-sulfate.

For additional problems, see W. B. Wood, J. H. Wilson, R. M. Benbow, and L. E. Hood, *Biochemistry: A Problems Approach* (Benjamin, 1974), ch. 8; I. M. Klotz, *Energy Changes in Biochemical Reactions* (Academic Press, 1967); and R. Montgomery and C. A. Swenson, *Quantitative Problems in the Biochemical Sciences,* 2nd ed. (Freeman, 1976), ch. 10.

CHAPTER 12

GLYCOLYSIS

We begin our consideration of the generation of metabolic energy with glycolysis, a nearly universal pathway in biological systems. *Glycolysis is the sequence of reactions that converts glucose into pyruvate with the concomitant production of ATP.* In aerobic organisms, glycolysis is the prelude to the citric acid cycle and the electron-transport chain, which together harvest most of the energy contained in glucose. Under aerobic conditions, pyruvate enters mitochondria, where it is completely oxidized to CO_2 and H_2O. If the supply of oxygen is insufficient, as in actively contracting muscle, pyruvate is converted into lactate. In some anaerobic organisms, such as yeast, pyruvate is transformed into ethanol. The formation of ethanol and lactate from glucose are examples of fermentations.

The elucidation of glycolysis has a rich history. Indeed, the development of biochemistry and the delineation of this central pathway went hand-in-hand. A key discovery was made by Hans Buchner and Eduard Buchner in 1897, quite by accident. They were interested in manufacturing cell-free extracts of yeast for possible therapeutic use. These extracts had to be preserved without using antiseptics such as phenol, and so they decided to try sucrose, a commonly used preservative in kitchen chemistry. They obtained a

Glycolysis—
Derived from the Greek words *glycos*, sugar (sweet), and *lysis*, dissolution.

Fermentation—
An ATP-generating process in which organic compounds act as both donors and acceptors of electrons. Fermentation can occur in the absence of O_2. Discovered by Pasteur, who described fermentation as "la vie sans l'air" (life without air).

startling result: sucrose was rapidly fermented into alcohol by the yeast juice. The significance of this finding was immense. *The Buchners demonstrated for the first time that fermentation could occur outside living cells.* The accepted view of their day, asserted by Louis Pasteur in 1860, was that fermentation is inextricably tied to living cells. The chance discovery of the Buchners refuted this vitalistic dogma and opened the door to modern biochemistry. Metabolism became chemistry.

The next important contribution was made by Arthur Harden and William Young in 1905. They added yeast juice to a solution of glucose and found that fermentation started almost immediately. However, the rate of fermentation soon decreased markedly unless inorganic phosphate was added. Furthermore, they found that the added inorganic phosphate disappeared in the course of fermentation, and so they inferred that *inorganic phosphate was incorporated into a sugar phosphate.* Harden and Young isolated a hexose diphosphate, which was later shown to be fructose 1,6-diphosphate. They also discovered that yeast juice contains two kinds of substances necessary for fermentation: "zymase" and "cozymase." They found that yeast juice lost its activity if it was dialyzed or heated to 50°C. However, the inactive dialyzed juice became active when it was mixed with inactive heated juice. Thus, activity depended on the presence of two kinds of substances: a heat-labile, nondialyzable component (called *zymase*) and a heat-stable, dialyzable fraction (called *cozymase*). We now know that "zymase" consists of a number of enzymes, whereas "cozymase" consists of metal ions, adenosine triphosphate (ATP), adenosine diphosphate (ADP), and coenzymes such as nicotinamide adenine dinucleotide (NAD^+).

Studies of muscle extracts carried out several years later showed that many of the reactions of lactic fermentation were the same as those of alcoholic fermentation. *This was an exciting discovery because it revealed an underlying unity in biochemistry.* The complete glycolytic pathway was elucidated by 1940, largely because of the contributions made by Gustav Embden, Otto Meyerhof, Carl Neuberg, Jacob Parnas, Otto Warburg, Gerty Cori, and Carl Cori. Glycolysis is sometimes called the Embden-Meyerhof pathway.

> *Enzyme—*
> A term coined by Friedrich Wilhelm Kühne in 1878 to designate catalytically active substances that had previously been called ferments. Derived from the Greek words *en,* in, and *zyme,* leaven.

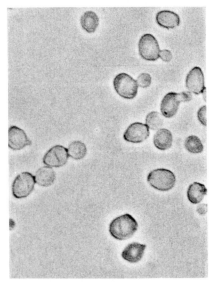

Figure 12-1
Light micrograph of yeast cells.
[Courtesy of Dr. Randy Schekman.]

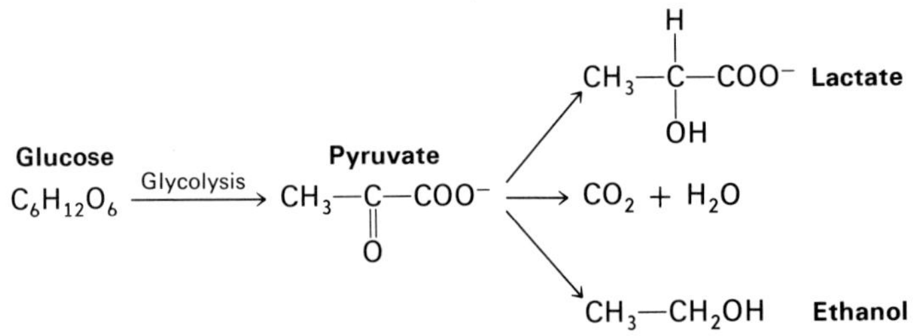

Figure 12-2
Some fates of glucose.

Let us consider the nomenclature and structure of monosaccharides, the simplest carbohydrates, before turning to the reactions of glycolysis. Monosaccharides are aldehydes or ketones that have two or more hydroxyl groups; their empirical formula is $(CH_2O)_n$. The simplest ones, for which $n = 3$, are glyceraldehyde and dihydroxyacetone. They are *trioses*. Glyceraldehyde is called an *aldose* because it contains an aldehyde group, whereas dihydroxyacetone is a *ketose* because it contains a keto group.

Glyceraldehyde has a single asymmetric carbon. Thus, there are two stereoisomers of this three-carbon aldose, named D-glyceraldehyde and L-glyceraldehyde. The prefixes D and L designate the absolute configuration.

Sugars with 4, 5, 6, and 7 carbon atoms are called *tetroses, pentoses, hexoses,* and *heptoses,* respectively. Their formulas and stereochemical relationships are given in the appendix to this chapter (p. 280). Two common hexoses are D-*glucose* and D-*fructose*. The prefixed symbol D means that the absolute configuration at the asymmetric carbon furthest from the aldehyde or keto group, namely C-5, is the same as that in D-glyceraldehyde. Note that glucose is an aldose, whereas fructose is a ketose.

The predominant forms of glucose and fructose in solution are not open-chain structures. Rather, the open-chain forms of glucose and fructose can cyclize into rings. In general, an aldehyde can react with an alcohol to form a *hemiacetal*. The C-1 aldehyde in the open-chain form of glucose reacts with the C-5 hydroxyl group to form an *intramolecular hemiacetal*. The resulting six-membered sugar ring is called *pyranose* because of its similarity to pyran.

D-Glyceraldehyde
(An aldose)

Dihydroxyacetone
(A ketose)

Figure 12-3
Absolute configuration of D-glyceraldehyde.

D-Glucose
(An aldose)

D-Fructose
(A ketose)

Aldehyde   **Alcohol**   **Hemiacetal**

D-Glucose
(Open-chain form)

α-D-Glucopyranose
(A ring form of glucose)

Similarly, a ketone can react with an alcohol to form a *hemiketal*. The C-2 keto group in the open-chain form of fructose reacts with the C-5 hydroxyl group to form an *intramolecular hemiketal*. This five-membered sugar ring is called *furanose* because of its similarity to furan.

Ketone   **Alcohol**   **Hemiketal**

D-Fructose ⇌ ⇌ **α-D-Fructofuranose**
(A ring form of fructose)

The structural formulas of glucopyranose and fructofuranose shown above and on page 257 are Haworth projections. In such a projection, the carbon atoms in the ring are not explicitly shown. The approximate plane of the ring is perpendicular to the plane of the paper, with the heavy line on the ring closest to the reader.

An additional asymmetric center is created when glucose cyclizes. Carbon-1, the carbonyl carbon atom in the open-chain form, becomes an asymmetric center in the ring form. Two ring structures can be formed: α-D-glucopyranose and β-D-glucopyranose (Figure 12-4). *The designation α means that the hydroxyl group attached to C-1 is below the plane of the ring; β means that it is above the plane of the ring.* The C-1 carbon is called the anomeric carbon atom, and so the α and β forms are *anomers*.

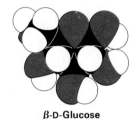

β-D-Glucose

Figure 12-4
Space-filling model of β-D-glucose.

α-D-Glucose   **β-D-Glucose**

α-D-Fructose   **β-D-Fructose**

Chair form of a pyranose
(e = equatorial substituent;
a = axial substituent)

β-D-Glucose

The same nomenclature applies to the ring form of fructose, except that α and β refer to the hydroxyl groups attached to C-2, the anomeric carbon atom in ketoses.

Alpha-D-glucopyranose, β-D-glucopyranose, and the open-chain form of glucose interconvert rapidly, as do the furanose and open-chain forms of fructose. The designations glucose and fructose will be used to refer to the equilibrium mixture of the open-chain and ring forms of these sugars.

The six-membered pyranose ring is not planar. Rather, the preferred conformation is a *chair form*. The substituents are of two types: *axial* and *equatorial*. In β-D-glucopyranose, all of the hydroxyl groups are equatorial.

Learning the sequence of events in a metabolic pathway is facilitated by having a firm grasp of the structures of the reactants and an understanding of the types of reactions taking place. The intermediates in glycolysis have either six carbons or three carbons. The *six-carbon units* are derivatives of *glucose* and *fructose*. The *three-carbon units* are derivatives of *dihydroxyacetone, glyceraldehyde, glycerate,* and *pyruvate*.

All intermediates in glycolysis between glucose and pyruvate are *phosphorylated*. The phosphoryl groups in these compounds are in either *ester* or *anhydride* linkage. Now let us look at some of the kinds of reactions that occur in glycolysis:

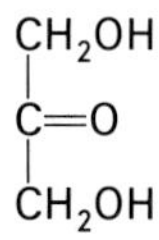

1. *Phosphoryl transfer.* A phosphoryl group is transferred from ATP to a glycolytic intermediate, or vice versa.

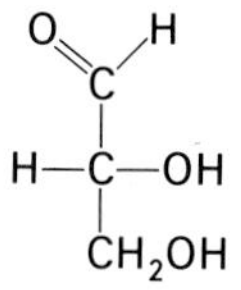

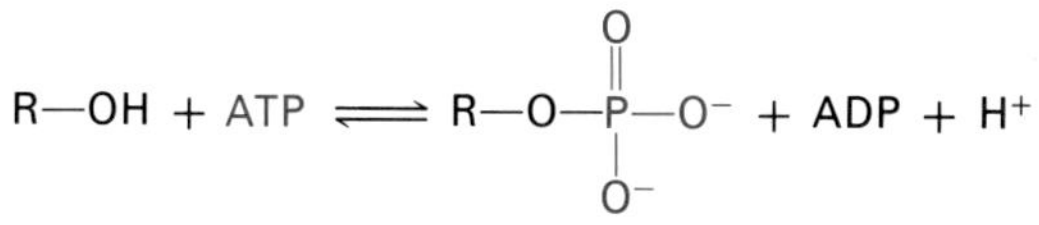

$$R\text{---}OH + ATP \rightleftharpoons R\text{---}O\text{---}\overset{\displaystyle O}{\underset{\displaystyle O^-}{\overset{\|}{P}}}\text{---}O^- + ADP + H^+$$

2. *Phosphoryl shift.* A phosphoryl group is shifted within a molecule from one oxygen atom to another.

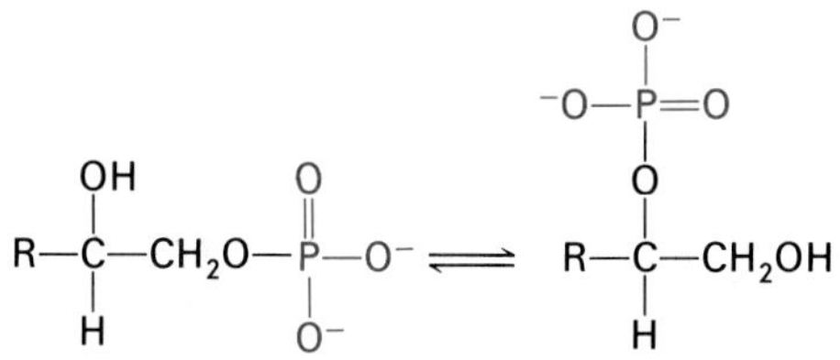

3. *Isomerization.* A ketose is converted into an aldose, or vice versa.

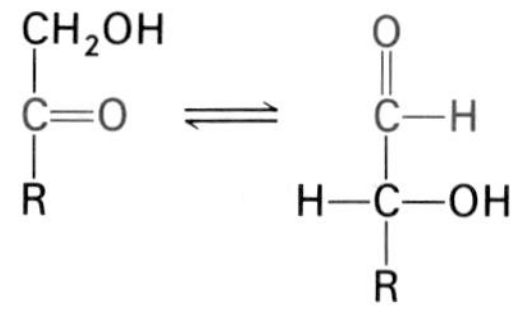

4. *Dehydration.* A molecule of water is eliminated.

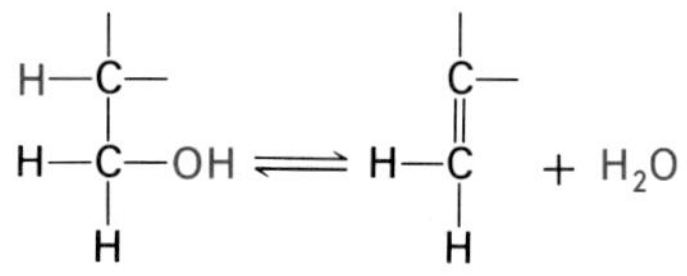

5. *Aldol cleavage.* A carbon–carbon bond is split in a reversal of an aldol condensation.

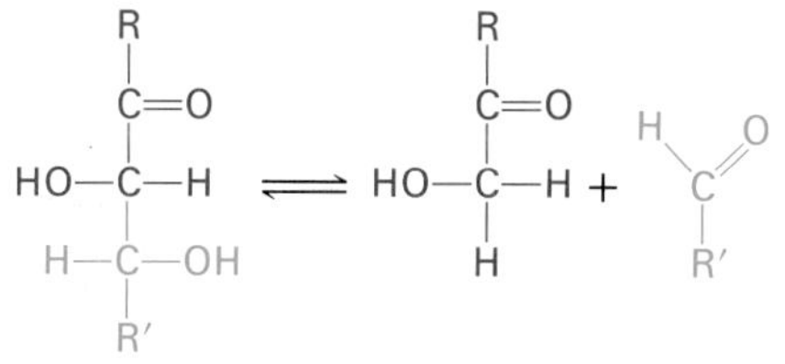

CH₂OH / C=O / CH₂OH — **Dihydroxyacetone**

Glyceraldehyde

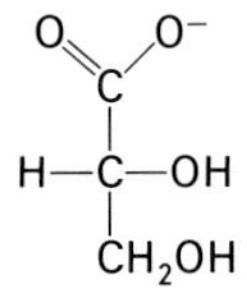

Glycerate

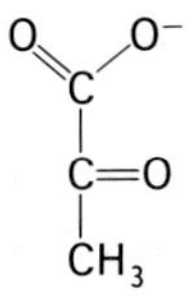

Pyruvate

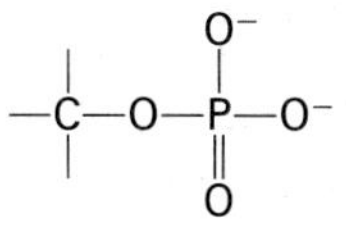

Ester

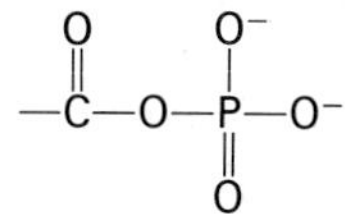

Anhydride

Ketose   **Aldose**

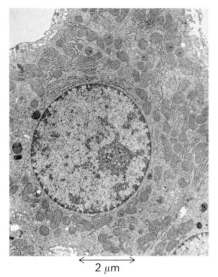

Figure 12-5
Electron micrograph of a liver cell.
Glycolysis occurs in the cytosol.
[Courtesy of Dr. Anne Hubbard.]

FORMATION OF FRUCTOSE 1,6-DIPHOSPHATE FROM GLUCOSE

We now start our journey down the glycolytic pathway. The reactions in this pathway take place in the cell cytosol. The first stage, which is the conversion of glucose into fructose 1,6-diphosphate, consists of three steps: a phosphorylation, an isomerization, and a second phosphorylation reaction. *The strategy of these initial steps in glycolysis is to form a compound that can be readily cleaved into phosphorylated three-carbon units.* Energy is subsequently extracted from the three-carbon units.

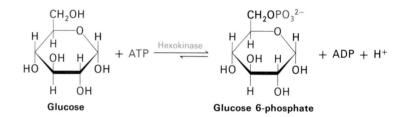

Glucose enters most cells by a specific carrier or active-transport system (p. 871). *Glucose inside cells has only one fate: it is phosphorylated by ATP to form glucose 6-phosphate.* The transfer of the phosphoryl group from ATP to the hydroxyl group on C-6 of glucose is catalyzed by *hexokinase*.

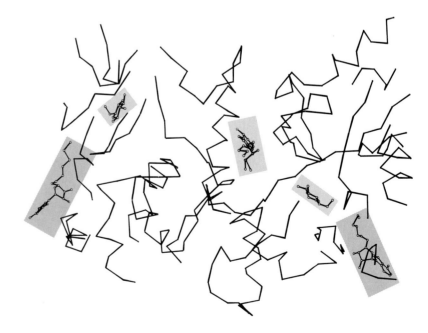

Figure 12-6
Schematic diagram of the α-carbon backbone of yeast hexokinase. The two identical subunits of this dimeric enzyme interact with each other nonsymmetrically. Glucose (shaded green) and ATP (shaded red) are bound to the catalytic site of each dimer. An additional ATP molecule (shaded blue) is bound at the interface between subunits. [Courtesy of Dr. Thomas Steitz.]

Phosphoryl transfer is a basic reaction in biochemistry. An enzyme that transfers a phosphoryl group from ATP to an acceptor is called a *kinase*. Hexokinase, then, is an enzyme that transfers a phosphoryl group from ATP to a variety of six-carbon sugars (*hexoses*). Hexokinase, like all other kinases, requires Mg^{2+} (or another divalent metal ion such as Mn^{2+}) for activity. The divalent metal ion forms a complex with ATP. The structures of two possible Mg^{2+}-ATP complexes are shown at the right.

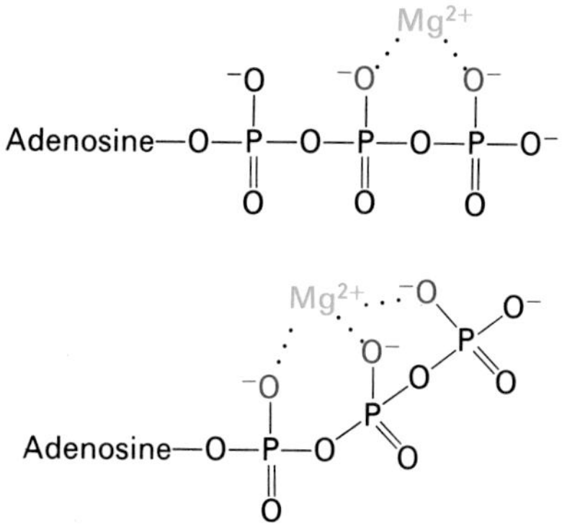

The next step in glycolysis is the *isomerization of glucose 6-phosphate to fructose 6-phosphate*. The *six-membered pyranose ring* of glucose 6-phosphate is converted into the *five-membered furanose ring* of fructose 6-phosphate. Recall that the open-chain form of glucose has an aldehyde group on C-1, whereas the open-chain form of fructose has a keto group on C-2. The aldehyde on C-1 reacts with the hydroxyl group on C-5 to form the pyranose ring, whereas the keto group on C-2 reacts with the C-5 hydroxyl to form the furanose ring. Thus, the isomerization of glucose 6-phosphate to fructose 6-phosphate is a *conversion of an aldose into a ketose*.

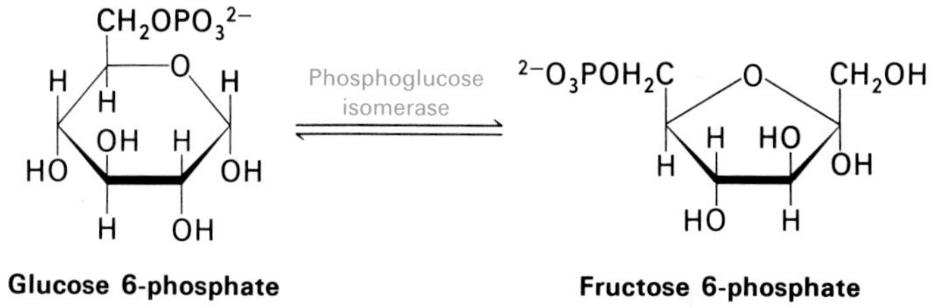

Figure 12-7
Modes of binding of Mg^{2+} to ATP.

The open-chain representations of these sugars show the essence of this reaction.

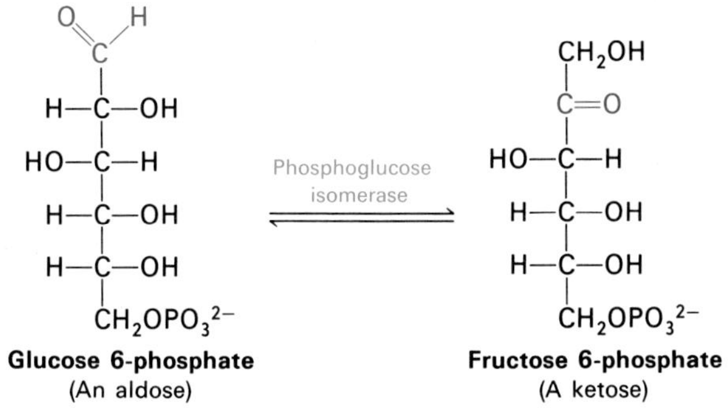

A second phosphorylation reaction follows the isomerization step. *Fructose 6-phosphate is phosphorylated by ATP to fructose 1,6-diphosphate.*

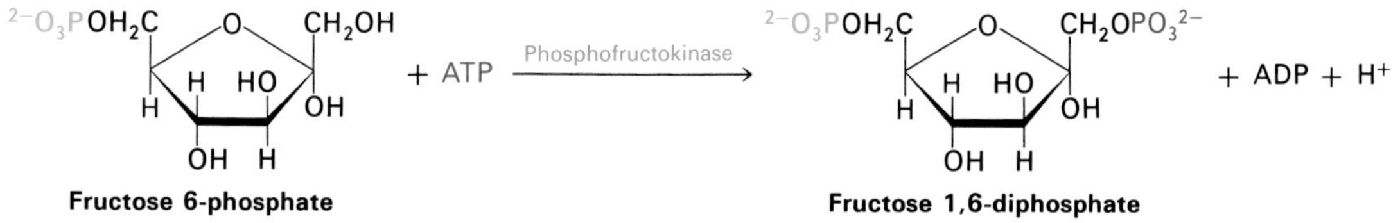

This reaction is catalyzed by *phosphofructokinase,* an allosteric enzyme. The pace of glycolysis is critically dependent on the level of activity of this enzyme. The catalytic activity of phosphofructokinase is allosterically controlled by ATP and several other metabolites (p. 266).

FORMATION OF GLYCERALDEHYDE 3-PHOSPHATE BY CLEAVAGE AND ISOMERIZATION

The second stage of glycolysis consists of four steps, starting with the splitting of fructose 1,6-diphosphate to yield *glyceraldehyde 3-phosphate* and *dihydroxyacetone phosphate.* The remaining steps in glycolysis involve three-carbon units rather than six-carbon units.

| Fructose 1,6-diphosphate | Dihydroxyacetone phosphate | Glyceraldehyde 3-phosphate |

This reaction is catalyzed by *aldolase.* This enzyme derives its name from the nature of the reverse reaction, which is an aldol condensation.

Glyceraldehyde 3-phosphate is on the direct pathway of glycolysis, but dihydroxyacetone phosphate is not. However, dihydroxyacetone phosphate can be readily converted into glyceraldehyde 3-phosphate. These compounds are isomers: dihydroxyacetone phosphate is a ketose, whereas glyceraldehyde 3-phosphate is an aldose. The isomerization of these three-carbon phosphorylated sugars is catalyzed by *triose phosphate isomerase.* This reaction is very fast and reversible. At equilibrium, 96% of the triose phosphate is dihydroxyacetone phosphate. However, the reaction proceeds readily from dihydroxyacetone phosphate to glyceraldehyde 3-phosphate because of efficient removal of this product.

Glyceraldehyde 3-phosphate (An aldose)   Dihydroxyacetone phosphate (A ketose)

Thus, two molecules of glyceraldehyde 3-phosphate are formed from one molecule of fructose 1,6-diphosphate by the sequential

Table 12-2 267
Reactions of glycolysis

Step	Reaction	Enzyme	Type*	$\Delta G°'$	ΔG
1	Glucose + ATP ⟶ glucose 6-phosphate + ADP + H⁺	Hexokinase	a	−4.0	−8.0
2	Glucose 6-phosphate ⇌ fructose 6-phosphate	Phosphoglucose isomerase	c	+0.4	−0.6
3	Fructose 6-phosphate + ATP ⟶ fructose 1,6-diphosphate + ADP + H⁺	Phosphofructokinase	a	−3.4	−5.3
4	Fructose 1,6-diphosphate ⇌ dihydroxyacetone phosphate + glyceraldehyde 3-phosphate	Aldolase	e	+5.7	−0.3
5	Dihydroxyacetone phosphate ⇌ glyceraldehyde 3-phosphate	Triose phosphate isomerase	c	+1.8	+0.6
6	Glyceraldehyde 3-phosphate + P_i + NAD⁺ ⇌ 1,3-diphosphoglycerate + NADH + H⁺	Glyceraldehyde 3-phosphate dehydrogenase	f	+1.5	−0.4
7	1,3-Diphosphoglycerate + ADP ⇌ 3-phosphoglycerate + ATP	Phosphoglycerate kinase	a	−4.5	+0.3
8	3-Phosphoglycerate ⇌ 2-phosphoglycerate	Phosphoglyceromutase	b	+1.1	+0.2
9	2-Phosphoglycerate ⇌ phosphoenolpyruvate + H_2O	Enolase	d	+0.4	−0.8
10	Phosphoenolpyruvate + ADP + H⁺ ⟶ pyruvate + ATP	Pyruvate kinase	a	−7.5	−4.0

Note: $\Delta G°'$ and ΔG are expressed in kcal/mol. ΔG, the actual free-energy change, has been calculated from $\Delta G°'$ and known concentrations of reactants under typical physiological conditions.

*Reaction type: (a) Phosphoryl transfer   (d) Dehydration
                (b) Phosphoryl shift   (e) Aldol cleavage
                (c) Isomerization   (f) Phosphorylation coupled to oxidation

catalyzed by hexokinase, phosphofructokinase, and pyruvate kinase are virtually irreversible, and so they would be expected to have regulatory as well as catalytic roles. In fact, all three enzymes are control sites.

Phosphofructokinase is the most important control element in glycolysis. This tetrameric enzyme is *inhibited by high levels of ATP,* which lowers the affinity of the enzyme for fructose 6-phosphate. A high concentration of ATP converts the hyperbolic binding curve for fructose 6-phosphate into a sigmoidal one (Figure 12-9). This allosteric effect is elicited by the binding of ATP to a highly specific regulatory site that is distinct from the catalytic site. The inhibitory action of ATP is reversed by AMP, and so *the activity of the enzyme increases when the ATP/AMP ratio is lowered.* In other words, *glycolysis is stimulated when the energy charge of the cell is low.* As already mentioned, glycolysis also furnishes the carbon skeleton for biosyntheses, and so phosphofructokinase might also be expected to be regulated by a signal indicating whether building blocks are abundant or scarce. Indeed, *phosphofructokinase is inhibited by citrate,* an early intermediate in the citric acid cycle (p. 284). A high level of citrate means that biosynthetic precursors are abundant and so additional glucose should not be degraded for this purpose. Citrate inhibits phosphofructokinase by enhancing the inhibitory effect of ATP.

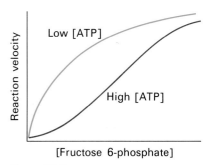

Figure 12-9
Allosteric regulation of phosphofructokinase. A high level of ATP inhibits the enzyme by decreasing its affinity for fructose 6-phosphate. AMP diminishes and citrate enhances the inhibitory effect of ATP.

Thus, phosphofructokinase is most active when the cell needs both energy and building blocks, as signalled by a low ATP/AMP ratio and a low level of citrate. The enzyme is moderately active when either energy or a carbon skeleton is needed. When both are abundant, the activity of phosphofructokinase is nearly switched off.

Hexokinase and pyruvate kinase also participate in regulating the rate of glycolysis. Pyruvate kinase from muscle and liver is allosterically inhibited by ATP, and so the conversion of phosphoenolpyruvate into pyruvate is blocked when the energy charge is high. Hexokinase is allosterically inhibited by glucose 6-phosphate. The level of fructose 6-phosphate increases when phosphofructokinase is blocked, and so there is a corresponding increase in the level of glucose 6-phosphate, which is in equilibrium with fructose 6-phosphate. *Hence, the inhibition of phosphofructokinase by a high ATP/AMP ratio or a high level of citrate leads to the inhibition of hexokinase.* In liver, glucose is phosphorylated into glucose 6-phosphate even when the glucose 6-phosphate level is high because of the presence of *glucokinase,* which differs from hexokinase. Glucokinase has a high K_M for glucose, and so it is effective only when glucose is abundant. The role of glucokinase is to provide glucose 6-phosphate for the synthesis of glycogen, a storage form of glucose (p. 357). The high K_M of glucokinase in the liver gives the brain and muscle first call on glucose when its supply is limited.

Why is phosphofructokinase rather than hexokinase the pacemaker of glycolysis? The reason becomes evident on noting that glucose 6-phosphate is not only a glycolytic intermediate. Glucose 6-phosphate can also be converted into glycogen or it can be oxidized by the pentose phosphate pathway (p. 333) to generate NADPH. The first irreversible reaction unique to the glycolytic pathway, called the *committed step,* is the phosphorylation of fructose 6-phosphate to fructose 1,6-diphosphate. Thus, it is highly appropriate for phosphofructokinase to be the primary control site in glycolysis. *In general, the enzyme catalyzing the committed step in a metabolic sequence is the most important control element in the pathway.*

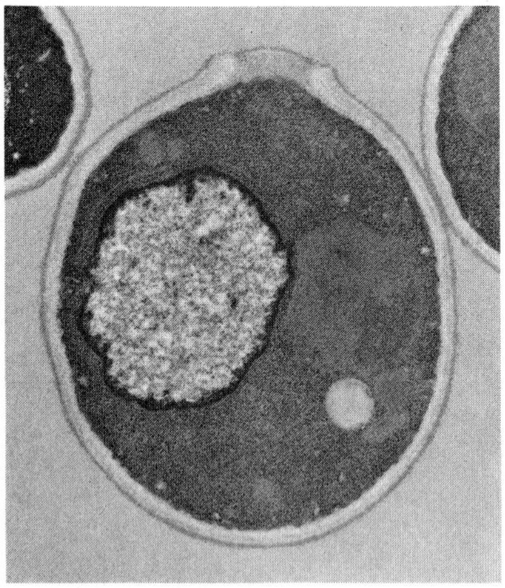

PYRUVATE CAN BE CONVERTED INTO ETHANOL, LACTATE, OR ACETYL COENZYME A

The sequence of reactions from glucose to pyruvate is very similar in all organisms and in all kinds of cells. In contrast, the fate of pyruvate in the generation of metabolic energy is variable. Three reactions of pyruvate are considered here.

1. *Ethanol* is formed from pyruvate in yeast and several other microorganisms. The first step is the decarboxylation of pyruvate:

$$\text{Pyruvate} + H^+ \longrightarrow \text{acetaldehyde} + CO_2$$

Figure 12-10
Electron micrograph of a yeast cell.
[Courtesy of Lynne Mercer.]

This reaction is catalyzed by pyruvate decarboxylase, which contains thiamine pyrophosphate as a coenzyme. Thiamine pyrophosphate is the coenzyme in a variety of decarboxylases (see p. 290 for a discussion of its mechanism of action).

The second step is the reduction of acetaldehyde to ethanol by NADH. *Alcohol dehydrogenase,* which contains a zinc ion at its active site, catalyzes this oxidation-reduction reaction.

$$\text{Acetaldehyde} + \text{NADH} + \text{H}^+ \Longrightarrow \text{ethanol} + \text{NAD}^+$$

The conversion of glucose into ethanol is called *alcoholic fermentation.* The net reaction of this anaerobic process is

$$\text{Glucose} + 2\,\text{P}_i + 2\,\text{ADP} + 2\,\text{H}^+ \longrightarrow$$
$$2\,\text{ethanol} + 2\,\text{CO}_2 + 2\,\text{ATP} + 2\,\text{H}_2\text{O}$$

It is important to note that NAD^+ and NADH do not appear in this equation. Acetaldehyde is reduced to ethanol so that NAD^+ is regenerated for use in the reaction catalyzed by glyceraldehyde 3-phosphate dehydrogenase. Thus, there is no net oxidation-reduction in the conversion of glucose into ethanol.

2. *Lactate* is normally formed from pyruvate in a variety of microorganisms. The reaction also occurs in the cells of higher organisms when the amount of oxygen is limiting, as in muscle during intense activity. The reduction of pyruvate by NADH to form lactate is catalyzed by *lactate dehydrogenase:*

The overall reaction in the conversion of glucose into lactate is

$$\text{Glucose} + 2\,\text{P}_i + 2\,\text{ADP} \longrightarrow 2\,\text{lactate} + 2\,\text{ATP} + 2\,\text{H}_2\text{O}$$

As in alcoholic fermentation, there is no net oxidation-reduction. The NADH formed in the oxidation of glyceraldehyde 3-phosphate is consumed in the reduction of pyruvate. *The regeneration of NAD^+ in the reduction of pyruvate to lactate or ethanol sustains the continued operation* *of glycolysis under anaerobic conditions.* If NAD^+ were not regenerated, glycolysis could not proceed beyond glyceraldehyde 3-phosphate, which means that no ATP would be generated. In effect, the formation of lactate buys time, as will be discussed in Chapter 15.

3. Only a small fraction of the energy of glucose is released in its anaerobic conversion into lactate (or ethanol). Much more energy can be extracted aerobically by means of the citric acid cycle and the electron-transport chain. The entry point to this oxidative pathway is *acetyl coenzyme A* (acetyl CoA), which is formed inside mitochondria by the oxidative decarboxylation of pyruvate:

$$Pyruvate + NAD^+ + CoA \longrightarrow acetyl\ CoA + CO_2 + NADH$$

This reaction, which is catalyzed by the pyruvate dehydrogenase complex, will be discussed in detail in the next chapter. The NAD^+ required for this reaction and for the oxidation of glyceraldehyde 3-phosphate is regenerated when NADH ultimately transfers its electrons to O_2 through the electron-transport chain in mitochondria.

THE BINDING SITE FOR NAD⁺ IS VERY SIMILAR IN A VARIETY OF DEHYDROGENASES

Lactate dehydrogenase from skeletal muscle, a 140-kdal tetramer, and alcohol dehydrogenase, an 84-kdal dimer, have quite different three-dimensional structures. However, their binding sites for NAD^+ are strikingly similar (Figure 12-11). The NAD^+ binding region is made up of four α helices and six strands of parallel β sheets. The conformations of NAD^+ bound to lactate dehydrogenase and to alcohol dehydrogenase also are nearly the same. The adenosine moiety of NAD^+ is bound in a hydrophobic crevice. In contrast, the nicotinamide unit is bound so that the reactive side of the ring is in a polar environment, whereas the other side makes contact with hydrophobic residues of the enzyme. The bound NAD^+ has an extended conformation (Figure 12-12), the adenine and nicotinamide rings being 14 Å apart. The three-dimensional structures of glyceraldehyde 3-phosphate dehydrogenase and malate dehydrogenase (a citric acid cycle enzyme, p. 288) are also known at high resolution. Their NAD^+ binding sites are very similar to those in lactate dehydrogenase and alcohol dehydrogenase. *It seems likely that the NAD^+ binding region common to these four enzymes exemplifies a fundamental structural motif of NAD^+-linked dehydrogenases.*

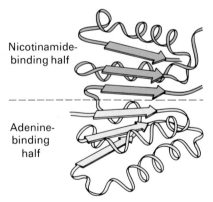

Nicotinamide-binding half

Adenine-binding half

Figure 12-11
Schematic diagram of the NAD^+ binding region in dehydrogenases. The nicotinamide-binding half (shaded green) is structurally similar to the adenine-binding half (shaded yellow) of the site. [After M. G. Rossmann, A. Liljas, C.-I. Brändén, and L. J. Banaszak. In *The Enzymes,* 3rd ed., vol. 10 (Academic Press, 1975), p. 68.]

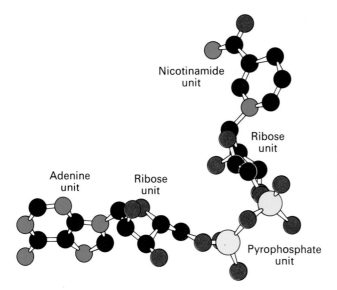

Nicotinamide
unit

Ribose
unit

Adenine
unit

Ribose
unit

Pyrophosphate
unit

Figure 12-12
Model of NAD$^+$. The conformation shown here is the one found in the complex of NAD$^+$ and lactate dehydrogenase.

GLUCOSE INDUCES A LARGE CONFORMATIONAL CHANGE IN HEXOKINASE

X-ray crystallographic studies of yeast hexokinase have revealed that the binding of glucose leads to a large conformational change in the enzyme. Hexokinase consists of two lobes, which come together when glucose is bound (Figure 12-13). Glucose induces a 12-degree rotation of one lobe with respect to the other, resulting in movements of the polypeptide backbone of as much as 8 Å. The cleft between the lobes closes and the bound glucose becomes surrounded by protein, except for its 6-hydroxymethyl group. The closing of the cleft in hexokinase is a striking example of the role of *induced fit* (p. 110) in enzyme action, as originally proposed by Koshland. The glucose-induced structural changes are likely to be significant in two ways. First, the environment around the glucose becomes much more nonpolar, which enhances phosphoryl transfer from ATP. Second, the embracing of glucose by hexokinase enables the enzyme to discriminate against H_2O as a substrate. If hexokinase were rigid, a water molecule occupying the binding site for the —CH$_2$OH of glucose would attack the α-phosphate of ATP. In other words, a rigid kinase would necessarily be an ATPase as well as a kinase. This undesirable activity is prevented by making hexokinase enzymatically active only when glucose closes the cleft. We see here how *enzyme flexibility contributes to specificity*. It is interesting to note that pyruvate kinase, phosphoglycerate kinase, and phosphofructokinase also contain clefts between lobes that close when substrate is bound. The *substrate-induced cleft closing* seen in hexokinase exemplifies an important feature of kinases in general.

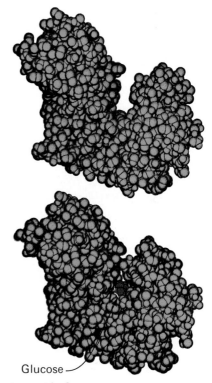

Glucose

Figure 12-13
The conformation of hexokinase changes markedly on binding glucose (shown in red). The two lobes of the enzyme come together and surround the substrate. [Courtesy of Dr. Thomas Steitz.]

ALDOLASE FORMS A SCHIFF BASE WITH DIHYDROXYACETONE PHOSPHATE

Let us now turn to aldolase, which catalyzes the condensation of dihydroxyacetone phosphate and glyceraldehyde 3-phosphate to form fructose 1,6-diphosphate. First, dihydroxyacetone phosphate forms a protonated Schiff base with a specific lysine residue in the active site of animal aldolases.

Figure 12-14
Photomicrograph of crystals of aldolase.
[Courtesy of Dr. David Eisenberg.]

This protonated Schiff base plays a critical role in catalysis because it promotes the formation of the enolate anion of dihydroxyacetone phosphate.

Glyceraldehyde 3-phosphate then adds to this enolate anion intermediate, which yields the protonated Schiff base of fructose 1,6-diphosphate.

This Schiff base is deprotonated and hydrolyzed to yield fructose 1,6-diphosphate and the regenerated enzyme.

The pathway for the cleavage of fructose 1,6-diphosphate is simply the reverse of the one given for its formation.

A THIOESTER IS FORMED IN THE OXIDATION OF GLYCERALDEHYDE 3-PHOSPHATE

A different kind of covalent enzyme-substrate intermediate is formed in glyceraldehyde 3-phosphate dehydrogenase. This enzyme catalyzes the oxidative phosphorylation of its aldehyde substrate.

$$\text{Glyceraldehyde 3-phosphate} + P_i + NAD^+ \longrightarrow$$
$$1,3\text{-DPG} + NADH + H^+$$

In the conversion of an aldehyde into an acyl phosphate, the aldehyde is first oxidized.

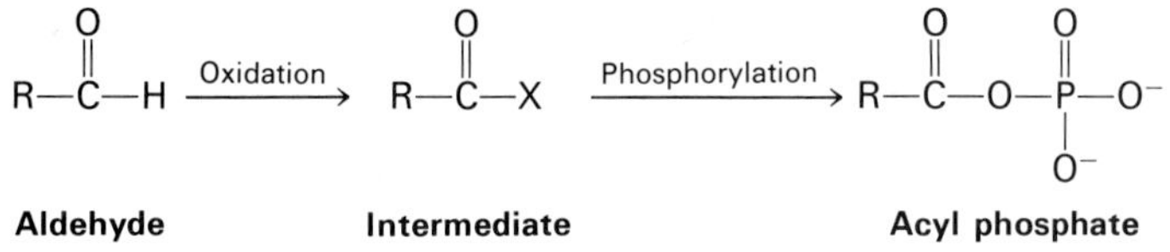

This requires the *removal of a hydride ion* ($:H^-$), which is a hydrogen nucleus and two electrons. There is a large barrier to the removal of a hydride ion from an aldehyde because of the dipolar character of the carbonyl group. The carbon atom of the carbonyl group already has a partial positive charge.

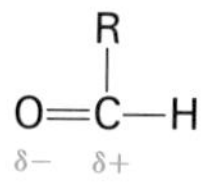

The removal of the hydride ion is facilitated by making the carbon atom less positively charged. This is accomplished by the *addition of a nucleophile,* represented by X^- in the following equation:

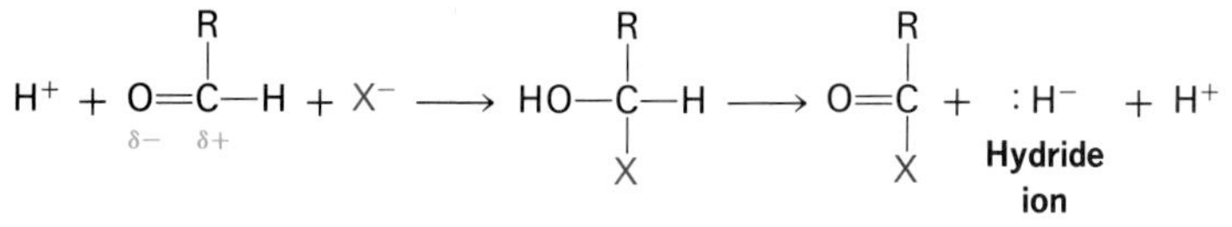

The hydride ion readily leaves the addition compound because its carbon atom no longer carries a large positive charge. Furthermore, some of the free energy of the oxidation is preserved in the acyl intermediate. Addition of orthophosphate to this acyl intermediate yields an acyl phosphate, which has a high group-transfer potential. This sequence of reactions is called a *substrate-level phosphorylation*.

$$R-\overset{O}{\underset{\|}{C}}-X + {}^-O-\overset{O}{\underset{\underset{OH}{|}}{\overset{\|}{P}}}-O^- \longrightarrow R-\overset{O}{\underset{\|}{C}}-O-\overset{O}{\underset{\underset{O^-}{|}}{\overset{\|}{P}}}-O^- + XH$$

Now let us see how glyceraldehyde 3-phosphate dehydrogenase carries out these steps (Figure 12-15). *The nucleophile X^- is the sulfhydryl group of a cysteine residue at the active site.* The aldehyde substrate reacts with the ionized form of this sulfhydryl group to form a hemithioacetal. The next step is the transfer of a hydride ion. *The acceptor for the hydride ion is a molecule of NAD^+ that is tightly bound to the enzyme.* The products of this reaction are the reduced coenzyme NADH and a thioester. *This thioester is an energy-rich intermediate,* corresponding to the acyl intermediate mentioned earlier. NADH dissociates from the enzyme, and NAD^+ again binds to the active site. Orthophosphate then attacks the thioester to form 1,3-DPG, an energy-rich phosphate. A crucial aspect of the formation of 1,3-DPG from 3-phosphoglyceraldehyde is that a thermodynamically unfavorable reaction, the formation of an acyl phosphate from a carboxylate, is driven by a thermodynamically favorable reaction, the oxidation of an aldehyde.

$$R-\overset{O}{\underset{\|}{C}}-H + NAD^+ + H_2O \rightleftharpoons R-\overset{O}{\underset{\|}{C}}-O^- + NADH + 2\,H^+$$

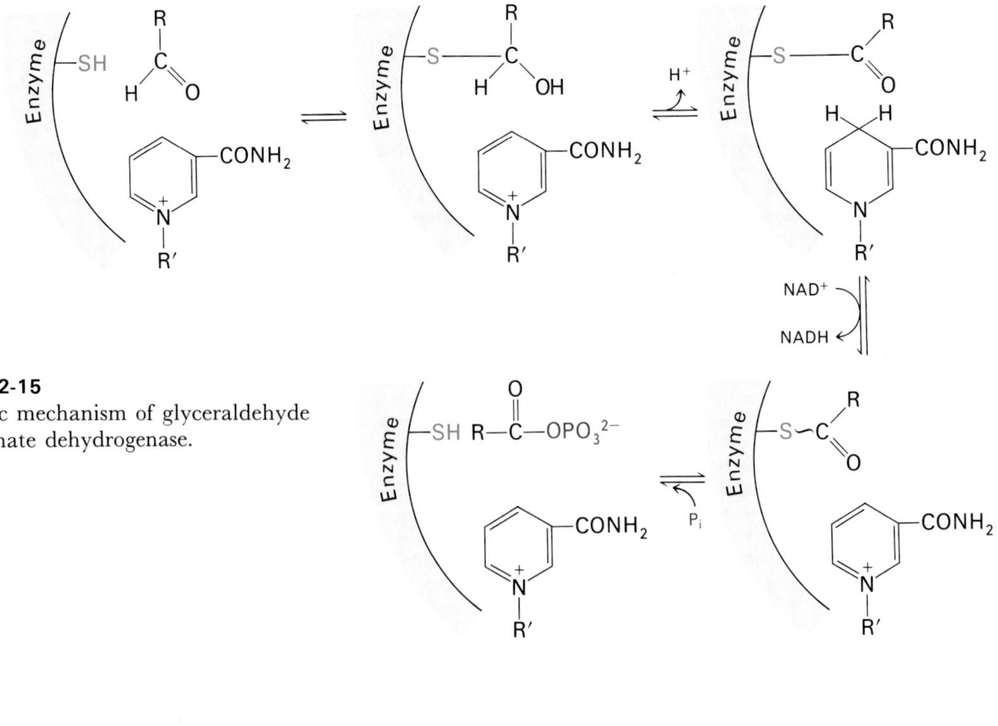

Figure 12-15
Catalytic mechanism of glyceraldehyde 3-phosphate dehydrogenase.

These two reactions are *coupled by the thioester intermediate,* which preserves much of the free energy released in the oxidation reaction. We see here the *use of a covalent enzyme-bound intermediate as a mechanism of energy coupling.*

ARSENATE, AN ANALOG OF PHOSPHATE, ACTS AS AN UNCOUPLER

Arsenate (AsO_4^{3-}) closely resembles P_i in structure and reactivity. In the reaction catalyzed by glyceraldehyde 3-phosphate dehydrogenase, arsenate can replace phosphate in attacking the energy-rich thioester intermediate. The product of this reaction, 1-arseno-3-phosphoglycerate is unstable, in contrast with 1,3-diphosphoglycerate. 1-Arseno-3-phosphoglycerate and other acyl arsenates are very rapidly and spontaneously hydrolyzed. Hence, the net reaction in the presence of arsenate is

Glyceraldehyde 3-phosphate + NAD$^+$ + H$_2$O $\longrightarrow$
3-phosphoglycerate + NADH + 2 H$^+$

Note that glycolysis proceeds in the presence of arsenate but that the ATP normally formed in the conversion of 1,3-diphosphoglycerate into 3-phosphoglycerate is lost. Thus, *arsenate uncouples oxidation and phosphorylation by forming a highly labile acyl arsenate.* One likely reason for the choice of phosphorus over arsenate in the evolution of biomolecules is the greater kinetic stability of its energy-rich compounds.

1-Arseno-3-phosphoglycerate

ENOL PHOSPHATES HAVE A HIGH GROUP-TRANSFER POTENTIAL

Because it is an *acyl phosphate,* 1,3-DPG has a high group-transfer potential. A different kind of high-energy phosphate compound is formed several steps later in glycolysis. Phosphoenolpyruvate, an *enol phosphate,* is formed by the dehydration of 2-phosphoglycerate. The $\Delta G^{\circ\prime}$ of hydrolysis of a phosphate ester of an ordinary alcohol is -3 kcal/mol, whereas that of phosphoenolpyruvate is -14.8 kcal/mol. Why does phosphoenolpyruvate have such a high phosphate group transfer potential? The answer is that the reaction does not stop with the enol formed upon transfer of the phosphoryl group. The enol undergoes a conversion into a ketone—namely, pyruvate.

Phosphoenolpyruvate

Phosphoenolpyruvate   Enolpyruvate   Pyruvate

The $\Delta G^{\circ\prime}$ of the enol $\longrightarrow$ ketone conversion is very large, of the order of -10 kcal/mol. This should be compared with the $\Delta G^{\circ\prime}$ of hydrolysis of phosphoenolpyruvate to enolpyruvate, which is about -3 kcal/mol. Thus, the *high phosphate group transfer potential of phosphoenolpyruvate arises from the large driving force of the subsequent enol $\longrightarrow$ ketone conversion.*

METABOLISM OF 2,3-DPG, A REGULATOR OF OXYGEN TRANSPORT

Recall that 2,3-diphosphoglycerate (2,3-DPG) is a regulator of oxygen transport in erythrocytes. It decreases the oxygen affinity of hemoglobin by stabilizing the deoxygenated form of hemoglobin (Chapter 4). Red blood cells have a high concentration of 2,3-DPG, typically 4 mM, in contrast with most other cells, which have only trace amounts. 2,3-DPG has a general role as a cofactor in the conversion of 3-phosphoglycerate into 2-phosphoglycerate by phosphoglycerate mutase.

The synthesis and degradation of 2,3-DPG are detours from the glycolytic pathway (Figure 12-16).

Table 12-3
Typical concentrations of glycolytic intermediates in erythrocytes

Intermediate	μM
Glucose	5000
Glucose 6-phosphate	83
Fructose 6-phosphate	14
Fructose 1,6-diphosphate	31
Dihydroxyacetone phosphate	138
Glyceraldehyde 3-phosphate	19
1,3-Diphosphoglycerate	1
2,3-Diphosphoglycerate	4000
3-Phosphoglycerate	118
2-Phosphoglycerate	30
Phosphoenolpyruvate	23
Pyruvate	51
Lactate	2900
ATP	1850
ADP	138
P_i	1000

After S. Minakami and H. Yoshikawa,
Biochem. Biophys. Res. Comm. 18(1965):345.

Glyceraldehyde 3-phosphate
$\parallel$
1,3-Diphosphoglycerate $\xrightarrow{\text{Mutase}}$
$\parallel$ 2,3-Diphosphoglycerate
3-Phosphoglycerate $\xleftarrow{\text{Phosphatase}}$
$\parallel$
2-Phosphoglycerate

Figure 12-16
Pathway for the synthesis and degradation of 2,3-diphosphoglycerate.

Diphosphoglycerate mutase converts 1,3-DPG into 2,3-DPG.

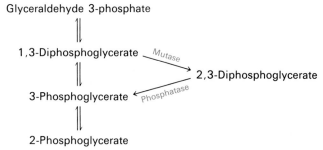

1,3-Diphosphoglycerate   2,3-Diphosphoglycerate

2,3-DPG is hydrolyzed to 3-phosphoglycerate by *2,3-diphosphoglycerate phosphatase.* A phosphatase is an enzyme that catalyzes the hydrolysis of a phosphate ester.

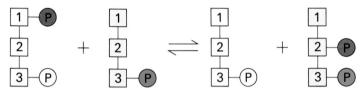

2,3-Diphosphoglycerate   **3-Phosphoglycerate**

This mutase reaction has an interesting mechanism. *3-Phosphoglycerate* is an obligatory participant although it does not appear in the overall stoichiometry. The mutase binds 1,3-DPG and 3-phosphoglycerate simultaneously. In this ternary complex, a phosphoryl group is transferred from the 1-position of 1,3-DPG to the 2-position of 3-phosphoglycerate (Figure 12-17).

Figure 12-17
Schematic representation of the participation of 3-phosphoglycerate in the conversion of 1,3-diphosphoglycerate into 2,3-diphosphoglycerate.

In this mutase reaction, 2,3-DPG is a potent competitive inhibitor of 1,3-DPG. Thus, the rate of synthesis of 2,3-DPG depends in part on its own concentration. Another controlling factor is the concentration of 1,3-DPG, because the enzyme is not usually saturated with 1,3-DPG. In contrast, the level of 3-phosphoglycerate in the red cell is nearly adequate to saturate the mutase. Thus, *the rate of synthesis of 2,3-DPG is controlled by the concentration of unbound 1,3-DPG and 2,3-DPG.*

DEFECTS IN RED-CELL GLYCOLYSIS ALTER OXYGEN TRANSPORT

Glycolysis in red cells and oxygen transport are linked by 2,3-DPG. This means that defects in glycolysis can affect oxygen transport. Indeed, the oxygen dissociation curves of blood from some patients with inherited defects of glycolysis in red blood cells are altered (Figure 12-18). In *hexokinase deficiency,* the concentration of glycolytic intermediates is low because the phosphorylation of glucose, the first step, is impaired. Hence, the concentration of 2,3-DPG is diminished in these red cells, and so their hemoglobin has an *abnormally high oxygen affinity.* Just the opposite is found in *pyruvate kinase*

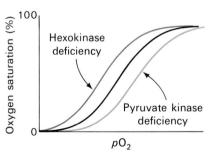

Figure 12-18
Oxygen dissociation curves of normal red cells (black line), red cells from a patient with hexokinase deficiency (red line), and red cells from a patient with pyruvate kinase deficiency (green line).

deficiency. The glycolytic intermediates are at an abnormally high concentration because the terminal step is blocked. Consequently, the level of 2,3-DPG is about twice the normal one, which leads to a *low oxygen affinity*. The abnormal oxygen dissociation curves in hexokinase and pyruvate kinase deficiency were puzzling until it was discovered that 2,3-DPG is a regulator of oxygen transport.

SUMMARY

Glycolysis is the set of reactions that converts glucose into pyruvate. In aerobic organisms, glycolysis is the prelude to the citric acid cycle and the electron-transport chain, where most of the free energy in glucose is harvested. The ten reactions of glycolysis occur in the cytosol. In the first stage, glucose is converted into fructose 1,6-diphosphate by a phosphorylation, an isomerization, and a second phosphorylation reaction. Two molecules of ATP are consumed per molecule of glucose in these reactions, which prepare for the net synthesis of ATP. In the second stage, fructose 1,6-diphosphate is cleaved by aldolase into dihydroxyacetone phosphate and glyceraldehyde 3-phosphate, which are readily interconvertible. Glyceraldehyde 3-phosphate is then oxidized and phosphorylated to form 1,3-DPG, an acyl phosphate with a high phosphate-transfer potential. 3-Phosphoglycerate is formed as an ATP is generated. In the last stage of glycolysis, phosphoenolpyruvate, a second intermediate with a high phosphate-transfer potential, is formed by a phosphoryl shift and a dehydration. An ATP is generated as phosphoenolpyruvate is converted into pyruvate. There is a net gain of two molecules of ATP in the formation of two molecules of pyruvate from one molecule of glucose.

The electron acceptor in the oxidation of glyceraldehyde 3-phosphate is NAD^+, which must be regenerated for glycolysis to proceed. In aerobic organisms, the NADH formed in glycolysis transfers its electrons to O_2 through the electron-transport chain, which thereby regenerates NAD^+. Under anaerobic conditions, NAD^+ is regenerated by the reduction of pyruvate to lactate. In some microorganisms, NAD^+ is normally regenerated by the synthesis of lactate or ethanol from pyruvate. These two processes are called fermentations.

The glycolytic pathway has a dual role: it degrades glucose to generate ATP and it provides building blocks for the synthesis of cellular components. The rate of conversion of glucose into pyruvate is regulated to meet these two major cellular needs. The reactions of glycolysis are readily reversible under physiological conditions except for the ones catalyzed by hexokinase, phosphofructokinase, and pyruvate kinase. Phosphofructokinase, the most important control element in glycolysis, is inhibited by high levels of ATP and citrate and is activated by AMP. Hence, phospho-fructokinase is active when the need arises for either energy or building blocks. Hexokinase is inhibited by glucose 6-phosphate, which accumulates when phosphofructokinase is inactive. Pyruvate kinase, the other control site, is allosterically inhibited by ATP, and so the conversion of phosphoenolpyruvate into pyruvate is blocked when the energy charge of the cell is high.

SELECTED READINGS

BOOKS AND GENERAL REVIEWS

Davison, E. A., 1967. *Carbohydrate Chemistry*. Holt, Rinehart, and Winston.

Pigman, W. W., and Horton, D., (eds.), 1972. *The Carbohydrates: Chemistry and Biochemistry*. Academic Press.

Dickens, F., Randle, P. J., and Whelan, W. J., 1968. *Carbohydrate Metabolism and Its Disorders*, vols. 1 and 2. Academic Press.

GLYCOLYTIC ENZYMES

Boyer, P. D., (ed.), 1972. *The Enzymes* (3rd ed.). Academic Press. [Volumes 5 through 9 contain authoritative reviews on each of the glycolytic enzymes. Alcohol dehydrogenase and lactate dehydrogenase are discussed in Volume 10.]

Blake, C. C. F., 1975. X-ray studies of glycolytic enzymes. *Essays Biochem.* 11:37–79.

Bennett, W. S., Jr., and Steitz, T. A., 1978. Glucose-induced conformational change in yeast hexokinase. *Proc. Nat. Acad. Sci.* 75:4848–4852.

Knowles, J. R., and Albery, W. J., 1977. Perfection in enzyme catalysis: the energetics of triosephosphate isomerase. *Acc. Chem. Res.* 10:105–111.

Biesecker, G., Harris, J. I., Thierry, J. C., Walker, J. E., and Wonacott, A. J., 1977. Sequence and structure of D-glyceraldehyde 3-phosphate dehydrogenase from *Bacillus stearothermophilus*. *Nature* 266:328–333.

Anderson, C. M., Zucker, F. H., and Steitz, T. A., 1979. Space-filling models of kinase clefts and conformation changes. *Science* 204:375–380. [A well-illustrated and lucid article showing that kinases generally have a deep cleft that closes on binding substrate.]

REGULATION OF GLYCOLYSIS

Newsholme, E. A., and Start, C., 1973. *Regulation in Metabolism*. Wiley. [Chapters 3 and 6 provide excellent accounts of the control of carbohydrate metabolism.]

Ottaway, J. H., and Mowbray, J., 1977. The role of compartmentation in the control of glycolysis. *Curr. Top. Cell Regul.* 12:108–195.

Clark, M. G., and Lardy, H. A., 1975. Regulation of intermediary carbohydrate metabolism. *Intern. Rev. Sci.* (*Biochem.*) 5:223–266.

MOLECULAR EVOLUTION

Rossman, M. G., Liljas, A., Bränden, C.-I., and Banaszak, L. J., 1975. Evolutionary and structural relationships among dehydrogenases. *In* Boyer, P. D., (ed.), *The Enzymes* (3rd ed.), vol. 11, pp. 61–102.

Levine, M., Muirhead, H., Stammers, D. K., and Stuart, D. I., 1978. Structure of pyruvate kinase and similarities with other enzymes: possible implications for protein taxonomy and evolution. *Nature* 271:626–630.

2,3-DIPHOSPHOGLYCERATE

Benesch, R. E., and Benesch, R., 1970. The reaction between diphosphoglycerate and hemoglobin. *Fed. Proc.* 29:1101–1104.

Rose, Z. B., 1970. Enzymes controlling 2,3-diphosphoglycerate in human erythrocytes. *Fed. Proc.* 29:1105–1111.

Delivoria-Papadopoulos, M., Oski, F. A., and Gottlieb, A. J., 1969. Oxygen-hemoglobin dissociation curves: effect of inherited enzyme defects of the red cell. *Science* 165:601–602.

HISTORICAL ASPECTS

Kalckar, H. M., (ed.)., 1969. *Biological Phosphorylations: Development of Concepts*. Prentice-Hall. [Contains many of the classical papers on glycolysis.]

Fruton, J. S., 1972. *Molecules and Life: Historical Essays on the Interplay of Chemistry and Biology*. Wiley-Interscience. [Includes a meticulously documented account of the elucidation of the nature of fermentation and how it led to enzyme chemistry.]

APPENDIX
Stereochemical Relations
of Some Sugars

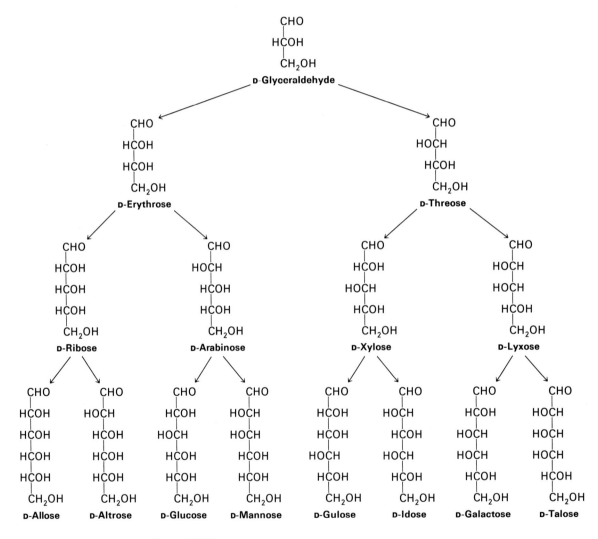

Figure 12-19
Stereochemical relations of D-aldoses containing three, four, five, and six carbon atoms.

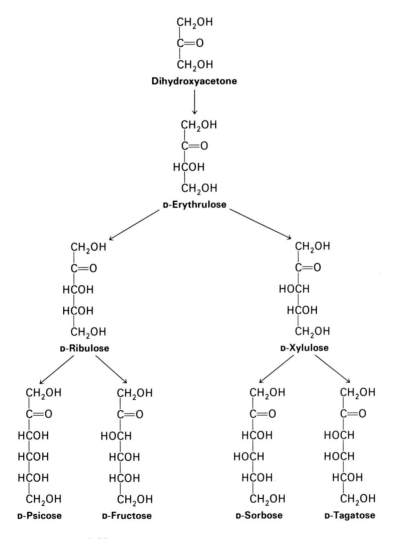

Figure 12-20
Stereochemical relations of D-ketoses containing three, four, five, and six carbon atoms.

PROBLEMS

1. Carbohydrates that differ only in the configuration about their carbonyl carbon atoms are *anomers. Epimers* differ in configuration about a specific carbon atom other than the carbonyl carbon atom. Indicate whether each of the following pairs of sugars consists of anomers, epimers, or an aldose-ketose pair:
   (a) D-glyceraldehyde and dihydroxyacetone.
   (b) D-glucose and D-mannose.
   (c) D-glucose and D-fructose.
   (d) α-D-glucose and β-D-glucose.
   (e) D-ribose and D-ribulose.
   (f) D-galactose and D-glucose.

2. Glucose labeled with ^{14}C at C-1 is incubated with the glycolytic enzymes and necessary cofactors. What is the distribution of ^{14}C in the pyruvate that is formed? (Assume that the interconversion of glyceraldehyde 3-phosphate and dihydroxyacetone phosphate is very rapid compared with the subsequent step.)

3. Write a balanced equation for the conversion of glucose into lactate.
   (a) Calculate the standard free-energy change of this reaction using the data given in Table 12-2 (p. 267) and the fact that $\Delta G^{\circ\prime}$ is -6 kcal for the reaction

   $$\text{Pyruvate} + \text{NADH} + \text{H}^+ \rightleftharpoons \text{lactate} + \text{NAD}^+$$

   (b) What is the free-energy change ($\Delta G'$, not $\Delta G^{\circ\prime}$) of this reaction when the concentrations of reactants are: glucose, 5 mM; lactate, .05 mM; ATP, 2 mM; ADP, 0.2 mM; and P_i, 1 mM?

4. What is the equilibrium ratio of phosphoenolpyruvate to pyruvate under standard conditions when [ATP]/[ADP] = 10?

5. What are the equilibrium concentrations of fructose 1,6-diphosphate, dihydroxyacetone phosphate, and glyceraldehyde 3-phosphate when 1 mM fructose 1,6-diphosphate is incubated with aldolase under standard conditions?

6. 3-Phosphoglycerate labeled uniformly with ^{14}C is incubated with 1,3-DPG labeled with ^{32}P at C-1. What is the radioisotope distribution of the 2,3-DPG that is formed on addition of DPG mutase?

7. Xylose has the same structure as glucose except that it has a hydrogen atom at C-6 in place of a hydroxymethyl group. The rate of ATP hydrolysis by hexokinase is markedly enhanced by the addition of xylose. Why?

8. The active site of phosphoglycerate mutase contains two histidine residues that are parallel to one another and about 4 Å apart. Propose a plausible catalytic mechanism.

For additional problems, see W. B. Wood, J. H. Wilson, R. M. Benbow, and L. E. Hood. *Biochemistry: A Problems Approach* (Benjamin, 1974), ch. 9.

CHAPTER 13

CITRIC ACID CYCLE

The preceding chapter considered the glycolytic pathway, in which glucose is converted into pyruvate. Under aerobic conditions, the next step in the generation of energy from glucose is the oxidative decarboxylation of pyruvate to form acetyl coenzyme A (acetyl CoA). This activated acetyl unit is then completely oxidized to CO_2 by the *citric acid cycle*, a series of reactions that is also known as the *tricarboxylic acid cycle* or the *Krebs cycle*. The citric acid cycle is the *final common pathway for the oxidation of fuel molecules*—amino acids, fatty acids, and carbohydrates. Most fuel molecules enter the cycle as acetyl CoA. The cycle also provides intermediates for biosyntheses. The reactions of the citric acid cycle occur inside mitochondria, in contrast with those of glycolysis, which occur in the cytosol.

FORMATION OF ACETYL COENZYME A FROM PYRUVATE

The oxidative decarboxylation of pyruvate to form acetyl CoA, which occurs in the mitochondrial matrix, is the link between glycolysis and the citric acid cycle:

$$\text{Pyruvate} + \text{CoA} + \text{NAD}^+ \longrightarrow \text{acetyl CoA} + CO_2 + \text{NADH}$$

This irreversible funneling of the product of glycolysis into the citric acid cycle is catalyzed by the *pyruvate dehydrogenase complex*. This

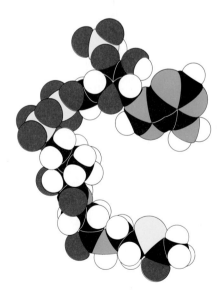

Figure 13-1
Model of acetyl CoA.

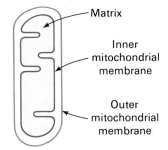

Figure 13-2
Schematic diagram of a mitochondrion. The oxidative decarboxylation of pyruvate and the sequence of reactions in the citric acid cycle occur within the mitochondrial matrix.

very large multienzyme complex is a highly integrated array of three kinds of enzymes, which will be discussed in detail later in this chapter (p. 294).

AN OVERVIEW OF THE CITRIC ACID CYCLE

The overall pattern of the citric acid cycle is shown in Figure 13-3. A four-carbon compound (oxaloacetate) condenses with a two-carbon acetyl unit to yield a six-carbon tricarboxylic acid (citrate). An isomer of citrate is then oxidatively decarboxylated. The resulting five-carbon compound (α-ketoglutarate) is oxidatively decarboxylated to yield a four-carbon compound (succinate). Oxaloacetate is then regenerated from succinate. Two carbon atoms enter the cycle as an acetyl unit and two carbon atoms leave the cycle in the form of two molecules of CO_2. An acetyl group is more reduced than CO_2, and so oxidation-reduction reactions must take place in the citric acid cycle. In fact, there are four such reactions. Three hydride ions (hence, six electrons) are transferred to three NAD^+ molecules, whereas one pair of hydrogen atoms (hence, two electons) is transferred to a flavin adenine dinucleotide (FAD) molecule. These electron carriers yield eleven molecules of adenosine triphosphate (ATP) when they are oxidized by O_2 in the electron-transport chain. In addition, one high-energy phosphate bond is formed in each round of the citric acid cycle itself.

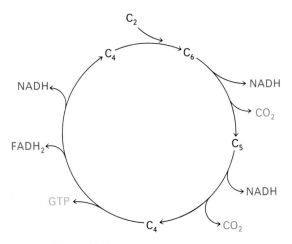

Figure 13-3
An overview of the citric acid cycle.

OXALOACETATE CONDENSES WITH ACETYL COENZYME A TO FORM CITRATE

The cycle starts with the joining of a four-carbon unit, oxaloacetate, and a two-carbon unit, the acetyl group of acetyl CoA. Oxaloacetate reacts with acetyl CoA and H_2O to yield citrate and CoA.

$$O=\overset{\underset{\displaystyle H_2C-COO^-}{|}}{\underset{}{C}}-COO^- + \overset{\underset{\displaystyle S-CoA}{|}}{\overset{\displaystyle O}{\overset{\|}{C}}}-CH_3 + H_2O \longrightarrow HO-\overset{\underset{\displaystyle CH_2-COO^-}{|}}{\overset{\displaystyle CH_2-COO^-}{|}}{C}-COO^- + HS-CoA + H^+$$

Oxaloacetate   Acetyl CoA   Citrate

This reaction, which is an aldol condensation followed by a hydrolysis, is catalyzed by *citrate synthetase* (originally called condensing enzyme). Oxaloacetate first condenses with acetyl CoA to form *citryl CoA,* which is then hydrolyzed to citrate and CoA. Hydrolysis of citryl CoA pulls the overall reaction far in the direction of the synthesis of citrate.

$$HO-\overset{\underset{\displaystyle CH_2-COO^-}{|}}{\underset{\displaystyle CH_2-\overset{\displaystyle O}{\overset{\|}{C}}-S-CoA}{|}}{C}-COO^-$$

Citryl CoA

CITRATE IS ISOMERIZED INTO ISOCITRATE

Citrate must be isomerized into isocitrate to enable the six-carbon unit to undergo oxidative decarboxylation. The isomerization of citrate is accomplished by a *dehydration* step followed by a *hydration* step. The result is an interchange of an H and OH. The enzyme catalyzing both steps is called *aconitase* because the presumed intermediate is cis-*aconitate.*

Citrate   cis-Aconitate   Isocitrate

ISOCITRATE IS OXIDIZED AND DECARBOXYLATED TO ALPHA-KETOGLUTARATE

We come now to the first of four oxidation-reduction reactions in the citric acid cycle. The oxidative decarboxylation of isocitrate is catalyzed by *isocitrate dehydrogenase:*

Isocitrate + NAD$^+$ $\rightleftharpoons$ α-ketoglutarate + CO$_2$ + NADH

The intermediate in this reaction is oxalosuccinate, which rapidly loses CO$_2$ while bound to the enzyme to give α-ketoglutarate.

Isocitrate   Oxalosuccinate α-Ketoglutarate

The rate of formation of α-ketoglutarate is important in determining the overall rate of the cycle, as will be discussed later (p. 301). Also noteworthy is that there are two kinds of isocitrate dehydrogenases. One is specific for NAD^+, the other for nicotinamide adenine dinucleotide phosphate ($NADP^+$). The NAD^+-specific enzyme, which is located in mitochondria, is the important one for the citric acid cycle. The $NADP^+$-specific enzyme, which is present both in mitochondria and in the cytoplasm, has a different metabolic role.

SUCCINYL COENZYME A IS FORMED BY THE OXIDATIVE DECARBOXYLATION OF ALPHA-KETOGLUTARATE

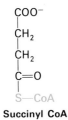

Succinyl CoA

The conversion of isocitrate into α-ketoglutarate is followed by a second oxidative decarboxylation reaction, the formation of succinyl CoA from α-ketoglutarate:

$$\alpha\text{-Ketoglutarate} + NAD^+ + CoA \rightleftharpoons$$
$$\text{succinyl CoA} + CO_2 + NADH$$

This reaction is catalyzed by the *α-ketoglutarate dehydrogenase complex*, an organized assembly consisting of three kinds of enzymes. *The mechanism of this reaction is very similar to that of the conversion of pyruvate into acetyl CoA.* The same cofactors are used: NAD^+, CoA, thiamine pyrophosphate, lipoamide, and FAD. Indeed, the pyruvate dehydrogenase complex and the α-ketoglutarate dehydrogenase complex have common structural features (p. 294).

A HIGH-ENERGY PHOSPHATE BOND IS GENERATED FROM SUCCINYL COENZYME A

Succinate

The succinyl thioester of CoA is an energy-rich bond. The $\Delta G^{\circ\prime}$ for hydrolysis of succinyl CoA is about -8 kcal/mol, which is comparable to that of ATP (-7.3 kcal/mol). *The cleavage of the thioester bond of succinyl CoA is coupled to the phosphorylation of guanosine diphosphate* (GDP):

$$\text{Succinyl CoA} + P_i + GDP \rightleftharpoons \text{succinate} + GTP + CoA$$

This readily reversible reaction ($\Delta G^{\circ\prime} = -0.8$ kcal/mol) is catalyzed by *succinyl CoA synthetase*. The phosphoryl group in guanosine triphosphate (GTP) is readily transferred to adenosine diphosphate (ADP) to form ATP, in a reaction catalyzed by *nucleoside diphosphokinase*.

$$GTP + ADP \rightleftharpoons GDP + ATP$$

The generation of a high-energy phosphate bond from succinyl CoA is an example of a *substrate-level phosphorylation*. In fact, this is the only reaction in the citric acid cycle that directly yields a high-energy phosphate bond. The contrasting process is *respiratory-chain* phosphorylation (also called *oxidative phosphorylation*), which is the formation of ATP coupled to the oxidation of NADH and $FADH_2$ by O_2. Substrate-level phosphorylation was previously encountered in two reactions of glycolysis: in the oxidation of glyceraldehyde 3-phosphate and in the conversion of phosphoenol pyruvate into pyruvate. Oxidative phosphorylation will be discussed in the next chapter.

REGENERATION OF OXALOACETATE BY OXIDATION OF SUCCINATE

Reactions of four-carbon compounds constitute the final stage of the citric acid cycle (Figure 13-4). Succinate is converted into oxalo-

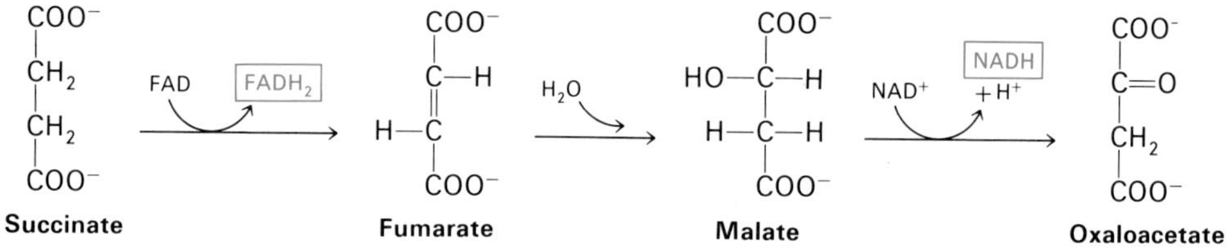

Figure 13-4
Final stage of the citric acid cycle: from succinate to oxaloacetate.

acetate in three steps: an oxidation, a hydration, and a second oxidation reaction. Oxaloacetate is thereby regenerated for another round of the cycle, concomitant with the trapping of energy in the form of $FADH_2$ and NADH.

Succinate is oxidized to fumarate by *succinate dehydrogenase*. The hydrogen acceptor is FAD rather than NAD^+, which is used in the other three oxidation reactions in the cycle. FAD is the hydrogen acceptor in this reaction because the free-energy change is insufficient to reduce NAD^+. FAD is nearly always the electron acceptor in reactions of this type. In succinate dehydrogenase, the isoalloxazine ring of FAD is covalently attached to a histidine side chain of the enzyme (denoted E-FAD).

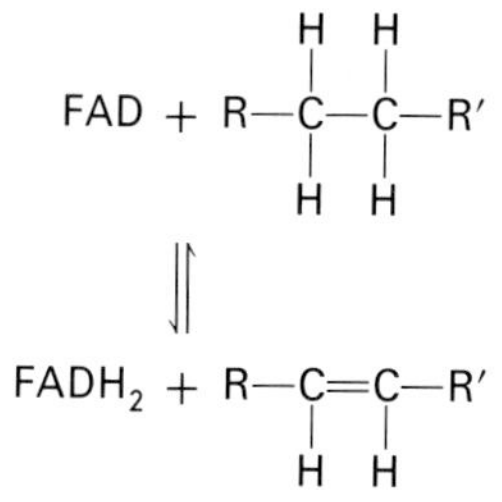

$$\text{E-FAD} + \text{succinate} \rightleftharpoons \text{E-FADH}_2 + \text{fumarate}$$

Succinate dehydrogenase contains four iron atoms and four inorganic sulfides in addition to the flavin. There is no heme in this enzyme. Rather, the iron atoms are probably bonded to the inorganic sulfides. Consequently, this type of protein is known as an *iron-sulfur* (FeS) *protein* or a *nonheme iron protein*. Iron-sulfur proteins play an important role in the electron-transport systems of mitochondria and chloroplasts (p. 312 and p. 439). Succinate dehydrogenase comprises a 70-kdal unit, which includes FAD and two FeS clusters, and a 27-kdal unit, which consists of a single FeS cluster. This enzyme differs from the others in the citric acid cycle in being an integral part of the inner mitochondrial membrane. In fact, *succinate dehydrogenase is directly linked to the electron-transport chain.* The $FADH_2$ produced by the oxidation of succinate does not dissociate from the enzyme, in contrast with NADH. Rather, two electrons from $FADH_2$ are transferred directly to the Fe^{3+} atoms of the enzyme. The ultimate acceptor of these electrons is molecular oxygen, as will be discussed in the next chapter.

The next step in the cycle is the hydration of fumarate to form L-malate. *Fumarase* catalyzes a stereospecific *trans* addition of H and OH, as shown by deuterium-labeling studies. The OH group adds to only one side of the double bond of fumarate; hence, only the L-isomer of malate is formed.

Finally, malate is oxidized to form oxaloacetate. This reaction is catalyzed by *malate dehydrogenase* and NAD^+ is again the hydrogen acceptor.

$$\text{Malate} + NAD^+ \rightleftharpoons \text{oxaloacetate} + NADH + H^+$$

STOICHIOMETRY OF THE CITRIC ACID CYCLE

The net reaction of the citric acid cycle is:

$$\text{Acetyl CoA} + 3\,NAD^+ + FAD + GDP + P_i + 2\,H_2O \longrightarrow$$
$$2\,CO_2 + 3\,NADH + FADH_2 + GTP + 2\,H^+ + CoA$$

The reactions that give this stoichiometry (Figure 13-5 and Table 13-1) are recapitulated on the next two pages.

Table 13-1
Citric acid cycle

Step	Reaction	Enzyme	Cofactor	Type*	$\Delta G°'$
1	Acetyl CoA + oxaloacetate + H_2O $\longrightarrow$ citrate + CoA + H^+	Citrate synthetase	CoA	a	-7.5
2	Citrate $\rightleftharpoons$ *cis*-aconitate + H_2O	Aconitase	Fe^{2+}	b	$+2.0$
3	*cis*-Aconitate + H_2O $\rightleftharpoons$ isocitrate	Aconitase	Fe^{2+}	c	-0.5
4	Isocitrate + NAD^+ $\rightleftharpoons$ α-ketoglutarate + CO_2 + NADH	Isocitrate dehydrogenase	NAD^+	d + e	-2.0
5	α-Ketoglutarate + NAD^+ + CoA $\rightleftharpoons$ succinyl CoA + CO_2 + NADH	α-Ketoglutarate dehydrogenase complex	NAD^+ CoA TPP Lipoic acid FAD	d + e	-7.2
6	Succinyl CoA + P_i + GDP $\rightleftharpoons$ succinate + GTP + CoA	Succinyl CoA synthetase	CoA	f	-0.8
7	Succinate + FAD (enzyme-bound) $\rightleftharpoons$ fumarate + $FADH_2$ (enzyme-bound)	Succinate dehydrogenase	FAD	e	~0
8	Fumarate + H_2O $\rightleftharpoons$ malate	Fumarase	None	c	-0.9
9	L-Malate + NAD^+ $\rightleftharpoons$ oxaloacetate + NADH + H^+	Malate dehydrogenase	NAD^+	e	$+7.1$

*Reaction type: (a) Condensation   (d) Decarboxylation
            (b) Dehydration   (e) Oxidation
            (c) Hydration   (f) Substrate-level phosphorylation

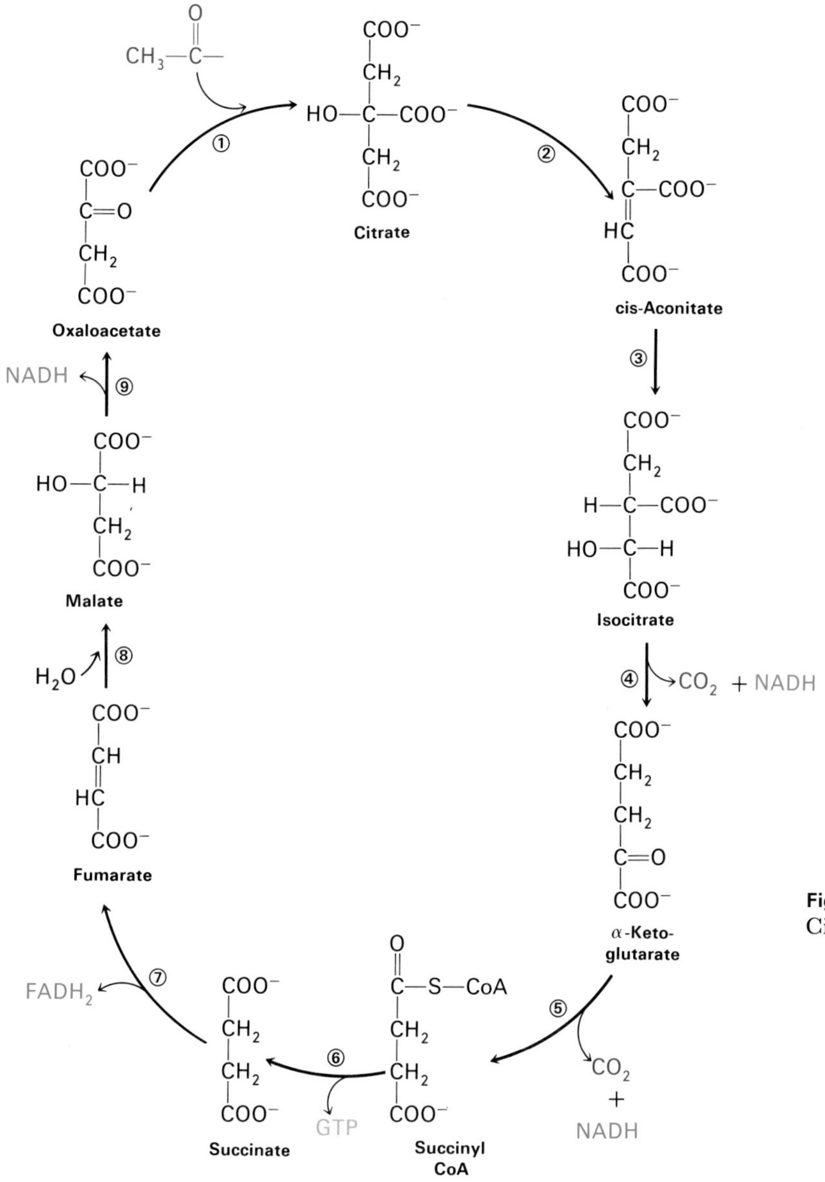

Figure 13-5
Citric acid cycle.

1. Two carbon atoms enter the cycle in the condensation of an acetyl unit (from acetyl CoA) with oxaloacetate. Two carbon atoms leave the cycle in the form of CO_2 in the successive decarboxylations catalyzed by isocitrate dehydrogenase and α-ketoglutarate dehydrogenase. As will be discussed shortly, the two carbon atoms that leave the cycle are different from the ones that entered in that round.

2. Four pairs of hydrogen atoms leave the cycle in the four oxidation reactions. Two NAD^+ molecules are reduced in the oxidative decarboxylations of isocitrate and α-ketoglutarate, one FAD molecule is reduced in the oxidation of succinate, and one NAD^+ molecule is reduced in the oxidation of malate.

3. One high-energy phosphate bond (in the form of GTP) is generated from the energy-rich thioester linkage in succinyl CoA.

4. Two water molecules are consumed: one in the synthesis of citrate by the hydrolysis of citryl CoA, the other in the hydration of fumarate.

Looking ahead, the NADH and $FADH_2$ formed in the citric acid cycle are oxidized by the electron-transport chain (Chapter 14). ATP is generated as electrons are transferred from these carriers to O_2, the ultimate acceptor. Three ATP are formed for each NADH in the mitochondrion, whereas two ATP are generated per $FADH_2$. Note that only one high-energy phosphate bond per acetyl unit is directly formed in the citric acid cycle. Eleven more high-energy phosphate bonds are generated when three NADH and one $FADH_2$ are oxidized by the electron-transport chain.

Molecular oxygen does not participate directly in the citric acid cycle. However, the cycle operates only under aerobic conditions because NAD^+ and FAD can be regenerated in the mitochondrion only by electron transfer to molecular oxygen. *Glycolysis has both an aerobic and an anaerobic mode, whereas the citric acid cycle is strictly aerobic.* Recall that glycolysis can proceed under anaerobic conditions because NAD^+ is regenerated in the conversion of pyruvate into lactate.

PYRUVATE DEHYDROGENASE COMPLEX: AN ORGANIZED ENZYME ASSEMBLY

We now turn to some reaction mechanisms. The oxidative decarboxylation of pyruvate to form acetyl CoA is catalyzed by the *pyruvate dehydrogenase complex,* an organized assembly of three kinds of enzymes (Table 13-2). The net reaction catalyzed by this complex is

$$\text{Pyruvate} + \text{CoA} + \text{NAD}^+ \longrightarrow \text{acetyl CoA} + \text{CO}_2 + \text{NADH}$$

The mechanism of this reaction is more complex than might be

Table 13-2
Pyruvate dehydrogenase complex of *E. coli*

Enzyme	Abbreviation	Number of chains	Prosthetic group	Reaction catalyzed
Pyruvate dehydrogenase component	A or E_1	24	TPP	Decarboxylation of pyruvate
Dihydrolipoyl transacetylase	B or E_2	12	Lipoamide	Oxidation of the C_2 unit and transfer to CoA
Dihydrolipoyl dehydrogenase	C or E_3	12	FAD	Regeneration of the oxidized form of lipoamide

suggested by its stoichiometry. *Thiamine pyrophosphate* (TPP), *lipo-amide,* and *FAD* serve as catalytic cofactors, in addition to CoA and NAD^+, the stoichiometric cofactors.

There are four steps in the conversion of pyruvate into acetyl CoA. First, pyruvate is *decarboxylated* after it combines with TPP. This reaction is catalyzed by the *pyruvate dehydrogenase component* of the multienzyme complex.

$$\text{Pyruvate} + \text{TPP} \longrightarrow \text{hydroxyethyl-TPP} + CO_2$$

A key feature of TPP, the prosthetic group of the pyruvate dehy-drogenase component, is that the carbon atom between the nitrogen and sulfur atoms in the thiazole ring is highly acidic.

**Thiamine pyrophosphate
(TPP)**

It ionizes to form a *carbanion,* which readily adds to the carbonyl group of pyruvate.

Pyruvate   **Carbanion of TPP**   **Addition compound**

The positively charged ring nitrogen of TPP then acts as an elec-tron sink to stabilize the formation of a negative charge, which is necessary for decarboxylation. Protonation then gives *hydroxyethyl thiamine pyrophosphate.*

Addition compound   **Resonance forms of ionized
hydroxyethyl-TPP**

Second, the hydroxyethyl group attached to TPP is *oxidized* to form an acetyl group and concomitantly *transferred to lipoamide.* The oxidant in this reaction is the disulfide group of lipoamide, which is

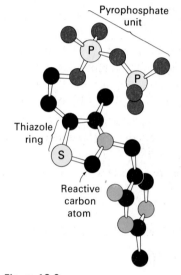

Figure 13-6
Model of thiamine pyrophosphate.

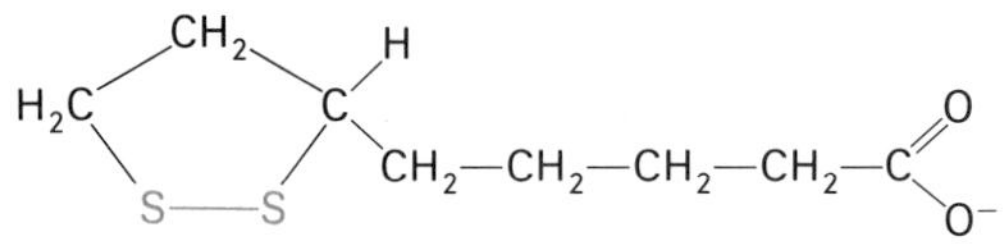

Lipoic acid
(Ionized form)

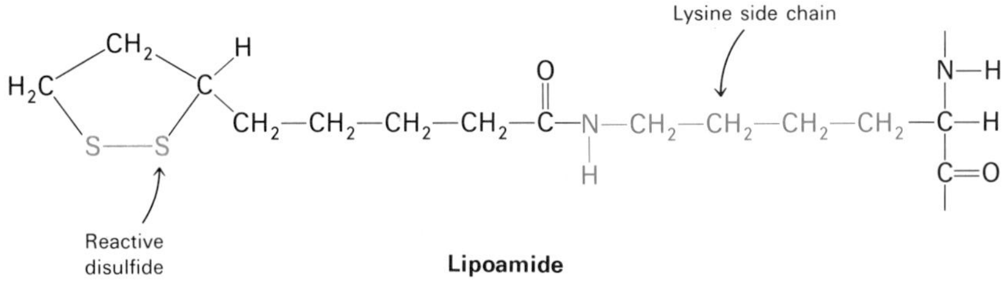

Lysine side chain

Reactive
disulfide

Lipoamide

Figure 13-7
Structures of lipoic acid and lipoamide. Lipoic acid is covalently attached to a specific lysine side chain of dihydrolipoyl transacetylase. Note that this prosthetic group is at the end of a long, flexible chain, which enables it to rotate from one active site to another in the enzyme complex.

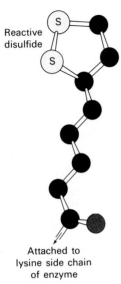

Reactive
disulfide

Attached to
lysine side chain
of enzyme

Figure 13-8
Model of the lipoyl part of lipoamide.

converted into the sulfhydryl form. This reaction, catalyzed by the *dihydrolipoyl transacetylase* part of the complex, yields *acetyllipoamide*.

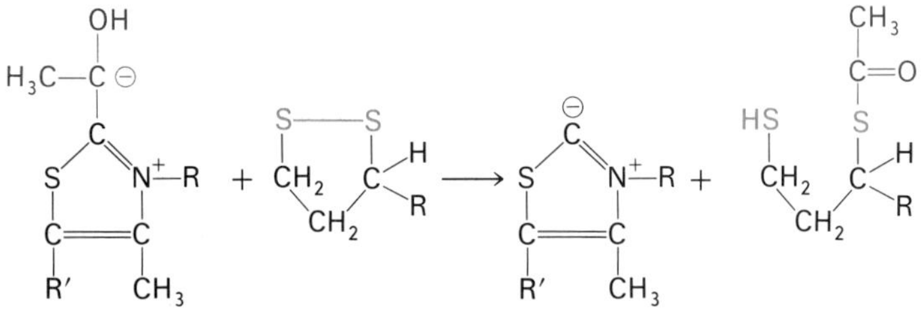

Hydroxyethyl-TPP   **Lipoamide**   **Carbanion**   **Acetyllipoamide**
(Ionized form)   **of TPP**

Third, *the acetyl group is transferred from acetyllipoamide to CoA to form acetyl CoA*. Dihydrolipoyl transacetylase also catalyzes this reaction. The energy-rich thioester bond is preserved as the acetyl group is transferred to CoA.

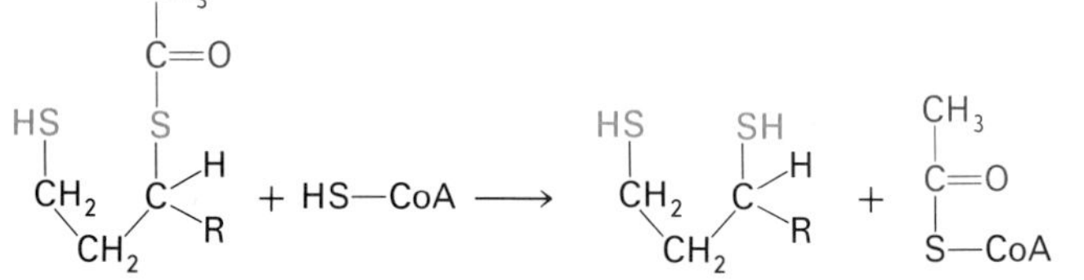

Acetyllipoamide   **Dihydrolipoamide**   **Acetyl CoA**

Fourth, *the oxidized form of lipoamide is regenerated* to complete the reaction. NAD^+ is the oxidant in this reaction, which is catalyzed by the *dihydrolipoyl dehydrogenase* part of the complex. FAD is the prosthetic group of this enzyme.

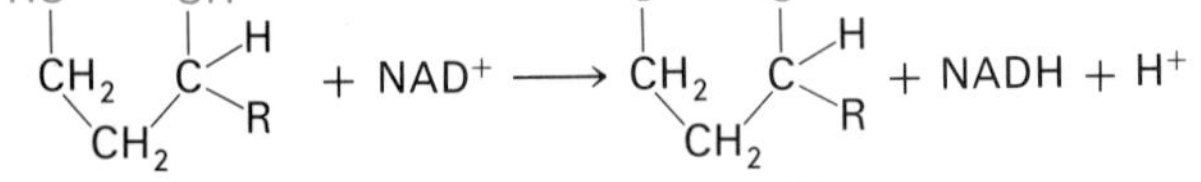

Dihydrolipoamide   **Lipoamide**

The studies of Lester Reed have been sources of insight into the structure and assembly of the pyruvate dehydrogenase complex. The enzyme complex from *E. coli* has been studied intensively. It has a mass of about 4,600 kdal and consists of 48 polypeptide chains. A polyhedral structure with a diameter of about 300 Å is evident in electron micrographs (Figure 13-9). *The transacetylase polypeptide chains are the core of the pyruvate dehydrogenase complex.* The pyruvate dehydrogenase units and lipoyl dehydrogenase units are bound to the outside of this transacetylase core (Figure 13-10).

The constituent polypeptide chains of the complex are held together by noncovalent forces. At alkaline pH, the complex dissociates into the pyruvate dehydrogenase component and a subcomplex of the other two enzymes. The transacetylase can then be separated from the dehydrogenase at neutral pH in the presence of urea. *These three enzymes spontaneously associate to form the pyruvate dehydrogenase complex* when they are mixed at neutral pH in the absence of urea, which suggests that the native enzyme complex may be formed by a self-assembly process.

The structural integration of three kinds of enzymes makes possible the coordinated catalysis of a complex reaction (Figure 13-11). All of the inter-

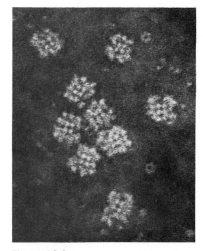

Figure 13-9
Electron micrograph of the pyruvate dehydrogenase complex from *E. coli*. [Courtesy of Dr. Lester Reed.]

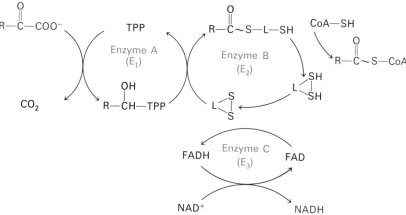

Figure 13-11
Summary of the reactions catalyzed by the pyruvate dehydrogenase complex. L refers to the lipoyl group.

Figure 13-10
Model of the pyruvate dehydrogenase complex from *E. coli*. [After a drawing kindly provided by Dr. Lester Reed.]

mediates in the oxidative decarboxylation of pyruvate are tightly bound to the complex. The close proximity of one enzyme to another *increases the overall reaction rate* and *minimizes side reactions.* The activated intermediates are transferred from one active site to another by the lipoamide prosthetic group of the transacetylase. The attachment of the lipoyl group to the ε-amino group of a lysine residue on the transacetylase provides a flexible arm for the reactive ring. This *14-Å molecular string* enables the lipoyl moiety of a transacetylase subunit to interact with the thiamine pyrophosphate unit of an adjacent pyruvate dehydrogenase subunit and with the flavin unit of an adjacent lipoyl dehydrogenase. Furthermore, the lipoyl moieties of the multienzyme complex can interact with each other, forming an *interacting network of reactive groups*. The net charge on the lipoyl moiety during its cycle of transformation is 0, -1, or -2, if the sulfhydryl groups are fully ionized. This change in net charge may provide the driving force for the directed movement of the lipoyl group.

VARIATION ON A MULTIENZYME THEME: THE ALPHA-KETOGLUTARATE DEHYDROGENASE COMPLEX

The oxidative decarboxylation of α-ketoglutarate closely resembles that of pyruvate:

$$\alpha\text{-Ketoglutarate} + \text{CoA} + \text{NAD}^+ \longrightarrow$$
$$\text{succinyl CoA} + \text{CO}_2 + \text{NADH}$$
$$\text{Pyruvate} + \text{CoA} + \text{NAD}^+ \longrightarrow$$
$$\text{acetyl CoA} + \text{CO}_2 + \text{NADH}$$

The same cofactors are participants: TPP, lipoamide, CoA, FAD, and NAD^+. In fact, *the oxidative decarboxylation of α-ketoglutarate is catalyzed by an enzyme complex that is structurally similar to the pyruvate dehydrogenase complex*. There are three kinds of enzymes in the α-ketoglutarate dehydrogenase complex: an α-ketoglutarate dehydrogenase component (A'), a transsuccinylase one (B'), and a dihydrolipoyl dehydrogenase one (C'). Again, A' binds to B', and B' binds to C', but A' does not bind directly to C'. Thus, *transsuccinylase (like transacetylase) is the core of the complex*.

The α-ketoglutarate dehydrogenase component (A') and transsuccinylase (B') are different from the corresponding enzymes (A and B) in the pyruvate dehydrogenase complex. However, *the dihydrolipoyl dehydrogenase parts (C and C') of the two complexes are similar*. Reconstitution experiments have shown that a complex formed from A, B, and C' is as active in the oxidative decarboxylation of pyruvate as one formed from A, B, and C. Similarly, C and C' are interchangeable in forming a synthetic complex that oxidatively decarboxylates α-ketoglutarate.

It was noted earlier that chymotrypsin, trypsin, thrombin, and elastase are homologous enzymes. Here we see that the pyruvate and α-ketoglutarate dehydrogenase complexes are *homologous enzyme assemblies*. The structural and mechanistic motifs giving coordinated catalysis in an entrée to the citric acid cycle are used again later in the cycle.

BERIBERI IS CAUSED BY A DEFICIENCY OF THIAMINE

"A certain very troublesome affliction, which attacks men, is called by the inhabitants Beri-beri (which means sheep). I believe those, whom this same disease attacks, with their knees shaking and the legs raised up, walk like sheep. It is a kind of paralysis, or rather Tremor: for it penetrates the motion and sensation of the hands and feet indeed sometimes of the whole body. . . ." Jacobus Bonitus, a Dutch physician, gave this description of beriberi in 1630, while working in Java.

Beriberi is caused by a dietary deficiency of thiamine (also called Vitamin B_1). The disease has been and continues to be a serious health problem in the Far East because rice, the major food, has a rather low content of thiamine. The problem is exacerbated if the rice is polished, because only the outer layer contains appreciable amounts of thiamine. Beriberi is also occasionally seen in alcoholics who are severely malnourished. The disease is characterized by neurologic and cardiac symptoms. Damage to the peripheral nervous system is expressed in terms of pain in the limbs, weakness of the musculature, and distorted skin sensation. The heart may be enlarged and the cardiac output inadequate.

How are these symptoms elicited by a deficiency of thiamine? TPP is the prosthetic group of three important enzymes: pyruvate dehydrogenase, α-ketoglutarate dehydrogenase, and transketolase. Transketolase transfers two-carbon units from one sugar to another; its role in the pentose phosphate pathway will be discussed later. The common feature of enzymatic reactions utilizing TPP is the transfer of an activated aldehyde unit. In beriberi, the levels of pyruvate and α-ketoglutarate in the blood are higher than normal. The increase in the level of pyruvate in the blood is especially pronounced after ingestion of glucose. A related finding is that the enzymatic activities of the pyruvate and α-ketoglutarate dehydrogenase complexes in vivo are abnormally low. Futhermore, the transketolase activity of red cells is low in beriberi. This enzymatic assay is useful in diagnosing the disease.

SYMMETRIC MOLECULES MAY REACT ASYMMETRICALLY

Let us follow the fate of a particular carbon atom in the citric acid cycle. Suppose that oxaloacetate were labeled with ^{14}C in the carboxyl carbon furthest from the keto group. Analysis of the α-ketoglutarate formed would show that none of the radioactive label had been lost. Decarboxylation of α-ketoglutarate would then yield succinate devoid of radioactivity. All of the label would be in the released CO_2.

Table 13-3
Commonly used radioisotopes

Isotope	Half-life
^3H	12.26 years
^{14}C	5730 years
^{22}Na	2.62 years
^{32}P	14.28 days
^{35}S	87.9 days
^{42}K	12.36 hours
^{45}Ca	163 days
^{59}Fe	45.6 days
^{125}I	60.2 days
^{203}Hg	46.9 days

Path 1

The finding that *all* of the label emerges in the CO_2 came as a surprise. Citrate is a symmetric molecule. Consequently, it was assumed that the two $-CH_2COO^-$ groups in it would react identically. Thus, for every citrate undergoing the reactions shown in Path 1, it was thought that another citrate molecule would react as shown in Path 2. If so, then only *half* of the label should have emerged in the CO_2.

Path 2
(Does not occur)

The interpretation of these experiments, which were carried out in 1941, was that citrate (or any other symmetric compound) could not be an intermediate in the formation of α-ketoglutarate because of the asymmetric fate of the label. This interpretation seemed compelling until Alexander Ogston incisively pointed out in 1948 that it is a fallacy to assume that the two identical groups of a symmetric molecule cannot be distinguished: "On the contrary, it is possible that *an asymmetric enzyme which attacks a symmetrical compound can distinguish between its identical groups* . . . the asymmetrical occurrence of isotope in a product cannot be taken as conclusive evidence against its arising from a symmetrical precursor."

Let us examine Ogston's assertion. For simplicity, consider a molecule in which two hydrogen atoms, a group X, and a different group Y are bonded to a tetrahedral carbon atom. Let us label one hydrogen A, the other B. Now suppose that an enzyme binds three groups of this substrate: X, Y, and H. Can H_A be distinguished from H_B? Figure 13-12 shows X, Y, and H_A bound to three points on the enzyme. Note that X, Y, and H_B cannot be bound to this active site; two of these three groups can be bound, but not all three. Thus, H_A *and* H_B *will have different fates.*

It should be noted that H_A and H_B are sterically not equivalent even though the molecule $CXYH_2$ is optically inactive. Similarly, the $-CH_2COO^-$ groups in citrate are sterically not equivalent even though citrate is optically inactive. *The symmetry rules that determine whether a compound has indistinguishable substituents are different from those that determine whether it is optically inactive:* (1) a molecule is optically inactive if it can be superimposed on its mirror image; (2) a molecule has indistinguishable substituents only if these groups can be brought into coincidence by a rotation that leaves the rest of the structure invariant.

Sterically nonequivalent groups such as H_A and H_B will almost always be

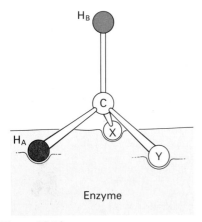

Figure 13-12
H_A and H_B are sterically not equivalent if the substrate $CXYH_2$ is bound to the enzyme at three points.

distinguished in enzymatic reactions. The essence of the differentiation of these groups is that the enzyme holds the substrate in a specific orientation. Attachment at three points, as depicted in Figure 13-12, is a readily visualized way of achieving a particular orientation of the substrate, but it is not the only means of doing so.

The terms chiral and prochiral are now extensively used to describe the stereochemistry of molecules. A *chiral* molecule has handedness and hence is optically active. A *prochiral* molecule, such as citrate or $CXYH_2$, lacks handedness and is optically inactive, but it can become chiral in one step. A prochiral molecule (such as $CXYH_AH_B$) is transformed into a chiral one ($CXYZH_B$) when one of its identical atoms or groups (H_A in this example) is replaced. The prefixes R and S are used to unambiguously designate the configuration of chiral and prochiral centers, as described in the appendix to this chapter (p. 305).

> *Chirality—*
> "I call any geometrical figure, or any group of points, *chiral,* and say it has *chirality,* if its image in a plane mirror, ideally realized, cannot be brought to coincide with itself."
>   LORD KELVIN (1893)
>
> Derived from the Greek word *cheir* for hand.

STEREOSPECIFIC TRANSFER OF HYDROGEN BY NAD⁺-DEHYDROGENASES

In the 1950s, Birgit Vennesland, Frank Westheimer, and their associates carried out some elegant experiments on the stereospecificity of hydrogen transfer by NAD^+-dehydrogenases. Ethanol labeled with two deuterium atoms at C-1 was the substrate in a reaction catalyzed by alcohol dehydrogenase. They found that the reduced coenzyme contained one atom of deuterium per molecule, whereas the other deuterium atom was in acetaldehyde. Thus, none of the deuterium was lost to the solvent, which shows that deuterium was directly transferred from the substrate to NAD^+.

$$H_3C-\overset{\overset{\displaystyle D}{|}}{\underset{\underset{\displaystyle D}{|}}{C}}-OH + NAD^+ \longrightarrow CH_3-C\overset{\diagup O}{\underset{\diagdown D}{}} + \begin{matrix}\text{Reduced NAD}\\ \text{containing}\\ \text{1 deuterium}\end{matrix} + H^+$$

The deuterated reduced coenzyme formed in this reaction was then used to reduce acetaldehyde. The striking result was that *all of the deuterium was transferred from the coenzyme to the substrate.* None remained in NAD^+.

$$CH_3-C\overset{\diagup O}{\underset{\diagdown H}{}} + H^+ + \begin{matrix}\text{Reduced NAD}\\ \text{containing}\\ \text{1 deuterium}\end{matrix} \longrightarrow \underset{\underset{\text{(Contains}}{\text{1 deuterium)}}}{CH_3CHDOH} + \underset{\underset{\text{(Contains}}{\text{no deuterium)}}}{NAD^+}$$

These reactions revealed that the transfer catalyzed by alcohol dehydrogenase is stereospecific. *The positions occupied by the two hydrogen atoms at C-4 in NADH are not equivalent.* One of them (H_A) is in front of the nicotinamide plane, the other (H_B) is behind. In other words, C-4 is a *prochiral center.* Alcohol dehydrogenase, a chiral rea-

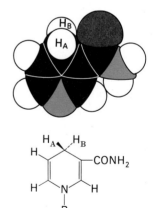

NADH

Figure 13-13
Model and formula of the nicotinamide moiety of NADH showing that H_A and H_B are on opposite sides of the ring. In the RS nomenclature, H_A is *pro-R* and H_B is *pro-S.*

gent, distinguishes between positions A and B at C-4. Deuterium is transferred from deuterated ethanol to position A only:

In the reverse reaction, the deuterium atom at position A is removed and directly transferred to acetaldehyde.

Some dehydrogenases, such as glyceraldehyde phosphate dehydrogenase, transfer hydrogen to position B. Thus, *there are two classes of NAD$^+$ (and NADP$^+$) dehydrogenases: A-stereospecific and B-stereospecific*. A comparison of the three-dimensional structures of NAD$^+$ dehydrogenases shows that the difference in stereospecificity of the A- and B-type enzymes arises from a 180-degree rotation of the nicotinamide ring with respect to the adjacent ribose unit. The result of this 180-degree flip is that opposite sides of the nicotinamide ring are exposed and hence reactive in A- and B-type dehydrogenases. As might be expected, all known dehydrogenases are stereospecific. When a dehydrogenase reacts with a range of substrates, the stereospecificity of hydrogen transfer is the same for all. The conservation of NAD$^+$-binding sites in the course of evolution is emphasized by the finding that the stereospecificity of a particular dehydrogenase is independent of species (e.g., yeast and horse alcohol dehydrogenase have the same stereospecificity). Furthermore, the stereospecificity of a particular reaction is the same for NAD$^+$ and NADP$^+$ in those enzymes that can use either coenzyme.

LETHAL SYNTHESIS: THE CONVERSION OF FLUOROACETATE INTO FLUOROCITRATE

The citric acid cycle is blocked in animals that have ingested the leaves of *Dichapetalum cymosum,* a poisonous South African plant. Within an hour, the level of citrate in many organs increases more than tenfold. The poisoned animals have convulsions and usually die a short time later. The toxic agent in these leaves is *fluoroacetate,* which has also been used as a rat poison. However, fluoroacetate itself does not affect purified enzymes of the citric acid cycle. Why, then, is the cycle blocked in vivo following the ingestion of fluoroacetate? The reason is that *fluoroacetate is enzymatically converted into fluorocitrate, a potent inhibitor of aconitase.* Fluoroacetate is activated to fluoroacetyl CoA, which then condenses with oxaloacetate to give fluorocitrate.

The active site of aconitase contains an Fe^{2+} ion, which normally chelates the hydroxyl oxygen and two carboxylate oxygens of

Fluoroacetate

Fluoroacetyl CoA

Fluorocitrate

citrate. In contrast, fluorocitrate binds so that its fluorine atom is chelated to the ferrous ion (Figure 13-14). The high electronegativity of fluorine enables it to interact strongly with Fe^{2+}, which results in an inhibited enzyme. It is important to note that the conversion of fluoroacetate into fluorocitrate is not a unique case of lethal synthesis. *A variety of chemicals, relatively harmless in themselves, can be enzymatically altered to form highly deleterious substances.* For example, some polycyclic aromatic hydrocarbons are enzymatically modified in vivo into potent mutagens and carcinogens (p. 475).

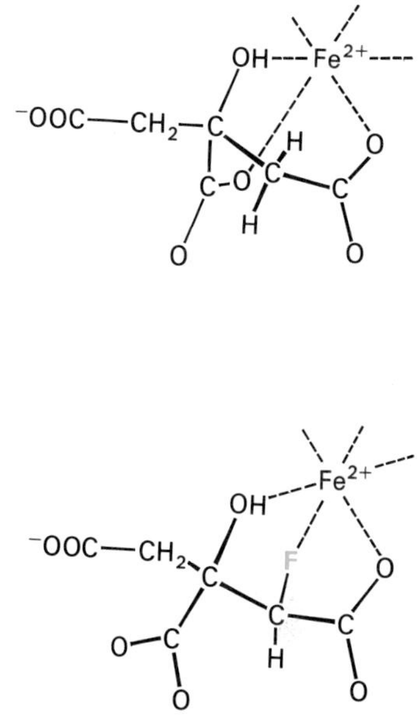

THE CITRIC ACID CYCLE IS A SOURCE OF BIOSYNTHETIC PRECURSORS

Thus far, discussion has focused on the citric acid cycle as the *major degradative pathway for the generation of ATP.* The citric acid cycle also *provides intermediates for biosyntheses* (Figure 13-15). For example, a majority of the carbon atoms in porphyrins come from *succinyl CoA.* Many of the amino acids are derived from *α-ketoglutarate and oxaloacetate.* The biosyntheses of these compounds will be discussed in subsequent chapters. The important point to note now is that *citric acid cycle intermediates must be replenished if any are drawn off for biosyntheses.* Suppose that oxaloacetate is converted into amino acids for protein synthesis. The citric acid cycle will cease to operate unless oxaloacetate is formed de novo because acetyl CoA cannot enter the cycle unless it condenses with oxaloacetate. How is oxaloacetate replenished? Mammals lack the enzymatic machinery to convert acetyl CoA into oxaloacetate or another citric acid cycle intermediate. Rather, oxaloacetate is formed by the carboxylation of pyruvate, a reaction catalyzed by pyruvate carboxylase (p. 348).

$$\text{Pyruvate} + CO_2 + \text{ATP} + H_2O \rightleftharpoons$$
$$\text{oxaloacetate} + \text{ADP} + P_i + 2\,H^+$$

Figure 13-14
Proposed modes of binding of citrate and fluorocitrate to the active site of aconitase: (A) binding of citrate resulting in catalysis; (B) binding of fluorocitrate resulting in inhibition. The fluorine atom of fluorocitrate is chelated to the ferrous ion at the active site. [After a drawing kindly provided by Dr. Jenny Glusker.]

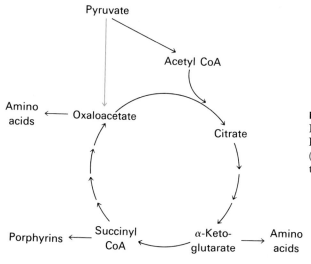

Figure 13-15
Biosynthetic roles of the citric acid cycle. Intermediates drawn off for biosyntheses (shown by red arrows) are replenished by the formation of oxaloacetate from pyruvate.

The carboxylation of pyruvate is an example of an *anaplerotic reaction* (from the Greek, to "fill up").

CONTROL OF THE PYRUVATE DEHYDROGENASE COMPLEX

The formation of acetyl CoA from pyruvate is a key irreversible step in metabolism because *animals are unable to convert acetyl CoA into glucose.* The oxidative decarboxylation of pyruvate to acetyl CoA commits the carbon atoms of glucose to two potential fates: oxidation to CO_2 by the citric acid cycle, with the concomitant generation of energy, or incorporation into lipid (p. 401). It would therefore seem essential that the activity of the pyruvate dehydrogenase complex be stringently regulated. In fact, this enzyme complex is controlled in three ways:

1. *Product inhibition.* Acetyl CoA and NADH, the products of the oxidation of pyruvate, inhibit the enzyme complex. Acetyl CoA inhibits the transacetylase component, whereas NADH inhibits the dihydrolipoyl dehydrogenase component. These inhibitory effects are reversed by CoA and NAD^+, respectively.

2. *Feedback regulation by nucleotides.* The activity of the enzyme complex is controlled by the *energy charge* (p. 251). Specifically, the pyruvate dehydrogenase component is inhibited by GTP and activated by AMP. Again, the activity of the complex is reduced when the cell is rich in immediately available energy.

3. *Regulation by covalent modification.* The complex becomes enzymatically inactive when a specific serine residue of the pyruvate dehydrogenase component is phosphorylated by ATP. Phosphorylation is enhanced by high ratios of ATP/ADP, acetyl CoA/CoA, and $NADH/NAD^+$, whereas it is inhibited by pyruvate. The enzyme complex becomes active again when the phosphoryl group is hydrolyzed by a specific phosphatase. Dephosphorylation is enhanced by a high level of pyruvate. *Covalent modification is an important mechanism for controlling enzymatic activity.* Phosphorylation and dephosphorylation as control reactions will be encountered again when glycogen synthesis and degradation are considered.

CONTROL OF THE CITRIC ACID CYCLE

The rate of the citric acid cycle is also precisely adjusted to meet the cell's needs for ATP. *The synthesis of citrate from oxaloacetate and acetyl CoA is an important control point in the cycle.* ATP is an allosteric inhibitor of citrate synthetase. The effect of ATP is to increase the K_M for acetyl CoA. Thus, as the level of ATP increases, less of this enzyme is saturated with acetyl CoA and so less citrate is formed.

A second control point is isocitrate dehydrogenase. This enzyme is allosterically stimulated by ADP, which enhances its affinity for substrates. The binding of isocitrate, NAD^+, Mg^{2+}, and ADP is mutually cooperative. In contrast, NADH inhibits isocitrate dehydrogenase by directly displacing NAD^+.

A third control site in the citric acid cycle is α-ketoglutarate dehydrogenase. Some aspects of its control are like those of the pyruvate dehydrogenase complex, as might be expected from their structural homology. Alpha-ketoglutarate dehydrogenase is inhibited by succinyl CoA and NADH, the products of the reaction it catalyzes. Also, α-ketoglutarate dehydrogenase is inhibited by a high energy charge. *In short, the funneling of two-carbon fragments into the citric acid cycle and the rate of the cycle are reduced when the cell has a high level of ATP.* This control is achieved by a variety of complementary mechanisms at several sites (Figure 13-16).

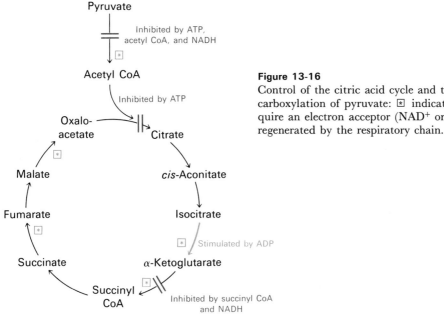

Figure 13-16
Control of the citric acid cycle and the oxidative decarboxylation of pyruvate: ▣ indicates steps that require an electron acceptor (NAD^+ or FAD) that is regenerated by the respiratory chain.

KREBS' DISCOVERY OF THE CYCLE

"I have often been asked how the work on the tricarboxylic acid cycle arose and developed. Was the concept perhaps due to a sudden inspiration and vision?" Hans Krebs' reply is that "it was of course nothing of the kind, but a very slow evolutionary process, extending over some five years beginning (as far as I am involved) in 1932." Krebs first studied the rate of oxidation of a variety of compounds by using kidney and liver slices. The substances chosen were ones that might be expected to be intermediates in the oxidation of foodstuffs. The idea behind these experiments was that such intermediates would be rapidly oxidized and thereby identified. The important finding was that *citrate,* succinate, fumarate, and acetate were very readily oxidized in various tissues.

A critical contribution was made by Albert Szent-Györgyi in

1935. He studied the oxidation of various substances by using suspensions of minced pigeon-breast muscle. This very active flight muscle has an exceptionally high rate of oxidation, which facilitated the experimental studies. Szent-Györgyi found that the addition of certain C_4-dicarboxylic acids increased the uptake of O_2 far more than could be caused by their direct oxidation. In other words, they produced a catalytic (rather than a stoichiometric) enhancement of O_2 uptake. *This catalytic stimulation of respiration was obtained with succinate, fumarate, and malate.*

The next breakthrough was the elucidation of the biological pathway of oxidation of citrate by Carl Martius and Franz Knoop in 1937. They showed that citrate is isomerized to isocitrate by way of *cis*-aconitate and that isocitrate is oxidatively decarboxylated to α-ketoglutarate. It had already been known that α-ketoglutarate can be oxidized to succinate. *Thus, their discovery revealed the pathway from citrate to succinate.* This finding came at a propitious moment, for it enabled Krebs to interpret the recent observation that *citrate catalytically enhanced* the respiration of minced pigeon-breast muscle.

Additional critical information came from the use of *malonate, a specific inhibitor of succinate dehydrogenase.* Malonate is a competitive inhibitor of this enzyme because it closely resembles succinate. It had been known for some time that *malonate poisons respiration.* Krebs reasoned that succinate dehydrogenase must therefore play a key role in respiration. This was reinforced by the finding that succinate accumulated when citrate was added to malonate-poisoned muscle. Furthermore, succinate also accumulated when fumarate was added to malonate-poisoned muscle. The first of these experiments showed that the pathway from citrate to succinate is physiologically significant. The second experiment revealed that there is a pathway from fumarate to succinate distinct from the reaction catalyzed by succinate dehydrogenase.

Krebs then discovered that citrate is readily formed by a muscle suspension if oxaloacetate is added. *The discovery of the synthesis of citrate from oxaloacetate and a substance derived from pyruvate or acetate enabled Krebs to formulate a complete scheme.* His postulated tricarboxylic acid cycle suddenly provided a coherent picture of how carbohydrates are oxidized. Many experimental facts, such as the catalytic enhancement of respiration by succinate and other intermediates, fell neatly into place. It is noteworthy that the citric acid cycle was not the only metabolic cycle, or the first, to be discovered by Krebs. Six years earlier, he had shown that urea is synthesized by a cyclic metabolic pathway called the ornithine cycle (Chapter 18). Thus, the concept of a cyclic metabolic pathway had already been fully recognized by Krebs when he pondered the data and designed the experiments that led to the proposal of the citric acid cycle.

COO^-
|
CH_2
|
CH_2
|
COO^-

Succinate

COO^-
|
CH_2
|
COO^-

Malonate

SUMMARY

The citric acid cycle is the final common pathway for the oxidation of fuel molecules. It also serves as a source of building blocks for biosyntheses. Most fuel molecules enter the cycle as acetyl CoA. The link between glycolysis and the citric acid cycle is the oxidative decarboxylation of pyruvate to form acetyl CoA. This reaction and those of the cycle occur inside mitochondria, in contrast with glycolysis, which occurs in the cytosol. The cycle starts with the condensation of oxaloacetate (C_4) and acetyl CoA (C_2) to give citrate (C_6), which is isomerized to isocitrate (C_6). Oxidative decarboxylation of this intermediate gives α-ketoglutarate (C_5). The second molecule of CO_2 comes off in the next reaction, in which α-ketoglutarate is oxidatively decarboxylated to succinyl CoA (C_4). The thioester bond of succinyl CoA is cleaved by P_i to yield succinate, and a high-energy phosphate bond in the form of GTP is concomitantly generated. Succinate is oxidized to fumarate (C_4), which is then hydrated to form malate (C_4). Finally, malate is oxidized to regenerate oxaloacetate (C_4). Thus, two carbon atoms from acetyl CoA enter the cycle, and two carbon atoms leave the cycle as CO_2 in the successive decarboxylations catalyzed by isocitrate dehydrogenase and α-ketoglutarate dehydrogenase. In the four oxidation-reductions in the cycle, three pairs of electrons are transferred to NAD^+ and one pair to FAD. These reduced electron carriers are subsequently oxidized by the electron-transport chain to generate eleven molecules of ATP. In addition, one high-energy phosphate bond is directly formed in the citric acid cycle. Hence, a total of twelve high-energy phosphate bonds are generated for each two-carbon fragment that is completely oxidized to H_2O and CO_2.

The citric acid cycle operates only under aerobic conditions because it requires a supply of NAD^+ and FAD. These electron acceptors are regenerated when NADH and $FADH_2$ transfer their electrons to O_2 through the electron-transport chain, with the concomitant production of ATP. Consequently, the rate of the citric acid cycle depends on the need for ATP. The regulation of three enzymes in the cycle is also important for control. A high energy charge diminishes the activities of citrate synthetase, isocitrate dehydrogenase, and α-ketoglutarate dehydrogenase. The irreversible formation of acetyl CoA from pyruvate is another important regulatory site. The activity of the pyruvate dehydrogenase complex is controlled by product inhibition, feedback regulation by nucleotides, and covalent modification. These mechanisms complement each other in reducing the rate of formation of acetyl CoA when the energy charge of the cell is high.

SELECTED READINGS

BOOKS AND GENERAL REVIEWS

Lowenstein, J. M., (ed.), 1969. *Citric Acid Cycle: Control and Compartmentation.* Marcel Dekker.

Goodwin, T. W., (ed.), 1968. *The Metabolic Roles of Citrate.* Academic Press.

Lowenstein, J. M., 1971. The pyruvate dehydrogenase complex and the citric acid cycle. *Compr. Biochem.* 18S:1–55.

REACTION MECHANISMS AND STEREOSPECIFICITY

Popják, G., 1970. Stereospecificity of enzymic reactions. *In* Boyer, P. D., (ed.), *The Enzymes* (3rd ed.), vol. 2, pp. 115–215. Academic Press. [An excellent review containing a discussion of the stereochemistry of the citric acid cycle.]

Ogston, A. G., 1948. Interpretation of experiments on metabolic processes using isotopic tracer elements. *Nature* 162:963.

Hirschmann, H., 1960. The nature of substrate asymmetry in stereoselective reactions. *J. Biol. Chem.* 235:2762–2767.

Bentley, R., 1969. *Molecular Asymmetry in Biology,* vols. 1 and 2. Academic Press. [These volumes contain a wealth of information about stereospecificity in biochemical reactions. Chapter 2 of Volume 1 discusses nomenclature and Chapter 4 discusses prochirality.]

FLUOROCITRATE AND ACONITASE

Peters, R., 1954. Biochemical light upon an ancient poison: a lethal synthesis. *Endeavor* 13:147–154. [An account of the conversion of fluoroacetate into fluorocitrate, an inhibitor of aconitase.]

Gribble, G. W., 1973. Fluoroacetate toxicity. *J. Chem. Ed.* 50:460–462.

Glusker, J. P., 1971. Aconitase. *In* Boyer, P. D., (ed.), *The Enzymes* (3rd ed.), vol. 5, pp. 413–439.

PYRUVATE AND α-KETOGLUTARATE DEHYDROGENASE COMPLEXES

Reed, L. J., 1974. Multienzyme complexes. *Acc. Chem. Res.* 7:40–46.

Bates, D. L., Danson, M. J., Hale, G., Hooper, E. A., and Perham, R. N., 1977. Self-assembly and catalytic activity of the pyruvate dehydrogenase multienzyme complex of *Escherichia coli. Nature* 268:313–316.

Collins, J. H., and Reed, L. J., 1977. Acyl group and electron pair relay system: a network of interacting lipoyl moieties in the pyruvate and α-ketoglutarate dehydrogenase complexes from *Escherichia coli. Proc. Nat. Acad. Sci.* 74:4223–4227.

Gutowski, J. A., and Lienhard, G. E., 1976. Transition state analogs for thiamine pyrophosphate-dependent enzymes. *J. Biol. Chem.* 251:2863–2866.

Randle, P. J., 1978. Pyruvate dehydrogenase complex: meticulous regulator of glucose disposal in animals. *Trends Biochem. Sci.* 3:217–219.

DeRosier, D. J., and Oliver, R. M., 1971. A low resolution electron density map of lipoyl transsuccinylase, the core of the α-ketoglutarate dehydrogenase complex. *Cold Spring Harbor Symp. Quant. Biol.* 36:199–203.

DISCOVERY OF THE CITRIC ACID CYCLE

Krebs, H. A., and Johnson, W. A., 1937. The role of citric acid in intermediate metabolism in animal tissues. *Enzymologia* 4:148–156.

Krebs, H. A., 1970. The history of the tricarboxylic acid cycle. *Perspect. Biol. Med.* 14:154–170.

APPENDIX
The RS Designation of Chirality

The absolute configuration of any chiral center can be unambiguously specified using the RS notation introduced in 1956 by Robert Cahn, Christopher Ingold, and Vladimir Prelog. Consider the chiral compound CHFClBr (Figure 13-17A). The first

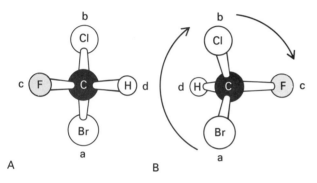

R configuration
(Clockwise)

Figure 13-17
The stereoisomer of CHFClBr shown here has an R configuration.

step in using the RS designation is to assign a *priority sequence* to the four substituents by applying the role that *an atom with a higher atomic number has a higher priority than an atom with a lower atomic number.* Hence, the priority sequence of these four substituents is (a) Br, (b) Cl, (c) F, and (d) H, with Br having the highest priority (a). The next step is to *orient the molecule so that the group of lowest priority (d) points away from the viewer.* For CHFClBr, this means that the molecule should be oriented so that H is away from us (behind the plane of the page in Figure 13-17B). *The final step is to ask whether the path from a to b to c under these conditions is clockwise or counterclockwise.* If it is clockwise (right-handed), the configuration is R (from the Latin *rectus,* right). If it

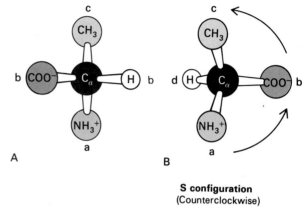

S configuration
(Counterclockwise)

Figure 13-18
L-Alanine has an S configuration.

is counterclockwise (left-handed), the configuration is S (from the Latin *sinister,* left). The configuration of the stereoisomer of CHFClBr shown in Figure 13-17 is R.

Now let us consider the RS designation of alanine (Figure 13-18A). The four atoms bonded to the α-carbon are N, C, C, and H. The priority sequence of the methyl carbon and the carboxyl carbon is determined by going outward to the next set of atoms. *It is useful to note the following priority sequence for biochemically important groups:* —SH (highest), —OR, —OH, —NHR, —NH$_2$, —COOR, —COOH, —CHO, —CH$_2$OH, —C$_6$H$_5$, —CH$_3$, —T, —D, —H (lowest). Thus, the priority sequence of the four groups attached to the α-carbon of alanine is (a) —NH$_3^+$, (b) —COO$^-$, (c) —CH$_3$, and (d) —H. The next step is to orient L-alanine so that its lowest priority group (—H) is behind the plane of the page (Figure 13-18B). The path from (a) —NH$_3^+$ to (b) —COO$^-$ to (c) —CH$_3$ is then counterclockwise (left-handed), and so L-alanine has an S configuration.

PROBLEMS

1. What is the fate of the radioactive label when each of the following compounds is added to a cell extract containing the enzymes and cofactors of the glycolytic pathway, the citric acid cycle, and the pyruvate dehydrogenase complex? (The ^{14}C label is denoted by an asterisk.)

(a)

$$H_3\overset{*}{C}-\overset{O}{\underset{\|}{C}}-COO^-$$

(b)

$$H_3C-C-\overset{*}{C}OO^-$$
$$\overset{\|}{O}$$

(c)

$$H_3C-\overset{O}{\underset{\|}{C}}-\underset{*}{COO^-}$$

(d)

$$H_3\overset{*}{C}-\overset{O}{\underset{\|}{C}}-S-CoA$$

(e) Glucose 6-phosphate labeled at C-1.

2. Is it possible to get *net synthesis* of oxaloacetate by adding acetyl CoA to an extract that contains the enzymes and cofactors of the citric acid cycle?

3. What are the relative concentrations of citrate, isocitrate, and *cis*-aconitate at equilibrium? (Use the data given in Table 13-1.)

4. What is the $\Delta G°'$ for the oxidation of the acetyl unit of acetyl CoA by the citric acid cycle?

5. A sample of deuterated reduced NAD was prepared by incubating H_3C-CD_2-OH and NAD^+ with alcohol dehydrogenase. This re-

duced coenzyme was added to a solution of 1,3-DPG and glyceraldehyde 3-phosphate dehydrogenase. The NAD^+ formed by this second reaction contained one atom of deuterium, whereas glyceraldehyde 3-phosphate, the other product, contained none. What does this experiment reveal about the stereospecificity of glyceraldehyde 3-phosphate dehydrogenase?

6. Thiamine thiazolone pyrophosphate binds to pyruvate dehydrogenase about 20,000 times as strongly as thiamine pyrophosphate, and it competitively inhibits the enzyme. Why?

TPP   **Thiazolone analog of TPP**

7. The oxidation of malate by NAD^+ to form oxaloacetate is a highly endergonic reaction under standard conditions $(\Delta G°' = +7\,\text{kcal/mol})$. The reaction proceeds readily from malate to oxaloacetate under physiological conditions because the steady-state concentrations of the products are low compared with those of the substrates. Assuming an $[NAD^+]/[NADH]$ ratio of 8 and a pH of 7, what is the lowest [malate]/[oxaloacetate] ratio at which oxaloacetate can be formed from malate?

For additional problems, see W. B. Wood, J. H. Wilson, R. M. Benbow, and L. E. Hood, *Biochemistry: A Problems Approach* (Benjamin, 1974), ch. 10.

CHAPTER 14

OXIDATIVE PHOSPHORYLATION

The NADH and $FADH_2$ formed in glycolysis, fatty acid oxidation, and the citric acid cycle are energy-rich molecules because each contains a pair of electrons that have a high transfer potential. When these electrons are transferred to molecular oxygen, a large amount of energy is liberated. This released energy can be used to generate ATP. *Oxidative phosphorylation is the process in which ATP is formed as electrons are transferred from NADH or $FADH_2$ to O_2 by a series of electron carriers.* This is the major source of ATP in aerobic organisms. For example, oxidative phosphorylation generates 32 of the 36 molecules of ATP that are formed when glucose is completely oxidized to CO_2 and H_2O. Some salient features of this process are:

1. Oxidative phosphorylation is carried out by *respiratory assemblies* that are located in the *inner membrane of mitochondria.* The citric acid cycle and the pathway of fatty acid oxidation, which supply most of the NADH and $FADH_2$, are in the adjacent mitochondrial matrix.

2. The oxidation of NADH yields 3 ATP, whereas the oxidation of $FADH_2$ yields 2 ATP. Oxidation and phosphorylation are coupled processes.

3. The respiratory assembly contains numerous electron carriers, such as the cytochromes. The step-by-step transfer of electrons from NADH or $FADH_2$ to O_2 through these carriers leads to the pump-

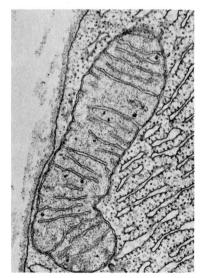

Figure 14-1
Electron micrograph of a mitochondrion. [Courtesy of Dr. George Palade.]

Respiration—
An ATP-generating process in which an inorganic compound (such as O_2) serves as the ultimate electron acceptor. The electron donor can be either an organic compound or an inorganic one.

ing of protons out of the mitochondrial matrix and to the generation of a membrane potential (proton-motive force). Protons are pumped by three kinds of electron-transfer complexes. ATP is synthesized when protons flow back to the mitochondrial matrix through an enzyme complex. Thus, *oxidation and phosphorylation are coupled by a proton gradient across the inner mitochondrial membrane.*

OXIDATIVE PHOSPHORYLATION OCCURS IN MITOCHONDRIA

Mitochondria are oval-shaped organelles, typically about 2 μm in length and 0.5 μm in diameter. Techniques for the isolation of mitochondria were devised in the late 1940s. Eugene Kennedy and Albert Lehninger then discovered that *mitochondria contain the respiratory assembly, the enzymes of the citric acid cycle, and the enzymes of fatty acid oxidation.* Electron microscopic studies by George Palade and Fritjof Sjöstrand revealed that mitochondria have two membrane systems: an *outer membrane* and an extensive, highly folded *inner membrane.* The inner membrane is folded into a series of internal ridges called *cristae.* Hence, there are two compartments in mitochondria: the *intermembrane space* between the outer and inner membranes, and the *matrix,* which is bounded by the inner membrane (Figure 14-2). *The*

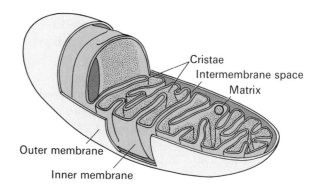

Figure 14-2
Diagram of a mitochondrion. [After *Biology of the Cell* by Stephen L. Wolfe. © 1972 by Wadsworth Publishing Company, Inc. Belmont, California 94002. Adapted by permission of the publisher.]

respiratory assembly is an integral part of the inner mitochondrial membrane, whereas most of the reactions of the citric acid cycle and fatty acid oxidation occur in the matrix.

The outer membrane is quite permeable to most small molecules and ions. In contrast, the inner membrane is intrinsically impermeable to nearly all ions and most uncharged molecules. There are specific protein carriers that transport molecules such as ADP and long-chain fatty acids across the inner mitochondrial membrane.

REDOX POTENTIALS AND FREE-ENERGY CHANGES

In oxidative phosphorylation, the *electron-transfer potential* of NADH or $FADH_2$ is converted into the *phosphate-transfer potential* of ATP. We need quantitative expressions for these forms of free energy. The measure of phosphate-transfer potential is already familiar to us: it is given by $\Delta G°'$ for hydrolysis of the phosphate compound. The corresponding expression for the electron-transfer potential is E_0', the redox potential (also called the oxidation-reduction potential).

The redox potential is an electrochemical concept. Consider a substance that can exist in an oxidized form X and a reduced form X^-. Such a pair is called a *redox couple* (Figure 14-3). The redox

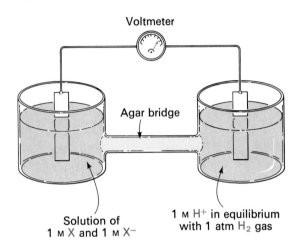

Voltmeter

Agar bridge

Solution of
1 M X and 1 M X^-

1 M H^+ in equilibrium
with 1 atm H_2 gas

Figure 14-3
Arrangement for the measurement of the standard oxidation-reduction potential of a redox couple.

potential of this couple can be determined by measuring the electromotive force generated by a sample *half-cell* with respect to a *standard reference half-cell*. The sample half-cell consists of an electrode immersed in a solution of 1 M oxidant (X) and 1 M reductant (X^-). The standard reference half-cell consists of an electrode immersed in a 1 M H^+ solution that is in equilibrium with H_2 gas at 1 atmosphere pressure. The electrodes are connected to a voltmeter, and electrical continuity between the half-cells is established by an agar bridge. Electrons then flow from one half-cell to the other. If the reaction proceeds in the direction

$$X^- + H^+ \longrightarrow X + \tfrac{1}{2}H_2$$

the reactions in the half-cells are

$$X^- \longrightarrow X + e^-$$
$$H^+ + e^- \longrightarrow \tfrac{1}{2}H_2$$

Thus, electrons flow from the sample half-cell to the standard reference half-cell, and consequently the sample-cell electrode is negative with respect to the standard-cell electrode. *The redox potential of the X : X^- couple is the observed voltage at the start of the experiment* (when X, X^-, and H^+ are 1 M). *The redox potential of the H^+ : H_2 couple is defined to be 0 V* (volts).

The meaning of the redox potential is now evident. A negative redox potential means that a substance has lower affinity for electrons than does H_2, as in the above example. A positive redox potential means that a substance has higher affinity for electrons than does H_2. These comparisons refer to standard conditions, namely 1 M oxidant, 1 M reductant, 1 M H^+, and 1 atmosphere H_2.

Thus, *a strong reducing agent (such as NADH) has a negative redox potential, whereas a strong oxidizing agent (such as O_2) has a positive redox potential.*

The redox potentials of many biologically important redox couples are known (Table 14-1). The free-energy change of an oxida-

Table 14-1
Standard oxidation-reduction potentials of some reactions

Oxidant	Reductant	n	E_0' (V)
α-Ketoglutarate	Succinate + CO_2	2	−0.67
Acetate	Acetaldehyde	2	−0.60
Ferredoxin (oxidized)	Ferredoxin (reduced)	1	−0.43
2 H^+	H_2	2	−0.42
NAD^+	NADH + H^+	2	−0.32
$NADP^+$	NADPH + H^+	2	−0.32
Lipoate (oxidized)	Lipoate (reduced)	2	−0.29
Glutathione (oxidized)	Glutathione (reduced)	2	−0.23
Acetaldehyde	Ethanol	2	−0.20
Pyruvate	Lactate	2	−0.19
Fumarate	Succinate	2	0.03
Cytochrome b (+3)	Cytochrome b (+2)	1	0.07
Dehydroascorbate	Ascorbate	2	0.08
Ubiquinone (oxidized)	Ubiquinone (reduced)	2	0.10
Cytochrome c (+3)	Cytochrome c (+2)	1	0.22
Fe (+3)	Fe (+2)	1	0.77
$\frac{1}{2} O_2$ + 2 H^+	H_2O	2	0.82

Note: E_0' is the standard oxidation-reduction potential (pH 7, 25°C) and *n* is the number of electrons transferred. E_0' refers to the partial reaction written as

$$\text{Oxidant} + e^- \longrightarrow \text{reductant}$$

tion-reduction reaction can be readily calculated from the difference in redox potentials of the reactants. For example, consider the reduction of pyruvate by NADH:

(a) Pyruvate + NADH + H^+ ⇌ lactate + NAD^+

The redox potential of the NAD^+:NADH couple is −0.32 V, whereas that of the pyruvate:lactate couple is −0.19 V. By convention, redox potentials refer to partial reactions written as follows: oxidant + e^- → reductant. Hence,

(b) Pyruvate + 2 H^+ + 2 e^- ⟶ lactate $\qquad E_0' = -0.19$ V
(c) NAD^+ + H^+ + 2 e^- ⟶ NADH $\qquad E_0' = -0.32$ V

Subtracting reaction c from reaction b yields the desired reaction (a) and a $\Delta E_0'$ of +0.13 volt. Now we can calculate the $\Delta G^{\circ\prime}$ for the reduction of pyruvate by NADH. The standard free-energy change $\Delta G^{\circ\prime}$ is related to the change in redox potential $\Delta E_0'$ by

$$\Delta G^{\circ\prime} = -n\text{F}\Delta E_0'$$

in which n is the number of electrons transferred, F is the caloric equivalent of the faraday (23.062 kcal V^{-1} mol^{-1}), $\Delta E_0'$ is in volts, and $\Delta G^{\circ\prime}$ is in kilocalories per mole. For the reduction of pyruvate, $n = 2$ and so

$$\Delta G^{\circ\prime} = -2 \times 23.062 \times 0.13$$
$$= -6 \, \text{kcal/mol}$$

Note that a *positive $\Delta E_0'$ signifies an exergonic reaction* under standard conditions.

THE SPAN OF THE RESPIRATORY CHAIN IS 1.14 VOLTS, WHICH CORRESPONDS TO 53 KCAL

The driving force of oxidative phosphorylation is the electron-transfer potential of NADH or $FADH_2$. Let us calculate the $\Delta E_0'$ and $\Delta G^{\circ\prime}$ associated with the oxidation of NADH by O_2. The pertinent partial reactions are

(a) $\frac{1}{2} O_2 + 2 \, H^+ + 2 \, e^- \rightleftharpoons H_2O$ $E_0' = +0.82 \, \text{V}$
(b) $NAD^+ + H^+ + 2 \, e^- \rightleftharpoons NADH$ $E_0' = -0.32 \, \text{V}$

Subtracting reaction b from reaction a yields

(c) $\frac{1}{2} O_2 + NADH + H^+ \rightleftharpoons H_2O + NAD^+$

$$\Delta E_0' = +1.14 \, \text{V}$$

The free energy of oxidation of this reaction is then given by

$$\Delta G^{\circ\prime} = -n\text{F}\Delta E_0' = -2 \times 23.062 \times 1.14$$
$$= -52.6 \, \text{kcal/mol}$$

FLAVIN, IRON-SULFUR, QUINONE, AND HEME GROUPS CARRY ELECTRONS FROM NADH TO O_2

Electrons are transferred from NADH to O_2 through a series of electron carriers: flavins, iron-sulfur complexes, quinones, and hemes (Figure 14-4). These electron carriers, except for the quinones, are prosthetic groups of proteins. The first reaction is the oxidation of NADH by *NADH-Q reductase* (also called *NADH dehydrogenase*), an enzyme consisting of at least sixteen polypeptide

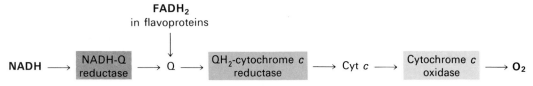

Figure 14-4
Sequence of electron carriers in the respiratory assembly. Protons are pumped by the three complexes shown in color.

chains. Two electrons are transferred from NADH to the *flavin mononucleotide* (FMN) prosthetic group of this enzyme to give the reduced form, $FMNH_2$.

$$NADH + H^+ + FMN \longrightarrow FMNH_2 + NAD^+$$

The electrons are then transferred from $FMNH_2$ to a series of *iron-sulfur complexes* (abbreviated as FeS), the second type of prosthetic

Flavin mononucleotide (FMN)

Reduced flavin mononucleotide ($FMNH_2$)

group in NADH-Q reductase. The iron is not part of a heme group, and so iron-sulfur proteins are also known as *nonheme iron proteins*. Recent studies have revealed that FeS complexes play a critical role in a wide range of redox reactions in biological systems. Three types of FeS centers are known (Figure 14-5). In the simplest kind, a single iron atom is tetrahedrally coordinated to the sulfhydryl groups of four cysteine residues of the protein. The second kind, denoted by (Fe_2S_2), contains two iron atoms and two inorganic sulfides, in addition to four cysteine residues. The third type of complex, designated as (Fe_4S_4), contains four iron atoms, four inorganic sulfides, and four cysteine residues. The iron atom in these FeS complexes can be in the Fe^{2+} (reduced) or Fe^{3+} (oxidized) state. NADH-Q reductase contains both the (Fe_2S_2) and (Fe_4S_4) types of complexes.

Electrons are then transferred from the iron-sulfur centers of NADH-Q reductase to *coenzyme Q* (abbreviated as Q).

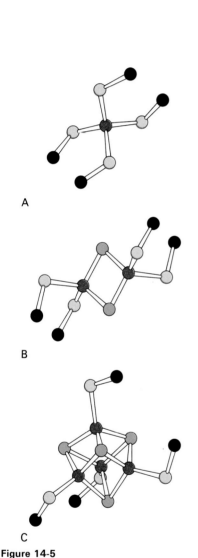

Figure 14-5
Molecular models of iron-sulfur complexes: (A) FeS center; (B) Fe_2S_2 center; (C) Fe_4S_4 center. Iron atoms are shown in red, cysteine sulfur atoms in yellow, and inorganic sulfur atoms in green. [Drawn from atomic coordinates of model compounds kindly provided by Jeremy Berg.]

Coenzyme Q is a quinone derivative with a long isoprenoid tail. It is also called *ubiquinone* because it is ubiquitous in biological systems. The number of isoprene units in Q depends on the species. The most common form in mammals contains ten isoprene units, and so it is designated Q_{10}.

Oxidized form of coenzyme Q_{10}
(Oxidized Q_{10})

Reduced form of coenzyme Q_{10}
(Reduced Q_{10})

The isoprenoid tail makes Q highly nonpolar, which enables it to diffuse rapidly in the hydrocarbon phase of the inner mitochondrial membrane. Coenzyme Q is the only electron carrier in the respiratory chain that is not tightly bound or covalently attached to a protein. In fact, *Q serves as a highly mobile carrier of electrons between the flavoproteins and the cytochromes of the electron-transport chain.*

Recall that $FADH_2$ is formed in the citric acid cycle in the oxidation of succinate to fumarate by succinate dehydrogenase. This enzyme is one of the two components of the *succinate-Q reductase complex,* the other being an FeS protein. This complex, like NADH-Q reductase, is an integral part of the inner mitochondrial membrane. The high-potential electrons of $FADH_2$ in the succinate dehydrogenase component are transferred to FeS centers in the complex and then to Q for entry into the electron-transport chain. Likewise, glycerol phosphate dehydrogenase (p. 322) and fatty acyl CoA dehydrogenase (p. 389) transfer their high-potential electrons to Q to form QH_2, the reduced state.

The electron carriers between QH_2 and O_2 are all *cytochromes,* apart from one FeS protein. The central role of the cytochromes in respiration was discovered in 1925 by David Keilin. *A cytochrome is an electron-transporting protein that contains a heme prosthetic group.* The iron atom in cytochromes alternates between a reduced ferrous ($+2$) state and an oxidized ferric ($+3$) state during electron transport. A heme group, like an FeS center, is a *one-electron carrier,* in contrast with NADH, flavins, and Q, which are two-electron carriers. Thus, a molecule of QH_2, the reduced form of the quinone, transfers its two high-potential electrons to two molecules of cytochrome b, the next member of the electron-transport chain. The free-radical semiquinone (designated QH·) may be an intermediate in these transfers.

There are five cytochromes between QH_2 and O_2 in the electron-transport chain. Cytochromes b and c_1, in addition to an FeS protein, are components of the *QH_2-cytochrome c reductase complex.* Cytochrome c transfers electrons from this complex to the *cytochrome c oxidase complex,* which contains cytochromes a and a_3. The redox potentials (electron affinities or oxidizing power) of these cytochromes increase sequentially.

$$QH_2 \longrightarrow \text{Cyt } b \longrightarrow \text{FeS} \longrightarrow \text{Cyt } c_1 \longrightarrow$$
$$\text{Cyt } c \longrightarrow \text{Cyt } a \longrightarrow \text{Cyt } a_3 \longrightarrow O_2$$

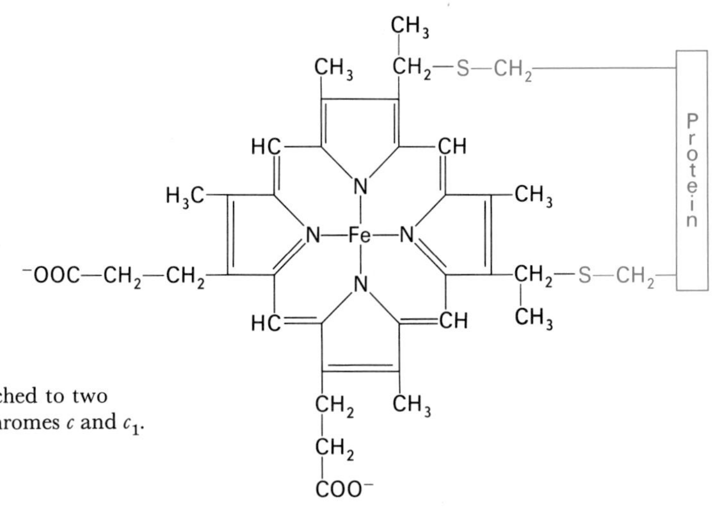

Figure 14-6
The heme is covalently attached to two
cysteine side chains in cytochromes c and c_1.

R—CH=CH$_2$
**Vinyl group
of the heme**

+

HS—CH$_2$—R′
**Cysteine residue
of the protein**

↓

R—CH$_2$—S—CH$_2$—R′
Thioether linkage

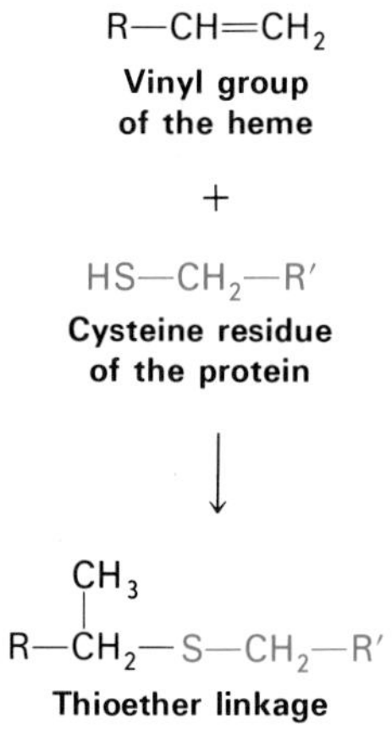

These cytochromes have distinctive structures and properties. The prosthetic group of cytochromes b, c_1, and c is iron-protoporphyrin IX, commonly called *heme,* which is the same prosthetic group as that in myoglobin and hemoglobin. In cytochrome b, the heme is not covalently bonded to the protein, whereas in cytochromes c and c_1, the heme is covalently attached to the protein by thioether linkages (Figure 14-6). These linkages are formed by the addition of the sulfhydryl groups of two cysteine residues to the vinyl groups of the heme.

Cytochromes a and a_3 have a different iron-porphyrin prosthetic group, called *heme a*. It differs from the heme in cytochromes c and c_1 in that a formyl group replaces one of the methyl groups, and a hydrocarbon chain replaces one of the vinyl groups. Cytochromes a and a_3 are the terminal members of the respiratory chain. They exist as a complex, which is sometimes called *cytochrome oxidase.*

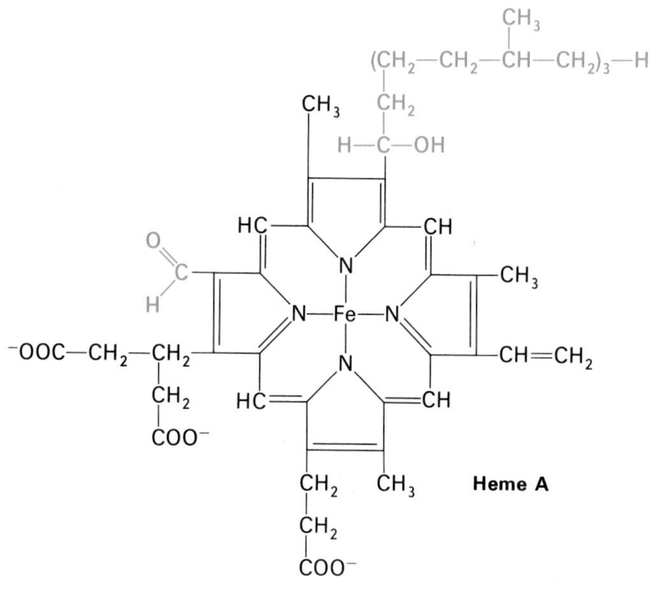

Heme A

The *QH₂-cytochrome* c *reductase complex* transfers electrons from QH_2 to cytochrome c, a water-soluble peripheral membrane protein (see pp. 326–329).

$$\begin{array}{c}
QH_2 \\ Q
\end{array}
\begin{array}{c}
Cyt\ b\ (+3) \\ Cyt\ b\ (+2)
\end{array}
\begin{array}{c}
FeS\ (+2) \\ FeS\ (+3)
\end{array}
\begin{array}{c}
Cyt\ c_1\ (+3) \\ Cyt\ c_1\ (+2)
\end{array}
\begin{array}{c}
Cyt\ c\ (+2) \\ Cyt\ c\ (+3)
\end{array}$$
$$\underline{QH_2\text{-cytochrome } c \text{ reductase}}$$

Reduced cytochrome c then transfers its electron to the *cytochrome* c *oxidase complex*. The role of cytochrome c is analogous to that of coenzyme Q: it is a mobile carrier of electrons between different complexes in the respiratory chain.

$$\begin{array}{c}
Cyt\ c\ (+2) \\ Cyt\ c\ (+3)
\end{array}
\begin{array}{c}
Cyt\ a\ (+3) \\ Cyt\ a\ (+2)
\end{array}
\begin{array}{c}
Cyt\ a_3\ (+2) \\ Cyt\ a_3\ (+3)
\end{array}
\begin{array}{c}
Cu\ (+2) \\ Cu\ (+1)
\end{array}
\begin{array}{c}
H_2O \\ O_2
\end{array}$$
$$\underline{Cytochrome\ c\ oxidase}$$

Electrons are transferred to the cytochrome a moiety of the complex, and then to cytochrome a_3, which contains copper. This copper atom alternates between a $+2$ oxidized form and a $+1$ reduced form as it transfers electrons from cytochrome a_3 to molecular oxygen. The formation of water is a four-electron process, whereas heme groups are one-electron carriers. It is not known how four electrons converge to reduce a molecule of O_2.

$$O_2 + 4\,H^+ + 4\,e^- \longrightarrow 2\,H_2O$$

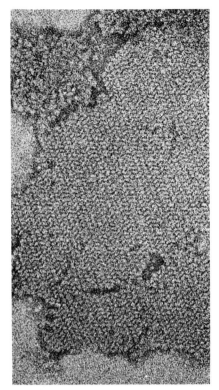

Figure 14-7
Electron micrograph of a two-dimensional crystalline array of cytochrome c oxidase. [Courtesy of Dr. Steven Fuller and Dr. Roderick Capaldi.]

OXIDATION AND PHOSPHORYLATION ARE COUPLED BY A PROTON GRADIENT

Thus far, consideration has been given to the flow of electrons from NADH to O_2, an exergonic process:

$$NADH + \tfrac{1}{2}O_2 + H^+ \rightleftharpoons H_2O + NAD^+$$
$$\Delta G^{\circ\prime} = -52.6\ \text{kcal/mol}$$

This free energy of oxidation is used to synthesize ATP.

$$ADP + P_i + H^+ \rightleftharpoons ATP + H_2O \qquad \Delta G^{\circ\prime} = +7.3\ \text{kcal/mol}$$

The synthesis of ATP is carried out by a molecular assembly in the inner mitochondrial membrane. This enzyme complex (p. 320) is called the *mitochondrial ATPase* because it was discovered through its catalysis of the hydrolytic reaction.

How is the oxidation of NADH coupled to the phosphorylation of ADP? It was first suggested that electron transfer leads to the formation of a covalent high-energy intermediate that serves as the precursor of ATP. This chemical-coupling hypothesis was based on the mechanism of substrate-level phosphorylation, as exemplified by the glyceraldehyde 3-phosphate dehydrogenase reaction, in which 1,3-DPG is the high-energy intermediate (p. 273). An alternative proposal was that the free energy of oxidation is trapped in an activated protein conformation, which then drives the synthesis of ATP. Investigators in many laboratories tried for several decades to isolate these putative energy-rich intermediates, but none were found.

A radically different mechanism, *the chemiosmotic hypothesis,* was postulated by Peter Mitchell in 1961. He proposed that electron transport and ATP synthesis are coupled by a proton gradient rather than by a covalent high-energy intermediate or an activated protein. According to this model, the transfer of electrons through the respiratory chain results in the pumping of protons from the matrix to the cytoplasmic side of the inner mitochondrial membrane. The H^+ concentration becomes higher on the cytoplasmic side, and a membrane potential with the cytoplasmic side positive is generated (Figure 14-8). This proton-motive force is postulated to

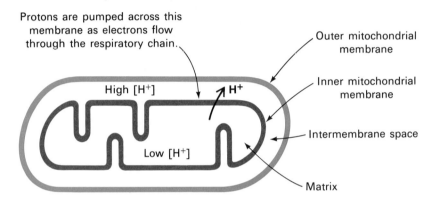

Figure 14-8
A proton gradient and a membrane potential across the inner mitochondrial membrane are generated by electron transfer through the respiratory chain.

drive the synthesis of ATP by the ATPase complex. In this model, the electron-transport chain interacts with the ATP-synthesizing complex only through the membrane potential. The model requires that the electron carriers in the respiratory chain and the ATPase be *vectorially organized*—that is, they must be oriented with respect to the two faces of the inner mitochondrial membrane. Furthermore, the inner mitochondrial membrane must be quite impermeable to protons, because a *closed compartment* is essential for the existence of a proton gradient. The essence of this proposed mechanism is that the *primary energy-conserving event is the movement of protons across the inner mitochondrial membrane.*

Mitchell's hypothesis that oxidation and phosphorylation are coupled by a proton gradient is now supported by a wealth of evidence:

1. A proton gradient across the inner mitochondrial membrane is generated during electron transport. The pH outside is 1.4 units lower than inside, and the membrane potential is 0.14 V, the outside being positive. The total electrochemical potential Δp (in volts) consists of a membrane-potential contribution ($\Delta\psi$) and a H^+ concentration-gradient contribution (ΔpH). In the following equation, R is the gas constant, T is the absolute temperature, and F is the caloric equivalent of the faraday.

$$\Delta p = \Delta \psi - \frac{RT}{F} \Delta pH = \Delta \psi - .06 \, \Delta pH$$
$$= .14 - .06(-1.4) = .224 \, V$$

This total proton-motive force of 0.224 V corresponds to a free energy of 5.2 kcal per mole of protons.

2. ATP is synthesized when a pH gradient is imposed on mitochondria or chloroplasts (p. 443) in the absence of electron transport.

3. A purple-membrane protein from halobacteria pumps protons when illuminated (p. 451). Synthetic vesicles containing this bacterial protein and a purified ATPase from beef-heart mitochondria synthesize ATP when illuminated. In this experiment, the purple-membrane protein replaces the respiratory chain, which shows that the respiratory chain and ATPase are biochemically separate systems, linked only by a proton gradient.

4. Both the respiratory chain (p. 319) and the ATPase (p. 320) are vectorially organized in the inner mitochondrial membrane.

5. A closed compartment is essential for oxidative phosphorylation. ATP synthesis coupled to electron transfer does not occur in soluble preparations or in membrane fragments lacking well-defined inside and outside compartments.

6. Substances that carry protons across the inner mitochondrial membrane dissipate the proton gradient and thereby uncouple oxidation and phosphorylation (p. 325).

A PROTON GRADIENT IS GENERATED AT THREE SITES

Protons are pumped at three sites as electrons flow through the respiratory chain from NADH to O_2 (Figure 14-9): *site 1* is the NADH-Q reductase complex; *site 2* is the QH_2-cytochrome c reductase complex; and *site 3* is the cytochrome c oxidase complex. The proton gradient generated at each site by the flow of a pair of electrons from NADH is used to synthesize one molecule of ATP. These sites have been identified by several experimental approaches:

1. *Comparison of the ATP yield from the oxidation of several substrates.* The oxidation of NADH yields three ATP, whereas the oxidation of succinate yields two ATP. The electrons from $FADH_2$ enter the electron-transport chain at coenzyme Q, which is at a lower energy level than phosphorylation site 1. Only one ATP is formed when ascorbate, an artificial substrate, is oxidized, because its electrons enter at cytochrome c, which is at a lower energy level than phosphorylation site 2. The *P : O ratio,* defined as the number of moles of inorganic phosphate incorporated into organic form per atom of oxygen consumed, is a frequently used index of oxidative phospho-

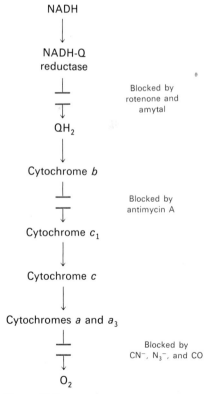

Figure 14-9
Sites of action of some inhibitors of electron transport.

rylation. The P : O ratios for the oxidation of NADH, succinate, and ascorbate are 3, 2, and 1, respectively.

2. *Thermodynamic estimates.* The $\Delta G^{\circ\prime}$ for electron transfer from NADH to the lowest-energy FeS center in NADH-Q reductase is -12 kcal/mol; from cytochromes b to c_1 in QH_2-cytochrome c reductase, -10 kcal/mol; and from cytochrome a to O_2 in cytochrome c oxidase, -24 kcal/mol. These oxidation-reduction reactions are sufficiently exergonic to drive the synthesis of ATP under standard conditions ($\Delta G^{\circ\prime} = -7.3 \text{ kcal/mol}$). The $\Delta G^{\circ\prime}$ values of the other electron-transfer reactions, namely those mediated by coenzyme Q and cytochrome c, are too small to sustain ATP synthesis.

3. *Specific inhibition of electron flow. Rotenone* and *amytal* specifically inhibit electron transfer within the NADH-Q reductase complex, and so they prevent the generation of a proton gradient at site 1 (see Figure 14-8). In contrast, these inhibitors do not interfere with the oxidation of succinate, because the electrons of this substrate enter the electron-transport chain beyond the block at coenzyme Q. *Antimycin A* inhibits electron flow between cytochromes b and c_1, which prevents ATP synthesis coupled to the generation of a proton gradient at site 2. This block can be bypassed by the addition of ascorbate, which directly reduces cytochrome c. Electrons then flow from cytochrome c to O_2, with the concomitant synthesis of ATP coupled to a proton gradient at site 3. Finally, electron flow can be blocked between the cytochrome oxidase complex and O_2 by CN^-, N_3^-, and *CO*. Cyanide and azide react with the ferric form of this carrier, whereas carbon monoxide inhibits the ferrous form. Phosphorylation coupled to the generation of a proton gradient at site 3 does not occur in the presence of these inhibitors because electron flow is blocked.

The sites of action of these inhibitors were revealed by the *crossover technique.* Britton Chance devised some elegant spectroscopic methods for determining the proportions of the oxidized and reduced forms of each carrier. This is feasible because each carrier has a characteristic absorption spectrum for its oxidized and reduced forms. The addition of an electron-transport inhibitor changes the proportion of the oxidized and reduced forms of each carrier. For example, addition of antimycin A causes the carriers between NADH and cytochrome b to become more reduced. In contrast, the carriers between cytochrome c and O_2 become more oxidized. Hence, it can be concluded that antimycin A inhibits the conversion of cytochrome b into c_1, because this step is the *crossover point.*

4. *Synthesis of vesicle systems containing a single site.* Each of the three proton-pumping sites has been reconstituted in synthetic phospholipid vesicles, together with the ATPase. Addition of an oxidizable substrate to these vesicles leads to the generation of a proton-motive force that is sufficient for the synthesis of one ATP per electron pair.

The generation of a proton gradient by the flow of electrons through the three energy-conserving sites in the respiratory chain requires that they be asymmetrically oriented. Furthermore, these three enzyme complexes must span the membrane so that protons can be pumped from the matrix side to the cytoplasmic side. Submitochondrial particles formed by the sonication of mitochondria (see Figure 14-10) have been very helpful in studying these important aspects of the electron-transport chain. The outer surface of submitochondrial particles corresponds to the matrix surface of the inner membrane of intact mitochondria. Thus, both surfaces of the inner mitochondrial membrane are experimentally accessible: the cytoplasmic surface in intact mitochondria, and the matrix surface in submitochondrial particles. The arrangement of the protein components of the respiratory chain in these preparations has been studied using proteolytic enzymes, specific antibodies, lectins, and nonpenetrating labeling reagents. For example, subunits II, V, and VI of cytochrome oxidase can be labeled only from the matrix side. In addition, cytochrome oxidase binds cytochrome c only on the cytoplasmic face and pumps protons in only one direction. These experiments and corresponding ones for NADH-Q reductase and QH_2-cytochrome c reductase (Table 14-2) have shown that *all three energy-conserving sites traverse the inner mitochondrial membrane and are asymmetrically oriented.*

Table 14-2
Components of the mitochondrial electron-transport chain

Enzyme complex	Mass (kdal)	Prosthetic group	Location of binding sites		
			M-side	Middle	C-side
NADH-Q reductase	850	FMN FeS	NADH	Q	
Succinate-Q reductase	97	FAD FeS	Succinate	Q	
QH$_2$-cytochrome c	280	Heme b Heme c$_1$ FeS		Q	Cyt c
Cytochrome c	13	Heme c			Cyt c_1 Cyt a
Cytochrome c oxidase	200	Heme a Heme a$_3$ Cu	O$_2$		Cyt c

Source: J. W. DePierre and L. Ernster, *Ann. Rev. Biochem.* 46(1977):215.
Note: All components are integral membrane proteins, except for cytochrome c, which is a water-soluble peripheral membrane protein, and coenzyme Q, which is a lipid-soluble quinone. M-side, middle, and C-side refer to the matrix side, the hydrocarbon core, and the cytosol side, respectively, of the inner mitochondrial membrane.

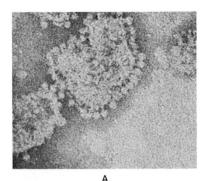

A

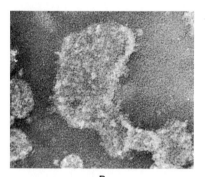

B

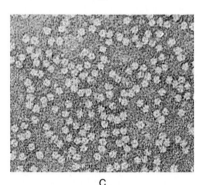

C

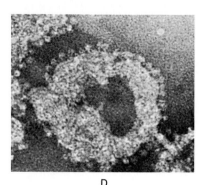

D

Where do the translocated protons come from? One possibility is that the pumped protons are those *directly* participating in the chemical reactions catalyzed by the three sites. For example, when two electrons are donated by NADH to the NADH-Q reductase complex, two H^+ are involved—one from NADH itself and one from the solvent. Alternatively, the translocated protons may arise *indirectly* by *conformational interactions* between the catalytic site and proton-binding sites in different regions of the enzyme complex. The stoichiometries of the electron-transfer reactions show that only one proton per electron, or two per energy-conserving site, can be pumped by the direct mechanism. However, H^+/site ratios of between 3 and 4 have been observed, which do not fit this direct mechanism. It seems likely that *electron transfer through each of the energy-conserving sites produces transmitted conformational changes that eject protons from the matrix to the cytoplasmic side of the membrane.* Recall that the protons involved in the Bohr effect in hemoglobin come from sites in the molecule distant from the heme group (p. 80).

ATP IS SYNTHESIZED AS PROTONS FLOW BACK TO THE MATRIX THROUGH A PROTON CHANNEL

We turn now to the utilization of the proton gradient to synthesize ATP. The enzyme catalyzing this process appears in electron micrographs of submitochondrial particles as spherical projections from the surface (Figure 14-10). In intact mitochondria, these projections are on the matrix side of the inner mitochondrial membrane. In 1960, Efraim Racker found that these knobs (which are 85 Å in diameter) can be removed by mechanical agitation. The stripped submitochondrial particles can transfer electons through their electron-transport chain, but they can no longer synthesize ATP. In contrast, the separated 85-Å spheres catalyze the hydrolysis of ATP. Most interesting, Racker found that the addition of these ATPase spheres to the stripped submitochondrial particles restored their capacity to synthesize ATP. These spheres are referred to as *coupling factor 1, or F_1. The physiological role of the F_1 unit is to catalyze the synthesis of ATP.* The ATPase activity exhibited by solubilized F_1 (in the absence of a proton gradient) is the reverse of its physiological reaction.

Figure 14-10
Electron micrographs of (A) a submitochondrial particle showing F_1 projections on its surface, (B) a submitochondrial particle treated with urea, which has removed the F_1 projections, (C) isolated F_1 units, and (D) a reconstituted submitochondrial particle formed by adding F_1 (a coupling factor) to stripped membranes. The particle shown in part B can transfer electrons to O_2, but it cannot form ATP. The reconstituted particle shown in part D carries out oxidative phosphorylation. [Courtesy of Dr. Efraim Racker.]

Table 14-3
Components of the mitochondrial ATP-synthesizing complex

Subunits	Mass (kdal)	Role	Location
F_1	360	Contains catalytic site for ATP synthesis	Spherical headpiece on matrix side
α	53		
β	50		
γ	33		
δ	17		
ϵ	7		
F_0	29	Contains proton channel	Transmembrane
	22		
	12		
	8		
F_1 inhibitor	10	Regulates proton flow and ATP synthesis	Stalk between F_0 and F_1
Oligomycin-sensitivity-conferring protein (OSCP)	18		
Fc_2 (F_6)	8		

Source: J. W. DePierre and L. Ernster, *Ann. Rev. Biochem.* 46(1977):216.

The F_1 unit, which has a mass of 360 kdal and contains five kinds of polypeptide chains (Table 14-3), is only part of the ATP-synthesizing machinery of mitochondria. The other major unit of this complex is F_0, a hydrophobic segment of four polypeptide chains that is anchored in the inner mitochondrial membrane. *F_0 is the proton channel of the complex.* The stalk between F_0 and F_1 includes several other proteins (Table 14-3). One of them renders the complex sensitive to *oligomycin*, an antibiotic that blocks ATP synthesis by interfering with the utilization of the proton gradient.

The flow of protons through the F_0 channel from the cytoplasmic to the matrix side of the membrane results in ATP synthesis by F_1. How is proton flow coupled to ATP synthesis? As in proton pumping, direct and indirect mechanisms have been suggested. One proposal is that the proton flux acts directly on the ATP-synthesizing site. In this scheme, the P_i is activated and simultaneously attacked by ADP to give ATP. Alternatively, the proton flux may be coupled to ATP synthesis by conformational changes transmitted through the enzyme complex.

ELECTRONS FROM CYTOPLASMIC NADH ENTER MITOCHONDRIA BY THE GLYCEROL PHOSPHATE SHUTTLE

Intact mitochondria are impermeable to NADH and NAD^+. How then does cytoplasmic NADH get oxidized by the respiratory chain? NADH is formed in glycolysis in the oxidation of glyceraldehyde 3-phosphate. NAD$^+$ must be regenerated for glycolysis to continue. The solution is that *electrons from NADH*, rather than NADH itself, are carried across the mitochondrial membrane. One carrier is *glycerol 3-phosphate*, which readily traverses the outer mitochondrial membrane. The first step in this shuttle (Figure 14-11) is the

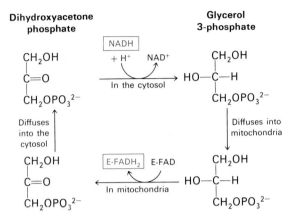

Figure 14-11
Glycerol phosphate shuttle.

transfer of electrons from NADH to dihydroxyacetone phosphate to form glycerol 3-phosphate. This reaction, catalyzed by glycerol 3-phosphate dehydrogenase, occurs in the cytosol. Glycerol 3-phosphate then enters mitochondria, where it is reoxidized to dihydroxyacetone phosphate by the FAD prosthetic group of a dehydrogenase that is bound to the inner mitochondrial membrane. This mitochondrial FAD-linked glycerol dehydrogenase is different from the NAD$^+$-linked glycerol dehydrogenase in the cytosol. The dihydroxyacetone phosphate formed in the oxidation of glycerol 3-phosphate then diffuses out of the mitochondria into the cytosol to complete the shuttle. The net reaction is

$$\text{NADH} + \text{H}^+ + \text{E-FAD} \longrightarrow \text{NAD}^+ + \text{E-FADH}_2$$

| Cyto- | Mito- | Cyto- | Mito- |
| plasmic | chondrial | plasmic | chondrial |

The reduced flavin inside the mitochondria transfers its electrons to the respiratory chain at the level of coenzyme Q. *Consequently, two rather than three ATP are formed when cytoplasmic NADH transported by the glycerol phosphate shuttle is oxidized by the respiratory chain.* At first glance, this shuttle may seem to waste one ATP per cycle. This lower yield arises because FAD rather than NAD$^+$ is the electron acceptor in mitochondrial glycerol 3-phosphate dehydrogenase. The use of FAD enables electrons from cytoplasmic NADH to be transported into mitochondria against an NADH concentration gradient. The price of this transport is one ATP per two electrons. This glycerol phosphate shuttle is especially prominent in insect flight muscle.

In heart and liver, electrons from cytoplasmic NADH are brought into mitochondria by the *malate-aspartate shuttle*, which is mediated by two membrane carriers and four enzymes. Electrons are transferred from NADH in the cytosol to malate, which traverses the inner mitochondrial membrane and is then reoxidized to form NADH in the mitochondrial matrix. Oxaloacetate does not readily cross the inner mitochondrial membrane, and so a transamination reaction is needed to form aspartate, which does traverse this barrier. The net reaction of the malate-aspartate shuttle is

$$\underset{\substack{\text{Cyto-}\\\text{plasmic}}}{\text{NADH}} + \underset{\substack{\text{Mito-}\\\text{chondrial}}}{\text{NAD}^+} \Longleftrightarrow \underset{\substack{\text{Cyto-}\\\text{plasmic}}}{\text{NAD}^+} + \underset{\substack{\text{Mito-}\\\text{chondrial}}}{\text{NADH}}$$

This shuttle, in contrast with the glycerol phosphate shuttle, is readily reversible. Consequently, NADH can be brought into mitochondria by the malate-aspartate shuttle only if the NADH/NAD^+ ratio is higher in the cytosol than in the mitochondrial matrix. No energy is consumed by the malate-aspartate shuttle in transferring electrons from NADH into the mitochondrial respiratory chain, and so three ATP are synthesized per NADH transferred.

THE ENTRY OF ADP INTO MITOCHONDRIA REQUIRES THE EXIT OF ATP

ATP and ADP do not diffuse freely across the inner mitochondrial membrane. Rather, a specific carrier enables these highly charged molecules to traverse this permeability barrier. An interesting property of this carrier is that the flows of ATP and ADP are coupled. *ADP enters the mitochondrial matrix only if ATP exits, and vice versa.* This coupled flow of ATP and ADP is an example of *facilitated exchange diffusion.* It is mediated by the *ATP-ADP translocase,* a dimer of identical 29-kdal subunits. This translocase is abundant in the inner mitochondrial membrane, comprising about 6% of the total protein. ADP on the cytoplasmic side of the membrane is transported in preference to ATP, which may account in part for the tenfold higher $[\text{ATP}]/[\text{ADP}][\text{P}_i]$ ratio on the cytoplasmic side relative to the matrix side. The coupled transport of ATP and ADP by the translocase is probably driven by the proton gradient across the inner mitochondrial membrane. The ATP-ADP translocase is specifically inhibited by very low concentrations of *atractyloside,* a plant glycoside, or of *bongkrekic acid,* an antibiotic from a mold. Oxidative phosphorylation stops soon after the addition of these inhibitors because the supply of ADP inside the mitochondrion cannot be replenished.

MITOCHONDRIA CONTAIN NUMEROUS TRANSPORT SYSTEMS FOR IONS AND METABOLITES

The ATP-ADP translocase is only one of many mitochondrial transport systems. The inner mitochondrial membrane contains a variety of carriers for ions and charged metabolites. For example, the *dicarboxylate carrier* mediates the facilitated exchange diffusion of malate, succinate, fumarate, and P_i. The *tricarboxylate carrier* exchanges OH^- for P_i. Pyruvate in the cytosol enters the mitochondrial matrix in exchange for OH^- by means of the *pyruvate carrier*. Glutamate and aspartate are exchanged by the *glutamate carrier*, which can also transport OH^-. Mitochondria also contain a *calcium ion transport system*. The proton-motive force generated by electron transfer rather than ATP is the immediate source of free energy for the accumulation of Ca^{2+} in the mitochondrial matrix.

THE COMPLETE OXIDATION OF GLUCOSE YIELDS 36 ATP

We can now calculate how many ATP are formed when glucose is completely oxidized (Table 14-4). The overall reaction is

$$Glucose + 36\,ADP + 36\,P_i + 36\,H^+ + 6\,O_2 \longrightarrow$$
$$6\,CO_2 + 36\,ATP + 42\,H_2O$$

The *P:O ratio is 3* because 36 ATP are formed and 12 atoms of oxygen are consumed. Most of the ATP, 32 out of 36, is generated by oxidative phosphorylation.

The overall efficiency of ATP generation is high. The oxidation of glucose yields 686 kcal under standard conditions:

$$Glucose + 6\,O_2 \longrightarrow 6\,CO_2 + 6\,H_2O \qquad \Delta G^{\circ\prime} = -686\,kcal$$

The free energy stored in 36 ATP is 263 kcal, because $\Delta G^{\circ\prime}$ for the hydrolysis of ATP is -7.3 kcal. Hence, the thermodynamic efficiency of ATP formation from glucose is 263/686, or 38%, under standard conditions.

The *respiratory quotient* (RQ), a frequently used index in metabolic studies of whole organisms, is defined as

$$RQ = \frac{\text{moles of } CO_2 \text{ produced}}{\text{moles of } O_2 \text{ consumed}}$$

The RQ is 1.0 for the complete oxidation of carbohydrate. For fats and proteins, it is about 0.71 and 0.80, respectively. Thus, the RQ can be used as an index of the relative utilization of carbohydrate, fat, and protein by an organism.

THE RATE OF OXIDATIVE PHOSPHORYLATION IS DETERMINED BY THE NEED FOR ATP

Under most physiological conditions, electron transport is tightly coupled to phosphorylation. *Electrons do not usually flow through the electron-transport chain to O_2 unless ADP is simultaneously phosphorylated to ATP.* Oxidative phosphorylation requires a supply of NADH (or other source of electrons at high potential), O_2, ADP, and P_i. The most important factor in determining the rate of oxidative phos-

Table 14-4
ATP yield from the complete oxidation of glucose

Reaction sequence	ATP yield per glucose
Glycolysis: glucose into pyruvate (in the cytosol)	
Phosphorylation of glucose	-1
Phosphorylation of fructose 6-phosphate	-1
Dephosphorylation of 2 molecules of 1,3-DPG	$+2$
Dephosphorylation of 2 molecules of phosphoenolpyruvate	$+2$
2 NADH are formed in the oxidation of 2 molecules of glyceraldehyde 3-phosphate	
Conversion of pyruvate into acetyl CoA (inside mitochondria)	
2 NADH are formed	
Citric acid cycle (inside mitochondria)	
2 molecules of guanosine triphosphate are formed from 2 molecules of succinyl CoA	$+2$
6 NADH are formed in the oxidation of 2 molecules each of isocitrate, α-ketoglutarate, and malate	
2 $FADH_2$ are formed in the oxidation of 2 molecules of succinate	
Oxidative phosphorylation (inside mitochondria)	
2 NADH formed in glycolysis; each yields 2 ATP (assuming transport of NADH by the glycerol phosphate shuttle)	$+4$
2 NADH formed in the oxidative decarboxylation of pyruvate; each yields 3 ATP	$+6$
2 $FADH_2$ formed in the citric acid cycle; each yields 2 ATP	$+4$
6 NADH formed in the citric acid cycle; each yields 3 ATP	$+18$
NET YIELD PER GLUCOSE	$\overline{+36}$

phorylation is the *level of ADP*. The rate of oxygen consumption of a tissue homogenate increases markedly when ADP is added and then returns to its initial value when the added ADP has been converted into ATP (Figure 14-12).

The regulation of the rate of oxidative phosphorylation by the ADP level is called *respiratory control*. The physiological significance of this regulatory mechanism is evident. The ADP level increases when ATP is consumed, and so oxidative phosphorylation is coupled to the utilization of ATP. *Electrons do not flow from fuel molecules to O_2 unless ATP needs to be synthesized.*

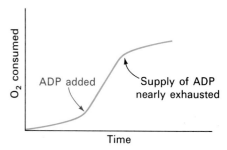

Figure 14-12
Respiratory control. Electrons are transferred to O_2 only if ADP is phosphorylated to ATP.

DINITROPHENOL UNCOUPLES OXIDATIVE PHOSPHORYLATION BY DISSIPATING THE PROTON GRADIENT

The tight coupling of electron transport and phosphorylation is disrupted by 2,4-dinitrophenol (DNP) and some other acidic aro-

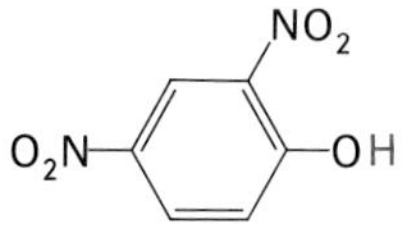

**2,4-Dinitrophenol
(DNP)**

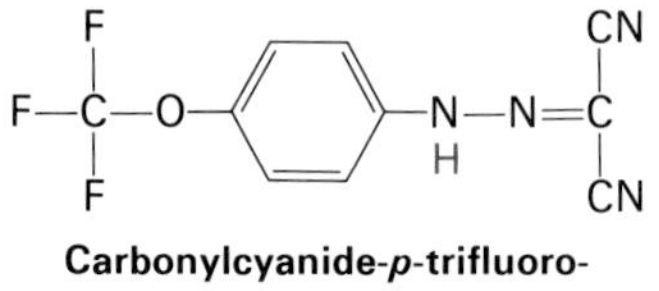

**Carbonylcyanide-*p*-trifluoro-
methoxyphenylhydrazone**

Figure 14-13
Formulas of two uncouplers of oxidative phosphorylation. These lipid-soluble substances can carry protons across the inner mitochondrial membrane. The dissociable proton is shown in red.

matic compounds (Figure 14-13). These substances carry protons across the inner mitochondrial membrane. In the presence of these uncouplers, electron transport from NADH to O_2 proceeds normally, but ATP is not formed by the mitochondrial ATPase because the proton-motive force across the inner mitochondrial is dissipated. The loss of respiratory control leads to increased oxygen consumption and oxidation of NADH. In contrast, DNP has no effect on substrate-level phosphorylations. DNP and other uncouplers are very useful tools in metabolic studies because of their specific effect on the respiratory chain.

The uncoupling of oxidative phosphorylation can be biologically useful. *It is a means of generating heat to maintain body temperature in hibernating animals, some newborn animals, and in mammals adapted to cold.* Brown adipose tissue, which is very rich in mitochondria, is specialized for this process of *thermogenesis*. Fatty acids act as uncouplers in brown adipose tissue. In turn, norepinephrine controls the release of fatty acids. Thus, the degree of uncoupling of oxidative phosphorylation in brown adipose tissue is under hormonal control. The mitochondria in this tissue can serve as ATP generators or as miniature furnaces.

There is an interesting case report of a thirty-eight-year-old woman who was incapable of performing prolonged physical work. Her basal metabolic rate was more than twice normal, but her thyroid function was normal. A muscle biopsy showed that her mitochondria were highly variable and atypical in structure. Biochemical studies then revealed that these mitochondria were not subject to respiratory control. NADH was oxidized irrespective of whether ADP was present. In other words, *oxidation and phosphorylation were not tightly coupled in these mitochondria.* Furthermore, their P:O ratio was below normal. Thus, in this patient, much of the energy of fuel molecules was converted into heat rather than ATP. The molecular defect in such mitochondria has not yet been elucidated.

THREE-DIMENSIONAL STRUCTURE OF CYTOCHROME *C*

Cytochrome *c* is the only electron-transport protein that can be separated from the inner mitochondrial membrane by gentle treatment. The solubility of this peripheral membrane protein in water has facilitated its purification and crystallization. Indeed, much more is known about the structure of cytochrome *c* than of any other electron-carrying protein.

Cytochrome *c* consists of a single polypeptide chain of 104 amino acid residues and a covalently attached heme group. The three-dimensional structures of the ferrous and ferric forms of cytochrome *c* have been elucidated at nearly atomic resolution by Richard Dickerson (Figure 14-14). The protein is roughly spherical, with a diameter of 34 Å. The heme group is surrounded by many tightly

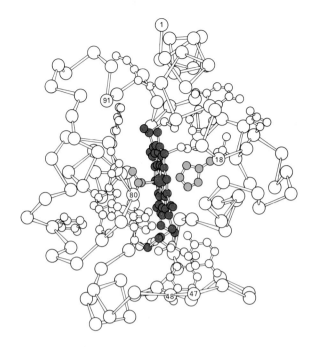

Figure 14-14
Three-dimensional structure of reduced cytochrome c from tuna. The heme group (red), methionine 80 (green), histidine 18 (green), and the α-carbon atoms are shown. [After T. Takano, O. B. Kallai, R. Swanson, and R. E. Dickerson, *J. Biol. Chem.* 248(1973):5244.]

packed hydrophobic side chains. The iron atom is bonded to the sulfur atom of a methionine residue and to the nitrogen atom of a histidine residue (Figure 14-15). The hydrophobic character of the heme environment makes the redox potential of cytochrome c more positive (corresponding to a higher electron affinity) than that of the same heme complex in an aqueous milieu. It is energetically more costly to remove an electron from the heme in cytochrome c than from a heme in water because the dielectric constant near the iron atom is lower in cytochrome c.

The overall structure of the molecule can be characterized as a shell one residue thick surrounding the heme. Hydrophobic side chains make up the innermost part of the shell. The main chain comes next, followed by charged side chains on the surface. There is very little α helix and no β pleated sheet. In essence, the polypeptide chain is wrapped around the heme. Residues 1 to 47 are on the histidine 18 side of the heme (called the right side), whereas 48 to 91 are on the methionine 80 side (called the left side). Residues 92 to 104 come back across the heme to the right side.

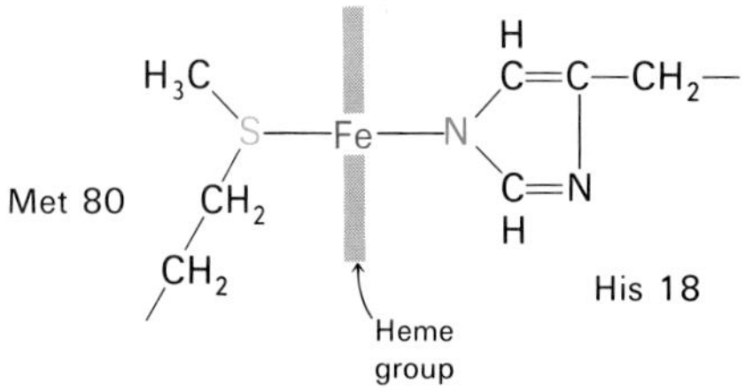

Figure 14-15
The iron atom of the heme group in cytochrome c is bonded to a methionine and a histidine side chain.

INTERACTION OF CYTOCHROME C WITH ITS REDUCTASE AND OXIDASE

As discussed previously, cytochrome c carries electrons from the QH_2-cytochrome c reductase complex (the second energy-conserving site) to the cytochrome c oxidase complex (the third energy-conserving site). How does cytochrome c interact with its reductase and then with its oxidase? An important clue comes from the distribution of charged residues on the surface of the protein. Cyto-

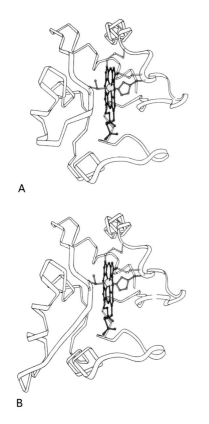

A

B

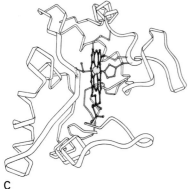

C

Figure 14-16
Conservation of the three-dimensional structure of cytochrome c in evolution is exemplified by the similarity in the conformation of (A) cytochrome c from tuna-heart mitochondria, (B) cytochrome c_2 from *Rhodospirillum rubrum,* a photosynthetic bacterium, and (C) cytochrome c_{550} from *Paracoccus dentrificans,* a denitrifying bacterium. [After F. R. Salamme. Reproduced, with permission, from the *Annual Review of Biochemistry*, Volume 46. © 1977 by Annual Reviews Inc.]

chrome c molecules from all species studied thus far have clusters of lysine side chains around the heme crevice on one face of the protein (the front of Figure 14-14). *This distribution of charges on the surface of cytochrome c almost certainly plays a role in the recognition and binding of the reductase and oxidase.* For example, the interaction of cytochrome c with cytochrome oxidase is impaired if lysine 13 is modified. Furthermore, polylysine competes with cytochrome c in binding to the oxidase and reductase.

How does cytochrome c accept an electron from its reductase, which it then donates to its oxidase? A priori, two kinds of mechanisms are possible. An electron could be transferred between heme groups in different proteins by a relay of aromatic side chains. Alternatively, an electron could be directly transferred from one heme to another. It is important to note that an electron carried by a heme is not confined to its iron atom. Rather, it is partially delocalized over the entire conjugated π-electron network of the heme. Hence, an electron can be transferred from one heme to another if their edges are sufficiently close (less than about 8 Å apart) and if their planes are nearly parallel. The direct mechanism for electron transfer seems more likely because the free energy required to form a free-radical anion of an aromatic side chain is very high. Furthermore, one of the edges of the heme group of cytochrome c (the front edge in Figure 14-14) is accessible for direct electron-transfer reactions.

THE CONFORMATION OF CYTOCHROME *C* HAS REMAINED ESSENTIALLY CONSTANT FOR A BILLION YEARS

Cytochrome c is present in all organisms that have mitochondrial respiratory chains: plants, animals, and eucaryotic microorganisms. This electron carrier evolved more than 1.5 billion years ago, before the divergence of plants and animals. The function of this protein has been conserved throughout this period as evidenced by the fact that *the cytochrome c of any eucaryotic species reacts in vitro with the cytochrome oxidase of any other species tested thus far.* For example, wheat-germ cytochrome c reacts with human cytochrome oxidase. A second criterion of the conservation of function is that the redox potentials of all cytochrome c molecules studied are close to $+0.25$ V. Third, the absorption spectra of cytochrome c molecules of many species are virtually indistinguishable. In fact, some procaryotic cytochromes, such as cytochrome c_2 from a photosynthetic bacterium and cytochrome c_{550} from a denitrifying bacterium, closely resemble cytochrome c from tuna-heart mitochondria (Figure 14-16).

The amino acid sequences of cytochrome c from more than eighty widely ranging eucaryotic species have been determined by Emil Smith, Emanuel Margoliash, and others. The striking finding is that *26 of 104 residues have been invariant for more than a billion and a half years of evolution.* The reasons for the constancy of many of these residues are evident now that the three-dimensional structure of the molecule is known. As might be expected, the heme ligands methionine 80 and histidine 18 are invariant, as are the two cysteines that are covalently bonded to the heme. A sequence of 11 residues from residue 70 to 80 is nearly the same in all cytochrome *c* molecules. Many hydrophobic residues in contact with the heme are invariant. Furthermore, most of the glycine residues in cytochrome *c* have been preserved. As discussed earlier (p. 190), glycine is important because it is small. The compact folding of the polypeptide chain requires the presence of glycine at certain sites. Several invariant lysine and arginine residues are located in the positively charged clusters on the surface of the molecule. One of these clusters interacts with cytochrome *c* reductase, and another with cytochrome oxidase.

POWER TRANSMISSION BY PROTON GRADIENTS: A CENTRAL MOTIF OF BIOENERGETICS

The major concept presented in this chapter is that mitochondrial electron transfer and ATP synthesis are linked by a transmembrane proton gradient. ATP synthesis in bacteria and chloroplasts (p. 443) also is driven by proton gradients. In fact, proton gradients power a variety of energy-requiring processes such as the active transport of Ca^{2+} by mitochondria, the entry of some amino acids and sugars into bacteria, the rotation of bacterial flagella, and the transfer of electrons from NADH to NADPH. Proton gradients can also be used to generate heat, as in hibernation. It is evident that *proton gradients are a central interconvertible currency of free energy in biological systems* (Figure 14-17).

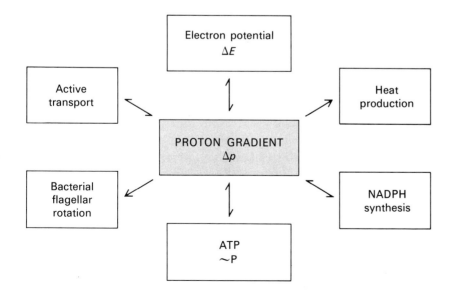

Figure 14-17
The proton gradient is an interconvertible form of free energy.

Why were proton gradients selected for this role in the course of evolution? Mitchell (1976) offers the following reasons:

> The transduction of energy by proticity is marvellously simple and effective because all that is required is a thin topologically closed insulating lipid membrane between two aqueous proton-conductor phases. A generator of proticity plugged through the membrane at any point can act as a source of protonic power that can be drawn upon by a suitable consumer unit plugged through the membrane at any other point, even when the producer and consumer units are quite far apart on a molecular scale. This chemiosmotic type of system for power transmission in biology is familiar to the electrophysiologist studying animal plasma membranes across which power is transmitted by gradients of Na^+ and K^+ ion activity, for example in nervous conduction and in intestinal nutrient absorption. But, because of the relatively high reactivity and mobility of the proton, which allows it to function at very low concentration, the use of proticity for power transmission in biology is unique in its ubiquitous occurrence and wide range of application. . . . It is instructive to remember that these micro-miniaturized *protonic* devices were invented by accident and developed by natural selection to a marvellous state of sophistication millions of years before the advent of modern technology led to the development of micro-miniaturized *electronic* devices. The elucidation of the detailed mechanisms of the natural protonic devices may thus provide intellectual nourishment and inspiration not only for the biochemist in search of knowledge, but also for the technologist in search of inventions.

SUMMARY

In oxidative phosphorylation, the synthesis of ATP is coupled to the flow of electrons from NADH or $FADH_2$ to O_2 by a proton gradient across the inner mitochondrial membrane. Electron flow through three asymmetrically oriented transmembrane complexes results in the pumping of protons out of the mitochondrial matrix and the generation of a membrane potential. ATP is synthesized when protons flow back to the matrix through a channel in an ATP-synthesizing complex, which is known as the mitochondrial ATPase. Oxidative phosphorylation exemplifies a fundamental theme of bioenergetics: the transmission of free energy by proton gradients.

The electron carriers in the respiratory assembly of the inner mitochondrial membrane are flavins, iron-sulfur complexes, quinones, and the heme groups of cytochromes. Electrons from NADH are transferred to the FMN prosthetic group of NADH-Q reductase, the first of three complexes. This reductase also contains FeS centers. The electrons emerge in QH_2, the reduced form of ubiquinone (Q). This highly mobile carrier transfers its electrons to QH_2-cytochrome c reductase, a complex that contains cytochromes b and c_1 and an FeS center. This second complex reduces cytochrome c, a water-soluble peripheral membrane protein. Cytochrome c, like Q, is a mobile carrier of electrons, which are then transferred to cytochrome c oxidase. This third complex contains cytochromes a and a_3. Cuprous ion in this oxidase transfers electrons to O_2, the ultimate acceptor, to form H_2O.

The flow of two electrons through each of these three complexes generates a proton gradient sufficient to synthesize one molecule of ATP. Hence, three ATP are formed per NADH oxidized, whereas only two ATP are made per $FADH_2$ oxidized because its electrons enter the chain at QH_2, after the first proton-pumping site. Also, only two ATP are generated in the oxidation of NADH formed in the cytosol because one ATP is lost when the glycerol phosphate shuttle carries these electrons into mitochondria. The entry of ADP into mitochondria is coupled to the exit of ATP by a process called facilitated exchange diffusion. Thirty-six ATP are generated when a molecule of glucose is completely oxidized to CO_2 and H_2O. Electron transport is normally tightly coupled to phosphorylation. NADH and $FADH_2$ are oxidized only if ADP is simultaneously phosphorylated to ATP. This coupling, called respiratory control, can be disrupted by uncouplers such as DNP, which dissipate the proton gradient by carrying protons across the inner mitochondrial membrane.

SELECTED READINGS

WHERE TO START

Hinkle, P. C., and McCarty, R. E., 1978. How cells make ATP. *Sci. Amer.* 238(3):104–123. [Available as *Sci. Amer.* Offprint 1383.]

Mitchell, P., 1979. Keilin's respiratory chain concept and its chemiosmotic consequences. *Science* 206:1148–1159. [Mitchell reviews the evolution of the chemiosmotic concept in this Nobel lecture.]

Stoeckenius, W., 1976. The purple membrane of salt-loving bacteria. *Sci. Amer.* 234(6):38–46. [Offprint 1340.]

Dickerson, R. E., 1972. The structure and history of an ancient protein. *Sci. Amer.* 226(4)58–72. [A beautifully illustrated account of the conformation and evolution of cytochrome c. Offprint 1245.]

Dickerson, R. E., 1980. Cytochrome c and the evolution of energy metabolism. *Sci. Amer.* 242(3):137–153.

BOOKS

Racker, E., 1976. *A New Look at Mechanisms in Bioenergetics.* Academic Press. [A lively, personal account of electron transport and phosphorylation.]

Quagliariello, E., Papa, S., Palmieri, F., Slater, E. C., and Siliprandi, N., (eds.), 1975. *Electron Transfer Chains and Oxidative Phosphorylation.* Academic Press.

Lovenberg, W., (ed.), 1977. *Iron-sulfur proteins,* vol. 3. Academic Press.

REVIEWS

Boyer, P. D., Chance, B., Ernster, L., Mitchell, P., Racker, E., and Slater, E. C., 1977. Oxidative phosphorylation and photophosphorylation. *Ann. Rev. Biochem.* 46:955–1026. [A series of review articles by leading investigators in this field. It is helpful to start with the one by Ernster.]

Fillingame, R., 1980. The proton-translocating pumps of oxidative phosphorylation. *Ann. Rev. Biochem.* 49: 1079–1114.

Downie, J. A., Gibson, F., and Cox, G. B., 1979. Membrane adenosine triphosphatases of prokaryotic cells. *Ann. Rev. Biochem.* 48:103–131.

DePierre, J. W., and Ernster, L., 1977. Enzyme topology of intracellular membranes. *Ann. Rev. Biochem.* 6:201–262. [A review of the membrane-sidedness of the respiratory chain, ATPase, and other proteins of the inner mitochondrial membrane.]

Salemme, F. R., 1977. Structure and function of cytochromes c. *Ann. Rev. Biochem.* 46:299–329.

Beinert, H., 1977. Iron-sulfur centers of the mitochondrial electron transfer system. *In* Lovenberg, W., (ed.),

332

Iron-sulfur proteins, vol. 3, pp. 61–100. Academic Press.

Sweeney, W. V., and Rabinowitz, J. C., 1980. Proteins containing 4Fe-4S clusters: an overview. *Ann. Rev. Biochem.* 49:139–161.

Crane, F. L., 1977. Hydroquinone dehydrogenases. *Ann. Rev. Biochem.* 46:439–469.

ARTICLES

Racker, E., and Stoeckenius, W., 1974. Reconstitution of purple membrane vesicles catalyzing light-driven proton uptake and adenosine triphosphate formation. *J. Biol. Chem.* 249:662–663.

Mitchell, P., 1976. Vectorial chemistry and the molecular mechanics of chemiosmotic coupling: power transmission by proticity. *Trans. Biochem. Soc.* 4:399–430.

Yoshida, M., Okamoto, H., Sone, N., Hirata, H., and Kagawa, Y., 1977. Reconstitution of thermostable ATPase capable of energy coupling from its purified subunits. *Proc. Nat. Acad. Sci.* 74:936–940.

Skulachev, V. P., 1977. Transmembrane electrochemical H$^+$-potential as a convertible energy source for the living cell. *FEBS (Fed. Eur. Biochem. Soc.) Lett.* 74:1–9.

Luft, R., Ikkos, D., Palmieri, G., Ernster, L., and Afzelius, B., 1962. A case of severe hypermetabolism of nonthyroid origin with a defect in the maintenance of mitochondrial respiratory control: a correlated clinical, biochemical, and morphological study. *J. Clin. Invest.* 41:1776–1804.

Erecinska, M., Wilson, D. F., and Nishiki, K., 1978. Homeostatic regulation of cellular energy metabolism: experimental characterization in vivo and fit to a model. *Amer. J. Physiol.* 234:82–89.

HISTORICAL ASPECTS

Racker, E., 1980. From Pasteur to Mitchell: a hundred years of bioenergetics. *Fed. Proc.* 39:210–215.

Keilin, D., 1966. *The History of Cell Respiration and Cytochromes.* Cambridge University Press.

Kalckar, H. M., (ed.), 1969. *Biological Phosphorylations: Development of Concepts.* Prentice-Hall. [A collection of classical papers on oxidative phosphorylation and other aspects of bioenergetics.]

Fruton, J. S., 1972. *Molecules and Life: Historical Essays on the Interplay of Chemistry and Biology.* Wiley-Interscience. [Includes an excellent account of cellular respiration, starting on page 262.]

PROBLEMS

1. What is the yield of ATP when each of the following substrates is completely oxidized by a cell homogenate? Assume that glycolysis, the citric acid cycle, and oxidative phosphorylation are fully active.
   (a) Pyruvate.
   (b) NADH.
   (c) Fructose 1,6-diphosphate.
   (d) Phosphoenolpyruvate.
   (e) Glucose.
   (f) Dihydroxyacetone phosphate.

2. (a) Write an equation for the oxidation of reduced glutathione (see p. 344) by O_2. Calculate $\Delta E_0'$ and $\Delta G°'$ for this reaction using the data in Table 14-1.
   (b) What is $\Delta E_0'$ and $\Delta G°'$ for the reduction of oxidized glutathione by NADPH?

3. What is the effect of each of the following inhibitors on electron transport and ATP formation by the respiratory chain?
   (a) Azide.
   (b) Atractyloside.
   (c) Rotenone.
   (d) DNP.
   (e) Carbon monoxide.
   (f) Antimycin A.

4. The addition of oligomycin to mitochondria markedly decreases both the rate of electron transfer from NADH to O_2 and the rate of ATP formation. The subsequent addition of DNP leads to an increase in the rate of electron transfer without changing the rate of ATP formation. What does oligomycin inhibit?

5. Compare the $\Delta G°'$ for the oxidation of succinate by NAD$^+$ and by FAD. Use the data given in Table 14-1, and assume that E_0' for the FAD/FADH$_2$ redox couple is nearly 0 V. Why is FAD rather than NAD$^+$ the electron acceptor in the reaction catalyzed by succinate dehydrogenase?

6. The immediate administration of nitrite is a highly effective treatment for cyanide poisoning. What is the basis for the action of this antidote? (Hint: Nitrite oxidizes ferrohemoglobin to ferrihemoglobin.)

7. For a proton-motive force of 0.2 V (matrix negative), what is the maximum [ATP]/[ADP][P$_i$] ratio compatible with ATP synthesis? In calculating this ratio, assume that the number of protons translocated per ATP formed is 2, 3, and 4 and that the temperature is 25°C.

CHAPTER 15

PENTOSE PHOSPHATE PATHWAY AND GLUCONEOGENESIS

The preceding chapters on glycolysis, the citric acid cycle, and oxidative phosphorylation were primarily concerned with the generation of ATP, starting with glucose as a fuel. We now turn to the generation of a different type of metabolic energy—reducing power. Some of the electrons and hydrogen atoms of fuel molecules must be conserved for biosynthetic purposes rather than transferred to O_2 to generate ATP. *The currency of readily available reducing power in cells is NADPH.* The phosphoryl group on C-2 of one of the ribose units of NADPH distinguishes it from NADH. As mentioned previously (p. 246), there is a *fundamental distinction between NADPH and NADH in most biochemical reactions. NADH is oxidized by the respiratory chain to generate ATP, whereas NADPH serves as a hydrogen and electron donor in reductive biosyntheses.* This chapter also deals with the synthesis of glucose from noncarbohydrate precursors, a process called *gluconeogenesis.*

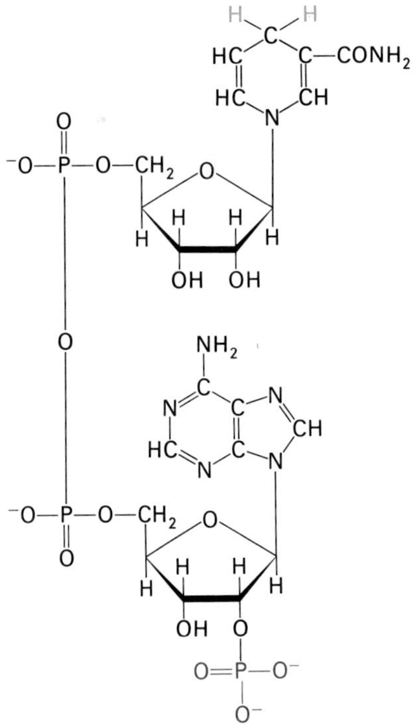

THE PENTOSE PHOSPHATE PATHWAY GENERATES NADPH AND SYNTHESIZES FIVE-CARBON SUGARS

In the pentose phosphate pathway, NADPH is generated as glucose 6-phosphate is oxidized to ribose 5-phosphate. This five-carbon

Reduced nicotinamide
adenine dinucleotide
phosphate (NADPH)

sugar and its derivatives are components of such important biomolecules as ATP, CoA, NAD$^+$, FAD, RNA, and DNA.

$$\text{Glucose 6-phosphate} + 2\,\text{NADP}^+ + H_2O \longrightarrow$$
$$\text{ribose 5-phosphate} + 2\,\text{NADPH} + 2\,H^+ + CO_2$$

The pentose phosphate pathway also catalyzes the interconversion of three-, four-, five-, six-, and seven-carbon sugars in a series of nonoxidative reactions. All of these reactions occur in the cytosol. In plants, part of the pentose phosphate pathway also participates in the formation of hexoses from CO_2 in photosynthesis (Chapter 19).

The pentose phosphate pathway is sometimes called *the pentose shunt, the hexose monophosphate pathway,* or *the phosphogluconate oxidative pathway.* The discovery of glucose 6-phosphate dehydrogenase, the first enzyme in the pathway, by Otto Warburg in 1931 led to its complete elucidation by Fritz Lipmann, Frank Dickens, Bernard Horecker, and Efraim Racker.

TWO NADPH ARE GENERATED IN THE CONVERSION OF GLUCOSE 6-PHOSPHATE INTO RIBULOSE 5-PHOSPHATE

The pentose phosphate pathway starts with the dehydrogenation of glucose 6-phosphate at C-1, a reaction catalyzed by *glucose 6-phosphate dehydrogenase* (Figure 15-1). This enzyme is highly specific for

Figure 15-1
Oxidative branch of the pentose phosphate pathway. These three reactions are catalyzed by glucose 6-phosphate dehydrogenase, lactonase, and 6-phosphogluconate dehydrogenase.

NADP$^+$; the K_M for NAD$^+$ is about a thousand times as great as that for NADP$^+$. The product is *6-phosphoglucono-δ-lactone,* which is an intramolecular ester between the C-1 carboxyl group and the C-5 hydroxyl group. The next step is the hydrolysis of 6-phosphoglucono-δ-lactone by a specific *lactonase* to give *6-phosphogluconate.* This six-carbon sugar is then oxidatively decarboxylated by *6-phosphogluconate dehydrogenase* to yield *ribulose 5-phosphate.* NADP$^+$ is again the electron acceptor.

The final step in the synthesis of ribose 5-phosphate is the isomerization of ribulose 5-phosphate by *phosphopentose isomerase.*

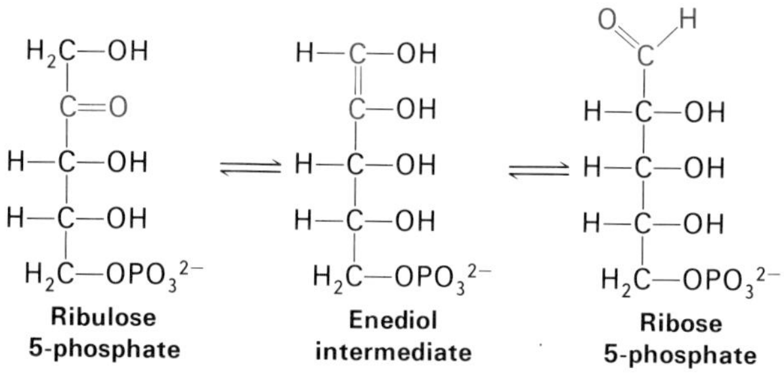

Ribulose 5-phosphate ⇌ Enediol intermediate ⇌ Ribose 5-phosphate

This reaction is similar to the glucose 6-phosphate ⇌ fructose 6-phosphate and to the dihydroxyacetone phosphate ⇌ glyceraldehyde 3-phosphate reactions in glycolysis. *These three ketose-aldose isomerizations proceed through an enediol intermediate.*

THE PENTOSE PHOSPHATE PATHWAY AND GLYCOLYSIS ARE LINKED BY TRANSKETOLASE AND TRANSALDOLASE

The preceding reactions yield two NADPH and one ribose 5-phosphate for each glucose 6-phosphate oxidized. However, many cells need much more NADPH for reductive biosyntheses than ribose 5-phosphate for incorporation into nucleotides and nucleic acids. Under these conditions, ribose 5-phosphate is converted into glyceraldehyde 3-phosphate and fructose 6-phosphate by *transketolase* and *transaldolase. These enzymes create a reversible link between the pentose phosphate pathway and glycolysis* by catalyzing these three reactions:

$$C_5 + C_5 \xrightleftharpoons{\text{transketolase}} C_3 + C_7$$

$$C_7 + C_3 \xrightleftharpoons{\text{transaldolase}} C_4 + C_6$$

$$C_5 + C_4 \xrightleftharpoons{\text{transketolase}} C_3 + C_6$$

The sum of these reactions is the *formation of two hexoses and one triose from three pentoses.*

The essence of these reactions is that *transketolase transfers a two-carbon unit, whereas transaldolase transfers a three-carbon unit.* The sugar that donates the two- or three-carbon unit is always a ketose, whereas the acceptor is always an aldose.

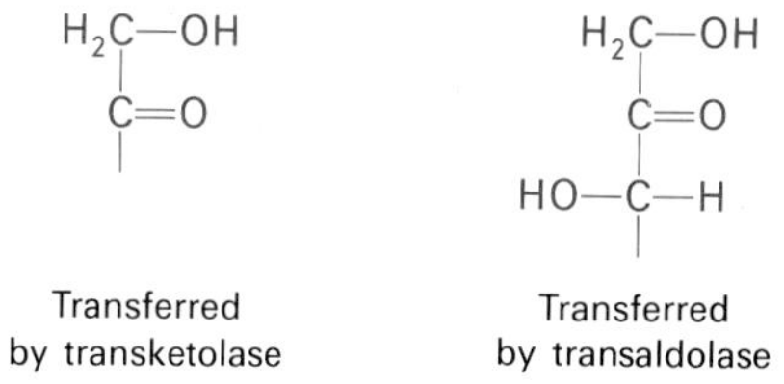

The first of the three reactions linking the pentose phosphate pathway and glycolysis is the formation of *glyceraldehyde 3-phosphate* and *sedoheptulose 7-phosphate* from two pentoses.

Transferred by transketolase

Transferred by transaldolase

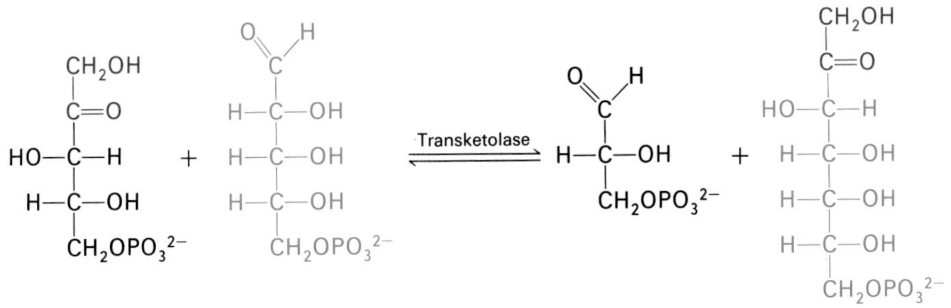

The structures below show the reactions discussed in the text.

Xylulose 5-phosphate + **Ribose 5-phosphate** ⇌ (Transketolase) **Glyceraldehyde 3-phosphate** + **Sedoheptulose 7-phosphate**

The donor of the two-carbon unit in this reaction is xylulose 5-phosphate, which is an epimer of ribulose 5-phosphate. A ketose is a substrate of transketolase only if its hydroxyl group at C-3 has the configuration of xylulose rather than ribulose. Ribulose 5-phosphate is converted into the appropriate epimer for the transketolase reaction by *phosphopentose epimerase*.

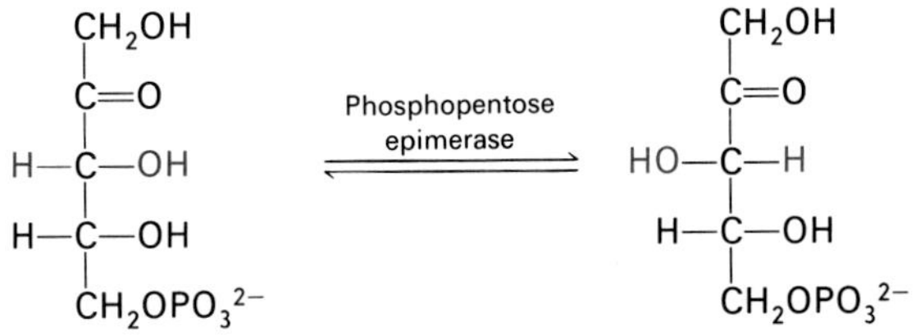

Ribulose 5-phosphate ⇌ (Phosphopentose epimerase) **Xylulose 5-phosphate**

Glyceraldehyde 3-phosphate and sedoheptulose 7-phosphate then react to form *fructose 6-phosphate* and *erythrose 4-phosphate*. This synthesis of a four-carbon sugar and a six-carbon sugar is catalyzed by *transaldolase*.

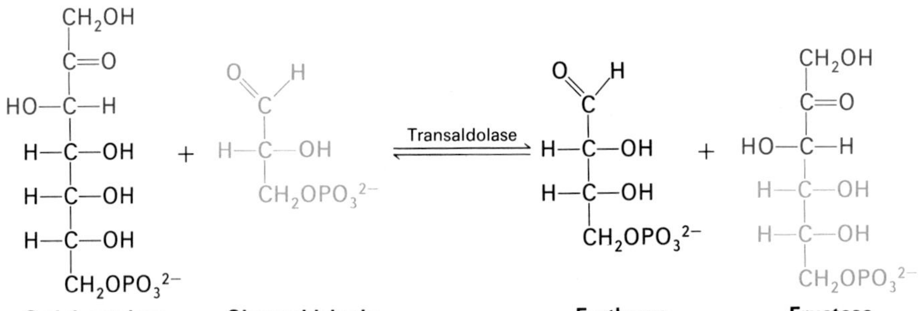

Sedoheptulose 7-phosphate + **Glyceraldehyde 3-phosphate** ⇌ (Transaldolase) **Erythrose 4-phosphate** + **Fructose 6-phosphate**

In the third reaction, transketolase catalyzes the synthesis of *fructose 6-phosphate* and *glyceraldehyde 3-phosphate* from erythrose 4-phosphate and xylulose 5-phosphate.

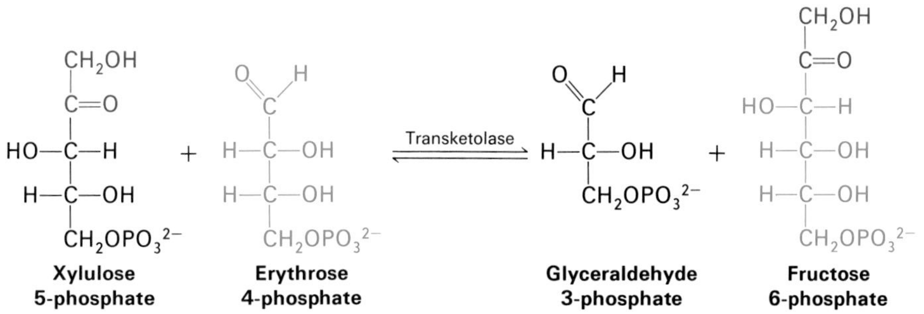

The chemical structures show:

Xylulose 5-phosphate + **Erythrose 4-phosphate** ⇌ (Transketolase) **Glyceraldehyde 3-phosphate** + **Fructose 6-phosphate**

The sum of these reactions is

2 Xylulose 5-phosphate + ribose 5-phosphate ⇌
2 fructose 6-phosphate + glyceraldehyde 3-phosphate

Xylulose 5-phosphate can be formed from ribose 5-phosphate by the sequential action of phosphopentose isomerase and phosphopentose epimerase, and so the net reaction starting from ribose 5-phosphate is

3 Ribose 5-phosphate ⇌
2 fructose 6-phosphate + glyceraldehyde 3-phosphate

Thus, excess ribose 5-phosphate formed by the pentose phosphate pathway can be quantitatively converted into glycolytic intermediates.

Table 15-1
Pentose phosphate pathway

Reaction	Enzyme
OXIDATIVE BRANCH	
Glucose 6-phosphate + $NADP^+$ ⇌ 6-phosphoglucono-δ-lactone + NADPH + H^+	Glucose 6-phosphate dehydrogenase
6-Phosphoglucono-δ-lactone + H_2O ⟶ 6-phosphogluconate + H^+	Lactonase
6-Phosphogluconate + $NADP^+$ ⟶ ribulose 5-phosphate + CO_2 + NADPH	6-Phosphogluconate dehydrogenase
NONOXIDATIVE BRANCH	
Ribulose 5-phosphate ⇌ ribose 5-phosphate	Phosphopentose isomerase
Ribulose 5-phosphate ⇌ xylulose 5-phosphate	Phosphopentose epimerase
Xylulose 5-phosphate + ribose 5-phosphate ⇌ sedoheptulose 7-phosphate + glyceraldehyde 3-phosphate	Transketolase
Sedoheptulose 7-phosphate + glyceraldehyde 3-phosphate ⇌ fructose 6-phosphate + erythrose 4-phosphate	Transaldolase
Xylulose 5-phosphate + erythrose 4-phosphate ⇌ fructose 6-phosphate + glyceraldehyde 3-phosphate	Transketolase

THE RATE OF THE PENTOSE PHOSPHATE PATHWAY IS CONTROLLED BY THE LEVEL OF NADP+

The first reaction in the oxidative branch of the pentose phosphate pathway, the dehydrogenation of glucose 6-phosphate, is essentially irreversible. In fact, this reaction is rate-limiting under physiological conditions and serves as the control site. The most important regulatory factor is the level of $NADP^+$, the electron acceptor in the oxidation of glucose 6-phosphate to 6-phosphogluconolactone. Also, NADPH competes with $NADP^+$ in binding to the enzyme, and ATP competes with glucose 6-phosphate. The ratio of $NADP^+$ to NADPH in the cytosol from the liver of a well-fed rat is about 0.014, several orders of magnitude lower than the ratio of NAD^+ to NADH, which is 700 under the same conditions. The marked effect of the $NADP^+$ level on the rate of the oxidative branch ensures that NADPH generation is tightly coupled to its utilization in reductive biosyntheses. The control of the nonoxidative branch of the pentose phosphate pathway has not yet been defined.

THE FLOW OF GLUCOSE 6-PHOSPHATE DEPENDS ON THE NEED FOR NADPH, RIBOSE 5-PHOSPHATE, AND ATP

Let us follow the fate of glucose 6-phosphate in four different situations:

1. *Much more ribose 5-phosphate than NADPH is required.* Most of the glucose 6-phosphate is converted into fructose 6-phosphate and glyceraldehyde 3-phosphate by the glycolytic pathway. Transaldolase and transketolase then convert two molecules of fructose 6-phosphate and one molecule of glyceraldehyde 3-phosphate into three molecules of ribose 5-phosphate by a reversal of the reactions described earlier. The stoichiometry of this mode (Figure 15-2A) is

$$5 \text{ Glucose 6-phosphate} + \text{ATP} \longrightarrow 6 \text{ ribose 5-phosphate} + \text{ADP} + \text{H}^+$$

2. *The needs for NADPH and ribose 5-phosphate are balanced.* The predominant reaction under these conditions is the formation of two NADPH and one ribose 5-phosphate from glucose 6-phosphate by the oxidative branch of the pentose phosphate pathway. The stoichiometry of this mode (Figure 15-2B) is

$$\text{Glucose 6-phosphate} + 2\,\text{NADP}^+ + \text{H}_2\text{O} \longrightarrow \text{ribose 5-phosphate} + 2\,\text{NADPH} + 2\,\text{H}^+ + \text{CO}_2$$

3. *Much more NADPH than ribose 5-phosphate is required; glucose 6-phosphate is completely oxidized to CO_2.* Three groups of reactions are active in this situation. First, two NADPH and one ribose 5-phosphate are formed by the oxidative branch of the pentose phosphate pathway. Then, ribose 5-phosphate is converted into fructose 6-phosphate and glyceraldehyde 3-phosphate by transketolase and transaldolase. Finally, glucose 6-phosphate is resynthesized from fructose 6-phosphate and glyceraldehyde 3-phosphate by the gluconeogenic pathway (discussed later in this chapter). The stoichiometries of these three sets of reactions (Figure 15-2C) are

$$6 \text{ Glucose 6-phosphate} + 12\,\text{NADP}^+ + 6\,\text{H}_2\text{O} \longrightarrow 6 \text{ ribose 5-phosphate} + 12\,\text{NADPH} + 12\,\text{H}^+ + 6\,\text{CO}_2$$

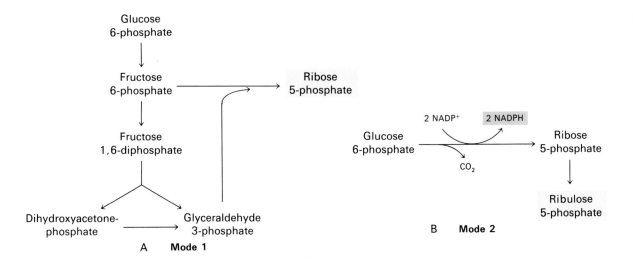

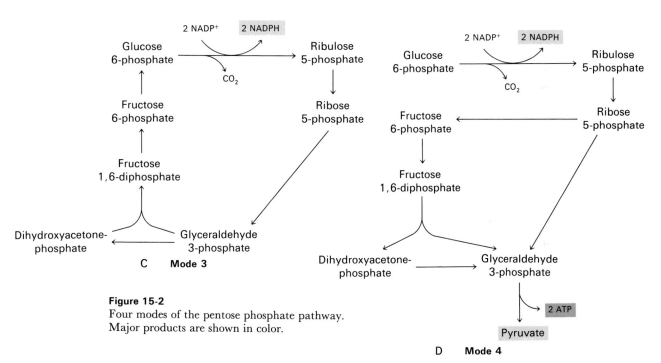

Figure 15-2
Four modes of the pentose phosphate pathway.
Major products are shown in color.

6 Ribose 5-phosphate $\longrightarrow$

        4 fructose 6-phosphate + 2 glyceraldehyde 3-phosphate

4 Fructose 6-phosphate + 2 glyceraldehyde 3-phosphate +

        $H_2O \longrightarrow$ 5 glucose 6-phosphate + P_i

The sum of these reactions is

Glucose 6-phosphate + 12 $NADP^+$ + 7 $H_2O \longrightarrow$

        6 CO_2 + 12 NADPH + 12 H^+ + P_i

*Thus, the equivalent of a glucose 6-phosphate can be completely oxidized to
CO_2 with the concomitant generation of NADPH.* The essence of these
reactions is that *the ribose 5-phosphate produced by the pentose phosphate
pathway is recycled into glucose 6-phosphate* by transketolase, transaldolase, and some of the enzymes of the gluconeogenic pathway.

4. Much more NADPH than ribose 5-phosphate is required; glucose 6-phosphate is converted into pyruvate. Alternatively, ribose 5-phosphate formed by the oxidative branch of the pentose phosphate pathway can be converted into pyruvate (Figure 15-2D). Fructose 6-phosphate and glyceraldehyde 3-phosphate derived from ribose 5-phosphate proceed down the glycolytic pathway rather than reverting to glucose 6-phosphate. In this mode, *ATP and NADPH are concomitantly generated, and five of the six carbons of glucose 6-phosphate emerge in pyruvate:*

$$3 \text{ Glucose 6-phosphate} + 6 \text{ NADP}^+ + 5 \text{ NAD}^+ + 5 \text{ P}_i + 8 \text{ ADP}$$
$$\longrightarrow 5 \text{ pyruvate} + 3 \text{ CO}_2 + 6 \text{ NADPH} + 5 \text{ NADH}$$
$$+ 8 \text{ ATP} + 2 \text{ H}_2\text{O} + 8 \text{ H}^+$$

Pyruvate formed by these reactions can be oxidized to generate more ATP or it can be used as a building block in a variety of biosyntheses.

THE PENTOSE PHOSPHATE PATHWAY IS MUCH MORE ACTIVE IN ADIPOSE TISSUE THAN IN MUSCLE

Radioactive-labeling experiments can provide estimates of how much glucose 6-phosphate is metabolized by the pentose phosphate pathway and how much is metabolized by the combined action of glycolysis and the citric acid cycle. One aliquot of a tissue homogenate is incubated with glucose labeled with ^{14}C at C-1, and another with glucose labeled with ^{14}C at C-6. The radioactivity of the CO_2 produced by the two samples is then compared. The rationale of this experiment is that only C-1 is decarboxylated by the pentose phosphate pathway, whereas C-1 and C-6 are decarboxylated to the same extent when glucose is metabolized by the glycolytic pathway, the pyruvate dehydrogenase complex, and the citric acid cycle. The reason for the equivalence of C-1 and C-6 in the latter set of reactions is that glyceraldehyde 3-phosphate and dihydroxyacetone phosphate are rapidly interconverted by triose phosphate isomerase.

This experimental approach has shown that *the activity of the pentose phosphate pathway is very low in skeletal muscle, whereas it is very high in adipose tissue.* These findings support the idea that a major role of the pentose phosphate pathway is to generate NADPH for reductive biosyntheses. Large amounts of NADPH are consumed by adipose tissue in the reductive synthesis of fatty acids from acetyl CoA (see Chapter 17).

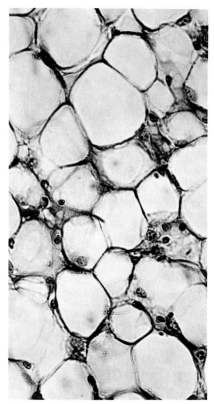

Figure 15-3
Light micrograph of adipose tissue.

TPP, THE PROSTHETIC GROUP OF TRANSKETOLASE, TRANSFERS ACTIVATED ALDEHYDES

Transketolase contains a tightly bound thiamine pyrophosphate (TPP) as its prosthetic group. This prosthetic group has been encountered before in the decarboxylation of pyruvate by the pyruvate dehydrogenase complex. The mechanism of catalysis of transketolase is similar in that an *activated aldehyde unit is transferred to an acceptor*. The acceptor in the transketolase reaction is an aldose, whereas in the pyruvate dehydrogenase reaction it is lipoamide. In both reactions, the site of addition of the keto substrate is the *thiazole ring* of the prosthetic group. The C-2 carbon atom is highly acidic and readily ionizes to give a *carbanion*. This carbanion adds to the carbonyl group of the ketose substrate (e.g., xylulose 5-phosphate, fructose 6-phosphate, and sedoheptulose 7-phosphate).

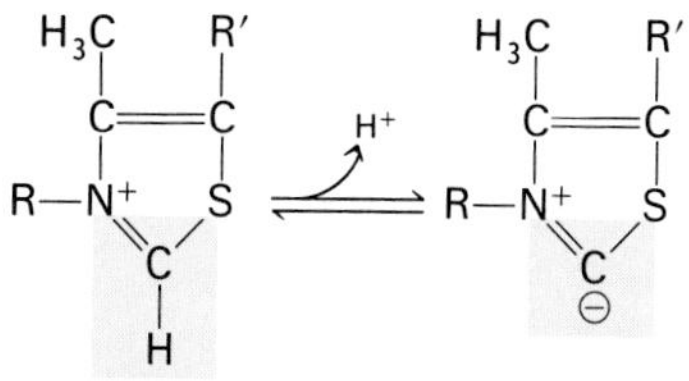

(TPP)
Thiamine
pyrophosphate

Carbanion

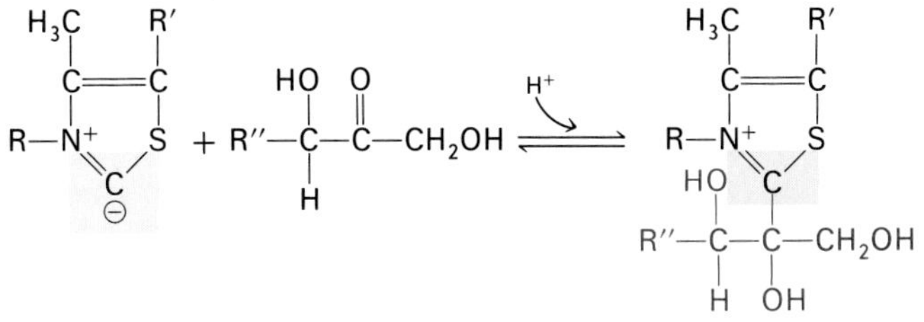

Carbanion Ketose substrate Addition compound

This addition compound loses its R—CHOH moiety to yield a negatively charged, *activated glycoaldehyde* unit. The positively charged nitrogen in the thiazole ring acts as an electron sink to promote the development of a negative charge on the activated intermediate.

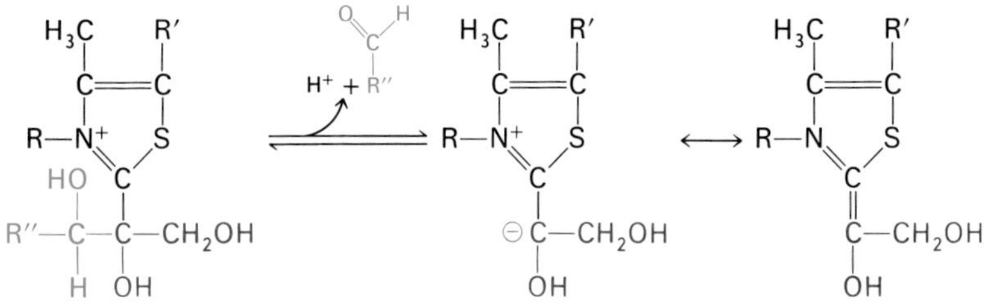

Addition
compound

Resonance forms of the
activated glycoaldehyde unit

The carbonyl group of a suitable aldehyde acceptor then condenses with the activated glycoaldehyde unit to form a new ketose, which is released from the enzyme.

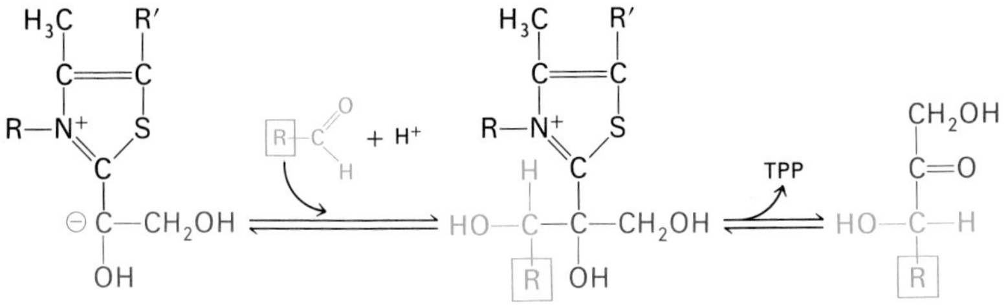

Activated
glycoaldehyde

Addition
compound

Ketose
product

TRANSKETOLASE DEFECTIVE IN TPP BINDING CAN CAUSE A NEUROPSYCHIATRIC DISORDER

The *Wernicke-Korsakoff syndrome,* a striking neuropsychiatric disorder, is caused by a lack of thiamine in the diet of susceptible persons. This disease is characterized by paralysis of eye movements, abnormal stance and gait, and markedly deranged mental function. In particular, a patient with this syndrome is disoriented and has a severely impaired memory. It has been known for some time that only a small minority of alcoholics and other chronically malnourished persons develop this disorder. Also, its incidence is much higher among Europeans than among non-Europeans on thiamine-deficient diets. These observations suggest that genetic factors may be important determinants of whether a thiamine-deficient person develops the Wernicke-Korsakoff syndrome. Recent studies of transketolase from cultured fibroblasts show that this is in fact the case. *Transketolase from patients with the Wernicke-Korsakoff syndrome binds thiamine pyrophosphate tenfold less avidly than does the enzyme from normal persons.* The other two thiamine-dependent enzymes, pyruvate dehydrogenase and α-ketoglutarate dehydrogenase, are normal in this disorder. The abnormality in transketolase becomes clinically evident only when the level of thiamine pyrophosphate is too low to saturate the enzyme. This is a clear-cut example of the interplay between genetic and environmental factors in the production of disease. The Wernicke-Korsakoff syndrome also provides a vivid demonstration of how a reduction in the activity of a single enzyme can have profound neurologic and behavioral consequences.

ACTIVATED DIHYDROXYACETONE IS CARRIED BY TRANSALDOLASE AS A SCHIFF BASE

Transaldolase transfers a three-carbon *dihydroxyacetone* unit from a ketose donor to an aldose acceptor. Transaldolase, in contrast with transketolase, does not contain a prosthetic group. Rather, *a Schiff base is formed between the carbonyl group of the ketose substrate and the ϵ-amino group of a lysine residue at the active site of the enzyme.* This kind of covalent enzyme-substrate (ES) intermediate is like that formed in fructose diphosphate aldolase in the glycolytic pathway.

Ketose substrate   Schiff base

The Schiff base becomes protonated, the bond between C-3 and C-4 is split, and an aldose is released.

Protonated
Schiff base   Carbanion   Aldose

The negative charge on the dihydroxyacetone moiety is stabilized by resonance. The positively charged nitrogen atom of the Schiff base acts as an electron sink. This nitrogen atom plays the same role in transaldolase as does the thiazole-ring nitrogen in transketolase.

The Schiff base between dihydroxyacetone and transaldolase is stable until a suitable aldose becomes bound. The carbanion of the dihydroxyacetone moiety then reacts with the carbonyl group of the aldose. The ketose product is released by hydrolysis of the Schiff base.

Resonance forms of the Schiff base carbanion

Carbanion   Aldose substrate   Schiff base   Ketose product

GLUCOSE 6-PHOSPHATE DEHYDROGENASE DEFICIENCY CAUSES A DRUG-INDUCED HEMOLYTIC ANEMIA

An antimalarial drug, pamaquine, was introduced in 1926. Most patients tolerated the drug well, but a few developed severe symptoms within a few days after therapy was started. The urine turned black, jaundice developed, and the hemoglobin content of the blood dropped sharply. In some cases, massive destruction of red blood cells caused death.

The basis of this *drug-induced hemolytic anemia* was elucidated in 1956. The primary defect is a *deficiency in glucose 6-phosphate dehydrogenase in red cells*. The pentose phosphate pathway is the only source of NADPH in red cells, and so the production of NADPH is diminished in glucose 6-phosphate dehydrogenase deficiency. The major role of NADPH in red cells is to reduce the disulfide form of *glutathione* to the sulfhydryl form. This reaction is catalyzed by *glutathione reductase*.

Reduced glutathione
(γ-Glutamylcysteinylglycine)

$$\begin{array}{c}\gamma\text{-Glu—Cys—Gly} \\ | \\ S \\ | \\ S \\ | \\ \gamma\text{-Glu—Cys—Gly}\end{array} + \text{NADPH} + \text{H}^+ \rightleftharpoons 2\ \begin{array}{c}\gamma\text{-Glu—Cys—Gly} \\ | \\ \text{SH}\end{array} + \text{NADP}^+$$

Oxidized glutathione   **Reduced glutathione**

The reduced form of glutathione, a tripeptide with a free sulfhydryl group, serves as a *sulfhydryl buffer* that maintains the cysteine residues of hemoglobin and other red-cell proteins in the reduced state. The ratio of the reduced form of glutathione (GSH) to the oxidized form (GSSG) is normally about 500. The reduced form also plays a role in detoxification by reacting with hydrogen peroxide and organic peroxides.

$$2\,\text{GSH} + \text{R—O—OH} \longrightarrow \text{GSSG} + \text{H}_2\text{O} + \text{ROH}$$

Reduced glutathione appears to be essential for maintaining normal red-cell structure and for keeping hemoglobin in the ferrous state. Cells with a lowered level of reduced glutathione are more susceptible to hemolysis for reasons that are not yet understood. Drugs such as pamaquine may distort the surface of red cells in the absence of reduced glutathione, which would make them more liable to destruction and removal by the spleen. These drugs also increase the rate of formation of toxic peroxides, which are normally eliminated by reaction with reduced glutathione.

Glucose 6-phosphate dehydrogenase deficiency is not a rare disease. It is inherited as a sex-linked trait. Female heterozygotes have two populations of red cells: one has normal enzymatic activity, whereas the other is deficient in glucose 6-phosphate dehydrogenase. The glucose 6-phosphate dehydrogenase in most other organs is specified by a different gene. The incidence of the most common (type A) deficiency of glucose 6-phosphate dehydrogenase, characterized by a tenfold reduction in enzymatic activity in red cells, is 11% among black Americans. This high frequency suggests that the deficiency may be advantageous under certain environmental conditions. Indeed, *glucose 6-phosphate dehydrogenase deficiency in red cells seems to protect a person from falciparum malaria,* because the parasites that cause this disease require the pentose phosphate pathway and reduced glutathione for optimal growth. Thus, glucose 6-phosphate dehydrogenase deficiency and sickle-cell trait are parallel mechanisms of protection against malaria, which accounts for their high gene frequencies in malaria-infested regions of the world.

The occurrence of glucose 6-phosphate dehydrogenase deficiency clearly demonstrates that *atypical reactions to drugs may have a genetic basis.* This inherited enzymatic deficiency is relatively benign until certain drugs are administered. We again see here the interplay of heredity and environment in the production of disease. Galactosemia, hereditary fructose intolerance, phenylketonuria, and succinylcholine sensitivity also illustrate this interaction in a striking way.

GLUTATHIONE REDUCTASE TRANSFERS ELECTRONS FROM NADPH TO OXIDIZED GLUTATHIONE BY WAY OF FAD

The regeneration of reduced glutathionine is catalyzed by glutathione reductase, a dimer of 50-kdal subunits. The electrons from NADPH are not directly transferred to the disulfide bond in oxidized glutathione. Rather, they are transferred from NADPH to a tightly bound flavin adenine dinucleotide (FAD), then to a disulfide bridge between two cysteine residues in the subunit, and finally to oxidized glutathione.

$$\text{NADPH} \longrightarrow \text{FAD} \longrightarrow \begin{array}{c} \text{Cys}_{46}-\text{S} \\ | \\ \text{Cys}_{41}-\text{S} \end{array} \longrightarrow \begin{array}{c} \text{G}-\text{S} \\ | \\ \text{G}-\text{S} \end{array}$$

Each subunit consists of three structural domains: an FAD-binding domain, an NADPH-binding domain, and an interface domain (Figure 15-4). The FAD domain and NADP$^+$ domain resemble

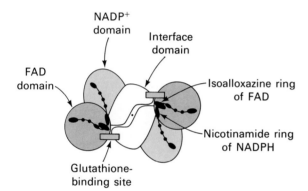

FAD
domain

NADP$^+$
domain

Interface
domain

Isoalloxazine ring
of FAD

Nicotinamide ring
of NADPH

Glutathione-
binding site

Figure 15-4
Schematic diagram of the domain structure of glutathione reductase. Each subunit in this dimeric enzyme consists of an NADP$^+$ domain, an FAD domain, and an interface domain. Glutathione is bound to the FAD domain of one subunit and the interface domain of another. [After G. E. Schultz, R. H. Schirmer, W. Sachsenheimer, and E. F. Pai, *Nature* 273(1978):123.]

each other and are similar to nucleotide-binding domains in other dehydrogenases. FAD and NADP$^+$ are bound in an extended form, with their isoalloxazine and nicotinamide rings next to each other (Figure 15-4). It is interesting to note that the binding site for oxidized glutathione is formed by the FAD domain of one subunit and the interface domain of the other subunit.

GLUCOSE CAN BE SYNTHESIZED FROM NONCARBOHYDRATE PRECURSORS

We now turn to the *synthesis of glucose from noncarbohydrate precursors,* a process called *gluconeogenesis.* This metabolic pathway is very important because certain tissues, such as the brain, are highly dependent on glucose as the primary fuel. The daily glucose requirement of the brain in a typical adult is about 120 g, which accounts for most of the 160 g of glucose needed by the whole body. The amount of glucose present in body fluids is about 20 g, and that readily available from glycogen, a storage form of glucose (p. 357), is approximately 190 g. Thus, the direct glucose reserves are sufficient to meet the needs for glucose for about a day. In a longer period of starvation, glucose must be formed from noncarbohydrate sources for survival. Gluconeogenesis is also important during periods of intense exercise.

The major noncarbohydrate precursors of glucose are *lactate, amino acids,* and *glycerol.* Lactate is formed by active skeletal muscle when the rate of glycolysis exceeds the metabolic rate of the citric acid cycle and the respiratory chain (p. 269). Amino acids are derived from proteins in the diet and, during starvation, from the breakdown of proteins in skeletal muscle (p. 553). The hydrolysis of triacylglycerols (p. 386) in fat cells yields glycerol and fatty acids. Glycerol is a precursor of glucose, but fatty acids cannot be converted into glucose in animals, for reasons that will be discussed later (p. 394). The *gluconeogenic pathway converts pyruvate into glucose.* The principal points of entry are pyruvate, oxaloacetate, and dihydroxyacetone phosphate (Figure 15-5).

The major site of gluconeogenesis is the *liver.* Gluconeogenesis also occurs in the cortex of the *kidney,* but the total amount of glucose formed there is about one-tenth of that formed in the liver

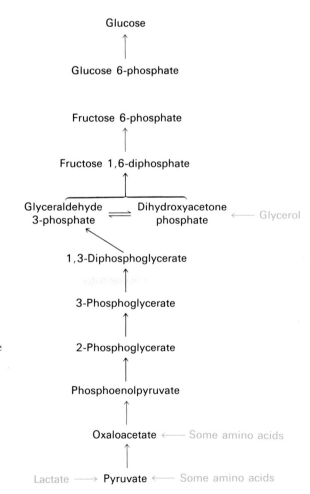

Figure 15-5
Pathway of gluconeogenesis. The distinctive reactions of this pathway are denoted by red arrows. The other reactions are common to glycolysis. The enzymes of gluconeogenesis are located in the cytosol, except for pyruvate carboxylase (in mitochondria), and glucose 6-phosphatase (attached to the endoplasmic reticulum). The entry points for lactate, glycerol, and amino acids are shown.

because of the kidney's smaller mass. Very little gluconeogenesis takes place in the brain, skeletal muscle, or heart muscle. Rather, *gluconeogenesis in the liver and kidney helps to maintain the glucose level in the blood so that brain and muscle can extract sufficient glucose from it to meet their metabolic demands.*

GLUCONEOGENESIS IS NOT A REVERSAL OF GLYCOLYSIS

In glycolysis, glucose is converted into pyruvate; in gluconeogenesis, pyruvate is converted into glucose. However, gluconeogenesis is not a reversal of glycolysis. A different pathway is required because the thermodynamic equilibrium of glycolysis lies far on the side of pyruvate formation. The actual ΔG for the formation of pyruvate from glucose is about -20 kcal/mol under typical cellular conditions (p. 267). Most of the decrease in free energy in glycolysis takes place in the three essentially irreversible steps catalyzed by hexokinase, phosphofructokinase, and pyruvate kinase.

$$\text{Glucose} + \text{ATP} \xrightarrow{\text{hexokinase}} \text{glucose 6-phosphate} + \text{ADP}$$

$$\text{Fructose 6-phosphate} + \text{ATP} \xrightarrow{\text{phosphofructokinase}}$$
$$\text{fructose 1,6-diphosphate} + \text{ADP}$$

$$\text{Phosphoenolpyruvate} + \text{ADP} \xrightarrow{\text{pyruvate kinase}} \text{pyruvate} + \text{ATP}$$

In gluconeogenesis, these virtually irreversible reactions of glycolysis are bypassed by the following new steps:

1. *Phosphoenolpyruvate is formed from pyruvate by way of oxaloacetate.* First, pyruvate is carboxylated to oxaloacetate at the expense of an ATP. Then, oxaloacetate is decarboxylated and phosphorylated to yield phosphoenolpyruvate, at the expense of a second high-energy phosphate bond.

$$\text{Pyruvate} + CO_2 + \text{ATP} + H_2O \rightleftharpoons$$
$$\text{oxaloacetate} + \text{ADP} + P_i + 2\,H^+$$

$$\text{Oxaloacetate} + \text{GTP} \rightleftharpoons \text{phosphoenolpyruvate} + \text{GDP} + CO_2$$

The first reaction is catalyzed by *pyruvate carboxylase,* and the second by *phosphoenolpyruvate carboxykinase.* The sum of these reactions is

$$\text{Pyruvate} + \text{ATP} + \text{GTP} + H_2O \rightleftharpoons$$
$$\text{phosphoenolpyruvate} + \text{ADP} + P_i + 2\,H^+$$

This pathway for the formation of phosphoenolpyruvate from pyruvate is thermodynamically feasible, because $\Delta G^{\circ\prime}$ is $+0.2$ kcal/mol in contrast with $+7.5$ kcal/mol for the reaction catalyzed by pyruvate kinase. This much more favorable $\Delta G^{\circ\prime}$ results from the input of an additional high-energy phosphate bond.

COO⁻
|
C=O
|
CH₃
Pyruvate

Pyruvate
carboxylase

COO⁻
|
C=O
|
CH₂
|
COO⁻
Oxaloacetate

Phosphoenol-
pyruvate
carboxykinase

COO⁻   O
| ‖
C—O—P—O⁻
‖ |
CH₂   O⁻
Phosphoenolpyruvate

2. Fructose 6-phosphate is formed from fructose 1,6-diphosphate by hydrolysis of the phosphate ester at C-1. Fructose 1,6-diphosphatase catalyzes this exergonic hydrolysis.

$$\text{Fructose 1,6-diphosphate} + H_2O \longrightarrow \text{fructose 6-phosphate} + P_i$$

3. Glucose is formed by hydrolysis of glucose 6-phosphate, a reaction catalyzed by glucose 6-phosphatase.

$$\text{Glucose 6-phosphate} + H_2O \longrightarrow \text{glucose} + P_i$$

Glucose 6-phosphatase is bound to the endoplasmic reticulum and acts on substrate located in the cytosol. This enzyme is not present in brain and muscle; hence, glucose does not leave these organs.

Table 15-2
Enzymatic differences between glycolysis and gluconeogenesis

Glycolysis	Gluconeogenesis
Hexokinase	Glucose 6-phosphatase
Phosphofructokinase	Fructose 1,6-diphosphatase
Pyruvate kinase	Pyruvate carboxylase
	Phosphoenolpyruvate carboxykinase

BIOTIN IS A MOBILE CARRIER OF ACTIVATED CO_2

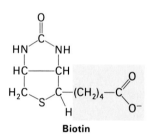

Biotin

The finding that mitochondria can form oxaloacetate from pyruvate led to the discovery of pyruvate carboxylase by Merton Utter in 1960. This enzyme is of especial interest because of its catalytic and allosteric properties. Pyruvate carboxylase contains a covalently attached prosthetic group, *biotin,* which serves as a *carrier of activated CO_2.* The carboxyl terminus of biotin is linked to the ϵ-amino group of a specified lysine residue by an amide bond.

Note that biotin is attached to pyruvate carboxylase by a *long, flexible chain* like that of lipoamide in the pyruvate dehydrogenase complex.

The carboxylation of pyruvate occurs in two stages:

$$\text{Biotin-enzyme} + \text{ATP} + \text{HCO}_3^- \xrightleftharpoons[\quad]{\substack{Mg^{2+} \\ \text{acetyl CoA}}}$$
$$CO_2 \sim \text{biotin-enzyme} + \text{ADP} + P_i$$

$$CO_2 \sim \text{biotin-enzyme} + \text{pyruvate} \xrightleftharpoons{\text{Mn}^{2+}} \text{biotin-enzyme} + \text{oxaloacetate}$$

The carboxyl group in the carboxybiotin-enzyme intermediate is bonded to the N-1 nitrogen atom of the biotin ring. The carboxyl group in this carboxybiotin intermediate is *activated*. The $\Delta G^{\circ\prime}$ for its cleavage

$$CO_2 \sim \text{biotin-enzyme} + H^+ \longrightarrow CO_2 + \text{biotin-enzyme}$$

is -4.7 kcal/mol, which enables carboxybiotin to transfer CO_2 to acceptors without the input of additional free energy.

The activated carboxyl group is then transferred from carboxybiotin to pyruvate to form oxaloacetate. The long, flexible link between biotin and the enzyme enables this prosthetic group to rotate from one active site of the enzyme (the ATP-bicarbonate site) to the other (the pyruvate site).

Carboxybiotin-enzyme intermediate

Pyruvate   **Oxaloacetate**

PYRUVATE CARBOXYLASE IS ACTIVATED BY ACETYL CoA

The activity of pyruvate carboxylase depends on the presence of acetyl CoA. *Biotin is not carboxylated unless acetyl CoA (or a closely related acyl CoA) is bound to the enzyme.* The second partial reaction is not affected by acetyl CoA. The allosteric activation of pyruvate carboxylase by acetyl CoA is an important physiological control mechanism. Oxaloacetate, the product of the pyruvate carboxylase reaction, is both a stoichiometric intermediate in gluconeogenesis and a catalytic intermediate in the citric acid cycle. *A high level of acetyl CoA signals the need for more oxaloacetate.* If there is a surplus of ATP, oxaloacetate will be consumed in gluconeogenesis. If there is a deficiency of ATP, oxaloacetate will enter the citric acid cycle upon condensing with acetyl CoA.

Thus, not only is pyruvate carboxylase important in gluconeogenesis, but it also plays a *critical role in maintaining the level of citric acid cycle intermediates.* These intermediates need to be replenished because they are consumed in some biosynthetic reactions, such as heme synthesis. This role of pyruvate carboxylase is termed *anaplerotic,* meaning to fill up.

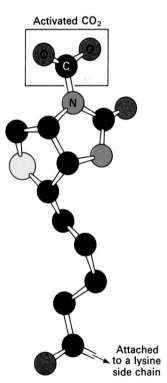

Figure 15-6
Molecular model of carboxybiotin.

OXALOACETATE IS SHUTTLED INTO THE CYTOSOL AND CONVERTED INTO PHOSPHOENOLPYRUVATE

Pyruvate carboxylase is a mitochondrial enzyme, whereas the other enzymes of gluconeogenesis are cytoplasmic. Oxaloacetate, the product of the pyruvate carboxylase reaction, is transported across the mitochondrial membrane in the form of *malate*. Oxaloacetate is reduced to malate inside the mitochondrion by an NADH-linked malate dehydrogenase. Malate is transported by a carrier across the mitochondrial membrane and is reoxidized to oxaloacetate by an NAD^+-linked malate dehydrogenase in the cytosol.

Oxaloacetate is simultaneously *decarboxylated* and *phosphorylated* by phosphoenolpyruvate carboxykinase in the cytosol.

Oxaloacetate   Phosphoenolpyruvate

The CO_2 that was added to pyruvate by pyruvate carboxylase comes off in this step. In fact, the phosphorylation reaction is made energetically feasible by the concomitant decarboxylation. *Decarboxylations often drive reactions that would otherwise be highly endergonic.* This device will be encountered again in fatty acid synthesis.

SIX HIGH-ENERGY PHOSPHATE BONDS ARE SPENT IN SYNTHESIZING GLUCOSE FROM PYRUVATE

The stoichiometry of gluconeogenesis is

$$2 \text{ Pyruvate} + 4\,\text{ATP} + 2\,\text{GTP} + 2\,\text{NADH} + 2\,H_2O \longrightarrow$$
$$\text{glucose} + 4\,\text{ADP} + 2\,\text{GDP} + 6\,P_i + 2\,NAD^+$$
$$\Delta G^{\circ\prime} = -9 \text{ kcal/mol}$$

In contrast, the stoichiometry for the reversal of glycolysis is

$$2 \text{ Pyruvate} + 2\,\text{ATP} + 2\,\text{NADH} + 2\,H_2O \longrightarrow$$
$$\text{glucose} + 2\,\text{ADP} + 2\,P_i + 2\,NAD^+$$
$$\Delta G^{\circ\prime} = +20 \text{ kcal/mol}$$

Note that *six* high-energy phosphate bonds are used to synthesize glucose from pyruvate in gluconeogenesis, whereas only *two* molecules of ATP are generated in glycolysis in the conversion of glucose into pyruvate. Thus, the extra price of gluconeogenesis is four high-energy phosphate bonds per glucose synthesized from pyruvate. The four extra high-energy phosphate bonds are needed to turn an energetically unfavorable process (the reversal of glycolysis,

$\Delta G^{\circ\prime} = +20 \, \text{kcal/mol}$) into a favorable one (gluconeogenesis, $\Delta G^{\circ\prime} = -9 \, \text{kcal/mol}$). Another way of looking at this energetic difference between glycolysis and gluconeogenesis is to recall that the input of an ATP equivalent changes the equilibrium constant of a reaction by a factor of about 10^8 (p. 243). Hence, the input of four additional high-energy bonds in gluconeogenesis changes the equilibrium constant by a factor of 10^{32}, which makes the conversion of pyruvate into glucose thermodynamically feasible.

GLUCONEOGENESIS AND GLYCOLYSIS ARE RECIPROCALLY REGULATED

Gluconeogenesis and glycolysis are coordinated so that one pathway is relatively inactive while the other is highly active. If both sets of reactions were highly active at the same time, the net result would be the hydrolysis of four $\sim\text{P}$ (two ATP plus two GTP) per reaction cycle. Both glycolysis and gluconeogenesis are highly exergonic under cellular conditions, and so there is no thermodynamic barrier to such cycling. Rather, the activities of the distinctive enzymes of each pathway are controlled so that both pathways are not highly active at the same time. For example, AMP stimulates phosphofructokinase (p. 267), whereas it inhibits fructose 1,6-diphosphatase. Citrate has the reverse effect on these enzymes. *Consequently, phosphorylation of fructose 6-phosphate, the rate-limiting step in glycolysis, is enhanced when the energy charge of the cell is low. Conversely, fructose 1,6-diphosphate is hydrolyzed and gluconeogenesis is stimulated when the energy charge is high and citric acid cycle intermediates are abundant.* Pyruvate kinase (p. 265) and pyruvate carboxylase (p. 349) are also reciprocally regulated. Fructose 1,6-diphosphate stimulates and ATP inhibits pyruvate kinase, whereas acetyl CoA stimulates and ADP inhibits pyruvate carboxylase. Hence, *pyruvate flows to phosphoenolpyruvate and gluconeogenesis is favored when the liver cell is rich in fuel molecules and ATP.*

SUBSTRATE CYCLES AMPLIFY METABOLIC SIGNALS AND PRODUCE HEAT

A pair of reactions such as the phosphorylation of fructose 6-phosphate to fructose 1,6-diphosphate and its hydrolysis back to fructose 6-phosphate is called a *substrate cycle*. As already mentioned, both reactions are not simultaneously fully active in most cells because of reciprocal allosteric controls. However, isotope labeling studies have shown that phosphorylation of fructose 1,6-diphosphate occurs during gluconeogenesis. A limited degree of cycling also occurs in other pairs of opposed irreversible reactions. This cycling was regarded as an imperfection in metabolic control, and so substrate cycles have sometimes been called *futile cycles*. However, it now

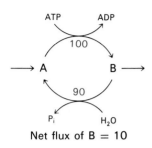

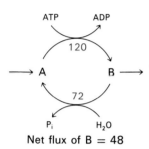

Figure 15-7
Example of an ATP-driven substrate cycle operating at two different rates. A small change in the rates of the two opposing reactions results in a large change in the *net* flux of product B.

seems more likely that substrate cycles are biologically important. One possibility is that *substrate cycles amplify metabolic signals.* Suppose that the rate of conversion of A into B is 100 and of B into A is 90, giving an initial net flux of 10. Assume that an allosteric effector increases the A → B rate by 20% to 120 and reciprocally decreases the B → A rate by 20% to 72. The new net flux is 48, and so a 20% change in the rates of the opposing reactions has led to a 480% increase in the net flux. In the example shown in Figure 15-7, this amplification is achieved by the hydrolysis of ATP.

The other potential biological role of substrate cycles is the *generation of heat produced by the hydrolysis of ATP.* A striking example is provided by bumblebees, which have to maintain a thoracic temperature of about 30°C in order to fly. A bumblebee is able to maintain this high thoracic temperature and forage for food even when the ambient temperature is only 10°C because phosphofructokinase and fructose diphosphatase in its flight muscle are simultaneously highly active. This fructose diphosphatase is not inhibited by AMP, which suggests that the enzyme is specially designed for the generation of heat. In contrast, the honeybee has almost no fructose diphosphatase activity in its flight muscle and consequently cannot fly when the ambient temperature is low. Excessive heat production caused by too high a rate of cycling between fructose 6-phosphate and fructose 1,6-diphosphate can occur. Halothane, an anesthetic, produces this condition, called *malignant hyperthermia,* in a susceptible strain of pigs.

LACTATE FORMED BY CONTRACTING MUSCLE IS CONVERTED INTO GLUCOSE BY THE LIVER

A major raw material of gluconeogenesis is lactate produced by active skeletal muscle. The rate of production of pyruvate by glycolysis exceeds the rate of oxidation of pyruvate by the citric acid cycle in contracting skeletal muscle under anaerobic conditions. Moreover, the rate of formation of NADH in glycolysis in active muscle is greater than the rate of its oxidation by the respiratory chain. Continued glycolysis depends on the availablility of NAD$^+$ for the oxidation of glyceraldehyde 3-phosphate. This is achieved by lactate dehydrogenase, which oxidizes NADH to NAD$^+$ as it reduces pyruvate to lactate.

Lactate is a dead end in metabolism. It must be converted back into pyruvate before it can be metabolized. The only purpose of the reduction of pyruvate to lactate is to regenerate NAD$^+$ so that gly-colysis can proceed in active skeletal muscle. *The formation of lactate buys time and shifts part of the metabolic burden from muscle to liver.*

The plasma membrane of most cells is highly permeable to lactate and pyruvate. Both substances diffuse out of active skeletal muscle into the blood and are carried to the liver. Much more lactate than pyruvate is carried because of the high NADH/NAD$^+$ ratio in contracting skeletal muscle. The lactate that enters the liver is oxidized to pyruvate, a reaction favored by the low NADH/NAD$^+$ in the cytosol of liver. Pyruvate is then converted into glucose by the gluconeogenic pathway in liver. Glucose then enters the blood and is taken up by skeletal muscle. Thus, *liver furnishes glucose to contracting skeletal muscle, which derives ATP from the glycolytic conversion of glucose into lactate. Glucose is then synthesized from lactate by the liver. These conversions constitute the Cori cycle* (Figure 15-8).

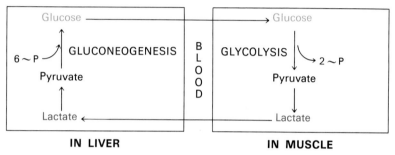

Figure 15-8
The Cori cycle. Lactate formed by active muscle is converted into glucose by the liver. This cycle shifts part of the metabolic burden of active muscle to the liver.

These conversions are facilitated by differences in the catalytic properties of lactate dehydrogenase enzymes in skeletal muscle and liver. Lactate dehydrogenase is a tetramer of 35-kdal subunits. There are two kinds of polypeptide chains called M and H, which can form five types of tetramers: M_4, M_3H, M_2H_2, M_1H_3, and H_4. These species are called *isoenzymes* (or isozymes). The M_4 isoenzyme has a much higher affinity for pyruvate than does the H_4 isoenzyme. The other isoenzymes have intermediate affinities. The principal isoenzyme in skeletal muscle is M_4, whereas the main one in liver and heart muscle is H_4. These isoenzymes have been studied intensively, but the reasons for the existence of multiple forms are an enigma.

SUMMARY

The pentose phosphate pathway generates NADPH and ribose 5-phosphate in the cytosol. NADPH is used in reductive biosyntheses, whereas ribose 5-phosphate is used in the synthesis of RNA, DNA, and nucleotide coenzymes. The pentose phosphate pathway starts with the dehydrogenation of glucose 6-phosphate to form a lactone, which is hydrolyzed to give 6-phosphogluconate and then oxidatively decarboxylated to yield ribulose 5-phosphate. $NADP^+$ is the electron acceptor in both of these oxidations. The last step is the isomerization of ribulose 5-phosphate (a ketose) to ribose 5-phosphate (an aldose). A different mode of the pathway is active when cells need much more NADPH than ribose 5-phosphate. Under these conditions, ribose 5-phosphate is converted into glyceraldehyde 3-phosphate and fructose 6-phosphate by transketolase and transaldolase. Transketolase contains TPP as its prosthetic group. These enzymes create a reversible link between the pentose phosphate pathway and glycolysis. Xylulose 5-phosphate, sedoheptulose 7-phosphate, and erythrose 4-phosphate are intermediates in these interconversions. In this way, twelve NADPH can be generated for each glucose 6-phosphate that is completely oxidized to CO_2. Only the nonoxidative branch of the pathway is active when much more ribose 5-phosphate than NADPH needs to be synthesized. Under these conditions, fructose 6-phosphate and glyceraldehyde 3-phosphate (formed by the glycolytic pathway) are converted into ribose 5-phosphate without the formation of NADPH. Alternatively, ribose 5-phosphate formed by the oxidative branch can be converted into pyruvate through fructose 6-phosphate and glyceraldehyde 3-phosphate. In this mode, ATP and NADPH are generated, and five of the six carbons of glucose 6-phosphate emerge in pyruvate. The interplay of the glycolytic and pentose phosphate pathways enables the levels of NADPH, of ATP, and of building blocks such as ribose 5-phosphate and pyruvate to be continuously adjusted to meet cellular needs.

Gluconeogenesis is the synthesis of glucose from noncarbohydrate sources, such as lactate, amino acids, and glycerol. Several of the reactions that convert pyruvate, a major entry point, into glucose are common to glycolysis. However, gluconeogenesis requires four new reactions to bypass the essential irreversibility of the corresponding reactions in glycolysis. Pyruvate is carboxylated to oxaloacetate in mitochondria, which in turn is decarboxylated and phosphorylated to phosphoenolpyruvate in the cytosol. Two high-energy phosphate bonds are consumed in these reactions, which are catalyzed by pyruvate carboxylase and phosphoenolpyruvate carboxykinase. Pyruvate carboxylase contains a biotin prosthetic group. The other distinctive reactions of gluconeogenesis are the hydrolyses of fructose 1,6-diphosphate and glucose 6-phosphate, which are catalyzed by specific phosphatases. Gluconeogenesis and glycolysis are reciprocally regulated so that one pathway is relatively inactive while the other is highly active.

SELECTED READINGS

WHERE TO START

Horecker, B. L., 1976. Unravelling the pentose phosphate pathway. *In* Kornberg, A., Cornudella, L., Horecker, B. L., and Oro, J., (eds.), *Reflections on Biochemistry,* pp. 65–72. Pergamon.

REVIEWS OF THE PENTOSE PHOSPHATE PATHWAY AND GLUCONEOGENESIS

Pontremoli, S., and Grazi, E., 1969. Hexose monophosphate oxidation. *Compr. Biochem.* 17:163–189.

Pontremoli, S., and Grazi, E., 1968. Gluconeogenesis. *In* Dickens, F., Randle, P. J., and Whelan, W. J., (eds.), *Carbohydrate Metabolism and Its Disorders,* vol. 1, pp. 259–295. Academic Press.

ENZYMES AND REACTION MECHANISMS

Horecker, B. L., 1964. Transketolase and transaldolase. *Compr. Biochem.* 15:48–70.

Scrutton, M. C., and Young, M. R., 1972. Pyruvate carboxylase. *In* Boyer, P. D., (ed.), *The Enzymes* (3rd ed.), vol. 6, pp. 1–35, Academic Press.

Wood, H. G., and Barden, R. E., 1977. Biotin enzymes. *Ann. Rev. Biochem.* 46:385–414.

Schultz, G. E., Schirmer, R. H., Sachsenheimer, W., and Pai, E. F., 1978. The structure of the flavoenzyme glutathione reductase. *Nature* 273:120–124.

REGULATION

Newsholme, E. A., and Start, C., 1973. *Regulation in Metabolism.* Wiley. [Chapter 6 deals with the control of glycolysis and gluconeogenesis in liver and kidney cortex.]

Soling, H. D., and Willms, B., (eds.), 1971. *Regulation of Gluconeogenesis.* Academic Press.

Newsholme, E. A., and Crabtree, B., 1976. Substrate cycles in metabolic regulation and in heat generation. *Biochem. Soc. Symp.* 41:61–109.

Katz, J., and Rognstad, R., 1976. Futile cycles in the metabolism of glucose. *Curr. Topics Cell Reg.* 10:238–287.

Nordlie, R. C., 1974. Metabolic regulation by multifunctional glucose-6-phosphatase. *Curr. Top. Cell Regul.* 8:33–111.

Ureta, T., 1978. The role of isozymes in metabolism: A model of metabolic pathways as the basis for the biological role of isozymes. *Curr. Top. Cell Regul.* 13:233–250.

GENETIC DISEASES

Beutler, E., 1978. Glucose 6-phosphate dehydrogenase deficiency. *In* Stanbury, J. B., Wyngaarden, J. B., and Fredrickson, D. S., (eds.), *The Metabolic Basis of Inherited Disease* (4th ed.), pp. 1430–1451. McGraw-Hill.

Luzzatto, L., Usanga, E. A., and Reddy, S., 1969. Glucose 6-phosphate dehydrogenase deficient red cells: resistance to infection by malarial parasites. *Science* 164:839–842.

Blass, J. P., and Gibson, G. E., 1977. Abnormality of a thiamine-requiring enzyme in patients with Wernicke-Korsakoff syndrome. *New Engl. J. Med.* 297:1367–1370.

PROBLEMS

1. What is the stoichiometry of the synthesis of ribose 5-phosphate from glucose 6-phosphate without the concomitant generation of NADPH? What is the stoichiometry of the synthesis of NADPH from glucose 6-phosphate without the concomitant formation of pentose sugars?

2. Glucose labeled with ^{14}C at C-6 is added to a solution containing the enzymes and cofactors of the oxidative branch of the pentose phosphate pathway. What is the fate of the radioactive label?

3. Which reaction in the citric acid cycle is most analogous to the oxidative decarboxylation of 6-phosphogluconate to ribulose 5-phosphate? What kind of enzyme-bound intermediate is formed in both reactions?

4. Ribose 5-phosphate labeled with ^{14}C at C-1 is added to a solution containing transketolase, transaldolase, phosphopentose epimerase, phosphopentose isomerase, and glyceraldehyde 3-phosphate. What is the distribution of the radioactive label in the erythrose 4-phosphate and fructose 6-phosphate that are formed in this reaction mixture?

5. Avidin, a 70-kdal protein in egg white, has a very high affinity for biotin. In fact, it is a highly specific inhibitor of biotin enzymes. Which of the following conversions would be blocked by the addition of avidin to a cell homogenate?
   (a) Glucose $\longrightarrow$ pyruvate.
   (b) Pyruvate $\longrightarrow$ glucose.
   (c) Oxaloacetate $\longrightarrow$ glucose.
   (d) Glucose $\longrightarrow$ ribose 5-phosphate.
   (e) Pyruvate $\longrightarrow$ oxaloacetate.
   (f) Ribose 5-phosphate $\longrightarrow$ glucose.

6. Design a chemical experiment to identify the lysine residue that forms a Schiff base at the active site of transaldolase.

For additional problems, see W. B. Wood, J. H. Wilson, R. M. Benbow, and L. E. Hood, *Biochemistry: A Problems Approach* (Benjamin, 1974), ch. 9.

CHAPTER 16

GLYCOGEN AND
DISACCHARIDE METABOLISM

Glycogen is a *readily mobilized storage form of glucose*. It is a very large, branched polymer of glucose residues (Figure 16-1). Most of the glucose residues in glycogen are linked by α-1,4-glycosidic bonds. The branches are created by α-1,6-glycosidic bonds, of which there is one in about ten residues.

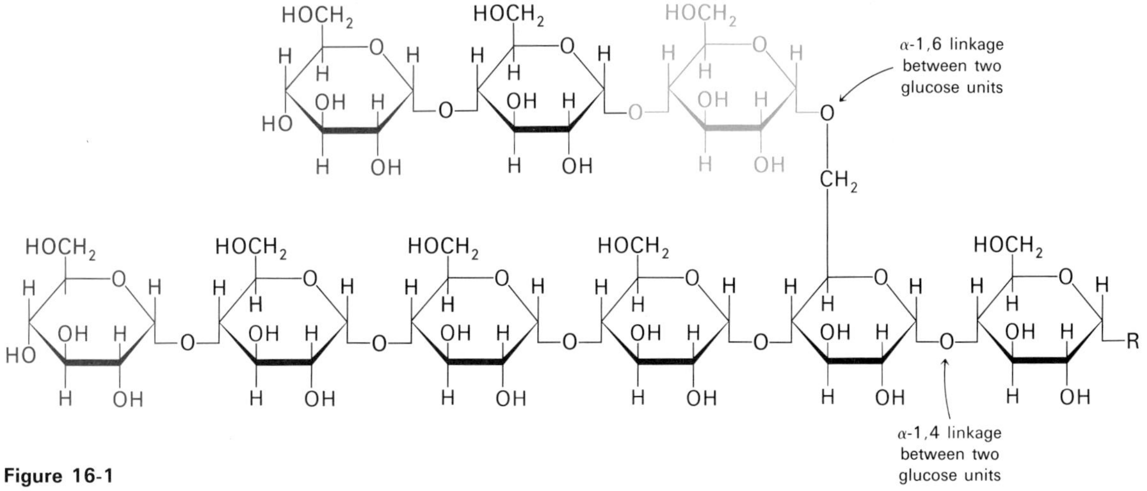

Figure 16-1
Structure of two outer branches of a glycogen particle. The residues at the non-reducing ends are shown in red. The residue that starts a branch is shown in green. The rest of the glycogen molecule is represented by R.

The presence of glycogen greatly increases the amount of glucose that is immediately available between meals and during muscular activity. The amount of glucose in the body fluids of an average 70-kg man has an energy content of only 40 kcal, whereas the total body glycogen has an energy content of more than 600 kcal, even after an overnight fast. The two major sites of glycogen storage are the liver and skeletal muscle. The concentration of glycogen is higher in the liver than in muscle, but more glycogen is stored in skeletal muscle because of its much greater mass. Glycogen is present in the cytosol in the form of granules with diameters ranging

Figure 16-2
Diagram of a cross section of a glycogen molecule. (Residues are differentiated by the same colors as in Figure 16-1.)

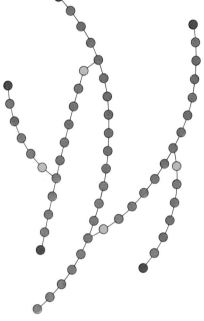

Figure 16-3
Electron micrograph of a liver cell. The dense particles in the cytoplasm are glycogen granules. [Courtesy of Dr. George Palade.]

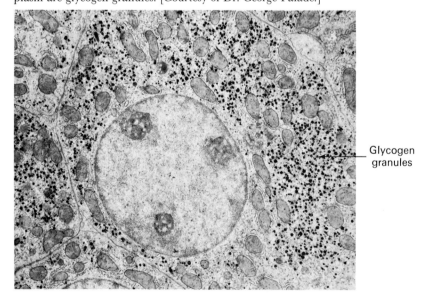

Glycogen
granules

from about 100 to 400 Å. This range in size reflects the fact that glycogen molecules do not have a unique size; a typical distribution is centered on a mass of several thousand kilodaltons. Glycogen granules have a dense appearance in electron micrographs (Figure 16-3). They contain the enzymes that catalyze the synthesis and degradation of glycogen and some of the enzymes that regulate these processes. However, a glycogen granule differs from a multi-enzyme complex (such as the pyruvate dehydrogenase complex) in that the bound enzymes are not present in defined stoichiometric ratios. Also, the degree of structural organization in a glycogen granule is less than in a multienzyme complex.

The synthesis and degradation of glycogen are considered here in some detail for several reasons. First, these processes are important because they *regulate the blood glucose level* and provide a *reservoir of glucose* for strenuous muscular activity. Second, the synthesis and degradation of glycogen occur by *different reaction pathways,* which illustrates an important principle of biochemistry. Third, the hormonal regulation of glycogen metabolism is mediated by mechanisms that are of general significance. The *role of cyclic adenosine monophosphate* (cyclic AMP) in the coordinated control of glycogen synthesis and breakdown is well understood and is a source of insight into the action of hormones in a variety of other systems. Fourth, a number of inherited enzyme defects resulting in impaired glycogen metabolism have been characterized. Some of these *glycogen storage diseases* are lethal in infancy, whereas others have a relatively mild clinical course. The final part of this chapter deals with the metabolism of the common disaccharides: *lactose, maltose, and sucrose.*

PHOSPHORYLASE CATALYZES THE PHOSPHOROLYTIC CLEAVAGE OF GLYCOGEN INTO GLUCOSE 1-PHOSPHATE

The pathway of glycogen breakdown was elucidated by the incisive studies of Carl Cori and Gerty Cori. They showed that glycogen is *cleaved by orthophosphate* to yield a new kind of phosphorylated sugar, which they identified as *glucose 1-phosphate*. The Coris also isolated and crystallized *glycogen phosphorylase,* the enzyme that catalyzes this reaction.

Glycogen + P_i $\rightleftharpoons$ glucose 1-phosphate + glycogen
(*n* residues)   (*n* − 1 residues)

Phosphorylase catalyzes the sequential removal of glycosyl residues from the nonreducing end of the glycogen molecule. The glycosidic linkage between C-1 of the terminal residue and C-4 of the adjacent one is split by orthophosphate. Specifically, the bond between the C-1 carbon atom and the glycosidic oxygen atom is cleaved by orthophosphate, and the α configuration at C-1 is retained.

Phosphorolysis—
The cleavage of a bond by orthophosphate (in contrast with *hydrolysis,* which refers to cleavage by water).

HOCH$_2$ HOCH$_2$ P$_i$ HOCH$_2$ HOCH$_2$

H   O   H   H   O   H   Phosphorylase   H   O   H   O   H

HO   OH   H   OH   H   OR   HO   OH   H   OPO$_3^{2-}$ +   HO   OH   H   OR

H   OH   H   OH   H   OH   H   OH

Glycogen   **Glucose 1-phosphate**   **Glycogen**
(*n* residues)   (*n* − 1 residues)

This reaction probably proceeds through a carbonium ion intermediate. The breaking of the C$_1$—O bond, the retention of configuration at C-1, and the likely participation of a carbonium ion intermediate are reminiscent of the lysozyme-catalyzed cleavage of chitin.

The reaction catalyzed by phosphorylase is readily reversible in vitro. At pH 6.8, the equilibrium ratio of orthophosphate to glucose 1-phosphate is 3.6. The $\Delta G°'$ for this reaction is small because a glycosidic bond is replaced by a phosphate ester bond that has a nearly equal transfer potential. However, phosphorolysis proceeds far in the direction of glycogen breakdown in vivo because the [P$_i$]/[glucose 1-phosphate] ratio is usually greater than 100.

The phosphorolytic cleavage of glycogen is energetically advantageous because the released sugar is phosphorylated. In contrast, a hydrolytic cleavage would yield glucose, which would have to be phosphorylated at the expense of an ATP to enter the glycolytic pathway. An additional advantage of phosphorolytic cleavage for muscle cells is that glucose 1-phosphate cannot diffuse out of the cell, whereas glucose can. The significance of the retention of phosphorylated sugars by muscle will be discussed shortly.

A DEBRANCHING ENZYME ALSO IS NEEDED FOR THE BREAKDOWN OF GLYCOGEN

Glycogen is degraded to a limited extent by phosphorylase alone. However, the α-1,6-glycosidic bonds at the branch points are not susceptible to cleavage by phosphorylase. Indeed, phosphorylase stops cleaving α-1,4 linkages when it reaches a terminal residue four away from a branch point. The action of phosphorylase on two outer branches of a glycogen particle is shown in Figure 16-4. Five α-1,4-glycosidic bonds on one branch and three on the other are cleaved by phosphorylase. Cleavage by phosphorylase stops at this stage because terminal residues *a* and *d* are four away from the branch point *h*. A new enzymatic activity is required at this stage. A *transferase shifts a block of three glycosyl residues from one outer branch to the other.* The α-1,4-glycosidic link between *c* and *z* is broken, and a new α-1,4 link between *c* and *d* is formed. This transfer exposes residue *z* to the action of a third degradative enzyme, an *α-1,6-glucosidase,* which is also known as the *debranching enzyme.* This enzyme hydrolyzes the α-1,6-glycosidic bond between residues *z* and *h*.

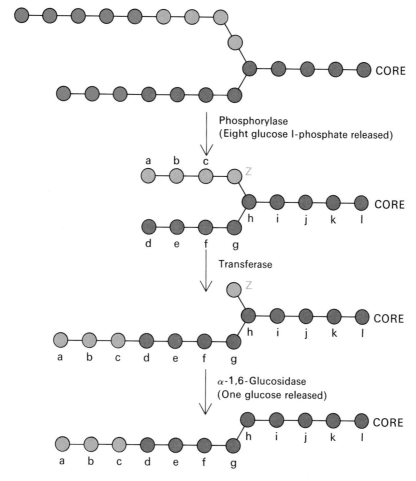

Figure 16-4
Steps in the degradation of glycogen.

Thus, the transferase and the debranching enzyme (α-1,6-gluco-sidase) convert the branched structure into a linear one, which paves the way for further cleavage by phosphorylase. The hydrolysis of z renders all of the residues a through l susceptible to phosphorylase. It is interesting to note that a single 160-kdal polypeptide chain contains both the transferase and the α-1,6-glucosidase active sites.

PHOSPHOGLUCOMUTASE CONVERTS GLUCOSE 1-PHOSPHATE INTO GLUCOSE 6-PHOSPHATE

The glucose 1-phosphate formed in the phosphorolytic cleavage of glycogen is converted into glucose 6-phosphate by *phosphoglucomutase*. The equilibrium mixture contains 95% glucose 6-phosphate. The catalytic site of an active enzyme molecule contains a phosphorylated serine residue. In catalysis, this phosphoryl group is probably transferred to the hydroxyl group at C-6 of glucose 1-phosphate to form glucose 1,6-diphosphate. This intermediate then transfers its C-1 phosphoryl group to the serine residue at the active site of the enzyme, which results in glucose 6-phosphate and regeneration of the phosphoenzyme.

The phosphoryl group on the mutase is slowly lost by hydrolysis. It is restored by phosphoryl transfer from glucose 1,6-diphosphate, which is formed from glucose 1-phosphate and ATP in a reaction catalyzed by phosphoglucokinase. These reactions are similar to those of *phosphoglyceromutase*, a glycolytic enzyme (p. 264). The role of 2,3-diphosphoglycerate (2,3-DPG) in the interconversion of 2-phosphoglycerate and 3-phosphoglycerate is similar to that of glucose 1,6-diphosphate in the interconversion of the phosphoglucoses. Furthermore, a phosphoenzyme intermediate participates in both reactions.

LIVER CONTAINS GLUCOSE 6-PHOSPHATASE, A HYDROLYTIC ENZYME ABSENT FROM MUSCLE

A major function of the liver is to maintain a relatively constant level of glucose in the blood. The liver releases glucose into the blood during muscular activity and in the intervals between meals. The released glucose is taken up primarily by the brain and by skeletal muscle. Phosphorylated glucose, in contrast with glucose, cannot readily diffuse out of cells. The liver contains a hydrolytic enzyme, *glucose 6-phosphatase,* that enables glucose to leave that organ. This enzyme is essential for gluconeogenesis (p. 348).

$$\text{Glucose 6-phosphate} + H_2O \longrightarrow \text{glucose} + P_i$$

Glucose 6-phosphatase is also present in the kidneys and intestine, but *it is absent from muscle and the brain.* Consequently, glucose 6-phosphate is retained by muscle and the brain, which need large amounts of this fuel for the generation of ATP. In contrast, glucose is not a major fuel for the liver. Rather, the liver stores and releases glucose primarily for the benefit of other tissues (see p. 548).

GLYCOGEN IS SYNTHESIZED AND DEGRADED BY DIFFERENT PATHWAYS

The reaction catalyzed by glycogen phosphorylase is readily reversed in vitro, because the $\Delta G^{\circ\prime}$ for the elongation of glycogen by glucose 1-phosphate is -0.5 kcal/mol. In fact, the Coris were able to synthesize glycogen from glucose 1-phosphate using phosphorylase and a branching enzyme. However, a number of subsequent experimental observations indicated that glycogen is synthesized in vivo by a different pathway. First, the reaction catalyzed by phosphorylase is at equilibrium when the $[P_i]/[\text{glucose 1-phosphate}]$ ratio is 3.6 at neutral pH, whereas this ratio in cells is usually greater than 100. Hence, the phosphorylase reaction in vivo must proceed in the direction of glycogen degradation. Second, hormones that lead to an increase in phosphorylase activity always elicit glycogen breakdown. Third, patients who lack muscle phosphorylase entirely (in McArdle's disease, see p. 376) are able to synthesize muscle glycogen.

In 1957, Luis Leloir and his coworkers showed that glycogen is synthesized by a different pathway. The glycosyl donor is uridine diphosphate glucose (UDP-glucose) rather than glucose 1-phosphate. *The synthetic reaction is not a reversal of the degradative reaction:*

Synthesis: $\text{Glycogen}_n + \text{UDP-glucose} \longrightarrow$

$$\text{glycogen}_{n+1} + \text{UDP}$$

Degradation: $\text{Glycogen}_{n+1} + P_i \longrightarrow$

$$\text{glycogen}_n + \text{glucose 1-phosphate}$$

We now know that *biosynthetic and degradative pathways in biological systems* are almost always distinct. Glycogen metabolism provided the first example of this important principle. *Separate pathways afford much greater flexibility, both in energetics and in control.* The cell is no longer at the mercy of mass action; glycogen can be synthesized despite a high ratio of orthophosphate to glucose 1-phosphate.

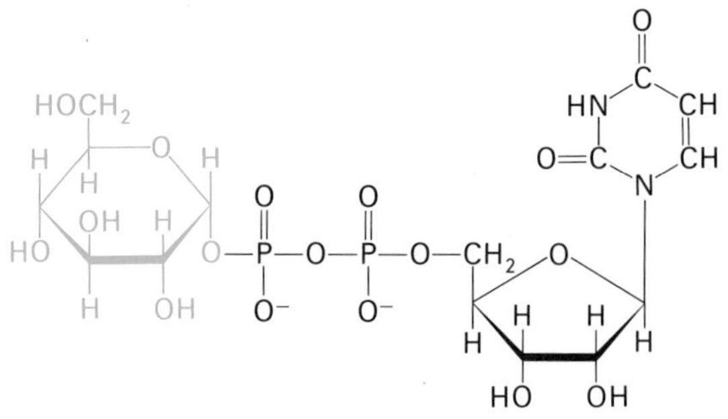

Uridine diphosphate glucose
(UDP-glucose)

UDP-GLUCOSE IS AN ACTIVATED FORM OF GLUCOSE

UDP-glucose, the glucose donor in the biosynthesis of glycogen, is an *activated form of glucose,* just as ATP and acetyl CoA are activated forms of orthophosphate and acetate, respectively. The C-1 carbon atom of the glucosyl unit of UDP-glucose is activated because its hydroxyl group is esterified to the diphosphate moiety of UDP.

UDP-glucose is synthesized from glucose 1-phosphate and uridine triphosphate (UTP) in a reaction catalyzed by *UDP-glucose pyrophosphorylase.* The pyrophosphate liberated in this reaction comes from the outer two phosphoryl residues of UTP.

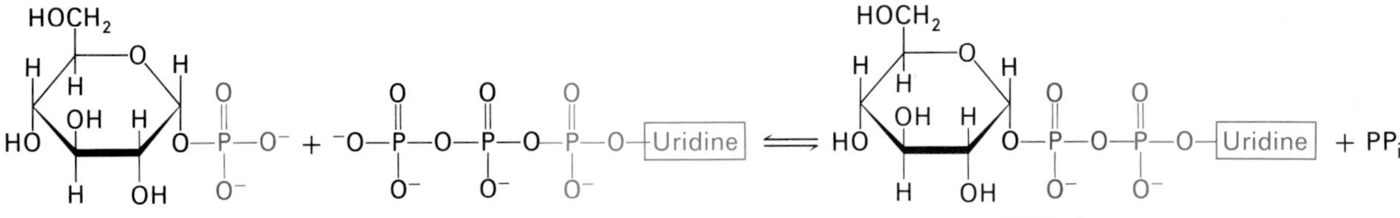

Glucose 1-phosphate

UDP-glucose

This reaction is readily reversible. However, pyrophosphate is rapidly hydrolyzed in vivo to orthophosphate by an inorganic pyrophosphatase. The essentially irreversible hydrolysis of pyrophosphate drives the synthesis of UDP-glucose.

$$\text{Glucose 1-phosphate} + \text{UTP} \rightleftharpoons \text{UDP-glucose} + \text{PP}_i$$
$$\text{PP}_i + \text{H}_2\text{O} \longrightarrow 2\,\text{P}_i$$

$$\text{Glucose 1-phosphate} + \text{UTP} + \text{H}_2\text{O} \longrightarrow \text{UDP-glucose} + 2\,\text{P}_i$$

The synthesis of UDP-glucose exemplifies a recurring theme in biochemistry: *many biosynthetic reactions are driven by the hydrolysis of pyrophosphate.* Another aspect of this reaction has broad significance. Nucleoside diphosphate sugars serve as glycosyl donors in the biosynthesis of many disaccharides and polysaccharides.

GLYCOGEN SYNTHETASE CATALYZES THE TRANSFER OF GLUCOSE FROM UDP-GLUCOSE TO A GROWING CHAIN

New glucosyl units are added to the nonreducing terminal residues of glycogen. The activated glucosyl unit of UDP-glucose is transferred to the hydroxyl group at a C-4 terminus of glycogen to form an α-1,4-glycosidic linkage. In this elongation reaction, UDP is displaced by this terminal hydroxyl group of the growing glycogen molecule. This reaction is catalyzed by *glycogen synthetase,* which can add glucosyl residues only if the polysaccharide chain already contains more than four residues. Thus, glycogen synthesis requires a *primer,* which is formed by a different synthetase.

A BRANCHING ENZYME FORMS ALPHA-1,6 LINKAGES

Glycogen synthetase catalyzes only the synthesis of α-1,4 linkages. Another enzyme is needed to form the α-1,6 linkages that make glycogen a branched polymer. *Branching is important because it increases the solubility of glycogen.* Furthermore, branching creates a large number of nonreducing terminal residues, which are the sites of action of glycogen phosphorylase and synthetase. Thus, *branching increases the rate of glycogen synthesis and degradation.*

Branching occurs after a number of glucosyl residues are joined in α-1,4 linkage by glycogen synthetase. A branch is created by the breakage of an α-1,4 link and the formation of an α-1,6 link, which is a different reaction from debranching. A block of residues, typically seven in number, is transferred to a more interior site. The *branching enzyme* that catalyzes this reaction is quite exacting. The block of seven or so residues must include the nonreducing terminus and come from a chain at least eleven residues long. In addition, the new branch point must be at least four residues away from a preexisting one.

GLYCOGEN IS A VERY EFFICIENT STORAGE FORM OF GLUCOSE

What is the cost of converting glucose 6-phosphate into glycogen and back into glucose 6-phosphate? The pertinent reactions have already been described, except for reaction 5 below, which is the regeneration of UTP. UDP is phosphorylated by ATP in a reaction catalyzed by *nucleoside diphosphokinase*.

(1) $\quad$ Glucose 6-phosphate $\longrightarrow$ glucose 1-phosphate

(2) $\quad$ Glucose 1-phosphate + UTP $\longrightarrow$ UDP-glucose + PP_i

(3) $\quad PP_i + H_2O \longrightarrow 2 P_i$

(4) $\quad$ UDP-glucose + glycogen$_n$ $\longrightarrow$ glycogen$_{n+1}$ + UDP

(5) $\quad$ UDP + ATP $\longrightarrow$ UTP + ADP

Sum: Glucose 6-phosphate + ATP + glycogen$_n$ + $H_2O \longrightarrow$
$\qquad$ glycogen$_{n+1}$ + ADP + 2 P_i

Thus, one high-energy phosphate bond is spent in incorporating glucose 6-phosphate into glycogen. The energy yield from the breakdown of glycogen is highly efficient. About 90% of the residues are phosphorolytically cleaved to glucose 1-phosphate, which is converted at no cost into glucose 6-phosphate. The other 10% are branch residues, which are hydrolytically cleaved. One ATP is then used to phosphorylate each of these glucose molecules to glucose 6-phosphate. The complete oxidation of glucose 6-phosphate yields thirty-seven molecules of ATP and storage consumes slightly more than one ATP per glucose 6-phosphate, and so *the overall efficiency of storage is nearly 97%.*

CYCLIC AMP IS CENTRAL TO THE COORDINATED CONTROL OF GLYCOGEN SYNTHESIS AND BREAKDOWN

The occurrence of separate pathways for the synthesis and degradation of glycogen means that they must be rigorously controlled. ATP would be wastefully hydrolyzed if both sets of reactions were fully active at the same time. In fact, *glycogen synthesis and degradation are coordinately controlled so that glycogen synthetase is nearly inactive when phosphorylase is fully active, and vice versa.*

Glycogen metabolism is profoundly affected by specific hormones. *Insulin*, a polypeptide hormone (p. 21), increases the capacity of the liver to synthesize glycogen. The mechanism of action of insulin is not yet understood. High levels of insulin in the blood signal the fed state, whereas low levels signal the fasted state (p. 549). Much more is known about the mode of action of *epinephrine* and *glucagon*, which have effects opposite to those of insulin. Muscular activity or its anticipation leads to the release of epinephrine by the adrenal medulla. *Epinephrine* markedly stimulates glycogen breakdown in muscle and, to a lesser extent, in liver. The liver is more responsive to *glucagon*, a polypeptide hormone that is secreted by the α cells of the pancreas when the blood sugar level is low. This hormone increases the blood sugar level by stimulating the breakdown of glycogen in the liver.

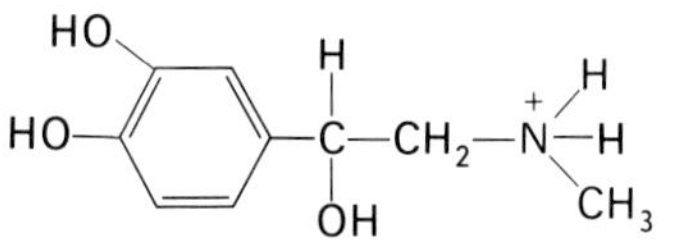

Epinephrine

$^+$H$_3$N-His-Ser-Glu-Gly-Thr-Phe-Thr-Ser-Asp-Tyr-   10

-Ser-Lys-Tyr-Leu-Asp-Ser-Arg-Arg-Ala-Gln-   20

-Asp-Phe-Val-Gln-Trp-Leu-Met-Asn-Thr-COO$^-$ 29

Glucagon

Earl Sutherland discovered that the action of epinephrine and glucagon metabolism is mediated by *cyclic AMP*. This discovery led to the recognition that cyclic AMP is ubiquitous in all forms of life and plays a key role in controlling biological processes (see Chapter 35). The synthesis of this regulatory molecule from ATP is catalyzed by *adenylate cyclase*, an enzyme associated with the plasma membrane. The synthesis of cyclic AMP is accelerated by the subsequent hydrolysis of pyrophosphate.

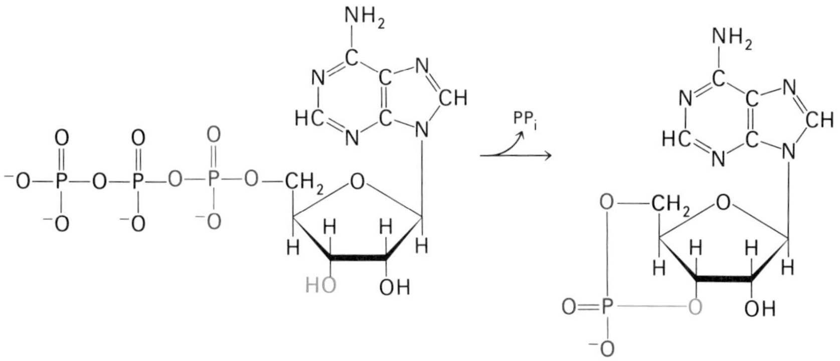

ATP   **Cyclic AMP**

Epinephrine and glucagon do not enter their target cells. Rather, they bind to the plasma membrane and stimulate adenylate cyclase (p. 842). The increased intracellular level of cyclic AMP triggers a series of reactions that activate phosphorylase and inhibit glycogen synthetase. We will now consider the structural basis of the control of these enzymes, and then turn to the reaction cascade that links cyclic AMP to these crucial enzymes of glycogen metabolism.

PHOSPHORYLASE IS ACTIVATED BY PHOSPHORYLATION OF A SPECIFIC SERINE RESIDUE

Skeletal muscle phosphorylase exists in two interconvertible forms: an *active* phosphorylase *a* and a usually *inactive* phosphorylase *b* (Figure 16-5). The enzyme is a dimer of 92-kdal subunits. Phosphorylase *b* is converted into phosphorylase *a* by the phosphorylation of a single serine residue (serine 14) in each subunit.

$$\boxed{\begin{array}{c}\text{Phosphorylase}\\ b\end{array}} + \text{ATP} \longrightarrow \boxed{\begin{array}{c}\text{Phosphorylase}\\ a\end{array}} + \text{ADP} + \text{H}^+$$

$$\underset{\text{Ser—OH}}{\qquad\qquad} \qquad\qquad \underset{\text{Ser—O—PO}_3^{2-}}{\qquad\qquad\qquad}$$

This covalent modification is catalyzed by a specific enzyme, *phosphorylase kinase,* which was discovered by Edmond Fischer and Edwin Krebs. Phosphorylase *a* is deactivated by a specific phosphatase that hydrolyzes the phosphoryl group attached to serine 14.

Muscle phosphorylase *b* is active only in the presence of high concentrations of AMP, which acts allosterically. AMP binds to the nucleotide binding site and alters the conformation of phosphorylase *b*. ATP acts as a negative allosteric effector by competing with AMP. Glucose 6-phosphate also inhibits phosphorylase *b*, presumably by binding to another site. Under most physiological conditions, *phosphorylase* b *is inactive because of the inhibitory effects of ATP and*

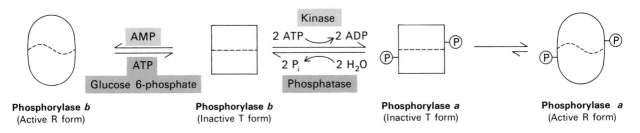

Figure 16-5
Schematic diagram of the control of glycogen phosphorylase in skeletal muscle. The enzyme can adopt a catalytically inactive T (tense) conformation or an active R (relaxed) conformation. The R ⇌ T equilibrium of phosphorylase *a* is far on the side of the active R state. In contrast, phosphorylase *b* is mostly in the inactive T state unless the level of AMP is high and the levels of ATP and glucose 6-phosphate are low. Under most physiological conditions, the proportion of active enzyme is determined by the rates of phosphorylation and dephosphorylation.

glucose 6-phosphate. In contrast, *phosphorylase* a *is fully active*, irrespective of the levels of AMP, ATP, and glucose 6-phosphate. The proportion of active enzyme is determined primarily by the rates of phosphorylation and dephosphorylation. In resting muscle, nearly all of the enzyme is in the inactive *b* form. As described in the next section, increased levels of epinephrine and electrical stimulation of muscle result in the formation of the active *a* form.

THREE-DIMENSIONAL STRUCTURE OF GLYCOGEN PHOSPHORYLASE

X-ray crystallographic studies of the *a* and *b* forms of glycogen phosphorylase are sources of insight into the catalytic and control mechanisms of this key enzyme in metabolism. The 841 amino acid residues of the monomeric subunit are compactly folded into three structural domains (Figure 16-6): an *amino-terminal domain* (310 residues), a *glycogen-binding domain* (160 residues), and a *carboxy-terminal domain* (371 residues). The *catalytic site* is located in a deep crevice

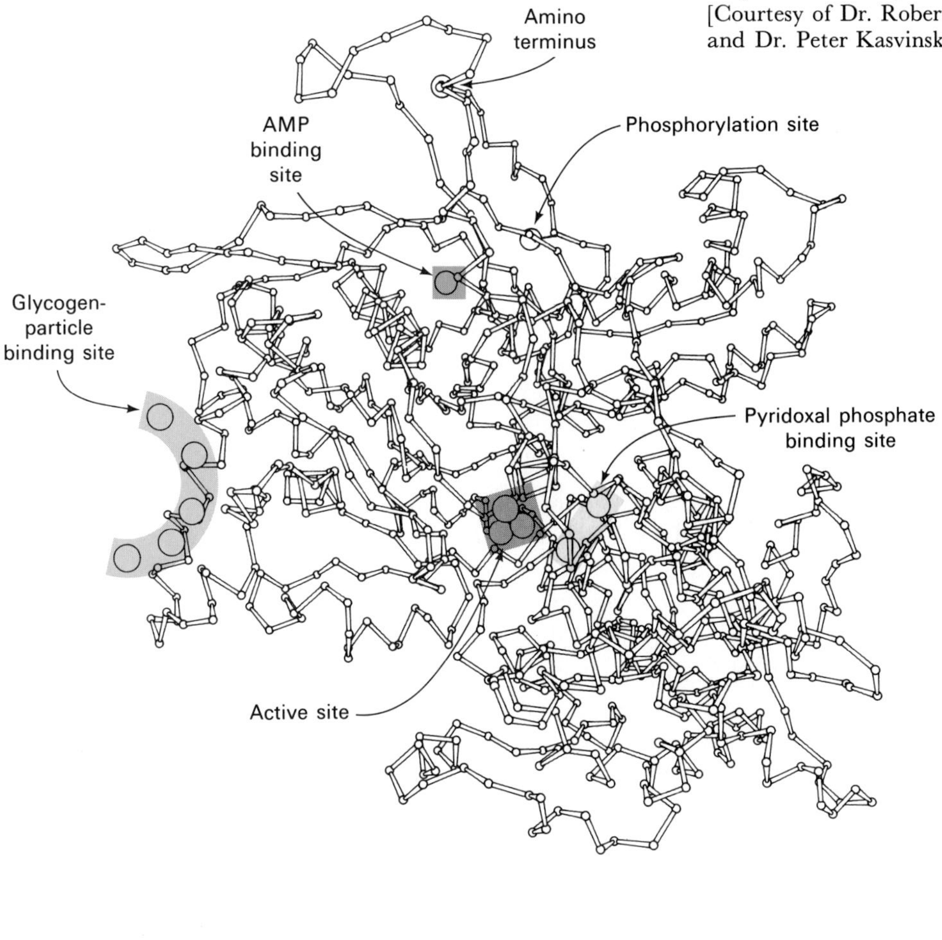

Figure 16-6
α-Carbon diagram of phosphorylase *a* showing the locations of the catalytic site, glycogen-particle binding site, allosteric sites, and phosphorylation site. [Courtesy of Dr. Robert Fletterick, Dr. Neil Madsen, and Dr. Peter Kasvinsky.]

formed by residues from each of these three domains. This shielding of the active site from the aqueous milieu is expected to favor phosphorolysis over hydrolysis. *Pyridoxal phosphate* (vitamin B_6), which is essential for enzymatic activity, is bound near the site for glucose 1-phosphate. The aldehyde group of this cofactor forms a Schiff-base linkage with a lysine side chain in the carboxy-terminal domain. The retention of enzymatic activity following reduction of this Schiff base with borohydride shows that the aldehyde group does not participate in catalysis, in contrast with other pyridoxal enzymes. However, the phosphoryl group of pyridoxal phosphate seems to directly participate in catalysis. Phosphorylase also contains a *glycogen-binding site* that is distinct from the catalytic site. This site, which is about 30 Å away from the catalytic site, is important for the attachment of the enzyme to the glycogen particle. The large separation between this attachment site and the catalytic site enables the enzyme to phosphorolyze many terminal residues without having to dissociate and reassociate after each round of catalysis.

In addition, phosphorylase contains at least two allosteric *control sites*. Glucose and nucleosides, which are allosteric inhibitors of liver phosphorylase *a* (p. 374), bind near the catalytic site. AMP, an allosteric activator of phosphorylase *b*, binds to a site near the interface between subunits, far from the catalytic site and the glycogen-binding site. Serine 14, the *site of phosphorylation* in the conversion of phosphorylase *b* into *a*, is also located near the subunit interface. In phosphorylase *a*, this phosphoryl group is hydrogen bonded to the side chain of arginine 69. In contrast, the region comprising the 19 amino-terminal residues of phosphorylase *b* does not have a well-defined structure. The flexibility of this region in the *b* form is reminiscent of the very floppy activation domain in trypsinogen, which adopts a well-defined conformation on conversion into trypsin (p. 168). Detailed studies of the structural changes elicited by phosphorylation and by allosteric effectors are in progress. It is already evident that glycogen phosphorylase is a highly sophisticated integrator of information about the energy metabolism of the cell.

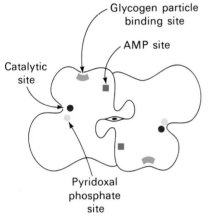

Figure 16-7
Schematic diagram of the phosphorylase *b* dimer viewed down the molecular twofold axis showing the domains and the binding sites for glucose-1-phosphate and maltotriose. [After I. T. Weber, L. N. Johnson, K. S. Wilson, D. G. R. Yeates, D. L. Wild, and J. A. Jenkins, *Nature* 274(1978):433.]

PHOSPHORYLASE KINASE ALSO IS ACTIVATED BY PHOSPHORYLATION

The activity of phosphorylase kinase also is regulated by covalent modification. Phosphorylase kinase, like phosphorylase, is converted from *a low-activity form into a high-activity one by phosphorylation*. The enzyme that catalyzes this activation reaction is a component of the hormone-cyclic-AMP system, which will be discussed shortly. Phosphorylase kinase can be partially activated in a different way by Ca^{2+} levels of the order of 10^{-7} M. This mode of activation of the kinase is biologically significant because muscle contraction is triggered by the release of Ca^{2+} (Chapter 34). Thus, *glycogen breakdown* *and muscle contraction are linked by a transient increase in the Ca^{2+} level in the cytoplasm.*

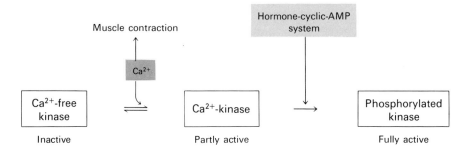

Muscle contraction

Hormone-cyclic-AMP system

Ca^{2+}

Ca^{2+}-free kinase	Ca^{2+}-kinase	Phosphorylated kinase
Inactive	Partly active	Fully active

GLYCOGEN SYNTHETASE IS INACTIVATED BY THE PHOSPHORYLATION OF A SPECIFIC SERINE RESIDUE

The synthesis of glycogen is closely coordinated with its degradation. The activity of glycogen synthetase, like that of phosphorylase, is regulated by covalent modification. Phosphorylation converts glycogen synthetase *a* into a usually inactive *b* form. The phosphorylated *b* form (formerly called the D or dependent form) requires a high level of glucose 6-phosphate for activity, whereas the dephosphorylated *a* form (formerly called the I or independent form) is active in its presence or absence. Thus, *phosphorylation has opposite effects on the enzymatic activities of glycogen synthetase and phosphorylase.*

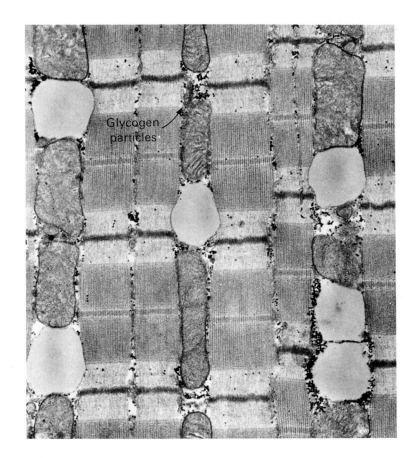

Glycogen particles

Figure 16-8
Electron micrograph showing the localization of glycogen particles in cardiac muscle. [Courtesy of Dr. Don W. Fawcett.]

A REACTION CASCADE CONTROLS THE PHOSPHORYLATION OF GLYCOGEN SYNTHETASE AND PHOSPHORYLASE

We turn now to the link between the hormones that affect glycogen metabolism and the phosphorylation reactions that determine the activities of glycogen synthetase and phosphorylase. The reaction sequence (Figure 16-9) is:

1. Epinephrine binds to the plasma membrane of the muscle cell and stimulates adenylate cyclase.

2. Adenylate cyclase in the plasma membrane catalyzes the formation of cyclic AMP from ATP.

3. The increased intracellular level of cyclic AMP activates a *protein kinase*. This kinase is inactive in the absence of cyclic AMP. The binding of cyclic AMP allosterically stimulates the protein kinase (see Chapter 35).

4. The cyclic-AMP–dependent protein kinase phosphorylates both phosphorylase kinase and glycogen synthetase. *The phosphorylation of both enzymes is the basis of the coordinated regulation of glycogen synthesis and breakdown.* Phosphorylation by this cyclic-AMP–dependent kinase *switches on phosphorylase* (by means of phosphorylase kinase) and simultaneously *switches off glycogen synthetase* (directly).

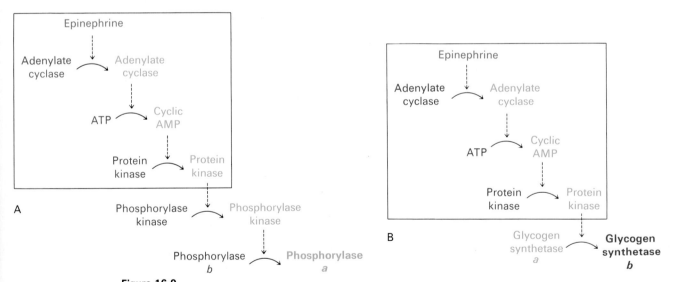

Figure 16-9
Reaction cascades for the control of glycogen metabolism: (A) glycogen degradation; (B) glycogen synthesis. Inactive forms are shown in red, and active ones in green. The sequence of reactions leading to the activation of the protein kinase is the same for the regulation of glycogen degradation and synthesis.

The changes in enzymatic activity produced by phosphorylation can be reversed by the hydrolytic removal of the phosphoryl group. For example, the conversion of phosphorylase *a* into *b* is catalyzed by *phosphorylase phosphatase*.

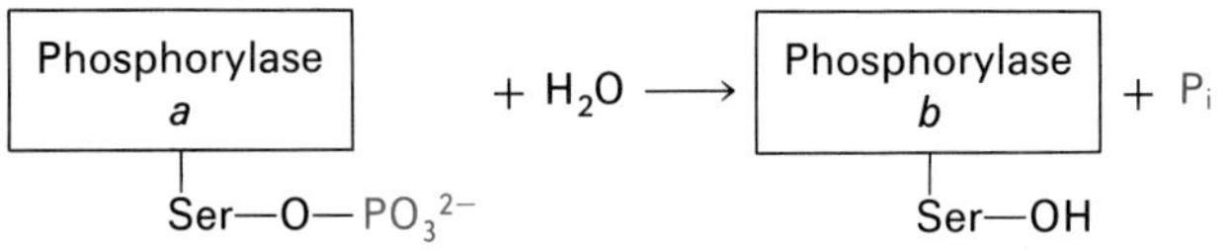

This enzyme also hydrolyzes the phosphoryl group on the active form of phosphorylase kinase, which renders this kinase inactive. Moreover, the same phosphatase also removes the phosphoryl group from glycogen synthetase *b* to convert it into the much more active *a* form. This is yet another molecular device for coordinating glycogen synthesis and breakdown. The activities of the phosphatases also appear to be regulated. For example, the combination of Ca^{2+} and Mg-ATP inhibits phosphorylase phosphatase, whereas it activates phosphorylase kinase. Muscle glycogen synthetase phosphatase is inhibited by glycogen. Thus, at high glycogen levels, muscle glycogen synthetase will remain in the phosphorylated *b* form.

The signal arising from cyclic AMP can also be switched off. The phosphodiester bond in cyclic AMP is hydrolyzed by a specific phosphodiesterase to form AMP, which does not activate the protein kinase. This highly exergonic reaction has a $\Delta G°'$ of -11.9 kcal/mol.

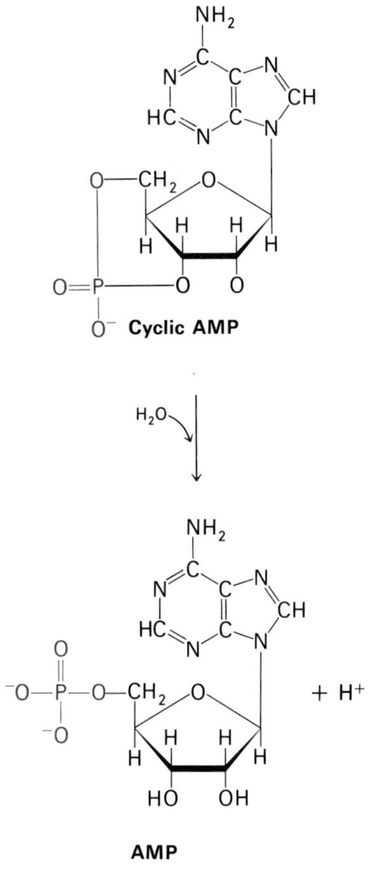

THE REACTION CASCADE AMPLIFIES THE HORMONAL SIGNAL

The enzymatic cascade in the control of glycogen metabolism is analogous to the proteolytic cascade in blood clotting (see p. 171). In both processes, the enzymatic cascade provides a *high degree of amplification*. In glycogen breakdown, there are three enzymatically catalyzed control stages, whereas in glycogen synthesis, there are two control stages. If glycogen phosphorylase and synthetase were directly regulated by the binding of epinephrine, more than a thousand times as much epinephrine would be required to elicit glycogen breakdown than is needed in the presence of the amplifying cascade.

Cyclic AMP

AMP

GLYCOGEN METABOLISM IN THE LIVER REGULATES THE BLOOD GLUCOSE LEVEL

The control of the synthesis and degradation of glycogen in the liver is central to the regulation of the blood glucose level. The concentration of glucose in the blood normally ranges from about 80 to 120 mg per 100 ml. The liver senses the concentration of glucose in the blood and takes up glucose if its level is above a threshold value or releases glucose if its level is below that value. The amount of liver phosphorylase *a* decreases rapidly when glucose is infused (Figure 16-10). After a lag period, the amount of glycogen synthetase *a* increases, which results in the synthesis of glycogen. Recent studies suggest that *phosphorylase* a *is the glucose sensor in liver cells*. The binding of glucose to phosphorylase *a* shifts the allosteric equilibrium from the R state to the T state (see Figure 16-5), which *exposes the phosphoryl group on serine 14 to hydrolysis by the phosphatase*. It is significant that the phosphatase binds tightly to phosphorylase *a* but acts catalytically only when glucose induces it into the T state.

How does glucose activate the synthetase? Recall that the same phosphatase acts on phosphorylase and glycogen synthetase. Phosphorylase *b*, in contrast with the *a* form, does not bind the phosphatase. Consequently, the conversion of phosphorylase *a* into *b* is accompanied by the *release of the phosphatase, which is then free to activate glycogen synthetase*. Removal of the phosphoryl group of inactive synthetase *b* converts it into the active *a* form. Initially, there are about ten phosphorylase *a* molecules per phosphatase. Hence, the activity of the synthetase begins to increase only after most of phosphorylase *a* is converted into the *b* form (Figure 16-10). This remarkable glucose-sensing system depends on three key elements: (1) communication between the allosteric site for glucose and the serine phosphate; (2) the use of the same phosphatase to inactivate phosphorylase and activate the synthetase, and (3) the binding of the phosphatase to phosphorylase *a* to prevent premature activation of the synthetase.

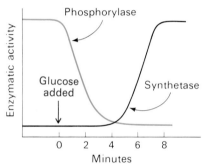

Figure 16-10
The infusion of glucose leads to the inactivation of phosphorylase, followed by the activation of glycogen synthetase. [After W. Stalmans, H. De Wulf, L. Hue, and H. G. Hers, *Eur. J. Biochem.* 41(1974):127.]

A VARIETY OF GENETICALLY DETERMINED GLYCOGEN STORAGE DISEASES ARE KNOWN

The first glycogen storage disease was described by Edgar von Gierke in 1929. A patient with this disease has a huge abdomen caused by a *massive enlargement of the liver*. There is a pronounced *hypoglycemia* between meals. Furthermore, the blood glucose level does not rise on administration of epinephrine and glucagon. An infant with this glycogen storage disease may have convulsions because of the low blood glucose level.

The enzymatic defect in von Gierke's disease was elucidated by the Coris in 1952. They found that *glucose 6-phosphatase was missing from the liver of a patient with this disease.* This was the first demonstration of an inherited deficiency of a liver enzyme. The liver glycogen is normal in structure but present in abnormally large amounts. The absence of glucose 6-phosphatase in the liver causes hypoglycemia because glucose cannot be formed from glucose 6-phosphate. This phosphorylated sugar does not leave the liver because it cannot traverse the plasma membrane. There is a compensatory increase in glycolysis in the liver, leading to a high level of lactate and pyruvate in the blood. Patients who have von Gierke's disease also have an increased dependence on fat metabolism.

A number of glycogen storage diseases have been characterized (Table 16-1). The Coris elucidated the biochemical defect in another glycogen storage disease (type III), which cannot be distinguished from von Gierke's disease (type I) by physical examination alone. In type III disease, the structure of liver and muscle glycogen

Table 16-1
Glycogen storage diseases

Type	Defective enzyme	Organ affected	Glycogen in the affected organ	Clinical features
I VON GIERKE'S DISEASE	Glucose 6-phosphatase	Liver and kidney	Increased amount; normal structure.	Massive enlargement of the liver. Failure to thrive. Severe hypoglycemia, ketosis, hyperuricemia, hyperlipemia.
II POMPE'S DISEASE	α-1,4-Glucosidase (lysosomal)	All organs	Massive increase in amount; normal structure.	Cardiorespiratory failure causes death, usually before age 2.
III CORI'S DISEASE	Amylo-1,6-glucosidase (debranching enzyme)	Muscle and liver	Increased amount; short outer branches.	Like Type I, but milder course.
IV ANDERSEN'S DISEASE	Branching enzyme (α-1,4 $\longrightarrow$ α-1,6)	Liver and spleen	Normal amount; very long outer branches.	Progressive cirrhosis of the liver. Liver failure causes death usually before age 2.
V McARDLE'S DISEASE	Phosphorylase	Muscle	Moderately increased amount; normal structure.	Limited ability to perform strenuous exercise because of painful muscle cramps. Otherwise patient is normal and well developed.
VI HERS' DISEASE	Phosphorylase	Liver	Increased amount.	Like Type I, but milder course.
VII	Phosphofructokinase	Muscle	Increased amount; normal structure.	Like Type V.
VIII	Phosphorylase kinase	Liver	Increased amount; normal structure.	Mild liver enlargement. Mild hypoglycemia.

Note: Types I through VII are inherited as autosomal recessives. Type VIII is sex-linked.

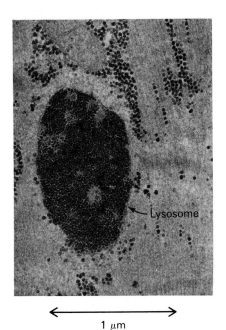

← 1 μm →

Figure 16-11
Electron micrograph of skeletal muscle from an infant with type II glycogen storage disease (Pompe's disease). The lysosomes are engorged with glycogen because of a deficiency in the α-1,4-glucosidase. This defect in the degradation of glycogen is confined to lysosomes. The amount of glycogen in the cytosol is normal. [From H. G. Hers and F. Van Hoof (eds.), *Lysosomes and Storage Diseases* (Academic Press, 1973), p. 205.]

is abnormal and the amount is markedly increased. Most striking, the outer branches of the glycogen are very short. *Patients having this type lack the debranching enzyme* (α-1,6-glucosidase), and so only the outermost branches of glycogen can be effectively utilized. Thus, only a small fraction of this abnormal glycogen is functionally active as an accessible store of glucose.

A defect in glycogen metabolism confined to muscle is found in McArdle's disease (type V). *Muscle phosphorylase activity is absent,* and the patient's capacity to perform strenuous exercise is limited because of painful muscle cramps. The patient is otherwise normal and well developed. Thus, effective utilization of muscle glycogen is not essential for life.

STARCH IS THE STORAGE POLYSACCHARIDE IN PLANTS

We turn now to the other common polysaccharides. The nutritional reservoir in plants is *starch*, of which there are two forms. *Amylose,* the unbranched type of starch, consists of glucose residues in α-1,4 linkage. *Amylopectin,* the branched form, has about one α-1,6 linkage per thirty α-1,4 linkages, and so it is like glycogen except for its lower degree of branching.

More than half of the carbohydrate ingested by humans is starch. Both amylopectin and amylose are rapidly hydrolyzed by *α-amylase,* which is secreted by the salivary glands and the pancreas. Alpha-amylase hydrolyzes internal α-1,4 linkages to yield *maltose, maltotriose,* and *α-dextrin.* Maltose consists of two glucose residues in α-1,4 linkage (Figure 16-12) and maltotriose consists of three such residues. Alpha-dextrin is made up of several glucose units joined by an α-1,6 linkage in addition to α-1,4 linkages. Maltose and maltotriose are hydrolyzed to glucose by *maltase,* whereas α-dextrin is hydrolyzed to glucose by *α-dextrinase.* There is a different kind of amylase in malt, called *β-amylase,* which hydrolyzes starch into maltose. Beta-amylase acts only on residues at the nonreducing terminus.

The other major polysaccharide of plants is *cellulose,* which serves a structural rather than a nutritional role. In fact, *cellulose is the most abundant organic compound in the biosphere,* containing more than half of all the organic carbon. Cellulose is an unbranched polymer of glucose residues joined by β-1,4 linkages. Mammals do not have cellulases and therefore cannot digest wood and vegetable fibers. However, some ruminants harbor cellulase-producing bacteria in their digestive tracts and thus can digest cellulose.

Dextran is another polysaccharide made up of glucose residues only, mainly in α-1,6 linkage. Occasional branches are formed by α-1,2, α-1,3, or α-1,4 linkages, depending on the species. Dextran is a storage polysaccharide in yeasts and bacteria. The exoskeletons of insects and crustacea contain *chitin,* which consists of *N*-acetylglucosamine in β-1,4 linkage. Thus, chitin is like cellulose except that the substituent at C-2 is an acetylated amino group rather than a hydroxyl group.

MALTOSE, SUCROSE, AND LACTOSE ARE THE COMMON DISACCHARIDES

The structures of the three common disaccharides are shown in Figure 16-12. As mentioned previously, *maltose* arises from the hydrolysis of starch and is then hydrolyzed to glucose by maltase. *Sucrose,* the common table sugar, is obtained commercially from cane or beet. The anomeric carbon atoms of a glucose and a fructose residue are in α-glycosidic linkage in sucrose. Consequently, sucrose has no reducing end group, in contrast with most other sugars. The hydrolysis of sucrose to glucose and fructose is catalyzed by *sucrase*. *Lactose* is the disaccharide of milk. It is not found elsewhere in appreciable amounts. Lactose is hydrolyzed to galactose and glucose by *lactase*. Lactase, sucrase, maltase, and α-dextrinase are bound to mucosal cells lining the small intestine.

UDP-sugars are the activated intermediates in the synthesis of sucrose and lactose, just as UDP-glucose is the glucosyl donor in the synthesis of glycogen. In fact, *nucleoside diphosphate sugars are the activated intermediates in nearly all syntheses of glycosidic linkages.* For example, whether cellulose is synthesized from adenosine diphosphoglucose (ADP-glucose), cytidine diphosphoglucose (CDP-glucose), or guanosine diphosphoglucose (GDP-glucose) depends on the kind of plant. *Sucrose* is synthesized by the transfer of glucose from UDP-glucose to fructose 6-phosphate to form sucrose 6-phosphate, which is then hydrolyzed to sucrose.

LACTOSE SYNTHESIS IS CONTROLLED BY A MODIFIER SUBUNIT

The control of lactose synthesis is especially interesting. Lactose synthetase consists of a catalytic subunit and a modifier subunit. The isolated catalytic subunit, called *galactosyl transferase,* is catalytically active in transferring galactose from UDP-galactose to *N*-acetylglucosamine to form *N*-acetyllactosamine.

$$\text{UDP-galactose} + N\text{-acetylglucosamine}$$

$$\Big\downarrow \text{catalytic subunit alone}$$

$$\text{UDP} + N\text{-acetyllactosamine}$$

The specificity of the catalytic subunit changes on binding *α-lactalbumin,* the modifier subunit. The resulting complex, called *lactose synthetase,* transfers galactose to glucose rather than to *N*-acetylglucosamine, which yields lactose.

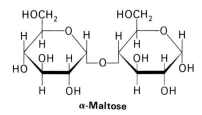

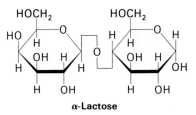

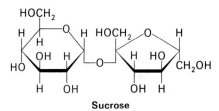

Figure 16-12
Structure of the common disaccharides: α-maltose, α-lactose, and sucrose. The α prefix refers to the configuration of the anomeric carbon atom at the reducing end of the disaccharide rather than to the configuration of the glycosidic linkage.

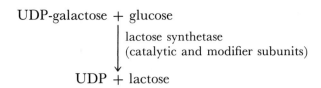

$$\text{UDP-galactose} + \text{glucose}$$

lactose synthetase
(catalytic and modifier subunits)

$$\text{UDP} + \text{lactose}$$

Galactosyl transferase is present in most tissues, where it participates in the synthesis of the carbohydrate moiety of glycoproteins. Lactose synthetase, in contrast, is found only in the mammary gland. During pregnancy, galactosyl transferase is synthesized and stored in the mammary gland, but there is little modifier subunit. At parturition, abrupt changes in the levels of some hormones elicit the synthesis of large amounts of the modifier subunit. The resulting lactose synthetase complex then forms large amounts of lactose. *The control of lactose synthesis illustrates that hormones may exert their physiological effects by switching enzymatic specificity.*

MOST ADULTS ARE INTOLERANT OF MILK BECAUSE THEY ARE DEFICIENT IN LACTASE

Nearly all infants and children are able to digest lactose. In contrast, a majority of adults in certain population groups are deficient in lactase, which makes them intolerant of milk. In a lactase-deficient adult, lactose accumulates in the lumen of the small intestine after ingestion of milk because there is no mechanism for the uptake of this disaccharide. The large osmotic effect of the unabsorbed lactose leads to an influx of fluid into the small intestine. Hence, the clinical symptoms of lactose intolerance are abdominal distention, nausea, cramping, pain, and a watery diarrhea. Lactase deficiency appears to be inherited as an autosomal recessive trait and is usually expressed in adolescence or young adulthood. The prevalence of lactase deficiency in human populations varies greatly. For example, 3% of Danes are deficient in lactase, compared with 97% of Thais. Human populations that do not consume milk in adulthood generally have a high incidence of lactase deficiency, which is also characteristic of other mammals. The capacity of humans to digest lactose in adulthood seems to have evolved since the domestication of cattle some ten thousand years ago.

ENTRY OF FRUCTOSE AND GALACTOSE INTO GLYCOLYSIS

Fructose comprises a significant portion of the carbohydrate in the diet. A typical daily intake is 100 grams, both as the free sugar and as a moiety of sucrose. A large part of the ingested fructose is metabolized by the liver, using the *fructose 1-phosphate pathway*. The first step in this pathway is the phosphorylation of fructose to fructose 1-phosphate by fructokinase. Fructose 1-phosphate is then split into

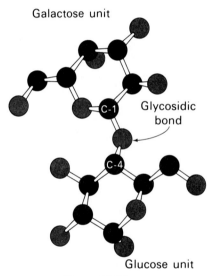

Galactose unit

Glycosidic
bond

Glucose unit

Figure 16-13
Model of lactose.

glyceraldehyde and dihydroxyacetone phosphate. This aldol cleavage is catalyzed by a specific fructose 1-phosphate aldolase. Glyceraldehyde is then phosphorylated to glyceraldehyde 3-phosphate by triose kinase so that it too can enter glycolysis. *Alternatively, fructose can be phosphorylated to fructose 6-phosphate by hexokinase.* However, the affinity of hexokinase for glucose is twenty times as high as it is for fructose. In the liver, the glucose level is high and so there is little phosphorylation of fructose to fructose 6-phosphate. However, in adipose tissue, the fructose level is much higher than that of glucose, and so the formation of fructose 6-phosphate is not competitively inhibited to an appreciable extent. Hence, most of the fructose in adipose tissue is metabolized through fructose 6-phosphate.

Galactose arises from the hydrolysis of lactose, the major carbohydrate of milk. Galactose is then converted into glucose 1-phosphate in four steps. The first reaction in the *galactose-glucose interconversion pathway* is the phosphorylation of galactose to galactose 1-phosphate by galactokinase.

$$\text{Galactose} + \text{ATP} \longrightarrow \text{galactose 1-phosphate} + \text{ADP} + \text{H}^+$$

UDP-galactose is then formed in an interchange reaction that is catalyzed by *galactose 1-phosphate uridyl transferase.*

Galactose 1-phosphate   UDP-glucose

UDP-galactose   Glucose 1-phosphate

Galactose is then epimerized to glucose while attached to UDP. The configuration of the hydroxyl group at C-4 is inverted by *UDP-galactose-4-epimerase,* which contains a tightly bound NAD$^+$. Fluorescence studies have shown that this NAD$^+$ is transiently reduced to NADH during catalysis. NAD$^+$ probably accepts the hydrogen atom attached to C-4 of the sugar, and a 4-keto sugar intermediate is formed. NADH then transfers its hydrogen to the other

$$\text{NAD}^+ \quad \begin{matrix} \text{HO} \\ | \diagup \\ \text{C} \\ | \diagdown \\ \text{H} \end{matrix} \quad \underset{\text{NADH}}{\overset{\text{H}^+}{\rightleftharpoons}} \quad \text{O}{=}\text{C}\diagdown \quad \rightleftharpoons \quad \text{NAD}^+ \quad \begin{matrix} \text{H} \\ | \diagup \\ \text{C} \\ | \diagdown \\ \text{HO} \end{matrix}$$

UDP-Galactose   **4-Keto**   **UDP-glucose**
(Only C-4 is shown.)   **intermediate**

side of C-4 to form the other epimer. The sum of the reactions catalyzed by galactokinase, the transferase, and the epimerase is

$$\text{Galactose} + \text{ATP} \longrightarrow \text{glucose 1-phosphate} + \text{ADP} + \text{H}^+$$

The reversible reaction catalyzed by the epimerase is important not only for the utilization of galactose as a fuel molecule. *The conversion of UDP-glucose into UDP-galactose is essential for the synthesis of galactosyl residues in complex polysaccharides and glycoproteins if the amount of galactose in the diet is inadequate to meet these needs.*

GALACTOSE IS HIGHLY TOXIC IF THE TRANSFERASE IS MISSING

The absence of galactose 1-phosphate uridyl transferase causes *galactosemia,* a severe disease that is inherited as an autosomal recessive. The metabolism of galactose in persons who have this disease is blocked at galactose 1-phosphate. Afflicted infants fail to thrive. Vomiting or diarrhea occurs when milk is consumed, and enlargement of the liver and jaundice are common. Furthermore, many galactosemics are mentally retarded. The blood galactose level is markedly elevated and galactose is found in the urine. In addition, the concentration of galactose 1-phosphate in red cells is elevated. The absence of the transferase in red blood cells is a definitive diagnostic criterion.

Galactosemia is treated by the exclusion of galactose from the diet. A galactose-free diet leads to a striking regression of virtually all of the clinical symptoms, except for mental retardation, which may not be reversible. Continued galactose intake may lead to death in some patients. *The damage in galactosemia is caused by an accumulation of toxic substances, rather than by the absence of an essential compound.* Patients are able to synthesize UDP-galactose from UDP-glucose because their epimerase activity is normal. One of the toxic substances is *galactitol,* which is formed by reduction of galactose. A high level of galactitol leads to the formation of cataracts in the lenses. Galactose 1-phosphate may be a precursor of some of the other toxic agents, or it may itself be toxic at high levels.

CH₂OH
|
H—C—OH
|
HO—C—H
|
HO—C—H
|
H—C—OH
|
CH₂OH

Galactitol

SUMMARY

Glycogen, a readily mobilized fuel store, is a branched polymer of glucose residues. Most of the glucose units in glycogen are linked by α-1,4-glycosidic bonds. At about every tenth residue, a branch is created by an α-1,6-glycosidic bond. Glycogen is present in large amounts in muscle and in the liver, where it is stored in the cytoplasm in the form of hydrated granules. Most of the glycogen molecule is degraded to glucose 1-phosphate by the action of phosphorylase. The glycosidic linkage between C-1 of a terminal residue and C-4 of the adjacent one is split by orthophosphate to give glucose 1-phosphate, which can be reversibly converted into glucose 6-phosphate. Branch points are degraded by the concerted action of two other enzymes, a transferase and an α-1,6-glucosidase. The latter enzyme (also known as the debranching enzyme) catalyzes the hydrolysis of α-1,6 linkages, which yields free glucose. Glycogen is synthesized by a different pathway. UDP-glucose, the activated intermediate in glycogen synthesis, is formed from glucose 1-phosphate and UTP. Glycogen synthetase catalyzes the transfer of glucose from UDP-glucose to the C-4 hydroxyl group of a terminal residue in the growing glycogen molecule. A branching enzyme converts some of the α-1,4 linkages into α-1,6 linkages.

Glycogen synthesis and degradation are coordinately controlled by an amplifying cascade of reactions. Glycogen synthetase is inactive when phosphorylase is active, and vice versa. Epinephrine and glucagon stimulate glycogen breakdown and inhibit its synthesis by increasing the intracellular level of cyclic AMP, which activates a protein kinase. Glycogen synthetase is then phosphorylated, which inactivates it. Phosphorylase is also phosphorylated but in this case the enzyme is converted into a more active form. The phosphoryl groups on these enzymes can be removed by the same phosphatase. Glycogen synthetase and phosphorylase are also regulated by noncovalent allosteric interactions. In fact, phosphorylase is a key part of the glucose-sensing system of liver cells.

The nutritional reservoir in plants is starch, which is like glycogen except that it is less branched. Starch is hydrolyzed to maltose, which consists of two glucose residues joined by an α-1,4 linkage. In turn, maltose is hydrolyzed to glucose. The other common disaccharides are sucrose, which is hydrolyzed to glucose and fructose, and lactose, which is hydrolyzed to galactose and glucose. Galactose and glucose are interconvertible when linked to UDP. UDP-galactose is the activated intermediate in the synthesis of lactose.

SELECTED READINGS

WHERE TO START

Cori, C. F., and Cori, G. T., 1947. Polysaccharide phosphorylase. In *Nobel Lectures: Physiology or Medicine (1942–1962)*, pp. 186–206. American Elsevier (1964).

Leloir, L. F., 1971. Two decades of research on the biosynthesis of saccharides. *Science* 172:1299–1302.

Kretchmer, N., 1972. Lactose and lactase. *Sci. Amer.* 227(4):70–78. [An interesting account of the population genetics of lactase deficiency. Available as *Sci. Amer.* Offprint 1259.]

REVIEWS AND BOOKS

Hers, H. G., 1976. The control of glycogen metabolism in the liver. *Ann. Rev. Biochem.* 45:167–190.

Fletterick, R. J., and Madsen, N. B., 1980. The structures and related functions of phosphorylase *a. Ann. Rev. Biochem.* 49:31–61.

Krebs, E. G., and Preiss, J., 1975. Regulatory mechanisms in glycogen metabolism. *In* Whelan, W. J., (ed.), *Biochemistry of Carbohydrates,* pp. 337–389. Butterworths.

Dickens, F., Randle, P. J., and Whelan, W. J., (eds.), 1968. *Carbohydrate Metabolism and Its Disorders,* vols. 1 and 2. Academic Press. [A valuable collection of review articles. Volume 1 contains reviews on glycogen metabolism in liver and muscle. Volume 2 contains reviews on glycogen-storage diseases and other disorders of glycogen metabolism.]

Newsholme, E. A., and Start, C., 1973. *Regulation in Metabolism.* Academic Press. [Chapter 4 deals with the regulation of glycogen metabolism.]

Stalmans, W., and Hers, H. G., 1973. Glycogen synthesis from UDPG. *In* Boyer, P. D., (ed.), *The Enzymes* (3rd ed.), vol. 9, pp. 310–361. Academic Press.

STRUCTURE OF PHOSPHORYLASE

Madsen, N. B., Kasvinsky, P. J., and Fletterick, R. J., 1978. Allosteric transitions of phosphorylase *a* and the regulation of glycogen metabolism. *J. Biol. Chem.* 253:9097–9101. [Proposed mechanism for the control of glycogen metabolism by glucose, based on x-ray crystallographic studies of allosteric transitions.]

Weber, I. T., Johnson, L. N., Wilson, K. S., Yeates, D. G. R., Wild, D. L., and Jenkins, J. A., 1978. Crystallographic studies on the activity of glycogen phosphorylase *b. Nature* 274:433–437.

INHERITED DISEASES

Howell, R. R., 1978. The glycogen storage diseases. *In* Stanbury, J. B., Wyngaarden, J. B., and Fredrickson, D. S., (eds.), *The Metabolic Basis of Inherited Disease* (4th ed.), pp. 137–159. McGraw-Hill.

Segal, S., 1978. Disorders of galactose metabolism. *In* Stanbury, J. B., Wyngaarden, J. B., and Fredrickson, D. S., (eds.), *The Metabolic Basis of Inherited Disease* (4th ed.), pp. 160–181.

PROBLEMS

1. Write a balanced equation for the formation of glycogen from galactose.

2. Write a balanced equation for the formation of glucose from fructose in the liver.

3. A sample of glycogen from a patient with liver disease is incubated with orthophosphate, phosphorylase, the transferase, and the debranching enzyme. The ratio of glucose 1-phosphate to glucose formed in this mixture is 100. What is the most likely enzymatic deficiency in this patient?

4. Suggest an explanation for the fact that the amount of glycogen in type-I glycogen-storage disease (von Gierke's disease) is increased.

5. Crystals of phosphorylase *a* grown in the presence of glucose shatter when a substrate such as glucose 1-phosphate is added. Why?

6. Muscle phosphorylase kinase is a very large enzyme (1,300 kdal) with the subunit structure $A_4B_4C_8$. Phosphorylation of the B subunits of this enzyme by the cyclic-AMP–dependent kinase is rapid, resulting in its activation. The A subunits of this enzyme are then phosphorylated at a slow rate, which makes both the A and B subunits susceptible to the action of a specific phosphatase. What might be the functional significance of the slow phosphorylation of the A subunits?

CHAPTER 17

FATTY ACID METABOLISM

We turn now from the metabolism of carbohydrates to that of fatty acids, a class of compounds containing a long hydrocarbon chain and a terminal carboxylate group. Fatty acids have two major physiological roles. First, *they are building blocks of phospholipids and glycolipids*. These amphipathic molecules are important components of biological membranes, as was discussed in Chapter 10. Second, *fatty acids are fuel molecules*. They are stored as *triacylglycerols*, which are uncharged esters of glycerol. Triacylglycerols are also called neutral fats or triglycerides.

$$H_3C-(CH_2)_7-\overset{\overset{\displaystyle H}{|}}{C}=\overset{\overset{\displaystyle H}{|}}{C}-(CH_2)_7-\overset{\overset{\displaystyle O}{||}}{C}-O-\overset{\displaystyle CH}{\underset{\displaystyle |}{\overset{\displaystyle |}{}}}$$

A triacylglycerol

NOMENCLATURE OF FATTY ACIDS

A brief comment on the nomenclature of fatty acids is appropriate before dealing with fatty acid metabolism. The systematic name for a fatty acid is derived from the name of its parent hydrocarbon by the substitution of *oic* for the final *e*. For example, the C_{18} saturated

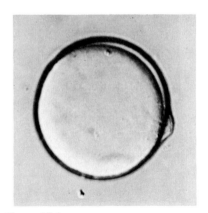

Figure 17-1
Photomicrograph of a fat cell. A large globule of fat is surrounded by a thin rim of cytoplasm and a bulging nucleus. [Courtesy of Dr. Pedro Cuatrecasas.]

fatty acid is called octadecanoic acid because the parent hydrocarbon is octadecane. A C_{18} fatty acid with one double bond is called octadec*enoic* acid; with two double bonds, octadeca*dienoic* acid; and with three double bonds, octadecatrienoic acid. The symbol $18:0$ denotes a C_{18} fatty acid with no double bonds, whereas $18:2$ signifies that there are two double bonds. Fatty acid carbon atoms are numbered starting at the carboxyl terminus:

$$H_3\underset{\omega}{C}-(CH_2)_n-\underset{\beta}{\overset{3}{C}}-\underset{\alpha}{\overset{2}{C}}-\overset{1}{C}\underset{OH}{\overset{O}{<}}$$

Carbon atoms 2 and 3 are often referred to as α and β, respectively. The methyl carbon atom at the distal end of the chain is called the ω carbon. The position of a double bond is represented by the symbol Δ followed by a superscript number. For example, *cis*-Δ^9 means that there is a *cis* double bond between carbon atoms 9 and 10; *trans*-Δ^2 means that there is a *trans* double bond between carbon atoms 2 and 3. Fatty acids are ionized at physiological pH, and so it is appropriate to refer to them according to their carboxylate form: for example, palmitate or hexadecanoate.

FATTY ACIDS VARY IN CHAIN LENGTH AND DEGREE OF UNSATURATION

Fatty acids in biological systems (Table 17-1) usually contain an even number of carbon atoms, typically between 14 and 24. The 16-

Table 17-1
Some naturally occurring fatty acids in animals

Number of carbons	Number of double bonds	Common name	Systematic name	Formula
12	0	Laurate	*n*-Dodecanoate	$CH_3(CH_2)_{10}COO^-$
14	0	Myristate	*n*-Tetradecanoate	$CH_3(CH_2)_{12}COO^-$
16	0	Palmitate	*n*-Hexadecanoate	$CH_3(CH_2)_{14}COO^-$
18	0	Stearate	*n*-Octadecanoate	$CH_3(CH_2)_{16}COO^-$
20	0	Arachidate	*n*-Eicosanoate	$CH_3(CH_2)_{18}COO^-$
22	0	Behenate	*n*-Docosanoate	$CH_3(CH_2)_{20}COO^-$
24	0	Lignocerate	*n*-Tetracosanoate	$CH_3(CH_2)_{22}COO^-$
16	1	Palmitoleate	*cis*-Δ^9-Hexadecenoate	$CH_3(CH_2)_5CH=CH(CH_2)_7COO^-$
18	1	Oleate	*cis*-Δ^9-Octadecenoate	$CH_3(CH_2)_7CH=CH(CH_2)_7COO^-$
18	2	Linoleate	*cis, cis*-Δ^9,Δ^{12}-Octadecadienoate	$CH_3(CH_2)_4(CH=CHCH_2)_2(CH_2)_6COO^-$
18	3	Linolenate	all *cis*-$\Delta^9,\Delta^{12},\Delta^{15}$-Octadecatrienoate	$CH_3CH_2(CH=CHCH_2)_3(CH_2)_6COO^-$
20	4	Arachidonate	all *cis*-$\Delta^5,\Delta^8,\Delta^{11},\Delta^{14}$-Eicosatetraenoate	$CH_3(CH_2)_4(CH=CHCH_2)_4(CH_2)_2COO^-$

and 18-carbon fatty acids are most common. The hydrocarbon chain is almost invariably unbranched in animal fatty acids. The alkyl chain may be saturated or it may contain one or more double bonds. The configuration of the double bonds in most unsaturated fatty acids is *cis*. The double bonds in polyunsaturated fatty acids are separated by at least one methylene group.

The properties of fatty acids and of lipids derived from them are markedly dependent on their chain length and on the degree to which they are unsaturated. Unsaturated fatty acids have a lower melting point than saturated fatty acids of the same length. For example, the melting point of stearic acid is 69.6°C, whereas that of oleic acid (which contains one *cis* double bond) is 13.4°C. The melting points of polyunsaturated fatty acids of the C_{18} series are even lower. Chain length also affects the melting point, as illustrated by the fact that the melting temperature of palmitic acid (C_{16}) is 6.5° lower than that of stearic acid (C_{18}). *Thus, short chain length and unsaturation enhance the fluidity of fatty acids and of their derivatives.* The significance of membrane fluidity has already been discussed (Chapter 10).

TRIACYLGLYCEROLS ARE HIGHLY CONCENTRATED ENERGY STORES

Triacylglycerols are highly concentrated stores of metabolic energy because they are *reduced* and *anhydrous*. The yield from the complete oxidation of fatty acids is about 9 kcal/g, in contrast with about 4 kcal/g for carbohydrates and proteins. The basis of this large difference in caloric yield is that fatty acids are much more highly reduced. Furthermore, triacylglycerols are very nonpolar and so they are stored in a nearly anhydrous form, whereas proteins and carbohydrates are much more polar and hence more highly hydrated. In fact, a gram of dry glycogen binds about two grams of water. *Consequently, a gram of nearly anhydrous fat stores more than six times as much energy as a gram of hydrated glycogen,* which is the reason that triacylglycerols rather than glycogen were selected in evolution as the major energy reservoir. Consider a typical 70-kg man who has fuel reserves of 100,000 kcal in triacylglycerols, 25,000 kcal in protein (mostly in muscle), 600 kcal in glycogen, and 40 kcal in glucose. Triacylglycerols constitute about 11 kg of his total body weight. If this amount of energy were stored in glycogen, his total body weight would be 55 kg greater.

In mammals, the major site of accumulation of triacylglycerols is the cytoplasm of *adipose cells* (*fat cells*). Droplets of triacylglycerol coalesce to form a large globule, which may occupy most of the cell volume. Adipose cells are specialized for the synthesis and storage of triacylglycerols and for their mobilization into fuel molecules that are transported to other tissues by the blood.

Figure 17-2
Scanning electron micrograph of a fat cell. [From *Tissues and Organs* by Richard G. Kessel and Randy H. Kardon. Copyright © 1979. W. H. Freeman and Company.]

TRIACYLGLYCEROLS ARE HYDROLYZED BY CYCLIC-AMP–REGULATED LIPASES

The initial event in the utilization of fat as an energy source is the hydrolysis of triacylglycerol by lipases.

Triacylglycerol   Glycerol   Fatty acids

The activity of adipose-cell lipase is regulated by hormones. Epinephrine, norepinephrine, glucagon, and adrenocorticotropic hormone stimulate the adenylate cyclase of adipose cells. The increased level of cyclic adenosine monophosphate (cyclic AMP) then stimulates a protein kinase, which activates the lipase by phosphorylating it. Thus, epinephrine, norepinephrine, glucagon, and adrenocorticotropic hormone cause lipolysis. *Cyclic AMP is a second messenger in the activation of lipolysis in adipose cells,* which is analogous to its role in the activation of glycogen breakdown (Chapter 16). In contrast, insulin inhibits lipolysis.

Glycerol formed by lipolysis is phosphorylated and reduced to dihydroxyacetone phosphate, which in turn is isomerized to glyceraldehyde 3-phosphate. This intermediate is on both the glycolytic and the gluconeogenic pathways. Hence, glycerol can be converted into pyruvate or glucose in the liver, which contains these enzymes. The reverse process can occur by the reduction of dihydroxyacetone phosphate to glycerol 3-phosphate. Hydrolysis by a phosphatase then gives glycerol. Thus, glycerol and glycolytic intermediates are readily interconvertible.

Glycerol   L-Glycerol 3-phosphate   Dihydroxyacetone phosphate

FATTY ACIDS ARE DEGRADED BY THE SEQUENTIAL REMOVAL OF TWO-CARBON UNITS

In 1904, Franz Knoop made a critical contribution to the elucidation of the mechanism of fatty acid oxidation. He fed dogs straight-chain fatty acids in which the ω-carbon atom was joined to a phenyl group. Knoop found that the urine of these dogs contained a derivative of phenylacetic acid when they were fed phenylbutyrate. In contrast, a derivative of benzoic acid was formed when they were fed phenylpropionate. In fact, phenylacetic acid was produced whenever a fatty acid containing an even number of carbon atoms was fed, whereas benzoic acid was formed whenever a fatty acid containing an odd number was fed (Figure 17-3). Knoop deduced from these findings that *fatty acids are degraded by oxidation at the β-carbon.* These experiments are a landmark in biochemistry because they were the first to use a synthetic label to elucidate reaction mechanisms. Deuterium and radioisotopes came into biochemistry several decades later.

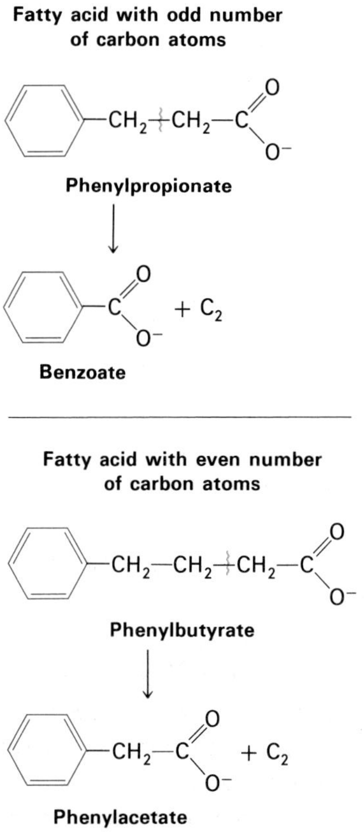

FATTY ACIDS ARE LINKED TO COENZYME A BEFORE THEY ARE OXIDIZED

Eugene Kennedy and Albert Lehninger showed in 1949 that fatty acids are oxidized in mitochondria. Subsequent work demonstrated that they are activated prior to their entry into the mitochondrial matrix. Adenosine triphosphate (ATP) drives the formation of a thioester linkage between the carboxyl group of a fatty acid and the sulfhydryl group of CoA. This activation reaction occurs on the mitochondrial membrane, where it is catalyzed by *acyl CoA synthetase* (also called fatty acid thiokinase).

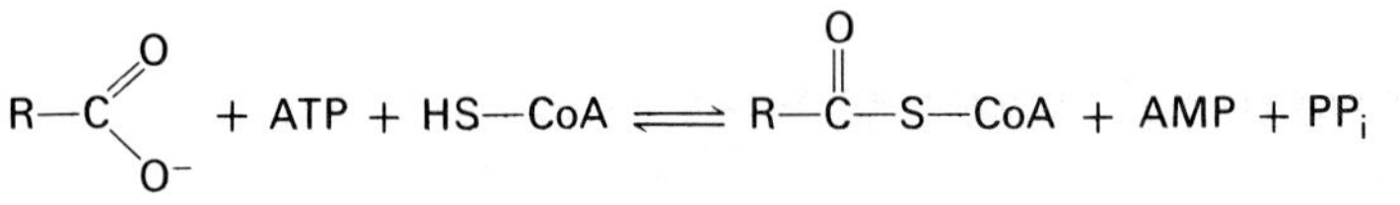

$$R-\overset{O}{\underset{O^-}{C}} \;+\; ATP \;+\; HS-CoA \;\rightleftharpoons\; R-\overset{O}{\overset{\|}{C}}-S-CoA \;+\; AMP \;+\; PP_i$$

Paul Berg showed that the activation of a fatty acid occurs in two steps. First, the fatty acid reacts with ATP to form an *acyl adenylate.* In this mixed anhydride, the carboxyl group of a fatty acid is bonded to the phosphoryl group of AMP. The other two phosphoryl groups of the ATP substrate are released as pyrophosphate. The sulfhydryl group of CoA then attacks the acyl adenylate, which is tightly bound to the enzyme, to form acyl CoA and AMP.

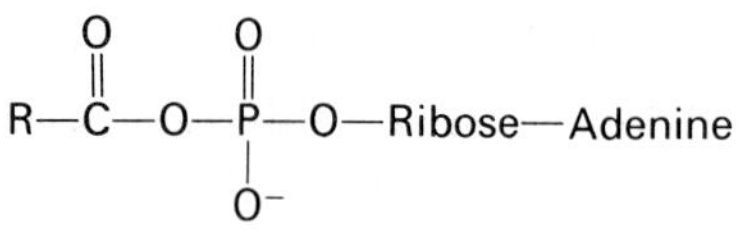

$$\underset{\text{Fatty acid}}{R-\overset{O}{\underset{O^-}{C}}} \;+\; ATP \;\rightleftharpoons\; \underset{\text{Acyl adenylate}}{R-\overset{O}{\overset{\|}{C}}-AMP} \;+\; PP_i$$

$$\underset{}{R-\overset{O}{\overset{\|}{C}}-AMP} \;+\; HS-CoA \;\rightleftharpoons\; \underset{\text{Acyl CoA}}{R-\overset{O}{\overset{\|}{C}}-S-CoA} \;+\; AMP$$

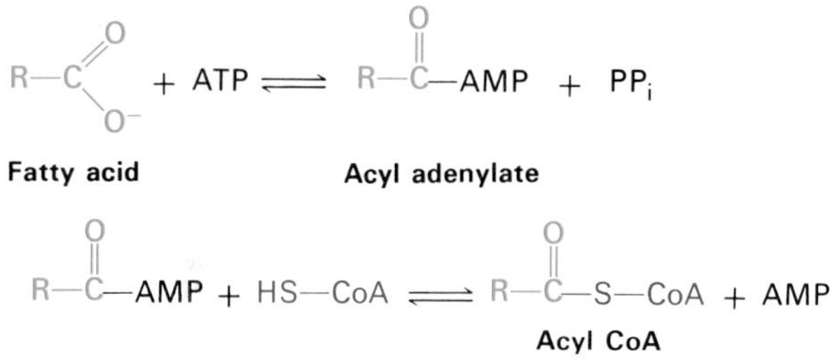

These partial reactions are freely reversible. In fact, the equilibrium constant for the sum of these reactions is close to 1.

$$R-COO^- \;+\; CoA \;+\; ATP \;\rightleftharpoons\; acyl\ CoA \;+\; AMP \;+\; PP_i$$

Fatty acid with odd number of carbon atoms

Phenylpropionate

Benzoate

Fatty acid with even number of carbon atoms

Phenylbutyrate

Phenylacetate

Figure 17-3
Knoop's experiment showing that fatty acids are degraded by the removal of two-carbon units.

$$R-\overset{O}{\overset{\|}{C}}-O-\overset{O}{\underset{O^-}{\overset{\|}{P}}}-O-Ribose-Adenine$$

**Acyl adenylate
(Acyl-AMP)**

One high-energy bond is broken (between PP_i and AMP) and one high-energy bond is formed (the thioester linkage in acyl CoA). How is this reaction driven forward? The answer is that pyrophosphate is rapidly hydrolyzed by a pyrophosphatase.

$$R\text{---}COO^- + CoA + ATP + H_2O \longrightarrow$$
$$\text{acyl CoA} + AMP + 2\,P_i + 2\,H^+$$

This makes the overall reaction irreversible because two high-energy bonds are consumed, whereas only one is formed. We see here another example of a recurring theme in biochemistry: *many biosynthetic reactions are made irreversible by the hydrolysis of inorganic pyrophosphate.*

There is another recurring motif in this activation reaction. The occurrence of an enzyme-bound acyl adenylate intermediate is not unique to the synthesis of acyl CoA. *Acyl adenylates are frequently formed when carboxyl groups are activated in biochemical reactions.* For example, amino acids are activated for protein synthesis by a similar mechanism.

CARNITINE CARRIES LONG-CHAIN ACTIVATED FATTY ACIDS INTO THE MITOCHONDRIAL MATRIX

Fatty acids are activated on the outer mitochondrial membrane, whereas they are oxidized in the mitochondrial matrix. Long-chain acyl CoA molecules do not readily traverse the inner mitochondrial membrane, and so a special transport mechanism is needed. Activated long-chain fatty acids are carried across the inner mitochondrial membrane by *carnitine.* The acyl group is transferred from the sulfur atom of CoA to the hydroxyl group of carnitine to form *acyl carnitine,* which diffuses across the inner mitochondrial membrane. On the matrix side of this membrane, the acyl group is transferred back to CoA, which is thermodynamically feasible because the *O*-acyl link in carnitine has a high group-transfer potential. These transacylation reactions are catalyzed by *fatty acyl CoA: carnitine fatty acid transferase.*

A defect in the transferase or a deficiency of carnitine might be expected to impair the oxidation of long-chain fatty acids. Such a disorder has in fact been found in identical twins who had had aching muscle cramps since early childhood. The aches were precipitated by fasting, exercise, or a high-fat diet; fatty acid oxidation is the major energy-yielding process in these three states. The enzymes of glycolysis and glycogenolysis were found to be normal.

Lipolysis of triacylglycerols was normal, as evidenced by a rise in the concentration of unesterified fatty acids in the plasma after fasting. Assay of a muscle biopsy showed that the long-chain acyl CoA synthetase was fully active. Furthermore, medium-chain (C_8 and C_{10}) fatty acids were normally metabolized. It is known that carnitine is not required for the permeation of medium-chain acyl CoAs into the mitochondrial matrix. This case report demonstrates in a striking way that *the impaired flow of a metabolite from one compartment of a cell to another can cause disease.*

ACETYL CoA, NADH, AND FADH₂ ARE GENERATED IN EACH ROUND OF FATTY ACID OXIDATION

A saturated acyl CoA is degraded by a recurring sequence of four reactions: oxidation linked to flavin adenine dinucleotide (FAD), hydration, oxidation linked to NAD^+, and thiolysis by CoA (Figure 17-4). The fatty acyl chain is shortened by two carbon atoms as a result of these reactions, and $FADH_2$, NADH, and acetyl CoA are generated. David Green, Severo Ochoa, and Feodor Lynen made important contributions to the elucidation of this series of reactions, which is called the *β-oxidation pathway.*

The first reaction in each round of degradation is the *oxidation* of acyl CoA by an acyl CoA dehydrogenase to give an enoyl CoA with a *trans* double bond between C-2 and C-3.

$$\text{Acyl CoA} + \text{E-FAD} \longrightarrow trans\text{-}\Delta^2\text{-enoyl CoA} + \text{E-FADH}_2$$

It is interesting to note that the dehydrogenation of acyl CoA is very similar to the dehydrogenation of succinate in the citric acid cycle. In fact, the first three reactions in each round of fatty acid degradation closely resemble the last steps in the citric acid cycle:

Acyl CoA $\longrightarrow$ enol CoA $\longrightarrow$ hydroxyacyl CoA $\longrightarrow$ ketoacyl CoA
Succinate $\longrightarrow$ fumarate $\longrightarrow$ malate $\longrightarrow$ oxaloacetate

The next step is the *hydration* of the double bond between C-2 and C-3 by enoyl CoA hydratase.

$$trans\text{-}\Delta^2\text{-Enoyl CoA} + \text{H}_2\text{O} \rightleftharpoons \text{L-3-hydroxyacyl CoA}$$

The hydration of the enoyl CoA is stereospecific, as are the hydrations of fumarate and aconitate. Only the L-isomer of 3-hydroxyacyl CoA is formed when the *trans*-Δ^2 double bond is hydrated. The enzyme also hydrates a *cis*-Δ^2 double bond, but the product then is the D-isomer. We will return to this point shortly in a discussion of the oxidation of unsaturated fatty acids.

The hydration of enoyl CoA is the prelude to the second *oxidation* reaction, which converts the hydroxyl group at C-3 into a keto group and generates NADH. This oxidation is catalyzed by L-3-hydroxyacyl CoA dehydrogenase, which is absolutely specific for the L-isomer of the hydroxyacyl substrate.

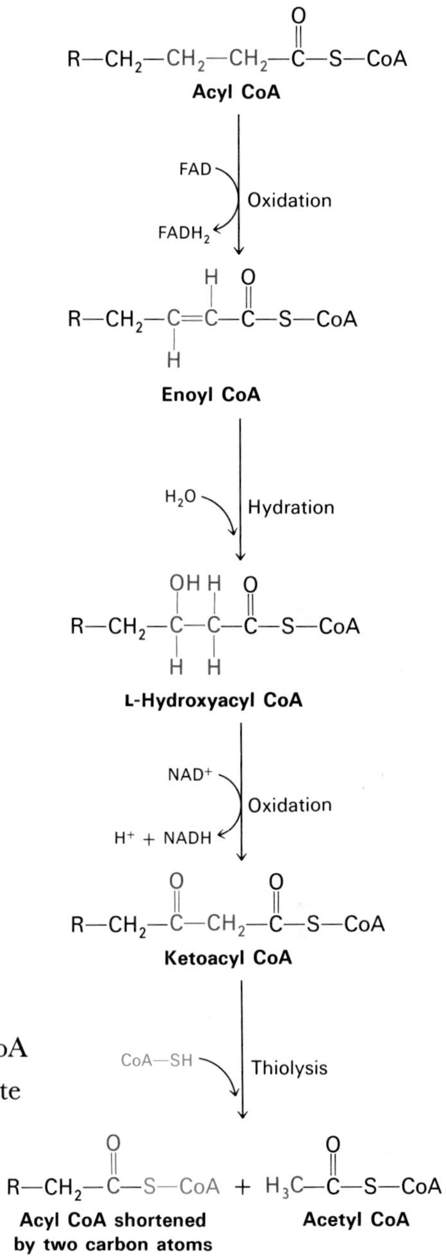

Figure 17-4
Reaction sequence in the degradation of fatty acids: oxidation, hydration, oxidation, and thiolysis.

$$\text{L-3-Hydroxyacyl CoA} + \text{NAD}^+ \rightleftharpoons$$
$$\text{3-ketoacyl CoA} + \text{NADH} + \text{H}^+$$

The preceding reactions have oxidized the methylene group at C-3 to a keto group. The final step is the *cleavage* of 3-ketoacyl CoA by the thiol group of a second molecule of CoA, which yields acetyl CoA and an acyl CoA shortened by two carbon atoms. This thiolytic cleavage is catalyzed by β-ketothiolase.

$$\underset{\text{(\textit{n} carbons)}}{\text{3-Ketoacyl CoA}} + \text{HS—CoA} \rightleftharpoons \text{acetyl CoA} + \underset{\text{(\textit{n} − 2 carbons)}}{\text{acyl CoA}}$$

The shortened acyl CoA then undergoes another cycle of oxidation, starting with the reaction catalyzed by acyl CoA dehydrogenase (Figure 17-5). Beta-ketothiolase, hydroxyacyl dehydrogenase, and enoyl CoA hydratase have broad specificity with respect to the length of the acyl group.

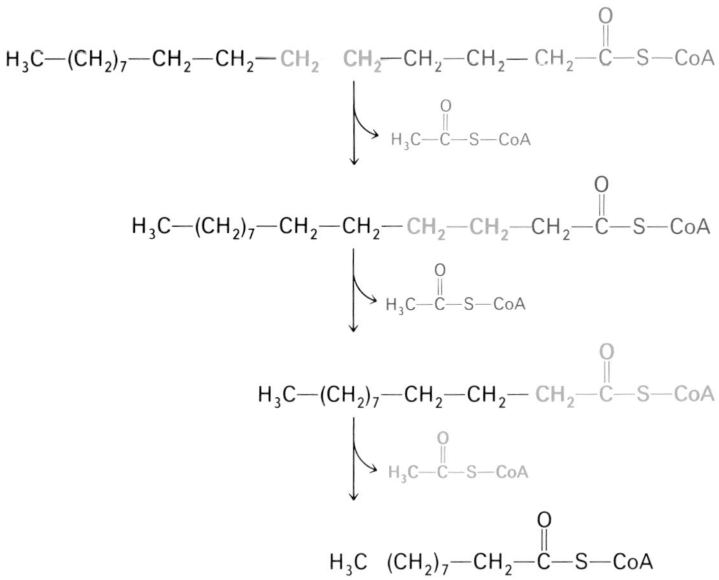

Figure 17-5
First three rounds in the degradation of palmitate. Two-carbon units are sequentially removed from the carboxyl end of the fatty acid.

Table 17-2
Principal reactions in fatty acid oxidation

Step	Reaction	Enzyme
1	Fatty acid + CoA + ATP $\rightleftharpoons$ acyl CoA + AMP + PP$_i$	Acyl CoA synthetase (also called fatty acid thiokinase; fatty acid: CoA ligase [AMP])
2	Carnitine + acyl CoA $\rightleftharpoons$ acyl carnitine + CoA	Fatty acyl CoA: carnitine fatty acid transferase
3	Acyl CoA + E-FAD $\longrightarrow$ $trans$-Δ^2-enoyl CoA + E-FADH$_2$	Acyl CoA dehydrogenases (several enzymes having different chain-length specificity)
4	$trans$-Δ^2-Enoyl CoA + H$_2$O $\rightleftharpoons$ L-3-hydroxyacyl CoA	Enoyl CoA hydratase (also called crotonase or 3-hydroxyacyl CoA hydrolyase)
5	L-3-Hydroxyacyl CoA + NAD$^+$ $\rightleftharpoons$ 3-ketoacyl CoA + NADH + H$^+$	L-3-Hydroxyacyl CoA dehydrogenase
6	3-Ketoacyl CoA + CoA $\rightleftharpoons$ acetyl CoA + acyl CoA (shortened by C$_2$)	β-Ketothiolase (also called thiolase)

We can calculate the energy yield derived from the oxidation of a fatty acid. In each reaction cycle, an acyl CoA is shortened by two carbons, and one $FADH_2$, NADH, and acetyl CoA are formed.

$$C_n\text{-acyl CoA} + FAD + NAD^+ + H_2O + CoA \longrightarrow$$
$$C_{n-2}\text{-acyl CoA} + FADH_2 + NADH + \text{acetyl CoA} + H^+$$

The degradation of palmitoyl CoA (C_{16}-acyl CoA) requires seven reaction cycles. In the seventh cycle, the C_4-ketoacyl CoA is thiolyzed to two molecules of acetyl CoA. Hence, the stoichiometry of oxidation of palmitoyl CoA is:

$$\text{Palmitoyl CoA} + 7\,FAD + 7\,NAD^+ + 7\,CoA + 7\,H_2O \longrightarrow$$
$$8\text{ acetyl CoA} + 7\,FADH_2 + 7\,NADH + 7\,H^+$$

Three ATP are generated when each of these NADH is oxidized by the respiratory chain, whereas two ATP are formed for each $FADH_2$, because their electrons enter the chain at the level of coenzyme Q. Recall that the oxidation of acetyl CoA by the citric acid cycle yields 12 ATP. Hence, the number of ATP formed in the oxidation of palmitoyl CoA is 14 from the 7 $FADH_2$, 21 from the 7 NADH, and 96 from the 8 molecules of acetyl CoA, which gives a total of 131. Two high-energy phosphate bonds are consumed in the activation of palmitate, in which ATP is split into AMP and $2\,P_i$. Thus, *the net yield from the complete oxidation of palmitate is 129 ATP.*

The efficiency of energy conservation in fatty acid oxidation can be estimated from the number of ATP formed and from the free energy of oxidation of palmitic acid to CO_2 and H_2O, as determined by calorimetry. The standard free energy of hydrolysis of 129 ATP is -940 kcal (129×-7.3 kcal). The standard free energy of oxidation of palmitic acid is $-2,340$ kcal. Hence, *the efficiency of energy conservation in fatty acid oxidation under standard conditions is about 40%,* a value similar to those of glycolysis, the citric acid cycle, and oxidative phosphorylation.

AN ISOMERASE AND AN EPIMERASE ARE REQUIRED FOR THE OXIDATION OF UNSATURATED FATTY ACIDS

We turn now to the oxidation of unsaturated fatty acids. Many of the reactions are the same as those for saturated fatty acids. In fact, only two additional enzymes—an isomerase and an epimerase—are needed to degrade a wide range of unsaturated fatty acids.

Consider the oxidation of palmitoleate. This C_{16} unsaturated fatty acid, which has one double bond between C-9 and C-10, is activated and transported across the inner mitochondrial membrane in the same way as palmitate. Palmitoleoyl CoA then undergoes three cycles of degradation, which are carried out by the same enzymes as in the oxidation of saturated fatty acids. However, the *cis*-Δ^3-enoyl CoA formed in the third round is not a substrate for

$$H_3C-(CH_2)_5-\overset{\overset{H}{|}}{\underset{4}{C}}=\overset{\overset{H}{|}}{\underset{3}{C}}-\overset{}{\underset{2}{CH_2}}-\overset{\overset{O}{\|}}{\underset{1}{C}}-S-CoA$$

cis-Δ^3-Enoyl CoA

Isomerase

$$H_3C-(CH_2)_5-\overset{}{\underset{4}{CH_2}}-\overset{\overset{H}{|}}{\underset{3}{C}}=\overset{\underset{H}{|}}{\underset{2}{C}}-\overset{\overset{O}{\|}}{\underset{1}{C}}-S-CoA$$

trans-Δ^2-Enoyl CoA

acyl CoA dehydrogenase. The presence of a double bond between C-3 and C-4 prevents the formation of another double bond between C-2 and C-3. This impasse is resolved by a new reaction that shifts the position and configuration of the cis-Δ^3 double bond. *An isomerase converts this double bond into a* trans-Δ^2 *double bond.* The subsequent reactions are those of the saturated fatty acid oxidation pathway, in which the trans-Δ^2-enoyl CoA is a regular substrate.

A second accessory enzyme is needed for the oxidation of polyunsaturated fatty acids. The activated C_{18} unsaturated fatty acid with cis-Δ^6 and cis-Δ^9 double bonds undergoes two rounds of degradation by the saturated fatty acid oxidation pathway. Then cis-Δ^2 Δ^5-enoyl CoA is hydrated by enoyl CoA hydratase, the same enzyme that hydrates trans-Δ^2 double bonds in the saturated pathway. However, hydration of a cis-Δ^2 double bond yields the D-isomer of 3-hydroxyacyl CoA, which is not a substrate for L-3-hydroxyacyl CoA dehydrogenase. This hurdle is overcome by an *epimerase that inverts the configuration of the hydroxyl group at C-3.*

$$H_3C-(CH_2)_7-\overset{H}{C}=\overset{H}{C}-CH_2-\overset{\overset{H}{|}}{\underset{\underset{OH}{|}}{C}}-CH_2-\overset{\overset{O}{\|}}{C}-S-CoA$$

D-3-Hydroxy-cis-Δ^5-enoyl CoA

Epimerase

$$H_3C-(CH_2)_7-\overset{H}{C}=\overset{H}{C}-CH_2-\overset{\overset{OH}{|}}{\underset{\underset{H}{|}}{C}}-CH_2-\overset{\overset{O}{\|}}{C}-S-CoA$$

L-3-Hydroxy-cis-Δ^5-enoyl CoA

ODD-CHAIN FATTY ACIDS YIELD PROPIONYL COENZYME A IN THE FINAL THIOLYSIS STEP

Fatty acids having an odd number of carbon atoms are minor species. They are oxidized in the same way as fatty acids having an even number, except that propionyl CoA and acetyl CoA, rather than two molecules of acetyl CoA, are produced in the final round of degradation. The activated three-carbon unit in propionyl CoA enters the citric acid cycle after it is converted into succinyl CoA. The pathway from propionyl CoA to succinyl CoA will be discussed in the next chapter (p. 418) because propionyl CoA is also formed in the oxidation of several amino acids.

$$H_3C-CH_2-\overset{\overset{O}{\|}}{C}-S-CoA$$

Propionyl CoA

KETONE BODIES ARE FORMED FROM ACETYL COENZYME A IF FAT BREAKDOWN PREDOMINATES

The acetyl CoA formed in fatty acid oxidation enters the citric acid cycle if fat and carbohydrate degradation are appropriately balanced. The molecular basis of the adage that *fats burn in the flame of carbohydrates* is now evident. The entry of acetyl CoA depends on the availability of oxaloacetate for the formation of citrate. However, if fat breakdown predominates, acetyl CoA undergoes a different fate. The reason is that the concentration of oxaloacetate is lowered if carbohydrate is not available or properly utilized. In fasting or in diabetes, oxaloacetate is used to form glucose and is thus unavailable for condensation with acetyl CoA. Under these conditions, acetyl CoA is diverted to the formation of acetoacetate and D-3-hydroxybutyrate. Acetoacetate, D-3-hydroxybutyrate, and acetone are sometimes referred to as *ketone bodies*.

Acetoacetate is formed from acetyl CoA in three steps (Figure 17-6). Two molecules of acetyl CoA condense to form acetoacetyl

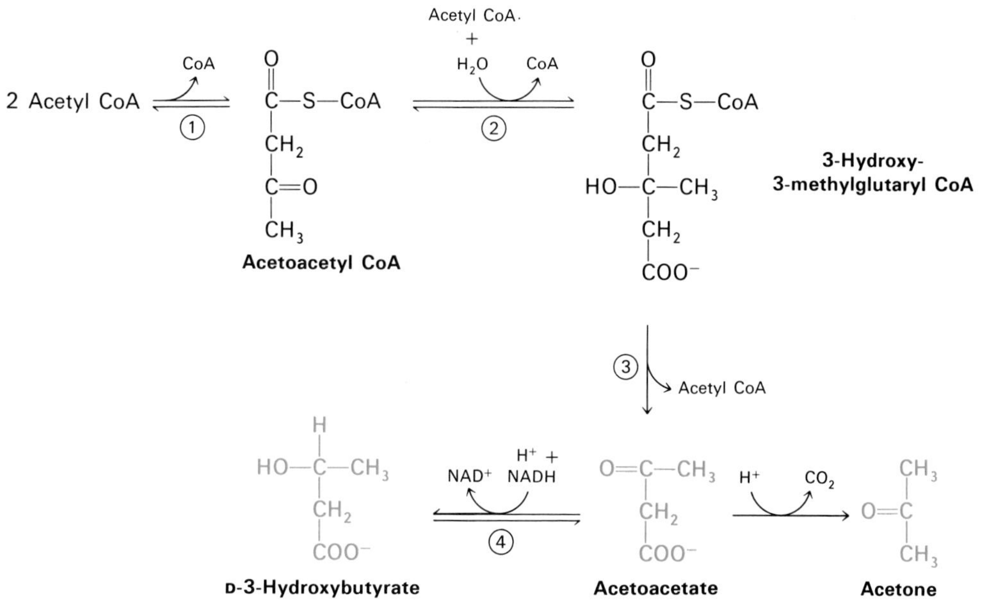

Figure 17-6
Formation of acetoacetate, D-3-hydroxybutyrate, and acetone from acetyl CoA. Enzymes catalyzing these reactions are: (1) 3-ketothiolase, (2) hydroxymethylglutaryl CoA synthetase, (3) hydroxymethylglutaryl CoA cleavage enzyme, and (4) D-3-hydroxybutyrate dehydrogenase. Acetoacetate spontaneously decarboxylates to form acetone.

CoA. This reaction, which is catalyzed by thiolase, is a reversal of the thiolysis step in the oxidation of fatty acids. Acetoacetyl CoA then reacts with acetyl CoA and water to give 3-hydroxy-3-methylglutaryl CoA and CoA. The unfavorable equilibrium in the formation of acetoacetyl CoA is compensated for by the favorable equilibrium of this reaction, which is due to the hydrolysis of a thioester linkage. 3-Hydroxy-3-methylglutaryl CoA is then cleaved to acetyl CoA and acetoacetate. The sum of these reactions is

$$2 \text{ Acetyl CoA} + \text{H}_2\text{O} \longrightarrow \text{acetoacetate} + 2 \text{ CoA} + \text{H}^+$$

3-Hydroxybutyrate is formed by the reduction of acetoacetate in the mitochondrial matrix. The ratio of hydroxybutyrate to acetoacetate depends on the NADH/NAD$^+$ ratio inside mitochondria.

Acetoacetate also undergoes a slow, spontaneous decarboxylation to acetone. The odor of acetone may be detected in the breath of a person who has a high level of acetoacetate in the blood.

ACETOACETATE IS A MAJOR FUEL IN SOME TISSUES

The major site of production of acetoacetate and 3-hydroxybutyrate is the liver. These substances diffuse from the liver mitochondria into the blood and are transported to peripheral tissues. Until a few years ago, these ketone bodies were regarded as degradation products of little physiological value. However, the studies of George Cahill and others have revealed that these derivatives of acetyl CoA are important molecules in energy metabolism. *Acetoacetate and 3-hydroxybutyrate are normal fuels of respiration and are quantitatively important as sources of energy.* Indeed, heart muscle and the renal cortex use acetoacetate in preference to glucose. In contrast, glucose is the major fuel for the brain in well-nourished persons on a balanced diet. However, the brain adapts to the utilization of acetoacetate during starvation and diabetes. In prolonged starvation, 75% of the fuel needs of the brain are met by acetoacetate.

Acetoacetate can be activated by the transfer of CoA from succinyl CoA in a reaction catalyzed by a specific CoA transferase. Acetoacetyl CoA is then cleaved by thiolase to yield two molecules of acetyl CoA, which can then enter the citric acid cycle. The liver can supply acetoacetate to other organs because it lacks the specific CoA transferase.

Acetoacetate can be regarded as a water-soluble, transportable form of acetyl units. Fatty acids are released by adipose tissue and converted into acetyl units by the liver, which then exports them as acetoacetate. As might be expected, acetoacetate also has a regulatory role. *High levels of acetoacetate in the blood signify an abundance of acetyl units and lead to a decrease in the rate of lipolysis in adipose tissue.*

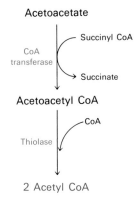

Figure 17-7
Utilization of acetoacetate as a fuel. Acetoacetate can be converted into two molecules of acetyl CoA, which then enter the citric acid cycle.

ANIMALS CANNOT CONVERT FATTY ACIDS INTO GLUCOSE

It is important to note that *animals are unable to convert fatty acids into glucose.* Specifically, acetyl CoA cannot be converted into pyruvate or oxaloacetate in animals. The two carbon atoms of the acetyl group of acetyl CoA enter the citric acid cycle, but two carbon atoms leave the cycle in the decarboxylations catalyzed by isocitrate dehydrogenase and α-ketoglutarate dehydrogenase. Consequently, oxaloacetate is regenerated but it is not formed de novo when the acetyl unit of acetyl CoA is oxidized by the citric acid cycle. In contrast, plants have two additional enzymes enabling them to convert the carbon atoms of acetyl CoA into glucose (see problem 6 on p. 406).

Fatty acid synthesis does not occur by a reversal of the degradative pathway. Rather, it consists of a new set of reactions, which once again exemplifies the principle that *synthetic and degradative pathways in biological systems are usually distinct.* Some important features of the pathway for the biosynthesis of fatty acids are:

1. Synthesis takes place in the *cytosol,* in contrast with degradation, which occurs in the mitochondrial matrix.

2. The intermediates in fatty acid synthesis are covalently linked to the sulfhydryl groups of an *acyl carrier protein* (ACP), whereas the intermediates in fatty acid breakdown are bonded to coenzyme A.

3. Many of the enzymes of fatty acid synthesis in higher organisms are organized into a *multienzyme complex* called the *fatty acid synthetase.* In contrast, the degradative enzymes do not seem to be associated.

4. The growing fatty acid chain is elongated by the *sequential addition of two-carbon units* derived from acetyl CoA. The activated donor of two-carbon units in the elongation step is *malonyl-ACP.* The elongation reaction is driven by the release of CO_2.

5. The reductant in fatty acid synthesis is *NADPH.*

6. Elongation by the fatty acid synthetase complex stops upon formation of *palmitate* (C_{16}). Further elongation and the insertion of double bonds are carried out by other enzyme systems.

THE FORMATION OF MALONYL COENZYME A IS THE COMMITTED STEP IN FATTY ACID SYNTHESIS

Salih Wakil's finding that bicarbonate is required for fatty acid biosynthesis was an important clue in the elucidation of this process. In fact, fatty acid synthesis starts with the carboxylation of acetyl CoA to *malonyl CoA.* This irreversible reaction is the committed step in fatty acid synthesis.

$$H_3C-\overset{O}{\overset{\|}{C}}-S-CoA + ATP + HCO_3^- \longrightarrow ^{-O}\overset{O}{\overset{\diagdown}{C}}-CH_2-\overset{O}{\overset{\|}{C}}-S-CoA + ADP + P_i + H^+$$

Acetyl CoA **Malonyl CoA**

The synthesis of malonyl CoA is catalyzed by *acetyl CoA carboxylase,* which contains a biotin prosthetic group. The carboxyl group of biotin is covalently attached to the ε-amino group of a lysine residue, as in pyruvate carboxylase (p. 348). Another similarity between acetyl CoA carboxylase and pyruvate carboxylase is that acetyl CoA is carboxylated in two stages. First, a carboxybiotin intermediate is formed at the expense of an ATP. The activated CO_2 group in this intermediate is then transferred to acetyl CoA to form malonyl CoA.

$$\text{Biotin-enzyme} + \text{ATP} + \text{HCO}_3^- \Longrightarrow$$
$$CO_2 \sim \text{biotin-enzyme} + \text{ADP} + \text{P}_i$$

$$CO_2 \sim \text{biotin-enzyme} + \text{acetyl CoA} \longrightarrow$$
$$\text{malonyl CoA} + \text{biotin-enzyme}$$

Substrates are bound to this enzyme and products are released in a specific sequence (Figure 17-8). Acetyl CoA carboxylase exemplifies

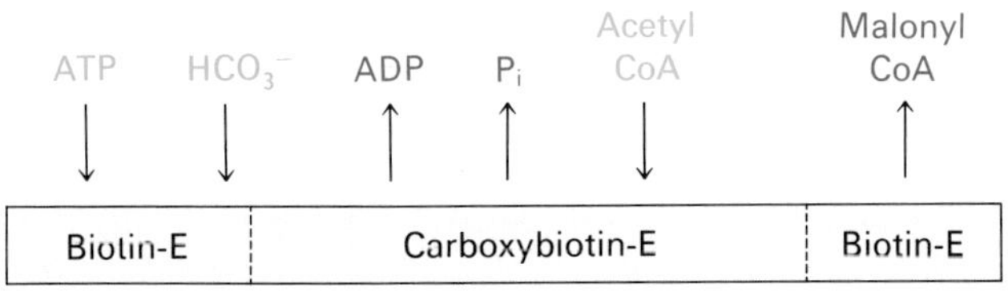

Figure 17-8
Reaction sequence of acetyl CoA carboxylase.

				Acetyl CoA		Malonyl CoA
ATP	HCO₃⁻	ADP	P_i			
↓	↓	↑	↑	↓		↑
Biotin-E		Carboxybiotin-E				Biotin-E

a *ping-pong reaction mechanism* in which one or more products are released before all the substrates are bound.

Acetyl CoA carboxylase from *E. coli* has been separated into subunits that catalyze partial reactions. Biotin is covalently attached to a small protein (22 kdal) called the *biotin carboxyl carrier protein.* The carboxylation of the biotin unit in this carrier protein is catalyzed by *biotin carboxylase,* a second subunit. The third component of the system is a *transcarboxylase,* which catalyzes the transfer of the activated CO_2 unit from carboxybiotin to acetyl CoA. The length and

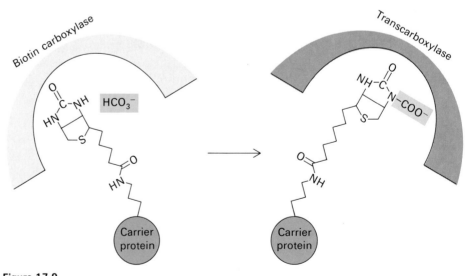

Figure 17-9
Schematic diagram showing the proposed movement of the biotin prosthetic group from the site where it acquires a carboxyl group from HCO_3^- to the site where it donates this group to acetyl CoA.

flexibility of the link between biotin and its carrier protein enable the activated carboxyl group to move from one active site to another in the enzyme complex (Figure 17-9), as in pyruvate carboxylase (p. 349).

In eucaryotes, acetyl CoA carboxylase exists as an enzymatically inactive protomer (450 kdal) or as an active, filamentous polymer (Figure 17-10). This interconversion is allosterically regulated, as might be expected because acetyl CoA carboxylase catalyzes the first committed step in fatty acid synthesis. *The key allosteric activator is citrate,* which shifts the enzyme equilibrium to the active filamentous form. The orientation of biotin with respect to its substrates is optimized in the filamentous form. In contrast, palmitoyl CoA shifts the equilibrium to the inactive protomer form. Thus, *palmitoyl CoA, the end product, inhibits the first committed step in the synthesis of fatty acids.* The control of acetyl CoA carboxylase in *E. coli* is quite different from that in eucaryotes. In bacteria, fatty acids are primarily precursors of phospholipids rather than of storage fuels, and so they are controlled differently. Citrate has no effect on the *E. coli* enzyme. Rather, the activity of its transcarboxylase component is regulated by guanine nucleotides that coordinate fatty acid synthesis with bacterial growth and division.

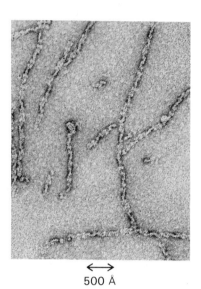

←——→
500 Å

Figure 17-10
Electron micrograph of the enzymatically active filamentous form of acetyl CoA carboxylase from chicken liver. [Courtesy of Dr. M. Daniel Lane.]

INTERMEDIATES IN FATTY ACID SYNTHESIS ARE ATTACHED TO AN ACP

P. Roy Vagelos discovered that the intermediates in fatty acid synthesis are linked to an acyl carrier protein. Specifically, they are linked to the sulfhydryl terminus of a phosphopantetheine group (Figure 17-11). In the degradation of fatty acids, this unit is part of CoA, whereas, in synthesis, it is attached to a serine residue of the ACP. This single polypeptide chain of 77 residues can be regarded as a giant prosthetic group, a "macro CoA."

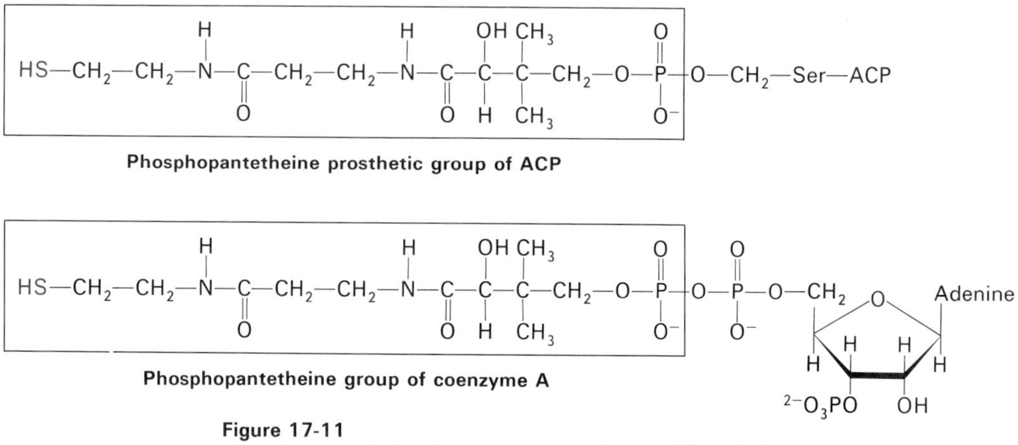

Phosphopantetheine prosthetic group of ACP

Phosphopantetheine group of coenzyme A

Figure 17-11
Phosphopantetheine is the reactive unit of ACP and CoA.

THE ELONGATION CYCLE IN FATTY ACID SYNTHESIS

The enzyme system that catalyzes the synthesis of saturated long-chain fatty acids from acetyl CoA, malonyl CoA, and NADPH is called the *fatty acid synthetase*. It is present in higher organisms as a multienzyme complex. In contrast, the constituent enzymes of bacterial fatty acid synthetases are dissociated when the cells are disrupted. The availability of these isolated enzymes has facilitated the elucidation of the steps in fatty acid synthesis (Table 17-3). In

Table 17-3
Principal reactions in fatty acid synthesis

Step	Reaction	Enzyme
1	Acetyl CoA + HCO_3^- + ATP $\longrightarrow$ malonyl CoA + ADP + P_i + H^+	Acetyl CoA carboxylase
2	Acetyl CoA + ACP $\rightleftharpoons$ acetyl-ACP + CoA	Acetyl transacylase
3	Malonyl CoA + ACP $\rightleftharpoons$ malonyl-ACP + CoA	Malonyl transacylase
4	Acetyl-ACP + malonyl-ACP $\longrightarrow$ acetoacetyl-ACP + ACP + CO_2	Acyl-malonyl-ACP condensing enzyme
5	Acetoacetyl-ACP + NADPH + H^+ $\rightleftharpoons$ D-3-hydroxybutyryl-ACP + $NADP^+$	β-Ketoacyl-ACP-reductase
6	D-3-Hydroxybutyryl-ACP $\rightleftharpoons$ crotonyl-ACP + H_2O	3-Hydroxyacyl-ACP-dehydratase
7	Crotonyl-ACP + NADPH + H^+ $\longrightarrow$ butyryl-ACP + $NADP^+$	Enoyl-ACP reductase

fact, the reactions leading to fatty acid synthesis in higher organisms are very much like those of bacteria.

The elongation phase of fatty acid synthesis starts with the formation of acetyl-ACP and malonyl-ACP. *Acetyl transacylase* and *malonyl transacylase* catalyze these reactions.

$$\text{Acetyl CoA} + \text{ACP} \rightleftharpoons \text{acetyl-ACP} + \text{CoA}$$
$$\text{Malonyl CoA} + \text{ACP} \rightleftharpoons \text{malonyl-ACP} + \text{CoA}$$

Malonyl transacylase is highly specific, whereas acetyl transacylase can transfer acyl groups other than the acetyl unit, though at a much slower rate. Fatty acids with an odd number of carbon atoms are synthesized starting with propionyl-ACP, which is formed from propionyl CoA by acetyl transacetylase.

Acetyl-ACP and malonyl-ACP react to form acetoacetyl-ACP. This condensation reaction is catalyzed by the *acyl-malonyl-ACP condensing enzyme.*

$$\text{Acetyl-ACP} + \text{malonyl-ACP} \longrightarrow$$
$$\text{acetoacetyl-ACP} + \text{ACP} + CO_2$$

In the condensation reaction, a four-carbon unit is formed from a two-carbon unit and a three-carbon unit, and CO_2 is released. Why isn't the four-carbon unit formed from two two-carbon units? In other words, why are the reactants acetyl-ACP and malonyl-ACP rather than two molecules of acetyl-ACP? The answer is that the equilibrium is highly unfavorable for the synthesis of acetoacetyl-ACP from two molecules of acetyl-ACP. In contrast, *the equilibrium is favorable if malonyl-ACP is a reactant because its decarboxylation contrib-*

utes *a substantial decrease in free energy*. In effect, the condensation reaction is driven by ATP, though ATP does not directly participate in the condensation reaction. Rather, ATP is used to form an energy-rich substrate in the carboxylation of acetyl CoA to malonyl CoA. The free energy stored in malonyl CoA in the carboxylation reaction is released in the decarboxylation accompanying the formation of acetoacetyl-ACP. Although HCO_3^- is required for fatty acid synthesis, its carbon atom does not appear in the product. *Rather, all of the carbon atoms of fatty acids containing an even number of them are derived from acetyl CoA.*

The next three steps in fatty acid synthesis reduce the keto group at C-3 to a methylene group (Figure 17-12). First, acetoacetyl-ACP is reduced to D-3-hydroxybutyryl-ACP. This reaction differs from the corresponding one in fatty acid degradation in two respects: (1) the D- rather than the L-epimer is formed; and (2) NADPH is the reducing agent, whereas NAD^+ is the oxidizing agent in β-oxidation. This difference exemplifies the general principle that *NADPH is consumed in biosynthetic reactions, whereas NADH is generated in energy-yielding reactions.* Then D-3-hydroxybutyryl-ACP is *dehydrated* to form crotonyl-ACP, which is a *trans-Δ^2-enoyl-ACP*. The final step in the cycle *reduces* crotonyl-ACP to butyryl-ACP.

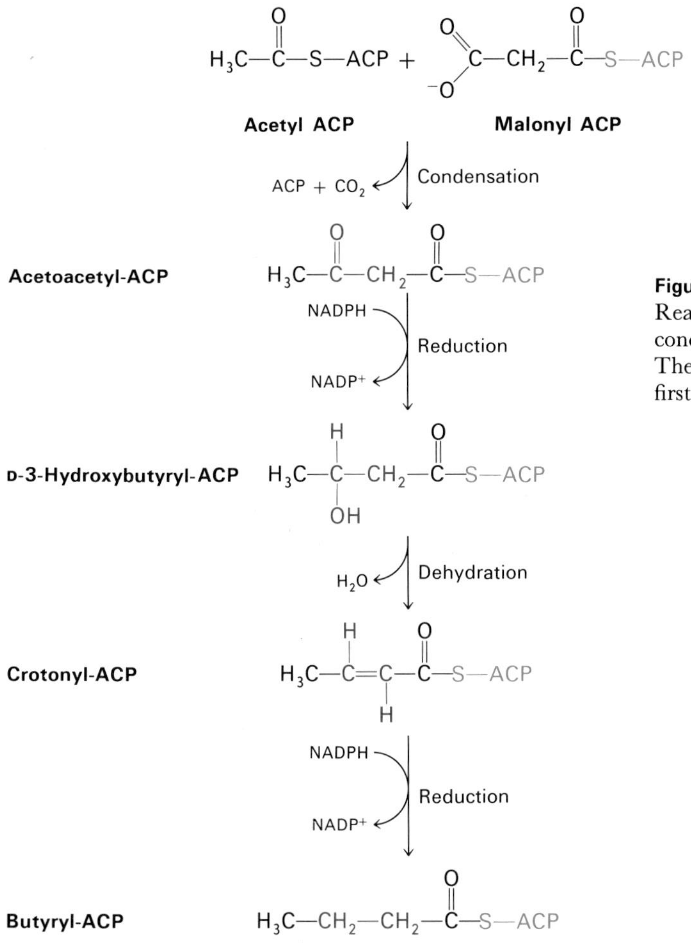

Figure 17-12
Reaction sequence in the synthesis of fatty acids: condensation, reduction, dehydration, and reduction. The intermediates shown here are produced in the first round of synthesis.

NADPH is again the reductant, whereas FAD^+ is the oxidant in the corresponding reaction in β-oxidation. These last three reactions—a reduction, a dehydration, and a second reduction—convert acetoacetyl-ACP into butyryl-ACP, which completes the first elongation cycle.

In the second round of fatty acid synthesis, butyryl-ACP condenses with malonyl-ACP to form a C_6-β-ketoacyl-ACP. This reaction is like the one in the first round, in which acetyl-ACP condenses with malonyl-ACP to form a C_4-β-ketoacyl-ACP. Reduction, dehydration, and a second reduction convert the C_6-β-ketoacyl-ACP into a C_6-acyl-ACP, which is ready for a third round of elongation. The elongation cycles continue until C_{16}-acyl-ACP is formed. This intermediate is not a substrate for the condensing enzyme. Rather, it is hydrolyzed to yield palmitate and ACP.

STOICHIOMETRY OF FATTY ACID SYNTHESIS

The stoichiometry of the synthesis of palmitate is

$$\text{Acetyl CoA} + 7 \text{ malonyl CoA} + 14 \text{ NADPH} + 7 \text{ H}^+ \longrightarrow$$
$$\text{palmitate} + 7 \text{ CO}_2 + 14 \text{ NADP}^+ + 8 \text{ CoA} + 6 \text{ H}_2\text{O}$$

The equation for the synthesis of the malonyl CoA used in the above reaction is

$$7 \text{ Acetyl CoA} + 7 \text{ CO}_2 + 7 \text{ ATP} \longrightarrow$$
$$7 \text{ malonyl CoA} + 7 \text{ ADP} + 7 \text{ P}_i + 7 \text{ H}^+$$

Hence, the overall stoichiometry for the synthesis of palmitate is

$$8 \text{ Acetyl CoA} + 7 \text{ ATP} + 14 \text{ NADPH} \longrightarrow$$
$$\text{palmitate} + 14 \text{ NADP}^+ + 8 \text{ CoA} + 6 \text{ H}_2\text{O} + 7 \text{ ADP} + 7 \text{ P}_i$$

FATTY ACIDS ARE SYNTHESIZED BY A MULTIENZYME COMPLEX IN EUCARYOTES

The fatty acid synthetases of eucaryotes, in contrast with those of bacteria, are well-defined multienzyme complexes. The one from yeast has a mass of 2,300 kdal and appears in electron micrographs as an ellipsoid with a length of 250 Å and a cross-sectional diameter of 210 Å (Figure 17-13). *It consists of only two kinds of polypeptide chains, each encoded by a single gene.* Subunit A (185 kdal) contains the acyl carrier protein, the condensing enzyme, and the β-ketoacyl reductase, whereas subunit B (175 kdal) contains acetyl transacylase, malonyl transacylase, β-hydroxyacyl dehydratase, and enoyl reductase. Mammalian fatty acid synthetase (400 kdal) also consists of two kinds of subunits, which are like the ones in yeast. In fact, *many eucaryotic multienzyme complexes consist of multifunctional proteins, in which*

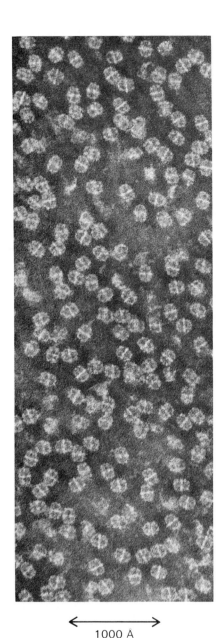

← 1000 Å →

Figure 17-13
Electron micrograph of the fatty acid synthetase complex from yeast. [Courtesy of Dr. Felix Wieland and Dr. Elmar A. Siess.]

different enzymes are linked covalently in a single polypeptide chain. An advantage of this arrangement is that the synthesis of different enzymes is coordinated. Furthermore, a multienzyme complex consisting of covalently joined enzymes is more stable than one formed by noncovalent associations.

Lynen proposed that the growing fatty acid chain is transferred from the ACP to the condensing enzyme and back again in each round of elongation. The first *translocation* makes room for the incoming malonyl unit, and the second occurs in the condensation step. It is interesting to note that there are analogous translocations in protein synthesis.

The flexibility and 20-Å maximal length of the phosphopantetheinyl moiety appear to be critical for the function of the multienzyme complex, because they enable the growing fatty acid chain to come into close contact with the active site of each enzyme in the complex. The enzyme subunits need not undergo large structural rearrangements to interact with the substrate. Instead, the substrate is on a long, flexible arm that can reach each of the active sites. Recall that biotin and lipoamide are also on long, flexible arms in their multienzyme complexes. The organized structure of the fatty acid synthetases of yeast and higher organisms enhances the efficiency of the overall process because intermediates are directly transferred from one active site to the next. The reactants are not diluted in the cytosol. Moreover, they do not have to find each other by random diffusion. Another advantage of such a multienzyme complex is that the covalently bound intermediates are sequestered and protected from competing reactions.

CITRATE CARRIES ACETYL GROUPS FROM MITOCHONDRIA TO THE CYTOSOL FOR FATTY ACID SYNTHESIS

The synthesis of palmitate requires the input of 8 molecules of acetyl CoA, 14 NADPH, and 7 ATP. Fatty acids are synthesized in the cytosol, whereas acetyl CoA is formed from pyruvate in mitochondria. Hence, acetyl CoA must be transferred from mitochondria to the cytosol for fatty acid synthesis. However, mitochondria are not readily permeable to acetyl CoA. Recall that carnitine carries only long-chain fatty acids. *The barrier to acetyl CoA is bypassed by citrate, which carries acetyl groups across the inner mitochondrial membrane.* Citrate is formed in the mitochondrial matrix by the condensation of acetyl CoA and oxaloacetate. It then diffuses to the cytosol, where it is cleaved by *citrate lyase:*

$$\text{Citrate} + \text{ATP} + \text{CoA} \longrightarrow \text{acetyl CoA} + \text{ADP} + \text{P}_i + \text{oxaloacetate}$$

Thus, acetyl CoA and oxaloacetate are transferred from mitochondria to the cytosol at the expense of an ATP.

Lyases— Enzymes catalyzing the cleavage of C—C, C—O, or C—N bonds by elimination. A double bond is formed in these reactions.

SOURCES OF NADPH FOR FATTY ACID SYNTHESIS

Oxaloacetate formed in the transfer of acetyl groups to the cytosol must now be returned to the mitochondria. The inner mitochondrial membrane is impermeable to oxaloacetate. Hence, a series of bypass reactions are needed. Most important, these reactions generate much of the NADPH needed for fatty acid synthesis. First, oxaloacetate is reduced to malate by NADH. This reaction is catalyzed by a *malate dehydrogenase* in the cytosol.

$$\text{Oxaloacetate} + \text{NADH} + \text{H}^+ \rightleftharpoons \text{malate} + \text{NAD}^+$$

Second, malate is oxidatively decarboxylated by an *NADP⁺-linked malate enzyme*. This reaction has not been mentioned before.

$$\text{Malate} + \text{NADP}^+ \longrightarrow \text{pyruvate} + \text{CO}_2 + \text{NADPH}$$

The pyruvate formed in this reaction readily diffuses into mitochondria, where it is carboxylated to oxaloacetate by pyruvate carboxylase.

$$\text{Pyruvate} + \text{CO}_2 + \text{ATP} + \text{H}_2\text{O} \longrightarrow$$
$$\text{oxaloacetate} + \text{ADP} + \text{P}_i + 2\,\text{H}^+$$

The sum of these three reactions is

$$\text{NADP}^+ + \text{NADH} + \text{ATP} + \text{H}_2\text{O} \longrightarrow$$
$$\text{NADPH} + \text{NAD}^+ + \text{ADP} + \text{P}_i + \text{H}^+$$

Thus, *one NADPH is generated for each acetyl CoA that is transferred from the mitochondria to the cytosol.* Hence, eight NADPH are formed when eight molecules of acetyl CoA are transferred to the cytosol for the synthesis of palmitate. *The additional six NADPH required for this process come from the pentose phosphate pathway.*

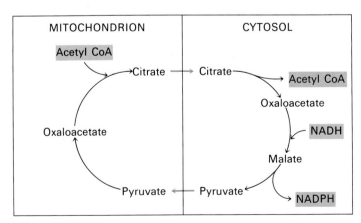

Figure 17-14
Acetyl CoA is transferred from mitochondria to the cytosol, and NADH is concomitantly converted into NADPH by this series of reactions.

ELONGATION AND UNSATURATION OF FATTY ACIDS ARE CARRIED OUT BY ACCESSORY ENZYME SYSTEMS

The major product of the fatty acid synthetase is palmitate. In eucaryotes, longer fatty acids are formed by elongation reactions catalyzed by enzyme systems associated with the *endoplasmic reticulum membrane* (also known as *microsomal systems*). Two-carbon units are added to the carboxyl end of both saturated and unsaturated fatty acids. Microsomal systems also introduce double bonds into long-chain acyl CoAs. For example, in the conversion of stearoyl CoA into oleoyl CoA, a *cis-*Δ^9 double bond is inserted by an oxidase that employs *molecular oxygen* and *NADH* (or *NADPH*):

$$\text{Stearoyl CoA} + \text{NADH} + \text{H}^+ + \text{O}_2 \longrightarrow$$
$$\text{oleoyl CoA} + \text{NAD}^+ + 2\,\text{H}_2\text{O}$$

A variety of unsaturated fatty acids can be formed from oleate by a combination of elongation and unsaturation reactions. For example, oleate can be elongated to a 20:1 *cis-*Δ^{11} fatty acid. Alternatively, a second double bond can be inserted to yield an 18:2 *cis-*Δ^6, Δ^9 fatty acid. Similarly, palmitate (16:0) can be oxidized to palmitoleate (16:1 *cis-*Δ^9), which can then be elongated to *cis*-vaccenate (18:1 *cis-*Δ^{11}).

Mammals lack the enzymes to introduce double bonds at carbon atoms beyond C-9 in the fatty acid chain. Hence, mammals cannot synthesize linoleate (18:2 *cis-*Δ^9, Δ^{12}) and linolenate (18:3 *cis-*Δ^9, Δ^{12}, Δ^{15}). *Linoleate and linolenate are the two essential fatty acids.* The term "essential" means that they must be supplied in the diet because they are required by the organism and cannot be endogenously synthesized. Linoleate and linolenate furnished by the diet are the starting points for the synthesis of a variety of the other unsaturated fatty acids. Unsaturated fatty acids in mammals are derived from either palmitoleate (16:1), oleate (18:1), linoleate (18:2), or linolenate (18:3). The number of methylene carbons between the ω-CH$_3$ group of an unsaturated fatty acid and the nearest double bond identifies its precursor:

Precursor	Formula
Linolenate	$CH_3-(CH_2)_1-CH=CH-R$
Linoleate	$CH_3-(CH_2)_4-CH=CH-R$
Palmitoleate	$CH_3-(CH_2)_5-CH=CH-R$
Oleate	$CH_3-(CH_2)_7-CH=CH-R$

CONTROL OF FATTY ACID SYNTHESIS

The synthesis of fatty acids is maximal when carbohydrate is abundant and the level of fatty acids is low. Both short-term and long-term control mechanisms are important. The concentration of *cit-*

rate in the cytosol is the most important short-term regulator of fatty acid synthesis. As already mentioned, citrate stimulates acetyl CoA carboxylase, the enzyme catalyzing the committed step in fatty acid synthesis. The level of citrate is high when both acetyl CoA and ATP are abundant. Recall that isocitrate dehydrogenase is inhibited by a high energy charge (p. 301). Hence, *a high level of citrate indicates that two-carbon units and ATP are available for fatty acid synthesis.* The effect of citrate on acetyl CoA carboxylase is antagonized by *palmitoyl CoA,* which is abundant when there is an excess of fatty acids. Palmitoyl CoA also inhibits the carrier that transports citrate from mitochondria to the cytosol. In addition, palmitoyl CoA inhibits NADPH generation by glucose 6-phosphate dehydrogenase.

Long-term control is mediated by changes in the rates of synthesis and degradation of the enzymes participating in fatty acid synthesis. This type of regulation is also known as *adaptive control.* Animals that have been fasted have striking increases in the amounts of acetyl CoA carboxylase and fatty acid synthetase in the liver within a few days of being fed high-carbohydrate, low-fat diets.

SUMMARY

Fatty acids are physiologically important both as components of phospholipids and glycolipids and as fuel molecules. They are stored in adipose tissue as triacylglycerols (neutral fat), which can be mobilized by the hydrolytic action of lipases that are under hormonal control. Fatty acids are activated to acyl CoAs, transported across the inner mitochondrial membrane by carnitine, and degraded in the mitochondrial matrix by a recurring sequence of four reactions: oxidation linked to FAD, hydration, oxidation linked to NAD^+, and thiolysis by CoA. The $FADH_2$ and NADH formed in the oxidation steps transfer their electrons to O_2 by means of the respiratory chain, whereas the acetyl CoA formed in the thiolysis step normally enters the citric acid cycle by condensing with oxaloacetate. When the concentration of oxaloacetate is insufficient, acetyl CoA gives rise to acetoacetate and 3-hydroxybutyrate, which are normal fuel molecules. In starvation and in diabetes, large amounts of acetoacetate, 3-hydroxybutyrate, and acetone (collectively known as ketone bodies) accumulate in the blood. Mammals are unable to convert fatty acids into glucose because there is no pathway for the net production of oxaloacetate, pyruvate, or other gluconeogenic intermediate from acetyl CoA.

Fatty acids are synthesized in the cytosol by a different pathway from that of β-oxidation. Synthesis starts with the carboxylation of acetyl CoA to malonyl CoA. This ATP-driven reaction is catalyzed by acetyl CoA carboxylase, a biotin-enzyme. Citrate allosterically stimulates this committed step in fatty acid synthesis. The intermediates in fatty acid synthesis are linked to an acyl carrier protein

(ACP), specifically to the sulfhydryl terminus of its phosphopante-theine prosthetic group. Acetyl-ACP is formed from acetyl CoA, and malonyl-ACP is formed from malonyl CoA. Acetyl-ACP and malonyl-ACP condense to form acetoacetyl-ACP, a reaction driven by the release of CO_2 from the activated malonyl unit. This is followed by a reduction, a dehydration, and a second reduction. NADPH is the reductant in these steps. The butyryl-ACP formed in this way is ready for a second round of elongation starting with the addition of a two-carbon unit from malonyl-ACP. Seven rounds of elongation yield palmitoyl-ACP, which is hydrolyzed to palmitate. The synthesis of palmitate requires eight molecules of acetyl CoA, fourteen NADPH, and seven ATP. In higher organisms, the enzymes carrying out fatty acid synthesis are organized into a multienzyme complex. The two kinds of polypeptide chains in this complex contain several covalently linked enzymes. A reaction cycle based on the cleavage of citrate carries acetyl groups from mitochondria to the cytosol and generates some of the required NADPH. The rest is formed by the pentose phosphate pathway. Fatty acids are elongated and unsaturated by enzyme systems in the endoplasmic reticulum membrane. Mammals lack the enzymes to introduce double bonds distal to C-9, and so they require linoleate and linolenate in their diets.

SELECTED READINGS

WHERE TO START

Lynen, F., 1972. The pathway from "activated acetic acid" to the terpenes and fatty acids. In *Nobel Lectures: Physiology or Medicine (1963–1970)*, pp. 103–138. American Elsevier (1973). [An account of the author's pioneering work on fatty acid degradation and synthesis and of the roles of acetyl CoA and biotin.]

GENERAL REVIEWS

Wakil, S. J., and Barnes, E. M., Jr., 1971. Fatty acid metabolism. *Compr. Biochem.* 18S:57–104.

Newsholme, E. A., and Start, C., 1973. *Regulation in Metabolism*. Wiley. [Chapters 4 and 7 deal with the regulation of fat metabolism.]

FATTY ACID OXIDATION

McGarry, J. D., and Foster, D. W., 1980. Regulation of hepatic fatty acid oxidation and ketone body production. *Ann. Rev. Biochem.* 49:395–420.

Bressler, R., 1970. Fatty acid oxidation. *Compr. Biochem.* 18:331–359.

Garland, P. B., Shepherd, D., Nicholls, D. G., Yates, D. W., and Light, P. A., 1969. Interactions between fatty acid oxidation and the tricarboxylic acid cycle. *In* Lowenstein, J. M., (ed.), *Citric Acid Cycle: Control and Compartmentation*, pp. 163–212. Dekker.

CARNITINE

Bremer, J., 1977. Carnitine and its role in fatty acid metabolism. *Trends Biochem. Sci.* 2:207–209.

Fritz, I. B., 1968. The metabolic consequences of the effects of carnitine on long-chain fatty acid oxidation. *In* Gran, F. C., (ed.), *Symposium on Cellular Compartmentalization and Control of Fatty Acid Metabolism*, pp. 39–63. Academic Press.

Engel, W. K., Vick, N. A., Glueck, C. J., and Levy, R. I., 1970. A skeletal-muscle disorder associated with intermittent symptoms and a possible defect of lipid metabolism. *N. Engl. J. Med.* 282:697–704.

FATTY ACID SYNTHESIS

Volpe, J. J., and Vagelos, P. R., 1976. Mechanism and regulation of biosynthesis of saturated fatty acids. *Physiol. Rev.* 56:339–417.

Bloch, K., and Vance, D., 1977. Control mechanisms in the synthesis of saturated fatty acids. *Ann. Rev. Biochem.* 46:263–298.

Lynen, F., 1972. Enzyme systems for fatty acid synthesis. *Biochem. Soc. Symp.* 35:5–26.

Lane, M. D., Moss, J., and Polakis, S. E., 1974. Acetyl coenzyme A carboxylase. *Curr. Top. Cell Regul.* 8:139–187.

406   Numa, S., and Yamashita, S., 1974. Regulation of lipogenesis in animal tissues. *Curr. Top. Cell Regul.* 8:197–239.

Utter, M. F., 1969. Metabolic roles of oxaloacetate. *In* Lowenstein, J. M., (ed.), *Citric Acid Cycle: Control and Compartmentation*, pp. 249–296. [Includes a discussion of the role of oxaloacetate in the transport of acetyl groups to the cytosol for fatty acid synthesis.]

ENZYME MECHANISMS

Wood, H. G., and Barden, R. E., 1977. Biotin enzymes. *Ann. Rev. Biochem.* 46:385–413.

Alberts, A. W., and Vagelos, P. R., 1972. Acyl-CoA carboxylases. *In* Boyer, P. D., (ed.), *The Enzymes* (3rd ed.), vol. 6, pp. 37–82. Academic Press.

Vagelos, P. R., 1973. Acyl group transfer (acyl carrier protein). *In* Boyer, P. D., (ed.), *The Enzymes* (3rd ed.), vol. 8, pp. 155–199.

PROBLEMS

1. Write a balanced equation for the conversion of glycerol into pyruvate. Which enzymes are required in addition to those of the glycolytic pathway?

2. Write a balanced equation for the conversion of stearate into acetoacetate.

3. Compare the following aspects of fatty acid oxidation and synthesis:

   (a) Site of the process.
   (b) Acyl carrier.
   (c) Reductants and oxidants.
   (d) Stereochemistry of the intermediates.
   (e) Direction of synthesis or degradation.
   (f) Organization of the enzyme system.

4. For each of the following unsaturated fatty acids, indicate whether the biosynthetic precursor in animals is palmitoleate, oleate, linoleate, or linolenate.

   (a) $18:1 \ cis \ \Delta^{11}$
   (b) $18:3 \ cis \ \Delta^{6}, \Delta^{9}, \Delta^{12}$
   (c) $20:2 \ cis \ \Delta^{11}, \Delta^{14}$
   (d) $20:3 \ cis \ \Delta^{5}, \Delta^{8}, \Delta^{11}$
   (e) $22:1 \ cis \ \Delta^{13}$
   (f) $22:6 \ cis \ \Delta^{4}, \Delta^{7}, \Delta^{10}, \Delta^{13}, \Delta^{16}, \Delta^{19}$

5. Consider a cell extract that actively synthesizes palmitate. Suppose that a fatty acid synthetase in this preparation forms one palmitate in about five minutes. A large amount of malonyl CoA labeled with ^{14}C in each carbon of its malonyl unit is suddenly added to this system, and fatty acid synthesis is stopped a minute later by altering the pH. The fatty acids in the supernatant are analyzed for radioactivity. Which carbon atom of the palmitate formed by this system is more radioactive—C-1 or C-14?

6. The citric acid cycle is modified in some plants and microorganisms. They contain isocitrase and malate synthetase, two enzymes absent from animals. Isocitrase catalyzes the aldol cleavage of isocitrate to succinate and glyoxylate ($OHC—COO^-$), whereas malate synthetase catalyzes the formation of malate from acetyl CoA, glyoxylate, and water.

   (a) If isocitrate dehydrogenase is absent and malate dehydrogenase is present, can oxaloacetate be formed from acetyl CoA in these organisms? If so, write a balanced equation for this conversion.

   (b) Can these organisms synthesize glucose from acetyl CoA?

CHAPTER 18

AMINO ACID DEGRADATION AND THE UREA CYCLE

Amino acids in excess of those needed for the synthesis of proteins and other biomolecules cannot be stored, in contrast with fatty acids and glucose, nor are they excreted. Rather, surplus amino acids are used as metabolic fuel. *The α-amino group is removed and the resulting carbon skeleton is converted into a major metabolic intermediate.* Most of the amino groups of surplus amino acids are converted into urea, whereas their carbon skeletons are transformed into acetyl CoA, acetoacetyl CoA, pyruvate, or one of the intermediates of the citric acid cycle. Hence *fatty acids, ketone bodies, and glucose can be formed from amino acids.*

α-AMINO GROUPS ARE CONVERTED INTO AMMONIUM ION BY OXIDATIVE DEAMINATION OF GLUTAMATE

The major site of amino acid degradation in mammals is the liver. The fate of the α-amino group will be considered first, followed by that of the carbon skeleton. The α-amino group of many amino acids is transferred to α-ketoglutarate to form *glutamate*, which is then oxidatively deaminated to yield NH_4^+.

Amino acid   Glutamate

Transaminases catalyze the transfer of an α-amino group from an α-amino acid to an α-keto acid. These enzymes are also called *aminotransferases*.

$$H-\underset{\boxed{R_1}}{\overset{NH_3^+}{\underset{|}{\overset{|}{C}}}}-COO^- + \underset{\textcircled{R_2}}{\overset{O}{\underset{|}{\overset{\parallel}{C}}}}-COO^- \rightleftharpoons \underset{\boxed{R_1}}{\overset{O}{\underset{|}{\overset{\parallel}{C}}}}-COO^- + H-\underset{\textcircled{R_2}}{\overset{NH_3^+}{\underset{|}{\overset{|}{C}}}}-COO^-$$

Glutamate transaminase, the most important of these enzymes, catalyzes the transfer of an amino group to α-ketoglutarate.

α-Amino acid + α-ketoglutarate $\rightleftharpoons$ α-keto acid + glutamate

Alanine transaminase, which is also prevalent in mammalian tissue, catalyzes the transfer of an amino group to pyruvate.

α-Amino acid + pyruvate $\rightleftharpoons$ α-keto acid + alanine

The alanine formed in this step can transfer its amino group to α-ketoglutarate to form glutamate. *These two transaminases funnel α-amino groups from a variety of amino acids into glutamate for conversion into NH_4^+.*

Ammonium ion is formed from glutamate by oxidative deamination. This reaction is catalyzed by *glutamate dehydrogenase,* which is unusual in being able to utilize either NAD^+ or $NADP^+$.

$$\underset{\textbf{Glutamate}}{H-\overset{NH_3^+}{\underset{|}{\overset{|}{C}}}-COO^-}_{\begin{subarray}{l}|\\CH_2\\|\\CH_2\\|\\COO^-\end{subarray}} + \underset{(or\ NADP^+)}{NAD^+} + H_2O \rightleftharpoons NH_4^+ + \underset{\textbf{α-Ketoglutarate}}{\overset{O}{\underset{|}{\overset{\parallel}{C}}}-COO^-}_{\begin{subarray}{l}|\\CH_2\\|\\CH_2\\|\\COO^-\end{subarray}} + \underset{(or\ NADPH)}{NADH} + H^+$$

The activity of glutamate dehydrogenase is allosterically regulated. The vertebrate enzyme consists of six identical subunits, which can polymerize further. Guanosine triphosphate (GTP) and adenosine triphosphate (ATP) are allosteric inhibitors, whereas guanosine diphosphate (GDP) and adenosine diphosphate (ADP) are allosteric activators. Hence, *a lowering of the energy charge accelerates the oxidation of amino acids.*

The sum of the reactions catalyzed by the transaminases and by glutamate dehydrogenase is

α-Amino acid + $\underset{(or\ NADP^+)}{NAD^+}$ + H_2O $\rightleftharpoons$

α-keto acid + NH_4^+ + $\underset{(or\ NADPH)}{NADH}$ + H^+

In terrestrial vertebrates, NH_4^+ is converted into urea, which is excreted. The synthesis of urea will be discussed shortly.

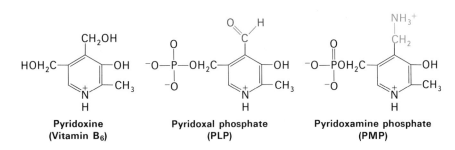

PYRIDOXAL PHOSPHATE, THE PROSTHETIC GROUP IN TRANSAMINASES, FORMS SCHIFF-BASE INTERMEDIATES

The prosthetic group of all transaminases is *pyridoxal phosphate* (PLP), which is derived from *pyridoxine* (Vitamin B$_6$). During transamination, pyridoxal phosphate is transiently converted into *pyridoxamine phosphate* (PMP).

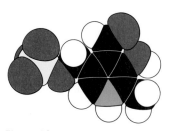

Figure 18-1
Space-filling model of pyridoxal phosphate (PLP).

PLP enzymes form covalent Schiff-base intermediates with their substrates. In the absence of substrate, the aldehyde group of PLP is in Schiff-base linkage with the *ε-amino group of a specific lysine residue at the active site.* A new Schiff-base linkage is formed on addition of an amino acid substrate. *The α-amino group of the amino acid substrate displaces the ε-NH₂ group of the active-site lysine.* The amino acid–PLP Schiff base that is formed remains tightly bound to the enzyme by noncovalent forces.

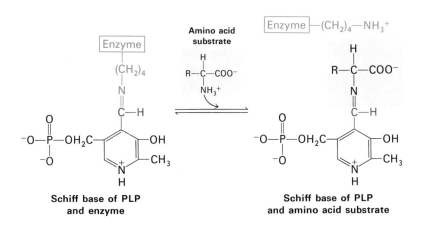

This Schiff base and the one between PLP and the active-site lysine are *aldimines*. During catalysis, the double bond in the amino acid–PLP Schiff-base linkage shifts position to form a *ketimine,* which is then hydrolyzed to *PMP* and an *α-keto acid.* A reaction mechanism proposed by Esmond Snell and Alexander Braunstein is shown in Figure 18-2. The active-site lysine or another suitably positioned

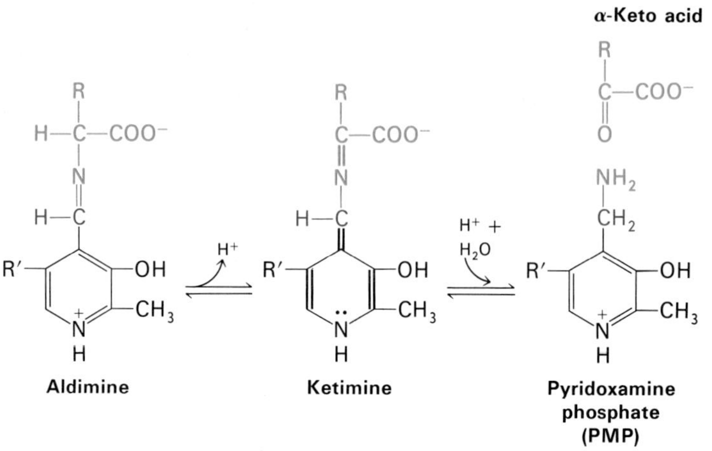

Figure 18-2
Proposed mechanism for transamination reactions.

basic group probably facilitates the aldimine-ketimine conversion by serving as an electron sink. These steps constitute half of the overall reaction:

$$\text{Amino acid}_1 + \text{E-PLP} \rightleftharpoons \alpha\text{-keto acid} + \text{E-PMP}$$

The second half of the overall reaction occurs by a reversal of the above pathway. A second α-keto acid reacts with the enzyme–pyridoxamine phosphate complex (E-PMP) to yield a second amino acid and regenerate the enzyme–pyridoxal phosphate complex (E-PLP).

$$\alpha\text{-Keto acid}_2 + \text{E-PMP} \rightleftharpoons \text{amino acid}_2 + \text{E-PLP}$$

The sum of these partial reactions is

$$\text{Amino acid}_1 + \alpha\text{-keto acid}_2 \rightleftharpoons \text{amino acid}_2 + \alpha\text{-keto acid}_1$$

The catalytic versatility of PLP enzymes is remarkable. Transamination is just one of a wide range of amino acid transformations that are catalyzed by PLP enzymes. The other reactions at the α-carbon atom of amino acids are decarboxylations, deaminations, racemizations, and aldol cleavages (Figure 18-3). In addition, PLP enzymes catalyze elimination and replacement reactions at the β-carbon atom (e.g., tryptophan synthetase, p. 498) and the γ-carbon atom (e.g., cystathionase, p. 495) of amino acid substrates. These reactions have the following features in common. First, a *Schiff base* is formed by the amino acid substrate (the amine component) and PLP (the carbonyl component). Second, PLP acts as an *electron sink* to stabilize catalytic intermediates that are negatively charged. The

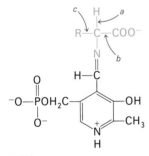

Figure 18-3
Pyridoxal phosphate enzymes labilize one of three bonds at the α-carbon atom of an amino acid substrate. For example, bond *a* is labilized by transaminases, bond *b* by decarboxylases, and bond *c* by aldolases (such as threonine aldolases). PLP enzymes also catalyze reactions at the β- and γ-carbon atoms of amino acids.

ring nitrogen of PLP attracts electrons from the amino acid substrate. In other words, PLP is an *electrophilic catalyst.* Third, the product Schiff base is then hydrolyzed.

SERINE AND THREONINE CAN BE DIRECTLY DEAMINATED

The α-amino groups of serine and threonine can be directly converted into NH_4^+ because each of these amino acids contains a hydroxyl group in its side chain. These direct deaminations are catalyzed by *serine dehydratase* and *threonine dehydratase,* in which PLP is the prosthetic group.

$$\text{Serine} \longrightarrow \text{pyruvate} + NH_4^+$$
$$\text{Threonine} \longrightarrow \alpha\text{-ketobutyrate} + NH_4^+$$

These enzymes are called dehydratases because dehydration precedes deamination. Serine loses a hydrogen atom from its α-carbon and a hydroxyl group from its β-carbon to yield amino-acrylate. This unstable compound reacts with H_2O to give pyruvate and NH_4^+.

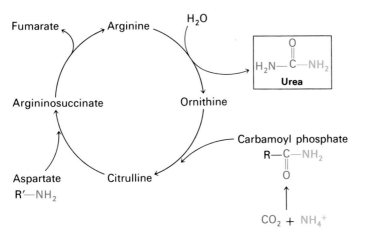

NH_4^+ IS CONVERTED INTO UREA IN MOST TERRESTRIAL VERTEBRATES AND THEN EXCRETED

Some of the NH_4^+ formed in the breakdown of amino acids is consumed in the biosynthesis of nitrogen compounds. In most terrestrial vertebrates, the excess NH_4^+ is converted into urea and then excreted. In birds and terrestrial reptiles, NH_4^+ is converted into

Figure 18-4
The urea cycle.

uric acid for excretion, whereas, in many aquatic animals, NH_4^+ itself is excreted. These three classes of organisms are called *ureotelic, uricotelic,* and *ammonotelic.*

In terrestrial vertebrates, urea is synthesized by the *urea cycle.* This series of reactions was proposed by Hans Krebs and Kurt Henseleit in 1932, five years before the elucidation of the citric acid cycle. In fact, the urea cycle was the first cyclic metabolic pathway to be discovered. One of the nitrogen atoms of the urea synthesized by this pathway comes from ammonia, whereas the other nitrogen atom comes from aspartate. The carbon atom of urea is derived from CO_2. *Ornithine* is the carrier of these carbon and nitrogen atoms in the urea cycle.

The immediate precursor of urea is *arginine,* which is hydrolyzed to urea and ornithine by *arginase.*

Arginine   **Urea**   **Ornithine**

The other reactions of the urea cycle result in the synthesis of arginine from ornithine. First, a carbamoyl group is transferred to ornithine to form *citrulline,* a reaction catalyzed by ornithine transcarbamoylase. The carbamoyl donor in this reaction is carbamoyl phosphate, which has a high transfer potential because of its anhydride bond.

Ornithine   **Carbamoyl phosphate**   **Citrulline**

Argininosuccinate synthetase then catalyzes the condensation of citrulline and aspartate. This synthesis of *argininosuccinate* is driven by the cleavage of ATP into AMP and pyrophosphate and by the subsequent hydrolysis of pyrophosphate.

Citrulline   **Aspartate**   **Argininosuccinate**

Finally, argininosuccinase cleaves argininosuccinate into *arginine* and fumarate. Note that the carbon skeleton of aspartate is preserved in these reactions, which transfer the amino group of aspartate to form arginine.

Argininosuccinate   **Arginine**   **Fumarate**

Carbamoyl phosphate is synthesized from NH_4^+, CO_2, ATP, and H_2O in a complex reaction that is catalyzed by *carbamoyl phosphate synthetase*. An unusual feature of this enzyme is that it requires *N*-acetylglutamate for activity.

Carbamoyl phosphate

The consumption of two molecules of ATP makes this synthesis of carbamoyl phosphate essentially irreversible.

THE UREA CYCLE IS LINKED TO THE CITRIC ACID CYCLE

The stoichiometry of urea synthesis is

$$CO_2 + NH_4^+ + 3\,ATP + aspartate + 2\,H_2O \longrightarrow$$
$$urea + 2\,ADP + 2\,P_i + AMP + PP_i + fumarate$$

Pyrophosphate is rapidly hydrolyzed, and so four high-energy phosphate bonds are used to synthesize one molecule of urea. *The synthesis of fumarate by the urea cycle is important because it links the urea cycle and the citric acid cycle* (Figure 18-5). Fumarate is hydrated to malate,

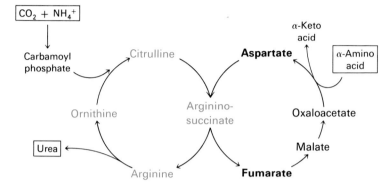

Figure 18-5
The urea cycle, the citric acid cycle, and the transamination of oxaloacetate are linked by fumarate and aspartate.

which is in turn oxidized to oxaloacetate. This key intermediate has several possible fates: (1) oxaloacetate can be transaminated to aspartate; (2) it can be converted into glucose by the gluconeogenic pathway; or (3) it can condense with acetyl CoA to form citrate.

The compartmentation of the urea cycle and its associated reactions is also noteworthy. The formation of NH_4^+ by glutamate dehydrogenase, its incorporation into carbamoyl phosphate, and the subsequent synthesis of citrulline occur in the mitochondrial matrix. In contrast, the next three reactions of the urea cycle, which lead to the formation of urea, take place in the cytosol.

INHERITED ENZYMATIC DEFECTS OF THE UREA CYCLE CAUSE HYPERAMMONEMIA

High levels of NH_4^+ are toxic to humans. The synthesis of urea in the liver is the major route of removal of NH_4^+. A complete block of any of the steps of the urea cycle is probably incompatible with life, because there is no known alternative pathway for the synthesis of urea. Inherited disorders caused by a partial block of any of the urea cycle reactions have been diagnosed. *The common finding is an elevated level of NH_4^+ in the blood (hyperammonemia).* A nearly total deficiency of any of the urea cycle enzymes results in coma and death shortly after birth. Partial deficiencies of these enzymes cause mental retardation, lethargy, and episodic vomiting. A low-protein diet leads to a lowering of the ammonium level in the blood and to clinical improvement in the milder forms of these inherited disorders.

Why are high levels of NH_4^+ toxic? A plausible explanation is that a high concentration of ammonium ion shifts the equilibrium of the reaction catalyzed by glutamate dehydrogenase toward the formation of glutamate, which results in the *depletion of α-ketogluta-*

rate. This reaction will be driven further by the subsequent incorporation of NH_4^+ into glutamate to form glutamine (p. 487). The depletion of α-ketoglutarate, a citric acid cycle intermediate, leads to a *decrease in the rate of formation of ATP.* The brain is highly vulnerable to decreases in the ATP level.

FATES OF THE CARBON ATOMS OF DEGRADED AMINO ACIDS

Thus far, we have considered a series of reactions that removes the α-amino group from amino acids and converts it into urea. We now turn to the fates of the resulting carbon skeletons. *The strategy of amino acid degradation is to form major metabolic intermediates that can be converted into glucose or be oxidized by the citric acid cycle.* In fact, the carbon skeletons of the diverse set of twenty amino acids are funneled into only seven molecules: *pyruvate, acetyl CoA, acetoacetyl CoA, α-ketoglutarate, succinyl CoA, fumarate, and oxaloacetate.* We see here another example of the remarkable economy of metabolic conversions.

Amino acids that are degraded to acetyl CoA or acetoacetyl CoA are termed *ketogenic* because they give rise to ketone bodies. In contrast, amino acids that are degraded to pyruvate, α-ketoglutarate, succinyl CoA, fumarate, or oxaloacetate are termed *glucogenic.* Net synthesis of glucose from these amino acids is feasible because these citric acid cycle intermediates and pyruvate can be converted into phosphoenolpyruvate and then into glucose (p. 346). Recall that mammals lack a pathway for the net synthesis of glucose from acetyl CoA or acetoacetyl CoA.

α-Ketoglutarate

NH_4^+ glutamate dehydrogenase

glutamate

NH_4^+ glutamine synthetase

glutamine

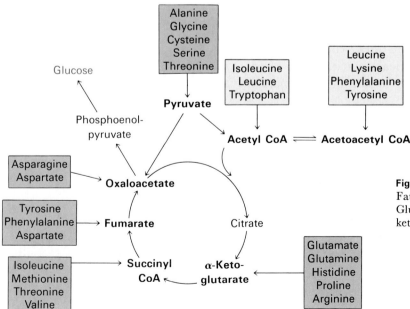

Figure 18-6
Fates of the carbon skeletons of amino acids. Glucogenic amino acids are shaded red, and ketogenic amino acids are shaded yellow.

Of the basic set of twenty amino acids, only leucine is purely ketogenic. Isoleucine, lysine, phenylalanine, tryptophan, and tyrosine are both ketogenic and glucogenic. Some of their carbon atoms emerge in acetyl CoA or acetoacetyl CoA, whereas others appear in potential precursors of glucose. The other fourteen amino acids are purely glucogenic.

THE C₃ FAMILY: ALANINE, SERINE, AND CYSTEINE ARE CONVERTED INTO PYRUVATE

Pyruvate is the entry point for the three-carbon amino acids: ala-

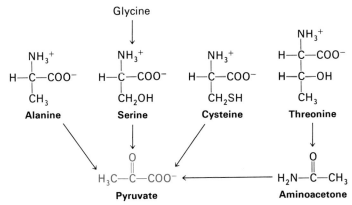

Figure 18-7
Pyruvate is the point of entry for alanine, serine, cysteine, glycine, and threonine.

nine, serine, and cysteine (Figure 18-7). The transamination of alanine directly yields pyruvate:

$$\text{Alanine} + \alpha\text{-ketoglutarate} \rightleftharpoons \text{pyruvate} + \text{glutamate}$$

As mentioned previously (p. 408), glutamate is then oxidatively deaminated, which yields NH_4^+ and regenerates α-ketoglutarate. The sum of these reactions is

$$\text{Alanine} + NAD^+ \longrightarrow \text{pyruvate} + NH_4^+ + NADH + H^+$$

Another simple reaction in the degradation of amino acids is the *deamination of serine to pyruvate* by serine dehydratase (p. 411).

$$\text{Serine} \longrightarrow \text{pyruvate} + NH_4^+$$

Cysteine can be converted into pyruvate by several pathways, with its sulfur atom emerging in H_2S, SO_3^{2-}, or SCN^-.

The carbon atoms of two other amino acids can be converted into pyruvate. *Glycine* can be converted into serine by enzymatic addition of a hydroxymethyl group (p. 491). *Threonine* can give rise to pyruvate by way of aminoacetone.

THE C$_4$ FAMILY: ASPARTATE AND ASPARAGINE ARE CONVERTED INTO OXALOACETATE

Aspartate, a four-carbon amino acid, is directly *transaminated to oxaloacetate,* a citric acid cycle intermediate:

$$\text{Aspartate} + \alpha\text{-ketoglutarate} \rightleftharpoons \text{oxaloacetate} + \text{glutamate}$$

Asparagine is hydrolyzed by *asparaginase* to NH_4^+ and aspartate, which is then transaminated.

Recall that aspartate can also be converted into *fumarate* by the urea cycle (p.414). Fumarate is also a point of entry for half of the carbon atoms of tyrosine and phenylalanine, as will be discussed shortly.

THE C$_5$ FAMILY: SEVERAL AMINO ACIDS ARE CONVERTED INTO α-KETOGLUTARATE THROUGH GLUTAMATE

The carbon skeletons of several five-carbon amino acids enter the citric acid cycle at *α-ketoglutarate.* These amino acids are converted into *glutamate,* which is then oxidatively deaminated by glutamate dehydrogenase to yield α-ketoglutarate (Figure 18-8).

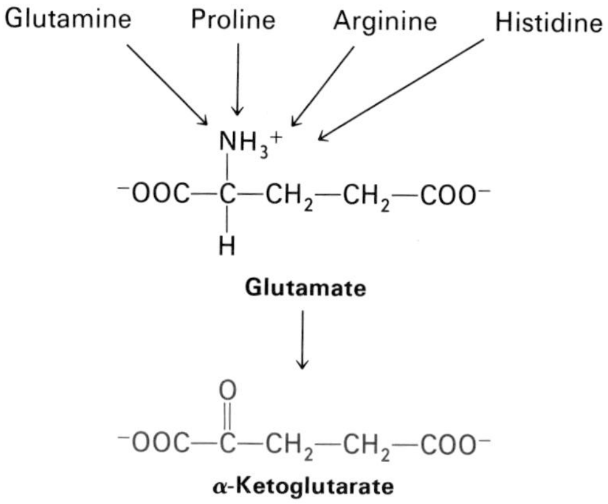

Figure 18-8
α-Ketoglutarate is the point of entry of several C$_5$ amino acids that are first converted into glutamate.

Histidine is converted into 4-imidazolone 5-propionate (Figure 18-9). The amide bond in the ring of this intermediate is hydrolyzed to the *N*-formimino derivative of glutamate, which is then converted into glutamate by transfer of its formimino group to tetrahydrofolate, a carrier of activated one-carbon units (see p. 491).

Glutamine is hydrolyzed to glutamate and NH_4^+ by glutaminase. *Proline* and *arginine* are converted into glutamate γ-semialdehyde, which is then oxidized to glutamate (Figure 18-10).

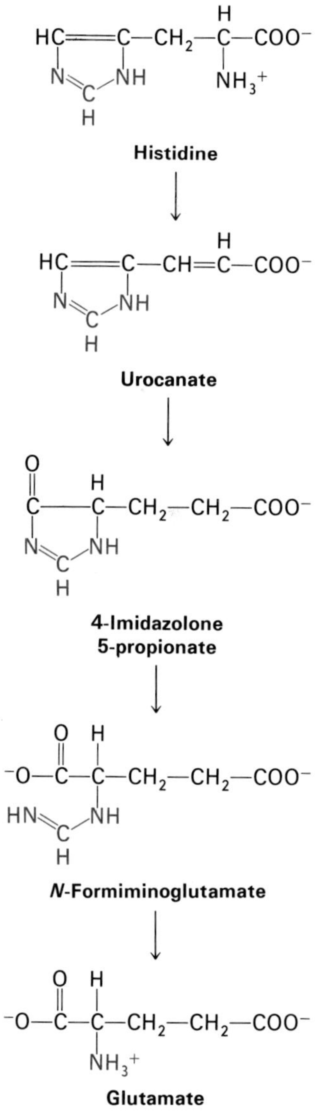

Figure 18-9
Conversion of histidine into glutamate.

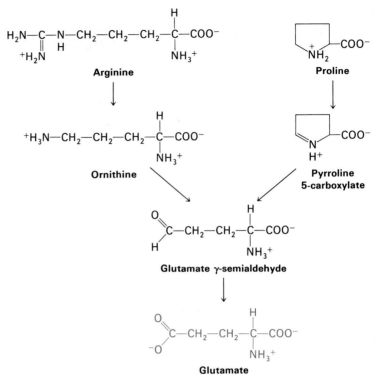

Figure 18-10
Conversion of proline and arginine into glutamate.

SUCCINYL COENZYME A IS A POINT OF ENTRY FOR SOME AMINO ACIDS

Succinyl CoA is the point of entry for some of the carbon atoms of methionine, isoleucine, threonine, and valine. Methylmalonyl CoA is an intermediate in the breakdown of these four amino acids (Figure 18-11).

The pathway from propionyl CoA to succinyl CoA is especially interesting. Propionyl CoA is carboxylated at the expense of an

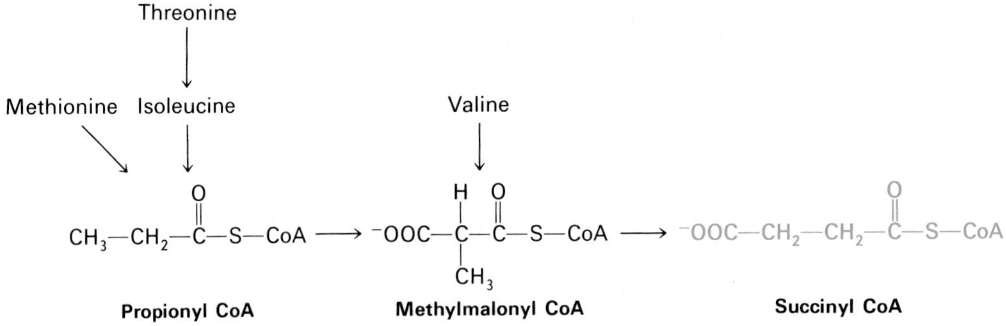

Figure 18-11
Conversion of methionine, isoleucine, threonine, and valine into succinyl CoA.

ATP to yield the D-isomer of methylmalonyl CoA. This carboxylation reaction is catalyzed by *propionyl CoA carboxylase,* a biotin enzyme that has a catalytic mechanism like that of acetyl CoA carboxylase and pyruvate carboxylase. The D-isomer of methylmalonyl CoA is racemized to the L-isomer, which is the substrate for the mutase enzyme that converts it into succinyl CoA.

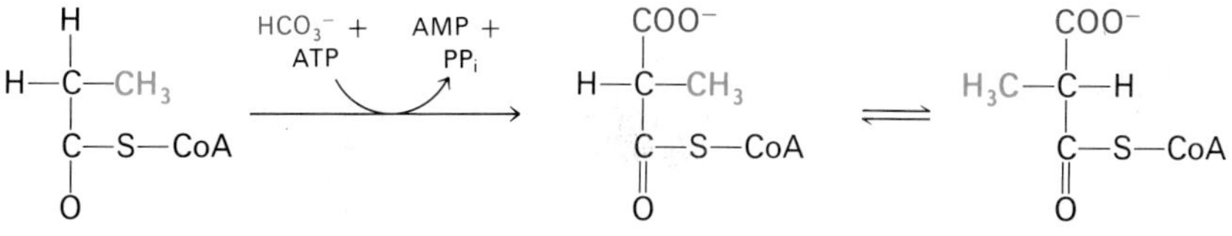

Propionyl CoA   D-Methylmalonyl CoA   L-Methylmalonyl CoA

Succinyl CoA is formed from L-methylmalonyl CoA by an *intramolecular rearrangement.* The —CO—S—CoA group migrates from C-2 to C-3 in exchange for a hydrogen atom. *This very unusual isomerization is catalyzed by methylmalonyl CoA mutase, one of the two mammalian enzymes known to contain a derivative of Vitamin B_{12} as its coenzyme.*

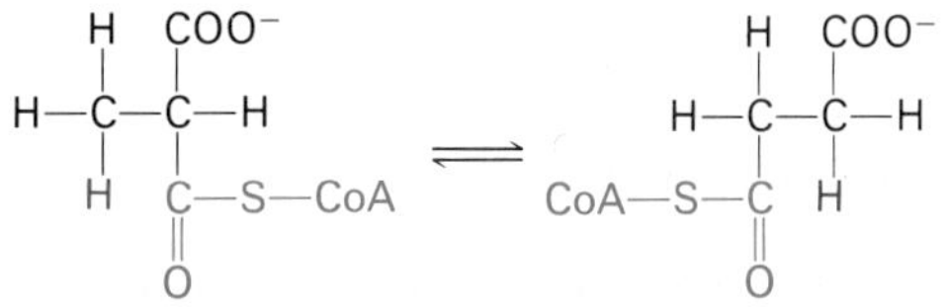

L-Methylmalonyl CoA   Succinyl CoA

This pathway from propionyl CoA to succinyl CoA also participates in the oxidation of *fatty acids that have an odd number of carbon atoms.* The final thiolytic cleavage of an odd-numbered acyl CoA yields acetyl CoA and propionyl CoA (p. 392).

COBALAMIN (VITAMIN B_{12}) ENZYMES CATALYZE REARRANGEMENTS AND METHYLATIONS

Cobalamin (Vitamin B_{12}) has been a challenging problem in biochemistry and medicine since the discovery by George Minot and William Murphy in 1926 that pernicious anemia can be treated by feeding the patient large amounts of liver. Cobalamin was purified and crystallized in 1948 and its complex three-dimensional structure was elucidated in 1956 by Dorothy Hodgkin. The core of cobalamin consists of a *corrin ring with a central cobalt atom* (Figure 18-12). The corrin ring, like a porphyrin, has *four pyrrole units.* Two of them (rings A and D) are directly bonded to each other, whereas the others are joined by methene bridges, as in porphyrins. The substituents on the pyrrole rings are methyl, propionamide, and acetamide groups.

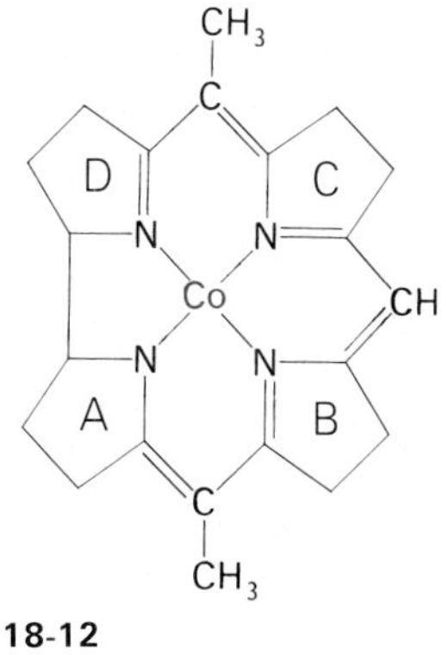

Figure 18-12
Corrin core of cobalamin. Substituents on the pyrroles and the other two cobalt ligands are not shown in this diagram.

A cobalt atom is bonded to the four pyrrole nitrogens. *The fifth substituent* (below the corrin plane in Figure 18-13) is a derivative of *dimethylbenzimidazole* that contains ribose 3-phosphate and amino-isopropanol. One of the nitrogen atoms of dimethylbenzimidazole

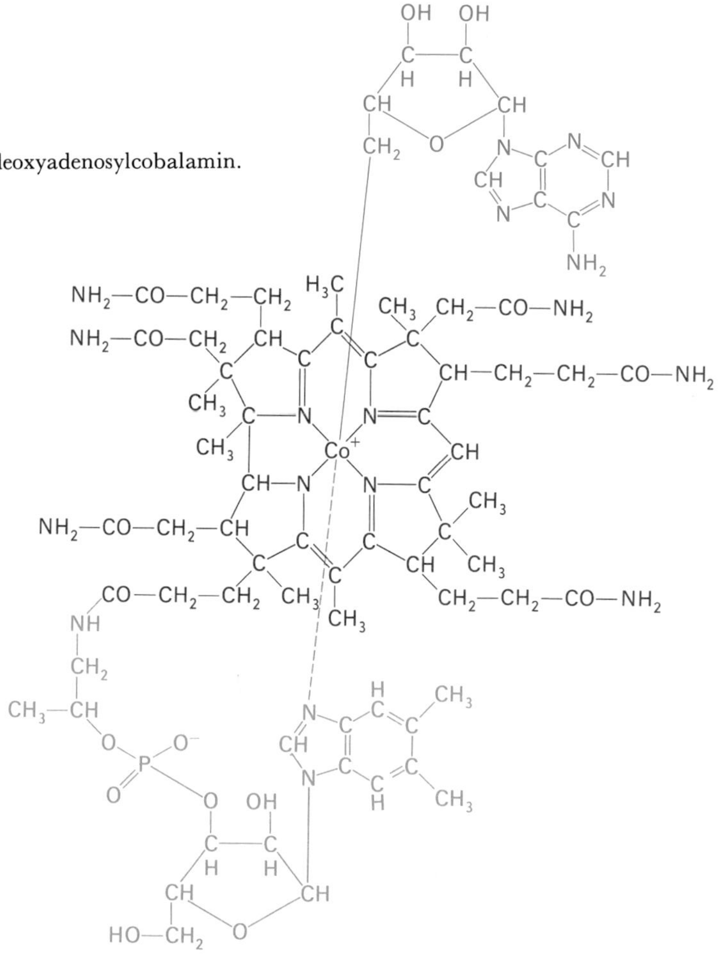

Figure 18-13
Structure of 5'-deoxyadenosylcobalamin.

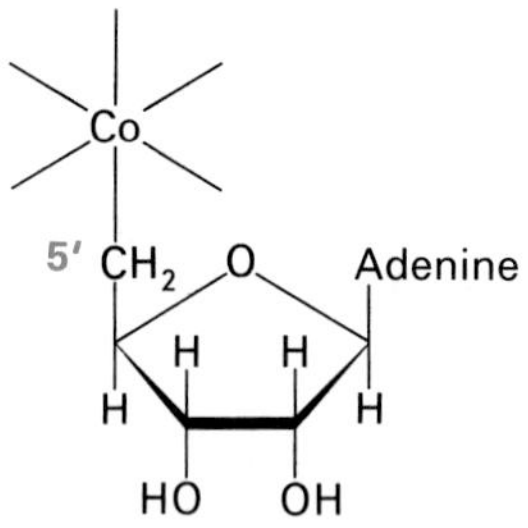

Figure 18-14
The 5' carbon of 5'-deoxyadenosine is coordinated to the cobalt atom in 5'-deoxyadenosylcobalamin. This is the only known example of a carbon–metal bond in a biomolecule.

is linked to cobalt. The amino group of aminoisopropanol is in amide linkage with a side chain of the D ring. The *sixth substituent* on the cobalt atom (located above the corrin plane in Figure 18-13) can be CN^-, $-CH_3$, OH^-, or deoxyadenosyl. The presence of the cyanide ion at the sixth coordination position in cyanocobalamin, the most common commercial form of the vitamin, results from the isolation procedure. Cyanide is not combined with cobalamin in vivo.

The cobalt atom in cobalamin can have a +1, +2, or +3 oxidation state. The cobalt atom is in the +3 state in hydroxocobalamin (where OH^- occupies the sixth coordination site). This form, called B_{12a} (Co^{3+}), is reduced to a divalent state called B_{12r} (Co^{2+}) by a flavo-protein reductase. The B_{12r} (Co^{2+}) form is reduced by a second flavoprotein reductase to B_{12s} (Co^+). NADH is the reductant in

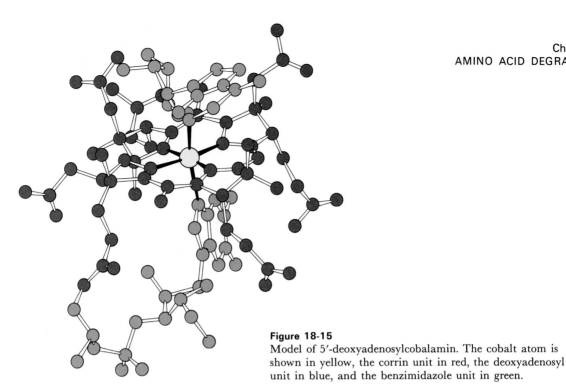

Figure 18-15
Model of 5′-deoxyadenosylcobalamin. The cobalt atom is shown in yellow, the corrin unit in red, the deoxyadenosyl unit in blue, and the benzimidazole unit in green.

both reactions. The B_{12s} form is the substrate for the final enzymatic reaction that yields the active coenzyme. The 5′-deoxyadenosyl group is transferred from ATP to Vitamin B_{12s} (Co^+) to give *5′-deoxyadenosylcobalamin, which is the coenzyme of methylmalonyl CoA mutase* (Figure 18-13).

$$B_{12a}(Co^{3+}) \longrightarrow B_{12r}(Co^{2+}) \longrightarrow B_{12s}(Co^+) \longrightarrow$$
$$5'\text{-deoxyadenosylcobalamin}$$

Cobalamin enzymes catalyze two types of reactions: (1) *rearrangements,* as exemplified by the conversion of L-methylmalonyl CoA into succinyl CoA; and (2) *methylations,* as in the synthesis of methionine (p. 494). The rearrangement reactions are interchanges of two groups attached to adjacent carbon atoms (Figure 18-16). A hydrogen atom migrates from one carbon atom to the next, and an R group concomitantly moves in the reverse direction. In these intramolecular rearrangements, the carbon–cobalt bond of 5′-deoxyadenosylcobalamin is cleaved. The substrate carbon atom that gave up its hydrogen atom becomes transiently bonded to the cobalt.

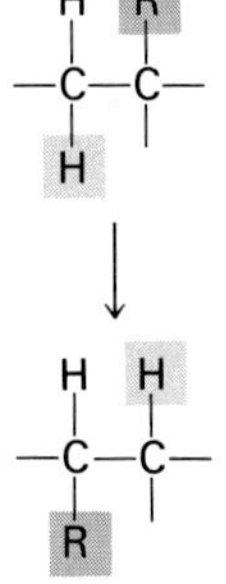

Figure 18-16
Rearrangement reaction catalyzed by cobalamin enzymes. The R group can be an amino group, a hydroxyl group, or a substituted carbon.

ABSORPTION OF COBALAMIN IS IMPAIRED IN PERNICIOUS ANEMIA

There is a specialized transport system for the absorption of cobalamin. The stomach secretes a glycoprotein called *intrinsic factor,* which binds cobalamin in the intestinal lumen. This complex is subsequently bound by a *specific receptor in the lining of the ileum*. The complex of cobalamin and intrinsic factor is then dissociated by a *releasing factor* and actively transported across the ileal membrane into the bloodstream. *Pernicious anemia is caused by a deficiency of intrinsic factor, which leads to impaired absorption of cobalamin.* This disease was originally treated by feeding patients large amounts of liver, a rich source of cobalamin, so that enough of the vitamin was absorbed even in the absence of intrinsic factor. The most reliable therapy is intramuscular injection of cobalamin at monthly intervals.

Animals and plants are unable to synthesize cobalamin. *This vitamin is unique in that it appears to be synthesized only by microorganisms,* particularly anaerobic bacteria. A normal person requires less than 10 μg of cobalamin per day. Nutritional deficiency of cobalamin is rare because this vitamin is found in virtually all animal tissues.

SEVERAL INHERITED DEFECTS OF METHYLMALONYL COENZYME A METABOLISM ARE KNOWN

Several *inherited disorders of methylmalonyl CoA metabolism* have recently been characterized. These disorders usually become evident in the first year of life, when the striking symptom is *acidosis*. The pH of arterial blood is about 7.0, rather than the normal value of 7.4. *Large amounts of methylmalonate appear in the urine of patients who have these disorders.* A normal person excretes less than 5 mg of methylmalonate per day, whereas a patient with defective methylmalonyl CoA metabolism may excrete 1 g or more. About half of the patients with methylmalonic aciduria improve markedly when large doses of cobalamin are administered parenterally. The arterial blood pH returns to normal, and there is a marked decrease in the excretion of methylmalonate. These responsive patients usually have a defect in the transferase that catalyzes the synthesis of deoxyadenosylcobalamin:

$$B_{12s}(Co^+) \;\xrightarrow{\;|\;|\;}\; \text{deoxyadenosylcobalamin}$$

In contrast, other patients with impaired methylmalonyl CoA metabolism do not respond to large doses of cobalamin. Some of them may have a defective methylmalonyl CoA mutase apoenzyme. This form of methylmalonic aciduria is frequently lethal.

LEUCINE IS DEGRADED TO ACETYL COENZYME A AND ACETOACETYL COENZYME A

As mentioned earlier, leucine is the only one of the basic set of twenty amino acids that is purely ketogenic. It is degraded by reactions that have already been encountered in fatty acid degradation and the citric acid cycle. First, leucine is transaminated to the corresponding α-keto acid, *α-ketoisocaproate*. This α-keto acid is then *oxidatively decarboxylated* to *isovaleryl CoA*. This reaction is analogous to the oxidative decarboxylation of pyruvate to acetyl CoA and of α-ketoglutarate to succinyl CoA.

Leucine   α-Ketoisocaproate   Isovaleryl CoA

Isovaleryl CoA is *dehydrogenated* to yield *β-methylcrotonyl CoA*. This oxidation is catalyzed by isovaleryl CoA dehydrogenase, in which the hydrogen acceptor is FAD, as in the analogous reaction in fatty acid oxidation that is catalyzed by acyl CoA dehydrogenase. *Beta-methylglutaconyl CoA* is formed by *carboxylation* of β-methylcrotonyl CoA at the expense of an ATP. As might be expected, the mechanism of carboxylation of β-methylcrotonyl CoA carboxylase is very similar to that of pyruvate carboxylase and acetyl CoA carboxylase. In fact, much of our present knowledge of the mechanism of biotin-dependent carboxylations comes from Feodor Lynen's pioneering work on this enzyme.

Isovaleryl CoA   β-Methylcrotonyl CoA   β-Methylglutaconyl CoA

Beta-methylglutaconyl CoA is then *hydrated* to form *β-hydroxy-β-methylglutaryl CoA,* which is cleaved to *acetyl CoA* and *acetoacetate.* This reaction has already been discussed in regard to the formation of ketone bodies from fatty acids (p. 393).

β-Methylglutaconyl CoA   β-Hydroxy-β-methylglutaryl CoA   Acetoacetate

It is interesting to note that many coenzymes participate in the degradation of leucine to acetyl CoA and acetoacetate: *PLP* in transamination; *TPP, lipoate, FAD, and NAD+* in oxidative decarboxylation; FAD again in dehydrogenation; and *biotin* in carboxylation. *Coenzyme A* is the acyl carrier in these reactions.

The degradative pathways of valine and isoleucine resemble that of leucine.

All three amino acids are initially transaminated to the corresponding α-keto acid, which is then oxidatively decarboxylated to yield a derivative of CoA. The subsequent reactions are like those of fatty acid oxidation. Isoleucine yields acetyl CoA and propionyl CoA, whereas valine yields methylmalonyl CoA. There is an inborn error of metabolism that affects the degradation of valine, isoleucine, and leucine. In *maple syrup urine disease,* the oxidative decarboxylation of these three amino acids is blocked. The amounts of leucine, isoleucine, and valine in blood and urine are markedly elevated, which results in a corresponding increase in the α-keto acids derived from these amino acids. The urine of patients having this disease has the odor of maple syrup, hence the name of the disease. Maple syrup urine disease is usually fatal unless the patient is placed on a diet low in valine, isoleucine, and leucine early in life.

PHENYLALANINE AND TYROSINE ARE DEGRADED BY OXYGENASES TO ACETOACETATE AND FUMARATE

The pathway for the degradation of phenylalanine and tyrosine has some very interesting features. This series of reactions shows how *molecular oxygen is used to break an aromatic ring.* The first step is the hydroxylation of phenylalanine to tyrosine, a reaction catalyzed by *phenylalanine hydroxylase.* This enzyme is called a *monooxygenase* (also called a *mixed-function oxygenase*) because one atom of O_2 appears in the product and the other in H_2O.

The reductant here is *tetrahydrobiopterin,* an electron carrier that has not been previously discussed. The oxidized form of this electron carrier is *dihydrobiopterin.*

Dihydrobiopterin
(Oxidized form)

Tetrahydrobiopterin
(Reduced form)

NADPH reduces dihydrobiopterin to regenerate tetrahydrobiopterin, a reaction catalyzed by dihydrobiopterin reductase. The sum of the reactions catalyzed by phenylalanine hydroxylase and dihydrobiopterin reductase is

Phenylalanine $+ \ O_2 \ + \ NADPH \ + \ H^+ \longrightarrow$
$$tyrosine \ + \ NADP^+ \ + \ H_2O$$

The next step is the transamination of tyrosine to p-*hydroxyphenylpyruvate* (Figure 18-17). This α-keto acid then reacts with O_2 to form *homogentisate*. The enzyme catalyzing this complex reaction, *p*-hydroxyphenylpyruvate hydroxylase, is called a *dioxygenase* because both atoms of O_2 become incorporated into the product. The aromatic ring of homogentisate is then cleaved by O_2, which yields *4-maleylacetoacetate*. This reaction is catalyzed by homogentisate oxidase, another dioxygenase. In fact, *nearly all cleavages of aromatic rings in biological systems are catalyzed by dioxygenases,* a class of enzymes discovered by Osamu Hayaishi. 4-Maleylacetoacetate is then isomerized to *4-fumarylacetoacetate,* which is finally hydrolyzed to *fumarate* and *acetoacetate.*

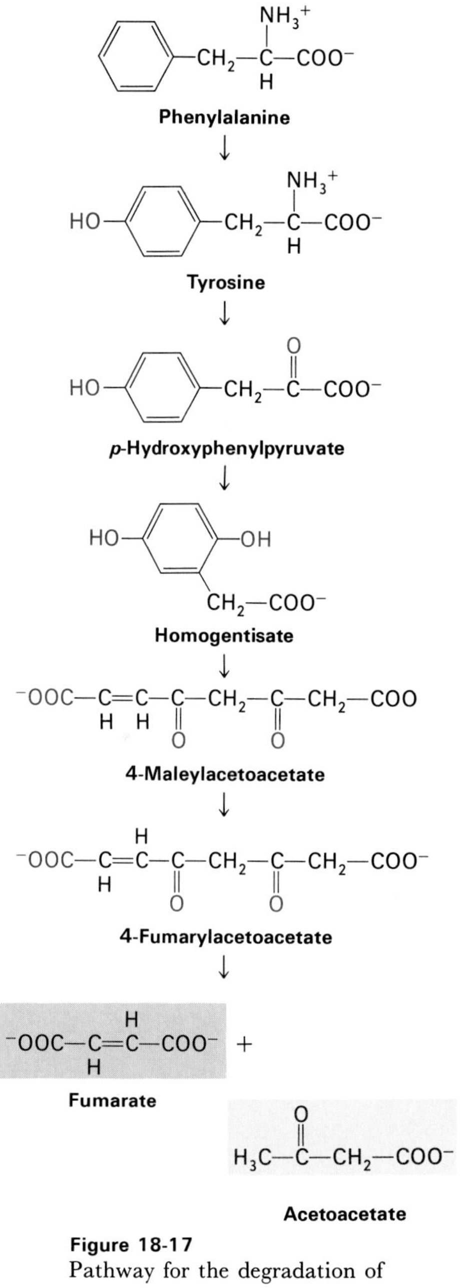

GARROD'S DISCOVERY OF INBORN ERRORS OF METABOLISM

Alcaptonuria is an inherited metabolic disorder caused by the absence of homogentisate oxidase. Homogentisate accumulates and is excreted in the urine, which turns dark on standing as homogentisate is oxidized and polymerized to a melaninlike substance. Alcaptonuria is a relatively benign condition, as described by Zacutus Lusitanus in 1649:

> The patient was a boy who passed black urine and who, at the age of fourteen years, was submitted to a drastic course of treatment which had for its aim the subduing of the fiery heat of his viscera, which was supposed to bring about the condition in question by charring and blackening his bile. Among the measures prescribed were bleedings, purgation, baths, a cold and watery diet and drugs galore. None of these had any obvious effect, and eventually the patient, who tired of the futile and superfluous therapy, resolved to let things take their natural course. None of the predicted evils ensued, he married, begat a large family, and lived a long and healthy life, always passing urine black as ink.

In 1902, Archibald Garrod showed that alcaptonuria is transmitted as a single recessive Mendelian trait. Furthermore, he recognized that homogentisate is a normal intermediate in the degradation of phenylalanine and tyrosine and that it accumulates in alcaptonuria because its degradation is blocked. He concluded that "the splitting of the benzene ring in normal metabolism is the work of a special enzyme, that in congenital alcaptonuria this enzyme is wanting." Garrod perceived the direct relationship between genes

Figure 18-17
Pathway for the degradation of phenylalanine and tyrosine.

and enzymes, and recognized the importance of chemical individuality. His book *Inborn Errors of Metabolism* was a most imaginative and important contribution to biology and medicine.

A BLOCK IN THE HYDROXYLATION OF PHENYLALANINE CAN LEAD TO SEVERE MENTAL RETARDATION

Phenylketonuria, an inborn error of phenylalanine metabolism, can have devastating effects, in contrast with alcaptonuria. Untreated individuals with phenylketonuria are nearly always *severely mentally retarded*. In fact, about 1% of patients in mental institutions have phenylketonuria. The weight of the brain of these individuals is below normal, myelination of their nerves is defective, and their reflexes are hyperactive. The life expectancy of untreated phenylketonurics is drastically shortened. Half are dead by age twenty, and three-quarters by age thirty.

Phenylketonuria is caused by an *absence or deficiency of phenylalanine hydroxylase* or, more rarely, of its tetrahydrobiopterin cofactor. Phenylalanine cannot be converted into tyrosine, and so there is an accumulaton of *phenylalanine in all body fluids*. Some fates of phenylalanine that are quantitatively insignificant in normal persons become prominent in phenylketonurics. The most evident of these is the transamination of phenylalanine to form *phenylpyruvate*. The disease acquired its name from the high levels of this phenylketone in urine. Phenyllactate, phenylacetate, and *o*-hydroxylphenylacetate are derived from phenylpyruvate. The α-amino group of glutamine forms an amide bond with the carboxyl group of phenylacetate, which yields phenylacetylglutamine.

There are many other derangements of amino acid metabolism associated with phenylketonuria, particularly among the aromatic compounds. Phenylketonurics have a lighter skin and hair color than their siblings. The hydroxylation of tyrosine is the first step in the formation of the pigment melanin. In phenylketonurics this reaction is competitively inhibited by the high levels of phenylalanine, and so less melanin is formed. *The biochemical basis of mental retardation in untreated phenylketonuria is an enigma.*

Phenylketonurics appear normal at birth but are severely defective by age one if untreated. The therapy for phenylketonuria is a *low phenylalanine diet*. The aim is to provide just enough phenylalanine to meet the needs for growth and replacement. Proteins that have an initially low content of phenylalanine, such as casein from milk, are hydrolyzed and phenylalanine is removed by adsorption. A low phenylalanine diet must be started very soon after birth to prevent irreversible brain damage. In one study, the average I.Q. of phenylketonurics treated within a few weeks after birth was 93; a control group of siblings treated starting at age one had an average I.Q. of 53.

Phenylpyruvate

Melanin—
A black pigment in skin and hair. From the Greek word *melan*, meaning black. This polymeric pigment is formed in granules called melanosomes that are rich in *tyrosinase*, a monoxygenase.

Early diagnosis of phenylketonuria is essential and has been accomplished by mass screening programs. In past years, the urine of newborns was assayed by the addition of $FeCl_3$, which gives an olive green color in the presence of phenylpyruvate. The phenylalanine level in the blood is now the preferred diagnostic criterion because it is more reliable. The incidence of phenylketonuria is about 1 in 20,000 newborns. The disease is inherited as an *autosomal recessive*. Heterozygotes, which comprise about 1.5% of a typical population, appear normal. Carriers of the phenylketonuric gene have a reduced level of phenylalanine hydroxylase, as reflected in an increased level of phenylalanine in the blood. However, these criteria are not absolute, because the blood levels of phenylalanine of carriers and normal persons overlap to some extent. The measurement of the kinetics of disappearance of intravenously administered phenylalanine is a more definitive test for the carrier state. It should be noted that a high blood level of phenylalanine in a pregnant woman can result in abnormal development of the fetus. This is a striking example of maternal-fetal relationships at the molecular level.

> Nature is nowhere accustomed more openly to display her secret mysteries than in cases where she shows traces of her workings apart from the beaten path; nor is there any better way to advance the proper practice of medicine than to give our minds to the discovery of the usual law of Nature by careful investigation of cases of rarer forms of disease.
>
> —WILLIAM HARVEY (1657)

SUMMARY

Surplus amino acids are used as metabolic fuel. The degradation of most surplus amino acids starts with the removal of their α-amino groups by transamination to an α-keto acid. Pyridoxal phosphate is the coenzyme in all transaminases. The α-amino groups funnel into α-ketoglutarate to form glutamate, which is then oxidatively deaminated by glutamate dehydrogenase to give NH_4^+ and α-ketoglutarate. NAD^+ or $NADP^+$ is the electron acceptor in this reaction. In terrestrial vertebrates, NH_4^+ is converted into urea by the urea cycle. Urea is formed by the hydrolysis of arginine. The subsequent reactions of the urea cycle synthesize arginine from ornithine, the other product of the hydrolysis reaction. First, ornithine is carbamoylated to citrulline by carbamoyl phosphate. Citrulline then condenses with aspartate to form argininosuccinate, which is cleaved to arginine and fumarate. The carbon atom and one nitrogen atom of urea come from carbamoyl phosphate, which is synthesized from CO_2, NH_4^+, and ATP. The other nitrogen atom of urea comes from aspartate. Four high-energy phosphate bonds are consumed in the synthesis of one molecule of urea.

The carbon atoms of degraded amino acids are converted into pyruvate, acetyl CoA, acetoacetate, or a citric acid cycle intermediate. Most amino acids are purely glucogenic, one is purely ketogenic, and a few are both ketogenic and glucogenic. Alanine, serine, cysteine, glycine, and threonine are degraded to pyruvate. Asparagine and aspartate are converted into oxaloacetate. Alpha-ketoglutarate is the point of entry for glutamate and four amino acids (glutamine, histidine, proline, and arginine) that can be converted into glutamate. Succinyl CoA is the point of entry for some of the carbon atoms of the four amino acids (methionine, isoleucine, threonine, and valine) that are degraded by way of methylmalonyl CoA. Deoxyadenosylcobalamin, a Vitamin B_{12} derivative, is required for the isomerization of methylmalonyl CoA to succinyl CoA. Leucine is degraded to acetoacetyl CoA and acetyl CoA. The aromatic rings of tyrosine and phenylalanine are degraded by oxygenases. Phenylalanine hydroxylase, a monooxygenase, uses tetrahydrobiopterin as the reductant. Some of the carbon atoms of phenylalanine and tyrosine are converted into fumarate, whereas others emerge in acetoacetate.

SELECTED READINGS

BOOKS ON AMINO ACID METABOLISM

Meister, A., 1965. *Biochemistry of the Amino Acids* (2nd ed.), vols. 1 and 2. Academic Press.

Bender, D. A., 1975. *Amino Acid Metabolism.* Wiley.

Grisolia, S., Báguena, R., and Mayor, F., (eds.), 1976. *The Urea Cycle.* Wiley.

REACTION MECHANISMS

Walsh, C., 1979. *Enzymatic Reaction Mechanisms.* Freeman. [Contains excellent accounts of the catalytic mechanisms of pyridoxal phosphate enzymes, cobalamin enzymes, and oxygenases.]

Snell, E. E., and DiMari, S. J., 1970. Schiff base intermediates in enzyme catalysis. *In* Boyer, P. D., (ed.), *The Enzymes* (3rd ed.), vol. 2, pp. 335–370. Academic Press.

Barker, H. A., 1972. Coenzymes B_{12}-dependent mutases causing carbon chain rearrangements. *In* Boyer, P. D., (ed.), *The Enzymes* (3rd ed.), vol. 6, pp. 509–537.

Abeles, R., and Dolphin, D., 1976. The Vitamin B_{12} coenzyme. *Acc. Chem. Res.* 9:114–120.

Corey, E. J., Cooper, N. J., and Green, M. L. H., 1977. Biochemical catalysis involving coenzyme B-12: a rational stepwise mechanistic interpretation of vicinal interchange rearrangements. *Proc. Nat. Acad. Sci. USA* 74:811–815.

INHERITED DISORDERS

Nyhan, W. L., (ed.), 1975. *Hereditable Disorders of Amino Acid Metabolism.* Wiley.

Stanbury, J. B., Wyngaarden, J. B., and Fredrickson, D. S., (eds.), 1978. *The Metabolic Basis of Inherited Disease* (4th ed.), McGraw-Hill. [Contains excellent articles on inherited disorders of amino acid metabolism, including ones by V. E. Shih on urea cycle disorders, L. E. Rosenberg on Vitamin B_{12} metabolism, and A. Y. Tourian and J. B. Sidbury on phenylketonuria.]

Garrod, A. E., 1909. *Inborn Errors in Metabolism.* Oxford University Press (reprinted in 1963 with a supplement by H. Harris).

PROCESS OF DISCOVERY

Childs, B., 1970. Sir Archibald Garrod's conception of chemical individuality: a modern appreciation. *New Engl. J. Med.* 282:71–78

Holmes, F. L., 1980. Hans Krebs and the discovery of the ornithine cycle. *Fed. Proc.* 39:216–225.

PROBLEMS

1. Name the α-keto acid that is formed by trans-amination of each of the following amino acids:
   (a) Alanine.
   (d) Leucine.
   (b) Aspartate.
   (e) Phenylalanine.
   (c) Glutamate.
   (f) Tyrosine.

2. Write a balanced equation for the conversion of aspartate into glucose by way of oxaloacetate. Cite the coenzymes that participate in these steps.

3. Write a balanced equation for the conversion of aspartate into oxaloacetate by way of fumarate.

4. Consider the mechanism of the conversion of L-methylmalonyl CoA into succinyl CoA by L-methylmalonyl CoA mutase.
   (a) Design an experiment to distinguish be-tween the migration of the COO^- group and that of the —CO—S—CoA group in this reaction.
   (b) In fact, it is the —CO—S—CoA group that migrates. Now design an experiment to distinguish between intermolecular and intra-molecular transfer of the —CO—S—CoA group.
   (c) What is the significance of the finding that no tritium is incorporated into succinyl CoA when the mutase reaction is carried out in tritiated water?

5. Pyridoxal phosphate stabilizes carbanionic intermediates by serving as an electron sink. Which other prosthetic group catalyzes reactions in this way?

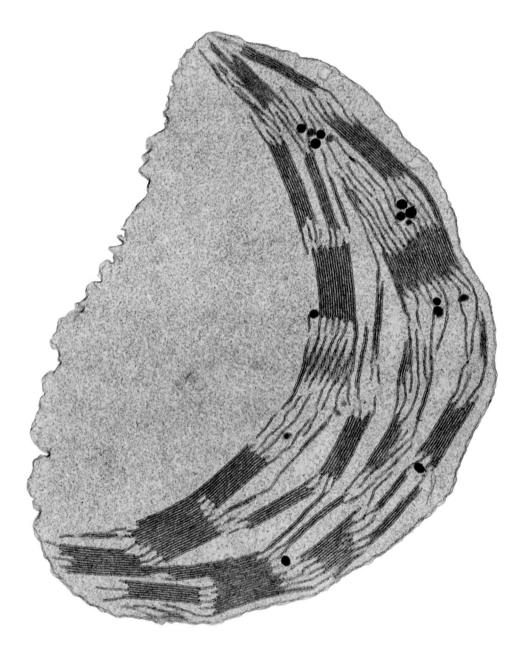

Electron micrograph of a whole chloroplast from a spinach leaf.
[Courtesy of Dr. Kenneth Miller.]

CHAPTER 19

PHOTOSYNTHESIS

All free energy consumed by biological systems arises from solar energy that is trapped by the process of photosynthesis. The basic equation of photosynthesis is simple—indeed deceptively so:

$$H_2O + CO_2 \xrightarrow{\text{light}} (CH_2O) + O_2$$

In this equation, (CH_2O) represents carbohydrate. The mechanism of photosynthesis is complex and requires the interplay of many macromolecules and small molecules. Photosynthesis in green plants occurs in chloroplasts, which are specialized organelles. The energy conversion apparatus is an integral part of the *thylakoid* membrane system in a chloroplast (Figure 19-1). The first step in photosynthesis is the absorption of light by a *chlorophyll molecule*. The energy is transferred from one chlorophyll to another until it reaches a chlorophyll with special properties at a site called the *reaction center*. Light is converted into chemically useful energy at two kinds of reaction centers. In fact, *the cooperation of two light reactions* is required for photosynthesis. One of them, called *photosystem I,* generates reducing power in the form of NADPH, whereas the other, called *photosystem II,* splits water to produce O_2 and generates a reductant. A proton gradient across the thylakoid membrane is generated when O_2 is evolved and when electrons flow through an electron-transport chain linking the two photosystems. The synthe-

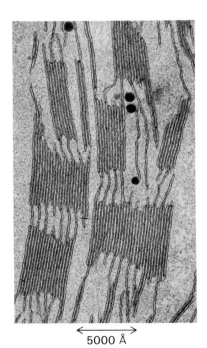

$\xleftarrow{\hspace{1cm}}$ 5000 Å $\xrightarrow{\hspace{1cm}}$

Figure 19-1
Electron micrograph of part of a chloroplast from a spinach leaf. The thylakoid membranes form stacks called grana. [Courtesy of Dr. Kenneth Miller.]

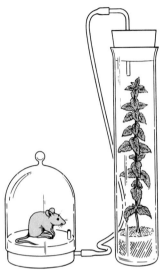

Figure 19-2
Priestley's classic experiment on photosynthesis. [After E.I. Rabinowitch. Photosynthesis. Copyright © 1948 by Scientific American, Inc. All rights reserved.]

sis of ATP is driven by this proton gradient, as in oxidative phosphorylation. ATP can also be generated without the concomitant formation of NADPH.

The NADPH and ATP generated by light are then used to reduce CO_2 to carbohydrate by a series of dark reactions called the *Calvin cycle*. These reactions occur in the soluble component of chloroplasts. The first step is the reaction of CO_2 with ribulose diphosphate to form two molecules of *3-phosphoglycerate*. Hexose is formed from 3-phosphoglycerate by the gluconeogenic pathway, and ribulose diphosphate is regenerated by the action of transketolase, aldolase, and several other enzymes. In one round of the cycle, 3 ATP and 2 NADPH take CO_2 to the level of a hexose phosphate.

DISCOVERY OF THE BASIC EQUATION OF PHOTOSYNTHESIS

Most of the basic equation of photosynthesis could have been written at the end of the eighteenth century. The production of oxygen in photosynthesis was discovered by Joseph Priestley in 1780. He found that plants could "restore air which has been injured by the burning of candles." He placed a sprig of mint in an inverted glass jar in a vessel of water and found several days later "that the air would neither extinguish a candle, nor was it all inconvenient to a mouse which I put into it." This great chemist of the eighteenth century was also a nonconformist English minister. Indeed, theology, philosophy, and politics were his primary interests. In 1791, Priestley was forced to leave England because of his sympathies with the French Revolution. He went to France and then to the United States, where he died in 1804, after a few quiet years high on the banks of the Susquehanna. A quite different fate was suffered by a contemporary of Priestley, who set the groundwork for the elucidation of the basic process of photosynthesis. Antoine Lavoisier devised methods for the handling of gases and discovered the concept of oxidation and the law of conservation of mass in chemical reactions. He was associated with the monarchy and was executed in 1794 by the French revolutionaries. The judge who pronounced sentence commented that "the Republic has no need for scientists."

The next major contribution to the elucidation of photosynthesis was made by Jan Ingenhousz, a Dutchman, who was court physician to the Austrian empress. Ingenhousz was a worldly man who liked to visit London. He once heard a discussion of Priestley's experiments on the restoration of air by plants and was so taken by it that he decided that he must do some experiments at the "earliest opportunity." This came six years later, when Ingenhousz rented a villa near London and spent a summer feverishly performing more than five hundred experiments. He discovered the role of light in photosynthesis:

I observed that plants not only have the faculty to correct bad air in six or ten days, by growing in it, as the experiments of Dr. Priestley indicate, but that they perform this important office in a complete manner in a few hours; that *this wonderful operation is by no means owing to the vegetation of the plant, but to the influence of light of the sun upon the plant.*

Ingenhousz was quick to publish his findings because he was concerned that someone else would beat him to it. By the end of the summer, he had published a book titled *Experiments upon Vegetables, Discovering Their Great Power of Purifying the Common Air in Sunshine and Injuring it in the Shade and at Night.*

Ingenhousz's fear of being scooped was justified. Similar experiments were being carried out in Geneva by Jean Senebier, a Swiss pastor. His distinctive contribution was to show that "fixed air"—namely, CO_2—is taken up in photosynthesis. The role of water in photosynthesis was demonstrated by Theodore de Saussure, also a Genevan. He showed that the sum of the weights of organic matter produced by plants and of the oxygen evolved is much more than the weight of CO_2 consumed. From Lavoisier's law of the conservation of mass, de Saussure concluded that another substance was utilized. The only inputs in his system were CO_2, water, and light. Hence, de Saussure concluded that the other reactant must be water.

The final contribution to the basic equation of photosynthesis came nearly a half-century later. Julius Robert Mayer, a German surgeon, discovered the law of conservation of energy in 1842. Mayer recognized that plants convert solar energy into chemical free energy:

The plants take in one form of power, light; and produce another power, chemical difference.

The amount of energy stored by photosynthesis is enormous. More than 10^{17} kcal of free energy is stored annually by photosynthesis on earth, which corresponds to the assimilation of more than 10^{10} tons of carbon into carbohydrate and other forms of organic matter.

CHLOROPHYLLS ARE THE PHOTORECEPTOR MOLECULES

Mayer stated, "Nature has put itself the problem of how to catch in flight light streaming to the earth and to store the most elusive of all powers in rigid form." What is the mechanism of trapping this most elusive of all powers? The first step is the absorption of light by a photoreceptor molecule. The principal photoreceptor in the chloroplasts of green plants is *chlorophyll* a, a substituted tetrapyrrole. The four nitrogen atoms of the pyrroles are coordinated to a magnesium atom. Thus, chlorophyll is a *magnesium porphyrin,* whereas heme is an iron porphyrin. The porphyrin ring structure of chlorophyll differs

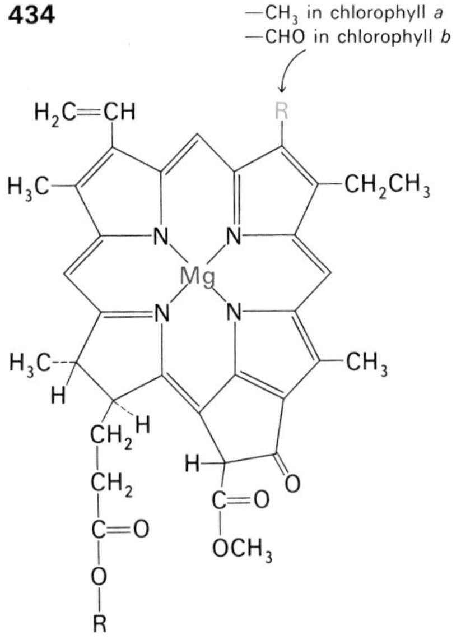

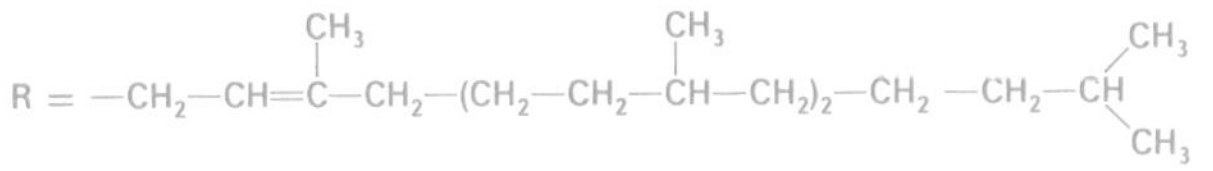

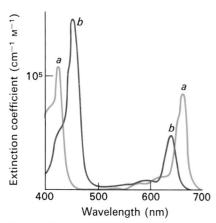

Figure 19-3
Formulas of chlorophylls *a* and *b*.

from that of heme in several ways: (1) one of the pyrroles is partially reduced; (2) a cyclopentanone ring is fused to one of the pyrroles; (3) both acid side chains in chlorophyll are esterified, whereas in heme they are free. One of the acid chains in chlorophyll is a methyl ester, whereas the other is an ester of phytol ($C_{20}H_{39}OH$). This long-chain alcohol consists of four isoprene units, which make it highly hydrophobic. *Chlorophyll* b differs from chlorophyll *a* in a substituent on one of the pyrroles: in chlorophyll *b*, there is a formyl group in place of a methyl group.

These chlorophylls are very effective photoreceptors because they contain networks of alternating single and double bonds. In other words, they are *polyenes*. They have very strong absorption bands in the visible region of the spectrum, where the solar output reaching the earth also is maximal. The peak extinction coefficients of chlorophylls *a* and *b* are higher than $10^5 \text{ cm}^{-1} \text{ M}^{-1}$, among the highest observed for organic compounds.

The absorption spectra of chlorophylls *a* and *b* are different (Figure 19-4). Light that is not appreciably absorbed by chlorophyll *a*—at 460 nm, for example—is captured by chlorophyll *b*, which has intense absorption at that wavelength. Thus, *these two kinds of chlorophyll complement each other in absorbing the incident sunlight.* There is still a large spectral region, from 500 to 600 nm, where the absorption of light is relatively weak. However, most plants need not capture light in that spectral region, because enough is absorbed in the blue and red parts of the spectrum.

Figure 19-4
Absorption spectra of chlorophylls *a* and *b*.

THE PRIMARY EVENTS OF PHOTOSYNTHESIS OCCUR IN A HIGHLY ORGANIZED MEMBRANE SYSTEM

Chloroplasts, the organelles of photosynthesis, are typically 5 μm long. Like a mitochondrion, a chloroplast has an outer membrane and an inner membrane with an intervening intermembrane space (Figure 19-5). The inner membrane surrounds a *stroma* containing

Figure 19-5
Diagram of a chloroplast. [After *Biology of the Cell* by Stephen L. Wolfe. © 1972 by Wadsworth Publishing Company, Inc., Belmont, California 94022. Adapted by permission of the publisher.]

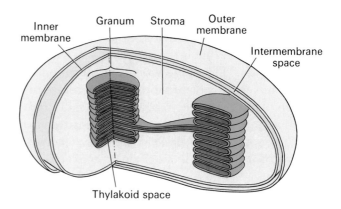

soluble enzymes and membranous structures called *thylakoids,* which are flattened sacs. A pile of these sacs is a granum. Different grana are connected by membrane regions called stroma lamellae. The thylakoid membranes separate the thylakoid space from the stroma space. Thus, chloroplasts have three different membranes (*outer, inner,* and *thylakoid membranes*) and three separate spaces (*intermembrane, stroma,* and *thylakoid spaces*). In developing chloroplasts, thylakoids arise from invaginations of the inner membrane, and so they are analogous to mitochondrial cristae.

The thylakoid membranes contain the chlorophyll molecules and the other components of the energy-transducing machinery. They have nearly equal amounts of lipids and proteins. The lipid composition is highly distinctive: about 40% of the total lipids are *galactolipids* and 4% are *sulfolipids,* whereas only 10% are phospholipids. The thylakoid membrane, like the inner mitochondrial membrane, is impermeable to most molecules and ions. The stroma contains the soluble enzymes that utilize the NADPH and ATP synthesized by the thylakoids to convert CO_2 into sugar. Chloroplasts contain their own DNA (p. 694) and the machinery for synthesizing RNA and protein. Thus, *a chloroplast is an organelle with considerable autonomy.* The inner membrane of a chloroplast, which contains translocators for a variety of compounds such as ATP and dicarboxylic acids, is the site of interaction between the chloroplast and the rest of the cell. The outer membrane of a chloroplast, like that of a mitochondrion, is highly permeable to small molecules and ions.

THE PHOTOSYNTHETIC UNIT: PHOTONS FUNNEL INTO A REACTION CENTER

When the rate of photosynthesis is measured as a function of the intensity of illumination, it increases linearly at low intensities and reaches a saturating value at high intensities (Figure 19-6). A saturating value is observed in strong light because chemical reactions that utilize the absorbed photons become rate-limiting. Thus, *photosynthesis can be separated into light reactions and dark reactions.* As will be discussed shortly, the light reactions generate NADPH and ATP, whereas the dark reactions use these energy-rich molecules to reduce CO_2.

In 1932, Robert Emerson and William Arnold measured the oxygen yield of photosynthesis when *Chlorella* cells were exposed to light flashes lasting a few microseconds. They expected to find that the yield per flash would increase with the flash intensity until each chlorophyll molecule absorbed a photon, which would then be used in dark reactions. Their experimental observation was entirely unexpected: a saturating light flash led to the production of only one molecule of O_2 per 2,500 chlorophyll molecules.

This experiment led to the concept of the *photosynthetic unit.* Hans Gaffron proposed that light is absorbed by hundreds of chlorophyll

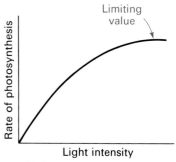

Figure 19-6
The rate of photosynthesis reaches a limiting value when the light intensity suffices to excite only a small fraction of the chlorophyll molecules.

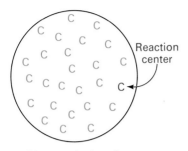

Photosynthetic unit

Figure 19-7
Diagram of a photosynthetic unit. The antenna chlorophyll molecules (denoted by green C's) transfer their excitation energy to a specialized chlorophyll at the reaction center (denoted by a red C).

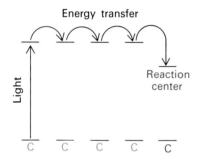

Figure 19-8
Diagram of the energy levels of the excited state of the antenna chlorophylls and of the reaction center.

molecules, which then transfer their excitation energy to a site at which chemical reactions occur (Figure 19-7). This site is called a *reaction center*. Thus, the function of most chlorophyll molecules in the photosynthetic unit is to absorb light. Only a small proportion of chlorophylls, those at reaction centers, mediate the transformation of light into chemical energy. The chlorophylls at the reaction center are chemically identical with the other chlorophylls in the photosynthetic unit, but their properties are different because of their unique environment. One difference is that the energy level of their excited state is lower than that of the other chlorophylls, which makes them energy traps. The energy absorbed by chlorophyll molecules hops around the photosynthetic unit until it reaches a reaction-center chlorophyll. The transfer of energy to the reaction center is very rapid, occurring in less than 10^{-10} sec.

THE OXYGEN EVOLVED IN PHOTOSYNTHESIS COMES FROM WATER

Let us turn now to the chemical changes in photosynthesis. The source of the evolved oxygen in green plants has important implications for the mechanism of photosynthesis. Comparative studies of photosynthesis in many organisms revealed the source as early as 1931. Some photosynthetic bacteria convert hydrogen sulfide into sulfur in the presence of light. Cornelis Van Niel perceived that the reactions for photosynthesis in green plants and green sulfur bacteria are very similar:

$$CO_2 + 2\,H_2O \xrightarrow{\text{light}} (CH_2O) + O_2 + H_2O$$
$$CO_2 + 2\,H_2S \xrightarrow{\text{light}} (CH_2O) + 2\,S + H_2O$$

The sulfur that is formed by the photosynthetic bacteria is analogous to the oxygen that is evolved in plants. Van Niel proposed a general formula for photosynthesis:

$$\underset{\substack{\text{Hydrogen} \\ \text{acceptor}}}{CO_2} + \underset{\substack{\text{Hydrogen} \\ \text{donor}}}{2\,H_2A} \xrightarrow{\text{light}} \underset{\substack{\text{Reduced} \\ \text{acceptor}}}{(CH_2O)} + \underset{\substack{\text{Dehydrogenated} \\ \text{donor}}}{2\,A} + H_2O$$

The hydrogen donor H_2A is H_2O in green plants and H_2S in the photosynthetic sulfur bacteria. Thus, photosynthesis in plants could be formulated as a reaction in which CO_2 is reduced by hydrogen derived from water. Oxygen evolution would then be the necessary consequence of this dehydrogenation process. The essence of this view of photosynthesis is that *water is split by light*.

The availability in 1941 of a heavy isotope of oxygen, namely ^{18}O, made it feasible to test this concept directly. In fact, ^{18}O appeared in the evolved oxygen when photosynthesis was carried out in water enriched in this isotope. This result confirmed the proposal that the O_2 formed in photosynthesis comes from water.

$$H_2{}^{18}O + CO_2 \xrightarrow{\text{light}} (CH_2O) + {}^{18}O_2$$

HILL REACTION: ILLUMINATED CHLOROPLASTS EVOLVE OXYGEN AND REDUCE AN ARTIFICIAL ELECTRON ACCEPTOR

In 1939, Robert Hill discovered that isolated chloroplasts evolve oxygen when they are illuminated in the presence of a suitable electron acceptor, such as ferricyanide. There is a concomitant reduction of ferricyanide to ferrocyanide. The Hill reaction is a landmark in the elucidation of the mechanism of photosynthesis for several reasons:

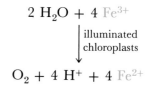

1. It dissected photosynthesis by showing that oxygen evolution can occur without the reduction of CO_2. Artificial electron acceptors such as ferricyanide can substitute for CO_2.

2. It confirmed that the evolved oxygen comes from water rather than from CO_2, because no CO_2 was present.

3. It showed that isolated chloroplasts can perform a significant partial reaction of photosynthesis.

4. It revealed that a primary event in photosynthesis is the *light-activated transfer of an electron from one substance to another against a chemical-potential gradient.* The reduction of ferric to ferrous ion by light is a conversion of light into chemical energy.

PHOTOSYNTHESIS REQUIRES THE INTERACTIONS OF TWO KINDS OF PHOTOSYSTEMS

A number of experimental observations led to the finding that there are two different photosystems in chloroplasts. The rate of photosynthesis was investigated as a function of the wavelength of the incident light. The photosynthetic rate divided by the number of quanta observed at each wavelength gives the relative quantum efficiency of the process. For a single kind of photoreceptor, the quantum efficiency is expected to be independent of wavelength over its entire absorption band. This is not the case in photosynthesis: the quantum efficiency of photosynthesis drops sharply at wavelengths longer than 680 nm, although chlorophyll still absorbs light in the range from 680 to 700 nm (Figure 19-9). However, the rate of photosynthesis using long-wavelength light can be enhanced by adding light of a shorter wavelength, such as 600 nm. The photosynthetic rate in the presence of both 600-nm and 700-nm light is

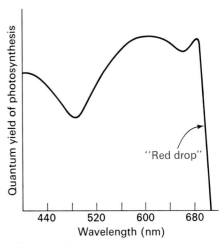

Figure 19-9
The quantum yield of photosynthesis drops abruptly when the excitation wavelength is greater than 680 nm.

greater than the sum of the rates obtained when the two wavelengths are given separately. These observations, called the red drop and the enhancement phenomenon, led Emerson to propose that *photosynthesis requires the interaction of two light reactions: both of them can be driven by light of less than 680 nm but only one of them by light of longer wavelength.*

ROLES OF THE TWO PHOTOSYSTEMS

Photosystem I, which can be excited by light of wavelength shorter than 700 nm, generates a strong reductant that leads to the formation of NADPH. In contrast, photosystem II, which requires light of shorter wavelength than 680 nm, produces a strong oxidant that leads to the formation of O_2. In addition, photosystem I produces a weak oxidant, whereas photosystem II produces a weak reductant. The interaction of these species results in the formation of ATP. This part of photosynthesis, which was discovered by Daniel Arnon, is called photosynthetic phosphorylation or *photophosphorylation.*

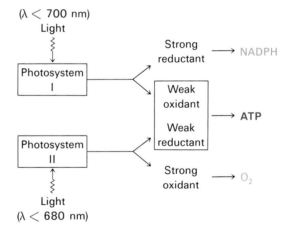

Figure 19-10
Interaction of photosystems I and II in photosynthesis.

Photosystems I and II are structurally distinct. When thylakoid membranes are treated with detergents, photosystem I particles are preferentially released. By density-gradient centrifugation, it is feasible to separate particles that have only photosystem I activity from others that are enriched in photosystem II activity. Most of the chlorophyll molecules are bound to specific proteins. A complex consisting of 14 chlorophyll *a* molecules bound to a 110-kdal protein has been isolated from photosystem I particles. A second type of complex, made up of photosystem II particles, contains 3 chlorophyll *a* and 3 chlorophyll *b* molecules bound to a 28-kdal protein. The best-characterized chlorophyll-protein complex, purified from green bacteria, consists of three 50-kdal subunits, each with seven bacteriochlorophyll molecules (Figure 19-11). *One role of the protein in these complexes is to maintain an optimal geometry for energy transfer between chlorophylls.*

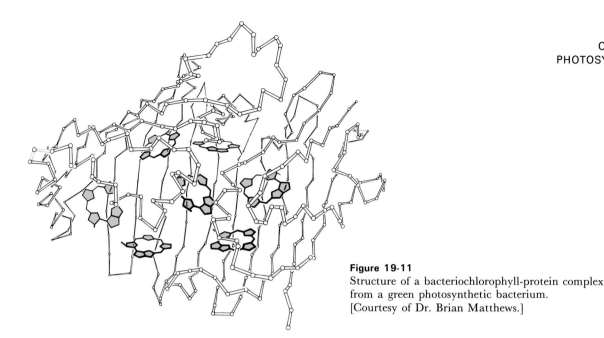

Figure 19-11
Structure of a bacteriochlorophyll-protein complex
from a green photosynthetic bacterium.
[Courtesy of Dr. Brian Matthews.]

PHOTOSYSTEM I GENERATES NADPH VIA REDUCED FERREDOXIN

The reaction center of photosystem I (Figure 19-12) is a chlorophyll *a* molecule in a unique environment. The absorption maximum is shifted from 680 to 700 nm. Consequently, this reaction center is called *P700* (P stands for pigment).

Many chlorophyll molecules absorb light and transfer the excitation energy to P700. When P700 is excited, it transfers an electron to *bound ferredoxin* (P430), the membrane-bound form of ferredoxin, an 11.6-kdal iron-sulfur protein of the Fe_4S_4 type. In the dark, P700 has an oxidation-reduction potential of $+0.4$ V. When P700 is excited by light, it has a different distribution of electrons, which changes its oxidation-reduction potential to about -0.6 V. Thus, *light pumps an electron in photosystem I from a potential of about $+0.4$ to -0.6 V.* The energy of a red photon, 1.8 eV (electron volts), is sufficient to elevate an electron by 1.0 V. The transfer of an excited electron from P700 to bound ferredoxin makes P700 electron deficient. This oxidized form of P700 must regain an electron before P700 can again function as a reaction center. The source of the restoring electrons will be discussed shortly.

Bound ferredoxin then transfers its electron to the soluble form of ferredoxin. The iron atom at the active site of ferredoxin is alternately oxidized and reduced. Reduced ferredoxin transfers its electron to $NADP^+$ to form NADPH. This reaction is catalyzed by *ferredoxin-NADP reductase,* which contains FAD as its prosthetic group. Note that the reduction of $NADP^+$ to NADPH is a two-electron process, whereas ferredoxin is a one-electron carrier. Hence, the

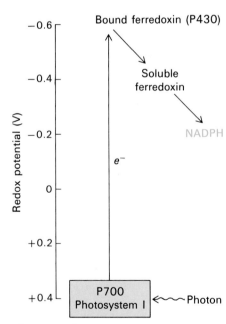

Figure 19-12
Formation of NADPH by photosystem I.

electrons from two reduced ferredoxins must converge to form one NADPH.

$$2 \text{ Ferredoxin}_{\text{reduced}} + H^+ + NADP^+ \xrightarrow{\text{ferredoxin-NADP reductase}}$$

$$2 \text{ ferredoxin}_{\text{oxidized}} + NADPH$$

PHOTOSYSTEM II GENERATES A STRONG OXIDANT THAT SPLITS WATER

Relatively little is known about the reaction center (P680) and the primary electron acceptor of photosystem II (Figure 19-13). The

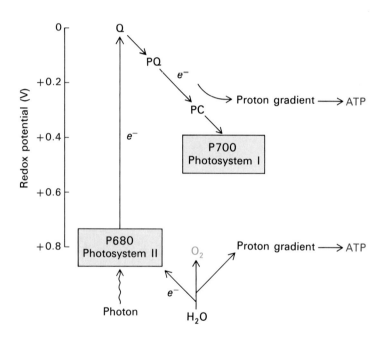

Figure 19-13
Excitation of photosystem II leads to the formation of O_2 and the transfer of an electron to photosystem I. A proton gradient across the thylakoid membrane is generated by these processes. (PQ refers to plastoquinone and PC to plastocyanin.)

redox potential of this reaction center is about $+0.8$ V. Light creates a very strong oxidant, Z^+ (the oxidized form of the reaction center or derived from it), and a weak reductant, Q^-, at this center. Z^+ extracts electrons from water to form O_2. Manganese plays an essential role in this process.

$$4 Z^+ + 2 H_2O \longrightarrow 4 Z + 4 H^+ + O_2$$

A PROTON GRADIENT IS FORMED AS ELECTRONS FLOW FROM PHOTOSYSTEM II TO I

The two photosystems are linked by a series of electron carriers that have two roles:

1. Electrons flow through this link from photosystem II to I.

Table 19-1
Composition of photosystems I and II

Photosystem I

~200	Antenna chlorophylls
~50	Carotenoids
1	P700 reaction center
1	Cytochrome c_{552} (f)
1	Plastocyanin
2	Cytochrome b_{563} (b_6)
1	Bound ferredoxin
1	Soluble ferredoxin
1	Ferredoxin-NADP reductase

Photosystem II

~200	Antenna chlorophylls
~50	Carotenoids
1	P680 reaction center
1	Primary electron donor (Z)
1	Primary electron acceptor (Q)
~4	Plastoquinone
6	Manganese atoms
2	Cytochrome b_{559}

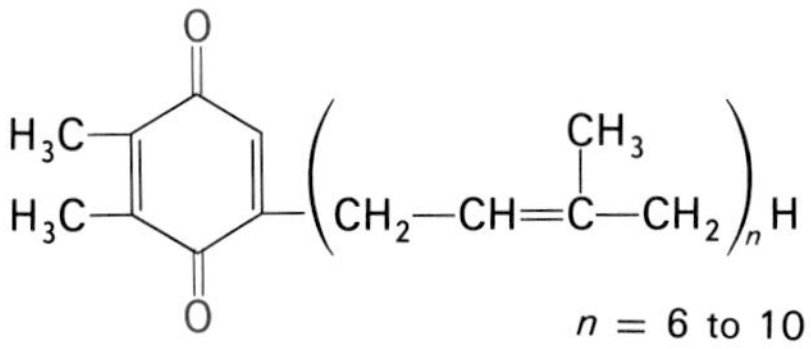

Plastoquinone

These electrons are needed to *regenerate the reduced form of P700,* the reaction center of photosystem I.

2. A *proton gradient* across the thylakoid membrane is formed as electrons flow through this link. Protons formed in the evolution of O_2 also contribute to this proton gradient. The thylakoid space becomes more acidic. This proton gradient drives the synthesis of ATP (p. 443).

The electron from the photoactivated reaction center of photosystem II is transferred to Q, which is a tightly bound molecule of *plastoquinone.* This quinone closely resembles ubiquinone (p. 312), a member of the electron-transport chain in mitochondria, and vitamin K (p. 249). The electron is then transferred from Q to a pool of *mobile plastoquinones* and then to *cytochrome* b_{559}. The next member of the electron-transport chain is *cytochrome* c_{552} (formerly called cytochrome *f*), which is an integral membrane protein, in contrast with cytochrome *c* in mitochondria. The final electron carrier from photosystem II to I is *plastocyanin,* a 10.5-kdal protein with a single *copper* atom, which is coordinated to four groups: a cysteine, a methionine, and two histidine side chains (Figure 19-14). This coordination geometry is distorted from the planar array characteristic of low-molecular-weight Cu^{2+} complexes. The localized strain at the Cu atom probably facilitates electron transfer as Cu alternates between +1 and +2 oxidation states. The Cu atom is near the surface of the plastocyanin molecule, shielded only by a histidine side chain, and

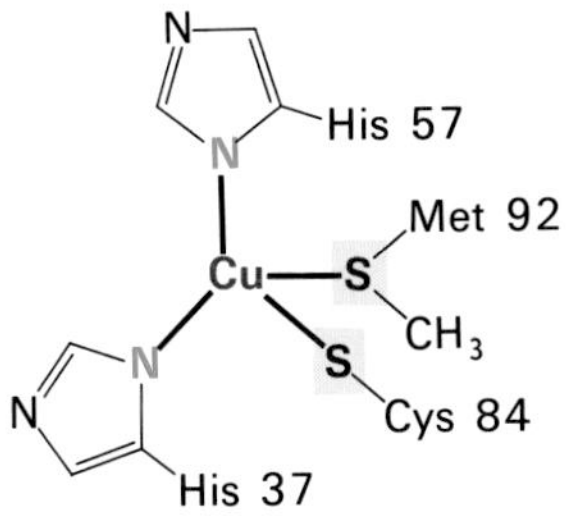

Figure 19-14
Structure of the coordinated copper ion in plastocyanin.

so electrons could be transferred to Cu by a direct mechanism (p. 328). Electron transfer from reduced plastocyanin to the oxidized form of P700 enables this reaction center to again serve as an electron donor to form NADPH when illuminated. The net reaction produced by the photoactivation of photosystems I and II is

$$2\,H_2O + 2\,NADP^+ \longrightarrow O_2 + 2\,NADPH + 2\,H^+$$

In other words, *light causes electrons to flow from H₂O to NADPH* (Figure 19-15).

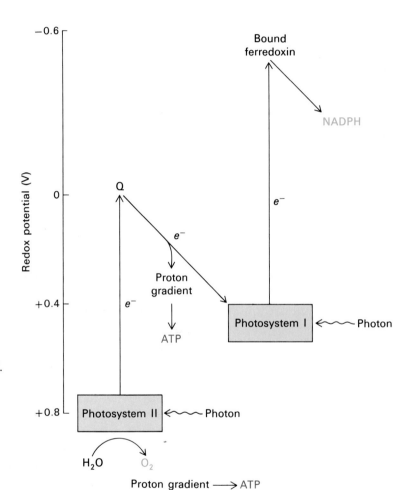

Figure 19-15
Electron flow in noncyclic photophosphorylation. There is net electron flow from H_2O to $NADP^+$ to form NADPH.

ATP CAN ALSO BE FORMED BY CYCLIC ELECTRON FLOW THROUGH PHOTOSYSTEM I

There is an alternate pathway for electrons arising from P700, the reaction center of photosystem I. The high-potential electron in bound ferredoxin can be transferred to cytochrome b_{563} rather than to $NADP^+$. This electron then flows back to the oxidized form of P700 through cytochrome c_{552} and plastocyanin. In other words,

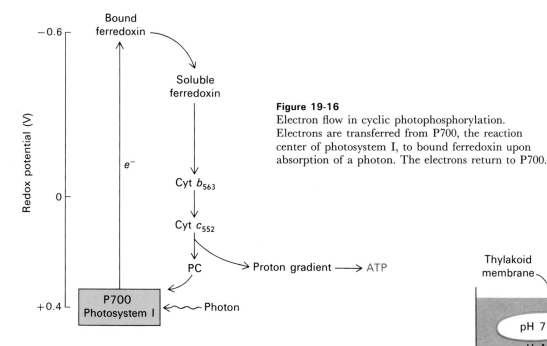

Figure 19-16
Electron flow in cyclic photophosphorylation.
Electrons are transferred from P700, the reaction
center of photosystem I, to bound ferredoxin upon
absorption of a photon. The electrons return to P700.

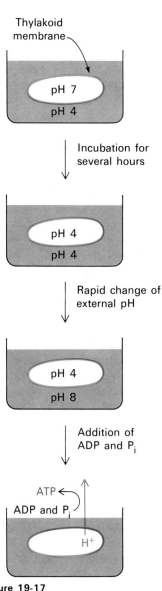

Figure 19-17
Synthesis of ATP by chloroplasts following the imposition of a pH gradient.

there is a cyclic flow of electrons. ATP is generated as electrons return to the reaction center via cytochrome b_{563} and plastocyanin. Hence, this process is called *cyclic photophosphorylation* (Figure 19-16). In this mode, *ATP is generated without the concomitant formation of NADPH*. Photosystem II does not participate in cyclic photophosphorylation, and so O_2 is not formed from H_2O during this process. Cyclic photophosphorylation is operative when there is insufficient $NADP^+$ to accept electrons from reduced ferredoxin. This condition exists when the ratio of NADPH to $NADP^+$ is high.

A PROTON GRADIENT ACROSS THE THYLAKOID MEMBRANE DRIVES THE SYNTHESIS OF ATP

In 1966, André Jagendorf showed that chloroplasts synthesize ATP in the dark when an artificial pH gradient is imposed across the thylakoid membrane. This transient pH gradient was obtained by soaking chloroplasts in a pH 4 buffer for several hours. These chloroplasts were then rapidly mixed with a pH 8 buffer containing ADP and P_i. The pH of the stroma suddenly increased to 8, whereas the pH of the thylakoid space remained at 4. *A burst of ATP synthesis then accompanied the disappearance of the pH gradient across the thylakoid membrane* (Figure 19-17). This revealing experiment provided strong support for the chemiosmotic hypothesis for ATP synthesis (p. 316).

Indeed, the mechanism of ATP synthesis in chloroplasts is very similar to that in mitochondria. *ATP formation is driven by a proton-motive force in both photophosphorylation and oxidative phosphorylation.* Furthermore, the enzyme assembly catalyzing ATP formation in

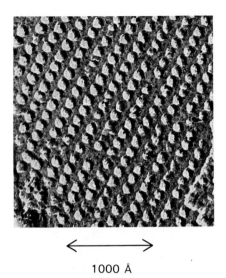

$\longleftrightarrow$
1000 Å

Figure 19-18
Freeze-fracture electron micrograph of spinach thylakoid membranes. A regular lattice of transmembrane particles is evident [Courtesy of Dr. Kenneth Miller.]

chloroplasts, called the *CF₁-CF₀ complex* (C stands for chloroplast and F for factor), is very similar to the mitochondrial F_1-F_0 complex. CF_1 catalyzes the formation of ATP from ADP and P_i. It contains pairs of five kinds of subunits and has a mass of 325 kdal. CF_0, which contains the proton channel, consists of three kinds of subunits. The knobs on the external surface of the thylakoid membrane are the CF_1 units of these ATP-synthesizing complexes.

Electron transfer through the asymmetrically oriented photosystems I and II produces a large proton gradient across the thylakoid membrane. *The thylakoid space becomes markedly acidic, with the pH approaching 4. The light-induced transmembrane proton gradient is about 3.5 pH units.* As discussed previously (p. 316), the proton-motive force, Δp, consists of a pH-gradient contribution and a membrane-potential contribution. In chloroplasts, nearly all of Δp arises from the pH gradient, whereas in mitochondria the contribution from the membrane potential is larger. The reason for this difference is that the thylakoid membrane is quite permeable to Cl^- and Mg^{2+}. The light-induced transfer of H^+ into the thylakoid space is accompanied by the transfer of either Cl^- in the same direction or Mg^{2+} (1 per 2 H^+) in the opposite direction. Consequently, electrical neutrality is maintained and no membrane potential is generated. The proton-motive force of about 0.2 V across the thylakoid membrane is equivalent to about 4.8 kcal per mole of protons. *About three protons flow through the CF_1-CF_0 complex per ATP synthesized, which corresponds to a free-energy input of 14.4 kcal per mole of ATP.* No ATP is synthesized if the pH gradient is less than two units because the driving force is then too small.

CF_1 is on the stromal surface of the thylakoid membrane, and so the newly synthesized ATP is released into the stromal space. Likewise, NADPH formed by photosystem I is released into the stromal space. Thus, *ATP and NADPH, the products of the light reactions of photosynthesis, are appropriately positioned for the subsequent dark reactions in which CO_2 is converted into carbohydrate.*

ELUCIDATION OF THE PATH OF CARBON BY RADIOACTIVE PULSE LABELING

In 1945, Melvin Calvin and his colleagues started a series of investigations that resulted in the elucidation of the dark reactions of photosynthesis. They used the unicellular green alga *Chlorella* in their work because it was easy to culture these organisms in a highly reproducible way. Their findings later proved to be pertinent to a wide variety of photosynthetic organisms ranging from photosynthetic bacteria to higher plants.

The aim of their work was to determine the pathway by which CO_2 becomes fixed into carbohydrate. The experimental strategy was to use radioactive ^{14}C to trace the fate of CO_2. Radioactive $^{14}CO_2$ was injected into an illuminated suspension of algae that had been carrying out photosynthesis with normal CO_2. The algae were killed after a preselected time by dropping the suspension into alcohol, which also stopped the enzymatic reactions.

The radioactive compounds in the algae were separated and identified by two-dimensional paper chromatography. The paper chromatogram was then pressed against photographic film, which became black where the paper contained a radioactive spot. In his Nobel Lecture, Calvin noted that their primary data resided "in the number, position, and intensity—that is, radioactivity—of the blackened areas. The paper ordinarily does not print out the names of these compounds, unfortunately, and our principal chore for the succeeding ten years was to properly label those blacked areas on the film."

CO$_2$ REACTS WITH RIBULOSE DIPHOSPHATE TO FORM TWO PHOSPHOGLYCERATES

The radiochromatogram after sixty seconds of illumination was so complex (Figure 19-19) that it was not feasible to detect the earliest intermediate in the fixation of CO_2. However, the pattern after only five seconds of illumination was much simpler. In fact, there was just one prominent radioactive spot, which proved to be *3-phosphoglycerate*.

The formation of 3-phosphoglycerate as the first detectable radioactive intermediate suggested that a two-carbon compound is the acceptor for the CO_2. This proved not to be so. The actual reaction sequence is more complex:

$$C_5 \xrightarrow{CO_2} C_6 \xrightarrow{H_2O} C_3 + C_3$$

The CO_2 molecule condenses with ribulose 1,5-diphosphate to form a transient six-carbon compound, which is rapidly hydrolyzed to two molecules of 3-phosphoglycerate. The overall reaction is highly exergonic ($\Delta G^{\circ\prime} = -12.4$ kcal/mol).

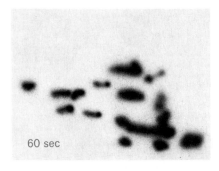

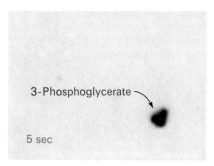

Figure 19-19
Radiochromatograms of illuminated suspensions of algae 5 sec and 60 sec after the injection of CO_2. [Courtesy of Dr. J.A. Bassham.]

Ribulose 1,5-diphosphate → Transient intermediate → Two molecules of 3-phosphoglycerate

This reaction is catalyzed by *ribulose 1,5-diphosphate carboxylase* (also called ribulose 1,5-*bis*phosphate carboxylase), which is located on the stromal surface of thylakoid membranes. This enzyme is very

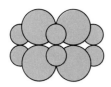

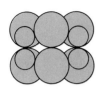

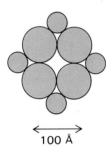

$\xleftarrow{\hspace{1cm}}$ 100 Å $\xrightarrow{\hspace{1cm}}$

Figure 19-20
Schematic diagrams of the structure of ribulose 1,5-diphosphate carboxylase. [After a drawing kindly provided by Dr. David Eisenberg.]

$C_6 + C_3 \xrightarrow{\text{transketolase}} C_4 + C_5$

$C_4 + C_3 \xrightarrow{\text{aldolase}} C_7$

$C_7 + C_3 \xrightarrow{\text{transketolase}} C_5 + C_5$

abundant in chloroplasts, comprising more than 16% of their total protein. In fact, ribulose 1,5-diphosphate carboxylase is probably the most abundant protein in the biosphere. It consists of eight large subunits (55 kdal) and eight small ones (15 kdal) arranged in two layers (Figure 19-20). The large subunits are catalytically active without the small subunits, which have a regulatory role. This enzyme is both an oxygenase (p. 450) and a carboxylase.

FORMATION OF FRUCTOSE 6-PHOSPHATE AND REGENERATION OF RIBULOSE 1,5-DIPHOSPHATE

The steps in the conversion of 3-phosphoglycerate into fructose 6-phosphate (Figure 19-21) are like those of the gluconeogenic pathway (p. 346), except that glyceraldehyde 3-phosphate dehydrogenase in chloroplasts is specific for NADPH, rather than NADH.

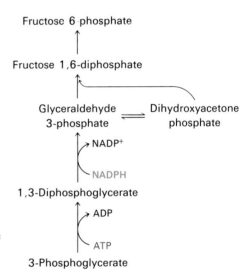

Figure 19-21
Pathway for the conversion of 3-phosphoglycerate into fructose 6-phosphate in chloroplasts.

These reactions bring CO_2 to the level of a hexose. The remaining task is to regenerate ribulose diphosphate, the acceptor of CO_2 in the first dark step. The problem is to construct a five-carbon sugar from six-carbon and three-carbon sugars. This is accomplished by reactions that are catalyzed by *transketolase* and *aldolase*. Transketolase also plays a role in the pentose phosphate pathway (p. 335). Recall that transketolase, a thiamine pyrophosphate (TPP) enzyme, transfers a two-carbon unit ($CH_2OH—CO—$) from a ketose to an aldose. Aldolase carries out an aldol condensation between dihydroxyacetone phosphate and an aldehyde. This enzyme is highly specific for dihydroxyacetone phosphate, but it accepts a wide variety of aldehydes. The specific reactions catalyzed by transketolase and aldolase in the Calvin cycle are:

Fructose 6-phosphate + glyceraldehyde 3-phosphate $\xrightarrow{\text{transketolase}}$
$$\text{xylulose 5-phosphate} + \text{erythrose 4-phosphate}$$

Erythrose 4-phosphate + dihydroxyacetone phosphate $\xrightarrow{\text{aldolase}}$
$$\text{sedoheptulose 1,7-diphosphate}$$

Sedoheptulose 7-phosphate + glyceraldehyde 3-phosphate $\xrightarrow{\text{transketolase}}$
$$\text{ribose 5-phosphate} + \text{xylulose 5-phosphate}$$

Four additional enzymes are needed for the dark reactions of photosynthesis. One, a *phosphatase,* hydrolyzes sedoheptulose 1,7-diphosphate to sedoheptulose 7-phosphate. A second, *phosphopentose epimerase,* converts xylulose 5-phosphate into ribulose 5-phosphate. A third, *phosphopentose isomerase,* converts ribose 5-phosphate into ribulose 5-phosphate. Recall that the second and third enzymes also participate in the pentose phosphate pathway (p. 337). The sum of the preceding reactions is:

Fructose 6-phosphate + 2 glyceraldehyde 3-phosphate +
$$\text{dihydroxyacetone phosphate} \longrightarrow \text{3 ribulose 5-phosphate}$$

Finally, the fourth, *phosphoribulose kinase,* catalyzes the phosphorylation of ribulose 5-phosphate to regenerate ribulose 1,5-diphosphate, the acceptor of CO_2.

Ribulose 5-phosphate + ATP $\longrightarrow$
$$\text{ribulose 1,5-diphosphate} + \text{ADP} + \text{H}^+$$

This series of reactions is called the *Calvin cycle* (Figure 19-22).

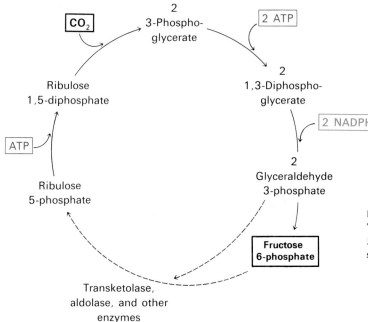

Figure 19-22
The Calvin cycle. The formation of ribulose 5-phosphate from three-carbon and six-carbon sugars is not explicitly shown in this diagram.

THREE ATP AND TWO NADPH BRING CO_2 TO THE LEVEL OF A HEXOSE

What is the energy expenditure for synthesizing a hexose? Six rounds of the Calvin cycle are required, because one carbon atom is reduced in each (Figure 19-22). Twelve ATP are expended in phosphorylating twelve molecules of 3-phosphoglycerate to 1,3-diphosphoglycerate, and twelve NADPH are consumed in reducing twelve molecules of 1,3-diphosphoglycerate to glyceraldehyde 3-phosphate. An additional six ATP are spent in regenerating ribulose 1,5-diphosphate.

We now write a balanced equation for the net reaction of the Calvin cycle:

$$6\,CO_2 + 18\,ATP + 12\,NADPH + 12\,H_2O \longrightarrow$$
$$C_6H_{12}O_6 + 18\,ADP + 18\,P_i + 12\,NADP^+ + 6\,H^+$$

Thus, three molecules of ATP and two of NADPH are consumed in converting CO_2 into a hexose such as glucose or fructose. The efficiency of photosynthesis can be estimated in the following way:

1. The $\Delta G^{\circ\prime}$ for the reduction of CO_2 to the level of hexose is 114 kcal/mol.

2. The reduction of $NADP^+$ is a two-electron process. Hence, the formation of two NADPH requires the pumping of four photons by photosystem I. The electrons given up by photosystem I are replenished by photosystem II, which needs to absorb an equal number of photons. Hence eight photons are needed to generate the required NADPH. The three molecules of ATP needed to convert CO_2 into a hexose are formed concurrently.

3. A mole of photons of 600-nm wavelength has an energy content of 47.6 kcal, and so the energy input of eight moles of photons is 381 kcal. Thus, the overall efficiency of photosynthesis under standard conditons is at least 114/381, or 30%.

REGULATION OF THE CALVIN CYCLE

A variety of control mechanisms assure that the Calvin cycle operates only when ATP and NADPH are produced by the light reactions of photosynthesis. The rate-limiting step in the Calvin cycle is the carboxylation of ribulose 1,5-diphosphate to form two molecules of 3-phosphoglycerate. *The activity of ribulose 1,5-diphosphate carboxylase is markedly enhanced by illumination in three ways:*

1. NADPH formed by photosystem I is an allosteric activator of the carboxylase.

2. The enzymatic rate increases markedly when the pH increases from 7 to 9. The light-induced acidification of the thylakoid space renders the stroma more alkaline, which activates the carboxylase.

3. The carboxylase is activated by Mg^{2+}. Recall that Mg^{2+} is released into the stroma as protons are pumped into the thylakoid space on illumination (p. 444).

TROPICAL PLANTS HAVE A C_4 PATHWAY THAT ACCELERATES PHOTOSYNTHESIS BY CONCENTRATING CO_2

Tropical plants, such as sugar cane, have an accessory pathway for transporting CO_2 to the site of the Calvin cycle in their photosynthetic cells. The first clue of the existence of this pathway came from studies showing that radioactivity from a pulse of $^{14}CO_2$ appeared initially in malate and aspartate, which are four-carbon compounds, rather than in 3-phosphoglycerate. The essence of this pathway, which was elucidated by M. D. Hatch and C. R. Slack, is that *C_4 compounds carry CO_2 from mesophyll cells, which are in contact with air, to bundle-sheath cells, which are the major sites of photosynthesis* (Figure 19-23). Decarboxylation of the C_4 compound in the bundle-sheath cell maintains a high concentration of CO_2 at the site of the Calvin cycle. The C_3 compound returns to the mesophyll cell for another round of carboxylation.

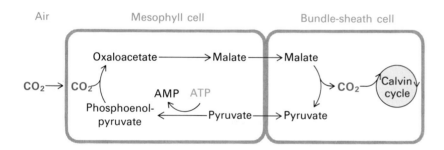

Figure 19-23
Schematic diagram of the essential features of the C_4 pathway.

This C_4 pathway (Figure 19-23) for the transport of CO_2 starts in the mesophyll cell with the condensation of CO_2 and phosphoenolpyruvate to form *oxaloacetate,* a reaction catalyzed by phosphoenolpyruvate carboxylase. In some species, oxaloacetate is converted into *malate* by an $NADP^+$-linked malate dehydrogenase. Malate goes into the bundle-sheath cell and is decarboxylated within the chloroplasts by an $NADP^+$-linked malate enzyme. The released CO_2 enters the Calvin cycle in the usual way by condensing with ribulose 1,5-diphosphate. Pyruvate formed in this decarboxylation reaction returns to the mesophyll cell. Finally, phosphoenolpyruvate is regenerated by the reaction of pyruvate with ATP and P_i. This unique reaction, catalyzed by pyruvate-P_i dikinase, is driven by the subsequent hydrolysis of PP_i. The net reaction of this C_4 pathway is

$$CO_2 \text{ (in mesophyll cell)} + ATP + H_2O \longrightarrow$$
$$CO_2 \text{ (in bundle-sheath cell)} + AMP + 2\,P_i$$

Thus, *two high-energy phosphate bonds are consumed in transporting CO_2 to the chloroplasts of the bundle-sheath cells.*

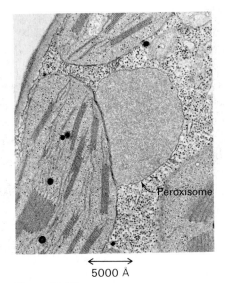

Ribulose 1,5-diphosphate

$$O_2$$

3-Phosphoglycerate

$$
\begin{array}{c}
COO^- \\
| \\
CH_2OPO_3{}^{2-}
\end{array}
$$

Phosphoglycolate

$$H_2O$$

$$P_i$$

$$
\begin{array}{c}
COO^- \\
| \\
CH_2OH
\end{array}
$$

Glycolate

$$O_2$$

$$H_2O_2$$

$$
\begin{array}{c}
COO^- \\
| \\
C \\
O^{\diagdown}\,^H
\end{array}
$$

Glyoxylate

Figure 19-24
Formation and breakdown of glycolate.

Peroxisome

5000 Å

Figure 19-25
Electron micrograph of a peroxisome in a plant cell. [Courtesy of Dr. Sue Ellen Frederick.]

When the C_4 pathway and the Calvin cycle operate together, the net reaction is

$$6\,CO_2 + 30\,ATP + 12\,NADPH + 12\,H_2O \longrightarrow$$
$$C_6H_{12}O_6 + 30\,ADP + 30\,P_i + 12\,NADPH^+ + 18\,H^+$$

Note that 30 ATP are consumed per hexose formed when the C_4 pathway delivers CO_2 to the Calvin cycle, in contrast with 18 ATP per hexose in the absence of the C_4 pathway. The high concentration of CO_2 in the bundle-sheath cells of C_4 plants, which is due to the expenditure of the additional 12 ATP, is critical for their rapid photosynthetic rate, because CO_2 is the limiting factor when light is abundant. A high CO_2 concentration also minimizes the energy loss caused by photorespiration, a process described in the next section.

GLYCOLATE IS THE MAJOR SUBSTRATE IN PHOTORESPIRATION

Illuminated plants consume O_2 and evolve CO_2 by a process called *photorespiration*, which is different from mitochondrial respiration. The biological role of photorespiration is an enigma. *Glycolate*, the major substrate of photorespiration, is derived from *phosphoglycolate*, which is formed by the oxygenation of ribulose 1,5-diphosphate (Figure 19-24). The reaction is catalyzed by ribulose 1,5-diphosphate carboxylase, which is an *oxygenase* as well as a carboxylase. In fact, oxygenation and carboxylation are competing reactions employing the same active site. The phosphoglycolate formed in this way is hydrolyzed to glycolate by a specific phosphatase. The subsequent metabolism of glycolate occurs in *peroxisomes* (also called *microbodies*) (Figure 19-25). Glycolate is oxidized to *glyoxylate* by glycolate oxidase. The H_2O_2 produced in this reaction is cleaved by catalase to H_2O and O_2. Transamination of glyoxylate then yields *glycine*. In mitochondria, serine can be formed from two molecules of glycine.

Photorespiration is a seemingly wasteful process in that organic carbon is converted into CO_2 without the production of ATP and NADH or other evident gain. Plants lacking a C_4 pathway lose between 25% and 50% of their fixed carbon by photorespiration. In contrast, *tropical plants with a C_4 pathway have little photorespiration because the oxygenation of ribulose 1,5-diphosphate is competitively inhibited by the high concentration of CO_2 in their bundle-sheath cells.* The oxygenase activity of ribulose 1,5-diphosphate carboxylase increases more rapidly with temperature than does the carboxylase activity. Hence, the C_4 pathway is especially important for minimizing photorespiration at higher temperatures. The geographical distribution of C_4 plants (which have the C_4 pathway) and C_3 plants (which lack this pathway) can now be understood in molecular terms. C_4 plants have the advantage in environments with high temperature and high illumination, and so they are prevalent in the tropics. C_3 plants, which consume 18 ATP per hexose formed in the absence of photorespiration (compared with 30 ATP for C_4 plants), are more efficient at temperatures of less than about 28°C, and so they predominate in temperate environments.

THE PURPLE-MEMBRANE PROTEIN OF HALOBACTERIA PUMPS PROTONS TO SYNTHESIZE ATP

Insight into the mechanism of photophosphorylation and oxidative phosphorylation has come from studies of the photosynthetic process in halobacteria, which use a light-absorbing group different from chlorophyll. These bacteria require high concentrations of NaCl for growth, the optimum being 4.3 M NaCl (ordinary seawater contains 0.6 M NaCl). Halobacteria are found in natural salt lakes and in areas where seawater is evaporated to produce salt, such as the one in San Francisco Bay near the airport. These bacteria contain a cell envelope consisting of a cell membrane surrounded by a glycoprotein cell wall. Walther Stoeckenius separated the cell membrane into yellow, red, and purple fractions. The yellow fraction mainly contains the walls of gas vacuoles, which enable the bacteria to adjust their depth in the water. The red fraction contains the respiratory chain and other enzymatic machinery for oxidative phosphorylation, in addition to a red screening pigment that protects the bacteria from the lethal effects of blue light. The purple fraction consists of a 26-kdal *purple-membrane protein* and forty associated lipids. This protein is also called *bacteriorhodopsin* because it contains a *retinal* chromophore, as does rhodopsin, a visual pigment in animal eyes (p. 897). The retinal group is joined to a specific lysine side chain of the protein by a protonated Schiff-base linkage.

In the presence of O_2, halobacteria synthesize ATP by oxidative phosphorylation. When oxygen is scarce, they switch to a photosynthetic mode (Figure 19-26). Stoeckenius and his collaborators showed that *the role of the purple-membrane protein is to pump protons from the inside to the outside of the cell when illuminated.* In fact, the purple-membrane protein traverses the cell membrane (Figure 10-32 on p. 228), as expected for a proton pump. Flash spectroscopic studies have demonstrated that light converts the retinal Schiff-base linkage from a protonated form with a 560-nm absorption band into a deprotonated form peaked at 412 nm. This reaction cycle takes a few milliseconds, and so several hundred H^+ per second can be pumped by each purple-membrane-protein molecule in bright light. The proton-motive force generated in this way is then used by the ATP-synthesizing assembly (F_0-F_1) in the spatially separate red-membrane region to phosphorylate ADP. The F_0-F_1 complex is oriented so that its globular F_1 unit faces the inside of the cell. The same ATP-synthesizing assembly is used to carry out oxidative phosphorylation and photo-

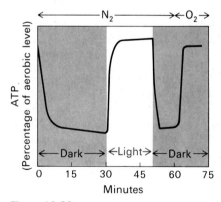

Figure 19-26
Experiment showing that halobacteria can synthesize ATP when illuminated in the absence of O_2. [After W. Stoeckenius. The purple membrane of salt-loving bacteria. Copyright © 1976 by Scientific American, Inc. All rights reserved.]

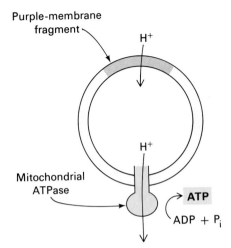

Figure 19-27
Reconstituted membrane vesicles containing the purple-membrane protein and a mitochondrial ATP-synthesizing complex form ATP on illumination.

phosphorylation. The proton-motive force generated by either process can also drive the active transport of ions and amino acids into halobacteria.

The availability of the purified purple-membrane fraction enabled Stoeckenius and Efraim Racker to carry out a highly informative reconstitution experiment. They formed synthetic phospholipid vesicles containing the purple-membrane protein from halobacteria and the F_1-F_0 ATP-synthesizing complex from beef-heart mitochondria (Figure 19-27). Illumination of these vesicles resulted in the formation of a proton gradient across the membrane and the generation of ATP from ADP and P_i. The synthesis of ATP was abolished by the addition of proton carriers that dissipated the proton gradient, which shows that the proton-motive force is essential. *These reconstituted vesicles, which contain only a proton pump and an ATP synthetase in a closed membrane that is impermeable to protons, are the simplest model system for chemiosmotic energy conversion.*

SUMMARY

The first step in photosynthesis is the absorption of light by chlorophyll molecules, which are organized into photosynthetic units in the thylakoid membranes of chloroplasts. The excitation energy is transferred from one chlorophyll molecule to another until it is trapped by a reaction center. The critical event at a reaction center is the light-activated transfer of an electron to an acceptor against a chemical-potential gradient. Photosynthesis in green plants requires the interaction of two light reactions. Photosystem I generates a strong reductant (bound ferredoxin) that leads to the formation of NADPH. Photosystem II produces a strong oxidant that forms O_2 from H_2O.

Electron flow through an electron-transport chain from photosystem II to I generates a proton gradient, which in turn leads to the synthesis of ATP. Thus, light causes the flow of electrons from H_2O to NADPH with the concomitant generation of ATP. Alternatively, a proton gradient and then ATP can be generated without the formation of NADPH by a process called cyclic photophosphorylation following the absorption of light by photosystem I. Photophosphorylation, like oxidative phosphorylation, is driven by a proton-motive force, and so a closed compartment is required.

The ATP and NADPH formed in the light reactions of photosynthesis are used to convert CO_2 into hexoses and other organic compounds. The dark phase of photosynthesis, called the Calvin cycle, starts with the reaction of CO_2 and ribulose 1,5-diphosphate to form two molecules of *3-phosphoglycerate*. The steps in the conversion of 3-phosphoglycerate into fructose 6-phosphate and glucose 6-phosphate are like those,of gluconeogenesis, except that glyceraldehyde 3-phosphate dehydrogenase in chloroplasts is specific for NADPH rather than NADH. Ribulose diphosphate is regenerated from fructose 6-phosphate, glyceraldehyde 3-phosphate, and dihydroxyacetone phosphate in a complex series of reactions. Several of the steps in the regeneration of ribulose 1,5-diphosphate are like those of the pentose phosphate pathway. Three ATP and two NADPH are consumed in converting CO_2 into a hexose. Four photons are absorbed by photosystem I and another four by photosystem II to generate two NADPH.

Tropical plants have an accessory pathway for concentrating CO_2 at the site of the Calvin cycle. This C_4 pathway enables tropical plants to take advantage of high light levels and minimize the oxygenation of ribulose 1,5-diphosphate, a competing reaction.

SELECTED READINGS

WHERE TO START

Hinkle, P. C., and McCarty, R. E., 1978. How cells make ATP. *Sci. Amer.* 238(3):104–123. [Contains a lucid discussion of photophosphorylation. Available as *Sci. Amer.* Offprint 1383.]

Stoeckenius, W., 1976. The purple membrane of salt-loving bacteria. *Sci. Amer.* 234(6):38–46. [Presentation of the discovery of a photosynthetic mechanism based on retinal rather than chlorophyll.]

Miller, K. R., 1979. The photosynthetic membrane. *Sci. Amer.* 241(4):102–113. [Offprint 1448.]

Bassham, J. A., 1962. The path of carbon in photosynthesis. *Sci. Amer.* 206(6):88–100. [Offprint 122.]

REVIEWS

Arnon, D. I., 1977. Photosynthesis 1950–75: changing concepts and perspectives. *In* Trebst, A., and Avron, M., (eds.), *Encyclopedia of Plant Physiology,* vol. 5, New Series, pp. 7–56. Springer-Verlag. [Emphasizes photophosphorylation and the role of ferredoxin.]

Halliwell, B., 1978. The chloroplast at work: a review of modern developments in our understanding of chloroplast metabolism. *Prog. Biophys. Mol. Biol.* 33:1–54. [Emphasizes the dark reactions of chloroplasts.]

BOOKS

Bonner, J., and Varner, J. E., (eds.), 1976. *Plant Biochemistry* (3rd ed.). Academic Press. [Contains articles by R. B. Park on the chloroplast, M. D. Hatch on the path of carbon in photosynthesis, and B. Kok on the path of energy in photosynthesis.]

Gregory, R. P. F., 1977. *Biochemistry of Photosynthesis* (2nd ed.), Wiley.

Govindjee, (ed.), 1975. *Bioenergetics of Photosynthesis.* Academic Press.

Olson, J. M., and Hind, G., (eds.). *Chlorophyll-Proteins, Reaction Centers, and Photosynthetic Membranes.* Brookhaven Symposium in Biology, vol. 28.

PRIMARY EVENTS IN PHOTOSYNTHESIS

Jagendorf, A. T., 1967. Acid-base transitions and phosphorylation by chloroplasts. *Fed. Proc.* 26:1361–1369.

Blankenship, R. E., and Parson, W. W., 1978. The photochemical electron transfer reactions of photosynthetic bacteria and plants. *Ann. Rev. Biochem.* 47:635–653.

Avron, M., 1977. Energy transduction in chloroplasts. *Ann. Rev. Biochem.* 46:143–155.

Radmer, R., and Kok, B., 1975. Energy capture in photosynthesis: photosystem II. *Ann. Rev. Biochem.* 44:409–433.

Junge, W., 1977. Membrane potentials in photosynthesis. *Ann. Rev. Plant Physiol.* 28:503–536.

Stoeckenius, W., Lozier, R. H., and Bogomolni, R. A., 1979. Bacteriorhodopsin and the purple membrane of halobacteria. *Biochem. Biophys. Acta* 505:215–278.

CARBON FIXATION IN PHOTOSYNTHESIS

Zelitch, I., 1975. Pathways of carbon fixation in green plants. *Ann. Rev. Biochem.* 44:123–145.

Kelly, G. J., Latzko, E., and Gibbs, M., 1976. Regulatory aspects of photosynthetic carbon metabolism. *Ann. Rev. Plant Physiol.* 27:181–205.

Jensen, R. G., and Bahr, J. T., 1977. Ribulose 1,5-biphosphate carboxylase-oxygenase. *Ann. Rev. Plant Physiol.* 28:379–400.

STRUCTURE OF PHOTOSYNTHETIC PROTEINS

Fenna, R. E., and Matthews, B. W., 1975. Chlorophyll arrangement in a bacteriochlorophyll protein from *Chlorobium limicola. Nature* 258:573–577.

Colman, P. M., Freeman, H. C., Guss, J. M., Murata, M., Norris, V. A., Ramshaw, J. A. M., and Venkatappa, M. P., 1978. X-ray crystal structure analysis of plastocyanin at 2.7 Å resolution. *Nature* 272:319–324.

Baker, T. S., Eisenberg, D., and Eiserling, F., 1977. Ribulose biphosphate carboxylase: a two-layered, square-shaped molecule of symmetry 422. *Science* 196:293–295.

PROBLEMS

1. Calculate the $\Delta E_0'$ and $\Delta G^{\circ\prime}$ for the reduction of $NADP^+$ by FRS.

2. Sedoheptulose 1,7-diphosphate is an intermediate in the Calvin cycle but not in the pentose phosphate pathway. What is the enzymatic basis of this difference?

3. Suppose that an illuminated suspension of *Chlorella* was actively carrying out photosynthesis when the light was suddenly switched off. How would the levels of 3-phosphoglycerate and ribulose 1,5-diphosphate change during the next minute?

4. Suppose that an illuminated suspension of *Chlorella* was actively carrying out photosynthesis in the presence of 1% CO_2 when the concentration of CO_2 was abruptly reduced to .003%. What effect would this have on the levels of 3-phosphoglycerate and ribulose 1,5-diphosphate during the next minute?

5. The photosynthetic apparatus of blue-green algae contains large amounts of phycoerythrin and phycocyanin in addition to chlorophyll *a*. Phycoerythrin absorbs maximally between 480 and 600 nm, whereas phycocyanin absorbs maximally near 620 nm. Suggest a role for these accessory photosynthetic pigments.

6. The Van Niel equation for photosynthesis in higher plants is

$$6\,CO_2 + 12\,H_2O \longrightarrow$$
$$C_6H_{12}O_6 + 6\,H_2O + 6\,O_2$$

(a) H_2O appears on both sides of this equation. Is this merely a formalism or does it express an important aspect of the mechanism of photosynthesis?

(b) The conventional equation for respiration is

$$C_6H_{12}O_6 + 6\,O_2 \longrightarrow 6\,CO_2 + 6\,H_2O$$

Is the Van Niel equation (in reverse) more revealing regarding the mechanism of respiration? In other words, does the combustion of glucose require the input of six molecules of H_2O?

7. Chloroplasts illuminated in the absence of ADP and P_i form ATP in a subsequent dark period on addition of ADP and P_i. The amount of ATP synthesized in the dark is markedly increased if pyridine is added before illumination. Why?

8. Dichlorophenyldimethylurea (DCMU), an herbicide, interferes with photophosphorylation and O_2 evolution in the absence of an artificial electron acceptor. However, it does not block the Hill reaction. Propose a site for the inhibitory action of DCMU.

For additional problems, see W. B. Wood, J. H. Wilson, R. J. Benbow, and L. E. Hood, *Biochemistry: A Problems Approach* (Benjamin, 1974), ch. 12.

On the facing page: Model of *S*-adenosylmethionine, a methyl donor in many biosyntheses.

PART III

BIOSYNTHESIS OF
MACROMOLECULAR
PRECURSORS

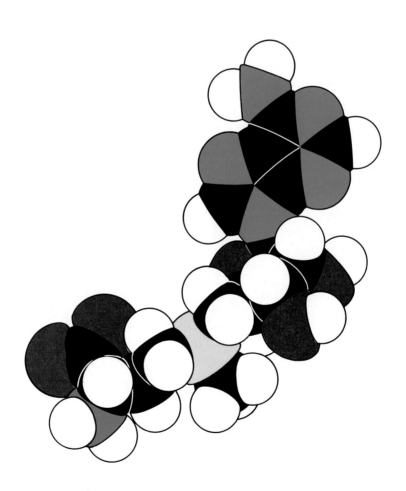

Polarized light micrograph of macrophages containing
birefringent crystals of cholesterol esters.
[Courtesy of Dr. Richard Anderson.]

BIOSYNTHESIS OF MEMBRANE LIPIDS
AND STEROID HORMONES

We turn now from the generation of metabolic energy to the biosynthesis of macromolecular precursors and related biomolecules. This chapter deals with the biosynthesis of phosphoglycerides, sphingolipids, and cholesterol, three important components of biological membranes (Chapter 10). The synthesis of triacylglycerols is also considered here because this pathway overlaps that of the phosphoglycerides; and the formation of steroid hormones is presented because they are derived from cholesterol. The biosynthesis of cholesterol exemplifies a fundamental mechanism for the assembly of extended carbon skeletons from five-carbon units.

Anabolism—
  Biosynthetic processes.
Catabolism—
  Degradative processes.
  Derived from the Greek
  words *ana*, up; *cata*, down;
  ballein, to throw.

PHOSPHATIDATE IS AN INTERMEDIATE IN THE SYNTHESIS OF PHOSPHOGLYCERIDES AND TRIACYLGLYCEROLS

Phosphatidate (diacylglycerol 3-phosphate) is a common intermediate in the synthesis of phosphoglycerides and triacylglycerols. The starting point is *glycerol 3-phosphate*, which is formed mainly by reduction of dihydroxyacetone phosphate and to a lesser extent by phosphorylation of glycerol. Glycerol 3-phosphate is acylated by acyl CoA to form *lysophosphatidate*, which is again acylated by acyl CoA to yield *phosphatidate*. These acylations are catalyzed by glycerol-phosphate acyl transferase.

Glycerol 3-phosphate — Lysophosphatidate — Phosphatidate

The pathways diverge at phosphatidate. In the synthesis of triacylglycerols, phosphatidate is hydrolyzed by a specific phosphatase to give a *diacylglycerol*. This intermediate is acylated to a *triacylglycerol* in a reaction that is catalyzed by diglyceride acyl transferase. These enzymes are associated in a membrane-bound *triacylglycerol synthetase complex*.

Phosphatidate — Diacylglycerol — Triacylglycerol

CDP-DIACYLGLYCEROL IS THE ACTIVATED INTERMEDIATE IN THE DE NOVO SYNTHESIS OF PHOSPHOGLYCERIDES

There are several routes for the synthesis of phosphoglycerides. The de novo pathway starts with the formation of *cytidine diphosphodiacylglycerol* (CDP-diacylglycerol) from phosphatidate and cytidine triphosphate (CTP). This reaction is driven forward by the hydrolysis of pyrophosphate.

Phosphatidate — CDP-diacylglycerol

The activated phosphatidyl unit then reacts with the hydroxyl group of a polar alcohol. The products are *phosphatidyl serine* and cytidine monophosphate (CMP) when the polar alcohol is serine.

$$H_2C—O—\overset{\overset{O}{\|}}{C}—R_1$$
$$R_2—\overset{\overset{}{\underset{O}{}}}{C}—O—\overset{}{C}—H$$

CDP-diacylglycerol   +   **Serine**

Phosphatidyl serine   +   **CMP**

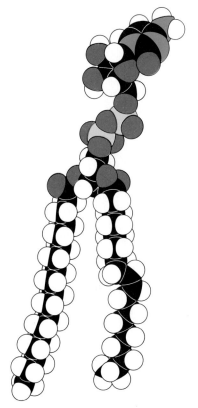

Figure 20-1
Space-filling model of CDP-diacylglycerol.

Thus, a cytidine nucleotide plays the same role in the synthesis of phosphoglycerides as does a uridine nucleotide in the formation of glycogen. In both biosyntheses, an activated intermediate (UDP-glucose or CDP-diacylglycerol) is formed from a phosphorylated substrate (glucose 1-phosphate or phosphatidate) and a nucleoside triphosphate (UTP or CTP). The activated intermediate then reacts with a hydroxyl group (the 4-OH terminus of glycogen or the hydroxyl side chain of serine).

PHOSPHATIDYL ETHANOLAMINE AND PHOSPHATIDYL CHOLINE CAN BE FORMED FROM PHOSPHATIDYL SERINE

Decarboxylation of phosphatidyl serine by a pyridoxal phosphate enzyme yields *phosphatidyl ethanolamine*. The amino group of this phosphoglyceride is then methylated three times to form *phosphatidyl choline*. S-*adenosylmethionine*, the methyl donor in this and many other methylations, will be discussed later (p. 493).

Phosphatidyl serine   **Phosphatidyl ethanolamine**   **Phosphatidyl choline**

PHOSPHOGLYCERIDES CAN ALSO BE SYNTHESIZED BY A SALVAGE PATHWAY

Phosphatidyl choline can also be synthesized by a pathway that makes use of choline. Hence, these reactions are called the *salvage pathway* (Figure 20-2). Choline is phosphorylated by ATP to *phos-*

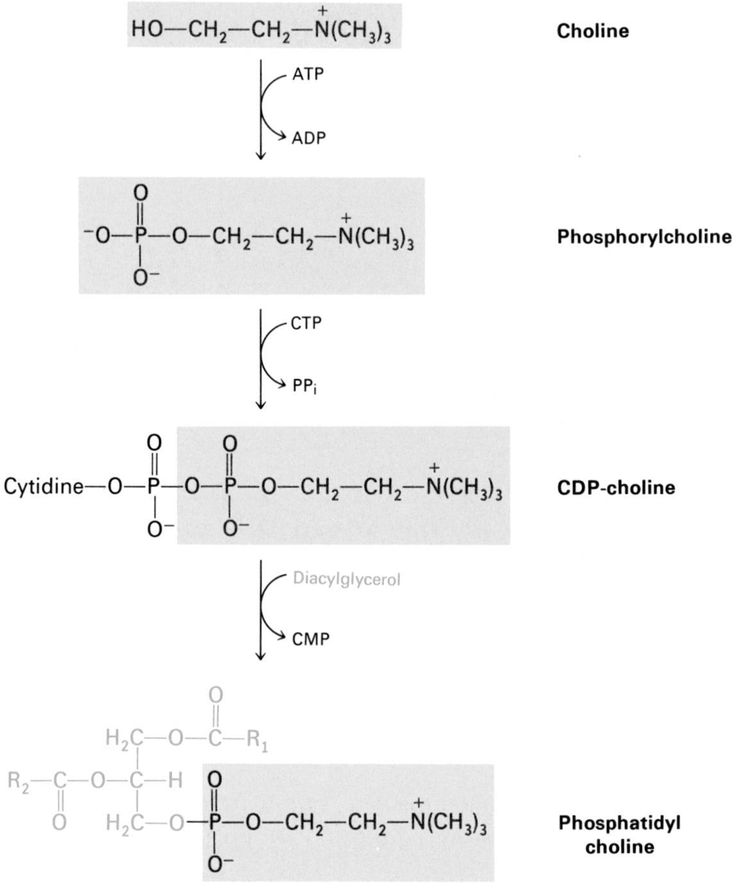

Figure 20-2
Synthesis of phosphatidyl choline by the salvage pathway.

phorylcholine, which then reacts with CTP to form *CDP-choline.* The phosphorylcholine unit of CDP-choline is then transferred to a diacylglycerol to form *phosphatidyl choline.* Note that the activated species in this salvage pathway is the cytidine derivative of phosphorylcholine rather than of phosphatidate.

Phosphatidyl ethanolamine can be synthesized from ethanolamine by analogous reactions. *Plasmalogens* are also formed by the salvage pathway. A plasmalogen differs from a phosphoglyceride in that the substituent at C-1 of the glycerol moiety is an α, β unsaturated ether. The final step in the synthesis of phosphatid*al* choline (the plasmalogen corresponding to phosphatid*yl* choline) is the transfer of phosphorylcholine from CDP-choline to an alkenyl glycerol ether.

Chapter 20
BIOSYNTHESIS OF LIPIDS

An alkenyl glycerol ether   **Phosphatidal choline**

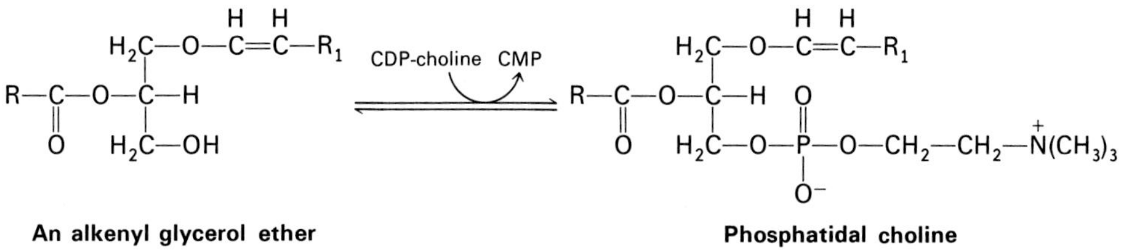

SEVERAL SPECIFIC PHOSPHOLIPASES HAVE BEEN ISOLATED

Enzymes can be used as highly specific biochemical reagents. For example, we have already seen that trypsin is an invaluable tool in studies of protein structure. Specific phospholipases are assuming comparable significance in studies of biological membranes and of their phospholipid constituents. Phospholipases are grouped according to their specificities. The bonds hydrolyzed by phospholipases A_1, A_2, C, and D are shown in Figure 20-3.

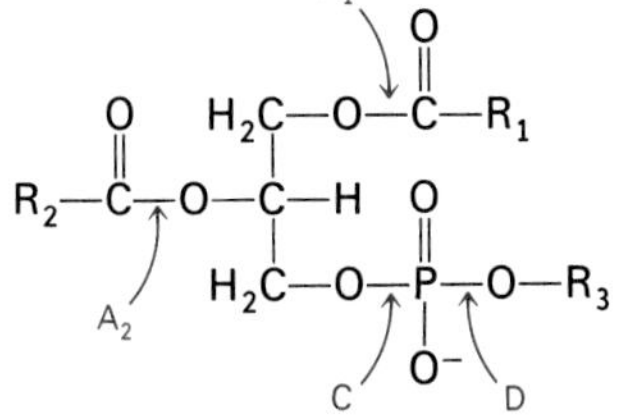

SYNTHESIS OF CERAMIDE, THE BASIC STRUCTURAL UNIT OF SPHINGOLIPIDS

We turn now from phosphoglycerides to sphingolipids. The backbone is *sphingosine,* a long-chain aliphatic amine, rather than glycerol. Palmitoyl CoA and serine condense to form dihydrosphingosine, which is then oxidized to sphingosine by a flavoprotein.

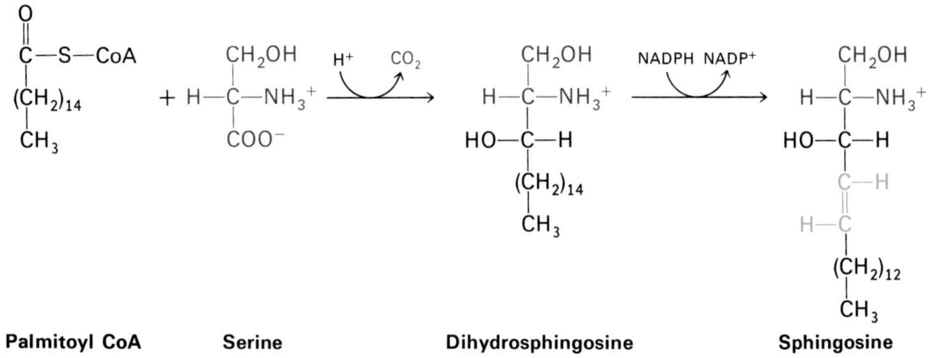

Figure 20-3
Specificity of phospholipases.

Palmitoyl CoA   **Serine**   **Dihydrosphingosine**   **Sphingosine**

The amino group of sphingosine is acylated in all sphingolipids (Figure 20-4). A long-chain acyl CoA reacts with sphingosine to form *ceramide* (*N*-acyl sphingosine). The terminal hydroxyl group is also substituted in sphingolipids. In *sphingomyelin,* the substituent is phosphorylcholine, which comes from CDP-choline. In a *cerebroside,* glucose or galactose is linked to the terminal hydroxyl group of

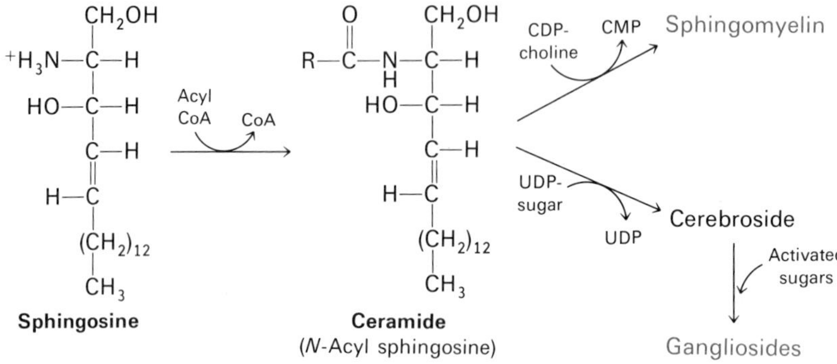

Figure 20-4
Ceramide is the precursor of sphingomyelin and of gangliosides.

ceramide. UDP-glucose or UDP-galactose is the sugar donor in the synthesis of cerebrosides. In a *ganglioside,* an oligosaccharide is linked to ceramide by a glucose residue (see Figure 20-5).

GANGLIOSIDES ARE CARBOHYDRATE-RICH SPHINGOLIPIDS THAT CONTAIN ACIDIC SUGARS

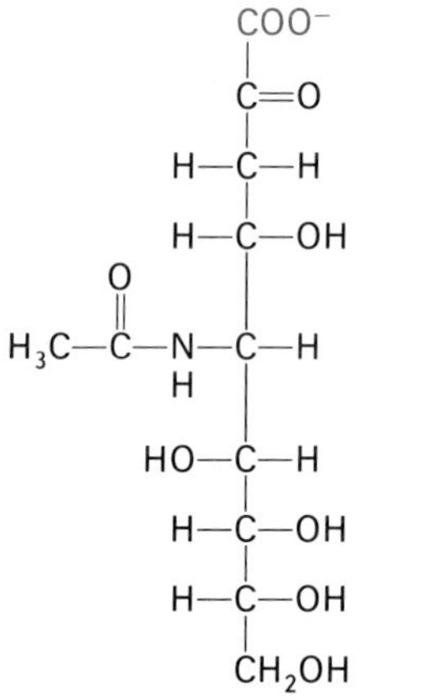

Gangliosides are the most complex sphingolipids. An *oligosaccharide chain* containing at least one acidic sugar is attached to ceramide. The acidic sugar is N-*acetylneuraminate* or N-glycolylneuraminate. These acidic sugars are called *sialic acids.* Their 9-carbon backbones are synthesized from phosphoenolpyruvate (a C_3 unit) and N-acetylmannosamine 6-phosphate (a C_6 unit).

Gangliosides are synthesized by the ordered, step-by-step addition of sugar residues to ceramide. UDP-glucose, UDP-galactose, and UDP-N-acetylgalactosamine are the activated donors of these sugar residues. The CMP derivative of N-acetylneuraminate is the activated donor of this acidic sugar. CMP-N-acetylneuraminate is synthesized from CTP and N-acetylneuraminate; C-1 is the activated carbon, as in the UDP-sugars. The structure of the resulting

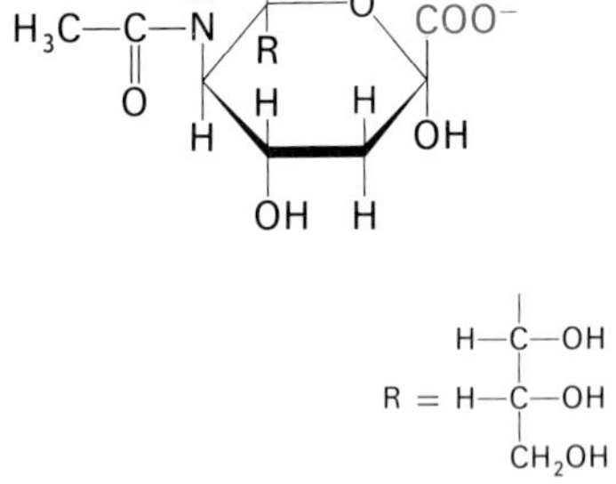

N-Acetylneuraminate
(Open-chain form)

N-Acetylneuraminate
(Pyranose form)

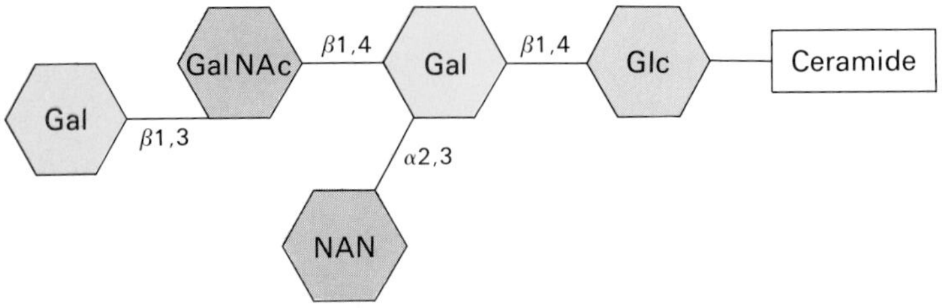

Figure 20-5
Structure of ganglioside G_{M1}. Abbreviations used: Gal, galactose; GalNAc, *N*-acetylgalactosamine; Glc, glucose; NAN, *N*-acetylneuraminate. The types of linkages between the sugars—such as $\beta1,4$—are noted at the bonds joining them.

ganglioside is determined by the specificity of the glycosyl transferases in the cell. More than fifteen different gangliosides have been characterized. The structure of ganglioside G_{M1} is shown in Figure 20-5.

TAY-SACHS DISEASE: AN INHERITED DISORDER OF GANGLIOSIDE BREAKDOWN

Gangliosides are found in highest concentration in the nervous system, particularly in gray matter, where they constitute 6% of the lipids. Gangliosides are continually synthesized and degraded by the sequential removal of their terminal sugars. The glycosyl hydrolases that catalyze these reactions are highly specific. Ganglioside breakdown occurs inside *lysosomes*. These organelles contain many types of degradative enzymes and are specialized for the orderly destruction of cellular components.

Disorders of ganglioside breakdown can have serious clinical consequences. In Tay-Sachs disease, the symptoms are usually evident before an affected infant is a year old. Weakness, retardation in development, and difficulty in eating are typical early symptoms. Blindness usually follows a few months later. Tay-Sachs disease is usually fatal before age three. The striking pathological changes in this disease occur in the nervous system, where the ganglion cells of the cerebral cortex and some other parts of the brain become enormously swollen. Also, there is a distinctive cherry-red spot in the retina.

The ganglioside content of the brain of an infant with Tay-Sachs disease is greatly elevated. Specifically, *the concentration of ganglioside G_{M2} is many times higher than normal.* The abnormally high level of this ganglioside is caused by a deficiency of the enzyme that removes its terminal *N*-acetylgalactosamine residue. The missing or deficient enzyme is a specific β-N-*acetylhexosaminidase*.

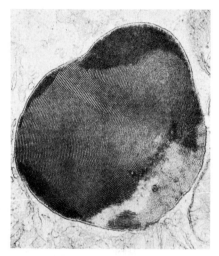

Figure 20-6
Electron micrograph of a lysosome.
[Courtesy of Dr. George Palade.]

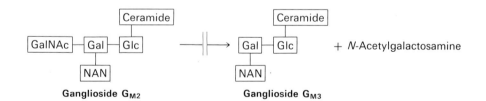

Ganglioside G_{M2} Ganglioside G_{M3} + *N*-Acetylgalactosamine

Tay-Sachs disease is inherited as an *autosomal recessive*. The carrier rate is 1/30 in Jewish Americans and 1/300 in non-Jewish Americans. Consequently, the incidence of the disease is about 100 times higher in Jewish Americans. Tay-Sachs disease can be diagnosed during fetal development. Amniotic fluid is obtained by *amniocentesis* and assayed for β-N-acetylhexosaminidase activity.

CHOLESTEROL IS SYNTHESIZED FROM ACETYL COENZYME A

Figure 20-7
Space-filling model of cholesterol.

We now turn to the synthesis of cholesterol, a steroid that modulates the fluidity of eucaryotic membranes (p. 227). Cholesterol is also the precursor of steroid hormones such as progesterone, testosterone, estradiol, and cortisol. An early and important clue concerning the synthesis of cholesterol came from the work of Konrad Bloch in the 1940s. Acetate isotopically labeled in its carbon atoms was prepared and fed to rats. The cholesterol that was synthesized by these rats contained the isotopic label, which showed that acetate is a precursor of cholesterol. In fact, *all twenty-seven carbon atoms of cholesterol are derived from acetyl CoA.* Further insight into the synthesis of cholesterol came from studies that utilized acetate labeled in either its methyl or its carboxyl carbon as the precursor. Degradation of cholesterol synthesized from acetate labeled in one of its carbons showed the origin of each atom of the cholesterol molecule (Figure 20-8). This pattern played a crucial role in the generation and testing of hypotheses concerning the pathway of cholesterol synthesis.

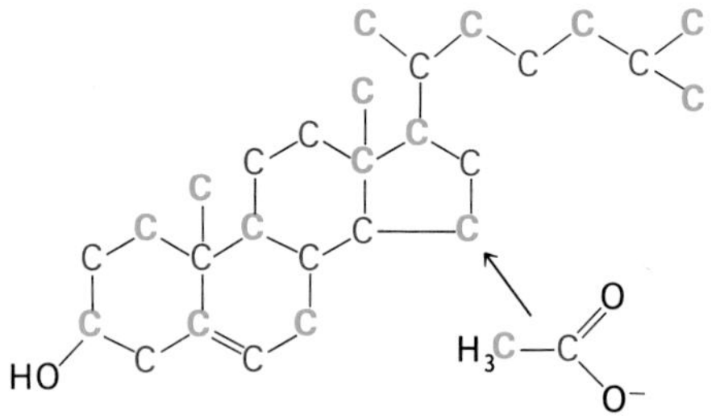

Figure 20-8
Isotope-labeling pattern of cholesterol synthesized from acetate labeled in its methyl (green) or carboxyl (red) carbon atom.

MEVALONATE AND SQUALENE ARE INTERMEDIATES IN THE SYNTHESIS OF CHOLESTEROL

The next major clue came from the discovery that *squalene,* a C_{30} hydrocarbon, is an intermediate in the synthesis of cholesterol. Squalene consists of six *isoprene* units.

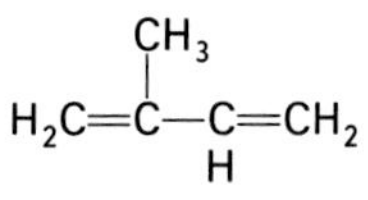

Isoprene

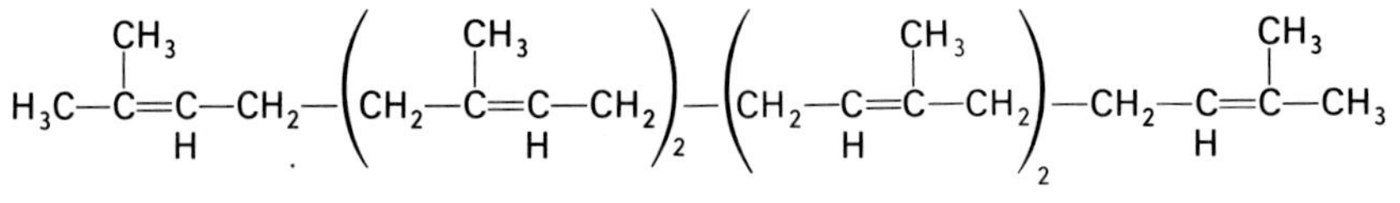

Squalene

This finding raised the question of how isoprene units are formed from acetate.

$$\text{Acetate} \longrightarrow [\text{isoprene}] \longrightarrow \text{squalene} \longrightarrow \text{cholesterol}$$
$$C_2 C_5 C_{30} C_{27}$$

The answer came unexpectedly from unrelated studies of bacterial mutants. It was found that mevalonate could substitute for acetate in meeting the nutritional needs of acetate-requiring mutants.

The discovery of mevalonate was a key step in the elucidation of the pathway of cholesterol biosynthesis, because it was soon recognized that this C_6 acid could decarboxylate to yield the postulated C_5 isoprene intermediate. Isotope-labeling studies then showed that mevalonate is indeed a precursor of squalene and that it could be formed from acetate. The activated isoprene intermediate proved to be *isopentenyl pyrophosphate,* which is formed by decarboxylation of a derivative of mevalonate. A pathway for the synthesis of cholesterol from acetate could then be outlined:

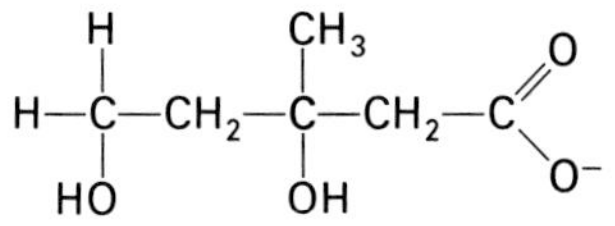

$$\text{Acetate} \longrightarrow \text{mevalonate} \longrightarrow$$
$$\quad C_2 \qquad\qquad C_6$$
$$\text{isopentenyl pyrophosphate} \longrightarrow \text{squalene} \longrightarrow \text{cholesterol}$$
$$\quad\quad C_5 \qquad\qquad\qquad\quad C_{30} \qquad\qquad C_{27}$$

Mevalonate

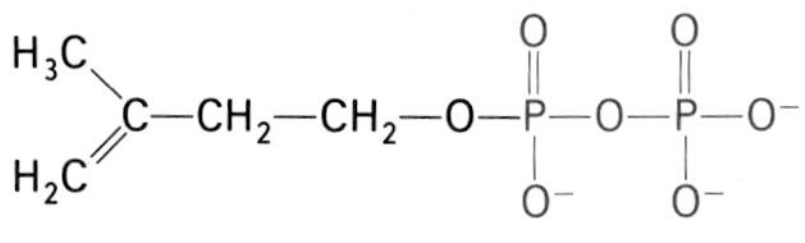

Isopentenyl pyrophosphate

SYNTHESIS OF ISOPENTENYL PYROPHOSPHATE, AN ACTIVATED INTERMEDIATE IN CHOLESTEROL FORMATION

The first stage in the synthesis of cholesterol is the formation of isopentenyl pyrophosphate from acetyl CoA. This set of reactions starts with the formation of 3-hydroxy-3-methylglutaryl CoA from acetyl CoA and acetoacetyl CoA. One of the fates of 3-hydroxy-3-methylglutaryl CoA, its cleavage to acetyl CoA and acetoacetate, was discussed previously in regard to the formation of ketone bodies (p. 393). Alternatively, 3-hydroxyl-3-methylglutaryl CoA can be reduced to mevalonate (Figure 20-9). 3-Hydroxy-3-methylglutaryl

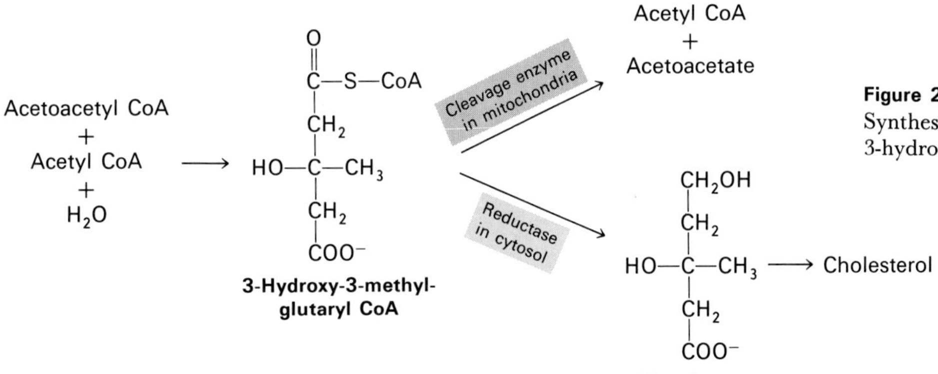

Figure 20-9
Synthesis and fates of 3-hydroxy-3-methylglutaryl CoA.

CoA is present both in the cytosol and in the mitochondria of liver cells. The mitochondrial pool of this intermediate is mainly a precursor of ketone bodies, whereas the cytoplasmic pool gives rise to mevalonate for the synthesis of cholesterol.

The synthesis of mevalonate is the committed step in cholesterol formation.

The enzyme catalyzing this irreversible step, 3-hydroxy-3-methyl-glutaryl CoA reductase, is an important control site in cholesterol biosynthesis, as will be discussed shortly.

$$\text{3-Hydroxy-3-methylglutaryl CoA} + 2\,\text{NADPH} + 2\,\text{H}^+ \longrightarrow$$
$$\text{mevalonate} + 2\,\text{NADP}^+ + \text{CoA}$$

Mevalonate is converted into 3-phospho-5-pyrophosphomevalonate by three consecutive phosphorylations. This labile intermediate loses CO_2 and P_i, which yields *3-isopentenyl pyrophosphate* (Figure 20-10).

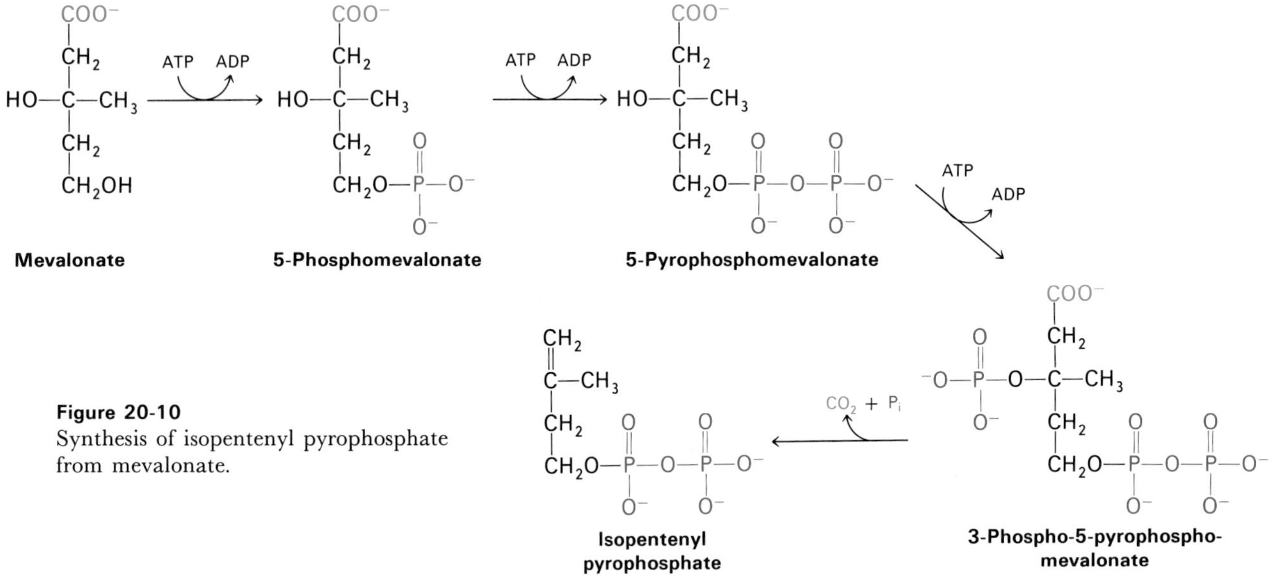

Figure 20-10
Synthesis of isopentenyl pyrophosphate from mevalonate.

SYNTHESIS OF SQUALENE FROM ISOPENTENYL PYROPHOSPHATE

Squalene is synthesized from isopentenyl pyrophosphate by the reaction sequence

$$C_5 \longrightarrow C_{10} \longrightarrow C_{15} \longrightarrow C_{30}$$

This stage in the synthesis of cholesterol (Figure 20-11) starts with the isomerization of *isopentenyl pyrophosphate* to *dimethylallyl pyrophosphate*.

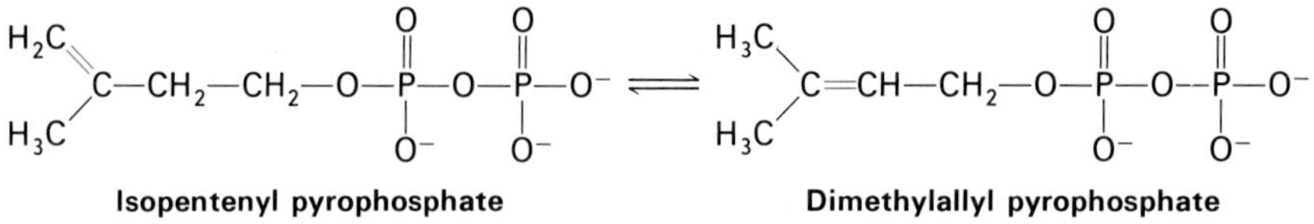

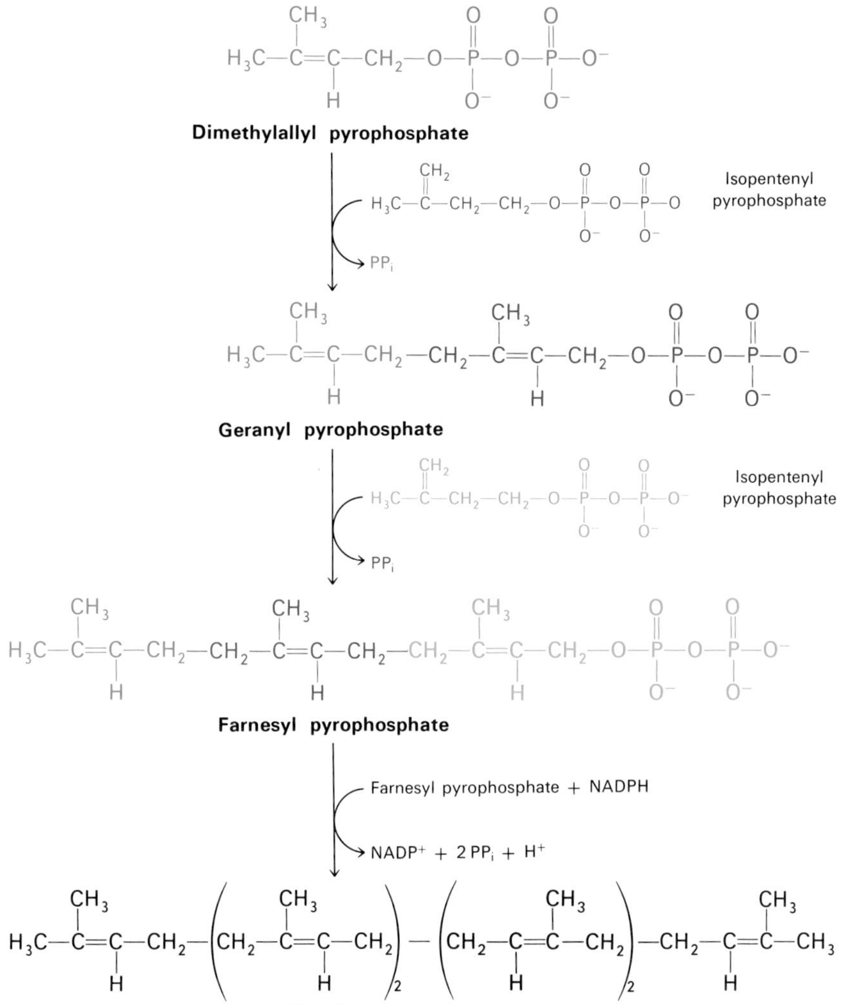

Dimethylallyl pyrophosphate

Isopentenyl pyrophosphate

PP$_i$

Geranyl pyrophosphate

Isopentenyl pyrophosphate

PP$_i$

Farnesyl pyrophosphate

Farnesyl pyrophosphate + NADPH

NADP$^+$ + 2 PP$_i$ + H$^+$

Squalene

Figure 20-11
Synthesis of squalene from dimethylallyl pyrophosphate, an isomer of isopentenyl pyrophosphate.

These two C$_5$ units condense to form *geranyl pyrophosphate* (C$_{10}$),
which condenses with another molecule of isopentenyl pyrophosphate to form *farnesyl pyrophosphate* (C$_{15}$). The last step in the synthesis of *squalene* is a reductive condensation of two molecules of
farnesyl pyrophosphate:

$$\text{2 Farnesyl pyrophosphate} + \text{NADPH} \longrightarrow$$
$$\text{C}_{15}$$

$$\text{squalene} + \text{2 PP}_i + \text{NADP}^+ + \text{H}^+$$
$$\text{C}_{30}$$

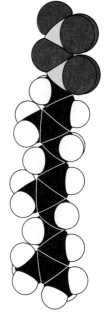

Figure 20-12
Space-filling model of
farnesyl pyrophosphate.

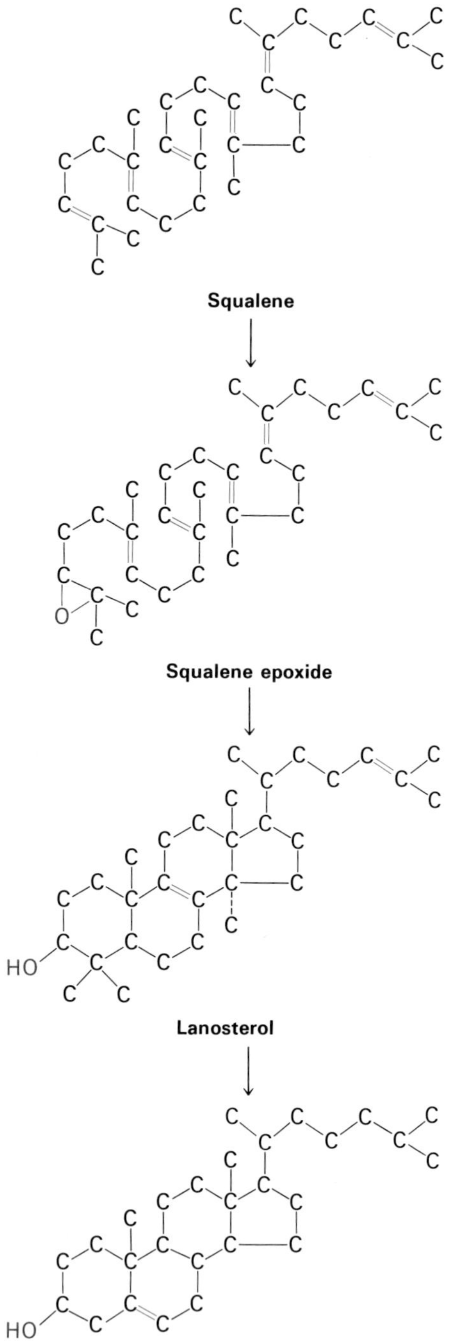

Squalene

Squalene epoxide

Lanosterol

Cholesterol

Figure 20-14
Synthesis of cholesterol from squalene.

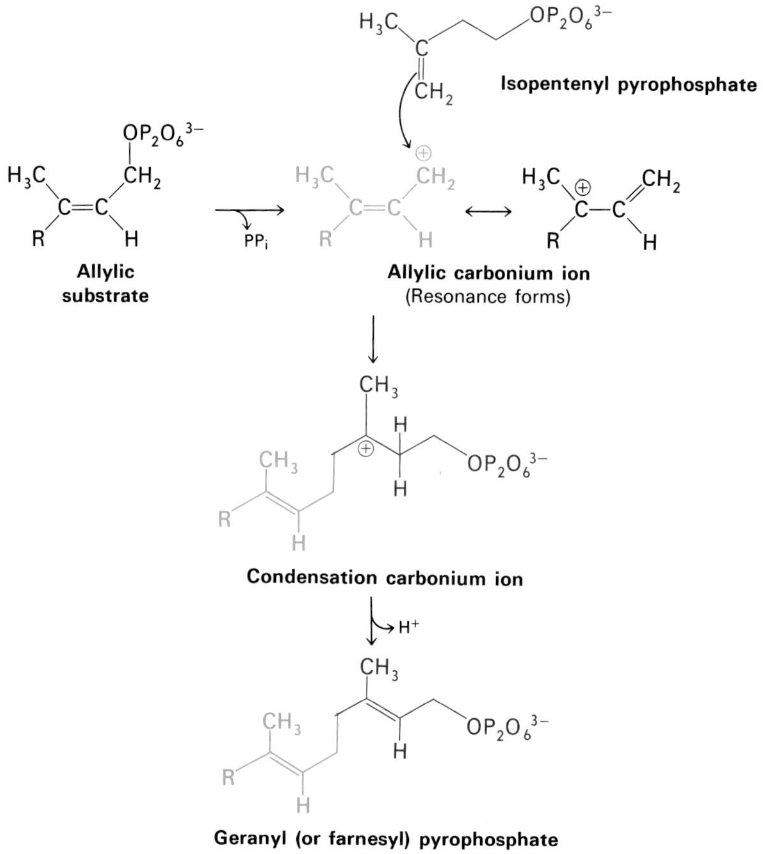

Isopentenyl pyrophosphate

Allylic substrate

Allylic carbonium ion
(Resonance forms)

Condensation carbonium ion

Geranyl (or farnesyl) pyrophosphate

Figure 20-13
Proposed mechanism for the head-to-tail joining of isopentenyl pyrophosphate to an allylic substrate (e.g., dimethylallyl pyrophosphate or geranyl pyrophosphate). The three steps are ionization, condensation, and elimination.

SQUALENE EPOXIDE CYCLIZES TO LANOSTEROL, WHICH IS CONVERTED INTO CHOLESTEROL

The final stage of cholesterol biosynthesis starts with the cyclization of squalene (Figure 20-14). This stage of cholesterol biosynthesis, in contrast with the preceding ones, *requires molecular oxygen. Squalene epoxide,* the reactive intermediate, is formed in a reaction that uses O_2 and NADPH. Squalene epoxide is then cyclized to *lanosterol* by a cyclase. There is a concerted movement of electrons through four double bonds and a migration of two methyl groups in this remarkable closure. Finally, lanosterol is converted into *cholesterol* by the removal of three methyl groups, the reduction of one double bond by NADPH, and the migration of the other double bond. It is interesting to note that cholesterol evolved only after the earth's atmosphere became aerobic. Cholesterol is ubiquitous in eucaryotes but absent from most procaryotes.

BILE SALTS DERIVED FROM CHOLESTEROL FACILITATE THE DIGESTION OF LIPIDS

Bile salts are polar derivatives of cholesterol. These compounds are highly effective *detergents* because they contain polar and nonpolar regions. Bile salts are synthesized in the liver, stored and concentrated in the gallbladder, and then released into the small intestine. Bile salts, the major constituent of bile, *solubilize dietary lipids.* The resulting increase in the surface area of the lipids has two consequences: it promotes their hydrolysis by lipases and facilitates their absorption. Bile salts are also the major breakdown products of cholesterol.

Cholesterol is converted into trihydroxycoprostanoate and then into *cholyl CoA,* the activated intermediate in the synthesis of most bile salts (Figure 20-15). The activated carboxyl carbon of cholyl

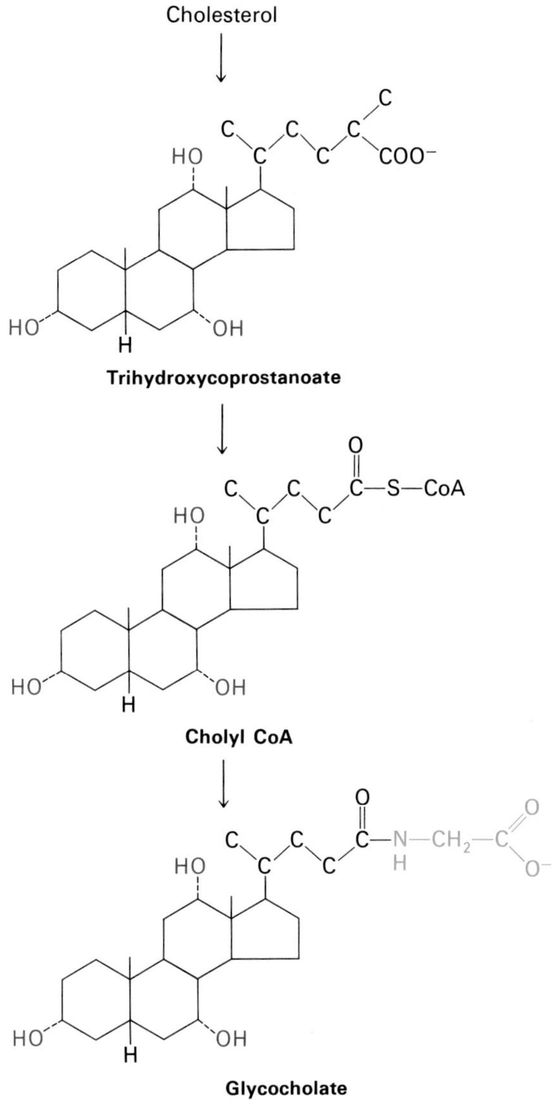

Figure 20-15
Synthesis of glycocholate, the major bile salt.

CoA then reacts with the amino group of glycine to form *glycocholate* or with the amino group of taurine (H_2N—CH_2—CH_2—SO_3^-) to form *taurocholate*. *Glycocholate is the major bile salt.*

CHOLESTEROL SYNTHESIS BY THE LIVER IS SUPPRESSED BY DIETARY CHOLESTEROL

Cholesterol can be obtained from the diet or it can be synthesized de novo. The major site of cholesterol synthesis in mammals is the liver. Appreciable amounts of cholesterol are also formed by the intestine. An adult on a low-cholesterol diet typically synthesizes about 800 mg of cholesterol per day. The rate of cholesterol formation by these organs is highly responsive to the amount of cholesterol absorbed from dietary sources. *This feedback regulation is mediated by changes in the activity of 3-hydroxy-3-methylglutaryl CoA reductase.* As discussed previously, this enzyme catalyzes the formation of mevalonate, which is the committed step in cholesterol biosynthesis. Dietary cholesterol suppresses the synthesis of the reductase in the liver and leads to the inactivation of existing enzyme molecules.

CHOLESTEROL AND OTHER LIPIDS ARE TRANSPORTED TO SPECIFIC TARGETS BY A SERIES OF LIPOPROTEINS

Cholesterol, triacylglycerols, and other lipids are transported in body fluids by a series of lipoproteins classified according to increasing density (Table 20-1): *chylomicrons, very low density lipoproteins* (VLDL), *low-density lipoproteins* (LDL), and *high-density lipoproteins* (HDL). These lipoproteins consist of a core of hydrophobic lipids surrounded by polar lipids and then by a shell of apoproteins. Eight types of apoproteins have been isolated and characterized thus far: apo A-I, A-II, B, C-I, C-II, C-III, D, and E. These complexes *solubilize highly hydrophobic lipids.* Furthermore, their protein components contain signals that *regulate the entry and exit of particular lipids at specific targets.*

Table 20-1
Plasma lipoproteins

Type	Density (g/cm^3)	Major polypeptides
Chylomicrons	<0.94	apo A, B, C
Very low density lipoproteins (VLDL)	0.940–1.006	apo B, C, E
Low-density lipoproteins (LDL)	1.006–1.063	apo B
High-density lipoproteins (HDL)	1.063–1.210	apo A

The plasma lipoproteins are synthesized and secreted by the liver and intestine. *Chylomicrons,* the largest of the lipoproteins, transport dietary triacylglycerols, cholesterol, and other lipids from the intestine to adipose tissue and the liver. Their density is very low (<0.94 g/cm^3) because they are rich in triacylglycerols and have a protein content of less than 2%. The triacylglycerols in chylomicrons are hydrolyzed within a few minutes by lipases located in the capillaries of adipose and other peripheral tissues. The cholesterol-rich residue, known as remnant particles, are taken up by the liver. *Very low density lipoproteins* are primarily synthesized by the liver. VLDL delivers endogenously synthesized triacylglycerols to adipose tissue. The residue is transformed into *low-density lipoproteins,* which are rich in cholesterol esters. Most of the cholesterol in LDL is esterified to linoleate, a polyunsaturated fatty acid. The role of LDL is to transport cholesterol to peripheral tissues and regulate de novo cholesterol synthesis at these sites, as will be discussed shortly. *High-density lipoproteins,* which are synthesized by the liver, are rich in phospholipids and cholesterol. One role of HDL is to transport cholesterol from peripheral tissues to the liver.

THE LOW-DENSITY LIPOPROTEIN RECEPTOR PLAYS A KEY ROLE IN CONTROLLING CHOLESTEROL METABOLISM

Cholesterol, a component of all eucaryotic plasma membranes, is essential for the growth and viability of cells in higher organisms. However, too much cholesterol can be lethal because of atherosclerosis resulting from the deposition of plaques of cholesterol esters. It is evident that cholesterol metabolism must be precisely regulated. The mode of control in the liver, the primary site of cholesterol synthesis, has already been discussed: dietary cholesterol reduces the activity and amount of 3-hydroxy-3-methylglutaryl CoA reductase, the enzyme catalyzing the committed step. Studies of cultured human fibroblasts by Michael Brown and Joseph Goldstein have been sources of insight into the control of cholesterol metabolism in nonhepatic cells. In general, cells outside the liver and intestine obtain cholesterol from the plasma rather than by synthesizing it de novo. Specifically, *their primary source of cholesterol is the low-density lipoprotein.* The steps in the uptake of cholesterol by the *LDL pathway* are:

1. LDL binds to a specific receptor on the plasma membrane of nonhepatic cells. The receptors for LDL are localized in specialized regions called *coated pits,* which contain *clathrin* (p. 717).

2. The receptor-LDL complex is internalized by *endocytosis*—that is, the plasma membrane in the vicinity of the complex invaginates and then fuses to form an endocytic vesicle (Figure 20-16).

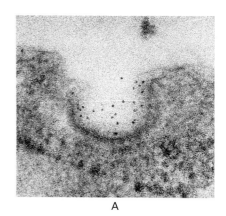

A

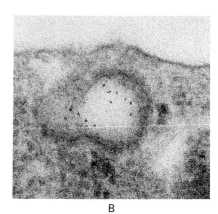

B

Figure 20-16
Endocytosis of LDL bound to its receptor on the surface of cultured human fibroblasts (LDL was visualized by conjugating it to ferritin): (A) electron micrograph showing LDL-ferritin (dark dots) bound to a coated-pit region of the cell surface; (B) this region invaginates and fuses to form an endocytic vesicle. [From R. G. W. Anderson, M. S. Brown, and J. L. Goldstein, *Cell* 10(1977):351. MIT Press.]

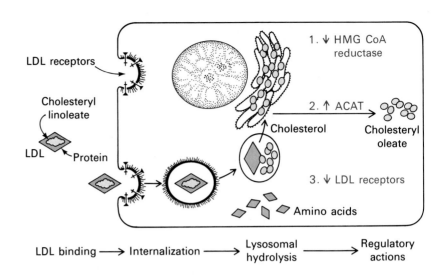

Figure 20-17
Steps in the low-density lipoprotein pathway in cultured human fibroblasts. HMG CoA reductase denotes 3-hydroxy-3-methylglutaryl CoA reductase, and ACAT denotes acyl CoA: cholesterol acyltransferase. [After a drawing kindly provided by Dr. Michael Brown and Dr. Joseph Goldstein.]

3. These vesicles, containing LDL, subsequently fuse with *lysosomes,* which carry a wide array of degradative enzymes. The protein component of the LDL is hydrolyzed to free amino acids. The cholesterol esters in the LDL are hydrolyzed by a lysosomal acid lipase.

4. *The released unesterified cholesterol can then be used for membrane biosynthesis.* Alternatively, it can be *reesterified for storage inside the cell.* In fact, free cholesterol activates acyl CoA-cholesterol acyl transferase (ACAT), the enzyme catalyzing this reaction. Reesterified cholesterol mainly contains oleate and palmitoleate, which are monounsaturated fatty acids, in contrast with the cholesterol esters in LDL, which are rich in linoleate, a polyunsaturated fatty acid.

The cholesterol content of cells that have an active LDL pathway is regulated in two ways. First, the *released cholesterol suppresses the formation of 3-hydroxy-3-methylglutaryl CoA reductase, which thereby inhibits the de novo synthesis of cholesterol.* Second, the LDL receptor is itself subject to feedback regulation. In fibroblasts, the half-life of the LDL receptor is about one day. *When cholesterol is abundant inside the cell, new LDL receptors are not synthesized, and so the uptake of additional cholesterol from plasma LDL is blocked.*

ABSENCE OF THE LDL RECEPTOR LEADS TO HYPERCHOLESTEROLEMIA AND PREMATURE ATHEROSCLEROSIS

The importance of the LDL receptor is highlighted by studies of *familial hypercholesterolemia.* The total concentration of cholesterol and LDL in the plasma is markedly elevated in this genetic disorder, which results from a mutation at a single autosomal locus.

The cholesterol level in the plasma of homozygotes is typically 680 mg/dl, compared with 300 mg/dl in heterozygotes and 175 mg/dl in normal individuals. *Cholesterol is deposited in various tissues because of the high concentration of LDL-cholesterol in the plasma.* Nodules of cholesterol called xanthomas are prominent in skin and tendons. More harmful is the deposition of cholesterol in arterial plaques, which produce atherosclerosis. In fact, *most homozygotes die of coronary artery disease in childhood.* The disease in heterozygotes has a more variable and milder clinical course. *The molecular defect in most cases of familial hypercholesterolemia is an absence or deficiency of functional receptors for LDL.* Homozygotes have almost no receptors for LDL, whereas heterozygotes have about half the normal number. Consequently, the entry of LDL into nonhepatic cells is impaired and so the plasma level of LDL is increased. Familial hypercholesteremia can also result from a defect in the internalization of the LDL-receptor complex.

NOMENCLATURE OF STEROIDS

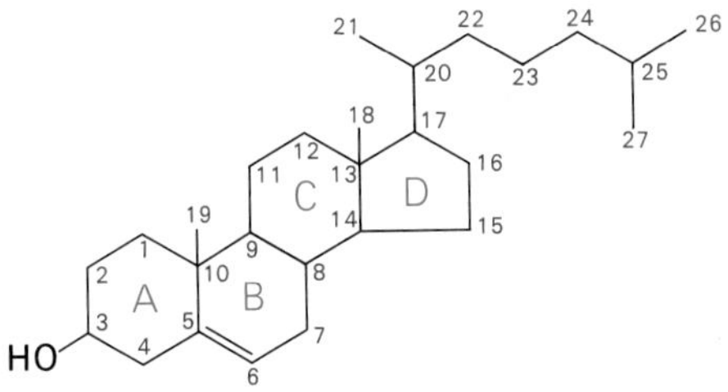

A few comments concerning steroid nomenclature are appropriate before turning to the synthesis of steroid hormones. Carbon atoms in steroids are numbered as shown in Figure 20-18 for cholesterol. The rings in steroids are named A, B, C, and D. Cholesterol contains two angular methyl groups: the C-19 methyl group is attached to C-10, and the C-18 methyl group is attached to C-13. A line above C-10 or C-13 denotes a methyl group. By definition, the C-18 and C-19 methyl groups of cholesterol are *above* the plane containing the four rings. A substituent that is above the plane is termed *β-oriented* and is shown by a *solid* bond. In contrast, a substituent that is below the plane is *α-oriented* and denoted by a *dashed* or dotted line.

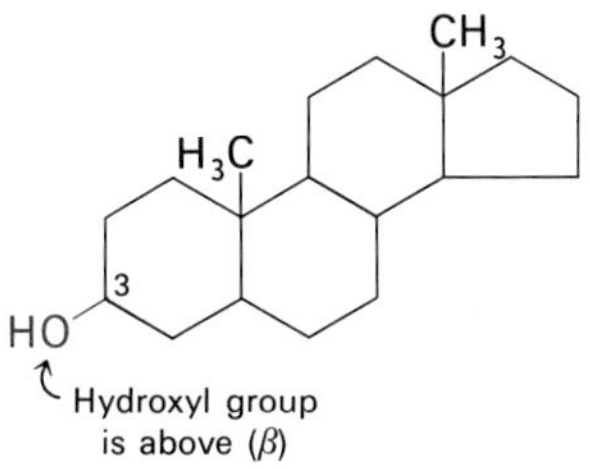

Figure 20-18
Numbering of the carbon atoms of cholesterol.

3β-Hydroxy — Hydroxyl group is above (β)

3α-Hydroxy — Hydroxyl group is below (α)

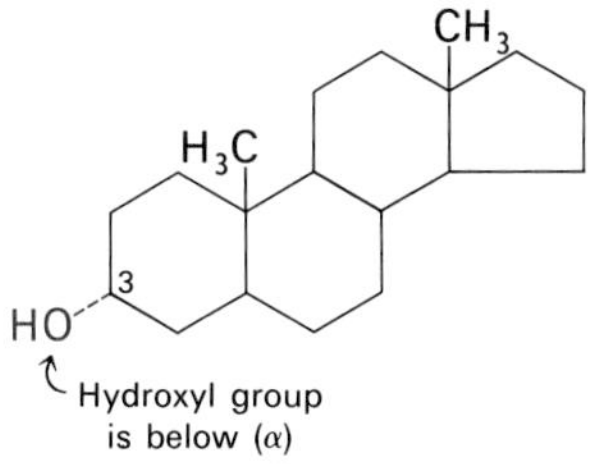

A hydrogen atom attached to C-5 can be α- or β-oriented. When this hydrogen atom is *α-oriented,* the A and B rings are fused in a *trans* conformation, whereas a *β-orientation* yields a *cis* fusion. The absence of a symbol for the C-5 hydrogen atom implies a *trans*

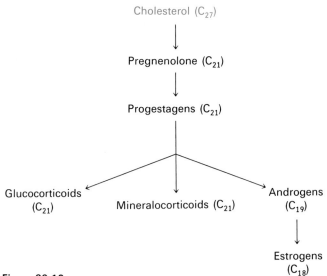

5β-Hydrogen
(A *cis* fusion)

5α-Hydrogen
(A *trans* fusion)

fusion. The C-5 hydrogen atom is α-oriented in all steroid hormones that contain a hydrogen atom in that position. In contrast, bile salts have a β-oriented hydrogen atom at C-5. Thus, *a cis fusion is characteristic of the bile salts, whereas a* trans *fusion is characteristic of all steroid hormones that possess a hydrogen atom at C-5.* A *trans* fusion yields a nearly planar structure, whereas a *cis* fusion gives a buckled structure.

STEROID HORMONES ARE DERIVED FROM CHOLESTEROL

Cholesterol is the precursor of the five major classes of steroid hormones: progestagens, glucocorticoids, mineralocorticoids, androgens, and estrogens (Figure 20-19). *Progesterone,* a *progestagen,* pre-

Cholesterol (C_{27})

↓

Pregnenolone (C_{21})

↓

Progestagens (C_{21})

Glucocorticoids (C_{21}) Mineralocorticoids (C_{21}) Androgens (C_{19})

↓

Estrogens (C_{18})

Figure 20-19
Biosynthetic relations of steroid hormones.

pares the lining of the uterus for implantation of an ovum. Progesterone is also essential for the maintenance of pregnancy. *Androgens* (such as *testosterone*) are responsible for the development of male secondary sex characteristics, whereas *estrogens* (such as *estrone*) are required for the development of female secondary sex characteristics. Estrogens also participate in the ovarian cycle. *Glucocorticoids* (such as *cortisol*) promote gluconeogenesis and the formation of glycogen and enhance the degradation of fat and protein. *Mineralocorticoids* (such as *aldosterone*) increase reabsorption of Na^+, Cl^-, and HCO_3^- by the kidney, which leads to an increase in blood volume and blood pressure. The major sites of synthesis of these classes of hormones are: progestagens, corpus luteum; estrogens, ovary; androgens, testis; glucocorticoids and mineralocorticoids, adrenal cortex.

Hydroxylation reactions play a very important role in the synthesis of cholesterol from squalene and in the conversion of cholesterol into steroid hormones and bile salts. All of these hydroxylations require *NADPH and O_2*. The oxygen atom of the incorporated hydroxyl group comes from O_2 rather than from H_2O, as shown by the use of ^{18}O-labeled O_2 and H_2O. One oxygen atom of the O_2 molecule goes into the substrate, whereas the other is reduced to water. The enzymes catalyzing these reactions are called *monooxygenases* (or mixed-function oxygenases). Recall that a monooxygenase also participates in the hydroxylation of phenylalanine (p. 424).

$$RH + O_2 + NADPH + H^+ \longrightarrow ROH + H_2O + NADP^+$$

Hydroxylation requires the activation of oxygen. In the synthesis of steroid hormones and bile salts, activation is accomplished by a specialized cytochrome called P_{450} because its complex with CO has an absorption maximum at 450 nm. Cytochrome P_{450} is the terminal component of an *electron-transport chain* that has been found in adrenal mitochondria and liver microsomes. The role of this assembly is hydroxylation rather than oxidative phosphorylation. NADPH transfers its high-potential electrons to a flavoprotein in this chain, which are then conveyed to *adrenodoxin*, a nonheme iron protein. Adrenodoxin transfers an electron to the oxidized form of cytochrome P_{450}. The reduced form of P_{450} then activates O_2.

The cytochrome P_{450} system is also important for the *detoxification of foreign substances* (xenobiotic compounds). For example, the hydroxylation of phenobarbital, a barbiturate, *increases its solubility* and *facilitates its excretion*. Likewise, polycyclic aromatic hydrocarbons are hydroxylated by the P_{450} system. The introduction of hydroxyl groups provides sites for conjugation with highly polar units (e.g., glucuronate or sulfate), which markedly increases the solubility of the modified aromatic molecule. However, the action of the P_{450} system is not always beneficial. Recent studies have shown that *most powerful carcinogens are converted in vivo into a chemically reactive form*. This process of *metabolic activation* is usually carried out by the P_{450} system.

PREGNENOLONE IS FORMED FROM CHOLESTEROL
BY CLEAVAGE OF ITS SIDE CHAIN

Steroid hormones contain 21 or fewer carbon atoms, whereas cholesterol contains 27. The first stage in the synthesis of steroid hormones is the removal of a C_6 unit from the side chain of cholesterol to form *pregnenolone*. The side chain of cholesterol is hydroxylated at

C-20 and then at C-22, followed by the cleavage of the bond between C-20 and C-22. The latter reaction is catalyzed by desmolase. All three reactions utilize NADPH and O_2.

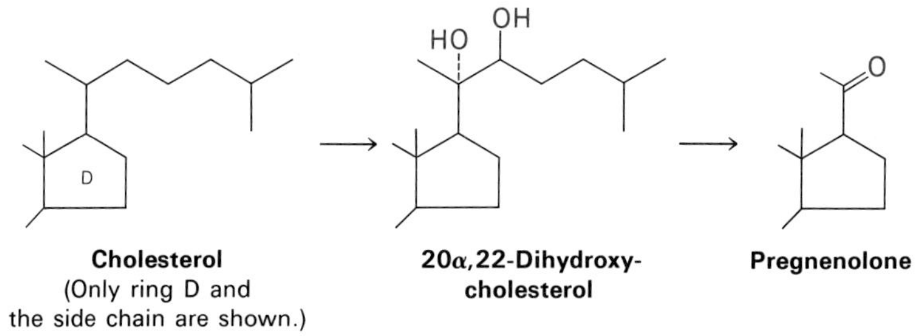

Cholesterol
(Only ring D and
the side chain are shown.)

**20α,22-Dihydroxy-
cholesterol**

Pregnenolone

Adrenocorticotropic hormone (ACTH, or corticotropin), a polypeptide synthesized by the anterior pituitary gland, stimulates the conversion of cholesterol into pregnenolone, which is the precursor of all steroid hormones.

SYNTHESIS OF PROGESTERONE AND CORTICOIDS

Progesterone is synthesized from pregnenolone in two steps. The 3-hydroxyl group of pregnenolone is oxidized to a 3-keto group, and the Δ^5 double bond is isomerized to a Δ^4 double bond. *Cortisol*, the

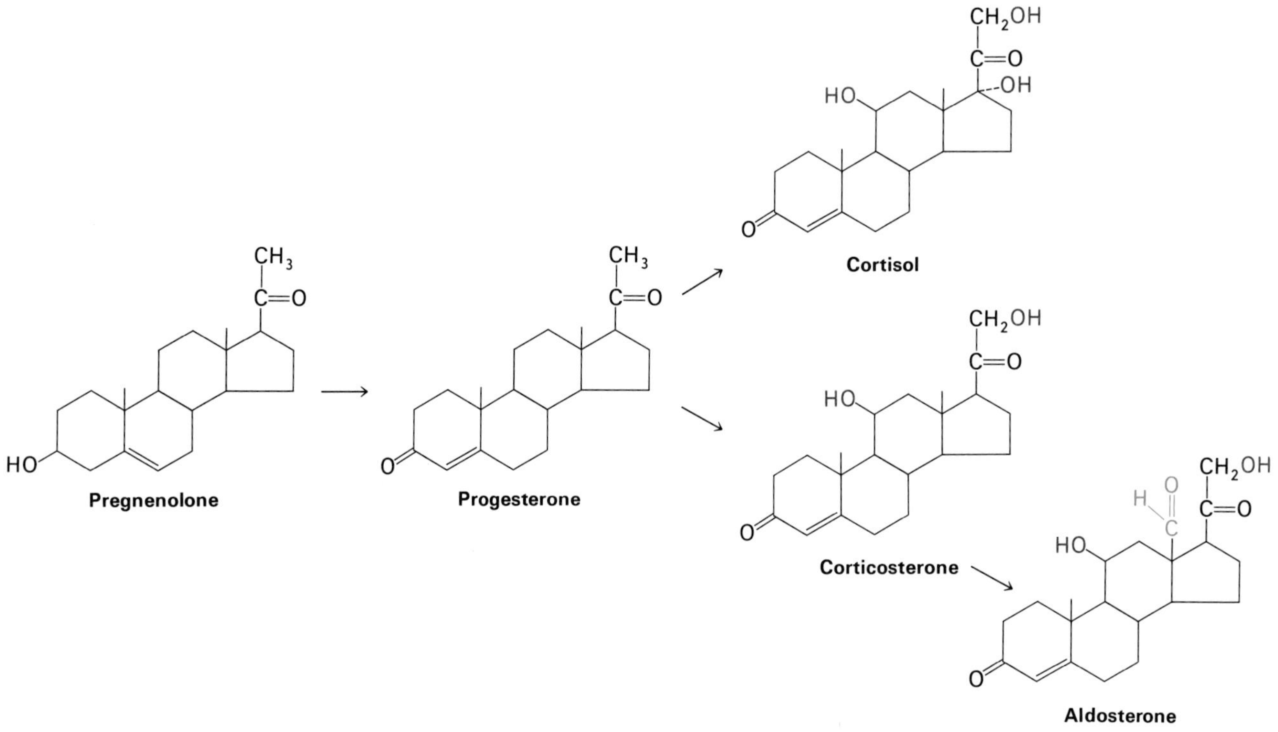

Pregnenolone

Progesterone

Cortisol

Corticosterone

Aldosterone

Figure 20-20
Synthesis of progesterone and corticoids.

major glucocorticoid, is synthesized from progesterone by hydroxylations at C-17, C-21, and C-11; C-17 must be hydroxylated before C-21, whereas hydroxylation at C-11 may occur at any stage. The enzymes catalyzing these hydroxylations are highly specific, as shown by some inherited disorders of steroid metabolism. The initial step in the synthesis of *aldosterone,* the major mineralocorticoid, is the hydroxylation of progesterone at C-21. The resulting deoxycorticosterone is hydroxylated at C-11. The C-18 angular methyl group is then oxidized to an aldehyde, which yields aldosterone.

SYNTHESIS OF ANDROGENS AND ESTROGENS

The synthesis of androgens (Figure 20-21) starts with the hydroxylation of progesterone at C-17. The side chain consisting of C-20 and C-21 is then cleaved to yield *androstenedione,* an androgen. *Testosterone,* another androgen, is formed by reduction of the 17-keto group of androstenedione. Androgens contain nineteen carbon atoms. Estrogens are synthesized from androgens by the loss of the C-19 angular methyl group and the formation of an aromatic A ring. These reactions require NADPH and O_2. *Estrone,* an estrogen, is derived from androstenedione, whereas *estradiol,* another estrogen, is formed from testosterone.

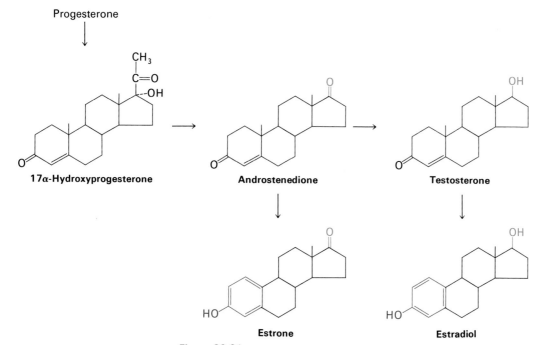

Figure 20-21
Synthesis of androgens and estrogens.

DEFICIENCY OF 21-HYDROXYLASE CAUSES VIRILIZATION AND ENLARGEMENT OF THE ADRENALS

The most common inherited disorder of steroid-hormone synthesis is a deficiency of *21-hydroxylase,* an enzyme needed for the synthesis of glucocorticoids and mineralocorticoids. The diminished production of glucocorticoids leads to an increased secretion of ACTH by the anterior pituitary gland. This response is an expression of a normal feedback mechanism that controls adrenal cortical activity. *The adrenal glands enlarge because of the high level of ACTH in the blood, and more pregnenolone is synthesized.* Consequently, the concentrations of progesterone and 17α-hydroxyprogesterone increase. In turn, there is a *marked increase in the amount of androgens,* because they are derived from 17α-hydroxyprogesterone.

The striking clinical finding in 21-hydroxylase deficiency is *virilization* caused by high levels of androgens. In affected females, virilization is usually evident at birth. Androgens secreted during the development of the female fetus produce a masculinization of the external genitalia. In males, the sexual organs appear normal at birth. Sexual precocity becomes apparent several months later. There is accelerated growth and very early bone maturation so that short stature is the typical final result. About half of the patients with 21-hydroxylase deficiency *persistently lose Na$^+$ in the urine.* They have very low levels of aldosterone, the principal mineralocorticoid. Loss of salt leads to dehydration and hypotension, which may lead to shock and sudden death.

Effective therapy is available for 21-hydroxylase deficiency. The administration of a glucocorticoid provides this needed hormone and concomitantly eliminates the excessive secretion of ACTH. Excessive formation of androgens is thereby stopped. A mineralocorticoid may also be given to patients who lose salt. Some of the symptoms of 21-hydroxylase deficiency are reversed if therapy is started in the first two years of life.

Several other inherited defects of steroid-hormone synthesis are known. The affected enzymes include 11-hydroxylase, 17-hydroxylase, 3β-dehydrogenase, and desmolase. All of these enzymatic lesions lead to a compensatory enlargement of the adrenal gland. Hence, the clinical term for this group of disorders is *congenital adrenal hyperplasia.* Like 21-hydroxylase deficiency, 11-hydroxylase deficiency is accompanied by virilization.

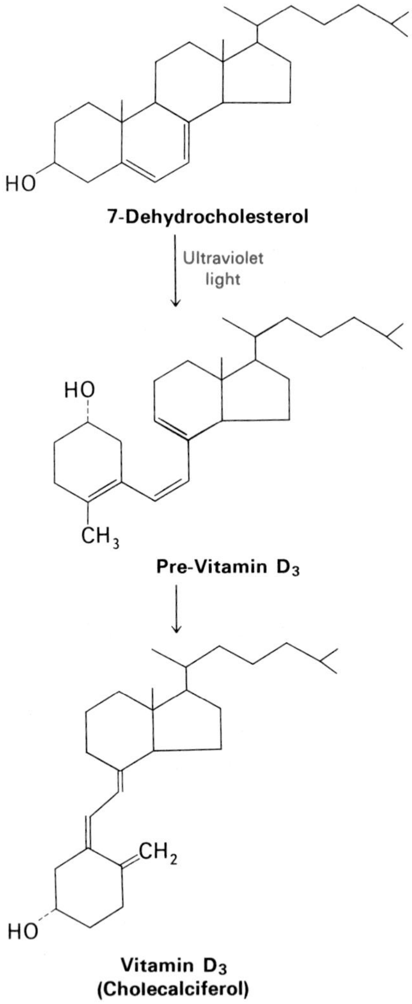

VITAMIN D IS DERIVED FROM CHOLESTEROL BY THE ACTION OF LIGHT

Cholesterol is also the precursor of Vitamin D, which plays an essential role in the control of calcium and phosphorus metabolism. *7-Dehydrocholesterol (pro-Vitamin D$_3$) is photolyzed by ultraviolet light to pre-Vitamin D$_3$, which spontaneously isomerizes to Vitamin D$_3$* (Figure 20-22). Vitamin D$_3$ (cholecalciferol) is converted into an active hor-

7-Dehydrocholesterol

Ultraviolet
light

Pre-Vitamin D$_3$

**Vitamin D$_3$
(Cholecalciferol)**

Figure 20-22
Conversion of 7-dehydrocholesterol into vitamin D (cholecalciferol). Vitamin D$_2$ (ergocalciferol) can be formed in a similar way starting with ergosterol, a plant sterol. Vitamin D$_2$ differs from D$_3$ in having a C-22–C-23 double bond and a C-24 methyl group.

mone by hydroxylation reactions occurring in the liver and kidneys. Vitamin D deficiency in childhood produces *rickets,* which is characterized by inadequate calcification of cartilage and bone. As described by Daniel Webster in 1645, in rickets,

> . . . the whole bony structure is as flexible as softened wax, so that the flaccid and enervated legs can hardly support the superposed weight of the body; hence the tibia, giving way beneath the overpowering weight of the frame, bend inwards; and for the same reason the legs are drawn together at their tops; and the back, by reason of the bending of the spine, sticks out in a hump in the lumbar region . . . the patients in their weakness cannot (in the most severe stages of the disease) bear to sit upright, much less stand. . . .

Rickets was so common that Webster referred to it as the "Children's disease of the English." We now know that rickets was prevalent in these children because there was little sunlight for many months of the year. Consequently, 7-dehydrocholesterol in the skin was not photolyzed to pre-Vitamin D_3. Furthermore, their diets provided little Vitamin D because most naturally occurring foods have a low content of this vitamin. Fish-liver oils are a notable exception and so cod-liver oil was used for many years as a rich source of Vitamin D. Today, the most reliable dietary sources of Vitamin D are fortified foods. In the United States, milk is fortified to a level of 400 international units per quart (10 μg per quart). The recommended daily intake of Vitamin D is 400 international units, irrespective of age. In adults, Vitamin D deficiency leads to softening and weakening of bones, a condition called *osteomalacia.* The occurrence of osteomalacia in Bedouin Arab women who are clothed so that only their eyes are exposed to sunlight is a striking reminder that Vitamin D is needed by adults as well as by children.

Rickets—
From the Old English word *wrickken,* to twist.

Osteomalacia—
From the Greek words *osteon,* bone, and *malakia,* softness.

FIVE-CARBON UNITS ARE JOINED TO FORM A WIDE VARIETY OF BIOMOLECULES

The synthesis of squalene (C_{30}) from isopentenyl pyrophosphate (C_5) exemplifies a fundamental mechanism for the assembly of carbon skeletons of biomolecules. *A remarkable array of compounds are formed from isopentenyl pyrophosphate, the basic five-carbon building block.* The fragrances of many plants arise from volatile C_{10} and C_{15} compounds, which are called *terpenes.* For example, myrcene ($C_{10}H_{16}$) from bay leaves consists of two isoprene units, as does limonene ($C_{10}H_{15}$) from lemon oil. Zingiberene ($C_{15}H_{24}$), from the oil of ginger, is made up of three isoprene units. Some terpenes, such as geraniol from geraniums and menthol from peppermint oil, are alcohols, and others, such as citronellal, are aldehydes. *Natural rubber* is a linear polymer of *cis*-isoprene units.

We have already encountered several molecules that contain isoprenoid side chains. The C_{30} hydrocarbon *side chain of Vitamin K_2,* a key molecule in clotting (p. 175), is built from six C_5 units. *Coenzyme*

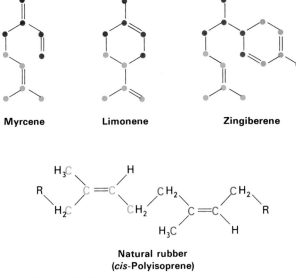

Myrcene   Limonene   Zingiberene

Natural rubber
(*cis*-Polyisoprene)

Figure 20-23
Formulas of some isoprenoids. The five-carbon
building blocks are shown in color.

"Perfumes, colors, and sounds
echo one another."
CHARLES BAUDELAIRE
Correspondances

Q_{10} in the mitochondrial respiratory chain (p. 313) has a side chain made up of ten isoprene units. Yet another example is the *phytol side chain* of *chlorophyll* (p. 434), which is formed from four isoprene units.

Isoprenoids can bring delight by their color as well as by their fragrance. Indeed, isoprenoids can be regarded as the sensual molecules! The color of tomatoes and carrots comes from *carotenoids*, specifically from *lycopene and β-carotene*, respectively. These compounds absorb light because they contain extended networks of single and double bonds—that is, they are *polyenes*. Their C_{40} carbon skeletons are built by the successive addition of C_5 units to form *geranylgeranyl pyrophosphate, a C_{20} intermediate*, which then condenses tail-to-tail with another molecule of geranylgeranyl pyrophosphate. This biosynthetic pathway is like that of squalene except that C_{20} rather than C_{15} units are assembled and condensed.

$$C_5 \longrightarrow C_{10} \longrightarrow C_{15} \longrightarrow C_{30} \text{ (squalene)}$$
$$C_5 \longrightarrow C_{10} \longrightarrow C_{15} \longrightarrow C_{20} \longrightarrow C_{40} \text{ (phytoene)}$$

Phytoene, the C_{40} condensation product, is dehydrated to yield lycopene. Cyclization of both ends of lycopene gives β-carotene (Figure 20-24). Carotenoids serve as light-harvesting molecules in photosynthetic assemblies and also play a role in protecting procaryotes from the deleterious effects of light. Carotenoids are also essential for vision. β-Carotene is the precursor of retinal, the chromophore in all known visual pigments (p. 897). *These examples illustrate the fundamental role of isopentenyl pyrophosphate in the assembly of extended carbon skeletons in biomolecules. It is also evident that isoprenoids are ubiquitous in nature and have diverse significant roles.*

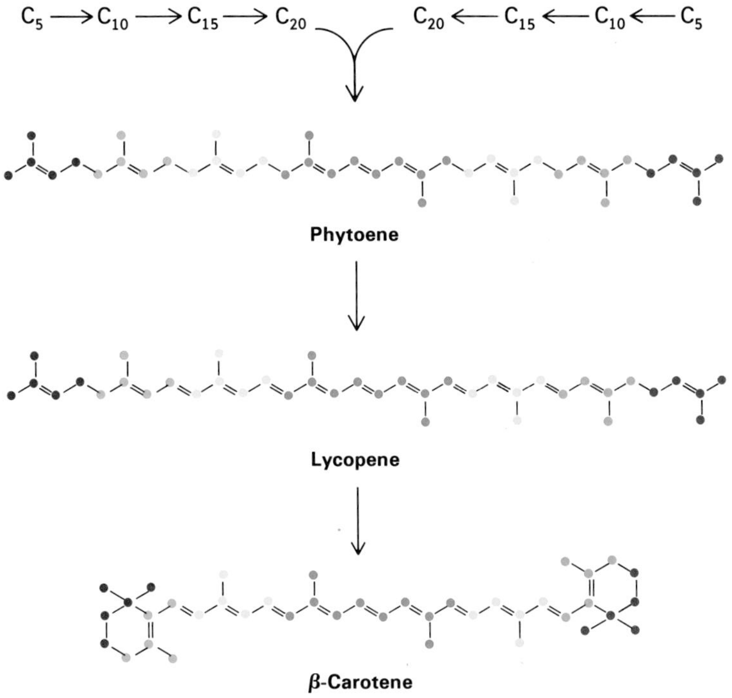

$$C_5 \longrightarrow C_{10} \longrightarrow C_{15} \longrightarrow C_{20} \qquad C_{20} \longleftarrow C_{15} \longleftarrow C_{10} \longleftarrow C_5$$

Phytoene

Lycopene

β-Carotene

Figure 20-24
Synthesis of C_{40} carotenoids: phytoene, lycopene, and β-carotene.

SUMMARY

Phosphatidate, an intermediate in the synthesis of phosphoglycerides and triacylglycerols, is formed by the acylation of glycerol 3-phosphate by acyl CoA. Hydrolysis of its phosphoryl group followed by acylation yields a triacylglycerol. CDP-diacylglycerol, the activated intermediate in the de novo synthesis of phosphoglycerides, is formed from phosphatidate and CTP. The activated phosphatidyl unit is then transferred to the hydroxyl group of a polar alcohol, such as serine, to form phosphatidyl serine. Decarboxylation of this phosphoglyceride yields phosphatidyl ethanolamine, which is methylated by S-adenosylmethionine to form phosphatidyl choline. This phosphoglyceride can also be synthesized by a salvage pathway that utilizes preformed choline. CDP-choline is the activated intermediate in this route. Sphingolipids are synthesized from ceramide, which is formed by the acylation of sphingosine. Gangliosides are sphingolipids that contain an oligosaccharide unit having at least one residue of N-acetylneuraminate or a related sialic acid. Gangliosides are synthesized by the step-by-step addition of activated sugars, such as UDP-glucose, to ceramide.

Cholesterol, a steroid component of eucaryotic membranes and a precursor of steroid hormones, is formed from acetyl CoA. The synthesis of mevalonate from 3-hydroxy-3-methylglutaryl CoA (derived from acetyl CoA and acetoacetyl CoA) is the committed step in the formation of cholesterol. Mevalonate is converted into isopentenyl pyrophosphate (C_5), which condenses with its isomer, dimethylallyl pyrophosphate (C_5), to form geranyl pyrophosphate (C_{10}). Addition of a second molecule of isopentenyl pyrophosphate yields farnesyl pyrophosphate (C_{15}), which condenses with itself to form squalene (C_{30}). This intermediate cyclizes to lanosterol (C_{30}), which is modified to yield cholesterol (C_{27}). The synthesis of cholesterol by the liver is regulated by changes in the amount and activity of 3-hydroxy-3-methylglutaryl CoA reductase, the enzyme catalyzing the committed step in its biosynthesis. The low-density lipoprotein receptor plays a critical role in controlling the entry of cholesterol into most cells outside the liver and intestine.

Five major classes of steroid hormones are derived from cholesterol: progestagens, glucocorticoids, mineralocorticoids, androgens, and estrogens. Hydroxylations by mixed-function oxidases that use NADPH and O_2 play an important role in the synthesis of steroid hormones and bile salts from cholesterol. Pregnenolone (C_{21}), a key intermediate in the synthesis of steroid hormones, is formed by scission of the side chain of cholesterol. Progesterone (C_{21}), synthesized from pregnenolone, is the precursor of cortisol and aldosterone. Cleavage of the side chain of progesterone yields androstenedione, an androgen (C_{19}). Estrogens (C_{18}) are synthesized from androgens by the loss of an angular methyl group and the formation of an aromatic A ring.

Vitamin D, which is important in the control of calcium and phosphorus metabolism, is formed from a derivative of cholesterol by the action of light. A remarkable array of biomolecules in addition to cholesterol and its derivatives are synthesized from isopentenyl pyrophosphate, the basic five-carbon building block. The hydrocarbon side chains of Vitamin K_2, coenzyme Q_{10}, and chlorophyll are examples of extended chains constructed from this activated C_5 unit.

SELECTED READINGS

WHERE TO START

Bloch, K., 1965. The biological synthesis of cholesterol. *Science* 150:19–28.

Brown, M. S., and Goldstein, J. L., 1976. Receptor-mediated control of cholesterol metabolism. *Science* 191:150–154.

BOOKS ON LIPID METABOLISM

Snyder, F., (ed.), 1977. *Lipid Metabolism in Mammals,* vols. 1 and 2. Plenum.

DeLuca, H. F., (ed.), 1978. *The Fat-Soluble Vitamins.* Plenum.

Wakil, S., (ed.), 1970. *Lipid Metabolism.* Academic Press.

Nes, W. R., and McKean, M. L., 1977. *Biochemistry of Steroids and Other Isopentanoids.* University Park Press.

ENZYMES AND REACTION MECHANISMS

Walsh, C., 1979. *Enzymatic Reaction Mechanisms.* Freeman. [Contains excellent discussions of reaction mechanisms in the biosynthesis of cholesterol and other compounds derived from isopentenyl pyrophosphate.]

Hayaishi, O., (ed.), 1974. *Molecular Mechanisms of Oxygen Activation.* Academic Press.

Holick, M. F., and Clark, M. B., 1978. The photobiogenesis and metabolism of vitamin D. *Fed. Proc.* 37:2567–2574.

McMurray, W. C., and Magee, W. L., 1972. Phospholipid metabolism. *Ann. Rev. Biochem.* 41:129–160.

Gatt, S., and Barenholz, Y., 1973. Enzymes of complex lipid metabolism. *Ann. Rev. Biochem.* 42:61–85.

Popják, G., and Cornforth, J. W., 1960. The biosynthesis of cholesterol. *Advan. Enzymol.* 22:281–335.

Bloch, K., 1976. On the evolution of a biosynthetic pathway. *In* Kornberg, A., Horecker, B. L., Cornudella, L., and Oro, J., (eds.), *Reflections on Biochemistry,* pp. 143–

150. Pergamon. [A stimulating discussion of the evolution of cholesterol and other sterols.]

LIPOPROTEINS

Scanu, A. M., 1977. Plasma lipoprotein structure. *Nature* 270:209–210.

Smith, L. C., Pownall, H. J., and Gotto, A. M., Jr., 1978. The plasma lipoproteins: structure and metabolism. *Ann. Rev. Biochem.* 47:751–777.

REGULATION

Brown, M. S., and Goldstein, J. L., 1974. Familial hypercholesterolemia: defective binding of lipoproteins to cultured fibroblasts associated with impaired regulation of 3-hydroxy-3-methylglutaryl coenzyme A reductase activity. *Proc. Nat. Acad. Sci.* 71:788–792.

Anderson, R. G. W., Goldstein, J. L., and Brown, M. S., 1977. A mutation that impairs the ability of lipoprotein receptors to localise in coated pits on the cell surface of human fibroblasts. *Nature* 270:695–699.

Goldstein, J. L., and Brown, M. S., 1977. The low-density lipoprotein pathway and its relation to atherosclerosis. *Ann. Rev. Biochem.* 46:897–930.

Dempsey, M. E., 1974. Regulation of steroid biosynthesis. *Ann. Rev. Biochem.* 43:967–990.

INHERITED DISEASES

Stanbury, J. B., Wyngaarden, J. B., and Fredrickson, D. S., (eds.), 1978. *The Metabolic Basis of Inherited Disease* (4th ed.). McGraw Hill. [Contains excellent articles on genetic disorders of sphingolipid, lipoprotein, cholesterol, and steroid metabolism.]

Dietschy, J. M., Gotto, A. M., Jr., and Ontko, J. A., (eds.), 1978. *Disturbances in Lipid and Lipoprotein Metabolism.* American Physiological Society.

PROBLEMS

1. Write a balanced equation for the synthesis of a triacylglycerol, starting from glycerol and fatty acids.

2. Write a balanced equation for the synthesis of phosphatidyl serine by the de novo pathway, starting from serine, glycerol, and fatty acids.

3. What is the activated reactant in each of these biosyntheses?
   (a) Phosphatidyl serine from serine.
   (b) Phosphatidyl ethanolamine from ethanolamine.
   (c) Ceramide from sphingosine.
   (d) Sphingomyelin from ceramide.
   (e) Cerebroside from ceramide.
   (f) Ganglioside G_{M1} from ganglioside G_{M2}.
   (g) Farnesyl pyrophosphate from geranyl pyrophosphate.

4. What is the distribution of isotopic labeling in cholesterol synthesized from each of these precursors?
   (a) Mevalonate labeled with ^{14}C in its carboxyl carbon atom.
   (b) Malonyl CoA labeled with ^{14}C in its carboxyl carbon atom.

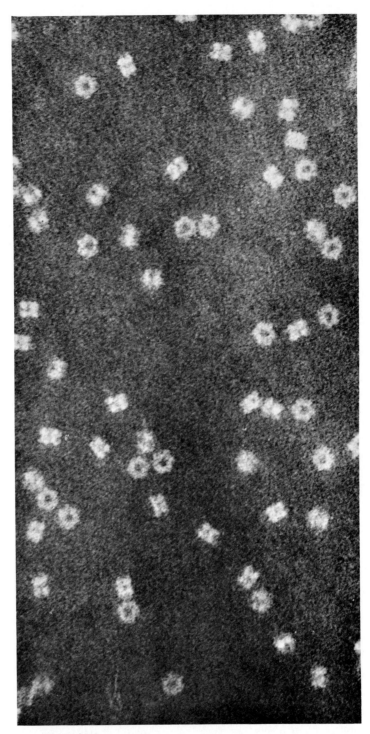

Figure 21-1
Electron micrograph of glutamine synthetase from *E. coli*.
This enzyme plays a key role in nitrogen metabolism.
[Courtesy of Dr. Earl Stadtman.]

CHAPTER 21

BIOSYNTHESIS OF
AMINO ACIDS AND HEME

This chapter deals with the biosynthesis of amino acids and some molecules derived from them. The flow of nitrogen into amino acids will be considered first. This process starts with the reduction of N_2 to NH_4^+ by nitrogen-fixing microorganisms. NH_4^+ is then assimilated into amino acids by way of glutamate and glutamine, the two pivotal molecules in nitrogen metabolism. Of the basic set of twenty amino acids, ten are synthesized from citric acid cycle and other major metabolic intermediates by quite simple reactions. We will consider these biosyntheses and examine the biosyntheses of the aromatic amino acids and of histidine as examples of amino acids formed by more complex routes. In fact, humans must obtain the latter group of ten amino acids from their diets, and so they are called essential amino acids. Two interesting carriers participate in these reactions: tetrahydrofolate, a highly versatile carrier of activated one-carbon units at three oxidation stages, and *S*-adenosylmethionine, the major methyl donor. The regulation of amino acid metabolism is another important area of inquiry. We will take a look at glutamine synthetase, which exemplifies some general principles. The final section of this chapter is concerned with the synthesis and degradation of heme.

MICROORGANISMS USE ATP AND A POWERFUL REDUCTANT TO CONVERT N_2 INTO NH_4^+

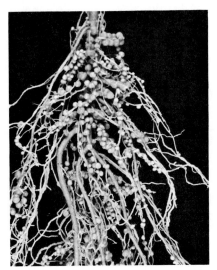

Figure 21-2
The nodules in the root system of the soybean are the sites of nitrogen fixation by *Rhizobium* bacteria. [Courtesy of Dr. Joe C. Burton, Nitragin Company, Inc.]

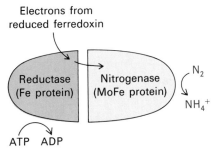

Figure 21-3
Schematic diagram of the nitrogenase complex. The reductase dissociates from the nitrogenase component before N_2 is converted into NH_4^+.

The nitrogen atoms of amino acids, purines, pyrimidines, and other biomolecules come from NH_4^+. Higher organisms are unable to convert N_2 into organic form. Rather, this conversion—called *nitrogen fixation*—is carried out by bacteria and blue-green algae. Some of these microorganisms—namely, the *Rhizobium* bacteria—invade the roots of leguminous plants and form root nodules, in which nitrogen fixation takes place (Figure 21-2). The relation between the bacteria and plant is symbiotic. The amount of N_2 fixed by microorganisms has been estimated to be about 2×10^{11} kg per year.

The $N \equiv N$ bond, which has a bond energy of 225 kcal/mol, is highly resistant to chemical attack. Indeed, Lavoisier named it "azote," meaning "without life," because it is quite unreactive. The industrial process for nitrogen fixation, devised by Fritz Haber in 1910 and currently used in fertilizer factories,

$$N_2 + 3\,H_2 \rightleftharpoons 2\,NH_3$$

is typically carried out over an iron catalyst at about 500°C and a pressure of 300 atm. It is not surprising, then, that the biological process of nitrogen fixation requires a complex enzyme. The *nitrogenase complex*, which carries out this process, consists of two kinds of protein components: a *reductase*, which provides electrons with high reducing power, and a *nitrogenase*, which uses these electrons to reduce N_2 to NH_4^+ (Figure 21-3). Each component is an *iron-sulfur protein*, in which iron is bonded to the sulfur atom of a cysteine residue and to inorganic sulfide (p. 312). The nitrogenase component of the complex also contains one or two *molybdenums* and so it has been known as the *MoFe protein*. It has the subunit structure $\alpha_2\beta_2$ and a mass of about 200 kdal. The reductase (also called the *Fe protein*) consists of two identical polypeptides and has a mass of about 65 kdal. In the nitrogenase complex, one or two Fe proteins are associated with a MoFe protein.

The conversion of N_2 into NH_4^+ by the nitrogenase complex *requires ATP and a powerful reductant*. In most nitrogen-fixing microorganisms, the source of high-potential electrons in this *six-electron reduction* is *reduced ferredoxin*, an electron carrier previously encountered in photosynthesis (p. 439). Whether reduced ferredoxin is then regenerated by photosynthetic or oxidative processes depends on the particular species. The stoichiometry of the reaction catalyzed by the nitrogenase complex is

$$N_2 + 6\,e^- + 12\,ATP + 12\,H_2O \longrightarrow$$
$$2\,NH_4^+ + 12\,ADP + 12\,P_i + 4\,H^+$$

Recent studies of nitrogenase suggest the following reaction sequence. First, reduced ferredoxin transfers its electrons to the reductase component of the complex. Second, ATP then binds to the reductase and shifts its redox potential from -0.29 to -0.40 V by altering its conformation. This enhancement of the reducing power of the reductase enables it to transfer its electrons to the nitrogenase component. Third, electrons are transferred, ATP is hydrolyzed, and the reductase dissociates from the nitrogenase component. Finally, N_2 bound to the nitrogenase component of the complex is reduced to NH_4^+.

Sources of energy for the chemical production of ammonia by the Haber process are becoming scarcer and more costly, and so there is much current interest in enhancing nitrogen fixation by microorganisms. One potential approach is to insert the genes for nitrogen fixation into nonleguminous plants, such as cereals. A difficulty that must be overcome is that the nitrogenase complex is exquisitely sensitive to inactivation by O_2. Leguminous plants maintain a very low concentration of free O_2 in their root nodules by binding O_2 to *leghemoglobin*. Another challenge that must be met in the formation of new nitrogen-fixing species is the requirement for a very high rate of ATP formation. In fact, nitrogen-fixing bacteria in the roots of pea plants consume nearly a fifth of all the ATP generated by the plant. A complementary approach is to increase the rate of nitrogen fixation by blue-green algae, which generate their own ATP by photosynthesis and thus are not dependent on an energy-yielding symbiotic relation.

NH_4^+ IS ASSIMILATED INTO AMINO ACIDS BY WAY OF GLUTAMATE AND GLUTAMINE

The next step in the assimilation of nitrogen into biomolecules is the entry of NH_4^+ into amino acids. *Glutamate* and *glutamine* play pivotal roles in this regard. The α-amino group of most amino acids comes from the α-amino group of glutamate by transamination. Glutamine, the other major nitrogen donor, contributes its side-chain nitrogen in the biosynthesis of a wide range of important compounds.

Glutamate is synthesized from NH_4^+ and α-ketoglutarate, a citric acid cycle intermediate, by the action of *glutamate dehydrogenase*. This enzyme has already been encountered in the degradation of amino acids (p. 408). In the biosynthetic direction, NADPH is the reductant, whereas NAD^+ is the oxidant in the catabolic direction.

$$NH_4^+ + \alpha\text{-ketoglutarate} + NADPH + H^+ \rightleftharpoons$$
$$\text{L-glutamate} + NADP^+ + H_2O$$

Ammonium ion is incorporated into glutamine by the action of *glutamine synthetase*. This amidation is driven by the hydrolysis of ATP.

Glutamate　　**Glutamine**

The regulation of glutamine synthetase plays a critical role in controlling nitrogen metabolism, as will be discussed shortly.

Glutamate dehydrogenase and glutamine synthetase are present in all organisms. Most procaryotes also contain *glutamate synthase,* which catalyzes the reductive amination of α-ketoglutarate. The nitrogen donor in this reaction is glutamine, and so two molecules of glutamate are formed.

$$\alpha\text{-Ketoglutarate} + \text{glutamine} + \text{NADPH} + \text{H}^+ \longrightarrow$$
$$2 \text{ glutamate} + \text{NADP}^+$$

When NH_4^+ *is limiting, most of the glutamate is made by the sequential action of glutamine synthetase and glutamate synthase.* The sum of these reactions is

$$\text{NH}_4^+ + \alpha\text{-ketoglutarate} + \text{NADPH} + \text{ATP} \longrightarrow$$
$$\text{L-glutamate} + \text{NADP}^+ + \text{ADP} + \text{P}_i$$

Note that this stoichiometry differs from that of the glutamate dehydrogenase reaction in that an ATP is hydrolyzed. Why is this more expensive pathway sometimes used by *E. coli?* The answer is that the K_M of glutamate dehydrogenase for NH_4^+ is high (~1 mM), and so this enzyme is not saturated when NH_4^+ is limiting. In contrast, glutamine synthetase has very high affinity for NH_4^+.

Table 21-1
Basic set of twenty amino acids

Nonessential	Essential
Alanine	Arginine
Asparagine	Histidine
Aspartate	Isoleucine
Cysteine	Leucine
Glutamate	Lysine
Glutamine	Methionine
Glycine	Phenylalanine
Proline	Threonine
Serine	Tryptophan
Tyrosine	Valine

AMINO ACIDS ARE SYNTHESIZED FROM CITRIC ACID CYCLE AND OTHER MAJOR METABOLIC INTERMEDIATES

Thus far, we have considered the conversion of N_2 into NH_4^+ and the assimilation of NH_4^+ into glutamate and glutamine. We turn now to the biosynthesis of the other amino acids. Bacteria such as *E. coli* can synthesize the entire basic set of twenty amino acids, whereas humans can make only half of them. The amino acids that must be supplied in the diet are called *essential,* whereas the others are termed *nonessential* (Table 21-1). *These designations refer to the needs of an organism under a particular set of conditions.* For example, enough arginine is synthesized by the urea cycle to meet the needs of an adult but not those of a growing child. A deficiency of even one amino acid results in a *negative nitrogen balance.* In this state, more protein is degraded than is synthesized, and so more nitrogen is excreted than is ingested.

The pathways for the biosynthesis of amino acids are diverse. However, they have an important common feature: *their carbon skeletons come from glycolytic, pentose phosphate pathway, or citric acid cycle intermediates.* A further simplification is that there are only *six biosynthetic families* (Figure 21-4).

The nonessential amino acids are synthesized by quite simple reactions, whereas the pathways for the formation of the essential amino acids are quite complex. For example, the nonessential amino acids *alanine* and *aspartate* are synthesized in a single step from pyruvate and oxaloacetate, respectively. Each acquires its amino group from glutamate in a transamination reaction in which pyridoxal phosphate is the cofactor (p. 410):

$$\text{Pyruvate} + \text{glutamate} \rightleftharpoons \text{alanine} + \alpha\text{-ketoglutarate}$$

$$\text{Oxaloacetate} + \text{glutamate} \rightleftharpoons \text{aspartate} + \alpha\text{-ketoglutarate}$$

Asparagine is then synthesized by the amidation of aspartate:

$$\text{Aspartate} + \text{NH}_4^+ + \text{ATP} \longrightarrow$$
$$\text{asparagine} + \text{AMP} + \text{PP}_i + \text{H}^+$$

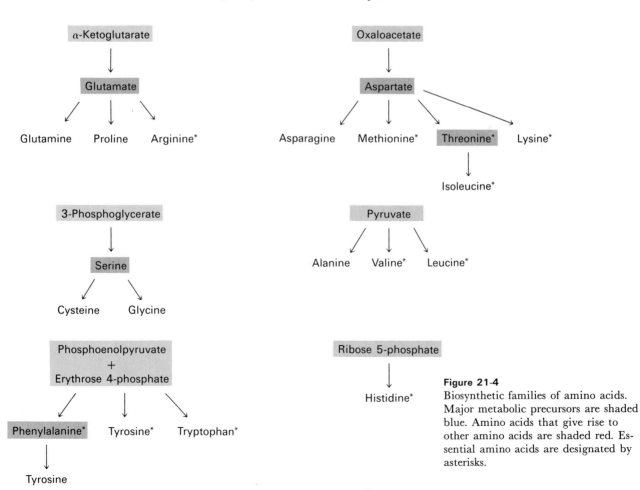

Figure 21-4
Biosynthetic families of amino acids. Major metabolic precursors are shaded blue. Amino acids that give rise to other amino acids are shaded red. Essential amino acids are designated by asterisks.

In mammals, the nitrogen donor in the synthesis of asparagine is glutamine rather than NH_4^+.

Another one-step synthesis of a nonessential amino acid is the hydroxylation of phenylalanine (an essential amino acid) to *tyrosine*, a reaction occurring in mammals.

$$\text{Phenylalanine} + O_2 + \text{NADPH} + H^+ \longrightarrow$$
$$\text{tyrosine} + \text{NADP}^+ + H_2O$$

This reaction is catalyzed by phenylalanine hydroxylase, a monooxygenase discussed previously (p. 424). It is noteworthy that tyrosine is an essential amino acid in individuals lacking this enzyme.

GLUTAMATE IS THE PRECURSOR OF GLUTAMINE AND PROLINE

The synthesis of glutamate by the reductive amination of α-ketoglutarate has already been discussed (p. 487), as has the conversion of glutamate into glutamine (p. 488). Glutamate is the precursor of one other nonessential amino acid, proline. First, the γ-carboxyl group of glutamate reacts with ATP to form an acyl phosphate. This mixed anhydride is then reduced by NADPH to an aldehyde. Glutamic γ-semialdehyde cyclizes with a loss of H_2O to give Δ'-pyrroline-5-carboxylate, which is reduced by NADPH to yield proline.

Glutamate   Glutamic-γ-semialdehyde   Δ'-Pyrroline-5-carboxylate   Proline

SERINE IS SYNTHESIZED FROM 3-PHOSPHOGLYCERATE

Serine is synthesized from 3-phosphoglycerate, an intermediate in glycolysis. The first step is an oxidation to 3-phosphohydroxypyruvate. This α-keto acid is transaminated to 3-phosphoserine, which is then hydrolyzed to yield serine.

3-Phosphoglycerate   3-Phosphohydroxy-pyruvate   3-Phosphoserine   Serine

Alternatively, hydrolysis of the phosphate group may precede oxidation and transamination:

3-Phosphoglycerate $\longrightarrow$ glycerate $\longrightarrow$
$$\text{hydroxypyruvate} \longrightarrow \text{serine}$$

Serine is the precursor of *glycine* and *cysteine*. In the formation of glycine, the side-chain β-carbon atom of serine is transferred to *tetrahydrofolate*, a carrier of one-carbon units that will be discussed shortly.

Serine + tetrahydrofolate $\rightleftharpoons$
$$\text{glycine} + \text{methylenetetrahydrofolate} + H_2O$$

This conversion is catalyzed by *serine transhydroxymethylase*, a pyridoxal phosphate (PLP) enzyme. The bond between the α- and β-carbon atoms of serine is labilized by the formation of a Schiff base between serine and PLP. The β-carbon atom of serine is then transferred to tetrahydrofolate. Glycine can also be formed from CO_2, NH_4^+, and methylenetetrahydrofolate in a reaction catalyzed by *glycine synthase*. The conversion of serine into cysteine requires the substitution of a sulfur atom derived from methionine for the side-chain oxygen atom. This reaction sequence will be presented after one-carbon metabolism has been considered.

TETRAHYDROFOLATE CARRIES ACTIVATED ONE-CARBON UNITS AT SEVERAL OXIDATION LEVELS

Tetrahydrofolate (also called tetrahydropteroylglutamate), a highly versatile carrier of activated one-carbon units, consists of three groups: a substituted pteridine, *p*-aminobenzoate, and glutamate. Mammals are unable to synthesize a pteridine ring. They obtain tetrahydrofolate from their diets or from microorganisms in their intestinal tracts.

Tetrahydrofolate

The one-carbon group carried by tetrahydrofolate is bonded to its N-5 or N-10 nitrogen atom (denoted as N^5 and N^{10}) or to both. This unit can exist in three oxidation states (Table 21-2). The most reduced form carries a *methyl* group, whereas the intermediate form carries a *methylene* group. The most oxidized forms carry a *methenyl*, *formyl*, or *formimino* group. The most oxidized one-carbon unit, CO_2,

Table 21-2
One-carbon groups carried by tetrahydrofolate

Oxidation state	Group	
Most reduced	—CH_3	Methyl
Intermediate	—CH_2—	Methylene
Most oxidized	—CHO	Formyl
	—CHNH	Formimino
	—CH=	Methenyl

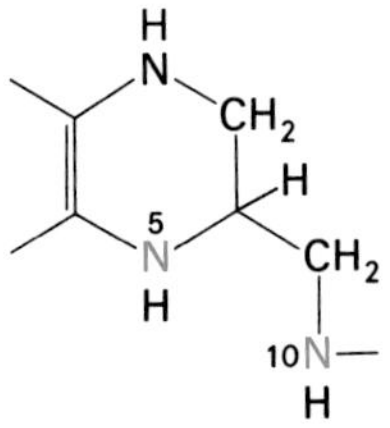

**Reactive part
of tetrahydrofolate**

is carried by biotin (p. 348) rather than by tetrahydrofolate.

These one-carbon units are interconvertible (Figure 21-5). N^5, N^{10}-*Methylene*tetrahydrofolate can be reduced to N^5-*methyl*tetrahydrofolate or oxidized to N^5-*methenyl*tetrahydrofolate. N^5, N^{10}-*Methenyl*tetrahydrofolate can be converted into N^5-*formimino*tetrahydrofolate and N^{10}-*formyl*tetrahydrofolate, which are at the same oxidation level. N^{10}-Formyltetrahydrofolate can also be synthesized from formate and ATP:

$$\text{Formate} + \text{ATP} + \text{tetrahydrofolate} \rightleftharpoons$$
$$N^{10}\text{-formyltetrahydrofolate} + \text{ADP} + \text{P}_i$$

These tetrahydrofolate derivatives serve as donors of one-carbon units in a variety of biosyntheses. Methionine is synthesized from homocysteine by transfer of the methyl group of N^5-methyltetrahydrofolate, as will be discussed shortly. Some of the carbon atoms of *purines* are derived

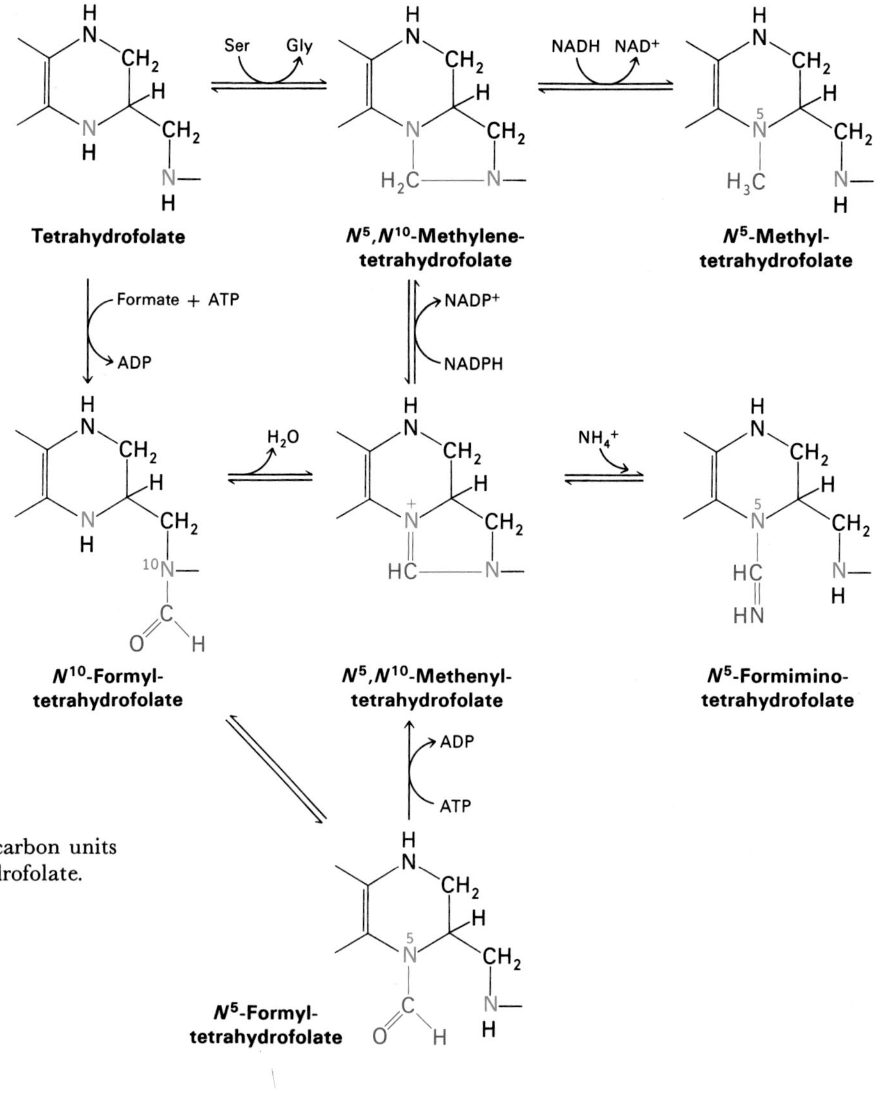

Figure 21-5
Conversions of one-carbon units
attached to tetrahydrofolate.

from the N^5, N^{10}-methenyl and the N^{10}-formyl derivatives of tetra-hydrofolate. The methyl group of *thymine,* a pyrimidine, comes from N^5, N^{10}-methylenetetrahydrofolate. This tetrahydrofolate deriva-tive also donates a one-carbon unit in the synthesis of *glycine* from CO_2 and NH_4^+, a reaction catalyzed by glycine synthase.

$$CO_2 + NH_4^+ + N^5, N^{10}\text{-methylenetetrahydrofolate} + NADH \rightleftharpoons$$
$$\text{glycine} + \text{tetrahydrofolate} + NAD^+$$

Thus, one-carbon units at each of the three oxidation levels are utilized in biosyntheses. In turn, *tetrahydrofolate serves as an acceptor of one-carbon units in degradative reactions.* The major source of one-carbon units is the conversion of serine into glycine, which yields N^5, N^{10}-methylenetetrahydrofolate, as previously mentioned. Serine can be derived from 3-phosphoglycerate (p. 490), and so *this pathway enables one-carbon units to be formed de novo from carbohydrate.* The breakdown of *histidine* yields N-formiminoglutamate, which transfers its formi-mino group to tetrahydrofolate to form the N^5-derivative.

S-ADENOSYLMETHIONINE IS THE MAJOR DONOR OF METHYL GROUPS

Tetrahydrofolate can carry a methyl group on its N^5-atom, but its transfer potential is not sufficiently high. Rather, the activated methyl donor in most biosyntheses is S-*adenosylmethionine,* which has already been encountered in the conversion of phosphatidyl etha-nolamine into phosphatidyl choline (p. 459). *S*-Adenosylmethio-nine is synthesized by the transfer of an adenosyl group from ATP to the sulfur atom of methionine. The methyl group of the methio-nine unit is activated by the positive charge on the adjacent sulfur atom, which makes it much more reactive than N^5-methyltetra-hydrofolate.

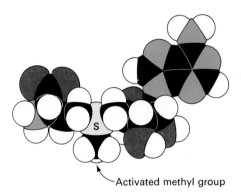

Activated methyl group

Figure 21-6
Space-filling model of *S*-adenosylmethionine.

Methionine + ATP ⟶ P_i + PP_i + **S-Adenosylmethionine**

The synthesis of *S*-adenosylmethionine is unusual in that the tri-phosphate group of ATP is split into pyrophosphate and orthophos-phate. Pyrophosphate is then hydrolyzed. Thus, all of the phospho-rus–oxygen bonds in ATP are split in this activation reaction, which markedly enhances the reactivity of the methyl group.

S-*Adenosylhomocysteine* is formed when the methyl group of S-adenosylmethionine is transferred to an acceptor such as phosphatidyl ethanolamine. S-Adenosylhomocysteine is then hydrolyzed to *homocysteine* and adenosine.

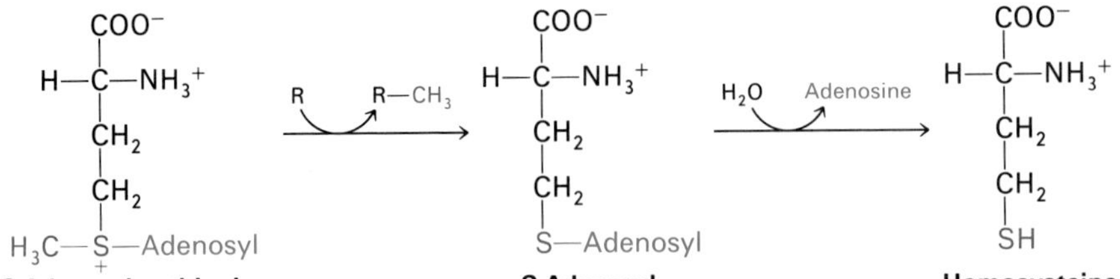

S-Adenosylmethionine → **S-Adenosyl-homocysteine** → **Homocysteine**

Methionine can be regenerated by the transfer of a methyl group from N^5-methyltetrahydrofolate, a reaction catalyzed by *homocysteine methyltransferase*.

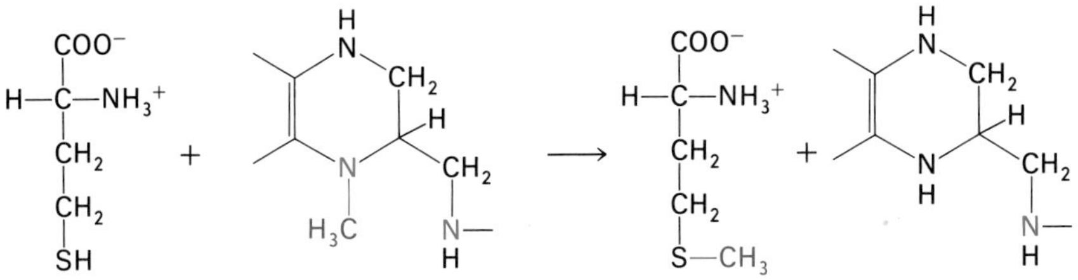

Homocysteine + **N^5-Methyltetrahydrofolate** → **Methionine** + **Tetrahydrofolate**

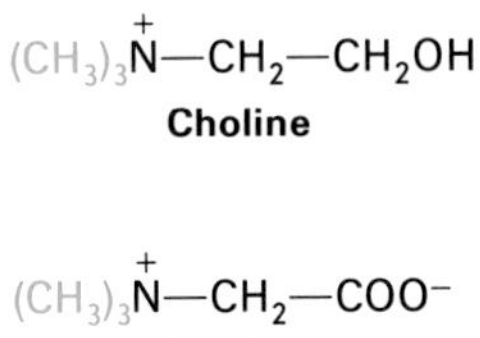

$(CH_3)_3\overset{+}{N}$—CH_2—CH_2OH
Choline

$(CH_3)_3\overset{+}{N}$—CH_2—COO^-
Betaine

This transfer of a methyl group is mediated by *methylcobalamin*, the coenzyme of homocysteine transmethylase. In fact, this reaction and the rearrangement of L-methylmalonyl CoA to succinyl CoA (p. 419) are the only two known Vitamin B_{12}-dependent reactions in mammals. Alternatively, homocysteine can be methylated to methionine by donors such as *betaine*, an oxidation product of choline.

These reactions constitute the *activated methyl cycle* (Figure 21-7). Methyl groups enter the cycle in the conversion of homocysteine into methionine and are then made highly reactive by the expenditure of 3 ~P. The high transfer potential of the methyl group in S-adenosylmethionine enables it to be transferred to a wide variety of acceptors, such as the amino group of the neurotransmitter norepinephrine (p. 895) and a glutamate residue of a regulatory protein in chemotaxis (p. 909).

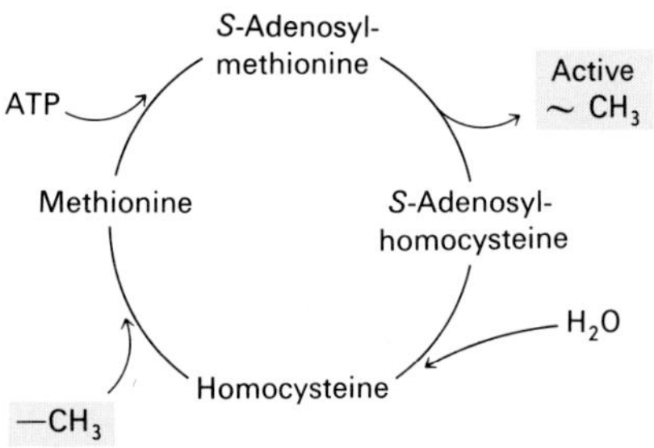

Figure 21-7
Activated methyl cycle.

CYSTEINE IS SYNTHESIZED FROM SERINE AND HOMOCYSTEINE

Homocysteine is an intermediate in the synthesis of cysteine, in addition to being a precursor of methionine in the activated methyl

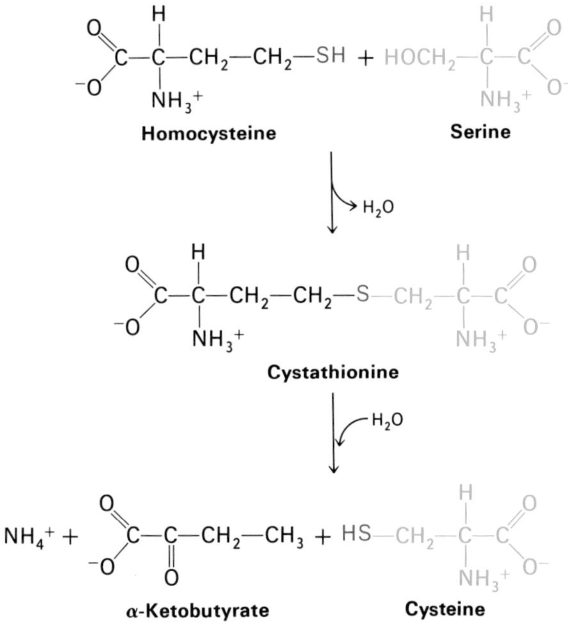

Figure 21-8
Synthesis of cysteine.

cycle. Serine and homocysteine condense to form *cystathionine* (Figure 21-8). This reaction is catalyzed by cystathionine synthetase, a PLP enzyme. Cystathionine is then deaminated and cleaved to cysteine and α-ketobutyrate by *cystathioninase,* another PLP enzyme. The net reaction is:

$$\text{Homocysteine} + \text{serine} \longrightarrow \text{cysteine} + \alpha\text{-ketobutyrate}$$

Note that the sulfur atom of cysteine is derived from homocysteine, whereas the carbon skeleton comes from serine.

This completes our consideration of the biosynthesis of the nonessential amino acids. The formation of tyrosine by the hydroxylation of phenylalanine was discussed earlier (p. 424).

SHIKIMATE AND CHORISMATE ARE INTERMEDIATES IN THE BIOSYNTHESIS OF AROMATIC AMINO ACIDS

We turn now to the biosynthesis of essential amino acids, which are formed by much more complex routes than are the nonessential amino acids. Two pathways have been selected for discussion here—those of the aromatic amino acids and of histidine.

Phenylalanine, tyrosine, and tryptophan are synthesized by a common pathway in *E. coli* (Figure 21-9). The initial step is the condensation of phosphoenolpyruvate (a glycolytic intermediate) and erythrose 4-phosphate (a pentose phosphate pathway intermediate). The resulting C_7 open-chain sugar loses its phosphoryl group and cyclizes to 5-dehydroquinate. Dehydration then yields 5-dehydroshikimate, which is reduced by NADPH to *shikimate* (Figure 21-10). A second molecule of phosphoenolpyruvate then condenses with 5-phosphoshikimate to give an intermediate that loses its phosphoryl group, which yields *chorismate.*

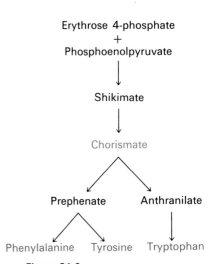

Figure 21-9
Pathway for the biosynthesis of aromatic amino acids in *E. coli.*

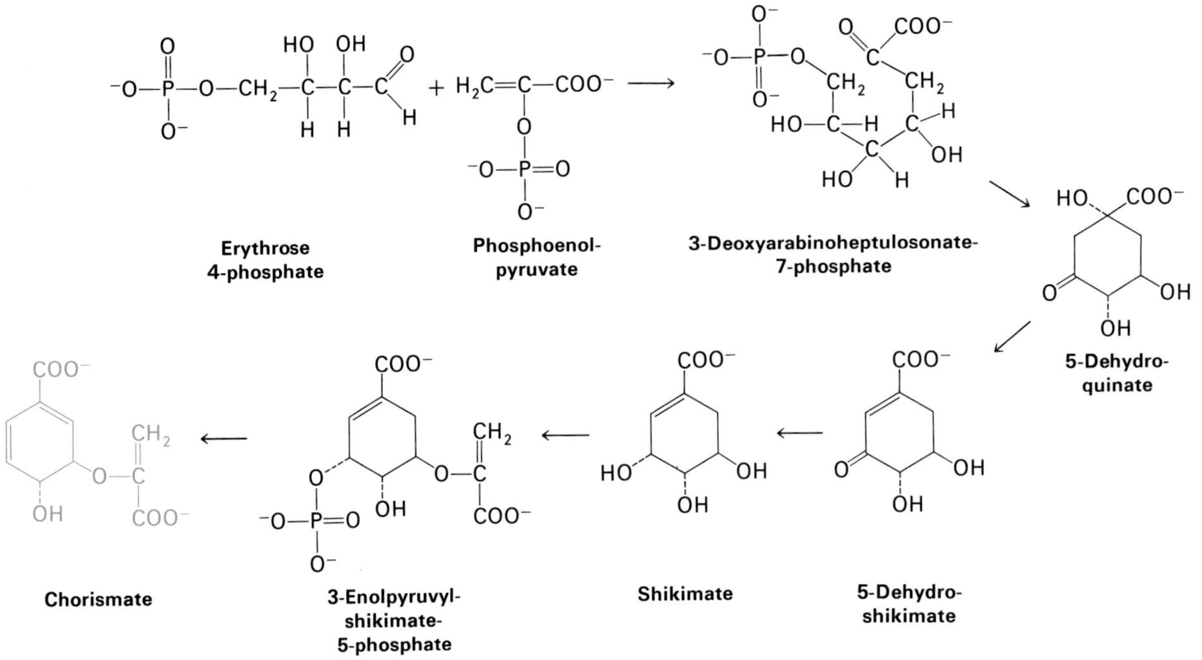

Figure 21-10
Synthesis of chorismate, an intermediate in the biosynthesis of phenylalanine, tyrosine, and tryptophan in *E. coli*.

The pathway bifurcates at chorismate. Let us first follow the *prephenate branch* (Figure 21-11). A mutase converts chorismate into prephenate, the immediate precursor of the aromatic ring of phenylalanine and tyrosine. Dehydration and decarboxylation yield *phenylpyruvate*. Alternatively, prephenate can be oxidatively decarboxylated to yield p-*hydroxyphenylpyruvate*. These α-keto acids are then transaminated to yield *phenylalanine* and *tyrosine,* respectively.

The branch starting with *anthranilate* leads to the synthesis of *tryptophan.* Chorismate acquires an amino group from the side chain of glutamine to form anthranilate. In fact, *glutamine serves as an amino donor in many biosynthetic reactions.* Anthranilate then condenses with *phosphoribosylpyrophosphate* (PRPP), *an activated form of ribose phosphate.* PRPP is also a key intermediate in the synthesis of histidine, purine nucleotides, and pyrimidine nucleotides (p. 514). The C-1 atom of ribose 5-phosphate becomes bonded to the nitrogen atom of anthranilate in a reaction that is driven by the hydrolysis of pyrophosphate.

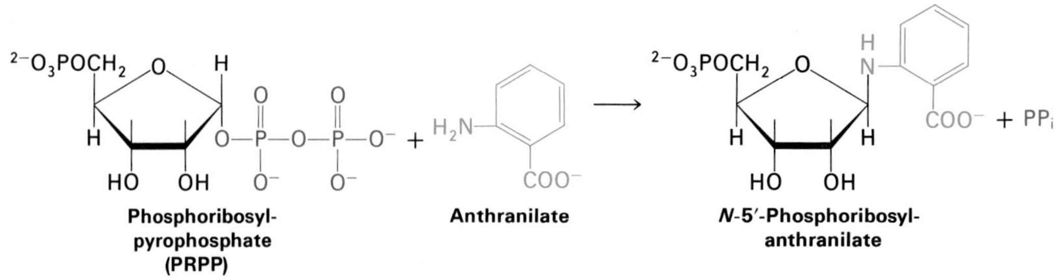

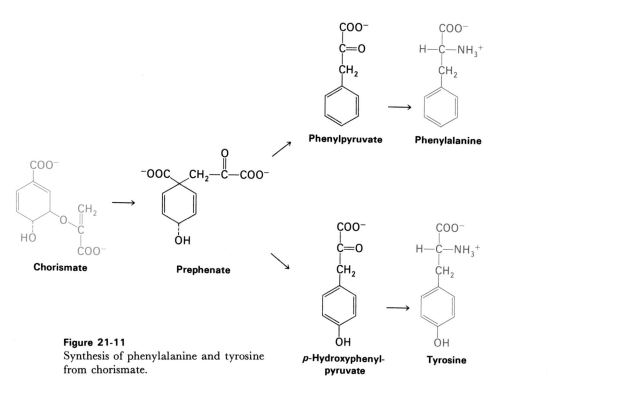

Figure 21-11
Synthesis of phenylalanine and tyrosine from chorismate.

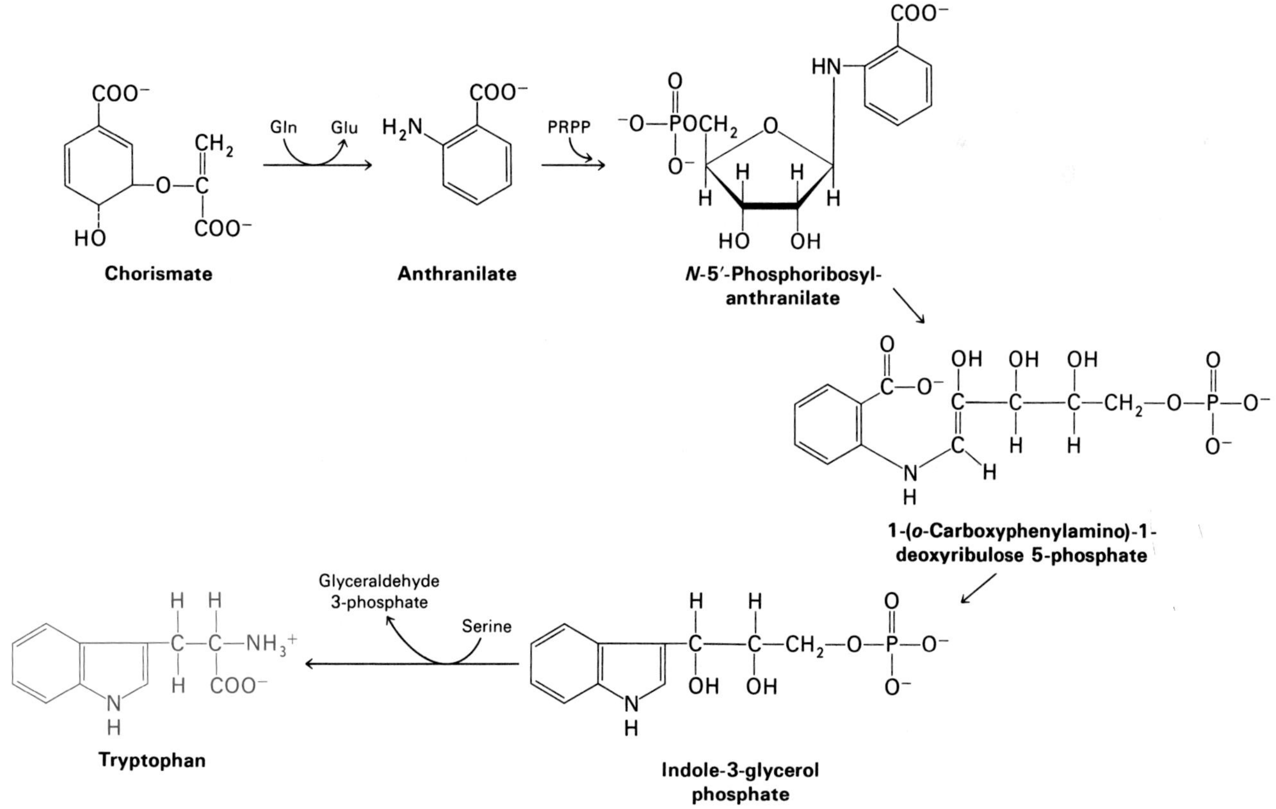

Figure 21-12
Synthesis of tryptophan from chorismate.

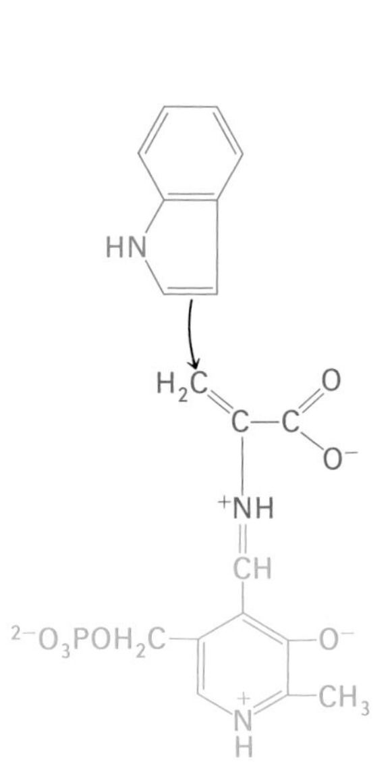

Figure 21-13
Proposed intermediate in the synthesis of tryptophan. Serine forms a Schiff base with PLP on a β chain, which is then dehydrated to give the Schiff base of aminoacrylate (shown in green). This enzyme-bound intermediate is attacked by indole, the product of the partial reaction catalyzed by the α subunit, to give tryptophan.

The ribose moiety of phosphoribosylanthranilate undergoes rearrangement (Figure 21-12) to yield 1-(o-carboxyphenylamino)-1-deoxyribulose 5-phosphate. This intermediate is dehydrated and decarboxylated to form *indole-3-glycerol phosphate*. Finally, indole-3-glycerol phosphate reacts with serine to form *tryptophan*. The glycerol phosphate side chain of indole-3-glycerol phosphate is replaced by the carbon skeleton and amino group of serine. This reaction is catalyzed by tryptophan synthetase.

Tryptophan synthetase of *E. coli* has the subunit structure $\alpha_2\beta_2$. The enzyme can be dissociated into two α subunits and a β_2 subunit. The isolated subunits catalyze partial reactions that lead to the synthesis of tryptophan:

$$\text{Indole-3-glycerol phosphate} \xrightarrow{\alpha \text{ subunit}}$$
$$\text{indole} + \text{glyceraldehyde 3-phosphate}$$

$$\text{Indole} + \text{serine} \xrightarrow{\beta_2 \text{ subunit}} \text{tryptophan} + \text{H}_2\text{O}$$

Each active site on the β_2 subunit contains a PLP prosthetic group. *The catalytic properties of the α and β_2 subunits are markedly altered on the formation of the $\alpha_2\beta_2$ complex.* The rates of the partial reactions are more than ten times as great for the $\alpha_2\beta_2$ complex as for the isolated subunits. Furthermore, the $\alpha_2\beta_2$ complex synthesizes tryptophan by a concerted mechanism. Indole formed by the first partial reaction reacts immediately with serine, so that indole is not released from the $\alpha_2\beta_2$ complex. Thus, the catalytic properties of a multisubunit enzyme can be altered by interactions between its subunits.

HISTIDINE IS SYNTHESIZED FROM ATP, PRPP, AND GLUTAMINE

The pathway for histidine biosynthesis in *E. coli* and *Salmonella* contains many complex and novel features (Figure 21-14). The reaction sequence starts with the condensation of ATP and PRPP, in which N-1 of the purine ring becomes bonded to C-1 of the ribose unit of PRPP. In fact, five carbon atoms of histidine come from PRPP. The adenine unit of ATP provides a nitrogen and a carbon atom of the imidazole ring of histidine. The other nitrogen atom of the imidazole ring comes from the side chain of glutamine. A noteworthy aspect of this pathway is that 5-aminoimidazole-4-carboxamide ribonucleotide, which is produced in the cleavage reaction that forms the imidazole ring, is an intermediate in purine biosynthesis (p. 515). Thus, histidine biosynthesis and purine biosynthesis are linked.

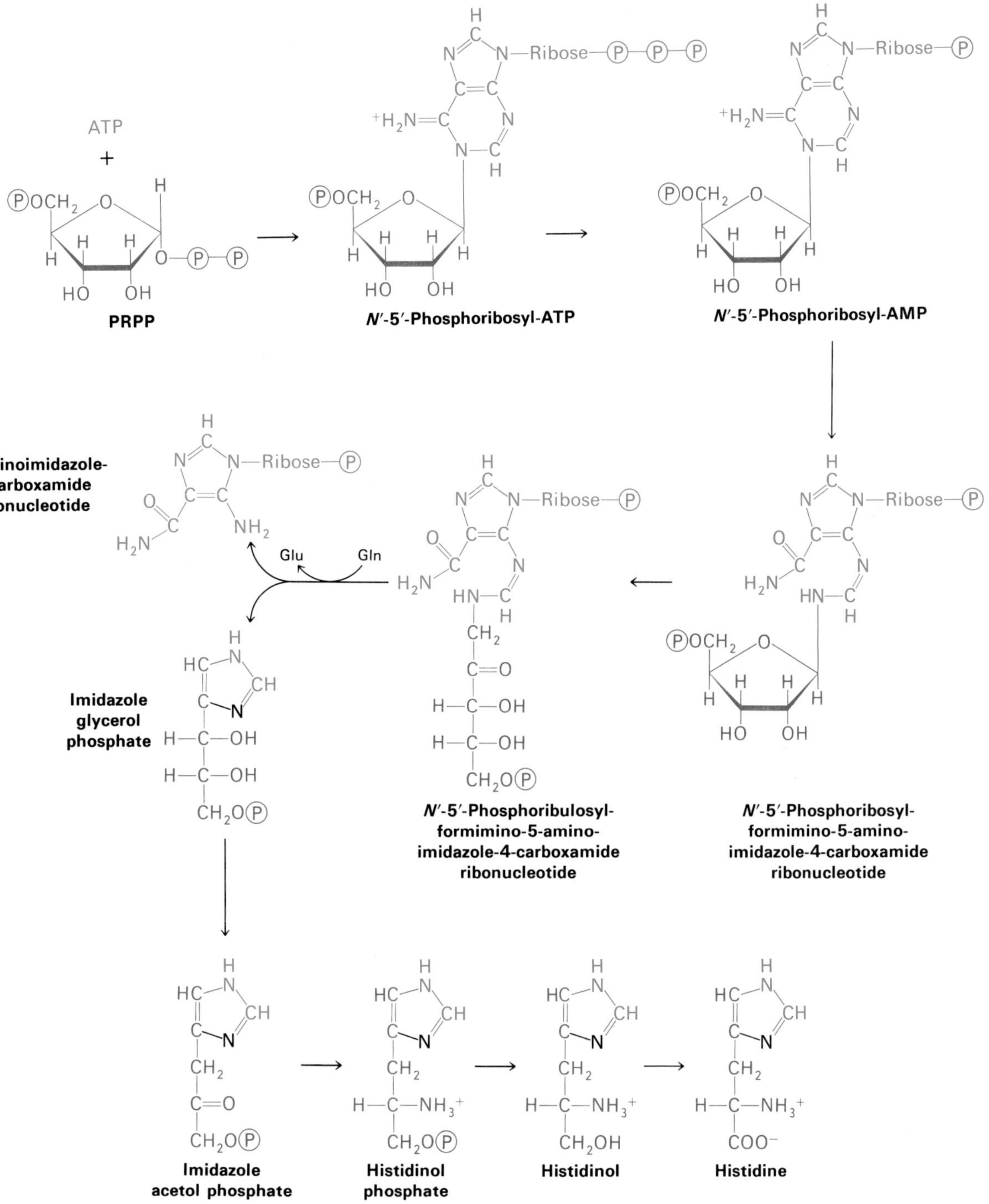

Figure 21-14
Pathway for the biosynthesis of histidine in *E. coli* and *Salmonella*
(Ⓟ denotes a phosphoryl group).

AMINO ACID BIOSYNTHESIS IS REGULATED BY FEEDBACK INHIBITION

The rate of synthesis of amino acids depends mainly on the *amounts* of the biosynthetic enzymes and on their enzymatic *activities.* We will now consider the control of enzymatic activity. The regulation of enzyme synthesis will be discussed in Chapter 28.

The first irreversible reaction in a biosynthetic pathway, called the committed step, is usually an important regulatory site. *The final product of the pathway (Z) often inhibits the enzyme that catalyzes the committed step (A → B).* This kind of control is essential for the conservation of building blocks and metabolic energy. The first example of this important principle of metabolic control came from studies of the biosynthesis of isoleucine in *E. coli.* The dehydration and deamination of threonine to α-ketobutyrate is the committed step in the synthesis of isoleucine. *Threonine deaminase,* the PLP enzyme that catalyzes this reaction, is allosterically inhibited by isoleucine.

Likewise, tryptophan inhibits the enzyme complex that catalyzes the first two steps in the conversion of chorismate into tryptophan.

Consider a branched biosynthetic pathway in which Y and Z are the final products.

Suppose that high levels of Y *or* Z completely inhibit the first common step (A → B). Then, high levels of Y would prevent the synthesis of Z even if there were a deficiency of Z. Such a regulatory scheme is obviously not optimal. In fact, several intricate control mechanisms have been found in branched biosynthetic pathways:

1. *Sequential feedback control.* The first common step (A → B) is not inhibited directly by Y or Z. Rather, these final products inhibit the reactions leading away from the point of branching: Y inhibits the C → D step, and Z inhibits the C → F step. In turn, high levels of

C inhibit the A → B step. Thus, the first common reaction is blocked only if both final products are present in excess.

Sequential feedback control regulates the synthesis of aromatic amino acids in *Bacillus subtilis*. The first divergent steps in the synthesis of phenylalanine, tyrosine, and tryptophan are inhibited by the respective final product. If all three are present in excess, chorismate and prephenate accumulate. These branch-point intermediates in turn inhibit the first common step in the overall pathway, which is the condensation of phosphoenolpyruvate and erythrose 4-phosphate.

2. *Enzyme multiplicity.* The distinguishing feature of this mechanism is that the first common step (A → B) is catalyzed by two different enzymes. One of them is inhibited by Y, and the other by Z. Thus, both Y and Z must be present at high levels to prevent the conversion of A into B completely. The other aspect of this control scheme is like that in sequential feedback control: Y inhibits the C → D step and Z inhibits the C → F step.

Differential inhibition of multiple enzymes controls a variety of biosynthetic pathways in microorganisms. In *E. coli,* the condensation of phosphoenolpyruvate and erythrose 4-phosphate is catalyzed by three different enzymes. One is inhibited by phenylalanine, another by tyrosine, and the third by tryptophan. Furthermore, there are two different mutases that convert chorismate into prephenate. One of them is inhibited by phenylalanine, the other by tyrosine.

3. *Concerted feedback control.* The first common step (A → B) is inhibited only if high levels of Y and Z are simultaneously present.

A high level of either product alone does not inhibit the A → B step. As in the two control schemes just discussed, Y inhibits the C → D step and Z inhibits the C → F step.

$$
\begin{array}{c}
\text{Inhibited} \\
\text{by Y}
\end{array}
$$

A —|—|—→ B —→ C

Inhibited by (Y + Z)

D —→ E —→ Y

F —→ G —→ Z

Inhibited by Z

An example of concerted feedback control is the inhibition of aspartyl kinase by threonine and lysine, the final products.

4. *Cumulative feedback control.* The first common step (A → B) is partially inhibited by each of the final products. Each final product acts independently of the others. Suppose that a high level of Y decreased the rate of the A → B step from 100 to 60 sec^{-1} and that Z alone decreased the rate from 100 to 40 sec^{-1}. Then, the rate of the A → B step in the presence of high levels of Y and Z would be 24 sec^{-1} ($0.6 \times 0.4 \times 100$ sec^{-1}).

THE ACTIVITY OF GLUTAMINE SYNTHETASE IS MODULATED BY ADENYLYLATION

The regulation of glutamine synthetase from *E. coli* is a striking example of *cumulative feedback inhibition.* Recall that glutamine is synthesized from glutamate, NH_4^+, and ATP (p. 488). Glutamine synthetase consists of twelve 50-kdal subunits arranged in two hexagonal rings that face each other (Figure 21-15). This enzyme is a key control element in intermediary metabolism because it regulates the flow of nitrogen, as shown by Earl Stadtman and his collaborators. The amide group of glutamine is a source of nitrogen in the biosyntheses of a variety of compounds such as tryptophan, histidine, carbamoyl phosphate, glucosamine 6-phosphate, CTP, and AMP. Glutamine synthetase is cumulatively inhibited by each of these final products of glutamine metabolism, as well as by alanine and glycine. There seem to be specific binding sites for each of these inhibitors. The enzymatic activity of glutamine synthetase is almost completely switched off when all eight final products are bound to the enzyme.

Another important and related feature of glutamine synthetase from *E. coli* is that its activity is altered by *reversible covalent modification* (Figure 21-16). This type of control was previously encountered in the synthesis and degradation of glycogen (p. 368). Phosphorylation activates glycogen phosphorylase and inactivates glycogen synthetase. The activity of glutamine synthetase is regulated in part by the covalent attachment of an *AMP unit* to the hydroxyl group of a

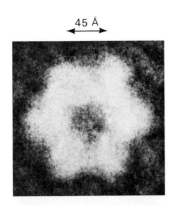

45 Å

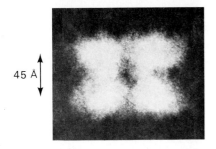

45 Å

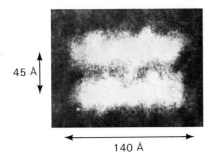

45 Å

140 Å

Figure 21-15
Three views of *E. coli* glutamine synthetase obtained by superposing several electron micrographs. The twelve subunits are arranged in two hexagonal rings that face each other. [Courtesy of Dr. Earl Stadtman.]

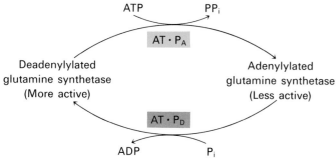

Figure 21-16
Control of the activity of glutamine synthetase by reversible covalent modification. Adenylylation is catalyzed by a complex of adenylyl transferase (AT) and one form of a regulatory protein (P_A). The same enzyme catalyzes deadenylylation when it is complexed with the other form (P_D) of the regulatory protein.

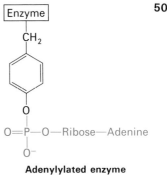

Adenylylated enzyme

specific tyrosine residue in each subunit. *This adenylylated enzyme is more susceptible to cumulative feedback inhibition than the deadenylylated form.* The covalently attached AMP unit can be removed from the adenylylated enzyme by phosphorolysis. An interesting feature of these reactions is that they are catalyzed by the same enzyme, *adenylyl transferase.* What then determines whether it acts to insert or remove an AMP unit? It turns out that the specificity of adenylyl transferase is controlled by a *regulatory protein* (designated P), which can exist in two forms, P_A and P_D. The complex of P_A and adenylyl transferase attaches AMP to glutamine synthetase, which thereby reduces its activity, whereas the complex of P_D and adenylyl transferase removes AMP. This brings us to another level of reversible covalent modification. P_A is converted into P_D by the attachment of uridine monophosphate (UMP), as shown in Figure 21-17. This reaction, which is catalyzed by a *uridyl transferase,* is stimulated by ATP and α-ketoglutarate, whereas it is inhibited by glutamine. In turn, the two UMP units on P_D can be enzymatically removed by hydrolysis.

The outcome of this *regulatory cascade* is that adenylylation is inhibited and deadenylylation is stimulated if the supply of activated nitrogen is low. Glutamine synthetase then becomes less susceptible to cumulative feedback inhibition, and the supply of glutamine consequently increases. Why is a cascade used to regulate this enzyme? One advantage of a cascade is that it *amplifies signals,* as in blood clotting (p. 171) and the control of glycogen metabolism (p. 373). Another likely reason is that the *potential for allosteric control is markedly increased because each enzyme in the cascade becomes an independent target for regulation.* The integration of nitrogen metabolism in a cell requires that a large number of input signals be detected and processed. There are limits to what a single protein can accomplish on its own—even a molecule as sentient as glutamine synthetase! The evolution of a cascade provided many more regulatory sites and made possible a finer tuning of the flow of nitrogen in the cell.

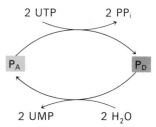

Figure 21-17
A higher level in the regulatory cascade of glutamine synthetase. P_A and P_D, the regulatory proteins that control the specificity of glutamine synthetase, are interconvertible. P_A is converted into P_D by uridylylation, which is reversed by hydrolysis. The enzymes catalyzing these reactions sense the concentrations of metabolic intermediates.

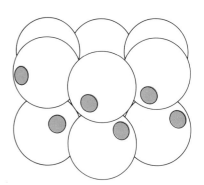

Figure 21-18
A model of glutamine synthetase showing the peripheral site of adenylylation. [After a drawing kindly provided by Dr. David Eisenberg.]

AMINO ACIDS ARE PRECURSORS OF A VARIETY OF BIOMOLECULES

Amino acids are the building blocks of proteins and peptides. They also serve as precursors of many kinds of small molecules that have important biological roles. Let us briefly survey some of the biomolecules that are derived from amino acids (Figure 21-19). *Purines* and

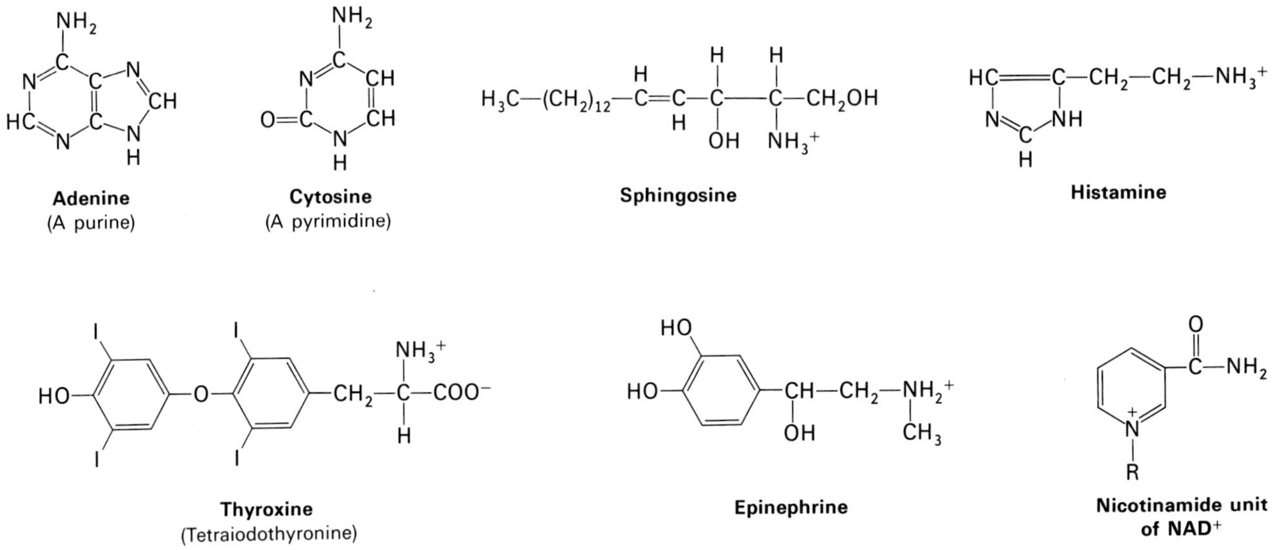

Adenine
(A purine)

Cytosine
(A pyrimidine)

Sphingosine

Histamine

Thyroxine
(Tetraiodothyronine)

Epinephrine

**Nicotinamide unit
of NAD+**

Figure 21-19
Biomolecules derived from amino acids.

pyrimidines are derived in part from amino acids. The biosynthesis of these precursors of DNA, RNA, and numerous coenzymes is discussed in detail in the next chapter. Six of the nine atoms of the purine ring and four of the six atoms of the pyrimidine ring are derived from amino acids. The reactive terminus of *sphingosine,* an intermediate in the synthesis of sphingolipids, comes from serine. *Histamine,* a potent vasodilator, is derived from histidine by decarboxylation. Tyrosine is a precursor of the hormones *thyroxine* (tetraiodothyronine) and *epinephrine* and of *melanin,* a polymeric pigment. The neurotransmitter *5-hydroxytryptamine (serotonin)* and the *nicotinamide ring* of NAD^+ are synthesized from tryptophan. Glutamine contributes the amide group of the nicotinamide moiety.

PORPHYRINS ARE SYNTHESIZED FROM GLYCINE AND SUCCINYL COENZYME A

The involvement of an amino acid in the biosynthesis of the porphyrin rings of hemes and chlorophylls was first revealed by isotope-labeling experiments carried out by David Shemin and his colleagues. In 1945, they showed that the nitrogen atoms of heme were labeled following the feeding of ^{15}N-glycine to human subjects, whereas ^{15}N-glutamate resulted in very little labeling. Using carbon-14, which had just become available, they discovered that in nucleated duck erythrocytes 8 of the carbon atoms of heme are derived from the α-carbon of glycine and none from the carboxyl carbon (Figure 21-20). Subsequent studies demonstrated that the other 26 carbon atoms of heme can arise from acetate. Moreover, the ^{14}C in methyl-labeled acetate emerged in 24 of these 26 carbons, whereas, the ^{14}C in carboxyl-labeled acetate appeared only in the other two. This highly distinctive labeling pattern led Shemin to propose that a heme precursor is formed by the condensation of glycine with an activated succinyl compound. In fact, *the first step in the biosynthesis of porphyrins is the condensation of glycine and succinyl CoA to form δ-aminolevulinate.*

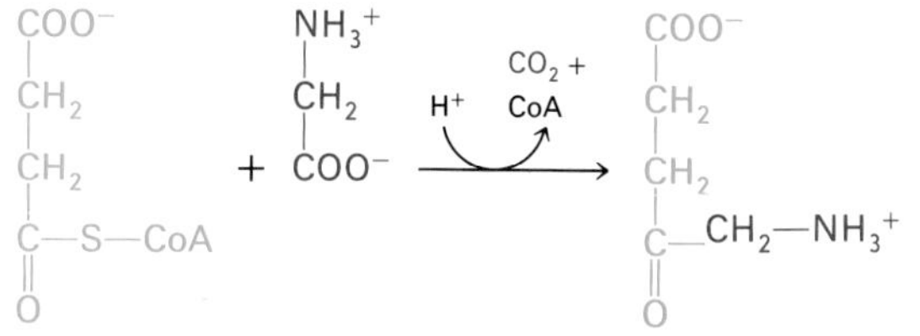

Succinyl CoA **Glycine** **δ-Aminolevulinate**

This reaction is catalyzed by δ-aminolevulinate synthetase, a PLP enzyme in mitochondria. As expected, this committed step in the biosynthesis of porphyrins is regulated. Two molecules of δ-aminolevulinate then condense to form *porphobilinogen.* This dehydration reaction is catalyzed by δ-aminolevulinate dehydrase.

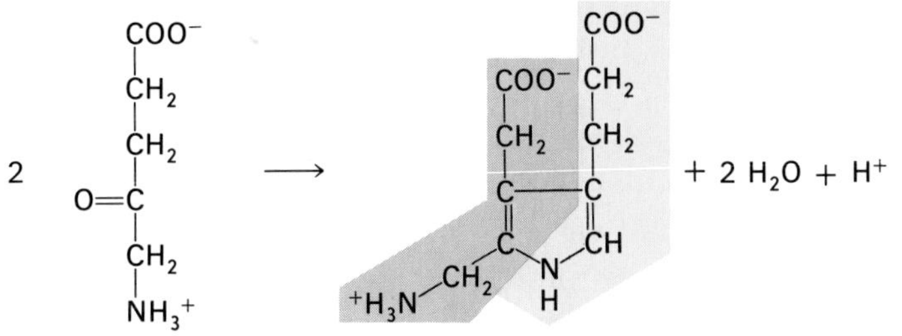

δ-Aminolevulinate **Porphobilinogen**

Four porphobilinogens condense head-to-tail to form a *linear tetrapyrrole,* which remains bound to the enzyme (Figure 21-21). An ammonium ion is released for each methylene bridge formed. This linear tetrapyrrole cyclizes by losing NH_4^+. The cyclic product is uroporphyrinogen III, which has an asymmetric arrangement of side chains. These reactions require a *synthetase* and a *cosynthetase.* In the presence of synthetase alone, uroporphyrinogen I, the symmetric isomer, is produced. The cosynthetase is essential for isomerizing

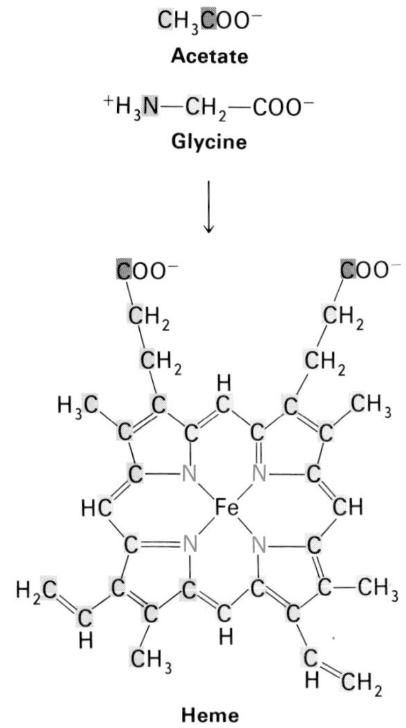

Figure 21-20
Labeling pattern of heme synthesized from glycine and acetate. The nitrogen atoms (blue) arise from the amino group of glycine. The origins of the carbon atoms are: yellow, from the α-carbon of glycine; green, mainly from the methyl carbon of acetate; and red, from the carboxyl carbon of acetate.

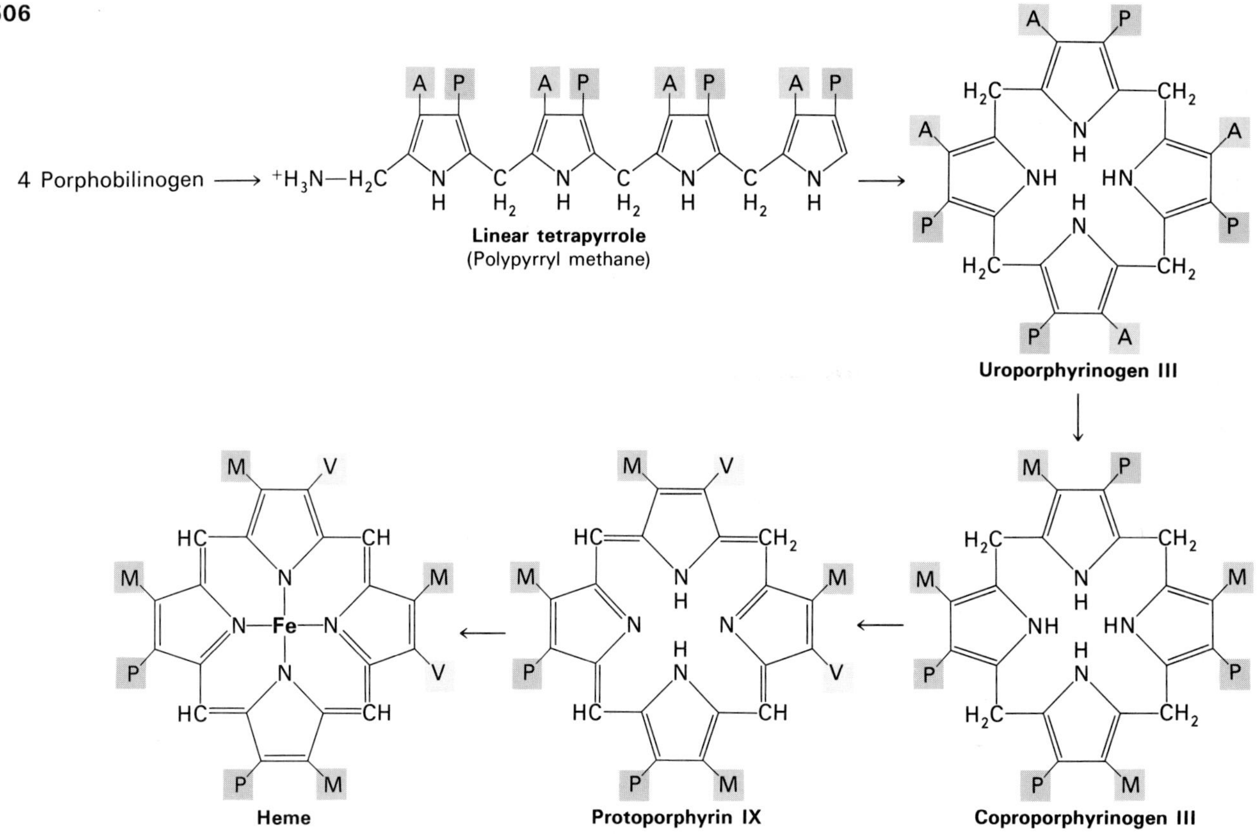

Figure 21-21
Pathway for the synthesis of heme from porphobilinogen. (Abbreviations: A, acetate; M, methyl; P, propionate, V, vinyl.)

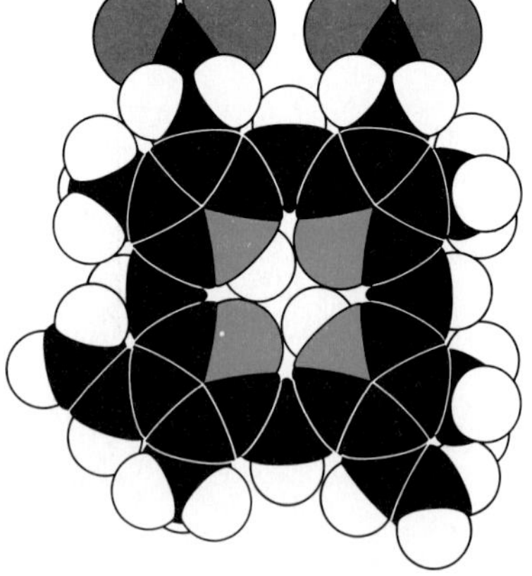

Figure 21-22
Space-filling model of protoporphyrin IX, the immediate precursor of heme.

one of the pyrrole rings to yield asymmetric uroporphyrinogen III.

The porphyrin skeleton is now formed. Subsequent reactions alter the side chains and the degree of saturation of the porphyrin ring (Figure 21-21). *Coproporphyrinogen III* is formed by decarboxylation of the acetate side chains. Unsaturation of the porphyrin ring and conversion of two of the propionate side chains into vinyl groups yield *protoporphyrin IX*. Chelation of iron finally gives *heme,* the prosthetic group of proteins such as myoglobin, hemoglobin, catalase, peroxidase, and cytochrome *c*. The insertion of the *ferrous* form of iron is catalyzed by *ferrochelatase.* Iron is transported in the plasma by *transferrin,* a protein that binds two ferric ions, and stored in tissues inside molecules of *ferritin.* The large internal cavity (~80 Å diameter) of ferritin can hold as many as 4500 ferric ions.

Several factors that regulate heme biosynthesis in animals have been elucidated. *Delta-aminolevulinate synthetase, the enzyme that catalyzes the committed step in this pathway, is feedback inhibited by heme,* as is δ-aminolevulinate dehydrase and ferrochelatase. Regulation also occurs at the level of enzyme synthesis. *Heme represses the synthesis of δ-aminolevulinate synthetase.* Recent studies suggest that the iron atom itself may be the active regulatory species.

Several inherited disorders of porphyrin metabolism are known. In *congenital erythropoietic porphyria,* there is a deficiency of uroporphyrinogen III cosynthetase, the isomerase that yields the asymmetric isomer on cyclization of the linear tetrapyrrole. The synthesis of the required amount of uroporphyrinogen III is accompanied by the formation of very large quantities of uroporphyrinogen I, the symmetric isomer devoid of a physiologic role. Uroporphyrin I, coproporphyrin I, and other symmetric derivatives also accumulate. Erythrocytes are prematurely destroyed in this disease, which is transmitted as an autosomal recessive. The urine of patients having this disease is red because of the excretion of large amounts of uroporphyrin I. Their teeth exhibit a strong red fluorescence under ultraviolet light because of the deposition of porphyrins. Furthermore, their *skin is usually very sensitive to light.*

Acute intermittent porphyria is a quite different disease. The liver, rather than the red cells, is affected, and the skin is not typically photosensitive. The activity of uroporphyrinogen synthetase is decreased in this disorder, and there is a compensatory increase in the level of δ-aminolevulinate synthetase. Consequently, the concentrations of δ-aminolevulinate and porphobilinogen in the liver are increased and so large amounts of these compounds are excreted in the urine. The disease is inherited as an *autosomal dominant.* The striking clinical symptoms are intermittent abdominal pain and neurologic disturbances. As its name implies, the disease is episodic in its clinical expression. Acute attacks are sometimes precipitated by drugs such as barbiturates and estrogens.

Uroporphyrinogen I

Uroporphyrinogen III

BILIVERDIN AND BILIRUBIN ARE INTERMEDIATES IN THE BREAKDOWN OF HEME

The normal human erythrocyte has a life span of about 120 days. Old cells are removed from the circulation and degraded by the spleen. The apoprotein of hemoglobin is hydrolyzed to its constituent amino acids. The first step in the degradation of the heme group to bilirubin (Figure 21-23) is the cleavage of its α-methene bridge to form *biliverdin,* a linear tetrapyrrole. This reaction is catalyzed by *heme oxygenase.* Two aspects of this reaction are noteworthy. First, this enzyme is a *monooxygenase:* O_2 and NADPH are required for the cleavage reaction. Second, a methene bridge carbon is released as *carbon monoxide.* This endogenous production of CO posed a special problem in the evolution of oxygen carriers (p. 55). The central methene bridge of biliverdin is then reduced by biliverdin reductase to form *bilirubin.* Again, the reductant is NADPH. The changing color of a bruise is a highly visible indication of these degradative reactions.

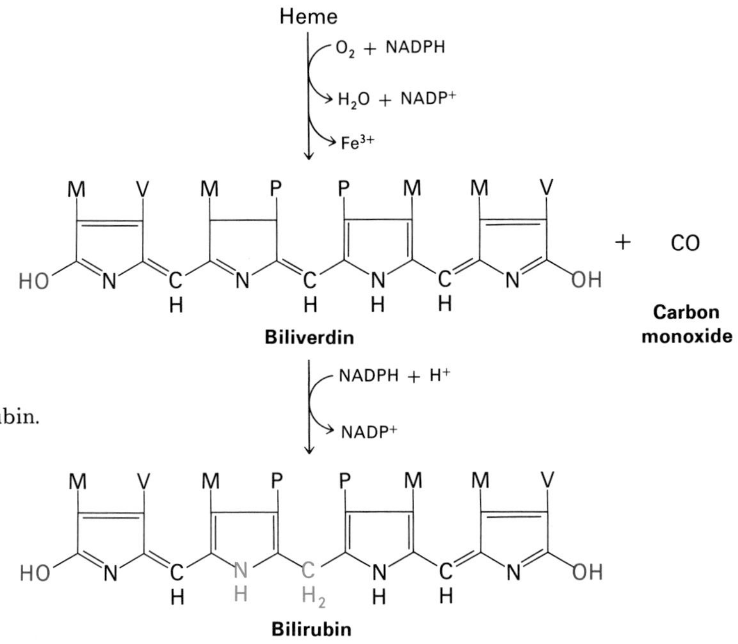

Heme

O_2 + NADPH

H_2O + NADP$^+$

Fe^{3+}

Biliverdin

+ CO

Carbon monoxide

NADPH + H$^+$

NADP$^+$

Bilirubin

Figure 21-23
Degradation of heme to bilirubin.

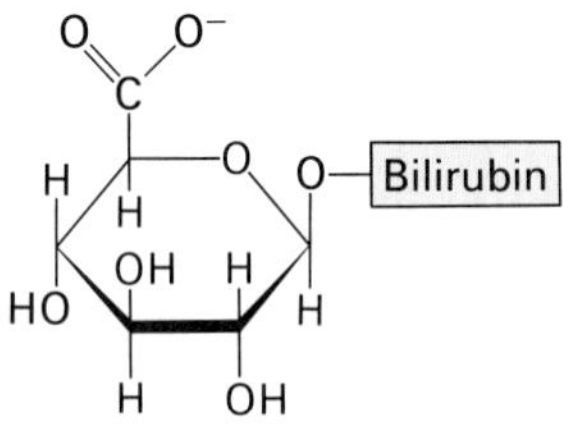

**A glucuronate unit
in bilirubin diglucuronide**

Bilirubin complexed to serum albumin is transported to the liver, where it is rendered more soluble by the attachment of sugar residues to its propionate side chains. The solubilizing sugar is *glucuronate*, which differs from glucose in having a COO$^-$ group at C-6 rather than a CH$_2$OH group. The conjugate of bilirubin and two glucuronates, called *bilirubin diglucuronide*, is secreted into bile. *UDP-glucuronate, derived from the oxidation of UDP-glucose, is the activated intermediate in the synthesis of bilirubin diglucuronide.* Thus, the iron atom of heme is recycled, whereas the organic moiety is converted into a soluble, open-chain form that is excreted.

SUMMARY

Microorganisms use ATP and a powerful reductant to convert N$_2$ into NH$_4^+$, which is consumed by higher organisms in the synthesis of amino acids, nucleotides, and other biomolecules. The major points of entry of NH$_4^+$ into intermediary metabolism are glutamine, glutamate, and carbamoyl phosphate. Humans can synthesize only half of the basic set of twenty amino acids. These amino acids are called nonessential, in contrast with the essential ones, which must be supplied in the diet. The pathways for the synthesis of nonessential amino acids are quite simple. Glutamate dehydrogenase catalyzes the reductive amination of α-ketoglutarate to glutamate. Alanine and aspartate are synthesized by transamination of pyruvate and oxaloacetate, respectively. Glutamine is synthesized from NH$_4^+$ and glutamate, and asparagine is synthesized similarly.

Proline is derived from glutamate. Serine, formed from 3-phosphoglycerate, is the precursor of glycine and cysteine. Tyrosine is synthesized by the hydroxylation of phenylalanine, an essential amino acid. The pathways for the biosynthesis of essential amino acids are much more complex than for the nonessential ones. Most of these pathways are regulated by feedback inhibition, in which the committed step is allosterically inhibited by the final product. The regulation of glutamine synthetase from *E. coli* provides a striking demonstration of cumulative feedback inhibition and of control by a cascade of reversible covalent modifications.

Tetrahydrofolate, a carrier of activated one-carbon units, plays an important role in amino acid and nucleotide metabolism. This coenzyme carries one-carbon units at three oxidation states, which are interconvertible: most reduced—methyl; intermediate—methylene; most oxidized—formyl, formimino, and methenyl. The major donor of activated methyl groups is *S*-adenosylmethionine, which is synthesized by the transfer of an adenosyl group from ATP to the sulfur atom of methionine. *S*-Adenosylhomocysteine is formed when the activated methyl group is transferred to an acceptor. It is hydrolyzed to adenosine and homocysteine, which is then methylated to methionine to complete the activated methyl cycle.

Amino acids are precursors of a variety of biomolecules. Porphyrins are synthesized from glycine and succinyl CoA, which condense to give δ-aminolevulinate. This intermediate condenses with itself to form porphobilinogen. Four porphobilinogens combine to form a linear tetrapyrrole, which cyclizes to form uroporphyrinogen III. Oxidation and side-chain modifications lead to the synthesis of protoporphyrin IX, which acquires an iron atom to form heme. Delta-aminolevulinate synthetase, the enzyme that catalyzes the committed step in this pathway, is feedback inhibited by heme.

SELECTED READINGS

BOOKS

Bender, D. A., 1975. *Amino Acid Metabolism.* Wiley.

Meister, A., 1965. *Biochemistry of the Amino Acids* (2nd ed.), vols. 1 and 2. Academic Press. [A comprehensive and authoritative treatise on amino acid metabolism.]

Prusiner, S., and Stadtman, E. R., (eds.), 1973. *The Enzymes of Glutamine Metabolism.* Academic Press. [A valuable collection of articles on the role of glutamine in nitrogen metabolism.]

NITROGEN FIXATION

Mortenson, L. E., and Thorneley, R. N. F., 1979. Structure and function of nitrogenase. *Ann. Rev. Biochem.* 48:387–318.

Hageman, R. V., and Burris, R. H., 1978. Nitrogenase and nitrogenase reductase associate and dissociate with each catalytic cycle. *Proc. Nat. Acad. Sci.* 75:2699–2702.

Wolff, T. E., Berg, J. M., Warrick, C., Hodgson, K. O.,

Holm, R. H., and Frankel, R. B., 1978. The molybdenum-iron-sulfur cluster complex $[Mo_2Fe_6S_9(SC_2H_5)_8]^{3-}$: a synthetic approach to the molybdenum site in nitrogenase. *J. Am. Chem. Soc.* 100:4630–4632.

ONE-CARBON METABOLISM

Walsh, C., 1979. *Enzymatic Reaction Mechanisms.* Freeman. [Chapter 25 provides an excellent account of reaction mechanisms in one-carbon metabolism.]
Ridley, W. P., Dizikes, L. J., and Wood, J. M., 1977. Biomethylation of toxic elements in the environment. *Science* 197:329–332.

AMINO ACID BIOSYNTHESIS AND REGULATION

Stadtman, E. R., and Ginsburg, A., 1974. The glutamine synthetase of *Escherichia coli:* structure and control. *In* Boyer, P. D., (ed.), *The Enzymes* (3rd ed.), vol. 10, pp. 755–807. Academic Press.
Rhee, S. G., Park, R., Chock, P. B., and Stadtman, E. R., 1978. Allosteric regulation of monocyclic interconvertible enzyme cascade systems: use of *Escherichia coli* glutamine synthetase as an experimental model. *Proc. Nat. Acad. Sci.* 75:3138–3142.

Umbarger, H. E., 1978. Amino acid biosynthesis and its regulation. *Ann. Rev. Biochem.* 47:533–606. [An excellent review of amino acid biosynthesis in bacteria.]
Yanofsky, C., and Crawford, I. P., 1972. Tryptophan synthetase. *In* Boyer, P. D., (ed.), *The Enzymes* (3rd ed.), vol. 7, pp. 1–31.

IRON AND HEME METABOLISM

Smith, K. M., (ed.), 1976. *Porphyrins and Metalloporphyrins.* Elsevier.
Porter, R., and Fitzsimons, D. W., (eds.), 1976. *Iron Metabolism.* Ciba Foundation Symposium, vol. 51 (New Series). Elsevier.
Maines, M. D., and Kappas, A., 1977. Metals as regulators of heme metabolism. *Science* 198:1215–1221.
Granick, S., and Beale, S. I., 1978. Hemes, chlorophylls, and related compounds: biosynthesis and metabolic regulation. *Advan. Enzymol.* 40:33–203. [An excellent review article.]
Meyer, U. A., and Schmid, R., 1978. The porphyrias. *In* Stanbury, J. B., Wyngaarden, J. B., and Fredrickson, D. S., (eds.), *The Metabolic Basis of Inherited Disease* (4th ed.), pp. 1166–1220. McGraw-Hill.

PROBLEMS

1. Write a balanced equation for the synthesis of alanine from glucose.

2. What are the intermediates in the flow of nitrogen from N_2 to heme?

3. What derivative of folate is a reactant in each of the following conversions?
   (a) Glycine $\rightarrow$ serine.
   (b) Histidine $\rightarrow$ glutamate.
   (c) Homocysteine $\rightarrow$ methionine.

4. In the reaction catalyzed by glutamine synthetase, an oxygen atom is transferred from the side chain of glutamate to orthophosphate, as shown by ^{18}O-labeling studies. Propose an interpretation for this finding.

5. Isovaleric acidemia is an inherited disorder of leucine metabolism caused by a deficiency of isovaleryl-CoA dehydrogenase. Many infants having this disease die in the first month of life. It has recently been reported that the administration of large amounts of glycine to two infants with this disease led to marked clinical improvement. What is the rationale for the use of glycine therapy?

6. Nitrogenase catalyzes the formation of HD in the presence of D_2, a strong reductant, and ATP. Furthermore, H-D exchange depends on the presence of N_2. Propose an intermediate in N_2 fixation that would account for these observations.

7. Blue-green algae form *heterocysts* when deprived of ammonia and nitrate. They lack nuclei and are attached to adjacent vegetative cells. Heterocysts have photosystem I activity but are entirely devoid of photosystem II activity. What is their role?

CHAPTER 22

BIOSYNTHESIS OF NUCLEOTIDES

This chapter deals with the biosynthesis of nucleotides. These compounds participate in nearly all biochemical processes:

1. They are the *activated precursors of DNA and RNA*.

2. Nucleotide derivatives are *activated intermediates in many biosyntheses*. For example, UDP-glucose and CDP-diacylglycerol are precursors of glycogen and phosphoglycerides, respectively.

3. ATP, an adenine nucleotide, is the *universal currency of energy* in biological systems.

4. Adenine nucleotides are *components of three major coenzymes:* NAD$^+$, FAD, and CoA.

5. Nucleotides are *metabolic regulators*. Cyclic AMP is a ubiquitous mediator of the action of many hormones. Covalent modifications introduced by ATP alter the activities of some enzymes, as exemplified by the phosphorylation of glycogen synthetase and the adenylylation of glutamine synthetase.

NOMENCLATURE OF BASES, NUCLEOSIDES, AND NUCLEOTIDES

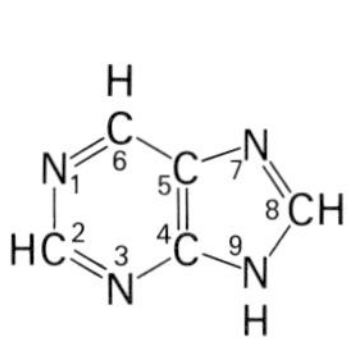

A nucleotide consists of a nitrogeneous base, a sugar, and one or more phosphate groups. The nitrogeneous base is a *purine* or *pyrimidine* derivative. The two major purines are *adenine* and *guanine,* and

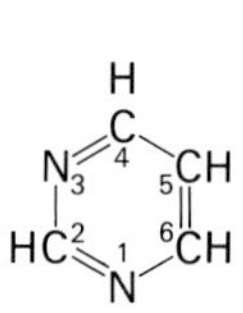

Purine

Pyrimidine

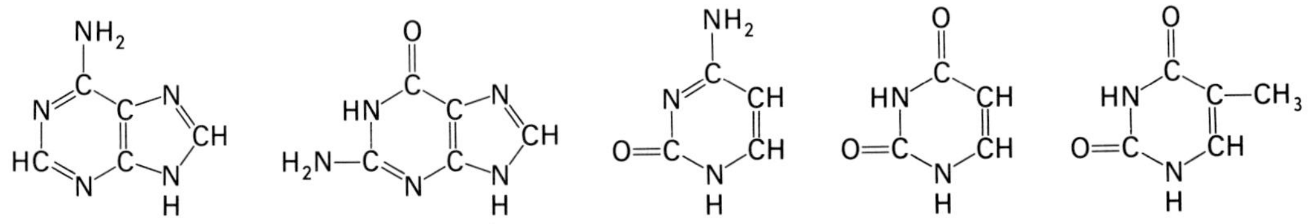

Adenine Guanine Cytosine Uracil Thymine

Figure 22-1
Structures of the major purines and pyrimidines.

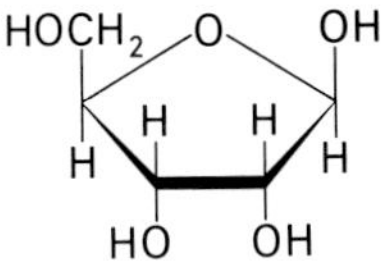

β-D-Ribose

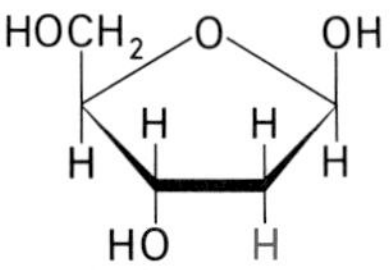

β-D-2-Deoxyribose

the three major pyrimidines are *cytosine, uracil,* and *thymine* (Figure 22-1).

A *nucleoside* consists of a purine or pyrimidine base linked to a pentose. The pentose is D-*ribose* or *2-deoxy-*D-*ribose.* In a nucleoside, the glycosidic C-1 carbon atom of the pentose is bonded to N-1 of the pyrimidine or N-9 of the purine base. The configuration of this *N*-glycosidic linkage is β in all naturally occurring nucleosides.

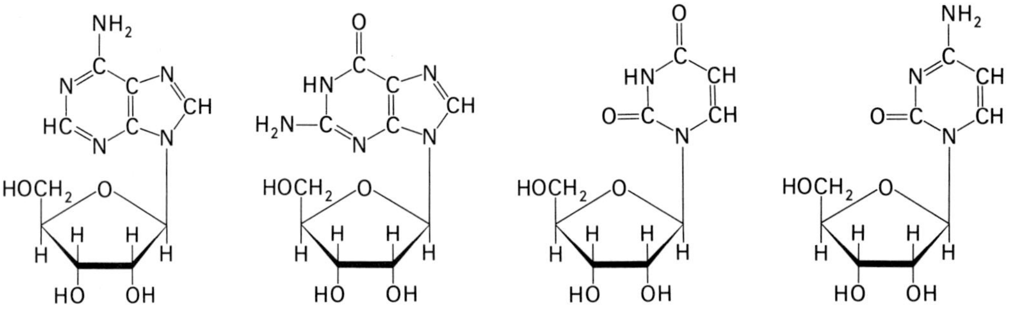

Adenosine Guanosine Uridine Cytidine

Figure 22-2
Structures of the major ribonucleosides.

In a ribonucleoside, the pentose is ribose, whereas in a deoxyribonucleoside it is deoxyribose. The major ribonucleosides are *adenosine, guanosine, uridine,* and *cytidine.* The major deoxyribonucleosides are *deoxyadenosine, deoxyguanosine, deoxythymidine,* and *deoxycytidine.*

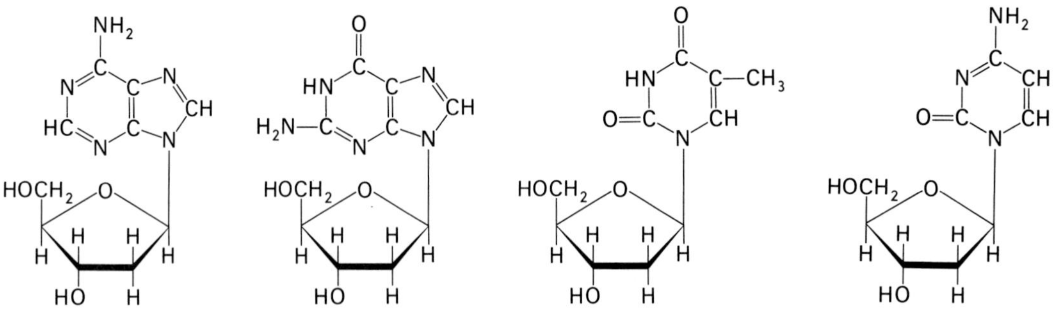

Deoxyadenosine Deoxyguanosine Deoxythymidine Deoxycytidine

Figure 22-3
Structures of the major deoxyribonucleosides.

Note that uracil is replaced by thymine, its methylated analog, in the deoxy series.

A *nucleotide* is a phosphate ester of a nucleoside. At least one of the hydroxyl groups of the pentose moiety of a nucleotide is esterified. The most common site of esterification is the hydroxyl group attached to C-5 of the pentose. Such a compound is called a *nucleoside 5′-phosphate* or a *5′-nucleotide*. A primed number designates an atom of the pentose, whereas an unprimed number designates an atom of the purine or pyrimidine ring. The type of pentose is denoted by the prefix in the terms *5′-ribonucleotide* and *5′-deoxyribonucleotide*.

The nucleotide derived by esterification of the 5′-hydroxyl group of adenosine is *adenosine 5′-phosphate*, which is often called *adenylate*. The name adenylic acid is sometimes used, but adenylate is preferred because the phosphate group is ionized at physiological pH. The standard abbreviation for this compound is AMP (for adenosine monophosphate). The common names of the other major 5′-ribonucleotides are *guanylate* (GMP), *uridylate* (UMP), and *cytidylate* (CMP). The major 5′-deoxyribonucleotides are called *deoxyadenylate* (dAMP), *deoxyguanylate* (dGMP), *deoxythymidylate* (dTMP), and *deoxycytidylate* (dCMP). The first letter in their abbreviations is a *d* to denote that they are 2′-deoxyribonucleotides.

In a *nucleoside 5′-diphosphate*, a diphosphate group is esterified to the 5′-hydroxyl of the pentose, whereas in a *nucleoside 5′-triphosphate*, a triphosphate group is linked to this hydroxyl. Thus, the adenine ribonucleotide series is called adenosine 5′-monophosphate (AMP), adenosine 5′-diphosphate (ADP), and adenosine 5′-triphosphate (ATP). The corresponding deoxyribonucleotides are deoxyadenosine 5′-monophosphate (dAMP), deoxyadenosine 5′-diphosphate (dADP), and deoxyadenosine 5′-triphosphate (dATP). The nomenclature of bases, nucleosides, and nucleotides is summarized in Table 22-1.

**Adenosine 5′-phosphate
(AMP)**

**Thymidine 5′-phosphate
(TMP)**

Table 22-1
Nomenclature of bases, nucleosides, and nucleotides

Base	Ribonucleoside	Ribonucleotide (5′-monophosphate)
Adenine (A)	Adenosine	Adenylate (AMP)
Guanine (G)	Guanosine	Guanylate (GMP)
Uracil (U)	Uridine	Uridylate (UMP)
Cytosine (C)	Cytidine	Cytidylate (CMP)

Base	Deoxyribonucleoside	Deoxyribonucleotide (5′-monophosphate)
Adenine (A)	Deoxyadenosine	Deoxyadenylate (dAMP)
Guanine (G)	Deoxyguanosine	Deoxyguanylate (dGMP)
Thymine (T)	Deoxythymidine	Deoxythymidylate (dTMP)
Cytosine (C)	Deoxycytidine	Deoxycytidylate (dCMP)

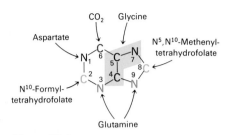

Figure 22-4
Origins of the atoms in the purine ring.

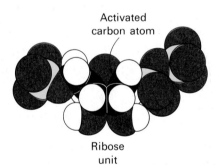

Activated
carbon atom

Ribose
unit

Figure 22-5
Space-filling model of 5-phosphoribo-syl-1-pyrophosphate (PRPP), the acti-vated donor of the sugar unit in the biosynthesis of nucleotides.

THE PURINE RING IS SYNTHESIZED FROM AMINO ACIDS, TETRAHYDROFOLATE DERIVATIVES, AND CO_2

The purine ring in purine nucleotides is assembled from a variety of precursors (Figure 22-4). *Glycine* provides C-4, C-5, and N-7. The N-1 atom comes from *aspartate*. The other two nitrogen atoms, N-3 and N-9, come from the amide group of the side chain of *glutamine*. Activated derivatives of *tetrahydrofolate* furnish C-2 and C-8, whereas CO_2 is the source of C-6.

PRPP IS THE DONOR OF THE RIBOSE PHOSPHATE MOIETY OF NUCLEOTIDES

The pathway of purine biosynthesis was elucidated in the 1950s by John Buchanan, G. Robert Greenberg, and others. The ribose phos-phate portion of purine and pyrimidine nucleotides comes from *5-phosphoribosyl-1-pyrophosphate* (PRPP), a key intermediate in the biosynthesis of histidine and tryptophan. PRPP is synthesized from ATP and ribose 5-phosphate, which is primarily formed by the pentose phosphate pathway (p. 333). The pyrophosphate group is transferred from ATP to C-1 of ribose 5-phosphate. PRPP has an α configuration.

Ribose 5-phosphate → **5-Phosphoribosyl-1-pyrophosphate (PRPP)**

THE PURINE RING IS ATTACHED TO RIBOSE PHOSPHATE DURING ITS ASSEMBLY

The committed step in the de novo synthesis of purine nucleotides is the formation of *5-phosphoribosylamine* from PRPP and glutamine. The amino group from the side chain of glutamine displaces the pyrophosphate group attached to C-1 of PRPP. The configuration at C-1 is inverted from α to β in this reaction. The resulting C–N glycosidic bond has the β configuration that is characteristic of naturally occurring nucleotides. This reaction is driven forward by the hydrolysis of pyrophosphate.

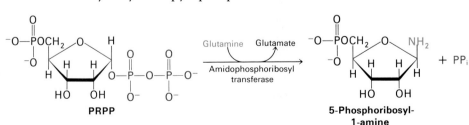

PRPP → **5-Phosphoribosyl-1-amine**

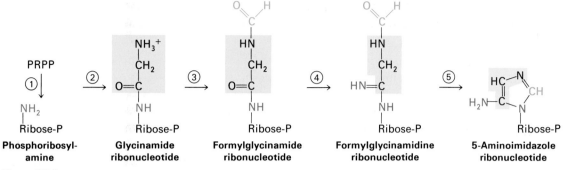

Figure 22-6
First stage of purine biosynthesis: formation of 5-aminoimidazole ribonucleotide from PRPP. The essence of these reactions is (1) displacement of PP_i by the side-chain amino group of glutamine, (2) addition of glycine, (3) formylation by methenyltetrahydrofolate, (4) transfer of a nitrogen atom from glutamine, and (5) dehydration and ring closure.

Glycine joins phosphoribosylamine to yield *glycinamide ribonucleotide* (Figure 22-6). An ATP is consumed in the formation of an amide bond between the carboxyl group of glycine and the amino group of phosphoribosylamine. The α-amino terminus of the glycine residue is then formylated by methenyltetrahydrofolate to give α-N-*formylglycinamide ribonucleotide*. The amide group in this compound is converted into an amidine group. The nitrogen atom is donated by the side chain of glutamine in a reaction that consumes an ATP. *Formylglycinamidine ribonucleotide then undergoes ring closure to form 5-aminoimidazole ribonucleotide.* This intermediate contains the complete five-membered ring of the purine skeleton.

The next phase in the synthesis of the purine skeleton, the formation of a six-membered ring, starts at this point (Figure 22-7). Three

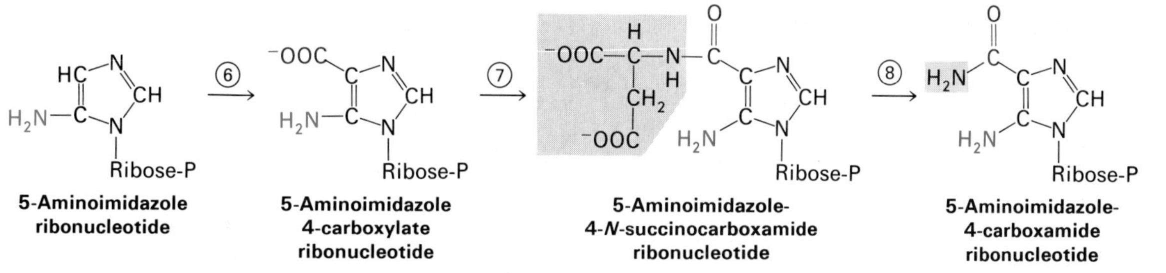

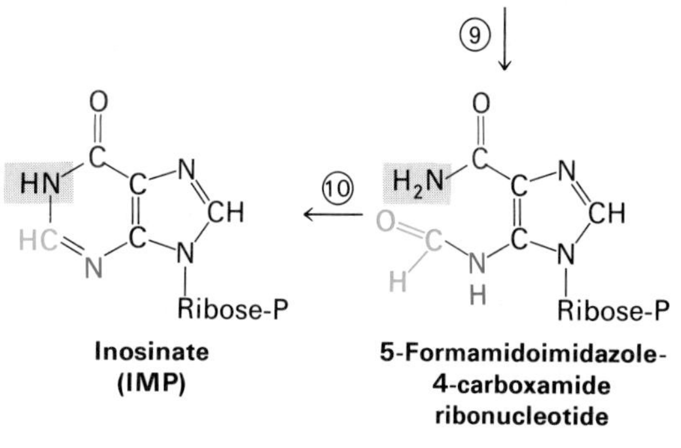

Figure 22-7
Second stage of purine biosynthesis: formation of inosinate from 5-aminoimidazole ribonucleotide. The essence of these reactions is (6) carboxylation, (7) addition of aspartate, (8) elimination of fumarate (leaving the amino group of aspartate), (9) formylation by N^{10}-formyltetrahydrofolate, and (10) dehydration and ring closure.

of the six atoms of this ring are already present in aminoimidazole ribonucleotide. The other three come from CO_2, aspartate, and formyltetrahydrofolate. The next carbon atom in the six-membered ring is introduced by the carboxylation of aminoimidazole ribonucleotide, yielding *5-aminoimidazole-4-carboxylate ribonucleotide.*

The amino group of aspartate then reacts with the carboxyl group of this intermediate to form *5-aminoimidazole-4-N-succinocarboxamide ribonucleotide.* An ATP is consumed in the formation of this amide bond. The carbon skeleton of the aspartate moiety comes off as fumarate in the next reaction, which yields *5-aminoimidazole-4-carboxamide ribonucleotide.* Note that the result of these two reactions is the conversion of a carboxylate into an amide. Thus, *aspartate contributes only its nitrogen atom to the purine ring.* The final atom of the purine ring is contributed by N^{10}-formyltetrahydrofolate. The resulting *5-formamidoimidazole-4-carboxamide ribonucleotide* undergoes dehydration and ring closure to form *inosinate* (IMP), which contains a complete purine ring. The purine-base part of inosinate is called *hypoxanthine.*

AMP AND GMP ARE FORMED FROM IMP

Inosinate is the precursor of AMP and GMP (Figure 22-8). *Adenylate* is synthesized from inosinate by the insertion of an amino group at C-6 in place of the carbonyl oxygen. Aspartate again contributes its amino group by addition of this amino acid followed by elimination of fumarate. GTP is the donor of a high-energy phosphate bond in the synthesis of *adenylosuccinate* from inosinate and aspartate. The removal of fumarate from adenylosuccinate and from 5-aminoimidazole-4-*N*-succinocarboxamide ribonucleotide is catalyzed by the same enzyme.

Figure 22-8
AMP and GMP are synthesized from IMP.

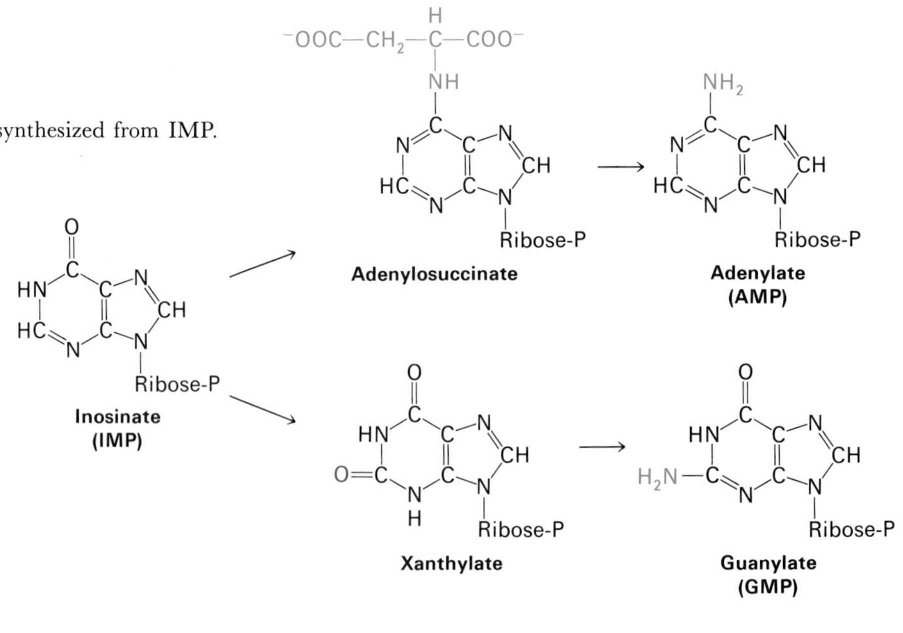

Guanylate (GMP) is synthesized by the oxidation of inosinate, followed by the insertion of an amino group at C-2. NAD^+ is the hydrogen acceptor in the oxidation of inosinate to xanthylate (XMP). The amino group in the side chain of glutamine is then transferred to xanthylate. Two high-energy phosphate bonds are consumed in this reaction, because ATP is cleaved into AMP and PP_i, which is subsequently hydrolyzed.

In the conversion of inosinate into adenylate and into guanylate, a carbonyl oxygen atom is replaced by an amino group. A similar change occurs in the synthesis of formylglycinamide ribonucleotide from its amide precursor (step 4 on p. 515), in the formation of CTP from UTP (p. 522), and in the conversion of citrulline into arginine in the urea cycle (p. 413). *The common mechanistic theme of these reactions is the conversion of the carbonyl oxygen into a derivative that can be readily displaced by an amino group.* The tautomeric form of the carbonyl group reacts with ATP (or GTP) to form a phosphoryl ester, which is nucleophilically attacked by an amine (Figure 22-9). Inor-

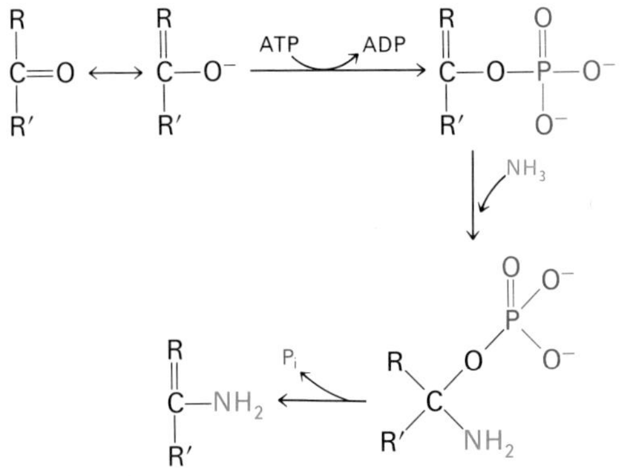

Figure 22-9
Reaction mechanism for the replacement of a carbonyl oxygen by an amino group.

ganic phosphate is then expelled from this tetrahedral adduct to complete the reaction. The attacking amine can be NH_3, the side-chain amide group of glutamine, or the α-amino group of aspartate. The leaving group in this class of reactions can be P_i, PP_i, or the AMP moiety. For example, PP_i is displaced by the amino group of glutamine in the synthesis of 5-phosphoribosyl-1-amine from PRPP (p. 514).

PURINE BASES CAN BE RECYCLED BY SALVAGE REACTIONS THAT UTILIZE PRPP

Free purine bases are formed by the hydrolytic degradation of nucleic acids and nucleotides. Purine nucleotides can be synthesized from these preformed bases by a *salvage reaction,* which is simpler and much less costly than the reactions of the *de novo pathway* discussed above. In the salvage reaction, the ribose phosphate moiety of PRPP is transferred to the purine to form the corresponding nucleotide:

PRPP → **Purine ribonucleotide**

There are two salvage enzymes with different specificities. *Adenine phosphoribosyl transferase* catalyzes the formation of adenylate:

$$\text{Adenine} + \text{PRPP} \longrightarrow \text{adenylate} + \text{PP}_i$$

whereas *hypoxanthine-guanine phosphoribosyl transferase* catalyzes the formation of inosinate and guanylate:

$$\text{Hypoxanthine} + \text{PRPP} \longrightarrow \text{inosinate} + \text{PP}_i$$

$$\text{Guanine} + \text{PRPP} \longrightarrow \text{guanylate} + \text{PP}_i$$

The versatile and efficient use of the purine ring is also evident in the biosynthesis of histidine. The six-membered portion of the purine ring of ATP contributes part of the imidazole ring of histidine (p. 499). The rest of the purine skeleton is not discarded. Rather, it is conserved in 5-aminoimidazole-4-carboxamide ribonucleotide, an intermediate in the de novo pathway of purine biosynthesis.

AMP AND GMP ARE FEEDBACK INHIBITORS OF PURINE NUCLEOTIDE BIOSYNTHESIS

The synthesis of purine nucleotides is controlled by feedback inhibition at several sites (Figure 22-10).

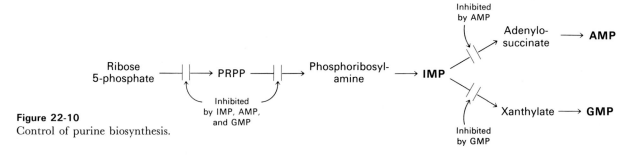

Figure 22-10
Control of purine biosynthesis.

1. Feedback inhibition of *5-phosphoribosyl-1-pyrophosphate synthetase* by purine nucleotides regulates the level of PRPP. This synthetase is inhibited by AMP, GMP, and IMP.

2. The committed step in purine nucleotide biosynthesis is the conversion of PRPP into phosphoribosylamine by transfer of the

side-chain amino group of glutamine. *Glutamine PRPP amidotransferase is feedback inhibited by many purine ribonucleotides.* It is noteworthy that AMP and GMP, the final products of the pathway, are synergistic in inhibiting this enzyme.

3. Inosinate is the branching point in the synthesis of AMP and GMP. *The reactions leading away from inosinate are sites of feedback inhibition.* AMP inhibits the conversion of inosinate into adenylosuccinate, its immediate precursor. Similarly, GMP inhibits the conversion of inosinate into xanthylate, its immediate precursor.

4. GTP is a substrate in the synthesis of AMP, whereas ATP is a substrate in the synthesis of GMP. This *reciprocal substrate relation* tends to balance the synthesis of adenine and guanine ribonucleotides.

THE PYRIMIDINE RING IS SYNTHESIZED FROM CARBAMOYL PHOSPHATE AND ASPARTATE

The pyrimidine ring is assembled first and then linked to ribose phosphate to form a pyrimidine nucleotide, in contrast with the reaction sequence in the de novo synthesis of purine nucleotides. PRPP is the donor of ribose phosphate in the synthesis of pyrimidine nucleotides, as well as of purine nucleotides. The precursors of the pyrimidine ring are carbamoyl phosphate and aspartate (Figure 22-11).

The synthesis of pyrimidines starts with the formation of *carbamoyl phosphate,* which is also an intermediate in the synthesis of urea (p. 412). The synthesis of this activated carbamoyl donor is compartmentalized in eucaryotes. Carbamoyl phosphate consumed in the·synthesis of pyrimidines is formed in the cytosol, whereas that used in the synthesis of urea is formed in mitochondria (p. 414). There are two distinct carbamoyl phosphate synthetases. Another noteworthy difference is that glutamine rather than NH_4^+ is the nitrogen donor in the cytosol synthesis of carbamoyl phosphate.

$$\text{Glutamine} + 2\,\text{ATP} + \text{HCO}_3^- \longrightarrow$$
$$\text{carbamoyl phosphate} + 2\,\text{ADP} + \text{P}_i + \text{glutamate}$$

The committed step in the biosynthesis of pyrimidines is the formation of N-carbamoylaspartate from aspartate and carbamoyl phosphate. This carbamoylation is catalyzed by *aspartate transcarbamoylase,* an especially interesting regulatory enzyme (see p. 522).

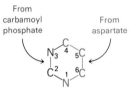

Figure 22-11
Origins of the atoms in the pyrimidine ring. C-2 and N-3 come from carbamoyl phosphate, whereas the other atoms of the ring come from aspartate.

Carbamoyl phosphate   **Aspartate**   *N*-**Carbamoylaspartate**

The pyrimidine ring is formed in the next reaction, in which carbamoylaspartate cyclizes with loss of water to yield *dihydroorotate*. *Orotate* is then formed by dehydrogenation of dihydroorotate.

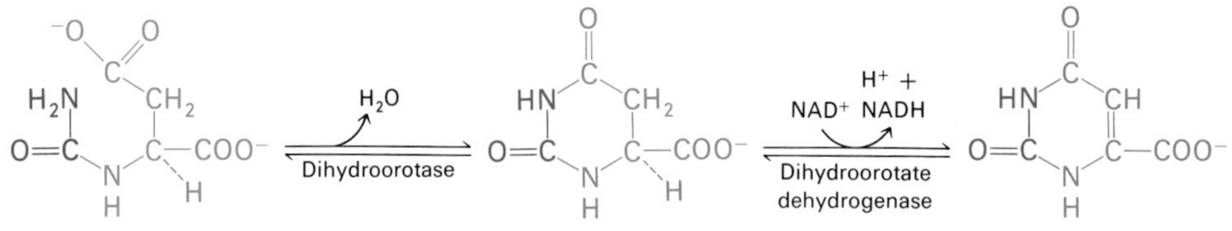

N-Carbamoylaspartate   Dihydroorotate   Orotate

OROTATE ACQUIRES A RIBOSE PHOSPHATE MOIETY FROM PRPP

The next step in the synthesis of pyrimidine nucleotides is the *acquisition of a ribose phosphate group*. Orotate (a free pyrimidine) reacts with PRPP to form *orotidylate* (a pyrimidine nucleotide). This reaction, which is catalyzed by orotidylate pyrophosphorylase, is driven forward by the hydrolysis of pyrophosphate. Orotidylate is then decarboxylated to yield *uridylate* (UMP), a major pyrimidine nucleotide.

Orotate   Orotidylate   Uridylate (UMP)

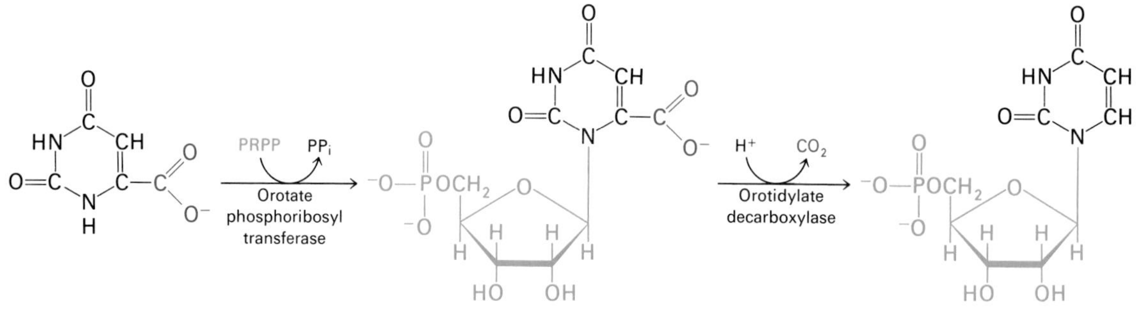

A SINGLE POLYPEPTIDE CHAIN CONTAINS THE FIRST THREE ENZYMES OF PYRIMIDINE BIOSYNTHESIS

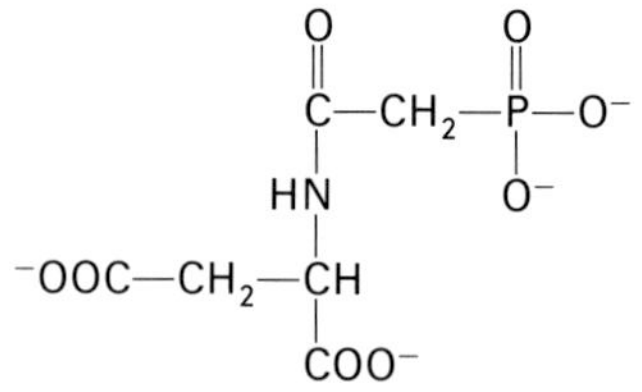

N-(Phosphonacetyl)-L-aspartate
(PALA)

Figure 22-12
Structure of PALA, a potent inhibitor of ATCase.

In *E. coli,* the six enzymes that synthesize UMP from simple precursors do not appear to be associated. In contrast, in higher organisms, several of these enzymes form a multienzyme complex. Large amounts of this complex were obtained by treating cultured mammalian cells with N-(*phosphonacetyl*)-L-*aspartate* (PALA), a potent inhibitor of aspartate transcarbamoylase (ATCase). PALA binds tightly to ATCase ($K_i = 10^{-8}$ M) because it has some of the structural features of the transition state in catalysis (Figure 22-12). The surviving cells overcame the inhibitory effect of PALA by synthesizing 100-fold more ATCase than do normal cells. The levels of carbamoyl phosphate synthetase and of dihydroorotase were also elevated 100-fold, whereas there was little change in the amounts of the enzymes catalyzing the subsequent steps in pyrimidine biosynthesis. These observations led to the finding that *carbamoyl phosphate synthetase, aspartate transcarbamoylase, and dihydroorotase are covalently joined on a single 200-kdal polypeptide chain.* Orotate phosphoribosyltransferase and orotidylate decarboxylase, the enzymes catalyzing the last two steps in pyrimidine biosynthesis, form another complex. They, too, may be covalently joined. Recall that the fatty acid synthetase complex in yeast consists of two kinds of polypeptide chains, each containing several enzymes (p. 400). *Covalent linkage of functionally related enzymes may be quite general in eucaryotes.* Such an arrangement would facilitate the assembly of a multienzyme complex. Another likely advantage of having several enzymes on a single polypeptide chain is that they would be synthesized in equimolar amounts.

NUCLEOSIDE MONO-, DI-, AND TRIPHOSPHATES ARE INTERCONVERTIBLE

The active forms of nucleotides in biosyntheses and energy conversions are the diphosphates and triphosphates. Nucleoside monophosphates are converted by specific *nucleoside monophosphate kinases* that utilize ATP as the phosphoryl donor. For example, UMP is phosphorylated by *UMP kinase.*

$$\text{UMP} + \text{ATP} \rightleftharpoons \text{UDP} + \text{ADP}$$

AMP, ADP, and ATP are interconverted by *adenylate kinase* (also called myokinase). The equilibrium constants of these reactions are close to 1.

$$\text{AMP} + \text{ATP} \rightleftharpoons \text{ADP} + \text{ADP}$$

Nucleoside diphosphates and triphosphates are interconverted by *nucleoside diphosphate kinase,* an enzyme that has broad specificity, in contrast with the monophosphate kinases. In the following equation, X and Y can be any of several ribonucleosides or deoxyribonucleosides.

$$\text{XDP} + \text{YTP} \rightleftharpoons \text{XTP} + \text{YDP}$$

For example,

$$\text{UDP} + \text{ATP} \rightleftharpoons \text{UTP} + \text{ADP}$$

CTP IS FORMED BY AMINATION OF UTP

Cytidine triphosphate (CTP) is derived from uridine triphosphate (UTP), the other major pyrimidine ribonucleotide. The carbonyl oxygen at C-4 is replaced by an amino group. In mammals, the side chain of glutamine is the amino donor, whereas NH_4^+ is used in this reaction in *E. coli*. An ATP is consumed in both amination reactions.

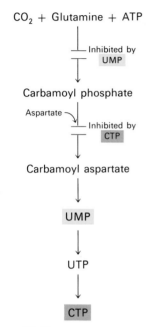

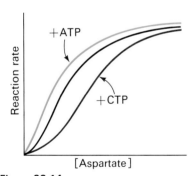

Figure 22-13
Control of pyrimidine biosynthesis.

Figure 22-14
Allosteric effects in aspartate transcarbamoylase. ATP is an activator, whereas CTP is an inhibitor. [After J. C. Gerhart, *Curr. Top. Cell Regul.* 2(1970):275.]

PYRIMIDINE NUCLEOTIDE BIOSYNTHESIS IS REGULATED BY FEEDBACK INHIBITION

The committed step in pyrimidine nucleotide biosynthesis in *E. coli* is the formation of *N*-carbamoylaspartate from aspartate and carbamoyl phosphate. *Aspartate transcarbamoylase, the enzyme that catalyzes this reaction, is feedback inhibited by CTP, the final product in the pathway.* A second control site is *carbamoyl phosphate synthetase*, which is feedback inhibited by UMP (Figure 22-13).

The allosteric properties of ATCase have been intensively investigated by John Gerhart and Howard Schachman. *The binding of carbamoyl phosphate and aspartate is cooperative*, as reflected in the sigmoidal dependence of the reaction velocity on substrate concentration (Figure 22-14). *CTP inhibits the enzyme by decreasing its affinity for substrates* without affecting its V_{max}. The extent of inhibition exerted by CTP, which may reach 90%, depends on the concentrations of the substrates. In contrast, *ATP is an activator of ATCase*. The affinity of the enzyme for its substrates is enhanced by ATP, whereas V_{max} is unaffected. Furthermore, the binding of ATP and CTP to the regulatory site of ATCase is competitive. High levels of ATP displace CTP from the enzyme so that it cannot exert its inhibitory effect.

The biological significance of the activation of ATCase by ATP is twofold. First, *it tends to equalize the rates of formation of purine and pyrimidine nucleotides.* Comparable quantities of these two types of nucleotides are needed for the synthesis of nucleic acids. Second, *activation by ATP signals its availability as a substrate* for some of the reactions of pyrimidine nucleotide biosynthesis, such as the synthesis of carbamoyl phosphate and the phosphorylations of UMP to UTP.

ASPARTATE TRANSCARBAMOYLASE CONSISTS OF SEPARABLE CATALYTIC AND REGULATORY SUBUNITS

The regulatory properties of ATCase vanish when the enzyme is treated with mercurials such as *p*-hydroxymercuribenzoate. ATP and CTP no longer have any effect on catalytic activity. Furthermore, the binding of substrates becomes noncooperative. However, the modified enzyme has full catalytic activity. This loss of regulatory properties with retention of enzymatic activity is called *desensitization.*

The desensitization of ATCase by mercurials is accompanied by its *dissociation into two kinds of subunits,* as shown by ultracentrifuge studies (Figure 22-15). The sedimentation coefficient of the native enzyme is 11.6S, whereas that of the dissociated subunits is 2.8S and 5.8S. These subunits can be readily separated by ion-exchange chromatography because they differ markedly in charge or by centrifugation in a sucrose density gradient because they differ in size. *p*-Hydroxymercuribenzoate, the dissociating agent, can be removed after the subunits are separated. The larger of the subunits, called the *catalytic subunit,* is catalytically active. However, the activity of the isolated catalytic subunit is not affected by ATP and CTP. The smaller of the subunits, called the *regulatory subunit,* is devoid of catalytic activity but contains specific binding sites for CTP and ATP. The catalytic subunit consists of three 34-kdal polypeptide chains, whereas the regulatory subunit is made up of two 17-kdal polypeptide chains.

The catalytic and regulatory subunits combine rapidly when they are mixed. The resulting complex has the same structure, R_6C_6, as that of the native enzyme.

$$3\,R_2 + 2\,C_3 \longrightarrow R_6C_6$$

Furthermore, *the reconstituted enzyme has the same allosteric properties as those of the native enzyme.*

X-ray crystallographic studies of ATCase are in progress in William Lipscomb's laboratory. An electron-density map at 3.0-Å resolution shows that the two catalytic trimers (C_3) are above and below an equatorial belt of three regulatory dimers (R_2) (Figure 22-16). A distinctive feature of the molecule is that it contains a large central cavity, which is accessible through several channels. It is interesting to note that the allosteric sites for CTP are far from the catalytic sites.

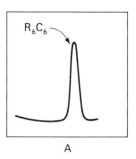

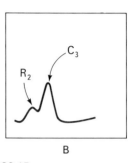

Figure 22-15
Sedimentation velocity patterns of (A) native ATCase and (B) the enzyme dissociated by a mercurial into regulatory and catalytic subunits. [After J. C. Gerhart and H. K. Schachman. *Biochemistry* 4(1965):1054.]

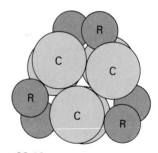

Figure 22-16
Arrangement of the catalytic (C, shown in blue) and regulatory (R, shown in red) subunits of aspartate transcarbamoylase. [After a drawing kindly provided by Dr. William Lipscomb.]

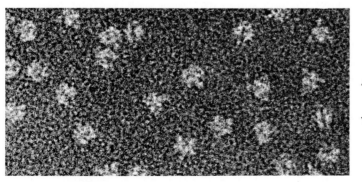

200 Å

Figure 22-17
Electron micrograph of aspartate transcarbamoylase. [Courtesy of Dr. Robley C. Williams.]

DEOXYRIBONUCLEOTIDES ARE SYNTHESIZED BY REDUCTION OF RIBONUCLEOSIDE DIPHOSPHATES

We turn now to the synthesis of deoxyribonucleotides. These precursors of DNA are formed by the reduction of ribonucleotides. The 2'-hydroxyl group on the ribose moiety is replaced by a hydrogen atom. In *E. coli* and in mammals, the substrates in this reaction are ribonucleoside *di*phosphates. The overall stoichiometry is

$$\text{Ribonucleoside diphosphate} + \text{NADPH} + \text{H}^+ \longrightarrow$$
$$\text{deoxyribonucleoside diphosphate} + \text{NADP}^+ + \text{H}_2\text{O}$$

The actual reaction mechanism is more complex than implied by this equation. Peter Reichard has shown that in *E. coli* the electrons from NADPH are transferred to the substrate through a series of sulfhydryl groups. *Ribonucleotide reductase* (also called ribonucleoside diphosphate reductase) catalyzes the final stage, which has the stoichiometry

$$\text{Ribonucleoside diphosphate} + \text{R} \begin{array}{c} \diagup \text{SH} \\ \diagdown \text{SH} \end{array} \longrightarrow$$

$$\text{deoxyribonucleoside diphosphate} + \text{R} \begin{array}{c} \diagup \text{S} \\ \mid \\ \diagdown \text{S} \end{array} + \text{H}_2\text{O}$$

This enzyme consists of two subunits, B1 (a 160-kdal dimer) and B2 (a 78-kdal dimer). The B1 subunit contains the binding sites for ribonucleotide substrates and for allosteric effectors. In addition, B1 contains sulfhydryls that serve as the immediate electron donors in the reduction of the ribose unit. B2, an iron-sulfur protein, participates in catalysis by forming an unusual free radical on the aromatic ring of a tyrosine residue. The B1 and B2 subunits together form the active sites of the enzyme (Figure 22-18).

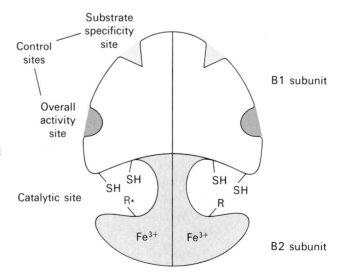

Figure 22-18
Model of ribonucleotide reductase from *E. coli*. [After L. Thelander and P. Reichard. *Ann. Rev. Biochem.* 48(1979):136. ©1979 by Annual Reviews Inc.]

How are electrons transferred from NADPH to the sulfhydryl groups at the catalytic site of ribonucleotide reductase? One carrier of reducing power is *thioredoxin,* a 12-kdal protein with two cysteine residues in close proximity (Figure 22-19). These sulfhydryls are oxidized to a disulfide in the reaction catalyzed by ribonucleotide reductase. In turn, reduced thioredoxin is regenerated by the reaction of NADPH with oxidized thioredoxin. This reaction is catalyzed by *thioredoxin reductase,* a flavoprotein.

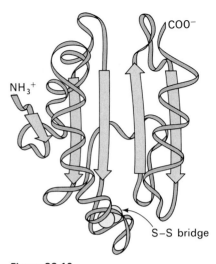

$$\text{Thioredoxin} \overset{\displaystyle S}{\underset{\displaystyle S}{\Big\langle}} + NADPH + H^+ \longrightarrow$$

$$\text{thioredoxin} \overset{\displaystyle SH}{\underset{\displaystyle SH}{\Big\langle}} + NADP^+$$

It was thought for some time that thioredoxin is the only carrier of reducing power to ribonucleotide reductase. However, a mutant of *E. coli* totally devoid of thioredoxin was found to form deoxyribonucleotides. This surprising observation led to the isolation of a second carrier system. The electron donor in this mutant proved to be *glutathione,* a cysteine-containing tripeptide. As discussed previously (p. 345), *glutathione reductase* catalyzes the reduction of oxidized glutathione (the disulfide form) by NADPH. In addition, *glutaredoxin,* a new protein, is needed to transfer the reducing power of glutathione to ribonucleotide reductase (Figure 22-20). The rela-

Figure 22-19
Schematic diagram of the main-chain conformation of oxidized thioredoxin from *E. coli.* The reactive disulfide is shown in yellow. [After a drawing kindly provided by Dr. Carl-Ivar Brändén.]

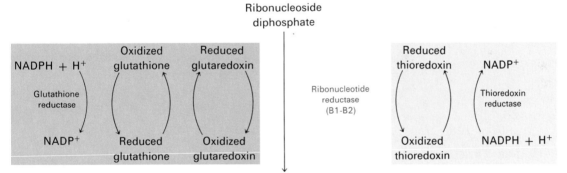

Figure 22-20
Ribonucleoside diphosphates are reduced to deoxyribonucleoside diphosphates by ribonucleotide reductase. Electrons are transferred from NADPH through a series of sulfhydryls. The thioredoxin system (shown in yellow) and the glutaredoxin system (shown in green) can serve as sources of reducing power.

tive contributions of the thioredoxin and glutaredoxin systems in ribonucleotide reduction by normal cells are not yet known.

The reduction of ribonucleotide diphosphates is precisely controlled by allosteric interactions. The B1 subunit of ribonucleotide reductase contains two types of allosteric sites: one of them controls the *overall activity* of the enzyme, whereas the other regulates *substrate specificity.* The overall catalytic activity of ribonucleotide reductase is diminished by the binding of dATP, which signals an abundance of deoxyribonucleotides. This feedback inhibition is reversed by the binding of ATP. The binding of dATP or ATP to the substrate-specificity control sites enhances the reduction of UDP and CDP, the pyrimidine nucleotides. The reduction of GDP is promoted by the binding of dTTP, which also inhibits the further reduction of pyrimidine ribonucleotides. The subsequent increase in the level of dGTP leads to a stimulation of ADP reduction. It is evident that ribonucleotide reductase has a variety of conformational states, each with different catalytic properties. This complex pattern of regulation provides the appropriate supply of the four deoxyribonucleotides needed for the synthesis of DNA.

DEOXYTHYMIDYLATE IS FORMED BY METHYLATION OF DEOXYURIDYLATE

Uracil is not a component of DNA. Rather, DNA contains *thymine*, the methylated analog of uracil. This finishing touch occurs at the level of the deoxyribonucleoside monophosphate: deoxyuridylate (dUMP) is methylated to deoxythymidylate (dTMP) by *thymidylate synthetase*. The methyl donor in this reaction is a tetrahydrofolate derivative rather than *S*-adenosylmethionine. Specifically, the methyl carbon comes from N^5,N^{10}-methylenetetrahydrofolate. Note that the methyl group inserted into deoxyuridylate is more reduced than the methylene group in this tetrahydrofolate derivative. What is the source of electrons for this reduction? The two electrons come from the tetrahydrofolate moiety itself in the form of a hydride ion (H^-), which is removed from the ring. This hydrogen becomes part of the methyl group of dTMP. In this reaction, tetrahydrofolate is oxidized to dihydrofolate. Thus N^5,N^{10}-methylenetetrahydrofolate serves both as an *electron donor* and as a *one-carbon donor* in the methylation reaction (Figure 22-21).

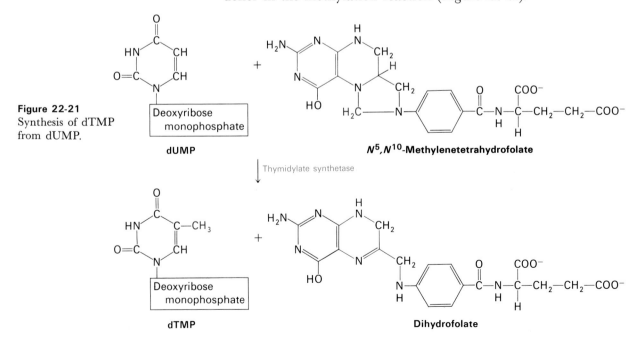

Figure 22-21
Synthesis of dTMP from dUMP.

Recall that one-carbon transfers occur at the level of tetrahydrofolate rather than dihydrofolate. Hence, tetrahydrofolate must be regenerated. This is accomplished by *dihydrofolate reductase,* utilizing NADPH as the reductant.

$$\text{Dihydrofolate} + \text{NADPH} + \text{H}^+ \longrightarrow \text{tetrahydrofolate} + \text{NADP}^+$$

SEVERAL ANTICANCER DRUGS BLOCK THE SYNTHESIS OF DEOXYTHYMIDYLATE

Rapidly dividing cells require an abundant supply of deoxythymidylate for the synthesis of DNA. The vulnerability of these cells to the inhibition of dTMP synthesis has been exploited in cancer chemotherapy (Figure 22-22). Thymidylate synthetase and dihy-

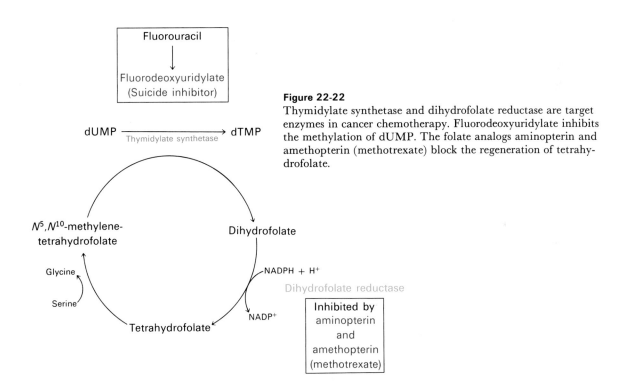

Figure 22-22
Thymidylate synthetase and dihydrofolate reductase are target enzymes in cancer chemotherapy. Fluorodeoxyuridylate inhibits the methylation of dUMP. The folate analogs aminopterin and amethopterin (methotrexate) block the regeneration of tetrahydrofolate.

drofolate reductase are choice target enzymes. *Fluorouracil* (or fluorodeoxyuridine), a clinically useful anticancer drug, is converted in vivo into *fluorodeoxyuridylate* (F-dUMP). *This analog of dUMP irreversibly inhibits thymidylate synthetase after acting as a normal substrate through part of the catalytic cycle.* First, a sulfhydryl group of the enzyme adds to C-6 of the bound F-dUMP. Methylenetetrahydrofolate then adds to C-5 of this intermediate. In the case of dUMP, a hydride ion is subsequently shifted to the methylene group of the folate, and a proton is taken away from C-5 of the bound nucleotide. However, for F-dUMP, F$^+$ cannot be abstracted by the enzyme and so catalysis is blocked at the stage of the covalent complex formed by F-dUMP, methylenetetrahydrofolate, and the sulfhydryl group of the enzyme (Figure 22-23).

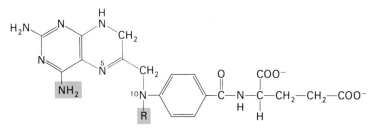

Figure 22-23
Thymidylate synthetase is irreversibly inhibited by fluorodeoxyuridylate (F-dUMP). This analog forms a covalent complex with both a sulfhydryl residue of the enzyme (shown in blue) and methylenetetrahydrofolate (shown in yellow).

The synthesis of dTMP can also be blocked by inhibiting the regeneration of tetrahydrofolate (Figure 22-22). Analogs of dihydrofolate, such as *aminopterin* and *amethopterin* (*methotrexate*) are potent competitive inhibitors ($K_i < 10^{-9}$ M) of dihydrofolate reductase (Figure 22-24). Amethopterin is a valuable drug in the treatment of acute leukemia and choriocarcinoma.

**Structure of aminopterin (R = H)
and amethopterin (R = CH$_3$)**

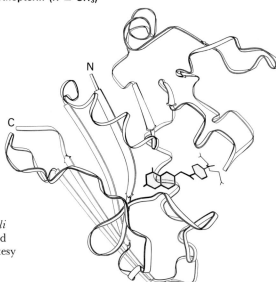

Figure 22-24
Three-dimensional structure of *E. coli* dihydrofolate reductase with a bound amethopterin (methotrexate). [Courtesy of Dr. Joseph Kraut.]

The *biosynthesis of nicotinamide adenine dinucleotide* (NAD⁺) starts with the formation of *nicotinate ribonucleotide* from nicotinate and PRPP. *Nicotinate* (also called *niacin*) is derived from tryptophan. Humans can synthesize the required amount of nicotinate if the supply of tryptophan in the diet is adequate. However, an exogenous supply of nicotinate is required if the dietary intake of tryptophan is low. *Pellagra* is a deficiency disease caused by a dietary insufficiency of tryptophan and nicotinate.

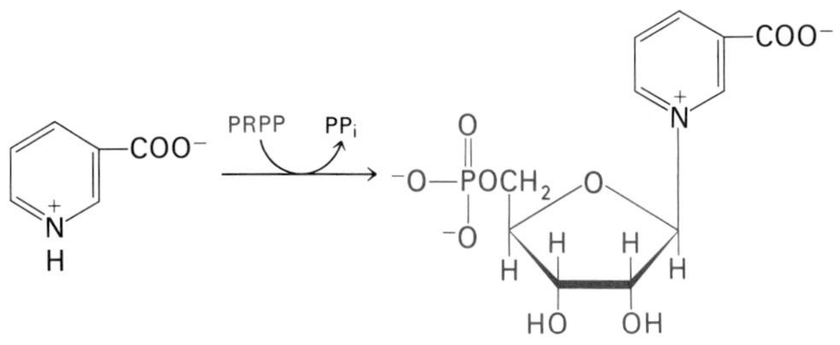

Nicotinate   Nicotinate ribonucleotide

An AMP moiety is transferred from ATP to nicotinate ribonucleotide to form *desamido-NAD⁺*. The final step is the transfer of the amide group of glutamine to the nicotinate carboxyl group to form NAD⁺ (Figure 22-25). NADP⁺ is derived from NAD⁺ by phospho-

Nicotinate
ribonucleotide   **Desamido-NAD⁺**   **NAD⁺**

Figure 22-25
Synthesis of NAD⁺ from nicotinate ribonucleotide.

rylation of the 2′-hydroxyl group of the adenine ribose moiety. This transfer of a phosphoryl group is catalyzed by NAD⁺ kinase.

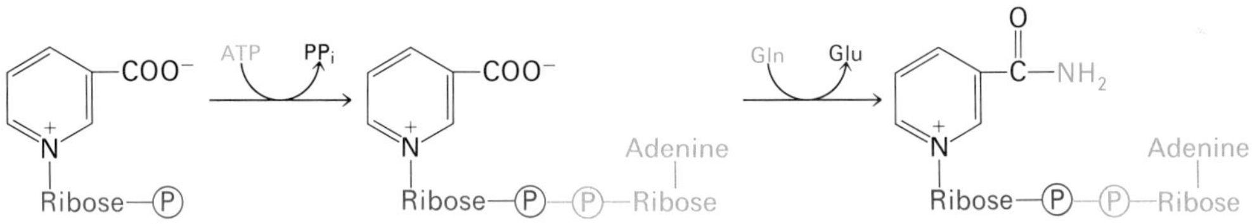

Flavin adenine dinucleotide (FAD) is synthesized from riboflavin and two molecules of ATP. Riboflavin is phosphorylated by ATP to give *riboflavin 5′-phosphate* (also called *flavin mononucleotide*). FAD is then formed by the transfer of an AMP moiety from a second molecule of ATP to riboflavin 5′-phosphate.

Riboflavin + ATP ⟶ riboflavin 5′-phosphate + ADP

Riboflavin 5′-phosphate + ATP ⇌

flavin adenine dinucleotide + PP$_i$

The *synthesis of coenzyme A* (CoA) in animals starts with the phosphorylation of *pantothenate* (Figure 22-26). Pantothenate is required in the diet of animals, whereas it is synthesized by plants and microorganisms. A peptide bond is formed between the carboxyl group of 4'-phosphopantothenate and the amino group of cysteine. The carboxyl group of the cysteine moiety is lost, which results in *4'-phosphopantotheine*. The AMP moiety of ATP is then transferred to this intermediate to form *dephosphocoenzyme A*. Finally, phosphorylation of its 3'-hydroxyl group yields coenzyme A.

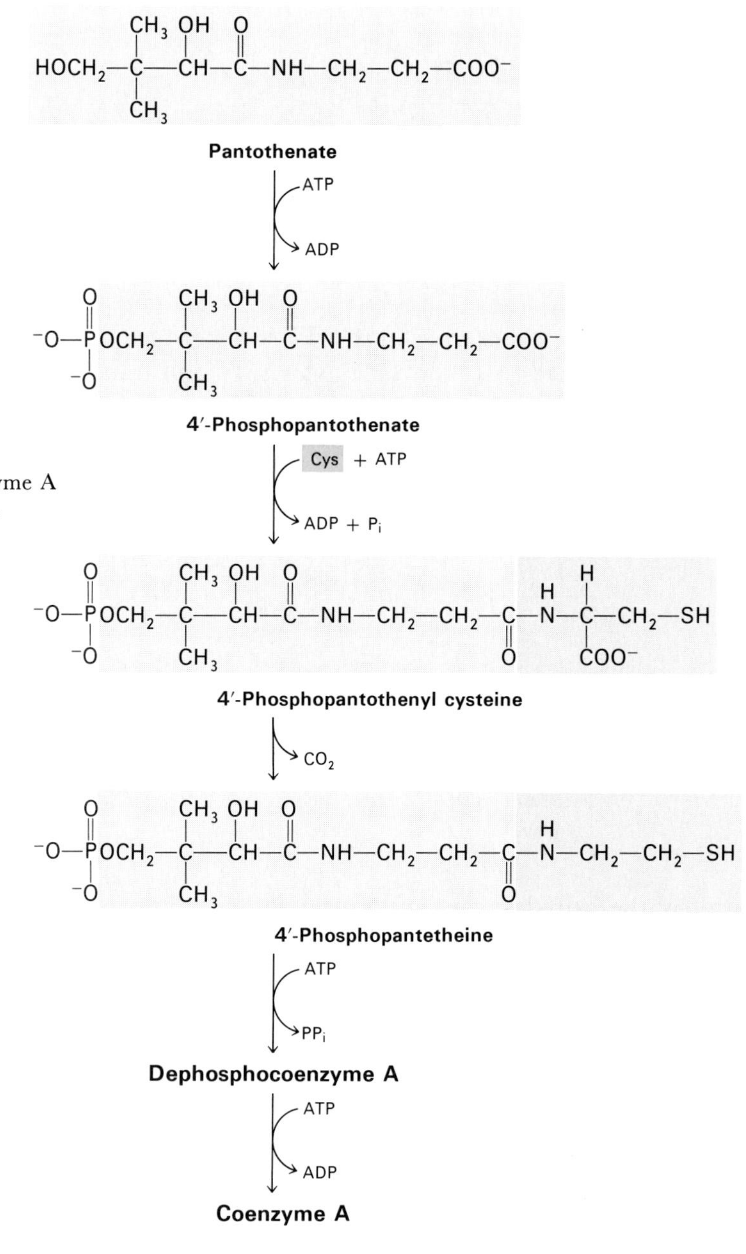

Figure 22-26
Synthesis of coenzyme A from pantothenate.

A common feature of the biosyntheses of NAD$^+$, FAD, and CoA is the *transfer of the AMP moiety of ATP to the phosphate group of a phosphorylated intermediate.* The pyrophosphate formed in this reaction is hydrolyzed to orthophosphate. This is a recurring motif in biochemistry: *biosynthetic reactions are frequently driven by the hydrolysis of the released pyrophosphate.*

PURINES ARE DEGRADED TO URATE IN HUMANS

The nucleotides of a cell undergo continuous turnover. Nucleotides are hydrolytically degraded to nucleosides by *nucleotidases.* Phosphorolytic cleavage of nucleosides to free bases and ribose 1-phosphate (or deoxyribose 1-phosphate) is catalyzed by *nucleoside phosphorylases.* Ribose 1-phosphate is isomerized by *phosphoribomutase* to ribose 5-phosphate, a substrate in the synthesis of PRPP. Some of the bases are reused to form nucleotides by salvage pathways.

The pathway for the degradation of AMP (Figure 22-27) includes an additional step. AMP is deaminated to IMP by *adenylate*

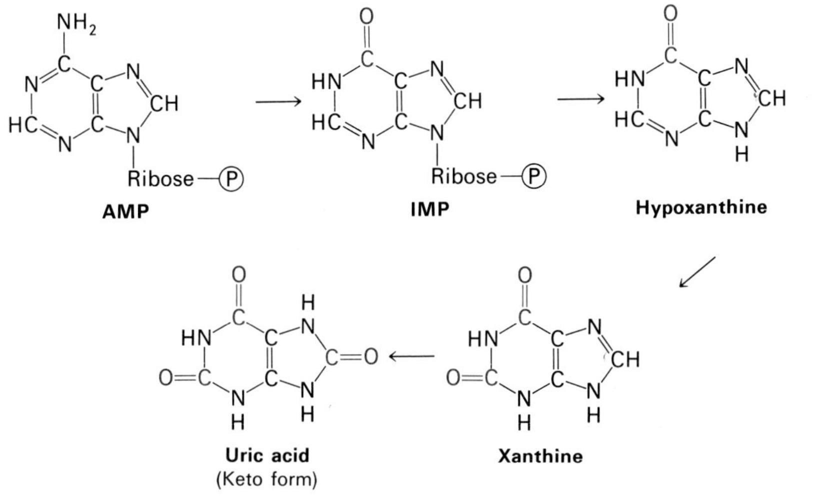

Figure 22-27
Degradation of AMP to uric acid.

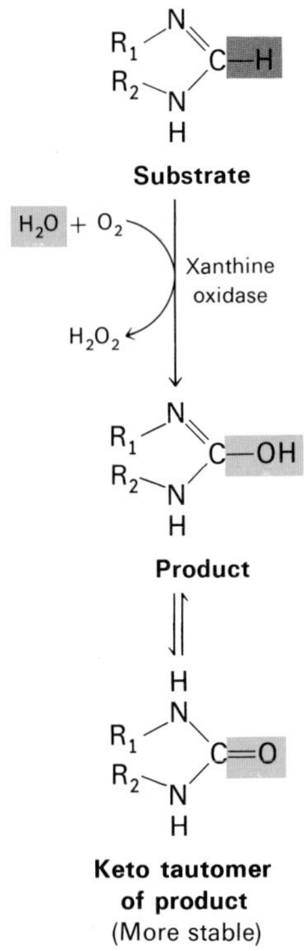

Figure 22-28
Oxidation reaction catalyzed by xanthine oxidase.

deaminase. The subsequent reactions leading to the free base hypoxanthine follow the general pattern. *Xanthine oxidase,* a molybdenum and iron-containing flavoprotein, oxidizes hypoxanthine to *xanthine* and then to *urate.* Molecular oxygen, the oxidant in both reactions, is reduced to H_2O_2, which is decomposed to H_2O and O_2 by catalase. Xanthine is also an intermediate in the formation of urate from guanine. *In humans, urate is the final product of purine degradation and is excreted in the urine.*

URATE IS FURTHER DEGRADED IN SOME ORGANISMS

The breakdown of purines proceeds further in some species (Figure 22-29). Mammals other than primates excrete *allantoin*, which is

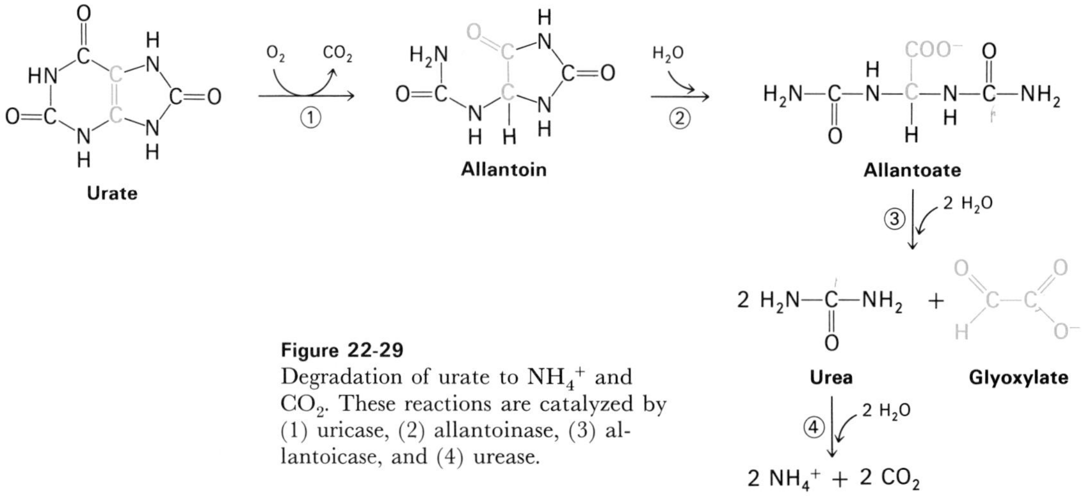

Figure 22-29
Degradation of urate to NH_4^+ and CO_2. These reactions are catalyzed by (1) uricase, (2) allantoinase, (3) allantoicase, and (4) urease.

formed by oxidation of urate. Teleost fish excrete *allantoate,* which is formed by hydration of allantoin. The degradation proceeds a step further in amphibians and most fish. Allantoate is hydrolyzed to two molecules of *urea* and one of *glyoxylate.* Finally, some marine invertebrates hydrolyze urea to NH_4^+ and CO_2. It seems likely that the enzymes catalyzing these reactions were progressively lost in the evolution of primates.

BIRDS AND TERRESTRIAL REPTILES EXCRETE URATE INSTEAD OF UREA TO CONSERVE WATER

In terrestrial reptiles and birds, urea is not the final product of amino nitrogen metabolism. These animals synthesize purines from their excess amino nitrogen and then degrade these purines to urate. *This circuitous path for the elimination of amino nitrogen serves a vital function: the conservation of water. Urate is the vehicle for the excretion of amino nitrogen because of its very low solubility at acid pH.* The pK_a of the most acidic group in uric acid is 5.4.

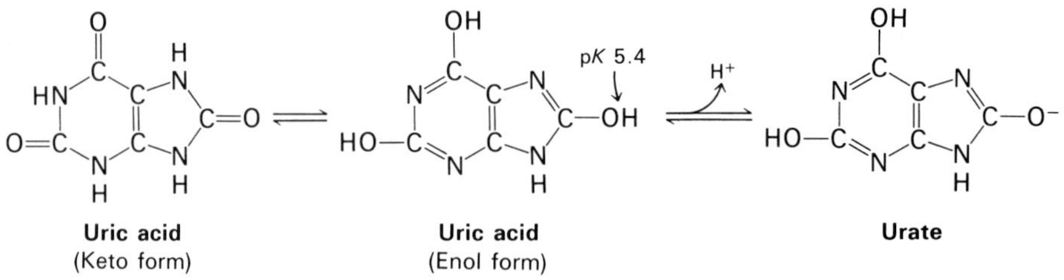

The acidic urine of birds and terrestrial reptiles consists of a paste of crystals of uric acid. Little water accompanies the excretion of these crystals. In contrast, the excretion of a comparable amount of highly soluble urea would be accompanied by a large efflux of water.

DEGRADATION OF PYRIMIDINES

The degradation of thymine (Figure 22-30) is illustrative of the breakdown of pyrimidines. Thymine is degraded to β-aminoisobutyrate, which is metabolized as though it were an amino acid. The amino group is removed by transamination to yield methylmalonate semialdehyde, which is converted into methylmalonyl CoA. The conversion of methylmalonyl CoA into succinyl CoA, the point of entry into the citric acid cycle, has already been discussed (p. 419).

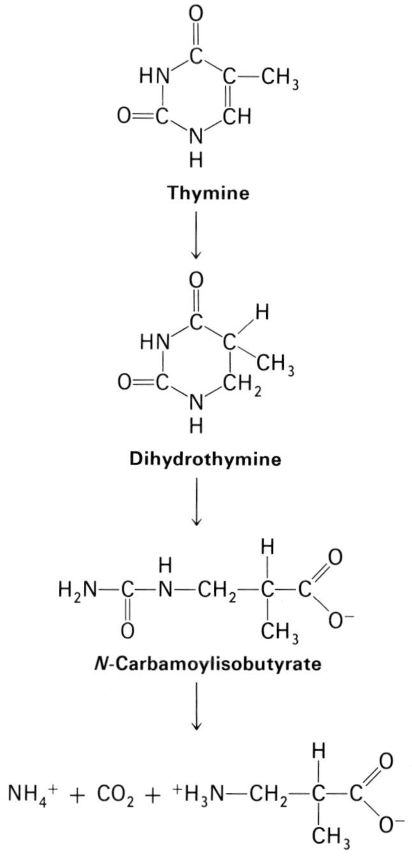

EXCESSIVE PRODUCTION OF URATE IS A CAUSE OF GOUT

Gout is a disease that affects the joints and leads to arthritis. The major biochemical feature of gout is an *elevated level of urate in the serum*. Inflammation of the joints is triggered by the *precipitation of sodium urate crystals*. Kidney disease may also occur because of the deposition of urate crystals in that organ. Gout primarily affects adult males. A vivid description of an acute attack of gout was given by Thomas Sydenham, an outstanding seventeenth-century English physician, who himself was afflicted with this disease:

> The victim goes to bed and sleeps in good health. About two o'clock in the morning he is awakened by a severe pain in the great toe; more rarely in the heel, ankle or instep. This pain is like that of a dislocation, and yet the parts feel as if cold water were poured over them. Then follow chills and shivers, and a little fever. The pain, which was at first moderate, becomes more intense. With its intensity the chills and shivers increase. After a time this comes to its height, accommodating itself to the bones and ligaments of the tarsus and metatarsus. Now it is a violent stretching and tearing of the ligaments—now it is a gnawing pain and now a pressure and tightening. So exquisite and lively meanwhile is the feeling of the part affected, that it cannot bear the weight of the bedclothes nor the jar of a person walking in the room. The night is passed in torture, sleeplessness, turning of the part affected, and perpetual change of posture; the tossing about of the body being as incessant as the pain of the tortured joint, and being worse as the fit comes on.

The biochemical lesion in most cases of gout has not been elucidated. It seems likely that gout is an expression of a variety of inborn errors of metabolism in which *excessive production of urate* is a common finding. Some patients with this abnormality have a par-

Figure 22-30
Degradation of thymine.

tial deficiency of *hypoxanthine-guanine phosphoribosyltransferase,* the enzyme that catalyzes the salvage synthesis of IMP and GMP.

$$\text{Hypoxanthine} + \text{PRPP} \longrightarrow\!\!\!\!\!\not\longmapsto \quad \text{IMP} \; + \text{PP}_i$$
$$\text{(or guanine)} \qquad\qquad\qquad \text{(or GMP)}$$

Deficiency of this enzyme leads to reduced synthesis of GMP and IMP by the salvage pathway and an increase in the level of PRPP. There is a *marked acceleration of purine biosynthesis by the de novo pathway.* A few patients with gout have an *abnormally high level of phosphoribosylpyrophosphate synthetase.* The allosteric control of this enzyme is impaired in these patients. This results in *excessive production of PRPP,* which in turn accelerates the rate of de novo synthesis of purines.

Allopurinol, an analog of hypoxanthine in which the positions of N7 and C8 are interchanged, is used to treat gout. The mechanism of action of allopurinol is very interesting: it acts *first as a substrate* and *then as an inhibitor* of xanthine oxidase. This enzyme hydroxylates allopurinol to *alloxanthine,* which then remains tightly bound to the active site. The molybdenum atom of xanthine oxidase is kept in the +4 oxidation state by the binding of alloxanthine instead of returning to the +6 oxidation state as it does in a normal catalytic cycle. This mode of action of allopurinol is an example of *suicide inhibition,* in which an enzyme converts a compound into a potent inhibitor that immediately inactivates the enzyme.

Allopurinol

Hypoxanthine

Allopurinol → (Xanthine oxidase) → **Alloxanthine**

The synthesis of urate from hypoxanthine and xanthine decreases soon after the administration of allopurinol.

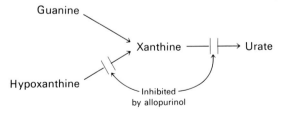

The serum concentrations of hypoxanthine and xanthine increase after administration of allopurinol, whereas that of urate drops. The formation of uric acid stones is virtually abolished by allopurinol, and there is some improvement in the arthritis. Also, *there is a decrease in the total rate of purine biosynthesis.* This inhibitory action of allopurinol depends on its reaction with PRPP to form the ribonucleotide. Consequently, the level of PRPP, the limiting substrate in the de novo synthesis of purines, is lowered. Furthermore, allopurinol ribonucleotide inhibits the conversion of PRPP into phosphoribosylamine by amidophosphoribosyltransferase.

LESCH-NYHAN SYNDROME: SELF-MUTILATION, MENTAL RETARDATION, AND EXCESSIVE PRODUCTION OF URATE

A nearly total absence of hypoxanthine-guanine phosphoribosyltransferase has devastating consequences. The most striking expression of this inborn error of metabolism, called the *Lesch-Nyhan syndrome,* is *compulsive self-destructive behavior.* At age two or three, children with this disease begin to bite their fingers and lips. The tendency to self-mutilate is so extreme that it is necessary to protect these patients by such measures as wrapping their hands in gauze. Those afflicted also tend to be aggressive toward others. *Mental deficiency and spasticity* are other characteristics of the Lesch-Nyhan syndrome. Elevated levels of urate in the serum lead to the formation of stones early in life, followed by the symptoms of *gout* years later. The disease is inherited as a sex-linked recessive.

The biochemical consequences of the virtual absence of hypoxanthine-guanine phosphoribosyltransferase are an *overproduction of urate* and an *elevated concentration of PRPP.* Also, there is a marked increase in the rate of purine biosynthesis by the de novo pathway. The relationships between the absence of the transferase and the bizarre neurologic signs are an enigma. The brain may be very dependent on the salvage pathway for the synthesis of IMP and GMP. The normal level of hypoxanthine-guanine phosphoribosyltransferase is higher in the brain than in any other tissue. In contrast, the activity of the amidotransferase that catalyzes the committed step in the de novo pathway is rather low in the brain. Allopurinol is effective in diminishing urate synthesis in the Lesch-Nyhan syndrome. However, it has no effect on the rate of de novo synthesis of purines and it fails to alleviate the neurologic expressions of the disease. Patients with the Lesch-Nyhan syndrome do not convert allopurinol into the ribonucleotide because they lack hypoxanthine-guanine phosphoribosyltransferase. Hence, the administration of allopurinol does not lower the level of PRPP in these individuals and so de novo purine synthesis is not diminished.

The Lesch-Nyhan syndrome demonstrates that the salvage pathway for the synthesis of IMP and GMP is not gratuitous. The salvage pathway evidently serves a critical role that is not yet fully understood. Furthermore, the interplay between the de novo and salvage pathways of purine synthesis remains to be elucidated. Moreover, the Lesch-Nyhan syndrome reveals that abnormal behavior such as self-mutilation and extreme hostility can be caused by the absence of a single enzyme. This finding has important implications for the future development of psychiatry.

SUMMARY

The purine ring is assembled from a variety of precursors: glutamine, glycine, aspartate, methenyltetrahydrofolate, N^{10}-formyltetrahydrofolate, and CO_2. The committed step in the de novo synthesis of purine nucleotides is the formation of 5-phosphoribosylamine from PRPP and glutamine. The purine ring is attached to ribose phosphate during its assembly. The addition of glycine, followed by formylation, amination, and ring closure, yields 5-aminoimidazole ribonucleotide. This intermediate contains the completed five-membered ring of the purine skeleton. The addition of CO_2, the nitrogen atom of aspartate and a formyl group, followed by ring closure, yields inosinate (IMP), a purine ribonucleotide. AMP and GMP are formed from IMP. Purine ribonucleotides can also be synthesized by a salvage pathway in which a preformed base reacts directly with PRPP. Feedback inhibition of 5-phosphoribosyl-1-pyrophosphate synthetase and of glutamine-PRPP amidotransferase by purine nucleotides is important in regulating their biosynthesis.

The pyrimidine ring is assembled first and then linked to ribose phosphate to form a pyrimidine nucleotide, in contrast with the sequence in the de novo synthesis of purine nucleotides. PRPP is again the donor of the ribose phosphate moiety. The synthesis of the pyrimidine ring starts with the formation of carbamoylaspartate from carbamoyl phosphate and aspartate, a reaction catalyzed by aspartate transcarbamoylase. Dehydration, cyclization, and oxidation yield orotate, which reacts with PRPP to give orotidylate. Decarboxylation of this pyrimidine nucleotide yields UMP. CTP is then formed by amination of UTP. Pyrimidine biosynthesis in *E. coli* is regulated by feedback inhibition of aspartate transcarbamoylase, the enzyme that catalyzes the committed step. CTP inhibits and ATP stimulates this enzyme. Aspartate transcarbamoylase consists of separable regulatory and catalytic subunits. In mammalian cells, a single polypeptide chain contains the first three enzymes of pyrimidine biosynthesis.

Deoxyribonucleotides, the precursors of DNA, are formed by the reduction of ribonucleoside diphosphates. These conversions are catalyzed by ribonucleotide reductase. Electrons are transferred from NADPH to the sulfhydryl groups at the catalytic sites of this enzyme by thioredoxin or glutaredoxin. dTMP is formed by the methylation of dUMP. The one-carbon and electron donor in this reaction is N^5,N^{10}-methylenetetrahydrofolate, which is converted into dihydrofolate. In turn, tetrahydrofolate is regenerated by the reduction of dihydrofolate. Dihydrofolate reductase, which catalyzes this reaction, is inhibited by folate analogs such as aminopterin and amethopterin (methotrexate). These compounds are used as antitumor drugs.

SELECTED READINGS

BOOKS AND REVIEWS

Kornberg, A., 1980. *DNA Replication.* Freeman. [Chapter 2 gives an excellent account of the biosynthesis of DNA precursors.]

Henderson, J. F., and Paterson, A. R. P., 1973. *Nucleotide Metabolism: An Introduction.* Academic Press.

Elliott, K., and Fitzsimons, D. W., (eds.), 1977. *Purine and Pyrimidine Metabolism.* Ciba Foundation Symposium 48.

Blakley, R. L., 1969. *The Biochemistry of Folic Acid and Related Pterdines.* North-Holland.

Walsh, C., 1979. *Enzymatic Reaction Mechanisms.* Freeman. [Several important reaction mechanisms in the biosynthesis of nucleotides are lucidly treated. See Chapter 5 for a discussion of amino transfers, Chapter 13 for xanthine oxidase, and Chapter 25 for one-carbon transfers.]

Kelley, W. N., and Weiner, I. M., (eds.), 1978. *Uric Acid.* Springer-Verlag. [A valuable collection of reviews on uric acid biochemistry and physiology, gout, and purine metabolism.]

Jones, M. E., 1980. Pyrimidine nucleotide biosynthesis in animals: genes, enzymes, and regulation of UMP biosynthesis. *Ann. Rev. Biochem.* 49:253–279.

ASPARTATE TRANSCARBAMOYLASE

Kantrowitz, E. V., Pastra-Landis, S. C., and Lipscomb, W. N., 1980. *E. coli* aspartate transcarbamylase. Part II: Structure and allosteric interactions. *Trends Biochem. Sci.* 5:150–153.

Gerhart, J. C., 1970. A discussion of the regulatory properties of aspartate transcarbamylase from *Escherichia coli. Curr. Top. Cell Regul.* 2:275–325.

Jacobson, G. R., and Stark, G. R., 1973. Aspartate transcarbamylases. *In* Boyer, P. D., (ed.), *The Enzymes* (3rd ed.), vol. 9, pp. 225–308. Academic Press.

Monaco, H. L., Crawford, J. L., and Lipscomb, W. N., 1978. Three-dimensional structures of aspartate carbamoyltransferase from *Escherichia coli* and of its complex with cytidine triphosphate. *Proc. Nat. Acad. Sci.* 75:5276–5280.

Hensley, P., and Schachman, H. K., 1979. Communication between dissimilar subunits in aspartate transcarbamylase: effect of inhibitor and activator on the conformation of the catalytic polypeptide chains. *Proc. Nat. Acad. Sci.* 76:3732–3736.

RIBONUCLEOSIDE DIPHOSPHATE REDUCTASE

Thelander, L., and Reichard, P., 1979. Reduction of ribonucleotides. *Ann. Rev. Biochem.* 48:133–158.

Holmgren, A., Soderberg, B.-O., Eklund, H., and Brändén, C.-I., 1975. Three-dimensional structure of *Escherichia coli* thioredoxin-S_2 to 2.8 Å resolution. *Proc. Nat. Acad. Sci.* 72:2305–2309.

Holmgren, A., 1976. Hydrogen donor system for *Escherichia coli* ribonucleoside-diphosphate reductase dependent upon glutathione. *Proc. Nat. Acad. Sci.* 73:2275–2279. [Finding of the new glutaredoxin system for reducing ribonucleotide reductase.]

OTHER ENZYMES

Coleman, P. F., Suttle, D. P., and Stark, G. R., 1977. Purification from hamster cells of the multifunctional protein that initiates de novo synthesis of pyrimidine nucleotides. *J. Biol. Chem.* 252:6379–6385.

Pagolotti, A. L., Jr., and Santi, D. V., 1977. The catalytic mechanism of thymidylate synthetase. *Bioorg. Chem.* 1:277–311.

Matthews, D. A., Alden, R. A., Bolin, J. T., Freer, S. T., Hamlin, R., Xuong, N., Kraut, J., Williams, M., Poe, M., and Hoogsteen, K., 1977. Dihydrofolate reductase: x-ray structure of the binary complex with methotrexate. *Science* 197:452–455.

Benkovic, S. J., 1980. On the mechanism of action of folate- and biopterin-requiring enzymes. *Ann. Rev. Biochem.* 49:227–251.

INBORN ERRORS

Stanbury, J. B., Wyngaarden, J. B., and Fredrickson, D. S., (eds.), 1978. *The Metabolic Basis of Inherited Diseases* (4th ed.), McGraw-Hill. [The chapters on inborn errors of purine and pyrimidine metabolism are excellent. See the articles on gout and the Lesch-Nyhan syndrome by J. B. Wyngaarden and W. B. Kelley, orotic aciduria by W. N. Kelly and L. H. Smith, Jr., and xanthinuria by J. B. Wyngaarden.]

Seegmiller, J. E., 1980. *In* Bundy, P. K., and Rosenberg, L. E., (eds.), *Metabolic Control and Disease* (8th ed.), pp. 777–937. Saunders. [An excellent presentation of diseases of purine and pyrimidine metabolism.]

PROBLEMS

1. Write a balanced equation for the synthesis of PRPP from glucose via the oxidative branch of the pentose phosphate pathway.

2. Write a balanced equation for the synthesis of orotate from glutamine, CO_2, and aspartate.

3. What is the activated reactant in the biosynthesis of each of these compounds?
   (a) Phosphoribosylamine.
   (b) Carbamoylaspartate.
   (c) Orotidylate (from orotate).
   (d) Nicotinate ribonucleotide.
   (e) Phosphoribosyl anthranilate.

4. Amidotransferases are inhibited by the antibiotic azaserine (O-diazoacetyl-L-serine), which is an analog of glutamine.

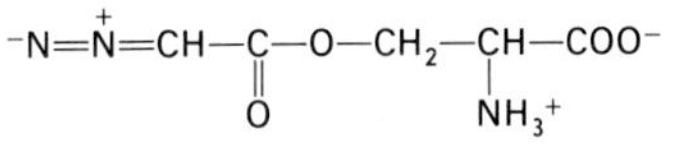

$$^-N{=}\overset{+}{N}{=}CH-\underset{\underset{O}{\|}}{C}-O-CH_2-\underset{\underset{NH_3^+}{|}}{CH}-COO^-$$

Azaserine

Which intermediates in purine biosynthesis would accumulate in cells treated with azaserine?

5. Write a balanced equation for the synthesis of dTMP from dUMP that is coupled to the conversion of serine into glycine.

6. Bacterial growth is inhibited by sulfanilamide and related sulfa drugs, and there is a concomitant accumulation of 5-aminoimidazole-4-carboxamide ribonucleotide. This inhibition is reversed by the addition of p-aminobenzoate.

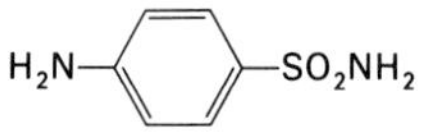

$$H_2N-\!\!\left\langle\right\rangle\!\!-SO_2NH_2$$

Sulfanilamide

Propose a mechanism for the inhibitory effect of sulfanilamide.

7. What are the major biosynthetic reactions that utilize PRPP?

8. Carbamoyl phosphate synthetase from *E. coli* consists of a small subunit (40 kdal) and a large one (130 kdal). The isolated small subunit has glutaminase activity, whereas the isolated large subunit can synthesize carbamoyl phosphate from NH_3 but not from glutamine. Furthermore, the isolated small subunit is an ATPase in the presence of bicarbonate. Propose a plausible reaction mechanism for the intact enzyme on the basis of the activities of its separated subunits.

For additional problems see W. B. Wood, J. H. Wilson, R. M. Benbow, and L. E. Hood, *Biochemistry: A Problems Approach* (Benjamin, 1974), ch. 15.

CHAPTER 23

INTEGRATION OF METABOLISM

How is the intricate network of reactions in metabolism coordinated to meet the needs of the whole organism? This chapter presents some of the principles underlying the integration of metabolism in mammals. It will start with a recapitulation of the strategy of metabolism and of recurring motifs in its regulation. The interplay of different pathways is then described in terms of the flow of molecules at three key crossroads: glucose 6-phosphate, pyruvate, and acetyl CoA. This view of major junctions is followed by a discussion of differences in the metabolic patterns of the brain, muscle, adipose tissue, and the liver. The major hormonal regulators of fuel metabolism—insulin, glucagon, epinephrine, and norepinephrine—are considered next. We then turn to a most important aspect of the integration of metabolism—the control of the level of glucose in the blood. The last part of the chapter deals with the remarkable metabolic adaptations in prolonged starvation.

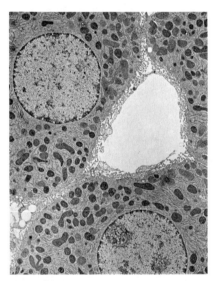

Figure 23-1
Electron micrograph of liver cells. The liver plays a key role in the integration of metabolism. [Courtesy of Dr. Ann Hubbard.]

STRATEGY OF METABOLISM: A RECAPITULATION

As discussed in Chapter 11, the basic strategy of metabolism is to form ATP, reducing power, and building blocks for biosyntheses. Let us briefly review these central themes:

1. *ATP is the universal currency of energy.* The high phosphoryl transfer potential of ATP enables it to serve as the energy source in muscle contraction, active transport, signal amplification, and

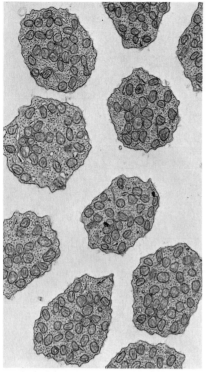

Figure 23-2
Electron micrograph showing numerous mitochondria in the inner segment of retinal rod cells. These photoreceptor cells generate large amounts of ATP and are highly dependent on a continuous supply of O_2. [Courtesy of Dr. Michael Hogan.]

biosyntheses. The hydrolysis of an ATP molecule changes the equilibrium ratio of products to reactants of a coupled reaction by a factor of about 10^8. Hence, a thermodynamically unfavorable reaction sequence can be made highly favorable by coupling it to the hydrolysis of a sufficient number of ATP molecules. For example, three molecules of ATP are consumed in the conversion of mevalonate into isopentenyl pyrophosphate, the activated five-carbon unit in the synthesis of cholesterol.

2. *ATP is generated by the oxidation of fuel molecules such as glucose, fatty acids, and amino acids.* The common intermediate in most of these oxidations is acetyl CoA. The acetyl unit is completely oxidized to CO_2 by the citric acid cycle with the concomitant formation of NADH and $FADH_2$. These electron carriers then transfer their high-potential electrons to the respiratory chain. The subsequent flow of electrons to O_2 leads to the pumping of protons across the inner mitochondrial membrane. This proton gradient is then used to synthesize ATP. Glycolysis is the other ATP-generating process but the amount formed is much smaller than in oxidative phosphorylation. The oxidation of glucose to pyruvate yields only two ATP, whereas thirty-six ATP are formed when glucose is completely oxidized to CO_2. However, glycolysis can proceed rapidly for a short time under anaerobic conditions, in contrast with oxidative phosphorylation, which requires a continuous supply of O_2.

3. *NADPH is the major electron donor in reductive biosyntheses.* In most biosyntheses, the products are more reduced than the precursors, and so reductive power is needed as well as ATP. The high-potential electrons required to drive these reactions are usually provided by NADPH. For example, in fatty acid biosynthesis, the keto group of an added two-carbon unit is reduced to a methylene group by the input of four electrons from two molecules of NADPH. The activation of O_2 by mixed-function oxygenases in hydroxylation reactions also illustrates the pervasive role of NADPH as a reductant. The pentose phosphate pathway supplies much of the required NADPH. Large amounts of this electron carrier are also formed by the malate enzyme in the transfer of acetyl CoA from mitochondria to the cytosol in fatty acid synthesis.

4. *Biomolecules are constructed from a relatively small set of building blocks.* The highly diverse molecules of life are synthesized from a much smaller number of precursors. The metabolic pathways that generate ATP and NADPH also provide building blocks for the biosynthesis of more complex molecules. For example, dihydroxyacetone phosphate formed in glycolysis gives rise to the glycerol backbone of phosphatidyl choline and other phosphoglycerides. Phosphoenolpyruvate, another glycolytic intermediate, provides part of the carbon skeleton of the aromatic amino acids. Acetyl CoA, the common intermediate in the breakdown of most fuels, supplies a two-carbon unit in a wide variety of biosyntheses. Succinyl CoA, formed in the citric acid cycle, is one of the precursors of

porphyrins. Ribose 5-phosphate, which is formed in addition to NADPH by the pentose phosphate pathway, is the source of the sugar unit of nucleotides. Many biosyntheses also require one-carbon units. Tetrahydrofolate is a source of these units at several oxidation levels. The formation of these derivatives and of S-adenosylmethionine, the major donor of methyl groups, is closely tied to amino acid metabolism. Thus, the central metabolic pathways have anabolic as well as catabolic roles.

5. *Biosynthetic and degradative pathways are almost always distinct.* For example, the pathway for the synthesis of fatty acids is different from that of its degradation. Likewise, glycogen is synthesized and degraded by different sets of reactions. This separation enables biosynthetic and degradative pathways to be thermodynamically favorable at all times. A biosynthetic pathway is made exergonic by coupling it to the hydrolysis of a sufficient number of ATP. For example, four more $\sim$P are spent in converting pyruvate into glucose in gluconeogenesis than are gained in converting glucose into pyruvate in glycolysis. The four additional $\sim$P assure that gluconeogenesis, like glycolysis, is highly exergonic under all cellular conditions. *The essential point is that the rates of metabolic pathways are governed more by the activities of key enzymes than by mass action.* The separation of biosynthetic and degradative pathways contributes greatly to the effectiveness of metabolic control.

RECURRING MOTIFS IN METABOLIC REGULATION

The complex network of reactions in a cell is precisely regulated and coordinated. Metabolism is controlled in several ways:

1. *Allosteric interactions.* The flow of molecules in most metabolic pathways is determined primarily by the amounts and activities of certain enzymes rather than by the availability of substrate. Essentially irreversible reactions are potential control sites. The first irreversible reaction in a pathway (the committed step) is usually an important control element. Enzymes catalyzing committed steps are allosterically regulated, as exemplified by phosphofructokinase in glycolysis and acetyl CoA carboxylase in fatty acid synthesis. Subsequent irreversible reactions in a pathway also may be controlled. Allosteric interactions enable such enzymes to detect diverse signals and to integrate this information.

2. *Covalent modification.* Some regulatory enzymes are controlled by covalent modification in addition to allosteric interactions. For example, the catalytic activity of glycogen phosphorylase is enhanced whereas that of glycogen synthetase is diminished by phosphorylation. These covalent modifications are catalyzed by specific enzymes. Another example is glutamine synthetase, which is rendered less active by the covalent insertion of an AMP unit. Again,

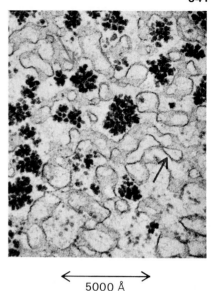

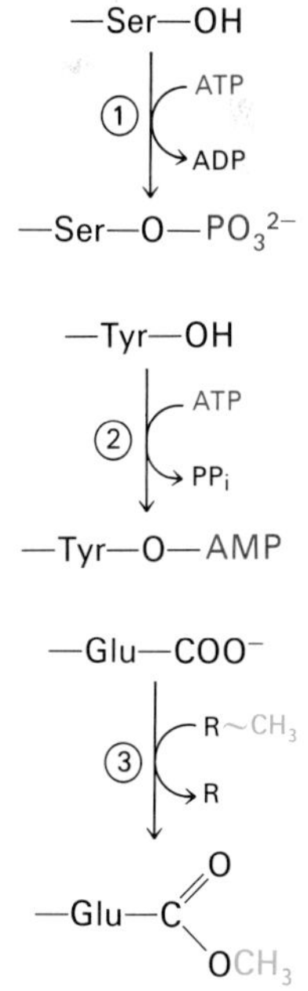

Figure 23-3
Electron micrograph of a portion of a liver cell. The green arrow points to a glycogen particle and the red arrow to the smooth endoplasmic reticulum. Hydroxylation reactions catalyzed by mixed-function oxygenases take place in the smooth endoplasmic reticulum. [Courtesy of Dr. George Palade.]

Figure 23-4
Examples of reversible covalent modifications of proteins:
(1) phosphorylation, (2) adenylylation, and (3) methylation.

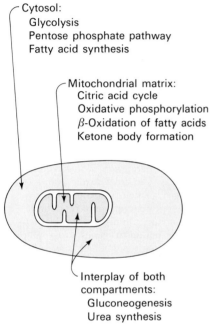

Cytosol:
Glycolysis
Pentose phosphate pathway
Fatty acid synthesis

Mitochondrial matrix:
Citric acid cycle
Oxidative phosphorylation
β-Oxidation of fatty acids
Ketone body formation

Interplay of both
compartments:
Gluconeogenesis
Urea synthesis

Figure 23-5
Compartmentation of the major
pathways of metabolism.

addition and removal of this modifying group is catalyzed by specific enzymes. Why is covalent modification used in addition to noncovalent allosteric control? Covalent modifications of key enzymes in metabolism are the final stage of amplifying cascades. Consequently, metabolic pathways can be rapidly switched on or off by very small triggering signals, as shown by the action of epinephrine in stimulating the breakdown of glycogen.

3. *Enzyme levels.* The amounts of enzymes, as well as their activities, are controlled. The rates of synthesis and degradation of some regulatory enzymes are subject to hormonal factors.

4. *Compartmentation.* The metabolic patterns of eucaryotic cells are markedly affected by the presence of compartments. Glycolysis, the pentose phosphate pathway, and fatty acid synthesis take place in the cytosol, whereas fatty acid oxidation, the citric acid cycle, and oxidative phosphorylation are carried out in mitochondria. Some processes, such as gluconeogenesis and urea synthesis, depend on the interplay of reactions that occur in both compartments. The fates of certain molecules depend on whether they are in the cytosol or in mitochondria, and so their flow across the inner mitochondrial membrane is often regulated. For example, fatty acids transported into mitochondria are rapidly degraded, in contrast with fatty acids in the cytosol, which are esterified or exported. Recall that long-chain fatty acids are transported into the mitochondrial matrix as esters of carnitine, a carrier that enables these molecules to traverse the inner mitochondrial membrane.

5. *Metabolic specializations of organs.* Regulation in higher eucaryotes is profoundly affected and enhanced by the existence of organs with different metabolic roles. These interactions will be discussed shortly.

MAJOR METABOLIC PATHWAYS AND CONTROL SITES

Let us review the roles of the major pathways of metabolism and the principal sites for their control:

1. *Glycolysis.* This sequence of reactions in the cytosol converts glucose into two molecules of pyruvate with the concomitant generation of two ATP and two NADH. The NAD$^+$ consumed in the reaction catalyzed by glyceraldehyde 3-phosphate dehydrogenase must be regenerated for glycolysis to proceed. Under anaerobic conditions, as in highly active skeletal muscle, this is accomplished by the reduction of pyruvate to lactate. Alternatively, under aerobic conditions, NAD$^+$ is regenerated by the transfer of electrons from NADH to O$_2$ through the electron-transport chain. Glycolysis serves two main purposes: it degrades glucose to generate ATP and it provides carbon skeletons for biosyntheses. The rate of conversion

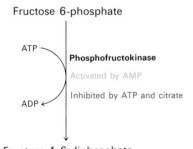

Fructose 6-phosphate

ATP

Phosphofructokinase

Activated by AMP

Inhibited by ATP and citrate

ADP

Fructose 1,6-diphosphate

Figure 23-6
Phosphofructokinase is the key enzyme
in the regulation of glycolysis.

of glucose into pyruvate is regulated to meet these dual needs. *Phosphofructokinase, which catalyzes the committed step in glycolysis, is the most important control site.* A high level of ATP inhibits phosphofructokinase. This inhibitory effect is enhanced by citrate and reversed by AMP. Thus, the rate of glycolysis depends on the need for ATP, as signalled by the ATP/AMP ratio, and on the need for building blocks, as signalled by the level of citrate.

2. *Citric acid cycle.* This final common pathway for the oxidation of fuel molecules—carbohydrates, amino acids, and fatty acids—takes place inside mitochondria. Most fuels enter the cycle as acetyl CoA. The complete oxidation of an acetyl unit generates one GTP, three NADH, and one $FADH_2$. The four pairs of electrons are then transferred to O_2 through the electron-transport chain, which results in the formation of a proton gradient that drives the synthesis of eleven ATP. NADH and $FADH_2$ are oxidized only if ADP is simultaneously phosphorylated to ATP. *This tight coupling, called respiratory control, ensures that the rate of the citric acid cycle matches the need for ATP.* An abundance of ATP also diminishes the activities of three enzymes in the cycle—citrate synthetase, isocitrate dehydrogenase, and α-ketoglutarate dehydrogenase. The citric acid cycle also has an anabolic role. It provides intermediates for biosyntheses, such as succinyl CoA, the source of part of the carbon skeleton of porphyrins.

3. *Pentose phosphate pathway.* This series of reactions, which takes place in the cytosol, has two purposes: the generation of NADPH for reductive biosyntheses and the formation of ribose 5-phosphate for the synthesis of nucleotides. Two NADPH are generated in the conversion of glucose 6-phosphate into ribose 5-phosphate. The dehydrogenation of glucose 6-phosphate is the committed step in this pathway. This reaction is controlled by the level of $NADP^+$, the electron acceptor. The extra phosphoryl group in NADPH is a tag that distinguishes it from NADH. This differentiation makes it possible to have both a high $NADPH/NADP^+$ ratio and a high $NAD^+/NADH$ ratio in the same compartment. Consequently, reductive biosyntheses and glycolysis can proceed simultaneously at a high rate.

4. *Gluconeogenesis.* Glucose can be synthesized by the liver and kidneys from noncarbohydrate precursors such as lactate, glycerol, and amino acids. The major entry point of this pathway is pyruvate, which is carboxylated to oxaloacetate in mitochondria. Oxaloacetate is then decarboxylated and phosphorylated in the cytosol to form phosphoenolpyruvate. The other distinctive reactions of gluconeogenesis are two hydrolytic steps that bypass the irreversible reactions of glycolysis. *Gluconeogenesis and glycolysis are usually reciprocally regulated so that one pathway is quiescent while the other is highly active.* For example, AMP inhibits and citrate activates fructose 1,6-diphosphatase, a key enzyme in gluconeogenesis, whereas these

Pasteur effect—
The inhibition of glycolysis by respiration, discovered by Louis Pasteur in studying fermentation by yeast.
The consumption of carbohydrate is about sevenfold lower under aerobic conditions than under anaerobic ones. The inhibition of phosphofructokinase by citrate and ATP accounts for much of the Pasteur effect.

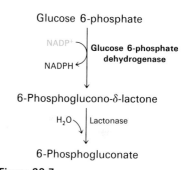

Figure 23-7
The dehydrogenation of glucose 6-phosphate is the committed step in the pentose phosphate pathway.

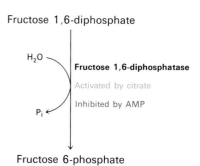

Figure 23-8
Fructose 1,6-diphosphatase is the key control site in gluconeogenesis.

molecules have opposite effects on phosphofructokinase, the pace-maker of glycolysis.

5. *Glycogen synthesis and degradation.* Glycogen, a readily mobilizable fuel store, is a branched polymer of glucose residues. The activated intermediate in its synthesis is UDP-glucose, which is formed from glucose 1-phosphate and UTP. Glycogen synthetase catalyzes the transfer of glucose from UDP-glucose to the terminal hydroxyl residue of a growing strand. Glycogen is degraded by a different pathway. Phosphorylase catalyzes the cleavage of glycogen by orthophosphate to give glucose 1-phosphate. *Glycogen synthesis and degradation are coordinately controlled by a hormone-triggered amplifying cascade so that the synthetase is inactive when phosphorylase is active, and vice versa.* These enzymes are regulated by phosphorylation and noncovalent allosteric interactions (p. 372).

6. *Fatty acid synthesis and degradation.* Fatty acids are synthesized in the cytosol by the addition of two-carbon units to a growing chain on an acyl carrier protein. Malonyl CoA, the activated intermediate, is formed by the carboxylation of acetyl CoA. Acetyl groups are carried from mitochondria to the cytosol by citrate. This shuttle provides part of the NADPH required for the reduction of the added acetyl unit, the rest coming from the pentose phosphate pathway. *Citrate stimulates acetyl CoA carboxylase, the enzyme catalyzing the committed step. When ATP and acetyl CoA are abundant, the level of citrate increases, which accelerates the rate of fatty acid synthesis.* Fatty acids are degraded by a different pathway in a different compartment. They are degraded to acetyl CoA in the mitochondrial matrix by β-oxidation. The acetyl CoA then enters the citric acid cycle if the supply of oxaloacetate is sufficient. Alternatively, acetyl CoA can give rise to ketone bodies. The $FADH_2$ and NADH formed in the β-oxidation pathway transfer their electrons to O_2 through the electron-transport chain. As for the citric acid cycle, β-oxidation can continue only if NAD^+ and FAD are regenerated. *Hence, the rate of fatty acid degradation is also coupled to the need for ATP.*

KEY JUNCTIONS: GLUCOSE 6-PHOSPHATE, PYRUVATE, AND ACETYL CoA

The factors governing the flow of molecules in metabolism can be further understood by examining three key crossroads: glucose 6-phosphate, pyruvate, and acetyl CoA. Each of these molecules has several contrasting fates:

1. *Glucose 6-phosphate.* Glucose entering a cell is rapidly phosphorylated to glucose 6-phosphate, which can be stored as glycogen, degraded by way of pyruvate, or converted into ribose 5-phosphate (Figure 23-10). Glycogen is formed when glucose 6-phosphate and ATP are abundant. In contrast, glucose 6-phosphate flows into the glycolytic pathway when ATP or carbon skeletons for

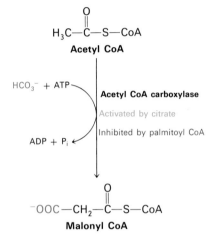

Figure 23-9
Acetyl CoA carboxylase is the key control site in fatty acid synthesis.

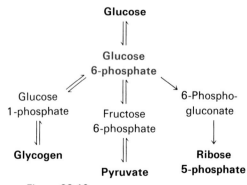

Figure 23-10
Metabolic fates of glucose 6-phosphate.

biosyntheses are required. Thus, the conversion of glucose 6-phosphate into pyruvate can be anabolic as well as catabolic. The third major fate of glucose 6-phosphate, to flow through the pentose phosphate pathway, provides NADPH for reductive biosyntheses and ribose 5-phosphate for the synthesis of nucleotides. The relative amounts of these two outputs can be adjusted over a very broad range by this highly flexible series of reactions, as discussed previously (p. 338). Glucose 6-phosphate can be formed by the mobilization of glycogen or it can be synthesized from pyruvate and other noncarbohydrate precursors by the gluconeogenic pathway. As will be discussed shortly, a low level of glucose in the blood stimulates both glycogenolysis and gluconeogenesis in the *liver and kidneys. These organs are distinctive in possessing glucose 6-phosphatase, which enables glucose to be released into the blood.*

2. *Pyruvate.* This three-carbon α-keto acid is another major metabolic junction (Figure 23-11). Pyruvate is derived primarily from

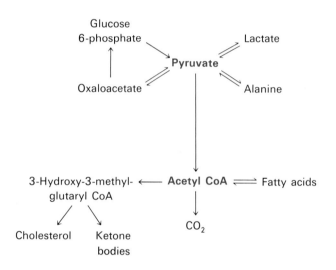

Figure 23-11
Metabolic fates of pyruvate and acetyl CoA.

glucose 6-phosphate, lactate, and alanine. Lactate is just the reduced form of pyruvate. The facile reduction of pyruvate catalyzed by lactate dehydrogenase serves to regenerate NAD^+, which enables glycolysis to proceed transiently under anaerobic conditions. Lactate formed in active tissues such as contracting muscle is subsequently oxidized back to pyruvate, primarily in the liver. The essence of this interconversion is that it buys time and shifts part of the metabolic burden of active muscle to the liver. Another readily reversible reaction in the cytosol is the transamination of pyruvate, an α-keto acid, to alanine, the corresponding amino acid. Several amino acids can enter the central metabolic pathways in this way. Conversely, several amino acids can be synthesized from carbohydrate precursors by this route. Thus, transamination is a major link between amino acid and carbohydrate metabolism. A third fate of pyruvate is its carboxylation to oxaloacetate inside mitochondria. This reaction and the subsequent conversion of oxaloacetate into phosphoenolpyruvate bypass an irreversible step of glycolysis and hence enable glucose to be synthesized from pyruvate. The carboxylation of pyruvate is also important for replenishing citric acid cycle intermediates. The activation of pyruvate carboxylase by acetyl CoA enhances the synthesis of oxaloacetate when the citric acid cycle is slowed by a paucity of this intermediate. On the other hand, oxaloacetate synthesized from pyruvate flows into the gluconeogenic pathway when the citric acid cycle is inhibited by an abundance of ATP. The fourth major fate of pyruvate is its oxidative decarboxylation to acetyl CoA. This irreversible reaction inside mitochondria is a decisive reaction in metabolism: it commits the carbon atoms of carbohydrates and amino acids to oxidation by the citric acid cycle or to the synthesis of lipids. The pyruvate dehydrogenase complex, which catalyzes this irreversible funneling, is stringently regulated by multiple allosteric interactions and covalent modifications. Pyruvate is rapidly converted into acetyl CoA only if ATP is needed or if two-carbon fragments are required for the synthesis of lipids.

3. *Acetyl CoA.* The major sources of this activated two-carbon unit are the oxidative decarboxylation of pyruvate and the β-oxidation of fatty acids (Figure 23-11). Acetyl CoA is also derived from ketogenic amino acids. The fate of acetyl CoA, in contrast with many molecules in metabolism, is quite restricted. The acetyl unit can be completely oxidized to CO_2 by the citric acid cycle. Alternatively, 3-hydroxy-3-methylglutaryl CoA can be formed from three molecules of acetyl CoA. This six-carbon unit is a precursor of cholesterol and of ketone bodies. A third major fate of acetyl CoA is its export to the cytosol in the form of citrate for the synthesis of fatty acids. It is important to stress again that acetyl CoA cannot be converted into pyruvate in mammals. Consequently, mammals are unable to convert lipid into carbohydrate.

METABOLIC PROFILES OF THE MAJOR ORGANS

The metabolic patterns of the brain, muscle, adipose tissue, and the liver are strikingly different. Let us consider how these organs differ in their use of fuels to meet their energy needs:

1. *Brain. Glucose is virtually the sole fuel for the human brain, except during prolonged starvation.* The brain lacks fuel stores and hence requires a continuous supply of glucose, which enters freely at all times. It consumes about 120 g daily, which corresponds to an energy input of about 420 kcal. The brain accounts for some 60% of the utilization of glucose by the whole body in the resting state. *During starvation, ketone bodies (acetoacetate and 3-hydroxybutyrate, its reduced counterpart) replace glucose as fuel for the brain.* Acetoacetate is activated by the transfer of CoA from succinyl CoA to give acetoacetyl CoA (Figure 23-12). Cleavage by thiolase then yields two mole-

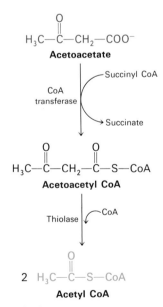

Figure 23-12
Entry of ketone bodies into the citric acid cycle.

cules of acetyl CoA, which enter the citric acid cycle. Fatty acids do not serve as fuel for the brain because they are bound to albumin and so they do not traverse the blood-brain barrier. In essence, *ketone bodies are transportable equivalents of fatty acids.* As will be discussed shortly, the shift in fuel usage from glucose to ketone bodies has the important effect of minimizing protein breakdown during starvation.

2. *Muscle. The major fuels for muscle are glucose, fatty acids, and ketone bodies.* Muscle differs from the brain in having a large store of glycogen (1,200 kcal). In fact, about three-fourths of all the glycogen in the body is stored in muscle (Table 23-1). The glycogen content of

Table 23-1
Fuel reserves in a typical 70-kg man

Organ	Available energy (kcal)		
	Glucose or glycogen	Triacylglycerols	Mobilizable proteins
Blood	60	45	0
Liver	400	450	400
Brain	8	0	0
Muscle	1,200	450	24,000
Adipose tissue	80	135,000	40

Source: After G. F. Cahill, Jr., *Clin. Endocrinol. Metab.* 5(1976):398.

muscle after a meal can be as high as 1%. This glycogen is readily converted into glucose 6-phosphate for use within muscle cells. Muscle, like the brain, lacks glucose 6-phosphatase, and so it does not export glucose. *Rather, muscle retains glucose, its preferred fuel for bursts of activity.* In actively contracting skeletal muscle, the rate of glycolysis far exceeds that of the citric acid cycle. Much of the

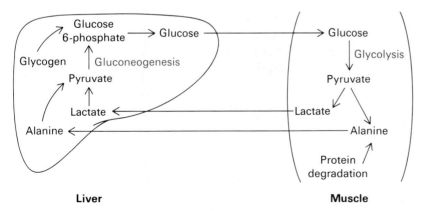

Liver　　　　　　　　　　　　　　　**Muscle**

Figure 23-13
Metabolic interchanges between muscle and the liver.

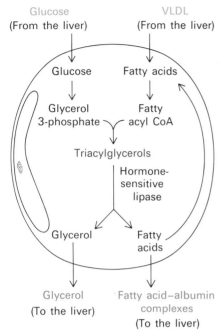

Glucose   VLDL
(From the liver) (From the liver)

Figure 23-14
Synthesis and degradation of triacyl-glycerols by adipose tissue. Fatty acids are delivered to adipose cells in the form of very low density lipoproteins (VLDL).

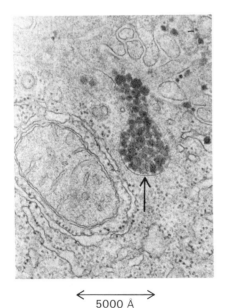

← 5000 Å →

Figure 23-15
Electron micrograph of part of a liver cell actively engaged in the synthesis and secretion of very low density lipo-protein (VLDL) particles. The arrow points to a vesicle that is releasing its content of VLDL particles. [Courtesy of Dr. George Palade.]

pyruvate formed under these conditions is reduced to lactate, which flows to the liver, where it is converted into glucose. These interchanges, known as the Cori cycle (p. 353), shift part of the metabolic burden of muscle to the liver. In addition, a large amount of alanine is formed in active muscle by the transamination of pyruvate. Alanine, like lactate, can be converted into glucose by the liver. The metabolic pattern of resting muscle is quite different. *In resting muscle, fatty acids are the major fuel.* Ketone bodies can also serve as fuel for heart muscle. In fact, heart muscle consumes acetoacetate in preference to glucose.

3. *Adipose tissue.* The triacylglycerols stored in adipose tissue are an enormous store of metabolic fuel. Their energy content is 135,000 kcal in a typical 70-kg man. Adipose tissue is specialized for the esterification of fatty acids and for their release from triacylglycerols. In humans, the liver is the major site of fatty acid synthesis, and so the principal biosynthetic task of adipose tissue is to activate these fatty acids and transfer the resulting CoA derivatives to glycerol. Glycerol 3-phosphate, a key intermediate in this biosynthesis (p. 458), comes from the reduction of dihydroxyacetone phosphate, which is formed from glucose by the glycolytic pathway. Adipose cells are unable to phosphorylate endogeneous glycerol because they lack the kinase. Hence, *adipose cells need glucose for the synthesis of triacylglycerols.* Triacylglycerols are hydrolyzed to fatty acids and glycerol by lipases. The release of the first fatty acid from a triacylglycerol, the rate-limiting step, is catalyzed by a hormone-sensitive lipase that is reversibly phosphorylated. As in glycogen metabolism, cyclic AMP is an intermediate in this hormone-triggered amplifying cascade. Triacylglycerols in adipose cells are continually being hydrolyzed and resynthesized. Glycerol derived from their hydrolysis is exported to the liver. Most of the fatty acids formed on hydrolysis are reesterified if glycerol 3-phosphate is abundant. In contrast, they are released into the plasma if glycerol 3-phosphate is scarce because of a paucity of glucose. Thus, the glucose level inside adipose cells is a major factor in determining whether fatty acids are released into the blood.

4. *Liver.* The metabolic activities of the liver are essential for providing fuel to the brain, muscle, and other peripheral organs. Most compounds absorbed by the intestine pass through the liver, which enables it to regulate the level of many metabolites in the blood. The liver can take up large amounts of glucose and convert it into glycogen. As much as 400 kcal can be stored in this way. The liver can release glucose into the blood by breaking down its store of glycogen and by carrying out gluconeogenesis. The main precursors of glucose are lactate and alanine from muscle, glycerol from adipose tissue, and glucogenic amino acids from the diet. The liver also plays a central role in the regulation of lipid metabolism. When fuels are abundant, fatty acids are synthesized by the liver, esterified, and then secreted into the blood in the form of very low density lipoprotein (VLDL) (Figure 23-15). This plasma lipoprotein is

the main source of fatty acids used by adipose tissue to synthesize triacylglycerols. However, in the fasting state, the liver converts fatty acids into ketone bodies. How does a liver cell choose between these contrasting paths? The choice depends on whether the fatty acid is to enter the mitochondrial matrix. Recall that long-chain fatty acids traverse the inner mitochondrial membrane only if they are esterified to carnitine (p. 388). The enzyme catalyzing the formation of acyl carnitine on the outer side of this membrane is inhibited by malonyl CoA, the committed intermediate in the synthesis of fatty acids. *Thus, when long-chain fatty acids are being synthesized, they are prevented from entering the mitochondrial matrix, the compartment of β-oxidation and ketone-body formation. Instead, these fatty acids become incorporated into triacylglycerols and phospholipids.* In contrast, the level of malonyl CoA is low when fuels are scarce. Under these conditions, fatty acids liberated from adipose tissues enter the mitochondrial matrix for conversion into ketone bodies. How does the liver meet its own energy needs? It prefers keto acids derived from the degradation of amino acids to glucose as its own fuel. In fact, the main purpose of glycolysis in the liver is to form building blocks for biosyntheses. Furthermore, the liver cannot use acetoacetate as a fuel because it lacks the transferase to activate it to acetyl CoA. Thus, the liver eschews the fuels it exports to muscle and the brain—an altruistic organ indeed!

Fatty acyl carnitine

HORMONAL REGULATORS OF FUEL METABOLISM

Hormones play a key role in integrating metabolism. In particular, *insulin, glucagon, epinephrine, and norepinephrine* have large effects on the storage and mobilization of fuels and on related facets of metabolism:

1. *Insulin.* This 5.8-kdal-protein hormone (pp. 21 and 846), which is secreted by the β-cells of the pancreas, is the most important regulator of fuel metabolism. *In essence, insulin signals the fed state: it stimulates the storage of fuels and the synthesis of proteins in a variety of ways.* Glycogen synthesis in both muscle and liver is stimulated, whereas gluconeogenesis by the liver is suppressed. Insulin accelerates glycolysis in the liver, which in turn increases the synthesis of fatty acids. The entry of glucose into muscle and adipose cells is promoted by insulin. The abundance of fatty acids and glucose in adipose tissue results in the synthesis and storage of triacylglycerols. The action of insulin also extends to amino acid and protein metabolism. Insulin promotes the uptake of branched-chain amino acids (valine, leucine, and isoleucine) by muscle, which favors a building up of muscle protein. Indeed, insulin has a general stimulating effect on protein synthesis. In addition, it inhibits the intracellular degradation of proteins.

2. *Glucagon.* This 3.5-kdal-polypeptide hormone (p. 367) is secreted by the α-cells of the pancreas in response to a *low blood sugar level. The main target organ of glucagon is the liver.* Glucagon stimulates

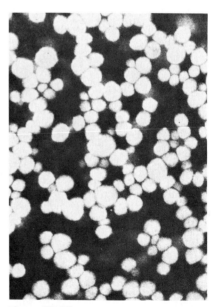

Figure 23-16
Electron micrograph of very low density lipoprotein particles (VLDL). These particles, which have diameters ranging between 300 and 800 Å, carry triacylglycerols from the liver to adipose tissue. [Courtesy of Dr. Robert Mahley.]

glycogen breakdown and inhibits glycogen synthesis by triggering the cAMP-mediated cascade that leads to the phosphorylation of phosphorylase and glycogen synthetase (p. 372). Glucagon also inhibits fatty acid synthesis by diminishing the production of pyruvate and by lowering the activity of acetyl CoA carboxylase. In addition, gluconeogenesis is enhanced by glucagon. The net result of these actions of glucagon is to *markedly increase the release of glucose by the liver.* In addition, glucagon raises the level of cyclic AMP in adipose cells, which promotes the breakdown of triacylglycerols.

3. *Epinephrine and norepinephrine.* These catecholamine hormones are secreted by the adrenal medulla and sympathetic nerve endings in response to a *low blood glucose level.* Like glucagon, they stimulate the mobilization of glycogen and triacylglycerols by triggering the cAMP-mediated cascade. They differ from glucagon in that their glycogenolytic effect is greater in muscle than in the liver. Another action of the catecholamines is to inhibit the uptake of glucose by muscle. Instead, fatty acids released from adipose tissue are used as fuel. Epinephrine also stimulates the secretion of glucagon and inhibits the secretion of insulin. *Thus, the catecholamines increase the amount of glucose released into the blood by the liver and decrease the utilization of glucose by muscle.*

THE BLOOD GLUCOSE LEVEL IS BUFFERED BY THE LIVER

The blood glucose level in a typical person after an overnight fast is 80 mg/100 ml (4.4 mM). The blood glucose level during the day normally ranges from about 80 mg/100 ml before meals to about 120 mg/100 ml after meals. How is the blood glucose level maintained relatively constant despite large changes in the input and utilization of glucose? The major control elements have already been discussed, and so they need only be brought together here. The blood glucose level is controlled primarily by the action of the liver, which can take up or release large amounts of glucose in response to hormonal signals and the level of glucose itself (Figure 23-17). After a carbohydrate-containing meal, the increased concentration of glucose in the blood leads to a higher level of glucose 6-phosphate in the liver because only then do the catalytic sites of glucokinase become filled with glucose. Recall that glucokinase, in contrast with hexokinase, has a high K_M for glucose ($\sim$10 mM, compared with a fasting blood glucose level of 4.4 mM) and is not inhibited by glucose 6-phosphate. *Consequently, glucose 6-phosphate is formed more rapidly by the liver as the blood glucose level rises.* The fate of glucose 6-phosphate is then largely controlled by the opposed effects of glucagon and insulin. Glucagon triggers the cAMP-mediated cascade (p. 372) that results in the breakdown of glycogen, whereas insulin antagonizes this action. *A high blood glucose level leads to a decreased secretion of glucagon and an increased secretion of insulin by the pancreas. Consequently, glycogen is rapidly synthesized when the blood glucose*

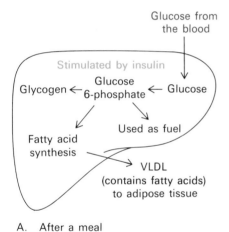

A. After a meal

B. After an overnight fast

Figure 23-17
Control of the blood glucose level by the liver.

level is high. These hormonal effects on glycogen synthesis and storage are reinforced by a direct action of glucose itself. As discussed in an earlier chapter (p. 374), *phosphorylase a is a glucose sensor in addition to being the enzyme that cleaves glycogen.* When the glucose level is high, the binding of glucose to phosphorylase *a* renders it susceptible to the action of a phosphatase that converts it into phosphorylase *b*, which is inactive in degrading glycogen. This conversion also releases the phosphatase, which enables it to activate glycogen synthetase. *Thus, glucose allosterically shifts the glycogen system from a degradative to a synthetic mode.*

The high insulin level in the fed state also promotes *the entry of glucose into muscle and adipose tissue.* The synthesis of glycogen by muscle as well as by liver is stimulated by insulin. In fact, muscle can store about three times as much glycogen as the liver because of its larger mass. The entry of glucose into adipose tissue provides glycerol 3-phosphate for the synthesis of triacylglycerols.

The blood glucose level begins to drop several hours after a meal, which leads to decreased insulin secretion and increased glucagon secretion. The effects just described are then reversed. The activation of the cAMP-mediated cascade results in a higher level of phosphorylase *a* and a lower level of glycogen synthetase *a*. The effect of hormones on this cascade is reinforced by the diminished binding of glucose to phosphorylase *a*, which makes it less susceptible to the hydrolytic action of the phosphatase. Instead, the phosphatase remains bound to phosphorylase *a* and so the synthetase stays in the inactive phosphorylated form. Consequently, there is a rapid mobilization of glycogen. The large amount of glucose formed by the hydrolysis of glucose 6-phosphate derived from glycogen is then released from the liver into the blood. The diminished utilization of glucose by muscle and adipose tissue also contributes to the maintenance of the blood glucose level. The entry of glucose into muscle and adipose tissue decreases because of the low insulin level. Both muscle and liver use fatty acids as fuel when the blood glucose level drops. *Thus, the blood glucose level is kept higher than about 80 mg/100 ml by three major factors: the mobilization of glycogen and release of glucose by the liver, the release of fatty acids by adipose tissue, and the shift in the fuel used from glucose to fatty acids by muscle and the liver.*

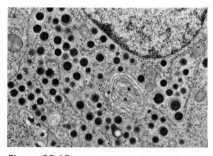

Figure 23-18
Electron micrograph of granules containing glucagon in an α-cell of the pancreas. [Courtesy of Dr. Arthur Like.]

METABOLIC ADAPTATIONS IN PROLONGED STARVATION MINIMIZE PROTEIN DEGRADATION

We turn now to metabolic adaptations in prolonged starvation. A typical 70-kg man has fuel reserves before the onset of starvation of some 1,600 kcal in glycogen, 24,000 kcal in mobilizable protein, and 135,000 kcal in triacylglycerols (Table 23-1). The energy need for a 24-hour period ranges from about 1,600 kcal in the basal state to 6,000 kcal, depending on the extent of activity. Thus, stored fuels suffice to meet caloric needs in starvation for one to three months. However, the carbohydrate reserves are exhausted in only a day.

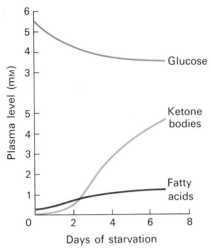

Figure 23-19
The plasma levels of fatty acids and ketone bodies increase in starvation, whereas that of glucose decreases.

Even under these conditions the blood glucose level is maintained above 50 mg/100 ml. The brain cannot tolerate appreciably lower glucose levels for even short periods. Hence, *the first priority of metabolism in starvation is to provide sufficient glucose to the brain and other tissues (e.g., red blood cells) that are absolutely dependent on this fuel.* However, precursors of glucose are not highly abundant. Most of the energy is stored in the fatty acyl moieties of triacylglycerols. Recall that fatty acids cannot be converted into glucose because acetyl CoA cannot be transformed into pyruvate (p. 394). The glycerol moiety of triacylglycerols can be converted into glucose but only a limited amount is available. The only other potential source of glucose is amino acids derived from the breakdown of proteins. Muscle is the largest potential source of amino acids during starvation. However, survival for most people depends on being able to move about, which requires a large muscle mass. *Thus, the second priority of metabolism in starvation is to preserve protein. This is accomplished by shifting the fuel being used from glucose to fatty acids and ketone bodies* (Figure 23-19).

The metabolic changes during the first day of starvation are like those after an overnight fast. The low blood sugar level leads to decreased secretion of insulin and increased secretion of glucagon. *The dominant metabolic processes are the mobilization of triacylglycerols in adipose tissue and gluconeogenesis by the liver. The liver obtains energy for its own needs by oxidizing fatty acids released from adipose tissue.* The concentrations of acetyl CoA and citrate consequently increase, which switches off glycolysis. The uptake of glucose by muscle is markedly diminished because of the low insulin level, whereas fatty acids enter freely. Consequently, *muscle also shifts from glucose to fatty acids for fuel.* The β-oxidation of fatty acids by muscle halts the conversion of pyruvate into acetyl CoA. Hence, pyruvate, lactate, and alanine are exported to the liver for conversion into glucose. Proteolysis of muscle protein provides some of these three-carbon precursors of glucose. Glycerol derived from the cleavage of triacylglycerols is another raw material for the synthesis of glucose by the liver.

The most important change after about three days of starvation is that large amounts of acetoacetate and 3-hydroxybutyrate (ketone bodies) are formed by the liver (Figure 23-20). Their synthesis from acetyl CoA increases markedly because the citric acid cycle is unable to oxidize all of the acetyl units generated by the degradation of fatty acids. Gluconeogenesis depletes the supply of oxaloacetate, which is essential for the entry of acetyl CoA into the citric acid cycle. Consequently, the liver produces large quantities of ketone bodies, which are released into the blood. At this time, the brain begins to consume appreciable amounts of acetoacetate in place of glucose. After three days of starvation, about a third of the energy needs of the brain are met by ketone bodies (Table 23-2). The heart also uses ketone bodies as fuel. These changes in fuel usage are referred to as *ketosis*.

Table 23-2
Fuel metabolism in starvation

Fuel exchanges and consumption	Amount formed or consumed in 24 hours (grams)	
	3rd day	40th day
Fuel use by the brain		
Glucose	100	40
Ketone bodies	50	100
All other use of glucose	50	40
Fuel mobilization		
Adipose-tissue lipolysis	180	180
Muscle-protein degradation	75	20
Fuel output of the liver		
Glucose	150	80
Ketone bodies	150	150

After several weeks of starvation, ketone bodies become the major fuel of the brain (Table 23-2). Only 40 grams of glucose is needed per day for the brain, compared with about 120 grams in the first day of starvation. *The effective conversion of fatty acids into ketone bodies by the liver and their use by the brain markedly diminishes the need for glucose. Hence, less muscle is degraded than in the first days of starvation.* The breakdown of 20 grams of muscle compared with 75 grams early in starvation is most important for survival. The duration of starvation compatible with life is mainly determined by the size of the triacylglycerol depot.

HUGE FAT DEPOTS ENABLE MIGRATORY BIRDS TO FLY LONG DISTANCES

Migratory birds provide another striking illustration of the biological value of triacylglycerols. Some small land birds fly in the autumn from their breeding grounds in New England to their wintering grounds in the West Indies and then back again in the spring. They fly nonstop over water for a distance of some 2400 kilometers. These birds sustain a velocity of 40 kilometers per hour for 60 hours. This remarkable feat is made possible by the existence of very large fat depots that are efficiently mobilized during the long flight. Birds that migrate for short distances or not at all are quite lean. They have fat indices of about 0.3; the fat index is defined as the ratio of the dry weight of total body fat to that of nonfat. In contrast, long-range migrants become moderately obese in preparing to travel over land and then become highly obese just before they begin their

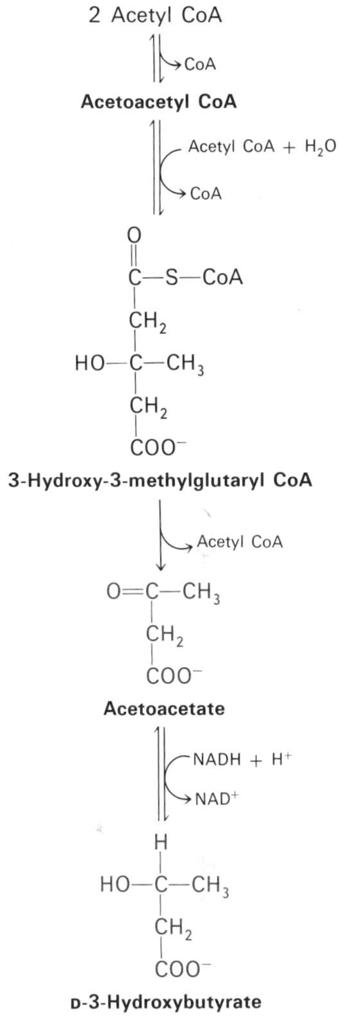

Figure 23-20
Synthesis of ketone bodies by the liver.

A ruby-throated hummingbird.

overseas journey. Indeed, their fat index is then close to 3. In the ruby-throated hummingbird, about 0.15 gram of triacylglycerols is accumulated each day per gram of body weight. The comparable weight gain in a human would be 10 kilograms per day. The accumulated fat in migratory birds is stored under the skin, in the abdominal cavity, in muscle, and in the liver. About two-thirds of this fat store is consumed in the long flight over water. The transition to the use of fatty acids and ketone bodies as fuels must be very rapid because almost no protein is degraded during the 60-hour flight. Oxidation of fat also provides these birds with the water needed to make up for losses through respiratory passages. The high efficiency of triacylglycerols as storage fuel is also noteworthy. Recall that triacylglycerols store six times as much energy as does glycogen because they are anhydrous and more highly reduced (p. 385). A migratory bird carrying the same amount of fuel in the form of glycogen would never get off the ground!

SUMMARY

The basic strategy of metabolism is to form ATP, reducing power, and building-blocks for biosyntheses. This complex network of reactions is controlled by allosteric interactions, reversible covalent modifications, changes in the amounts of enzymes, compartmentation, and interactions between metabolically distinctive organs. The enzyme catalyzing the committed step in a pathway is usually the most important control site, as exemplified by phosphofructokinase in glycolysis and acetyl CoA carboxylase in fatty acid synthesis. Opposing pathways such as gluconeogenesis and glycolysis are reciprocally regulated so that one pathway is usually quiescent while the other is highly active. Another pair of opposed reaction sequences, glycogen synthesis and degradation, are coordinately controlled by a hormone-triggered amplifying cascade that leads to the phosphorylation of glycogen synthetase and phosphorylase. The role of compartmentation in control is illustrated by the contrasting fates of fatty acids in the cytosol and the mitochondrial matrix.

The metabolic patterns of the brain, muscle, adipose tissue, and the liver are very different. Glucose is essentially the sole fuel for the brain in a well-fed person. During starvation, ketone bodies (acetoacetate and 3-hydroxybutyrate) become the predominant fuel of the brain. Muscle uses glucose, fatty acids, and ketone bodies as fuel and it synthesizes glycogen as a fuel reserve for its own needs. Adipose tissue is specialized for the synthesis, storage, and mobilization of triacylglycerols. The diverse metabolic activities of the liver support the other organs. The liver can rapidly mobilize glycogen and carry out gluconeogenesis to meet the glucose needs of other organs. The liver plays a central role in the regulation of lipid metabolism. When fuels are abundant, fatty acids are synthesized, esterified, and sent from the liver to adipose tissue in the form of very low density lipoprotein (VLDL). In the fasting state, however, fatty acids are converted into ketone bodies by the liver. The activities of these organs are integrated by hormones. Insulin signals the fed state: it stimulates the formation of glycogen and triacylglycerols and the synthesis of proteins. In contrast, glucagon signals a low blood glucose level: it stimulates glycogen breakdown and gluconeogenesis by the liver and triacylglycerol hydrolysis by adipose tissue. The effects of epinephrine and norepinephrine on fuels are like those of glucagon, except that muscle rather than the liver is their primary target.

The blood glucose level in a well-fed person typically ranges from 80 mg/100 ml to 120 mg/100 ml. After a meal, the rise in the blood glucose level leads to increased secretion of insulin and decreased secretion of glucagon. Consequently, glycogen is synthesized in muscle and the liver. The increased entry of glucose into adipose tissue provides glycerol 3-phosphate for the synthesis of triacylglycerols. These effects are reversed when the blood glucose level drops several hours later. Glucose is then formed by the degradation of glycogen and by the gluconeogenic pathway, and fatty acids are released by the hydrolysis of triacylglycerols. Liver and muscle then use fatty acids instead of glucose to meet their own energy needs so that glucose is conserved for use by the brain and other tissues that are highly dependent on it. The metabolic adaptations in starvation are designed to minimize protein degradation. Large amounts of ketone bodies are formed by the liver from fatty acids and released into the blood within a few days after the onset of starvation. After several weeks of starvation, ketone bodies become the major fuel of the brain. The diminished need for glucose decreases the rate of muscle breakdown, and so the likelihood of survival is enhanced.

SELECTED READINGS

WHERE TO START

Newsholme, E. A., and Start, C., 1973. *Regulation in Metabolism.* Wiley. [An excellent account of the control of carbohydrate and fat metabolism.]

Stalmans, W., 1976. The role of the liver in the homeostasis of blood glucose. *Curr. Top. Cell Regul.* 11:51–97.

Cahill, G. F., Jr., 1976. Starvation in man. *Clin. Endocrinol. Metab.* 5:397–415.

Newsholme, E. A., 1980. A possible metabolic basis for the control of body weight. *New Eng. J. Med.* 302:400–405. [Presentation of the hypothesis that obesity results from diminished cycling of substrate through pairs of opposed reactions, such as the ones catalyzed by phosphofructokinase and fructose 1,6-diphosphatase.]

BOOKS

Bondy, P. K., and Rosenberg, L. E., (eds.), 1980. *Metabolic Control and Disease.* Saunders.

Lundquist, F., and Tygstrup, N., (eds.), 1974. *Regulation of Hepatic Metabolism.* Munksgaard. [Contains articles on the integrative role of the liver.]

Howald, H., and Poortmans, J. R., (eds.), 1975. *Metabolic Adaptation to Prolonged Physical Exercise.* Birkhauser Verlag, Basel.

Atkinson, D. E., 1977. *Cellular Energy Metabolism and Its Regulation.* Academic Press.

CARBOHYDRATE METABOLISM

Felig, P., 1980. Disorders of carbohydrate metabolism. *In* Bondy, P. K., and Rosenberg, L. E., (eds.), *Metabolic Control and Disease.* Saunders. [An excellent review of the control of carbohydrate metabolism and its relationship to protein and lipid metabolism.]

Esmann, V., (ed.), 1978. *Regulatory Mechanisms of Carbohydrate Metabolism.* Pergamon Press.

Cahill, G. F., Jr., and Owen, O. E., 1968. Some observations on carbohydrate metabolism in man. *In* Dickens, F., Randle, P. J., and Whelan, W. J., (eds.), *Carbohydrate Metabolism and Its Disorders,* vol. 1, pp. 497–522. Academic Press.

GLUCONEOGENESIS

Snell, K., 1979. Alanine as a gluconeogenic carrier. *Trends Biochem. Sci.* 4:124–128.

Hanson, R., and Mehlman, M., (ed.), 1976. *Gluconeogenesis.* Wiley.

AMINO ACID METABOLISM

Felig, P., 1975. Amino acid metabolism in man. *Ann. Rev. Biochem.* 44:933–955.

Dice, J. F., Walker, C. D., Byrne, B., and Cardiel, A. 1978. General characteristics of protein degradation in diabetes and starvation. *Proc. Nat. Acad. Sci.* 75:2093–2097.

LIPID METABOLISM

McGarry, J. D., and Foster, D. W., 1980. Regulation of hepatic fatty acid oxidation and ketone body production. *Ann. Rev. Biochem.* 49:395–420.

McGarry, J. D., Takabayashi, Y., and Foster, D. W., 1978. The role of malonyl CoA in the coordination of fatty acid synthesis and oxidation in isolated rat hepatocytes. *J. Biol. Chem.* 253:8294–8300.

Williamson, D. H., 1979. Recent developments in ketone-body metabolism. *Biochem. Soc. Trans.* 7:1313–1321.

Odum, E. P., 1965. Adipose tissue in migratory birds. *In* Renold, A. E., and Cahill, G. F., (eds.), *Handbook of Physiology,* Section 5, *Adipose Tissue,* pp. 37–43. American Physiological Society.

Taylor, S. I., and Jungas, R. L., 1974. Regulation of lipogenesis in adipose tissue. *Arch. Biochem. Biophys.* 164:12–19.

McGarry, J. D., and Foster, D. W., 1977. Hormonal control of ketogenesis. *Arch. Intern. Med.* 137:495–501.

Grey, N. J., Karl, I., and Kipnis, D. M., 1975. Physiologic mechanisms in the development of starvation ketosis in man. *Diabetes* 24:10–16.

SPECTROSCOPY OF CELLS AND TISSUES

Shulman, R. G., Brown, T. R., Ugurbil, K., Ogawa, S., Cohen, S. M., den Hollander, J. A., 1979. Cellular applications of ^{31}P and ^{13}C nuclear magnetic resonance. *Science* 205:160–166. [Review of a highly informative new method for following metabolic events in intact cells and tissues.]

McLaughlin, A. C., Takeda, H., and Chance, B., 1979. Rapid ATP assays in perfused mouse liver by ^{31}P NMR. *Proc. Nat. Acad. Sci.* 76:5445–5449. [A promising approach for monitoring the metabolism of an intact organ.]

On the facing page: Model of a pair of deoxyribonucleotide units of a DNA double helix. Adenine is hydrogen bonded to thymine in this base pair.

PART IV

INFORMATION

storage, transmission, and expression of genetic information

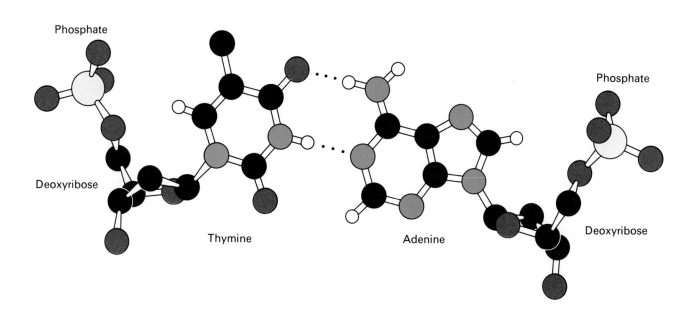

Phosphate

Phosphate

Deoxyribose

Deoxyribose

Thymine

Adenine

DNA: GENETIC ROLE, STRUCTURE, AND REPLICATION

Deoxyribonucleic acid (DNA) is the molecule of heredity. DNA is a very long, threadlike macromolecule made up of a large number of deoxyribonucleotides. *The purine and pyrimidine bases of DNA carry genetic information, whereas the sugar and phosphate groups perform a structural role.* This chapter deals with the structure, genetic role, and replication of DNA. The emphasis here is on DNA in procaryotes because they are simpler and exemplify principles that are common to all organisms. Eucaryotic chromosomes will be considered in Chapter 29.

COVALENT STRUCTURE AND NOMENCLATURE OF DNA

The *backbone* of DNA, which is constant throughout the molecule, consists of deoxyriboses linked by phosphodiester bridges. The 3'-hydroxyl of the sugar moiety of one deoxyribonucleotide is joined to the 5'-hydroxyl of the adjacent sugar by a phosphodiester bond.

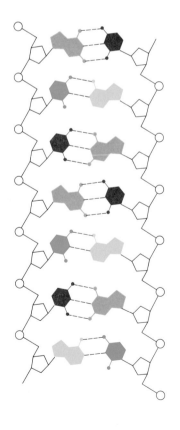

Figure 24-1
Schematic diagram of the structure of DNA. The sugar-phosphate backbone is shown in black, whereas the purine and pyrimidine bases are shown in color. [After A. Kornberg. The synthesis of DNA. Copyright © 1968 by Scientific American, Inc. All rights reserved.]

Adenine
(A)

Guanine
(G)

Thymine
(T)

Cytosine
(C)

The *variable* part of DNA is its *sequence of bases*. DNA contains four kinds of bases. The two purines are *adenine* (A) and *guanine* (G). The two pyrimidines are *thymine* (T) and *cytosine* (C). The structure of a DNA chain is shown in Figure 24-2.

The structure of a DNA chain can be concisely represented in the following way. The symbols for the four principal deoxyribonucleosides are

A   G   C   T

The bold line refers to the sugar, whereas A, G, C, and T represent the bases. The Ⓟ within the diagonal line in the diagram below denotes a phosphodiester bond. This diagonal line joins the end of one bold line and the middle of another. These junctions refer to the 5′-OH and 3′-OH, respectively. In this example, the symbol Ⓟ indicates that deoxyadenylate is linked to deoxycytidine by a phosphodiester bridge. The 3′-OH of deoxyadenylate is joined to the 5′-OH of deoxycytidine by a phosphoryl group.

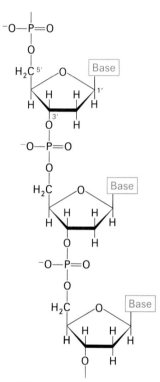

Figure 24-2
Structure of part of a DNA chain.

Now suppose that deoxyguanylate becomes linked to the deoxycytidine unit in this dinucleotide. The resulting trinucleotide can be represented by

An abbreviated notation for this trinucleotide is pApCpG or ACG.

A DNA chain has polarity. One end of the chain has a 5′-OH group and the other a 3′-OH group that is not linked to another nucleotide. By convention, the symbol ACG means that the unlinked

5′-OH group is on deoxyadenosine, whereas the unlinked 3′-OH group is on deoxyguanosine. Thus, *the base sequence is written in the 5′ → 3′ direction.* Recall that the amino acid sequence of a protein is written in the amino → carboxyl direction. Note that ACG and GCA refer to different compounds, just as Glu-Phe-Ala differs from Ala-Phe-Glu.

TRANSFORMATION OF PNEUMOCOCCI BY DNA REVEALED THAT GENES ARE MADE OF DNA

The pneumococcus bacterium played an important part in the discovery of the genetic role of DNA. A pneumococcus is normally surrounded by a slimy, glistening polysaccharide capsule. This outer layer is essential for the pathogenicity of the bacterium, which causes pneumonia in humans and other susceptible mammals. Mutants devoid of a polysaccharide coat are not pathogenic. The normal bacterium is referred to as the S form (because it forms smooth colonies), whereas mutants without capsules are called R forms (because they form rough colonies). One group of R mutants lacks the dehydrogenase enzyme that converts UDP-glucose into UDP-glucuronate. This enzyme is required for the synthesis of the capsular polysaccharide, which in one type of pneumococcus consists of an alternating sequence of glucose and glucuronate residues.

| Glucuronate unit | Glucose unit | Glucuronate unit | Glucose unit |

In 1928, Fred Griffith discovered that a nonpathogenic R mutant could be *transformed* into the pathogenic S form in the following way. He injected mice with a mixture of live R and heat-killed S pneumococci. The striking finding was that this mixture was lethal to the mice, whereas either live R or heat-killed S pneumococci alone were not. The blood of the dead mice contained live S pneumococci. Thus, the heat-killed S pneumococci had somehow transformed live R pneumococci into live S pneumococci. This change was permanent: the transformed pneumococci yielded pathogenic progeny of the S form. It was then found that this R → S transformation can occur in vitro. Some of the cells in a growing culture of the R form were transformed into the S form by the addition of a *cell-free extract* of heat-killed S pneumococci. This finding set the stage for the elucidation of the chemical nature of the "transforming principle."

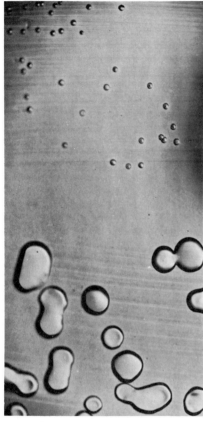

Figure 24-3
Transformation of nonpathogenic R pneumococci (small colonies) to pathogenic S pneumococci (large glistening colonies) by DNA from heat-killed S pneumococci. [From O. T. Avery, C. M. MacLeod, and M. McCarty. *J. Exp. Med.* 79(1944):158.]

The cell-free extract of heat-killed S pneumococci was fractionated and the transforming activity of its components assayed (Figure 24-3). In 1944, Oswald Avery, Colin MacLeod, and Maclyn McCarty published their discovery that *"a nucleic acid of the deoxyribose type is the fundamental unit of the transforming principle of Pneumococcus Type III."* The experimental basis for their conclusion was: (1) the purified, highly active transforming principle gave an elemental chemical analysis that agreed closely with that calculated for DNA; (2) the optical, ultracentrifugal, diffusive, and electrophoretic properties of the purified material were like those of DNA; (3) there was no loss of transforming activity upon extraction of protein or lipid; (4) trypsin and chymotrypsin did not affect transforming activity; (5) ribonuclease (known to digest ribonucleic acid) had no effect on the transforming principle; and (6) in contrast, transforming activity was lost following the addition of deoxyribonuclease.

This work is a landmark in the development of biochemistry. Until 1944, it was generally assumed that chromosomal proteins carry genetic information and that DNA plays a secondary role. This prevailing view was decisively shattered by the rigorously documented finding that *purified DNA has genetic specificity.* Avery gave a vivid description of this research and of its implications in a letter that he wrote in 1943 to his brother, a medical microbiologist at another university (Figure 24-4).

Further support for the genetic role of DNA came from the studies of a virus that infects *E. coli.* The T2 bacteriophage consists of a core of DNA surrounded by a protein coat. In 1951, Roger Herriott suggested that "the virus may act like a little hypodermic needle full of transforming principles; the virus as such never enters the cell; only the tail contacts the host and perhaps enzymatically cuts a small hole through the outer membrane and then the nucleic acid of the virus head flows into the cell." This idea was tested by Alfred Hershey and Martha Chase in the following way. Phage DNA was labeled with the radioisotope ^{32}P, whereas the protein coat was labeled with ^{35}S. These labels are highly specific because DNA does not contain sulfur and the protein coat is devoid of phosphorus. A sample of an *E. coli* culture was infected with labeled phage, which became attached to the bacteria during a short incubation period. The suspension was spun for a few minutes in a Waring Blendor at 10,000 rpm. This treatment subjected the phage-infected cells to very strong shearing forces, which severed the connections between the viruses and bacteria. The resulting suspension was centrifuged at a speed sufficient to throw the bacteria to the bottom of the tube. Thus, the pellet contained the infected bacteria, whereas the supernatant contained smaller particles. These fractions were analyzed for ^{32}P and ^{35}S to determine the location of the phage DNA and the

For the past two years, first with MacLeod and now with Dr. McCarty, I have been trying to find out what is the chemical nature of the substance in the bacterial extract which induces this specific change. The crude extract of Type III is full of capsular polysaccharide, C (somatic) carbohydrate, nucleoproteins, free nucleic acids of both the yeast and thymus type, lipids, and other cell constituents. Try to find in the complex mixtures the active principle! Try to isolate and chemically identify the particular substance that will by itself, when brought into contact with the R cell derived from Type II, cause it to elaborate Type III capsular polysaccharide and to acquire all the aristocratic distinctions of the same specific type of cells as that from which the extract was prepared! Some job, full of headaches and heartbreaks. But at last perhaps we have it.

. . . if we prove to be right—and of course that is a big if—then it means that both the chemical nature of the inducing stimulus is known and the chemical structure of the substance produced is also known, the former being thymus nucleic acid, the latter Type III polysaccharide, and both are thereafter reduplicated in the daughter cells and after innumerable transfers without further addition of the inducing agent and the same active and specific transforming substance can be recovered far in excess of the amount originally used to induce the reaction. Sounds like a virus—may be a gene. But with mechanisms I am not now concerned. One step at a time and the first step is what is the chemical nature of the transforming principle? Some one else can work out the rest. Of course the problem bristles with implications. It touches the biochemistry of the thymus type of nucleic acids which are known to constitute the major part of chromosomes but have been thought to be alike regardless of origin and species. It touches genetics, enzyme chemistry, cell metabolism and carbohydrate synthesis. But today it takes a lot of well documented evidence to convince anyone that the sodium salt of deoxyribose nucleic acid, protein free, could possibly be endowed with such biologically active and specific properties and that is the evidence we are now trying to get. It is lots of fun to blow bubbles but it is wiser to prick them yourself before someone else tries to.

Figure 24-4
Part of a letter from Oswald Avery to his brother Roy, written in May 1943. [From R. D. Hotchkiss. In *Phage and the Origins of Molecular Biology*, J. Cairns, G. S. Stent, and J. D. Watson, eds. (Cold Spring Harbor Laboratory, 1966), pp. 185–186.]

protein coat. The results of these experiments were:

1. Most of the phage DNA was found in the bacteria.

2. Most of the phage protein was found in the supernatant.

3. The blender treatment had almost no effect on the competence of the infected bacteria to produce progeny virus.

Additional experiments showed that less than 1% of the ^{35}S was transferred from the parental phage to the progeny phage. In contrast, 30% of the parental ^{32}P appeared in the progeny. These sim-

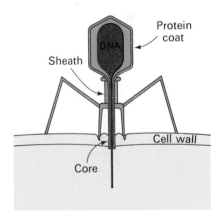

Figure 24-5
Diagram of a T2 bacteriophage injecting its DNA into a bacterial cell. [After W. B. Wood and R. S. Edgar. Building a bacterial virus. Copyright © 1967 by Scientific American, Inc. All rights reserved.]

ple, incisive experiments led to the conclusion that "*a physical separation of the phage T2 into genetic and non-genetic parts is possible. . . .* The sulfur-containing protein of resting phage particles is confined to a protective coat that is responsible for the adsorption of bacteria, and functions as an instrument for the injection of the phage DNA into the cell. This protein probably has no function in the growth of intracellular phage. The DNA has some function. Further chemical inferences should not be drawn from the experiments presented."

The cautious tone of this conclusion did not detract from its impact. The genetic role of DNA soon became a generally accepted fact. The experiments of Hershey and Chase strongly reinforced what Avery, MacLeod, and McCarty had found eight years earlier in a different system. Additional support came from studies of the DNA content of single cells, which showed that in a given species *the DNA content is the same for all cells that have a diploid set of chromosomes. Haploid cells were found to have half as much DNA.*

THE GENES OF SOME VIRUSES ARE MADE OF RNA

Genes in all procaryotic and eucaryotic organisms are made of DNA. In viruses, genes are made of either DNA or RNA. Tobacco mosaic virus, which infects the leaves of tobacco plants, is one of the best-characterized RNA viruses. It consists of a single molecule of RNA surrounded by a protein coat of 2,130 identical subunits (see Chapter 30 for a discussion of its structure and assembly). The protein can be separated from the RNA by treatment of the virus with phenol. *The isolated viral RNA is infective, whereas the viral protein is not.* Synthetic hybrid virus particles provide additional evidence that the genetic specificity of the virus resides exclusively in its RNA. There are a variety of strains of tobacco mosaic virus. A synthetic hybrid virus was prepared from the RNA of strain 1 and the protein of strain 2. Another was prepared from the RNA of strain 2 and the protein of strain 1. After infection, *the progeny virus always consisted of RNA and protein corresponding to the specificity of the RNA in the infecting hybrid virus.*

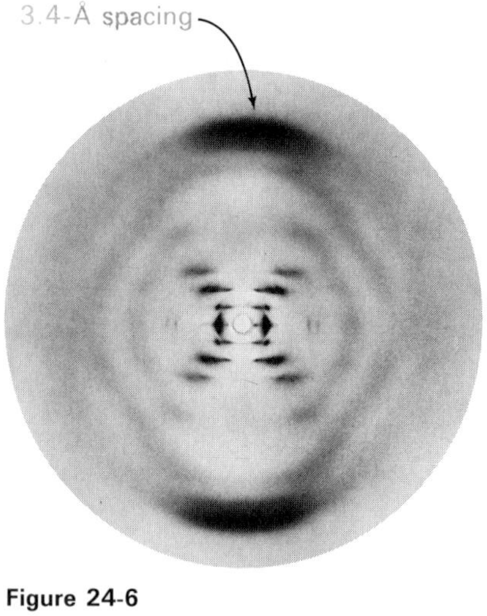

THE WATSON-CRICK DNA DOUBLE HELIX

In 1953, James Watson and Francis Crick deduced the three-dimensional structure of DNA and immediately inferred its mechanism of replication. This brilliant accomplishment ranks as one of the most significant in the history of biology because it led the way to an understanding of gene function in molecular terms. Watson and Crick analyzed x-ray diffraction photographs of DNA fibers taken by Rosalind Franklin and Maurice Wilkins and derived a

3.4-Å spacing

Figure 24-6
X-ray diffraction photograph of a hydrated DNA fiber (form B). The central cross is diagnostic of a helical structure. The strong arcs on the meridian arise from the stack of base pairs, which are 3.4 Å apart. [Courtesy of Dr. Maurice Wilkins.]

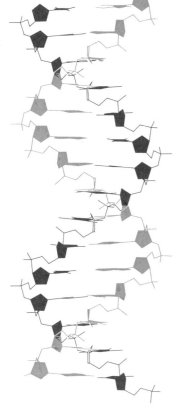

Figure 24-7
Skeletal model of double-helical DNA.
The structure repeats at intervals of
34 Å, which corresponds to ten residues
on each chain.

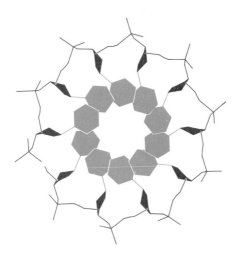

Figure 24-8
Diagram of one of the strands of a
DNA double helix, viewed down the
helix axis. The bases (all pyrimidines
here) are inside, whereas the sugar-
phosphate backbone is outside. The
tenfold symmetry is evident. The bases
are shown in blue and the sugars in
red.

structural model that has proved to be essentially correct. The important features of their model DNA are:

1. Two helical polynucleotide chains are coiled around a common axis. The chains run in opposite directions (Figure 24-7).

2. The purine and pyrimidine bases are on the inside of the
helix, whereas the phosphate and deoxyribose units are on the outside (Figure 24-8). The planes of the bases are perpendicular to the
helix axis. The planes of the sugars are nearly at right angles to
those of the bases.

3. The diameter of the helix is 20 Å. Adjacent bases are separated by 3.4 Å along the helix axis and related by a rotation of 36
degrees. Hence, the helical structure repeats after ten residues on
each chain; that is, at intervals of 34 Å.

4. The two chains are held together by hydrogen bonds between
pairs of bases. Adenine is always paired with thymine. Guanine is
always paired with cytosine (Figures 24-9 and 24-10).

5. The sequence of bases along a polynucleotide chain is not
restricted in any way. *The precise sequence of bases carries the genetic
information.*

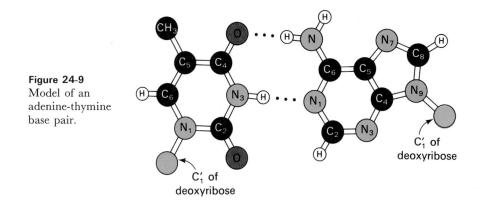

Figure 24-9
Model of an adenine-thymine base pair.

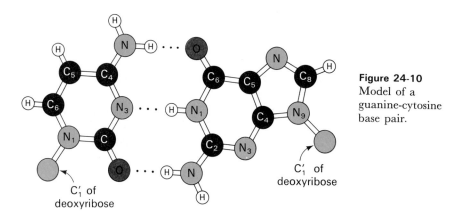

Figure 24-10
Model of a guanine-cytosine base pair.

Figure 24-11
Superposition of an A-T base pair (shown in yellow) on a G-C base pair (shown in blue). Note that the positions of the glycosidic bonds and of the C-1′ atom of deoxyribose (shown in green) are almost identical for the two base pairs.

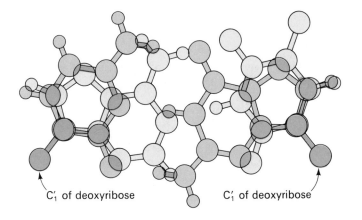

The most important aspect of the DNA double helix is the specificity of the pairing of bases. Watson and Crick deduced that adenine must pair with thymine, and guanine with cytosine, because of steric and hydrogen-bonding factors. The steric restriction is imposed by the regular helical nature of the sugar-phosphate backbone of each polynucleotide chain. The glycosidic bonds that are attached to a bonded pair of bases are always 10.85 Å apart (Figure 24-11). A purine-pyrimidine base pair fits perfectly in this space. In contrast, there is insufficient room for two purines. There is more than enough space for two pyrimidines, but they would be too far apart to form hydrogen bonds. Hence, one member of a base pair in a DNA helix must always be a purine and the other a pyrimidine because of steric factors. The base pairing is further restricted by hydrogen-bonding requirements. The hydrogen atoms in the purine and pyrimidine bases have well-defined positions. Adenine cannot pair with cytosine because there would be two hydrogens near one of the bonding positions and none at the other. Likewise, guanine cannot pair with thymine. In contrast, adenine forms two hydrogen bonds with thymine, whereas guanine forms three with cytosine (Figures 24-9 and 24-10). The orientations and distances of these hydrogen bonds are optimal for achieving strong interaction between the bases.

This base-pairing scheme was strongly supported by the results of earlier studies of the base compositions of DNAs from different species. In 1950, Erwin Chargaff found that the *ratios of adenine to thymine and of guanine to cytosine were nearly 1.0 in all species studied.* The meaning of these equivalences was not evident until the Watson-Crick model was proposed. Only then could it be seen that they reflect an essential facet of DNA structure and function—the specificity of base pairing.

THE COMPLEMENTARY CHAINS ACT AS TEMPLATES FOR EACH OTHER IN DNA REPLICATION

The double helical model of DNA immediately suggested a mechanism for the replication of DNA. Watson and Crick (1953*b*) published their hypothesis a month after they had presented their structural model in a beautifully simple and lucid paper:

> . . . If the actual order of the bases on one of the pairs of chains were given, one could write down the exact order of the bases on the other one, because of the specific pairing. Thus one chain is, as it were, the complement of the other, and it is this feature which suggests how the deoxyribonucleic acid molecule might duplicate itself.
>
> Previous discussions of self-duplication have usually involved the concept of a template, or mould. Either the template was supposed to copy itself directly or it was to produce a "negative," which in its turn was to act as a template and produce the original "positive" once again. In no case has it been explained in detail how it would do this in terms of atoms and molecules.
>
> Now our model for deoxyribonucleic acid is, in effect, a *pair* of templates, each of which is complementary to the other. We imagine that prior to duplication the hydrogen bonds are broken, and the two chains unwind and separate. Each chain then acts as a template for the formation onto itself of a new companion chain, so that eventually we shall have *two* pairs of chains, where we only had one before. Moreover, the sequence of the pairs of bases will have been duplicated exactly.

Forms of DNA—
A-DNA
Double-helical DNA containing about 11 residues per turn. The planes of the base pairs in this right-handed helix are tilted 20 degrees away from the perpendicular to the helix axis. Formed by the dehydration of B-DNA.

B-DNA
The classical Watson-Crick double helix containing about 10 residues per turn. The planes of the base pairs in this right-handed helix are perpendicular to the helix axis.

Z-DNA
A left-handed double helical form of DNA containing about 12 residues per turn. Proposed by Alexander Rich and co-workers on the basis of the crystal structure of d(CG)$_3$.

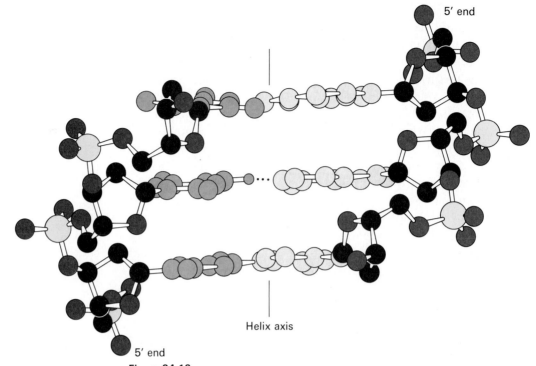

5′ end

Helix axis

5′ end

Figure 24-12
Model of a double-helical DNA molecule showing three base pairs.
Note that the two strands run in opposite directions.

DNA REPLICATION IS SEMICONSERVATIVE

Watson and Crick proposed that one of the strands of each daughter
DNA molecule is newly synthesized, whereas the other is derived
from the parent DNA molecule. This distribution of parental atoms
is called *semiconservative.* A critical test of this hypothesis was carried
out by Matthew Meselson and Franklin Stahl. The parent DNA
was labeled with ^{15}N, the heavy isotope of nitrogen, to make it
denser than ordinary DNA. This was accomplished by growing
E. coli for many generations in a medium that contained ^{15}NH$_4$Cl
as the sole nitrogen source. The bacteria were abruptly transferred
to a medium that contained ^{14}N, the ordinary isotope of nitrogen.
The question asked was: What is the distribution of ^{14}N and ^{15}N in
the DNA molecules after successive rounds of replication?

The distribution of ^{14}N and ^{15}N was revealed by the newly developed technique of *density-gradient equilibrium sedimentation.* A small
amount of DNA was dissolved in a concentrated solution of cesium
chloride having a density close to that of the DNA ($\sim$1.7 g cm^{-3}).
This solution was centrifuged until it was nearly at equilibrium.
The opposing processes of sedimentation and diffusion created a
gradient in the concentration of cesium chloride across the centri-

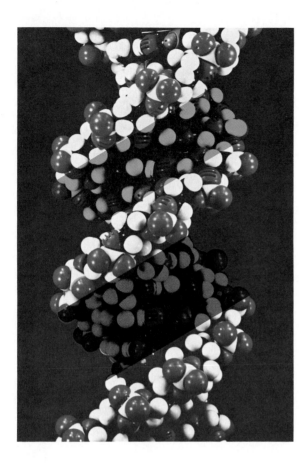

Figure 24-13
Space-filling model of a DNA double
helix. Two kinds of grooves are evident: a major groove (red) and a
minor groove (blue). Note that part of
each base is accessible for interaction
with other molecules. [Courtesy of Dr.
Sung-Hou Kim.]

fuge cell. The result was a stable density gradient, ranging from
1.66 to 1.76 g cm^{-3}. The DNA molecules in this density gradient
were driven by centrifugal force into the region where the solution
density was equal to their own buoyant density. High-molecular-
weight DNA yielded a sharp band that was detected by its absorption of ultraviolet light. The ^{14}N DNA and ^{15}N DNA molecules
were clearly resolved in the density gradient because they differ in
density by about 1% (Figure 24-14).

A B

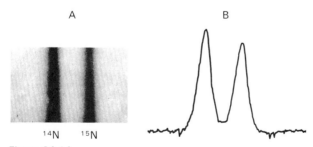

^{14}N ^{15}N

Figure 24-14
Resolution of ^{14}N DNA and ^{15}N DNA by density-gradient
centrifugation: (A) ultraviolet absorption photograph of a
centrifuge cell; (B) densitometric tracing of the absorption
photograph. [From M. Meselson and F. W. Stahl. *Proc. Nat.
Acad. Sci.* 44(1958):671.]

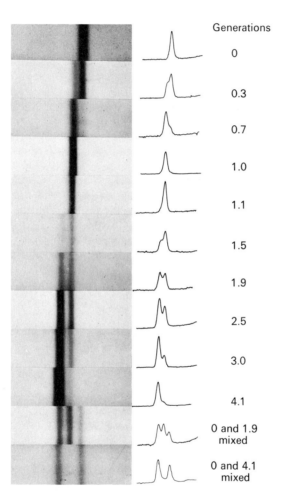

Generations

0

0.3

0.7

1.0

1.1

1.5

1.9

2.5

3.0

4.1

0 and 1.9
mixed

0 and 4.1
mixed

Figure 24-15
Demonstration of semiconservative replication
in *E. coli* by density-gradient centrifugation.
The position of a band of DNA depends on its
content of ^{14}N and ^{15}N. After 1.0 generation, all
of the DNA molecules are hybrids containing
equal amounts of ^{14}N and ^{15}N. No parental
DNA (^{15}N) is left after 1.0 generation. [From
M. Meselson and F. W. Stahl. *Proc. Nat. Acad.
Sci.* 44(1958):671.]

Original parent molecule

First generation daughter molecules

Second generation daughter molecules

Figure 24-16
Schematic diagram of semiconservative
replication. Parental DNA is shown in
green and newly synthesized DNA in
red. [After M. Meselson and F. W.
Stahl. *Proc. Nat. Acad. Sci.* 44(1958):671.]

DNA was extracted from the bacteria at various times after they were transferred from a ^{15}N to a ^{14}N medium. Analysis of these samples by the density-gradient technique showed that there was a single band of DNA after one generation (Figure 24-15). The density of this band was precisely halfway between those of ^{14}N DNA and ^{15}N DNA. *The absence of ^{15}N DNA indicated that parental DNA was not preserved as an intact unit on replication.* The absence of ^{14}N DNA indicated that all of the daughter DNA molecules derived some of their atoms from the parent DNA. This proportion had to be one-half, because the density of the hybrid DNA band was halfway between those of ^{14}N DNA and ^{15}N DNA.

After two generations, there were equal amounts of two bands of DNA. One was hybrid DNA, the other was ^{14}N DNA. Meselson and Stahl concluded from these incisive experiments *"that the nitrogen of a DNA molecule is divided equally between two physically continuous subunits; that, following duplication, each daughter molecule receives one of these; and that the subunits are conserved through many duplications."* Their results agreed perfectly with the Watson-Crick model for DNA replication (Figure 24-16).

A FEW VIRUSES HAVE SINGLE-STRANDED DNA FOR A PART OF THEIR LIFE CYCLES

DNA is not always double-stranded. Robert Sinsheimer discovered that the DNA in *φX174, a small virus that infects* E. coli, *is single stranded.* Several experimental results led to this unexpected conclusion. First, the base ratios of φX174 DNA do not conform to the rule that [A] = [T] and [G] = [C]. Second, a solution of φX174 DNA is much less viscous than a solution of the same concentration of *E. coli* DNA. The hydrodynamic properties of φX174 DNA are like those of a randomly coiled polymer. In contrast, the DNA double helix behaves hydrodynamically as a quite rigid rod. Third, the amino groups of the bases of φX174 DNA react readily with formaldehyde, whereas the bases in double helical DNA are virtually inaccessible to this reagent.

The finding of this single-stranded DNA raised doubts concerning the universality of the semiconservative replicative scheme proposed by Watson and Crick. However, it was soon shown that φX174 DNA is single stranded for only a part of the life cycle of the virus. Sinsheimer found that infected *E. coli* cells contain a *double-stranded form of φX174 DNA.* This double-helical DNA is called the *replicative form* because it serves as the template for the synthesis of the DNA of the progeny virus. Viruses that contain single-stranded RNA also replicate by means of a double-stranded replicative form. The detailed mechanisms of these processes will be discussed in Chapter 30. The important point now is that these studies reinforced the generality of the Watson-Crick scheme for replication. *Double-helical DNA (or RNA) is the replicative form of all known genes.*

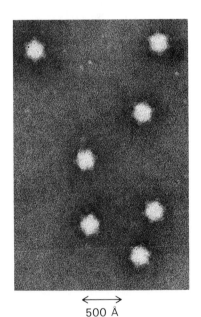

←——→
500 Å

Figure 24-17
Electron micrograph of φX174 virus particles. [Courtesy of Dr. Robley Williams.]

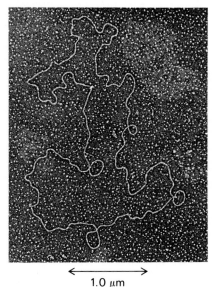

← 1.0 μm →

Figure 24-18
Electron micrograph of a DNA molecule from λ bacteriophage (RF II form). [Courtesy of Dr. Thomas Broker.]

Let us consider some properties of DNA before turning to its mechanism of replication, which is an intricate enzymatic process. A striking characteristic of naturally occurring DNA molecules is their length. *The* E. coli *chromosome is a single molecule of double-helical DNA consisting of four million base pairs.* The mass of this DNA molecule is 2.6×10^6 kdal. It has a *highly asymmetric shape:* its contour length is 14×10^6 Å, whereas its diameter is 20 Å. The 1.4-mm contour length of this DNA molecule corresponds to a macroscopic dimension, whereas its width of 20 Å is on the atomic scale. Bruno Zimm found that the largest chromosome of *Drosophila melanogaster* contains a single DNA molecule of 6.2×10^7 base pairs, which has a contour length of 2.1 cm. Such highly asymmetric DNA molecules are very susceptible to cleavage by shearing forces. They are easily degraded into segments whose masses are one thousand times less than that of the original molecule unless special precautions are taken in their handling.

The DNA molecules of many bacteria and viruses have been directly visualized by electron microscopy (Figure 24-18). The dimensions of some of these DNA molecules are given in Table 24-1.

Table 24-1
Sizes of DNA molecules

Organism	Base pairs (in thousands, or kb)	Contour length (μm)
Viruses		
Polyoma or SV40	5.1	1.7
λ phage	48.6	17
T2 phage	166	56
Vaccinia	190	65
Bacteria		
Mycoplasma	760	260
E. coli	4,000	1,360
Eucaryotes		
Yeast	13,500	4,600
Drosophila	165,000	56,000
Human	2,900,000	990,000

Source: After A. Kornberg, *DNA Replication* (W.H. Freeman and Company, 1980), p. 20.

Kilobase (kb)—

A unit of length equal to 1000 base pairs of a double-stranded nucleic acid molecule (or 1000 bases of a single-stranded molecule).

One kilobase of double-stranded DNA has a contour length of 0.34 μm and a mass of about 660 kdal.

It should be noted that even the smallest DNA molecules are highly elongate. The DNA from polyoma virus, for example, consists of 5,100 base pairs and has a contour length of 1.7 μm (17,000 Å). Recall that hemoglobin has a diameter of 65 Å and that collagen, one of the longest proteins, has a length of 3,000 Å. This comparison emphasizes the remarkable length and asymmetry of DNA molecules.

The two strands of a DNA helix readily come apart when the hydrogen bonds between its paired bases are disrupted. This can be accomplished by heating a solution of DNA or by adding acid or alkali to ionize its bases. The unwinding of the double helix is called *melting* because it occurs abruptly at a certain temperature. The *melting temperature* (T_m) is defined as the temperature at which half of the helical structure is lost. The abruptness of the transition indicates that the DNA double helix is a *highly cooperative structure*. The melting of DNA is readily monitored by measuring its absorbance at 260 nm. The unstacking of the base pairs results in increased absorbance, an effect called *hyperchromism* (Figure 24-19).

The melting temperature of a DNA molecule depends markedly on its base composition. DNA molecules rich in GC base pairs have a higher T_m than those having an abundance of AT base pairs (Figure 24–20). In fact, the T_m of DNA from many species varies

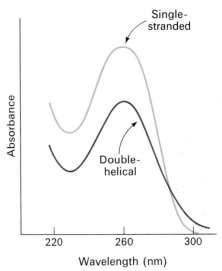

Figure 24-19
The absorbance at 260 nm of a DNA solution increases when the double helix is melted into single strands.

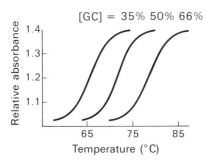

Figure 24-20
DNA melting curves. The absorbance at 260 nm (relative to that at 25°C) is plotted as a function of temperature. The T_m is 69°C for *E. coli* DNA (50% GC pairs) and 76°C for *P. aeruginosa* DNA (68% GC pairs).

linearly with GC content, rising from 77° to 100°C as the fraction of GC pairs increases from 20% to 78%. GC base pairs are more stable than AT pairs because their bases are held together by three hydrogen bonds rather than by two. Hence, *the AT-rich regions of DNA are the first to melt* (Figure 24-21). As will be discussed later, the double helix is melted in vivo by the action of specific proteins. Some of them drive the unwinding of DNA by hydrolyzing ATP (p. 585).

Separated complementary strands of DNA spontaneously reassociate to form a double helix when the temperature is lowered below T_m. This renaturation process is sometimes called *annealing*. The rate of reassociation depends on the concentration of complementary sequences (p. 696). The facility with which double helices can be melted and then reassociated is crucial for the biological functions of DNA.

SOME DNA MOLECULES ARE CIRCULAR

Electron microscopy has shown that intact DNA molecules from many sources are circular (see Figure 24-18). The finding that *E. coli* has a circular chromosome was anticipated by genetic studies that

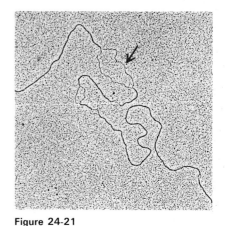

Figure 24-21
Electron micrograph of a DNA molecule partially unwound by alkali. Single-stranded regions appear as loops that stain less intensely than double-stranded segments. These unwound regions are rich in AT base pairs. One of them is marked by an arrow. [From R. B. Inman and M. Schnos. *J. Mol. Biol.* 49(1970):93.]

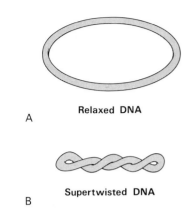

Relaxed DNA

A

Supertwisted DNA

B

Figure 24-22
Schematic diagram of (A) relaxed
DNA and (B) supertwisted DNA.

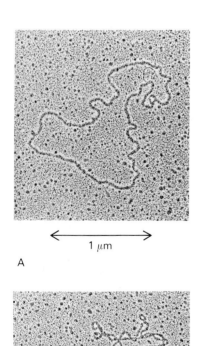

$\xleftarrow{\hspace{1cm}}$ 1 μm $\xrightarrow{\hspace{1cm}}$

A

B

Figure 24-23
Electron micrographs of mitochondrial
DNA: (A) relaxed circular form;
(B) supertwisted circular form.
[Courtesy of Dr. David Clayton.]

revealed that the *gene-linkage map of this bacterium is circular.* The term
"circular" refers to the continuity of the DNA chain, not to its
geometrical form. DNA molecules in vivo necessarily have a very
compact shape. Note that the contour length of the *E. coli* chromosome is about a thousand times as long as the greatest diameter of
the bacterium.

Not all DNA molecules are circular. DNA from the T7 bacteriophage, for example, is *linear.* The DNA molecules of some viruses,
such as the λ bacteriophage, *interconvert between linear and circular
forms.* The linear form is present inside the virus particle, whereas
the circular form is present in the host cell (see p. 739).

CIRCULAR DOUBLE-HELICAL DNA MOLECULES CAN BE SUPERTWISTED

A new property appears in the conversion of a linear DNA duplex
into a closed circular molecule. Jerome Vinograd discovered that
the axis of the double helix can itself be twisted to form a *superhelix,*
which can be right-handed or left-handed (Figure 24-22). The
terms supertwisted, superhelical, and supercoiled are synonymous.
A circular DNA without any superhelical turns is known as a *relaxed
molecule.* Energy is required to convert a relaxed DNA molecule into
a supertwisted one. For example, the formation of fifteen superhelical turns in SV40 DNA (which has a contour length of 1.7 μm) costs
about 100 kcal/mol. The strain energy stored in a supertwisted
DNA is nearly proportional to the square of the number of superhelical turns.

Supertwisting is likely to be biologically important for two reasons. The first is that *a supertwisted DNA has a more compact shape than
its relaxed counterpart* (Figure 24-23). Hence supertwisting may play a
role in the packaging of DNA. Second, supertwisting can influence
the degree of unwinding of the double helix and thereby affect its
interactions with other molecules. Specifically, *negative supertwisting
can result in an unwinding of the double helix.* It is interesting to note that
nearly all naturally occurring circular DNA molecules are negatively supertwisted.

An important characteristic of a closed circular DNA is its linking number (L), which is defined as the number of turns made by
one strand about the other and which must be an integer. The
degree of twisting of the double helix (T) and the extent of supertwisting (W) are reciprocally related by the expression

$$L = W + T$$

The linking number, a topological property, can change only if one
or both strands of the circular DNA are nicked. In fact, enzymes
catalyzing changes in L have been isolated. The catalytic action of
these *topoisomerases* can be readily monitored by gel electrophoresis

because supertwisted DNA is more compact and hence more mobile than is relaxed DNA (Figure 24-24).

DISCOVERY OF A DNA POLYMERASE

We turn now to the molecular events in the replication of DNA. The search for an enzyme that synthesizes DNA was initiated by Arthur Kornberg and his associates in 1955. This search soon proved fruitful, largely because three appropriate choices were made in the design of their experiments:

1. What are the *activated precursors* of DNA? They correctly deduced that *deoxyribonucleoside 5′-triphosphates* are the activated intermediates in DNA synthesis. This inference was based on two clues. First, pathways of purine and pyrimidine biosynthesis lead to nucleoside 5′-phosphates rather than to nucleoside 3′-phosphates. Second, ATP is the activated intermediate in the synthesis of a pyrophosphate bond in coenzymes such as NAD^+, FAD, and CoA.

2. What is the criterion of *DNA synthesis?* It was expected that the net amount of DNA synthesis might be very small in the initial experiments, especially where nucleases are prevalent. Hence, a sensitive assay was essential. This was accomplished by using *radioactive precursor nucleotides*. The incorporation of these precursors into DNA was detected by measuring the radioactivity of an *acid-precipitate of the incubation mixture*. The basis of this technique is that DNA is precipitated by acids, such as trichloroacetic acid, whereas precursor nucleotides stay in solution.

3. Which *kinds of cells* should be analyzed? *E. coli* bacteria were used after initial experiments with animal-cell extracts were negative. *E. coli* was chosen because it has a generation time of only twenty minutes and can be harvested in large quantities. As expected, this bacterium is a choice source of enzymes that synthesize DNA.

An extract of *E. coli* was incubated with radioactive deoxythymidine 5′-triphosphate. The level of radioactivity in this ^{14}C-labeled precursor was one million counts per minute. The acid-precipitate of this incubation mixture contained just fifty counts. Only a few picomoles of DNA was synthesized, but it was a start. Kornberg wrote: "Although the amount of nucleotide incorporated into nucleic acid was miniscule, it was nonetheless significantly above the level of background noise. Through this tiny crack we tried to drive a wedge. The hammer was enzyme purification, a technique that had matured during the elucidation of alcoholic fermentation."

This new enzyme was named DNA polymerase. It is now called *DNA polymerase I* because other DNA polymerases have since been isolated. After a decade of effort in Kornberg's laboratory, DNA

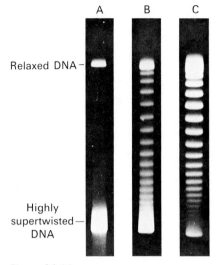

Figure 24-24
Gel patterns showing the relaxation of supertwisted SV40 viral DNA. Part A is a highly negatively supertwisted DNA. Incubation of the DNA with a topoisomerase for (B) 5 minutes and (C) 30 minutes leads to a series of bands that have less supertwisting. [From W. Keller, *Proc. Nat. Acad. Sci.* 72(1975):2553.]

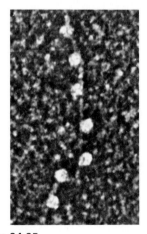

Figure 24-25
Electron micrograph of DNA polymerase molecules (spheres) bound to a DNA molecule (thin thread). [Courtesy of Dr. Jack Griffith.]

polymerase I was purified to homogeneity and characterized in detail. An appreciation of the magnitude of the task can be gained by noting that it took one hundred kilograms of *E. coli* cells to produce five hundred milligrams of pure enzyme.

DNA polymerase I is a 109-kdal single polypeptide chain. It catalyzes the *step-by-step addition of deoxyribonucleotide units to a DNA chain:*

$$(DNA)_{n \text{ residues}} + dNTP \rightleftharpoons (DNA)_{n+1} + PP_i$$

DNA polymerase I requires the following components to synthesize a chain of DNA:

1. All four *deoxyribonucleoside 5'-triphosphates*—dATP, dGTP, dTTP, and dCTP—must be present. The abbreviation dNTP will be used to refer to these deoxyribonucleoside triphosphates. Mg^{2+} is also required.

2. DNA polymerase I adds deoxyribonucleotides to the 3'-hydroxyl terminus of a preexisting DNA (or RNA) strand. In other words, a *primer* chain with a free 3'-OH group is required.

3. A DNA *template* is essential. The template can be single- or double-stranded DNA. Double-stranded DNA is an effective template only if its sugar-phosphate backbone is broken at one or more sites.

The chain-elongation reaction catalyzed by DNA polymerase occurs by means of *a nucleophilic attack of the 3'-OH terminus of the primer on the innermost phosphorus atom of the incoming deoxyribonucleoside triphosphate.* A phosphodiester bridge is formed and pyrophosphate is concomitantly released (Figure 24-26). The subsequent hydrolysis of pyrophosphate drives the polymerization forward. This displacement of the overall equilibrium could not occur if the activated

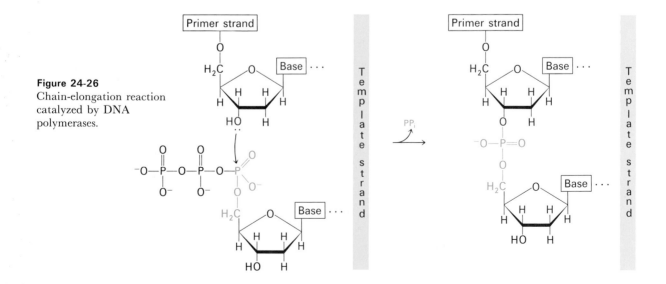

Figure 24-26
Chain-elongation reaction catalyzed by DNA polymerases.

intermediates were nucleoside diphosphates. We see here a compelling reason for the prevalence of nucleoside triphosphates rather than diphosphates as activated precursors in biosynthetic reactions. *The elongation of the DNA chain proceeds in the 5' → 3' direction.* About ten nucleotides are added per second per molecule of DNA polymerase I. Polymerization is *processive*—that is, many nucleotides are added without release of the enzyme from the template.

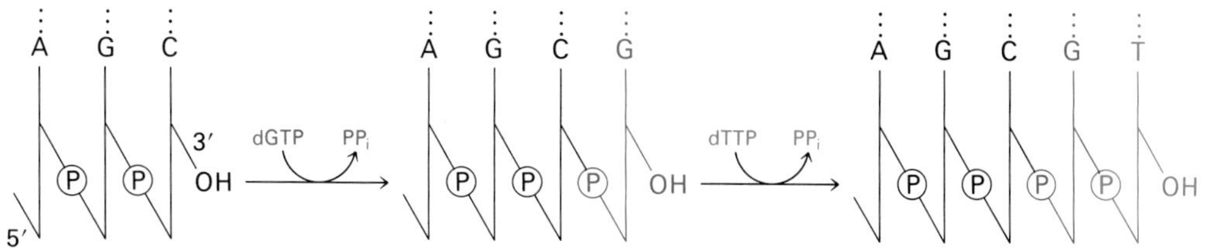

Figure 24-27
DNA polymerases catalyze elongation of DNA chains in the 5' → 3' direction.

DNA POLYMERASE TAKES INSTRUCTIONS FROM A TEMPLATE

DNA polymerase catalyzes the formation of a phosphodiester bond only if the base on the incoming nucleotide is complementary to the base on the template strand. The probability of making a covalent link is very low unless the incoming base forms a Watson-Crick type of base pair with the base on the template strand. Thus, DNA polymerase is a *template-directed enzyme*. In fact, it was the first enzyme of this type to be discovered.

Several kinds of experiments have shown that DNA polymerase takes instructions from a template:

1. The earliest evidence was the finding that *appreciable synthesis of DNA occurs only if all four deoxyribonucleoside triphosphates and a DNA template are present.*

2. DNA polymerase can incorporate certain base analogs. For example, uracil or 5-bromouracil can substitute for thymine. Hypoxanthine can substitute for guanine. The pattern of substitution is highly specific. *The rule is that a base analog can be incorporated only if it can form a Watson-Crick type of base pair with the complement of the base that it replaces.* Thus, hypoxanthine can substitute for guanine (but not for A, T, or C) because it can form an appropriate base pair with cytosine.

3. The *base composition* of newly synthesized DNA depends on the nature of the template, not on the relative proportions of the four precursor nucleotides. The product DNA has the same base composition as the double-helical template DNA. This result indicates that both strands of the template DNA are replicated by DNA polymerase.

4. The most compelling evidence comes from the finding that φX174 DNA replicated in vitro by DNA polymerase I is fully infective. Hence, the error rate of this enzyme must be very low.

DNA POLYMERASE I CORRECTS MISTAKES IN DNA

An additional enzymatic property of DNA polymerase is its capacity to hydrolyze DNA chains under certain conditions. DNA polymerase can hydrolyze DNA progressively from the 3′-hydroxyl terminus of a DNA chain. The products are mononucleotides. Thus, *DNA polymerase I is a 3′ → 5′ exonuclease* (Figure 24-28). The nucleotide that is removed must have a free 3′-OH terminus, and it must not be part of a double helix. Is this exonuclease activity an undesirable side effect of the enzyme or does it contribute to the biological action of DNA polymerase? Experiments using chemically synthesized polynucleotides with a mismatched residue at the primer terminus have indicated that the *3′ → 5′ nuclease activity has an editing function in polymerization.* Consider the polymer shown in Figure 24-28 in which a sequence of dT residues forms a double helix with a longer polymer of dA. At the 3′ terminus of this poly dT sequence there is a single dC residue that is not hydrogen bonded because it is not complementary to dA. Upon addition of DNA polymerase and dTTP, this mismatched dC residue was hydrolyzed before dT residues were added. Experiments with a variety of synthetic polymers have shown that *DNA polymerase I always removes mismatched residues at the primer terminus before proceeding with polymerization.* There is little or no hydrolysis if the terminus is properly matched and activated precursors are on hand. Polymerization prevents hydrolysis from the 3′ end.

It seems likely that DNA replication is very accurate because *base pairs are checked twice.* Polymerization does not usually occur unless the base pair fits into a double helix. However, if an error is made at this stage, it can be corrected before the next nucleotide is added. In effect, DNA polymerase I examines the result of each polymerization it catalyzes before proceeding to the next.

DNA polymerase I can also hydrolyze DNA starting from the 5′ end of a chain. This *5′ → 3′ nuclease* activity (Figure 24-29) is very different from the 3′ → 5′ exonuclease action discussed above. First, the cleaved bond must be in a double-helical region. Second, cleavage can occur at the terminal phosphodiester bond or at a bond several residues away from the 5′ terminus (which can bear a free hydroxyl group or be phosphorylated). Third, 5′ → 3′ nuclease activity is enhanced by concomitant DNA synthesis. Fourth, the active site for the 5′ → 3′ nuclease activity is clearly separate from the active sites for polymerization and 3′ → 5′ hydrolysis. DNA polymerase I can be split by proteases into a 36-kdal fragment with all of the original 5′ → 3′ exonuclease activity and a 75-kdal fragment with all of the polymerase and 3′ → 5′ exonuclease activities.

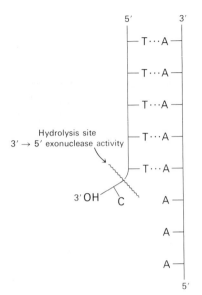

Figure 24-28
3′ → 5′ exonuclease action of DNA polymerase I.

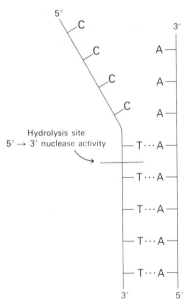

Figure 24-29
5′ → 3′ exonuclease action of DNA polymerase I.

Thus, *DNA polymerase I contains at least two distinct enzymes in one polypeptide chain.* The $5' \rightarrow 3'$ exonuclease complements the $3' \rightarrow 5'$ exonuclease activity by correcting errors of a different type. For example, the $5' \rightarrow 3'$ exonuclease participates in the excision of pyrimidine dimers that are formed following exposure of DNA to ultraviolet light (p. 588). Moreover, the $5' \rightarrow 3'$ exonuclease activity plays a key role in DNA replication itself (p. 584).

DNA LIGASE CONNECTS DNA FRAGMENTS

DNA polymerase I can add deoxyribonucleotides to a primer chain, but it cannot catalyze the joining of two DNA chains or the closure of a single DNA chain. The finding of circular DNA indicated that such an enzyme must exist. In 1967, researchers in several laboratories simultaneously discovered *DNA ligase,* an enzyme that catalyzes the formation of a phosphodiester bond between two DNA chains (Figure 24-30). This enzyme requires a *free OH group at*

DNA strand—3'—OH + ⁻O—P—O—5'—DNA strand

DNA Ligase
ATP (or NAD⁺)

DNA strand—3'—O—P—O—5'—DNA strand

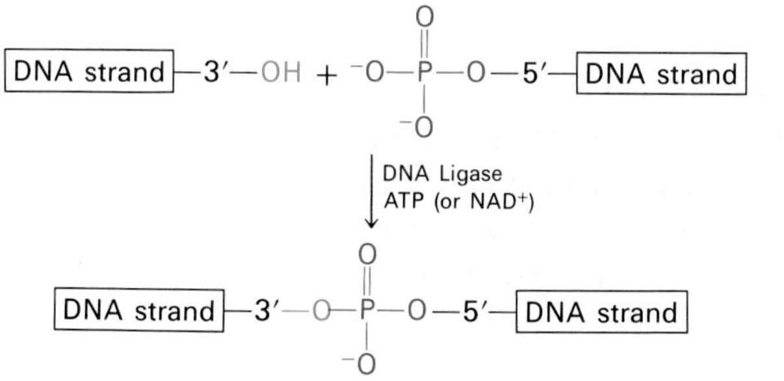

Figure 24-30
DNA ligase catalyzes the joining of two DNA chains that are part of a double-helical molecule.

the 3' end of one DNA chain and a phosphate group at the *5' end* of the other. The formation of a phosphodiester bond between these groups is an endergonic reaction. Hence, *an energy source is required for the joining reaction.* In *E. coli* and other bacteria, NAD^+ serves this role, whereas, in animal cells and bacteriophage, the reaction is driven by *ATP.*

One other aspect of DNA ligase merits attention. DNA ligase cannot link two molecules of single-stranded DNA. Rather, *the DNA chains joined by DNA ligase must be part of a double-helical DNA molecule.* Studies of model systems suggest that DNA ligase forms a phosphodiester bridge only if there are at least several base pairs in the region of this link. In fact, DNA ligase closes nicks in the backbone of double-helical DNA. This joining process is essential for the *normal synthesis of DNA, for repair of damaged DNA,* and for the *splicing of DNA chains in genetic recombination.*

Let us look at the mechanism of this reaction, which was elucidated by I. Robert Lehman. ATP or NAD^+ reacts with DNA ligase to form a *covalent enzyme-AMP complex* in which the AMP is linked to the ε-amino group of a lysine residue of the enzyme through a phosphoamide bond (Figure 24-31). The AMP moiety activates the

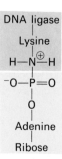

Figure 24-31
Covalent enzyme-AMP intermediate.

phosphate group at the 5′ terminus of the DNA. The final step is a *nucleophilic attack by the 3′-OH group on this activated phosphorus atom.* A phosphodiester bond is formed and AMP is released. This sequence of reactions is driven by the hydrolysis of the pyrophosphate that was released in the formation of the enzyme-adenylate complex. Thus, *two high-energy phosphate bonds are spent in forming a phosphodiester bridge in the DNA backbone if ATP is the energy source.*

Figure 24-32
Mechanism of the reaction catalyzed by DNA ligase.

$$E + ATP \text{ (or } NAD^+) \rightleftharpoons E\text{-}AMP + PP_i \text{ (or NMN)}$$

$$E\text{-}AMP + \text{(P)}—5'\text{-}DNA \rightleftharpoons E + AMP—\text{(P)}—5'\text{-}DNA$$

$$DNA\text{-}3'\text{-}OH + AMP—\text{(P)}—5'\text{-}DNA \rightleftharpoons DNA\text{-}3'\text{-}O—\text{(P)}—5'\text{-}DNA + AMP$$

$$DNA\text{-}3'\text{-}OH + \text{(P)}—5'\text{-}DNA + ATP \text{ (or } NAD^+) \rightleftharpoons DNA\text{-}3'\text{-}O—\text{(P)}—5'\text{-}DNA + AMP + PP_i \text{ (or NMN)}$$

DISCOVERY OF DNA POLYMERASES II AND III

We have seen that DNA polymerase I can synthesize and repair DNA in vitro. Does it perform these functions in vivo? This question is pertinent because an enzyme may or may not carry out the same reaction in vivo as it does in vitro. The reaction conditions inside a cell may be different and other enzymes may be present. In fact, *E. coli* contains at least two other DNA polymerases, named II and III, which were found some fifteen years after the discovery of DNA polymerase I. Why this lag? The reason is that the presence of polymerases II and III was masked by the high levels of activity due to polymerase I.

The breakthrough came in 1969, when Paula DeLucia and John Cairns isolated a mutant of *E. coli* that had from 0.5% to 1.0% of the normal polymerizing activity of DNA polymerase I in its extract. This mutant (named polA 1) multiplied at the same rate as the parent strain. Furthermore, a variety of phages replicated as well in polA 1 as in the parent strain. In contrast, this mutant was much more readily killed by ultraviolet light than was the parent strain. Also, polA 1 was more susceptible to the lethal effects of methylmethanesulfonate, a chemical mutagen. DeLucia and Cairns inferred that DNA replication was normal in their polA 1 mutant, but that DNA repair was markedly impaired. They suggested that a polymerase different from DNA polymerase I might be essential for DNA synthesis.

The low background level of polymerase I in the polA 1 mutant facilitated the discovery of new DNA polymerases. Two such enzymes were soon isolated and characterized in several laboratories. *DNA polymerases II and III are like polymerase I in these respects:*

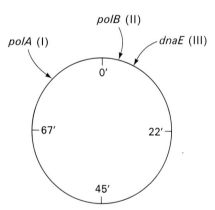

Figure 24-33
Locations of the structural genes for the three DNA polymerases in *E. coli.* DNA polymerase I maps at the *polA* locus, II at *polB,* and III at *dnaE.*

1. They catalyze a template-directed synthesis of DNA from deoxyribonucleoside triphosphate precursors.

2. A primer with a free 3'-OH group is required.

3. Synthesis is in the $5' \to 3'$ direction.

4. They possess $3' \to 5'$ exonuclease activity. DNA polymerase III, but not II, is also a $5' \to 3'$ exonuclease.

These polymerases differ in their template preferences. Polymerases II and III act optimally on double-stranded DNA templates that have short gaps. In contrast, extensive single-stranded regions near double-helical regions are preferred templates for polymerase I. The maximal catalytic rates of these enzymes in vitro also differ: 10 nucleotides per second are added by polymerase I, 0.5 per second by II, and 150 per second by III. DNA polymerase III is associated with several other proteins when it is physiologically active. This multisubunit complex is called *DNA polymerase III holoenzyme.*

What are the roles of these polymerases in vivo? As will be discussed shortly, a multienzyme complex containing polymerase III synthesizes most of the new DNA, whereas polymerase I erases the primer and fills gaps. The role of DNA polymerase II is not yet known. Recent biochemical and genetic studies have revealed that more than ten proteins, in addition to DNA polymerases I and III and DNA ligase, are required for DNA replication in *E. coli.* Before turning to the interplay of these proteins with DNA, let us take a view of replication at the level of the whole chromosome.

PARENTAL DNA IS UNWOUND AND NEW DNA IS SYNTHESIZED AT REPLICATION FORKS

DNA in the midst of replication has been visualized by autoradiography and by electron microscopy. In *autoradiography,* an image is formed by the decay of a suitable radioisotope, such as tritium. The emitted electrons affect the silver grains of a photographic emulsion. Black spots appear when the film is developed. The resolution of autoradiography is relatively low, of the order of several hundred angstroms. However, the technique is attractive because only what is labeled can be seen. DNA is made visible in autoradiographs by the incorporation of tritiated thymine or thymidine.

Autoradiographs and electron micrographs show the form of replicating DNA from *E. coli* to be a closed circle with an inner loop (Figure 24-34). Such forms are called theta structures because of their resemblance to the Greek letter θ. The theta structures show that the *DNA molecule maintains its circular form while it is being replicated.* The resolution of the technique does not suffice to display free ends. However, it is clear that long stretches of single-stranded

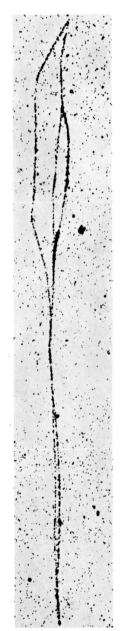

Figure 24-34
Autoradiograph of replicating DNA from *E. coli.* [Courtesy of Dr. John Cairns.]

582

Figure 24-35
Schematic diagram of the circular chromosome of *E. coli* during replication. The green strands in this theta structure depict parental DNA and the red strands represent newly synthesized DNA.

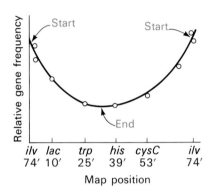

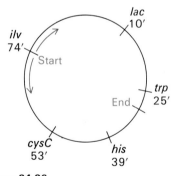

Figure 24-36
Relative amounts of different genes under conditions of rapid DNA synthesis in *E. coli* as a function of position on the genetic map.

DNA are absent. Thus, these pictures rule out a replication mechanism in which the parental DNA strands unwind completely before serving as templates for the synthesis of new DNA. Rather, *the synthesis of new DNA is closely coupled to the unwinding of parental DNA.* A site of simultaneous unwinding and synthesis is called a *replication fork.*

DNA REPLICATION STARTS AT A UNIQUE ORIGIN AND PROCEEDS SEQUENTIALLY IN OPPOSITE DIRECTIONS

Does DNA replication in *E. coli* start anywhere in the chromosome or is there a specific initiation site? DNA replication is a rigorously controlled process, and so a specific initiation site seems much more likely a priori. In fact, DNA replication in *E. coli* starts at a unique origin as shown by studies of the relative amounts of different genes under conditions of rapid DNA synthesis. Consider two genes, *a* and *b*. Suppose that *a* is located near the starting point of replication, whereas *b* is near the finish. Then, gene *a* will be replicated well ahead of *b*. In a rapidly growing culture, there will be nearly two *a* genes for each *b* gene. In contrast, if DNA replication starts randomly, there will be equal amounts of *a* and *b*. The relative frequencies of a number of genes were determined by hybridization, a technique that will be discussed later (p. 602). The results of these experiments clearly showed that the relative gene frequency does depend on the map position, as shown in Figure 24-36. These data revealed that:

1. *Replication starts at a unique site,* near the *ilv* gene, which is located at 74′ on the standard genetic map of *E. coli.*

2. Replication proceeds simultaneously in the two opposite directions, at about the same velocity. In other words, *there are two replication forks: one moves clockwise, the other counterclockwise.*

3. The two replication forks meet near *trp* (25′ on the map), a point diametrically opposite the origin of replication.

Further evidence for the bidirectionality of replication of *E. coli* DNA comes from autoradiography. Replication was initiated in a medium containing *moderately* radioactive tritiated thymine. After a few minutes incubation, the bacteria were transferred to a medium containing *highly* radioactive tritiated thymidine. Two levels of radioactivity were used to create two kinds of grain tracks in the autoradiographs: a low-density track corresponding to the DNA initially synthesized and a high-density track corresponding to the DNA synthesized later. If replication were unidirectional, tracks with a low grain density at one end and a high grain density at the other end would be seen. On the other hand, if replication were bidirectional, the middle of a track would have a low density (Figure 24-37). The autoradiographic patterns (Figure 24-38) provided a vivid answer. All of the grain tracks were denser on both ends than in the middle, indicating that *replication of the* E. coli *chromosome is bidirectional.*

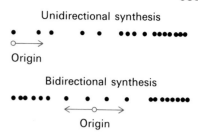

Unidirectional synthesis

Origin

Bidirectional synthesis

Origin

Figure 24-37
Autoradiographic pattern expected for unidirectional replication and bidirectional replication when bacteria are transferred from a medium containing moderately radioactive thymine to one containing highly radioactive thymine.

Figure 24-38
Autoradiograph of *E. coli* DNA during replication (under conditions described in Figure 24-37). The observed grain pattern indicates that replication is bidirectional. [From D. M. Prescott and P. L. Kuempel. *Proc. Nat. Acad. Sci.* 69(1972):2842.]

ONE STRAND OF DNA IS SYNTHESIZED DISCONTINUOUSLY

Let us return to the interplay of molecules in replication. At a replication fork, both strands of parental DNA serve as templates for the synthesis of new DNA. Recall that the parental strands are antiparallel. Hence, the *overall* direction of DNA synthesis must be $5' \rightarrow 3'$ for one daughter strand and $3' \rightarrow 5'$ for the other (Figure 24-39). However, all known DNA polymerases synthesize DNA in the $5' \rightarrow 3'$ direction but not in the $3' \rightarrow 5'$ direction. How then does one of the daughter DNA strands *appear at low resolution* to grow in the $3' \rightarrow 5'$ direction?

This dilemma was resolved by Reiji Okazaki, who found that a *large proportion of newly synthesized DNA exists as small fragments.* These units of about a thousand nucleotides (called Okazaki fragments) are present momentarily in the vicinity of the replication fork. As replication proceeds, these fragments become covalently joined by DNA ligase to form one of the daughter strands (Figure 24-40). The other new strand is synthesized continuously, or nearly so. The strand formed from Okazaki fragments is termed the *lagging strand,*

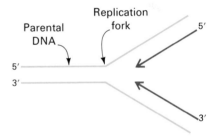

Replication fork

Parental DNA

Figure 24-39
At *low resolution,* the *apparent* direction of DNA replication is $5' \rightarrow 3'$ for one daughter strand and $3' \rightarrow 5'$ for the other. Actually, both strands are synthesized in the $5' \rightarrow 3'$ direction, as shown in Figure 24-40.

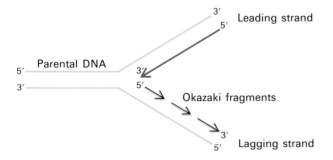

Leading strand

Parental DNA

Okazaki fragments

Lagging strand

Figure 24-40
Schematic diagram of a replication fork. Both strands of DNA are synthesized in the $5' \rightarrow 3'$ direction. The leading strand is synthesized continuously, whereas the lagging strand is synthesized in the form of short fragments (Okazaki fragments).

whereas the one synthesized with few or no interruptions is the *leading strand*. Both the Okazaki fragments and the leading strand are synthesized in the $5' \rightarrow 3'$ direction. *The discontinuous assembly of the lagging strand enables $5' \rightarrow 3'$ polymerization at the atomic level to give rise to overall growth in the $3' \rightarrow 5'$ direction.*

DNA SYNTHESIS IS PRIMED BY RNA

How does DNA synthesis begin? Recall that all these DNA polymerases require a primer with a free 3′-OH group for the initiation of DNA synthesis. What is the primer for the synthesis of the leading strand and the fragments? An important clue came from the observation that RNA synthesis is essential for the initiation of DNA synthesis. This finding suggested that RNA might prime the synthesis of DNA because it was already known that RNA polymerases can start chains de novo. It was then shown that *nascent DNA is covalently linked to a short stretch of RNA, which, in fact serves as the primer. Thus, RNA primes the synthesis of DNA.*

It seems likely that DNA replication in *E. coli* is carried out in the following way:

1. A specific RNA polymerase (called *primase*) synthesizes a short stretch of RNA (~ 10 nucleotides) that is complementary to one of the DNA template strands. Primase, in contrast with DNA polymerases, does not require a primer for the initiation of polynucleotide synthesis.

2. The 3′-hydroxyl group of the terminal ribonucleotide of this RNA chain serves as the primer for the synthesis of new DNA by DNA polymerase III holoenzyme. Most of the new DNA is synthesized by this multisubunit complex.

3. The RNA portion of this RNA-DNA hybrid is then hydrolyzed by DNA polymerase I.

4. Removal of RNA from the nascent chains leaves substantial gaps between the DNA fragments. These gaps are filled by DNA polymerase I, which is well suited for synthesizing DNA in response to instructions from a single-stranded template.

Recent studies have shown that the action of primase is preceded by the formation of a *prepriming intermediate* consisting of at least five proteins. One of them, the dnaB protein, propels itself along the DNA by hydrolyzing ATP. The dnaB protein may serve as a recognition signal to activate primase. Specificity for initiating a round of replication may reside in proteins that precisely deliver the dnaB protein to *ilv*, the origin of replication on the *E. coli* chromosome. The timing of DNA replication is critical because it must be coordi-

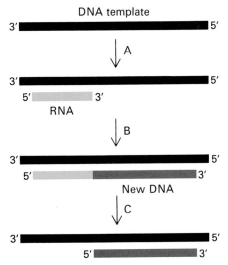

Figure 24-41
Initiation of DNA synthesis: (A) primase synthesizes a short complementary stretch of RNA; (B) this RNA serves as the primer for the synthesis of new DNA; (C) the RNA part of the new chain is hydrolyzed, which leaves a gap that is then filled.

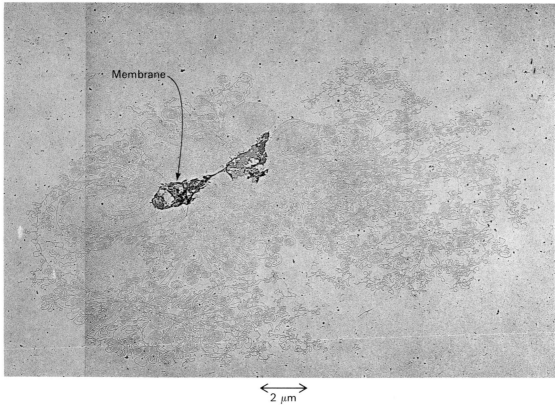

2 μm

Figure 24-42
Electron micrograph of the chromosome of *E. coli* attached to two fragments
of the cell membrane. This single molecule of double-helical DNA is intact
and supercoiled. [From H. Delius and A. Worcel, *J. Mol. Biol.* 82(1974):108.]

nated with cell division. Indeed, bacterial chromosomes are associated with an infolding of the cell membrane (Figure 24-42).

ATP HYDROLYSIS DRIVES THE UNWINDING OF PARENTAL DNA AT REPLICATION FORKS BY THE REP PROTEIN

In 1953, Watson and Crick noted that the "difficulty of untwisting
is a formidable one." Recent studies have revealed that the parental
double helix at the replication fork in *E. coli* is actively unwound by
an enzyme, the rep protein. The unwinding of parental DNA is
powered by the hydrolysis of ATP and so the rep protein is a
helicase. About two ATP are consumed for each base pair that is
separated. Each of the separated strands of parental DNA then
interacts with several molecules of the *single-stranded DNA binding
protein* (SS-binding protein). The role of the SS-binding protein is to
stabilize the single-stranded DNA formed by the action of helicase
so that the unwound region can serve as a template. SS-binding
proteins are also known as helix-destabilizing (HD), or melting,
proteins.

DNA GYRASE PUMPS NEGATIVE SUPERTWISTS INTO PARENTAL DNA TO PROMOTE UNWINDING

A topological problem arises in the unwinding of a closed circular DNA molecule. Specifically, unwinding of the double helix leads to the formation of positive supertwists in a closed molecule. The parental DNA at a replication fork rotates as rapidly as 100 revolutions per second, which is more than 100 times as fast as the rotation of a phonograph record. The positive supertwists imparted by this rotation must be relieved if unwinding is to proceed. In other words, a *molecular swivel* is needed. Martin Gellert has recently discovered that this function is performed by *DNA gyrase. This topoisomerase removes positive supertwists by nicking and then sealing phosphodiester bonds in the DNA backbone.* No ATP is needed to catalyze this thermodynamically favorable relaxation of the tertiary structure of DNA. In addition, *DNA gyrase can actively introduce negative supertwists into closed circular DNA at the expense of ATP hydrolysis* (Figure 24-43). These negative supertwists promote parental strand separation at the replication fork.

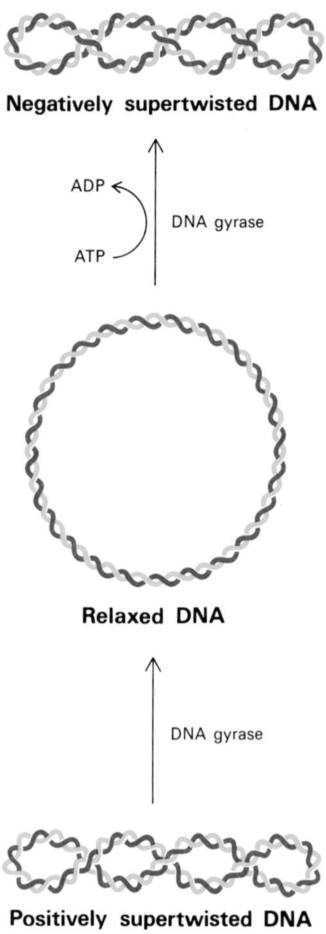

Negatively supertwisted DNA

ADP
ATP

DNA gyrase

Relaxed DNA

DNA gyrase

Positively supertwisted DNA

Figure 24-43
Catalytic activities of DNA gyrase.

THE COMPLEXITY OF THE REPLICATION APPARATUS MAY BE NECESSARY TO ASSURE VERY HIGH FIDELITY

Current knowledge of the molecular events in DNA replication in *E. coli* is summarized in Figure 24-44 and Table 24-2. A striking feature of this process is the intricate interplay of many proteins. Genetic analyses suggest that at least fifteen proteins directly participate in DNA replication. Why is DNA replication so complex? In particular, why does DNA synthesis start with an RNA primer that

Figure 24-44
Schematic diagram of the enzymatic events at a replication fork of *E. coli.* Enzymes shaded blue catalyze chain initiation, elongation, and ligation. [After *DNA Replication* by A. Kornberg. W. H. Freeman and Company. Copyright © 1980.]

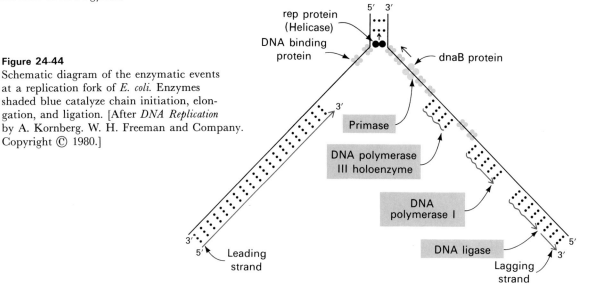

Table 24-2
Replication proteins of *E. coli*

587

Chapter 24
DNA

Protein	Role	Size (kdal)	Molecules per cell
dnaB protein	Enables primase to initiate	~250	20
Primase	Synthesizes RNA primers	60	100
rep protein	Unwinds the double helix	65	50
SS-binding	Stabilizes single-stranded regions	74	300
DNA gyrase	Introduces superhelical twists	400	
DNA polymerase III holoenzyme	Synthesizes DNA	>300	20
DNA polymerase I	Erases primer and fills gaps	109	300
DNA ligase	Joins the ends of DNA	74	300

is subsequently erased? An RNA primer would be unnecessary if DNA polymerases could start chains de novo. However, such a property would be incompatible with the very high fidelity of DNA polymerases. Recall that DNA polymerases test the correctness of the preceding base pair before forming a new phosphodiester bond. This editing function markedly decreases the error frequency. In contrast, RNA polymerases can start chains de novo because they do not examine the preceding base pair. Consequently, their error rates are orders of magnitude higher than those of DNA polymerases. The ingenious solution that has evolved is to *start DNA synthesis with a low-fidelity stretch of polynucleotide but mark it "temporary" by placing ribonucleotides in it*. These short RNA primer sequences are then excised by DNA polymerase I and replaced with a high-fidelity DNA sequence. It seems likely that much of the complexity of DNA replication is imposed by the need for very high accuracy. Genetic analyses indicate that the error rate is of the order of 1 per 10^9 to 10^{10} base pairs copied.

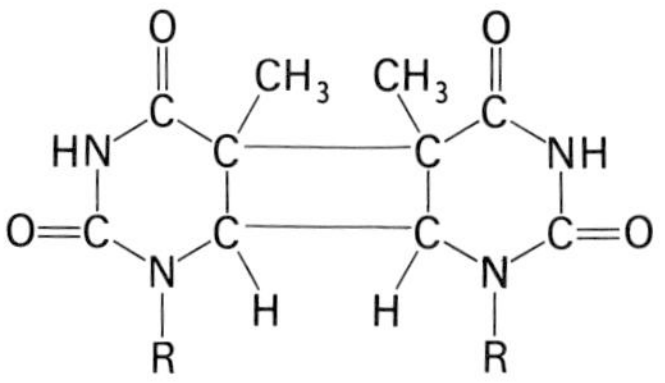

Thymine dimer

LESIONS IN DNA ARE CONTINUALLY BEING REPAIRED

DNA is damaged by a variety of chemical and physical agents, and so all cells possess mechanisms for repairing these lesions. Bases can be altered or lost, phosphodiester bonds in the backbone can be broken, and strands can become cross-linked. These lesions are produced by ionizing radiation, ultraviolet light, and a variety of chemicals. Much of the damage sustained by DNA can be repaired because genetic information is stored in both strands of the double helix, so that information lost by one strand can be retrieved from the other.

One of the best understood repair mechanisms is the excision of a *pyrimidine dimer* (Figure 24-45), *which is formed on exposure of DNA to ultraviolet light*. Adjacent pyrimidine residues on a DNA strand can become covalently linked under these conditions. Such a pyrimi-

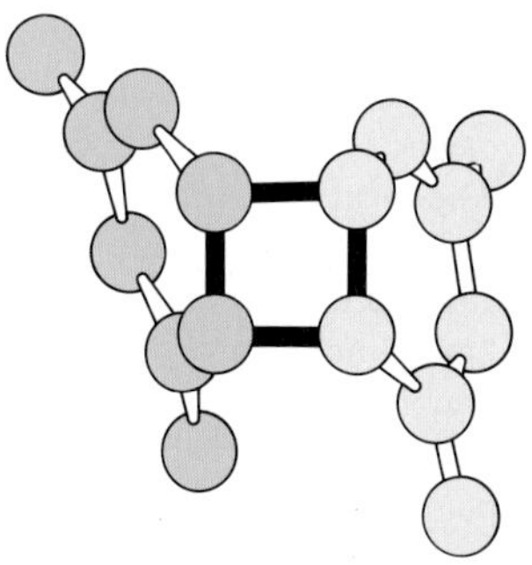

Figure 24-45
Model of a dimer of uracil produced by ultraviolet irradiation. A thymine dimer has nearly the same structure.

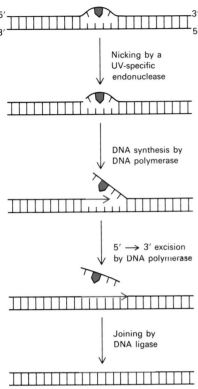

Figure 24-46
Repair of a region of DNA containing a thymine dimer by the sequential action of a specific endonuclease, a DNA polymerase, and a DNA ligase. The thymine dimer is shown in blue and the new region of DNA in red. [After P. C. Hanawalt, *Endeavor* 31(1972):83.]

Xeroderma—
From the Greek words for dry skin.
The term was first used by F. Hebra and M. Kaposi in 1874 to describe the "parchment skin" and abnormal pigmentation seen in one of their patients.

dine dimer cannot fit into a double helix and so replication and gene expression are blocked until the lesion is removed. Four enzymatic activities are essential for this repair process (Figure 24-46). First, a *UV-specific endonuclease* detects the distorted region and makes a nick near the dimer, usually on its 5′ side. The segment containing the dimer peels away, which allows *DNA polymerase I* (or a similar kind of polymerase) to carry out repair synthesis in the 5′ → 3′ direction. The 3′ end of the nicked strand is the primer, and the intact complementary strand is the template. Then, the pyrimidine dimer region is excised by the 5′ → 3′ *exonuclease* activity of DNA polymerase. Finally, the newly synthesized stretch of DNA and the original portion of the DNA chain are joined by *DNA ligase.* Alternatively, the pyrimidine dimer can be photochemically reversed. Nearly all cells contain a *photoreactivating enzyme* that recognizes the dimer and cleaves it to its original bases on absorbing blue light.

A SKIN CANCER IS CAUSED BY DEFECTIVE REPAIR OF DNA IN XERODERMA PIGMENTOSUM

Xeroderma pigmentosum is a rare skin disease in humans. It is genetically transmitted as an autosomal recessive trait. The skin in an affected homozygote is extremely sensitive to sunlight or ultraviolet light. In infancy, severe changes in the skin become evident and worsen with time. The skin becomes dry and there is a marked atrophy of the dermis. Keratoses appear, the eyelids become scarred, and the cornea ulcerates. Skin cancer usually develops at several sites. Many patients die before age thirty from metastases of these malignant skin tumors.

Ultraviolet light produces pyrimidine dimers in human DNA, as it does in *E. coli* DNA. Furthermore, the repair mechanisms seem to be similar. Studies of skin fibroblasts from patients with xeroderma pigmentosum have revealed a biochemical defect in one form of this disease. In normal fibroblasts, half of the pyrimidine dimers produced by ultraviolet radiation are excised in less than 24 hours. In contrast, there is almost no excision of dimers in this amount of time in fibroblasts derived from patients with xeroderma pigmentosum. Which step in repair is blocked? The answer came from studies of the molecular weights of DNA strands derived from UV-irradiated fibroblasts. In normal cells, there is a marked reduction of the molecular weight of single-stranded DNA within a few hours after radiation. This reduction in molecular weight is caused by the first step in the repair process, namely cleavage of a DNA chain near the pyrimidine dimer. In contrast, a reduction of molecular weight is not evident after UV-irradiation of xeroderma pigmentosum cells. Thus, *this skin disease can be caused by a defect in the endonuclease that hydrolyzes the DNA backbone near a pyrimidine dimer. The drastic clinical consequences of this enzymatic defect emphasize the critical importance of DNA repair processes.*

DNA CONTAINS THYMINE INSTEAD OF URACIL TO PERMIT REPAIR OF DEAMINATED CYTOSINE

Cytosine in DNA spontaneously deaminates at a perceptible rate to form uracil. The deamination of cytosine is potentially mutagenic because uracil pairs with adenine and so one of the daughter strands will contain an AU base pair rather than the original GC base pair.

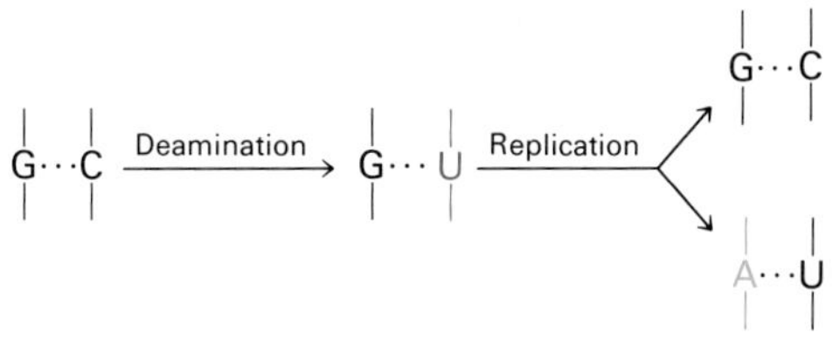

This mutation is prevented by a repair system that recognizes uracil to be foreign to DNA (Figure 24-47). First, a *uracil-DNA glycosidase* hydrolyzes the glycosidic bond between the uracil and deoxyribose moieties. At this stage, the DNA backbone is intact, but a base is missing. A specific endonuclease then recognizes this defect and nicks the backbone adjacent to the missing base. DNA polymerase I excises the residual deoxyribose phosphate unit and inserts cytosine, as dictated by the presence of guanine on the undamaged complementary strand. Finally, the repaired strand is sealed by DNA ligase.

For many years, the presence in DNA of thymine rather than uracil was an enigma. Both bases pair with adenine. The only difference between them is a methyl group in thymine in place of the C-5 hydrogen in uracil. Why is a methylated base employed in DNA and not in RNA? Recall that the methylation of deoxyuridylate to form deoxythymidylate is energetically expensive (p. 526). The recent discovery of an active repair system to correct the deamination of cytosine provides a convincing solution to this puzzle. Uracil-DNA glycosidase does not remove thymine from DNA. *Thus, the methyl group on thymine is a tag that distinguishes it from deaminated cytosine.* If this tag were absent, uracil correctly in place would be hard to distinguish from uracil formed by deamination. The defect would persist unnoticed, and so a GC base pair would necessarily be mutated to AU in one of the daughter DNA molecules. This mutation is prevented by a repair system that searches for uracil and leaves thymine alone. It seems likely that *thymine is used instead of uracil in DNA to enhance the fidelity of the genetic message.* In contrast, RNA is not repaired and so uracil is used in RNA because it is a less-expensive building block.

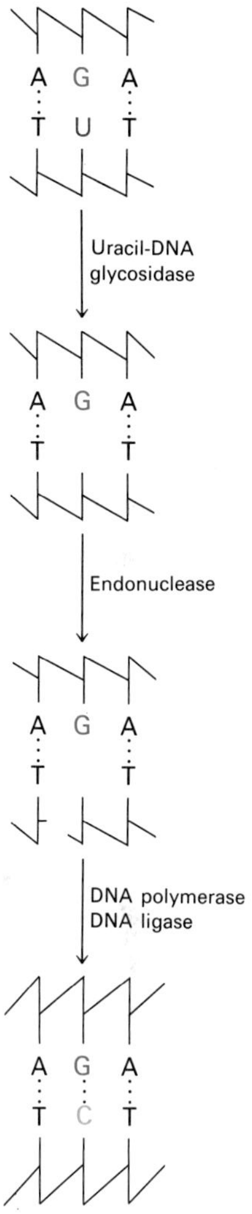

Figure 24-47
Uracil residues in DNA are excised and replaced by cytosine, the original base.

RESTRICTION ENDONUCLEASES HAVE REVOLUTIONIZED THE ANALYSIS OF DNA

Restriction enzymes are endonucleases that recognize specific base sequences in double-helical DNA and cleave both strands of the duplex. These exquisitely precise scalpels are marvelous gifts of nature to biochemists. Restriction enzymes have made possible experiments that were not even dreamed about just a few years ago. They are indispensable tools for analyzing chromosome structure, sequencing very long DNA molecules, isolating genes, and creating new DNA molecules that can be cloned. This development was pioneered by Werner Arber, Hamilton Smith, and Daniel Nathans.

Restriction endonucleases are found in a wide variety of procaryotes. Their biological role is to cleave foreign DNA molecules. The cell's own DNA is not degraded because the site recognized by its restriction enzyme is methylated. The interplay between restriction and modification will be considered in a later chapter (p. 732). The important point here is that many restriction enzymes recognize specific sequences of four to six base pairs and hydrolyze a phosphodiester bond in each strand in this region. A striking characteristic of these cleavage sites is that they possess *twofold rotational symmetry*. In other words, the sequence of base pairs is *palindromic*. The cleavage sites are symmetrically positioned with respect to the twofold axis. For example, the sequence recognized by a restriction enzyme from *Streptomyces achromogenes* is

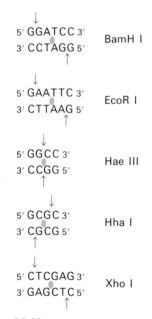

5′ GGATCC 3′   BamH I
3′ CCTAGG 5′

5′ GAATTC 3′   EcoR I
3′ CTTAAG 5′

5′ GGCC 3′   Hae III
3′ CCGG 5′

5′ GCGC 3′   Hha I
3′ CGCG 5′

5′ CTCGAG 3′   Xho I
3′ GAGCTC 5′

Figure 24-48
Specificities of some restriction endonucleases. The base-pair sequences that are recognized by these enzymes contain a twofold axis of symmetry. The two strands in these regions are related by a 180° rotation around the axis marked by the green symbol. The cleavage sites are denoted by red arrows. The abbreviated name of each restriction enzyme is given at the right of the sequence that it recognizes.

The CG phosphodiester bond in each strand distal to the symmetry axis is cleaved by this endonuclease. More than eighty restriction enzymes have been purified and characterized. Their names consist of a three letter abbreviation for the host organism (e.g., Eco for *E. coli,* Hin for *Haemophilus influenzae,* Hae for *H. aegyptius*) followed by a strain designation and a Roman numeral. The specificities of several of these enzymes are shown in Figure 24-48. Note that the cuts may be staggered or even.

Restriction enzymes are being used to cleave DNA molecules into specific fragments that are more readily analyzed and manipulated than the parent molecule. For example, the 5.1-kb circular duplex DNA of SV40 virus is cleaved at one site by EcoRI, four sites by Hpa, and eleven sites by Hind. A piece of DNA produced by the action of one restriction enzyme can be specifically cleaved into smaller fragments by another restriction enzyme. *Chromosomes can be mapped by using a series of restriction enzymes* (p. 760). *Furthermore, the pattern of restriction fragments can serve as a fingerprint of the DNA molecule.* Small differences between related DNA molecules can be readily detected because their restriction fragments can be separated and displayed by gel electrophoresis. For a particular gel, the electrophoretic mobility of a DNA fragment is inversely proportional to the logarithm of the number of base pairs up to a certain limit. Polyacrylamide gels are used to separate DNA molecules containing as many as 1000 base pairs, whereas more porous agarose gels are used to resolve larger molecules. Bands of radioactive DNA can be visualized by autoradiography. Alternatively, a gel can be stained with ethidium bromide, which exhibits an intense orange fluorescence on binding to double-helical DNA. A band containing 50 nanograms of DNA can readily be seen (Figure 24-49). Another important feature of these gels is their high resolving power.

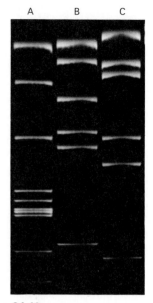

Figure 24-49
Gel electrophoresis pattern showing the fragments produced by cleaving SV40 DNA with each of three restriction enzymes. These fragments were made fluorescent by staining the gel with ethidium bromide. [Courtesy of Dr. Jeffrey Sklar.]

DNA CAN BE RAPIDLY SEQUENCED BY SPECIFIC CHEMICAL CLEAVAGE

The analysis of DNA structure and its relationship to gene expression has also been markedly facilitated by the development of powerful techniques for the sequencing of DNA molecules. The *chemical cleavage method* devised by Allan Maxam and Walter Gilbert starts with a DNA that is labeled at one end of one strand with ^{32}P. Polynucleotide kinase is usually used to insert ^{32}P at the 5' hydroxyl terminus. The labeled DNA is preferentially broken at one of the four nucleotides. The conditions are chosen so that an average of one break is made per chain. Partial cleavage at each base produces a nested set of radioactive fragments extending from the ^{32}P label to each of the positions of that base. For example, if the sequence is

$$5'-^{32}\text{P-GCTACGTA-}3'$$

the *radioactive* fragments produced by specific cleavage on the 5' side of each of the four bases would be

Cleavage at A:	^{32}P-GCT
	^{32}P-GCTACGT
Cleavage at G:	^{32}P-GCTAC
Cleavage at C:	^{32}P-G
	^{32}P-GCTA
Cleavage at T:	^{32}P-GC
	^{32}P-GCTACG

These fragments are then separated by polyacrylamide-gel electrophoresis, which can resolve DNA molecules differing in length by just one nucleotide. The next step is to look at an autoradiograph of this gel. As shown in Figure 24-50, the lowest band is in the C lane, the next one is in the T lane, followed by one in the A lane. Hence, the sequence of the first three nucleotides is 5' CTA 3'. Reading all seven bands in ascending order gives the sequence 5' CTACGTA 3'.

Cleavage at

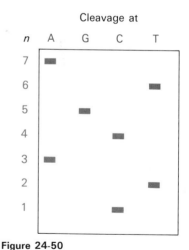

Figure 24-50
Schematic diagram of a gel showing the radioactive fragments formed by specific cleavage of 5'-^{32}P-GCTACGTA-3' at each of the four bases (A, G, C, and T lanes). The number of nucleotides (n) in a fragment is shown on the left. The base sequence is read directly in ascending order.

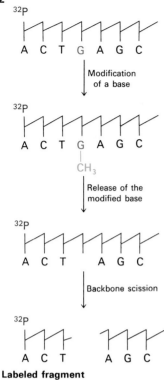

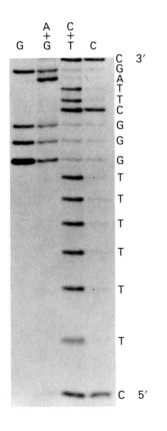

Figure 24-51
Strategy of the chemical-cleavage method for the sequencing of DNA.

Figure 24-52
Autoradiograph of a gel showing labeled fragments produced by chemical cleavage. The 5' end of the DNA strand was labeled with ^{32}P. The shortest nucleotide is at the bottom of the gel. Hence, the base sequence is 5'-CTTTTTTGGGCTTAGC-3'. [Courtesy of Dr. David Dressler.]

Thus, the autoradiograph of a gel produced from four different chemical cleavages displays a pattern of bands from which the sequence can be read directly.

How is the DNA specifically cleaved? Reagents that modify and then remove a base from its sugar are used (Figure 24-51). Purines are damaged by dimethylsulfate, which methylates guanine at N-7 and adenine at N-3. The glycosidic bond of a methylated purine is readily broken by heating at neutral pH, which leaves the sugar without a base. The backbone is then cleaved and the sugar is eliminated by heating in alkali. When the resulting end-labeled fragments are resolved on a polyacrylamide gel, a pattern of dark and light bands is seen on the autoradiograph. The dark bands correspond to fragments formed by breakage at guanine, because this base is methylated much more rapidly than is adenine. In contrast, the glycosidic bond of methylated adenosine is less stable than that of methylated guanosine. Hence, treatment with dilute acid after methylation preferentially releases adenine (A + G). Comparison of the A + G lane of a gel with a parallel G lane reveals whether cleavage occurred at A or G (Figure 24-52). Cytosine and thymine are split by *hydrazine.* The backbone is then cleaved by *piperidine,* which displaces the products of the hydrazine reaction and catalyzes elimination of the phosphates. These reactions give a series of bands of about equal intensity from cleavages at cytosines and thymines (C + T lane). These pyrimidines are distinguished by carrying out hydrazinolysis in the presence of 2 M NaCl, which suppresses the reaction with thymine (C lane). The DNA sequence is then read from the autoradiograph of the gel by comparing the G, A + G, C + T, and C lanes (Figure 24-52). The shortest fragment has the highest electrophoretic mobility and so the 5' end of the sequence starts at the bottom of the gel. Sequences of more than 150 bases can be readily determined by using several gels to resolve all of the labeled fragments.

THE ENTIRE BASE SEQUENCE OF φX174 DNA WAS DETERMINED BY ENZYMATIC REPLICATION TECHNIQUES

DNA can also be sequenced by generating fragments through the *controlled interruption of enzymatic replication,* a method developed by Fred Sanger and his associates. DNA polymerase I is used to copy a particular sequence of a single-stranded DNA. The synthesis is primed by a complementary fragment, which may be obtained from a restriction enzyme digest. In addition to the four radioactive deoxyribonucleoside triphosphates, the incubation mixture contains a *2′,3′-dideoxy analog* of one of them. The incorporation of this analog blocks further growth of the new chain because it lacks a 3′-hydroxyl terminus to form the next phosphodiester bond. Hence, fragments are produced in which the dideoxy analog is at the 3′ end (Figure 24-53). The four sets of *chain-terminated fragments* are then separated by gel electrophoresis and the base sequence of the new DNA is read from the autoradiograph of the gel. Sequences of as many as 200 bases can be determined starting with subpicomole amounts of purified DNA.

The complete sequence of the 5,375 bases in φX174 DNA was determined by Sanger by controlled enzymatic replication. The elucidation of the sequence of a complete DNA genome is a landmark in molecular biology. The wealth of information derived from this remarkable accomplishment will be considered in a later chapter (p. 632). It is interesting to note that Sanger's determination of the entire base sequence of a chromosome came a quarter century after his elucidation of the amino acid sequence of a protein.

SUMMARY

DNA is the molecule of heredity in all procaryotic and eucaryotic organisms. In viruses, the genetic material is either DNA or RNA. All cellular DNA consists of two very long, helical polynucleotide chains coiled around a common axis. The two strands of the double helix run in opposite directions. The sugar-phosphate backbone of each strand is on the outside of the double helix, whereas the purine and pyrimidine bases are inside. The two chains are held together by hydrogen bonds between pairs of bases. Adenine (A) is always paired with thymine (T), and guanine (G) is always paired with cytosine (C). Hence, one strand of a double helix is the complement of the other. Genetic information is encoded in the precise sequence of bases along a strand. Most DNA molecules are circular. The axis of the double helix of a circular DNA can itself be twisted to form a superhelix. Supertwisted DNA has a more compact shape than relaxed DNA.

In the replication of DNA, the two strands of a double helix unwind and separate as new chains are synthesized. Each parent strand acts as a template for the formation of a new complementary strand. Thus, the replication of DNA is semiconservative—each daughter molecule receives one strand from the parent DNA molecule. The replication of DNA is a complex process carried out by many proteins, including three kinds of DNA polymerases and a DNA ligase. The activated precursors in the synthesis of DNA are the four deoxyribonucleoside 5′-triphosphates. The new strand is

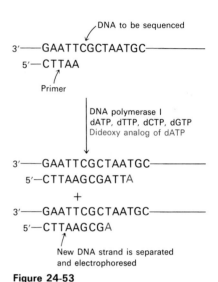

2′,3′-Dideoxy analog

Figure 24-53
Strategy of the chain-termination method for the sequencing of DNA. Fragments are produced by adding the 2′,3′-dideoxy analog of a dNTP to each of four polymerization mixtures. For example, the addition of the dideoxy analog of dATP results in fragments ending in A (shown in red).

synthesized in the $5' \rightarrow 3'$ direction by a nucleophilic attack by the 3'-hydroxyl terminus of the primer strand on the innermost phosphorus atom of the incoming deoxyribonucleoside triphosphate. Most important, DNA polymerases catalyze the formation of a phosphodiester bond only if the base on the incoming nucleotide is complementary to the base on the template strand. In other words, DNA polymerases are template-directed enzymes. DNA polymerases I, II, and III also have a $3' \rightarrow 5'$ exonuclease activity that enhances the fidelity of replication by removing mismatched residues. In addition, DNA polymerases I and III have a $5' \rightarrow 3'$ nuclease activity that is essential for the repair of DNA and for DNA replication.

DNA replication in *E. coli* starts at a unique origin and proceeds sequentially in opposite directions. At the replication fork (where two strands of DNA become four), both strands of parental DNA serve as templates for the synthesis of new DNA. The rep protein hydrolyzes ATP to unwind parental DNA. Unwinding is also made possible by DNA gyrase, which acts as a molecular swivel and pumps negative supertwists into parental DNA. One strand of DNA (the leading strand) is synthesized continuously, whereas the other strand (the lagging strand) is synthesized in the form of 1-kb fragments. The discontinuous assembly of the lagging strand enables $5' \rightarrow 3'$ polymerization at the atomic level to give rise to overall growth of this strand in the $3' \rightarrow 5'$ direction. The synthesis of new DNA is preceded by the synthesis of RNA, which serves as the primer. The RNA part of the nascent RNA-DNA chain is then hydrolyzed and replaced by DNA. Nascent DNA fragments that are base-paired to the same template strand then become linked by DNA ligase in a reaction driven by NAD^+.

Lesions in DNA produced by ionizing radiation, ultraviolet light, and a variety of chemical agents are continually being repaired. For example, pyrimidine dimers induced by ultraviolet light are excised by a specific endonuclease. Uracil formed by spontaneous deamination of cytosine in DNA is removed by a glycosidase that distinguishes between uracil and thymine.

Restriction enzymes recognize specific sequences with twofold symmetry and hydrolyze a phosphodiester bond in each DNA strand in such regions. These endonucleases can be used to cleave DNA molecules into specific fragments that can be readily analyzed. The development of rapid sequencing techniques based on specific fragmentation and gel-electrophoretic separation has also revolutionized the analysis of DNA structure.

SELECTED READINGS

WHERE TO START

Kornberg, A., 1979. Aspects of DNA replication. *Cold Spring Harbor Symp. Quant. Biol.* 43:1–9.

Alberts, B., and Sternglanz, R., 1977. Recent excitement in the DNA replication problem. *Nature* 269:655–661.

Fiddes, J. C., 1977. The nucleotide sequence of a viral DNA. *Sci. Amer.* 237(6):54–67. [Available as *Sci. Amer.* Offprint 1374.]

Hamilton, H. O., 1979. Nucleotide sequence specificity of restriction endonucleases. *Science* 205:455–622.

Nathans, D., 1979. Restriction endonucleases, simian virus 40, and the new genetics. *Science* 206:903–909.

Bauer, W. R., Crick, F. H. C., and White, J. H., 1980. Supercoiled DNA. *Sci. Amer.* 243(1):118–133. [A lucid account of the topology of DNA superhelices. Offprint 1474.]

BOOKS

Kornberg, A., 1980. *DNA Replication.* Freeman. (An outstanding and highly readable book.)

Cold Spring Harbor Laboratory, 1979. *Replication and Recombination.* Cold Spring Harbor Symposia of Quantitative Biology, vol. 43. (A most informative collection of articles on the replication, repair, and recombination of DNA.)

Bloomfield, V. A., Crothers, D. M., and Tinoco, I., Jr., 1974. *Physical Chemistry of Nucleic Acids.* Harper & Row.

Davidson, J. N., 1977. *The Biochemistry of the Nucleic Acids.* Academic Press.

Ts'o, P. O. P., 1974. *Basic Principles in Nucleic Acid Chemistry,* vol. 1 and 2. Academic Press.

DISCOVERY OF THE MAJOR CONCEPTS

Avery, O. T., MacLeod, C. M., and McCarty, M., 1944. Studies on the chemical nature of the substance inducing transformation of pneumococcal types. Induction of transformation by a deoxyribonucleic acid fraction isolated from Pneumococcus Type III. *J. Exp. Med.* 79:137–158.

Hershey, A. D., and Chase, M., 1952. Independent functions of viral protein and nucleic acid in growth of bacteriophage. *J. Gen. Physiol.* 36:39–56.

Watson, J. D., and Crick, F. H. C., 1953a. Molecular structure of nucleic acid. A structure for deoxyribose nucleic acid. *Nature.* 171:737–738.

Watson, J. D., and Crick, F. H. C., 1953b. Genetic implications of the structure of deoxyribonucleic acid. *Nature.* 171:964–967.

Kornberg, A., 1960. Biologic synthesis of deoxyribonucleic acid. *Science.* 131:1503–1508.

Meselson, M., and Stahl, F. W., 1958. The replication of DNA in *Escherichia coli. Proc. Nat. Acad. Sci.* 44:671–682.

Taylor, J. H., (ed.), 1965. *Selected Papers on Molecular Genetics.* Academic Press. (Contains the classic papers listed above.)

DNA SEQUENCING

Sanger, F., Nicklen, S., and Coulson, A. R., 1977. DNA sequencing with chain-terminating inhibitors. *Proc. Nat. Acad. Sci.* 74:5463–5467.

Maxam, A. M., and Gilbert, W., 1977. A new method for sequencing DNA. *Proc. Nat. Acad. Sci.* 74:560–564.

DNA REPLICATION PROTEINS

Wickner, S. H., 1978. DNA replication proteins of *Escherichia coli. Ann. Rev. Biochem.* 47:1163–1191.

Lehman, I. R., 1974. DNA ligase: structure, mechanism, and function. *Science* 186:790–797.

McHenry, C., and Kornberg, A., 1977. DNA polymerase III holoenzyme of *Escherichia coli:* purification and resolution into subunits. *J. Biol. Chem.* 252:6478–6484.

Rowen, L., and Kornberg, A., 1978. Primase, the dna G protein of *Escherichia coli:* an enzyme which starts DNA chains. *J. Biol. Chem.* 253:758–764.

Kornberg, A., Scott, J. F., and Bertsch, L. L., 1978. ATP utilization by *rep* protein in the catalytic separation of DNA strands at a replicating fork. *J. Biol. Chem.* 253:3298–3304.

Cozzarelli, N. R., 1977. The mechanism of action of inhibitors of DNA synthesis. *Ann. Rev. Biochem.* 46:641–668.

ORIGIN AND DIRECTION OF DNA REPLICATION

Prescott, D. M., and Kuempel, P. L., 1972. Bidirectional replication of the chromosome in *E. coli. Proc. Nat. Acad. Sci.* 69:2842–2845.

Bird, R. E., Lourarn, J., Martuscelli, J., and Caro, L., 1972. Origin and sequence of chromosome replication in *E. coli. J. Mol. Biol.* 70:549–566.

Ogawa, T., and Okazaki, T., 1980. Discontinuous DNA replication. *Ann. Rev. Biochem.* 49:421–458.

DNA STRUCTURE

Cantor, C. R., and Schimmel, P. R., 1980. *Biophysical Chemistry.* Freeman. [Chapters 3 and 6 in Part I and Chapters 22, 23, and 24 in Part III give an excellent account of the conformation of nucleic acids.]

SUPERHELICAL DNA

Bauer, W. R., 1978. Structure and reactions of closed duplex DNA. *Ann. Rev. Biophys. Bioeng.* 7:287–313.

Crick, F. H. C., 1976. Linking numbers and nucleosomes. *Proc. Nat. Acad. Sci.* 73:2639–2643. [A lucid introduction to linkage, writhing, and twist of superhelical DNA.]

596

Mizuuchi, K., O'Dea, M. H., and Gellert, M., 1978. DNA gyrase: subunit structure and ATPase activity of the purified enzyme. *Proc. Nat. Acad. Sci.* 75:5960–5963.

Cozzarelli, N. R., 1980. DNA gyrase and the supercoiling of DNA. *Science* 207:953–960.

DNA REPAIR

Hanawalt, P. C., Cooper, P. K., Ganesan, A. K., and Smith, C. A., 1979. DNA repair in bacteria and mammalian cells. *Ann. Rev. Biochem.* 48:783–836.

Lindahl, T., Ljungquist, S., Siegert, W., Nyberg, B., and Sperens, B., 1977. DNA *N*-glycosidases. Properties of uracil-DNA glycosidase from *Escherichia coli*. *J. Biol. Chem.* 252:3286–3294.

Cleaver, J. E., 1978. Xeroderma pigmentosum. *In* Stanbury, J. B., Wyngaarden, J. B., and Fredrickson, D. S., (eds.), *The Metabolic Basis of Inherited Disease* (4th ed.), pp. 1072–1095. McGraw-Hill.

REMINISCENCES AND HISTORICAL ACCOUNTS

Cairns, J., Stent, G. S., and Watson, J. D., (eds.), 1966. *Phage and the Origins of Molecular Biology.* Cold Spring Harbor Laboratory. [A fascinating collection of reminiscences by some of the architects of molecular biology.]

Watson, J. D., 1968. *The Double Helix.* Atheneum. [A lively, personal account of the discovery of the structure of DNA and its biological implications.]

Olby, R., 1974. *The Path to the Double Helix.* University of Washington Press.

Portugal, F. H., and Cohen, J. S., 1977. *A Century of DNA: A History of the Discovery of the Structure and Function of the Genetic Substance.* MIT Press.

Judson, H., 1979. *The Eighth Day of Creation.* Simon and Schuster.

PROBLEMS

1. Write the complementary sequence (in the standard $5' \rightarrow 3'$ notation) for:
   (a) GATCAA.
   (b) TCGAAC.
   (c) ACGCGT.
   (d) TACCAT.

2. The composition (in mole fraction units) of one of the strands of a double-helical DNA is [A] = 0.30 and [G] = 0.24.
   (a) What can you say about [T] and [C] for the same strand?
   (b) What can you say about [A], [G], [T], and [C] of the complementary strand?

3. The DNA of a deletion mutant of λ bacteriophage has a contour length of 15 μm instead of 17 μm. How many base pairs are missing from this mutant?

4. What result would Meselson and Stahl have obtained if the replication of DNA were conservative? Give the expected distribution of DNA molecules after 1.0 and 2.0 generations for conservative replication (i.e., the parental double helix stays together).

5. Whether the joining of two DNA chains by known DNA ligases is driven by NAD+ or ATP depends on the species. Suppose that a new DNA ligase requiring a different energy donor is found. Propose a plausible substitute for NAD+ or ATP in this reaction.

6. DNA ligase from *E. coli.* relaxes supertwisted circular DNA in the presence of AMP but not in its absence. What is the mechanism of this reaction and why is it dependent on AMP?

7. An autoradiograph of a gel containing four lanes of DNA fragments produced by chemical cleavage is shown in Figure 24-54. The DNA contained a ^{32}P label at its 5' end. What is its sequence?

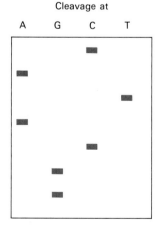

Figure 24-54
Schematic diagram of a gel showing the radioactive fragments formed by specific cleavage of an oligonucleotide.

For additional problems, see W. B. Wood, J. H. Wilson, R. M. Benbow, and L. E. Hood, *Biochemistry: A Problems Approach* (Benjamin, 1974), chs. 16 and 17.

CHAPTER 25

MESSENGER RNA
AND TRANSCRIPTION

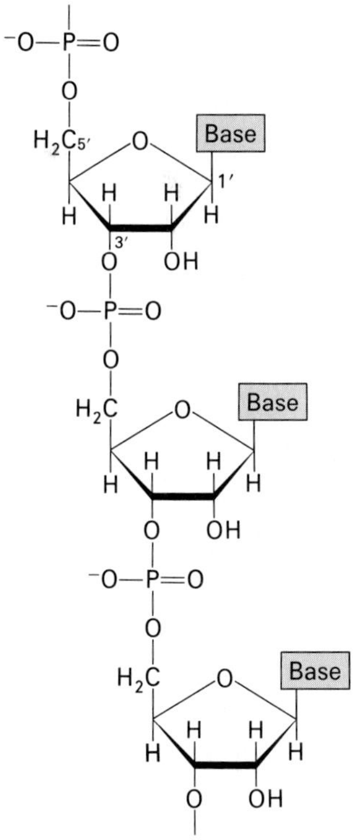

We turn now to the expression of the genetic information contained in DNA. Genes act by determining the kinds of proteins that are made by cells. However, DNA is not the direct template for protein synthesis. Rather, the templates for protein synthesis are RNA (ribonucleic acid) molecules. The flow of genetic information in normal cells is

$$\text{DNA} \xrightarrow[\text{transcription}]{} \text{RNA} \xrightarrow[\text{translation}]{} \text{protein}$$

In this chapter, evidence is presented that *certain RNA molecules are the information-carrying intermediates in protein synthesis, whereas other RNA molecules are part of the machinery of protein synthesis.* The synthesis of RNA according to instructions given by a DNA template is then considered. This process of *transcription* is followed by *translation,* in which RNA templates specify the synthesis of proteins, as will be discussed in Chapter 27.

The participation of RNA in protein synthesis seemed likely by 1940, several years before it was shown that DNA is the genetic material. Torbjörn Caspersson ascertained the distribution of RNA and DNA in single eucaryotic cells by light microscopy, making use

Figure 25-1
Structure of part of an RNA chain.

of the fact that both RNA and DNA strongly absorb 260-nm ultra-violet light, whereas only DNA is intensely stained by the Feulgen reagent. He found that almost all of the DNA is in the nucleus, whereas most of the RNA is in the cytoplasm. Jean Brachet came to the same conclusion by separating cells into nuclear and cytoplasmic fractions and chemically analyzing them for DNA and RNA. Furthermore, he showed that the RNA in the cytoplasmic fraction is located in small particles, in association with protein. These particles consisting of RNA and protein were shown to be the sites of protein synthesis. They are now called *ribosomes*.

STRUCTURE OF RNA

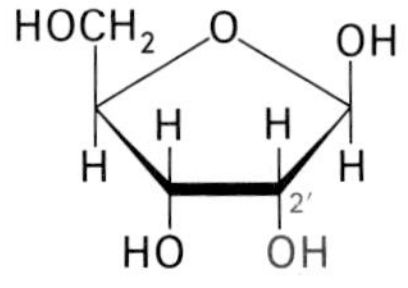

RNA, like DNA, is a long, unbranched macromolecule consisting of nucleotides joined by $3' \rightarrow 5'$ phosphodiester bonds. The number of nucleotides in RNA ranges from as few as seventy-five to many thousands. The covalent structure of RNA differs from that of DNA in two ways. As indicated by its name, the sugar unit in RNA is *ribose* rather than deoxyribose. Ribose contains a $2'$-hydroxyl group not present in deoxyribose. The other difference is that one of the four major bases in RNA is *uracil* (U) instead of thymine. Uracil, like thymine, can form a base pair with adenine. However, it lacks the methyl group present in thymine.

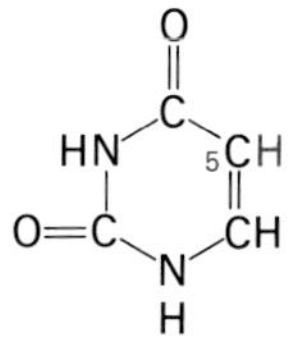

RNA molecules are single stranded, except in some viruses. Consequently, an RNA molecule need not have complementary base ratios. In fact, the proportion of adenine differs from that of uracil, and the proportion of guanine differs from that of cytosine, in most RNA molecules. However, RNA molecules do contain regions of double-helical structure that are produced by the formation of hairpin loops (Figure 25-2). In these regions, A pairs with U and G pairs

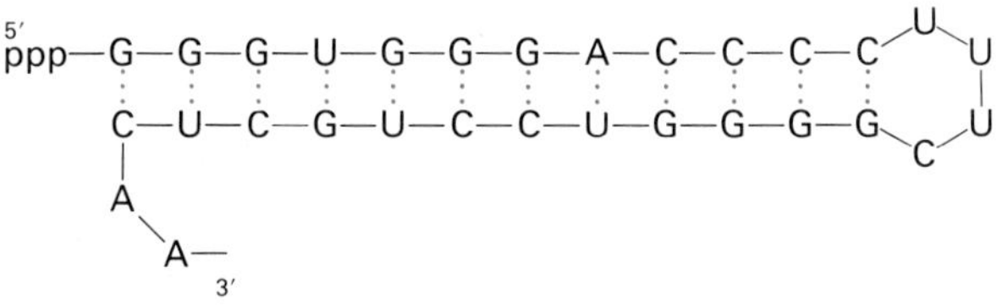

Ribose

Uracil

Figure 25-2
RNA can fold back on itself to form double-helical regions.

with C. G can also form a base pair with U, but it is less strong than the GC base pair (p. 650). The base pairing in RNA hairpins is frequently imperfect. Some of the apposing bases may not be complementary, and one or more bases along a single strand may be looped out to facilitate the pairing of the others. The proportion of helical regions in different kinds of RNA varies over a wide range; a value of 50% is typical.

CELLS CONTAIN THREE TYPES OF RNA: RIBOSOMAL, TRANSFER, AND MESSENGER

Cells contain three kinds of RNA (Table 25-1). *Messenger RNA* (mRNA) is the template for protein synthesis. There is an mRNA

Table 25-1
RNA molecules in *E. coli*

Type	Relative amount (%)	Sedimentation coefficient (S)	Mass (kdal)	Number of nucleotides
Ribosomal RNA (rRNA)	80	23	1.2×10^3	3,700
		16	0.55×10^3	1,700
		5	3.6×10^1	120
Transfer RNA (tRNA)	15	4	2.5×10^1	75
Messenger RNA (mRNA)	5	Heterogeneous		

molecule corresponding to each gene or group of genes that is being expressed. Consequently, mRNA is a very heterogeneous class of molecules. In *E. coli,* the average length of an mRNA molecule is about 1.2 kb. *Transfer RNA* (tRNA) carries amino acids in an activated form to the ribosome for peptide-bond formation, in a sequence determined by the mRNA template. There is at least one kind of tRNA for each of the twenty amino acids. Transfer RNA consists of about seventy-five nucleotides (25 kdal), which makes it the smallest of the RNA molecules. *Ribosomal RNA* (rRNA) is the major component of ribosomes, but its precise role in protein synthesis is not yet known. In *E. coli,* there are three kinds of rRNA, called 23S, 16S, and 5S RNA because of their sedimentation behavior. One molecule of each of these species of rRNA is present in each ribosome. Ribosomal RNA is the most abundant of the three types of RNA. Transfer RNA comes next, followed by messenger RNA, which comprises only 5% of the total RNA.

> *Svedberg unit (S)—*
> The sedimentation coefficients of biological macromolecules are expressed in Svedberg units (abbreviated as S). 1 S = 10^{-13} sec. Named after The Svedberg, the Swedish inventor of the ultracentrifuge.

FORMULATION OF THE CONCEPT OF MESSENGER RNA

The concept of mRNA was formulated by Francois Jacob and Jacques Monod in a classic paper published in 1961. Because proteins are synthesized in the cytoplasm rather than in the nucleus of eucaryotic cells, it was evident that there must be a chemical intermediate specified by the genes, which they called the structural messenger. What is the nature of this intermediate? An important clue came from their studies of the control of protein synthesis in *E. coli* (Chapter 28). Certain enzymes in *E. coli,* such as those that participate in the uptake and utilization of lactose, are inducible— that is, the amount of these enzymes increases more than a thousandfold if an inducer (such as isopropylthiogalactoside) is present.

The kinetics of induction were very revealing. The addition of an inducer elicited maximal synthesis of the lactose enzymes within a few minutes. Furthermore, the removal of the inducer resulted in the cessation of the synthesis of these enzymes in an equally short time. These experimental findings were incompatible with the presence of stable templates for the formation of these enzymes. Hence, Jacob and Monod surmised that *the messenger must be a very short-lived intermediate.* They then proposed that the messenger should have the following properties:

1. The messenger should be a polynucleotide.

2. The base composition of the messenger should reflect the base composition of the DNA that specifies it.

3. The messenger should be very heterogeneous in size because genes (or groups of genes) vary in length. They correctly assumed that three nucleotides code for one amino acid and calculated that the molecular weight of a messenger should be at least a half million.

4. The messenger should be transiently associated with ribosomes, the sites of protein synthesis.

5. The messenger should be synthesized and degraded very rapidly. A corollary is that base analogs such as 5-fluorouracil should be incorporated into the messenger within a few minutes.

It was apparent to Jacob and Monod that none of the known RNA fractions at that time met these criteria. Ribosomal RNA, then generally assumed to be the template for protein synthesis, was too homogeneous in size. Also, its base composition was similar in species that had very different DNA base ratios. Furthermore, rRNA was stable and it incorporated 5-fluorouracil at a slow rate. Transfer RNA also seemed an unlikely candidate for the same reasons. In addition, it was too small. However, there were suggestions in the literature of a third class of RNA that appeared to meet the above criteria for the messenger. In *E. coli* infected with T2 bacteriophage, there was a new RNA fraction of appropriate size that had a very short half-life. Most interesting, the base composition of this new RNA fraction was like that of the viral DNA rather than like that of *E. coli* DNA.

EXPERIMENTAL EVIDENCE FOR MESSENGER RNA, THE INFORMATIONAL INTERMEDIATE IN PROTEIN SYNTHESIS

The hypothesis of a short-lived messenger RNA as the information-carrying intermediate in protein synthesis was tested shortly after the concept was formulated. Sydney Brenner, Francois Jacob,

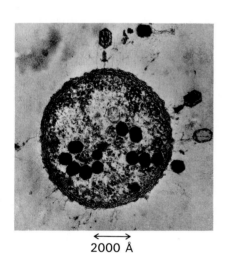

2000 Å

Figure 25-3
Electron micrograph of an *E. coli* cell infected by T2 viruses. [Courtesy of Dr. Lee Simon.]

and Matthew Meselson carried out experiments on *E. coli* infected with T2 bacteriophage. Nearly all of the proteins made after infection are genetically determined by the phage. The synthesis of these proteins is not accompanied by net synthesis of RNA. However, a minor RNA fraction with a short half-life appears soon after infection. In fact, as mentioned earlier, this RNA fraction has a nucleotide composition like that of the phage DNA. This system appeared to be optimal for a test of the messenger hypothesis because there was a sudden switch in the kinds of proteins synthesized. Another attractive feature was that rRNA and tRNA are not synthesized after infection.

How did this switch in the kinds of proteins made after infection occur? One possibility was that the phage DNA specifies a new set of ribosomes. In this model, genes control the synthesis of specialized ribosomes, and each ribosome can make only one kind of protein. An alternative model, the one proposed by Jacob and Monod, was that ribosomes are nonspecialized structures that receive genetic information from the gene in the form of an unstable messenger RNA. The experiments of Brenner, Jacob, and Meselson were designed to determine whether new ribosomes are synthesized after infection or whether new RNA joins preexisting ribosomes.

Bacteria were grown in a medium containing heavy isotopes (^{15}N and ^{13}C), infected with phage, and then immediately transferred to a medium containing light isotopes (^{14}N and ^{12}C). Constituents synthesized before and after infection could be separated by density-gradient centrifugation because their densities differed ("heavy" and "light," respectively). New RNA was labeled by the radioisotope ^{32}P- or ^{14}C-uracil and new protein by ^{35}S. These experiments showed that

1. *Ribosomes were not synthesized after infection,* as evidenced by the absence of "light" ribosomes.

2. RNA was synthesized after infection. Most of the radioactively labeled RNA emerged in the "heavy" ribosome peak. Thus, *most of the new RNA was associated with preexisting ribosomes.* Additional experiments showed that this new RNA turns over rapidly during the growth of phage.

3. The radioisotope ^{35}S appeared transiently in the "heavy" ribosome peak, which showed that *new proteins were synthesized in preexisting ribosomes.*

These experiments led to the conclusion that *ribosomes are nonspecialized structures which synthesize, at a given time, the protein dictated by the messenger they happen to contain.* Studies of uninfected bacterial cells also showed that messenger RNA is the information-carrying link between gene and protein. In a very short time, the concept of messenger RNA became a central facet of molecular biology.

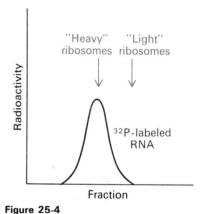

Figure 25-4
Density-gradient centrifugation of ribosomes and RNA of normal and T2 phage-infected bacteria. ^{32}P-labeled RNA synthesized *after* infection banded with ribosomes formed prior to infection ("heavy" ribosomes). Ribosomes were not synthesized after infection.

HYBRIDIZATION STUDIES SHOWED THAT MESSENGER RNA IS COMPLEMENTARY TO ITS DNA TEMPLATE

In 1961, Sol Spiegelman developed a new technique called *hybridization* to answer the following question: Is the base sequence of the RNA synthesized after infection with T2 phage complementary to the base sequence of T2 DNA? It was known from the work of Julius Marmur and Paul Doty that heating double-helical DNA above its melting temperature results in the formation of single strands. These strands reassociate to form a double-helical structure with biological activity if the mixture is cooled slowly. They also found that double-helical molecules are formed only from strands derived from the same or from closely related organisms. This observation suggested to Spiegelman that a double-stranded DNA-RNA hybrid might be formed from a mixture of single-stranded DNA and RNA if their base sequences were complementary. The experimental design was

1. The RNA synthesized after infection of *E. coli* with T2 phage (T2 mRNA) was labeled with ^{32}P. T2 DNA labeled with ^{3}H was prepared in a separate experiment.

2. A mixture of T2 mRNA and T2 DNA was heated to 100°C, which melted the double-helical DNA into single strands. This solution of single-stranded RNA and DNA was slowly cooled to room temperature.

3. The cooled mixture was analyzed by density-gradient centrifugation. The samples were centrifuged for several days in swinging bucket rotors. The plastic sample tubes were then punctured at the bottom and drops were collected for analysis.

Three bands were found (Figure 25-6). The densest of these was single-stranded RNA. A second band contained double-helical DNA. A third band consisting of double-stranded DNA-RNA hybrid molecules was present near the DNA band. Thus, T2 mRNA formed a hybrid with T2 DNA. In contrast, T2 mRNA did not hybridize with DNA derived from a variety of bacteria and unrelated viruses, even if their base ratios were like those of T2 DNA. Subsequent experiments showed that the mRNA fraction of uninfected cells hybridized with the DNA derived from that particular

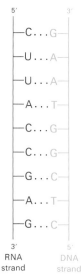

Figure 25-5
An RNA-DNA hybrid can be formed if the RNA and DNA have complementary sequences.

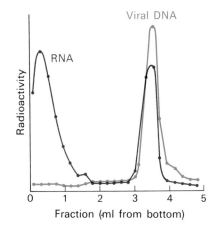

Figure 25-6
The RNA produced after *E. coli* has been infected with T2 phage is complementary to the viral DNA. In this hybridization experiment, RNA was labeled with ^{32}P, whereas T2 DNA was labeled with ^{3}H. The distribution of radioactivity in this cesium chloride density gradient shows that much of the RNA synthesized after infection bands with the T2 DNA. [After S. Spiegelman. Hybrid nucleic acids. Copyright © 1964 by Scientific American, Inc. All rights reserved.]

organism but not from unrelated ones. These incisive experiments revealed that *the base sequence of mRNA is complementary to that of its DNA template.* Furthermore, a powerful tool for tracing the flow of genetic information in cells and for determining whether two nucleic acid molecules are similar was developed.

RIBOSOMAL RNA AND TRANSFER RNA ARE ALSO SYNTHESIZED ON DNA TEMPLATES

The hybridization technique was then used to determine whether rRNA and tRNA are also synthesized on DNA templates. The formation of RNA-DNA hybrids was detected by a filter assay rather than by density-gradient centrifugation because it is simpler, more sensitive, and much more rapid. Single-stranded RNA passes through a nitrocellulose filter, whereas double-helical DNA and RNA-DNA hybrids are retained by this filter. RNA from *E. coli* labeled with ^{32}P was added to unlabeled *E. coli* DNA. This mixture was heated, slowly cooled, and then filtered through nitrocellulose. The radioactivity retained on the filter was counted. The results were unequivocal: *RNA-DNA hybrids were formed with both rRNA (5S, 16S, and 23S) and tRNA, which shows that complementary sequences for these RNA molecules are present in the* E. coli *genome.*

ALL CELLULAR RNA IS SYNTHESIZED BY RNA POLYMERASE

The concept of mRNA stimulated the search for an enzyme that synthesizes RNA according to instructions given by a DNA template. The experimental strategy was like the one used in the search for DNA polymerase I. In 1960, Jerard Hurwitz and Samuel Weiss independently discovered such an enzyme, which they named *RNA polymerase*. The enzyme from *E. coli* requires the following components for the synthesis of RNA:

1. *Template.* The preferred template is *double-stranded DNA.* Single-stranded DNA can also serve as a template. RNA, whether single or double stranded, is not an effective template, nor are RNA-DNA hybrids.

2. *Activated precursors.* All four *ribonucleoside triphosphates*—ATP, GTP, UTP, and CTP—are required.

3. *Divalent metal ion.* Mg^{2+} or Mn^{2+} are effective. Mg^{2+} meets this requirement in vivo.

RNA polymerase catalyzes the initiation and elongation of RNA chains. The reaction catalyzed by this enzyme is

$$(\text{RNA})_{n \text{ residues}} + \text{ribonucleoside triphosphate} \Longleftrightarrow$$

$$(\text{RNA})_{n+1 \text{ residues}} + \text{PP}_i$$

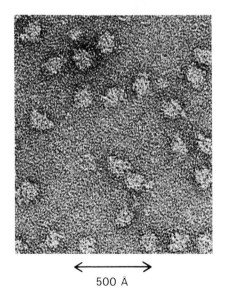

$$\longleftrightarrow$$
500 Å

Figure 25-7
Electron micrograph of RNA polymerase from *E. coli.* [Courtesy of Dr. Robley Williams and Dr. Michael Chamberlin.]

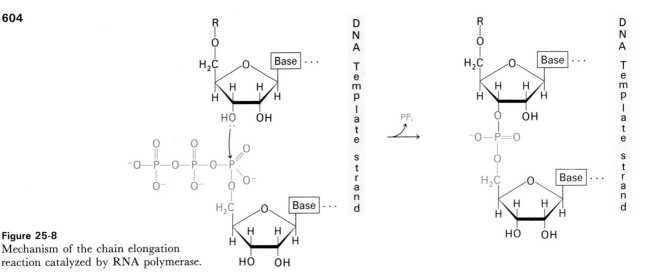

Figure 25-8
Mechanism of the chain elongation reaction catalyzed by RNA polymerase.

The synthesis of RNA is like that of DNA in several respects (Figure 25-8). First, the direction of synthesis is *5′ → 3′*, as will be discussed shortly. Second, the mechanism of elongation appears to be similar. There is a nucleophilic attack by the 3′-OH group at the terminus of the growing chain on the innermost nucleotidyl phosphate of the incoming nucleoside triphosphate. Third, the synthesis is driven forward by the hydrolysis of pyrophosphate.

RNA synthesis differs from that of DNA in several important ways. First, RNA polymerase does not require a primer. Second, the DNA template is fully conserved in RNA synthesis, whereas it is semiconserved in DNA synthesis. Third, RNA polymerase has no known nuclease activities.

All three types of cellular RNA—mRNA, tRNA, and rRNA—are synthesized in *E. coli* by the same RNA polymerase according to instructions given by a DNA template. In mammalian cells, there is a division of labor among several different kinds of RNA polymerases. It is also significant that some viruses code for RNA synthesizing enzymes that are very different from those of their host cells, as exemplified by the RNA polymerase specified by T7, a DNA phage, and by the RNA replicase that is encoded by Qβ, an RNA phage. The Qβ replicase is an RNA-dependent RNA polymerase because it takes instructions from an RNA template rather than from a DNA template (Chapter 30). In contrast, the cellular enzymes that synthesize RNA are *DNA-dependent RNA polymerases*.

RNA POLYMERASE TAKES INSTRUCTIONS FROM A DNA TEMPLATE

RNA polymerase, like the DNA polymerases described in the preceding chapter, takes instructions from a DNA template. The earliest evidence was provided by the finding that the *base composition* of the newly synthesized RNA is the complement of that of the DNA template strand. If poly (dT), a synthetic polydeoxyribonucleotide containing only thymidylate residues, is used as the template, then only one ribonucleoside triphosphate (ATP) is incorporated into the new polyribonucleotide chain. The product is polyriboadenylate (abbreviated as poly rA). If the alternating copolymer poly d(A-T) is used as the template for RNA polymerase, then UTP and ATP are incorporated and the product is poly r(A-U). The base composition of RNA synthesized using single-stranded φX174 DNA as template further demonstrates the complementary relationship between RNA product and DNA template (Table 25-2). Hybridization experiments also revealed that RNA synthesized by RNA polymerase is complementary to its DNA template. The strongest evidence for the fidelity of transcription comes from base-sequence studies showing that the RNA sequence is the precise complement of the DNA template sequence (Figure 25-9).

Table 25-2
Base composition of RNA synthesized from a viral DNA template

DNA template (*plus strand of φX174*)		RNA product	
A	0.25	0.25	U
T	0.33	0.32	A
G	0.24	0.23	C
C	0.18	0.20	G

5′-GCGGCGACGCGCAGUUAAUCCCACAGCCGCCAGUUCCGCUGGCGGCAUUUU-3′   **mRNA**

3′-CGCCGCTGCGCGTCAATTAGGGTGTCGGCGGTCAAGGCGACCGCCGTAAAA-5′   **DNA**

5′-GCGGCGACGCGCAGTTAATCCCACAGCCGCCAGTTCCGCTGGCGGCATTTT-3′

Figure 25-9
The base sequence of mRNA is the complement of that of the DNA template strand. The sequence shown here is from the tryptophan operon.

USUALLY ONLY ONE STRAND OF DNA IS TRANSCRIBED IN A PARTICULAR REGION OF THE GENOME

Are both strands of a double-helical DNA template transcribed or is transcription restricted to one of them? A priori, it seems unlikely that both strands of a particular region of DNA could code for functional proteins. Hybridization studies of *E. coli* infected with the φX174 phage provided one of the earliest answers to this question. The φX174 phage particles contain a single strand of DNA called the plus strand. After the plus strand enters the bacterium, synthesis of a complementary minus strand of DNA yields a circular, double-stranded DNA molecule termed the replicative form (RF). This RF DNA directs the synthesis of mRNA, which in turn specifies the synthesis of phage proteins. Radioactive phage mRNA was prepared by adding ^{32}P-phosphate to the medium shortly after *E. coli* was infected with φX174. This labeled mRNA was isolated. Also, RF DNA was separated into its constituent plus and minus strands, which is feasible because they differ in base composition and hence in buoyant density. A hybridization assay was then carried out to determine whether the newly synthesized mRNA was complementary to the plus strand or to the minus strand or to both strands. A decisive result was obtained: only the minus strand of the

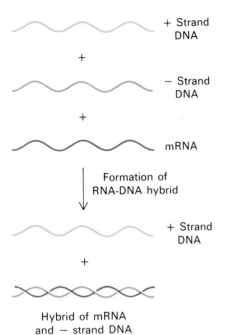

+ Strand
DNA

+

− Strand
DNA

+

mRNA

Formation of
RNA-DNA hybrid

+ Strand
DNA

+

Hybrid of mRNA
and − strand DNA

Figure 25-10
Only one strand of the RF DNA of
φX174 serves as a template for transcription, as shown by this hybridization experiment.

RF DNA formed a hybrid with the labeled mRNA. Thus, *only one
strand of the RF DNA of φX174 serves as a template for transcription in
vivo.*

Likewise, only one strand of the DNA of viruses such as T7, SP8,
and α is transcribed in vivo. The situation for a virus such as the T4
or λ phage is more complex. In part of its genome, one strand serves
as the template, whereas, in a different region, the other strand may
be the template. This change in template between one group of
genes and another also occurs in *E. coli.*

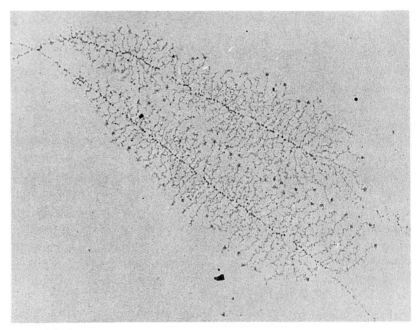

Figure 25-11
Electron micrograph showing the process of transcription. [From O. L.
Miller, Jr., and Barbara A. Hamkalo. Visualization of RNA synthesis on chro-
mosomes. *Int. Rev. Cytol.* 33(1972):1.]

RNA POLYMERASE FROM *E. coli* CONSISTS OF SUBUNITS

RNA polymerase from *E. coli* is a very large and complex enzyme.
The mass of the whole enzyme is nearly 500 kdal. RNA polymerase
is made up of subunits (Table 25-3), as can be shown by dissociating
the enzyme in a concentrated urea solution. The subunit composi-
tion of the entire enzyme, called a *holoenzyme,* is $\alpha_2\beta\beta'\sigma$. As will be
discussed shortly, the σ subunit dissociates from the enzyme follow-
ing the initiation of RNA synthesis. RNA polymerase without the
sigma subunit is called the *core enzyme* ($\alpha_2\beta\beta'$). The catalytic site of
RNA polymerase is contained in the $\alpha_2\beta\beta'$ core. The β' subunit has
been implicated in the binding of the DNA template, and the β

Table 25-3
Subunits of RNA polymerase
from *E. coli*

Subunit	Number	Mass (kdal)
α	2	40
β	1	155
β'	1	165
σ	1	95

subunit in the binding of ribonucleoside triphosphate substrates. The σ subunit of the holoenzyme participates in the selection of initiation sites for transcription.

The synthesis of DNA by *E. coli* RNA polymerase takes place in three stages: (1) initiation, (2) elongation, and (3) termination, which are discussed below.

TRANSCRIPTION IS INITIATED AT PROMOTER SITES ON THE DNA TEMPLATE

Transcription starts at specific sites, called *promoters,* on the DNA template. How does RNA polymerase find these sites? One approach to this question is to isolate and sequence fragments that are protected by RNA polymerase holoenzyme from digestion by pancreatic deoxyribonuclease. Another approach is to analyze certain mutants with enhanced or diminished rates of transcription of particular genes. Such studies have shown that promoter sites consist of about forty base pairs, which correspond to a span of DNA of about 140 Å. Analyses of a variety of promoters in *E. coli* and phage DNA suggest that a sequence of seven base pairs centered about ten bases before the start of the mRNA is a key part of the recognition signal (Figure 25-13). A second recognition site, located about thirty-five bases away from the start of the mRNA, has also been implicated in the initial binding of RNA polymerase.

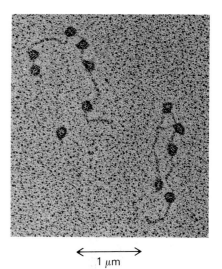

Figure 25-12
Electron micrograph of RNA polymerase holoenzyme bound to several promoter sites on a fragment of T7 phage DNA. [From R. C. Williams. *Proc. Nat. Acad. Sci.* 74(1977):2313.]

mRNA starts here
↓

	5′														3′
(A)	C G T A T G T T G T G T G G A
(B)	G C T A T G G T T A T T T C A
(C)	G T T A A C T A G T A C G C A
(D)	G T G A T A C T G A G C A C A
(E)	G T T T T C A T G C C T C C A
(F)	T T A T A A T G G T T A C A A

Figure 25-13
Promoter sequences from the (A) *lac,* (B) *gal,* and (C) *trp* operons of *E. coli,* (D) lambda phage, (E) φX174 phage, and (F) SV40 virus. The presumed region of homology, known as the Pribnow sequence, is shown in green.

THE SIGMA SUBUNIT ENABLES RNA POLYMERASE TO RECOGNIZE PROMOTER SITES

The $\alpha_2\beta\beta'$ core of RNA polymerase is unable to start transcription at promoter sites. Rather, *the $\alpha_2\beta\beta'\sigma$ holoenzyme is essential for specific initiation.* The role of the sigma subunit was discovered in the following way. RNA polymerase purified by chromatography on a phosphocellulose column was found to be nearly inactive when assayed with T4 DNA as template but still quite active with calf-thymus DNA as template. In contrast, RNA polymerase purified by centrifugation on a glycerol gradient was highly active with both templates. This observation suggested that a factor might be miss-

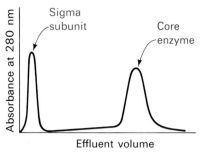

Figure 25-14
Resolution of RNA polymerase into the σ subunit and the core enzyme ($\alpha_2\beta\beta'$) on a phosphocellulose column.

ing from the phosphocellulose-purified RNA polymerase. This was indeed the case (Figure 25-14). The activity of the phosphocellulose-purified enzyme could be markedly enhanced by the addition of another fraction of the effluent of this column. This stimulatory fraction, called the sigma factor, had no catalytic activity by itself. Subsequent experiments showed that the phosphocellulose-purified enzyme lacked the sigma subunit, whereas the glycerol-gradient–purified enzyme retained this subunit. Thus, the enzyme that was fully active with the T4 DNA template was the $\alpha_2\beta\beta'\sigma$ holoenzyme, whereas the $\alpha_2\beta\beta'$ core enzyme was ineffective in transcribing this DNA. The addition of σ to the core enzyme led to the reconstitution of the fully active holoenzyme. The core enzyme was able to transcribe calf-thymus DNA because this template contains many nicks. RNA synthesized in vitro by the core polymerase does not correspond to in vivo transcripts. For example, the core enzyme transcribes both strands of phage DNA templates, whereas the holoenzyme asymmetrically transcribes one strand, as it does in vivo.

The sigma subunit contributes to specific initiation by *decreasing the affinity of RNA polymerase for general regions of DNA by a factor of 10^4*. In addition, *sigma activates RNA polymerase to recognize promoter sequences*. The sigma subunit also participates in *opening the DNA double helix* so that one of its strands can serve as a template. In fact, nearly one turn of the DNA helix is unwound by the binding of RNA polymerase holoenzyme. The stage is then set for the formation of the first phosphodiester bond of the new RNA chain.

The sigma subunit dissociates from the holoenzyme after the new RNA chain is started. The core polymerase continues to transcribe the DNA template. Thus, *the role of the holoenzyme is selection and initiation, whereas that of the core enzyme is elongation*. The dissociated sigma subunit then joins another core polymerase to initiate another round of transcription.

The starting of new RNA chains is controlled in a variety of ways. Some promoter sequences stimulate a high rate of initiation, whereas others are less effective. The rate of initiation is also regulated by proteins that bind to DNA within promoter sites or next to them. Repressors block transcription by interfering with the binding of RNA polymerase, whereas positive regulatory factors enhance initiation by facilitating the binding of RNA polymerase. These important control mechanisms will be discussed in detail in Chapter 28.

RNA CHAINS START WITH pppG OR pppA

Most newly synthesized RNA chains carry a tag that reveals how RNA chains are started. The 5′ end of a new RNA chain is highly distinctive: *the molecule starts with either pppG or pppA*. In contrast with DNA synthesis, primer is not needed. *RNA chains can be formed de novo*.

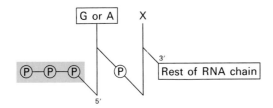

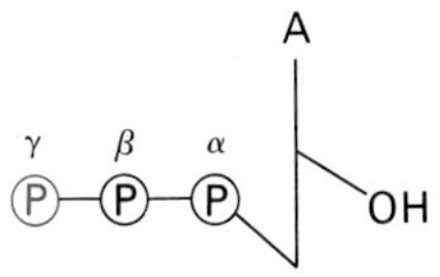

^{32}P γ-labeled ATP

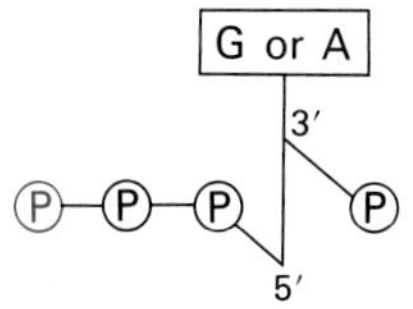

**Radioactive nucleoside
tetraphosphate**

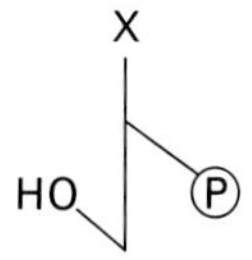

**Nucleoside
2′-(3′-) monophosphates**

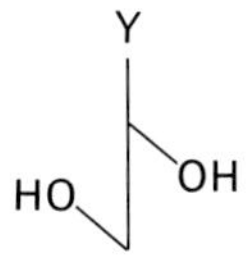

Nucleoside

This telltale tag at the 5′ terminus was discovered in two ways. It was found that RNA chains incorporated ^{32}P when the incubation mixture contained ^{32}P γ-labeled ATP. Clearly, the incorporated label had to be in a terminal position because only the α phosphorus atom of a nucleoside triphosphate can become part of internal phosphodiester bridges in the RNA. Furthermore, alkaline hydrolysis of newly synthesized RNA yielded three kinds of products: nucleosides, nucleoside 2′- (or 3′-) monophosphates, and nucleoside tetraphosphates (Figure 25-15). When ^{32}P γ-labeled ATP was used as a substrate, adenosine 3′-phosphate-5′-triphosphate was formed. The γ-phosphorus atom of this nucleoside tetraphosphate was labeled. The corresponding result was obtained when ^{32}P γ-labeled GTP was used. However, radioactivity was not incorporated when γ-labeled UTP or CTP were used. *Thus, a new RNA chain has a triphosphate group at its 5′ terminus and a free hydroxyl group at its 3′ terminus.*

RNA CHAINS ARE SYNTHESIZED IN THE 5′ → 3′ DIRECTION

Is RNA synthesized in the 5′ → 3′ direction or in the 3′ → 5′ direction? Two contrasting mechanisms of chain growth are depicted in Figure 25-16. For 5′ → 3′ growth, the triphosphate terminus is in-

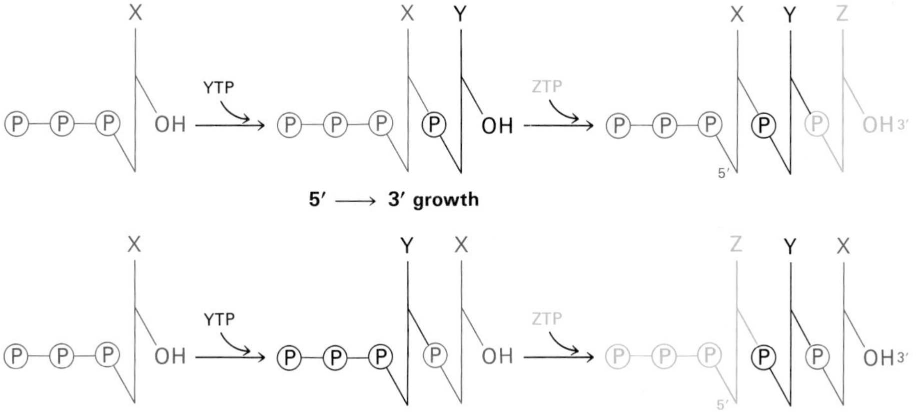

Figure 25-15
Alkaline hydrolysis products from an RNA chain synthesized with γ-labeled ^{32}P ATP.

5′ ⟶ 3′ growth

3′ ⟶ 5′ growth

Figure 25-16
Location of ^{32}P expected for 5′ → 3′ growth and for 3′ → 5′ growth. The observed location of label shows that RNA chains are synthesized in the 5′ → 3′ direction.

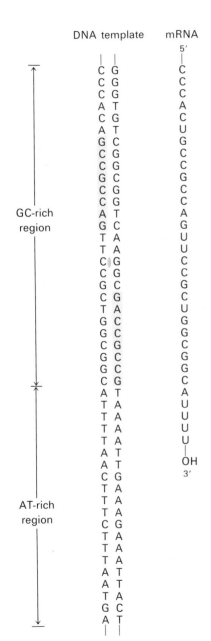

Figure 25-17
DNA sequence corresponding to the 3′ end of *trp* mRNA from *E. coli*. A two-fold axis of symmetry (marked by the green symbol) relates the base pairs that are shaded yellow.

serted at the start of the synthesis of a chain, whereas in 3′ → 5′ growth, the triphosphate terminus comes from the last residue that is incorporated. The kinetics of incorporation of radioactivity into RNA synthesized from ^{32}P γ-labeled GTP (or ATP) distinguished between these alternatives. When ^{32}P γ-labeled GTP was used as a substrate, the ratio of ^{32}P incorporation to total nucleotide incorporation was highest shortly after the components were mixed and then decreased progressively with time. Furthermore, the total radioactivity of the RNA already labeled was not reduced by the subsequent addition of a large excess of nonradioactive GTP to the incubation mixture. Thus, ^{32}P entered the RNA molecule at the start of its synthesis rather than at the last step. *Hence, the growth of an RNA chain is in the 5′ → 3′ direction, as in DNA synthesis.*

The RNA polymerase core moves along the DNA template strand in the 3′ → 5′ direction because the template strand is antiparallel to the newly synthesized RNA strand. The same RNA polymerase molecule synthesizes an entire transcript—in other words, transcription is *processive.* The transcribed region of DNA regains its double-helical conformation as the next section of DNA unwinds. The maximal rate of elongation is about 50 nucleotides per second.

It is important to note that RNA polymerase lacks nuclease activity. *RNA polymerase, in contrast with DNA polymerase, does not edit the nascent polynucleotide chain. In turn, the fidelity of transcription is much lower than that of replication.* The error rate of RNA synthesis is of the order of one mistake per 10^4 or 10^5, which is about 10^5 times as high as that in DNA synthesis. The much lower fidelity of RNA synthesis can be tolerated because a cell synthesizes many RNA transcripts of a gene.

THE DNA TEMPLATE CONTAINS STOP SIGNALS FOR TRANSCRIPTION

The termination of transcription is as precisely controlled as its initiation. The DNA template contains stop signals, which are being deciphered by comparing a variety of base sequences of these regions. A common characteristic is that *a GC-rich region before the termination site is followed by an AT-rich sequence.* The most distinctive feature of termination sequences is the *twofold symmetry of their GC-rich region* (Figure 25-17). Hence, the RNA transcript of this region is self-complementary, and so it can base pair to form a hairpin structure (Figure 25-18). In addition, *the nascent RNA chain ends with several U residues,* which are specified by the series of A bases in the AT-rich region of the DNA template. One or more of these structural features cause RNA polymerase to pause when it encounters such a signal. At some termination sites, the nascent RNA chain is released without the assistance of another protein. At other sites, the participation of the *rho protein* is essential for chain termination.

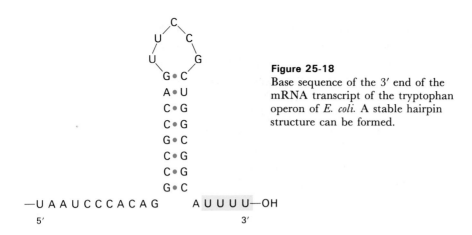

Figure 25-18
Base sequence of the 3′ end of the mRNA transcript of the tryptophan operon of *E. coli*. A stable hairpin structure can be formed.

THE RHO PROTEIN PARTICIPATES IN THE TERMINATION OF TRANSCRIPTION

Evidence for the participation of the rho protein in termination at some sites came from the finding that RNA molecules synthesized in vitro in its presence are shorter than those made in its absence. For example, RNA synthesized from fd phage DNA in the presence of rho has a sedimentation coefficient of 10S, whereas 23S RNA is made in its absence. Additional information about the action of rho was obtained by adding this termination factor to the incubation mixture at various times after the initiation of RNA synthesis. Spe-

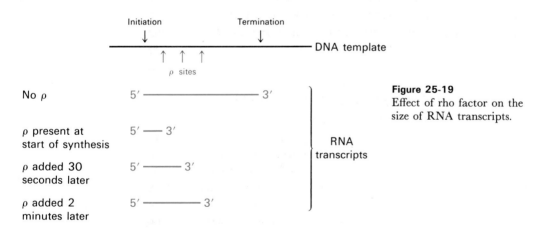

Figure 25-19
Effect of rho factor on the size of RNA transcripts.

cies with sedimentation coefficients of 13S, 17S, and 23S were obtained when rho was added a few seconds, two minutes, and ten minutes, respectively, after initiation. This result suggests that the template contains at least three termination sites that respond to rho (yielding 10S, 13S, and 17S RNA) and one termination site that does not require rho (yielding 23S RNA). Thus, specific termination can occur in the absence of rho. However, rho detects additional termination signals that are not recognized by RNA polymerase alone.

How does the rho protein provoke termination of RNA synthesis at certain stop signals? One possibility is that rho, a 200-kdal tetramer, attaches to RNA and moves toward an RNA polymerase molecule that is paused at a termination site. According to this model, rho displaces RNA polymerase from the 3'-OH end of RNA, which releases the RNA transcript. It is interesting to note that rho hydrolyzes ATP in terminating transcription.

As might be expected, the termination of transcription of certain genes is controlled. As will be described later, bacteriophage λ synthesizes *antitermination proteins* that allow certain genes to be transcribed and expressed (p. 680). In *E. coli*, specialized termination signals called *attenuators* are regulated to meet the nutritional needs of the cell (p. 677).

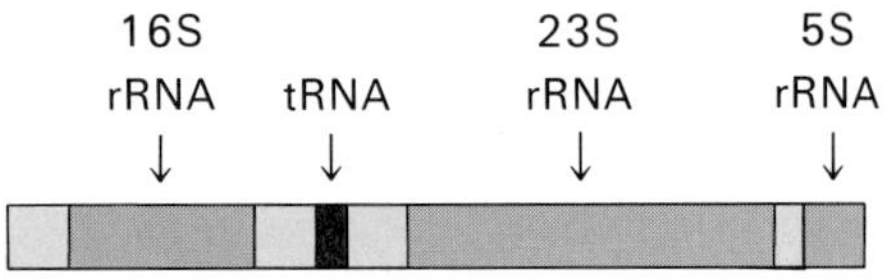

Figure 25-20
Cleavage of this primary transcript produces 5S, 16S, and 23S rRNA molecules and a tRNA molecule. Spacer regions are shown in yellow.

MANY RNA MOLECULES ARE CLEAVED AND CHEMICALLY MODIFIED AFTER TRANSCRIPTION

The formation of functionally active RNA molecules continues after transcription is completed. *In procaryotes, transfer RNA molecules and ribosomal RNA molecules are generated by the cleavage and chemical modification of certain nascent RNA chains.* For example, in *E. coli*, three kinds of ribosomal RNA molecules and a transfer RNA molecule are excised from a primary RNA transcript that also contains spacer regions (Figure 25-20). Other transcripts contain arrays of several kinds of tRNAs or of several copies of the same tRNA. The nucleases that cleave and trim these precursors of rRNA and tRNA are highly precise. For example, ribonuclease P generates the correct 5'-terminus of all tRNA molecules in *E. coli*. Ribonuclease III excises 5S, 16S, and 23S rRNA precursors from the primary transcript by cleaving specific sites in double-helical hairpin regions. *In contrast, mRNA molecules in procaryotes undergo little or no modification.* In fact, many of them are translated while they are being transcribed. However, some viral mRNAs (e.g., T7 phage mRNA) are cleaved by ribonuclease III before they are translated.

A second type of processing is the *addition of nucleotides to the termini of some RNA chains.* For example, CCA is added to the 3' end of tRNA molecules that do not already possess this terminal sequence. In eucaryotes, a long stretch of poly A is added to the 3' end and a methylated G nucleotide (called a cap) is added to the 5' end of most mRNA molecules (p. 706).

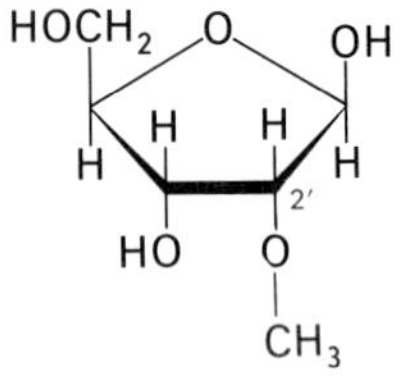

Modifications of bases and ribose units constitute a third type of processing. In eucaryotes, the 2'-hydroxyl group of about one of a hundred ribose units in rRNA is methylated enzymatically by *S*-adenosylmethionine. In bacteria, bases rather than ribose units in rRNA are methylated. Especially interesting are the unusual bases found in all tRNA molecules (p. 646). They are formed by enzymatic modification of a standard ribonucleotide in a tRNA precur-

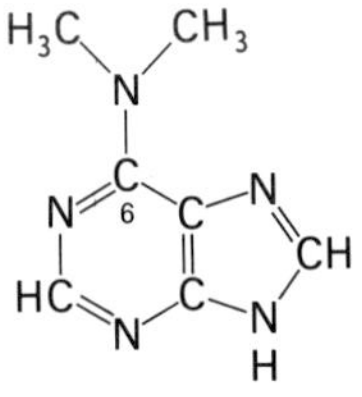

2'-*O*-Methylribose

6-Dimethyladenine

sor. For example, pseudouridylate and ribothymidylate are formed by modification of uridylate residues after transcription.

In eucaryotes, all RNA transcripts are extensively processed. The cleavage and modification of rRNA and tRNA precursors resemble those occurring in procaryotes. A striking difference is that *mRNA in eucaryotes is derived from the cleavage and splicing of sequences from a large transcript,* as will be discussed in a subsequent chapter (p. 705). In eucaryotes, transcription and translation take place in different compartments of the cell. *It seems likely that RNA processing plays a key role in regulating the transport of mRNA, rRNA, and tRNA from the nucleus to the cytosol.*

ANTIBIOTIC INHIBITORS OF TRANSCRIPTION: RIFAMYCIN AND ACTINOMYCIN

Antibiotics are interesting molecules because many of them are highly specific inhibitors of biological processes. Actinomycin and rifamycin are two antibiotics that inhibit transcription in quite different ways. Rifamycin, which is derived from *Streptomyces,* and rifampicin, a semisynthetic derivative, *specifically inhibit the initiation of RNA synthesis.* The binding of RNA polymerase to the DNA template is not blocked by this antibiotic. Rather, rifampicin interferes with the formation of the first phosphodiester bond in the RNA chain. In contrast, chain elongation is not appreciably affected by this antibiotic. This high degree of selectivity in its inhibitory action makes rifampicin a very useful experimental tool. For example, it can be used to block the initiation of new RNA chains without interfering with the transcription of chains that are already being synthesized. The site of action of rifampicin appears to be the *β subunit of RNA polymerase.* Mutants of *E. coli* that are resistant to rifampicin (called *rif-r mutants*) have been isolated. In some of these mutants, the *β* subunit has an altered electrophoretic mobility.

Actinomycin D, a polypeptide-containing antibiotic from *Streptomyces,* inhibits transcription by an entirely different mechanism from that of rifampicin. *Actinomycin D binds tightly to double-helical DNA and thereby prevents it from being an effective template for RNA synthesis.* It consists of two identical cyclic peptides joined to a phenoxazone ring system (Figure 25-22). The composition of these cyclic peptides is unusual in that sarcosine, methylvaline, and D-valine are present. Also, there is an ester bond between the hydroxyl group of threonine and the carboxyl group of methylvaline.

Actinomycin D binds tightly to double-helical DNA but not to single-stranded DNA or RNA, double-stranded RNA, or RNA-DNA hybrids. Furthermore, the binding of actinomycin to DNA is markedly enhanced by the *presence of guanine residues.* Spectroscopic and hydrodynamic studies of complexes of actinomycin D and DNA have suggested that the phenoxazone ring of actinomycin

Ribothymidine

Pseudouridine

Rifamycin B

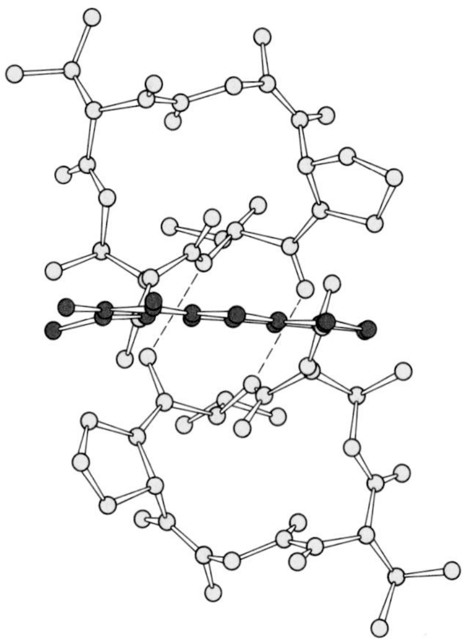

Figure 25-21
Model of actinomycin D. The phenoxazone ring is shown in red and the cyclic peptides in yellow. [After a drawing kindly provided by Dr. Henry Sobell.]

L-Methylvaline

Sarcosine

L-Proline

D-Valine

L-Threonine

Phenoxazone ring

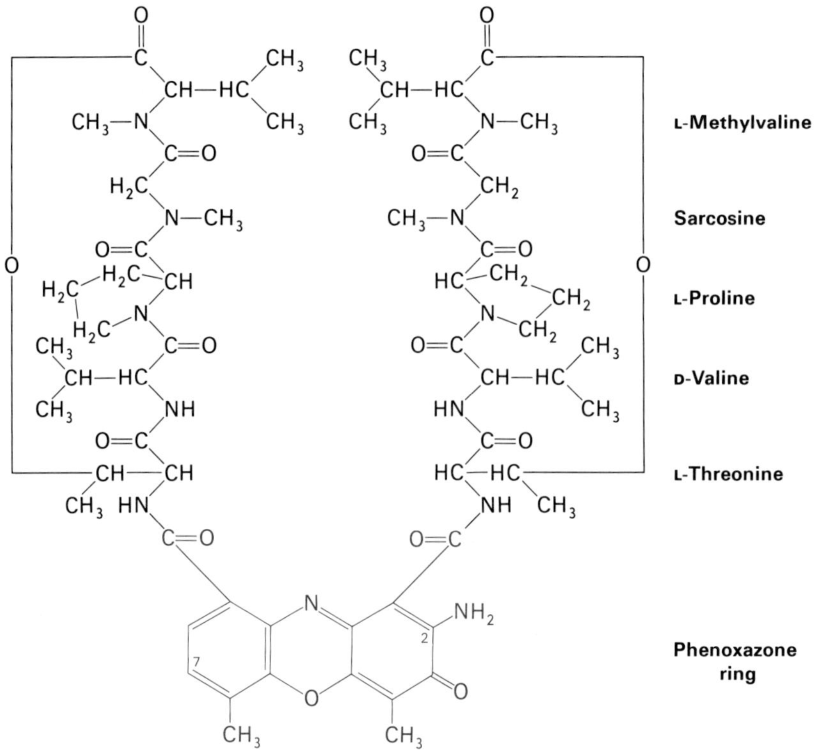

Actinomycin D

Figure 25-22
Structure of actinomycin D.

slips in between neighboring base pairs in DNA. This mode of binding is called *intercalation* (Figure 25-23). At low concentrations, *actinomycin D inhibits transcription without appreciably affecting DNA replication. Also, protein synthesis is not directly inhibited by actinomycin.* For these reasons, actinomycin D has been extensively used as a highly specific inhibitor of the formation of new RNA in both procaryotic and eucaryotic cells.

Figure 25-23
Proposed structure of the complex of actinomycin D and DNA. The phenoxazone ring of actinomycin D (shown in red) is intercalated between two GC base pairs of the DNA (shown in blue). The cyclic peptides of actinomycin D (shown in yellow) bind to the narrow groove of the DNA helix. There are several hydrogen bonds between actinomycin D and the adjacent guanines. The axis of symmetry relating the subunits on actinomycin coincides with the axis of symmetry relating the sugar-phosphate backbone and base sequence on the DNA helix. [After a drawing kindly provided by Dr. Henry Sobell.]

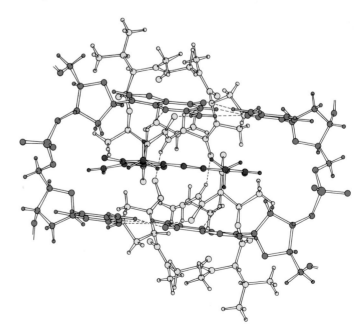

The structure of a crystalline complex of one actinomycin molecule and two deoxyguanosine molecules has recently been solved at atomic resolution (see Figure 25-23). The phenoxazone ring of actinomycin is sandwiched between the two guanine rings in this complex. One of the cyclic peptides is located above the phenoxazone ring, the other below. Each of these peptides forms a strong hydrogen bond with the 2-amino group of a guanine residue. There are numerous favorable van der Waals interactions between the antibiotic and the nucleosides. A significant feature of this complex is that it is nearly symmetric. There is a twofold axis of symmetry along the line joining the central O and N atoms of the phenoxazone ring. The conformation of this complex suggested that *actinomycin recognizes the base sequence GpC in DNA.* Note that, if the sequence along one strand of DNA is 5' GpC 3', it must be 3' CpG 5' along the complementary strand. It seems likely that actinomycin intercalates between two GC base pairs in DNA and interacts with the G residues in much the same way as it does in its complex with deoxyguanosine. In this model, the cyclic peptide units are located in the minor groove of the DNA helix. *A key aspect of this model is that the symmetry of actinomycin matches the symmetry of a specific base sequence in DNA.*

POWERFUL METHODS FOR THE SEQUENCING OF RNA HAVE BEEN DEVELOPED

RNA molecules, like DNA molecules, can now be rapidly and directly sequenced by taking advantage of the high resolving power of gel electrophoresis. The 3' or 5' end of an RNA chain is first labeled with a radioactive group. The labeled chain is then partially digested at one of its four bases to produce a nested set of fragments. Specific (or selective) cleavage can be accomplished enzymatically with *base-specific endonucleases* (Table 25-4). For example, T_1 ribonu-

Table 25-4
Enzymes used in RNA sequencing

Enzyme	Type	Substrate	Products
Ribonuclease T_1	Endonuclease (and exonuclease)	—XpGpYpZ—	—XpGp + YpZ—
Pancreatic ribonuclease	Endonuclease (and exonuclease)	—XpUpYpZ—	—XpUp + YpZ—
		—XpCpYpZ—	—XpCp + YpZ—
Ribonuclease U_2	Endonuclease (and exonuclease)	—XpGpYpZ—	—XpGp + YpZ—
		—XpApYpZ—	—XpAp + YpZ—
Ribonuclease I (from *P. polycephalum*)	Endonuclease (and exonuclease) $N = A,G,U$ but not C	—XpNpYpZ—	—XpNp + YpZ
Alkaline phosphatase (from *E. coli*)	Phosphatase	—YpZp	—YpZ + P_i
		pXpY—	XpY— + P_i
Bovine-spleen phosphodiesterase	Exonuclease	XpYpZ—	Xp + YpZ—
Snake-venom phosphodiesterase	Exonuclease	—XpYpZ	—XpY + pZ
Polynucleotide kinase	Kinase	ATP + XpYp—	pXpYp— + ADP

cleave hydrolyzes RNA chains on the 3′ side of a G residue. The RNA chain can also be specifically fragmented by *chemically modifying one of the four bases.* The backbone is then split at the site of the damaged base. The four sets of fragments are resolved on a gel and the base sequence of the RNA is then directly read from the autoradiograph (Figure 25-24), as for DNA (p. 591). RNA molecules consisting of 100 to 200 bases can be readily sequenced by these approaches.

A different strategy is to *sequence a complementary DNA molecule* rather than the RNA itself. How is a complementary piece of DNA obtained? One method is to fragment a large DNA by using a restriction endonuclease. The restriction fragment containing a complementary sequence is then identified by *hybridizing* it with the RNA of interest. Alternatively, complementary DNA can be enzymatically synthesized from an RNA template by a *reverse transcriptase* from a tumor virus (p. 743). In fact, the complete sequence of the 576 nucleotides in the mRNA for the β chain of human hemoglobin was determined in this way.

SUMMARY

The flow of genetic information in normal cells is from DNA to RNA to protein. The synthesis of RNA from a DNA template is called transcription, whereas the synthesis of a protein from an RNA template is termed translation. There are three kinds of RNA molecules: messenger RNA (mRNA), transfer RNA (tRNA), and ribosomal RNA (rRNA). These RNA molecules are single stranded. Transfer RNA and ribosomal RNA contain extensive double-helical regions that arise from the folding of the chain into hairpins. The smallest RNA molecules are the tRNAs, which contain about 75 nucleotides, whereas the largest ones are among the mRNAs, which may have more than 5000 nucleotides. Long RNA chains can be sequenced by cleaving them and separating the fragments by gel electrophoresis.

All cellular RNA is synthesized by RNA polymerase according to instructions given by a DNA template. The activated intermediates are ribonucleoside triphosphates. The direction of RNA synthesis is 5′ → 3′, like that of DNA synthesis. RNA polymerase differs from DNA polymerase in not requiring a primer and in not having nuclease activities. Another difference is that the DNA template is fully conserved in RNA synthesis, whereas it is semiconserved in DNA synthesis. RNA polymerase from *E. coli* has the subunit structure $\alpha_2\beta\beta'\sigma$ and a mass of nearly a half million daltons. The core enzyme consists of $\alpha_2\beta\beta'$. The sigma subunit enhances the rate and specificity of transcription by enabling RNA polymerase to recognize start signals on the DNA template. The binding of RNA polymerase to these promoter regions leads to a local unwinding of the

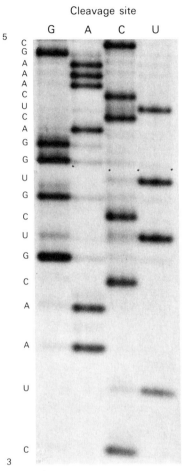

Cleavage site

Figure 25-24
Autoradiograph of fragments of 5S RNA from yeast. The 3′ end is radioactive. The four lanes correspond to cleavage at G, A, C, and U residues. The sequence shown here is

5′-CGAAACUCAGGUGCUGCAAUC-3′

[From D. A. Peattie. *Proc. Nat. Acad. Sci.* 76(1979):1762.]

DNA template, which sets the stage for the formation of the first phosphodiester bond. RNA chains start with pppG or pppA. The sigma subunit dissociates from the holoenzyme following initiation of the new chain. The termination of transcription is as precisely controlled as its initiation. The DNA template contains stop signals for transcription. Some of them are recognized by a protein called rho. Transcription can be blocked by specific inhibitors. For example, the antibiotic rifampicin inhibits the initiation of RNA synthesis, whereas actinomycin D blocks the elongation of RNA. Many RNA molecules are cleaved and chemically modified (e.g., methylated) after transcription. In procaryotes, tRNA and rRNA molecules are extensively processed, whereas mRNAs undergo little modification. In eucaryotes, all RNA transcripts are extensively modified.

SELECTED READINGS

WHERE TO START

Miller, O. L., Jr., 1973. The visualization of genes in action. *Sci. Amer.* 228(3):34–42. [Available as *Sci. Amer.* Offprint 1267.]

Spiegelman, S., 1964. Hybrid nucleic acids. *Sci. Amer.* 210(5):48–56. [Offprint 183.]

Chamberlin, M. J., 1976. RNA polymerase: an overview. *In* Losick, R., and Chamberlin, M., (eds.), *RNA Polymerase,* pp. 17–67. Cold Spring Harbor Laboratory.

DISCOVERY OF MESSENGER RNA

Jacob, F., and Monod, J., 1961. Genetic regulatory mechanisms in the synthesis of proteins. *J. Mol. Biol.* 3:318–356.

Brenner, S., Jacob, F., and Meselson, M., 1961. An unstable intermediate carrying information from genes to ribosomes for protein synthesis. *Nature.* 190:576–581.

Hall, B. D., and Spiegelman, S., 1961. Sequence complementarity of T2-DNA and T2-specific RNA. *Proc. Nat. Acad. Sci.* 47:137–146.

RNA POLYMERASE

Losick, R., and Chamberlin, M., (eds.), 1976. *RNA Polymerase.* Cold Spring Harbor Laboratory. [Contains many excellent articles.]

Chamberlin, M. J., 1974. The selectivity of transcription. *Ann. Rev. Biochem.* 43:721–776.

Travers, A., 1976. RNA polymerase specificity and the control of growth. *Nature.* 263:641–646.

INITIATION OF TRANSCRIPTION

Hayashi, M., Hayashi, M. N., and Spiegelman, S., 1963. Restriction of in vivo genetic transcription to one of the complementary strands of DNA. *Proc. Nat. Acad. Sci.* 50:664–672.

Taylor, K., Hradecna, Z., and Szybalski, W., 1967. Asymmetric distribution of the transcribing regions on the complementary strands of coliphage λ DNA. *Proc. Nat. Acad. Sci.* 57:1618–1625.

Burgess, R. R., Travers, A. A., Dunn, J. J., and Bautz, E. K. F., 1969. Factor stimulating transcription by RNA polymerase. *Nature.* 221:43–46. [Discovery of the sigma factor.]

Gilbert, W., 1976. Starting and stopping sequences for the RNA polymerase. *In* Losick, R., and Chamberlin, M., (eds.), *RNA Polymerase,* pp. 193–205. Cold Spring Harbor Laboratory.

Reznikoff, W. S., and Abelson, J. N., 1978. The *lac* promoter. *In* Miller, J. H., and Reznikoff, W. S., (eds.), *The Operon,* pp. 221–243. Cold Spring Harbor Laboratory. [Contains a comparison of 29 promoter sequences.]

TERMINATION OF TRANSCRIPTION

Adhya, S., and Gottesman, M., 1978. Control of transcription termination. *Ann. Rev. Biochem.* 47:967–996.

Das, A., Merril, C., and Adhya, S., 1978. Interaction of RNA polymerase and rho in transcription termination: coupled ATPase. *Proc. Nat. Acad. Sci.* 75:4828–4832.

618

Wu, A. M., and Platt, T., 1978. Transcription termination: nucleotide sequence at 3' end of tryptophan operon in *Escherichia coli*. *Proc. Nat. Acad. Sci.* 75:5442–5446.

ANTIBIOTIC INHIBITORS OF RNA SYNTHESIS

Goldberg, I. H., and Friedman, P. A., 1971. Antibiotics and nucleic acids. *Ann. Rev. Biochem.* 40:775–810.

Sobell, H. M., 1974. How actinomycin binds to DNA. *Sci. Amer.* 231(2)82–91. [Offprint 1303.]

PROCESSING OF RNA

Perry, R. P., 1976. Processing of RNA. *Ann. Rev. Biochem.* 45:605–630.

Steitz, J. A., and Young, R. A., 1978. Complementary sequences 1700 nucleotides apart form a ribonuclease III cleavage site in *Escherichia coli* ribosomal precursor RNA. *Proc. Nat. Acad. Sci.* 75:3593–3597.

Altman, S., 1978. Biosynthesis of tRNA. *In* Altman, S., (ed.), *Transfer RNA,* pp. 48–77. MIT Press.

SEQUENCING OF RNA

Peattie, D. A., 1979. Direct chemical method for sequencing RNA. *Proc. Nat. Acad. Sci.* 76:1760–1764.

Simoncsits, A., Brownlee, G. G., Brown, R. S., Rubin, J. R., and Guilley, H., 1977. New rapid gel sequencing method for RNA. *Nature.* 269:833–836.

PROBLEMS

1. Compare DNA polymerase I and RNA polymerase from *E. coli* in regard to each of the following features:
   (a) Subunit structure.
   (b) Activated precursors.
   (c) Direction of chain elongation.
   (d) Nuclease activities.
   (e) Conservation of the template.
   (f) Need for a primer.
   (g) Energetics of the elongation reaction.

2. Write the sequence of the mRNA molecule synthesized from a DNA template strand having the sequence

   5'-ATCGTACCGTTA-3'

3. RNA is readily hydrolyzed by alkali, whereas DNA is not. Why?

4. How does cordycepin (3'-deoxyadenosine) block the synthesis of RNA?

5. What are the expected products of partial digestion of the oligoribonucleotide

   5'-pGCAGUACUGUC-3'

   by each of the following enzymes?
   (a) Pancreatic ribonuclease.
   (b) Ribonuclease T_1.
   (c) Ribonuclease U_2.
   (d) Physarum ribonuclease I.

6. Extensive hydrolysis of an oligoribonucleotide by pancreatic ribonuclease yields Cp, 2 Up, AGCp, and GAAUp. Hydrolysis of the same oligonucleotide by ribonuclease T_1 gives AAUp, UAGp, and CCUGp? What is its sequence?

For additional problems, see W. B. Wood, J. H. Wilson, R. M. Benbow, and L. E. Hood, *Biochemistry: A Problems Approach* (Benjamin, 1974), chs. 16 and 17.

CHAPTER 26

THE GENETIC CODE AND GENE-PROTEIN RELATIONS

In the preceding chapter, the role of messenger RNA as the information-carrying intermediate between the gene and its polypeptide product was described. In this chapter, the consideration of the flow of information from gene to protein is continued. The central theme here is the genetic code, which is the relationship between the sequence of bases in DNA (or its mRNA transcript) and the sequence of amino acids in a protein. The code, which is the same in all organisms, is beautiful in its simplicity. Three bases, called a codon, specify an amino acid. Codons are read sequentially by transfer RNA (tRNA) molecules, which serve as adaptors in protein synthesis. The complete elucidation of the genetic code in the 1960s is one of the outstanding accomplishments of modern biology.

TRANSFER RNA IS THE ADAPTOR MOLECULE IN PROTEIN SYNTHESIS

We have seen that mRNA is the template for protein synthesis. How does it direct amino acids to become joined in the correct sequence? In 1958, Francis Crick wrote:

> One's first naive idea is that the RNA will take up a configuration capable of forming twenty different "cavities," one for the side chain of each of the twenty amino acids. If this were so, one might expect to be able to play the problem backwards—that is, to find the configuration of RNA by trying to form such cavities. All attempts to do this have failed, and on physical-chemical grounds the idea does not seem in the least plausible.

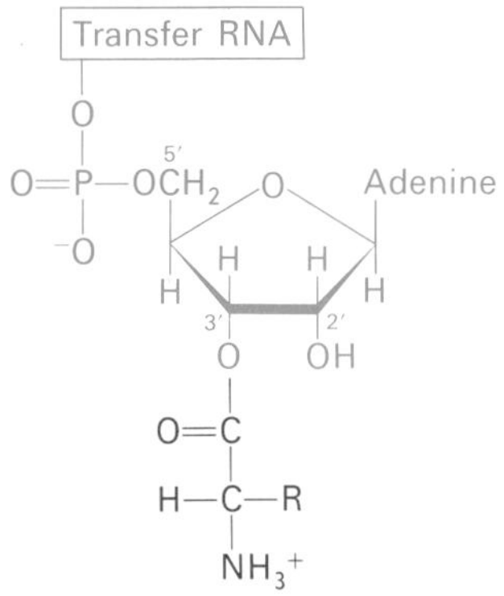

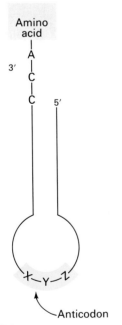

Figure 26-1

Mode of attachment of an amino acid (shown in red) to a tRNA molecule. The amino acid is esterified to the 3′-hydroxyl group of the terminal adenosine of tRNA. A tRNA having an attached amino acid is an aminoacyl-tRNA or a "charged" tRNA, whereas a tRNA without an attached amino acid is "uncharged."

Figure 26-2

Symbolic diagram of an aminoacyl-tRNA showing the amino acid attachment site and the anticodon, which is the template recognition site.

He observed that RNA does not have the knobby hydrophobic surfaces to distinguish valine from leucine and isoleucine, nor does it have properly positioned charged groups to discriminate between positively and negatively charged amino acid side chains. Crick then proposed an entirely different mechanism for recognition of the mRNA template:

> . . . RNA presents mainly a sequence of sites where hydrogen bonding could occur. One would expect, therefore, that whatever went onto the template in a *specific* way did so by forming hydrogen bonds. It is therefore a natural hypothesis that the amino acid is carried to the template by an adaptor molecule, and that the adaptor is the part which actually fits onto the RNA. In its simplest form one would require twenty adaptors, one for each amino acid.

This highly innovative hypothesis soon became an established fact. *The adaptor in protein synthesis is tRNA.* The structure and reactions of these remarkable adaptor molecules will be considered in detail in the next chapter. For the moment it suffices to note that tRNA contains an *amino acid attachment site* and a *template recognition site* (Figures 26-1 and 26-2). A tRNA molecule carries a specific amino acid in an activated form to the site of protein synthesis. The carboxyl group of this amino acid is esterified to the 3′- (or 2′-) hydroxyl group of the ribose unit at the 3′ end of the tRNA chain. The esterified amino acid may migrate between the 2′- and 3′-hydroxyl groups during protein synthesis. The joining of an amino acid to a tRNA to form an aminoacyl-tRNA is catalyzed by a specific enzyme called an aminoacyl-tRNA synthetase (or activating enzyme). This esterification reaction is driven by ATP. There is at least one specific synthetase for each of the twenty amino acids. The template recognition site on tRNA is a sequence of three bases called the *anticodon* (Figure 26-2). The anticodon on tRNA recognizes a complementary sequence of three bases on mRNA, called the *codon*.

AMINO ACIDS ARE CODED BY GROUPS OF THREE BASES STARTING FROM A FIXED POINT

The *genetic code* is the relationship between the sequence of bases in DNA (or its RNA transcripts) and the sequence of amino acids in proteins. Experiments by Francis Crick, Sydney Brenner, and others established the following features of the genetic code by 1961:

1. *What is the coding ratio?* A single base can specify only four kinds of amino acids because there are four kinds of bases in DNA. Sixteen kinds of amino acids can be specified by two bases $(4 \times 4 = 16)$, whereas sixty-four kinds of amino acids can be determined by three bases $(4 \times 4 \times 4 = 64)$. Proteins are built from a basic set of twenty amino acids, and so it was evident from this simple calculation that three or more bases are probably needed to

specify one amino acid. Genetic experiments then showed that *a group of three bases in fact codes for one amino acid*. This group of bases is called a *codon*.

2. *Is the code nonoverlapping or overlapping?* In a nonoverlapping triplet code, each group of three bases specifies only one amino acid, whereas in a completely overlapping triplet code, ABC specifies the first amino acid, BCD the second, CDE the third, and so forth.

These alternatives were distinguished by studies of the sequence of amino acids in mutants. Suppose that the base C is mutated to C'. In a nonoverlapping code, only amino acid 1 will be changed. In a completely overlapping code, amino acids 1, 2, and 3 will be altered by a mutation of C to C'. Studies of the sequence of amino acids in the coat protein of mutants of tobacco mosaic virus showed that only a single amino acid was usually altered. Also, recall from the discussion of abnormal hemoglobins (Chapter 5) that only one amino acid is changed in most mutants. Hence, it was concluded that the *genetic code is nonoverlapping*.

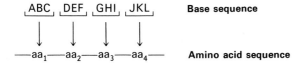

3. *How is the correct group of three bases read?* One possibility a priori is that one of the four bases (denoted as Q) serves as a "comma" between groups of three bases:

$$\ldots \text{QABCQDEFQGHIQJKLQ} \ldots$$

This turned out not to be the case. Rather, *the sequence of bases is read sequentially from a fixed starting point*. There are no commas. Suppose that a mutation deletes base G:

Start   G (deleted)
↓
ABC ¦ DEF ¦ HIJ ¦ KLM ¦ NOP

aa₁—aa₂—aa₃—aa₄—aa₅
  Normal   Altered

The first two amino acids in this polypeptide chain will be normal, but the rest of this base sequence will be read incorrectly because the *reading frame has been shifted* by the deletion of G. Now suppose that a base Z has been added between F and G:

Start   Insertion
↓
ABC ¦ DEF ¦ ZGH ¦ IJK ¦ LMN

aa₁—aa₂—aa₃—aa₄—aa₅
  Normal   Altered

Start
↓
ABC ¦ DEF ¦ GHI ¦ JKL ¦ MNO

aa₁—aa₂—aa₃—aa₄—aa₅

This addition mutation also disrupts the reading frame starting at the codon for amino acid 3. In fact, genetic studies of recombinants of addition and deletion mutants revealed many of the features of the genetic code.

4. As mentioned earlier, there are sixty-four possible base triplets and twenty amino acids. Is there just one triplet for each of the twenty amino acids or are some amino acids coded by more than one triplet? Genetic studies showed that most of the sixty-four triplets do code for amino acids. Subsequent biochemical studies demonstrated that sixty-one of the sixty-four triplets specify particular amino acids. Thus, *for most amino acids, there is more than one code word.* In other words, the genetic code is *degenerate.*

DECIPHERING THE GENETIC CODE: SYNTHETIC RNA CAN SERVE AS MESSENGER

What is the relationship between sixty-four kinds of code words and the twenty kinds of amino acids? In principle, this question can be directly answered by comparing the amino acid sequence of a protein with the corresponding base sequence of its gene or mRNA. However, this approach was not experimentally feasible in 1961 because the base sequences of genes and mRNA molecules were entirely unknown. The breaking of the genetic code then did not seem imminent, but the situation suddenly changed. Marshall Nirenberg discovered that *the addition of polyuridylate (poly U) to a cell-free protein-synthesizing system led to the synthesis of polyphenylalanine.* Poly U evidently served as a messenger RNA. The first code word was deciphered: UUU codes for phenylalanine. This remarkable experiment pointed the way to the complete elucidation of the genetic code.

Let us consider this landmark experiment in more detail. The two essential components were a cell-free system that actively synthesizes protein and a synthetic polyribonucleotide that serves as the messenger RNA. A cell-free protein-synthesizing system was obtained from *E. coli* in the following way. Bacterial cells were gently broken open by grinding them with finely powdered alumina to yield a cell sap. Cell-wall and cell-membrane fragments were then removed by centrifugation. The resulting extract contained DNA, mRNA, tRNA, ribosomes, enzymes, and other cell constituents. Protein was synthesized by this cell-free system when it was supplemented with ATP, GTP, and amino acids. At least one of the added amino acids was radioactive so that its incorporation into protein could be detected. This mixture was incubated at 37°C for about an hour. Trichloroacetic acid was then added to terminate the reaction and precipitate the proteins, leaving the free amino acids in the supernatant. The precipitate was washed and its radioactivity was then counted to determine how much labeled amino acid was incorporated into newly synthesized protein. A crucial feature of this system is that protein synthesis can be halted by the addition of deoxyribonuclease, which destroys the template for the synthesis of new mRNA. The mRNA present at the time of addition of deoxyribonuclease is labile, and so protein synthesis stops within a few minutes. Nirenberg then found that protein synthesis resumed on addition of a crude fraction of mRNA. Thus, here was a *cell-free protein-synthesizing system that was responsive to the addition of mRNA.*

The other critical component in this experiment was a synthetic polyribonucleotide—namely, poly U. Poly U was synthesized by using *polynucleotide phosphorylase,* an enzyme discovered in 1955 by Marianne Grunberg-Manago and Severo Ochoa. This enzyme catalyzes the synthesis of polyribonucleotides from ribonucleoside diphosphates:

$$(RNA)_n + \text{ribonucleoside diphosphate} \rightleftharpoons (RNA)_{n+1} + P_i$$

The reactions catalyzed by this enzyme and by RNA polymerase are very different. In this reaction, the activated substrates are ribonucleoside diphosphates rather than triphosphates. Orthophosphate is a product instead of pyrophosphate. Hence, this reaction cannot be driven to the right by the hydrolysis of pyrophosphate. Indeed, the equilibrium in vivo is in the direction of RNA degradation, not synthesis. A critical difference is that *polynucleotide phosphorylase does not utilize a template.* The RNA synthesized by this enzyme has a composition dictated by the ratios of the ribonucleotides present in the incubation mixture and a sequence that is essentially random. For this reason, polynucleotide phosphorylase proved to be a most valuable experimental tool in deciphering the genetic code. Thus, poly U was synthesized by incubating high concentrations of UDP in the presence of this enzyme. Copolymers of two ribonucleotides, say U and A, with a random sequence were prepared by incubating UDP and ADP with this enzyme.

A variety of synthetic ribonucleotides were added to this cell-free protein-synthesizing system and the incorporation of ^{14}C-labeled L-phenylalanine was measured. The results were striking:

Polyribonucleotide added	[14]C counts per minute
None	44
Poly A	50
Poly C	38
Poly U	39,800

The same experiment was carried out with a different ^{14}C-labeled amino acid in each incubation mixture. It was found that *poly A led to the synthesis of polylysine* and that *poly C led to the synthesis of polyproline.* Thus, three code words were deciphered.

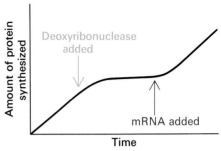

Figure 26-3
Protein synthesis in a cell-free system stops a few minutes after the addition of deoxyribonuclease and resumes following the addition of mRNA.

Code word	Amino acid
UUU	Phenylalanine
AAA	Lysine
CCC	Proline

The code word GGG could not be deciphered in the same way because poly G does not act as a template, probably because it forms a triple-stranded helical structure. Polyribonucleotides with extensive regions of ordered structure are ineffective templates for protein synthesis.

CODON COMPOSITIONS FOR MANY AMINO ACIDS WERE DETERMINED BY USING COPOLYMERS AS TEMPLATES

Polyribonucleotides consisting of two kinds of bases were used as templates to further elucidate the genetic code. For example, a random copolymer of U and G contains eight different triplets: UUU, UUG, UGU, GUU, UGG, GUG, GGU, and GGG. The expected incidence of these triplets can be readily calculated from the mole proportions of U and G in the copolymer, which were 0.76 and 0.24, respectively (Table 26-1). The extent of incorporation of various amino acids elicited by this template is given in Table 26-2. Phenylalanine was incorporated to the greatest extent, as expected, because the triplet UUU was the most prevalent one. Valine, leu-

Table 26-1
Expected incidence of triplets in a random copolymer of U (0.76) and G (0.24)

Triplet	Probability	Relative incidence
UUU	$.76 \times .76 \times .76 = .439$	100
UUG	$.76 \times .76 \times .24 = .139$	31.6
UGU	$.76 \times .24 \times .76 = .139$	31.6
GUU	$.24 \times .76 \times .76 = .139$	31.6
UGG	$.76 \times .24 \times .24 = .0438$	10.0
GUG	$.24 \times .76 \times .24 = .0438$	10.0
GGU	$.24 \times .24 \times .76 = .0438$	10.0
GGG	$.24 \times .24 \times .24 = .0138$	3.1

Table 26-2
Amino acid incorporation stimulated by a random copolymer of U (0.76) and G (0.24)

Amino acid	Relative amount incorporated	Inferred codon composition
Phenylalanine	100	UUU
Valine	37	2U, 1G
Leucine	36	2U, 1G
Cysteine	35	2U, 1G
Tryptophan	14	1U, 2G
Glycine	12	1U, 2G

cine, and cysteine came next. The extent to which they were incorporated was slightly more than a third of that of phenylalanine, which corresponds with the calculated incidence of triplets that contain two U and one G. The incorporation of other amino acids was very low. Thus, *it was concluded that valine, leucine, and cysteine are specified by codons that contain 2U and 1G, whereas tryptophan and glycine are specified by codons that contain 1U and 2G.*

The same type of experiment was carried out with other random copolymers—namely, UA, UC, AC, AG, and CG, as well as UGC, AGC, UAC, and UAG. The *compositions* of codons corresponding to each of the twenty amino acids were deduced in this way in the laboratories of Nirenberg and Ochoa.

TRINUCLEOTIDES PROMOTE THE BINDING OF SPECIFIC
TRANSFER RNA MOLECULES TO RIBOSOMES

The use of mixed copolymers as templates revealed the *composition but not the sequence of codons* corresponding to particular amino acids (except for UUU, AAA, and CCC). As described earlier, valine is coded by a triplet with 2U and 1G. Is it UUG, UGU, or GUU? The answer to this question was provided by two entirely different experimental approaches: (1) the use of synthetic polyribonucleotides with an ordered sequence, and (2) the codon-dependent binding of specific tRNA molecules to ribosomes.

In 1964, Nirenberg discovered that *trinucleotides promote the binding of specific tRNA molecules to ribosomes in the absence of protein synthesis.* For example, the addition of pUpUpU led to the binding of phenylalanine tRNA, whereas pApApA markedly enhanced the binding of lysine tRNA, as did pCpCpC for proline tRNA. Dinucleotides were ineffective in stimulating the binding of tRNA to ribosomes. These studies showed that a *trinucleotide (like a triplet in mRNA) specifically binds the particular tRNA molecule for which it is the code word.* A simple and rapid binding assay was devised: tRNA molecules bound to ribosomes are retained by a cellulose nitrate filter, whereas unbound tRNA passes through the filter. The kind of tRNA retained by the filter was identified by using tRNA molecules charged with a particular ^{14}C-labeled amino acid.

All sixty-four trinucleotides were synthesized by organic chemical or enzymatic techniques. For each trinucleotide, the binding of tRNA molecules corresponding to all twenty amino acids was assayed. For example, it was found that pUpUpG stimulated the binding of leucine tRNA only, pUpGpU stimulated the binding of cysteine tRNA only, and pGpUpU stimulated the binding of valine tRNA only. Hence, it was concluded that the codons UUG, UGU, and GUU correspond to leucine, cysteine, and valine, respectively. For a few codons, no tRNA was strongly bound, whereas, for a few others, more than one kind of tRNA was bound. For most codons, clear-cut binding results were obtained. In fact, about fifty codons were deciphered by this simple and elegant experimental approach.

COPOLYMERS WITH A DEFINED SEQUENCE WERE
ALSO INSTRUMENTAL IN BREAKING THE CODE

At about the same time, H. Gobind Khorana succeeded in synthesizing polyribonucleotides with a defined repeating sequence. A variety of copolymers with two, three, and four kinds of bases were synthesized by a combination of organic chemical and enzymatic techniques. Let us consider the strategy for the synthesis of poly (GUA), for example. This ordered copolymer has the sequence

GUAGUAGUAGUAGUAGUAGUAGUA. . .

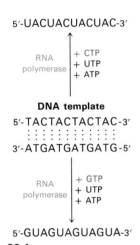

5′-UACUACUACUAC-3′

RNA polymerase | + CTP
+ UTP
+ ATP

DNA template
5′-TACTACTACTAC-3′
┊┊┊┊┊┊┊┊┊┊┊┊
3′-ATGATGATGATG-5′

RNA polymerase | + GTP
+ UTP
+ ATP

5′-GUAGUAGUAGUA-3′

Figure 26-4
The strand of this double-helical DNA template to be transcribed is determined by the choice of ribonucleoside triphosphates in the incubation mixture.

First, Khorana synthesized by organic chemical methods two complementary deoxyribonucleotides, each with nine residues: d(TAC)$_3$ and d(GTA)$_3$. These two oligonucleotides then served as templates for the synthesis of long DNA chains from the four deoxyribonucleoside triphosphates by DNA polymerase I. Either oligonucleotide alone was not an effective template. If both were present, d(TAC)$_3$ was the template for the enzymatic synthesis of poly d(GTA), whereas d(GTA)$_3$ was the template for the synthesis of poly d(TAC). These long complementary DNA chains formed a double-helical structure. The next step was to obtain long polyribonucleotide chains with a sequence corresponding to poly d(TAC) and poly d(GTA). This was accomplished by using the poly d(TAC):poly d(GTA) duplex as the template for synthesis by RNA polymerase. The DNA strand to be transcribed could be selected by adding three appropriate ribonucleoside triphosphates. When GTP, UTP, and ATP were added to the incubation mixture, the polyribonucleotide product synthesized from the poly d(TAC) template strand was poly (GUA). The other strand was not transcribed because CTP, one of the required substrates, was missing. When CTP, UTP, and ATP were added, poly (UAC) was synthesized from the other template strand. Thus, *organic synthesis, followed by template-directed syntheses carried out by DNA polymerase and then RNA polymerase, yielded two long polyribonucleotides having defined repeating sequences* (Figure 26-4).

These regular copolymers were used as templates in the cell-free protein-synthesizing system. Let us examine some of the results. A copolymer consisting of an alternating sequence of two bases P and Q

PQP ┊ QPQ ┊ PQP ┊ QPQ ┊ PQP ┊ . . .

contains two codons, PQP and QPQ. The polypeptide product should therefore contain an alternating sequence of two kinds of amino acids (abbreviated as aa$_1$ and aa$_2$):

aa$_1$—aa$_2$—aa$_1$—aa$_2$—aa$_1$—

Whether aa$_1$ or aa$_2$ is amino-terminal in the polypeptide product depends on whether the reading frame starts at P or Q. When poly (UG) was the template, a polypeptide with an alternating sequence of cysteine and valine was synthesized.

UGU ┊ GUG ┊ UGU ┊ GUG ┊ UGU ┊ GUG

Cys—Val—Cys—Val—Cys—Val

This result unequivocally confirmed the triplet nature of the code and showed that either UGU or GUG codes for cysteine and that the other of these two triplets codes for valine. When considered together with tRNA binding data, it was evident that UGU codes for cysteine and GUG codes for valine. The polypeptides synthesized in response to several other alternating copolymers of two bases were

Template	Product	Codon assignments	
Poly (UC)	Poly (Ser-Leu)	UCU ⟶ Ser	CUC ⟶ Leu
Poly (AG)	Poly (Arg-Gln)	AGA ⟶ Arg	GAG ⟶ Gln
Poly (AC)	Poly (Thr-His)	ACA ⟶ Thr	CAC ⟶ His

Now let us consider a template consisting of three bases in a repeating sequence, poly (PQR). If the reading frame starts at P, the resulting polypeptide should contain only one kind of amino acid, the one coded by the triplet PQR:

Start
↓
PQR | PQR | PQR | PQR | PQR | PQR . . .

aa_1—aa_1—aa_1—aa_1—aa_1—aa_1

However, if the reading frame starts at Q, the polypeptide synthesized should contain a different amino acid, the one coded by the triplet QRP:

Start
↓
P | QRP | QRP | QRP | QRP | QRP | QRP . . .

aa_2—aa_2—aa_2—aa_2—aa_2—aa_2

If the reading frame starts at R, a third kind of polypeptide should be formed, containing only the amino acid coded by the triplet RPQ:

Start
↓
PQ | RPQ | RPQ | RPQ | RPQ | RPQ | RPQ . . .

aa_3—aa_3—aa_3—aa_3—aa_3—aa_3

Thus, the expected products are three different homopolypeptides. In fact, this was observed for most templates consisting of repeating sequences of three nucleotides. For example, poly (UUC) led to the synthesis of polyphenylalanine, polyserine, and polyleucine. This result, taken together with the outcome of other experiments, showed that UUC codes for phenylalanine, UCU codes for serine, and CUU codes for leucine. The polypeptides synthesized in response to other templates of this type are given in Table 26-3. Note that poly (GUA) and poly (GAU) each elicited the synthesis of two rather than three homopolypeptides. The reason will be evident shortly.

Khorana also synthesized several copolymers with repeating tetranucleotides, such as poly (UAUC). This template led to the

Table 26-3
Homopolypeptides synthesized using messengers containing repeating trinucleotide sequences

Messenger	Homopolypeptides synthesized
Poly (UUC)	Phe; Ser; Leu
Poly (AAG)	Lys; Glu; Arg
Poly (UUG)	Cys; Leu; Val
Poly (CCA)	Gln; Thr; Asn
Poly (GUA)	Val; Ser
Poly (UAC)	Tyr; Thr; Leu
Poly (AUC)	Ile; Ser; His
Poly (GAU)	Met; Asp

synthesis of a polypeptide with the repeating sequence Tyr-Leu-Ser-Ile, irrespective of the reading frame:

$$\text{UAU | CUA | UCU | AUC | UAU | CUA | UCU | AUC}$$

$$\text{Tyr—Leu—Ser—Ile—Tyr—Leu—Ser—Ile}$$

Four codon assignments were deduced from this result.

A very different result was obtained when poly (GUAA) was the template. The only products were dipeptides and tripeptides. Why not longer chains? The reason is that one of the triplets present in this polymer—namely, UAA—codes not for an amino acid but rather for the termination of protein synthesis:

$$\text{GUA | AGU | AAG | UAA | GUA | A . . .}$$

$$\text{Val—Ser—Lys—Stop}$$

Only di- and tripeptides were formed when poly (AUAG) was the template because UAG is a second signal for chain termination:

$$\text{AUA | GAU | AGA | UAG | AUA | G . . .}$$

$$\text{Ile—Asp—Arg—Stop}$$

Now let us look again at Table 26-3. Two rather than three homopolypeptides were synthesized with poly (GUA) as a template because the third reading frame corresponds to the sequence

$$\text{G | UAG | UAG | UAG . . .}$$

$$\text{Stop—Stop—Stop}$$

which is a repeating sequence of termination signals. What about poly (GAU)? Again, only two homopolypeptides were synthesized with this template, because the third reading frame corresponds to

$$\text{GA | UGA | UGA | UGA | U . . .}$$

$$\text{Stop—Stop—Stop}$$

UGA is yet another signal for chain termination. In fact, *UGA, UAA, and UGA are the only three codons that do not specify an amino acid.*

Khorana's synthesis of polynucleotides with a defined sequence was an outstanding accomplishment. The use of these polymers as templates for protein synthesis, together with Nirenberg's studies of the trinucleotide-stimulated binding of tRNA to ribosomes, resulted in the complete elucidation of the genetic code by 1966, an outcome that would have been deemed a quixotic dream only six years earlier. The powerful synthetic methods designed by Khorana also made possible another feat: the total synthesis of a DNA molecule corresponding in sequence to a transfer RNA molecule.

All sixty-four codons have been deciphered (Table 26-4). Sixty-one triplets correspond to particular amino acids, whereas three code for chain termination. Because there are twenty amino acids and sixty-one triplets that code for them, it is evident that the code is highly *degenerate*. In other words, *many amino acids are designated by more than one triplet*. Only tryptophan and methionine are coded by just one triplet. The other eighteen amino acids are coded by two or more. Indeed, leucine, arginine, and serine are specified by six codons each. Under normal physiological conditions, *the code is not ambiguous:* a codon designates only one amino acid.

Codons that specify the same amino acid are called *synonyms*. For example, CAU and CAC are synonyms for histidine. Note that synonyms are not distributed haphazardly throughout the table of the genetic code (Table 26-4). An amino acid specified by two or more synonyms occupies a single box (unless there are more than four synonyms). The amino acids in a box are specified by codons that have the same first two bases but differ in the third base, as exemplified by GUU, GUC, GUA, and GUG. Thus, *most synonyms differ only in the last base of the triplet*. Inspection of the code shows that XYC and XYU always code for the same amino acid, whereas

Table 26-4
The genetic code

First position (5′ end)	Second position				Third position (3′ end)
	U	C	A	G	
U	Phe	Ser	Tyr	Cys	U
	Phe	Ser	Tyr	Cys	C
	Leu	Ser	Stop	Stop	A
	Leu	Ser	Stop	Trp	G
C	Leu	Pro	His	Arg	U
	Leu	Pro	His	Arg	C
	Leu	Pro	Gln	Arg	A
	Leu	Pro	Gln	Arg	G
A	Ile	Thr	Asn	Ser	U
	Ile	Thr	Asn	Ser	C
	Ile	Thr	Lys	Arg	A
	Met	Thr	Lys	Arg	G
G	Val	Ala	Asp	Gly	U
	Val	Ala	Asp	Gly	C
	Val	Ala	Glu	Gly	A
	Val	Ala	Glu	Gly	G

Note: Given the position of the bases in a codon, it is possible to find the corresponding amino acid. For example, the codon 5′ AUG 3′ on mRNA specifies methionine, whereas CAU specifies histidine. UAA, UAG, and UGA are termination signals. AUG is part of the initiation signal, in addition to coding for internal methionines.

XYG and XYA usually code for the same amino acid. The structural basis for these equivalences of codons will become evident when we consider the nature of the anticodons of tRNA molecules (p. 650).

What is the biological significance of the extensive degeneracy of the genetic code? One possibility is that *degeneracy minimizes the deleterious effects of mutations.* If the code were not degenerate, then twenty codons would designate amino acids and forty-four would lead to chain termination. The probability of mutating to chain termination would therefore be much higher with a nondegenerate code than with the actual code. It is important to recognize that chain-termination mutations usually lead to inactive proteins, whereas substitutions of one amino acid for another are usually relatively harmless. *Degeneracy of the code may also be significant in permitting DNA base composition to vary over a wide range without altering the amino acid sequence of the proteins encoded by the DNA.* The $[G] + [C]$ content of bacterial DNA ranges from less than 30% to more than 70%. DNA molecules with quite different $[G] + [C]$ contents could code for the same proteins if different synonyms were consistently used.

START AND STOP SIGNALS FOR PROTEIN SYNTHESIS

It has already been mentioned that *UAA, UAG, and UGA designate chain termination.* These codons are read not by tRNA molecules but rather by specific proteins called release factors. The start signal for protein synthesis is more complex. Polypeptide chains in bacteria start with a modified amino acid—namely, formylmethionine (fMet). There is a specific tRNA that carries fMet. This fMet-tRNA recognizes the codon AUG (or, less frequently, GUG). However, AUG is also the codon for an internal methionine, and GUG is the codon for an internal valine. This means that the signal for the first amino acid in the polypeptide chain must be more complex than for all subsequent ones. AUG (or GUG) is *part of the initiation signal.* A signal preceding AUG (or GUG) determines whether it is to be read as a chain initiation signal or as a codon for an internal methionine (or valine) residue. The mechanisms of initiation and termination of protein synthesis will be discussed in the next chapter (p. 656).

**Formylmethionine
(fMet)**

THE GENETIC CODE IS UNIVERSAL

As mentioned earlier, the genetic code was deciphered by studies of trinucleotides and synthetic mRNA templates in cell-free systems derived from bacteria. Two questions arise: Is the genetic code the same in vivo as in vitro? Is the genetic code the same in all organisms? A definitive answer to these questions has come from analyses of mutations in viruses, bacteria, and higher organisms. A large number of amino acid substitutions arising from mutations of the genes for hemoglobin in humans, the coat protein in tobacco mosaic virus (TMV), and the α chain of tryptophan synthetase in *E. coli* are known. *Nearly all of the amino acid substitutions can be accounted for by a change of a single base* (Table 26-5). *This is a stringent test of the correctness of the entire genetic code and of its universality.*

Table 26-5
Mutations in human hemoglobin, *E. coli* tryptophan synthetase, and tobacco mosaic virus (TMV) coat protein

Protein	Amino acid substitution	Inferred codon change
Hemoglobin	Glu ⟶ Val*	GAA ⟶ GUA
Hemoglobin	Glu ⟶ Lys*	GAA ⟶ AAA
Hemoglobin	Glu ⟶ Gly	GAA ⟶ GGA
Tryptophan synthetase	Gly ⟶ Arg	GGA ⟶ AGA
Tryptophan synthetase	Gly ⟶ Glu	GGA ⟶ GAA
Tryptophan synthetase	Glu ⟶ Ala	GAA ⟶ GCA
TMV coat protein	Leu ⟶ Phe	CUU ⟶ UUU
TMV coat protein	Glu ⟶ Gly	GAA ⟶ GGA
TMV coat protein	Pro ⟶ Ser	CCC ⟶ UCC

*The substitutions Glu → Val and Glu → Lys occur at position 6 in the β chain of human hemoglobin S and C, respectively.

Why has the code remained invariant through billions of years of evolution? Consider the effect of a mutation that would alter the reading of mRNA. Such a mutation would change the amino acid sequence of most, if not all, of the proteins synthesized by that particular organism. Many of these changes would undoubtedly be deleterious, and so there would be strong selection against such a mutation.

THE BASE SEQUENCE OF A GENE AND THE AMINO ACID SEQUENCE OF ITS POLYPEPTIDE PRODUCT ARE COLINEAR

We turn now to relationships between genes and proteins. High-resolution genetic-mapping studies carried out by Seymour Benzer revealed that genes are unbranched structures. This important result was in harmony with the established fact that DNA consists of a linear sequence of base pairs. Polypeptide chains also have unbranched structures. The following question was therefore asked in the early 1960s: *Is there a linear correspondence between a gene and its polypeptide product?*

Charles Yanofsky's approach to this problem was to use mutants of *E. coli* that produced an altered enzyme molecule. Numerous mutants of the α chain of tryptophan synthetase were isolated and their positions on the genetic map for the α chain were determined

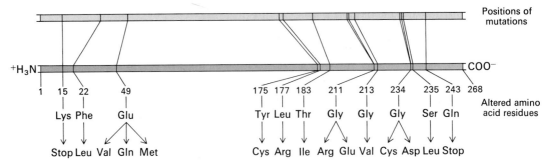

Figure 26-5
Colinearity of the gene and amino acid sequence of the α chain of tryptophan synthetase. The locations of mutations in DNA (shown in yellow) were determined by genetic-mapping techniques. The positions of the altered amino acids in the amino acid sequence (shown in blue) are in the same order as the corresponding mutations. [After C. Yanofsky. Gene structure and protein structure. Copyright © 1967 by Scientific American, Inc. All rights reserved.]

by carrying out recombination studies with a transducing phage. Some of these mutants were very close to each other on the genetic map, whereas others were far apart within the same gene. The next major task was to determine the location of the amino acid substitution for each of these ten mutants. First, the amino acid sequence of the 168 residues of the wild-type α chain was determined. Then, the site and kind of amino acid substitution in each mutant was established by fingerprinting. *The order of the mutants on the genetic map was the same as the order of the corresponding changes in amino acid sequence of the polypeptide product* (Figure 26-5). In other words, *the gene for the α chain and its polypeptide product are colinear.*

SOME VIRAL DNA SEQUENCES CODE FOR MORE THAN ONE PROTEIN

The observation that φX174 DNA codes for more proteins than expected from its nucleotide content posed an enigma. How does this virus encode more than 2,000 amino acid residues though it contains only 5,375 bases? The answer came from the complete base sequence (p. 593), which revealed that φX174 DNA contains several *overlapping genes*. Some regions of its mRNA transcript are translated in *different reading frames*, which gives rise to proteins with different amino acid sequences (Figure 26-6). For example, the same 300 nucleotides code for the gene E protein and for most of the gene D protein. An even more extreme example is provided by G4, a related bacteriophage, which contains short stretches of DNA that code for three different proteins. *These viruses use overlapping genes to pack more information into their small DNA molecules.* However, a price is paid in achieving such genetic economy. The amino acid sequences encoded by overlapping genes are highly constrained.

	C	
Leu	T	
	G	Trp
	G	
Asp	A	
	C	Thr
	T	
Phe	T	
	T	Leu
	G	
Val	T	
	G	Trp
	G	
Gly	G	
	A	Asp
	T	
Tyr	A	
	C	Thr
	C	
Pro	C	
	T	Leu
	C	
Arg	G	
	C	Ala
	T	
Phe	T	
	T	Phe
	C	

Figure 26-6
Overlapping genes in φX174 DNA. The two amino acid sequences specified by different reading frames are shown next to the base sequence.

Hence, overlapping genes are probably used extensively only when the amount of DNA in a genome is sharply limited, as in a virus with a well-defined protein shell.

EUCARYOTIC GENES ARE MOSAICS OF TRANSLATED AND UNTRANSLATED DNA SEQUENCES

In bacteria, polypeptide chains are encoded by a continuous array of triplet codons in DNA. It was assumed for many years that genes in higher organisms are also continuous. This view was unexpectedly shattered in 1977, when investigators in several laboratories discovered that several genes are *discontinuous*. For example, the gene for the β chain of hemoglobin is interrupted within its amino acid coding sequence by a long *noncoding intervening sequence* of 550 base pairs and a short one of 120 base pairs. Thus, the β-*globin gene is split into three coding sequences.*

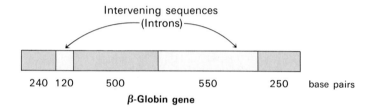

β-Globin gene

This unexpected structure was revealed by electron-microscopic studies of hybrids formed between β-globin mRNA and a segment of mouse DNA containing a β-globin gene (Figure 26-7). The DNA duplex is partially denatured to allow mRNA to hybridize to the complementary strand of DNA. The single-stranded region of DNA then loops out and appears in electron micrographs as a thin line, in contrast with double-standed DNA or DNA-RNA hybrid regions, which have a thick appearance. If the β-globin gene were continuous, a single loop would be seen. However, three loops are

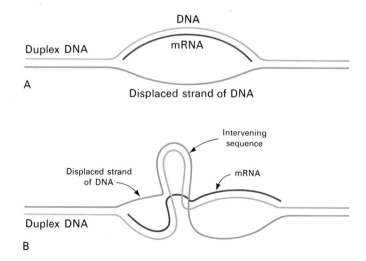

Figure 26-7
Detection of intervening sequences by electron microscopy. An mRNA molecule (shown in red) is hybridized to genomic DNA containing the corresponding gene. (A) A single loop of single-stranded DNA (shown in blue) is seen if the gene is continuous. (B) Two loops of single-stranded DNA (blue) and a loop of double-stranded DNA (blue and green) are seen if the gene contains an intervening sequence.

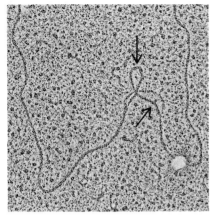

Figure 26-8
Electron micrograph of a hybrid between β-globin mRNA and a fragment of genomic DNA containing the β-globin gene. The thick loops of double-helical DNA that buckle out of the structure are the intervening sequences in DNA that are absent from mRNA (as in part B of Figure 26-7). The upper arrow points to the large intervening sequence, and the lower arrow to the small one. [Courtesy of Dr. Philip Leder.]

clearly evident in electron micrographs of these hybrids (Figure 26-8), which indicates that the gene is interrupted by at least one stretch of DNA that is absent from the mRNA. Additional evidence for intervening sequences came from restriction endonuclease mapping of the β-globin gene and of a reverse transcript of β-globin mRNA. The large difference between these maps showed that the genomic DNA contains untranslated sequences between the coding regions and defined the locations of these sequences.

At what stage in gene expression are intervening sequences removed? Newly synthesized RNA chains isolated from nuclei are much larger than the mRNA molecules derived from them. In fact, the primary transcript of the β-globin gene contains two untranslated regions. *These intervening sequences in the 15S primary transcript are excised and the coding sequences are simultaneously ligated by a precise splicing enzyme to form the mature 9S mRNA* (Figure 26-9). The coding

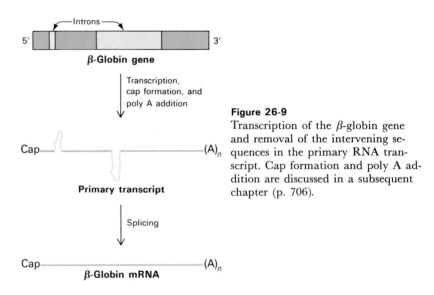

Figure 26-9
Transcription of the β-globin gene and removal of the intervening sequences in the primary RNA transcript. Cap formation and poly A addition are discussed in a subsequent chapter (p. 706).

sequences of split genes are called *exons* (for expressed regions), whereas their untranslated intervening sequences are known as *introns*.

Another split eucaryotic gene is the one for ovalbumin in chickens, which is made up of eight exons separated by seven long introns (Figure 26-10). Even more striking is the conalbumin gene,

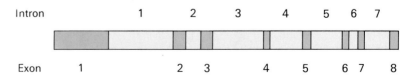

Figure 26-10
Structure of the chick ovalbumin gene. The introns (noncoding regions) are shown in yellow and the exons (translated regions) in blue.

which contains no fewer than seventeen exons. A common feature in the expression of these genes is that their exons are ordered in the same sequence in mRNA as in DNA. Thus, *split genes, like contiguous genes, are colinear with their polypeptide products.*

All avian and mammalian genes mapped thus far, except for the histone genes (p. 700), are split. Why do nearly all genes in higher eucaryotes contain intervening sequences? One possibility is that split genes reflect the evolutionary process. Exons may correspond to folding units or functional domains that were brought together to generate proteins with novel properties. Another possibility is that the excision of intervening sequences controls the flow of mRNA from the nucleus to the cytosol. According to this hypothesis, the splicing of the primary transcript plays a key role in determining which proteins are made by a cell. The discovery of split genes in higher organisms has clearly opened up fascinating areas of inquiry that are likely to have a major effect on our understanding of cell growth and differentiation.

MUTATIONS ARE PRODUCED BY CHANGES IN THE BASE SEQUENCE OF DNA

There are several types of mutations: (1) the *substitution* of one base pair for another or of several pairs; (2) the *deletion* of one or more base pairs; and (3) the *insertion* of one or more base pairs (Figure 26-11). *The substitution of one base pair for another is the most common type of mutation.* For T4 bacteriophage, the spontaneous mutation rate has been estimated to be 1.7×10^{-8} per base per replication. The corresponding values for *E. coli* and *Drosophila melanogaster* are 4×10^{-10} and 7×10^{-11}, respectively.

There are two kinds of single base-pair substitutions. A *transition* is the replacement of one purine by another purine or of one pyrimidine by another pyrimidine. In contrast, a *transversion* is the replacement of a purine by a pyrimidine or of a pyrimidine by a purine.

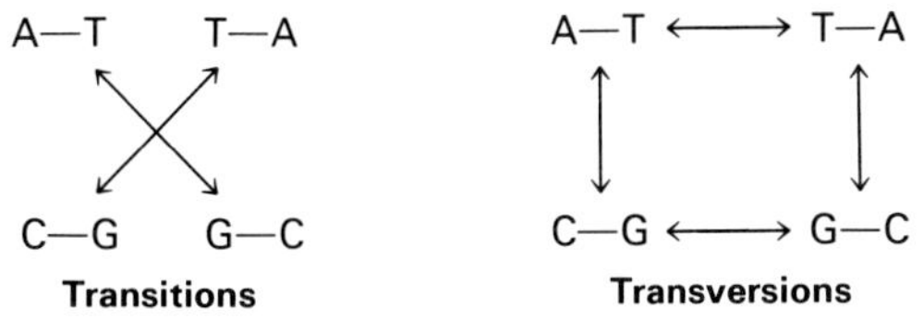

A mechanism for the spontaneous occurrence of transitions was suggested by Watson and Crick in their classical paper on the DNA double helix. They noted that some of the hydrogen atoms on each of the four bases can change their locations. *These tautomeric forms, which have a low probability of occurrence, can form base pairs other than*

There have been no problems in St. Louis created by pre-trial publicity because the newspapers have exorcised sound judgment and restraint in what they have printed.
—*St. Louis Globe-Democrat.*
Substitution

In analyzing the pictures relayed by the Mariner 10 spacecraft en route to Mercury, the scientists said the planet becomes much more dramatic looking when viewed through an ultra-violent lens.
—*Boston Globe.*
Insertion

"I can speak just as good nglish as you," Gorbulove corrected in a merry voice.
—*Seattle Times.*
Deletion

Tomorrow: "Give Baby Time to Learn to Swallow Solid Food." etaoin-oshrdlucmfwypvbgkq
—*Youngstown (Ohio) Vindicator.*
Nonsense

Figure 26-11
Types of mutations, as illustrated by typographical errors. [After S. Benzer. *Harvey Lectures,* Ser. 56 (Academic Press, 1961), p. 3.]

A—T   T—A   A—T ⟷ T—A

C—G   G—C   C—G ⟷ G—C

Transitions   **Transversions**

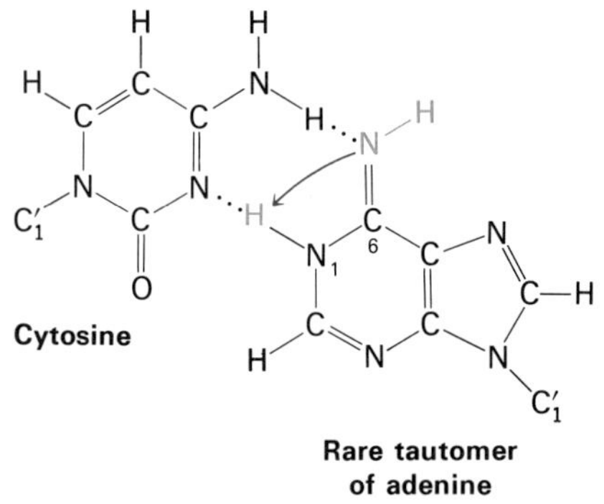

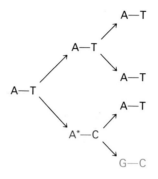

Cytosine

**Rare tautomer
of adenine**

Figure 26-12
The rare tautomer of adenine pairs
with cytosine instead of thymine. This
tautomer is formed by the shift of a
proton from the 6-amino group to N-1.

A-T and G-C. For example, the imino tautomer of adenine can pair
with cytosine (Figure 26-12). In the next round of replication, this
adenine will probably assume its normal tautomeric state, and so it
will pair with thymine, whereas cytosine will pair with guanine.
Thus, one of the daughter DNA molecules will contain a G-C base
pair in place of an A-T base pair (Figure 26-13).

$$A—T \to A—T \to A—T$$
$$A—T \to A—T \to A—T$$
$$A—T \to A^*—C \to A—T$$
$$A^*—C \to G—C$$

Figure 26-13
The pairing of the rare tautomer of
adenine (A*) with cytosine leads to a
G-C base pair in the next generation.

Mutations can also be caused by *defective DNA polymerases.* There
is a mutant of *E. coli* that has a mutation rate that is a thousand
times as high as normal, probably owing to an altered polymerase.
A distinctive feature of the mutations in this organism is that nearly
all of them are transversions of A-T to C-G.

SOME CHEMICAL MUTAGENS ARE QUITE SPECIFIC

Base analogs such as 5-bromouracil and 2-aminopurine can be incorporated into DNA. They lead to transitions as a consequence of
altered base pairing in the course of their incorporation into DNA
or in a subsequent round of replication (Figure 26-14). *5-Bromo-
uracil,* an analog of thymine, normally pairs with adenine. However,
the proportion of the enol tautomer of 5-bromouracil is higher than
that of thymine, presumably because of the highly electronegative
character of the bromine atom compared with a methyl group at
C-5. The enol form of 5-bromouracil pairs with guanine, which
causes transitions of A-T $\leftrightarrow$ G-C. *2-Aminopurine* normally pairs with
thymine. In contrast with adenine, the normal tautomer of 2-ami-
nopurine can form a single hydrogen bond with cytosine. Hence,
2-aminopurine can produce transitions of A-T $\leftrightarrow$ G-C.

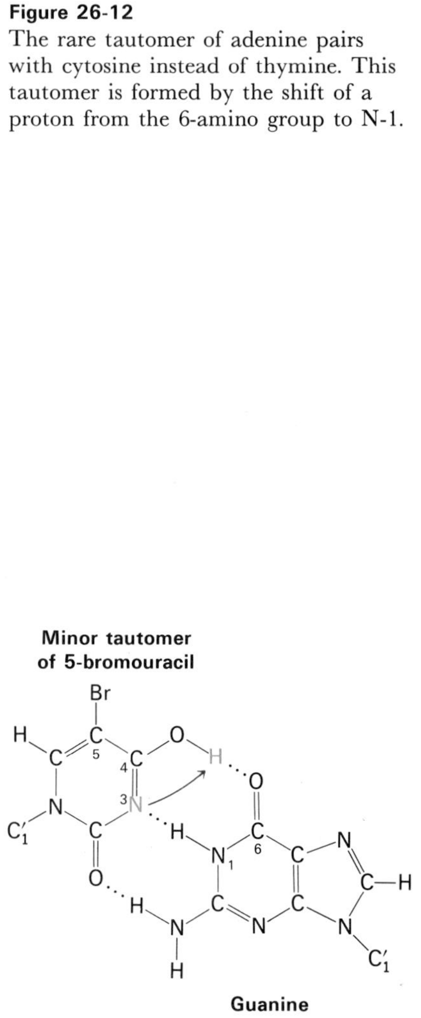

Other mutagens act by chemically modifying the bases of DNA. For example, nitrous acid reacts with bases that contain amino groups.
Adenine is oxidatively deaminated to hypoxanthine, cytosine to
uracil, and guanine to xanthine. Hypoxanthine pairs with cytosine
rather than with thymine. Uracil pairs with adenine rather than
with guanine. Xanthine, like guanine, pairs with cytosine. Consequently, nitrous acid causes transitions of A-T $\leftrightarrow$ G-C. *Hydroxylamine* (NH_2OH) is a highly specific mutagen. It reacts almost exclu-

**Minor tautomer
of 5-bromouracil**

Guanine

Figure 26-14
5-Bromouracil, an analog of thymine,
occasionally pairs with guanine instead
of adenine. The presence of bromine at
C-5 increases the proportion of the rare
tautomer formed by the shift of a proton from N-3 to the C-4 oxygen atom.

sively with cytosine to give a derivative that pairs with adenine rather than with guanine. Hydroxylamine produces a unidirectional transition of C-G to A-T.

A different kind of mutation is produced by flat aromatic molecules such as the acridines. These compounds *intercalate* in DNA—that is, they slip in between adjacent base pairs in the DNA double helix (p. 614). Intercalators appear to stabilize a shifted pairing of bases in a repetitive sequence such as CGCGCGCG. Consequently, *they lead to the insertion or deletion of one or more base pairs*. The effect of such mutations is to alter the reading frame, unless an integral multiple of three base pairs is inserted or deleted. In fact, it was the analysis of such mutants that revealed the triplet nature of the genetic code.

MANY POTENTIAL CARCINOGENS CAN BE DETECTED BY THEIR MUTAGENIC ACTION ON BACTERIA

Many human cancers are caused by exposure to toxic chemicals. These chemical carcinogens are usually mutagenic, which suggests that *damage to DNA is a fundamental event in both carcinogenesis and mutagenesis.* It is important to identify these compounds and ascertain their potency so that human exposure to them can be minimized. Bruce Ames has devised a simple and sensitive test for detecting chemical mutagens. A thin layer of agar containing about 10^9 bacteria of a specially constructed tester strain of *Salmonella* is placed on a petri dish. These bacteria are unable to grow in the absence of histidine because of a mutation in one of their genes for the biosynthesis of this amino acid. The addition of a mutagen to the center of a plate results in many new mutations. A small proportion of them reverse the original mutation so that histidine is synthesized. These *revertants* multiply in the absence of exogenous histidine and appear as discrete colonies after the plate is incubated at 37°C for two days (Figure 26-15). For example, 0.5 μg of 2-aminoanthracene gives 11,000 revertant colonies compared with only 30 spontaneous revertants in the absence of this mutagen. A series of concentrations of a chemical can readily be tested to generate a dose-response curve. These relationships are usually linear, which suggests that there is no threshold concentration for mutagenesis.

Some of the tester strains are responsive to *base-pair substitutions,* whereas others detect *deletions or additions of base pairs (frame shifts).* The sensitivity of these specially designed strains has been enhanced by genetically deleting their excision-repair systems. Also, they lack the lipopolysaccharide barrier that normally coats the surface of *Salmonella.* The absence of this barrier facilitates the entry of potential mutagens. Another important feature of this detection system is the addition of a *mammalian liver homogenate.* Recall that

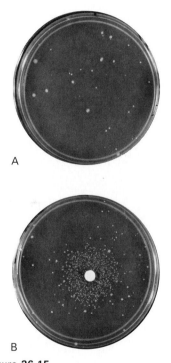

A

B

Figure 26-15
Salmonella test for mutagens: (A) petri plate containing about 10^9 bacteria that cannot synthesize histidine; (B) the addition of a filter-paper disc containing a mutagen, which leads to a large number of revertants that can synthesize histidine. These revertants appear as a ring of colonies around the disc. The small number of visible colonies in part A are spontaneous revertants. [From B. N. Ames, J. McCann, and E. Yamasaki. *Mutation Res.* 31(1975):347.]

some potential carcinogens are converted into their active forms by enzyme systems in the liver or other mammalian tissues (p. 475). Bacteria lack these enzymes, and so the test plate requires a few milligrams of a liver homogenate to activate this group of mutagens. For example, a compound with a reactive side chain (enabling it to form a covalent bond with DNA) and an aromatic region (allowing it to intercalate) is far more mutagenic than one with just the aromatic region.

The *Salmonella* test is now being extensively used to help evaluate the mutagenic and carcinogenic risks of a large number of chemicals. This rapid and inexpensive bacterial assay for mutagenicity complements epidemiologic surveys and animal tests that are necessarily slower, more laborious, and far more expensive. The *Salmonella* test for mutagenicity is an outgrowth of studies of gene-protein relationships in bacteria. We see here a striking example of how fundamental research in molecular biology can directly lead to important advances in public health.

SUMMARY

The base sequence of a gene is colinear with the amino acid sequence of its polypeptide product. The genetic code is the relationship between the sequence of bases in DNA (or its RNA transcript) and the sequence of amino acids in proteins. Amino acids are coded by groups of three bases (called codons) starting from a fixed point. Sixty-one of the sixty-four codons specify particular amino acids, whereas the other three codons (UAA, UAG, and UGA) are signals for chain termination. Thus, for most amino acids there is more than one code word. In other words, the code is degenerate. Codons specifying the same amino acid are called synonyms. Most synonyms differ only in the last base of the triplet. Some viral DNA sequences code for more than one protein because their transcripts are translated in different reading frames.

The genetic code, which is the same in all organisms, was deciphered by three kinds of experiments. First, synthetic polyribonucleotides (formed by polynucleotide phosphorylase) were used as mRNAs in cell-free protein-synthesizing systems, following the discovery that poly U codes for polyphenylalanine. Second, a trinucleotide (like a codon in mRNA) specifically binds the particular tRNA for which it is a code word. All sixty-four trinucleotides were synthesized and tested for their capacity to stimulate the binding of specific tRNAs to ribosomes. Third, synthetic polyribonucleotides with a defined repeating sequence were used as mRNAs. For example, poly (UAUC) led to the synthesis of poly (Tyr-Leu-Ser-Ile).

Most genes in higher eucaryotes are discontinuous. Coding sequences (exons) in these split genes are separated by intervening sequences (introns), which are removed in the conversion of the primary transcript into mRNA. For example, the β-globin gene in mammals contains two introns. Split genes, like continuous genes, are colinear with their polypeptide products.

Mutations are caused by changes in the base sequence of DNA. The major types are substitutions, deletions, and insertions. The substitution of one base pair for another is the most common type of mutation. A transition is the replacement of one purine by another purine or of one pyrimidine by another pyrimidine. A transversion is the replacement of a purine by a pyrimidine or vice versa. Mutations may be produced by the spontaneous tautomerization of a base or by the action of base analogs (e.g., 5-bromouracil) or other chemical mutagens (e.g., nitrous acid). Many potential carcinogens can be detected by their mutagenic action on bacteria.

SELECTED READINGS

WHERE TO START

Yanofsky, C., 1967. Gene structure and protein structure. *Sci. Amer.* 216(5):80–94. [Available as *Sci. Amer.* Offprint 1074. Presents the evidence for colinearity.]

Nirenberg, M. W., 1963. The genetic code II. *Sci. Amer.* 208(3):80–94. [Offprint 153. Describes the use of synthetic polyribonucleotides to decipher the code.]

Crick, F. H. C., 1966. The genetic code III. *Sci. Amer.* 215(4):55–62. [Offprint 1052. A view of the code when it was almost completely elucidated.]

Kourilsky, P., and Chambon, P., 1978. The ovalbumin gene: an amazing gene in eight pieces. *Trends Biochem. Sci.* 3:244–247.

THE ADAPTOR (tRNA) HYPOTHESIS

Crick, F. H. C., 1958. On protein synthesis. *Symp. Soc. Exp. Biol.* 12:138–163. [A brilliant anticipatory view of the problem of protein synthesis. The adaptor hypothesis is presented in this article.]

THE GENETIC CODE

Crick, F. H. C., Barnett, L., Brenner, S., and Watts-Tobin, R. J., 1961. General nature of the genetic code for proteins. *Nature* 192:1227–1232.

Khorana, H. G., 1968. Nucleic acid synthesis in the study of the genetic code. In *Nobel Lectures: Physiology or Medicine* (1963–1970), pp. 341–369. American Elsevier (1973).

Nirenberg, M., 1968. The genetic code. In *Nobel Lectures: Physiology or Medicine* (1963–1970), pp. 372–395. American Elsevier (1973).

Garen, A., 1968. Sense and nonsense in the genetic code. *Science* 160:149–159.

Woese, C. R., 1967. *The Genetic Code.* Harper & Row.

Cold Spring Harbor Laboratory, 1966. *The Genetic Code* (Cold Spring Harbor Symposia on Quantitative Biology, vol. 31). [An outstanding collection of papers establishing the major features of the code.]

OVERLAPPING GENES

Barrell, B. G., Air, G. M., and Hutchinson, C. A., III, 1976. Overlapping genes in bacteriophage φX174. *Nature* 264:34–40.

COLINEARITY OF GENE AND PROTEIN

Yanofsky, C., Carlton, B. C., Guest, J. R., Helinski, D. R., and Henning, U., 1964. On the colinearity of gene structure and protein structure.. *Proc. Nat. Acad. Sci.* 51:266–272.

Sarabhai, A. S., Stretton, O. W., Brenner, S., and Bolle, A., 1964. Colinearity of gene with polypeptide chain. *Nature* 201:13–17.

SPLIT GENES AND INTERVENING SEQUENCES

Crick, F., 1979. Split genes and RNA splicing. *Science* 204:264–271.

Jeffreys, A. J., and Flavell, R. A., 1977. The rabbit β-globin gene contains a large insert in the coding sequence. *Cell* 12:1097–1108.

Breathnack, R., Mandel, J. L., and Chambon, P., 1977. Ovalbumin gene is split in chicken DNA. *Nature* 270:314–319.

Cochet, M., Gannon, F., Hen, R., Maroteaux, L., Perrin, F., and Chambon, P., 1979. Organisation and sequence studies of the 17-piece chicken conalbumin gene. *Nature* 282:567–574.

640   Tilghman, S. M., Tiemeier, D. C., Seidman, J. G., Peterlin, B. M., Sullivan, M., Maijel, J. V., and Leder, P., 1978. Intervening sequence of DNA identified in the structural portion of a mouse β-globin gene. *Proc. Nat. Acad. Sci.* 75:725–729.

MUTAGENESIS

Hayes, W., 1968. *The Genetics of Bacteria and Their Viruses* (2nd ed.). Wiley. [Chapter 13 on the nature of mutations is lucid and concise.]

Drake, J. W., and Baltz, R. H., 1976. The biochemistry of mutagenesis. *Ann. Rev. Biochem.* 45:11–38.

Drake, J. W., 1970. *The Molecular Basis of Mutation.* Holden-Day.

DETECTION OF CARCINOGENS

Ames, B. N., 1979. Identifying environmental chemicals causing mutations and cancer. *Science* 204:587–593.

Devoret, R., 1979. Bacterial tests for potential carcinogens. *Sci. Amer.* 241(2):40–49.

McCann, J., Spingarn, N. E., Kobori, J., and Ames, B. N., 1975. Detection of carcinogens as mutagens: bacterial tester strains with R factor plasmids. *Proc. Nat. Acad. Sci.* 72:979–983.

PROBLEMS

1. What amino acid sequence is encoded by the following base sequence of an mRNA molecule? Assume that the reading frame starts at the 5' end.

    5'-UUGCCUAGUGAUUGGAUG-3'

2. What is the sequence of the polypeptide formed on addition of poly (UUAC) to a cell-free protein-synthesizing system?

3. Suppose that the single-stranded RNA from tobacco mosaic virus was treated with a chemical mutagen, that mutants were obtained having serine or leucine instead of proline at a specific position, and that further treatment of these mutants with the same mutagen yielded phenylalanine at this position.

    (a) What are the plausible codon assignments for these four amino acids?
    (b) Was the mutagen 5-bromouracil, nitrous acid, or an acridine dye?

4. A protein chemist told a molecular geneticist that he had found a new mutant hemoglobin in which aspartate replaced lysine. The molecular geneticist expressed surprise and sent his friend scurrying back to the laboratory.
    (a) Why was the molecular geneticist dubious about the reported amino acid substitution?
    (b) Which amino acid substitutions would have been more palatable to the molecular geneticist?

5. The amino acid sequences of part of lysozyme from wild-type T4 bacteriophage and a mutant are

    Wild-type
    -Thr-Lys-Ser-Pro-Ser-Leu-Asn-Ala-Ala-Lys-
    -Thr-Lys-Val-His-His-Leu-Met-Ala-Ala-Lys-
    Mutant

    (a) Could this mutant have arisen by the change of a single base pair in T4 DNA? If not, how might this mutant have been produced?
    (b) What is the base sequence of the mRNA that codes for the five amino acids in the wild-type that are different from those in the mutant?

6. The RNA transcript of a region of G4 phage DNA contains the sequence 5'-AAAUGAGGA-3'. This region encodes three different proteins. What are their amino acid sequences?

For additional problems, see W. B. Wood, J. H. Wilson, R. M. Benbow, and L. E. Hood, *Biochemistry: A Problems Approach* (Benjamin, 1974), ch. 19.

CHAPTER 27

PROTEIN SYNTHESIS

From the preceding chapter, we know that the sequence of amino acids in a protein is determined by the sequence of codons in mRNA as read by tRNA molecules. We turn now to the mechanism of protein synthesis, a process called *translation* because the language consisting of the four base-pairing letters of nucleic acids is turned into that comprising the twenty letters of proteins. As might be expected, translation is a more complex process than either the replication or the transcription of DNA, which take place within the framework of a common base-pairing language. In fact, translation necessitates the coordinated interplay of more than a hundred kinds of macromolecules. Transfer RNA molecules, activating enzymes, soluble factors, and mRNA are required in addition to ribosomes.

Let us consider the overall process of protein synthesis before it is described in more detail. A protein is synthesized in the amino-to-carboxyl direction by the sequential addition of amino acids to the carboxyl end of the growing peptide chain. The activated precursors are aminoacyl-tRNAs, in which the carboxyl group of an amino acid is joined to the 3′ terminus of a tRNA. The linking of an amino acid to its corresponding tRNA is catalyzed by an aminoacyl-tRNA synthetase. This activation reaction, which is analogous to the activation of fatty acids, is driven by ATP. For each amino acid, there is at least one kind of tRNA and activating enzyme. There are three stages in protein synthesis: initiation, elongation, and termination. *Initiation* results in the binding of the initiator tRNA to the start signal of mRNA. The initiator tRNA occupies the P (peptidyl) site on the ribosome, which is one of two binding sites for tRNA. *Elongation* starts with the binding of an aminoacyl-tRNA to the other tRNA site on the ribosome, which is called the A (aminoacyl) site. A peptide bond is then formed between the amino group of the incoming aminoacyl-tRNA and the carboxyl group of the fMet carried by the initiator tRNA. The resulting dipeptidyl-tRNA is then translocated from the A site to the P site, whereas the other tRNA molecule leaves the ribosome. The binding of amino-acyl-tRNA, the movement of peptidyl-tRNA from the A site to the P site, and the associated movement of the mRNA to the next codon require the hydrolysis of GTP. An aminoacyl-tRNA then binds to the vacant A site to start another round of elongation, which proceeds as described above. *Termination* occurs when a stop signal on the mRNA is read by a protein release factor, which leads to the release of the completed polypeptide chain from the ribosome. This chapter will deal primarily with protein synthesis in *E. coli,* where the process is best understood. Some differences between protein synthesis in procaryotes and eucaryotes will be considered in a later chapter (p. 709).

AMINO ACIDS ARE ACTIVATED AND LINKED TO TRANSFER RNA BY SPECIFIC SYNTHETASES

The formation of a peptide bond between the amino group of one amino acid and the carboxyl group of another is thermodynamically unfavorable. This thermodynamic barrier is overcome by activating the carboxyl group of the precursor amino acids. *The activated intermediates in protein synthesis are amino acid esters,* in which the carboxyl group of an amino acid is linked to either the 2'- or the 3'-hydroxyl group of the ribose unit at the 3' end of tRNA. The aminoacyl group can migrate rapidly between the 2'- and 3'-hydroxyl groups. This activated intermediate is called an *aminoacyl-tRNA* (Figure 27-1).

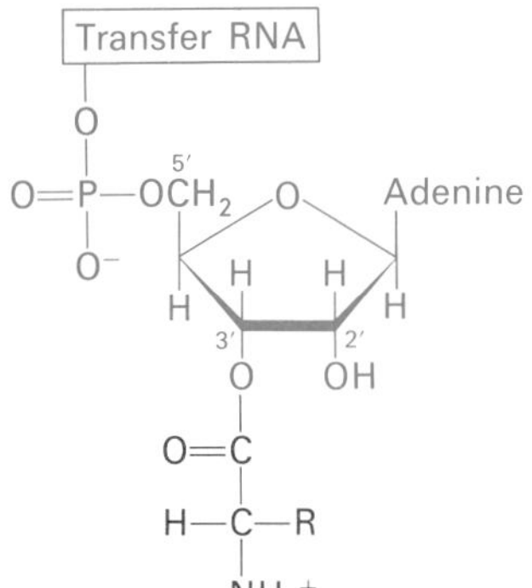

The attachment of an amino acid to a tRNA is important not only because this activates its carboxyl group so that it can form a peptide, but also because amino acids by themselves cannot recognize the codons on mRNA. Rather, amino acids are carried to the ribosomes by specific tRNAs, which do recognize codons on mRNA and thereby act as adaptor molecules.

In 1957, Paul Zamecnik and Mahlon Hoagland discovered that the activation of amino acids and their subsequent linkage to tRNAs are catalyzed by specific *aminoacyl-tRNA synthetases,* which are also called *activating enzymes.* For some synthetases, the first step is the formation of an *aminoacyl-adenylate* from an amino acid and ATP. This activated species is a mixed anhydride in which the carboxyl group of the amino acid is linked to the phosphoryl group of AMP. For other synthetases, the reaction of ATP, amino acid,

Figure 27-1
An amino acid is esterified to the 2'- or the 3'-hydroxyl group of the terminal adenosine in an aminoacyl-tRNA.

and tRNA occurs without a detectable aminoacyl-adenylate intermediate.

$$^{+}H_3N-\underset{\underset{R}{|}}{\overset{\overset{H}{|}}{C}}-\underset{\overset{|}{O^-}}{\overset{\overset{O}{\|}}{C}} + ATP \rightleftharpoons {}^{+}H_3N-\underset{\underset{R}{|}}{\overset{\overset{H}{|}}{C}}-\underset{}{\overset{\overset{O}{\|}}{C}}-O-\underset{\underset{O^-}{|}}{\overset{\overset{O}{\|}}{P}}-O-Ribose-Adenine + PP_i$$

Aminoacyl-adenylate
(Aminoacyl-AMP)

The next step is the transfer of the aminoacyl group of aminoacyl-AMP to a tRNA molecule to form *aminoacyl-tRNA,* the activated intermediate in protein synthesis. Whether the aminoacyl
group is transferred to the 2'-hydroxyl or the 3'-hydroxyl group of
the ribose unit at the 3' end of tRNA depends on the particular
species. The activated amino acid can migrate very rapidly between the 2'- and 3'- hydroxyl groups.

Aminoacyl-AMP + tRNA $\rightleftharpoons$ aminoacyl-tRNA + AMP

The sum of these activation and transfer steps is

Amino acid + ATP + tRNA $\rightleftharpoons$ aminoacyl-tRNA + AMP + PP$_i$

The $\Delta G°'$ of this reaction is close to 0, because the free energy of
hydrolysis of the ester bond of aminoacyl-tRNA is similar to that of
the terminal phosphoryl group of ATP. What then drives the synthesis of aminoacyl-tRNA? As expected, the reaction is driven by
the hydrolysis of pyrophosphate. The sum of these three reactions is
highly exergonic:

Amino acid + ATP + tRNA + H$_2$O $\rightleftharpoons$ aminoacyl-tRNA + AMP + 2 P$_i$

Thus, *two high-energy phosphate bonds are consumed in the synthesis of an
aminoacyl-tRNA.* One of them is consumed in forming the ester linkage of aminoacyl-tRNA, whereas the other is consumed in driving
the reaction forward.

The activation and transfer steps for a particular amino acid are
catalyzed by the same aminoacyl-tRNA synthetase. In fact, *the aminoacyl-AMP intermediate does not dissociate from the synthetase.* Rather, it
is tightly bound to the active site of the enzyme by noncovalent
interactions. The aminoacyl-AMP is normally a transient intermediate in the synthesis of aminoacyl-tRNA, but it is quite stable and
readily isolated if tRNA is absent from the reaction mixture.

An acyl-adenylate intermediate has already been encountered in
fatty acid activation (p. 387). It is interesting to note that Paul Berg
first discovered this intermediate in fatty acid activation, and then
recognized that it is also formed in amino acid activation. The
major difference between these reactions is that the acceptor of the
acyl group is CoA in the former and tRNA in the latter. The ener-

getics of these biosyntheses is very similar: both are made irreversible by the hydrolysis of pyrophosphate.

There is at least one aminoacyl-tRNA synthetase for each amino acid. These enzymes differ in size, subunit structure, and amino acid composition (Table 27-1).

Table 27-1
Properties of some aminoacyl-tRNA synthetases

Source	Amino acid specificity	Mass (kdal)	Subunit structure
E. coli	Histidine	85	α_2
E. coli	Isoleucine	114	One chain
E. coli	Lysine	104	α_2
E. coli	Glycine	227	$\alpha_2\beta_2$
Yeast	Lysine	138	α_2
Yeast	Phenylalanine	270	$\alpha_2\beta_2$
Beef pancreas	Tryptophan	108	α_2

THE FIDELITY OF PROTEIN SYNTHESIS DEPENDS ON THE HIGH SPECIFICITY OF AMINOACYL-tRNA SYNTHETASES

The correct translation of the genetic message depends on the high degree of specificity of aminoacyl-tRNA synthetases. *These enzymes are highly selective in their recognition of the amino acid to be activated and of the prospective tRNA acceptor.* As will be discussed shortly, tRNA molecules that accept different amino acids have different base sequences, and so they can be readily recognized by the synthetases. A much more demanding task for these enzymes is to discriminate between similar amino acids. For example, the only difference between isoleucine and valine is that isoleucine contains an additional methylene group (Figure 27-2). The additional binding energy contributed by this extra —CH_2— group favors the activation of isoleucine over valine by a factor of about 200. The concentration of valine in vivo is about five times that of isoleucine, and so valine would be mistakenly incorporated in place of isoleucine 1 in 40 times. However, the observed error frequency in vivo is only 1 in 3,000, indicating that there must be a subsequent editing step to enhance fidelity. In fact, the synthetase corrects its own errors. Valine that is mistakenly activated is not transferred to the tRNA specific for isoleucine. Instead, *this tRNA promotes the hydrolysis of valine-AMP, which thereby prevents its erroneous incorporation into proteins.* Furthermore, this hydrolytic reaction frees the synthetase for the activation and transfer of isoleucine, the correct amino acid. How does the synthetase avoid hydrolyzing isoleucine-AMP, the desired intermediate? Most likely, the hydrolytic site is just large enough to accommodate valine-AMP, but too small to allow the entry of isoleucine-AMP.

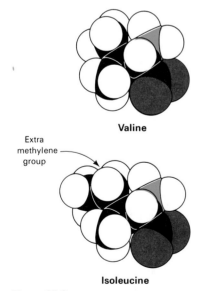

Valine

Extra methylene group

Isoleucine

Figure 27-2
Molecular models of valine and isoleucine. The extra methylene group in isoleucine is marked. The synthetases specific for these amino acids are highly discerning.

A number of other aminoacyl-tRNA synthetases also contain hydrolytic sites in addition to synthetic sites. These sites probably function as a *double sieve* to assure very high fidelity. The synthetic site rejects amino acids that are *larger* than the correct one, whereas the hydrolytic site destroys activated intermediates that are *smaller* than the correct species. It is evident that *the high fidelity of protein synthesis is critically dependent on the hydrolytic proofreading action of many aminoacyl-tRNA synthetases.*

TRANSFER RNA MOLECULES HAVE A COMMON DESIGN

The base sequence of a transfer RNA molecule was first determined by Robert Holley in 1965, the culmination of seven years of effort. Indeed, this study of yeast alanine tRNA provided the first complete sequence of a nucleic acid. This work was a source of insight into the biological activity of a tRNA molecule. In elucidating the base sequence, Holley also created techniques for the sequencing of nucleic acids in general. The sequence of yeast alanine tRNA is shown in Figure 27-4. The molecule is a single chain of seventy-six ribonucleotides. The 5' terminus is phosphorylated (pG), whereas

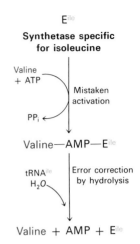

Figure 27-3
Proofreading by hydrolysis.

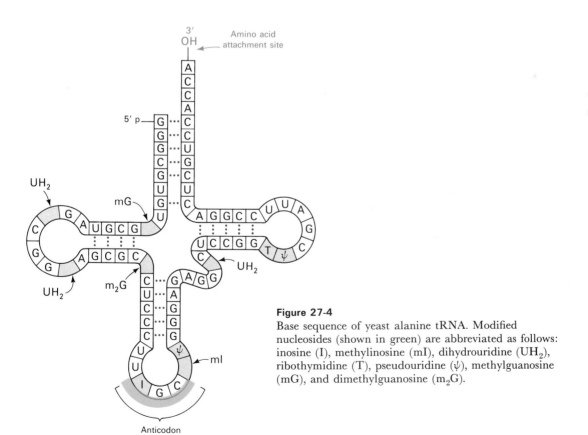

Figure 27-4
Base sequence of yeast alanine tRNA. Modified nucleosides (shown in green) are abbreviated as follows: inosine (I), methylinosine (mI), dihydrouridine (UH$_2$), ribothymidine (T), pseudouridine (ψ), methylguanosine (mG), and dimethylguanosine (m$_2$G).

Anticodon

$$-\overset{3'}{C}-G-\overset{5'}{I}-$$
$$-\underset{5'}{G}-C-\underset{3'}{C}-$$

Codon

the 3′ terminus has a free hydroxyl group. A distinctive feature of the molecule is its high content of bases other than A, U, G, and C. There are nine unusual nucleosides in this tRNA: inosine, pseudouridine, dihydrouridine, ribothymidine, and methylated derivatives of guanosine and inosine. The *amino acid attachment site* is the 3′-hydroxyl group of the adenosine residue at the 3′ terminus of the molecule. The sequence IGC in the middle of the molecule is the *anticodon*. It is complementary to GCC, one of the codons for alanine.

The sequences of several other tRNA molecules were determined a short time later. More than seventy sequences are now known. The striking finding is that all of these sequences can be written in a cloverleaf pattern in which about half the residues are base paired. Hence, *tRNA molecules have many common structural features.* This finding is not unexpected, because all tRNA molecules must be able to interact in nearly the same way with ribosomes and mRNAs. Specifically, all tRNA molecules must fit into the A and P sites on the ribosomes and must interact with the enzyme that catalyzes peptide-bond formation.

All transfer RNA molecules have these features in common:

1. They are single chains containing between *73 and 93 ribonucleotides* (about 25 kdal) each.

2. They contain *many unusual bases,* typically between 7 and 15 per molecule. Many of these unusual bases are methylated or dimethylated derivates of A, U, C, and G that are formed by enzymatic modification of a precursor tRNA (p. 612). The role of these unusual bases is uncertain. One possibility is that methylation prevents the formation of certain base pairs, thereby rendering some of the bases accessible for other interactions. Also, methylation imparts a hydrophobic character to some regions of tRNAs, which may be important for their interaction with the synthetases and with ribosomal proteins.

3. The 5′ end of tRNAs is phosphorylated. The 5′ terminal residue is usually pG.

4. The base sequence at the 3′ end of tRNAs is CCA. The activated amino acid is attached to the 3′-hydroxyl group of the terminal adenosine.

5. About half of the nucleotides in tRNAs are base paired to form double helices. Five groups of bases are not base paired: the 3′ *CCA terminal region;* the *TψC loop,* which acquired its name from the sequence ribothymine-pseudouracil-cytosine; the *"extra arm,"* which contains a variable number of residues; the *DHU loop,* which contains several dihydrouracil residues; and the *anticodon loop.*

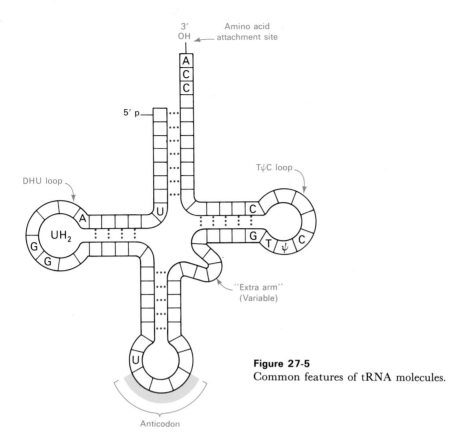

Figure 27-5
Common features of tRNA molecules.

6. The *anticodon loop* consists of seven bases, with the following sequence:

5'   3'
—Pyrimidine—Pyrimidine—X—Y—Z— Modified — Variable —
                        ‾‾‾‾‾   purine   base
                       Anticodon

TRANSFER RNA IS L-SHAPED

The three-dimensional structure of a tRNA molecule is now known in atomic detail as a result of x-ray crystallographic studies carried out in the laboratories of Alexander Rich and Aaron Klug. Their independent x-ray analyses of yeast phenylalanine tRNA have provided a wealth of structural information:

1. The molecule is *L-shaped* (Figure 27-6).

2. There are *two segments of double helix*. Each of these helices contains about ten base pairs, which correspond to one turn of helix. The helical segments are perpendicular to each other, which gives the molecule its L-shape. The base pairing given in the cloverleaf model, which was based on sequence studies, is correct.

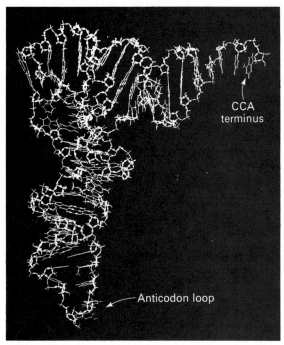

Figure 27-6
Photograph of a skeletal model of yeast phenylalanine tRNA based on an electron-density map at 3-Å resolution. [Courtesy of Dr. Sung-Hou Kim.]

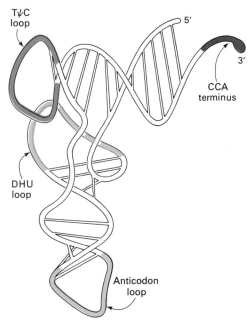

Figure 27-7
Schematic diagram of the three-dimensional structure of yeast phenylalanine tRNA. [After a drawing kindly provided by Dr. Sung-Hou Kim.]

3. Most of the bases in the nonhelical regions participate in unusual hydrogen-bonding interactions. These *tertiary interactions* are between bases that are not usually complementary (e.g., G-G, A-A, and A-C). Moreover, the ribose-phosphate backbone interacts with some bases and even with another region of the backbone itself. The 2′—OH groups of the ribose units act as hydrogen bond donors or acceptors in many of these interactions. In addition, a large proportion of the bases are stacked. These hydrophobic interactions between adjacent aromatic rings play a major role in stabilizing the molecular architecture.

4. The CCA terminus containing the *amino acid attachment site* is at one end of the L. The other end of the L is the *anticodon loop*. Thus, *the amino acid in aminoacyl-tRNA is far from the anticodon* (about 80 Å). The DHU and TψC loops form the corner of the L.

5. The CCA terminus and the adjacent helical region do not interact strongly with the rest of the molecule. This part of the molecule may change conformation during amino acid activation and during protein synthesis on the ribosome.

The determination of the three-dimensional structure of tRNA is an important step toward elucidating the process of translation in precise molecular terms. This achievement seemed only a remote possibility not many years ago. The thoughts and efforts of investigators in this field are now turning to even more demanding goals, such as solving the three-dimensional structure of tRNA-synthetase complexes and even of tRNA bound to mRNA and ribosomal proteins.

THE CODON IS RECOGNIZED BY THE ANTICODON RATHER THAN BY THE ACTIVATED AMINO ACID

It has already been mentioned that the anticodon on tRNA is the recognition site for the codon on mRNA and that recognition occurs by base pairing. Does the amino acid attached to the tRNA play a role in this recognition process? This question was answered in the following way. First, cysteine was attached to its cognate tRNA (denoted by tRNA^{Cys}). The attached cysteine unit was then converted into alanine by reacting Cys-tRNA^{Cys} with Raney nickel, which removed the sulfur atom from the activated cysteine residue without affecting its linkage to tRNA.

Thus, a *hybrid aminoacyl-tRNA* was produced in which alanine was covalently attached to a tRNA specific for cysteine. Does this hybrid tRNA recognize the codon for alanine or for cysteine? The answer was provided by using this hybrid tRNA in a cell-free protein-synthesizing system. The template was a random copolymer of U and G in the ratio of 5:1, which normally leads to the incorporation of cysteine (UGU) but not alanine (GCX). However, alanine was incorporated into a polypeptide when Ala-tRNA^{Cys} was added to the incubation mixture. In other words, alanine was incorporated as though it were cysteine because it was attached to the tRNA specific for cysteine. Thus, *codon recognition does not depend on the amino acid that is attached to tRNA*.

The same result was obtained when mRNA for hemoglobin served as the template. [14]C alanyl-tRNA^{Cys} was used as the hybrid aminoacyl-tRNA. The only radioactive tryptic peptide (i.e., containing ^{14}C-alanine) was one that normally contained cysteine but not alanine. On the other hand, no radioactivity was found in any of the peptides that normally contained alanine but not cysteine. This experiment and the preceding one unequivocally demonstrated that the amino acid in aminoacyl-tRNA does not play a role in selecting a codon.

A TRANSFER RNA MOLECULE MAY RECOGNIZE MORE THAN ONE CODON BECAUSE OF WOBBLE

What are the rules that govern the recognition of a codon by the anticodon of a tRNA? A simple hypothesis is that each of the bases of the codon forms a Watson-Crick type of base pair with a complementary base on the anticodon. The codon and anticodon would then be lined up in an antiparallel fashion. In the accompanying diagram, the prime denotes the complementary base. Thus X and X′ would be either A and U (or U and A) or G and C (or C and G). A specific prediction of this model is that a particular anticodon can recognize only one codon. The facts are otherwise. *Some pure tRNA molecules can recognize more than one codon.* For example, the yeast alanine tRNA studied by Holley binds to *three* codons: GCU, GCC, and GCA. The first two bases of these codons are the same, whereas the third is different. Could it be that the recognition of the third base of a codon is sometimes less discriminating than of the other two? The pattern of degeneracy of the genetic code indicates that this might be so. XYU and XYC always code for the same amino acid, whereas XYA and XYG usually do. Crick surmised from these data that the steric criteria for pairing of the third base might be less stringent than for the other two. Models of various base pairs were built to determine which ones are similar to the standard A-U and G-C base pairs with regard to the distance and angle between the glycosidic bonds. Inosine was included in this study because it appeared in several anticodons. Assuming some steric freedom ("wobble") in the pairing of the third base of the codon, the combinations shown in Table 27-2 seemed plausible.

The wobble hypothesis is now firmly established. The anticodons of tRNAs of known sequence bind to the codons predicted by this hypothesis. For example, the anticodon of yeast alanine tRNA is IGC. This tRNA recognizes the codons GCU, GCC, and GCA.

Anticodon

$$-\overset{3'}{X'}-Y'-\overset{5'}{Z'}-$$
$$-\underset{5'}{X}-Y-\underset{3'}{Z}-$$

Codon

Table 27-2
Allowed pairings at the third codon base according to the wobble hypothesis

First anticodon base	Third codon base
C	G
A	U
U	A or G
G	U or C
I	U, C, or A

Anticodon

$$-\overset{3'}{C}-G-\overset{5'}{I}-$$
$$-\underset{5'}{G}-C-\underset{3'}{U}-$$

Codon

Anticodon

$$-\overset{3'}{C}-G-\overset{5'}{I}-$$
$$-\underset{5'}{G}-C-\underset{3'}{C}-$$

Codon

Anticodon

$$-\overset{3'}{C}-G-\overset{5'}{I}-$$
$$-\underset{5'}{G}-C-\underset{3'}{A}-$$

Codon

Thus, I pairs with U, C, or A, as predicted. Phenylalanine tRNA, which has the anticodon GAA, recognizes the codons UUU and UUC but not UUA and UUG.

Anticodon

$$-\overset{3'}{A}-A-\overset{5'}{G}-$$
$$-\underset{5'}{U}-U-\underset{3'}{U}-$$

Codon

Anticodon

$$-\overset{3'}{A}-A-\overset{5'}{G}-$$
$$-\underset{5'}{U}-U-\underset{3'}{C}-$$

Codon

Thus, G pairs with either U or C in the third position of the codon, as predicted by the wobble hypothesis.

Two generalizations concerning the codon-anticodon interaction can be made:

1. The first two bases of a codon pair in the standard way. Recognition is precise. Hence, *codons that differ in either of their first two bases must be recognized by different tRNAs.* For example, both UUA and CUA code for leucine but are read by different tRNAs.

2. The first base of an anticodon determines whether a particular tRNA molecule reads one, two or three kinds of codons: C or A (1 codon), U or G (2 codons), or I (3 codons). Thus, *part of the degeneracy of the genetic code arises from imprecision (wobble) in the pairing of the third base of the codon.* We see here a strong reason for the frequent appearance of inosine, one of the unusual nucleosides, in anticodons. Inosine maximizes the number of codons that can be read by a particular tRNA molecule (Figure 27-8).

MUTANT TRANSFER RNA MOLECULES CAN SUPPRESS OTHER MUTATIONS

Further insight into the process of codon recognition has come from studies of mutant tRNAs in which a single base in the anticodon is changed. In effect, these spontaneous events in nature are the converse of the ones in which the amino acid attached to tRNA was chemically modified in vitro. These mutant tRNAs were discovered in an interesting way. Geneticists had known for many years that the deleterious effects of some mutations can be reversed by a second mutation. Suppose that an enzyme becomes inactive because of a mutation from GCU (alanine) to GAU (aspartate). What kind of mutations might restore the activity of this enzyme?

1. A reverse mutation of the same base, A → C, would give the original code word.

2. A different mutation, A → U, would give valine (GUU), which might result in a partially or fully active enzyme.

3. A mutation at a different site in the same gene might reverse the effects of the first mutation. Such an altered protein would differ from the wild-type by two amino acids.

4. A mutation in a *different gene* might overcome the deleterious effects of the first mutation. Such a mutation is called an *intergenic suppressor.*

The mode of action of intergenic suppressors was a puzzle for a long time. We now know from genetic and biochemical studies that

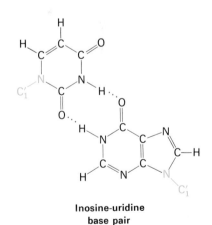

Inosine-cytidine
base pair

Inosine-adenosine
base pair

Inosine-uridine
base pair

Figure 27-8
Inosine can base-pair with cytosine, adenine, or uracil because of wobble.

most of these suppressors act by altering the reading of mRNA. For example, consider a mutation that leads to the appearance of the codon UAG, which is a termination signal. The result is an incomplete polypeptide chain. Such mutations are called *nonsense mutations* because incomplete polypeptides are usually inactive. The UAG nonsense mutation can be suppressed by mutations in several different genes. One of these suppressors reads UAG as a codon for tyrosine, which may lead to the synthesis of a functional protein rather than to an incomplete polypeptide chain (Figure 27-9). Why does this

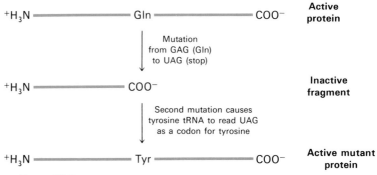

Figure 27-9
Suppression of a chain-terminating mutation by a second mutation in a tRNA molecule.

Anticodon

$$^{3'} \quad \quad ^{5'}$$
—A—U—C— Mutated base in the anticodon of a tyrosine tRNA (normally 3′ AUG 5′)

—U—A—G— } This chain-termination codon is read as tyrosine.
$$^{5'} \quad \quad ^{3'}$$

Codon

mutant tRNA insert tyrosine in response to the UAG codon? Tyrosine tRNA normally recognizes the codons UAC and UAU. This mutant tRNA is identical with the normal tyrosine tRNA except for a single base change in its anticodon: GUA → CUA. The mutation G → C in the first base of its anticodon changes its recognition properties. The altered anticodon can recognize only UAG, as predicted by the wobble hypothesis.

This kind of suppressor mutation is more likely to be selected if the tRNA undergoing mutation is not essential. In other words, there must be another tRNA species that recognizes the same codons as the tRNA undergoing mutation. In fact, in *E. coli*, there are two different tRNAs that normally recognize both UAC and UAU. The species present in a small amount is the one that becomes altered. The normal role of this minor species of tyrosine tRNA is uncertain. The occurrence of this suppressor mutation raises a second question. If the mutation to UAG is read as tyrosine rather than as a stop signal, what happens to normal chain termination? Surprisingly, most polypeptide chains terminate normally in the suppressor mutant, possibly because the termination signal may be more than just UAG. Indeed, some messages are known to end with two different stop codons. Also, suppression is not completely efficient.

Other kinds of mutant tRNAs have been identified. *Missense suppressors* alter the reading of mRNA so that one amino acid is inserted instead of another at certain codons (e.g., glycine instead of arginine). *Frameshift suppressors* are especially interesting. One of them has an extra base in its anticodon loop. Consequently, it reads *four* nucleotides as a codon rather than three. In particular, UUUC rather than UUU is read as the codon for phenylalanine. A mutation caused by the insertion of an extra base can be suppressed by this altered tRNA.

RIBOSOMES, THE ORGANELLES OF PROTEIN SYNTHESIS, CONSIST OF LARGE AND SMALL SUBUNITS

We turn now to the mechanism of protein synthesis. This complex process takes place in ribosomes, which can be regarded as the organelles of protein synthesis, just as mitochondria are the organelles of oxidative phosphorylation. A ribosome is a highly specialized and complex structure, approximately 200 Å in diameter. The best-characterized ribosomes are those of *E. coli*. The *E. coli* ribosome has a mass of 2,500 kdal and a sedimentation coefficient of 70S. It can be dissociated into a *large subunit (50S)* and a *small subunit (30S)*. These subunits can be further split into their constituent proteins and RNAs. The 30S subunit contains twenty-one proteins and a 16S RNA molecule. The 50S subunit contains about thirty-four proteins and two RNA molecules, a 23S species and a 5S species. About two-thirds of the mass of an *E. coli* ribosome is RNA, whereas the other third is protein.

The ribosomes in the cytoplasm of eucaryotic cells are somewhat larger than those of bacteria. An intact eucaryotic ribosome has a

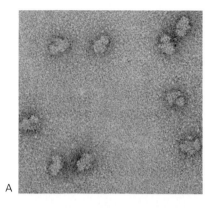

A

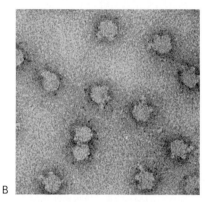

B

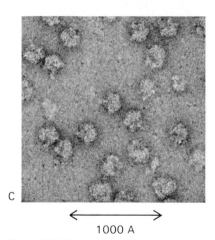

C

←――――→
1000 A

Figure 27-10
Electron micrographs of (A) 30S subunits, (B) 50S subunits, and (C) 70S ribosomes. [Courtesy of Dr. James Lake.]

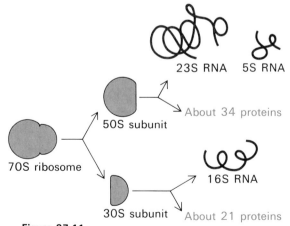

23S RNA   5S RNA

About 34 proteins

50S subunit

70S ribosome

16S RNA

30S subunit   About 21 proteins

Figure 27-11
Ribosomes can be dissociated into about fifty-five proteins and three RNA molecules.

sedimentation coefficient of 80S. Like a bacterial ribosome, it dissociates into a large subunit (60S) and a small subunit (40S). The small subunit contains a single 18S RNA molecule, whereas the large subunit contains three RNA molecules (28S, 7S, and 5S). The ribosomes in mitochondria and chloroplasts are different from those in the cytoplasm of eucaryotes. They are more like 70S than 80S particles. In fact, there are many similarities between protein synthesis in mitochondria, chloroplasts, and bacteria.

RIBOSOMES CAN BE RECONSTRUCTED FROM THEIR CONSTITUENT PROTEINS AND RNA MOLECULES

The 30S ribosomal subunit can be reconstituted from a mixture of 16S RNA and its twenty-one constituent proteins. The spontaneous reassembly of these components to form a fully functional 30S subunit was first achieved by Masayasu Nomura in 1968. The 50S subunit was reconstituted several years later. The significance of these experiments is twofold. First, they demonstrate that all of the information needed for the correct assembly of this organelle is contained in the structure of its components. Nonribosomal factors are not needed. Thus, *the formation of a ribosome in vitro is a self-assembly process.* Second, reconstitution can be used to ascertain whether a particular component is essential for the assembly of the ribosome or for a specific function. For example, the component responsible for sensitivity to the antibiotic streptomycin was identified in this way (p. 662). The following conclusions have emerged from studies of the reconstitution of the 30S subunit:

1. The 16S RNA is essential for its assembly and function. The requirement is quite specific because 16S RNA from yeast cannot substitute for 16S RNA from *E. coli.* The role of ribosomal RNA molecules is a challenging area of inquiry.

2. Reconstitution was performed with one protein omitted at a time to see whether an intact 30S particle can be formed from the other twenty proteins, in addition to 16S RNA. At least six proteins were found to be essential for the assembly of a 30S particle.

3. Most of the twenty-one proteins were needed (in addition to 16S RNA) for the reassembly of a functionally active 30S particle. Thus, *the 30S subunit is a cooperative functional entity.*

PROTEINS ARE SYNTHESIZED IN THE AMINO-TO-CARBOXYL DIRECTION

One of the first questions asked about the mechanism of protein synthesis was whether proteins are synthesized in the amino-to-carboxyl direction or in the reverse direction. The pulse-labeling

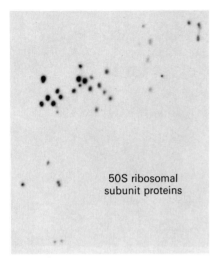

Figure 27-12
Separation of the proteins of the 50S ribosomal subunit by two-dimensional electrophoresis on a polyacrylamide gel. [Courtesy of Dr. Charles Cantor.]

Figure 27-13
Polyacrylamide-gel electrophoresis patterns of the native 30S ribosomal subunit (left) and a reconstituted 30S subunit (right). All of the proteins present in the native subunit are present in the reconstituted one. [Courtesy of Dr. Masayasu Nomura.]

studies of Howard Dintzis provided a clear-cut answer. Reticulocytes that were actively synthesizing hemoglobin were exposed to ^{3}H-leucine for periods shorter than required to synthesize a complete chain. The released hemoglobin was separated into α and β chains and then treated with trypsin. The specific radioactivity of these peptides—that is, the amount of radioactivity in them relative to their total content of leucine—was determined. There was more radioactivity in the peptides near the carboxyl end than in those near the amino end. In fact, there was a *gradient of increasing radioactivity from the amino to the carboxyl end of each chain* (Figure 27-14). The carboxyl end was the most heavily labeled because it was the last to be synthesized. Hence, *the direction of chain growth is from the amino to the carboxyl end.*

MESSENGER RNA IS TRANSLATED IN THE 5′ → 3′ DIRECTION

The direction of reading of mRNA was determined by using the synthetic polynucleotide

$$\overset{5'}{A}-A-A-(A-A-A)_n-A-A-\overset{3'}{C}$$

as the template in a cell-free protein-synthesizing system. AAA codes for lysine, whereas AAC codes for asparagine. The polypeptide product was

$$^+H_3N-Lys-(Lys)_n-Asn-C\overset{\displaystyle O}{\underset{\displaystyle O^-}{\diagdown}}$$

Because asparagine was the carboxyl-terminal residue, the codon AAC was the last to be read. Hence, *the direction of translation is 5′ → 3′.* Recall that mRNA is also synthesized in the 5′ → 3′ direction (p. 609). This means that mRNA can be translated as it is being synthesized. In fact, in *E. coli,* the 5′ end of mRNA interacts with ribosomes very soon after it is synthesized (Figure 27-15). Thus, *translation is closely coupled to transcription.*

SEVERAL RIBOSOMES SIMULTANEOUSLY TRANSLATE A MESSENGER RNA MOLECULE

Many ribosomes can simultaneously translate an mRNA molecule. This markedly increases the efficiency of utilization of the mRNA. The group of ribosomes bound to an mRNA molecule is called a *polyribosome* or a *polysome.* The ribosomes in this unit operate independently, each synthesizing a complete polypeptide chain. The maximum density of ribosomes on mRNA is about one ribosome per eighty nucleotides. Polyribosomes synthesizing hemoglobin (which contains about 145 amino acids per chain, or 500 nucleo-

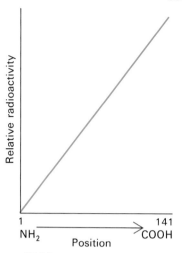

Figure 27-14
Distribution of ^{3}H-leucine in the α chain of hemoglobin synthesized after exposure of reticulocytes to a pulse of tritiated leucine. The higher radioactivity of the carboxyl end relative to the amino end indicates that the carboxyl end was synthesized last.

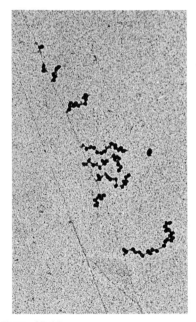

Figure 27-15
Transcription of a section of the DNA of *E. coli* and translation of the nascent mRNA. Only part of the chromosome is being transcribed. [From O. L. Miller, Jr., Barbara A. Hamkalo, and C. A. Thomas, Jr. Visualization of bacterial genes in action. *Science* 169(1970):392.]

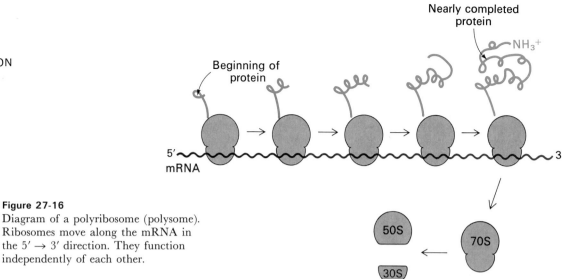

Figure 27-16
Diagram of a polyribosome (polysome).
Ribosomes move along the mRNA in
the 5′ → 3′ direction. They function
independently of each other.

tides per mRNA) typically consist of five ribosomes. Ribosomes closest to the 5′ end of the messenger have the shortest polypeptide chains, whereas those nearest the 3′ end have almost finished chains. Ribosomes dissociate into 30S and 50S subunits after the polypeptide product is released.

PROTEIN SYNTHESIS IN BACTERIA IS INITIATED BY FORMYLMETHIONINE TRANSFER RNA

How does protein synthesis start? The simplest possibility a priori is that the entire mRNA is translated so that no special start signal is needed. However, the experimental fact is that translation does not begin immediately at the 5′ terminus of mRNA. Indeed, the first translated codon is nearly always more than 25 nucleotides away from the 5′ end. Furthermore, many mRNA molecules in procaryotes are *polycistronic*—that is, they code for two or more polypeptide chains. For example, a single mRNA molecule about 7,000 nucleotides long specifies five enzymes in the biosynthetic pathway for tryptophan in *E. coli*. Each of these five proteins has its own start and stop signals on the mRNA. In fact, *all known bacterial mRNA molecules contain signals that define the beginning and end of each encoded polypeptide chain.*

A clue to the mechanism of chain initiation was provided by the finding that nearly half of the amino-terminal residues of proteins in *E. coli* are methionine, yet this amino acid is uncommon at other positions of the polypeptide chain. Furthermore, the amino terminus of nascent proteins is usually modified, which suggests that a derivative of methionine participates in initiation. In fact, *protein synthesis in bacteria starts with formylmethionine* (fMet). There is a special tRNA that brings formylmethionine to the ribosome to initiate protein synthesis. This *initiator tRNA* (abbreviated as *tRNA$_f$*) is different from the one that inserts methionine in internal positions (abbreviated as tRNA$_m$). The subscript f indicates that methionine

tRNA$_f$

↓

$^+H_3N-\underset{\underset{\underset{\underset{CH_3}{|}}{\overset{|}{S}}}{\overset{|}{\underset{\overset{|}{CH_2}}{CH_2}}}}{\overset{H}{\underset{|}{C}}}-\overset{O}{\overset{||}{C}}-O-tRNA_f$

Methionyl-tRNA$_f$
(Met-tRNA$_f$)

↓

$\underset{H}{\overset{O}{\overset{||}{C}}}-\underset{H}{\overset{H}{\underset{|}{N}}}-\underset{\underset{\underset{\underset{CH_3}{|}}{\overset{|}{S}}}{\overset{|}{\underset{\overset{|}{CH_2}}{CH_2}}}}{\overset{H}{\underset{|}{C}}}-\overset{O}{\overset{||}{C}}-O-tRNA_f$

Formylmethionyl-tRNA$_f$
(fMet-tRNA$_f$)

Figure 27-17
Formation of formylmethionyl-tRNA$_f$.

attached to the initiator tRNA can be formylated, whereas it cannot be formylated when attached to $tRNA_m$.

Methionine is linked to these two kinds of tRNAs by the same aminoacyl-tRNA synthetase. A specific enzyme then formylates the amino group of methionine that is attached to $tRNA_f$. The activated formyl donor in this reaction is N^{10}-formyltetrahydrofolate. It is significant that free methionine and methionyl-$tRNA_m$ are not substrates for this transformylase.

THE START SIGNAL IS AUG (OR GUG) PRECEDED BY SEVERAL BASES THAT CAN PAIR WITH 16S rRNA

What is the structural basis for the discrimination between AUG codons? The first step toward answering this question was the isolation of initiator regions from a number of mRNAs. This was accomplished by digesting mRNA-ribosome complexes (formed under conditions of chain initiation but not elongation) with pancreatic ribonuclease. In each case, a sequence of about thirty nucleotides was protected from digestion. As expected, each of these initiator regions displays an AUG (or GUG) codon (Figure 27-18). In addi-

AGCAC GAGGGG AAAUCUG AUG GAACGCUAC   *E. coli trpA*

UUUGGAU GGAG UGAAACG AUG GCGAUUGCA   *E. coli araB*

GGUAACC CAGGUAACAACC AUG CGAGUGUUG   *E. coli thrA*

CAAUUCAG GGUGG UGAAU GUG AAACCAGUA   *E. coli lacI*

AAUCUUG GAGGC UUUUUUU AUG GUUCGUUCU   φX174 phage A protein

UAAC UAAGGA UGAAAUGC AUG UCUAAGACA   Qβ phage replicase

UCC UAGGAGG UUUGACCU AUG CGAGCUUUU   R17 phage A protein

AUGUAC UAAGGAGG UUGU AUG GAACAACGC   λ phage *cro*

Pairs with 16S rRNA   Pairs with initiator tRNA

Figure 27-18
Sequences of initiation sites for protein synthesis in some bacterial and viral RNA molecules.

tion, each initiator region contains a purine-rich sequence centered about ten nucleotides on the 5′ side of the initiator codon. The role of this purine-rich region became evident when the sequence of 16S RNA was elucidated. The 3′ end of this ribosomal RNA contains a sequence of several bases that is complementary to the purine-rich region in the initiator sites of mRNA. In fact, a complex of the 3′ end of 16S RNA and the initiator region of mRNA has been isolated from an enzymatic digest of the initiation complex (Figure

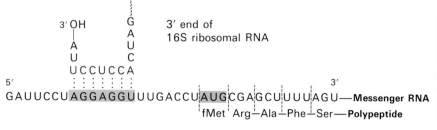

Figure 27-19
Base pairing between the purine-rich region (blue) in the initiator region of an mRNA and the 3′ end of 16S rRNA (red). The AUG codon (green) defines the start of the polypeptide chain. The mRNA shown here codes for the A protein of R17 phage.

30S initiation complex

70S initiation complex

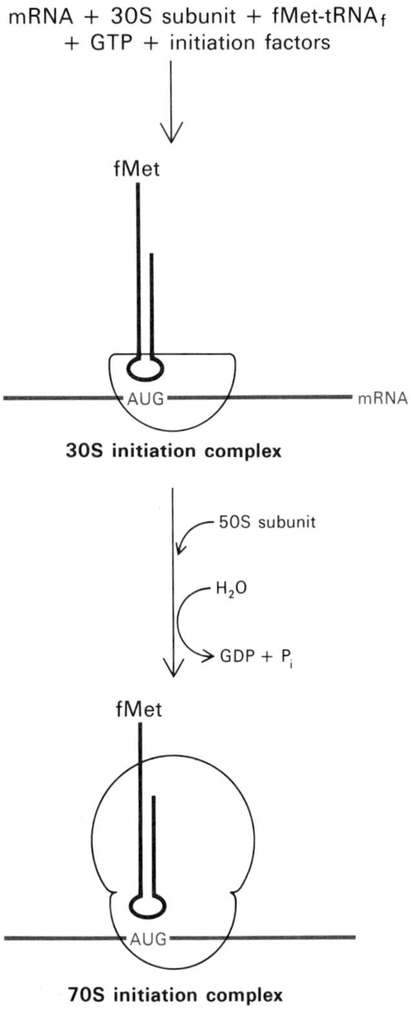

Figure 27-20
Initiation phase of protein synthesis: formation of a 30S initiation complex, followed by a 70S initiation complex.

27-19). The sequences of more than seventy known initiator sites show that the number of base pairs between mRNA and 16S rRNA ranges from three to nine. An interesting possibility is that the efficiency of initiation is modulated by the strength of this interaction. *Thus, two kinds of interactions determine where protein synthesis starts: the pairing of mRNA bases with the 3' end of 16S rRNA and with the formylmethionine initiator tRNA.*

FORMATION OF THE 70S INITIATION COMPLEX PLACES FORMYLMETHIONINE tRNA IN THE P SITE

Protein synthesis starts with the association of mRNA, a 30S ribosomal subunit, and formylmethionyl-tRNA$_f$ to form a *30S initiation complex* (Figure 27-20). The formation of this complex requires GTP and three protein factors called IF-1, IF-2, and IF-3. One of these *initiation factors,* IF-3, mediates the binding of mRNA to a 30S subunit complex. IF-3 also prevents the 50S and 30S subunits from coming together to form a dead-end 70S complex devoid of mRNA. IF-1 and IF-2 enhance the binding of initiator tRNA to the mRNA-30S subunit complex.

A 50S ribosomal subunit then joins a 30S initiation complex to form a 70S *initiation complex.* The bound GTP is hydrolyzed in this step. The 70S initiation complex (Figure 27-20) is ready for the elongation phase of protein synthesis. The fMet-tRNA$_f$ molecule occupies the P (peptidyl) site on the ribosome. The other site for a tRNA molecule on the ribosome, the A (aminoacyl) site, is empty. The existence of distinct P and A sites was inferred from studies of puromycin, an antibiotic that will be discussed later (p. 663). The important point now is that fMet-tRNA$_f$ is positioned so that its anticodon pairs with the initiating AUG (or GUG) codon on mRNA. Thus, the reading frame is defined by specific interactions of the ribosome and fMet-tRNA$_f$ with mRNA. In particular, recall that a purine-rich region on the 5' side of the initiating codon base-pairs with the 3' end of 16S RNA, a component of the 30S subunit. The delineation of this intricate pathway to properly position the initiator tRNA raises an interesting question. How were poly U and other synthetic polypeptides without start signals translated in the studies leading to the elucidation of the genetic code (p. 623)? The answer is that nonspecific initiation fortunately occurred because the concentration of Mg^{2+} in the reaction mixture was higher than it is in vivo.

ELONGATION FACTOR Tu DELIVERS AMINOACYL-tRNAs TO THE A SITE OF THE RIBOSOME

The elongation cycle in protein synthesis consists of three steps: (1) *binding of aminoacyl-tRNA (codon recognition),* (2) *peptide-bond formation,* and (3) *translocation.* The cycle begins with the insertion of an aminoacyl-tRNA into the empty A site on the ribosome. The particular species inserted depends on the mRNA codon that is positioned in the A site. The complementary aminoacyl-tRNA is delivered to the A site by a protein *elongation factor* called *EF-Tu*. GTP bound to EF-Tu is hydrolyzed as the aminoacyl-tRNA is precisely positioned on the ribosome. GDP remains tightly associated with EF-Tu until it is displaced by *EF-Ts*, another elongation factor. The resulting complex of Tu and Ts is then split by the binding of GTP to give a Tu-GTP complex, which is ready for another round of protein synthesis. This cyclic association and release of proteins (Figure 27-21) is accelerated and given directionality by the hydrolysis of GTP.

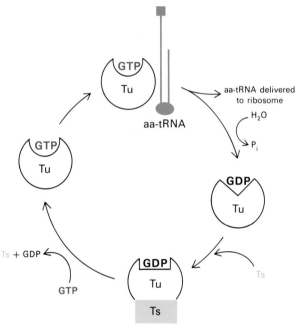

Figure 27-21
Reaction cycle of elongation factor Tu.

It is important to note that *EF-Tu does not interact with fMet-tRNA_f*. Hence, this initiator tRNA is not delivered to the A site. In contrast, met-tRNA$_m$, like all other aminoacyl-tRNAs, binds to EF-Tu. These findings account for the fact that *internal AUG codons are not read by the initiator tRNA.*

FORMATION OF A PEPTIDE BOND IS FOLLOWED BY TRANSLOCATION

We now have a complex in which aminoacyl-tRNA occupies the A site and fMet-tRNA occupies the P site. The stage is set for the *formation of a peptide bond* (Figure 27-22). This reaction is catalyzed by *peptidyl transferase,* an enzyme that is an integral part of the 50S

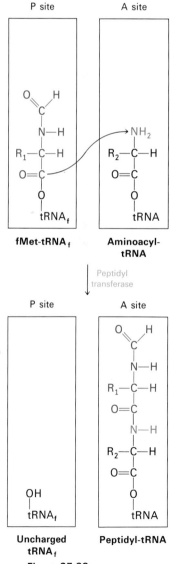

Figure 27-22
Peptide-bond formation.

subunit. The activated formylmethionine unit of fMet-tRNA$_f$ (in the P site) is transferred to the amino group of the aminoacyl-tRNA (in the A site) to form a dipeptidyl-tRNA.

An uncharged tRNA occupies the P site, whereas a dipeptidyl-tRNA occupies the A site, following the formation of a peptide bond. The next phase of the elongation cycle is *translocation*. Three movements occur: the uncharged tRNA leaves the P site, the peptidyl-tRNA moves from the A site to the P site, and mRNA moves a distance of three nucleotides. The result is that the next codon is positioned for reading by the incoming aminoacyl-tRNA. Translocation requires a third elongation factor, *EF-G* (also called *translocase*). The GTP bound to EF-G is hydrolyzed during translocation. Hydrolysis of GTP enables EF-G to be released from the ribosome so that it can act catalytically. *We see here yet another example of directed movement powered by the hydrolysis of a nucleoside triphosphate.* After translocation, the A site is empty, ready to bind an aminoacyl-tRNA to start another round of elongation (Figure 27-23).

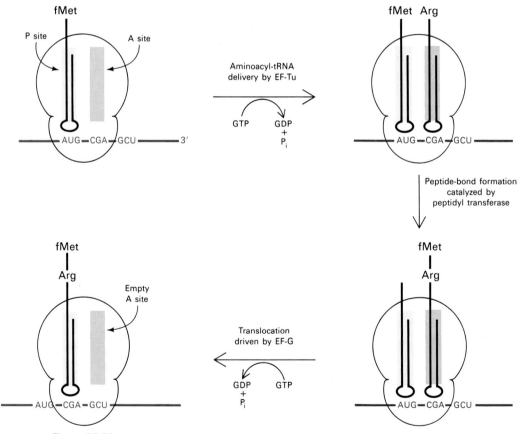

Figure 27-23
Elongation phase of protein synthesis: binding of aminoacyl-tRNA, peptide-bond formation, and translocation.

Aminoacyl-tRNA does not normally bind to the A site of a ribosome if the codon is UAA, UGA, or UAG. Normal cells do not contain tRNAs with anticodons complementary to these stop signals. Instead, these stop signals are recognized by release factors, which are proteins. One of these release factors, RF 1, recognizes UAA or UAG. The other release factor, RF 2, recognizes UAA or UGA. Thus, proteins can recognize trinucleotide sequences with high specificity.

The binding of a release factor to a termination codon in the A site somehow activates peptidyl transferase so that it hydrolyzes the bond between the polypeptide and the tRNA in the P site. The specificity of peptidyl transferase is altered by the release factor so that water rather than an amino group is the acceptor of the activated peptidyl moiety. The polypeptide chain then leaves the ribosome. The 70S ribosome then dissociates into 30S and 50S subunits as the prelude to the synthesis of another protein molecule.

MANY PROTEINS ARE MODIFIED
FOLLOWING TRANSLATION

Many of the polypeptides formed by the translation of mRNA are not the final products. Proteins may be subsequently modified in a variety of ways:

1. The formyl group at the amino terminus of bacterial proteins is hydrolyzed by a *deformylase.* One or more amino-terminal residues may be removed by *aminopeptidases.* In both procaryotes and eucaryotes, the terminal methionine is sometimes hydrolyzed while the rest of the polypeptide chain is being synthesized.

2. *Disulfide bonds* may be formed by the oxidation of two cysteine residues.

3. Certain amino acid side chains may be specifically modified. For example, some proline and lysine residues in collagen are *hydroxylated.* Glycoproteins are formed by the *attachment of sugars* to the side chains of asparagine, serine, and threonine. Some proteins are *phosphorylated. Prosthetic groups,* such as lipoate, are covalently attached to some enzymes.

4. Polypeptide chains may be specifically *cleaved,* as in the conversion of procollagen into collagen and of proinsulin into insulin. Translation of the mRNA of polio virus yields a very long polypeptide chain, which is hydrolyzed to form several proteins.

STREPTOMYCIN INHIBITS INITIATION AND CAUSES THE MISREADING OF MESSENGER RNA

We have already seen that antibiotics are very useful tools in biochemistry because many of them are highly specific in their action. For example, rifampicin is a potent inhibitor of the initiation of RNA synthesis (p. 613). Many antibiotic inhibitors of protein synthesis are known, and the mechanism of action of some of them has been determined. *Streptomycin*, a highly basic trisaccharide, interferes with the binding of formylmethionyl tRNA to ribosomes and thereby prevents the correct initiation of protein synthesis. Strepto-

Table 27-3
Antibiotic inhibitors of protein synthesis

Antibiotic	Action
Streptomycin	Inhibits initiation and causes misreading of mRNA (procaryotes)
Tetracycline	Binds to the 30S subunit and inhibits binding of aminoacyl-tRNAs (procaryotes)
Chloramphenicol	Inhibits the peptidyl transferase activity of the 50S ribosomal subunit (procaryotes)
Cycloheximide	Inhibits the peptidyl transferase activity of the 60S ribosomal subunit (eucaryotes)
Erythromycin	Binds to the 50S subunit and inhibits translocation (procaryotes)
Puromycin	Causes premature chain termination by acting as an analog of aminoacyl-tRNA (procaryotes and eucaryotes)

mycin also leads to a misreading of mRNA. If poly U is the template, isoleucine (AUU) is incorporated in addition to phenylalanine (UUU). The site of action of streptomycin in the ribosome has been determined by the reconstruction of ribosomal components from streptomycin-sensitive and streptomycin-resistant bacteria. These bacterial strains are related by mutation of a single gene. How do their ribosomes differ? Because ribosomes can be dissociated and assembled in vitro, it was feasible to determine whether the determinant of streptomycin sensitivity resides in the 50S or 30S ribosomal subunit. Hybrid ribosomes consisting of 50S subunits from resistant bacteria and 30S subunits from sensitive bacteria were found to be sensitive, whereas ribosomes prepared from the reverse combination were resistant. This experiment showed that the determinant of streptomycin sensitivity is located in the 30S subunit. The next experiment demonstrated that a protein in the 30S subunit rather than its 16S RNA molecule specifies streptomycin sensitivity. Finally, an extensive series of experiments revealed that a single protein in the 30S subunit, namely protein S12, is the determinant of streptomycin sensitivity.

Streptomycin

PUROMYCIN CAUSES PREMATURE CHAIN TERMINATION BECAUSE IT MIMICS AMINOACYL-TRANSFER RNA

The antibiotic puromycin inhibits protein synthesis by releasing nascent polypeptide chains before their synthesis is completed. Puromycin is an analog of the terminal aminoacyl-adenosine portion of aminoacyl-tRNA (Figure 27-24). It binds to the A site on the

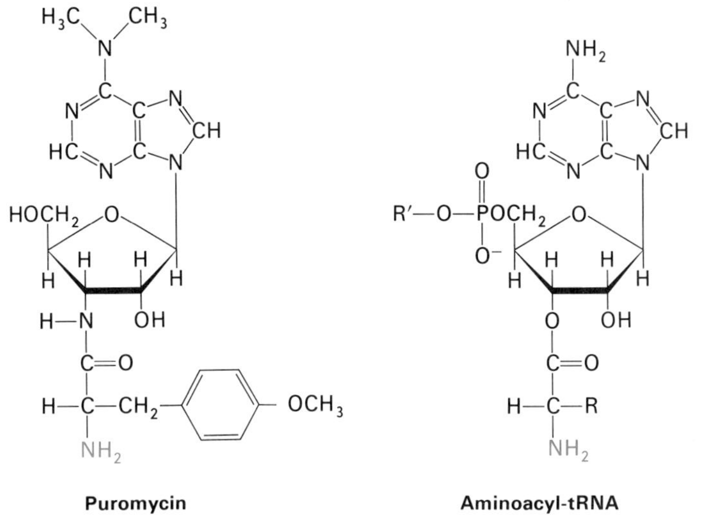

Puromycin   **Aminoacyl-tRNA**

Figure 27-24
Puromycin resembles the aminoacyl terminus of an aminoacyl-tRNA.

ribosome and inhibits the entry of aminoacyl-tRNA. Furthermore, puromycin contains an α-amino group. This amino group, like the one on aminoacyl-tRNA, forms a peptide bond with the carboxyl group of the growing peptide chain in a reaction that is catalyzed by peptidyl transferase. The product is a peptide having a covalently attached puromycin residue at its carboxyl end. Peptidyl puromycin then dissociates from the ribosome. Puromycin has been used to ascertain the functional state of ribosomes. In fact, the concept of A and P sites resulted from the use of puromycin to ascertain the location of peptidyl tRNA. When peptidyl tRNA is in the A site (before translocation), it cannot react with puromycin.

SOME SHORT PEPTIDES ARE NOT SYNTHESIZED BY RIBOSOMES

We turn now to a different mechanism of peptide-bond formation in biological systems. As an example, let us consider the biosynthesis of gramicidin S, a cyclic peptide antibiotic made up of two identical pentapeptides joined head-to-tail. This antibiotic is produced by certain strains of *Bacillus brevis,* a spore-forming bacterium.

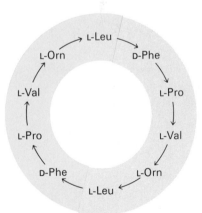

Gramicidin S

Figure 27-25
Amino acid sequence of gramicidin S, a cyclic peptide made of two identical pentapeptide units. The arrows denote the polarity of the polypeptide chain (amino to carboxyl direction).

The biosynthesis of gramicidin S does not depend on the presence of ribosomes or mRNA. A much simpler synthetic apparatus consisting only of two enzymes, E_I and E_{II}, is required. D-Phenylalanine is activated by E_{II} (100 kdal), whereas the other four amino acids of the pentapeptide unit are activated by E_I (180 kdal). Both enzymes also participate in peptide-bond formation.

In this system, amino acids are activated by the formation of enzyme-bound *thioesters*. The activated amino acid is attached to a sulfhydryl group of E_I or E_{II} rather than to the 3′ terminal hydroxyl group of a tRNA.

$$\text{Amino acid} + \text{ATP} \rightleftharpoons \text{Aminoacyl-AMP} + PP_i$$

$$\text{Aminoacyl-AMP} + \text{E—SH} \rightleftharpoons \text{E—S} \sim \overset{\overset{\displaystyle O}{\|}}{C} - \overset{\overset{\displaystyle H}{|}}{\underset{\underset{\displaystyle R}{|}}{C}} - NH_3^+ + \text{AMP}$$

Thioester

L-Proline, L-valine, L-ornithine, and L-leucine form thioester linkages with specific sulfhydryls of E_I when they are incubated in the presence of ATP. Likewise, D-phenylalanine forms a thioester bond with E_{II} in the presence of ATP.

Peptide synthesis in this system is initiated by the interaction of E_I and E_{II}. The D-phenylalanine residue on E_{II} is transferred to the imino group of the L-proline residue on E_I to form a dipeptide.

$$E_{II}\text{—S}\sim\overset{\overset{\displaystyle O}{\|}}{C}\text{—Phe—NH}_2 + E_I\text{—S}\sim\overset{\overset{\displaystyle O}{\|}}{C}\text{—Pro—NH} \rightleftharpoons E_{II}\text{—SH} + E_I\text{—S}\sim\overset{\overset{\displaystyle O}{\|}}{C}\text{—Pro—Phe—NH}_2$$

Activated dipeptide

The subsequent reactions require only E_I. The activated carbonyl group of the prolyl residue in the dipeptide reacts with the amino group of the valine residue on the same enzyme to yield a tripeptide. This process occurs again with ornithine and then with leucine to give an enzyme-bound pentapeptide. The growing peptide is transferred to a different sulfhydryl each time a peptide bond is formed.

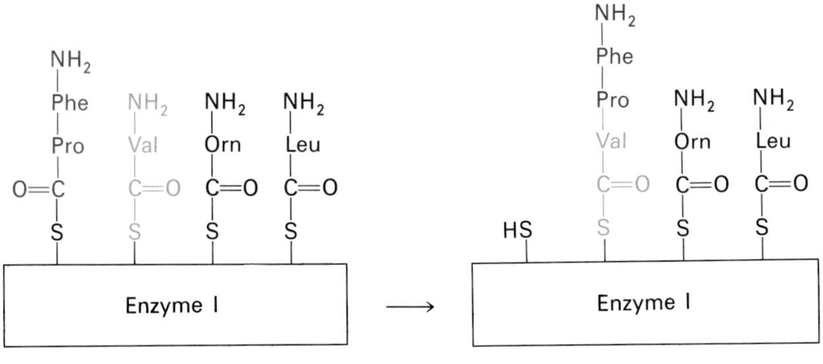

Finally, the activated pentapeptides attached to two different E_I molecules react with each other to form cyclic gramicidin S.

$$H_2N-Phe-Pro-Val-Orn-Leu-\overset{\overset{O}{\|}}{C}-S-\boxed{Enzyme\ I}$$

$$\boxed{Enzyme\ I}-S-\overset{\overset{}{\underset{\|}{C}}}{\underset{O}{}}-Leu-Orn-Val-Pro-Phe-NH_2$$

Two features of this biosynthetic pathway are noteworthy:

1. The amino acid sequence of gramicidin S is determined by the spatial arrangement and specificity of the enzymes in E_I and E_{II}. There is at least one protein subunit for each peptide bond formed. Hence, *this mode of synthesis is uneconomical compared with the ribosomal mechanism. Consequently, a peptide containing more than about fifteen residues is not synthesized by this mechanism.*

2. *The synthesis of gramicidin S resembles fatty acid synthesis,* because the activated intermediates in both processes are thioesters. Furthermore, E_I contains a covalently attached phosphopantetheine residue. This thiol probably carries the growing peptide chain from one site to the next in E_I. Fritz Lipmann suggested that *antibiotic polypeptide synthesis may be a surviving relic of a primitive mechanism of protein synthesis used early in evolution.* Ribosomal protein synthesis may have evolved from fatty acid synthesis.

SUMMARY

Protein synthesis (translation) requires the coordinated interplay of more than a hundred macromolecules, which include mRNA, tRNAs, activating enzymes, and protein factors, in addition to ribosomes. Protein synthesis starts with the ATP-driven activation of amino acids by aminoacyl-tRNA synthetases (activating enzymes), which link the carboxyl group of an amino acid to the 2'- or 3'-hydroxyl group of the adenosine unit at the 3' end of tRNA. There is at least one specific activating enzyme for each amino acid. Also, there is at least one specific tRNA for each amino acid. Transfer RNAs of different specificity have a common structural design. They are single RNA chains of about eighty nucleotides, containing some modified (e.g., methylated) derivatives of the standard bases. The base sequences of all known tRNAs can be written in a cloverleaf pattern in which about half of the nucleotides are base paired. X-ray crystallographic studies have shown that tRNA is L-shaped. The amino acid attachment site at the 3' CCA terminus is at one end of the L, whereas the anticodon is 80 Å away at the other end. Messenger RNA recognizes the anticodon of a tRNA rather than the attached amino acid. A codon on mRNA forms base pairs with the anticodon on tRNA. Some tRNAs recognize more than one codon because pairing of the third base of a codon is less discriminating than that of the other two (the wobble hypothesis). Protein synthesis takes place on ribosomes, which consist of large and small subunits, each about two-thirds RNA and one-third protein by weight. In *E. coli,* the 70S ribosome (2,500 kdal) is made up of 30S and 50S subunits.

There are three phases of protein synthesis: initiation, elongation, and termination. Messenger RNA, formylmethionyl-tRNA$_f$, and a 30S ribosomal subunit come together to form a 30S initiation complex. The start signal on mRNA is AUG (or GUG) preceded by a purine-rich sequence that can base-pair with 16S rRNA. A 50S ribosomal subunit then joins this complex to form a 70S initiation complex, which is ready for the next phase. The elongation cycle consists of the binding of aminoacyl-tRNA (codon recognition), peptide-bond formation, and translocation. The direction of chain growth is from the amino to the carboxyl end. Protein synthesis is terminated by release factors, which recognize the termination codons UAA, UGA, and UAG and lead to the hydrolysis of the bond between the polypeptide and tRNA. GTP is hydrolyzed in the formation of the 70S initiation complex, in the binding of aminoacyl-tRNA to the ribosome, and in the translocation step. Various steps in protein synthesis are specifically inhibited by toxins and antibiotics. Peptide antibiotics and other short polypeptides are synthesized by a nonribosomal pathway that resembles fatty acid synthesis.

SELECTED READINGS

WHERE TO START

Miller, O. L., Jr., 1973. The visualization of genes in action. *Sci. Amer.* 228(3):34–42. [Available as *Sci. Amer.* Offprint 1267.]

Rich, A., and Kim, S. H., 1978. The three-dimensional structure of transfer RNA. *Sci. Amer.* 238(1):52–62. [Offprint 1377.]

Clark, B. F. C., and Marcker, K. A., 1968. How proteins start. *Sci. Amer.* 218(1):36–42. [Offprint 1092.]

BOOKS ON PROTEIN SYNTHESIS

Weissbach, H., and Pestka, S., (eds.), 1977. *Molecular Mechanisms of Protein Biosynthesis.* Academic Press. [Contains articles on synthetases, ribosome structure and function, initiation, elongation, translocation, termination, and antibiotic inhibitors of protein synthesis.]

Cold Spring Harbor Laboratory, 1969. *Mechanisms of Protein Biosynthesis* (Cold Spring Harbor Symposium on Quantitative Biology, vol. 34).

RIBOSOMES

Nomura, M., Tissieres, A., and Lengyel, P., (eds.), 1975. *Ribosomes.* Cold Spring Harbor Laboratory. [Contains excellent reviews on the structure, function, and assembly of ribosomes.]

Brimacombe, R., Stoffler, G., and Wittmann, H. G., 1978. Ribosome structure. *Ann. Rev. Biochem.* 47:217–249.

Wittman, H. G., 1977. Structure and function of *Escherichia coli* ribosomes. *Fed. Proc.* 36:2025–2080.

Nomura, M., 1973. Assembly of bacterial ribosomes. *Science.* 179:864–873.

TRANSFER RNA

Schimmel, P., Söll, D., and Abelson, J., (eds.), 1979. *Transfer RNA.* Cold Spring Harbor Laboratory. [Part 1 deals with structure, properties, and recognition, and Part 2 with biosynthesis and genetic aspects of tRNA. An authoritative and valuable treatise.]

Kim, S.-H., 1978. Three-dimensional structure of transfer RNA and its functional implications. *Advan. Enzymol.* 46:279–315.

Jack, A., Ladner, J. E., and Klug, A., 1976. Crystallographic refinement of yeast phenylalanine transfer RNA at 2.5 Å resolution. *J. Mol. Biol.* 108:619–649.

Sussman, J. L., and Kim, S.-H., 1976. Three-dimensional structure of a transfer RNA in two crystal forms. *Science* 192:853–858.

AMINOACYL-tRNA SYNTHETASES

Schimmel, P. R., and Söll, D., 1979. Amino acyl-tRNA synthetases: general features and recognition of transfer RNAs. *Ann. Rev. Biochem.* 48:601–648.

INITIATION, ELONGATION, AND TERMINATION

Dintzis, H. M., 1961. Assembly of the peptide chains of hemoglobin. *Proc. Nat. Acad. Sci.* 47:247–261. [These important pulse-labeling experiments showed that proteins are synthesized from the amino to the carboxyl end.]

Steitz, J. A., 1979. Genetic signals and nucleotide sequences in messenger RNA. *In* Goldberger, R. F., (ed.), *Biological Regulation and Development,* vol. 1, pp. 349–399. Plenum.

Shine, J., and Dalgarno, L., 1974. The 3′-terminal sequence of *Escherichia coli* 16S ribosomal RNA: complementarity to nonsense triplets and ribosome binding sites. *Proc. Nat. Acad. Sci.* 71:1342–1346.

Steitz, J. A., and Jakes, K., 1975. How ribosomes select initiator regions in mRNA: base pair formation between the 3′-terminus of 16S rRNA and the mRNA during initiation of protein synthesis in *Escherichia coli. Proc. Nat. Acad. Sci.* 72:4734–4738.

Gupta, S. L., Waterson, J., Sopori, M., Weissman, S. M., and Lengyel, P., 1971. Movement of the ribosome along the messenger ribonucleic acid during protein synthesis. *Biochemistry* 10:4410–4421.

Kaziro, Y., 1978. The role of GTP in polypeptide chain elongation. *Biochim. Biophys. Acta* 505:95–127.

FIDELITY OF TRANSLATION

Fersht, A., and Dingwall, C., 1979. Evidence for the double-sieve editing mechanism in protein biosynthesis. *Biochemistry.* 18:2627–2631.

ANTIBIOTICS AND TOXINS

Pestka, S., (1971). Inhibitors of ribosome function. *Ann. Rev. Microbiol.* 25:487–562.

Jimenez, A., 1976. Inhibitors of translation. *Trends Biochem. Sci.* 1:28–29.

Tai, P.-C., Wallace, B. J., and Davis, B. D., 1978. Streptomycin causes misreading of natural messenger by interacting with ribosomes after initiation. *Proc. Nat. Acad. Sci.* 75:275–279.

NONRIBOSOMAL PEPTIDE SYNTHESIS

Lipmann, F., 1971. Attempts to map a process evolution of peptide biosynthesis. *Science* 173:875–884.

Perlman, D., and Bodanszky, M., 1971. Biosynthesis of peptide antibiotics. *Ann. Rev. Biochem.* 40:449–464.

Lipmann, F., 1973. Nonribosomal polypeptide synthesis on polyenzyme templates. *Acc. Chem. Res.* 6:361–367.

PROBLEMS

1. The formation of Ile-tRNA proceeds through an enzyme-bound Ile-AMP intermediate. Predict whether ^{32}P-labeled ATP is formed from ^{32}PP$_i$ when each of the following sets of components is incubated with the specific activating enzyme:
(a) ATP and ^{32}PP$_i$.
(b) tRNA, ATP, and ^{32}PP$_i$.
(c) Isoleucine, ATP, and ^{32}PP$_i$.

2. Ribosomes were isolated from bacteria grown in a "heavy" medium (^{13}C and ^{15}N) and from bacteria grown in a "light" medium (^{12}C and ^{14}N). These 70S ribosomes were added to an in vitro system actively engaged in protein synthesis. An aliquot removed several hours later was analyzed by density-gradient centrifugation. How many bands of 70S ribosomes would you expect to see in the density gradient?

3. How many high-energy phosphate bonds are consumed in the synthesis of a 200-residue protein, starting from amino acids?

4. There are two basic mechanisms for the elongation of biomolecules (Figure 27-26). In type 1, the activating group (labeled X) is released from the growing chain. In type 2, the activating group is released from the incoming unit as it is added to the growing chain. Indicate whether each of the following biosyntheses occurs by means of a type 1 or a type 2 mechanism:
(a) Glycogen synthesis.
(b) Fatty acid synthesis.
(c) $C_5 \rightarrow C_{10} \rightarrow C_{15}$ in cholesterol synthesis.
(d) DNA synthesis.
(e) RNA synthesis.
(f) Protein synthesis.

5. Mutations that produce termination codons are called nonsense mutations. These mutations can be suppressed by altered tRNAs. For example, the UGA codon is translated into tryptophan by a mutant tRNA. What is the most likely base change in this mutant tRNA?

6. Design an affinity-labeling reagent for one of the tRNA binding sites in *E. coli* ribosomes. How would you synthesize such a reagent?

7. An mRNA transcript of a T7 phage gene contains the base sequence

$$\downarrow$$

5′-AACUGCACGAGGUAACACAAGAUGGCU-3′

Predict the effect of a mutation that changes the G marked by an arrow to A.

8. What is a common feature of proofreading in protein synthesis and DNA synthesis?

Type 1   **Type 2**

Figure 27-26
Two modes of elongation.

For additional problems, see W. B. Wood, J. H. Wilson, R. M. Benbow, and L. E. Hood, *Biochemistry: A Problems Approach* (Benjamin, 1974), ch. 18.

CHAPTER 28

CONTROL OF GENE EXPRESSION

We have already seen that the activity of many proteins is regulated by a variety of mechanisms, such as proteolytic activation, allosteric interaction, and covalent modification. This chapter deals with the *control of the rate of protein synthesis,* which is also critical in determining the pattern of cellular processes. In bacteria, gene activity is regulated primarily at the level of *transcription* rather than translation. We will focus on the lactose and tryptophan operons of *E. coli* and on regulatory aspects of the bacteriophage lambda because much is known about the molecular basis of control of these systems. Furthermore, intensive study of these systems has revealed some general principles of gene regulation in procaryotes and viruses. Gene expression in eucaryotes is controlled in different ways, as will be discussed in the next chapter.

BETA-GALACTOSIDASE IS AN INDUCIBLE ENZYME

E. coli can use lactose as its sole source of carbon. An essential enzyme in the metabolism of this sugar is *β-galactosidase,* which hydrolyzes lactose to galactose and glucose (Figure 28-1). An *E. coli* cell growing on lactose contains several thousand molecules of β-galactosidase. In contrast, the number of β-galactosidase molecules per cell is fewer than ten if *E. coli* is grown on other sources of carbon, such as glucose or glycerol. Lactose induces a large increase in the amount of β-galactosidase in *E. coli* by eliciting the synthesis of new enzyme molecules rather than by activating a proenzyme

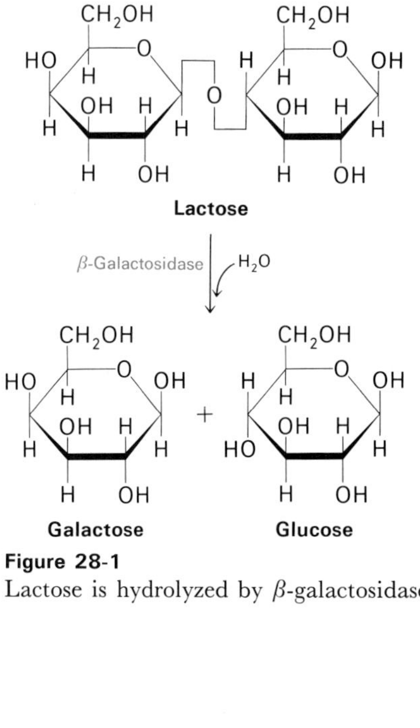

Figure 28-1
Lactose is hydrolyzed by β-galactosidase.

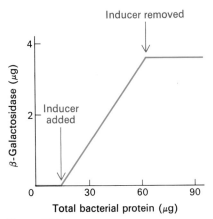

Figure 28-2
The increase in the amount of β-galactosidase parallels the increase in the number of cells in a growing culture of *E. coli*. The slope of this plot indicates that 6.6% of the protein synthesized is β-galactosidase.

(Figure 28-2). Hence, β-galactosidase is an inducible enzyme. Two other proteins are synthesized in concert with β-galactosidase—namely, *galactoside permease* and *thiogalactoside transacetylase*. The permease is required for the transport of lactose across the bacterial cell membrane, whereas the transacetylase is not essential for lactose metabolism. The physiologic role of the transacetylase is not yet known. In vitro, it catalyzes the transfer of an acetyl group from acetyl CoA to the C-6 hydroxyl group of a thiogalactoside.

The physiologic inducer is *allolactose,* which is formed from lactose by transglycosylation. The synthesis of allolactose is catalyzed by the few β-galactosidase molecules that are present prior to induction. A study of the nature of inducers showed that some β-galactosides are inducers without being substrates of β-galactosidase, whereas other compounds are substrates without being inducers. For example, *isopropylthiogalactoside* (IPTG) is a nonmetabolizable inducer (also called a gratuitous inducer).

1,6-Allolactose

IPTG

DISCOVERY OF A REGULATORY GENE

An important clue concerning the nature of the induction process was provided by the finding that the amounts of the permease and the transacetylase increased in direct proportion to that of β-galactosidase for all inducers tested. Further insight came from studies of mutants, which showed that β-galactosidase, the permease, and the transacetylase are encoded by three contiguous genes, called *z, y,* and *a,* respectively. Mutants defective in only one of these proteins were isolated. For example, $z^-y^+a^+$ denotes a mutant lacking β-galactosidase but having normal amounts of the permease and the transacetylase. A most interesting class of mutants affecting all three proteins was then isolated. These *constitutive mutants* synthesize large amounts of β-galactosidase, the permease, and the transacetylase whether or not inducer is present. François Jacob and Jacques Monod deduced that *the rate of synthesis of these three proteins is governed by a common element that is different from the genes that specify their structure.* The gene for this common regulatory element was named *i.* Wild-type inducible bacteria have the genotype $i^+z^+y^+a^+$, whereas the constitutive lactose mutants have the genotype $i^-z^+y^+a^+$.

How does the i^+ gene affect the rate of synthesis of the proteins encoded by the *z, y,* and *a* genes? The simplest hypothesis was that the i^+ gene determines the synthesis of a cytoplasmic substance called a *repressor,* which is missing or inactive in i^-. This idea was tested in an ingenious series of genetic experiments involving partially diploid bacteria that contained two sets of genes for the lactose region. One set was on the bacterial chromosome whereas the other was on an F′ sex factor, introduced by conjugation. For example, an i^+z^-/Fi^-z^+ diploid was isolated. In this diploid, i^+z^- is on the chromosome, whereas i^-z^+ is on the episome. Is this diploid inducible or constitutive for β-galactosidase? In other words, does i^+ on the bacterial chromosome repress the expression of z^+ on the episome? The experimental result was clear-cut: *the diploid was inducible rather than constitutive.* The same result was obtained for the diploid i^-z^+/Fi^+z^-. Hence, *a diffusible repressor is specified by the* i$^+$ *gene.*

AN OPERON IS A COORDINATED UNIT OF GENETIC EXPRESSION

These experiments led Jacob and Monod to propose the *operon model* for the regulation of protein synthesis. The genetic elements of this model are a *regulator gene,* an *operator gene,* and a set of *structural genes* (Figure 28-3). The regulator gene produces a *repressor* that can inter-

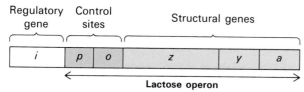

Figure 28-3
Map of the lactose operon and its regulatory gene. (This map is not drawn to scale: the p and o sites are actually much smaller than the other genes.)

act with the operator gene. Subsequent work revealed that the repressor is a protein. The operator gene is adjacent to the structural genes it controls. The binding of the repressor to the operator gene prevents the transcription of the structural genes. The operator gene and its associated structural genes are called an *operon.* For the lactose operon, the i gene is the regulator gene, the o gene is the operator gene, and the $z, y,$ and a genes are the structural genes. In addition, there is a *promoter site* (denoted by p) for the binding of RNA polymerase. This site for the initiation of transcription is next to the operator gene. An *inducer* such as IPTG binds to the repressor, which prevents it from interacting with the operator gene. The $z, y,$ and a genes can then be transcribed to give a single mRNA molecule that codes for all three proteins (Figure 28-4). An mRNA molecule coding for more than one protein is known as a *polycistronic (or polygenic) transcript.*

A

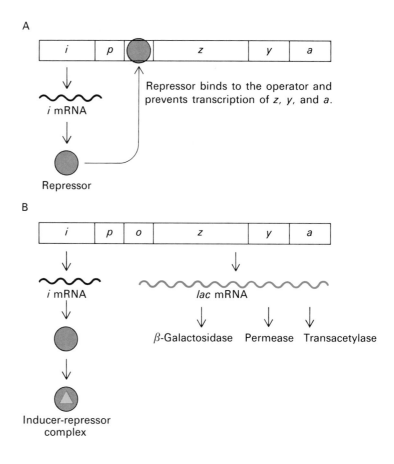

Repressor binds to the operator and prevents transcription of *z*, *y*, and *a*.

i mRNA

Repressor

B

i mRNA

lac mRNA

β-Galactosidase   Permease   Transacetylase

Inducer-repressor
complex

Figure 28-4
Diagram of the lactose operon in the (A) repressed and (B) induced states.

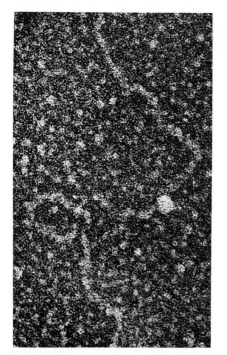

Figure 28-5
Electron micrograph of the *lac* repressor bound to DNA containing the *lac* operator. [Courtesy of Dr. Jack Griffith.]

THE *lac* REPRESSOR IS A TETRAMERIC PROTEIN

The repressor of the lactose operon (called the *lac* repressor) was isolated on the basis of its binding of IPTG. Walter Gilbert and Benno Müller-Hill showed that the *lac* repressor is a protein that binds to DNA carrying the *lac* operon but not to other DNA molecules. As expected, IPTG prevents the binding of *lac* repressor to *lac* operator DNA. A wild-type *E. coli* cell contains only about ten molecules of the *lac* repressor. It is difficult to purify the repressor from these cells because it constitutes only .001% of the total protein. However, much larger amounts of the *lac* repressor are made in i^{sq} mutants, which probably have a more efficient promoter for the *i* gene. The amount of *lac* repressor is increased further by using transducing phages that carry the *lac* region. Such an infected *E. coli* cell contains about 20,000 repressors (about 2% of the total protein), which makes it a choice starting material for the purification of *lac* repressor.

The repressor is a tetramer of identical 37-kdal subunits, each with one binding site for an inducer. The dissociation constant for IPTG is about 10^{-6} M. The repressor binds very tightly and rapidly to the operator. The dissociation constant of the repressor-operator complex is about 10^{-13} M. This high affinity is essential because only a few repressor molecules are present in a wild-type *E. coli* cell.

The rate constant for association, 7×10^9 M^{-1} sec^{-1}, is strikingly high, which suggests that the repressor finds the operator site by diffusing along a DNA molecule (a one-dimensional search) rather than by encountering it from the aqueous medium (a three-dimensional search).

THE *lac* OPERATOR HAS A SYMMETRIC BASE SEQUENCE

The availability of pure *lac* repressor has made it feasible to isolate the *lac* operator and determine its base sequence. Gilbert and his associates sonicated the DNA of a phage carrying the *lac* region into fragments that were approximately 1000 base pairs long. The *lac* repressor was added to the mixture of fragments, which was then filtered through a cellulose nitrate membrane. DNA fragments without bound *lac* repressor passed through this filter, whereas DNA-repressor complexes bound tightly to it. The bound DNA was released by adding IPTG. These released DNA fragments were then treated with pancreatic deoxyribonuclease in the presence of repressor. The rationale of this step is that the operator region is protected from digestion by deoxyribonuclease when it is complexed to the repressor (Figure 28-6). The base sequence of these operator fragments was then elucidated by determining the sequence of its RNA transcript. The base sequence of this region is very interesting: a total of 28 base pairs are related by a twofold axis of symmetry (Figure 28-7).

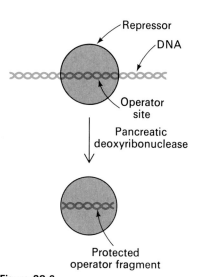

Figure 28-6
The *lac* repressor protects the *lac* operator from digestion by pancreatic deoxyribonuclease.

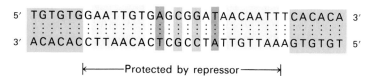

Figure 28-7
Nucleotide sequence of the *lac* operator. The symmetrically related regions are shown in color.

The symmetry of the repressor probably matches that of the operator. This principle of recognition also applies to termination signals for transcription (p. 676) and to the binding of actinomycin D, a peptide antibiotic, to DNA (p. 614).

CYCLIC AMP STIMULATES THE TRANSCRIPTION OF SEVERAL INDUCIBLE CATABOLIC OPERONS

It has long been known that *E. coli* grown on glucose have very low levels of catabolic enzymes, such as β-galactosidase, galactokinase, arabinose isomerase, and tryptophanase. Clearly, it would be wasteful to synthesize these enzymes when glucose is abundant. The molecular basis of this inhibitory effect of glucose, called *catabolite repression*, has been elucidated. A key clue was provided by the

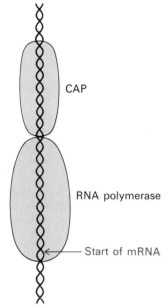

Figure 28-8
Schematic diagram of CAP and RNA polymerase on the DNA template. The locations of these proteins were inferred from nuclease digestion studies.

observation that glucose lowers the concentration of cyclic AMP in *E. coli.* It was then found that exogenous cyclic AMP can relieve the repression exerted by glucose. Subsequent biochemical and genetic studies revealed that *cyclic AMP stimulates the initiation of transcription of many inducible operons.*

Cyclic AMP synthesized in the absence of glucose binds to CAP (the *catabolite gene activator protein*), a dimer of 22-kdal subunits. *The complex of CAP and cyclic AMP, but not CAP alone, stimulates transcription by binding to certain promoter sites.* In the *lac* operon, CAP binds next to the site for RNA polymerase, as shown by deoxyribonuclease digestion studies. Specifically, CAP protects nucleotides -87 to -49 from digestion, whereas RNA polymerase protects nucleotides -48 to $+5$ (Figure 28-8). In this numbering system, the first transcribed nucleotide is $+1$. The DNA base sequence recognized by CAP contains a twofold axis of symmetry, a recurring theme in protein-DNA interactions. How does CAP stimulate the initiation of *lac* mRNA synthesis by a factor of 50? *The contiguous, nonoverlapping arrangement of the binding sites for CAP and RNA polymerase suggests that the binding of CAP to DNA creates an additional interaction site for RNA polymerase.* In contrast, the *lac* repressor binds to nucleotides -3 to $+21$, which significantly overlaps the RNA polymerase site from nucleotides -48 to $+5$. Thus, *the repressor prevents initiation by blocking the entry of RNA polymerase.* It seems likely that the cyclic-AMP–CAP complex acts similarly at other inducible operons. Thus, *inducible operons are controlled by complementary mechanisms that use cyclic AMP and specific inducers as the signal molecules.*

DIFFERENT FORMS OF THE SAME PROTEIN ACTIVATE AND INHIBIT TRANSCRIPTION OF THE ARABINOSE OPERON

Bacteria can use arabinose as a fuel by converting it into xylulose 5-phosphate, a pentose phosphate pathway intermediate (p. 337), by the sequential action of arabinose isomerase, ribulokinase, and ribulose 5-phosphate epimerase. These enzymes are encoded by the *araA, araB,* and *araD* genes, respectively. These structural genes, together with a promoter (*araI*) and an operator (*araO*), constitute the arabinose operon (Figure 28-9). This operon is regulated by *araC*, which is adjacent to the operator site. *The arabinose operon, like*

Figure 28-9
Map of the arabinose operon and its regulatory gene.

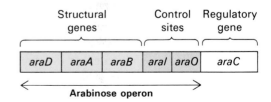

the lactose operon, can be activated by the complex of CAP and cyclic AMP. Indeed, the arabinose operon was the first shown to be subject to positive as well as negative regulation. The product of *araC* is a protein that can have two functionally different conformations, called P1 and P2. In the absence of arabinose, this protein acts as a repressor (P1). The P1 form binds to the operator, which prevents transcription of the operon. Arabinose removes P1 from the operator and shifts the conformational equilibrium to P2, the activator form. The P2 form, together with the complex of CAP and cyclic AMP, then binds to the promoter site, which enables RNA polymerase to initiate transcription. Thus, *the same regulatory protein serves a positive and a negative control function.*

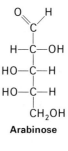

Arabinose

BOTH AN ATTENUATOR AND AN OPERATOR CONTROL TRANSCRIPTION OF THE TRYPTOPHAN OPERON

A new control element was discovered by Charles Yanofsky and his colleagues in their studies of the tryptophan operon of *E. coli*. The 7-kb mRNA transcript of this operon codes for the five enzymes that convert chorismate into tryptophan (p. 497). These five proteins are synthesized sequentially, coordinately, and in equimolar amounts by translation of this polycistronic *trp* mRNA. Translation

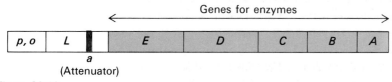

Figure 28-10
Diagram of the *trp* operon showing the promoter (*p*), operator (*o*), and attenuator (*a*) control sites and the genes for the leader sequence (*L*) and enzymes of the tryptophan pathway (*E, D, C, B,* and *A*).

takes place prior to the completion of transcription. *Trp* mRNA is synthesized in about four minutes and then rapidly degraded. The short lifetime of *trp* mRNA, which is only about three minutes, enables bacteria to respond quickly to their changing needs for tryptophan. In fact, *E. coli* can vary the rate of production of its biosynthetic enzymes for tryptophan over a 700-fold range.

How is this regulation accomplished? One level of control is achieved by the interaction of a specific repressor with the *trp* operator site on the DNA. The *trp* repressor is a 58-kdal protein encoded by the *trpR* gene, which is far from the *trp* operon. *A complex of this repressor and tryptophan binds tightly to the operator, whereas the repressor alone does not.* In other words, tryptophan is a *corepressor.* The target for the tryptophan-repressor complex is a DNA sequence with twofold symmetry (Figure 28-11). Again, symmetry plays an important

Start
↓
−21   −10   +1
CGAACTAGTTAACTAGTACGCAAG
┆┆┆┆┆┆┆┆┆┆:◉:┆┆┆┆┆┆┆┆┆┆┆┆
GCTTGATCAATTGATCATGCGTTC

Figure 28-11
Base sequence of the *trp* operator. The twofold axis of symmetry is denoted by the green symbol. The base pair labeled +1 is the start of the transcribed part of the operon.

role in the interaction of a protein with DNA. This operator site overlaps the promoter site for the initiation of transcription. Hence, *binding of the* trp *repressor to the operator prevents RNA polymerase from binding to the* trp *promoter and so the* trp *genes are not transcribed.*

For some time it was thought that end-product inhibition of the catalytic activity of the first enzyme complex in the tryptophan pathway (p. 500) and repressor-operator inhibition of transcription accounted for most of the regulation of tryptophan biosynthesis. This view was abruptly altered by the unexpected finding that certain mutants with *deletions* between the operator and the gene for the first enzyme (*trpE*) in the operon showed *increased* production of *trp* mRNA. Furthermore, analysis of the sequence of the 5′ end of *trp* mRNA revealed the presence of a *leader sequence* of 162 nucleotides before the initiation codon of *trpE.* It was then found that deletion mutants with enhanced *trp* mRNA levels mapped in this leader region, some 30 to 60 nucleotides before the start of *trpE.* The next striking observation was that a transcript containing only the first 130 nucleotides of the leader was produced when the tryptophan level was high, whereas a 7000-nucleotide *trp* mRNA including the entire leader sequence was synthesized when tryptophan was scarce. Hence, Yanofsky concluded that transcription of the *trp* operon must be regulated by a *controlled termination site,* called an *attenuator,* that is located between the operator and the gene for the first enzyme in the pathway. This physiologically regulated termination site, like termination sites at the ends of some operons (p. 610), contains a GC-rich sequence followed by an AT-rich one. Each of these regions in the attenuator contains a twofold axis of symmetry (Figure 28-12). Moreover, the terminated leader transcript ends with a series of U.

Figure 28-12
Base sequence of the *trp* attenuator site. Base pairs in the GC-rich region related by a twofold axis of symmetry are shown in blue, and those in the AT-rich region are shown in yellow.

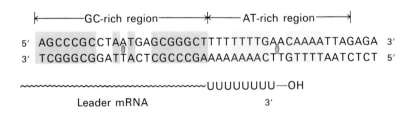

The *attenuator site* complements the operator site in regulating transcription of *trp* genes. When tryptophan is plentiful, initiation of transcription is blocked by the binding of the tryptophan-repressor complex to the operator. As the level of tryptophan in the cell decreases, repression is lifted and transcription begins. However, some of the RNA polymerase molecules fall off the template at the attenuator site, whereas others continue to synthesize the entire *trp* message. *The proportion of RNA polymerase molecules that proceed past the attenuator site increases as tryptophan becomes scarcer.*

ATTENUATION IS MEDIATED BY TRANSLATION OF THE LEADER mRNA

How does the attenuator site in the *trp* operon sense the level of tryptophan in the cell? An important clue came from the finding that part of the leader mRNA is translated. The presence of tryptophan residues at positions 11 and 12 of the fourteen-residue leader polypeptide (Figure 28-13) is highly significant. When tryptophan

Met - Lys - Ala - Ile - Phe - Val - Leu - Lys - Gly - Trp - Trp - Arg - Thr - Ser - Stop

∿AUG AAA GCA AUU UUC GUA CUG AAA GGU UGG UGG CGC ACU UCC UGA∿

Figure 28-13
Amino acid sequence of the *trp* leader peptide and the base sequence of the corresponding leader mRNA.

is abundant, the complete leader is synthesized. *However, when tryptophan is scarce, the ribosome stalls at the tandem UGG codons because of a paucity of tryptophanyl tRNA. The stalled ribosome somehow alters the structure of the mRNA so that RNA polymerase transcribes the operon beyond the attenuator site.* A key aspect of this control mechanism is that translation and transcription are closely coupled. The ribosome translating the *trp* leader mRNA follows closely behind the RNA polymerase molecule that is transcribing the DNA template. Recent studies suggest that the stalled ribosome switches the secondary structure of the mRNA from a base-paired arrangement that favors termination of transcription to a very different one that allows RNA polymerase to read through the attenuator site (Figure 28-14). We are

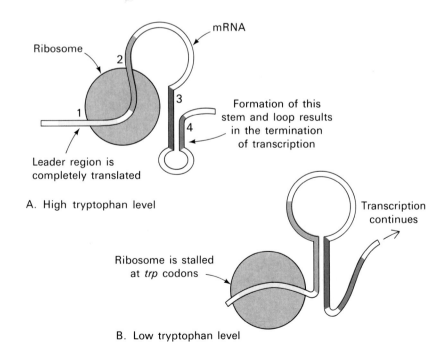

A. High tryptophan level

B. Low tryptophan level

Figure 28-14
Model for attenuation in the *E. coli trp* operon. When tryptophan is abundant (A), the leader region (segment 1) of the *trp* mRNA is fully translated. Segment 2 interacts with the ribosome, which enables segments 3 and 4 to base-pair. This base-paired region somehow signals RNA polymerase to terminate transcription. In contrast, when tryptophan is scarce (B), segments 3 and 4 do not interact because the ribosome is stalled at the *trp* codons of segment 1. Segment 2 interacts with 3 instead of being drawn into the ribosome and so segments 3 and 4 cannot pair. Consequently, transcription continues. [After D. L. Oxender, G. Zurawski, and C. Yanofsky. *Proc. Nat. Acad. Sci.* 76(1979):5524.]

beginning to see that nucleic acid molecules, like protein molecules, adopt alternative conformations that are regulated and have far-reaching physiological consequences.

THE ATTENUATOR SITE IN THE HISTIDINE OPERON CONTAINS SEVEN HISTIDINE CODONS IN A ROW

Two other operons for the biosynthesis of amino acids in *E. coli* are now known to have attenuator sites. The phenylalanine operon and the histidine operon, like the tryptophan operon, contain regulated termination sites preceding the first gene for an enzyme. Again, a leader region before the termination site is translated. The amino acid sequence of the leader peptide from the phenylalanine operon is striking: seven of the fifteen residues are phenylalanine (Figure 28-15). Even more remarkable, the leader peptide from the histi-

A

Met - Lys - His - Ile - Pro - Phe - Phe - Phe - Ala - Phe - Phe - Phe - Thr - Phe - Pro - Stop

5′ AUG AAA CAC AUA CCG UUU UUC UUC GCA UUC UUU UUU ACC UCC CCC UGA   3′

B

Met - Thr - Arg - Val - Gln - Phe - Lys - His - His - His - His - His - His - His - Pro - Asp -

5′ AUG ACA CGC GUU CAA UUU AAA CAC CAC CAU CAU CAC CAU CAU CCU GAC   3′

Figure 28-15
Amino acid sequence of the leader peptide and base sequence of the corresponding portion of mRNA from (A) the phenylalanine operon and (B) the histidine operon.

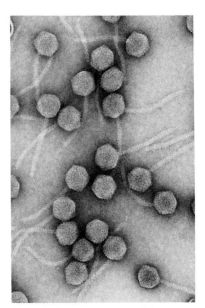

Figure 28-16
Electron micrograph of λ phages.
[Courtesy of Dr. A. Dale Kaiser.]

dine operon contains *seven histidine residues in a row. Clearly, these leader mRNAs are designed to sense the level of phenylalanine and histidine.* If the corresponding charged tRNA is scarce, translation of the leader is arrested. As discussed earlier for the *trp* operon, a stalled ribosome is thought to switch the mRNA conformation to a base-paired arrangement that allows RNA polymerase to read through the attenuator site. The presence of seven consecutive codons for histidine in the histidine operon leader mRNA markedly enhances the sensitivity of this detection system. In fact, a 15% decrease in the level of histidyl tRNA leads to a threefold increase in the number of mRNA molecules transcribed from this operon.

REPRESSORS AND ACTIVATORS DETERMINE THE DEVELOPMENT OF TEMPERATE PHAGE

We turn now to the role of repressors and activators of transcription in regulating the life cycle of *lambda* (λ) *bacteriophage.* The mature virus particle consists of a linear double-helical DNA molecule (48 kb) surrounded by a protein coat. Two developmental pathways are open to this virus: it can destroy its host or it can become part of its host (hence the name *temperate*). In the *lytic pathway,* the viral functions are fully expressed, which leads to the lysis of the bacterium and the production of a burst of about 100 progeny virus particles. Alternatively, λ can enter the *lysogenic pathway,* in which its DNA becomes covalently inserted into the host-cell DNA at a specific site. This recombination process involving a circular λ DNA molecule will be discussed later (p. 739). Most of the phage functions are switched off when its DNA is integrated in the host DNA. The viral DNA in this state is called a *prophage;* a host cell containing a prophage is called a *lysogenic bacterium.* The prophage replicates as part of the host chromosome in a lysogenic bacterium, usually for many generations. The lytic functions of λ are dormant but not lost in the lysogenic state (Figure 28-17). A variety of agents that interfere with DNA replication in the host induce the prophage to undergo lytic development.

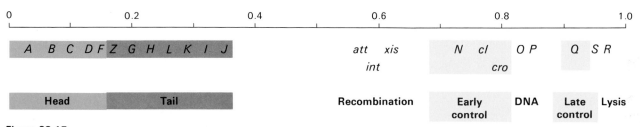

Figure 28-17
Genetic map of λ phage. Only some of the genes are shown here. The linear duplex DNA is converted into the circular form following entry into the bacterial cell.

Let us first consider the pattern of gene expression of λ in the lytic pathway. The goal of producing a large number of progeny is achieved by the *sequential transcription of viral genes.* Proteins needed for DNA replication and recombination are made first, followed by the synthesis of the head and tail proteins of the virus particle and of the proteins required for lysis of the host cell. Timing is critical; premature destruction of the host would of course be disadvantageous to the virus. There are three stages of gene expression in the lytic pathway: immediate-early, delayed-early, and late (Figure 28-18). In the *immediate-early stage,* RNA synthesis starts at two pro-

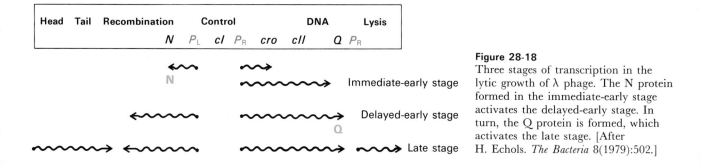

Figure 28-18
Three stages of transcription in the lytic growth of λ phage. The N protein formed in the immediate-early stage activates the delayed-early stage. In turn, the Q protein is formed, which activates the late stage. [After H. Echols. *The Bacteria* 8(1979):502.]

moter sites, P_L and P_R. One of these transcripts is the template for the N protein, which has a critical regulatory role. In the absence of N protein, the immediate-early transcripts end at either of two termination sites. The N protein antagonizes the termination of transcription at these sites and thereby permits further expression of λ genes. The N protein turns on the *delayed-early stage*. Proteins necessary for the replication of λ DNA and for recombination are made at this time. In addition, the *Q* gene is transcribed in the delayed-early stage. The Q protein is another critical regulator of gene expression in λ. The Q protein is required for the *late stage*. The genes for proteins needed for phage head and tail formation and for the lysis of the host are transcribed in the late stage. The Q protein, like the N protein, antagonizes the termination of transcription. *In short, the sequential regulation of lytic development is effected by two positive regulatory proteins, encoded by genes N and Q, which act by allowing transcription to proceed beyond several termination sites.*

There are three stages in the lysogenic cycle: *establishment, maintenance,* and *release.* The establishment of the prophage state requires the integration of the viral DNA into the host DNA and the inactivation of the lytic functions of the virus. These processes are complex and not fully understood. In contrast, the maintenance of the prophage state is relatively simple. A. Dale Kaiser showed that *only the cI gene is expressed in the prophage.* This gene codes for the λ repressor, which binds to two operator regions, O_L and O_R (Figure 28-19).

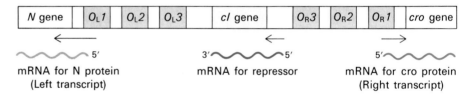

Figure 28-19
Diagram of the O_L and O_R operator regions and the adjacent genes. $O_L 1$ and $O_R 1$ have the highest affinity for the λ repressor. The repressor gene is *cI*. The left transcript starts with the *N* gene, whereas the right transcript starts with the *cro* gene.

Binding of the λ repressor to O_L directly prevents the leftward transcription of the immediate-early genes. In particular, the N protein is not synthesized and so the lytic pathway is blocked. By binding to O_R, the λ repressor prevents the rightward expression of the *cro* and *Q* genes. Indeed, the binding of the λ repressor to O_L and O_R silences the whole λ genome except for the *cI* gene, which codes for the λ repressor. As will be discussed shortly, the λ repressor itself controls the *cI* gene, thereby regulating its own level. Inactivation of the λ repressor enables lytic genes to be transcribed. The prophage is then excised from the host chromosome and the lytic functions are expressed.

The λ repressor has been isolated and studied in detail by Mark Ptashne. The 26-kdal monomer is in equilibrium with oligomers, which are the forms that bind to DNA. Two operator regions, O_L and O_R, are recognized by the same λ repressor. The *cI* gene that codes for the repressor is located between O_L and O_R (see Figure 28-19). Each of these operators contains three binding sites for the λ repressor. Nuclease-digestion studies have shown that the binding sites are seventeen base pairs long and that they are separated from each other by AT-rich regions that are from three to seven base pairs in length. The base sequences of these sites for the λ repressor are similar but not identical. The sequence recognized is 5'-TATCACCGC-3' or a similar one. These operator sites, like the *lac* operator, exhibit partial twofold symmetry.

The strongest binding site for repressor in both O_L and O_R is the one closest to the start of the first structural gene in the operon. The promoter site for the *N* gene lies within O_L and the promoter site for the *cro* gene lies within O_R. As in the lactose and arabinose operons, the binding of repressor to these operators blocks the binding of RNA polymerase to the corresponding promoter and so transcription is not initiated. The binding of λ repressor to two sites in O_L or O_R more effectively excludes RNA polymerase than binding to just one operator site.

THE LAMBDA REPRESSOR REGULATES ITS OWN SYNTHESIS

The number of λ repressor molecules in a lysogenized *E. coli* cell is precisely regulated. Too little repressor, even transiently, would put the cell on the lytic pathway. On the other hand, too much repressor would make it difficult for the phage to emerge should the bacterial environment become inhospitable. What controls the level of λ repressor? Recent studies indicate that this repressor regulates its own synthesis. Specifically, *repressor bound to $O_R 3$, the operator site closest to the* cI *gene, turns off transcription of this gene* (Figure 28-20). In contrast, *the binding of this repressor to $O_R 1$ enhances transcription of* cI. Recall that the affinity of $O_R 1$ for the λ repressor is higher than that of $O_R 3$. Thus, *transcription of* cI *is enhanced by low levels of the λ repressor, whereas it is inhibited by high levels of the same protein.* In other words, the expression of the *cI* gene is self-regulated.

The feedback circuit just described will tend to maintain the level of repressor so that the rest of the phage genome is not expressed. How then does the phage ever emerge from lysogeny? The critical trigger is a reduction in the number of λ repressors just sufficient to enable the *cro* gene to be transcribed. The newly synthesized cro protein then binds to $O_R 3$ to prevent the transcription

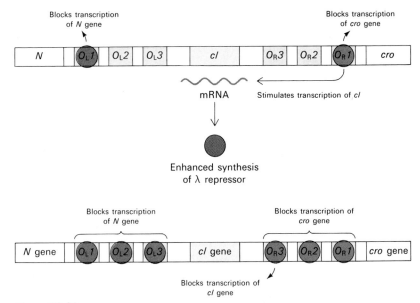

Figure 28-20
Self-regulation of the level of λ repressor: (A) the repressor binds to O_R1 when its level is low, which stimulates the transcription of $cI;$ (B) as the level of the λ repressor increases, it binds to O_R3 and inhibits further transcription of the cI gene.

of the cI gene. The important point is that O_R3 has a higher affinity for the cro protein than does O_R1. Thus, *low levels of the cro protein repress the synthesis of λ repressor without turning off the synthesis of the cro protein itself.* In this way, the λ repressor is prevented from regaining the upper hand. At this point, a chain of events leading to lysis is irreversibly set in motion. Thus, a subtle interplay between just a few proteins and operator sites determines the developmental pathway of this phage. It will be interesting to see whether regulatory motifs such as *multiple operator sites with differential affinities for proteins* are generally important in controlling procaryotic development.

SUMMARY

Cells regulate the amounts of proteins that are synthesized in a variety of ways. In *E. coli,* gene expression is regulated primarily at the level of transcription rather than translation. Many genes are organized into operons, which are coordinated units of genetic expression. An operon consists of control sites (an operator and a promoter) and a set of structural genes. Outside the operon, there is a regulator gene coding for a protein that can interact with the operator site. The lactose operator has a symmetric base sequence. In general, symmetry plays a central role in the recognition of specific sites on DNA by proteins. The *lac* operon is induced by β-galactosides such as allolactose and IPTG. Binding of an inducer to the *lac* repressor results in its displacement from the operator. RNA polymerase can then move through the operator to transcribe the *lac* operon. The tryptophan operon is repressed by tryptophan, which binds to the specific repressor and thereby enables it to interact with the operator. The effect is to switch off the transcription of the genes that code for the biosynthesis of tryptophan.

Several operons for the biosynthesis of amino acids, including the *trp* operon, are also controlled by an attenuator site. Transcription terminates at this site if the amino acid end product is abundant. Attenuation is mediated by translation of a leader mRNA. Broad control is exerted by cyclic AMP, which binds to a specific protein (CAP). This complex interacts with the promoter sites of several inducible operons and stimulates the initiation of transcription. The level of cyclic AMP is high only if glucose is scarce. Thus, glucose indirectly represses the synthesis of a variety of catabolic enzymes.

Bacteriophage λ can multiply and destroy its host (the lytic pathway) or its DNA can become integrated with that of its host (the lysogenic pathway). Three groups of genes are sequentially transcribed in the lytic pathway. The N protein formed in the immediate-early stage activates the transcription of the delayed-early genes, which in turn yields the Q protein, the activator for the late stage of transcription. The lysogenic pathway is maintained by the λ repressor, which is encoded by the *cI* gene. The repressor binds to the O_L and O_R operators and prevents the transcription of the immediate-early genes. The presence of multiple binding sites within these operators enables the λ repressor to regulate its own synthesis.

SELECTED READINGS

WHERE TO START

Jacob, F., and Monod, J., 1961. Genetic regulatory mechanisms in the synthesis of proteins. *J. Mol. Biol.* 3:318–356. [The operon model and the concept of messenger RNA were proposed in this superb paper.]

Ptashne, M., and Gilbert, W., 1970. Genetic repressors. *Sci. Amer.* 222(6):36–44. [Available as *Sci. Amer.* Offprint 1179.]

Maniatis, T., and Ptashne, M., 1976. A DNA operator-repressor system. *Sci. Amer.* 234(1):64–76. [Offprint 1333.]

LACTOSE OPERON

Miller, J. H., and Reznikoff, W. S., (eds.), 1978. *The Operon.* Cold Spring Harbor Laboratory. [An excellent collection of articles about regulation in *E. coli* and λ phage. The lactose, tryptophan, arabinose, histidine, and galactose operons are discussed in detail.]

Gilbert, W., and Muller-Hill, B., 1966. Isolation of the *lac* repressor. *Proc. Nat. Acad. Sci.* 56:1891–1898.

Dickson, R., Abelson, J., Barnes, W., and Reznikoff, W. 1975. Genetic regulation: the *lac* control region. *Science* 187:27–35.

ARABINOSE OPERON

Wilcox, G., Meuris, P., Bass, P., and Englesberg, E., 1974. Regulation of the arabinose operon in vitro. *J. Biol. Chem* 249:2946–2952.

Hirsh, J., and Schleif, R., 1976. Electron microscopy of gene regulation: the L-arabinose operon. *Proc. Nat. Acad. Sci.* 73:1518–1522.

TRYPTOPHAN AND HISTIDINE OPERONS

Platt, T., 1978. Regulation of gene expression in the tryptophan operon of *Escherichia coli. In* Miller, J. H., and Reznikoff, W. S., (eds.), *The Operon*, pp. 213–302. Cold Spring Harbor Laboratory. [A lucid introduction to the *trp* operon.]

684   Oxender, D. L., Zurawski, G., and Yanofsky, C., 1979. Attenuation in the *Escherichia coli* tryptophan operon: the role of RNA secondary structure involving the Trp codon region. *Proc. Nat. Acad. Sci.* 76:5524–5528.

Bertrand, K., Korn, L., Lee, F., Platt, T., Squires, C. L., Squires, C., and Yanofsky, C., 1975. New features of the regulation of the tryptophan operon. *Science* 189:22–26. [An account of the discovery of attenuation of transcription.]

Barnes, W. M., 1978. DNA sequence from the histidine operon control region: seven histidine codons in a row. *Proc. Nat. Acad. Sci.* 75:4281–4285.

Stephens, J. C., Artz, S. W., and Ames, B. N., 1975. Guanosine 5′-diphosphate 3′-diphosphate (ppGpp): positive effector for histidine operon transcription and general signal for amino acid deficiency. *Proc. Nat. Acad. Sci.* 72:4389–4393.

CYCLIC AMP AND CATABOLITE REPRESSION

Pastan, I., and Adhya, S., 1976. Cyclic adenosine 3′,5′-monophosphate in *Escherichia coli*. *Bacteriol. Rev.* 40:527–551.

Zubay, G., Schwartz, D., and Beckwith, J., 1970. Mechanism of activation of catabolite-sensitive genes: a positive control system. *Proc. Nat. Acad. Sci.* 66:104–110.

CONTROL OF TRANSCRIPTION IN λ PHAGE

Ptashne, M., Backman, K., Humayun, M. Z., Jeffrey, A., Maurer, R., Meyer, B., and Sauer, R. T., 1976. Autoregulation and function of a repressor in bacteriophage lambda. *Science* 194:156–161.

Johnson, A., Meyer, B. J., and Ptashne, M., 1978. Mechanism of action of the *cro* protein of bacteriophage λ. *Proc. Nat. Acad. Sci.* 75:1783–1787.

Ptashne, M., Jeffrey, A., Johnson, A. D., Maurer, R., Meyer, B. J., Pabo, C. O., Roberts, T. M., and Sauer, R. T., 1980. How the λ repressor and *cro* work. *Cell* 19:1–11.

Hershey, A. D., (ed.), 1971. *The Bacteriophage Lambda.* Cold Spring Harbor Laboratory. [Contains a wealth of information about λ.]

Reichardt, L., and Kaiser, A. D., 1971. Control of λ repressor synthesis. *Proc. Nat. Acad. Sci.* 68:2185–2189.

PROBLEMS

1. What is the effect of each of these mutations?
   (a) Deletion of the *lac* regulator gene.
   (b) Deletion of the *trp* regulator gene.
   (c) Deletion of the *ara* regulator gene.
   (d) Deletion of the *cI* gene of λ.
   (e) Deletion of the *N* gene of λ.

2. Superrepressed mutants (i^s) of the *lac* operon behave as noninducible mutants. The i^s gene is dominant over i^+ in partial diploids. What is the molecular nature of this mutation likely to be?

3. A mutant of *E. coli* synthesizes large amounts of β-galactosidase whether or not inducer is present. A partial diploid formed from this mutant and $Fi^+o^+z^-$ also synthesizes large amounts of β-galactosidase whether or not inducer is present. What kind of mutation might give these results?

4. A mutant unable to grow on galactose, lactose, arabinose, and several other carbon sources is isolated from a wild-type culture. The cyclic AMP level in this mutant is normal. What kind of mutation might give these results?

5. An *E. coli* cell bearing a λ prophage is immune to lytic infection by λ. Why?

6. Transcription of *cI* can be initiated at p_{RE}, the promoter for repressor establishment, or at p_{RM}, the promoter for repressor maintenance. The p_{RM} transcript begins at its 5′ end with a AUG codon for the λ repressor, whereas in the p_{RE} transcript this initiation codon is preceded by a sequence that is complementary to the 3′ end of 16S rRNA. Initiation at the p_{RE} site requires phage-encoded proteins that are not expressed in the lysogenic state.
   (a) Which transcript will be translated more efficiently?
   (b) What is the likely physiological significance of this difference?

For additional problems, see L. E. Hood, J. H. Wilson, and W. B. Wood, *Molecular Biology of Eucaryotic Cells* (Benjamin, 1975), ch. 1.

CHAPTER 29

EUCARYOTIC CHROMOSOMES AND GENE EXPRESSION

Eucaryotes contain far more genetic information than do procaryotes. For example, a human cell contains about a thousand times as much DNA as an *E. coli* cell and about a hundred thousand times as much DNA as a λ phage. This abundance of DNA endows eucaryotes with potentialities that are absent from procaryotes. A second difference is that the DNA of higher forms is associated with basic proteins called *histones,* whereas the DNA of lower forms is not. These basic proteins are a device for making DNA more compact so that a contour length of many centimeters can be accommodated within a few micrometers. The distinctive morphology of eucaryotic chromosomes, which is evident under the light microscope, indicates that they have a much higher degree of structural organization than do procaryotic genomes. Moreover, the shape of eucaryotic chromosomes changes markedly during the cell cycle. Another major difference—indeed, the one that sets eucaryotes apart from procaryotes—is that eucaryotic chromosomes are bounded by a *nuclear membrane.* This membrane and other internal ones are absent in procaryotes. An important consequence is that *transcription and translation are separated in space and time in eucaryotes,* whereas they are closely coupled in procaryotes. Primary transcripts are extensively modified, cleaved, and spliced in the nuclei of higher organisms. Only a small fraction of the RNAs synthesized in the nucleus emerge in the cytosol as mRNAs. Clearly, gene expression is far richer and more complex in eucaryotes than in pro-

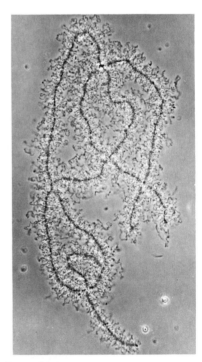

Figure 29-1
Phase-contrast light micrograph of a lampbrush chromosome from an oocyte. [Courtesy of Dr. Joseph Gall.]

caryotes. This field is advancing rapidly because eucaryotic genes can be isolated, cloned, sequenced, and expressed in well-defined systems. There is excitement in the air and a feeling that we may be on the threshold of unraveling a fundamental problem in biology, the mechanism of cell differentiation.

A EUCARYOTIC CHROMOSOME CONTAINS A SINGLE MOLECULE OF DOUBLE-HELICAL DNA

Does a chromosome contain one long molecule of DNA? This important question was difficult to answer for many years because very large DNA molecules are exquisitely sensitive to degradation by shearing forces. Bruno Zimm circumvented this problem by using the *viscoelastic technique,* which provides a measure of the size of the *largest DNA molecules in a mixture.* DNA molecules are stretched out by flow and then allowed to recoil to their normal form. The half time for recoiling depends on molecular weight. Cells from fruit flies were lysed in the measuring chamber to avoid the breakage of DNA that would be caused by transferring samples. Nucleases were inactivated by incubating the sample in detergent at 65°C. In addition, pronase was added to digest proteins bound to DNA.

The observed mass of the largest DNA molecules in this mixture was 41×10^9 daltons, which agreed closely with the known value of the DNA content of the largest chromosome in *Drosophila melanogaster,* which is 43×10^9 daltons. Excellent agreement also was obtained for a translocation mutant of the largest chromosome, which has an extra piece of DNA, giving it a DNA content of 59×10^9 daltons. The measured mass of this DNA molecule was 58×10^9 daltons. Radioautographs of *D. melanogaster* DNA (Figure 29-2) provide confirming evidence for the existence of very long DNA molecules. These studies reveal that *a Drosophila chromosome contains a single uninterrupted molecule of DNA.* Furthermore, this DNA molecule is *linear and unbranched.*

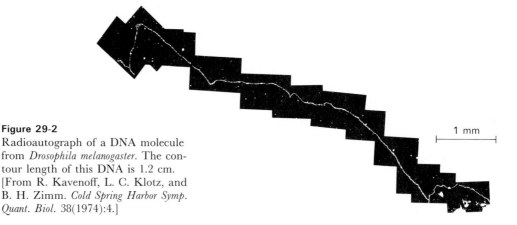

Figure 29-2
Radioautograph of a DNA molecule from *Drosophila melanogaster.* The contour length of this DNA is 1.2 cm. [From R. Kavenoff, L. C. Klotz, and B. H. Zimm. *Cold Spring Harbor Symp. Quant. Biol.* 38(1974):4.]

1 mm

EUCARYOTIC DNA IS TIGHTLY BOUND TO BASIC PROTEINS CALLED HISTONES

The DNA in eucaryotic chromosomes is not bare. Rather, eucaryotic DNA is tightly bound to a group of small, basic proteins called *histones*. In fact, histones comprise about half of the mass of eucaryotic chromosomes, the other half being DNA. This nucleoprotein chromosomal material is termed *chromatin*. Histones can be dissociated from DNA by treating chromatin with salt or dilute acid. The resulting mixture can then be fractionated by ion-exchange chromatography. The five types of histones are called H1, H2A, H2B, H3, and H4. They range in mass from about 11 to 21 kdal (Table 29-1). A striking feature of histones is their *high content of positively charged side chains:* about one in four residues is either lysine or arginine.

Table 29-1
Types of histones

Type	Lys/Arg ratio	Number of residues	Mass (kdal)	Location
H1	20.0	215	21.0	Linker
H2A	1.25	129	14.5	Core
H2B	2.5	125	13.8	Core
H3	0.72	135	15.3	Core
H4	0.79	102	11.3	Core

Each type of histone can exist in a variety of forms because of *posttranslational modifications* of certain side chains. For example, lysine 16 in H4 is often acetylated. Histones can also be methylated, ADP-ribosylated, and phosphorylated. The modulation of the charge, hydrogen-bonding capabilities, and shape of histones by these covalent modifications may be important in regulating the availability of DNA for replication and transcription.

THE AMINO ACID SEQUENCES OF H3 AND H4 ARE NEARLY THE SAME IN ALL PLANTS AND ANIMALS

Emil Smith and Robert DeLange have shown that the amino acid sequences of H4 from pea seedlings and calf thymus differ at only two sites out of 102 residues. The changes at these two sites are quite small: valine in place of isoleucine, and lysine in place of arginine. Thus, *the amino acid sequence of H4 has remained nearly constant in the 1.2 × 10^9 years following the divergence of plants and animals.* Likewise, H3 has changed little throughout this very long evolutionary period. The amino acid sequences of H3 from pea seedlings and calf thymus differ at just four positions. It is interesting to compare the rate of change of these histones in the course of evolution with those of

Ser -Gly-Arg-Gly -Lys -Gly-Gly-Lys -Gly-Leu-   10
Gly -Lys -Gly -Gly -Ala -Lys -Arg-His -Arg-Lys -   20
Val -Leu-Arg-Asp-Asn-Ile -Gln-Gly -Ile -Thr -   30
Lys -Pro-Ala -Ile -Arg-Arg-Leu-Ala -Arg-Arg-   40
Gly -Gly-Val -Lys -Arg-Ile -Ser-Gly -Leu-Ile -   50
Tyr -Glu-Glu-Thr -Arg-Gly -Val -Leu-Lys -Val -   60
Phe-Leu-Glu-Asn-Val -Ile -Arg-Asp-Ala -Leu-   70
Thr -Tyr -Thr -Glu -His -Ala -Lys -Arg-Lys -Thr -   80
Val -Thr -Ala -Met-Asp-Val -Val -Tyr -Ala -Leu-   90
Lys -Arg-Gln-Gly -Arg-Thr -Leu-Tyr -Gly -Phe-   100
Gly -Gly   102

Figure 29-3
Amino acid sequence of histone H4 from calf thymus. Several residues are modified. The α-amino group is acetylated, as is the ε-amino group of Lys 16. The ε-amino group of Lys 20 is methylated or dimethylated. Histone H4 from pea seedlings has the same amino acid sequence except that residue 60 is isoleucine and residue 77 is arginine.

other proteins. A useful index is the *unit evolutionary period*, which is the time required for the amino acid sequence to change by 1% after two evolutionary lines have diverged. The times for H3 and H4 of 300 million years and 600 million years, respectively, are much longer than those of other proteins studied thus far. For example, the unit evolutionary period is 20 million years for cytochrome *c*, 6 million years for hemoglobin, and 1 million years for fibrinopeptides. The remarkably conserved structure of H3 and H4 strongly suggests that they have a critical role that was established early in the evolution of eucaryotes and has remained nearly invariant since then.

NUCLEOSOMES ARE THE REPEATING UNITS OF CHROMATIN

How do histones interact with DNA to form the chromatin fiber? In 1974, Roger Kornberg proposed on the basis of several lines of evidence that *chromatin is made up of repeating units, each consisting of 200 base pairs of DNA and of two each of H2A, H2B, H3, and H4.* These repeating units are now known as *nucleosomes.* Most of the DNA is wound around the outside of a core of histones. The remainder of the DNA, called the linker, joins adjacent nucleosomes and contributes to the flexibility of the chromatin fiber. Thus, *a chromatin fiber is considered to be a flexibly jointed chain of nucleosomes,* rather like beads on a string.

This model for the structure of chromatin is supported by a wide range of experimental findings:

1. *Electron microscopy.* Linear arrays of 100-Å–diameter spheres that are connected by a thin strand are seen in electron micrographs of chromatin (Figure 29-4). The degree of extension of the chromatin fiber depends on the procedure used to prepare the specimen for electron microscopy. Other preparative methods yield electron micrographs in which the 100-Å spheres are more compactly arranged. Thus, electron microscopy directly supports the notion that chromatin is a chain of roughly spherical particles that are separated by flexible regions.

2. *X-ray and neutron diffraction.* A 100-Å repetition is also seen in x-ray diffraction patterns of chromatin fibers. Neutron diffraction studies indicate that the DNA is located near the outside of the nucleosome.

3. *Nuclease digestion.* Free DNA in solution can be cleaved at any of its phosphodiester bonds by pancreatic deoxyribonuclease I (DNase I) or micrococcal nuclease. In contrast, DNA in chromatin is protected from digestion except at a few sites. The digestion pattern of chromatin is striking in its simplicity: it consists of a ladder

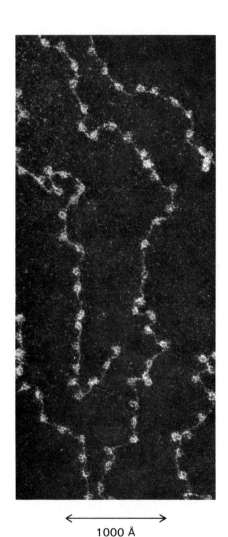

←——————→
1000 Å

Figure 29-4
Electron micrograph of chromatin. The beadlike particles have diameters of nearly 100 Å. [Courtesy of Dr. Ada Olins and Dr. Donald Olins.]

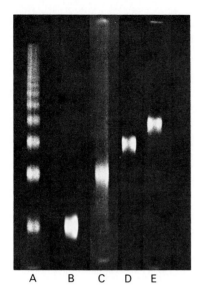

Figure 29-5
Gel-electrophoresis patterns of defined lengths of chromatin produced by limited micrococcal nuclease digestion. Part A shows an unfractionated digest. Fractionation by centrifugation on a sucrose gradient yielded a (B) monomer, (C) dimer, (D) trimer, and (E) tetramer. [From J. T. Finch, M. Noll, and R. D. Kornberg. *Proc. Nat. Acad. Sci.* 72(1975):3321.]

A   B   C   D   E

of discrete bands (Figure 29-5). The DNA contents of these fragments are multiples of a basic unit of about 200 base pairs. Electron micrographs show that the number of spherical particles in a fragment of chromatin is equal to the number of 200 base-pair units (Figure 29-6). For example, a fragment with 600 base pairs of DNA consists of three 100-Å–diameter particles. Thus, a bead seen in an electron micrograph corresponds to a nucleosome defined by nuclease digestion.

4. *Reconstitution.* A chromatinlike fiber can be formed in vitro by adding histones to DNA from adenovirus or simian virus 40 (SV40). The amount of DNA associated with a nucleosome in these reconstituted systems is close to 200 base pairs. Furthermore, equimolar amounts of H2A, H2B, H3, and H4 are required to form nucleosomes. Characteristic beads are not formed if any of the four histones is absent from the reconstitution mixture. On the other hand, H1 is not required, which goes along with the finding that H1 is not present in the nucleosomes of all eucaryotic cells. X-ray diffraction studies have also shown that H2A, H2B, H3, and H4 must be added to DNA to regain the diffraction pattern of chromatin.

A NUCLEOSOME CORE CONSISTS OF 140 BASE PAIRS OF DNA WOUND AROUND A HISTONE OCTAMER

The DNA content of nucleosomes from different organisms and cell types ranges from about 160 to 240 base pairs (Table 29-2). What is the structural basis of these differences? Again, nucleases have proved to be highly informative probes. Nucleosomes can be further digested with micrococcal nuclease to yield a *core particle* that contains 140 base pairs of DNA, irrespective of the initial DNA

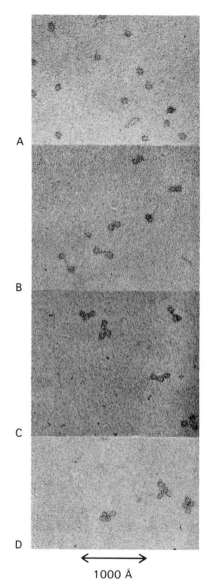

A

B

C

D

←——→
1000 Å

Figure 29-6
Electron micrographs of (A) monomer, (B) dimer, (C) trimer, and (D) tetramer nucleosomes produced as described in Figure 29-5. [From J. T. Finch, M. Noll, and R. D. Kornberg. *Proc. Nat. Acad. Sci.* 72(1975):3321.]

Table 29-2
DNA content of nucleosomes

Cell type	Number of base pairs
Yeast	165
HeLa	183
Rat bone marrow	192
Rat liver	196
Rat kidney	196
Chick oviduct	196
Chicken erythrocyte	207
Sea urchin gastrula	218
Sea urchin sperm	241

Figure 29-7
Electron micrograph of a crystal of nucleosome cores. Centers of adjacent nucleosome cores in this hexagonal array are 100 Å apart. [From J. T. Finch, L. C. Lutter, D. Rhodes, R. S. Brown, B. Rushton, M. Levitt, and A. Klug. *Nature* 269(1977):31.]

content of the particular nucleosome. The nucleosome core is probably nearly the same in all eucaryotes. *This core consists of 140 base pairs of DNA bound to a histone octamer (two each of H2A, H2B, H3, and H4).*

Nucleosome cores have been crystallized and are now being analyzed by electron microscopy (Figure 29-7) and x-ray diffraction. Crystallization of these particles demonstrates that the nucleosome cores of chromatin are quite homogenous. Aaron Klug and John Finch have found that the core is a flat particle of dimensions $110 \times 110 \times 55$ Å and that it is made up of two layers. The 140 base pairs of DNA are wound on the outside of the core to form $1\frac{3}{4}$ turns of a left-handed superhelix with a pitch of about 28 Å (Figure 29-8).

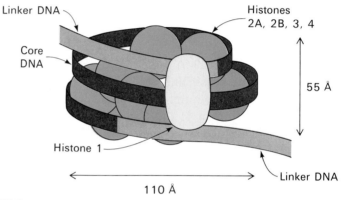

Figure 29-8
Schematic diagram of a nucleosome. The DNA double helix (shown in red) is wound around an octamer of histones (two molecules each of 2A, 2B, 3, and 4, shown in blue). Histone 1 (shown in yellow) binds to the outside of this core particle and to the linker DNA. [After *DNA Replication* by A. Kornberg. W. H. Freeman and Company. Copyright © 1980.]

As mentioned earlier, H1 is not invariably present in nucleosomes. The amino acid sequence of H1 is the most variable of the five histones. It also differs from the other histones in its stoichiometry, which is one per nucleosome. Furthermore, H1 can be readily dissociated from nucleosomes, which suggests that it has a peripheral location rather than being part of the core. In fact, H1 is released from nucleosomes when the DNA is trimmed from 160 to 140 base pairs. Thus, H1 is almost certainly on the outside of the nucleosome, near the linker DNA. H1 may serve as a bridge between different nucleosomes, which would render chromatin even more compact.

A NUCLEOSOME IS THE FIRST STAGE IN THE CONDENSATION OF DNA

The winding of DNA around a nucleosome core contributes to the packing of DNA by decreasing its linear extent. A 200-base-pair stretch of DNA would have a length of about 680 Å in solution. In contrast, this amount of DNA fits within the 100-Å diameter of the nucleosome. Thus, the *packing ratio* (degree of condensation) of the nucleosome is about 7. How does this value compare with the degree of condensation of DNA in chromosomes? Human metaphase chromosomes, which are highly condensed, contain a total of 5.3×10^9 base pairs, corresponding to a contour length of 180 cm. This DNA is packed into forty-six cylinders whose total length is 200 μm. Thus, the packing ratio of DNA in these metaphase chromosomes is about 10^4. DNA in interphase nuclei, where the chromatin is more dispersed, has a packing ratio of about 10^2 to 10^3. Clearly, the nucleosome is just the first step in the compaction of DNA.

What is the next level of organization of DNA? One possibility is that the nucleosomes themselves form a helical array. For example, a postulated solenoidal model of chromatin with a 360-Å diameter (Figure 29-9) would have a packing ratio of about 40. The folding of such solenoids into loops would provide additional condensation. It seems likely that a series of nonhistone proteins stabilizes higher-order structures of chromosomes. For example, metaphase chromosomes stripped of histones display a central protein scaffold that is surrounded by many very long loops of DNA (Figure 29-10). These DNA loops are probably bound to the scaffold. This arrangement may be important for moving chromosomes in mitosis and meiosis.

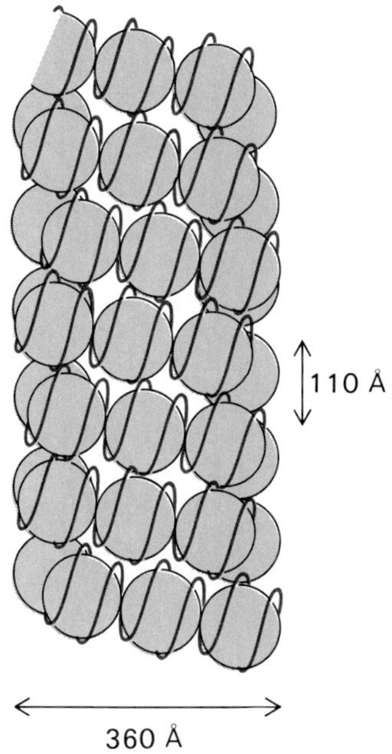

110 Å

360 Å

Figure 29-9
A proposed solenoidal model of chromatin consisting of six nucleosomes (shown in blue) per turn of helix. The DNA double helix (shown in red) is wound around each nucleosome. [After J. T. Finch and A. Klug. *Proc. Nat. Acad. Sci.* 73(1976):1900.]

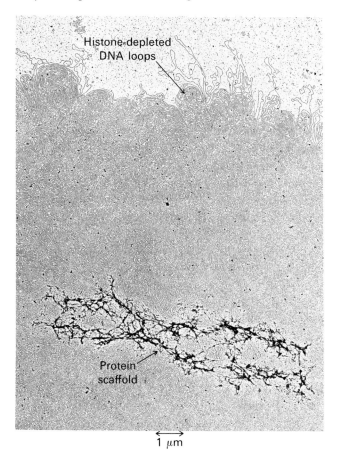

Histone-depleted DNA loops

Protein scaffold

1 μm

Figure 29-10
Electron micrograph showing histone-depleted DNA attached to a central protein scaffold. Histones were removed from metaphase chromosomes of HeLa cells by treating them with polyanions. [Courtesy of Dr. Ulrich Laemmli.]

EUCARYOTIC DNA IS REPLICATED BIDIRECTIONALLY FROM MANY ORIGINS

Eucaryotic DNA, like all other DNA molecules, is *replicated semiconservatively*. Newly replicated DNA is also distributed semiconservatively between sister chromatids of metaphase chromosomes (Figure 29-11). Electron microscopic studies have also shown that

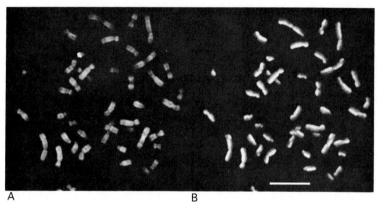

A   B

Figure 29-11
Fluorescence micrograph of human male metaphase chromosomes from cells that replicated twice in a medium containing the thymidine analog 5-bromodeoxyuridine (BrdU). Chromosomes were first stained with quinacrine (A), destained, and then restained with the *bis*benzimidazole dye 33258 Hoechst (B). The fluorescence of this latter dye is quenched by BrdU, and so sister chromatids containing DNA with only one polynucleotide chain substituted with BrdU fluoresce more brightly than those with BrdU in both polynucleotide chains. (The bar is 10 μm long.) [Courtesy of Dr. Samuel Latt.]

eucaryotic DNA is replicated *bidirectionally from many origins*. The use of many initiation points is necessary for rapid replication because of the great length of eucaryotic DNAs. For example, the largest chromosome in *Drosophila* has a length of about 2.1 cm, or 62,000 kb. A DNA replication fork in *Drosophila* moves at the rate of 2.6 kb/min, compared with about 16 kb/min in *E. coli*. The replication of the largest *Drosophila* chromosome would take more than sixteen days if there were only one origin for replication. The actual replication time of less than three minutes is achieved by *the cooperative action of more than 6,000 replication forks per DNA molecule*. A DNA molecule from the cleavage nuclei of *Drosophila* exhibits a serial array of replicated regions, or "eye forms" (Figure 29-12). The activation of each initiation point generates two diverging replication forks. The eye forms expanding in both directions merge to form the two daughter DNA molecules (Figure 29-13). An eye form within an eye form has not been observed, which indicates that an origin cannot be reactivated until after the entire DNA molecule is replicated.

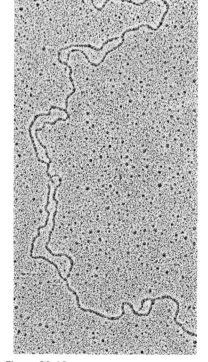

Figure 29-12
Electron micrograph of replicating chromosomal DNA from cleavage nuclei of *Drosophila*. The eye forms are the newly replicated regions. [Courtesy of Dr. David Hogness.]

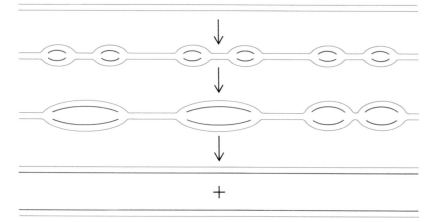

Figure 29-13
Schematic diagram of the replication of a eucaryotic chromosome. The parental DNA is shown in blue, and the newly replicated DNA in red.

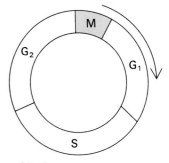

Figure 29-14
Phases in the life cycle of a eucaryotic cell: M (mitosis); G_1 (gap 1, prior to DNA synthesis); S (period of DNA synthesis); and G_2 (gap 2, between DNA synthesis and mitosis). The length of these periods depends on the cell type and conditions of growth. Mitosis is usually the shortest phase.

Eucaryotic cells contain three types of DNA polymerases (Table 29-3). The α polymerase plays a major role in chromosome replication, whereas the β enzyme participates in the repair of DNA. The

Table 29-3
Eucaryotic DNA polymerases

Type	Location	Major role	Mass (kdal)
α	Nucleus	Replication of nuclear DNA	140
β	Nucleus	Repair of nuclear DNA	40
γ	Mitochondria	Replication of mitochondrial DNA	150

amount of the α polymerase increases more than tenfold when quiescent cells begin to divide rapidly. The γ polymerase is responsible for the replication of mitochondrial DNA. These enzymes, like procaryotic DNA polymerases, use deoxyribonucleoside triphosphates as activated intermediates and carry out template-directed elongation of a primer in the $5' \rightarrow 3'$ direction. In contrast, these eucaryotic DNA polymerases lack nuclease activity. It seems likely that proofreading is performed by nucleases that are associated with these polymerases in multienzyme complexes.

NEW HISTONES FORM NEW NUCLEOSOMES ON THE LAGGING DAUGHTER DNA DUPLEX

At replication forks, the overall direction of DNA synthesis is $5' \rightarrow 3'$ for one daughter strand and $3' \rightarrow 5'$ for the other. As in the replication of procaryotic DNA (p. 583), this is accomplished by

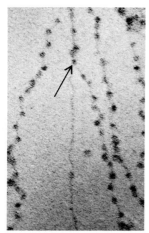

Figure 29-15
An asymmetric replication fork generated by DNA synthesis in the presence of cycloheximide, which blocks the formation of new histones. One of the daughter duplexes is beaded, whereas the other is bare, which shows that parental histones are bound to only one daughter duplex. [From D. Riley and H. Weintraub. *Proc. Nat. Acad. Sci.* 76(1979):331.]

having one strand, the leading strand, synthesized continuously, and the other, the lagging strand, discontinuously. How are old and new histones distributed? This question was answered by carrying out DNA synthesis in the presence of cycloheximide, an inhibitor of protein synthesis. Under these conditions, DNA synthesis continues for about 15 minutes. Half of the newly synthesized DNA is completely degraded by DNase I, whereas the other half is split into 200 base-pair fragments. This experiment, as well as density-labeling studies, suggested that parental histones are associated with one of the daughter DNA duplexes, the other being bare because of the absence of new histones. This interpretation is directly supported by electron micrographs showing that one of the daughter duplexes at a replication fork is beaded, whereas the other is naked (Figure 29-15). In other words, parental histones segregate conservatively during replication. This arrangement indicates that histones do not dissociate from DNA during replication. Rather, *old histones stay with the DNA duplex containing the leading strand, whereas new histones assemble on the DNA duplex containing the lagging strand.* A likely reason for this difference between the daughter DNA molecules is that histones bind much more strongly to double-stranded than to single-stranded DNA. Old histones probably do not follow the lagging duplex because it contains single-stranded regions prior to the joining of its Okazaki fragments.

MITOCHONDRIA AND CHLOROPLASTS CONTAIN THEIR OWN DNA

Not all of the genetic information of eucaryotic cells is encoded by their nuclear chromosomal DNA. Genetic studies of yeast led to the discovery of a mitochondrial genome distinct from the nuclear one. In 1949, Boris Ephrussi found that some mutants of baker's yeast are unable to carry out oxidative phosphorylation. These respiration-deficient mutants grow slowly by fermentation. They are called *petites* because they form very small colonies. Genetic analyses led to the surprising finding that petite mutations segregate independently of the nucleus, which suggested that mitochondria have their own genomes. Several years later, mitochondria were in fact found to contain DNA. Furthermore, mitochondrial DNA from a petite strain has a different buoyant density from that of wild-type yeast, which shows that a large part of the mitochondrial genome is altered in this mutant. It was subsequently found that chloroplasts in photosynthetic eucaryotes likewise contain DNA that is replicated, transcribed, and translated.

Mitochondrial DNA from animal cells is a circular duplex with a contour length of about 5 μm, corresponding to about 15 kb. Yeast mitochondrial DNA is about five times as long and chloroplast DNA is typically ten times as long. DNA molecules from mitochondria and chloroplasts are devoid of histones. They are relatively small, comparable in size to viral genomes. The best-characterized mitochondrial genome is that of yeast, which codes for about ten proteins, two ribosomal RNA molecules, and some twenty-six species of transfer RNA. The molecules specified by the mitochondrial DNA and synthesized within the organelle comprise only about 5% of the total mitochondrial protein. Thus, most of the proteins in a mitochondrion are encoded by the nuclear genome. However, the genetic contribution of mitochondrial DNA is essential. For example, three of the seven subunits of cytochrome oxidase and three of the ten subunits of the ATPase in the inner mitochondrial membrane are encoded by the mitochondrial genome. The existence of separate genomes poses some special problems. How is replication of mitochondrial DNA coordinated with chromosome duplication and cell division? How do proteins synthesized in the cytosol enter mitochondria and interact with mitochondrial gene products? Most enigmatic, why do mitochondria have their own genomes when 95% of their proteins are specified by the nuclear genome? The answers to these intriguing questions are not yet known.

EUCARYOTIC DNA CONTAINS MANY REPEATED BASE SEQUENCES

Studies of the kinetics of reassociation of thermally denatured DNA by Roy Britten and his associates have revealed that eucaryotic DNA, in contrast with procaryotic DNA, contains many repeated base sequences. In these experiments, DNA was sheared to small fragments and then denatured by heating the solution above the melting temperature (T_m) of the DNA. This solution of single-stranded DNA was then cooled to about 25°C below the T_m, which is optimal for the reassociation of complementary strands to form double-helical DNA. The kinetics of reassociation can be measured in a variety of ways. One technique is to follow the absorbance of the solution at 260 nm (p. 573). At this wavelength, the extinction coefficient of double-stranded DNA is about 40% less than that of single-stranded DNA; this decrease is called *hypochromicity*. Another experimental approach is based on the fact that *double-stranded DNA binds to hydroxyapatite (calcium phosphate) columns, whereas single-stranded DNA passes through.* An attractive feature of this technique is that large amounts of DNA can be fractionated on the basis of their rates of reassociation following thermal denaturation.

The observed kinetics of reassociation of DNA from *E. coli* or T4 phage followed the time course expected for the bimolecular reaction

$$S + S' \xrightarrow{k} D$$

in which S and S' are complementary single-stranded molecules, D is the reassociated double helix, and k is the rate constant for associ-

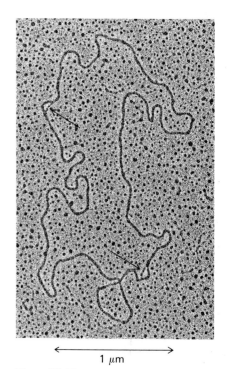

Figure 29-16
Electron microscopy of a mitochondrial DNA molecule containing two genomes joined head-to-tail to form a circle. Replication of this DNA molecule has just started. The arrows point to two loops 180 degrees apart on the circle. These displacement loops (D loops) contain newly synthesized DNA. The thinner line at each loop is a displaced single-stranded region of parental DNA. [Courtesy of Dr. David Clayton.]

ation. For such a reaction, the fraction f of single-stranded molecules decreases with time according to the expression

$$f = \frac{1}{1 + kC_0t}$$

in which C_0 is the total concentration of DNA (expressed in moles per liter of nucleotides) and t is time (sec). For a particular DNA and a particular set of experimental conditions (e.g., ionic strength, temperature, size of the DNA fragments), f depends only on C_0t, the product of DNA concentration and time. It is convenient to depict the kinetics of reassociation by plotting f versus the logarithm of C_0t. Such a C_0t *curve* has a sigmoidal shape (Figure 29-17).

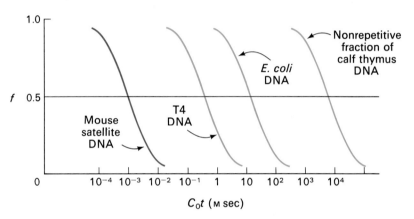

Figure 29-17
These C_0t curves depict the kinetics of reassociation of several kinds of thermally denatured DNA. The fraction of single-stranded molecules is plotted as a function of C_0t. The rapid reassociation of mouse satellite DNA shows that it contains very many repeated sequences. [After R. J. Britten and D. E. Kohne. *Science* 161(1968):530.]

An informative index of this plot is $C_0t_{0.5}$, the value of C_0t corresponding to reassociation of half of the DNA ($f = 0.5$). For *E. coli* DNA, $C_0t_{0.5}$ is about 9 M sec, whereas for T4 phage, $C_0t_{0.5}$ is 0.3 M sec. These values indicate that *E. coli* DNA reassociates about thirty times as slowly as T4 phage DNA. The reason is that *E. coli* DNA is larger and so the number of kinds of fragments in a sheared sample is greater than for T4 DNA. Hence, *the concentration of complementary fragments is lower in a solution of sheared* E. coli *DNA compared with one of T4 DNA* (containing the same concentration of nucleotides) *and thus its rate of reassociation is slower.* Studies of a variety of procaryotic DNA molecules showed that $C_0t_{0.5}$ is directly related to the size of the genome.

An unexpected result was obtained when mouse DNA was studied in this way. Mammalian genomes are about three orders of magnitude larger than those of *E. coli*, and so a $C_0t_{0.5}$ value of the order of 10^4 M sec seemed likely. A 10^{-4} M solution of DNA having such a $C_0t_{0.5}$ value would take 10^8 sec (about 3 years) to reassociate halfway. It came as a surprise to find that 10% of the mouse DNA was half-reassociated in a few seconds. Indeed, *this fraction of mouse DNA reassociates more rapidly than even the smallest viral DNAs, which indicates that it contains many repeated sequences.* An analysis of the C_0t curve suggested that this fraction of mouse DNA contains *of the order of a million copies of a repeating sequence of about 300 base pairs.* About 20% of the mouse DNA renatured at an intermediate rate, which was interpreted to indicate that this fraction contains between 10^3 and 10^4 copies of certain sequences. The other 70% of mouse DNA renatures very slowly. The $C_0 t_{0.5}$ value of this fraction shows that it consists of unique or nearly unique base sequences.

All eucaryotes studied thus far, perhaps except yeast, contain repetitive DNA, whereas procaryotes do not. For example, 30% of human DNA consists of sequences repeated at least twenty times. The proportion of highly repetitive, moderately repetitive, and single-copy DNA varies from one species to another.

HIGHLY REPETITIVE DNA (SATELLITE DNA) IS LOCALIZED AT CENTROMERES

Many highly repetitive DNAs can be isolated by density-gradient centrifugation because they have distinctive buoyant densities. For example, *Drosophila virilis* DNA exhibits a main band and three less-dense satellite bands. These satellite bands consist exclusively of highly repetitive DNA. In fact, each consists of a repeating heptanucleotide sequence:

5′—ACAAACT—3′	5′—ATAAACT—3′	5′—ACAAATT—3′
3′—TGTTTGA—5′	3′—TATTTGA—5′	3′—TGTTTAA—5′
Satellite I	**Satellite II**	**Satellite III**

The role of highly repetitive DNA is not known. However, the chromosomal localization of this fraction has been determined by the technique of *in situ hybridization,* which was devised by Joseph Gall and Mary Lou Pardue. Cells were immobilized under a thin layer of agar and treated with alkali to denature the DNA. This preparation was then incubated with tritium-labeled RNA transcribed in vitro using purified mouse satellite DNA as the template. Hybrids of this radioactive RNA and chromosomal regions containing satellite DNA were detected by autoradiography. A clear-cut result was obtained: *mouse satellite DNA is confined to centromeric regions.* The location of these sequences and the absence of corresponding RNA suggests that satellite sequences participate in aligning chromosomes during mitosis and meiosis.

GENES FOR RIBOSOMAL RNA ARE TANDEMLY REPEATED SEVERAL HUNDRED TIMES

The genes coding for ribosomal RNA molecules are highly distinctive in two respects. First, they are *tandemly repeated.* Nearly all eucaryotes have more than 100 copies of these genes. Second, most rRNA genes are located in specialized regions of chromosomes that are associated with *nucleoli.* Mutants lacking nucleoli synthesize very little rRNA and hence are not viable. Much is known about

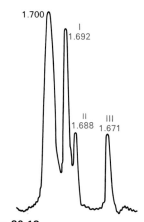

Figure 29-18
Three prominent bands of satellite DNA (ρ = 1.692, 1.688, and 1.671) are evident in this sedimentation pattern. DNA from *Drosophila virilis* was centrifuged to equilibrium in neutral CsCl. [After J. G. Gall, E. H. Cohen, and D. D. Atherton. *Cold Spring Harbor Symp. Quant. Biol.* 38(1974):417.]

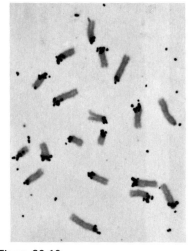

Figure 29-19
Autoradiograph of a mouse cell showing the localization of the satellite DNA. [Courtesy of Dr. Joseph Gall.]

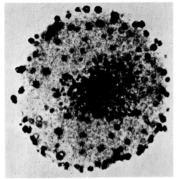

Figure 29-20
Micrograph of an isolated nucleus from a *Xenopus* oocyte stained to show the hundreds of nucleoli that are formed in the amplification of the ribosomal RNA genes. [From D. D. Brown and I. B. Dawid. *Science* 160(1968):272.]

these genes, largely because of the work of Max Birnstiel, Donald Brown, Oscar Miller, and their associates. The genes for the four kinds of ribosomal RNA—18S, 5.8S, 28S, and 5S rRNA—have been purified from the DNA of the African clawed toads *Xenopus laevis* and *Xenopus mulleri*. These creatures were chosen because their oocytes contain especially large amounts of these genes.

The genes for 18S, 5.8S, and 28S rRNA are clustered and tandemly repeated. In situ hybridization studies showed that these genes are located in nucleoli. The repeating unit consists of a *gene for a 40S precursor RNA* and a *spacer* (Figure 29-21). The 40S (8 kb) precursor is modified and cleaved to give the mature 18S, 5.8S, and 28S rRNAs (p. 705). The spacer is not transcribed (Figure 29-22). Somatic cells of toads contain about 500 copies of this repeating unit, all in a tandem array. In oogenesis, a striking selective amplification of these genes takes place. They are replicated several thousand times to form some 2×10^6 copies. Indeed, the genes for these rRNA molecules comprise nearly 75% of the total DNA in the oocyte. The amplified DNA is present as extrachromosomal circles that are bound to the very many new nucleoli. This *selective gene amplification* enables the oocyte to amass the 10^{12} ribosomes needed for the very rapid protein synthesis that occurs during cell cleav-

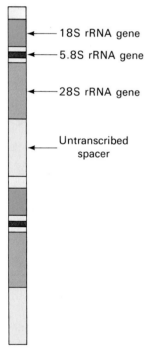

Figure 29-21
Organization of the genes for the 40S precursor of 18S, 5.8S, and 28S rRNAs in *Xenopus*. The tandemly repeated genes are separated by untranscribed spacers. The repeating unit is about 13 kb long.

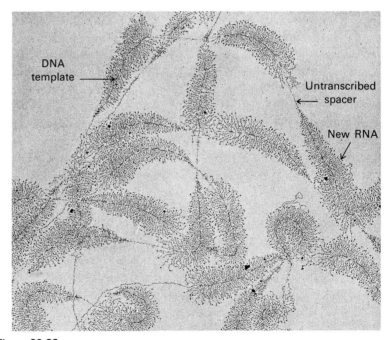

Figure 29-22
The tandem array of genes for 18S, 5.8S, and 28S rRNA is evident in this electron micrograph of nucleolar DNA. The dense axial fiber is DNA. The fine lateral fibers are newly synthesized RNA molecules with bound proteins. The tip of each arrowhead of the RNA molecules corresponds to an initiation point for transcription. The bare regions between arrowheads are the untranscribed spacers. [From O. L. Miller, Jr., and B. R. Beatty. Portrait of a gene. *J. Cell Physiol.* 74(supp. 1, 1969):225.]

ages. In the absence of gene amplification, it would take several centuries to assemble 10^{12} ribosomes!

The gene for the smallest ribosomal RNA molecule, 5S rRNA, is also tandemly repeated. Both somatic cells and oocytes contain about 24,000 copies of this gene, which codes for a 120-nucleotide RNA molecule. These genes are clustered at the ends of most of the toad chromosomes. Again, there is an untranscribed spacer region between the genes for this rRNA (Figure 29-23). Indeed, this spacer is several times as long as the gene. Curiously, it contains an untranscribed pseudogene with a base sequence that is similar to that of the gene itself. The role of the spacer in this and other eucaryotic genes is an enigma.

HISTONE GENES ARE CLUSTERED AND TANDEMLY REPEATED MANY TIMES

What is the arrangement of genes that code for proteins? The histone genes were among the first to be isolated and characterized because of the abundance of histone mRNA in rapidly dividing sea urchin embryos. These marine invertebrates develop from a zygote to a 1,000-cell blastula in about 10 hours, and so large amounts of histones must be synthesized for the assembly of new chromatin. Indeed, more than a quarter of the protein synthesized during early embryogenesis are histones. Moreover, histone mRNAs comprise about 70% of the messengers synthesized at this stage, and so it was relatively easy to isolate them. The kinetics of hybridization of histone mRNA with sea urchin DNA was then measured to determine the number of copies of histone genes. *The rate of hybridization was several hundred times as fast as would be given by single-copy sequences, which showed that the histone genes are highly reiterated.* The number of copies of the histone genes ranges from 300 to 1,000 in several species of sea urchin. In other organisms, the reiteration frequency is lower (Table 29-4). The number of copies of histone genes in an organism seems to be correlated with its need for the rapid synthesis of histone mRNAs.

Table 29-4
Reiteration frequency of the histone genes

Species	Number of copies
Sea urchin	300–1,000
Fruit fly (*Drosophila melanogaster*)	110
Clawed toad (*Xenopus laevis*)	20–50
Mouse	10–20
Chicken	10
Human	30–40

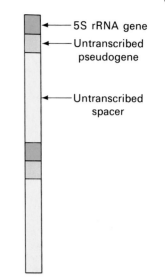

Figure 29-23
Organization of the genes for the 5S rRNA of *Xenopus laevis*. These tandemly repeated genes are also separated by untranscribed spacers. The repeating unit contains about 750 base pairs.

Figure 29-24
Light micrograph of a sea urchin embryo at the four-cell stage. [Courtesy of Dr. Annamma Spudich.]

How are the highly reiterated histone genes arranged in the genome? A fraction of DNA enriched in histone genes was prepared by taking advantage of the fact that these genes have a higher GC content (55%) than most of the sea urchin DNA (42%) and hence have a higher buoyant density. Cleavage of this enriched DNA by a restriction endonuclease yielded 7-kb fragments that hybridized to histone mRNAs. These fragments were then cloned in *E. coli* and analyzed by digestion with restriction enzymes and by electron microscopy. These studies revealed that *the genes for the five major histones are clustered in a basic 7-kb repeating unit* (Figure 29-25). The

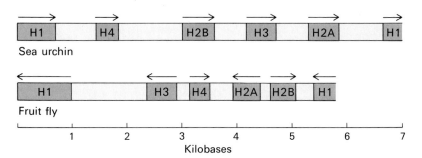

Figure 29-25
Map of the histone gene cluster in a sea urchin (*Strongylocentrotus purpuratus*) and a fruit fly (*Drosophila melanogaster*). Coding regions are shown in blue and spacers in yellow. The arrows denote the direction of transcription.

five coding regions of this repeating unit alternate with five spacers. The coding regions are not interrupted by intervening sequences, in contrast with other eucaryotic genes described earlier (p. 633). *This cluster of five histone genes is tandemly repeated many times.* The repetitions are very similar but not identical, which fits the finding that H1, H2A, and H2B are groups of closely related proteins rather than unique species. Different sets of histone genes are expressed in different tissues and at different times in development.

What is the functional significance of this arrangement of genes? Their repetition is undoubtedly important for the rapid synthesis of large amounts of histone mRNA. The reason for the tandem nature of their reiteration is not at all obvious. Clustering of the genes for the five types of histones may be important for coordinating their syntheses so that equimolar amounts of H2A, H2B, H3, and H4 and half as much H1 are produced.

MANY MAJOR PROTEINS ARE ENCODED BY SINGLE-COPY GENES

We have seen that the genes for ribosomal RNAs and histones are reiterated many times. Are there a large number of copies of other genes that code for abundant products? Donald Brown explored this question by isolating the gene for silk fibroin from the silkworm *Bombyx mori*. This system was chosen because the mRNA for silk fibroin has distinctive chemical characteristics. It is the template for repeating amino acid sequences that are rich in glycine.

Furthermore, very large amounts of silk fibroin are synthesized by a single type of giant cell at a particular stage of larval development. Fibroin mRNA was purified on the basis of its large size (9.1 kb) and its high content of G compared with that of rRNA and other RNA molecules. The gene for silk fibroin was partially purified by also taking advantage of its unusually high buoyant density. Hybridization of purified fibroin mRNA with DNA then revealed that *there is only one gene for silk fibroin per haploid genome.*

This result has important implications concerning gene expression and differentiation in eucaryotes. It shows that large amounts of a particular protein can be synthesized even if there is only one copy of its gene. *The single gene for silk fibroin is the template for the synthesis of 10^4 mRNA molecules, which are stable for days. Each mRNA is the template for the synthesis of 10^5 proteins. Thus, a single gene is sufficient for the synthesis of 10^9 protein molecules in four days.* It now appears that the reiterated genes for ribosomal RNAs and histones are exceptions rather than the common pattern. *The single gene found for silk fibroin is much more typical of genes that code for proteins, even abundant ones.* For example, reticulocytes contain one or just a few copies of the genes for the subunits of hemoglobin (p. 709). Likewise, large amounts of ovalbumin, the major protein in egg white, are synthesized by the oviduct of the laying hen, which contains only one copy of this gene per haploid genome. However, the possibility of selective gene amplification as a means of enhancing the synthesis of a specific protein by a particular cell type should be kept in mind. Some cultured tumor cells become resistant to analogs of folic acid by synthesizing two hundred times as much dihydrofolate reductase as does a sensitive cell. This increased level is a consequence of the *selective amplification* of the gene for dihydrofolate reductase. Clearly, eucaryotic genomes are highly dynamic systems.

Figure 29-26
A silkworm. [Courtesy of Dr. Karen Sprague.]

MOST SINGLE-COPY GENES ARE SEPARATED BY REPETITIVE SEQUENCES

About 70% of the DNA of a wide range of eucaryotes consists of unique sequences. What is the arrangement of these single-copy genes with respect to repetitive sequences? Analyses of eucaryotic chromosomes by a variety of experimental approaches have shown that *single-copy sequences are usually interspersed with moderately repetitive sequences that are typically three hundred base pairs long.* There are several thousand different moderately repetitive sequences. They recur several hundred times in the genome and comprise about 20% of the total DNA. The function of the interspersed moderately repetitive sequences of DNA is unknown. They may serve as sites for the binding of specific regulatory macromolecules, which could control the transcription of adjacent single-copy structural genes.

Table 29-5
Split eucaryotic genes

Gene	Number of introns
Ovalbumin	7
Ovomucoid	6
Conalbumin	17
α-Globin	2
β-Globin	2
δ-Globin	2
Immunoglobulin L-chain	2
Immunoglobulin H-chain	4

NEARLY ALL GENES CODING FOR PROTEINS IN HIGHER EUCARYOTES ARE SPLIT

The occurrence in eucaryotes of repetitive sequences and their interspersion with single-copy genes have no known counterparts in procaryotes. *Another striking difference in genome organization is the presence in higher eucaryotes of intervening sequences in nearly all genes that code for proteins.* As discussed earlier (p. 634), these intervening sequences (*introns*) are transcribed along with the coding sequences (*exons*) and then removed in the formation of the mature mRNA. The number of introns in split genes characterized thus far ranges from two to seventeen (Table 29-5 and Figure 29-27). Some of these introns are

Figure 29-27
Map of the conalbumin gene. The intervening sequences (introns), which are transcribed but absent from the mature mRNA, are shown in yellow.

long. For example, the mouse β-globin gene contains a 550-base-pair intron, which is longer than its exons. Two segments of an immunoglobulin gene are separated by an even larger intron, which is 1,250 base pairs long. In some genes, the total length of the introns exceeds that of the exons. For example, the gene for ovalbumin contains about 7,700 base pairs, whereas its mRNA is only 1,859 bases long. *In general, the intervening sequences of split genes are longer than their expressed sequences.* It is interesting to note that the base sequences of introns change more rapidly in evolution than those of exons. The base sequences of the smaller intron in the β-globin gene of mouse and rabbit are quite different. However, the lengths of these introns are similar and their positions in the gene are identical. Most significant, the junctions of introns and exons, which are the splicing sites, are conserved.

Some split genes for RNA molecules have also been found. For example, a transfer RNA gene in yeast contains a 14 base-pair intron next to the region that codes for the anticodon loop of the mature tRNA. A mitochondrial gene for a ribosomal RNA is also split. However, many genes for tRNA and rRNA are continuous. Furthermore, the genes for histones in sea urchins and fruit flies do not appear to contain intervening sequences. The split genes for proteins characterized thus far come from birds and mammals. It will be interesting to see whether lower eucaryotes also have many split genes.

The discovery of split genes, which came as a complete surprise, has raised some fascinating questions. Do split genes reflect the formation in evolution of new proteins from segments of old ones?

Do intervening sequences participate in the control of gene expression? Clearly, highly significant new areas of inquiry have been opened. Information about one important aspect of split genes, the excision of intervening sequences from the primary RNA transcript, is rapidly being acquired and will be discussed shortly.

RNA IN EUCARYOTIC CELLS IS SYNTHESIZED BY THREE TYPES OF RNA POLYMERASES

We turn now to transcription. RNA in eucaryotic cells is synthesized by three types of RNA polymerases that differ in their template specificity, localization, and susceptibility to inhibitors. The type-I polymerase is located in nucleoli, where it transcribes the tandem array of genes for 18S, 5.8S, and 28S ribosomal RNA. The other RNA molecules, 5S rRNA and all of the tRNAs, are synthesized by a different polymerase, type III, which is located in the nucleoplasm rather than in nucleoli. The large precursor RNA molecules that give rise to messenger RNAs are synthesized by RNA polymerase II, which is also located in the nucleoplasm. In contrast, all cellular RNA in procaryotes is synthesized by a single kind of RNA polymerase. A common feature of the eucaryotic RNA polymerases is that they are made up of two large and several small subunits.

Table 29-6
Eucaryotic RNA polymerases

Type	Localization	Cellular transcripts	Viral transcripts
I	Nucleolus	18S, 5.8S, and 28S rRNA	None identified
II	Nucleoplasm	mRNA precursors and hnRNA	mRNA precursors
III	Nucleoplasm	tRNA and 5S rRNA	Small RNAs

α-AMANITIN, A MUSHROOM POISON, IS A POTENT INHIBITOR OF RNA POLYMERASE II

More than a hundred deaths result each year in the world from the ingestion of poisonous mushrooms, particularly *Amanita phalloides,* which is also called the death cup or the destroying angel. One of the toxins in this mushroom is *α-amanitin,* a cyclic octapeptide that contains several unusual amino acids. α-Amanitin binds very tightly ($K = 10^{-8}$ M) to RNA polymerase II and thereby blocks the formation of precursors of mRNA. The type-III polymerase is inhibited by higher concentrations of α-amanitin (10^{-6} M), whereas the type-I enzyme is insensitive to this toxin. α-Amanitin blocks the elongation step in the synthesis of RNA.

Figure 29-28
Amanita phalloides, a poisonous mushroom that contains α-amanitin. [After G. Lincoff and D. H. Mitchel, *Toxic and Hallucinogenic Mushroom Poisoning* (Van Nostrand Reinhold, 1977), p. 30.]

$$\alpha\text{-Amanitin}$$
(Inhibitor of RNA polymerase II)

SPECIFIC GENES CAN BE ACTIVATED FOR TRANSCRIPTION

Gene expression in eucaryotes, as in procaryotes, is controlled to a large extent by the pattern of transcription. Some of the most direct evidence has come from studies of the chromosomes of developing insects. The giant *polytene chromosomes* from *Drosophila* salivary glands contain more than a thousand aligned DNA molecules that have not yet segregated. Each polytene chromosome has a characteristic series of bands (chromomeres) that can be seen under a light microscope. In the development of a larva into a pupa, certain bands become transiently enlarged (puffed) because the DNA in that region is converted from a condensed form into a dispersed one (Figure 29-29). *Puffs* correspond to transcriptionally active regions. A striking finding is that puffing can be induced in vitro in isolated salivary glands by *ecdysone,* an insect steroid hormone. Specific bands enlarge and then contract in a precise temporal sequence and there are concomitant changes in the population of mRNAs being synthesized.

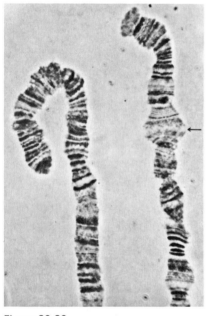

Figure 29-29
Puff formation in a polytene chromosome during development. The arrow marks a prominent puff in a *Drosophila* chromosome. [Courtesy of Dr. Joseph Gall.]

THREE KINDS OF RIBOSOMAL RNAs ARE FORMED BY THE PROCESSING OF A SINGLE PRIMARY TRANSCRIPT

The RNA molecules synthesized by the three kinds of RNA polymerases are called *primary transcripts.* These nascent RNAs are extensively modified in the nucleus before they enter the cytosol as mature rRNA, tRNA, and mRNA molecules. The formation of ribosomal RNAs from a primary transcript has been studied in detail and is the best-understood processing pathway. As described earlier, the genes for 18S, 5.8S, and 28S rRNA are clustered in a unit that is tandemly repeated many times. In fact, this cluster of three genes is transcribed by RNA polymerase I in nucleoli to give a 45S RNA (Figure 29-30). This 13-kb precursor is enzymatically modified and cleaved to yield the mature 18S, 5.8S, and 28S rRNAs

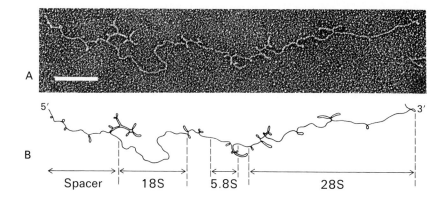

A

B

Spacer | 18S | 5.8S | 28S

Figure 29-30
The 45S precursor of ribosomal RNA in HeLa cells has a highly distinctive pattern of hairpin loops when spread and examined by electron microscopy. This pattern makes it feasible to map the linear arrangement of 28S and 18S RNA molecules derived from this precursor. Part A is an electron micrograph of the 45S precursor (bar = 2000 Å). Part B is a tracing of the molecule shown in part A. [From P. K. Wellauer and I. B. Dawid. *Proc. Nat. Acad. Sci.* 70(1973):2828.]

(Figure 29-31). About a hundred nucleotides are methylated, nearly all on the 2'-hydroxyl groups of their ribose units. The highly specific methylation sites are conserved in the rRNAs of eucaryotes as distant as yeast and fruit flies. In addition, more than a hundred uridines in the precursor RNA are isomerized to pseudouridines. Many ribosomal proteins are associated with these RNAs and their precursors during the maturation process. Indeed, interaction of the RNA with ribosomal proteins may render certain sites susceptible to the action of nucleases.

MESSENGER RNAs ARE SELECTIVELY GENERATED FROM LARGE NUCLEAR RNA PRECURSORS (hnRNA)

The formation of ribosomal RNAs in eucaryotes is similar to that of its procaryotic counterparts (p. 612). In contrast, there are some important differences between eucaryotic and procaryotic mRNAs:

1. *Primary transcripts in eucaryotes are not used directly as mRNA.* Rather, they are extensively processed before being transported from the nucleus to the cytosol. *Translation and transcription are spatially and temporally separated in eucaryotes, whereas they are closely coupled in procaryotes.*

2. Primary transcripts in eucaryotes range in size from about 2 to 20 kb, and so they have been called *heterogeneous nuclear RNA* (hnRNA). These primary transcripts are usually several times as long as the mRNAs derived from them. Extensive splicing and cleavage take place in the generation of eucaryotic mRNAs. It is not known whether hnRNA has a role apart from being a precursor of mRNA.

3. Eucaryotic mRNAs contain modified nucleotide *caps* at their 5' ends. Most of them also have long *poly A tails* at their 3' ends.

4. Eucaryotic mRNAs are *monocistronic*—that is, they are templates for the synthesis of a single polypeptide chain. In contrast, many procaryotic mRNAs are polycistronic (e.g., *lac* mRNA is the template for three polypeptide chains).

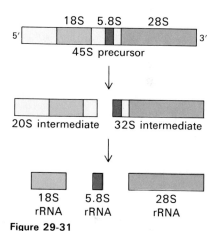

18S 5.8S 28S

5' ─── 45S precursor ─── 3'

20S intermediate   32S intermediate

18S rRNA   5.8S rRNA   28S rRNA

Figure 29-31
Formation of mammalian ribosomal RNAs from a primary transcript. Spacer regions are shown in yellow.

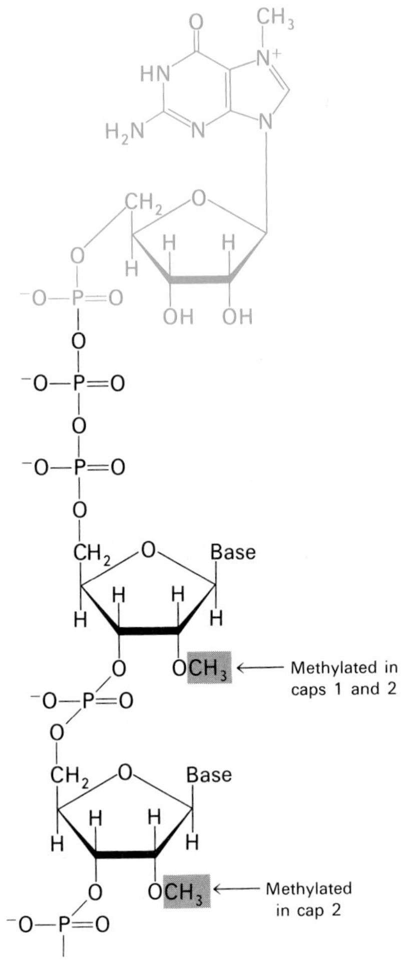

Figure 29-32
Structure of caps at the 5′ end of eucaryotic mRNAs. All caps contain 7-methylguanylate (shown in green) attached by a pyrophosphate linkage to the 5′ end. None of the riboses are methylated in cap 0, one is methylated in cap 1, and two are methylated in cap 2.

5. Not only does the population of mRNA molecules in a eucaryotic cell depend on the rates of transcription of particular genes, but eucaryotic cells have another level of decision-making as well: *whether to degrade and discard a primary transcript or process it into a mature mRNA and transport it into the cytosol.*

EUCARYOTIC mRNAs CONTAIN 5′ CAPS AND USUALLY HAVE 3′ POLY A TAILS

The 5′ end of all known eucaryotic mRNAs, but not of any tRNAs or rRNAs, is modified in a special way. 7-Methylguanosine is joined to mRNA by an unusual 5′–5′ pyrophosphate linkage (Figure 29-32). This highly distinctive structure, called a *cap*, is acquired by the primary transcript. The 5′ triphosphate end of the nascent chain is hydrolyzed to a diphosphate, to which a guanylate unit is transferred from GTP. The N-7 nitrogen of this terminal guanine is then methylated by *S*-adenosylmethionine to form *cap 0*. The adjacent riboses may be methylated to form *cap 1* and *cap 2* (Figure 29-32). These caps contribute to the stability of mRNAs by protecting their 5′ ends from phosphatases and nucleases. In addition, caps enhance the translation of mRNA by eucaryotic protein-synthesizing systems.

Most eucaryotic mRNAs also have *poly A tails* at their 3′ ends. A poly A polymerase adds some 150 to 200 nucleotides to primary transcripts that have a 3′ terminal GC and an AAUAA sequence about 20 residues away. Studies of mRNAs injected into *Xenopus* eggs have shown that *a poly A tail enhances the stability of an mRNA but is not required for its translation.* The absence of a poly A tail in histone mRNAs also shows that it is not essential for the transport of mRNA from the nucleus to the cytosol.

SPLICING ENZYMES ARE VERY PRECISE IN REMOVING INTRONS FROM PRIMARY TRANSCRIPTS OF SPLIT GENES

In the splicing of RNA, two phosphodiester bonds are broken and a new one is formed. The endonuclease and ligase activities probably reside in the same enzyme complex, and the splicing reaction is likely to be concerted. Splicing enzymes are now being isolated. It is evident that they must be highly precise in their action. A one-nucleotide slippage in a splice point would shift the reading frame on the 3′ side of the splice, which would yield an entirely different amino acid sequence. How do splicing enzymes recognize their targets? The base sequences of a number of intron-exon junctions of RNA transcripts for proteins are now known (Table 29-7). These junctions have a common structural motif: *the base sequence of an intron begins with GU and ends with AG.* It is noteworthy that the same splicing signal is found in chicken, rabbits, mice, and SV40 virus, which infects monkey and human cells.

Table 29-7
Base sequences of splice points in transcripts containing introns

Gene region	Exon	Intron	Exon
Ovalbumin intron 2	U A A G G U G A	∿∿∿∿∿∿∿∿∿	A C A G G U U G
Ovalbumin intron 3	U C A G G U A C	∿∿∿∿∿∿∿∿∿	U C A G U C U G
β-Globin intron 1	G C A G G U U G	∿∿∿∿∿∿∿∿∿	U C A G G C U G
β-Globin intron 2	C A G G G U G A	∿∿∿∿∿∿∿∿∿	A C A G U C U C
Immunoglobulin λ₁ intron 1	U C A G G U C A	∿∿∿∿∿∿∿∿∿	G C A G G G G C
SV40 virus early T-antigen	U A A G G U A A	∿∿∿∿∿∿∿∿∿	U U A G A U U C

The sequences surrounding the splice points in transfer RNA molecules are quite different, which suggests that there are at least two splicing enzymes, one for the production of mRNA, and the other for tRNA. The splicing reaction for tRNA appears to be very similar in species that are evolutionarily far apart. Cloned genes for a yeast tRNA were microinjected into *Xenopus* oocytes to determine whether a gene from a unicellular eucaryote can be transcribed and processed in an amphibian. The striking result is that the yeast gene is transcribed and properly processed in *Xenopus,* even though the genes for this tRNA are quite different in the two species. In particular, a 14-nucleotide intervening sequence was properly spliced out (Figure 29-33). Thus, *the specificity of splicing enzymes has been conserved over a very long evolutionary period.*

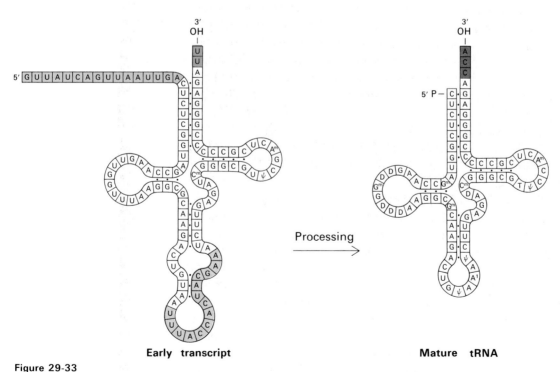

Figure 29-33
Processing of a precursor of yeast tyrosine tRNA. A 14-nucleotide intervening sequence (shown in yellow) is removed and a number of bases are modified. The 92-nucleotide product is derived from a 108-nucleotide primary transcript by cleavage of a 5′ leader and addition of CCA to the 3′ end.

THE BASE SEQUENCES OF SEVERAL MESSENGER RNAs ARE NOW KNOWN

A variety of eucaryotic mRNAs have been purified, and the base sequences of several have been determined. The starting point is a cell rich in the mRNA of interest. For example, reticulocytes are abundant in globin mRNAs, chicken oviduct in ovalbumin mRNA, plasma cells in immunoglobulin mRNAs, and sea urchin embryos in histone mRNAs. The isolation of a particular mRNA begins with the removal of protein and DNA from a cell extract. Most mRNAs can then be separated from other RNA species by taking advantage of their poly A tails. These mRNAs bind strongly to columns containing covalently attached poly U or poly T. The mRNAs eluted from such an affinity column can then be separated according to size by gel electrophoresis or sedimentation. Alternatively, a particular mRNA can be isolated by immunoprecipitation. For example, the addition of antibody specific for ovalbumin to a protein-synthesizing extract leads to the precipitation of polysomes containing ovalbumin mRNA. An mRNA preparation can be assayed by adding an aliquot to a cell-free protein-synthesizing system that requires exogenous messenger. Wheat germ extracts are useful in this regard.

The availability of a purified mRNA opens the way to many interesting experiments. First, the number of genes of a particular kind in the genome can be determined by hybridizing the corresponding purified mRNA (or a complementary DNA copy of it) with the chromosomal DNA. Second, DNA fragments containing a particular gene can be identified by hybridizing them with the mRNA. These fragments can then be cloned (p. 766) to obtain large amounts of the gene and its neighbors on the chromosome. Third, intervening sequences in the gene can be detected by electron microscopy (p. 633). Fourth, the mRNA can be sequenced to identify its regulatory signals.

The sequence of ovalbumin mRNA has recently been determined. This mRNA contains 1,859 nucleotides between its *5' cap* and *3' poly A tail* (Figure 29-34). *The 1,158 nucleotides that code for the protein are flanked on either side by untranslated sequences,* a short one on the 5' side and a long one on the 3' side. The 64-nucleotide untranslated region on the 5' side, in addition to the cap and the AUG start codon, interacts with the protein-synthesizing machinery and participates in the initiation of protein synthesis. A provocative finding is that it contains a sequence that is complementary to the 3' end of 18S rRNA. Recall that in procaryotes the initiator sequence on the mRNA pairs with the 3' end of 16S rRNA (p. 657). The role of the very long untranslated sequence on the 3' side is unknown. A distinctive feature of this 637-nucleotide region is that it contains many short repeating sequences. The poly A tail in ovalbumin mRNA and other mRNAs of known sequence is immediately preceded by GC, and some 20 bases before, by AAUAAA.

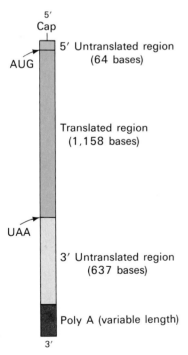

Figure 29-34
Schematic diagram of the organization of ovalbumin mRNA from chicken.

THE EUCARYOTIC RIBOSOME (80S) IS MADE UP OF A SMALL (40S) AND A LARGE (60S) SUBUNIT

The machinery of protein synthesis in eucaryotes is analogous to that of procaryotes, but the constituent proteins and RNAs are different and there are more of them in eucaryotes. The ribosomes in the cytosol of eucaryotic cells are somewhat larger than those of bacteria. Eucaryotic ribosomes (Figure 29-35) have a sedimentation coefficient of 80S rather than 70S. Like bacterial ribosomes, they dissociate into large subunits (60S) and small subunits (40S). The 40S subunit contains an 18S RNA molecule and about thirty proteins. The other three ribosomal RNA species—5S, 5.8S, and 28S—are located in the 60S subunit, which also contains about forty-five proteins.

The initiation, elongation, and termination phases of translation are generally similar in eucaryotes and procaryotes. However, some of the reaction mechanisms differ in detail. For example, *eucaryotes use a special tRNA for initiation (called tRNA$_i^{Met}$), but it is not formylated.* The sensitivity of eucaryotes and procaryotes to some inhibitors of translation also differs. *Cycloheximide* inhibits elongation in eucaryotes only, whereas erythromycin blocks this step in procaryotes only. It is interesting to note that *ribosomes in mitochondria and chloroplasts are more like bacterial ribosomes than the other ribosomes in the adjoining cytosol.* Furthermore, mitochondria and chloroplasts use a formylated initiator tRNA and are sensitive to most inhibitors that selectively block translation in procaryotes.

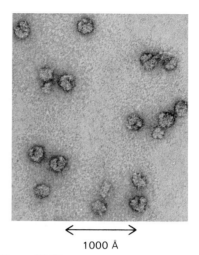

1000 Å

Figure 29-35
Electron micrograph of eucaryotic 80S ribosomes. [Courtesy of Dr. Miloslav Boublik.]

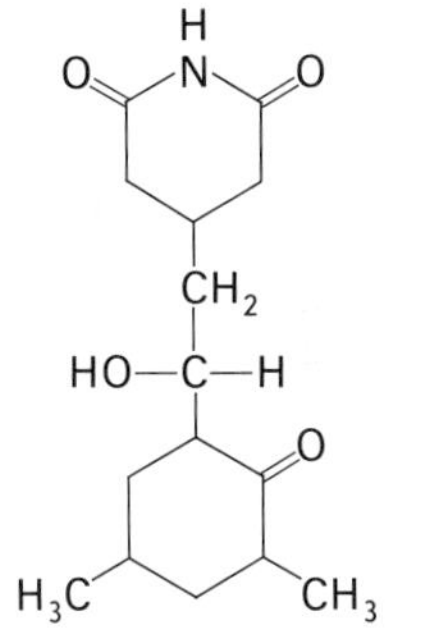

THALASSEMIAS ARE GENETIC DISORDERS OF HEMOGLOBIN SYNTHESIS

The study of hemoglobin has contributed a great deal to our understanding of protein structure and function (Chapters 4 and 5). Likewise, investigations of the genes for hemoglobin and their mode of expression are proving to be sources of insight into eucaryotic gene function in general. In development, embryonic hemoglobins are replaced by fetal hemoglobin (Hb F, $\alpha_2\gamma_2$) and then by adult hemoglobin (Hb A, $\alpha_2\beta_2$). Adults also produce a small amount of Hb A_2, which has the subunit structure $\alpha_2\delta_2$. Recall that Hb F has a higher oxygen affinity than does Hb A because it binds diphosphoglycerate less tightly (p. 73). This higher oxygen affinity optimizes transfer of oxygen from the maternal to the fetal circulation. Hb F is actually a mixture of two species, one with glycine and the other with alanine at residue 136 of the γ chain. These subunits are designated $^G\gamma$ and $^A\gamma$, respectively.

All of the hemoglobin genes have been mapped. There are two closely linked α genes per haploid genome. These genes appear to be identical except in rare mutants. The other globin genes are clustered in the sequence $^G\gamma$-$^A\gamma$-δ-β on a different chromosome (Fig-

Figure 29-36
Structure of cycloheximide, an inhibitor of the elongation step of protein synthesis in eucaryotes but not procaryotes.

Figure 29-37
Map of the human γ-, δ-, and β-globin genes.

ure 29-37). It will be interesting to learn why these functionally related genes are next to each other on a chromosome. One possibility is that the switch from the γ genes to the δ and β genes in development depends on their proximity. The linkage of these genes may also reflect their evolutionary history. The γ, δ, and β genes probably have a common ancestral gene, which became tandemly repeated and then diverged.

The *thalassemias* are a group of hereditary anemias in which *the rate of synthesis of one of the chains of hemoglobin is diminished*. The disease was named thalassemia, derived from the Greek word for the sea, because many patients who have it are of Mediterranean ancestry. Thalassemia major and minor refer to the homozygous and heterozygous states, respectively. The prefix α or β designates which chain is synthesized at an abnormally slow rate. Studies of globin mRNA and cloned globin DNA are beginning to provide an understanding of these disorders. The molecular defect in several thalassemias has been delineated:

1. *Gene deletion.* In some α-thalassemias, one or both α-globin genes are deleted.

2. *Instability of mRNA.* In hemoglobin Constant Spring, the α chain has 172 residues instead of 141 because of a mutation from UAA (stop) to CAA (glutamine). Translation of part of what is normally a noncoding region somehow renders the altered mRNA susceptible to the action of nucleases.

3. *Defective chain initiation.* In some β-thalassemias, initiation of translation is abnormally slow, probably because of a defect in the 5′ noncoding region.

4. *Premature chain termination.* A β-thalassemia arises from a single base change from AAG (lysine) to UAG (stop) at residue 17.

5. *Underproduction of mRNA.* In many β-thalassemias, the β-globin gene is present, but very little β-globin mRNA is produced. The cause of this underproduction is being actively studied. An intriguing possibility is that intervening sequences are not properly excised in some β-thalassemias.

TRANSLATION IS REGULATED BY A PROTEIN KINASE CASCADE THAT INACTIVATES AN INITIATION FACTOR

Extracts of reticulocytes synthesize hemoglobin subunits at a rapid rate until the supply of heme is depleted. In the absence of heme, protein synthesis halts because of the abrupt formation of an enzymatic inhibitor of protein synthesis. In fact, this inhibitor is a protein kinase. The target of this kinase is eIF-2, an initiation factor that binds GTP and delivers met-tRNA$_f$ to the 40S ribosomal subunit. Inactivation of eIF-2 by phosphorylation leads to a halt in the

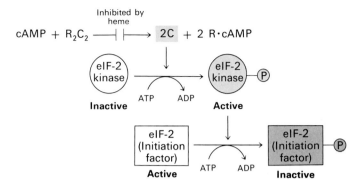

Figure 29-38
Phosphorylation cascade leading to the inactivation of initiation factor eIF-2.

initiation of protein synthesis. How does heme regulate the activity of this kinase? The effect is mediated by another kinase (Figure 29-38). The eIF-2 kinase that modifies the initiation factor itself exists in two forms, an inactive dephosphorylated state and an active phosphorylated state. The phosphorylation of eIF-2 kinase is catalyzed by a cyclic AMP-dependent kinase, which contains two regulatory (R) subunits and two catalytic (C) ones. The inactive R_2C_2 complex is dissociated by cyclic AMP into two catalytically active C subunits and two R subunits. *Heme blocks this dissociation and so the two kinases in the pathway do not become activated. Consequently, eIF-2 does not become phosphorylated, and so it remains active in initiating protein synthesis. This protein kinase cascade is reminiscent of the control of glycogen metabolism* (p. 372). Another similarity is that the regulatory effects of these kinases are reversed by specific phosphatases.

DIPHTHERIA TOXIN BLOCKS PROTEIN SYNTHESIS IN EUCARYOTES BY INHIBITING TRANSLOCATION

Diphtheria was a major cause of death in childhood prior to the advent of effective immunization. The lethal effects of this disease are due mainly to a toxin by *Corynebacterium diphtheriae,* a bacterium that grows in the upper respiratory tract. The structural gene for the toxin comes from a lysogenic phage that is harbored by some strains of *C. diphtheriae.* A few micrograms of this 61-kdal toxin is usually lethal in an unimmunized person because it inhibits protein synthesis. Specifically, diphtheria toxin blocks the elongation phase of protein synthesis in eucaryotes by inactivating the elongation factor necessary for translocation. This factor in eucaryotes, called elongation factor 2 (EF-2), or *translocase,* has a role analogous to that of EF-G in bacteria. The translocase is required for the GTP-driven movement of peptidyl tRNA from the A site to the P site and for the associated movement of messenger RNA immediately following the formation of a peptide bond. The mechanism of inactivation of the translocase by diphtheria toxin is especially interesting. *The toxin catalyzes the covalent modification of the translocase.* NAD^+ donates its adenosine diphosphate ribose moiety (ADPR) to the translocase, and nicotinamide is released.

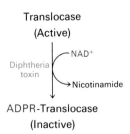

Diphtheria toxin is a bipartite molecule. It can be split into two pieces, a 21-kdal A fragment and a 40-kdal B fragment. *The B domain binds to the surface of susceptible cells, whereas the A domain catalyzes the ADP-ribosylation of translocase.* Specifically, the B domain binds to ganglioside G_{MI} on the plasma membrane, which enables the catalytic domain to enter the cell. In fact, the bound toxin is cleaved so that the B fragment remains on the cell surface, whereas the hydrophilic A fragment is transported into the cytosol. It is interesting to note that a variety of other toxins, such as cholera toxin (p. 846), also consist of a cell-surface binding domain and a catalytic domain that inactivates an important cell component.

RIBOSOMES BOUND TO THE ENDOPLASMIC RETICULUM SYNTHESIZE SECRETORY AND MEMBRANE PROTEINS

In eucaryotic cells, some ribosomes are free in the cytosol, whereas others are bound to an extensive membrane system called the *endoplasmic reticulum* (ER). The region that binds ribosomes is called the *rough ER* because of its studded appearance (Figure 29-39), in contrast with the smooth ER, which is devoid of ribosomes. Cells secreting large amounts of protein, such as pancreatic acinar cells, have a highly developed rough ER. In fact, all known secretory proteins are synthesized by ribosomes bound to the ER. Ribosomes attached to this membrane system also synthesize many of the proteins of the plasma membrane and of organelles such as lysosomes.

Membrane-bound ribosomes in the oocytes of hibernating lizards form crystalline sheets (Figure 29-40). These ordered arrays are now being analyzed by three-dimensional image-reconstruction techniques. A low-resolution map shows that both the large (60S) and small (40S) subunits of the ribosome lie close to the membrane surface. The large subunit has a protrusion extending into the membrane (Figure 29-41). The rough ER (but not the smooth ER) contains two transmembrane proteins called *ribophorins* that interact specifically with the large ribosomal subunit.

Three fundamental questions arise concerning the synthesis and fate of proteins formed by ribosomes attached to the rough ER:

1. *Are there two classes of ribosomes—one free in the cytosol, the other membrane bound—or are all ribosomes intrinsically the same?* If there is only one class of ribosomes, what determines whether a particular ribosome is free or bound to the rough ER?

2. *How does the nascent polypeptide chain emerging from a membrane-bound ribosome traverse the permeability barrier of the rough ER?* For example, secretory proteins such as the pancreatic zymogens (p. 158) are located inside the lumen of the ER soon after being synthesized.

3. *What determines the destination of a protein that is synthesized by a membrane-bound ribosome?* Some of these proteins are exported,

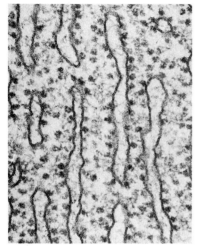

Figure 29-39
Electron micrograph of the rough endoplasmic reticulum. [Courtesy of Dr. George Palade.]

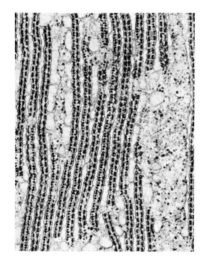

Figure 29-40
Electron micrograph of crystalline sheets of membrane-bound ribosomes in the oocytes of hibernating lizards. [Courtesy of Dr. Nigel Unwin.]

whereas others are directed to an organelle inside the cell. Proteins synthesized by the rough ER also emerge as integral membrane proteins in the plasma membrane and intracellular membranes.

SIGNAL SEQUENCES ENABLE SECRETORY PROTEINS TO CROSS THE ENDOPLASMIC RETICULUM MEMBRANE

Studies of the protein-synthesizing activities of ribosomes in cell-free systems provided the answer to the first of these questions. Free ribosomes from the cytosol were isolated and then added to rough ER membranes that had been stripped of their ribosomes. This reconstituted system actively synthesized secretory proteins when presented with the appropriate mRNAs and other soluble factors. Likewise, ribosomes derived from the rough ER were fully active in synthesizing proteins that are normally released into the cytosol. Furthermore, there were no detectable structural differences between free ribosomes and ribosomes derived from the rough ER. *Thus, membrane-bound ribosomes and free ribosomes are intrinsically the same. Whether a particular ribosome is free or attached to the rough ER depends only on the kind of protein it is making.*

What then is the marker on a nascent protein that determines whether its associated ribosome is to be free in the cytosol or bound to the ER membrane? In 1970, David Sabatini and Günter Blobel postulated that *the signal for attachment is provided by a sequence of amino acid residues near the amino-terminus of the nascent polypeptide chain.* This *signal hypothesis* (Figure 29-42) was soon supported by the finding of Cesar Milstein and George Brownlee that an immunoglobulin chain synthesized in vitro by free ribosomes contained an amino-terminal sequence of twenty residues that was absent from the mature protein synthesized in vivo. Blobel then found that all of the major secretory proteins of the pancreas contain amino-terminal extensions of some twenty residues when synthesized in vitro by free ribosomes. The signal sequences of many secretory proteins are now

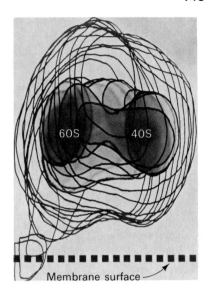

Figure 29-41
Low-resolution image of a membrane-bound ribosome. This image was reconstructed from a series of electron micrographs of ordered arrays of ribosomes at different tilt angles. [After a diagram kindly provided by Dr. Nigel Unwin.]

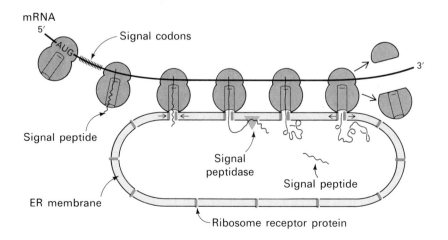

Figure 29-42
Signal hypothesis for the biosynthesis of secretory and membrane proteins. In this model, the amino-terminal sequence (shown in red) of the nascent polypeptide chain binds the ribosome to the ER membrane. The signal sequence is then excised by a peptidase on the luminal side of the ER. [After G. Blobel. In *International Cell Biology*, B. R. Brinkley and K. R. Porter, eds. (Rockefeller University Press, 1977), p. 318.]

known. They range in length from about fifteen to thirty residues and contain a high proportion of nonpolar residues (Figure 29-43).

H$_3^+$N-Met-Arg-Ser-Leu-Leu-Ile-Leu-Val-Leu-Cys-
-Phe-Leu-Pro-Leu-Ala-Ala-Leu-Gly⌇Gly-Lys-
Prelysozyme

Figure 29-43
Signal sequences of two secretory proteins. Hydrophobic residues are shaded yellow. The cleavage site is marked in red.

H$_3^+$N-Met-Lys-Trp-Val-Thr-Phe-Leu-Leu-Leu-Leu-
-Phe-Ile-Ser-Gly-Ser-Ala-Phe-Ser⌇Arg-
Preproalbumin

Hydrophobic signal sequences probably adopt conformations that are recognized by channel proteins in the ER membrane. *It seems likely that the nascent polypeptide chain is actively threaded through a tunnel in the ER membrane as it is being synthesized.* The signal sequence is then cleaved by a peptidase on the luminal side of the ER (see Figure 29-42). Nascent chains destined to become integral membrane proteins probably contain specific sequences that block the transfer of the polypeptide across the ER membrane before the carboxyl-terminal end is reached. In contrast, the entire polypeptide chain is transported across the ER membrane in the case of secretory proteins.

A key feature of the signal sequence mechanism is that transfer of the polypeptide across the ER membrane is coupled to translation. *However, some proteins can cross membranes after their synthesis is finished.* For example, most mitochondrial and chloroplast proteins are encoded by nuclear genes and synthesized by free ribosomes. These proteins are released into the cytosol and then traverse the organelle membrane. Transport of these proteins is clearly *posttranslational* rather than cotranslational. It is interesting to note that these mitochondrial and chloroplast proteins, like secretory proteins, contain N-terminal sequences that are removed soon after their emergence through the membrane.

GLYCOPROTEINS ACQUIRE THEIR CORE SUGARS FROM DOLICHOL DONORS IN THE ENDOPLASMIC RETICULUM

Nearly all proteins synthesized by ribosomes bound to the ER acquire covalently attached carbohydrate units. In contrast, soluble proteins synthesized by free ribosomes in the cytosol are almost always devoid of carbohydrate. As mentioned earlier, sugar units may orient glycoproteins in membranes (p. 222). In addition, carbohydrate groups may participate in determining the destination of a glycoprotein. A typical glycoprotein contains one or a few oligosaccharide units linked to asparagine side chains by *N*-glycosidic bonds. Less common is the attachment of sugars to serine and threonine side chains by *O*-glycosidic bonds. The sugar directly bonded

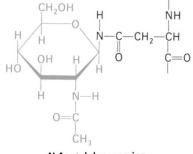

N-Acetylglucosamine
linked to an
asparagine residue
by an *N*-glycosidic bond

to asparagine residues is invariably *N*-acetylglucosamine, whereas *N*-acetylgalactosamine is the sugar joined to serine and threonine residues.

Many kinds of oligosaccharides are found in glycoproteins (Figure 29-44). However, a unifying pattern underlies this diversity: *carbohydrate units attached to asparagine residues (N-linked) have a common*

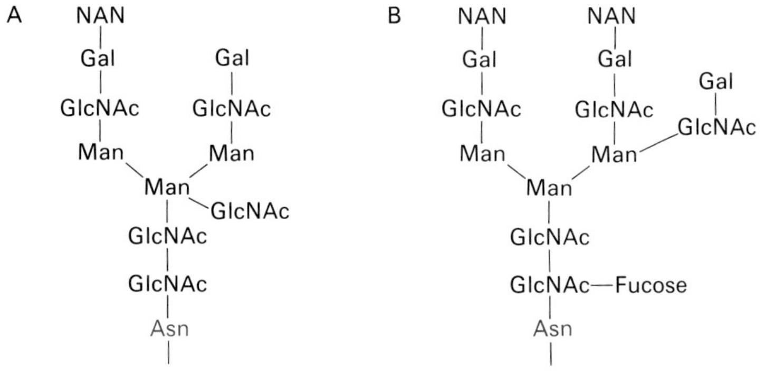

N-Acetylgalactosamine
linked to a
serine residue
by an *O*-glycosidic bond

Figure 29-44
Structure of an asparagine-linked oligosaccharide unit in (A) human immunoglobulin and (B) porcine thyroglobulin.

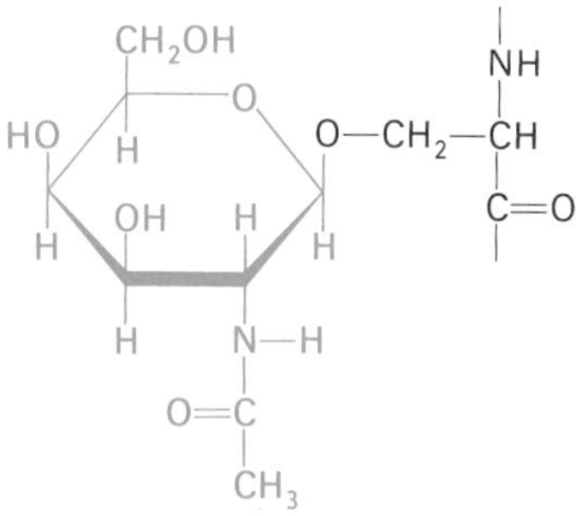

inner-core structure. This unifying motif is an expression of the mode of biosynthesis of oligosaccharide units of glycoproteins. *A common oligosaccharide block* (**Figure 29-45**) *is transferred from an activated lipid carrier to the growing polypeptide chain on the luminal side of the ER mem-*

Abbreviations for sugars—

Fuc	fucose
Gal	galactose
Glc	glucose
GlcNAc	*N*-acetylglucosamine
Man	mannose
NAN	*N*-acetylneuraminidate (sialic acid)

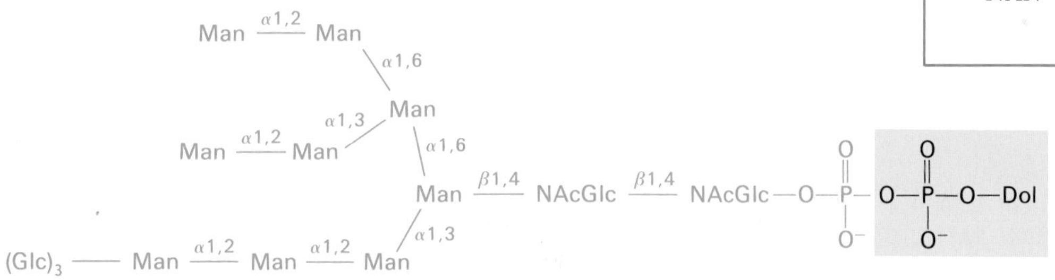

Figure 29-45
Structure of the activated core oligosaccharide (shown in green). The carrier (shown in yellow) is dolichol phosphate.

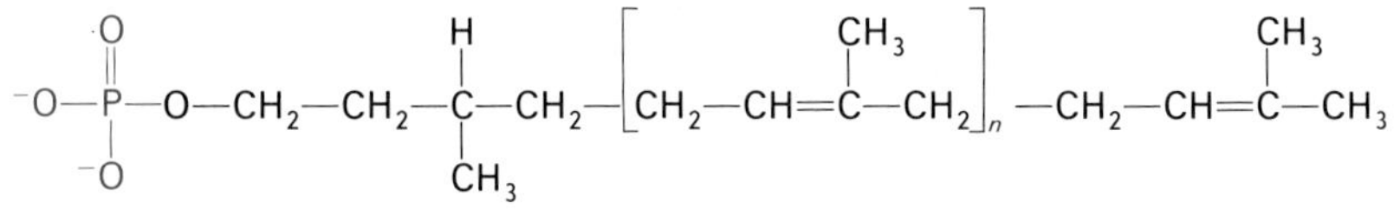

brane. The carrier is *dolichol phosphate,* a very long chain lipid containing some twenty isoprene (C_5) units. The terminal phosphoryl group of this highly hydrophobic carrier is the site of attachment of the activated oligosaccharide.

Dolichol phosphate
($n = 15$ to 19)

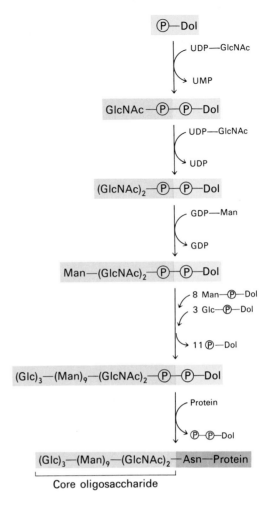

Figure 29-46
An activated core oligosaccharide is built by the sequential addition of single sugar units. This block is then transferred to an asparagine side chain of a nascent protein in the ER lumen.

Dolichol phosphate acquires an oligosaccharide unit consisting of two *N*-acetylglucosamines, nine mannoses, and three glucoses by the sequential addition of monosaccharides (Figure 29-46). The activated sugar donors in these reactions are derivatives of UDP, GDP, and dolichol. A series of specific transferases catalyze the synthesis of the activated core oligosaccharide, which is then transferred en bloc to a specific asparagine residue of the growing polypeptide chain. The activated oligosaccharide and the specific transferase are located on the luminal side of the ER, which accounts for the fact that proteins in the cytosol are not glycosylated. An asparagine residue can accept the oligosaccharide only if it is part of an Asn-X-Ser or Asn-X-Thr sequence. Two of the three glucose residues in the attached oligosaccharide are rapidly trimmed from most glycoproteins while they are still in the ER.

Dolichol pyrophosphate released in the transfer of the oligosaccharide to the protein is recycled to dolichol phosphate by the action of a phosphatase. This conversion is blocked by *bacitracin*, an antibiotic (p. 778). Another interesting antibiotic inhibitor is *tunicamycin*, a hydrophobic analog of UDP-*N*-acetylglucosamine.

Tunicamycin blocks the addition of *N*-acetylglucosamine to dolichol phosphate, the first step in the formation of the core oligosaccharide.

GLYCOPROTEINS ARE MODIFIED AND SORTED IN THE GOLGI APPARATUS

Proteins in the lumen of the ER and in the ER membrane are transported to the *Golgi apparatus,* which is a stack of flattened membranous sacs (Figure 29-47). The core oligosaccharides of glycopro-

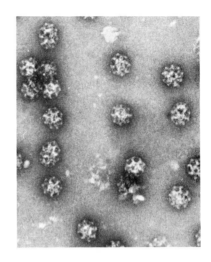

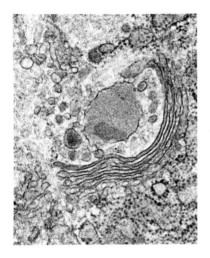

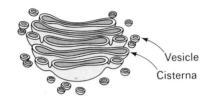

Figure 29-47
Electron micrograph (left) and schematic diagram (above) of the Golgi apparatus. Glycoproteins are modified, sorted, and packaged in this organelle. [Electron micrograph courtesy of Lynne Mercer.]

Figure 29-48
Electron micrograph of coated vesicles (top). Interpretive diagram of a coated vesicle (bottom). [Courtesy of Barbara Pearse.]

teins are trimmed and new sugars are added in this organelle. The Golgi apparatus also sorts and packages glycoproteins for transport to a variety of cellular locations. Glycoproteins are delivered from the ER to the convex face of the Golgi apparatus by 500-Å–diameter vesicles that have bristlelike coats around them (Figure 29-48). This coat is a polyhedral lattice of 180-kdal *clathrin* subunits (Figure 29-49). These *coated vesicles* bud from the ER and then fuse with the Golgi apparatus. They also carry membrane proteins from the concave face of the Golgi apparatus to lysosomes, the plasma membrane, and other cellular destinations. Moreover, coated vesicles bring proteins and lipids from the plasma membrane to internal membranes. *Thus, clathrin plays a key role in the transfer of membrane from one cellular location to another.* It is important to note that membrane asymmetry is preserved in these transport processes. The luminal sides of the coated vesicle membrane and of the Golgi membrane correspond to the luminal side of the ER membrane. When a coated vesicle fuses with the plasma membrane, its luminal surface becomes the extracellular surface of the plasma membrane. Hence,

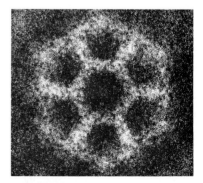

Figure 29-49
Image of a reassembled coated vesicle. This picture was obtained by superposing many electron micrographs. [From M. P. Woodward and T. F. Roth, *J. Supramol. Struct.* 11(1979):240.]

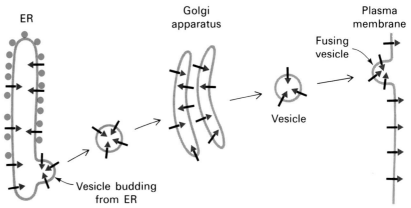

Figure 29-50
Asymmetry of cell membranes. The luminal faces of the ER and other organelles correspond to the extracellular face of the plasma membrane.

the luminal sides of the ER membrane and of other organelle membranes correspond to the external face of the plasma membrane (Figure 29-50). For this reason, the carbohydrate groups of glycoproteins in the plasma membrane are always on its extracellular surface (p. 222).

The oligosaccharide units of glycoproteins are remodeled in the Golgi apparatus. The one remaining glucose and several mannose residues of the core are trimmed. The formation of the carbohydrate unit of some glycoproteins is complete at this stage. The more complex oligosaccharides of other glycoproteins are constructed by the sequential addition of sugars to the trimmed core. For example, UDP-N-acetylglucosamine, UDP-galactose, and CMP-neuraminidate are the activated donors in the synthesis of a terminal trisaccharide unit that is found in some glycoproteins (Figure 29-51). This final stage of biosynthesis in the Golgi apparatus is called *terminal glycosylation* to distinguish it from *core glycosylation*, which takes place in the ER.

How are glycoproteins sorted in the Golgi apparatus and sent to their ultimate destinations? The answer to this intriguing question is not yet known, but an informative clue is provided by *I-cell disease* (also called mucolipidosis II). This lysosomal storage disease, which is inherited as an autosomal recessive trait, is characterized by severe psychomotor retardation and skeletal deformities. Lysosomes in the connective tissues of patients who have I-cell disease contain large inclusions (hence the name of the disease) of undigested glycosaminoglycans and glycolipids. *These inclusions are present because at least eight enzymes required for their degradation are missing from the affected lysosomes.* In contrast, very high levels of these enzymes are present in the blood and urine of these patients. Thus, active enzymes are synthesized, but they are exported instead of being sequestered in lysosomes. In other words, *a whole series of enzymes is mislocated in I-cell disease.* These enzymes lack mannose 6-phosphate, which is nor-

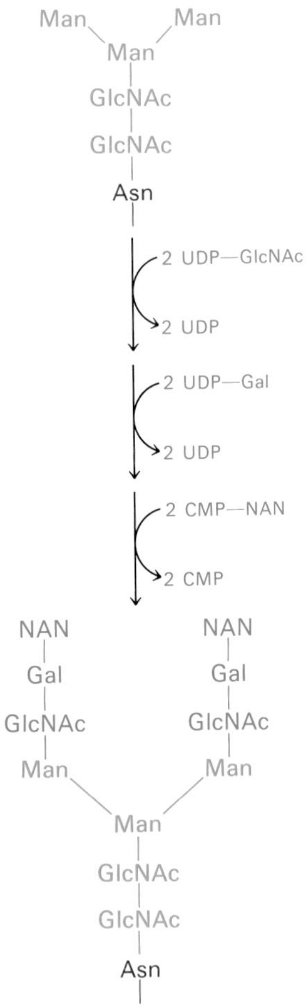

Figure 29-51
Final stage in the synthesis of a neuraminidate-containing carbohydrate unit of a glycoprotein (human transferrin). This terminal glycosylation takes place in the Golgi apparatus.

mally present. *Hence, it seems likely that mannose 6-phosphate is the marker that normally directs many hydrolytic enzymes from the Golgi apparatus to lysosomes.* It will be interesting to see whether the carbohydrate units of other glycoproteins participate in the direction of intracellular traffic.

SUMMARY

A eucaryotic chromosome contains a single molecule of double-helical DNA, which is at least one hundred times as large as DNA molecules in procaryotes. DNA is tightly bound to basic proteins, called histones. The chromatin fiber is a flexibly jointed chain of nucleosomes. The core of this repeating unit consists of 140 base pairs of DNA wound around a histone octamer made up of two each of histones 2A, 2B, 3, and 4. Nucleosome cores are joined by linker DNA, typically 60 base pairs long, which binds one molecule of H1. A sevenfold reduction in the length of DNA is achieved by nucleosomes, which are the first stage in the condensation of DNA. Eucaryotic DNA is replicated semiconservatively and semidiscontinuously from several thousand origins. A large number of initiation points are necessary because of the great length of eucaryotic DNA compared with procaryotic DNA.

The kinetics of reassociation of thermally denatured DNA shows that eucaryotic DNA, unlike procaryotic DNA, contains many repeated base sequences. *Highly repetitive DNA* (satellite DNA) is usually localized near centromeres. Some satellite DNAs have a heptanucleotide sequence repeated more than ten thousand times. Another category is *moderately repetitive DNA*. Tens to hundreds of copies of the genes for ribosomal RNAs and histones are present in many eucaryotes. Most of the moderately repetitive DNA probably has a regulatory role. The third category is *single-copy DNA,* consisting of sequences that occur only once (or just a few times) in a haploid genome. Many major proteins, such as silk fibroin, hemoglobin, and ovalbumin, are encoded by single-copy genes. They give rise to large amounts of stable mRNA. Single-copy sequences are usually interspersed with moderately repetitive sequences. Higher eucaryotes are also distinctive in having intervening sequences in many genes that code for proteins.

RNA in eucaryotic cells is synthesized by three types of RNA polymerases. These primary transcripts are extensively modified before they are transported from the nucleus to the cytosol as mature rRNA, tRNA, and mRNA molecules. Messenger RNAs are derived from much longer primary transcripts (hnRNA) by cleavage and splicing. A eucaryotic messenger has a 5′ cap and usually has a long 3′ poly A tail. Its single translated region is flanked by noncoding sequences, which can be quite long. The population of mRNAs in a eucaryotic cell depends on the selective processing of the primary transcripts as well as on the rates of transcription of particular genes. Control at the level of translation seems to be less important in determining which proteins are synthesized by a particular cell. The 80S eucaryotic ribosome consists of 40S and 60S subunits. As in procaryotes, there is a special initiator tRNA, but it is not formylated. Initiation in eucaryotes is controlled by a protein kinase cascade that inactivates an initiation factor (eIF-2). Mitochondria and chloroplasts have their own DNA, which is devoid of histones. Protein synthesis in these organelles resembles that of bacteria.

Secretory and membrane proteins are synthesized by ribosomes bound to the endoplasmic reticulum. Many of these proteins contain a highly hydrophobic amino-terminal sequence which acts as a signal for the attachment of the ribosome to the ER membrane. The nascent polypeptide chain is then actively threaded through the ER membrane and the signal sequence is cleaved on the luminal side. Nearly all proteins synthesized by membrane-bound ribosomes acquire one or more covalently attached carbohydrate units. A common oligosaccharide block is transferred in the ER from an activated dolichol donor to a specific asparagine side chain. Glycoproteins are transferred from the ER to the Golgi apparatus in coated vesicles. These clathrin-coated vesicles participate in a wide variety of membrane interchanges. Trimming and terminal glycosylation of the oligosaccharide units take place in the Golgi apparatus. In addition, glycoproteins are sorted there and sent to their destinations. Carbohydrate units may be important in this sorting process.

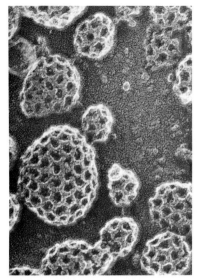

Figure 29-52
Electron micrograph of coated vesicles on the plasma membrane of a cell. [Courtesy of Dr. John Heuser.]

SELECTED READINGS

WHERE TO START

Crick, F., 1979. Split genes and RNA splicing. *Science* 204:264–271.

DeRobertis, E. M., and Gurdon, J. B., 1979. Gene transplantation and the analysis of development. *Sci. Amer.* 241(6):74–82. [Available as Offprint 1454.]

Beerman, W., and Clever, U., 1974. Chromosomal puffs. *Sci. Amer.* 210(4):50–58. [Offprint 180.]

Palade, G., 1975. Intracellular aspects of the process of protein synthesis. *Science* 189:347–358. [A lucid and highly informative Nobel Lecture.]

Rothman, J. E., and Lodish, H. F., 1979. The assembly of cell membranes. *Sci. Amer.* 240(1):48–63. [An excellent discussion of membrane asymmetry.]

CHROMATIN

Cold Spring Harbor Symposia, 1978. *Chromatin.* Cold Spring Harbor Symposia on Quantitative Biology, vol. 42.

McGhee, J. D., and Felsenfeld, G., 1980. Nucleosome structure. *Ann. Rev. Biochem.* 49:1115–1155.

Kornberg, R. D., 1977. Structure of chromatin. *Ann. Rev. Biochem.* 46:931–954.

Kornberg, R. D., 1974. Chromatin structure: a repeating unit of histones and DNA. *Science* 184:868–871.

Sperling, R., and Wachtel, E. J., in press. The histones. *Advan. Protein Chem.*

DNA REPLICATION

Kornberg, A., 1980. *DNA Replication.* Freeman. [Chapter 6 deals with eucaryotic DNA polymerase and replication.]

DePamphilis, M. L., and Wassarman, P. M., 1980. Replication of eucaryotic chromosomes: a close-up of the replication fork. *Ann. Rev. Biochem.* 49:627–666.

Kriegstein, H. J., and Hogness, D. S., 1974. The mechanism of DNA replication in *Drosophila* chromosomes: structure of replication forks and evidence for bidirectionality. *Proc. Nat. Acad. Sci.* 71:135–139.

Britten, R., and Kohn, D., 1968. Repeated segments of DNA. *Sci. Amer.* 222(4):24–31. [Offprint 1173.]

Davidson, E. H., and Britten, R. J., 1979. Regulation of gene expression: possible role of repetitive sequences. *Science* 204:1052–1059.

Long, E. O., and Dawid, I. B., 1980. Repeated genes in eucaryotes. *Ann. Rev. Biochem.* 49:727–766.

Schimke, R. T., Kaufman, R. J., Alt, F. W., and Kellems, R. F., 1978. Gene amplification and drug resistance in cultured murine cells. *Science* 202:1051–1055.

Kedes, L. H., 1979. Histone genes and histone messengers. *Ann. Rev. Biochem.* 48:837–870.

Tzagoloff, A., Macino, G., and Sebald, W., 1979. Mitochondrial genes and translation products. *Ann. Rev. Biochem.* 48:419–441.

SPLIT GENES, TRANSCRIPTION, AND RNA PROCESSING

Darnell, J. E., Jr., 1979. Transcription units for mRNA production in eukaryotic cells and their DNA viruses. *Prog. Nucleic Acid Res. Mol. Biol.* 22:327–353.

Abelson, J., 1979. RNA processing and the intervening sequence problem. *Ann. Rev. Biochem.* 48:1035–1069.

Royal, A., Garapin, A., Cami, B., Perrin, F., Mandel, J. L., LeMeur, M., Bregegegre, F., Gannon, F., LePennec, J. P., Chambon, P., and Kourilsky, P., 1979. The ovalbumin gene region: common features in the organisation of three genes expressed in chicken oviduct under hormonal control. *Nature* 279:438–445.

DeRobertis, E. M., and Olson, M. V., 1979. Transcription and processing of cloned yeast tyrosine tRNA genes microinjected into frog oocytes. *Nature* 278:137–143.

Shatkin, A. J., 1976. Capping of eucaryotic mRNAs. *Cell* 9:645–653.

TRANSLATION

Wool, I. G., 1979. Structure and function of eukaryotic ribosomes. *Ann. Rev. Biochem.* 48:719–754.

Revel, M., and Groner, Y., 1978. Post-transcriptional and translational controls of gene expression in eukaryotes. *Ann. Rev. Biochem.* 47:1079–1126.

Pappenheimer, A. M., Jr., 1977. Diphtheria toxin. *Ann. Rev. Biochem.* 46:69–94.

Weatherall, D. J., and Clegg, J. B., 1979. Recent developments in the molecular genetics of human hemoglobin. *Cell* 12:467–479.

Leder, A., Miller, H. I., Hamer, D. H., Seidman, J. G., Norman, B., Sullivan, M., and Leder, P., 1978. Comparison of cloned mouse α- and β-globin genes: conservation of intervening sequence locations and extragenic homology. *Proc. Nat. Acad. Sci.* 75:6187–6191.

Weatherall, D. J., 1978. The thalassemias. *In* Stanbury, J. B., Wyngaarden, J. B., and Fredrickson, D. S., (eds.), *The Metabolic Basis of Inherited Disease* (4th ed.), pp. 1508–1523. McGraw-Hill.

SYNTHESIS OF MEMBRANE PROTEINS AND SECRETED PROTEINS

Blobel, G., Walter, P., Chang, G. N., Goldman, B. M., Erickson, A. H., Lingappa, V. R., 1979. Translocation of proteins across membranes: the signal hypothesis and beyond. *Symp. Soc. Exp. Biol.* 33:9–36.

Unwin, P. N. T., 1977. Three-dimensional model of membrane-bound ribosomes obtained by electron microscopy. *Nature* 269:118–122.

Rothman, J. E., and Lodish, H. F., 1977. Synchronised transmembrane insertion and glycosylation of a nascent membrane protein. *Nature* 269:775–780.

Waechter, C. J., and Lennartz, W. J., 1976. The role of polyprenol-linked sugars in glycoprotein synthesis. *Ann. Rev. Biochem.* 45:95–112.

Kornfeld, R., and Kornfeld, S., 1976. Comparative aspects of glycoprotein structure. *Ann. Rev. Biochem.* 45:217–237.

Robbins, P. W., Hubbard, S. C., Turco, S. J., and Wirth, D. F., 1977. Proposal for a common oligosaccharide intermediate in the synthesis of membrane glycoproteins. *Cell* 12:893–900.

Pearse, B. M. F., 1980. Coated vesicles. *Trends Biochem. Sci.* 5:131–134. [A concise review of the structure and function of clathrin.]

Rothman, J. E., and Fine, R. E., 1980. Coated vesicles transport newly synthesized membrane glycoproteins from endoplasmic reticulum to plasma membrane in two successive stages. *Proc. Nat. Acad. Sci.* 77:780–784.

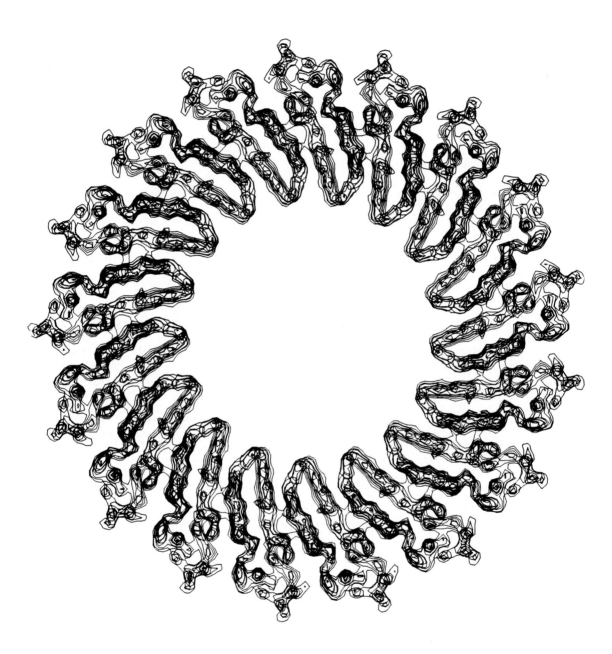

Electron-density map of a protein disc
from tobacco mosaic virus.
[Courtesy of Dr. Aaron Klug.]

CHAPTER 30

VIRUSES

Viruses are packets of infectious nucleic acid surrounded by protective coats. They are the most efficient of the self-reproducing intracellular parasites. Viruses are unable to generate metabolic energy or to synthesize proteins. They also differ from cells in having either DNA or RNA, but not both. The nucleic acid is single stranded in some viruses, and double stranded in others. Viruses range in complexity from $Q\beta$, an RNA phage having only 4 genes, to the poxviruses, which have about 250 genes. The completed extracellular product of virus multiplication is called a *virion* (or virus particle). In a virion, the viral nucleic acid is covered by a protein *capsid,* which protects it from enzymatic attack and mechanical breakage and delivers it to a susceptible host. In some of the more complex animal viruses, the capsid is surrounded by an *envelope* that contains lipid and glycoprotein.

We have already seen that the study of viruses has had a profound effect on the development of molecular biology, as in the discovery of messenger RNA. There is much current interest in viruses for several reasons. First, viral multiplication is a *model for cellular development* because it requires the sequential expression of genes and the assembly of macromolecules to form highly ordered structures. The relatively small number of viral genes, the rapid rate of replication, and the feasibility of genetic analysis contribute to the value of viruses as models. Second, viruses are sources of insight into *evolutionary processes* and molecular aspects of *host-parasite relations.* Third, *some viruses cause cancers in experimental animals.* The possible role of viruses in human cancers is now being intensively studied.

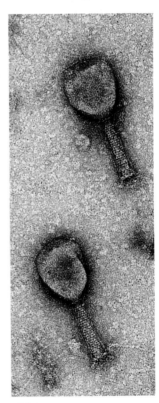

Figure 30-1
Electron micrograph of T4 virions from an infected cell. [Courtesy of Dr. Jonathan King, Dr. Yoshiko Kikuchi, and Dr. Elaine Lenk.]

Table 30-1
Types of viruses

Nucleic acid	Representative virus	Approximate number of genes
Single-stranded DNA	φX174 phage	5
Double-stranded DNA	Polyoma virus	6
	Adenovirus 2	30
	T4 phage	150
	Vaccinia poxvirus	240
Single-stranded RNA	Qβ phage	3
	Rous sarcoma virus	4
	Tobacco mosaic virus	6
	Poliovirus	8
	Influenza virus	12
Double-stranded RNA	Reovirus	22

Source: B. D. Davis, R. Dulbecco, H. N. Eisen, H. S. Ginsberg, and W. B. Wood, Jr. *Microbiology* (Harper & Row, 1973).

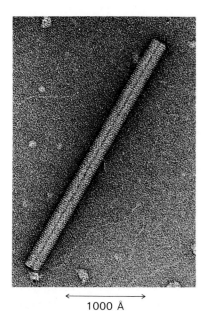

← 1000 Å →

Figure 30-2
Electron micrograph of a tobacco mosaic virus (TMV) particle. [Courtesy of Dr. Robley Williams.]

THE COAT OF A SMALL VIRUS CONSISTS OF MANY IDENTICAL PROTEIN SUBUNITS

The total number of amino acids in the coat of a virus always exceeds the number of nucleotides in its genome. For example, the protein coat of a tobacco mosaic virus (TMV) particle contains about 340,000 amino acid residues, whereas its RNA contains only about 6,400 nucleotides. In 1957, Francis Crick and James Watson noted that the protein coat of a virus cannot be one large molecule or an assembly of very many different small proteins because the amount of viral nucleic acid is much too small to code for such a large number of amino acid residues. On the other hand, the protein coat cannot be reduced in size if all the nucleic acid is to be covered. Viruses circumvent their genetic poverty by having coats made up of a *large number of one or a few kinds of protein subunits*. For example, the coat of TMV consists of 2,130 identical protein subunits (each 158 residues long).

A viral coat can be built from identical subunits in only a limited number of ways. Stability is achieved by forming a maximum number of bonds and by using the same kinds of contacts again and again. The resulting structure must be symmetric. Two kinds of arrangements of the protein coat are most likely: *a cylindrical shell having helical symmetry and a spherical shell having icosahedral symmetry* (Figure 30-3). In fact, *all small viruses are either rods or spheres* (or a combination of these shapes). The rules governing the structure of spherical viruses were deduced by Donald Caspar and Aaron Klug, who received some of their inspiration from the architectural designs of Buckminster Fuller.

Figure 30-3
Model of an icosahedron.

The simplest and best understood viral assembly process is that of TMV. This rod-shaped virus is 3,000 Å long and 180 Å in diameter (Figure 30-4) and has a mass of about 40,000 kdal. The 2,130 identical subunits in the protein coat are closely packed in a helical array around a single-stranded RNA molecule consisting of 6,390 nucleotides. The RNA is deeply buried in the protein, which renders it insusceptible to attack by ribonucleases. Each protein subunit interacts with three nucleotides. The dissociated RNA is very labile, whereas intact TMV remains infective for decades.

The subunits of TMV are not covalently bonded to one another. TMV can be dissociated into protein and RNA by agents such as concentrated acetic acid. In 1955, Heinz Fraenkel-Conrat and Robley Williams showed that the *dissociated coat subunits and RNA of TMV spontaneously reassemble under suitable conditions into virus particles that are indistinguishable from the original TMV in structure and infectivity.* This was the first example of the self-assembly of an active biological structure. *A self-assembly process is one in which the components associate spontaneously under appropriate environmental conditions to form a specific structure.*

←——100 Å——→

Figure 30-4
Model of a part of TMV, showing the helical array of protein subunits around a single-stranded RNA molecule. [After A. Klug and D. L. D. Caspar, *Advan. Virus Res.* 7(1960):274.]

PROTEIN DISCS ADD TO A LOOP OF RNA IN THE ASSEMBLY OF TMV

A priori, the simplest mechanism for the assembly of TMV would be the step-by-step addition of single protein subunits to the RNA. The difficulty with such a mechanism, however, is that nucleation would be very slow. About 17 coat subunits would have to add to a flexible RNA molecule before the complex could close on itself by forming a turn of helix and thereby acquire stability. This problem can be solved by adding a complex of many subunits to the RNA rather than one subunit at a time. In fact, the coat protein readily forms a *two-layered disc* consisting of 34 subunits. Each layer of a disc is a ring of 17 subunits, which is nearly the same as the number of subunits ($16\frac{1}{3}$) in a turn of the TMV helix. Klug and his associates solved the three-dimensional structure of the disc (Figure 30-5) and showed that it is a key intermediate in the assembly of TMV. *A critical property of the disc is that its subunits can slide over each other to form a two-turn helix, called a lockwasher* (Figure 30-6).

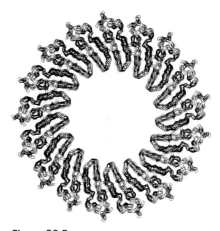

Figure 30-5
Electron-density map of a TMV protein disc. A 6 Å thick slice is shown here. [Courtesy of Dr. Aaron Klug.]

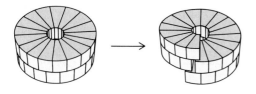

Figure 30-6
Schematic diagram showing the conversion of a TMV protein disc into the helical lock-washer form. [After A. Klug, *Fed. Proc.* 31(1972):40.]

Figure 30-7
Initiation region in TMV RNA for the assembly of the virus particle.

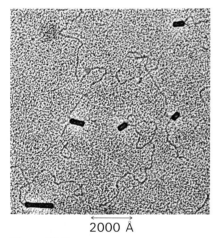

2000 Å

Figure 30-8
Electron micrograph of partially reconstituted TMV particles. Two RNA tails emerge from each growing virion. [From G. Lebeurier, A. Nicholaieff, and K. E. Richards, *Proc. Nat. Acad. Sci.* 74(1977):150.]

A disc interacts much more rapidly with TMV RNA than with foreign RNAs. Hence, it seemed likely that TMV RNA contains a base sequence that is specifically recognized by the disc and serves to initiate assembly. This initiation region was isolated by adding a few discs to TMV RNA to coat the initiation region and then digesting the rest of the RNA with a nuclease. The protected fragment contains a common core of about 65 nucleotides that binds very tightly and specifically to discs. The *base sequence of this initiation region strongly suggests that it forms a hairpin structure with a base-paired stem and a loop* (Figure 30-7). Most interesting, the loop contains G at every third position. This repeating triplet pattern matches the stoichiometry of three nucleotides per coat subunit in the virus. Hence, it seems likely that the loop binds the first disc to start the assembly of the virus.

Surprisingly, the initiation loop is far from either end of the RNA. Indeed, this starting point for assembly is about 5,300 nucleotides from the 5′ end and about 1,000 nucleotides from the 3′ end of the RNA. Another unexpected finding is that two tails of RNA emerge from the *same* side of a growing TMV particle (Figure 30-8). The length of the 3′ tail is rather constant throughout most of the assembly process, whereas the 5′ tail becomes shorter as the virus particle becomes longer.

A plausible model for the formation of TMV particles is shown in Figure 30-9. Assembly starts with the insertion of the initiation loop into the central hole of the two-layered protein disc. The loop binds to the first turn of the disc and the adjoining base-paired stem opens. This interaction transforms the disc into the helical lockwasher form and traps the viral RNA. The viral helix is now started. Another disc then adds to the newly formed loop of RNA that protrudes from the central hole. A new loop is formed after the addition of each disc by the drawing up of the 5′ tail through the central hole of the growing virus particle. Finally, the 3′ tail is coated in a way not yet determined.

The *two-layered disc markedly enhances the specificity of coating, in addition to assuring rapid nucleation.* A disc can bind a sequence of many nucleotides, whereas a single subunit can interact with only three. Consequently, the disc is much more discriminating than a single subunit could be in selecting TMV RNA instead of host mRNA for coating. Another important feature of discs is that they do not form helices devoid of RNA under physiological conditions. Two carboxyl groups in each subunit play a critical role in this regard. At neutral pH, both carboxyl groups are ionized in the helix form, but only one is ionized in the disc. The disc form is favored because of electrostatic repulsion between these closely spaced carboxylates in the helix form. The binding of RNA to the helix form provides enough free energy to overcome the electrostatic repulsion of the carboxylates. Thus, *the carboxylates are a negative switch to prevent the formation of a helix devoid of RNA.*

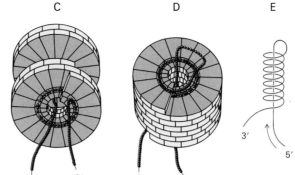

Figure 30-9
Model for the assembly of TMV: (A) the initiation region of the RNA loops into the central hole of the protein disc and transforms it into (B) the helical lock-washer form; (C) additional discs add to the looped end of the RNA; (D) one of the RNA tails is continually pulled through the central hole to interact with incoming discs; (E) schematic diagram of RNA in a partially assembled virus. The direction of RNA movement is denoted by the arrow. [After P. J. G. Butler and A. Klug. The assembly of a virus. Copyright © 1978 by Scientific American, Inc. All rights reserved.]

INFECTION BY T4 PHAGE PROFOUNDLY ALTERS MACROMOLECULAR SYNTHESIS IN *E. coli*

The T4 bacteriophage is a much more complex virus than TMV. Its double-stranded DNA contains about 165 genes, compared with 6 in TMV. However, much is known about the structure, multiplication, and assembly of T4 because it has been subjected to intensive genetic and biochemical analysis. The T4 virion consists of a *head,* a *tail,* and six *tail fibers* (Figure 30-10). The DNA molecule is tightly packed inside an icosahedral protein coat to form the head of the virus. The tail is made up of two coaxial hollow tubes that are connected to the head by a short neck. In the tail, a contractile sheath surrounds a central core through which the DNA is injected into the bacterial host. The tail terminates in a baseplate that has six short spikes and gives off six long, slender fibers.

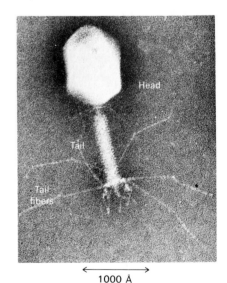

← 1000 Å →

Figure 30-10
Electron micrograph of a T4 phage. [From R. C. Williams and H. W. Fisher, *An Electron Micrographic Atlas of Viruses,* 1974. Courtesy of Charles C Thomas, Publisher, Springfield, Illinois.]

Table 30-2
Genes in T4 phage

Type	Number
Metabolic, essential	22
Metabolic, nonessential	60
Particle assembly, structural proteins	40
Particle assembly, other proteins	13
Total number of identified genes	135

Source: W. B. Wood and H. R. Revel, *Bacteriol Rev.* 40(1976):860.

NH$_2$

5-Hydroxymethylcytosine

CH$_2$OH

α-Glucosylated derivative
of 5-hydroxymethylcytosine

The tips of the tail fibers bind to a specific site on *E. coli.* An ATP-driven contraction of the tail sheath pulls the phage head toward the baseplate and tail fibers, which causes the central core to penetrate the cell wall but not the cell membrane. The naked T4 DNA then penetrates the cell membrane. *A few minutes later, all cellular, DNA, RNA, and protein synthesis stops, and the synthesis of viral macromolecules begins.* In other words, the infecting virus takes over the synthesizing machinery of the cell and substitutes its genes for the bacterial genes.

The T4 DNA contains three sets of genes that are transcribed at different times after infection: *immediate-early, delayed-early,* and *late.* The early genes are transcribed and translated before T4 DNA is synthesized. Some proteins coded by these genes are responsible for switching off the synthesis of cellular macromolecules. *Soon after infection, the host-cell DNA is degraded by a deoxyribonuclease that is specified by a T4 early gene. This enzyme does not hydrolyze T4 DNA because it lacks clusters of cytosines.* T4 DNA contains *5-hydroxymethylcytosine* (HMC) instead of cytosine. Furthermore, some HMC residues in T4 DNA are *glucosylated.*

The DNA of T4 bacteriophage contains these derivatives of cytosine because of the action of several phage-specified enzymes that are synthesized in the early phase of infection. One of them hydrolyzes dCTP to dCMP to prevent the incorporation of dCTP into T4 DNA. Then, dCMP is hydroxymethylated by a second enzyme to form 5-hydroxymethylcytidylate (dHMCMP). A third enzyme converts dHMCMP into the triphosphate, which is a substrate for DNA polymerases. Finally, some HMC bases in DNA are glycosylated by a fourth enzyme.

The synthesis of the *late proteins* is coupled to the replication of T4 DNA. The capsid proteins and a lysozyme are formed at this stage. The lysozyme digests the bacterial cell wall and causes its rupture when the assembly of progeny virions has been completed. About two hundred new virus particles emerge approximately twenty minutes after infection.

SCAFFOLD PROTEINS AND PROTEASES PARTICIPATE IN THE ORDERED ASSEMBLY OF THE T4 PHAGE

The assembly of T4 is a far more intricate process than that of TMV. The T4 capsid has a much higher degree of structural organization and contains some forty kinds of proteins. An additional thirteen proteins participate in the construction of this virus. Insight into the mechanism of assembly of T4 has come from the combined use of genetic, biochemical, and electron-microscopic techniques. The studies of William Wood and Robert Edgar on mutants of T4 that are defective in assembly have revealed that:

1. *There are three major pathways in the construction of the virus.* They lead independently to the head, the tail, and the tail fibers (Figure

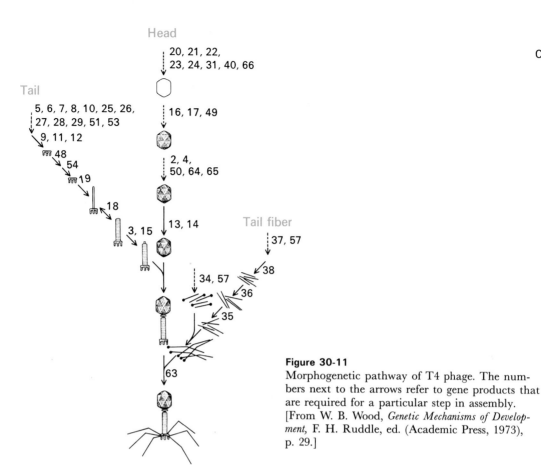

Figure 30-11
Morphogenetic pathway of T4 phage. The numbers next to the arrows refer to gene products that are required for a particular step in assembly. [From W. B. Wood, *Genetic Mechanisms of Development*, F. H. Ruddle, ed. (Academic Press, 1973), p. 29.]

30-11). A block in the formation of one of these components does not affect the synthesis of the other two.

2. *A strict sequential order* is followed in each of these pathways. All of the capsid proteins are synthesized simultaneously during the latter half of the infectious cycle. Hence, the sequential nature of the assembly of the head, tail, and tail fibers is improved by the structural features of the intermediates themselves. *None of the associations in a pathway occur at a significant rate unless the preceding one has taken place.* Some of the binding energy at each step may be used to lower the activation energy for the formation of the next complex, which thereby increases its rate.

3. The head and tail must be completed before they can combine. Then, completed tail fibers attach to the baseplate. Again, the strict sequential order assures that only complete virus particles are formed.

The construction of T4 virions does not occur by self-assembly alone. Rather, *scaffold proteins* and *proteases* play essential roles at certain stages of the assembly process. For example, three proteins that do not become part of the finished structure are needed for the formation of the central plug in the baseplate of the tail. These scaffold proteins serve as transient templates to promote the associ-

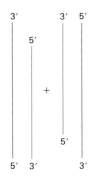

Figure 30-12
The 5′ ends of newly synthesized linear DNA molecules are incomplete. Parental DNA strands are shown in red and daughter ones in green.

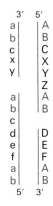

Figure 30-13
Concatameric intermediate in the replication of linear DNA duplexes. Incomplete two-stranded DNA molecules associate through their complementary single-stranded ends (AB to ab). The gaps are then filled.

ation of the constituents of the plug. Proteases have a prominent role in the assembly of the head. The 45-kdal major protein of the head, called *gp 23** (gp stands for gene product) is derived from gp 23, the 55-kdal precursor. Cleavage takes place when the head is partly assembled, which suggests that it may be the trigger for the entry of DNA. Three other head proteins are also known to be cleaved during assembly. *Thus, T4 is constructed by a combination of self-assembly, scaffold-assisted assembly, and enzyme-directed assembly.*

T4 DNA IS REPLICATED BY MEANS OF A CONCATAMERIC INTERMEDIATE

A special problem is encountered in replicating linear DNA molecules such as T4 DNA. The 5′ ends of nascent daughter DNA strands are incomplete because the RNA primer has been erased but not replaced by DNA (Figure 30-12). Recall that DNA polymerases are unable to synthesize in the 3′ → 5′ direction or start chains de novo (p. 584). This problem does not arise in the replication of circular DNA molecules because the 3′ end of a new strand serves as a primer to complete the synthesis of the daughter strand. How do T4 and other viruses with linear DNA genomes resolve this dilemma? An important clue came from the finding that these linear DNA molecules have *redundant ends*—that is, the base sequence at the left end of the DNA is precisely repeated at the right end:

5′ A B C D E——— V W X Y Z A B C 3′
3′ a b c d e——— v w x y z a b c 5′

Furthermore, long *concatamers* are formed in the replication of these DNA molecules. These findings suggested a mechanism for the completion of the 5′ ends of daughter strands. Nascent duplexes will rapidly associate because their single-stranded tails are complementary to each other (Figure 30-13). This complementarity arises from the redundancy in base sequence of the ends. The 3′ end of one duplex in the concatamer then serves as the primer to fill the gap at the 5′ end of the adjoining duplex in the repeating chain of new DNA molecules.

T4 DNA IS INSERTED INTO A PREFORMED HEAD

How is a unit-length DNA molecule formed and packaged inside a phage head? The problem is formidable: the DNA has a contour length of 56 μm, yet it must fit inside a head with a long axis of only 0.1 μm. Furthermore, the volume of the DNA ($1.8 \times 10^{-4} \mu m^3$) is not much less than that of the head ($2.5 \times 10^{-4} \mu m^3$). A priori, the DNA could be introduced into a preformed head or, alternatively, a head could be assembled around a core of condensed DNA. The

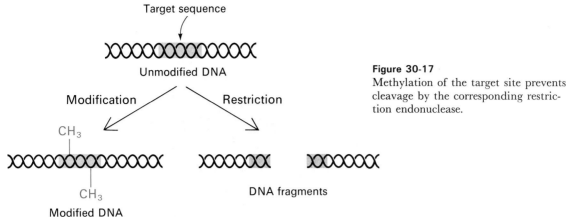

Figure 30-17
Methylation of the target site prevents cleavage by the corresponding restriction endonuclease.

Two types of restriction-modification systems are found in bacteria. In type-I systems, the methylase and nuclease activities are associated with a large multisubunit complex. For example, the *E. coli* B and K complexes consist of three kinds of polypeptide chains. The α chain has the endonuclease activity, the β chain has the methylase activity, and the γ chain has the recognition site for DNA. The type-I enzymes require *S*-adenosylmethionine and ATP for both their nuclease and their methylase activities. *Type-I enzymes cleave unmodified DNA at random sites a thousand base pairs or more to the 5′ side of the recognition site and concomitantly hydrolyze ATP.* In contrast, methylases and nucleases are separate in type-II systems. *S*-Adenosylmethionine is the methyl donor in the modification reaction, but it does not participate in DNA cleavage. Another difference is that type-II nucleases and methylases do not require ATP. Most striking, *type-II cleavages are highly specific.* As discussed in an earlier chapter (p. 590), many of these enzymes recognize a specific sequence of four to six base pairs and hydrolyze a single phosphodiester bond in each strand in this region. The distinctive characteristic of these cleavage sites is that they possess *twofold rotational symmetry* (Figure 30-18). Restriction enzymes are indispensable tools for analyzing DNA structure (p. 590) and constructing new DNA molecules (p. 761).

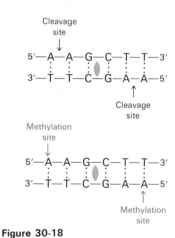

Figure 30-18
Specificity of the *Hind III* restriction endonuclease from *Hemophilus influenza* and of the corresponding methylase. The green symbol denotes the twofold axis of symmetry.

STRATEGIES FOR THE REPLICATION OF RNA VIRUSES

A special problem arises in the replication of RNA viruses because uninfected host cells lack enzymes to synthesize RNA according to instructions given by an RNA template. Consequently, RNA viruses must contain genetic information for the synthesis of an *RNA-directed RNA polymerase* (also called an *RNA replicase* or an *RNA synthetase*) or for an *RNA-directed DNA polymerase* (also called a *reverse transcriptase*). It is informative to classify RNA viruses according to the relation between their virion RNA and mRNA. By convention, mRNA is defined as (+) RNA and its complement as (−) RNA. Four pathways of replication and transcription of RNA viruses are known (Figure 30-19). Class 1 viruses (e.g., poliovirus) are *positive-*

Figure 30-19
Modes of gene expression of RNA viruses. [After D. Baltimore, *Bacteriol. Rev.* 35(1971):326.]

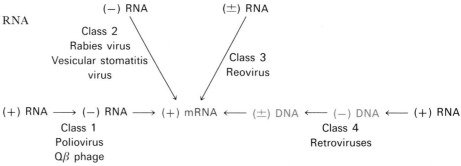

strand RNA viruses. They synthesize (−) RNA, which then serves as the template for the formation of (+) mRNA. Class 2 viruses (e.g., rabies virus) are *negative-strand RNA viruses* in which virion (−) RNA is the template for the synthesis of (+) mRNA. Class 3 viruses (e.g., reovirus) are *double-strand RNA viruses* in which the virion (±) RNA directs the asymmetric synthesis of (+) mRNA. Class 4 viruses, the most unusual, are the *retroviruses* (e.g., Rous sarcoma virus). They express the genetic information in their virion (+) RNA through a DNA intermediate that serves as the template for the synthesis of (+) RNA. Thus, the flow of information in retroviruses is from RNA to DNA and then back to RNA.

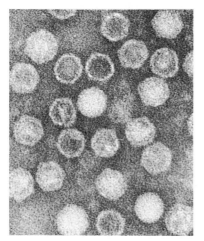

Figure 30-20
Electron micrograph of poliovirus particles. [Courtesy of Dr. John Finch.]

Picornaviruses—
A group of small (*pico*) RNA (*rna*) viruses. One single-stranded (+) RNA molecule is surrounded by an icosahedral protein shell (270 Å diameter). Poliovirus, rhinovirus (which causes the common cold), and the virus that causes foot-in-mouth disease in cattle are picornaviruses.

POLIOVIRUS PROTEINS ARE FORMED BY MULTIPLE CLEAVAGES OF A GIANT PRECURSOR

Poliovirus consists of a 7.5-kb single-stranded (+) RNA in an icosahedral capsid. This virion RNA molecule acts as a messenger on entering the cytoplasm of its host cell. It is translated by host ribosomes to give capsid proteins and a special RNA polymerase that takes instructions from an RNA template (an *RNA replicase*). The virion (+) RNA then directs this RNA replicase to synthesize (−) strands (Figure 30-21). In turn, the (−) RNA acts as the template for the synthesis of many (+) strands, which serve as messenger or are encapsidated to form new virions.

A striking feature of the expression of poliovirus genes is that the virion (+) RNA is the messenger for the synthesis of a continuous polypeptide chain of more than two thousand residues. David Baltimore showed that *this giant polypeptide is cleaved by host proteases to give seven proteins:* four coat proteins, an RNA replicase, and two proteins of unknown function (Figure 30-22). The nascent polypeptide chain is split into three pieces, which are further cleaved. In particular, two of the coat proteins are formed from a precursor in the final stage in the assembly of the virion.

A

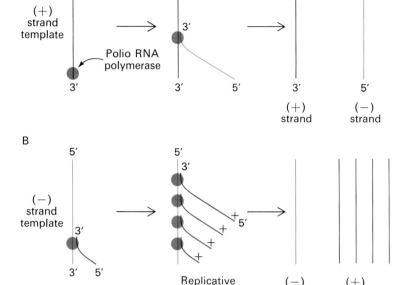

Figure 30-21
Replication of poliovirus RNA.

B

Why does poliovirus synthesize its proteins by this seemingly complex pathway? It appears to have little choice. Recall that *in eucaryotic cells an mRNA molecule can be translated into only one polypeptide chain for reasons that are not yet evident.* In contrast, procaryotic mRNAs are often polycistronic (e.g., *lac* mRNA). Thus, poliovirus cleaves a polyprotein to overcome a limitation imposed by its animal host cell.

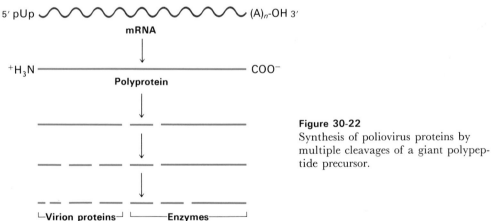

Figure 30-22
Synthesis of poliovirus proteins by multiple cleavages of a giant polypeptide precursor.

VESICULAR STOMATITIS VIRUS (VSV) TRANSCRIBES FIVE MONOCISTRONIC mRNAs FROM ITS GENOMIC RNA

Vesicular stomatitis virus (VSV), a mild pathogen of cattle, and rabies virus display a second mode of gene expression. Their virions contain a single-stranded ($-$) RNA molecule, which does not serve as messenger. *Hence the first step in its expression is the synthesis of* ($+$)

Rhabdoviruses—
  Bullet-shaped viruses, such as vesicular stomatitis virus and rabies virus. From *rhabdo,* the Greek word for rod.

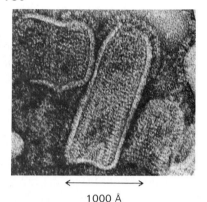

Figure 30-23
Electron micrograph of vesicular stomatitis virus. [From R. C. Williams and H. W. Fisher. *An Electron Micrographic Atlas of Viruses,* 1974. Courtesy of Charles C Thomas, Publisher, Springfield, Illinois.]

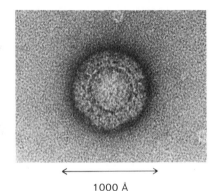

Figure 30-24
Electron micrograph of reovirus. [From R. C. Williams and H. W. Fisher. *An Electron Micrographic Atlas of Viruses,* 1974. Courtesy of Charles C Thomas, Publisher, Springfield, Illinois.]

Reovirus—
A double-stranded RNA virus isolated from the respiratory and intestinal tracts of humans and other mammals but not known to cause disease. The prefix *reo* is an acronym for respiratory enteric orphan, a virus in search of a disease.

---

mRNA. Uninfected cells lack an RNA replicase, and so these viruses must carry this enzyme in their virions and deliver it to the infected cell. In fact, two of the five proteins in the VSV virion carry out RNA replication. The 200-kdal L (large) and 45-kdal NS (nonstructural) proteins, though present in small amounts, are essential for infectivity.

The genomic RNA is complexed to many copies of N, the 50-kdal nucleocapsid protein, which is the major protein in the virion. VSV is enveloped by a lipid bilayer, which is acquired from the plasma membrane of the host in the process of budding out of the cell (Figure 30-23). The spikes in this membrane come from the G (glycoprotein) protein, a 65-kdal species encoded by the virus. The 29-kdal matrix protein, called M, lies between the envelope and nucleocapsid. *These five VSV proteins are formed by the translation of five (+) mRNAs rather than by cleavage of a polyprotein.* The same RNA replicase also synthesizes a long (+) RNA that contains all of the genetic information of the virus. This complete (+) RNA in turn serves as the template for the synthesis of (−) RNA, which is packaged to form new virions.

THE GENOME OF REOVIRUS CONSISTS OF TEN DIFFERENT DOUBLE-STRANDED RNA MOLECULES

Reovirus, a double-stranded RNA virus that infects mammalian cells, exemplifies a third type of viral genetic system. The core of the virion contains *ten different double-stranded (±) RNA molecules* that are associated with proteins. On entering the host, the virion loses its outer icosahedral shell, which is made up of three kinds of proteins. Removal of this shell activates an RNA polymerase that is carried in the core of the virion. This RNA-directed polymerase transcribes the ten (±) RNA molecules in entirety, so that the (+) mRNAs formed have the same length as the genome segments. The (±) RNA template is transcribed *asymmetrically* and *conservatively*—that is, only (+) RNA is formed, and the original (±) RNA duplex is not disrupted. The 5′ ends of these mRNAs are capped by enzymes in the core and then extruded through channels in the core (Figure 30-25). Thus, the core is a highly organized assembly for the synthesis of mRNA. *Each of the ten mRNAs is translated into one protein.* A complete set of ten (+) RNA molecules then joins some viral proteins to form a precore on which ten (−) strands are synthesized.

Why is the genome of reovirus segmented? As mentioned earlier, animal viruses cannot have polycistronic mRNAs. Poliovirus circumvents this limitation by cleaving a giant protein, whereas vesicular stomatitis virus transcribes its virion RNA into short mRNAs, each corresponding to one protein. *The stratagem used by reovirus is to have a separate chromosome for each kind of protein synthesized.*

In the fourth pathway for the expression of genetic information in RNA viruses, a DNA intermediate becomes integrated into the host genome. This more complex genetic system, which is found in the retroviruses (RNA tumor viruses), will be considered later in this chapter (p. 744).

SMALL RNA PHAGES CONTAIN OVERLAPPING GENES

RNA phages such as R17 (MS2, F2) and $Q\beta$ are very simple viruses. They have a regular polyhedral shape and a diameter of about 200 Å. The capsid of these closely related phages contains 180 molecules of the coat protein (14 kdal) and 1 molecule of the A (maturation) protein (38 kdal). The single-stranded (+) RNA also codes for a subunit of the replicase. It was thought until recently that these small RNA viruses contain only three genes. However, the finding of a phage mutant that formed normal virions but was unable to lyse the host cell led to a search for another virus-encoded protein. Indeed, these RNA phages do contain a fourth gene, which specifies a protein needed to lyse the bacterial host. The gene for this lysis protein overlaps the genes for the coat protein and replicase subunit (Figure 30-26). *These small RNA phages, like the small DNA phage φX174 (p. 632), use overlapping genes to pack more information into their small genomes.* The (+) RNA molecule of the virion serves both as a messenger for the synthesis of the four proteins and as a template for the synthesis of (−) RNA. Then, (−) RNA is used as the template for producing many copies of (+) RNA. Thus, the genetic system of these phages is like that of poliovirus.

The replicase that synthesizes the (+) and (−) strands of phage RNA is a very interesting enzyme. It is highly specific for the homologous phage RNA. Hence, host RNA molecules do not compete with phage RNA for replication. *The $Q\beta$ replicase consists of four subunits, of which only one is encoded by $Q\beta$ RNA. The other three subunits of the replicase are host proteins that have been diverted by the phage for its own ends.* Two of them are elongation factors for protein synthesis, EF Tu and EF Ts, and the third is a component of the 30S ribosomal subunit. Thus, $Q\beta$ creates a highly specific enzyme in a very economical way.

Regulatory mechanisms assure that replication and translation are appropriately timed. The (+) RNA is both a messenger for protein synthesis and a template for the synthesis of (−) RNA. It would be undesirable for both processes to occur simultaneously on the same (+) RNA because ribosomes travelling in the 5′ to 3′ direction would collide with the replicase moving in the 3′ to 5′ direction. This does not occur because $Q\beta$ replicase strongly inhibits the binding of ribosomes to (+) RNA until an adequate number of (−) RNA have been synthesized.

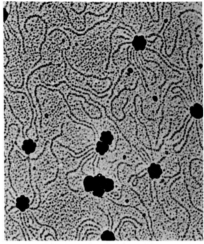

Figure 30-25
Synthesis of mRNA by reovirus cores. The mRNA molecules appear as threads emerging from the dark cores. [From N. M. Bartlett, S. G. Gillies, S. Bullivant, and A. R. Bellamy. *J. Virol.* 14(1974):324.]

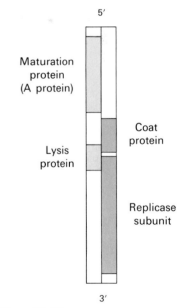

Figure 30-26
Overlapping genes in R17 and $Q\beta$ RNA. One of the reading frames is shown in yellow and the other in blue.

The four phage proteins are synthesized in different amounts. The coat protein, which is synthesized throughout infection, is the major product of translation. One reason is that ribosomes bind much more strongly to the initiation site for the coat cistron than to the other initiation sites on the (+) RNA. Furthermore, the coat protein inhibits translation of the replicase gene by blocking its initiation site. Thus, the coat protein is a specific translational repressor. The A protein is translated only from unfinished nascent (+) RNA molecules because its initiation site is blocked by base pairing in finished RNA molecules. The secondary structure of the completed RNA molecule permits only small amounts of the A protein to be made.

DARWINIAN EVOLUTION OF PHAGE RNA OUTSIDE A CELL

The purification of Qβ RNA and Qβ replicase free from contaminating nucleases enabled Sol Spiegelman to examine evolutionary events outside a living cell. One question asked was: *What happens to RNA molecules if the only demand made on them is that they multiply as rapidly as possible?* Molecules of Qβ RNA, Qβ replicase, and ribonucleoside triphosphates were incubated for twenty minutes, an incubation period that facilitates the selection of mutant RNA molecules that are rapidly replicated. An aliquot was diluted into a standard reaction mixture containing Qβ replicase and ribonucleoside triphosphates. A series of seventy-five transfers was carried out, and the RNA products were analyzed. The length of incubation was reduced periodically because the RNA molecules replicated more rapidly as the experiment proceeded. *The striking outcome was that the RNA molecules after seventy-five transfers ("generations") were only 12% as long as the original Qβ RNA.* Nucleotides not essential for replication were lost because the resulting shortened molecule was replicated more rapidly. The main constraint in this experiment was that the mutants had to retain the initiation sequence recognized by the Qβ replicase.

LYSOGENIC PHAGES CAN INSERT THEIR DNA INTO THE HOST-CELL DNA

Some bacteriophages have a choice of life styles: they can multiply and lyse an infected cell (*lytic pathway*) or their DNA can join the infected cell, retaining the capacity for multiplication and lysis (*lysogenic pathway*). Viruses that do not always kill their hosts are called *temperate* or *moderate*. The best-understood temperate virus is *lambda* (λ), which was discussed previously regarding its control of transcription (p. 679). Recall that the λ repressor binds to two sets of

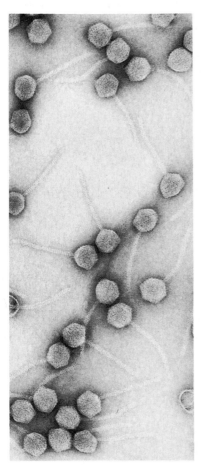

Figure 30-27
Electron micrograph of λ phage.
[Courtesy of Dr. A. Dale Kaiser.]

operator sites (the O_L and O_R regions) and that it regulates its own synthesis.

The DNA in the λ virion is a linear 48-kb duplex. The 5' end of each strand of this linear DNA molecule contains a single-stranded sequence of twelve nucleotides. These sequences are called *cohesive ends* because they are mutually complementary and can base-pair to each other. In fact, the cohesive ends come together following infection. The 5' phosphate terminus of each strand is then adjacent to its own 3' hydroxyl end. A host DNA ligase seals these two gaps to yield a *circular λ DNA molecule* (Figure 30-28).

This circular λ DNA molecule can be replicated by the interaction of λ proteins with the DNA replication machinery of the host. Alternatively, the λ DNA circle can be inserted into the bacterial chromosome by a single reciprocal recombination between specific 15-base-pair loci of both the λ and the *E. coli* DNA. The attachment site for λ in the *E. coli* DNA, called *attλ*, is located between *galE* and *bioA*, genes in the galactose and biotin operons. The base sequence of *attλ* is symbolized by B-B' (B for bacterial). The specific attachment site on λ, called *att*, is located next to the genes *int* (for integrate) and *xis* (for excise). The base sequence of *att* is symbolized by P-P' (P for phage). The int protein recognizes the P-P' sequence in the phage DNA and the B-B' sequence in *E. coli* DNA. A reciprocal transfer then occurs: P joins B' and B joins P'. This scheme (Figures 30-29 and 30-30) was originally proposed by Allan Campbell on the basis of genetic evidence.

The λ DNA is now part of the *E. coli* DNA molecule. This form of λ is called the *prophage,* and the *E. coli* cell containing the prophage is called a *lysogenic bacterium.* The prophage is stable in the absence of the xis protein. Transcription of the *xis* gene is blocked by the λ

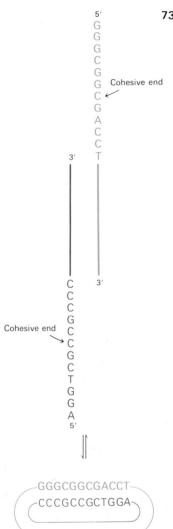

Figure 30-28
Conversion of linear λ DNA into the circular form.

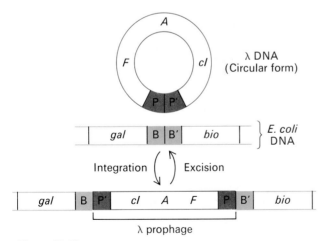

Figure 30-29
Schematic diagram of reciprocal recombination of λ DNA and *E. coli* DNA.

$$B{-}B' + P{-}P'$$
$$int \left(\ \right) int + xis$$
$$B{-}P' + P{-}B'$$

Figure 30-30
Integration and excision of λ DNA. This equation is incomplete because some of the reactants have not yet been identified.

repressor (p. 679). When repression is released, xis and int proteins together catalyze the breaking of the B-P′ and P-B′ sequences and again a reciprocal transfer occurs (Figure 30-30): P rejoins P′ and B rejoins B′, which recreates a circular molecule of λ DNA and a nonlysogenic *E. coli* chromosome. A key feature of this recombination system is that the int protein alone cannot recognize the two new sequences at the ends of the prophage (B-P′ and P-B′), which makes it stable. Thus, insertion occurs when only the int protein is present, whereas excision occurs when both the int and xis proteins are present.

RETROVIRUSES AND SOME DNA VIRUSES CAN INDUCE CANCER IN SUSCEPTIBLE HOSTS

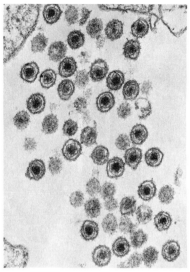

Figure 30-31
Electron micrograph of avian sarcoma virus (Rous sarcoma virus), a retrovirus. The densely stained virions are near the surface of an infected chicken cell. [Courtesy of Dr. Samuel Dales.]

In 1911, Peyton Rous prepared a cell-free filtrate from a connective-tissue tumor that appeared spontaneously in a hen and injected it into normal chickens. The striking result was that the recipients developed highly malignant tumors of the same kind, called *sarcomas.* Rous also found that the tumor-causing agent in the filtrate, now known as *Rous sarcoma virus* (RSV) or *avian sarcoma virus* (ASV) (Figure 30-31), could be propagated by serial passage through chickens. Avian sarcoma virus is a member of a group of *RNA tumor viruses* (oncogenic RNA viruses). These viruses contain (+)RNA in their virions and propagate through a double-helical DNA intermediate. Hence, they are known as *retroviruses.* In fact, retroviruses are the only RNA viruses that can produce cancer. Malignant tumors can also be induced by a variety of DNA viruses. *Simian virus 40* (*SV40*) and *polyoma virus,* which belong to the papovavirus group, are the most intensively studied *oncogenic DNA viruses* (Figure 30-32). There is much interest in these RNA and DNA viruses because they contain only four or five genes. Just one or two viral genes are directly involved in inducing cancer and so investigators hope to unravel their mode of action.

Tissue-culture systems have been developed for the study of cancer at the molecular level. Appropriate animal cells become permanently *transformed*—that is, rendered cancerlike—by injection with an oncogenic virus. Transformed cells differ from normal cells in their growth patterns and in the nature of their cell surfaces (Table 30-3). The most striking change is that *transformed cells grow continuously and chaotically,* without regard for their neighbors. Furthermore, transformed cells contain *viral-specific DNA that is integrated into the host genome,* which accounts for the fact that transformation is a heritable alteration. The abnormal characteristics of transformed cells are perpetuated in cultures derived from transformed colonies. Furthermore, some *transformed cells from tissue culture grow into a cancer when they are injected in sufficient number into an appropriate host.*

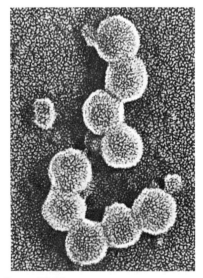

Figure 30-32
Electron micrograph of SV40, a DNA tumor virus. [Courtesy of Dr. Jack Griffith.]

Table 30-3
Altered properties of cells transformed by DNA or RNA tumor viruses

Growth patterns
    Form tumors when injected into susceptible animals

    Grow to much higher densities

    Growth becomes unoriented and cells detach from the substratum

    Protease activators are secreted into the medium, making the cells more invasive

    Requirement for serum factors decreases

Surface properties
    New viral-specific antigens appear

    Fibronectin, an external surface protein, disappears

    Ganglioside content decreases

    Fetal antigens become exposed

    Rate of transport of nutrients increases

    Agglutinability by plant lectins increases

Signs of the presence of tumor virus
    Virus DNA sequences are present

    Virus mRNAs are present

    Virus-specific antigens are detectable

Source: J. Tooze, ed. *The Molecular Biology of Tumour Viruses* (Cold Spring Harbor Laboratory, 1973), p. 351.

SV40 AND POLYOMA VIRUS CAN PRODUCTIVELY INFECT OR TRANSFORM HOST CELLS

SV40 and polyoma virus contain a small circular DNA duplex inside an icosahedral shell. In some cells (called *permissive hosts*), these viruses go through a *lytic cycle,* which results in the production of many new virions (Figure 30-33). Productively infected cells are

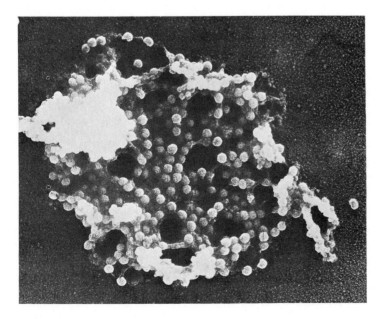

Figure 30-33
Electron micrograph of a fragment of the nuclear membrane of an SV40-infected cell showing nuclear pores and many imbedded virions. [Courtesy of Dr. Jack Griffith.]

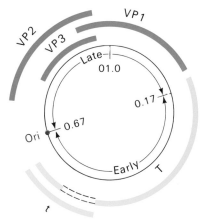

Figure 30-34
Genetic map of SV40 DNA, which contains 5,243 base pairs. The early region (transcribed counterclockwise) is shown in yellow, and the late region (transcribed clockwise) in blue. The origin of replication is marked in red.

killed by these viruses. In other types of cells (called *nonpermissive hosts*), some of the steps in viral expression are blocked for unknown reasons. No progeny virus are formed but a small proportion of the cells, of the order of 1 in 10^5, are *transformed* following integration of the viral DNA in the host genome.

The entire sequence of the 5,243 base pairs of SV40 DNA has been elucidated and many facets of its replication and transcription are understood. Half of the DNA is transcribed early in infection, whereas the other half is expressed later in infection, concomitant with the synthesis of viral DNA (Figure 30-34). In fact, the origin of DNA replication corresponds to the starting point of the early and late regions. The *early region*, which is transcribed in the counterclockwise direction, codes for the T antigen (A protein), which is essential for initiating DNA replication. A second immunologically distinctive protein, the small t antigen, is also specified by the early region. An intervening sequence is excised from the primary transcript in the synthesis of mRNA for the T antigen. Thus, SV40 takes advantage of the splicing apparatus in the cell nucleus. The base sequence of the origin of replication is also noteworthy. Two 13-base-pair sequences are related by a twofold axis of symmetry, and an adjoining region is rich in AT:

CAGAGGCCGAGGCGGC C TC GGCCTC TGCATAAATAAAAAAAAT TA

GTC TC CGGCT C CGCC GGAGCCGGAGACGTATT TATT TTTTT TAAT

T antigen binds to this site.

Transcription of the *late region*, which proceeds clockwise from the origin of replication (Figure 30-34), leads to the synthesis of three capsid proteins: VP1, VP2, and VP3. Again, intervening sequences are removed from a primary transcript. Three mRNAs appear to be produced by different splicing reactions. The amino acid sequence of VP3 overlaps the carboxy-terminal 70% of VP2. Furthermore, an overlap region of 22 nucleotides is read in one frame for VP2 and VP3 and in a different frame for VP1. Thus, *the limited amount of*

Table 30-4
Proteins encoded by SV40

Protein	Mass (kdal)	mRNA	Role
T antigen (A protein)	94	Early	Necessary for initiation of DNA replication and for transformation
t antigen	21	Early	Unknown
VP1	40	Late	Major capsid protein
VP2	39	Late	Minor capsid protein
VP3	27	Late	Minor capsid protein

genetic information in SV40 is exploited very fully. Another example of genetic economy is that SV40 does not synthesize its own proteins to package its DNA. Rather, host-cell histones become bound to the newly synthesized DNA (Figure 30-35). This supercoiled complex is then encapsidated by VP1, VP2, and VP3. Finally, progeny virions are released by lysis of the host cell, which is killed in the process.

In nonpermissive cells, the early region of SV40 is expressed, but DNA replication and transcription of the late region do not occur. *A small fraction of these cells become transformed by the integration of the SV40 genome into the host DNA.* In contrast with λ phage DNA integration, that of SV40 DNA does not appear to involve specific sites on either the viral or the host DNA. Both viral antigens (T and t) are required for transformation and probably also for the maintenance of the transformed state. Injection of these transformed cells into susceptible animals leads to the rapid production of tumors. A major goal of current research on SV40 is to find out how expression of the early region of the integrated form of this virus makes a cell cancerous.

RETROVIRUSES CONTAIN A REVERSE TRANSCRIPTASE THAT SYNTHESIZES DOUBLE-HELICAL DNA FROM (+) RNA

Retroviruses, the other class of tumor-producing viruses, contain a (+) RNA genome inside an icosahedral shell. This spherical nucleoprotein core is surrounded by an envelope consisting of viral-encoded glycoprotein molecules in a lipid bilayer derived from the plasma membrane of the host. Retroviruses are typically 1000 Å in diameter (see Figure 30-31).

In 1964, Howard Temin observed that infection by RNA tumor viruses such as avian sarcoma virus is blocked by inhibitors of DNA synthesis. Inhibitors such as amethopterin, 5-fluorodeoxyuridine, and cytosine arabinoside are effective if they are present during the first twelve hours following the introduction of viruses. This finding suggested that *DNA synthesis is required for the growth of RNA tumor viruses.* Furthermore, the production of progeny virus particles is inhibited by actinomycin D. This antibiotic is known to inhibit the synthesis of RNA from DNA templates (p. 613). Hence, *transcription of DNA seemed to be essential for the multiplication of RNA tumor viruses.* These unexpected results led Temin to propose that *a DNA provirus is an intermediate in the replication and oncogenic action of RNA tumor viruses.*

$$\begin{array}{ccc}
\text{RNA tumor} & \longrightarrow \quad \text{DNA} & \longrightarrow \quad \text{RNA tumor} \\
\text{virus} & \text{provirus} & \text{virus}
\end{array}$$

Temin's bold hypothesis that genetic information can flow from RNA to DNA was initially greeted with little enthusiasm by most investigators. It required the existence of a theretofore unknown

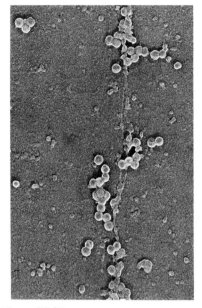

Figure 30-35
Electron micrograph of SV40 tumor virus particles developing in association with the host chromosome. [Courtesy of Dr. Jack Griffith.]

enzyme—one that synthesizes DNA according to instructions given by an RNA template (an *RNA-directed DNA polymerase*). In 1970, Temin and Baltimore independently discovered such an enzyme, which is called *reverse transcriptase,* in the virions of some RNA tumor viruses. All RNA tumor viruses subsequently studied have a reverse transcriptase, and so they are now known as *retroviruses.*

The life cycle of a typical retrovirus starts when infecting virions bind to specific receptors on the surface of the host and enter the cell. The viral ($+$) RNA is uncoated in the cytosol. Reverse transcriptase brought in by the virus particle then synthesizes a ($-$) DNA strand. This enzyme proceeds to digest the genomic RNA strand in the RNA-DNA hybrid. The ($-$) DNA then directs reverse transcriptase to synthesize a ($+$) DNA strand. *Thus, reverse transcriptase carries out a sequence of three reactions: RNA-directed DNA synthesis, hydrolysis of RNA, and DNA-directed DNA synthesis* (Figure 30-36).

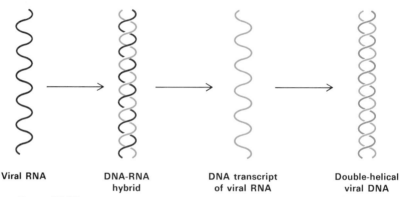

Viral RNA **DNA-RNA hybrid** **DNA transcript of viral RNA** **Double-helical viral DNA**

Figure 30-36
Synthesis of DNA from an RNA template by reverse transcriptase. The primer is not shown here.

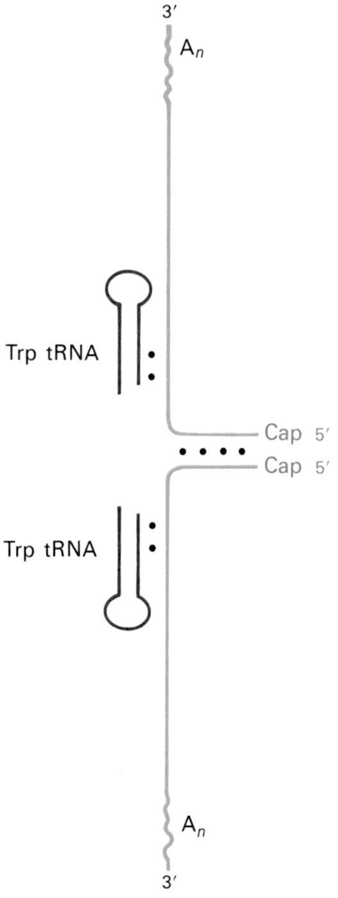

Reverse transcriptase, like other DNA polymerases, synthesizes DNA in the 5′ to 3′ direction and is unable to initiate chains de novo. How then is viral DNA synthesis primed? Initiation is accomplished very economically: the ($+$) RNA viral genome contains a noncovalently bound transfer RNA (trp tRNA in avian sarcoma virus) that was acquired from the host during the preceding round of infection. *The 3′ OH of this base-paired tRNA acts as the primer for DNA synthesis.* How is the entire ($+$) RNA strand replicated? Recall that a special problem is encountered in the replication of any *linear* DNA (p. 730). Retroviruses have met this challenge in an ingenious way. Their genomes consist of *two* molecules of ($+$) RNA rather than one (Figure 30-37). These molecules are hydrogen bonded to each other near their 5′ ends. Furthermore, the ($+$) RNA contains the same sequence at the 5′ and 3′ termini. This terminal redundancy is likely to be essential for the replication process, as with T4 phage (p. 730).

Figure 30-37
Schematic diagram of the genome of avian sarcoma virus. Two identical ($+$) RNA molecules are noncovalently bonded. The 5′ end is capped, and the 3′ end contains a poly A tail. A tRNA molecule that serves as primer is base paired to each RNA molecule.

RETROVIRUS DNA IS TRANSCRIBED ONLY WHEN IT IS INTEGRATED IN THE HOST GENOME

The viral DNA duplex becomes circular and enters the nucleus. Transcription of retroviral DNA occurs only after it has been integrated into the host-cell DNA. Thus, *integration is an obligatory step in the life cycle of retroviruses.* In contrast, integration and productive infection are alternative pathways for oncogenic DNA viruses. Another difference is that the frequency of integration of retroviral DNA is very high, as might be expected in view of its central role in productive infection.

The 10-kb genome of avian sarcoma virus contains four genes (Figure 30-38). Three of them—*gag, pol,* and *env* are essential for

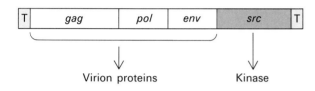

Figure 30-38
Genetic map of avian sarcoma virus. The genome is 10 kb long. T denotes the terminally redundant sequences.

productive infection. The *gag* gene specifies a 76-kdal polyprotein that is cleaved to form the four proteins of the viral core. *Pol* codes for the reverse transcriptase, which consists of an α and a β subunit. The 65-kdal α subunit is a proteolytic fragment of the 90-kdal β chain. *Env* specifies the glycoprotein of the viral envelope, which is important in the attachment of the virus to the surface of the host. The fourth gene, *src* (for sarcoma), is not needed for viral multiplication but is required for transformation, as will be discussed shortly.

The different viral mRNAs are probably spliced out of a 10-kb primary transcript. They are then transported into the cytosol, where they are translated. Genomic RNA and viral proteins migrate to the plasma membrane and become incorporated into it. A portion of the altered membrane then buds to form a new virus particle. Thus, *productive infection by retroviruses, in contrast with oncogenic DNA viruses, is not lytic. Retroviruses do not usually kill their hosts.* Retroviral DNA stays in the genome of the infected cell and continues to be expressed. Furthermore, the integrated viral DNA is replicated along with the host DNA, and so *the viral genome is inherited by the daughter cells.*

A KINASE ENCODED BY THE *src* GENE OF AVIAN SARCOMA VIRUS MEDIATES TRANSFORMATION

Studies of some *temperature-sensitive mutants* have been sources of insight into the mechanism of transformation by avian sarcoma virus. At high temperature, these mutants multiply normally but do not transform their hosts, whereas, at low temperature, both processes are normal. Furthermore, fibroblasts transformed by these mutants at low temperature revert to normal when the temperature is raised. Indeed, the shift from transformed to normal state is completely reversible. Analyses of these mutants have shown that only one viral protein is involved, the product of the *src* gene, which acquired its name because of *its ability to direct the synthesis of a sarcoma-producing protein.* It is noteworthy that viral DNA from these mutants remains integrated in the host genome at high temperature. Thus, *integration by itself does not lead to transformation. The* src *gene must also be expressed.*

What is the nature of the *src*-gene product? Recent studies indicate that this 60-kdal protein is a *kinase* that phosphorylates proteins. The targets of this kinase are now being studied to delineate the pathway of transformation. It is remarkable that a single small protein can profoundly alter the growth patterns of a cell and render it cancerous.

DOUBLE-STRANDED RNA INHIBITS PROTEIN SYNTHESIS IN INTERFERON-TREATED CELLS

The resistance of animal cells to many viruses is markedly enhanced by *interferons,* a group of small proteins. *They are synthesized and secreted by vertebrate cells* following a *virus infection.* Double-stranded RNA molecules are particularly effective in stimulating the formation of interferon. *Interferons bind to the plasma membrane of other cells in the organism and induce an antiviral state in them.* These cells acquire resistance to a broad spectrum of viruses. In contrast, immunity conferred by an antibody is highly specific. Interferons are very potent—as little as 10^{-11} M has a significant antiviral effect.

Interferon leads to an antiviral state by enhancing the production of three enzymes: an oligonucleotide synthetase, an endonuclease, and a kinase. These three enzymes are quiescent until the sensitized cell is infected by a virus or exposed to double-stranded RNA. *Activation of these enzymes blocks protein synthesis in two distinct ways* (Figure 30-40). The protein kinase phosphorylates an initiation factor in protein synthesis, which renders it inactive. Recall that this initiation factor is also the target of a regulatory cascade in reticulocytes (p. 711). The other major effect of double-stranded RNA in interferon-sensitized cells is to stimulate the degradation of mRNA. An endonuclease is activated by *oligoadenylates* called 2,5 A. Double-stranded RNA stimulates the synthetase that forms this endonuclease activator. It will be very interesting to see whether these pathways have a regulatory role in uninfected cells in addition to serving as a defense against viruses.

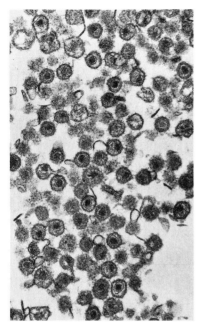

Figure 30-39
Electron micrograph of particles of avian myeloblastosis virus. Injection of this RNA tumor virus into newborn chicks produces a fatal leukemia within three weeks. [Courtesy of Dr. Ursula Heine, National Cancer Institute.]

2,5-A
(Endonuclease activator)

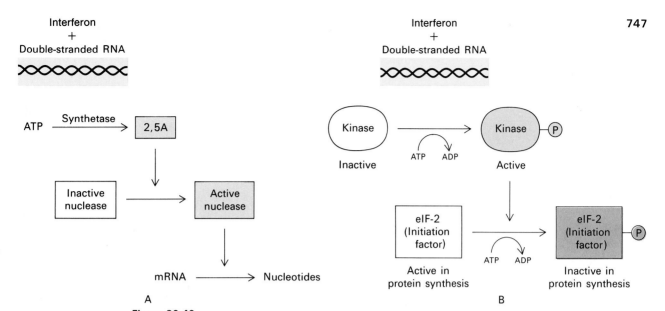

Figure 30-40
Double-stranded RNA (A) stimulates mRNA degradation and (B) inhibits the initiation of protein synthesis in interferon-treated cells. [After P. J. Farrell, G. C. Sen, M. F. Dubois, L. Ratner, E. Slattery, and P. Lengyel. *Proc. Nat. Acad. Sci.* 75(1978):5896.]

SUMMARY

Viruses are packets of infectious nucleic acid surrounded by a protective coat. They contain either DNA or RNA, which can be single stranded or double stranded. The simplest viruses have only 3 genes, whereas the most complex ones have about 250 genes. Most viral coats consist of a large number of one or a few kinds of protein subunits because viruses have a very limited amount of genetic information. The two principal arrangements are a cylindrical shell with helical symmetry and a spherical shell with icosahedral symmetry. Tobacco mosaic virus, which consists of a helical array of 2,130 identical subunits around a single-stranded RNA molecule, is formed by a self-assembly process in which protein discs add to a loop of RNA. Tomato bushy stunt virus can form an icosahedral shell because the domains of its coat protein are hinged. The assembly of T4 phage is a much more complex process that involves approximately fifty proteins. Three major pathways lead independently to the head, the tail, and the tail fibers. There is a stringent sequential order in the formation of T4, which occurs by a combination of self-assembly and enzyme-directed assembly. Scaffold proteins are also important in the assembly of T4.

Uninfected cells cannot synthesize RNA from an RNA template. Hence, a common feature of RNA viruses (except for the RNA tumor viruses) is that they specify an RNA-directed RNA polymerase. The RNA in the virion of some RNA viruses (e.g., $Q\beta$ phage) serves as a messenger on infection. For other RNA viruses (e.g., reovirus), the genome RNA must first be transcribed following infection. This is accomplished with an enzyme that is carried in the virion. A variety of strategies are employed by RNA viruses to synthesize proteins. Poliovirus proteins are formed by cleavage of a giant precursor. Reovirus contains a separate chromosome for each protein it synthesizes, whereas vesicular stomatitis virus transcribes several monocistronic mRNAs from a single genomic RNA.

Not all viruses destroy their hosts. The temperate phage λ can assume a dormant prophage form in which its DNA is integrated into the bacterial chromosome by reciprocal recombination. This prophage retains the capacity for multiplication and lysis, which occur when it is excised from the host DNA. Likewise, the DNA tumor viruses SV40 and polyoma can multiply and lyse their hosts or their DNA can become integrated into the host-cell DNA. Retroviruses can propagate only by integrating into the host-cell DNA. These viruses contain a reverse transcriptase that synthesizes duplex DNA from genomic RNA. Productive infection by retroviruses, in contrast with oncogenic DNA viruses, is not lytic. Cells containing SV40, polyoma, or retroviral DNA in their genomes have abnormal growth patterns and cell surfaces. They cause tumors when injected into susceptible animals. These viruses are attractive models for the study of the molecular basis of cancer because they have only about five genes. A kinase encoded by the *src* gene of avian sarcoma virus has been implicated in transformation.

The resistance of animal cells to many viruses is markedly enhanced by interferon, a protein synthesized and secreted by cells following a viral infection. Interferon enhances the production of three enzymes that block protein synthesis by inhibiting an initiation factor and by degrading mRNA.

> "The study of biology is partly an exercise in natural esthetics. We derive much of our pleasure as biologists from the continuing realization of how economical, elegant, and intelligent are the accidents of evolution that have been maintained by selection. A virologist is among the luckiest of biologists because he can see into his chosen pet down to the details of all of its molecules. The virologist sees how an extreme parasite functions using just the most fundamental aspects of biological behavior . . ."
>
> DAVID BALTIMORE
> Nobel Lecture (1976)
> © The Nobel Foundation, 1976

SELECTED READINGS

WHERE TO START

Butler, P. J. G., and Klug, A., 1978. The assembly of a virus. *Sci. Amer.* 239(5):62–69. [Mode of assembly of tobacco mosaic virus. Available as *Sci. Amer.* Offprint 1412.]

Bishop, J. M., 1980. The molecular biology of RNA tumor viruses: a physician's guide. *New Engl. J. Med.* 303:675–682.

Campbell, A. M., 1976. How viruses insert their DNA into the DNA of the host cell. *Sci. Amer.* 235(6):102–113. [Offprint 1347.]

Temin, H. M., 1972. RNA-directed DNA synthesis. *Sci. Amer.* 226(1):24–33. [Offprint 1239.]

Nathans, D., 1979. Restriction endonucleases, simian virus 40, and the new genetics. *Science* 206:903–909.

BOOKS

Luria, S. E., Darnell, J. E., Jr., Baltimore, D., and Campbell, A., 1978. *General Virology* (3rd ed.). Wiley. [A highly readable and interesting introduction to viruses.]

Tooze, J., (ed.), 1980. *The Molecular Biology of Tumour Viruses.* (2nd ed.) Cold Spring Harbor Laboratory.

Williams, R. C., and Fisher, H. W., 1974. *An Electron Micrographic Atlas of Viruses.* Thomas. [Contains superb electron micrographs and a very interesting accompanying text.]

VIRUS STRUCTURE

Bloomer, A. C., Champness, J. N., Bricogne, G., Staden, R., and Klug, A., 1978. Protein disk of tobacco mosaic virus at 2.8 Å resolution showing the interactions within and between subunits. *Nature* 276:362–368.

Stubbs, G., Warren, S., and Holmes, K., 1977. Structure of RNA and RNA binding site in tobacco mosaic virus from 4-Å map calculated from X-ray fibre diagrams. *Nature* 267:216–221.

Harrison, S. C., Olson, A. J., Schutt, C. E., Winkler, F. K., and Bricogne, G., 1978. Tomato bushy stunt virus at 2.9 Å resolution. *Nature* 276:368–373.

Harrison, S. C., 1978. Structure of simple viruses: specificity and flexibility in protein assemblies. *Trends Biochem. Sci.* 3:3–6.

Crick, F. H. C., and Watson, J. D., 1957. Virus structure: general principles. *In* Wolstenholme, G. E. W., (ed.), *Ciba Foundation Symposium on the Nature of Viruses,* pp. 5–13.

Caspar, D. L. D., and Klug, A., 1962. Physical principles in the construction of regular viruses. *Cold Spring Harbor Symp. Quant. Biol.* 27:1–24. [The basic theory of the structure of spherical viruses is presented in this classic paper.]

VIRUS ASSEMBLY

Lebeurier, G., Nicolaieff, A., and Richards, K. E., 1977.

Inside-out model for self-assembly of tobacco mosaic virus. *Proc. Nat. Acad. Sci.* 74:149–153.

Butler, P. J. G., Finch, J. T., and Zimmern, D., 1977. Configuration of tobacco mosaic virus RNA during virus assembly. *Nature* 265:217–219.

Wood, W. B., 1978. Bacteriophage T4 assembly and the morphogenesis of subcellular structures. *Harvey Lectures* 73:203–223.

Casjens, S., and King, J., 1975. Virus assembly. *Ann. Rev. Biochem.* 44:555–611.

RESTRICTION AND MODIFICATION

Smith, D. H., 1979. Nucleotide sequence specificity of restriction endonucleases. *Science* 205:455–462.

Arber, W., 1979. Promotion and limitation of genetic exchange. *Science* 205:361–365.

INTERFERON

Friedman, R. M., 1977. Antiviral activity of interferon. *Bacteriol. Rev.* 41:543–567.

Farrell, P. J., Sen, G. C., Dubois, M. F., Ratner, L., Slattery, E., and Lengyel, P., 1978. Interferon action: two distinct pathways for inhibition of protein synthesis by double-stranded RNA. *Proc. Nat. Acad. Sci.* 75:5893–5897.

Burke, D. C., 1977. The status of interferon. *Sci. Amer.* 236(4):42–50. [Offprint 1356.]

EXTRACELLULAR EVOLUTION OF Qβ RNA

Mills, D. R., Peterson, R. L., and Spiegelman, S., 1967. An extracellular Darwinian experiment with a self-duplicating nucleic acid molecule. *Proc. Nat. Acad. Sci.* 58:217–224.

LYSOGENY

Lwoff, A., 1966. The prophage and I. *In* Cairns, J., Stent, G. S., and Watson, J. D., (eds.), *Phage and the Origins of Molecular Biology,* pp. 88–99. Cold Spring Harbor Laboratory. [A delightful account of the discovery of the basis of lysogeny.]

TUMOR VIRUSES

Baltimore, D., 1976. Viruses, polymerases, and cancer. *Science* 192:632–636.

Temin, H., 1976. The DNA provirus hypothesis: the establishment and implications of RNA-directed DNA synthesis. *Science* 192:1075–1080.

Dulbecco, R., 1976. From the molecular biology of oncogenic DNA viruses to cancer. *Science* 192:437–440.

Bishop, J. M., 1978. Retroviruses. *Ann. Rev. Biochem.* 47:35–88.

Reddy, V. B., Thimmappaya, B., Dhar, R., Subramanian, K. N., Zain, B. S., Pan, J., Ghosh, P. K., Celma, M. L., and Weissman, S. M., 1978. The genome of simian virus 40. *Science* 200:494–502.

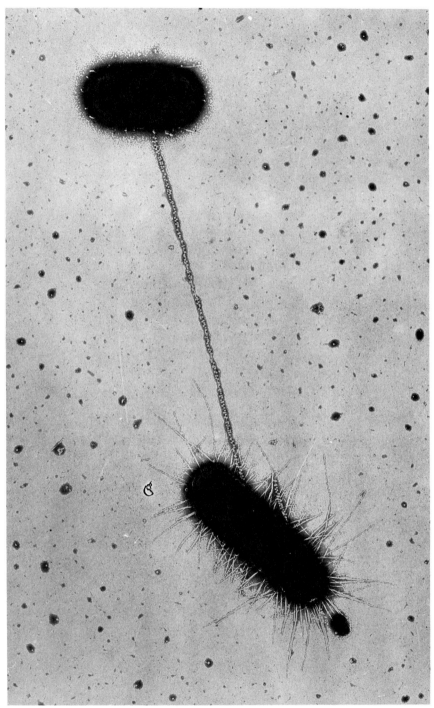

Figure 31-1
Transmission of genetic information from one *E. coli* cell to another. This electron micrograph shows two *E. coli* joined by a pilus during conjugation. DNA is transmitted through the pilus from the donor to the acceptor cell. [Courtesy of Dr. Charles Brinton and Dr. Judith Carnahan.]

CHAPTER 31

GENE REARRANGEMENTS:
RECOMBINATION, TRANSPOSITION,
AND CLONING

The theme of this chapter is the formation of new arrangements of genes by the movement of large blocks of DNA. The process of *genetic recombination,* in which a new DNA molecule is formed by the breakage and reunion of DNA strands, will be considered first. Genetic recombination is markedly enhanced by extensive sequence homology between the interacting DNA molecules. Intermediates in recombination have been isolated and an enzyme that catalyzes the interchange of DNA strands has recently been characterized. We will then consider *transposition,* which refers to the movement of a gene from one chromosome to another or from one site to a different one on the same chromosome. Transposition, in contrast with general recombination, does not require extensive sequence homology. In procaryotes, mobility is conferred by the presence of *insertion sequences* that enable unrelated pieces of DNA to be joined. Recombination and transposition have played important roles in evolution by *generating novel genomes.* The final part of this chapter deals with the *construction of new combinations of genes in the laboratory* and their expression in host cells. Genes can be covalently inserted into the DNA of plasmids and viruses by using restriction endonucleases and DNA ligase. These *recombinant DNA molecules* can be replicated and expressed in suitable host cells. The value of gene cloning and its prospective impact will also be discussed. The pace of research on recombination, transposition, and cloning is exceptionally rapid. These studies are providing powerful new methods for analyzing genomes and are contributing to a deeper understanding of their evolution and expression.

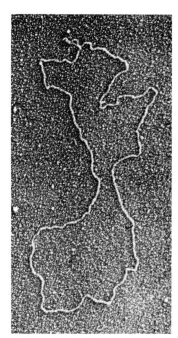

Figure 31-2
Electron micrograph of a circular DNA molecule containing genes that confer resistance to several antibiotics. Infectious drug resistance is a consequence of the transmission of such R (resistance) factor plasmids. [Courtesy of Dr. Stanley Cohen.]

GENETIC RECOMBINATION IS MEDIATED BY THE BREAKING AND JOINING OF DNA STRANDS

In genetic recombination, a DNA molecule with a base sequence derived partly from one parental DNA molecule and partly from another is formed. Studies of *E. coli* infected with a mixture of ^{32}P-labeled and bromouracil-labeled T4 phages were sources of insight into the molecular nature of recombinant molecules. The buoyant density of bromouracil-labeled DNA is appreciably higher than that of ^{32}P-labeled DNA, and so the parental molecules could be separated from each other and from recombinants by centrifuging them in a CsCl gradient. Analyses of these gradients showed that recombinant DNA molecules containing both ^{32}P and bromouracil were formed after infection with this mixture of phages. The nature of these hybrids depended on whether DNA synthesis occurred during their formation. In the absence of DNA synthesis, the ^{32}P-DNA in the recombinant was not covalently joined to the bromouracil-DNA. The hybrid could be dissociated into light and heavy components by heating above the melting temperature of the duplex. The parental units in this hybrid are held together by base pairing, and so this intermediate is called a *lap-joint* (Figure 31-3). However, if DNA synthesis occurred, the gaps in the lap-joint intermediate were filled by DNA polymerase I and the ends were joined by DNA ligase. The resulting recombinant molecule could not be separated into light and heavy components because these pieces of DNA became covalently linked. *These experiments revealed that single-stranded regions of DNA are intermediates in genetic recombination and suggested that enzymes are intimately involved in this process.*

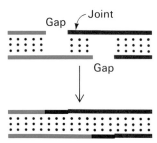

Figure 31-3
Lap-joint intermediate in the recombination of T4 DNA in infected bacteria. The ^{32}P-labeled DNA is shown in red and the bromouracil-labeled DNA in blue.

HOMOLOGOUS DNA STRANDS PAIR TO FORM A DUPLEX INTERMEDIATE IN GENETIC RECOMBINATION

Intermediates in genetic recombination have been visualized by electron microscopy. The DNA molecules chosen for this study were plasmids in *E. coli.* As will be discussed shortly (p. 755), these small circular DNA duplexes replicate autonomously within the bacterial cell. The number of plasmids in a bacterium increases from about twenty to one thousand when chloramphenicol is present. This antibiotic inhibitor of protein synthesis prevents replication of the bacterial chromosome but does not inhibit plasmid replication. Thus, the bacterial cell becomes filled with plasmid molecules, which can recombine. Electron microscopy of the plasmids isolated from these cells shows that about a quarter of them are dimers having the shape of a figure eight (Figure 31-4A). These dimers were then cleaved by the EcoR1 restriction endonuclease, which cuts the monomer plasmid at a unique site. If the dimers were interlocked monomers or double-length circles, unit-size rods would be formed by this procedure. On the other hand, a structure

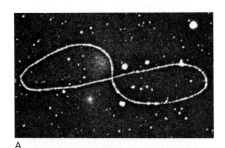

A

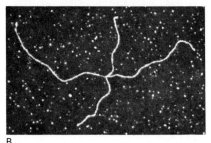

B

Figure 31-4
Electron micrographs of DNA molecules undergoing recombination: (A) a figure-eight intermediate consisting of two DNA molecules; (B) cleavage by a restriction endonuclease yields a chi form. [From H. Potter and D. Dressler. *Proc. Nat. Acad. Sci.* 76(1979):1086.]

with four arms resembling the Greek letter χ would be generated if two plasmid circles were covalently joined at a point of homology. In fact, nearly all of the figure eights were converted into chi forms (Figure 31-4B), which strongly suggests that *figure-eight forms are intermediates in recombination.* This inference is supported by the finding that figure eights are not formed in some mutants of *E. coli* that are defective in recombination.

What is the nature of the crossover region of the two genomes in the figure-eight forms? The contact point in the chi forms always divides the structure into pairs of equal-length arms. This means that *the genomes are joined at a region of homology* (Figure 31-5). There

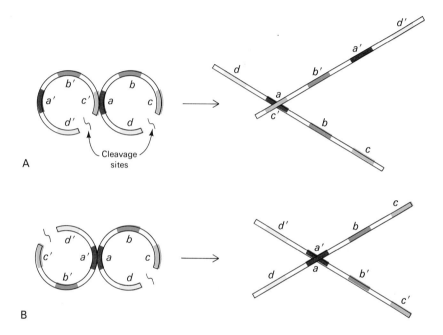

Figure 31-5
Expected cleavage patterns for two DNA molecules that are joined at (A) nonhomologous sites and (B) homologous sites. The observed symmetry of chi forms (as in Figure 31-4B) shows that DNA molecules in figure-eight forms are joined at homologous sites.

would be no special relations between the sizes of the four arms if the plasmids were joined at unrelated sequences. Furthermore, the contact point occurs with nearly equal probability along the entire plasmid, which shows that *pairing can occur at many locations.* The strand connections at the crossover region have been visualized by selectively denaturing it. Electron micrographs show that *four duplexes emerge from a ring of connecting single strands at the junction be-*

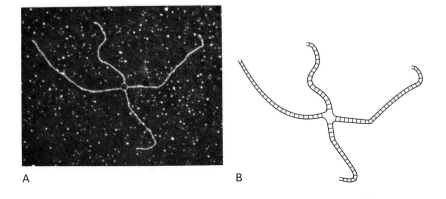

Figure 31-6
A. Electron micrograph of a chi form. B. Interpretive diagram. The region of homology (which is rich in AT base pairs) was selectively denatured by formamide to show the strand connections in the crossover. [From H. Potter and D. Dressler. *Cold Spring Harbor Symp. Quant. Biol.* 43(1979):973.]

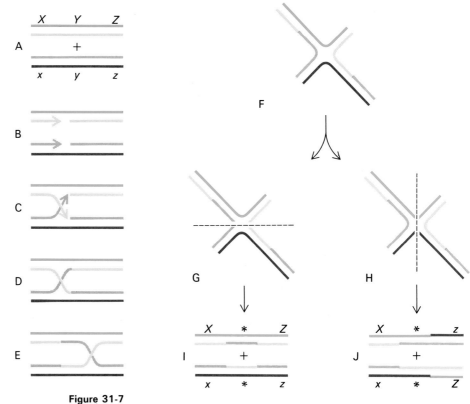

Figure 31-7
A model of genetic recombination proposed by Robin Holliday. One parental duplex is shown in green and the other in red. The darker strand of each duplex is the (+) strand. *X*, *Y*, and *Z* denote three genes; *x*, *y*, and *z* are alleles. The parental DNA molecules become covalently linked in parts C and D. Part F is an alternative representation of part E. Note that part F can be cleaved along the horizontal or the vertical axis. Rejoining of strands in parts G and H gives two different sets of recombinants (I and J). Regions containing one strand from each parental duplex are labeled with an asterisk. [After H. Potter and D. Dressler. *Cold Spring Harbor Symp. Quant. Biol.* 43(1979):973.]

tween the two genomes (Figure 31-6). This intermediate can then be cleaved and ligated to generate two different sets of recombinant molecules (Figure 31-7).

THE recA PROTEIN CATALYZES THE ATP-DRIVEN EXCHANGE OF DNA STRANDS IN GENETIC RECOMBINATION

The process discussed thus far is called *general genetic recombination* because exchange can occur between any pair of homologous sequences on the parental DNA molecules. In *E. coli*, general recombination depends on the *rec* genes. In rec⁻ cells, the bacterial DNA is unable to recombine with exogeneous DNA molecules. Three *rec* genes have been identified: *recA*, *recB*, and *recC*. The recB protein (140 kdal) and the recC protein (128 kdal) are the two subunits of a nuclease that unwinds double-helical DNA and fragments one of the unwound chains and then the other. These pieces, which contain several hundred nucleotides, are further digested by the recBC enzyme, which also acts as an exonuclease. The unwinding and nucleolytic activities of this enzyme complex depend on the hydrolysis of ATP. *It seems likely that the role of the recBC protein in recombination is to generate single-stranded DNA, which can then invade a duplex DNA molecule.*

How does single-stranded DNA find a homologous sequence in a duplex, pair with the complementary strand, and displace the other strand? The *recA protein (40 kdal) has recently been shown to catalyze this strand-assimilation reaction, which is driven by the hydrolysis of ATP.* The product of this reaction, consisting of a duplex and a displaced single-stranded loop, has the shape of the letter D, and so it is called a *D-loop structure* (Figure 31-8). The formation of a D loop is enhanced by a protein that specifically binds single-stranded DNA. This *single-stranded DNA binding protein,* which is also important in DNA replication (p. 585), stabilizes the single-stranded DNA formed by the recBC nuclease and stimulates the recA-catalyzed assimilation of this strand into a homologous duplex.

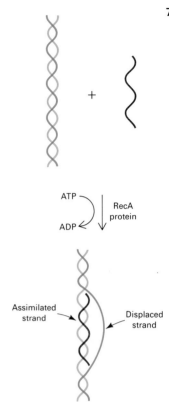

D-loop intermediate

Figure 31-8
The pairing of a single-stranded DNA molecule (shown in red) with the complementary strand (shown in green) of a duplex is catalyzed by the recA protein. The resulting structure is called a D loop.

BACTERIA CONTAIN PLASMIDS AND OTHER MOBILE GENETIC ELEMENTS

General genetic recombination generates new combinations of specific *alleles,* but it does not readily change the arrangement of entire *loci.* In other words, *ABC'D'E'* can be readily formed by homologous recombination of *ABCDE* with *A'B'C'D'E',* whereas *ABXYZCDE* and *ABE* cannot. These larger-scale genetic rearrangements can be mediated by *mobile genetic elements,* which are also called *translocatable elements* (Table 31-1). *Plasmids* are an important

Table 31-1
Mobile genetic elements in *Escherichia coli*

Type	Size (kb)	Characteristics
Plasmids		
F (fertility) factor	93	Confers maleness; transmissible by conjugation.
F' factors	> 100	Carry *E. coli* genes in addition to F-factor genes
R (resistance) factors	4 to 117	Carry drug-resistance genes; some also carry genes for conjugation.
Colicinogenic factors	6 to 141	Carry genes for colicin (toxin) production; some also carry genes for conjugation.
Lysogenic phages		
Lambda	48	A small proportion carry *E. coli* genes (*gal* or *bio*) in addition to viral genes.
Mu	38	All mu carry a small piece of the *E. coli* genome.
Insertion sequences (IS)	0.8 to 1.4	Genes flanked by a pair of IS are translocatable within a cell.

class of mobile genetic elements. They are circular duplex DNA molecules (Figure 31-9) ranging in size from two to several hundred kilobases. Plasmids carry genes for the inactivation of antibiotics, metabolism of natural products, and production of toxins. In effect, *plasmids are accessory chromosomes.* They differ from the bacterial chromosome in being dispensable under certain conditions. Also, *plasmids can replicate autonomously of the host chromosome.* An *E. coli* cell typically contains about twenty copies of the smaller chromosomes and one or two copies of the larger one.

THE F FACTOR ENABLES BACTERIA TO DONATE GENES TO RECIPIENTS BY CONJUGATION

Some plasmids enable bacteria to transfer genetic material to each other by forming a direct cell-cell contact. This process, called *conjugation,* was discovered in 1946 by Joshua Lederberg and Edward Tatum. During the conjugation of *E. coli,* one partner (the male) is the genetic donor and the other (the female) is the genetic recipient. Male bacteria contain special appendages called *sex pili* on their surfaces, whereas female cells contain receptor sites that bind pili. The male and female cells become joined by a pilus (see Figure 31-1), which retracts to enable the cells to establish direct contact for the passage of DNA (Figure 31-10). A male bacterium contains a plasmid called *F factor* (fertility factor), which contains genes for the formation of the sex pili and other components required for conjugation. One strand of the F-factor plasmid gets nicked and the duplex unwinds in conjugation (Figure 31-11). The 5' end of the nicked strand enters the recipient cell and a complementary strand is synthesized to form a closed, circular duplex. *The presence of an F-*

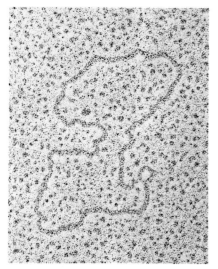

Figure 31-9
Electron micrograph of a small R-factor plasmid. [Courtesy of Dr. Jack Griffith.]

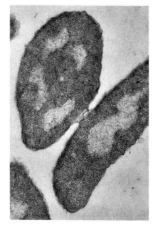

Figure 31-10
Electron micrograph of two *E. coli* in contact during conjugation. [Courtesy of Dr. Lucien Caro.]

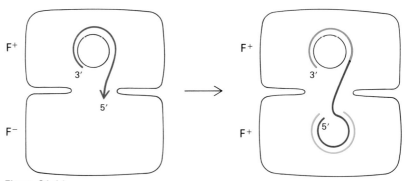

Figure 31-11
Proposed model for strand transfer in conjugation. The supercoiled R factor in the F⁺ donor unwinds after being nicked. Transfer of one strand of the R factor to the F⁻ recipient is coupled to replication of that strand of the donor. A complementary strand is then synthesized in the recipient. [After G. J. Warren, A. J. Twigg, and D. J. Sherratt. *Nature* 274(1978):260.]

factor plasmid in the recipient cell (originally F⁻) converts it into a male (*F⁺*). A male cell can spontaneously lose the F factor and thereby revert to F^-.

An F-factor plasmid can become integrated into the bacterial chromosome. Integration occurs by crossing-over at one of a number of sites on the bacterial chromosome. The frequency of integration is about 10^{-5} per generation. Bacteria harboring F factors in their chromosomes are called *Hfr cells* because they exhibit a high frequency of recombination. Hfr cells, like F^+ cells, are donors in conjugation (Figure 31-12). The difference between them is that *an Hfr cell donates the entire bacterial chromosome (including the integrated F factor)*, whereas an F^+ cell donates only the F factor. The entire chromosome is transferred in about 90 minutes in an Hfr × F^- cross. The order of entry of the donated genes into the recipient depends on the site of integration of the F factor and on its polarity. Hence, it is feasible to ascertain the order of genes in the donor's chromosome by interrupting the conjugation at different times and determining which markers have been transferred. The donated chromosome can recombine with the recipient's chromosome. The frequency of recombination is highest for genes that first entered the recipient because they are present for the longest time. Thus, genetic maps can be constructed by determining the *time of entry* and the *frequency of recombination* of markers contributed by the donor.

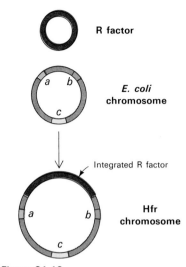

Figure 31-12
Schematic diagram showing the formation of an Hfr cell by integration of the F factor into the *E. coli* chromosome.

An Hfr cell reverts to the F^+ state by the excision of the F factor from its chromosome. This reversal of the integration of the F factor also has a frequency of about 10^{-5} per generation. In a small proportion of the revertants, crossing-over occurs at a site different from that of entry, which results in the *formation of a plasmid that contains chromosomal genes in addition to the F factor genes* (Figure 31-13). Such a plasmid is called an *F' factor,* the prime denoting the presence of chromosomal genes. Conjugation of an F' cell with an F^- cell results in the transfer of these chromosomal genes from the donor to the recipient cell, which then becomes diploid for these genes.

Thus, bacteria have mechanisms for transferring entire chromosomes and sets of genes from one cell to another. *The F factor can be regarded as a specially designed vector for the interchange of genetic material.* It is interesting to note that *lysogenic phages can also mediate the exchange of host genes.* For example, λ-phage DNA can be inserted between the *gal* and *bio* genes on the *E. coli* chromosome (p. 739). Excision is usually but not invariably precise. In about 1 of 10^5 virions, λ DNA contains either the *gal* operon or the *bio* gene. Infection by such phages, which are called λgal or λbio, results in the introduction of these *E. coli* genes in addition to λ genes. A related phage, called φ80, inserts near the *trp* operon and can carry *trp* genes from one infected cell to another. Bacteriophage μ, which inserts nearly anywhere on the *E. coli* chromosome, invariably emerges with a piece of the bacterial chromosome. *These transducing*

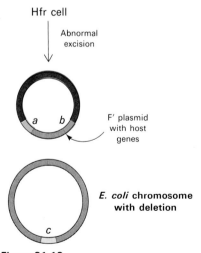

Figure 31-13
Abnormal excision leads to the formation of an F' plasmid, which contains part of the *E. coli* chromosome.

phages, like the F factor, are mobile genetic elements that promote the exchange of bacterial genes. Transduction probably accelerates bacterial evolution.

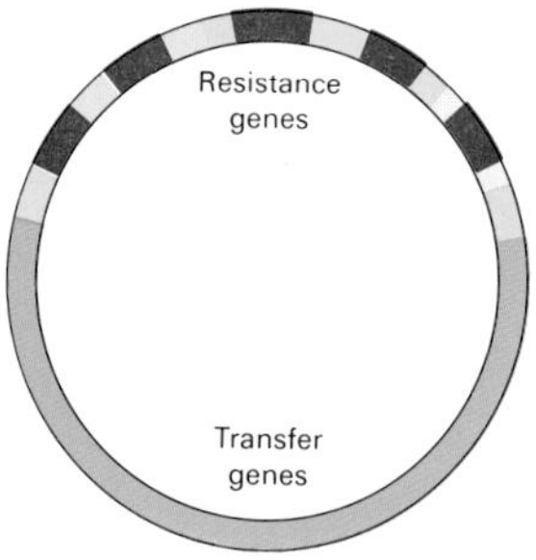

R-FACTOR PLASMIDS MAKE BACTERIA RESISTANT TO ANTIBIOTICS

A striking example of very rapid bacterial evolution took place in 1955 during an epidemic of bacterial dysentery. A strain of *Shigella dysenteriae* became simultaneously resistant to chloramphenicol, streptomycin, sulfanilamide, and tetracycline. Multiple-drug resistance of this kind is now common among many pathogenic microorganisms. The genes conferring resistance to multiple antibiotics are linked together on *R-factor (resistance factor) plasmids*. The larger of these plasmids contain a *resistance transfer factor* (RTF) in addition to several r-*genes* (Figure 31-14). The RTF region enables the plasmid to be transmitted to other bacteria by conjugation. In fact, the genes in the RTF region closely resemble those in F factors. The r-genes code for enzymes that inactivate specific drugs. R factors with an RTF region can be transmitted between different bacterial species in mixed cultures. Thus, *multiple-drug resistance can be infectious.*

The small R-factor plasmids are devoid of an RTF region and usually confer resistance to a single antibiotic. For example, the 8.2-kb R-pSC101 plasmid carries a gene for tetracycline resistance, but it cannot be transmitted by conjugation. However, this r-gene can join another plasmid with a different gene for drug resistance (Figure 31-15). A transmissible R plasmid is formed when these r-genes integrate into a plasmid that contains an RTF region. *Thus, complex R-factor plasmids are built from highly mobile modules that confer resistance to individual drugs.* These translocatable genetic elements are now called *transposons.*

INSERTION ELEMENTS CAN JOIN UNRELATED GENES

What is the structural basis of the high mobility of transposons? Electron-microscopic and nucleotide-sequence studies have revealed that *the sequence at one end of a transposon is repeated at the other end.* For example, the termini of Tn3, a transposon that codes for ampicillin resistance, consist of inverted repetitions of thirty-eight base pairs. The flanking sequences on the recipient are noninverted repetitions of a short sequence (typically five or nine base pairs) that was present once before insertion of the transposon (Figure 31-16). There is no homology between these flanking sequences on the recipient and the terminal sequences of the transposon. Fur-

Figure 31-14
Schematic diagram of an R factor. The *RTF* genes (for conjugation and replication) are shown in green, and the r-genes (for drug resistance) in red. Insertion sequences are shown in yellow.

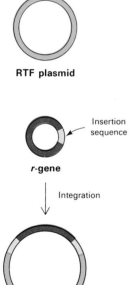

RTF plasmid

Insertion sequence

r-gene

Integration

Infectious R factor

Figure 31-15
An infectious R factor is formed when an r-gene joins an RTF plasmid.

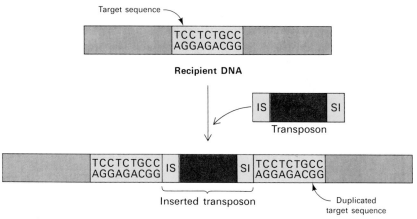

Figure 31-16
Transposition of a gene (red) bounded by insertion sequences (IS, yellow). The target site on the recipient (blue) is duplicated during transposition. One end of the transposon is an inverted repetition of the other end.

thermore, the *rec* genes of *E. coli* do not participate in the insertion process, which appears to be quite different from general genetic recombination. The termini of transposons probably act as *insertion elements* by directing the action of nucleases and other proteins involved in their integration. The important point is that *a transposon need not be homologous to the recipient DNA because the specificity of insertion is primarily determined by protein-DNA interactions rather than by base pairing.*

The smallest mobile genetic elements are *insertion sequences* (IS), which have a length of about 1 kb. Insertion sequences, in contrast with transposons, do not carry any genes. However, they markedly affect the expression of nearby genes. Insertion sequences usually block the transcription of distal genes in a transcription unit. They can also act as new promoters. Furthermore, insertion sequences enhance chromosomal rearrangements such as deletions and inversions. Several copies of four different insertion sequences (IS1, IS2, IS3, and IS4) have been found in the *E. coli* chromosome. Moreover, the terminal sequences of some transposons are identical with one of these insertion sequences. *It seems likely that a transposon is formed when a gene becomes bounded by a pair of insertion sequences.*

We have seen that plasmids and phages can exchange blocks of genes with bacterial chromosomes. Plasmids and phages can also recombine with each other. A tetracycline-resistance element has been shown to move from an R-factor plasmid to a *Salmonella* phage and then to the *Salmonella* chromosome, and from there to λ phage and then to the *trp* operon of *E. coli* and back to λ. This remarkable journey emphasizes the highly mobile nature of procaryotic genes. It will be very interesting to see whether genes in eucaryotes also are highly mobile.

NEW GENOMES CAN BE CONSTRUCTED IN THE LABORATORY AND CLONED IN HOST CELLS

The recent development of recombinant DNA technology is a major breakthrough in molecular biology. *New combinations of unrelated genes can be constructed in the laboratory.* These novel genomes can then be introduced into suitable cells and *amplified* many times by the DNA synthesizing machinery of the host. Some of these inserted genes are also transcribed and translated in their new setting. The major steps in the cloning of DNA are (Figure 31-17):

1. *Construction of a recombinant molecule.* A DNA fragment of interest is covalently joined to a DNA *vector* (*vehicle*). The essential feature of a vector is that it can replicate autonomously in an appropriate host. For example, plasmids and λ phage are choice vectors

Chimeric DNA—
A recombinant DNA molecule containing unrelated genes.
From *chimera*, a mythological creature with the head of a lion, the body of a goat, and the tail of a serpent.

". . . a thing of immortal make, not human, lion-fronted and snake behind, a goat in the middle, and snorting out the breath of the terrible flame of bright fire. . . ."

Iliad (6.179)

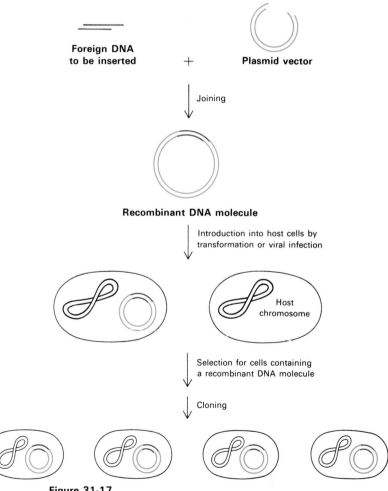

Figure 31-17
Synthesis and cloning of recombinant DNA molecules.

for cloning in *E. coli.* As will be described shortly, DNA molecules can be spliced by ligating fragments with cohesive single-stranded ends or with blunt ends.

2. *Introduction into host cells.* Most bacterial and eucaryotic cells take up naked DNA molecules from the medium. The efficiency of uptake is low (of the order of 1 of 10^6 DNA molecules), but an appreciable proportion of cells can be transformed under appropriate experimental conditions. Alternatively, target cells can be infected with reassembled virions harboring the recombinant DNA molecule. In this synthetic viral genome, a gene of interest replaces a segment of viral DNA that is not essential for replication.

3. *Selection.* The next step is to determine which cells harbor the recombinant DNA molecule containing the gene of interest. The desired clones can be selected on the basis of the presence of the vector or of the inserted gene itself. For example, some plasmid vectors confer resistance to an antibiotic. Another approach is to determine which cells bind RNA that is complementary to the gene of interest or synthesize protein encoded by it. Clones containing recombinant DNA are stable for at least several hundred generations.

The cloning of recombinant DNA has already greatly enhanced our understanding of chromosome structure and gene expression. Many inserted genes have been amplified by cloning, which has resulted in the production of large amounts of DNA for sequence and electron-microscopic analyses. Large quantities of otherwise scarce proteins also have been synthesized by these clones. Recombinant DNA techniques are also being used to dissect complex genomes and analyze the regulation of their expression.

RESTRICTION ENZYMES AND DNA LIGASE ARE ESSENTIAL TOOLS IN FORMING RECOMBINANT DNA MOLECULES

DNA molecules can be readily spliced in vitro by taking advantage of restriction endonucleases (p. 590), DNA ligases (p. 579), and other highly specific enzymes that act on DNA. The vector in a recombinant DNA experiment can be prepared for splicing by cleaving it at a single specific site with a restriction endonuclease. For example, the small plasmid pSC101 can be split at a unique site by the EcoR1 restriction enzyme. The staggered cuts made by this enzyme generate *complementary single-stranded ends (cohesive ends)*. Now suppose that the DNA fragment to be inserted into this plasmid were formed by cleaving a large DNA molecule with the EcoR1 endonuclease. The single-stranded ends of this fragment would then be complementary to those of the cut plasmid. The DNA

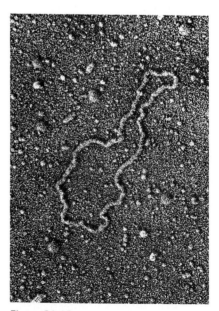

Figure 31-18
Electron micrograph of pSC101, a plasmid vector used to clone DNA. [Courtesy of Dr. Stanley Cohen.]

fragment and the cut plasmid can be annealed and then joined by DNA ligase (Figure 31-19). Bacteria are then exposed to this mix-

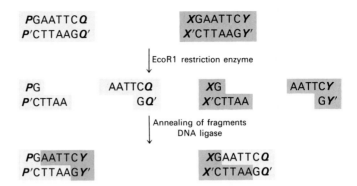

Figure 31-19
Joining of DNA molecules by the cohesive-end method. One of the parental DNA molecules (shown in green) carries genes *P* and *Q* separated by a restriction site, and the other (shown in red) carries *X* and *Y*. One of the recombinant molecules contains *P* and *Y*, and the other contains *Q* and *X*.

ture of DNA molecules. The small proportion of bacteria harboring the added plasmid are selected by taking advantage of the fact that pSC101 confers resistance to tetracycline. A different selection procedure is then used to determine which clones contain a DNA insert within the plasmid.

A second method for joining two unrelated DNA molecules requires the addition of poly dA tails to both 3′ ends of one molecule and poly dT tails to both 3′ ends of the other molecule (Figure 31-20). These homopolymeric extensions are synthesized by *terminal*

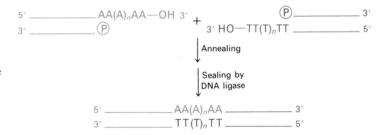

Figure 31-20
Joining of DNA molecules by the poly-dA–poly-dT tail method.

deoxynucleotidyl transferase (terminal transferase), an enzyme that adds nucleotides to the 3′ hydroxyl group of a DNA chain. This transferase, in contrast with DNA polymerase, does not take instructions from a template, and so tails made up of just one kind of nucleotide (typically 100 residues long) can be added. *The poly A tails on one DNA molecule are annealed to the poly T tails on the other.* The lengths of the tails are not precisely defined and so the gaps are filled by DNA polymerase I before the strands are joined by DNA ligase. Likewise, poly dG and poly dC tails can be used to anneal different DNA molecules.

A third approach for joining DNA molecules combines the advantages of the cohesive-end method with the generality of the dA-dT method. *A chemically synthesized linker* (from six to ten base pairs) that is susceptible to cleavage by a restriction enzyme is covalently joined to the ends of a DNA fragment or vector. The 5′ ends of the decameric linker and of the DNA molecule are phosphorylated by polynucleotide kinase and joined by T4 ligase, which can form a covalent bond between blunt-ended (flush-ended) DNA molecules. Cohesive ends are generated when these terminal extensions are cut by an appropriate restriction enzyme (Figure 31-21). Thus, cohesive ends corresponding to the specificity of a particular restriction enzyme can be fashioned on virtually any DNA molecule.

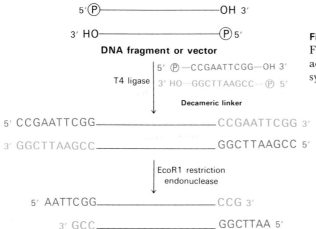

Figure 31-21
Formation of cohesive ends by the addition and cleavage of a chemically synthesized linker.

PLASMIDS AND LAMBDA PHAGE ARE CHOICE VECTORS FOR THE CLONING OF DNA IN BACTERIA

New vectors are being constructed to enhance the delivery of recombinant DNA and facilitate the selection of bacteria harboring them. For example, the plasmid pBR322 contains genes for resistance to tetracycline and ampicillin (a penicillinlike antibiotic). This plasmid can be cleaved at single sites by five different endonucleases (Figure 31-22). Insertion of DNA at the EcoR1 restriction site does not alter either gene for antibiotic resistance. However, insertion at the HindIII, SaI, or BamHI restriction sites leads to the inactivation of the gene for tetracycline resistance, an effect called *insertional inactivation.* Cells containing pBR322 with a DNA insert are resistant to ampicillin but sensitive to tetracycline and so they can be readily selected. Cells that failed to take up the vector are sensitive to both antibiotics, whereas cells containing pBR322 without a DNA insert are resistant to both.

Another group of vectors is derived from ColE1, a plasmid that codes for colicin E, a protein toxin that kills certain strains of *E. coli.* An advantage of using these *colicinogenic plasmids* as vectors is that

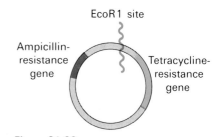

Figure 31-22
Genetic map of pBR322, a plasmid with two genes for antibiotic resistance.

their replication is not stringently controlled. About twenty-five copies of ColE1 are usually present in an infected cell. Treatment with chloramphenicol blocks protein synthesis and the replication of the host chromosome. However, the replication of ColE1 continues under these conditions. In fact, a thousand copies of ColE1 can accumulate in a chloramphenicol-treated cell, so that this plasmid accounts for nearly half of the cellular DNA.

Lambda phage is another choice vector. Large segments of its 48-kb DNA are not essential for lytic infection or integration and can be replaced by foreign DNA. Mutant λ phages designed for the cloning of DNA have been constructed. One of the mutants, called λgt-λβ, contains two EcoR1 cleavage sites instead of the five normally present (Figure 31-24). After cleavage, the middle segment of

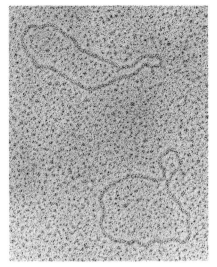

Figure 31-23
Electron micrograph of ColE1 plasmids. [Courtesy of Dr. Jack Griffith.]

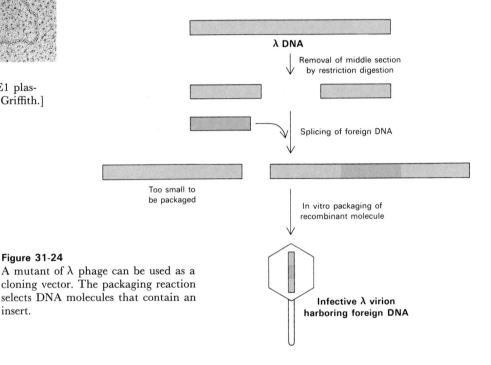

λ DNA

Removal of middle section by restriction digestion

Splicing of foreign DNA

Too small to be packaged

In vitro packaging of recombinant molecule

Infective λ virion harboring foreign DNA

Figure 31-24
A mutant of λ phage can be used as a cloning vector. The packaging reaction selects DNA molecules that contain an insert.

this λ DNA molecule can be removed. The two remaining pieces of DNA have a combined length equal to 72% of a unit genome. This amount of DNA is too little to be packaged in a λ head. The range of lengths that can be readily packaged is from 75% to 105% of a unit genome. However, *a suitably long DNA insert (say, 10 kb) between the two ends of λ DNA enables such a recombinant DNA molecule (93% in length) to be encapsidated.* Nearly all infective λ particles formed in this way will contain an inserted piece of foreign DNA. Another advantage of using these virions as vectors is that they enter bacteria with a much higher frequency than do plasmids. Recombinant DNA molecules can now be packaged in vitro to form infectious λ virions. A variety of λ mutants have been constructed for use as cloning vectors. Some of them can serve as vehicles for DNA inserts as large as 40 kb.

SPECIFIC EUCARYOTIC GENES CAN BE CLONED STARTING WITH A DIGEST OF GENOMIC DNA

As discussed in an earlier chapter (p. 685), the analysis of eucaryotic genomes is a formidable problem. A 1-kb gene is 2.5×10^{-4} of the *E. coli* genome but only 3.4×10^{-7} of a mammalian genome. Recombinant DNA techniques now make it feasible to insert a eucaryotic gene into *E. coli*, which greatly reduces the complexity of the problem. The experiment begins with the partial digestion of a eucaryotic genome to randomly generate fragments having a mean length of about 20 kb (Figure 31-25). Synthetic linkers are attached to the ends of these fragments, cohesive ends are formed, and the fragments are then joined to a vector such as λ phage DNA. The in vitro packaging of DNA into virions selects for recombinant DNA molecules that contain a large insert. *E. coli* are then infected with these recombinant phage. The resulting lysate contains fragments of eucaryotic DNA housed in phage and amplified about a millionfold. In effect, this lysate is a *cloned library of eucaryotic DNA*.

Such a library must then be screened to identify phage clones containing a particular eucaryotic gene. A calculation shows that only 1 of 180,000 clones will carry a single-copy eucaryotic gene. Thus, a very rapid and efficient screening process is essential. This is accomplished by hybridization. *The presence of a specific DNA sequence in a single λ plaque can be detected by using a radioactive complementary DNA or RNA molecule as a probe and visualizing its binding by autoradiography.* A million clones can be screened per day in this way. Thus, a clone containing a specific eucaryotic gene can be readily identified and isolated if its RNA transcript is available in relatively pure form.

EUCARYOTIC GENES CAN BE TRANSCRIBED AND TRANSLATED IN BACTERIAL CELLS

Recombinant DNA molecules carrying bacterial genes are often expressed in *E. coli*. For example, the cloning of the tryptophan operon of *E. coli* carried by a ColE1 plasmid vector resulted in the high-level production of the five biosynthetic enzymes specified by this operon. The amounts of these enzymes were about twenty times as high as those in a normal *E. coli* cell because of the presence of multiple copies of the recombinant plasmid. DNA from yeast, a simple eucaryote, can also be expressed in bacteria. In one study, the host cell was an *E. coli* mutant that required histidine because it lacked imidazole glycerol phosphate dehydrogenase. Infection of this mutant with λ phage containing a segment of yeast DNA yielded several bacterial clones that no longer required exogenous histidine. The inserted piece of yeast DNA supplied the missing

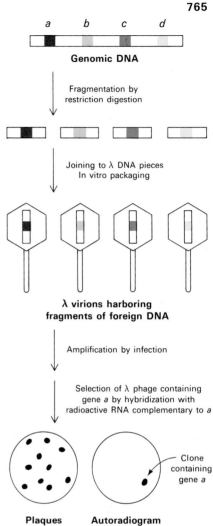

Figure 31-25
Strategy for cloning a specific eucaryotic gene starting with a digest of the whole genome.

gene, which was expressed by the transcriptional and translational machinery of the cell.

Can mammalian genes be expressed in bacteria? This question was explored by introducing a rat insulin gene into *E. coli* (Figure 31–26). The starting point was an insulinoma, a pancreatic tumor that secretes high levels of insulin. This tumor is rich in mRNA for the synthesis of preproinsulin, a precursor of the active hormone (p. 848). Double-stranded cDNA was prepared by reverse transcription of this mRNA and inserted into a plasmid vector. Why was

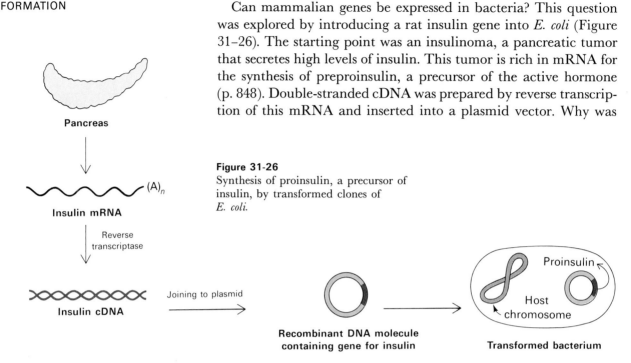

Figure 31-26
Synthesis of proinsulin, a precursor of insulin, by transformed clones of *E. coli.*

cDNA—
Complementary DNA synthesized by reverse transcriptase from an RNA template.

cDNA rather than genomic DNA introduced into *E. coli?* As discussed previously, many eucaryotic genes contain intervening sequences that are excised from primary transcripts (p. 702). Bacteria are probably unable to remove these sequences, and so it seems desirable to transform them with a DNA segment that is complementary to the mature mRNA. In fact, several bacterial clones transformed with cDNA for insulin were found to synthesize small amounts (about 100 copies per cell) of a precursor of insulin. Another important recent finding is that a DNA sequence coding for chicken ovalbumin is expressed in *E. coli.* About 1.5% of the protein synthesized by these transformed bacteria is full-length (43-kdal) ovalbumin. *It is evident that genes for eucaryotic proteins can be expressed by bacteria.*

A CHEMICALLY SYNTHESIZED GENE FOR SOMATOSTATIN, A PEPTIDE HORMONE, IS EXPRESSED IN *E. coli*

Recent advances in the chemical synthesis of specific DNA sequences have enhanced the scope and power of recombinant DNA technology. Genes with nearly any desired nucleotide sequence can be synthesized de novo and inserted into a vector for delivery into

E. coli. A nice example of this approach is provided by the synthesis of a gene for *somatostatin,* a 14-residue peptide (Figure 31-27) found in hypothalamic extracts. Somatostatin inhibits the secretion of growth hormone, insulin, and glucagon. A DNA molecule coding for this peptide was synthesized by joining eight oligonucleotide blocks. This gene was fused to a β-galactosidase gene on a plasmid vector. Transformation of *E. coli* led to the synthesis of a hybrid protein in which somatostatin was linked to β-galactosidase. The peptide bond between these moieties of the chimeric protein was split in vitro by cyanogen bromide (p. 25). The inserted gene contained a codon for methionine before the first codon of somatostatin for this reason. The carboxyl end of somatostatin was free because two stop codons followed the codon for the last residue of the peptide (Figure 31-27). The chimeric protein constituted an appreciable proportion (about 3%) of the total cellular protein. Moreover, somatostatin produced in this way was biologically active. *Thus, a functional polypeptide can be produced by cloning a chemically synthesized gene.*

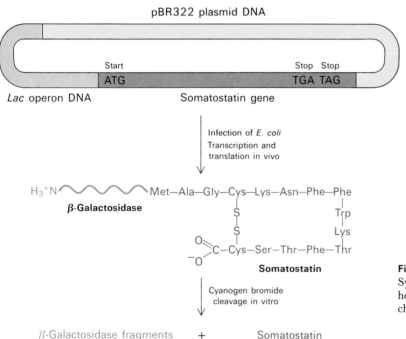

Figure 31-27
Synthesis of somatostatin, a peptide hormone, by *E. coli* transformed by a chemically synthesized gene.

IMPACT OF GENE CLONING

Recombinant DNA technology has opened new vistas in molecular biology. The amplification of genes by cloning in bacteria has provided abundant quantities of DNA for electron-microscopic and sequencing studies. Specific regions of DNA that control gene mobility, DNA replication, and transcription are being analyzed. New insights are emerging, as exemplified by the discovery of intervening sequences in many eucaryotic genes. Complex chromosomes are rapidly being mapped and dissected into units that can be manipulated and explored. The pace of research has been greatly accelerated. Gene cloning is also becoming important as a means of obtaining large quantities of specific proteins. For example, the amount of DNA ligase produced by *E. coli* can be increased 500-fold by recombinant DNA techniques. The synthesis of eucaryotic peptides and proteins by transformed bacteria is most promising. Hormones such as insulin and antiviral agents such as interferon may soon be produced by bacteria. A new pharmacology that is likely to have a profound effect on medicine is beginning to emerge. The potentiality of gene cloning for increasing agricultural yields is also being explored. Eucaryotic genes are now being introduced into eucaryotic cells as well as into bacteria. For example, SV40 virus has been used as a vector to bring the rabbit β-globin gene into monkey-kidney cells. These infected cells synthesize substantial amounts of rabbit β-globin protein. This experimental approach is a promising means of elucidating regulatory mechanisms in eucaryotic gene expression.

SUMMARY

In genetic recombination, a new DNA molecule is formed by the breakage and reunion of DNA strands. Any pair of homologous sequences on the parental DNA molecules can be exchanged, and so this process is called general genetic recombination. In *E. coli,* general recombination is mediated by the action of the *rec* genes. The recBC protein, a nuclease, generates single-stranded DNA, which can invade a duplex DNA molecule. This strand-assimilation reaction is catalyzed by the recA protein, which concomitantly hydrolyzes ATP.

Genetic rearrangements of nonhomologous loci are mediated by mobile genetic elements called transposons (translocatable elements). Plasmids, which are small circular DNA duplexes, are an important class of mobile genetic elements. These accessory chromosomes carry genes for the inactivation of antibiotics, metabolism of natural products, and production of toxins. Plasmids can replicate autonomously or become integrated into the host chromosome. The F-factor (fertility-factor) plasmid enables bacteria to donate genes to other bacteria by conjugation, a process mediated by direct physical contact between the donor (F^+) and recipient (F^-) cells. R-factor plasmids confer resistance to antibiotics. The larger of these plasmids contain a resistance transfer factor (RTF) that enables the plasmid to be transmitted to other bacteria by conjugation. Several genes that confer resistance to antibiotics can become linked to an RTF region. Hence, multiple-drug resistance can be infectious. Transposons are highly mobile because they are bounded by insertion sequences. These terminal sequences direct the action of proteins that mediate their integration into DNA. A transposon need not be homologous to the recipient DNA.

New combinations of unrelated genes can be constructed in the laboratory and cloned in host cells. First, a recombinant DNA molecule is synthesized by joining a DNA fragment to a DNA vector that can replicate autonomously in an appropriate host. Lambda phage, SV40 virus, and plasmids are choice vectors. Restriction enzymes and DNA ligase are essential tools in forming new DNA molecules. Cohesive ends, homopolymeric tails, and chemically synthesized linkers are three means of bringing DNA molecules together. Recombinant DNA molecules can be introduced into host cells by infecting them with reassembled viruses or exposing them to naked DNA molecules. The final step is to select cells that harbor the desired recombinant DNA molecule by testing for a distinctive property of a newly introduced gene (e.g., drug resistance). Specific eucaryotic genes can be cloned in *E. coli* starting with a digest of genomic DNA. Many eucaryotic genes can be transcribed and translated in bacterial cells. Gene cloning is a powerful technique for analyzing the architecture and expression of complex genes. Recombinant DNA technology is also a very promising method for synthesizing large amounts of scarce genes and proteins.

SELECTED READINGS

WHERE TO START

Gilbert, W., and Villa-Komaroff, L., 1980. Useful proteins from recombinant bacteria. *Sci. Amer.* 242(4):74–94. [Available as *Sci. Amer.* Offprint 1466.]

Cohen, S. N., and Shapiro, J. A., 1980. Transposable genetic elements. *Sci. Amer.* 242(2):40–49. [Offprint 1460.]

Abelson, J., and Butz, E., (eds.), 1980. Recombinant DNA. *Science* 209:1317–1438. [This highly informative issue of *Science* contains many fine articles on the structure and movements of eucaryotic genes and on the expression of cloned genes.]

Clowes, R. C., 1973. The molecule of infectious drug resistance. *Sci. Amer.* 228(4):18–27. [Offprint 1269.]

Cohen, S. N., 1975. The manipulation of genes. *Sci. Amer.* 233(1):24–33. [Offprint 1324. This article and several others of interest are reprinted in D. Freifelder, (ed.), *Recombinant DNA* (Freeman, 1978).]

Stanier, R. Y., Adelberg, E. A., and Ingraham, J. L., 1976. *The Microbial World* (4th ed.). Prentice-Hall.

[Chapter 15 on recombination, conjugation, and transduction is outstanding.]

GENETIC RECOMBINATION

Stahl, F. W., 1979. *Genetic Recombination.* Freeman.

Potter, H., and Dressler, D., 1978. In vitro system from *Escherichia coli* that catalyzes generalized genetic recombination. *Proc. Nat. Acad. Sci.* 75:3698–3702.

McEntee, K., Weinstock, G. M., and Lehman, I. R., 1979. Initiation of general recombination catalyzed in vitro by the recA protein of *E. coli. Proc. Nat. Acad. Sci.* 76:2615–2619.

Cunningham, R. P., Shibata, T., DasGupta, C., and Radding, C. M., 1979. Single strands induce recA protein to unwind duplex DNA for homologous pairing. *Nature* 281:191–195.

Radding, C. M., 1978. Genetic recombination: strand transfer and mismatch repair. *Ann. Rev. Biochem.* 47:847–880.

Calos, M. P., and Miller, J. H., 1980. Transposable elements. *Cell* 20:579–595.

Cohen, S. N., 1976. Transposable genetic elements and plasmid evolution. *Nature* 263:731–738.

Kleckner, N., 1977. Translocatable elements in procaryotes. *Cell* 11:11–23.

Hicks, J., Strathern, J. N., and Klar, A. J. S., 1979. Transposable mating type genes in *Saccharomyces cerevisiae*. *Nature* 282:478–483.

Gill, R. E., Heffron, F., and Falkow, S., 1979. Identification of the protein encoded by the transposable element Tn3 which is required for its transposition. *Nature* 282:797–801.

Chou, J., Lemaux, P. G., Casadaban, M. J., and Cohen, S. N., 1979. Transposition protein of Tn3: identification and characterisation of an essential repressor-controlled gene product. *Nature* 282:801–806.

Bukhari, A. I., Shapiro, J. A., and Adhya, S. L., (eds.), 1977. *DNA Insertion Elements and Episomes.* Cold Spring Harbor Laboratory.

Broda, P., 1979. *Plasmids.* Freeman.

CONSTRUCTION AND CLONING OF RECOMBINANT GENES

Setlow, J. K., and Hollaender, A., (eds.), 1979. *Genetic Engineering: Principles and Methods.* Plenum.

Scott, W. A., and Werner, R., 1977. *Molecular Cloning of Recombinant DNA.* Academic Press.

Helinski, D. R., 1978. Plasmids as vehicles for gene cloning: impact on basic and applied research. *Trends Biochem. Sci.* 3:10–14.

Cameron, J. R., Panesenko, S. M., Lehman, I. R., and Davis, R. W., 1975. In vivo construction of bacteriophage λ carrying segments of the *Escherichia coli* chromosome: selection of hybrids containing the gene for DNA ligase. *Proc. Nat. Acad. Sci.* 72:3416–3420.

Collins, J., and Hohn, B., 1978. Cosmids: a type of plasmid gene-cloning vector that is packageable in vitro in bacteriophage heads. *Proc. Nat. Acad. Sci.* 75:4242–4246.

Maniatis, T., Hardison, R. C., Lacy, E., Lauer, J. O'Connell, C., Quon, D., Sim, G. K., and Efstratiadis, A., 1978. The isolation of structural genes from libraries of eucaryotic DNA. *Cell* 15:687–701.

EXPRESSION OF CLONED GENES

Villa-Komaroff, L., Efstratiadis, A., Broome, S., Lomedico, P., Tizard, R., Naber, S. P., Chick, W. L., and Gilbert, W., 1978. A bacterial clone synthesizing proinsulin. *Proc. Nat. Acad. Sci.* 75:3727–3731.

Burrell, C. J., Mackory, P., Greenaway, P. J., Hofschneider, P. H., and Murray, K., 1979. Expression in *Escherichia coli* of hepatitis B virus DNA sequences cloned in plasmid pBR322. *Nature* 279:43–47.

Mulligan, R. C., Howard, B. H., and Berg, P., 1979. Synthesis of rabbit β-globin in cultured monkey kidney cells following infection with a SV40 β-globin recombinant genome. *Nature* 277:108–114.

CHEMICAL SYNTHESIS OF GENES

Khorana, H. G., 1979. Total synthesis of a gene. *Science* 203:614–625.

Itakura, K., Hirose, T., Crea, R., Riggs, A. D., Heyneker, H. L., Bolivar, F., and Boyer, H. W., 1977. Expression in *Escherichia coli* of a chemically synthesized gene for the hormone somatostatin. *Science* 198:1056–1063.

Crea, R., Kraszewski, A., Hirose, T., and Itakura, K., 1978. Chemical synthesis of genes for human insulin. *Proc. Nat. Acad. Sci.* 75:5765–5769.

On the facing page: Scanning electron micrograph of retinal rod cells. A rod cell can be excited by a single photon. [Courtesy of Dr. Deric Bownds.]

PART V

MOLECULAR PHYSIOLOGY

interaction of information, conformation,
and metabolism in physiological processes

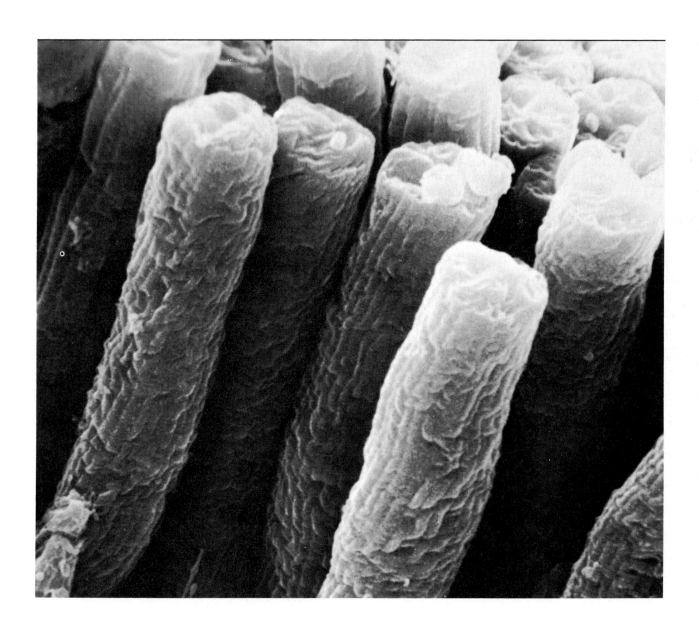

CHAPTER 32

BACTERIAL CELL ENVELOPES

Bacterial cells, in contrast with animal cells, are surrounded by cell walls, which confer mechanical support and give them their distinctive shape. The cell membrane in a bacterium cannot by itself withstand the high osmotic pressure inside the cell, which may reach 20 atmospheres because of the high internal concentration of metabolites. A bacterial cell lyses in ordinary media unless it is surrounded by a cell wall.

Bacterial cell walls are of considerable medical importance. The cell wall and associated substances are largely responsible for the *virulence of bacteria.* Indeed, the symptoms of numerous bacterial diseases can be produced by injecting isolated bacterial cell walls into experimental animals. Furthermore, the *characteristic antigens* of bacteria are located on their cell walls. Animals acquire immunity to some bacteria following the injection of extracts of bacterial cell walls. Finally, bacteria can be killed by preventing the synthesis of their cell walls. Indeed, this is the mechanism of action of *penicillin* and of some other antibiotics.

For more than a half-century, bacteria have been classified as being either gram-negative or gram-positive, in accord with their reaction with the Gram stain. The basis of this empirical distinction is now understood. These two classes of bacteria have different

Figure 32-1
Electron micrograph of the isolated cell wall of *Bacillus licheniformis.* [Courtesy of Dr. Nathan Sharon.]

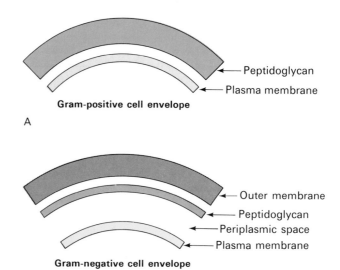

Figure 32-2
Schematic diagrams showing the differences between the cell envelopes of (A) gram-positive and (B) gram-negative bacteria.

kinds of cell envelopes (Figure 32-2). The plasma membrane of *gram-positive bacteria* is surrounded by a thick cell wall, typically 250 Å wide, composed of peptidoglycan and teichoic acid. *Gram-negative bacteria* have a more complex cell envelope. Their plasma membrane is surrounded by a 30-Å-wide peptidoglycan wall, which in turn is covered by an 80-Å outer membrane that is a mosaic of protein, lipid, and lipopolysaccharide.

THE CELL WALL IS AN ENORMOUS BAG-SHAPED MACROMOLECULE

Let us consider the structure and biosynthesis of the cell wall of *Staphylococcus aureus,* a gram-positive bacterium that causes pus-producing infections of tissues such as skin, bone, and lung. The cell-wall macromolecule is called a *peptidoglycan* because it consists of peptide and sugar units. In the peptidoglycan, *linear polysaccharide chains are cross-linked by short peptides.* The peptidoglycan is extensively cross-linked, which gives rise to a single enormous bag-shaped macromolecule. The isolated cell wall has the same shape as the bacterium it had formerly enveloped.

The peptidoglycan consists of three repeating units (Figure 32-3): (1) a *disaccharide* of *N*-acetylglucosamine (NAG) and *N*-acetylmuramic acid (NAM) in β-1,4-glycosidic linkage; (2) a *tetrapeptide* of L-alanine, D-glutamine, L-lysine, and D-alanine; and (3) a *pentaglycine bridge peptide*. The tetrapeptide is unusual in two respects. It contains D-amino acids, which are never found in proteins. Moreover, the D-glutamine residue forms a peptide linkage at its side-chain γ-carboxyl group.

In the intact peptidoglycan, NAG and NAM alternate in sequence to form a linear polysaccharide chain. The pentaglycine

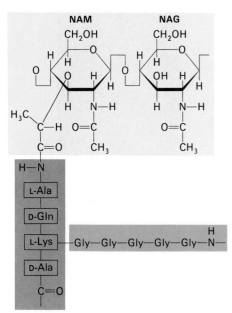

Figure 32-3
Basic structural unit of the peptidoglycan of *Staphylococcus aureus.*

peptide cross-links NAM residues on different polysaccharide strands. The amino group of (Gly)$_5$ forms a peptide bond with the carboxyl group of D-alanine, whereas the carboxyl group of (Gly)$_5$ forms a peptide bond with the side-chain ϵ-amino group of L-lysine.

STAGES IN THE SYNTHESIS OF THE PEPTIDOGLYCAN

There are five stages in the synthesis of the peptidoglycan:

1. *A peptide unit is built on NAM while the sugar is attached to uridine diphosphate.*

2. *The NAM-peptide unit is transferred to a carrier lipid.* A special problem is encountered in cell-wall synthesis. The final polymeric product lies outside the permeability barrier of the cell, whereas the precursors are formed inside the cell. UDP is too polar to get across the membrane. In contrast, the carrier lipid can function as a transmembrane shuttle because it is highly nonpolar.

3. *NAG and the pentaglycine bridge are added to the NAM-peptide unit* while it is attached to the carrier lipid.

4. *The disaccharide peptide unit is transferred from the carrier lipid to a growing polysaccharide chain.*

5. *Different polysaccharide strands are cross-linked by a transpeptidation reaction involving the pentaglycine bridges.*

SYNTHESIS OF A UDP-SUGAR-PEPTIDE UNIT

The biosynthesis of the peptidoglycan starts with the formation of activated sugars. *Uridine diphosphate-N-acetylglucosamine* (UDP-NAG) is synthesized from *N*-acetylglucosamine 1-phosphate and UTP in a reaction that is driven forward by the hydrolysis of PP$_i$ (Figure 32-5). *Uridine diphosphate-N-acetylmuramic acid* (UDP-NAM) is

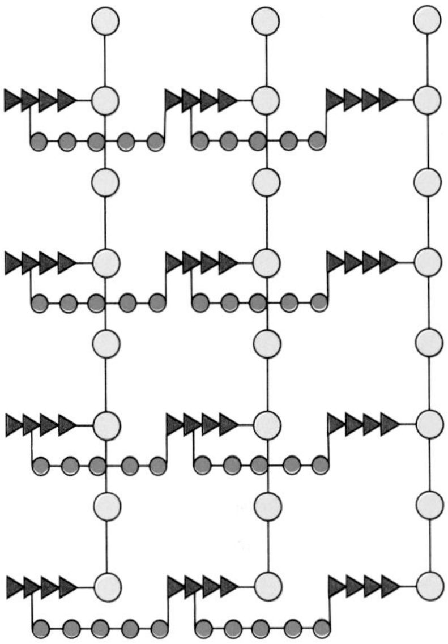

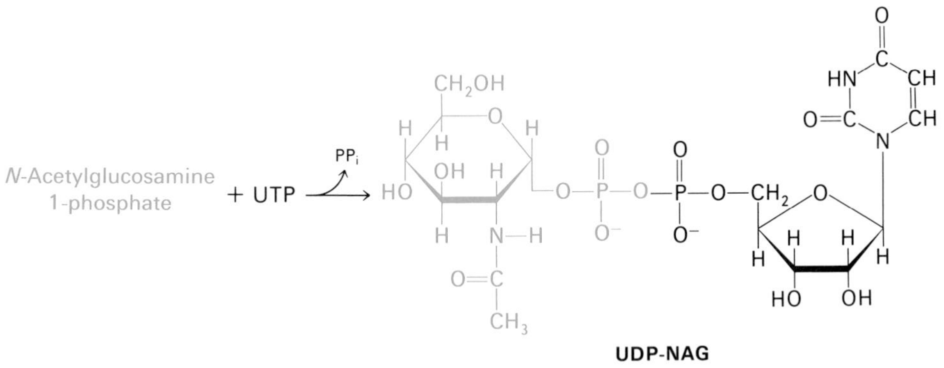

Figure 32-4
Schematic diagram of the peptidoglycan. The sugars are shown in yellow, the tetrapeptides in red, and the pentaglycine bridges in blue. The cell wall is a single, enormous bag-shaped macromolecule because of extensive cross-linking.

Figure 32-5
Synthesis of UDP-NAG.

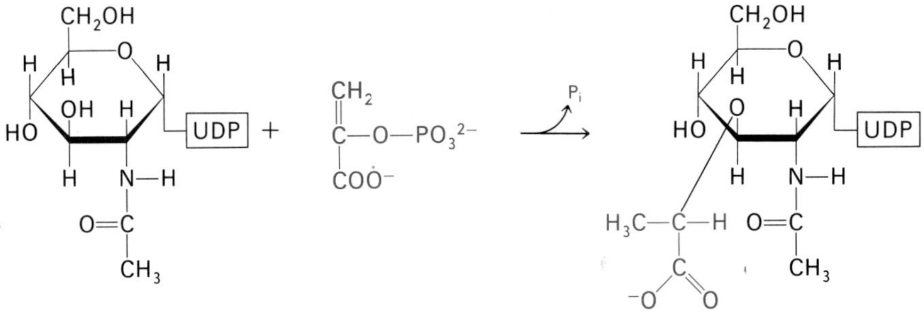

Figure 32-6
Synthesis of UDP-NAM.

UDP-NAG   Phosphoenolpyruvate   UDP-NAM

formed from UDP-NAG and phosphoenolpyruvate (Figure 32-6).

Growth of the *peptide chain* starts with the formation of a peptide bond between the amino group of L-alanine and the carboxyl group of the *N*-acetylmuramic acid moiety of UDP-NAM. D-Glutamate, L-lysine, and the dipeptide D-alanyl-D-alanine are successively attached (Figure 32-7). The D-amino acids are formed from their

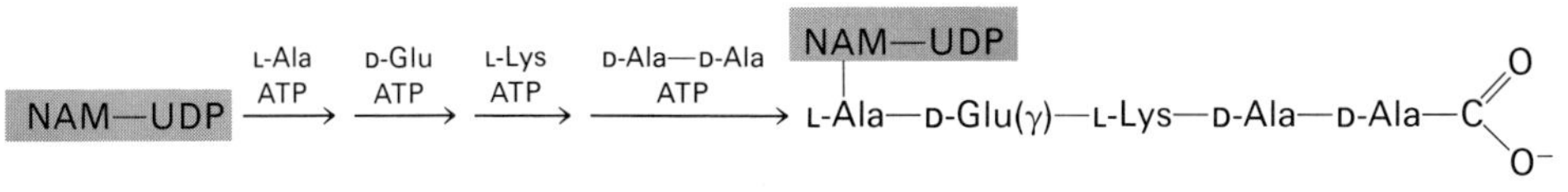

Figure 32-7
Synthesis of a pentapeptide attached to UDP-NAM.

L-isomer by the action of *racemases* that contain pyridoxal phosphate prosthetic groups. The formation of each of these peptide bonds requires an ATP. *These peptide-bond syntheses are catalyzed by special enzymes, not by the ribosomal pathway for protein synthesis.* The information concerning the amino acid sequence of the peptide is inherent in the specificities of the enzymes rather than in the base sequence of a messenger RNA. A peptide of this kind could not be synthesized by the usual mechanisms of protein synthesis because it contains D-amino acids and a γ-peptide bond.

TRANSFER OF THE SUGAR-PEPTIDE UNIT TO A CARRIER LIPID

The activated sugar-peptide unit is then transferred from UDP to a *carrier lipid* (Figure 32-8). This phosphorylated long-chain alcohol

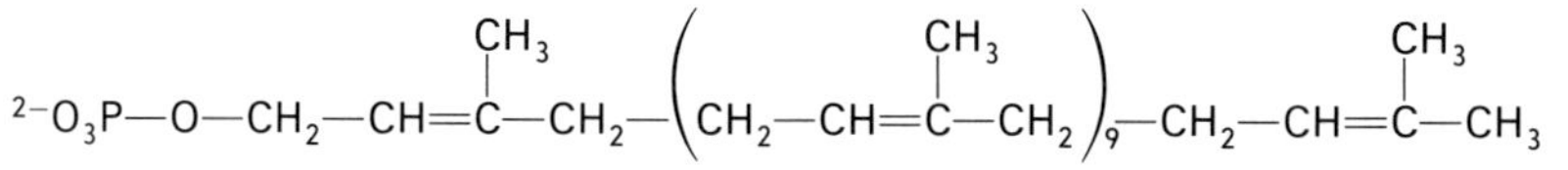

Figure 32-8
Structure of the carrier lipid in cell-wall biosynthesis.

contains eleven isoprene units, which make it very hydrophobic, in contrast with UDP. It seems likely that the hydrophobic character of the C_{55} carrier lipid enables the nascent sugar-peptide unit to traverse the permeability barrier imposed by the cell membrane.

NAM—O—P(=O)(O⁻)—O—P(=O)(O⁻)—O—Uridine + ⁻O—P(=O)(⁻O)—O—[Carrier lipid] ⇌ NAM—O—P(=O)(O⁻)—O—P(=O)(O⁻)—O—[Carrier lipid] + UMP

Recall that another highly hydrophobic carrier, dolichol phosphate, participates in the synthesis of the core oligosaccharide of eucaryotic glycoproteins (p. 715).

SYNTHESIS OF A DISACCHARIDE-PEPTIDE UNIT ATTACHED TO A CARRIER LIPID

The next step is the addition of NAG to the NAM moiety of the sugar-peptide and carrier-lipid unit. UDP-NAG, the activated sugar donor, reacts with C-4 of the NAM unit to form a β-1,4-glycosidic linkage between these sugars (Figure 32-9). The free α-

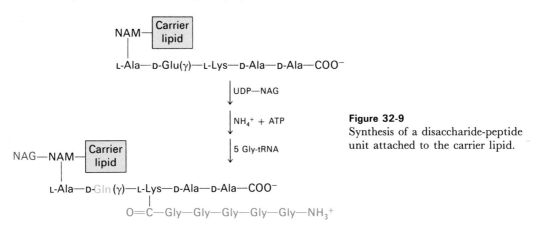

Figure 32-9
Synthesis of a disaccharide-peptide unit attached to the carrier lipid.

carboxyl group of D-glutamic acid in the peptide is amidated by NH_4^+ in a reaction that is driven by the hydrolysis of ATP. A *pentaglycine-bridge* unit is then added to the ϵ-amino group of the peptide lysine residue. This chain of five glycine residues is formed by the *successive transfer of single glycine residues from glycyl-tRNA*. This is a unique reaction in that tRNA otherwise serves as an amino acid donor only in protein synthesis on ribosomes. The construction of the basic structural unit of the cell wall is now complete.

TRANSFER OF THE DISACCHARIDE-PEPTIDE UNIT TO A GROWING POLYSACCHARIDE CHAIN

The disaccharide-peptide unit is transferred from the carrier lipid to the nonreducing end of a growing polysaccharide chain. The C-1 carbon of the NAM moiety is activated because it is linked to the carrier by a pyrophosphate bond. The C-1 carbon of the activated NAM unit reacts with C-4 of the NAG terminus of a growing polysaccharide chain to form a β-1,4-glycosidic bond (Figure 32-10).

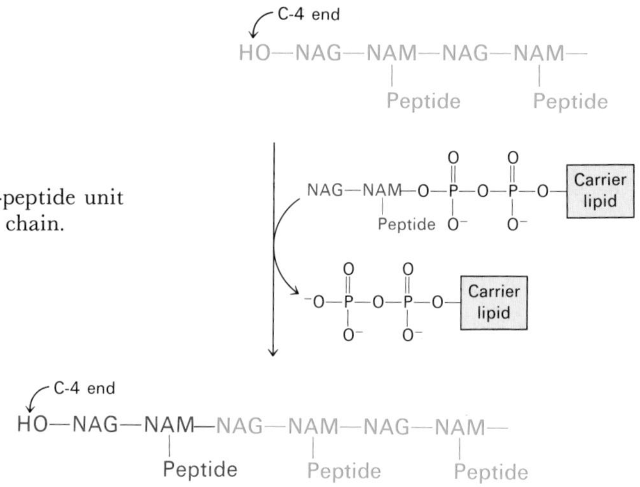

Figure 32-10
Transfer of the disaccharide-peptide unit to a growing polysaccharide chain.

The lipid carrier is released in the pyrophosphate form. It is then hydrolyzed to the monophosphate form by a specific phosphatase. This dephosphorylation step is essential for the regeneration of the species that accepts the sugar-peptide unit from UDP. Bacitracin, a peptide antibiotic, inhibits cell-wall biosynthesis by blocking this step.

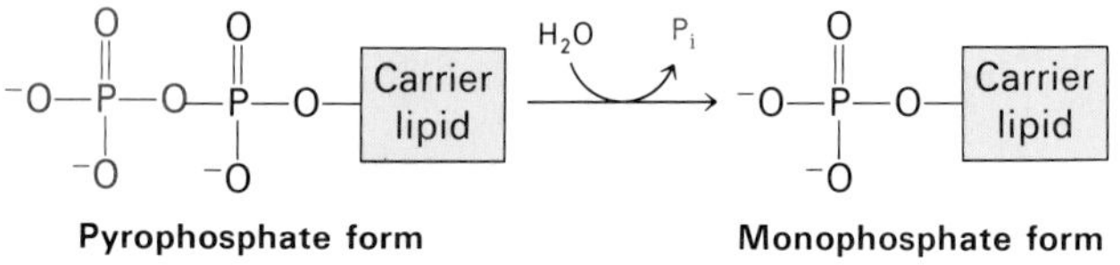

POLYSACCHARIDE CHAINS ARE CROSS-LINKED BY TRANSPEPTIDATION

Different polysaccharide chains are cross-linked by transpeptidation to form one enormous bag-shaped molecule. *The amino group at the end of one pentaglycine bridge attacks the peptide bond between the D-Ala-D-Ala residues in another peptide unit* (Figure 32-11). A peptide bond is formed between glycine and one of the D-alanine residues,

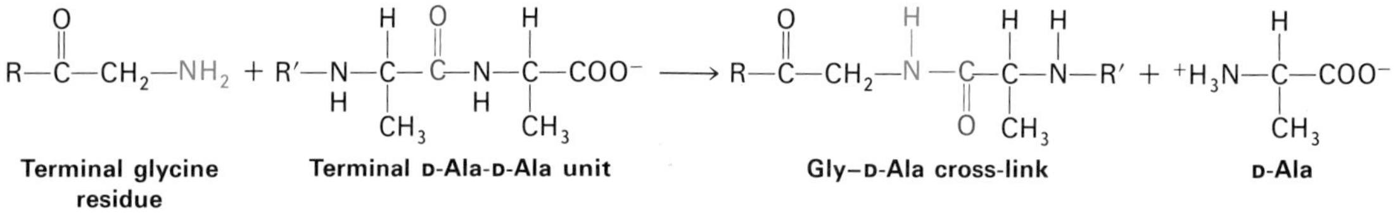

Figure 32-11
The amino group of the pentaglycine bridge attacks the peptide bond between two D-Ala residues to form a cross-link.

whereas the other D-alanine residue is released. This reaction is catalyzed by *glycopeptide transpeptidase*. Note that ATP is not consumed in the synthesis of this cross-link. This transpeptidation reaction uses the free energy already present in the D-Ala-D-Ala bond. This unusual means of forming a peptide bond may be necessary because the reaction occurs outside the cell membrane, beyond the reach of ATP. Recall that a peptide bond is formed in the cross-linking of fibrin by a transamidation reaction, which does not use ATP (p. 174).

TEICHOIC ACID COVERS THE PEPTIDOGLYCAN IN GRAM-POSITIVE BACTERIA

The surface of gram-positive bacteria consists of teichoic acid, which is a polymer of glycerol (or another sugar such as ribitol) linked by phosphodiester bridges (Figure 32-12). The available hydroxyl groups are esterified to alanine or to sugars such as glucose. Teichoic acid is attached to the NAG-NAM backbone of the peptidoglycan by a phosphodiester bond. The teichoic acid chains are elongated by the transfer of glycerol phosphate from CDP-glycerol to the free terminal hydroxyl group of the chain. Glucose is added to the hydroxyl groups of the teichoic acid backbone by reaction with UDP-glucose.

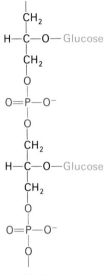

Figure 32-12
Structure of teichoic acid.

PENICILLIN KILLS GROWING BACTERIA BY INHIBITING CELL-WALL SYNTHESIS

Penicillin was discovered by Alexander Fleming in 1928, the result of a chance observation:

> While working with staphylococcus variants a number of culture-plates were set aside on the laboratory bench and examined from time to time. In the examinations these plates were necessarily exposed to the air and they became contaminated with various micro-organisms. It was noticed that around a large colony of a contaminating mould the staphylococcus colonies became transparent and were obviously undergoing lysis.
>
> Subcultures of this mould were made and experiments conducted with a view to ascertain something of the properties of the bacteriolytic substance which had evidently been formed in the mould culture and which had diffused into the surrounding medium. It was found that broth in which the mould had been grown at room temperature for one or two weeks had acquired marked inhibitory, bacteriocidal and bacteriolytic properties to many of the more common pathogenic bacteria.

Fleming was further encouraged by the finding that the extract from the *Penicillium* mold had no apparent toxicity when injected into animals. However, he tried to concentrate and purify the active antibiotic but found that "penicillin is easily destroyed, and to all intents and purposes we failed. We were bacteriologists—not chemists—and our relatively simple procedures were unavailing."

Ten years passed before penicillin was studied again. Howard Florey, a pathologist, and Ernst Chain, a biochemist, carried out a series of incisive studies that led to the isolation, chemical characterization, and clinical use of this antibiotic. Penicillin consists of a thiazolidine ring fused to a *β-lactam ring*, to which a variable group (R) is attached by a peptide bond. In *benzyl penicillin*, for example, R is a benzyl group (Figures 32-13 and 32-14). This struc-

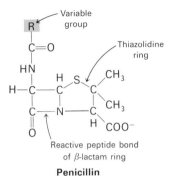

Figure 32-13
The reactive site of penicillin is the peptide bond of its β-lactam ring. The R group is variable. In benzyl penicillin, R is a benzyl group.

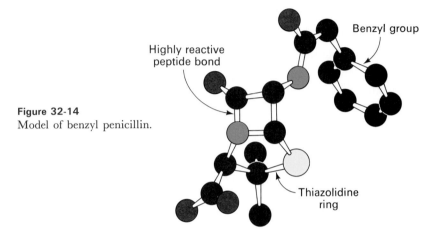

Figure 32-14
Model of benzyl penicillin.

ture can undergo a variety of rearrangements, which accounts for the instability first encountered by Fleming. In particular, the β-lactam ring is very labile. Indeed, this property is closely tied to the antibiotic activity of penicillin, as will be discussed shortly.

In 1957, Joshua Lederberg showed that bacteria ordinarily susceptible to penicillin could be grown in its presence if a hypertonic medium were used. The organisms obtained in this way, called *protoplasts,* are devoid of a cell wall, and consequently lyse when transferred to a normal medium. Thus, it was inferred that penicillin interfered with bacterial cell-wall synthesis. In 1965, James Park and Jack Strominger independently deduced that penicillin blocks the last step in bacterial cell-wall biosynthesis, namely the cross-linking reaction.

PENICILLIN BLOCKS CELL-WALL SYNTHESIS BY INHIBITING TRANSPEPTIDATION

Penicillin inhibits the transpeptidase that cross-links different peptidoglycan strands.

$$\text{R—Gly—NH}_2 + \text{R}'\text{—D-Ala—D-Ala—C}\overset{\text{O}}{\underset{\text{O}^-}{}} \quad \xrightarrow[\text{by penicillin}]{\text{Inhibited}} \quad \text{R}'\text{—D-Ala—Gly—R} + \text{D-Alanine}$$

The transpeptidase normally forms an *acyl intermediate* with the penultimate D-alanine residue of the peptide (Figure 32-15). This

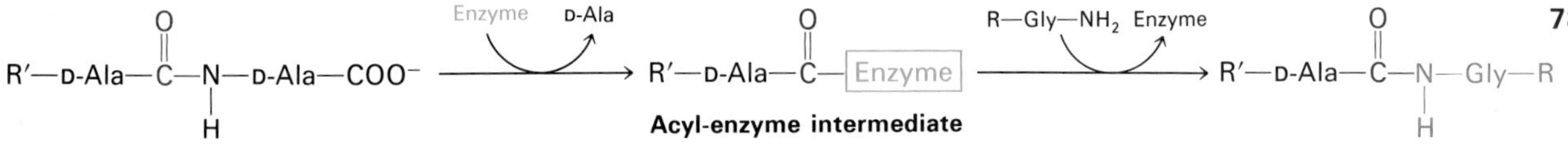

$$R'—\text{D-Ala}—\overset{\overset{\displaystyle O}{\|}}{C}—\overset{\overset{\displaystyle }{\underset{\underset{\displaystyle H}{|}}{N}}}—\text{D-Ala}—COO^- \xrightarrow{\quad \text{Enzyme} \quad \text{D-Ala} \quad} R'—\text{D-Ala}—\overset{\overset{\displaystyle O}{\|}}{C}—\boxed{\text{Enzyme}} \xrightarrow{\quad R—\text{Gly}—NH_2 \quad \text{Enzyme} \quad} R'—\text{D-Ala}—\overset{\overset{\displaystyle O}{\|}}{C}—\overset{\overset{\displaystyle }{\underset{\underset{\displaystyle H}{|}}{N}}}—\text{Gly}—R$$

Acyl-enzyme intermediate

Figure 32-15
An acyl-enzyme intermediate is formed in the transpeptidation reaction.

acyl-enzyme intermediate then reacts with the amino group of the terminal glycine in another peptide. Recent studies have shown that penicillin inhibits the transpeptidase by forming a covalent bond with a serine residue at the active site of the enzyme (Figure 32-16). *This penicillinoyl-enzyme complex does not undergo deacylation.*

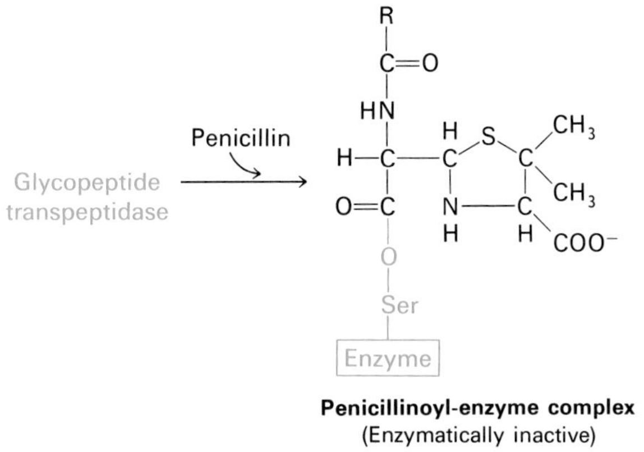

Glycopeptide transpeptidase $\xrightarrow{\quad \text{Penicillin} \quad}$

Figure 32-16
Formation of a penicillinoyl-enzyme complex, which is indefinitely stable.

Penicillinoyl-enzyme complex
(Enzymatically inactive)

Hence, the transpeptidase is irreversibly inhibited by penicillin.

Why is penicillin such an effective inhibitor of the transpeptidase? Molecular models have revealed that *penicillin resembles acyl-D-Ala-D-Ala, one of the substrates of this enzyme* (Figure 32-17). Moreover, the four-membered β-lactam ring of penicillin is strained, which makes this peptide bond highly reactive. Indeed, penicillin probably has the structure of the transition state of the normal substrate. In other words, penicillin is a *transition-state analog.*

It is of interest to compare penicillin with other substances that irreversibly inhibit enzymes. The stable, inactive *penicillinoyl-enzyme complex* is analogous to the *phosphoryl-enzyme complex* formed by the reaction of organic fluorophosphates with acetylcholinesterase or the serine proteases. Penicillin is far more specific, however, than the fluorophosphate inhibitors. The reactivity of penicillin is precisely targeted because of its structural resemblance to the D-Ala-D-Ala terminus of nascent strands of peptidoglycan. The low toxicity of penicillin, which is essential for its therapeutic effectiveness, is a consequence of its exquisite specificity. There is no known human enzyme that recognizes D-Ala-D-Ala, and so penicillin does not interfere with our own enzymatic machinery.

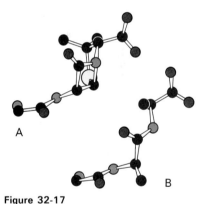

Figure 32-17
The conformation of penicillin in the vicinity of its reactive peptide bond (A) resembles the postulated conformation of the transition state of R—D-Ala—D-Ala (B) in the transpeptidation reaction. [After B. Lee. *J. Mol. Biol.* 61(1971):464.]

SOME BACTERIA ARE RESISTANT TO PENICILLIN BECAUSE THEY SYNTHESIZE A PENICILLIN-DESTROYING ENZYME

Some bacteria secrete *penicillinase,* an enzyme that cleaves the amide bond in the β-lactam ring of penicillin to form penicilloic acid, which is inactive as an antibiotic:

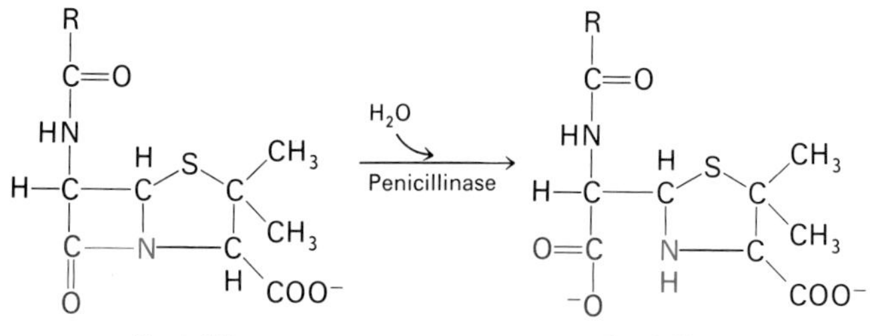

A variety of related penicillinases having a mass of about 30 kdal and a high turnover number (typically 10^3 sec^{-1}) have been isolated. The activity of the enzyme depends markedly on the nature of the R group attached to the β-lactam ring of penicillin. Semisynthetic penicillins having R groups that render them unsusceptible to penicillinase are clinically important.

The gene for penicillinase is located on various plasmids (p. 758). These extrachromosomal genetic elements are readily lost and acquired by some bacterial species. The enzyme is inducible in some bacterial strains. *Penicillinase probably evolved as a detoxification mechanism,* because it is found only in organisms that possess a peptidoglycan wall. The enzyme probably does not have a function other than to destroy penicillin, in view of the fact that loss of the gene has no deleterious effect in the absence of the antibiotic. Furthermore, there is a good correlation between the degree of resistance to penicillin and the amount of penicillinase activity.

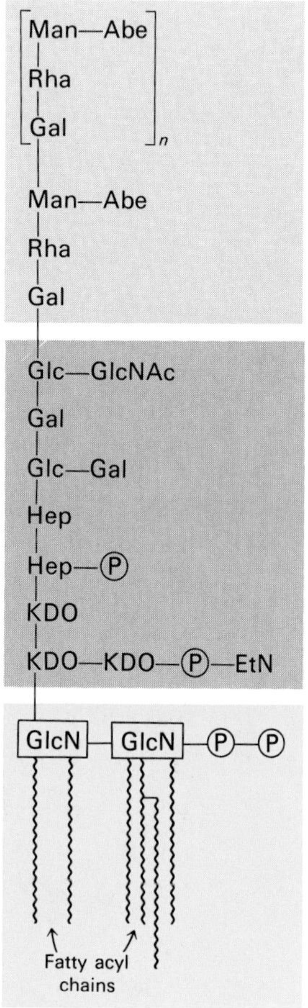

A LIPOPOLYSACCHARIDE-RICH OUTER MEMBRANE SURROUNDS GRAM-NEGATIVE BACTERIA

As mentioned earlier, the cell envelopes of gram-negative bacteria are more complex than those of gram-positive ones. The peptidoglycan layer of gram-negative cells (such as *Escherichia coli* and *Salmonella typhimurium*) is surrounded by an *outer membrane* that contains phospholipids, proteins, and lipopolysaccharides. This outer membrane, like the plasma membrane, has a bilayer arrangement. Thus, gram-negative cells have two membranes, whereas gram-positive cells have only one. A gram-negative bacterium is distinctive in having an aqueous compartment between its plasma membrane and peptidoglycan layer. This *periplasmic space* contains many proteins that bind and transport sugars and other nutrients (p. 905).

Lipopolysaccharides (LPS) in the outer membrane are highly unusual molecules consisting of three regions: *lipid A, core oligosaccharides,* and *O side chain* (Figure 32-18). The lipid-A moiety is the

Figure 32-18
A lipopolysaccharide molecule consists of three regions: lipid A, core oligosaccharide, and O side chain. The particular sequence of sugars shown here is from *Salmonella typhimurium.* Abbreviations used: Abe, abequose; EtN, ethanolamine; Gal, galactose; Glc, glucose; GlcN, glucosamine; GlcNAc, *N*-acetylglucosamine; Hep, heptulose; KDO, 2-keto-3-deoxyoctonate; Man, mannose; Rha, L-rhamnose.

hydrophobic part of this large (10-kdal) amphipathic molecule. It contains six saturated fatty acid chains that are linked to two glucosamine residues. These fatty acyl chains constitute about half of the outer leaflet of the outer membrane. They are absent from the inner leaflet, which contains phospholipids instead (Figure 32-19). The

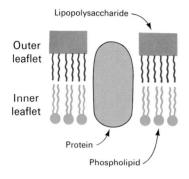

Figure 32-19
Lipopolysaccharide molecules are located in the outer leaflet, whereas phospholipids are positioned in the inner leaflet of the outer membrane.

core oligosaccharide region of the lipopolysaccharide comes next. The ten sugar units in this region project outward, followed by the O side chain, which is made up of many repeating tetrasaccharide units. These two regions contain several sugars that are rarely present elsewhere in nature: 2-keto-3-deoxyoctonate (KDO), an eight-carbon sugar; heptose, a seven-carbon sugar; and L-rhamnose and abequose, six-carbon sugars with $-CH_3$ instead of $-CH_2OH$ at C-6 (Figure 32-20). The core oligosaccharide and O side chain, in

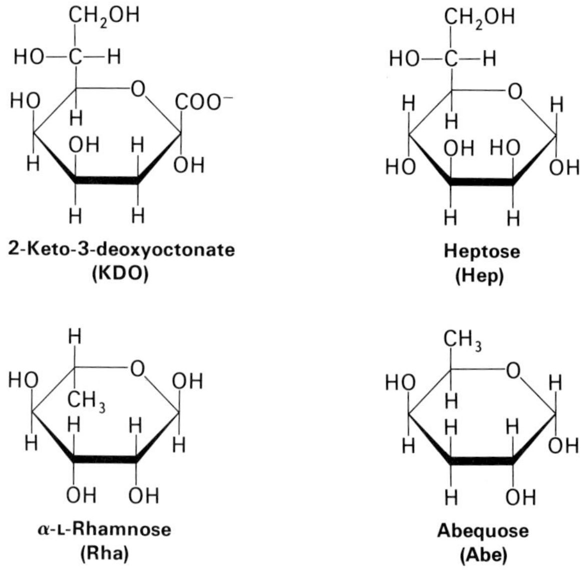

2-Keto-3-deoxyoctonate (KDO)

Heptose (Hep)

α-L-Rhamnose (Rha)

Abequose (Abe)

Figure 32-20
Formulas of some unusual sugars in lipopolysaccharides.

contrast with the fatty acyl part of lipid A, are highly hydrophilic. The lipopolysaccharide molecule is negatively charged because several of its sugars are phosphorylated.

Lipopolysaccharides are synthesized in the plasma membrane and are then transferred to the outer membrane. Lipid A is synthesized first, and the

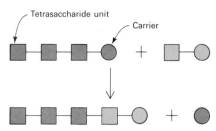

Figure 32-21
Mode of elongation of the O side chain.

core polysaccharide is then formed by the sequential addition of sugars from activated donors such as UDP-glucose and UDP-galactose. The O side chain is assembled in a different way. Its repeating tetrasaccharide unit is synthesized on a carrier lipid, the same C_{55}-isoprenoid as is used in the synthesis of peptidoglycans (p. 776). A growing chain of such units is elongated by the transfer of an oligosaccharide from one carrier to a single repeating unit on another carrier (Figure 32-21). This mode of elongation is analogous to that of fatty acid synthesis (p. 399) and protein synthesis (p. 659). Finally, the O side chain is added to the tip of the core oligosaccharide. It seems likely that the repeating tetrasaccharide of the O side chain is built on the inner leaflet of the plasma membrane and is then transferred to the outer leaflet, where it is added to the growing chain. The completed lipopolysaccharide molecule is probably translocated from the plasma membrane to the outer membrane through adhesions between these structures.

THE DIVERSITY OF O SIDE CHAINS AIDS GRAM-NEGATIVE BACTERIA IN EVADING HOST DEFENSES

Lipopolysaccharides contain highly unusual sugars and distinctive linkages. What are the roles of these baroque molecules? Mutants that are defective in the synthesis of lipopolysaccharides have provided clues about their function. The lipid-A region and the adjoining KDO sugars of the core oligosaccharide are probably indispensable for viability. The saturated fatty acyl chains contribute to the *barrier* role of the outer membrane. Periplasmic proteins are kept in and most noxious molecules are kept out. For example, penicillin does not readily enter gram-negative microorganisms. The lipid-A region may also confer *rigidity* on the outer membrane. In contrast, the O side chains are not essential for viability. For example, *E. coli* strain K12, a favorite of many biochemists, has no O side chains. However, gram-negative microorganisms in natural habitats almost always possess O side chains. The outer coat of polysaccharide makes the bacterial surface very hydrophilic, which renders it less susceptible to phagocytosis by host cells. Furthermore, the O side chains of lipopolysaccharides are highly diverse; the structure shown in Figure 32-18 is only one of many that have been found. Gram-negative bacteria can mutate rapidly to alter the nature of their O side chains. A host population that has not been exposed to the new surface structure will initially have a low level of antibody directed against the novel O side chain. *Thus, varying the O side chain is a way of staying one step ahead of the host's defense system.*

The genetic information for the alteration of O side chains sometimes comes from a temperate phage harbored by the gram-negative bacterium. For example, phage P22 contributes a gene for an enzyme that adds glucose to the repeating tetrasaccharide unit of the O side chain (Figure 32-22). This kind of alteration is called *phage conversion*. We see here a striking example of the interplay of

```
  |
Man—Abe
  |
Rha
  |
Gal
  |
  |  Phage
  ↓  P22
  |
Man—Abe
  |
Rha
  |
Gal—Glc
  |
```

Figure 32-22
Structural change in the O side chain of a *Salmonella* bacterium produced by infection with P22, a temperate phage.

bacterial and viral genes and of the evolutionary advantage of their symbiotic relation.

PORIN FORMS CHANNELS THAT ENABLE SMALL POLAR MOLECULES TO CROSS THE OUTER MEMBRANE

The outer membrane contains a large number ($\sim 10^5$) of molecules of *porin*, a 37-kdal transmembrane protein. Trimers of porin form channels that enable small polar molecules to rapidly diffuse across the outer membrane (Figure 32-23). *The channel has a diameter of about*

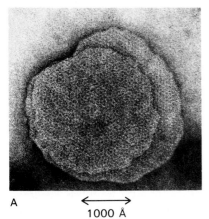

A

← 1000 Å →

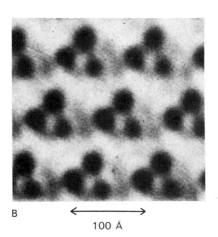

B

← 100 Å →

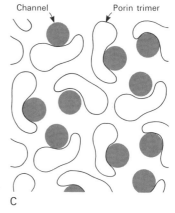

Channel   Porin trimer

C

Figure 32-23
A. Electron micrograph of negatively stained arrays of porin channels from *E. coli.*

B. A processed and filtered image of the electron micrograph shown in part A.

C. Interpretive diagram showing a low-resolution projection of the porin channels. [From A. C. Steven, B. Ten Heggeler, R. Müller, J. Kistler, and J. P. Rosenbusch, *J. Cell Biol.* 72(1977):292.]

10 Å and so molecules with a mass greater than about 600 daltons are excluded. Hydrophobic molecules, irrespective of size, do not readily traverse the porin channel, which suggests that it is filled with water and is lined by polar groups. Thus, the porin channel is well suited for the permeation of small polar metabolites such as monosaccharides. Certain larger polar molecules traverse the outer membrane by way of specific transport systems, as exemplified by the *maltose transport system.* Maltose (a disaccharide) and maltotriose diffuse too slowly through the porin channel, and so a special pathway is needed. Also, there are specific transport systems for vitamin B_{12} and iron chelates. Some phages enter these bacteria by binding to these specific transporters. For example, λ phage takes advantage of the maltose receptor.

The most abundant protein in the outer membrane ($\sim 7 \times 10^5$ copies) is a small *lipoprotein* containing only fifty-eight amino acid residues. This highly α helical protein contains three covalently attached fatty acids, all joined to the amino-terminal cysteine. In about a third of the lipoprotein molecules, the ε-amino group of the carboxyl-terminal lysine is linked to a carboxyl group of the underlying peptidoglycan. *Thus, the lipoprotein ties the outer membrane to the peptidoglycan layer and thereby contributes to the mechanical stability of the cell envelope.*

NASCENT OUTER-MEMBRANE PROTEINS CONTAIN SIGNAL SEQUENCES THAT ARE CLEAVED

A protein synthesized by a gram-negative microorganism can be located in the cytosol, plasma membrane, periplasmic space, outer membrane, or extracellular medium. How is a particular molecule targeted to one of these five potential destinations? A highly informative finding in this regard is that the lipoprotein of the outer membrane is synthesized as a *prolipoprotein* containing an extra twenty residues at the amino-terminal end (Figure 32-24). Likewise, other proteins destined for the plasma membrane or more external sites are distinguished from those retained in the cytosol by the presence of *signal sequences.* Ribosomes synthesizing proteins destined for transport out of the cytosol bind to the plasma membrane because of the presence of these highly hydrophobic amino-terminal extensions. Thus, the initial part of the secretory process of procaryotes is very much like that of eucaryotes (p. 713). The major difference at this stage is that ribosomes synthesizing cell-envelope proteins are bound to the *plasma membrane in procaryotes* (Figure 32-25), whereas their counterparts are attached to the *endoplasmic reticulum in eucaryotes* (p. 712). As in eucaryotes, signal sequences in procaryotes are cleaved by peptidases soon after the nascent chains cross the membrane barrier.

The signal-sequence hypothesis is strongly supported by studies of bacterial mutants. The maltose-binding protein is a 38-kdal molecule that participates in the uptake of this sugar. This protein is normally located in the periplasmic space. However, a mutation altering the amino-terminal end of its precursor changes the cellular location of the mature protein. The replacement of a hydrophobic amino acid in its signal sequence by a charged residue results in the accumulation of the maltose-binding protein in the cytosol. *Thus, a change of a single residue can alter the localization of a protein from the periplasmic space to the cytosol.* Conversely, can a protein normally

H₃⁺N-Met-Lys-Ala-Thr-Lys-Leu-
-Val-Leu-Gly-Ala-Val-Ile-
-Leu-Gly-Ser-Thr-Leu-Leu-
-Ala-Gly⌡Cys-Ser-Ser-Asn
Cleavage
Site

Figure 32-24
Amino acid sequence of the amino-terminal region of prolipoprotein. This signal sequence contains many hydrophobic residues (shown in yellow).

Figure 32-25
Procaryotic proteins destined for locations other than the cytosol are synthesized by ribosomes bound to the plasma membrane. A signal sequence (red) on the nascent chain directs the ribosome to the plasma membrane and enables the protein to be translocated across this membrane.

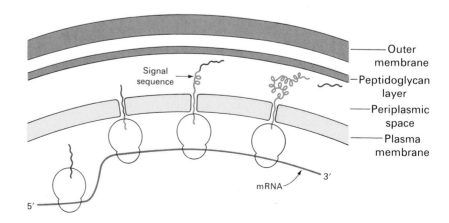

Outer membrane
Peptidoglycan layer
Periplasmic space
Plasma membrane
Signal sequence
mRNA
3'
5'

destined for the cytosol be misdirected to the outer membrane? The amino-terminal portion of the gene for a maltose-transport protein (also the λ-phage receptor) in the outer membrane was fused to the β-galactosidase gene. The chimeric protein encoded by this gene emerged in the outer membrane rather than in the cytosol, the usual location of β-galactosidase. This experiment suggests that *the address of an envelope protein is specified by the amino-terminal sequence of the nascent chain*. It is evident that procaryotic cells, like eucaryotic ones, can translocate proteins to specific targets. The molecular basis of this sorting process is now a challenging area of inquiry.

SUMMARY

Bacterial cells are surrounded by cell walls that protect them from osmotic lysis. The cell wall in gram-positive bacteria is typically 250 Å thick, consisting of peptidoglycan and teichoic acid, whereas the cell surface in gram-negative bacteria is more complex. The cell wall is one enormous bag-shaped macromolecule. The peptidoglycan in *S. aureus* consists of three repeating units: NAG-NAM, a tetrapeptide of D and L residues, and a pentaglycine chain. The pentaglycine unit cross-links different polysaccharide strands. UDP-NAG and UDP-NAM are the activated sugars in the biosynthesis of the peptidoglycan, which takes place in five stages. First, a peptide unit is built on NAM while it is attached to UDP. Second, the NAM-peptide is transferred to a carrier lipid. Third, NAG and the pentaglycine bridge are added to the NAM-peptide unit. Fourth, the NAG-NAM-peptide unit is transferred from the carrier lipid to a growing polysaccharide chain. Fifth, the amino group of a pentaglycine bridge on one strand attacks a terminal D-Ala-D-Ala peptide bond of another strand to form a cross-link.

Penicillin kills bacteria by inhibiting the synthesis of cell walls. Penicillin resembles acyl-D-Ala-D-Ala and contains a highly reactive peptide bond. It irreversibly inhibits glycopeptide transpeptidase and thereby blocks the cross-linking of the peptidoglycan. Some bacteria are resistant to penicillin because they contain penicillinase, which hydrolyzes penicillin to an inactive form.

A gram-negative cell, in contrast with a gram-positive one, has an outer membrane beyond its peptidoglycan layer. It also has a periplasmic space between its plasma membrane and peptidoglycan layer. The outer membrane contains highly distinctive lipopolysaccharides. These amphipathic molecules are made up of a lipid-A region anchored to the outer leaflet of the outer membrane and of core-oligosaccharide and O-side-chain regions, which project outward. The O side chains, which contain unusual sugars, are assembled by the polymerization of tetrasaccharide units on a carrier lipid. The diversity of O side chains helps gram-negative bacteria evade the immunological defenses of their hosts. Small polar molecules, but not hydrophobic ones, diffuse through aqueous channels in the outer membrane that are formed by porin, a transmembrane protein. Specific transport systems enable larger molecules such as maltose and vitamin B_{12} to traverse the outer membrane. The other major protein of this membrane is a small lipoprotein that ties the outer membrane to the underlying peptidoglycan. Cell-envelope proteins are synthesized by ribosomes that are attached to the plasma membrane. The precursors of these proteins contain amino-terminal signal sequences that are cleaved following translocation.

SELECTED READINGS

WHERE TO START

Nikaido, H., and Nakae, T., 1979. The outer membrane of gram-negative bacteria. *Advan. Microbiol. Physiol.* 14:163–250.

DiRienzo, J. M., Nakamura, K., and Inouye, M., 1978. The outer membrane proteins of gram-negative bacteria: biosynthesis, assembly, and functions. *Ann. Rev. Biochem.* 47:481–532.

Sharon, N., 1969. The bacterial cell wall. *Sci. Amer.* 220(5):92–98.

BACTERIAL CELL ENVELOPES

Leive, L., (ed.), 1973. *Bacterial Membranes and Walls.* Marcel Dekker. [An excellent collection of critical reviews. The article by M. M. Ghuysen and G. D. Shockman on the biosynthesis of peptidoglycan and the one by H. Nikaido on the biosynthesis of the cell envelope of gram-negative organisms are most pertinent to this chapter.]

Nikaido, H., 1979. Permeability of the outer membrane of bacteria. *Angew. Chem.* (Int. Ed.) 18:337–350.

Hussey, H., and Baddiley, J., 1976. Biosynthesis of bacterial cell walls. *In* Martonosi, A., (ed.), *The Enzymes of Biological Membranes,* vol. 2, pp. 227–326. Plenum.

Osborn, M. J., 1971. The role of membranes in the synthesis of macromolecules. *In* Rothfield, L. I., (ed.), *Structure and Function of Biological Membranes,* pp. 343–400. Academic Press.

Steven, A. C., ten Heggeler, B., Müller, R., Kistler, J., and Rosenbusch, J. P., 1977. Ultrastructure of a periodic protein layer in the outer membrane of *Escherichia coli. J. Cell. Biol.* 72:292–301.

PENICILLIN AND BACITRACIN

Fleming, A., 1929. On the antibacterial action of cultures of a penicillinium, with special reference to their use in the isolation of *B. influenzae. Brit. J. Exp. Pathol.* 10:226–236.

Chain, E., 1971. Thirty years of penicillin therapy. *Proc. Roy. Soc. London* (B) 179:293–319.

Strominger, J. L., Blumberg, P. M., Suginaka, H., Umbreit, J., and Wickus, G. G., 1971. How penicillin kills bacteria: progress and problems. *Proc. Roy. Soc. London* (B) 179:369–383.

Yocum, R. R., Waxman, D. J., Rasmussen, J. R., and Strominger, J. L., 1979. Mechanisms of penicillin action: penicillin and substrate bind covalently to the same active site serine in two bacterial D-alanine carboxypeptidases. *Proc. Nat. Acad. Sci.* 76:2730–2734.

Stone, K. J., and Strominger, J. L., 1971. Mechanism of action of bacitracin: complexation with metal ion and C_{55}-isoprenyl pyrophosphate. *Proc. Nat. Acad. Sci.* 68:3223–3227.

Boyd, D. B., 1977. Transition state structures of a dipeptide related to the mode of action of β-lactam antibiotics. *Proc. Nat. Acad. Sci.* 74:5239–5243.

SIGNAL SEQUENCES AND PROTEIN LOCALIZATION

Chang, C. N., Model, P., and Blobel, G., 1979. Membrane biogenesis: cotranslational integration of the bacteriophage f1 coat protein into an *Escherichia coli* membrane fraction. *Proc. Nat. Acad. Sci.* 75:4891–4895.

Inouye, S., Wang, S., Sekizawa, J., Halejoua, S., and Inouye, M., 1977. Amino acid sequence for the peptide extension on the prolipoprotein of the *Escherichia coli* outer membrane. *Proc. Nat. Acad. Sci.* 74:1004–1008.

Bedouelle, H., Bassford, P. J., Jr., Fowler, A. V., Zabin, I., Beckwith, J., and Hofnung, M., 1980. Mutations which alter the function of the signal sequence of the maltose binding protein of *Escherichia coli. Nature* 285:78–81.

Emr, S., Hedgpeth, J., Clement, J.-M., Silhavy, T. J., and Hofnung, M., 1980. Sequence analysis of mutations that prevent export of λ receptor, an *Escherichia coli* outer membrane protein. *Nature* 285:82–91.

Bassford, P., and Beckwith, J., 1979. *Escherichia coli* mutants accumulating the precursor of a secreted protein in the cytoplasm. *Nature* 277:538–541.

Emr, S. D., Schwartz, M. and Silhary, T. J., 1978. Mutations altering the cellular localization of the phage λ receptor, an *Escherichia coli* outer membrane protein. *Proc. Nat. Acad. Sci.* 75:5802–5806.

Lin, J. J. C., Kanazawa, H., Ozols, J., and Wu, H. C., 1978. An *Escherichia coli* mutant with an amino acid alteration within the signal sequence of outer membrane prolipoprotein. *Proc. Nat. Acad. Sci.* 75:4891–4895.

CHAPTER 33

IMMUNOGLOBULINS

By the turn of the century, many important features of the immune response had been discovered. An awareness of the protection conferred was fully appreciated much earlier, as evidenced by Thucydides' account of the plague that struck Athens in 430 B.C.:

> All speculation as to its origins and its causes, if causes can be found adequate to produce so great a disturbance, I leave to other writers . . . for myself, I shall simply set down its nature, and explain the symptoms by which it may be recognized by the student, if it should ever break out again. This I can the better do, as I had the disease myself, and watched its operation in the case of others. . . .
>
> Yet it was with those who had recovered from the disease that the sick and the dying found most compassion. These knew what it was from experience, and had now no fear for themselves; *for the same man was never attacked twice*—never at least fatally. And such persons not only received the congratulations of others, but themselves also, in the elation of the moment, *half entertained the vain hope that they were for the future safe from any disease whatsoever*. . . .

In characterizing the hope of protection against all disease a vain one, Thucydides clearly appreciated the *specificity* of the protection afforded by the immune system.

BASIC DEFINITIONS

An *antibody* (or *immunoglobulin*) is a protein synthesized by an animal in response to the presence of a foreign substance. The antibody has specific affinity for the foreign material that elicited its synthesis.

Figure 33-1
Electron micrograph of a crystal of antibody molecules. [From L. W. Labaw and D. R. Davies. *J. Ultrastruct. Res.* 40(1972):349.]

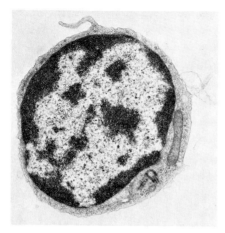

Figure 33-2
Electron micrograph of a B-lymphocyte.
[Courtesy of Lynne Mercer.]

An *antigen* (or *immunogen*) is a foreign macromolecule capable of eliciting antibody formation. Proteins, polysaccharides, and nucleic acids are usually effective antigens. The specificity of an antibody is directed against a particular site on an antigen called the *antigenic determinant*.

Small foreign molecules do not stimulate antibody formation. However, they can elicit the formation of specific antibody if they are attached to macromolecules. The macromolecule is then the *carrier* of the attached chemical group, which is called a *haptenic determinant*. The small foreign molecule by itself is called a *hapten*.

Antibody molecules are secreted by *plasma cells*, which are derived from *B-lymphocytes*. The other class of lymphocytes, *T-lymphocytes*, mediates the cellular immune response, which complements the humoral immune response carried out by soluble antibody molecules. This chapter will deal with the humoral immune response because much more is known about its molecular basis. However, it is important to note that antibody molecules are just one component of a vast integrated network of molecules and cells that recognizes foreign substances and eliminates them.

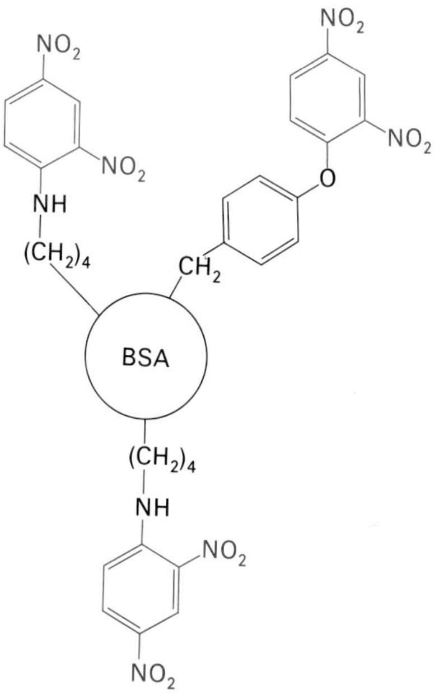

SYNTHESIS OF A SPECIFIC ANTIBODY FOLLOWING EXPOSURE TO AN ANTIGEN

Animals can make specific antibodies against virtually any foreign chemical group. The dinitrophenyl group (DNP) is particularly effective in eliciting antibody formation and therefore has been used extensively as a haptenic determinant. Anti-DNP antibody can be obtained in the following way:

1. DNP groups are covalently attached to a carrier protein such as bovine serum albumin (BSA) by reaction of fluorodinitrobenzene with lysine side chains and other nucleophiles of this protein (Figure 33-3).

2. DNP-BSA, the immunogen (antigen), is injected into a rabbit. The level of anti-DNP antibody in the serum of the rabbit starts to rise a few days later (Figure 33-4). These *early antibody molecules*

Figure 33-3
Dinitrophenylated bovine serum albumin
(DNP-BSA) is an effective immunogen.

Figure 33-4
Kinetics of the appearance of immunoglobulins
M and G (IgM and IgG) in the serum
following immunization.

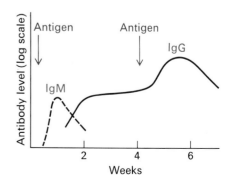

belong to the *immunoglobulin M* (IgM) class and have a mass of nearly 1000 kdal.

3. Approximately ten days after the injection of immunogen, the amount of immunoglobulin M decreases and there is a concurrent increase in the amount of anti-DNP antibody of a different class, called *immunoglobulin G* (IgG), which has a mass of 150 kdal.

4. The level of anti-DNP antibody of the immunoglobulin G class reaches a plateau approximately three weeks after the injection of immunogen. A booster dose of DNP-BSA given at that time produces a further increase in the level of anti-DNP antibody in the rabbit's serum.

5. Blood is drawn from the immunized rabbit. The resulting serum (called an *antiserum* because it is obtained after immunization) may contain as much as 1 mg of anti-DNP antibody per milliliter. The DNP group is particularly effective in eliciting the formation of large amounts of specific antibody. Nearly all of this antibody is of the immunoglobulin G class, the principal one in serum.

6. The next step is to separate anti-DNP antibody from antibodies of other specificities and from other serum proteins. The anti-DNP antibody differs from other proteins in the antiserum in its very high affinity for the DNP group. Consequently, it can be separated by *affinity chromatography*. A column consisting of dinitrophenyl groups covalently attached to an insoluble carbohydrate matrix is formed. The antiserum is applied to this DNP column, which is then washed with buffer. Most proteins in the antiserum flow through because they have little or no affinity for the DNP group or for the carbohydrate support. In contrast, anti-DNP antibody binds tightly to the column. It is released by adding a high concentration of dinitrophenol, which binds to the antibody and thereby displaces it from the DNP groups of the insoluble carbohydrate matrix. A soluble complex of dinitrophenol and anti-DNP antibody emerges from the column. The dinitrophenol is removed by dialysis or by ion-exchange chromatography, which yields a preparation of purified antibody.

THE COMBINING SITES OF ANTIBODIES ARE LIKE THE ACTIVE SITES OF ENZYMES

The *combining sites* of antibodies (i.e., their antigen-binding sites) resemble the active sites of enzymes in several ways:

1. The binding constants for haptens have been determined by equilibrium dialysis and spectroscopic techniques. For example, the binding of a colored hapten such as a DNP derivative quenches the fluorescence of the tryptophan residues of the antibody. The

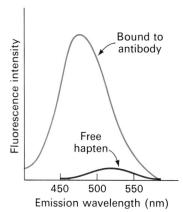

Figure 33-5
The fluorescence properties of ε-dansyllysine change markedly when this hapten binds to antidansyl antibody. The decrease in the wavelength of the emission maximum and the increase in quantum yield indicate that the hapten is bound to a nonpolar site.

extent of quenching is a measure of the saturation of the antibody combining sites. *Binding constants for most haptens range from* $K = 10^{-4}$ *to* 10^{-10} M. Thus the standard free energies of binding typically fall between about -6 and -15 kcal/mol, which is the same range as that for enzyme-substrate and enzyme-coenzyme complexes. Furthermore, *the binding forces in hapten-antibody complexes are like those in enzyme-substrate complexes.* Weak, noncovalent interactions of the electrostatic, hydrogen bond, and van der Waals types combine to give strong and specific binding.

2. An antibody specific for dextran, a polysaccharide of glucose residues, was tested for its binding of oligomers. The full binding affinity was obtained when the oligomer contained six glucose residues. An interesting comparison here is with lysozyme, which also has space for six sugar residues in its active-site cleft. Thus, *the binding site of this antidextran antibody is about 25 Å long.*

3. The spectroscopic properties of some haptens provide information about the polarity of the antibody combining site. For example, certain naphthalenes exhibit a weak yellow fluorescence when they are in a highly polar environment (such as water) and an intense blue fluorescence when they are in a markedly nonpolar environment (such as hexane). An intense blue fluorescence is observed when such a naphthalene hapten is bound to specific antibody, indicating that water is largely excluded from the combining site. In general, antibody combining sites are *nonpolar niches,* which enhances the binding of antigens for the reasons discussed previously in regard to enzyme-substrate interactions (p. 109).

4. The fit of hapten is rather precise. *Specificity is high but certainly not absolute.* The hapten is firmly held to the combining site, so that it has little rotational freedom. The rate constant for the binding of a variety of haptens is about 10^8 M^{-1} sec^{-1}. This high rate constant indicates that the binding step is a diffusion-controlled reaction. Large-scale conformational rearrangements do not seem to accompany the binding of hapten.

AN ANTIBODY PREPARATION OF A GIVEN SPECIFICITY IS NORMALLY HETEROGENEOUS

Antibodies differ from enzymes in a very important way. *Most normal antibodies of a given specificity, such as anti-DNP antibody, are not a single molecular species.* Analyses of the binding of DNP haptens to a preparation of anti-DNP antibody reveal a wide range of binding affinities. Some antibody molecules bind DNP with a K of 10^{-6} M, whereas others have a K of 10^{-10} M. In contrast, most enzymes have a single binding constant for a particular substrate or coenzyme. Furthermore, a large number of bands are evident on electrophoresis of anti-DNP antibody or other antibodies of a particular specificity. On the other hand, an enzyme displays a single electrophoretic band or a few discrete bands (e.g., the isoenzymes of lactate dehydrogenase).

The heterogeneity of antibodies of a given specificity is an intrinsic feature of the immune response. What is the source of this heterogeneity? An important finding is that *antibodies produced by a single cell are homogeneous.* However, different cells produce different kinds of antibodies. The anti-DNP antibody in serum is the product of many different antibody-producing cells, which accounts for its heterogeneity.

IMMUNOGLOBULIN G CAN BE ENZYMATICALLY CLEAVED INTO ACTIVE FRAGMENTS

Immunoglobulin G has a mass of 150 kdal. A fruitful approach in studying such a large protein is to split it into fragments that retain activity. In 1959, Rodney Porter showed that immunoglobulin G can be cleaved into three active 50-kdal fragments by the limited proteolytic action of the enzyme papain. Two of these fragments

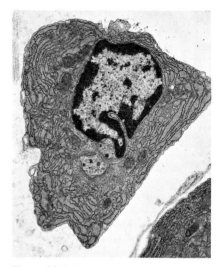

Figure 33-6
Electron micrograph of a plasma cell. The rough endoplasmic reticulum of this secretory cell is highly developed. [Courtesy of Lynne Mercer.]

IgG
(150 kdal)

$\downarrow$ Papain

2 F$_{ab}$ + F$_c$
(50 kdal each) (50 kdal)

Figure 33-7
Proteolytic cleavage of immunoglobulin G (IgG).

bind antigen. They are called F$_{ab}$ (*ab* stands for antigen-binding, *F* for fragment). *Each F$_{ab}$ contains one combining site for hapten or antigen,* which has the same binding affinity for hapten as does the whole molecule. However, *F$_{ab}$ does not form a precipitate with antigen* because it is univalent (i.e., contains only one combining site), in contrast with an intact immunoglobulin G antibody molecule, which contains two *identical* antigen-binding sites. Immunoglobulin G can form a precipitate with an antigen that contains more than one antigenic determinant. An extended lattice is then formed, in which each antibody molecule cross-links two or more antigens, and vice versa (Figure 33-8). The amount of precipitate formed is largest when the antibody and antigen are present in equivalent amounts.

The other fragment (also 50 kdal), called F$_c$, does not bind antigen, but it has other important biological activities. This fragment can cross the placental membrane, whereas F$_{ab}$ cannot. Thus, the capacity of immunoglobulin G to traverse the placental membrane and reach the fetal circulation depends on the presence of a *specific placental transmission site on F$_c$.* There is also a *site for the binding of complement* on F$_c$. The binding of an antibody to an antigenic determinant on the surface of a foreign cell often leads to the lysis of that cell. This sequel of the combination of antibody and antigen is mediated by a group of proteins known collectively as complement.

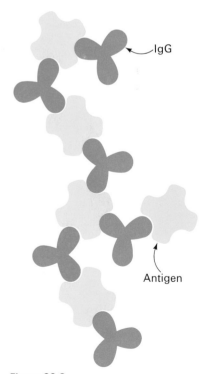

Figure 33-8
Schematic diagram of a lattice formed by the cross-linking of IgG (shown in blue) and an antigen (shown in yellow).

The binding of one of these proteins to the antigen-antibody complex triggers the reaction sequence that lyses the foreign cell. The F_c fragment derived its name from the fact that it is readily *crystallizable*. This finding provided the first hint that F_c is homogeneous, in contrast with F_{ab}, which usually cannot be crystallized because it is heterogeneous.

IMMUNOGLOBULIN G IS MADE UP OF L AND H CHAINS

Another breakthrough occurred in 1959, when Gerald Edelman showed that immunoglobulin G consists of two kinds of polypeptide chains. The disulfide bonds of the molecule were reduced by mercaptoethanol, and noncovalent interactions between the constituent chains were disrupted by 6 M urea. Chromatographic analysis of the resulting mixture revealed that immunoglobulin G consists of 25- and 50-kdal polypeptide chains, which are called the *light* (L) and *heavy* (H) chains, respectively.

Porter then found milder conditions, which made it possible to reconstitute immunoglobulin G from two H and two L chains. These studies led him to propose a model (Figure 33-9) for the subunit structure of immunoglobulin G in which each L chain is bonded to an H chain by a disulfide bond. The H chains are bonded to each other by at least one disulfide bond. Thus, *the subunit structure of immunoglobulin G is L_2H_2.*

The fragments produced by papain digestion were then related to this model. Papain cleaves the H chains on the carboxyl-terminal side of the disulfide bond that links the L and H chains. Thus, F_{ab} consists of the entire L chain and the amino-terminal half of the H chain, whereas F_c consists of the carboxyl-terminal halves of both H chains. The portion of the H chain contained in F_{ab} is called F_d.

IMMUNOGLOBULIN G IS A FLEXIBLE Y-SHAPED MOLECULE

Electron-microscopic studies by Robin Valentine and Michael Green showed that immunoglobulin G has the shape of the letter Y and that the hapten binds near the ends of the F_{ab} units. Furthermore, the F_c and the two F_{ab} units of the intact antibody are joined by a hinge that allows rapid variation in the angle between the F_{ab} units through a wide angular range (Figure 33-10). This kind of mobility, called *segmental flexibility*, enhances the formation of antibody-antigen complexes by enabling both combining sites on an antibody to bind a multivalent antigen. The distance between combining sites at the tips of the F_{ab} units can be adjusted to match the distance between specific determinants on the antigen (e.g., a virus with many repeating subunits). The affinity of an antibody for a multivalent antigen is typically 10^4 times as high when both com-

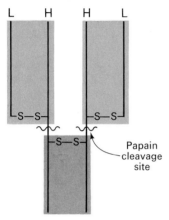

Figure 33-9
The subunit structure of IgG is L_2H_2. The F_{ab} units obtained by digestion with papain are shaded blue, whereas the F_c unit is shaded red.

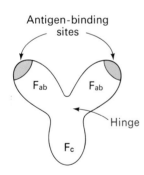

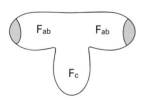

Figure 33-10
Immunoglobulin G has the shape of the letter Y. It contains a hinge that gives it segmental flexibility.

bining sites are involved as when only one binds the antigen. Segmental flexibility may also play a role in the transmission of information from the F_{ab} units to the F_c unit.

ARE ANTIBODIES FORMED BY SELECTION OR INSTRUCTION?

Enzymes acquired their specificities in millions of years of evolution. A specific antibody appears in the serum of an animal only a few weeks after it is exposed to any foreign determinant. How is a specific antibody formed in such a short time?

1. The *instructive theory,* which was proposed by Linus Pauling in 1940, postulated that the *antigen acts as a template that directs the folding of the nascent antibody chain.* In this model, antibody molecules of a given amino acid sequence have the potential for forming combining sites of many different specificities. The particular one formed depends on the antigen present at the time of folding. The instructive theory predicts that a specific antibody cannot be formed in the absence of the corresponding antigen.

2. The *selective theory,* which was originated by Macfarlane Burnet and Niels Jerne in the 1950s, postulates that the antigen controls only the amount of specific antibody produced. In this model, *the combining site of the specific antibody is completely determined before it ever encounters antigen.*

DEMISE OF THE INSTRUCTIVE THEORY

The instructive theory predicted that an antibody will lose its specificity if it is unfolded and then refolded in the absence of antigen. This prediction contrasted with the experimental results obtained in Christian Anfinsen's laboratory on the refolding of ribonuclease. Recall that unfolded ribonuclease spontaneously assumes the three-dimensional structure, specificity, and catalytic activity of the native enzyme upon removal of the denaturing agent (p. 34). A substrate need not be present during the refolding of ribonuclease. This important experiment on ribonuclease provided the impetus for devising a similar test for antibodies (Figure 33-11). An F_{ab} fragment from anti-DNP antibody rather than an intact antibody molecule was used because its smaller size simplified the experiment. The F_{ab} fragment was unfolded in 7 M guanidine hydrochloride, a highly effective denaturing agent. This denatured F_{ab} molecule had no detectable affinity for DNP haptens. Hydrodynamic and optical techniques indicated that the conformation of the denatured F_{ab} approached that of a random coil. The guanidine hydrochloride was then removed by dialysis, and the denatured F_{ab}

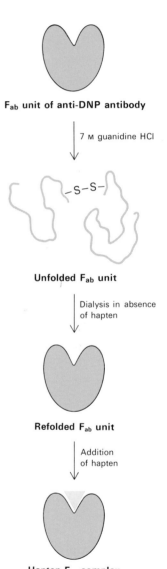

Figure 33-11
The F_{ab} unit of anti-DNP antibody binds DNP after being unfolded and refolded in the absence of this hapten. This experiment showed that specificity is inherent in the amino acid sequence.

refolded in *the absence of DNP hapten.* The striking result was that the refolded F_{ab} had a high affinity for DNP haptens. Because hapten was absent during refolding, it was evident that *the specificity of the combining site is inherent in the amino acid sequence of the antibody.* This experimental result is consistent with the selective theory but contradicts a critical prediction of the instructive theory. Furthermore, it was found that *antibody-producing cells can synthesize large amounts of antibody in the absence of antigen.* The essence of the selective theory—that the *antigen influences the amount of specific antibody produced but not its amino acid sequence or three-dimensional structure* —is now firmly established.

MYELOMA AND HYBRIDOMA IMMUNOGLOBULINS ARE HOMOGENEOUS

How is a specific antibody made before the arrival of an antigen? For many years, the heterogeneity of antibody preparations was an impediment to solving this problem because of the difficulty of analyzing a mixture of antibody molecules. This hurdle has been overcome by taking advantage of *multiple myeloma,* a malignant disorder of antibody-producing cells. In this cancer, transformation of a single lymphocyte or plasma cell results in uncontrolled cell division. Consequently, a very large number of cells of a single kind are produced. They are a *clone* because they are descended from the same cell and have identical properties. Large amounts of immunoglobulin of a single kind are secreted by these tumors. *Myeloma immunoglobulins have a normal structure and are typical of normal immunoglobulins, each being a homogeneous example of the large number of different antibodies present in an individual.* Myelomas also occur in mice. These tumors can be transplanted to other mice, where they proliferate. Furthermore, these antibody-producing tumors synthesize the same kind of homogeneous antibody generation after generation. In fact, homogeneous myeloma immunoglobulins have contributed to a major breakthrough in molecular immunology.

Cesar Milstein and Georges Köhler discovered that large amounts of homogeneous antibody of nearly any desired specificity can be obtained by *fusing an antibody-producing cell with a myeloma cell.* A mouse is immunized with an antigen and its spleen is removed several weeks later. A mixture of immune cells from this spleen is fused in vitro with myeloma cells. Hybrid cells can be selected by culturing the cells in a medium that does not sustain the growth of either parental line. Some of these hybrid cells acquire the neoplastic characteristics of myeloma cells and also synthesize antibody of the desired specificity. These *hybridoma cells* and their progeny indefinitely produce large amounts of the homogeneous antibody specified by the parent cell from the spleen. Many kinds of such *monoclonal antibodies* are now being produced for use as highly specific analytical reagents. For example, monoclonal antibodies directed against a particular drug or hormone can be used to assay very small amounts of such a molecule in body fluids.

EACH L AND H CHAIN CONSISTS OF A VARIABLE REGION AND A CONSTANT REGION

An abnormal protein appears in large amounts in the urine of many patients who have multiple myeloma. This protein attracted the attention of Henry Bence-Jones in 1847 because of its unusual solubility properties: it precipitates on being heated to 50°C and becomes soluble again on boiling. The *Bence-Jones protein* is now known to be a *dimer of the L chains* of a patient's myeloma immunoglobulin, which makes it a very suitable starting material for amino acid sequence studies.

The first complete amino acid sequences of myeloma L chains, which contain about 214 amino acids, were determined in 1965. These studies showed that Bence-Jones proteins from different patients have different amino acid sequences. Most striking, these differences in sequence are confined to the amino-terminal half of the polypeptide chain. Each Bence-Jones protein studied has a unique amino acid sequence from positions 1 to 108. In contrast, the sequences of a number of Bence-Jones proteins are the same starting at position 109. Thus, *the L chain consists of a variable region (residues 1 to 108) and a constant region (residues 109 to 214)* (Figure 33-12). This remarkable finding had no precedent in protein chemistry.

Not all L chains have the same constant region. In fact, there are two types of L chains, called kappa (κ) and lambda (λ), which exhibit many similarities. For example, each consists of 214 residues and contains an amino-terminal variable half and a carboxyl-terminal constant half. Each half of the κ and λ chains contains an intrachain disulfide loop. Both types have a carboxyl-terminal cysteine residue, which forms a disulfide bond with the H chain. Furthermore, the κ and λ chains have identical amino acid residues at 40% of the positions, which is nearly the same degree of homology exhibited by the α and β chains of human hemoglobin.

All κ chains have the same constant region except for residue 191, which is either leucine or valine. This *allotypic* difference is inherited in a classical Mendelian manner. The λ chains also exhibit variability at position 191, which is either lysine or arginine.

The H chain consists of 446 amino acid residues. A comparison of the sequences of H chains from different myeloma immunoglobulins shows that all of the differences in sequence are located in the 108 residues at the amino-terminal end. Thus, *the heavy chain, like the light chain, consists of a variable region* (V) *and a constant region* (C). The variable region of the heavy chain has the same length as that of the light chain, whereas the constant region is about three times as long (Figure 33-13).

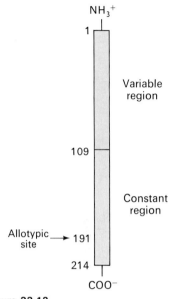

Figure 33-12
The light chain of an immunoglobulin consists of a variable region and a constant region.

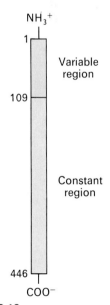

Figure 33-13
The heavy chain of IgG also consists of a variable region and a constant region.

THE ANTIGEN-BINDING SITE IS FORMED BY HYPER-VARIABLE REGIONS OF THE L AND H CHAINS

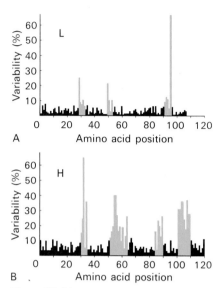

Figure 33-14
Hypervariable regions in (A) light and (B) heavy chains. The degree of variability in amino acid sequence is plotted as a function of amino acid position. [After J. D. Capra and A. B. Edmundson. The antibody combining site. Copyright © 1976 by Scientific American, Inc. All rights reserved.]

As mentioned earlier, the antigen-binding fragment, F_{ab}, consists of the entire L chain and part of the H chain. What are the relative contributions of the two chains to binding activity? An early clue came from a comparison of the amino acid sequences of many myeloma proteins, which revealed that the variable regions of the L and H chains are not uniformly variable. Rather, three segments in the L chain and four in the H chain display far more variability than do other residues in the variable regions of these chains (Figure 33-14). In 1970, Elwin Kabat proposed that these *hypervariable segments* form the antigen-binding site and that antibody specificity is determined by the nature of their amino acid residues.

Affinity-labeling studies were then carried out to test this hypothesis. Haptens containing reactive groups can be readily synthesized. The reactive hapten binds specifically to the antibody combining site and then forms a covalent bond with a suitable neighboring residue on the protein. For example, anti-DNP antibody can be labeled at its combining sites by a nitrophenyldiazonium hapten.

$$O_2N-\langle\text{ring}\rangle-N_2^+ \qquad \textit{p-}\textbf{Nitrophenyldiazonium hapten}$$

The diazonium group is highly reactive and can form a diazo bond with tyrosine, histidine, and lysine residues. Antibody modified in this way contains the label in both its L and its H chains. In fact, the labeled residues are located in the hypervariable regions of these chains. Thus, *hypervariable residues of both the L and the H chains form the antigen-binding site.* Subsequent x-ray crystallographic analyses have provided the most direct evidence in support of this conclusion, as will be discussed shortly.

THE VARIABLE AND CONSTANT REGIONS HAVE DIFFERENT ROLES

The amino-terminal parts of the light and heavy chains contain the variable regions, which are responsible for the binding of antigen. Differences in amino acid sequence yield combining sites that have different specificities. The rest of each chain contains the constant region, which mediates effector functions such as the binding of complement and the transfer of antibodies across the placental membrane. The same effector functions are mediated by antibodies of very different specificity. Thus, *antibodies consist of regions of unique sequence that confer antigen-binding specificity and regions of constant sequence that mediate common effector functions.* A striking structural feature of immunoglobulins is that the variable and constant regions are sharply demarcated in the amino acid sequences of the light and heavy chains.

The complete amino acid sequence of an immunoglobulin molecule was elucidated by Gerald Edelman in 1968. A most interesting feature of this sequence is the periodic location of intrachain disulfide bonds in both the light and the heavy chains (Figure 33-15).

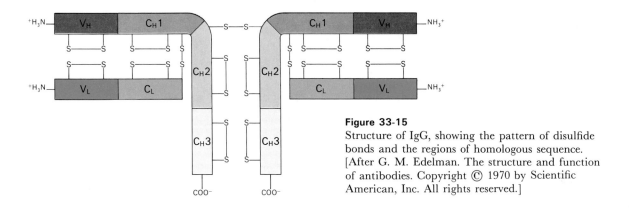

Figure 33-15
Structure of IgG, showing the pattern of disulfide bonds and the regions of homologous sequence. [After G. M. Edelman. The structure and function of antibodies. Copyright © 1970 by Scientific American, Inc. All rights reserved.]

Furthermore, there are many similarities in the amino acid sequences of different parts of the immunoglobulin molecule. The variable region of the light chain (V_L) is homologous with the variable region of the heavy chain (V_H). In addition, the constant region of the heavy chain (C_H) consists of equal thirds (C_H1, C_H2, and C_H3) that are similar in sequence. Furthermore, the constant region of the light chain (C_L) is homologous with the three domains of the constant region of the heavy chain.

The homologies evident in the amino acid sequences suggested that immunoglobulins are folded into compact domains, each consisting of a homologous region containing at least one active site that serves a distinctive molecular function (Figure 33-16). It also

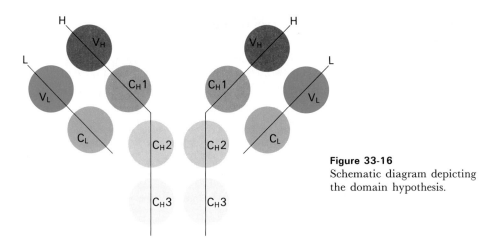

Figure 33-16
Schematic diagram depicting the domain hypothesis.

seemed likely that these domains would resemble each other in tertiary structure because they have similar amino acid sequences. X-ray crystallographic studies have verified the domain hypothesis. The three-dimensional structure of the $F_{ab'}$ fragment (which is nearly the same as the F_{ab} fragment) of a human immunoglobulin has been solved at 2.0-Å resolution by Roberto Poljak. This fragment consists of a *tetrahedral array of four globular subunits—V_L, V_H, C_L, and C_H1—which are strikingly similar in three-dimensional structure* (Figure 33-17). A common structural feature of these domains is the

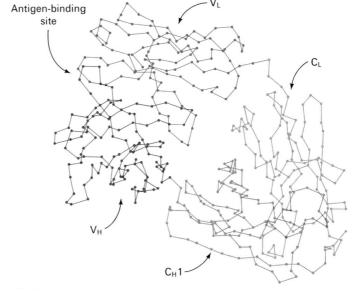

Figure 33-17
An α-carbon diagram of the $F_{ab'}$ unit. The L chain is shown in blue and the H chain in red. [From R. L. Poljak, L. M. Amzel, H. P. Avey, B. L. Chen, R. P. Phizackerley, and F. Saul. *Proc. Nat. Acad. Sci.* 70(1973):3306.]

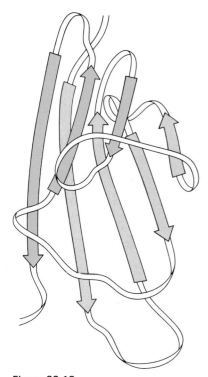

Figure 33-18
The basic structural motif of domains in immunoglobulins consists of two sheets of antiparallel β strands (shown in green and yellow). [After J. D. Capra and A. B. Edmundson. The antibody combining site. Copyright © 1976 by Scientific American, Inc. All rights reserved.]

presence of two broad sheets of antiparallel β-strands (Figure 33-18). Many hydrophobic side chains are tightly packed between these sheets. This recurring structural motif is called the *immunoglobulin fold*.

The three-dimensional structure of an intact immunoglobulin G molecule has been solved by David Davies. A 6-Å electron-density map shows that this antibody molecule is T-shaped (Figure 33-19). The domains in the F_c unit, like those in the F_{ab} units, display the characteristic immunoglobulin fold. The carbohydrate moiety covalently attached to an asparagine residue of the C_H2 domain is located between the C_H1 and C_H2 domains (Figure 33-19). Thus, the carbohydrate forms a large part of the interface between the F_{ab} and F_c units. A comparison of the relative orientations of adjacent domains in a variety of immunoglobulins and their fragments shows that the polypeptide chain between domains is quite flexible.

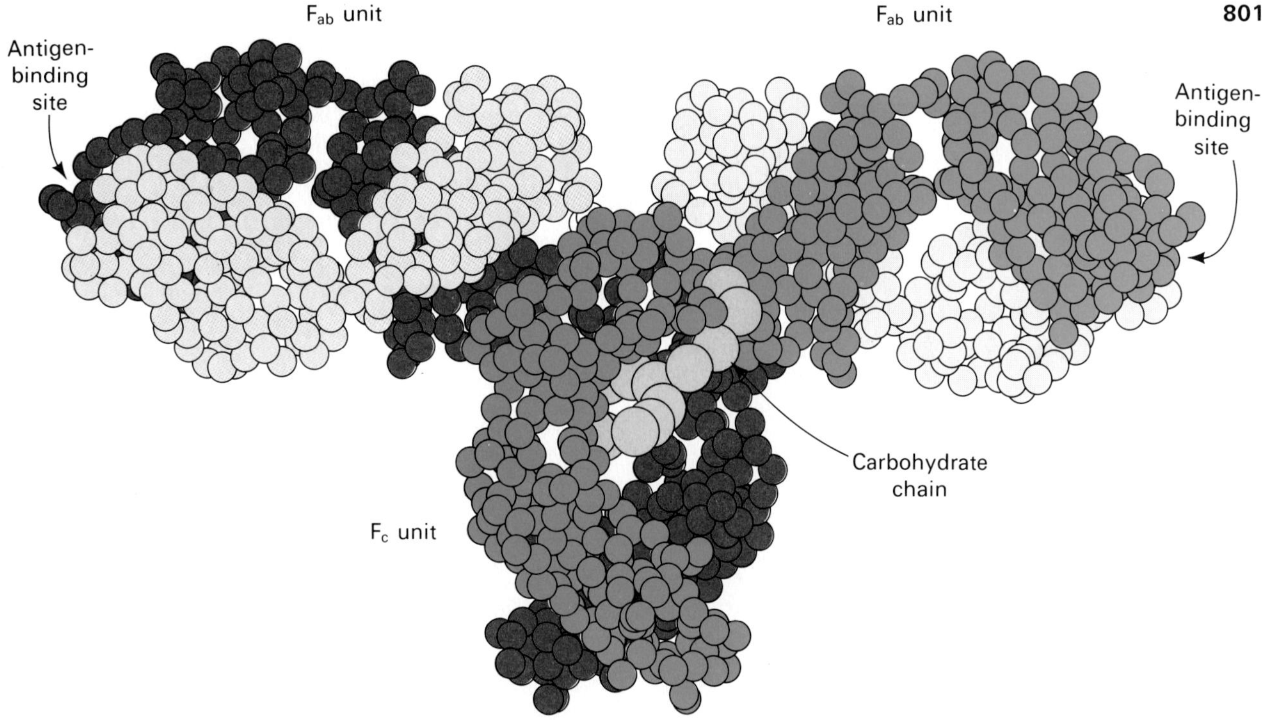

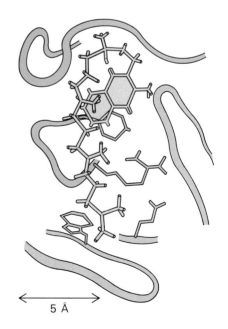

Figure 33-19
Schematic drawing of the three-dimensional structure of an IgG molecule.
Each amino acid residue is represented by a small sphere. One of the H
chains is shown in dark red, the other in dark blue. One of the L chains is
shown in light red, the other in light blue. A carbohydrate unit attached to a
C_H2 domain is shown in yellow. [After E. W. Silverton, M. A. Navia, and
D. R. Davies. *Proc. Nat. Acad. Sci.* 74(1977):5142.]

X-RAY ANALYSES OF ANTIBODY COMBINING SITES
HAVE REVEALED HOW SOME HAPTENS BIND

X-ray crystallographic studies of complexes of haptens and F_{ab} fragments are sources of insight into the nature of antibody combining
sites and the structural basis of their specificity. As predicted from
comparisons of amino acid sequences and affinity-labeling studies,
*the combining site for antigen is formed by residues from the hypervariable
segments of both the L and the H chains.* The mode of binding of a
vitamin K_1 derivative exemplifies this important structural feature
(Figure 33-20). Likewise, phosphorylcholine binds to a cavity that is

Figure 33-20
Mode of binding of a derivative of vitamin K_1 (shown in yellow) to an antibody combining site. Amino acid residues from the hypervariable segments of
the H chain are shown in red, and those from the L chain in blue. [After
I. M. Amzel, R. Poljak, F. Saul, J. Varga, and F. Richards. *Proc. Nat. Acad. Sci.*
71(1974):1427.]

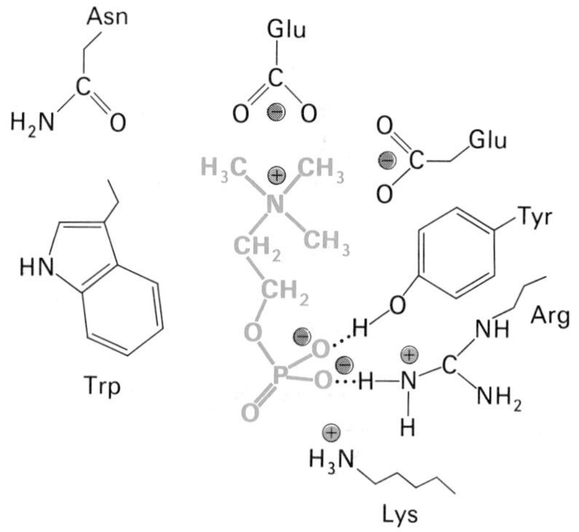

Figure 33-21
Mode of binding of phosphorylcholine (shown in green) to an antibody combining site. [After E. A. Padlan, D. R. Davies, S. Rudikoff, and M. Potter. *Immunochemistry* 13(1976):945.]

lined with residues from five hypervariable segments, two from the L chain and three from the H chain (Figure 33-21). This hapten interacts most strongly with residues from the H chain. The positively charged trimethylammonium group of phosphorylcholine is buried inside the wedge-shaped cavity, where it interacts electrostatically with two negatively charged glutamate side chains. The negatively charged phosphate group of the hapten binds to the positively charged guanidinium group of an arginine residue at the mouth of the crevice. The phosphate group is also hydrogen bonded to the hydroxyl group of a tyrosine residue and to the amino group of the arginine side chain. Numerous van der Waals interactions, such as the ones including a tryptophan side chain (Figure 33-21), also contribute to the stability of this complex.

DIFFERENT CLASSES OF IMMUNOGLOBULINS HAVE DISTINCT BIOLOGICAL ACTIVITIES

Thus far, we have focused on immunoglobulin G. There are in fact five classes of immunoglobulins (Table 33-1), each consisting of heavy and light chains. The constant regions of the heavy chains differ from one class to another, whereas the light chains are the same, either λ or κ. The heavy chains in immunoglobulin G are called γ chains, whereas those in immunoglobulins A, M, D, and E are called α, μ, δ, and ϵ, respectively. These distinctive heavy chains give the five classes of immunoglobulins different biological characteristics. As mentioned previously, *immunoglobulin M* (IgM) is the first class of antibodies to appear in the serum after injection of an antigen, whereas *immunoglobulin G* (IgG) is the principal antibody in the serum. For many years, IgG was called γ-globulin, because of its electrophoretic and solubility characteristics. *Immunoglobulin A* (IgA) is the major class of antibodies in external secretions, such as saliva, tears, bronchial mucus, and intestinal mucus. Thus, IgA serves as a first line of defense against bacterial and viral antigens.

Table 33-1
Properties of immunoglobulin classes

Class	Serum concentration (mg/ml)	Mass (kdal)	Sedimentation coefficient(s)	Light chains	Heavy chains	Chain structure
IgG	12	150	7	κ or λ	γ	$\kappa_2\gamma_2$ or $\lambda_2\gamma_2$
IgA	3	180–500	7, 10, 13	κ or λ	α	$(\kappa_2\alpha_2)_n$ or $(\lambda_2\alpha_2)_n$
IgM	1	950	18–20	κ or λ	μ	$(\kappa_2\mu_2)_5$ or $(\lambda_2\mu_2)_5$
IgD	0.1	175	7	κ or λ	δ	$\kappa_2\delta_2$ or $\lambda_2\delta_2$
IgE	0.001	200	8	κ or λ	ϵ	$\kappa_2\epsilon_2$ or $\lambda_2\epsilon_2$

Note: $n = 1$, 2, or 3. IgM and oligomers of IgA also contain J chains that join immunoglobulin molecules. IgA in secretions has an additional secretory piece.

The benefits conferred by immunoglobulins D (IgD) and E (IgE) are not yet known. However, the harmful effects of *immunoglobulin E* in mediating allergic reactions are clearly established.

ANTIBODY MOLECULES EVOLVED BY GENE DUPLICATION AND DIVERSIFICATION

The structural homologies between domains in immunoglobulins have important evolutionary implications. The similarities in the amino acid sequences of chymotrypsin, trypsin, elastase, and thrombin indicate that they evolved from a common ancestor (p. 166). Likewise, *the internal homologies in the immunoglobulin molecule suggest that antibodies evolved by a process of gene duplication and subsequent diversification.* An ancestral gene may have coded for a primitive antibody having a length of about 108 residues. Such an ancestral gene probably duplicated and then diverged to give rise to different kinds of variable and constant regions. Allen Edmundson has proposed that a dimer of L chains served as a primitive antibody (Figure 33-22).

VARIABLE AND CONSTANT REGIONS ARE ENCODED BY SEPARATE GENES THAT BECOME JOINED

How are immunoglobulin genes organized and expressed? The discovery of distinct variable and constant regions in the L and H chains raised the possibility that immunoglobulin genes, as well as their polypeptide products, have an unusual architecture. As mentioned earlier, the constant regions of κ chains are identical except at position 191, which is either leucine or valine. Pedigree analyses showed that these two variants (allotypes) are inherited in a classical Mendelian manner, which strongly implied that there is only one gene for the constant region of these chains. In contrast, genetic analyses showed that their variable regions are encoded by multiple

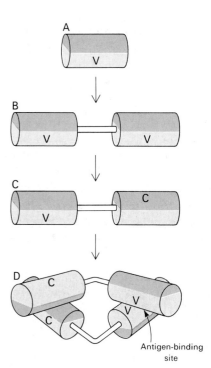

Figure 33-22
Proposed stages in the evolution of a primitive antibody. The gene coding for (A) a single polypeptide chain with about 108 residues duplicated to give (B) two identical domains joined by a short linear polypeptide chain. Divergent evolution of the duplicated genes altered the amino acid compositions of the two domains to give (C) variable and constant regions rotated with respect to each other. Association of these molecules yielded (D) a primitive F_{ab} unit with a combining site for antigen. [From J. D. Capra and A. B. Edmundson. The antibody combining site. Copyright © 1976 by Scientific American, Inc. All rights reserved.]

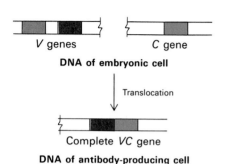

V genes C gene

DNA of embryonic cell

Translocation

Complete VC gene

DNA of antibody-producing cell

Figure 33-23
A V gene is translocated near a C gene in the differentiation of an antibody-producing cell.

genes. In 1965, William Dreyer and Claude Bennett proposed that multiple V genes are separate from a single C gene in the germ line. According to their model, one of these V genes becomes joined to the C gene during differentiation of the antibody-producing cell. It was known then that DNA can be spliced as in the integration of λ phage DNA.

A critical test of this translocation hypothesis had to await the isolation of pure immunoglobulin mRNA and the development of techniques for analyzing complex mammalian genomes. The use of restriction enzymes to split large chromosomes into specific pieces of DNA that can be readily analyzed was particularly important in this regard. The distribution of V and C genes on these restriction fragments could then be assayed by hybridizing them with mRNA segments specific for either the variable or the constant region. In 1976, Susumu Tonegawa found in this way that V *and C genes are far apart in embryonic (germ-line) DNA but are closely associated in the DNA of antibody-producing cells. Thus, immunoglobulin genes are translocated during the differentiation of lymphocytes* (Figure 33-23).

HOW ARE DIVERSE ANTIBODY SPECIFICITIES GENERATED?

An animal can synthesize large amounts of specific antibody against virtually any foreign determinant within a few weeks after being exposed to it. The number of different kinds of antibodies that can be made by an animal is very large, probably more than a million. We have seen that the structural basis of antibody specificity resides in the amino acid sequences of the variable regions of the light and heavy chains. This brings us to the key question: *How are different variable-region sequences generated?* More specifically, we can ask:

1. *When* is diversity generated? During the life of an individual animal (somatic) or in the course of evolution (genetic)?

2. *How* is diversity generated? By a random evolutionary process, by somatic recombination, or by somatic hypermutation?

The wealth of amino acid sequence data on myeloma immunoglobulins stimulated immunologists to formulate models for the generation of diversity. Three very different kinds of mechanisms have been proposed:

1. *Germ-line hypothesis.* In this model, diversity is already present in the germ line, in which there are a very large number ($>10^4$) of genes for the variable regions of antibodies. For each unique amino acid sequence in the variable regions, there is a corresponding piece of DNA in the germ line. Thus, this hypothesis proposes that diversity arose in the course of evolution by random mutation and selection.

2. *Somatic-recombination hypothesis.* In this model, there are only a small number of genes, of the order of 100, for the variable regions of antibodies. These genes, which are similar but not identical, undergo extensive recombination in the antibody-producing cell throughout the lifetime of an individual. Crossing-over is assumed to be intrachromosomal. A plausible calculation suggests that recombination among ten genes could generate 10^6 different amino acid sequences.

3. *Somatic-hypermutation hypothesis.* In this model, diversity is created by point mutations in a single gene for the variable region of a particular subclass. An extraordinarily high mutation rate would be needed to produce the required diversity during the lifetime of an individual. One proposal is that this is achieved in the antibody-producing cell (or its precursor) by a DNA repair mechanism that is designed to make mistakes.

THERE ARE SEVERAL HUNDRED GENES FOR THE VARIABLE REGIONS OF L AND H CHAINS

The revolution in DNA technology that has taken place in recent years has made the generation of antibody diversity experimentally accessible. In particular, the development of gene cloning and of methods for rapidly sequencing long stretches of DNA has been especially important. The use of these techniques has provided in a short time structural information needed to answer two key questions: How many variable-region genes are there in the germ line? Do their base sequences change during the differentiation of antibody-producing cells?

The answer to the first question is that there are *several hundred genes for the variable regions of κ light chains (V_κ) and of heavy chains (V_H).* If there are 300 V_κ genes and 300 V_H genes, a maximum of 9×10^4 different specificities could be generated by the combinatorial association (300×300) of L and H chains encoded by these genes. This calculated upper limit of 9×10^4 falls far short of the number of different kinds of antibodies that an animal can produce, which is generally agreed to be significantly more than 10^6. The disparity is even greater for λ light chains, which are known to be encoded by fewer than ten V_λ genes. In short, *the number of variable-region genes in the germ line is simply too small to be the sole source of antibody diversity. Clearly, some of this diversity must be generated during the lifetime of an animal in the differentiation of its lymphocytes.*

DISCOVERY OF *J* (JOINING) GENES, AN ADDITIONAL SOURCE OF ANTIBODY DIVERSITY

The next advance in the elucidation of how diversity is generated came from the determination of the base sequences of cloned im-

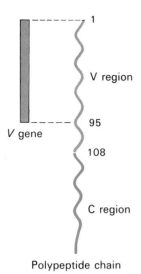

Figure 33-24
A *V* gene isolated from embryonic cells is foreshortened. It does not encode the last thirteen amino acid residues of the variable (V) region of the polypeptide chain.

munoglobulin genes from embryonic cells and myeloma cells. These highly informative studies were carried out by Tonegawa, Philip Leder, and Leroy Hood. The first surprise was that V *genes in embryonic cells do not encode the entire variable region of L and H chains.* The germ-line *V* gene stops at the codon for amino acid residue 95 rather than 108, which is the end of the variable region of the polypeptide chain (Figure 33-24). Where is the DNA that encodes the last thirteen residues of the variable region? In embryonic cells, this stretch of DNA is located in an unexpected place: near the *C* gene. It is called the *J* gene because it joins the *V* and *C* genes in a differentiated cell. In fact, *a tandem array of several* J *genes is located near the* C *gene in embryonic cells* (Figure 33-25). In the differentiation of an

Figure 33-25
A tandem array of *J* genes, which encode part of the last hypervariable segment of the variable region, is located near the *C* gene.

antibody-producing cell, a *V* gene is translocated to a site near the *C* gene by intrachromosomal recombination. Specifically, *a* V *gene becomes spliced to a* J *gene to form a complete gene for the variable region.* Each of these genes contains a short palindromic sequence next to the site of recombination. These palindromes may be recognition elements in this recombination process.

The *J* genes are important contributors to antibody diversity because they encode part of the last hypervariable segment of the L and H chains. In forming a continuous V_κ gene, any of several hundred *V* genes can become linked to any of five *J* genes. For example, 1,500 kinds of continuous *V* genes can be formed by the combinatorial joining of 300 incomplete *V* genes and 5 *J* genes. Thus, *somatic recombination of these gene segments amplifies the diversity already present in the germ line.*

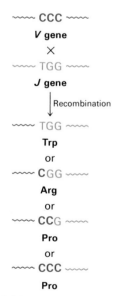

Figure 33-26
Imprecision in the exact site of splicing of a *V* gene to a *J* gene gives rise to additional diversity.

ALTERNATIVE FRAMES FOR THE JOINING OF *V* AND *J* GENES ALSO CONTRIBUTE TO DIVERSITY

The other surprising finding was that a set of five *J* genes yielded more than five amino acid sequences for the corresponding region of the light chain. Analyses of amino acid and base sequence data then showed that recombination of *V* and *J* genes is not absolutely precise. Recombination between these genes can occur at one of several bases near the codon for residue 95 (Figure 33-26). Thus, *additional diversity is generated by allowing* V *and* J *genes to become spliced in slightly different joining frames.* The immune system clearly takes delight in sweet disorder!

Only one member of an allelic pair of genes is expressed by a particular lymphocyte. Hence, all of the antigen-binding sites produced by a single cell are the same. The structural basis of this selective gene expression, known as *allelic exclusion*, is being elucidated. Analyses of restriction fragments have shown that an incomplete *V* gene becomes correctly joined to a *J* gene on only one of a pair of homologous chromosomes. Only the properly recombined immunoglobulin gene is expressed.

mRNAs FOR L AND H CHAINS ARE FORMED BY THE SPLICING OF PRIMARY TRANSCRIPTS

The mRNA for a κ light chain contains about 1,250 bases (Figure 33-27). Like other eucaryotic mRNAs, it contains a poly A tail at its

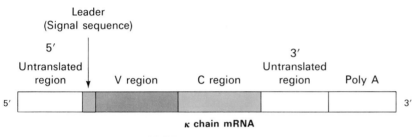

Figure 33-27
Structure of mRNA for an L chain.

3′ end and untranslated sequences at both its 5′ and its 3′ ends. A *leader sequence* near the 5′ end of the κ-chain mRNA codes for the hydrophobic N-terminal region of the nascent chain. This *signal sequence* (Figure 33-28) directs the ribosome to the endoplasmic reticulum and enables the nascent immunoglobulin chain to traverse this membrane and enter the lumen of the ER (p. 713). A peptidase then cleaves this signal sequence on the luminal side of the ER. The variable and constant regions of the light chain are encoded by a contiguous sequence on the mRNA that immediately follows the leader sequence.

The primary transcript that gives rise to this mRNA contains two intervening sequences (Figure 33-29). One of them separates the leader sequence from the start of the variable-region sequence, and

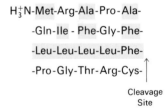

Figure 33-28
Signal sequence of a nascent L chain. Hydrophobic residues are shown in yellow.

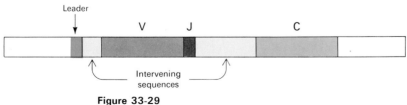

Figure 33-29
Primary transcript of an L chain mRNA.

the other separates the distal end of the joining-region sequence from the start of the constant-region sequence. *These intervening sequences are removed in the processing of the primary transcript.* It is interesting to note that the J gene carries information for two kinds of splicing, one at the DNA level for the fusion of V and J, and the other at the RNA level for the joining of J and C.

What is the structure of the gene for the heavy chain? Recall that an H chain consists of four domains: V_H, C_H1, C_H2, and C_H3. A DNA fragment containing the constant region of an IgG heavy chain was recently cloned. Electron-microscopic studies showed

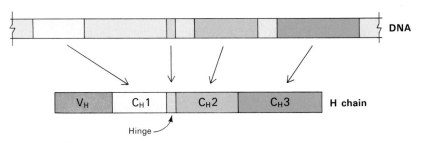

Figure 33-30
The three domains of the constant (C) region of a heavy chain and the hinge region are encoded by separate gene segments.

that C_H1, C_H2, and C_H3 are encoded by separate DNA segments (Figure 33-30). Furthermore, the hinge between C_H1 and C_H2 is encoded by yet another piece of DNA. Thus, *the domain construction of immunoglobulins (p. 799) is a reflection of their underlying gene architecture. Splicing enabled organisms in the course of evolution to create new proteins by assembling DNA segments that code for different functional domains.*

DIFFERENT CLASSES OF ANTIBODIES ARE FORMED BY THE HOPPING OF V_H GENES

As mentioned earlier, there are five classes of immunoglobulins. An antibody-producing cell first synthesizes IgM and then makes either IgG, IgA, IgD, or IgE of the *same specificity.* In this switch, the light chain is unchanged. Furthermore, the variable region of the heavy chain stays the same. *Only the constant region of the heavy chain changes,* and so this step in the differentiation of an antibody-producing cell is called C_H switching (Figure 33-31).

In embryonic mouse cells, the genes for the constant regions of the μ, γ, and α heavy chains—called C_μ, C_γ, and C_α—are next to each other (Figure 33-32). In fact, there are four genes for the constant regions of γ

J_H genes	C_μ	$C_{\gamma3}$	$C_{\gamma1}$	$C_{\gamma2b}$	$C_{\gamma2a}$	C_α

Figure 33-32
The genes for the constant regions of μ, γ, and α chains are next to each other. The positions of the C_δ and C_ε genes are not yet known.

V	C_μ

μ chain of IgM

C_H switching

V	C_α

α chain of IgA

or

V	C_γ

γ chain of IgG

or

V	C_δ

δ chain of IgD

or

V	C_ε

ε chain of IgE

Figure 33-31
Synthesis of different classes of immunoglobulins. The V_H region is first associated with C_μ and then with another C region to form an H chain of a different class.

chains, which agrees nicely with genetic analyses showing that there are four subclasses of IgG. A tandem array of J genes for the last hypervariable segment of the variable region is located next to the C_μ gene. *A complete gene for the heavy chains of IgM antibody is formed by the translocation of a V_H gene to a J_H gene* (Figure 33-33). This joining brings V_H, J_H, and C_μ together to form a functional gene. The intervening sequences between the leader and start of the variable region, between the end of J_H and the start of C_μ, and within C_μ are excised in the processing of the primary transcript to form mRNA for the μ chain.

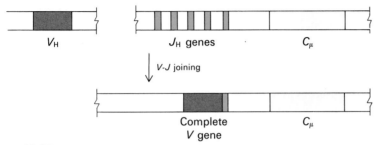

Figure 33-33
The joining of V_H and J_H creates a functional gene for a μ chain.

Analyses of restriction digests of DNA from embryonic and myeloma cells have revealed that C_H switching takes place at the level of DNA instead of RNA. *In the switch from IgM to IgA, for example, the V_H-J_H gene moves from a site near the C_μ gene to one near the C_α gene* (Figure 33-34). In this recombination, the genes between C_μ and C_α loop out and are lost. A palindromic sequence may align the DNA segments undergoing recombination. The translocation of the entire V_H-J_H gene accounts for the fact that the antigen-combining specificity of IgA produced by a particular cell is the same as that of IgM synthesized at an earlier stage of its development. It is not known how a cell chooses one of several C_H genes as the target in this translocation. *The biological significance of C_H switching is that a whole recognition domain (the variable domain) is shifted from the early constant region (C_μ) to one of several other constant regions that mediate different effector functions.*

DIVERSITY IS GENERATED BY SOMATIC RECOMBINATION OF MANY GERM-LINE GENES AND BY SOMATIC MUTATION

Let us recapitulate the sources of antibody diversity. The germ line contains a rather large repertoire of variable-region genes. For κ light chains, there are several hundred (say 300) V genes and five J genes. There are at least three frames for the joining of V and J. Hence a total of $300 \times 5 \times 3$, or 4,500, kinds of complete V_κ genes can be formed by the combinatorial associations of V and J. A comparable number of kinds of heavy chains are generated in a

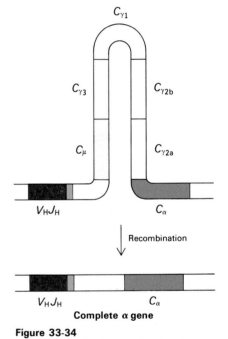

Figure 33-34
Structural basis of C_H switching. The V_H-J_H gene moves from its position near C_μ to one near C_α by intrachromosomal recombination.

similar way. The association of 4,500 L chains with 4,500 H chains would yield 4,500 × 4,500, or 2×10^7, different antibody specificities. This calculated number is sufficiently large to account for the remarkable range of antibodies that can be synthesized by an animal.

As mentioned earlier, there are far fewer V_λ genes than V_κ genes. In fact, mice seem to have only two V_λ genes. However, many more V_λ amino acid sequences are known. It seems likely that somatic mutation gives rise to much of the diversity of λ light chains. *Thus, nature draws on each of the three proposed sources of diversity* (p. 804)—*a germ-line repertoire, somatic recombination, and somatic mutation—to form the rich variety of antibodies that are needed to protect an organism from foreign incursions.*

CLONAL SELECTION THEORY OF ANTIBODY FORMATION

A unifying model of the immune response is provided by the *clonal selection theory,* which was developed by Niels Jerne, Macfarlane Burnet, David Talmage, and Joshua Lederberg in the 1950s. The essential features of this theory, now firmly established, are:

1. An antibody-producing cell has a unique base sequence in its DNA that determines the amino acid sequence of its immunoglobulin chains. The specificity of an antibody is determined by its amino acid sequence.

2. A single cell produces antibody of a single kind. *The commitment to synthesize a particular kind of antibody is made before the cell ever encounters antigen.*

3. Each cell, as it begins to mature, produces small amounts of specific antibody. Some of this antibody becomes bound to the surface of the cell. An immature cell is killed if it encounters the antigen corresponding to its antibody. Hence, an animal does not usually make antibody against its own macromolecules—it is *self-tolerant.* Such antibody-producing cells were eliminated during fetal life. In contrast, *a mature cell is stimulated if it encounters antigen. Large amounts of antibody are then synthesized and the cell is triggered to divide.* The descendants of this cell are a *clone.* They have the same genetic makeup as that of the cell that was initially triggered by its encounter with antigen, and therefore all of them make antibody of the same specificity.

4. A clone tends to persist after the disappearance of the antigen. These cells retain their capacity to be stimulated by antigen if it reappears, which provides for *immunological memory.*

Specific antibody molecules are located on the surfaces of anti-body-producing cells. The precursor cells are *lymphocytes,* which are very slow to divide unless stimulated by antigen. They then become *plasma cells,* which are very active in the synthesis and secretion of antibodies. If a population of lymphocytes is passed through a column containing a covalently bound dinitrophenyl (DNP) group, a small proportion remains bound. These bound lymphocytes synthesize anti-DNP antibody and contain antibody molecules of this specificity on their surfaces. The other lymphocytes do not make anti-DNP antibody.

How are lymphocytes triggered by exposure to specific antigen? Antibody molecules on lymphocyte surfaces are randomly distributed in the absence of antigen. The addition of antigen has a striking effect: the surface antibody molecules with their bound antigens come together to form a cap. Following this redistribution, the antibody molecules on the surface are engulfed into the interior of the cell. *Cap formation* is an energy-requiring process that involves contractile activity. The formation of a surface antibody-antigen lattice leads to cap formation, which may in turn stimulate cell division. The antigen must be multivalent (i.e., contain more than one specific determinant), so that it can cross-link antibody molecules on the lymphocyte surface. The link between these events on the cell surface and mitotic activity in the cell nucleus is a challenging area of inquiry.

Figure 33-35
Scanning electron micrograph of sheep erythrocytes bound to a human T-lymphocyte (center). The surface of this lymphocyte contains receptors that are specific for a component of the sheep erythrocyte. [Courtesy of Dr. Jerry Thornthwaite, Dr. Robert Leif, and Dr. Marilyn Cayer.]

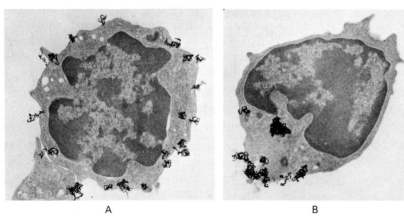

A   B

Figure 33-36
Radiographs showing the capping of immunoglobulins on the surface of a B-lymphocyte. The grains correspond to ^{125}I-labeled anti-immunoglobulin antibody which was added to mouse-spleen B-lymphocytes. (A) At 4°C, the immunoglobulin molecules are uniformly distributed on the cell surface. (B) On incubation at 37°C, the label becomes concentrated into a cap at one pole of of the cell and is then internalized by endocytosis. [Courtesy of Dr. Emil Unanue.]

BIOLOGICAL SIGNIFICANCE OF CLONAL SELECTION

The essence of the clonal selection theory, as expressed by Burnet, is that

> No combining site is in any evolutionary sense adapted to a particular antigenic determinant. The pattern of the combining site is there and if it happens to fit, in the sense that the affinity of adsorption to a given antigenic determinant is above a certain value, immunologically significant reaction will be initiated.

The selection of a preexisting pattern that is randomly generated is not a new theme in biology. Indeed, it is the heart of Darwin's theory of evolution. It is interesting to note that instructive theories have preceded selective theories: Lamarck's before Darwin's, antigen-directed folding before clonal selection.

Jerne has suggested that we should consider the possibility that selective mechanisms play a role in the operations of the nervous system, such as memory. In fact, the instructive and selective theories of learning have been stated long ago. Locke held that the brain could be likened to white paper, devoid of all characteristics, on which experience paints with almost endless variety. In contrast, Socrates asserted that "all learning consists of being reminded of what is preexisting in the brain." I hope that this chapter has seemed familiar to you!

SUMMARY

An antibody is a protein synthesized by an animal in response to a foreign macromolecule, called an antigen or immunogen, for which it has high affinity. Small foreign molecules (haptens) elicit the formation of specific antibody if they are attached to a macromolecule. Antibody synthesis occurs by selection rather than by instruction. An antigen binds to the surface of lymphocytes already committed to make antibodies specific for that antigen. The combination of antigen and surface receptors triggers cell division and the synthesis of large amounts of specific antibody. Antibodies directed against a specific determinant are usually heterogeneous, because they are the products of many antibody-producing cells. The antibodies produced by a single cell or by a clone are homogeneous. There are five classes of antibodies in the serum, of which immunoglobulin G is the principal one. Immunoglobulin M is the first class to appear in the serum following exposure to an antigen, whereas immunoglobulin A is the major class in external secretions. The roles of immunoglobulins D and E are not known.

Antibodies consist of light (L) and heavy (H) chains. Immunoglobulin G, which has the subunit structure L_2H_2, contains two binding sites for antigen. Immunoglobulin G can be enzymatically cleaved into two F_{ab} fragments, which bind antigen but do not form a precipitate, and one F_c fragment, which mediates effector functions such as the binding of complement. A comparison of the amino acid sequences of myeloma immunoglobulins has revealed that L and H chains consist of a variable (V) region (an amino-terminal sequence of about 108 residues) and a constant (C) region. Antigen-binding sites are formed by residues from the hypervariable segments of the variable regions of both the L and the H chains. Antibody molecules are folded into compact domains of about 108 residues that have homologous sequences. The constant-region domains, which mediate effector functions, probably evolved by duplication and diversification of an antigen-binding (variable) domain.

Variable and constant regions are encoded by separate genes that become joined during differentiation. There are several hundred genes for the variable regions of L and H chains. A complete gene for the variable region of a light chain is formed by the recombination of an incomplete V gene with one of several J (joining) genes that code for part of the last hypervariable segment. This tandem array of J genes is located near the C gene. Intervening sequences are removed from the primary transcript. The resulting mRNA codes for an N-terminal hydrophobic signal sequence that is cleaved from the nascent chain in addition to coding for the variable and constant regions of the immunoglobulin chain. The complete gene for the μ heavy chains of immunoglobulin M is also formed by the joining of V and J genes. A heavy chain of a different class (such as IgG) is then formed when this complete V_H gene is translocated to a different C gene (C_γ). This gene rearrangement is called C_H switching. The diversity of κ light chains and heavy chains arises from the somatic recombination of a rather large repertoire of germ-line variable- and joining-region genes. Different frames for the joining of these genes also contribute to their diversity. Somatic mutation is likely to be important in generating diverse λ light chains. The association of several thousand kinds of L chains with an equal number of H chains could account for the remarkable range of antibodies synthesized by an animal.

SELECTED READINGS

WHERE TO START

Capra, J. D., and Edmundson, A. B., 1977. The antibody combining site. *Sci. Amer.* 236(1):50–59. [Available as *Sci. Amer.* Offprint 1350.]

Milstein, C., 1980. Monoclonal antibodies. *Sci. Amer.* 243(4):66–74.

Porter, R. R., 1973. Structural studies of immunoglobulins. *Science* 180:713–716.

Edelman, G. M., 1973. Antibody structure and molecular immunology. *Science* 180:830–840.

BOOKS

Hood, L. E., Weissman, I. L., and Wood, W. B., 1978. *Immunology.* Benjamin. [A lucid, highly readable introduction to immunology.]

Kabat, E. A., 1976. *Structural Concepts in Immunology and Immunochemistry,* (2nd ed.). Holt, Rinehart, and Winston. [An excellent account of immunoglobulin structure and its relation to antigen-binding specificity.]

Mishell, B. B., and Shiigi, S. M., (eds.), 1980. *Selected Methods in Cellular Immunology.* Freeman. [A very useful guide to experimental procedures in immunology.]

STRUCTURE OF IMMUNOGLOBULINS

Amzel, L. M., and Poljak, R. J., 1979. Three-dimensional structure of immunoglobulins. *Ann. Rev. Biochem.* 48:961–967.

Silverton, E. W., Navia, M. A., and Davies, D. R., 1977. Three-dimensional structure of an intact human immunoglobulin. *Proc. Nat. Acad. Sci.* 74:5140–5144.

Valentine, R. C., and Green, N. M., 1967. Electron microscopy of an antibody-hapten complex. *J. Mol. Biol.* 27:615–617.

GENE ARCHITECTURE

Tonegawa, S., Maxam, A. M., Tizard, R., Bernard, O., and Gilbert, W., 1978. Sequence of a mouse germ-line gene for a variable region of an immunoglobulin light chain. *Proc. Nat. Acad. Sci.* 75:1485–1489.

Sakano, H., Rogers, J. H., Hüppi, K., Brack, C., Traunecker, A., Maki, R., Wall, R., and Tonegawa, S., 1979. Domains and the hinge region of an immunoglobulin heavy chain are encoded in separate DNA segments. *Nature* 277:627–633.

GENERATION OF DIVERSITY

Seidman, J. G., Leder, A., Nau, M., Norman, B., and Leder, P., 1978. Antibody diversity. *Science* 202:11–17.

Sakano, H., Hüppi, K., Heinrich, G., and Tonegawa, S., 1979. Sequences at the somatic recombination sites of immunoglobulin light-chain genes. *Nature* 280:288–294.

Seidman, J. G., Max, E. E., and Leder, P., 1979. A κ-immunoglobulin gene is formed by site-specific recombination without further somatic mutation. *Nature* 280:370–375.

Schilling, J., Clevinger, B., Davie, J. M., and Hood, L., 1980. Amino acid sequence of homogeneous antibodies to dextran and DNA rearrangements in heavy chain V-region gene segments. *Nature* 283:35–40.

Kataoka, T., Kawakami, T., Takahashi, N., and Honjo, T., 1980. Rearrangement of immunoglobulin γ1-chain gene and mechanism for heavy-chain class switch. *Proc. Nat. Acad. Sci.* 77:919–923.

Davis, M. M., Calame, K., Early, P. W., Livant, D. L., Joho, R., Weissman, I. L., and Hood, L., 1980. An immunoglobulin heavy-chain gene is formed by at least two recombinational events. *Nature* 283:733–739.

EFFECTOR FUNCTIONS OF IMMUNOGLOBULINS

Porter, R. R., and Reid, K. B. M., 1978. The biochemistry of complement. *Nature* 275:699–704.

Metzger, H., 1978. The IgE-mast cell system as a paradigm for the study of antibody mechanisms. *Immunol. Rev.* 41:186–199.

CLONAL SELECTION AND CELLULAR IMMUNOLOGY

Burnet, F. M., 1959. *The Clonal Selection Theory of Acquired Immunity.* Cambridge University Press.

Jerne, N. K., 1967. Antibodies and learning: selection versus instruction. *In* Quarton, G. C., Melnechuk, T., and Schmitt, F. O., (eds.), *The Neurosciences: A Study Program.* Rockefeller University Press.

Cunningham, B. A., 1977. The structure and function of histocompatability antigens. *Sci. Amer.* 237(4):96–107. [Offprint 1369.]

Raff, M. C., 1976. Cell-surface immunology. *Sci. Amer.* 234(5):30–39. [Offprint 1338.]

Jerne, N. K., 1973. The immune system. *Sci. Amer.* 229(1):52–60. [Offprint 1276.]

CHAPTER 34

MUSCLE CONTRACTION
AND CELL MOTILITY

How is chemical-bond energy transformed into coordinated motion? This is one of the most challenging questions in molecular biology today. Directed motion occurs in the movement of chromosomes in cell division, in the injection of bacteriophage DNA into a bacterium, in the beating of cilia and flagella, in the active transport of molecules, in the movement of mRNA in protein synthesis, and, most obviously, in the contraction of muscle. This chapter deals mainly with the structural basis of contraction in vertebrate striated muscle, which is the best understood of these processes. This contractile system consists of interdigitating protein filaments that slide past each other. The energy for contraction comes from the hydrolysis of ATP. The contraction of striated muscle is controlled by the concentration of Ca^{2+}, which is regulated by the sarcoplasmic reticulum, a specialized membrane system that sequesters Ca^{2+} in the resting state and releases it upon the arrival of a nerve impulse.

MUSCLE IS MADE UP OF INTERACTING THICK AND THIN PROTEIN FILAMENTS

Vertebrate muscle that is under voluntary control has a *striated* appearance when examined under the light microscope (Figure 34-1). It consists of cells, which are surrounded by an electrically excitable membrane called the *sarcolemma*. A muscle cell contains many parallel *myofibrils*, each about 1 μm in diameter. The myofi-

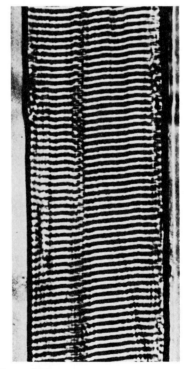

Figure 34-1
Phase-contrast light micrograph of a skeletal muscle fiber about 50 μm in diameter. The A bands are dark and the I bands are light. [Courtesy of Dr. Hugh Huxley.]

brils are immersed in the *sarcoplasm,* which is the intracellular fluid. The sarcoplasm contains glycogen, ATP, phosphocreatine, and glycolytic enzymes. Many regularly spaced mitochondria are found in the more active muscles.

An electron micrograph of a longitudinal section of a myofibril displays a wealth of structural detail (Figures 34-2 and 34-3). The functional unit, called a *sarcomere,* repeats every 2.3 μm (23,000 Å) along the fibril axis. The dark *A band* and the light *I band* alternate regularly. The central region of the A band, termed the *H zone,* is less dense than the rest of the band. A dark *M line* is found in the middle of the H zone. The I band is bisected by a very dense, narrow *Z line.*

Figure 34-2
Electron micrograph of a longitudinal section of a skeletal muscle fiber. [Courtesy of Dr. Hugh Huxley.]

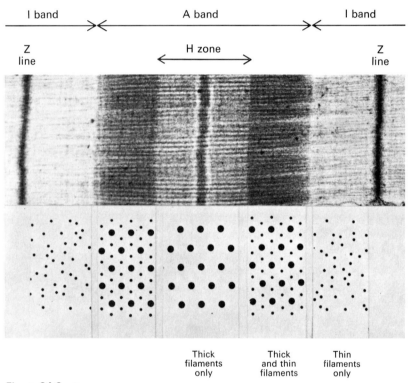

Figure 34-3
Electron micrograph of a longitudinal section of a skeletal muscle myofibril. Schematic diagrams of cross sections are shown below the micrograph. [Courtesy of Dr. Hugh Huxley.]

The underlying molecular plan of a sarcomere is revealed by electron micrographs of cross sections of a myofibril, which show that there are *two kinds of interacting protein filaments.* The *thick filaments* have diameters of about 150 Å, whereas the *thin filaments* have diameters of about 70 Å. The thick filament primarily contains *myosin.* The thin filament contains *actin, tropomyosin,* and *troponin.*

Alpha-actinin is present in the Z line, whereas an M-protein is located in the M line.

The I band consists of only thin filaments, whereas only thick filaments are found in the H zone of the A band. Both types of filaments are present in the other parts of the A band. A regular hexagonal array is evident in cross sections, which show that each thin filament has three neighboring thick filaments and each thick filament is encircled by six thin filaments (Figure 34-3). The thick and thin filaments interact by *cross-bridges,* which are domains of myosin molecules. Cross-bridges emerge at regular intervals from the thick filaments and bridge a gap of 130 Å between the surfaces of thick and thin filaments (Figures 34-4 and 34-5). In fact, the contractile force is generated by the interaction of myosin cross-bridges with actin units in thin filaments.

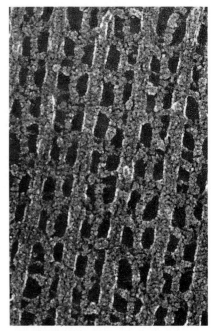

Figure 34-4
Electron micrograph showing cross-bridges between thick and thin filaments. [Courtesy of Dr. John Heuser.]

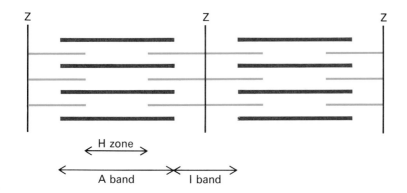

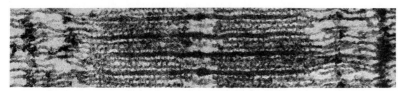

Figure 34-5
Schematic diagram of the structure of striated muscle, showing overlapping arrays of thick and thin filaments, and a corresponding electron micrograph of a very thin longitudinal section. [Courtesy of Dr. Hugh Huxley.]

THICK AND THIN FILAMENTS SLIDE PAST EACH OTHER AS MUSCLE CONTRACTS

Muscle shortens by as much as a third of its original length as it contracts. What causes this shortening? In the 1950s, two groups of investigators independently proposed a *sliding-filament model* on the basis of x-ray, light-microscopic, and electron-microscopic studies.

Figure 34-6
Sliding filament model. [From H. E. Huxley. The mechanism of muscular contraction. Copyright © 1965 by Scientific American, Inc. All rights reserved.]

The essential features of this model (Figure 34-6), which was proposed by Andrew Huxley and R. Niedergerke and by Hugh Huxley and Jean Hanson, are:

1. The lengths of the thick and thin filaments do not change during muscle contraction.

2. Instead, the length of the sarcomere decreases during contraction because the two types of filaments overlap more. *The thick and thin filaments slide past each other in contraction.*

3. The force of contraction is generated by a process that actively moves one type of filament past the neighboring filaments of the other type.

This model is supported by measurements of the lengths of the A and I bands and the H zone in stretched, resting, and contracted muscle (Figure 34-6). The length of the A band is constant, which means that the thick filaments do not change size. The distance between the Z line and the adjacent edge of the H zone is also constant, which indicates that the thin filaments do not change size. In contrast, the size of the H zone and also of the I band decreases on contraction, because the thick and thin filaments overlap more.

MYOSIN FORMS THICK FILAMENTS, HYDROLYZES ATP, AND BINDS ACTIN

Myosin has three important biological activities. First, myosin molecules spontaneously assemble into filaments in solutions of physiologic ionic strength and pH. In fact, the *thick filament* consists mainly of myosin molecules. Second, myosin is an enzyme. Vladimir Engelhardt and Militsa Lyubimova discovered in 1939 that myosin is an *ATPase.*

$$ATP + H_2O \rightleftharpoons ADP + P_i + H^+$$

This reaction is the immediate source of the free energy that drives muscle contraction. Third, myosin binds to the polymerized form of *actin*, the major constituent of the thin filament. Indeed, this interaction is critical for the generation of the force that translates the thick and thin filaments relative to each other.

Myosin is a very large molecule (500 kdal). It contains two identical major chains (200 kdal each) and four light chains (about 20 kdal each). Electron micrographs show that myosin consists of a double-headed globular region joined to a very long rod (Figures 34-7 and 34-8). The rod is a two-stranded α-helical coiled coil.

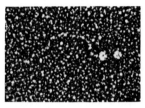

Figure 34-7
Electron micrograph of a myosin molecule. [Courtesy of Dr. Susan Lowey.]

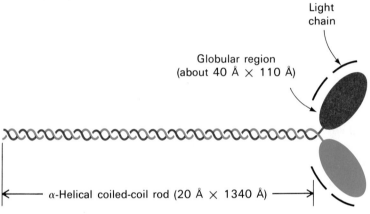

Light chain

Globular region
(about 40 Å × 110 Å)

α-Helical coiled-coil rod (20 Å × 1340 Å)

Figure 34-8
Schematic diagram of a myosin molecule.

MYOSIN CAN BE CLEAVED INTO ACTIVE FRAGMENTS

Myosin can be enzymatically split into fragments that retain some of the activities of the intact molecule. The fruitfulness of such an experimental approach was evident in the discussion of immunoglobulins (p. 793). In fact, the cleavage of myosin into active fragments came first. In 1953, Andrew Szent-Györgyi showed that myosin is split by trypsin into two fragments, called *light meromyosin* and *heavy meromyosin* (Figure 34-9).

Light meromyosin (LMM), like myosin, forms filaments. However, it lacks ATPase activity and does not combine with actin. Electron micrographs show that LMM is a rod, which accounts for the high viscosity of solutions of LMM. The optical rotatory prop-

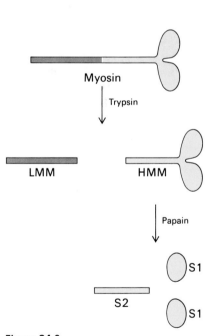

Myosin

Trypsin

LMM   HMM

Papain

S1

S2   S1

Figure 34-9
Enzymatic cleavage of myosin. The four light chains are not shown in this diagram.

erties of LMM point to an α-helix content of 90%. Furthermore, x-ray diffraction patterns of LMM fibers display a strong reflection at 5.1 Å, which is characteristic of an α-helical coiled coil. The composite picture is that *LMM is a two-stranded α-helical rod for its entire length of 850 Å.*

Heavy meromyosin (HMM) has entirely different properties. HMM catalyzes the hydrolysis of ATP and binds to actin, but it does not form filaments. HMM consists of a rod attached to a double-headed globular region (Figure 34-9). It can be split further into two globular subfragments (called S1) and one rod-shaped subfragment (called S2). *Each S1 fragment contains an ATPase active site and a binding site for actin.* Furthermore, the light chains of myosin are bound to the S1 fragments. *Light chains may modulate the ATPase activities of myosin.* For example, the control of smooth muscle contraction by Ca^{2+} is mediated by the light chains.

ACTIN FORMS FILAMENTS THAT BIND MYOSIN

Actin is the major constituent of thin filaments. In solutions of low ionic strength, actin is a 42-kdal monomer called *G-actin* because of its globular shape. As the ionic strength is increased to the physiologic level, G-actin polymerizes into a fibrous form called *F-actin,* which closely resembles thin filaments. An F-actin fiber looks in electron micrographs like two strings of beads wound around each other (Figure 34-10). X-ray diffraction patterns show that *F-actin is a helix of actin monomers.* The helix has a diameter of about 70 Å. The structure repeats at intervals of 360 Å along the helix axis (Figure 34-11).

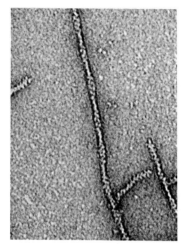

Figure 34-10
Electron micrograph of purified F-actin filaments. [Courtesy of Dr. James Spudich.]

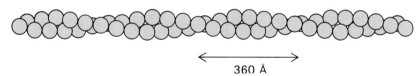

360 Å

Figure 34-11
Schematic diagram of the helical array of actin monomers in F-actin. [After a diagram kindly provided by Dr. James Spudich.]

A complex called *actomyosin* is formed when a solution of actin is added to a solution of myosin. The formation of this complex is accompanied by a large increase in the viscosity of the solution. In the 1940s, Albert Szent-Györgyi showed that this increase in viscosity is reversed by the addition of ATP. This observation revealed that *ATP dissociates actomyosin* to actin and myosin. Szent-Györgyi also prepared threads of actomyosin in which the molecules were oriented by flow. A striking result was obtained when the threads were immersed in a solution containing ATP, K^+, and Mg^{2+}. The *actomyosin threads contracted*, whereas threads formed from myosin alone did not. *These incisive experiments suggested that the force of muscle contraction arises from an interaction of myosin, actin, and ATP.*

ACTIN ENHANCES THE ATPase ACTIVITY OF MYOSIN

The ATPase activity of myosin is markedly enhanced by stoichiometric amounts of F-actin. The turnover number increases 200 times, from 0.05 sec^{-1} to 10 sec^{-1}. The hydrolysis of ATP by myosin alone is rapid, but the release of the products ADP and P_i is slow. Actin increases the turnover number of myosin by binding to the myosin-ADP-P_i complex and accelerating the release of ADP and P_i (Figure 34-12). Actomyosin then binds ATP, which leads to the

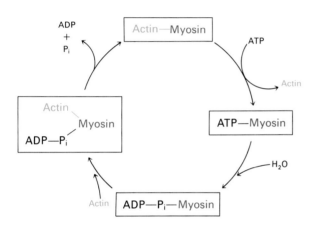

Figure 34-12
The hydrolysis of ATP drives the cyclic association and dissociation of actin and myosin.

dissociation of actin and myosin. The resulting ATP-myosin complex is then ready for another round of catalysis. These reactions require Mg^{2+}. An important feature of this reaction cycle, which was proposed by Edwin Taylor on the basis of rapid-kinetic studies, is that actin has high affinity for myosin and for the myosin-ADP-P_i complex but it has low affinity for the myosin-ATP complex. *Consequently, actin alternately binds to myosin and is released from myosin as ATP is hydrolyzed.* As will be discussed shortly, this ATP-driven change in the interaction of myosin and actin is the essence of force generation in muscle contraction.

THE THICK AND THIN FILAMENTS HAVE DIRECTION

The cyclic formation and dissociation of the complex between myosin and actin produces coherent motion because these molecules are components of highly ordered assemblies. The organization of myo-

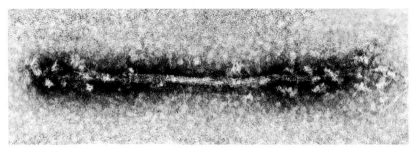

Figure 34-13
Electron micrograph of a reconstituted thick filament. The projections on either side of the bare zone are the cross-bridges. [Courtesy of Dr. Hugh Huxley.]

Figure 34-14
Electron micrograph of filaments of F-actin decorated with the S1 moiety of HMM. The "arrowheads" all point the same way. [Courtesy of Dr. James Spudich.]

sin molecules in the thick filament has been elucidated by Hugh Huxley through studies of intact filaments dissociated from muscle and of synthetic filaments formed from purified myosin molecules. A *dissociated thick filament* has a diameter of 160 Å and a length of 1.5 μm (15,000 Å). *Cross-bridges* emerge from this filament in a regular helical array, except for a 1500-Å *bare region* midway along its length (Figure 34-13).

The same structural features are evident in *synthetic thick filaments,* which are produced by lowering the ionic strength of a solution of myosin. The shortest synthetic filament is about 3000 Å long and contains a central bare region 1500 Å long. Because the size of the bare zone is the same in longer synthetic filaments, it is evident that the thick filament grows by the addition of molecules parallel to those already assembled. *The myosin molecules on one side of the bare zone point one way, whereas those on the other side point in the opposite direction. Thus, a thick filament is inherently bipolar.*

LMM, like myosin, aggregates in solutions of low ionic strength to form filaments having an axial period of 430 Å, which is the same as that of the cross-bridges in the intact thick filament. However, filaments formed from LMM are smooth, which indicates that the *cross-bridges arise from the HMM portion of myosin, whereas the LMM unit of myosin forms the backbone of the thick filament.*

Thin filaments also have direction. If myosin (or HMM or S1) is added to thin filaments or to F-actin, an arrowhead pattern appears in electron micrographs (Figure 34-14). These structures are picturesquely called *decorated filaments.* The arrows on both strands of a decorated filament always point the same way for its entire length. Thus, *thin filaments have inherent directionality, with both strands pointing the same way.* A number of preparations of thin filaments from homogenized muscle contain many filaments still joined to the Z line. Decoration of these filaments with heavy meromyosin reveals that the arrows on all filaments point away from the Z line. Thus, *all the thin filaments on one side of a Z line have the same orientation, whereas those on the opposite side have the reverse polarity.*

THE POLARITY OF THICK AND THIN FILAMENTS
REVERSES IN THE MIDDLE OF A SARCOMERE

The structural polarity of both the thick and the thin filaments is crucial for coherent motion. The sliding force developed by the interaction of individual actin and myosin units adds up because the interacting sites have the same relative orientation. Further-more, *the absolute direction of the sites reverses half-way between Z lines* (Figure 34-15). Consequently, the two thin filaments that interact with a thick filament are drawn toward each other, which shortens the distance between Z lines (Figure 34-16).

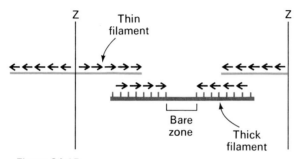

Figure 34-15
The polarity of the thick and thin filaments reverses halfway between the Z lines.

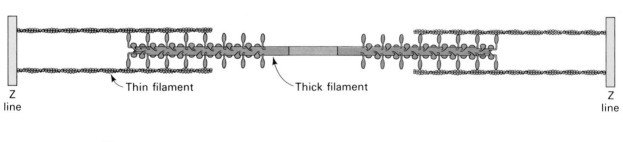

Figure 34-16
Schematic diagram showing the interaction of thick and thin filaments in skeletal muscle contraction. [After a diagram kindly provided by Dr. James Spudich.]

THE POWER STROKE COMES FROM THE TILTING OF A MYOSIN S1 HEAD COMPLEXED TO ACTIN

The contractile force is generated by the cyclic formation and dissociation of complexes between the S1 heads of myosin and actin. Repeated cycles of attachment, pulling, and detachment take place during even a single contraction. The mechanism of force generation is now being investigated by biochemical, electron-microscopic, and x-ray–diffraction techniques. The picture emerging from these studies is that S1 heads are detached from thin filaments in resting muscle (Figure 34-17A). In this state, S1 heads form an ordered helical array around the thick filament. When muscle is stimulated, S1 heads move away from the thick filament and attach to actin units on thin filaments (Figure 34-17B). The orientation of an S1 head changes in the next step so that its long axis is at an angle of about 45 degrees to the thin filament axis (Figure 34-17C). *This postulated tilting of an S1 head is likely to be the power stroke of muscle contraction.* This change in angle of the S1 domain is transmitted by the S2 unit of myosin to the thick filament. Consequently, the thick filament is pulled a distance of some 75 Å relative to the thin filament. In the final step, an S1 head is released from the thin filament (Figure 34-17D).

Now let us correlate these proposed structural states in the generation of the contractile force with intermediates in the hydrolysis of ATP. In the resting state, myosin contains tightly bound ADP and P_i (Figure 34-17A). On stimulation, the S1 head becomes attached to the thin filament in a perpendicular orientation (Figure 34-17B). ADP and P_i bound to S1 are then released, and the S1 head adopts a tilted orientation (Figure 34-17C). Thus, the power stroke is driven by the release of tightly bound ADP and P_i. The S1 head is then dissociated from the thin filament by the binding of ATP (Figure 34-17D). Consequently, the detached S1 unit returns to a perpendicular orientation near the thick filament. Finally, hydrolysis of ATP by S1 not combined with actin completes the cycle.

The myosin molecule contains two kinds of hinges that enable its S1 heads to reversibly attach and detach from actin and to change their orientation when bound. One type of hinge is located between each S1 head and the S2 rod, and the other is positioned between S2 and the LMM unit of myosin (Figure 34-18). These hinges are flexible regions of polypeptide chain, vulnerable to cleavage by proteolytic enzymes. Indeed, the enzymatic splitting of myosin into LMM, S2, and S1 pieces is an expression of the fact that *myosin is constructed of domains joined by hinges.* The role of the S2 domain is to transmit tension from S1 bound to a thin filament to the LMM domain that forms part of the thick filament. The hinge between S1 and S2 enables S1 to interact with actin in one way when ADP and P_i are bound and in a different way when they are released. The

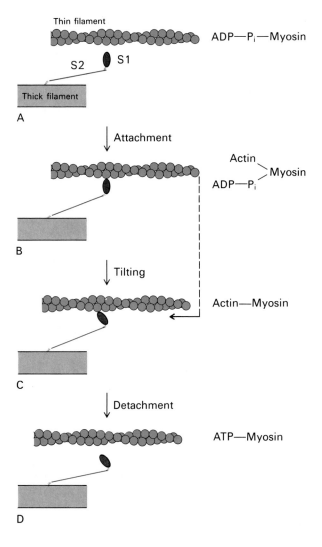

A

↓ Attachment

B

↓ Tilting

C

↓ Detachment

D

Thin filament

ADP—P$_i$—Myosin

S2 · S1

Thick filament

Actin
            ⟩Myosin
ADP—P$_i$

Actin—Myosin

ATP—Myosin

Figure 34-17
Proposed mechanism for the generation
of force by the interaction of the S1
units of a myosin filament with the
actin filament. The thick filament
moves relative to the thin filament
when the S1 unit bound to actin
changes its tilt (in the transition from
B to C).

Figure 34-18
Two kinds of hinges in myosin—one
between S1 and S2 and the other between S2 and the LMM region—are
important in allowing the S1 unit to
alter its contacts with actin during the
power stroke.

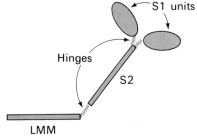

S1 units

Hinges

S2

LMM

other hinge, between S2 and LMM, allows considerable variation
in the position of S1 relative to the thick filament, which permits S1
to interact precisely with actin. Hence, tension can be generated
over a wide range of lateral spacings of the thick and thin filaments.
Thus, *segmental flexibility* plays a critical role in muscle contraction,
as it does in the action of antibody molecules (p. 794).

TROPONIN AND TROPOMYOSIN MEDIATE CALCIUM ION REGULATION OF MUSCLE CONTRACTION

The physiologic regulator of muscle contraction is Ca^{2+}. Setsuro Ebashi discovered that the effect of Ca^{2+} on the interaction of actin and myosin is mediated by *tropomyosin* and the *troponin complex,* which are located in the thin filament and constitute about a third of its mass. Tropomyosin is a two-stranded α-helical rod. This highly elongated 70-kdal protein is aligned nearly parallel to the long axis of the thin filament (Figure 34-19). Troponin is a complex of three polypeptide chains: TnC (18 kdal), TnI (24 kdal), and TnT (37 kdal). TnC binds calcium ions, TnI binds to actin, and TnT binds to tropomyosin. The troponin complex is located in the thin filaments at intervals of 385 Å, a period set by the length of tropomyosin. A troponin complex bound to a molecule of tropomyosin regulates the activity of about seven actin monomers.

The interaction of actin and myosin is inhibited by troponin and tropomyosin in the absence of Ca^{2+}. Tropomyosin sterically blocks the S1 binding sites on actin units in an inhibited thin filament. Nerve excitation triggers the release of Ca^{2+} by the sarcoplasmic reticulum, as will be discussed shortly. The released Ca^{2+} binds to the TnC component of troponin and causes conformational changes that are transmitted to tropomyosin and then to actin. Specifically, tropomyosin moves toward the center of the long helical groove of the thin filament. Consequently, the S1 heads of myosin molecules can then interact with actin units of the thin filament. Contractile force is generated and ATP is concomitantly hydrolyzed until Ca^{2+} is removed and tropomyosin again blocks access of the S1 heads. Thus, *Ca^{2+} controls muscle contraction by an allosteric mechanism in which the flow of information is*

$$Ca^{2+} \longrightarrow \text{troponin} \longrightarrow \text{tropomyosin} \longrightarrow \text{actin} \longrightarrow \text{myosin}$$

THE FLOW OF CALCIUM IONS IS CONTROLLED BY THE SARCOPLASMIC RETICULUM

The outer membrane of a muscle fiber becomes depolarized following the arrival of a nerve impulse at the end plate, which is the junction between nerve and muscle. The depolarization of the outer membrane is transmitted to the interior of the muscle fiber by the T tubules (T stands for their transverse orientation). The T tubules are in close proximity to a network of extremely fine channels called the *sarcoplasmic reticulum*, which is a reservoir of Ca^{2+} (Figure 34-20).

In the resting state, Ca^{2+} is sequestered in the sarcoplasmic reticulum by a Ca^{2+} active-transport system (p. 868). This ATP-driven pump lowers the concentration of Ca^{2+} in the cytoplasm of resting muscle to less

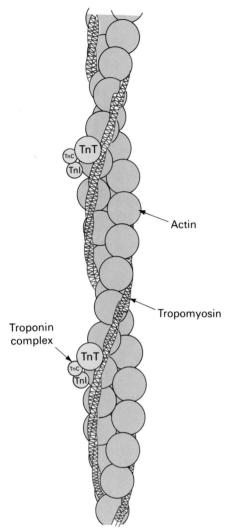

Figure 34-19
Proposed model of a thin filament in the resting state. The tropomyosin double helix (red) blocks the sites on actin (blue) that bind myosin S1 units in the contracting state. The binding of Ca^{2+} to the TnC component of the troponin complex (yellow) activates contraction by displacing the tropomyosin helix, which in turn exposes the S1 binding sites on actin. [After C. Cohen. Protein switch of muscle contraction. Copyright © 1975 by Scientific American, Inc. All rights reserved.]

amount of actin that is very similar to muscle actin. It forms thin filaments and interacts with myosin. Most interesting, *slime-mold actin* reacts with S1 heads from *vertebrate skeletal muscle* to give decorated filaments like those formed from vertebrate actin and myosin. In fact, slime-mold actin differs from rabbit-muscle actin at only 17 of 375 amino acid residues. Thus, *actin is a highly conserved, ancient protein of eucaryotes.* Likewise, myosin has been isolated from this slime mold and from many other eucaryotic cells. *There is greater variation in the properties of myosin molecules derived from different sources than in those of their actin molecules.* For example, myosins from many nonmuscle cells do not readily form thick filaments of the kind seen in skeletal muscle. However, nonmuscle myosins do form short bipolar filaments.

THE CELLULAR DISTRIBUTION OF MICROFILAMENTS CAN BE SEEN BY IMMUNOFLUORESCENCE MICROSCOPY

Most eucaryotic cells are rich in actin, which typically comprises 10% of the total cell protein. Indeed, actin is the most abundant protein in many kinds of cells. The myosin content of nonmuscle cells is usually about tenfold lower than their actin content. Hence, the ratio of actin to myosin is much higher in nonmuscle cells than in muscle.

Some of the actin in nonmuscle cells is polymerized into *microfilaments* (Figure 34-22), which resemble the thin filaments of muscle. *These 70-Å–diameter filaments are evident in electron micrographs of nearly*

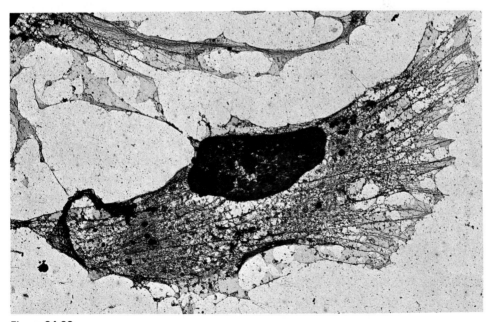

Figure 34-22
Electron micrograph showing many microfilaments in cytoskeletons of chick-embryo fibroblasts. [Courtesy of Dr. Susan Brown.]

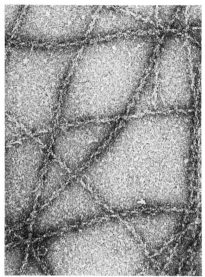

Figure 34-23
Electron micrograph of F-actin from *Dictyostelium* decorated with *Dictyostelium* myosin S1 heads. The arrowhead pattern is like that of skeletal muscle F-actin decorated with muscle S1. [Courtesy of Dr. James Spudich.]

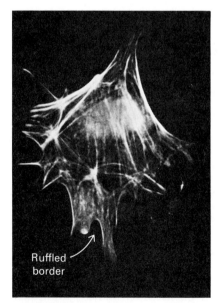

Ruffled
border

Figure 34-25
The distribution of actin filaments is more diffuse in a moving cell, particularly in its ruffled borders. [Courtesy of Dr. Elias Lazarides.]

all eucaryotic cells. Regular arrays of arrowheads are seen when microfilaments are decorated with S1 heads, which shows that they are made up of actin units and can interact with myosin (Figure 34-23).

The cellular distribution of microfilaments can also be displayed by *immunofluorescence microscopy.* Antibodies specific for actin are covalently labeled with a fluorescent tag such as fluorescein and then added to cultured cells. The resulting fluorescence micrograph shows a remarkable array of filaments (Figure 34-24). Some of these

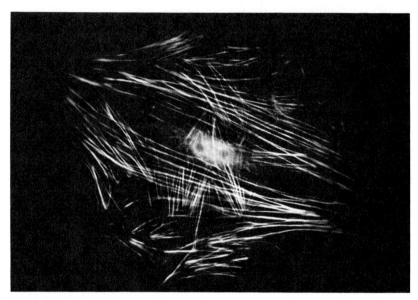

Figure 34-24
Immunofluorescence micrograph of a resting cell in a culture dish that was stained with a fluorescent antibody to actin. Long bundles of microfilaments are evident. [Courtesy of Dr. Elias Lazarides.]

fluorescent bundles span the entire length of the cell. In quiescent cells attached to a solid support, many of these actin filaments are aligned parallel to the long axis of the cell. In contrast, ruffled borders exhibit a diffuse fluorescence, which comes from actin monomers or short actin filaments (Figure 34-25). The distribution of other contractile proteins also has been investigated by immunofluorescence microscopy. *Tropomyosin* is also found in microfilaments (Figure 34-26), whereas troponin seems to be absent. Calcium ion is important in regulating nonmuscle motility as well as muscle contraction, but these processes are controlled differently. Little is yet known about the distribution of myosin. *α-Actinin,* which binds actin at the Z line of myofibrils, is also present in nonmuscle cells. α-Actinin is especially prominent in ruffled borders and at other sites where actin filaments undergo rapid assembly and disassembly. One possible role of α-actinin is to bridge actin filaments in the cytoskeleton and other motile assemblies.

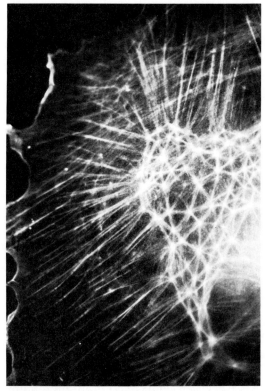

Figure 34-26
Geodesic domelike structures are seen in cells
stained with fluorescent antibody to actin.
[Courtesy of Dr. Elias Lazarides.]

MEMBRANE-ANCHORED ACTIN FILAMENTS MEDIATE THE CONTRACTION OF INTESTINAL MICROVILLI

The microvilli of intestinal epithelial cells (Figure 34-27) constitute
one of the best-characterized contractile systems in nonmuscle cells.
This array of cytoplasmic extensions on the apical surface of intestinal epithelium is often called a *brush border* because of its appearance
in light micrographs. Each microvillus contains a bundle of actin
filaments that are attached to the plasma membrane at the pro-
truding end. These filaments are also linked to the plasma membrane by threadlike connections at many points along the length of
the microvillus. The core of actin filaments extends below the surface of the cell into a region called the terminal web. The plasma
membrane of isolated brush borders can be removed by treating
them with a detergent, which leaves the rest of the structure intact.
The striking finding is that this demembranated brush border can
be induced to contract by the addition of Ca^{2+} and ATP. Specifically, the actin filaments plunge rapidly into the terminal web.

What is the structural basis of this movement? Decoration of the
actin filaments in the microvilli with myosin S1 heads shows that

Figure 34-27
Electron micrograph of the brush bor-
der of intestinal epithelium. The mi-
crovilli contain actin filaments. [From
M. S. Mooseker and L. G. Tilney,
J. Cell Biol. 67(1975):725.]

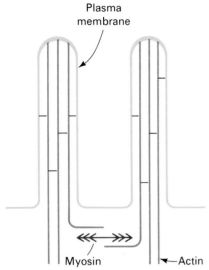

Plasma
membrane

Myosin ◀—Actin

Figure 34-28
Proposed organization of actin and
myosin in microvilli of the intestinal
brush border. [After M. S. Mooseker
and L. G. Tilney, *J. Cell Biol.*
67(1975):725.]

Cytochalasin B

all of them have the same polarity (Figure 34-28). The arrowheads point away from the tip of microvillus, just as arrowheads on decorated thin filaments in muscle point away from the Z line. In other words, the polarity of the actin filaments is such that the microvillus would be drawn into the cell on contraction. The other pertinent structural finding is that thick filaments containing myosin are present in the terminal web. It seems likely that bipolar myosin filaments in the terminal web interact with actin filaments from adjacent microvilli (Figure 34-28). In this model, actin filaments slide relative to the myosin filament, and so filament bundles move toward the center of the cell. Hence, the microvilli shorten, which may aid the absorption of nutrients into the cell.

CYTOCHALASIN AND PHALLOIDIN INHIBIT MOVEMENTS THAT REQUIRE TURNOVER OF ACTIN FILAMENTS

Cytochalasin B, an alkaloid from fungi, alters the shape of eucaryotic cells and inhibits many of their movements. For example, the ruffling of fibroblasts, the outgrowth of axons from ganglia, the retraction of blood clots by platelets, and the division of fertilized sea urchin eggs are inhibited by this alkaloid. Electron-microscopic studies suggested that cytochalasin acts on microfilaments because they disappear from cells treated with this alkaloid. In fact, *cytochalasin interferes with the assembly of actin filaments by interacting specifically with one end of the filament.* The inhibitory action of cytochalasin emphasizes the dynamic nature of microfilaments.

The importance of continual assembly and disassembly of microfilaments in cell motility is also underscored by studies of the mechanism of action of *phalloidin.* This toxic agent comes from *Amanita phalloides,* a highly poisonous mushroom that also contains α-amanitin, an inhibitor of RNA polymerase (p. 703). Phalloidin, like cytochalasin, blocks cell movements that require the turnover of microfilaments. *Phalloidin binds to actin units in microfilaments and prevents their depolymerization.* Thus, microfilaments are in effect locked by phalloidin.

MICROTUBULES PARTICIPATE IN CELL MOVEMENTS AND FORM PART OF THE CYTOSKELETON

The role of microfilaments in cell movements has been considered thus far. Now we turn to *microtubules,* a different class of fibrous elements that have architectural and contractile roles in nearly all eucaryotic cells (Figure 34-29). Microtubules are hollow cylindrical structures built from two kinds of similar 55-kdal subunits, *α- and β-tubulin.* Microtubules have an outer diameter of about 240 Å,

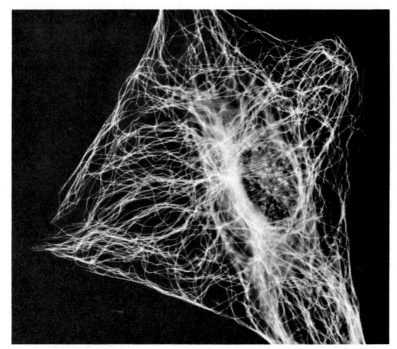

Figure 34-29
Immunofluorescence micrograph showing the distribution of
microtubules in a fibroblast. [Courtesy of Dr. Klaus Weber.]

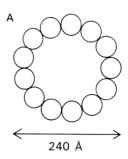

240 Å

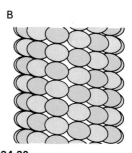

Figure 34-30
Helical pattern of tubulin subunits in
a microtubule: (A) cross-sectional view
showing the arrangement of thirteen
protofilaments; (B) surface lattice of α
and β subunits. [After J. A. Snyder
and J. R. McIntosh. *Ann. Rev. Biochem.*
45(1976):706.]

which clearly distinguishes them from microfilaments (70-Å diameter) and from intermediate filaments (100-Å–diameter structures
that serve as connectors). The rigid wall of a microtubule is made
up of a helical array of alternating α- and β-tubulin subunits (Figure 34-30). The repeating unit is an $\alpha\beta$ dimer, and the structure can
be regarded as being made of thirteen protofilaments that run parallel to the long axis of the microtubule.

Assembly and disassembly take place at opposite ends of microtubules. *Colchicine,* an alkaloid from the autumn crocus, inhibits cell
processes that depend on microtubule function by blocking their
polymerization. For example, dividing cells are arrested at metaphase by the action of colchicine because microtubules are essential
for moving chromosomes. Colchicine also inhibits the directed
movement of many kinds of particles in cells, and so secretory processes are often blocked by this compound. Colchicine has been used
for several centuries to treat acute attacks of gout.

Colchicine

THE BEATING OF CILIA AND FLAGELLA IS PRODUCED
BY THE DYNEIN-INDUCED SLIDING OF MICROTUBULES

Microtubules are major components of eucaryotic cilia and flagella.
These hairlike organelles protrude from the surfaces of many cells.

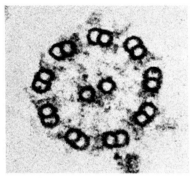

Figure 34-31
Electron micrograph of a cross section of a flagellar axoneme. Nine outer microtubule doublets surround two singlets. [Courtesy of Dr. Joel Rosenbaum.]

Motile cilia act as oars to move a stream of liquid parallel to the cell surface. For example, the coordinated beating of cilia lining the respiratory passages serves to sweep out foreign particles. Flagella propel free cells such as sperm and protozoa. Electron-microscopic studies have revealed that cilia and flagella from nearly all eucaryotes have the same fundamental design: a bundle of fibers called an *axoneme* is surrounded by a membrane that is continuous with the plasma membrane. In fact, the fibers in an axoneme are microtubules. A peripheral group of nine pairs of microtubules surrounds two singlet microtubules (Figure 34-31). This recurring motif is known as a *9 + 2 array*. The two central microtubules have diameters of 240 Å. Each of the nine outer doublets looks like the figure eight, 370 Å × 250 Å (Figure 34-32). One of the pair, *sub-*

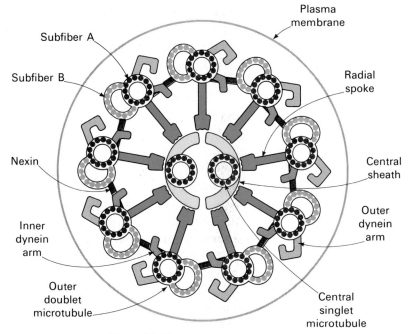

Figure 34-32
Schematic diagram of the structure of an axoneme.

fiber A, is smaller and is joined to the central sheath of the cilium by radial spokes. Two *arms* emerge from each subfiber A. All the arms point in the same direction in a given cilium. Treatment of cilia with detergent and then with a high salt concentration removes their outer membrane and solubilizes an ATPase called *dynein.* The outer fibers retain their ninefold cylindrical arrangement but are devoid of arms. The arms can be restored by the addition of dynein under suitable ionic conditions. Thus, *the dynein arms of subfibers A contain the ATPase activity.*

How does the splitting of ATP by dynein cause cilia and flagella to beat? Peter Satir and Ian Gibbons have shown that the *outer doublets of the axoneme slide past each other to produce bending*. The force between adjacent doublets is generated by *dynein cross-bridges*. It seems likely that the dynein arms of one doublet walk along the adjacent doublet as ATP is being hydrolyzed, in a manner analogous to the movement of myosin cross-bridges on an actin filament in skeletal muscle. In an intact cilium, the radial spokes resist this sliding motion, which instead is converted into a local bending.

Axonemes devoid of dynein have been found in a group of patients who have chronic pulmonary disorders. Bjorn Afzelius has shown that the cilia in their respiratory tracts are immotile. Males with this genetic defect are also infertile because their sperm, too, are unable to move. Defective radial spokes have recently been shown to be another cause of immotile cilia and flagella. These clinical observations provide further support for the proposed mechanism of beating of these organelles.

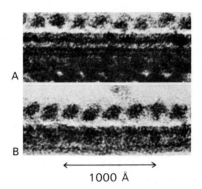

Figure 34-33
Dynein arms project periodically in a linear array along microtubules. Electron micrographs of (A) an intact axoneme and (B) a microtubule reconstituted from tubulin and dynein. [From L. T. Haimo, B. R. Telzer, and J. L. Rosenbaum, *Proc. Nat. Acad. Sci.* 76(1979):5760.]

SUMMARY

Vertebrate striated muscle consists of two kinds of interacting protein filaments. The thick filaments contain myosin, whereas the thin filaments contain actin, tropomyosin, and troponin. The hydrolysis of ATP by actomyosin drives the sliding of these filaments past each other. Myosin is a very large protein (500 kdal) composed of two major chains and four light chains. The major chains are folded so that myosin contains two globular regions (S1 heads) attached to a long α-helical rod. The globular regions and part of the rod form the cross-bridges that interact with actin to generate the contractile force. The rest of the myosin molecule forms the backbone of the thick filament. Actin, the major component of thin filaments, is a globular protein (42 kdal) that polymerizes to form 70-Å–diameter filaments. The thin and thick filaments have direction, which reverses half-way between the Z lines. ATP drives the cyclic formation and dissociation of complexes between the myosin cross-bridges of thick filaments and the actin units of thin filaments, which shortens the distance between Z lines. The power stroke comes from the tilting of a myosin S1 head complexed to actin. Hinges between domains of the myosin molecule are important in the generation of the contractile force. Muscle contraction is regulated by Ca^{2+}, whose effect is mediated by troponin and tropomyosin. The interaction of actin and myosin is inhibited by these proteins when the level of Ca^{2+} is low. Nerve excitation triggers the release of Ca^{2+} by the sarcoplasmic reticulum. Calcium ion binds to troponin, which initiates a series of conformational changes that allow actin and myosin to interact.

Actin and myosin are ancient proteins, as shown by their presence in slime molds. In fact, these proteins have contractile roles in nearly all eucaryotic cells. Actin, which is particularly abundant, forms microfilaments that have diameters of 70 Å. Microfilaments participate in a wide range of cellular movements, as exemplified by cell migration in development, clot retraction by blood platelets, and the movement of macrophages to injured tissues. Contraction of microvilli of the intestinal brush border is mediated by the interaction of actin filaments with bipolar myosin filaments that are smaller and less prevalent than the ones in muscle. Cytochalasin and phalloidin inhibit cell movements requiring the turnover of actin filaments. Cytochalasin interferes with assembly, whereas phalloidin blocks disassembly of microfilaments.

Eucaryotic cells also contain microtubules, which have architectural and contractile roles. These 240-Å–diameter hollow fibers are built from tubulin subunits. A eucaryotic cilium or flagellum contains nine microtubule doublets around two singlets. The outer doublets are cross-linked by dynein, which is an ATPase. The dynein-induced sliding of adjacent microtubule doublets produces a local bending of the cilium or flagellum and results in its beating. Colchicine inhibits movements mediated by microtubules by blocking their polymerization.

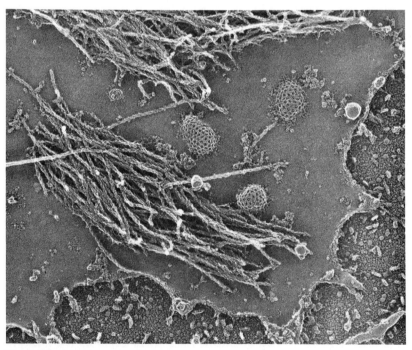

Figure 34-34
An electron microscopic view of the inner surface of the plasma membrane of a fibroblast. Bundles of actin filaments (decorated with S1) and coated pits are prominent. [Courtesy of Dr. John Heuser.]

SELECTED READINGS

WHERE TO START

Huxley, H. E., 1965. The mechanism of muscular contraction. *Sci. Amer.* 213(6):18–27. [Available as *Sci. Amer.* Offprint 1026.]

Cohen, C., 1975. The protein switch of muscle contraction. *Sci. Amer.* 233(5):36–45. [Offprint 1329.]

Murray, J. M., and Weber, A., 1974. The cooperative action of muscle proteins. *Sci. Amer.* 230(2):59–69. [Offprint 1290.]

Lazarides, E., and Revel, J. P., 1979. The molecular basis of cell movement. *Sci. Amer.* 240(5):100–113. [Offprint 1427.]

Satir, P., 1974. How cilia move. *Sci. Amer.* 231(4):44–52. [Offprint 1304.]

Dustin, P., 1980. Microtubules. *Sci. Amer.* 243(2):66–76. [Offprint 1477.]

BOOKS

Cold Spring Harbor Laboratory, 1972. *The Mechanism of Muscle Contraction* (Cold Spring Harbor Symposia on Quantitative Biology, vol. 37). [An outstanding volume that contains articles on the structure of muscle proteins, the generation of the contractile force, regulatory systems, the sarcoplasmic reticulum, myogenesis, and contractile proteins in nonmuscle tissue.]

Goldman, R., Pollard, T., and Rosenbaum, J., (eds.), 1976. *Cell Motility.* Cold Spring Harbor Laboratory. [Contains many excellent articles on microfilaments and microtubules and their roles in cell motility.]

Inoue, S., and Stephens, R. E., (eds.), 1975. *Molecules and Cell Movement.* Raven Press.

Hatano, S., Ishikawa, H., and Sato, H., (eds.), 1979. *Cell Motility: Molecules and Organization.* University Park Press.

Sugi, H., and Pollack, G. H., (eds.), 1979. *Cross-bridge Mechanism in Muscle Contraction.* University Park Press.

MUSCLE CONTRACTION

Huxley, H. E., 1971. The structural basis of muscle contraction. *Proc. Roy. Soc.* London (B) 178:131–149.

Huxley, A. F., 1975. The origin of force in skeletal muscle. *Ciba Found. Symp.* 31:271–290.

Harrington, W. F., 1979. Contractile proteins of muscle. *In* Neurath, H., and Hill, R. L., (eds.), *The Proteins* (3rd ed.), vol. 4, pp. 245–409. Academic Press.

Adelstein, R. S., and Eisenberg, E., 1980. Regulation and kinetics of the actin-myosin-ATP interaction. *Ann. Rev. Biochem.* 49:921–956.

Hill, T. L., 1977. *Free Energy Transduction in Biology.* Academic Press. [A lucid discussion of the thermodynamics of muscle contraction is given in Chapter 5.]

Huxley, H. E., Farugi, A. R., Bordas, J., Koch, M. H. J., and Milch, J. R., 1980. The use of synchrotron radiation in time-resolved x-ray diffraction studies of myo-sin layer-line reflections during muscle contraction. *Nature* 284:140–143.

Adelstein, R. S., (ed.), 1980. Symposium on phosphorylation of muscle contractile proteins. *Fed. Proc.* 39:1544–1573.

McCubbin, W. D., and Kay, C. M., 1980. Calcium-induced conformational changes in the troponin-tropomyosin complexes of skeletal and cardiac muscle and their roles in the regulation of contraction-relaxation. *Acc. Chem. Res.* 13:185–192.

ACTIN AND MYOSIN IN CELL MOTILITY

Korn, E. D., 1978. Biochemistry of actomyosin-dependent cell motility (a review). *Proc. Nat. Acad. Sci.* 75:588–599.

Clarke, M., and Spudich, J. A., 1977. Nonmuscle contractile proteins: the role of actin and myosin in cell motility and shape determination. *Ann. Rev. Biochem.* 46:797–822.

Tilney, L. G., 1979. Actin, motility, and membranes. *In* Cone, R. A., and Dowling, J. E., (eds.), *Membrane Transduction Mechanisms,* pp. 163–186. Raven Press.

Lin, S., Lin, D. C., and Flanagan, M. D., 1978. Specificity of the effects of cytochalasin B on transport and motile processes. *Proc. Nat. Acad. Sci.* 75:329–333.

Brown, S. S., and Spudich, J. A., 1979. Cytochalasin inhibits the rate of elongation of actin filament fragments. *J. Cell Biol.* 83:657–662.

MICROTUBULES

Kirschner, M. W., 1978. Microtubules: assembly and nucleation. *Int. Rev. Cytol.* 54:1–71.

Margolis, R. L., and Wilson, L., 1978. Opposite end assembly and disassembly of microtubules at steady-state in vitro. *Cell* 13:1–8.

Amos, L. A., and Baker, T. S., 1979. The three-dimensional structure of tubulin protofilaments. *Nature* 279:607–612.

Summers, K. E., and Gibbons, I. R., 1971. ATP-induced sliding of tubules in trypsin-treated flagella of sea-urchin sperm. *Proc. Nat. Acad. Sci.* 68:3092–3096. [These ingenious experiments revealed that the bending waves of normal flagella are the result of ATP-induced shearing forces between adjacent doublet-tubules.]

Haimo, L. T., Telzer, B. R., and Rosenbaum, J. L., 1979. Dynein binds to and crossbridges cytoplasmic microtubules. *Proc. Nat. Acad. Sci.* 76:5759–5763.

Sturgess, J. M., Chao, J., Wong, J., Aspin, N., and Turner, J. A. P., 1979. Cilia with defective radial spokes: a cause of human respiratory disease. *New Engl. J. Med.* 300:53–56.

INTERMEDIATE FILAMENTS

Lazarides, E., 1980. Intermediate filaments as mechanical integrators of cellular space. *Nature* 283:249–256.

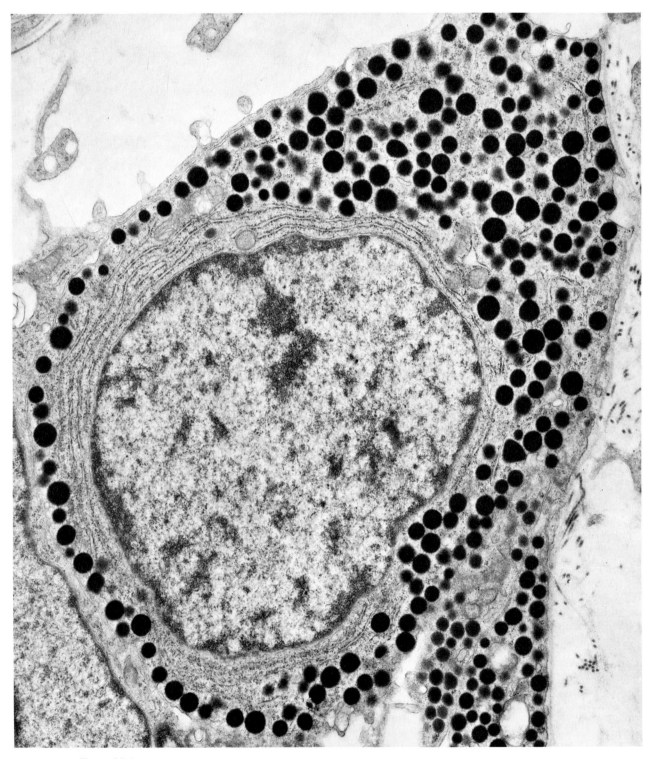

Figure 35-1
Electron micrograph of a somatotrope cell in the pituitary gland. Growth hormone (somatotropin) is stored in the prominent dense granules. [Courtesy of Lynne Mercer.]

CHAPTER 35

HORMONE ACTION

Hormones are chemical messengers that coordinate the activities of different cells in multicellular organisms. The term hormone was first used in 1904 by William Bayliss and Ernest Starling to describe the action of secretin, a molecule secreted by the duodenum that stimulates the subsequent flow of pancreatic juice. Very fruitful concepts emerged from this work—namely, that (1) hormones are molecules synthesized by specific tissues (*glands*); (2) they are secreted directly into the blood, which carries them to their sites of action; and (3) they specifically alter the activities of certain responsive tissues (target organs or *target cells*).

Hormones are chemically diverse. Some hormones, such as epinephrine and thyroxine, are *small molecules derived from amino acids*. Others, such as oxytocin, insulin, and thyroid-stimulating hormone, are *polypeptides* or *proteins*. A third group of hormones consists of the *steroids*, which are derived from cholesterol. The molecular bases of the actions of some hormones are now being elucidated. *Hormones can exert their specific effects in three ways:* (1) *by influencing the rate of synthesis of enzymes and other proteins;* (2) *by affecting the rate of enzymatic catalysis; and* (3) *by altering the permeability of cell membranes*. It is interesting to note that no known hormone is an enzyme or coenzyme. Rather, *hormones act by regulating preexisting processes*.

DISCOVERY OF CYCLIC AMP, A MEDIATOR OF THE ACTION OF MANY HORMONES

A major breakthrough in the elucidation of the mechanism of action of hormones was made by Earl Sutherland as an outgrowth of studies started in the 1950s. The initial aim was to determine how epinephrine and glucagon elicit the breakdown of glycogen and the production of glucose by the liver. Sutherland chose this system because, first, the effects of these hormones on glycogen breakdown are large and reproducible. Second, they occur within a few minutes. Third, liver slices are easy to prepare in large quantities. Fourth, much was known about the biochemistry of glycogen breakdown (Chapter 16). In fact, Sutherland started these studies on the mechanism of action of epinephrine and glucagon in the laboratory of Carl Cori and Gerty Cori.

One of the first steps was to identify the enzymatic reaction in the conversion of glycogen into glucose that was enhanced by these hormones. Measurements of the labeled intermediates following incubation of liver slices in the presence of $^{32}P_i$ showed that phosphorylase, rather than phosphoglucomutase or glucose 6-phosphatase, was rate limiting. Furthermore, epinephrine and glucagon led to an increase in the activity of phosphorylase. However, the nature of the activation process was not yet evident. Sutherland then found an enzyme that catalyzed the inactivation of active phosphorylase. This inactivating enzyme proved to be a phosphatase, which suggested that phosphorylase might be activated by being phosphorylated. This hypothesis was tested by incubating liver slices in the presence of $^{32}P_i$. Indeed, the rate of incorporation of ^{32}P into phosphorylase was found to be increased by glucagon and epinephrine, in direct proportion to their effects on glycogen breakdown. These studies revealed that *phosphorylase is activated by phosphorylation and inactivated by dephosphorylation. This was the first example of the regulation of an enzyme by covalent modification.*

The activation of phosphorylase by hormones in a preparation of broken liver cells was then studied. The exciting finding was that

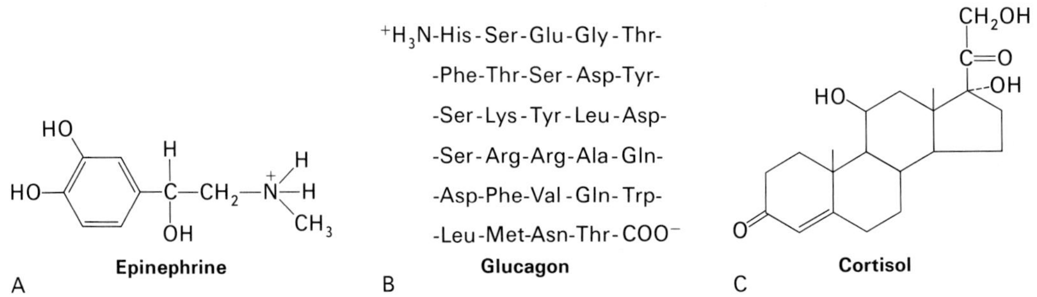

+H₃N-His-Ser-Glu-Gly-Thr-
-Phe-Thr-Ser-Asp-Tyr-
-Ser-Lys-Tyr-Leu-Asp-
-Ser-Arg-Arg-Ala-Gln-
-Asp-Phe-Val-Gln-Trp-
-Leu-Met-Asn-Thr-COO⁻

A **Epinephrine**   B **Glucagon**   C **Cortisol**

Figure 35-2
Examples of the three chemical classes of hormones: (A) epinephrine, an amino acid derivative; (B) glucagon, a polypeptide; (C) cortisol, a steroid.

the addition of epinephrine and glucagon resulted in the activation of phosphorylase, as in liver slices. *This observation of hormone action in a cell-free homogenate is a landmark in biochemistry.* Specific hormone effects had not previously been observed in cell-free systems. Some biologists had felt that hormones could act only on intact target cells. It is interesting to note that a similar attitude had been shattered a half-century earlier when the Buchners showed that fermentation can occur in cell-free extracts of yeast. However, in contrast with the glycolytic enzymes, not all of the components of the epinephrine and glucagon response system are soluble. Experiments showed that the response to these hormones is lost if the liver-cell homogenate is centrifuged. Hence, an essential part of this hormone response system must be located in a particulate fraction. In fact, the hormonal response was restored by adding back the particulate fraction to the original supernatant.

The role of the particulate fraction, which contains the plasma membrane, became evident from the following experiment. When the particulate fraction was incubated with epinephrine and glucagon, a heat-stable factor was produced. This factor activated phosphorylase when added to the supernatant fraction. Thus, *the hormonal response was separated into two parts: the interaction of the hormones with the particulate fraction to produce a heat-stable factor, followed by its effect on the supernatant to activate phosphorylase.* The next challenge was to identify this heat-stable intermediate of which only small amounts were available. Chemical analysis showed that it was an adenine ribonucleotide, but its properties were unusual. Sutherland wrote Leon Heppel about this molecule in the hope that he might be able to help in elucidating its structure. At the same time, David Lipkin wrote Heppel describing a new nucleotide that was produced by treating ATP with barium hydroxide. Heppel surmised that Lipkin and Sutherland were studying the same molecule, and he put them in touch with each other. Indeed, both investigators were studying the same molecule, which turned out to be adenosine 3′, 5′-monophosphate, now commonly referred to as *cyclic AMP* or *cAMP*. Another dividend of this chance encounter was that large amounts of cyclic AMP could be prepared for biochemical studies. Furthermore, the laboratory synthesis of cyclic AMP from ATP and barium hydroxide suggested a plausible route for its biosynthesis.

Adenosine 3′,5′-monophosphate (Cyclic AMP)

CYCLIC AMP IS SYNTHESIZED BY ADENYLATE CYCLASE AND DEGRADED BY A PHOSPHODIESTERASE

Cyclic AMP is formed from ATP by the action of *adenylate cyclase,* a membrane-bound enzyme.

$$ATP \xrightarrow{Mg^{2+}} \text{cyclic AMP} + PP_i + H^+$$

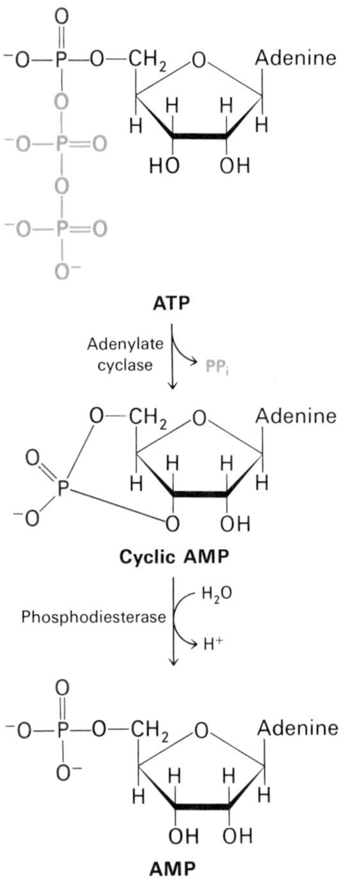

Figure 35-3
Enzyme-catalyzed synthesis and degradation of cyclic AMP.

This reaction is slightly endergonic, having a $\Delta G^{\circ\prime}$ of about 1.6 kcal/mol. The synthesis of cyclic AMP is driven by the subsequent hydrolysis of pyrophosphate. Cyclic AMP is destroyed by a specific *phosphodiesterase*, which hydrolyzes it to AMP.

$$\text{Cyclic AMP} + H_2O \xrightarrow{Mg^{2+}} \text{AMP} + H^+$$

This reaction is highly exergonic, having a $\Delta G^{\circ\prime}$ of about -12 kcal/mol. Cyclic AMP is a very stable compound in the absence of the phosphodiesterase.

CYCLIC AMP IS A SECOND MESSENGER IN THE ACTION OF MANY HORMONES

Sutherland's work led to the concept that cyclic AMP is a *second messenger* in the action of some hormones. The first messenger is the hormone itself. The essential features of this concept are:

1. Cells contain receptors for hormones in the plasma membrane.

2. The combination of a hormone with its specific receptor in the plasma membrane leads to the stimulation of adenylate cyclase, which is also bound to the plasma membrane.

3. The increased activity of adenylate cyclase increases the amount of cyclic AMP inside the cell.

4. Cyclic AMP then acts inside the cell to alter the rate of one or more processes.

An important feature of the second-messenger model is that the hormone need not enter the cell. Its impact is made at the cell membrane. The biological effects of the hormone are mediated inside the cell by cyclic AMP rather than by the hormone itself. A number of experimental criteria have been used to test the validity of this concept:

1. Adenylate cyclase in a target cell should be stimulated by hormones that affect that cell. Hormones that do not elicit a characteristic biological response in such a cell should not elevate its cyclase level.

2. The concentration of cyclic AMP in target cells should change in direct proportion to their biological response to hormonal stimulation, in terms of both kinetics and dependence on hormone concentration.

3. Inhibitors of the phosphodiesterase, such as theophylline and caffeine, should act synergistically with hormones that use cyclic AMP as a second messenger.

Caffeine
(1,3,7-Trimethylxanthine)

Theophylline
(1,3-Dimethylxanthine)

4. The biological effects of the hormone should be mimicked by the addition of cyclic AMP or a related compound to the target cells. (In practice, cyclic AMP cannot be readily used in this way because it penetrates cells poorly. However, less-polar derivatives of cyclic AMP, such as dibutyryl cyclic AMP, do enter cells and are active.)

Experiments based on these criteria have shown that *cyclic AMP is a second messenger for many hormones in addition to epinephrine and glucagon* (Table 35-1). Cyclic AMP affects an exceptionally wide range of cellular processes. For example, it enhances the degradation of storage fuels, increases the secretion of hydrochloric acid by the gastric mucosa, leads to the dispersion of melanin pigment granules, and diminishes the aggregation of blood platelets.

Table 35-1
Hormones using cyclic AMP as a second messenger

Calcitonin
Chorionic gonadotropin
Corticotropin
Epinephrine
Follicle-stimulating hormone
Glucagon
Luteinizing hormone
Lipotropin
Melanocyte-stimulating hormone
Norepinephrine
Parathyroid hormone
Thyroid-stimulating hormone
Vasopressin

A GUANYL-NUCLEOTIDE-BINDING PROTEIN COUPLES HORMONE RECEPTORS TO ADENYLATE CYCLASE

How does the binding of a hormone such as epinephrine or glucagon to a specific receptor lead to the activation of an adenylate cyclase molecule? The binding sites for these hormones are located on the extracellular surface of the plasma membrane, whereas the catalytic sites of adenylate cyclase face the cytosol. In fact, *hormone-binding sites and catalytic sites are located on different proteins,* which can be separated by centrifuging a detergent solution of the plasma membrane (Figure 35-4). Adenylate cyclase (185 kdal) is a large integral membrane protein, as is the receptor for epinephrine (75 kdal). This hormone receptor is also known as the *β-adrenergic receptor* because it binds a spectrum of pharmacologically active compounds.

Adenylate cyclase is not directly activated by the binding of hormone to a specific receptor. Instead, the effect is mediated by a third protein, which is called the *G protein* because it binds guanyl nucleotides. There are two forms of this 42-kdal regulatory protein. The GTP complex of this protein activates adenylate cyclase, whereas its complex with GDP does not. The G protein is converted from the inactive GDP form into the active GTP form by the exchange of GTP for bound GDP. This GTP-GDP exchange is cata-

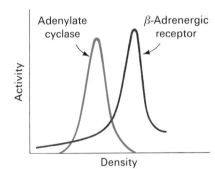

Figure 35-4
Separation of adenylate cyclase from the β-adrenergic receptor by centrifugation of detergent-solubilized plasma membrane on a sucrose density gradient. [After T. Haga, K. Haga, and A. G. Gilman, *J. Biol. Chem.* 252(1977): 5776.]

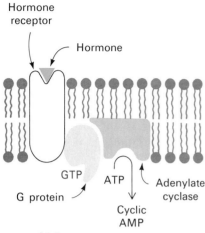

Hormone
receptor

Hormone

GTP   ATP

G protein

Adenylate
cyclase

Cyclic
AMP

Figure 35-5
Flow of information in the activation
of adenylate cyclase by the binding of
a hormone to its specific receptor. The
G protein plays a critical role in both
activation and deactivation of adenylate cyclase.

lyzed by the hormone-receptor complex but not by the unoccupied receptor. Thus, *the flow of information is from the hormone receptor to the G protein and then to adenylate cyclase* (Figure 35-5).

How is the activation of adenylate cyclase switched off? The G protein possesses yet another property that enables it to serve as the information-carrying intermediate between hormone receptors and adenylate cyclase. GTP bound to the G protein is slowly hydrolyzed to GDP. In other words, *the G protein is a GTPase. Thus, this regulatory protein has a built-in device for deactivation.* The proportion of G protein in the GTP state, and hence of adenylate cyclase in the active form, depends on the rate of exchange of GTP for GDP compared with the rate of hydrolysis of bound GTP. The rate of GTP-GDP exchange is markedly enhanced by the binding of the G protein to *occupied* hormone receptors. Consequently, nearly all of the G protein is in the GDP form and nearly all of the adenylate cyclase, in turn, is inactive when the hormone level is low. In some types of cells, the activation of adenylate cyclase also depends on the level of Ca^{2+}. The Ca^{2+} complex of *calmodulin,* a 17-kdal protein, is required in addition to the GTP form of the G protein for the activation of adenylate cyclase in those cells. The regulatory circuits that control adenylate cyclase are now beginning to be elucidated.

CYCLIC AMP STIMULATES PROTEIN KINASES

How does cyclic AMP influence so many cellular processes? Is there a common denominator for many of these effects? Indeed there is, and again the answer has come from studies of the control of glycogen metabolism, which has been described in a lighter vein as the metabolic birthplace of cyclic AMP. Edwin Krebs and Donal Walsh discovered that *cyclic AMP activates a protein kinase* in skeletal muscle. This protein kinase phosphorylates both glycogen synthetase (rendering it inactive) and phosphorylase kinase (rendering it active). In this way, cyclic AMP stimulates glycogen breakdown and stops glycogen synthesis in muscle (p. 372). A similar mechanism exists in liver. In fact, *all known effects of cyclic AMP result from the activation of protein kinases.* All cells studied thus far contain protein kinases that are activated by cyclic AMP at concentrations of the order of 10^{-8} M. These kinases modulate the activities of different proteins in different cells by phosphorylating them.

The mechanism of activation of the protein kinase in muscle by cyclic AMP is interesting. This enzyme consists of two kinds of subunits: a 49-kdal regulatory (R) subunit, which can bind cyclic AMP, and a 38-kdal catalytic (C) subunit. In the absence of cyclic AMP, the regulatory and catalytic subunits form an R_2C_2 complex that is enzymatically inactive. The binding of cyclic AMP to each of the regulatory subunits leads to the dissociation of the R_2C_2 complex into an R_2 subunit and two C subunits. These free catalytic subunits are then enzymatically active. Thus, *the binding of cyclic AMP to the regulatory subunit relieves its inhibition of the catalytic subunit. Cyclic AMP acts as an allosteric effector.* The presence of distinct regulatory and catalytic subunits in the protein kinase is reminiscent of the subunit structure of aspartate transcarbamoylase (p. 523).

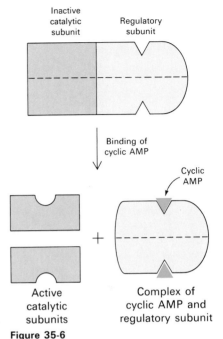

Figure 35-6
Cyclic AMP activates protein kinases by dissociating the complex of regulatory and catalytic subunits.

CYCLIC AMP IS AN ANCIENT HUNGER SIGNAL

As discussed previously, cyclic AMP has a regulatory role in bacteria, where it stimulates the transcription of certain genes (p. 673). It is evident that *cyclic AMP has a long evolutionary history as a regulatory molecule.* In bacteria, cyclic AMP is a hunger signal. It signifies an absence of glucose and leads to a synthesis of enzymes that can exploit other energy sources. In some mammalian cells, such as liver and muscle, cyclic AMP retains its ancient role as a hunger signal. However, it acts by stimulating a protein kinase rather than by enhancing the transcription of certain genes. Another important difference is that cyclic AMP has become a second messenger in higher organisms, where it participates in intercellular rather than intracellular communication.

Why was cyclic AMP chosen in the course of evolution to be a second messenger? Three factors seem important:

1. Cyclic AMP is derived from ATP, a ubiquitous molecule, in a simple reaction driven forward by the subsequent hydrolysis of pyrophosphate.

2. Though derived from a molecule that is at the center of metabolic transformations, cyclic AMP itself is not on a major metabolic pathway. It is used only as an integrator of metabolism, not as a biosynthetic precursor or intermediate in energy production. Hence, its concentration can be independently controlled. Furthermore, it is stable unless hydrolyzed by a specific phosphodiesterase.

3. Cyclic AMP has a sufficient number of functional groups so that it can bind tightly and specifically to receptor proteins, such as the regulatory subunit of the protein kinase in muscle, and elicit allosteric effects.

It is important to note that the *hormonal signal is greatly amplified by the use of cyclic AMP as a second messenger.* The concentration of many hormones in the blood is of the order of 10^{-10} M. In a stimulated target cell, the concentration of cyclic AMP is much higher. Many cyclic AMP molecules are synthesized by a single activated adenylate cyclase. The phosphorylation of many protein molecules by a protein kinase activated by cyclic AMP provides further amplification. The enzymatic cascade in the control of glycogen metabolism illustrates how small stimuli can trigger major changes in cells.

Figure 35-7
Space-filling model of cyclic AMP.

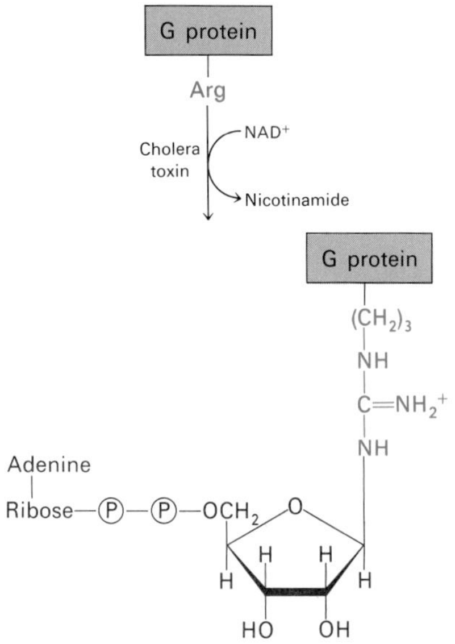

Figure 35-8
Cholera toxin catalyzes the ADP-ribosylation of the G protein, a regulator of the activity of adenylate cyclase.

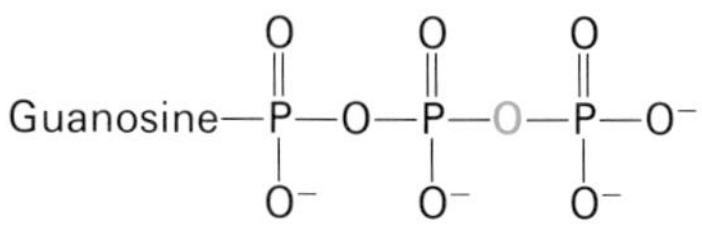

Guanosine triphosphate (GTP)

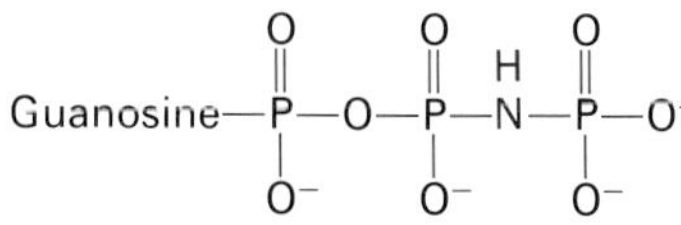

Guanosine 5'-[β,γ-imido]triphosphate (Gpp[NH]p)

CHOLERA TOXIN STIMULATES ADENYLATE CYCLASE BY INHIBITING THE GTPase ACTIVITY OF THE G PROTEIN

The direct participation of cyclic AMP in a disease process has been clearly established in cholera. This potentially lethal disease is caused by *Vibrio cholerae,* a gram-negative bacterium that is propelled by a single flagellum. The striking clinical feature of this disease is a massive diarrhea. Several liters of body fluid may be lost within a few hours, which leads to shock and death if the fluids are not replaced. The diarrhea is caused by a bacterial toxin rather than by a direct action of the bacteria themselves. *Cholera toxin (also called choleragen) increases the adenylate cyclase activity of the mucosa of the small intestine, which in turn raises the level of cyclic AMP in these cells.* The abnormally high level of cyclic AMP stimulates active transport of ions by these intestinal epithelial cells, which results in a very large efflux of Na^+ and water into the gut.

Cholera toxin, an 87-kdal protein, consists of an A_1 peptide linked by a disulfide bond to an A_2 peptide and five B peptides. This toxin enters cells by interacting with a G_{M1} ganglioside (p. 462) on the cell surface. This carbohydrate-rich sphingolipid is recognized by the B chains of the toxin. After gaining entry, the 23-kdal A_1 subunit covalently modifies the G protein that controls the activity of adenylate cyclase. Specifically, the A_1 subunit of the toxin catalyzes the transfer of an ADP-ribose unit from NAD^+ to an arginine side chain of the G protein (Figure 35-8). This ADP-ribosylation blocks the GTPase action of the G protein. In other words, the built-in device for deactivation (see Figure 35-5) is absent in the modified G protein. *In effect, the G protein is locked into the GTP form, and so adenylate cyclase stays persistently activated in the absence of hormone.* A similar effect can be obtained in vitro by the addition of guanylyl imidodiphosphate, a nonhydrolyzable analog of GTP, to unmodified G protein. This analog, like GTP, activates the G protein, but it cannot be converted into GDP. Consequently, adenylate cyclase is rendered permanently active by the binding of guanylyl imidodiphosphate to unmodified G protein, as it is by the binding of GTP to G protein that has lost its GTPase activity. The action of cholera toxin provides further evidence for the importance of the GTPase activity of the G protein. Recall that diphtheria toxin also exerts its harmful effects by ADP-ribosylating a protein with GTPase activity (p. 711).

INSULIN STIMULATES ANABOLIC PROCESSES AND INHIBITS CATABOLIC PROCESSES

We now turn to insulin, a polypeptide hormone that plays a key role in the integration of fuel metabolism, as was discussed in an earlier chapter (p. 549). A unifying characteristic is that insulin promotes anabolic processes and inhibits catabolic ones in muscle, liver, and adipose tissue. Specifically, insulin increases the rate of synthesis of glycogen, fatty acids, and proteins, and also stimulates glycolysis. An important action of the hormone is that *it promotes the entry of glucose, some other sugars, and amino acids into muscle and fat cells.* Hence, the level of glucose in the blood is lowered by insulin (called the *hypoglycemic effect*). Insulin inhibits catabolic processes such as the breakdown of glycogen and fat. It also decreases gluconeogenesis by lowering the level of enzymes such as pyruvate carboxylase and fructose 1,6-diphosphatase. Many of the effects of insulin are contrary to those elicited by epinephrine and glucagon. In essence, *epinephrine and glucagon signal that glucose is scarce, whereas insulin signals that glucose is abundant.*

PREPROINSULIN AND PROINSULIN ARE THE PRECURSORS OF THE ACTIVE HORMONE

Frederick Sanger showed that bovine insulin consists of two chains—an A chain of 21 residues and a B chain of 30 residues—which are covalently joined by two disulfide links (Figure 35-9). The same pattern exists in insulin molecules from many different species, including humans. How is this two-chain protein synthesized? A priori, one possibility is that the A and B chains are synthesized separately and that correct disulfide pairings are formed after the two chains associate in a specific way because of noncovalent interactions. This proposed mechanism was tested by determining whether active hormones having the correct disulfide pairings can be formed in vitro from reduced A and B chains. Anfinsen's demonstration that ribonuclease, a one-chain protein, can efficiently refold into an active enzyme after it is reduced and unfolded (p. 34) provided the stimulus for this experiment. The outcome with insulin was very different: less than 4% of the molecules regained the correct disulfide pairings. This low yield for the refolding of insulin was like that obtained for chymotrypsin, a three-chain enzyme connected by two interchain disulfide bonds (p. 159). In contrast, chymotrypsinogen, the one-chain precursor, refolded to a much greater degree. The contrasting behavior of chymotrypsin and chymotrypsinogen suggested that insulin did not refold properly because part of the necessary information was missing. These studies raised the possibility that insulin, like chymotrypsin, is derived from a single polypeptide chain.

A breakthrough came from Donald Steiner's study of an islet-cell adenoma of the pancreas. This rare human tumor produces large amounts of insulin. Slices of this pancreatic tumor were incubated with tritiated leucine and then analyzed. A new, highly radioactive protein was found (Figure 35-10). Subsequent studies revealed that

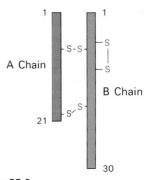

Figure 35-9
Chain structure and disulfide pairing in insulin.

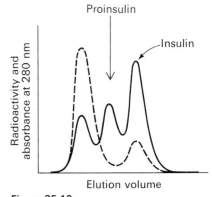

Figure 35-10
The appearance of a highly radioactive protein following pulse-labeling of an islet-cell adenoma led to the discovery of proinsulin. (The solid line refers to radioactivity; the dashed line to absorbance at 280 nm.)

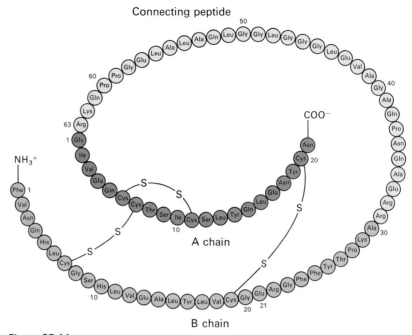

Figure 35-11
Amino acid sequence of porcine proinsulin: the A chain of insulin is shown in
red, the B chain in blue, and the connecting peptide (C-peptide) in yellow.
[After R. E. Chance, R. M. Ellis, and W. W. Bromer. *Science* 161(1968):165.
Copyright 1968 by the American Association for the Advancement of Science.]

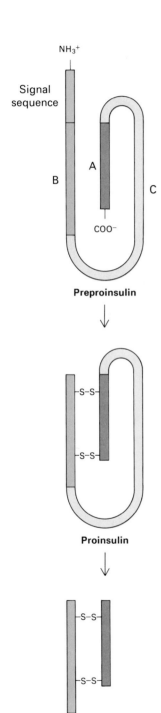

Figure 35-12
Enzymatic conversion of preproinsulin
into proinsulin and then into insulin.

this new protein, called *proinsulin,* is the *biosynthetic precursor of insulin.*

Proinsulin is a single polypeptide chain containing a sequence of
about thirty amino acid residues that is absent from insulin (Figure
35-11). This *connecting peptide (C-peptide)* joins the carboxyl end of
the B chain and the amino terminus of the A chain of the future
insulin molecule. As expected, *proinsulin forms correct disulfide pairings
when its disulfide bonds are reduced and then reoxidized.*

Recent studies of the biosynthesis of insulin have revealed that
proinsulin is not the earliest form of the hormone. The nascent
polypeptide chain in the lumen of the rough endoplasmic reticulum, called *preproinsulin,* contains an amino-terminal sequence of
sixteen residues that is absent from proinsulin. It seems likely that
this amino-terminal hydrophobic region serves as a signal sequence
that directs the nascent chain to the endoplasmic reticulum
(p. 713). The formation of proinsulin from preproinsulin takes
place in the luminal space of the endoplasmic reticulum, very soon
after the polypeptide chain traverses this membrane. Proinsulin is
then transported to the Golgi apparatus, where proteolysis of its
connecting peptide begins. The formation of insulin from proinsulin, which is not active as a hormone, continues in the storage granules. The connecting peptides in proinsulins from different species
have common structural features. In particular, the connecting
peptide in all proinsulins studied thus far contains Arg-Arg at the

amino end and Lys-Arg at the carboxyl end. A proteolytic enzyme similar to trypsin hydrolyzes the polypeptide chain at these positively charged residues. The insulin molecules in mature storage granules are secreted when the membrane of the granule fuses with the plasma membrane of the cell.

THREE-DIMENSIONAL STRUCTURE OF INSULIN

X-ray crystallographic analyses carried out in the laboratory of Dorothy Crowfoot Hodgkin revealed the three-dimensional structure of porcine insulin at a resolution of 1.9 Å. This accomplishment was completed approximately thirty-six years after Hodgkin obtained the first x-ray diffraction pattern of a protein crystal (pepsin) when she was a graduate student in the laboratory of John Bernal. In the intervening years, Hodgkin solved the structures of such biologically important molecules as cholesterol, penicillin, and Vitamin B_{12}. *Insulin has a compact three-dimensional structure* (Figure 35-14). Only the amino and carboxyl termini of the B chain are extended away from the rest of the protein. The A chain is nestled between these extended arms of the B chain. There is a nonpolar core, consisting of buried aliphatic side chains from both the A and the B chains. Insulin is also stabilized by several salt links and by hydrogen bonds between groups on the A and B chains, in addition to its two interchain disulfide bonds. The conformation of proinsulin is not yet known. However, spectroscopic studies suggest that proinsulin closely resembles insulin in its three-dimensional structure. Furthermore, proinsulin cocrystallizes with insulin.

INSULIN RECEPTORS ARE LOCATED IN THE PLASMA MEMBRANE OF TARGET CELLS

Insulin binds very tightly to specific receptors in the plasma membrane of target cells, as shown by using insulin made highly radioactive by the covalent insertion of ^{125}I. The dissociation constant of the insulin-receptor complex is about 10^{-10} M. Such tight binding is necessary because the insulin level in the blood is low (of the order of 10^{-10} M). The rate constant for association is very high (about 10^7 M^{-1} sec^{-1}), approaching the value for a diffusion-controlled reaction. A fat cell contains only 10^4 insulin receptors, which corresponds to a density of about *one receptor per square micrometer of plasma membrane*. In other words, there is just one insulin receptor per 10^6 phospholipid molecules.

The insulin receptor has been extracted from the plasma membrane by the addition of nonionic detergents, such as Triton X-100. The solubilized receptor has the same binding properties as the receptor in the intact plasma membrane. The solubilized receptor has been purified by affinity chromatography, a separation proce-

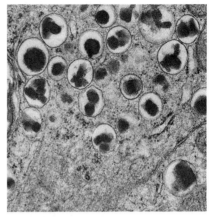

Figure 35-13
Electron micrograph of storage granules containing insulin in a β cell of the pancreas. Small crystals of insulin form within these granules because they contain high concentrations of the hormone. [Courtesy of Dr. Arthur Like.]

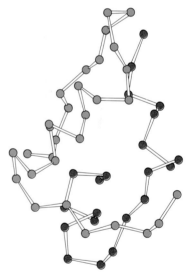

Figure 35-14
Alpha-carbon diagram of insulin: the A chain is shown in red, the B chain in blue.

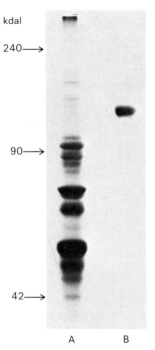

kdal

240→

90→

42→

A   B

Figure 35-15
Gel electrophoresis pattern of (A) an extract of liver plasma membrane and (B) purified insulin receptor. Electrophoresis was carried out under denaturing conditions. [After S. Jacobs, E. Hazum, Y. Schechter, and P. Cuatrecasas. *Proc. Nat. Acad. Sci.* 76(1979):4919.]

Diabetes—
    Named for the excessive urination in the disease. Aretaeus, a Cappadocian physician of the second century A.D., wrote: "The epithet diabetes has been assigned to the disorder, being something like passing of water by a siphon." He perceptively characterized diabetes as "being a melting-down of the flesh and limbs into urine."

Mellitus—
    From Latin, meaning sweetened with honey. Refers to the presence of sugar in the urine of patients having the disease.

dure that exploits the specific binding properties of the receptor. The insulin receptor binds tightly to an affinity column consisting of insulin covalently attached to agarose, whereas other molecules in the solubilized membrane preparation do not. The insulin receptor is then eluted by an acidic urea solution, which denatures the receptor and dissociates it from the insulin-agarose column. Fortunately, the purified receptor can then be renatured by removing the urea and raising the pH of the solution. Pedro Cuatrecasas has achieved a 250,000-fold purification of the insulin receptor by these procedures. The insulin-binding subunit of this receptor is a 135-kdal glycoprotein (Figure 35-15).

The insulin receptor comprises only about $4 \times 10^{-3}\%$ of the membrane protein of a liver homogenate. This means that 500 g of protein present in the homogenate derived from the livers of two hundred rats has to be fractionated to obtain 1 mg of pure receptor protein. Obtaining enough receptor for amino acid sequence and x-ray crystallographic studies, and for reconstitution experiments, will be a formidable but not impossible task.

It seems likely that many of the rapid effects of insulin, such as the activation of amino acid and glucose transport across the plasma membrane, are triggered by its interaction with receptors in the plasma membrane. Another early effect of insulin is the phosphorylation of a single protein in the 40S ribosomal subunit. These hormonal effects are not mediated by changes in the level of cyclic AMP. The intracellular messengers in the action of insulin have not yet been identified.

INSULIN DEFICIENCY CAUSES DIABETES

Diabetes mellitus is a complex disease that affects several hundred million people. Diabetes is characterized by an elevated level of glucose in the blood and in the urine. Glucose is excreted in the urine when the blood glucose level exceeds the reabsorptive capacity of the renal tubules. Water accompanies the excreted glucose, and so an untreated diabetic in the acute phase of the disease is hungry and thirsty. The loss of glucose depletes the carbohydrate stores, which leads to the breakdown of fat and protein. The mobilization of fats results in the formation of large amounts of acetyl CoA. Ketone bodies (acetoacetate, acetone, and hydroxybutyrate) are formed when acetyl CoA cannot enter the citric acid cycle because there is insufficient oxaloacetate. Recall that animals cannot synthesize oxaloacetate from acetyl CoA. Rather, oxaloacetate is derived from glucose and some amino acids (p. 546). The excretion of ketone bodies impairs the acid-base balance and causes further dehydration, which may lead to coma and death in the acute phase of the disease in an untreated diabetic.

Diabetes mellitus arises from a deficiency of insulin. In most cases, the underlying causes are unknown. Diabetes can be experimentally produced in an animal by surgical removal of much of the pancreas or by chemical destruction of the β cells. These insulin-producing cells can be selectively destroyed by administration of alloxan. Administration of anti-insulin antibody also leads to diabetes. Further evidence that insulin deficiency plays an important role in diabetes is provided by the fact that acute diabetes is rapidly reversed by the administration of insulin.

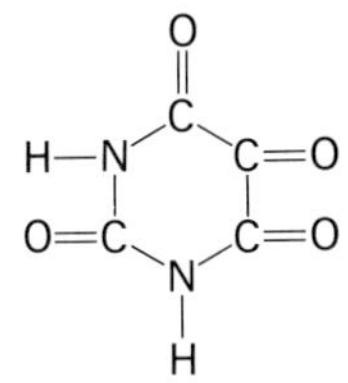

Why is a diabetic deficient in insulin relative to the needs of his tissues? *It is now evident that many kinds of underlying molecular defects can produce the clinical entity called diabetes mellitus.* A start has been made in identifying the causes of diabetes:

1. *Defective conversion of proinsulin into insulin.* Mutations that alter residues at the junction of the A chain (or B chain) and connecting peptide of proinsulin can interfere with its conversion into insulin. These patients have a high plasma level of proinsulin, which is inactive as a hormone.

2. *Structurally abnormal insulin molecules.* Another class of mutations results in the substitution of an amino acid residue in a critical region of the insulin molecule. For example, the replacement of leucine for phenylalanine near the carboxyl terminus of the B chain yields an insulin that is only a tenth as active as the normal hormone. It is interesting to note that this region of the insulin molecule has remained invariant in evolution, from the primitive hagfish to humans.

3. *Abnormal receptors for insulin.* Some patients secrete normal insulin molecules, but they do not bind in the usual way to their target cells. The receptors for insulin in the plasma membrane are defective in these diabetics.

4. *Defective coupling of insulin receptors.* The insulin molecules secreted by these patients are normal. Furthermore, their target cells contain the usual number of receptors for insulin. The binding characteristics of these receptors are also normal. It seems likely that the defect in these diabetics is intracellular. One possibility is that the insulin-receptor complex of these patients does not couple with the next component in the transmission of the hormonal signal.

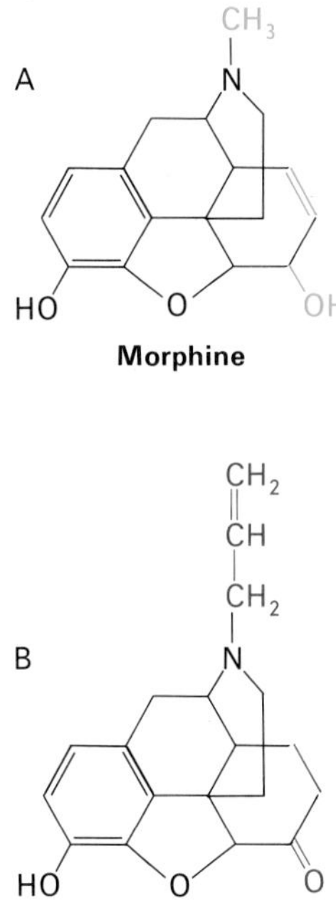

ENDORPHINS ARE BRAIN PEPTIDES THAT ACT LIKE OPIATES

Opiates such as morphine have been used for centuries to relieve pain. In 1680, Thomas Sydenham wrote, "Among the remedies which it has pleased Almighty God to give to man to relieve his sufferings, none is so universal and so efficacious as opium." Why do the brains of vertebrates contain receptors for alkaloids derived

Alloxan
(2,4,5,6-Tetraoxypyrimidine)

A

Morphine

B

Naloxone

Figure 35-16
Structures of (A) morphine, an opiate, and (B) naloxone, an antagonist of morphine.

^+H_3N-Tyr-Gly-Gly-Phe-**Met-COO**$^-$

Methionine enkephalin

A **(Met-enkephalin)**

^+H_3N-Tyr-Gly-Gly-Phe-**Leu-COO**$^-$

Leucine enkephalin

B **(Leu-enkephalin)**

^+H_3N-Tyr-Gly-Gly-Phe-Leu-Met-

-Thr-Ser-Glu-Lys-Ser-

-Gln-Thr-Pro-Leu-Val-

-Thr-Leu-Phe-Lys-Asn-

-Ala-Ile - Val-Lys-Asn-

-Ala-His-Lys-Lys-Gly-

-Gln-COO$^-$

C **β-Endorphin**

Figure 35-17
Amino acid sequences of (A) methionine enkephalin, (B) leucine enkephalin, and (C) β-endorphin. The common tetrapeptide sequence is shown in blue.

from the juice of poppy seeds? Neuropharmacologists surmised that opiate receptors serve to detect endogeneous regulators of pain perception rather than to respond to a plant product. According to this view, morphine exerts its pharmacological effects by mimicking molecules that are normally present in the bodies of vertebrates. The breakthrough came in 1975, when John Hughes isolated two peptides with opiatelike activity from pig brains. These related pentapeptides, called *methionine enkephalin* and *leucine enkephalin,* are abundant in certain nerve terminals. It seems likely that they participate in the integration of sensory information pertaining to pain.

A year later, Roger Guillemin isolated longer peptides, called *endorphins,* from the intermediate lobe of the pituitary gland. Endorphins are about as potent on a molar basis as morphine in relieving pain. Laboratory animals respond in a remarkable way to the injection of endorphins into the ventricles of the brain. For example, β-endorphin induces a profound analgesia of the whole body for several hours. Body temperature is lowered during this interval. Moreover, the animal becomes stuporous and assumes a stretched-out posture. These effects of endorphins disappear within a few hours and the animal again behaves normally. Another striking finding is that the actions of endorphins are reversed a few seconds after administering *naloxone* (Figure 35-16), a known antagonist of morphine. The behavioral effects induced by the endorphins suggest that these peptides may normally participate in regulating emotional responses. Many of the experimental methods needed to test this hypothesis have already been developed. For example, very small quantities of peptides such as the endorphins can be specifically assayed by the technique of *radioimmunoassay,* which combines the sensitivity of radioactive techniques with the specificity of immunoglobulin binding. We see here the beginning of a new and promising area of neurobiology and neuropsychiatry.

CLEAVAGE OF PRO-OPIOCORTIN YIELDS SEVERAL PEPTIDE HORMONES

Though β-endorphin was a newly discovered hormone, its amino acid sequence looked familiar. β-Endorphin has the same sequence as the carboxyl-terminal region of *β-lipotropin,* a hormone isolated from the pituitary gland by Choh Li. In fact, β-endorphin is formed in vivo by the proteolytic cleavage of β-lipotropin contained in storage granules. An even larger protein encompassing the sequences of both corticotropin and β-lipotropin was then discovered. This 29-kdal prohormone is called *pro-opiocortin* because it is the precursor of an opiate hormone and of corticotropin (Figure 35-18). *Corticotropin* (also called adrenocorticotropin, or ACTH) promotes the growth of the adrenal cortex and stimulates it to synthesize a set of steroid hormones. Pro-opiocortin also gives rise to two melano-

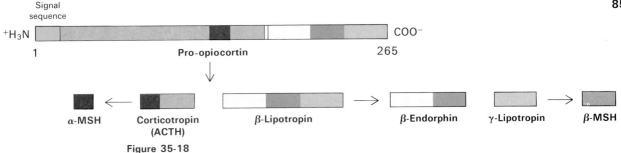

Figure 35-18
Pro-opiocortin is the biosynthetic precursor of several peptide hormones.

cyte-stimulating hormones (MSH). One of them, α-MSH, is a fragment of corticotropin and the other, β-MSH, comes from β-lipotropin (Figure 35-18). The amino-terminal half of pro-opiocortin may be a source of yet other hormones. Clearly, *pro-opiocortin is a cornucopia of peptide hormones.* This prohormone seems to be made of four homologous regions, which probably evolved by successive gene duplications.

The junctions between prospective active hormones in pro-opiocortin contain *pairs of basic residues* (Lys-Arg, Arg-Arg, or Lys-Lys). It is interesting to note that pairs of basic residues are also present at the boundaries of the connecting peptide in proinsulin (see Figure 35-11) and in proparathyroid hormone. Pairs of basic residues may mark prospective cleavage sites of prohormones in general.

PROSTAGLANDINS ARE MODULATORS OF HORMONE ACTION

We now turn to the prostaglandins, a group of fatty acids that affect a wide variety of physiological processes. These compounds were discovered in the 1930s but have become prominent only recently, largely because of the pioneering work of Sune Bergstrom. *A prostaglandin is a twenty-carbon fatty acid that contains a five-carbon ring.* The major classes are designated PGA, PGB, PGE, and PGF, followed by a subscript, which denotes the number of carbon–carbon double bonds outside the ring. *Prostaglandins seem to modulate the action of hormones rather than act as hormones themselves.* They often alter the activities of the cells in which they are synthesized. The nature of these effects may vary from one type of cell to another, in contrast with the uniformity in the action of a hormone.

The mechanism of the effect of prostaglandin PGE_1 on the breakdown of fat in adipose tissue has been studied in detail. Lipolysis is stimulated by hormones such as epinephrine, glucagon, corticotropin, and thyroid-stimulating hormone. PGE_1 at a concentration of 10^{-8} M strongly inhibits the lipolytic effects of these

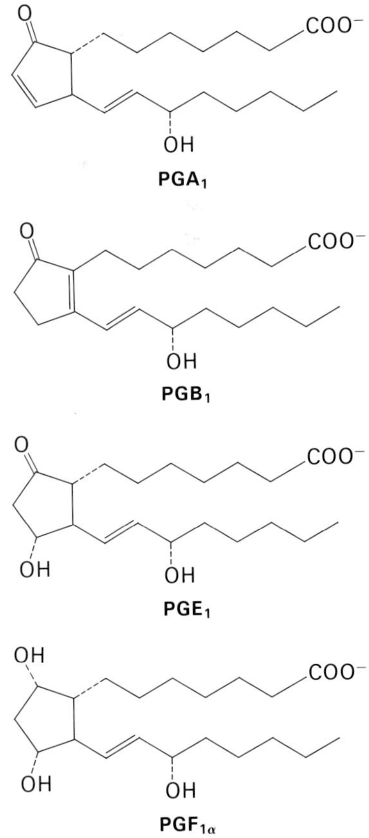

Figure 35-19
Structures of some prostaglandins.

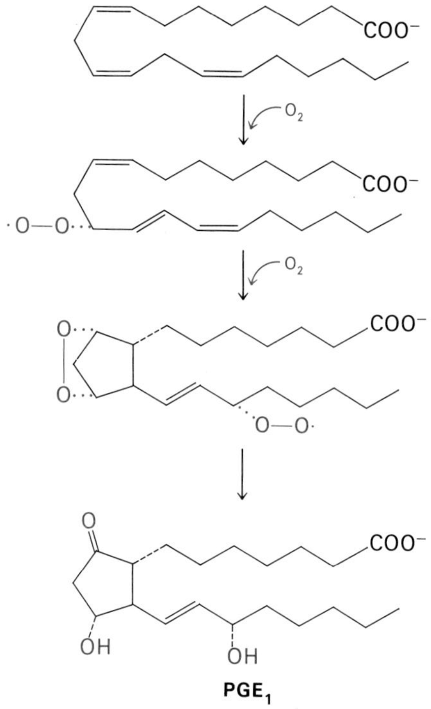

Figure 35-20
Synthesis of PGE$_1$ starting from cis-Δ^8,Δ^{11},Δ^{14}-eicosotrienoate.

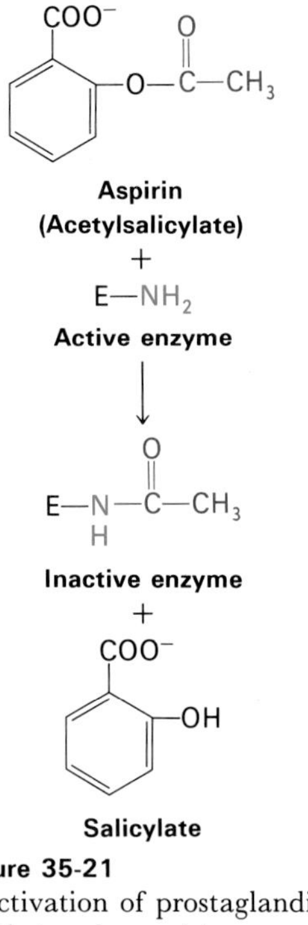

Figure 35-21
Inactivation of prostaglandin synthetase by aspirin.

hormones. A related finding is that PGE$_1$ prevents the rise of the intracellular level of cyclic AMP that is elicited by these hormones. However, PGE$_1$ does not inhibit lipolysis caused by the addition of dibutyryl cyclic AMP. Hence, *PGE$_1$ inhibits adenyl cyclase in fat cells.* In other cells, prostaglandins may have the opposite effect on cyclic AMP levels.

The molecular basis of many of the major actions of prostaglandins is not yet known. Some of the effects of prostaglandins are the stimulation of inflammation, the regulation of blood flow to particular organs, the control of ion transport across some membranes, and the modulation of synaptic transmission. There is much clinical interest in prostaglandins. For example, prostaglandins may participate in parturition. Infusion of PGE$_2$ induces delivery within a few hours. Also, prostaglandins are being evaluated for potential use as contraceptives. PGF$_{2\alpha}$ reduces the secretion of progesterone, a hormone needed for implantation of a fertilized ovum in the uterus.

PROSTAGLANDINS ARE SYNTHESIZED FROM UNSATURATED FATTY ACIDS

Prostaglandins are synthesized in membranes from C$_{20}$ fatty acids that contain at least three double bonds. These polyunsaturated fatty acids are required in the diets of mammals (p. 403). The precursors of prostaglandins are released from membrane phospholipids by phospholipases. For example, the biosynthesis of PGE$_1$ starts from Δ^8,Δ^{11},Δ^{14},-eicosatrienoate. A cyclopentane ring is formed and three oxygen atoms are introduced by *prostaglandin synthetase* (also called prostaglandin cyclooxygenase). As expected, all three oxygen atoms come from molecular oxygen (Figure 35-20). The heme-containing dioxygenase catalyzing these reactions is bound to the smooth endoplasmic reticulum.

John Vane discovered that *aspirin inhibits the biosynthesis of prostaglandins by inactivating prostaglandin synthetase.* Specifically, aspirin (acetylsalicylate) inhibits the oxygenase activity of this enzyme by acetylating the terminal amino group of one of its subunits (Figure 35-21). Prostaglandins enhance inflammatory effects, whereas aspirin diminishes them. This pharmacologic activity of aspirin is likely to be due to its inhibition of the biosynthesis of prostaglandins.

STEROID HORMONES ACTIVATE SPECIFIC GENES

The *primary effect* of steroid hormones such as estradiol, progesterone, and cortisone is on *gene expression* rather than on enzyme activities or transport processes. In contrast with epinephrine, these hormones must enter their target cells to exert their effects.

Furthermore, their primary site of action is in the *cell nucleus* rather than on the plasma membrane. The full impact of these steroids is achieved in hours rather than minutes because their biological effects depend on the *synthesis of new proteins*. Actinomycin D inhibits the action of these steroid hormones, which implies that the *synthesis of new messenger RNA is required for their action*.

The first step in the stimulation of uterine growth by 17β-estradiol is the binding of this hormone to a specific receptor in the cytoplasm of the uterine cell. The binding is very tight ($K \sim 10^{-9}$ M). This hormone-receptor complex then migrates to the nucleus of the cell. *The binding of estradiol to the receptor markedly enhances its affinity for DNA.* Furthermore, the receptor acquires a second subunit after binding estradiol. This is reflected in an increase in its sedimentation coefficient from 4S to 5S. It is not yet known whether the binding sites for the estradiol receptor on DNA are specified by the base sequence or by chromosomal proteins. The interaction of the hormone-receptor complex with DNA is highly specific. Furthermore, the activation of the receptor by the binding of hormone is also quite specific. Estrone binds to the receptor, but this complex does not interact with DNA. The failure of the estrone-receptor complex to bind to DNA agrees with the fact that estrone does not stimulate uterine growth.

The interactions of estradiol probably exemplify a general pattern for steroid hormones. For example, dexamethasone, a glucocorticoid steroid hormone, also binds to a specific receptor in the cytoplasm of hepatoma cells in tissue culture. The receptor undergoes a conformational change on binding the hormone and then migrates to the cell nucleus, where it complexes to specific sites on DNA. The number of binding sites on DNA for this receptor is estimated to be 1 per 10^6 base pairs. This finding is consistent with the observation that this hormone influences the transcription of a rather small number of genes. The mechanism of this highly selective control of transcription by steroid hormones is an intriguing problem.

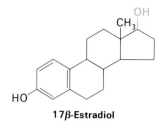

17β-Estradiol

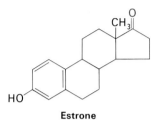

Estrone

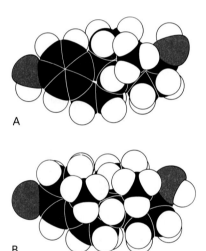

A

B

Figure 35-22
Space-filling models of (A) 17β-estradiol and (B) testosterone.

GROWTH-FACTOR PROTEINS SUCH AS NGF AND EGF STIMULATE TARGET CELLS TO PROLIFERATE

How is the growth of eucaryotic cells controlled? The recent isolation of growth-factor proteins that specifically stimulate target cells is a significant start in answering this challenging and important question. Rita Levi-Montalcini has discovered that *nerve growth factor (NGF) plays a critical role in the development of sympathetic neurons and of certain sensory neurons in vertebrates.* NGF stimulates these cells to divide and differentiate. The outgrowth of neurites from a ganglion

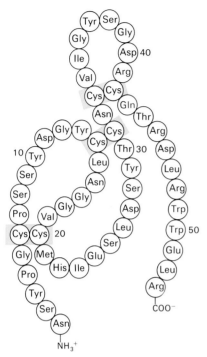

Figure 35-24
Amino acid sequence of epidermal growth factor (EGF). [After C. R. Savage, Jr., J. H. Hash, and S. Cohen. *J. Biol. Chem.* 248(1973):7669.]

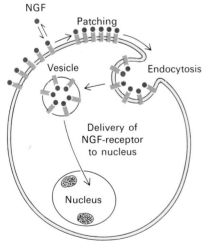

Figure 35-25
Internalization of NGF-receptor complexes. [After a drawing kindly provided by Dr. Eric Shooter.]

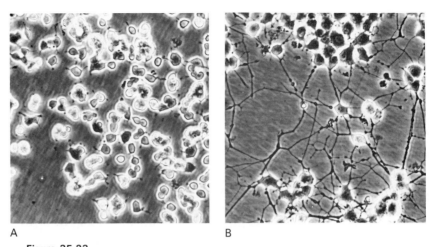

Figure 35-23
Nerve growth factor (NGF) induces neurites to grow out of cultured nerve cells: (A) no NGF added; (B) with NGF. [Courtesy of Bruce Yankner, Christiana Richter-Landsberg, and Dr. Eric Shooter.]

in culture is a sensitive assay for this growth factor (Figure 35-23). The biologically active NGF molecule consists of two identical 13-kdal polypeptide chains. This dimer (called the β subunit) is stored in the submaxillary gland, a site of synthesis, in the form of a 130-kdal complex that has the subunit structure $\alpha_2\gamma_2\beta$. The γ subunit is a proteolytic enzyme, whereas the α subunit inhibits this protease. NGF is synthesized as a prohormone containing the α and β subunits, which is then cleaved by the γ protease. The amino acid sequence of NGF resembles that of insulin. It is noteworthy that insulin has a stimulatory effect on the growth of most cells in addition to markedly enhancing anabolic processes in muscle, liver, and adipose tissue. The genes for NSF and insulin may have a common ancestor.

Epidermal growth factor (EGF) has also been purified and characterized. This 6-kdal polypeptide (Figure 35-24) stimulates the growth of epidermal and epithelial cells. EGF binds tightly to receptors on the plasma membrane of target cells. The dissociation constant of the EGF-receptor complex is about 10^{-10} M. Several minutes later, clusters of EGF-receptor complexes are internalized. The sites of endocytosis are coated-pit regions that contain clathrin (p. 717). Vesicles containing EGF then fuse with lysosomes. Likewise, NGF-receptor complexes are brought into the cell by endocytosis (Figure 35-25). It will be interesting to find out whether internalization of these complexes is essential for conveying the stimulatory signal to the nucleus of target cells.

Hormones are a group of diverse molecules that integrate the activities of different cells in multicellular organisms. Cyclic AMP plays a critical role in the action of many hormones (e.g., epinephrine and glucagon) by serving as a second messenger inside the target cell. Such hormones bind to specific receptors on the plasma membrane of the target cell and stimulate adenylate cyclase. This membrane-bound enzyme catalyzes the synthesis of cyclic AMP from ATP. A guanyl-nucleotide-binding protein (G protein) couples hormone receptors to adenylate cyclase. The activation of adenylate cyclase leads to an increased amount of cyclic AMP inside the cell. Cyclic AMP then activates a protein kinase, which phosphorylates one or more proteins. For example, the phosphorylation of glycogen synthetase and phosphorylase kinase in muscle and the liver results in decreased synthesis and enhanced degradation of glycogen. This reaction cascade amplifies the initial hormonal stimulus. Cyclic AMP has a long evolutionary history as a regulatory molecule, starting as a hunger signal. Cholera toxin catalyzes the ADP-ribosylation of the G protein, which leads to a persistent activation of adenylate cyclase in intestinal epithelial cells. Consequently there is a large efflux of Na^+ and water into the gut in cholera.

Insulin promotes anabolic processes (e.g., the synthesis of glycogen, fatty acids, and proteins) and inhibits catabolic ones (e.g., the breakdown of glycogen and fat). Insulin has a striking effect on membrane transport. Specifically, this hormone stimulates the entry of glucose and amino acids into muscle and fat cells. Preproinsulin and proinsulin are single-chain biosynthetic precursors of the active hormone. Preproinsulin contains a signal sequence, which is cleaved in the lumen of the rough endoplasmic reticulum. Insulin is then formed from proinsulin by proteolysis of the peptide that connects the A and B chains of the hormone. This processing takes place in the Golgi apparatus and in storage granules. Insulin is recognized by specific receptors located in the plasma membrane of target cells. Diabetes mellitus arises from a deficiency of insulin relative to the needs of the affected person. The disease, which has many underlying causes, is characterized by an elevated level of glucose in the blood and urine. Defective conversion of proinsulin into insulin, structurally abnormal insulin molecules, and defective cell-surface receptors for the hormone have been found in a small proportion of diabetics.

Endorphins and enkephalins are brain peptides that act like opiates such as morphine. β-Endorphin is derived from pro-opiocortin, a prohormone that is the source of other potent peptides such as corticotropin (ACTH), β-lipotropin, and melanocyte-stimulating hormones (α- and β-MSH). The junctions between prospective hormones in pro-opiocortin, as in proinsulin, contain pairs of basic residues. Small proteins also serve as growth hormones, as exemplified by nerve growth factor and epidermal growth factor. These proteins stimulate target cells to divide and differentiate. They act on a time scale of hours and days, as do steroid hormones, such as estradiol, progesterone, and cortisone. The major effects of the steroids are on gene expression. Estradiol enters its target cell and binds to a specific receptor in the cytosol. A second subunit joins this complex, which then enters the nucleus and binds to specific sites on DNA. Prostaglandins, which are synthesized from C_{20} polyunsaturated fatty acids, modulate the actions of hormones instead of acting as hormones themselves. Aspirin inhibits the synthesis of prostaglandins by covalently modifying the enzyme that catalyzes the insertion of oxygen.

SELECTED READINGS

WHERE TO START

Pastan, I., 1972. Cyclic AMP. *Sci. Amer.* 227(2):97–105. [Available as *Sci. Amer.* Offprint 1256.]

Rubenstein, E., 1980. Diseases caused by impaired communication among cells. *Sci. Amer.* 242(3):102–121. [Offprint 1463.]

Sutherland, E. W., 1972. Studies on the mechanism of hormone action. *Science* 177:401–408.

Synder, S. H., 1977. Opiate receptors and internal opiates. *Sci. Amer.* 236(3):44–56. [Offprint 1354.]

Guillemin, R., 1978. Peptides in the brain: the new endocrinology of the neuron. *Science* 202:390–402.

HORMONE RECEPTORS

Baxter, J. D., and MacLeod, K. M., 1980. Molecular basis for hormone action. *In* Bondy, P. K., and Rosenberg, L. E., (eds.), *Metabolic Control and Disease,* pp. 104–160. Saunders.

Bradshaw, R. A., and Frazier, W. A., 1977. Hormone receptors as regulators of hormone action. *Curr. Top. Cell Regul.* 12:1–35.

Kahn, C. R., 1976. Membrane receptors for hormones and neurotransmitters. *J. Cell Biol.* 70:261–286.

Goldstein, J. L., Anderson, R. G. W., and Brown, M. S., 1979. Coated pits, coated vesicles, and receptor-mediated endocytosis. *Nature* 279:679–685.

CYCLIC AMP AND ADENYLATE CYCLASE

Ross, E. M., and Gilman, A. G., 1980. Biochemical properties of hormone-sensitive adenylate cyclase. *Ann. Rev. Biochem.* 49:533–564.

Rodbell, M., 1980. The role of hormone receptors and GTP-regulatory proteins in membrane transduction. *Nature* 284:17–22.

Helmreich, E. J. M., Zenner, H. P., Pfeuffer, T., and Cori, C. F., 1976. Signal transfer from hormone receptor to adenylate cyclase. *Curr. Top. Cell Regul.* 10:41–87.

Greengard, P., 1978. Phosphorylated proteins as physiological effectors. *Science* 199:146–152.

Means, A. R., and Dedham, J. R., 1980. Calmodulin: an intracellular calcium receptor. *Nature* 285:73–77.

CHOLERA TOXIN

Moss, J., and Vaughan, M., 1979. Activation of adenylate cyclase by choleragen. *Ann. Rev. Biochem.* 48:581–600.

Hirschorn, N., and Greenough, W. B., III, 1971. Cholera. *Sci. Amer.* 225(2):15–21.

INSULIN AND DIABETES

Czech, M. P., 1977. Molecular basis of insulin action. *Ann. Rev. Biochem.* 46:359–384.

Steiner, D. F., 1977. Insulin today. *Diabetes* 26:322–340.

Renold, A. E., Mintz, D. H., Muller, W. A., and Cahill, G. F., Jr., 1978. Diabetes mellitus. *In* Stanbury, J. B., Wyngaarden, J. B., and Fredrickson, D. S., (eds.), *The Metabolic Basis of Inherited Disease* (4th ed.). McGraw-Hill.

Tager, H., Given, B., Baldwin, D., Mako, M., Markese, J., Rubenstein, A., Olefsky, J., Kobayashi, M., Kolterman, O., and Poucher, R., 1979. A structurally abnormal insulin causing human diabetes. *Nature* 281:122–125.

ENDORPHINS AND OPIATES

Guillemin, R., 1977. Endorphins: brain peptides that act like opiates. *New Engl. J. Med.* 296:226–228.

Nakanishi, S., Inoue, A., Kita, T., Nakamura, M., Chang, A. C. Y., Cohen, S. N., and Numa, S., 1979. Nucleotide sequence of cloned cDNA for bovine corticotropin-β-lipotropin precursor. *Nature* 278:423–427.

Synder, S. H., 1977. Opiate receptors in the brain. *New Engl. J. Med.* 296:266–271.

GROWTH FACTORS

Haigler, H. T., and Cohen, S., 1979. Epidermal growth factor: interactions with cellular receptors. *Trends Biochem. Sci.* 4:132–134.

Greene, L. A., and Shooter, E. M., 1980. The nerve growth factor: biochemistry, synthesis, and mechanism of action. *Ann. Rev. Neurosci.* 3:353–402.

Levi-Montalcini, R., and Calissano, P., 1979. The nerve-growth factor. *Sci. Amer.* 240(6):68–77.

STEROID HORMONES AND VITAMIN D

Yamamoto, K. R., and Alberts, B. M., 1976. Steroid receptors: elements for modulation of eukaryotic transcription. *Ann. Rev. Biochem.* 45:721–746.

DeLuca, H. F., 1974. Vitamin D: the vitamin and the hormone. *Fed. Proc.* 33:2211–2219.

Haussler, M. R., and McCain, T. A., 1977. Basic and clinical concepts related to vitamin D metabolism and action. *New Engl. J. Med.* 297:974–984; 1041–1049.

PROSTAGLANDINS

Samuelsson, B., Granstrom, E., Green, K., Hamberg, M., and Hammarstrom, S., 1975. Prostaglandins. *Ann. Rev. Biochem.* 44:669–695.

Roth, G. J., and Siok, C. J., 1978. Acetylation of the NH_2-terminal serine of prostaglandin synthetase by aspirin. *J. Biol. Chem.* 253:3782–3784.

Vane, J. R., 1971. Inhibition of prostaglandin synthesis as a mechanism of action for aspirin-like drugs. *Nat. New Biol.* 231:232–235.

RADIOIMMUNOASSAY OF HORMONES

Yalow, R. S., 1978. Radioimmunoassay: a probe for the fine structure of biologic systems. *Science* 200:1236–1246.

PROBLEM SET

Hood, L. E., Wilson, J. H., and Wood, W. B., 1975. *Molecular Biology of Eucaryotic Cells,* ch. 6. Benjamin.

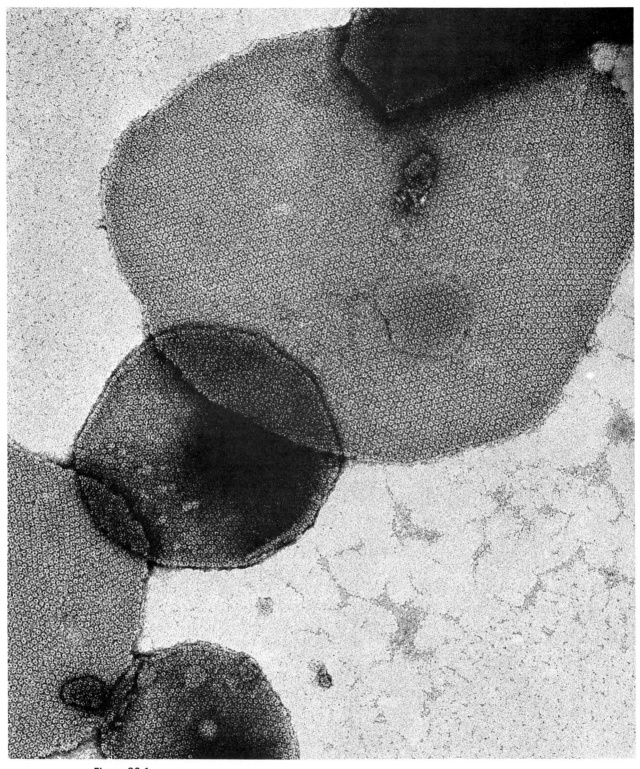

Figure 36-1
Electron micrograph of negatively stained junctions from liver cells. These 15-Å–diameter channels allow ions and small molecules to flow between the interiors of adjoining cells. [From E. L. Hertzberg and N. B. Gilula, *J. Biol. Chem.* 254(1979):2143.]

CHAPTER 36

MEMBRANE TRANSPORT

Biological membranes are highly selective permeability barriers. The flow of molecules and ions between a cell and its environment is precisely regulated by specific transport systems. Transport processes have several important roles:

1. They *regulate cell volume* and *maintain the intracellular pH and ionic composition within a narrow range* to provide a favorable environment for enzyme activity.

2. They extract and *concentrate metabolic fuels and building blocks* from the environment and *extrude toxic substances.*

3. They generate ionic gradients that are essential for the *excitability* of nerve and muscle.

The molecular mechanisms of many transport processes are now being elucidated. In this chapter, several bacterial and animal transport systems that carry ions, sugars, and amino acids across biological membranes are considered. Transport antibiotics produced by microorganisms are also discussed because analyses of their structures shed light on how transport systems discriminate between ions such as Na^+ and K^+. The last part of this chapter deals with channels that connect the interiors of apposed cells. These conduits (Figure 36-1) play an important role in intercellular communication.

DISTINCTION BETWEEN PASSIVE AND ACTIVE TRANSPORT

Whether a transport process is passive or active depends on the change in free energy of the transported species. Consider an uncharged solute molecule. The free-energy change in transporting this species from side 1 where it is present at a concentration of c_1 to side 2 where it is present at concentration c_2 is

$$\Delta G = RT \log_e \frac{c_2}{c_1} = 2.303 \, RT \log_{10} \frac{c_2}{c_1}$$

For a charged species, the electrical potential across the membrane must also be considered. The sum of the concentration and electrical terms is called the *electrochemical potential.* The free-energy change is then given by

$$\Delta G = RT \log_e \frac{c_2}{c_1} + Z \mathrm{F} \, \Delta V$$

in which Z is the electrical charge of the transported species, ΔV is the potential in volts across the membrane, and F is the faraday (23.062 kcal $\mathrm{V^{-1} \, mol^{-1}}$).

A transport process must be active when ΔG is positive, whereas it can be passive when ΔG is negative. *Active transport requires a coupled input of free energy, whereas passive transport can occur spontaneously.* For example, consider the transport of an uncharged molecule from $c_1 = 10^{-3}$ mM to $c_2 = 10^{-1}$ mM.

$$\Delta G = 2.3 \, RT \log_{10} \frac{10^{-1}}{10^{-3}}$$

$$= 2.3 \times 1.98 \times 298 \times 2$$

$$= +2.7 \, \mathrm{kcal/mol}$$

At 25°C (298 K), ΔG is $+2.7$ kcal/mol, which indicates that this transport process is active and thus requires an input of free energy. This transport process could be driven, for example, by the hydrolysis of ATP, which yields -7.3 kcal/mol under standard conditions.

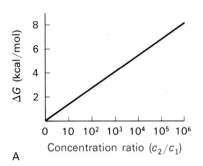

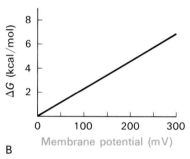

Figure 36-2
Free-energy change in transporting (A) an uncharged solute from a compartment at concentration c_1 to one at c_2 and (B) a singly charged species across a membrane to the side having the same charge as that of the transported ion. Note that the free-energy change imposed by a membrane potential of 59 mV is equivalent to that imposed by a concentration ratio of 10 for a singly charged ion at 25°C.

DISCOVERY OF THE ACTIVE-TRANSPORT SYSTEM FOR SODIUM AND POTASSIUM IONS

Most animal cells have a high concentration of K^+ and a low concentration of Na^+ relative to the external medium. These ionic gradients are generated by a specific transport system that is called the *Na+-K+ pump* because the movement of these ions is linked. The active transport of Na^+ and K^+ is of great physiological significance. Indeed, more than a third of the ATP consumed by a resting animal is used to pump these ions. The Na^+-K^+ gradient in animal cells controls cell volume, renders nerve and muscle cells electrically excitable, and drives the active transport of sugars and amino acids (p. 869).

In 1957, Jens Skou discovered an enzyme that hydrolyzes ATP only if Na^+ and K^+ are present in addition to Mg^{2+}. This enzyme is called a *Na^+-K^+ ATPase.*

$$ATP + H_2O \xrightarrow{Na^+, \; K^+, \; Mg^{2+}} ADP + P_i + H^+$$

Skou surmised that the Na^+-K^+ ATPase might be an integral part of the Na^+-K^+ pump and that the splitting of ATP could provide the energy needed for the active transport of Na^+ and K^+. Several lines of evidence have shown that the *Na^+-K^+ ATPase is in fact part of the Na^+-K^+ pump:*

1. The Na^+-K^+ ATPase is present wherever Na^+ and K^+ are actively transported. The level of enzymatic activity is quantitatively correlated with the extent of ion transport. For example, nerve cells are rich in both the Na^+-K^+ ATPase and the pump, whereas red blood cells have low levels of both.

2. Both the Na^+-K^+ ATPase and the pump are tightly associated with the plasma membrane.

3. Both the Na^+-K^+ ATPase and the pump are oriented in the same way in the plasma membrane.

4. Variations of the concentrations of Na^+ and K^+ have parallel effects on the ATPase activity and on the rate of transport of these ions.

5. Both the Na^+-K^+ ATPase and the pump are specifically inhibited by cardiotonic steroids. The concentration of inhibitor that causes half-maximal inhibition is the same for both processes.

6. The pump can be reversed under suitable ionic conditions so that ATP is synthesized from ADP and P_i.

BOTH THE ENZYME AND THE PUMP ARE ORIENTED IN THE MEMBRANE

Studies of the Na^+-K^+ pump in erythrocyte ghosts have revealed the orientation of this ATPase and pump. In a hypotonic salt solution, an erythrocyte swells and holes appear in its membrane. Hemoglobin diffuses out to leave a pale cell (hence, a ghost). The inside of the swollen erythrocyte can be equilibrated with the external medium. If the external medium is made isotonic, the membrane again becomes a permeability barrier. Thus, the molecular and ionic composition inside a ghost can be controlled by resealing it in an appropriate medium. Transport and enzymatic studies of

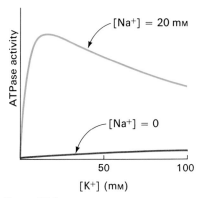

Figure 36-3
The Na^+-K^+ ATPase, a component of the Na^+-K^+ pump, hydrolyzes ATP only if both Na^+ and K^+ are present, in addition to Mg^{2+}.

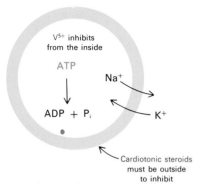

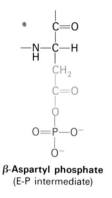

Figure 36-4
The Na⁺-K⁺ pump is oriented in the plasma membrane.

β-Aspartyl phosphate
(E-P intermediate)

such erythrocyte ghosts have shown that the Na^+-K^+ pump is oriented in the following way (Figure 36-4):

1. The Na^+ ion must be inside, whereas K^+ must be outside to activate the ATPase and to be transported across the membrane.

2. ATP is an effective substrate for the ATPase and the pump only when it is located inside the cell.

3. Cardiotonic steroids inhibit the pump and the ATPase only when they are located outside the cell.

4. Vanadate inhibits the pump and the ATPase only when it is located inside the cell.

ATP TRANSIENTLY PHOSPHORYLATES THE SODIUM-POTASSIUM PUMP

How does ATP drive the active transport of Na^+ and K^+? An important clue was provided by the finding that the Na^+-K^+ ATPase is phosphorylated by ATP in the presence of Na^+ and Mg^{2+}. The site of phosphorylation is the side chain of a specific aspartate residue. This phosphorylated intermediate (E-P) is then hydrolyzed if K^+ is present. The phosphorylation reaction does not require K^+, whereas the dephosphorylation reaction does not require Na^+ or Mg^{2+}.

$$E + ATP \underset{}{\overset{Na^+,\ Mg^{2+}}{\rightleftharpoons}} E-P + ADP$$
$$E-P + H_2O \xrightarrow{K^+} E + P_i$$

The Na^+-dependent phosphorylation and K^+-dependent dephosphorylation are not the only critical reactions. The pump also interconverts between at least two different conformations, denoted by E_1 and E_2. At least four conformational states—E_1, E_1-P, E_2-P, and E_2—participate in the transport of Na^+ and K^+ and concomitant hydrolysis of ATP (Figure 36-5). *Three Na⁺ and two K⁺ are transported*

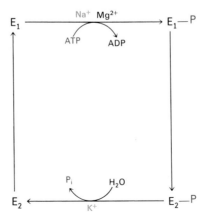

Figure 36-5
Conformational cycle of the Na^+-K^+ ATPase.

per ATP hydrolyzed. Hence, the pump generates an electric current across the plasma membrane. In other words, the Na$^+$-K$^+$ ATPase pump is *electrogenic.* The maximal turnover number of the ATPase is about 100 sec^{-1}.

The Na$^+$-K$^+$ ATPase is inhibited by nanomolar concentrations of vanadate ion (V^{5+}). This pentavalent ion locks the protein in the E$_2$ form. Vanadate is a transition-state analog for phosphoryl-group hydrolysis because it can adopt a bipyramidal structure like that made by phosphate (Figure 36-6).

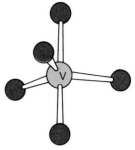

Figure 36-6
Structure of a vanadate ion (V^{5+}). The bipyramidal array of ligands around this ion is like that around a phosphorus atom during hydrolysis of a phosphoryl group.

ION TRANSPORT AND ATP HYDROLYSIS ARE TIGHTLY COUPLED

An important feature of the pump is that *ATP is not hydrolyzed unless Na$^+$ and K$^+$ are transported.* In other words, the system is coupled so that the energy stored in ATP is not dissipated. Tight coupling is a general characteristic of biological assemblies that mediate energy conversion. Recall that electrons do not normally flow through the mitochondrial electron-transport chain unless ATP is concomitantly generated (p. 324). The obligatory coupling of ATP hydrolysis and muscle contraction is another example of this principle.

The Na$^+$-K$^+$ pump can be reversed so that it synthesizes ATP. Net synthesis of ATP from ADP and P$_i$ occurs when the ionic gradients are very steep. This is achieved by incubating red cells in a medium containing a much higher than normal concentration of Na$^+$ and a much lower than normal concentration of K$^+$.

THE SODIUM-POTASSIUM PUMP IS AN OLIGOMERIC TRANSMEMBRANE PROTEIN

The Na$^+$-K$^+$ ATPase is a 270-kdal $\alpha_2\beta_2$ tetramer. The large α subunit (95 kdal) contains a site for the hydrolysis of ATP and a site for the binding of cardiotonic steroid inhibitors. The small β subunit (40 kdal) contains carbohydrate groups. Cross-links can be readily formed between the α subunits or between an α and a β subunit but not between the β subunits. This finding suggests that the α chains are in contact and that the β chains are far apart. As mentioned earlier, the ATP-hydrolysis site is on the cytosol side of the membrane, whereas the steroid-inhibitor site is on its extracellular face. Hence, each α subunit traverses the plasma membrane (Figure 36-7). The carbohydrate chains of the β subunits are on the extracellular side of the plasma membrane, as has been found for all other membrane glycoproteins (p. 222).

It is interesting to note that this enzyme complex has one binding site for steroid inhibitors, one phosphorylation site, and three binding sites for Na$^+$. How can an $\alpha_2\beta_2$ tetramer have an *odd* number of

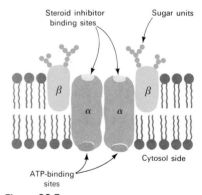

Figure 36-7
Schematic diagram of the subunit structure and orientation of the Na$^+$-K$^+$ pump.

binding sites? One possibility is that these binding sites are located at the interfaces of subunits. Recall that the $\alpha_2\beta_2$ hemoglobin tetramer has in its central cavity a single binding site for diphosphoglycerate (p. 78). Alternatively, the two $\alpha\beta$ halves of the enzyme may interact so that binding at one of a pair of potential sites precludes binding at the other. In fact, a number of oligomeric proteins display half-of-sites reactivity.

A MODEL FOR THE MECHANISM OF THE SODIUM-POTASSIUM PUMP

How does the phosphorylation and dephosphorylation of the ATPase result in the transport of Na^+ and K^+ across the membrane? Not enough is yet known about the structure of this pump to formulate a detailed mechanism of how it acts. Nevertheless, it is instructive to consider a simple model for a pump that was proposed by Oleg Jardetzky. In this model, a protein must fulfill three structural conditions to function as a pump:

1. It must contain a cavity large enough to admit a small molecule or ion.

2. It must be able to assume two conformations, such that the cavity is open to the inside in one form and to the outside in the other.

3. The affinity for the transported species must be different in the two conformations.

Let us apply this model to Na^+ and K^+ transport (Figure 36-8).

Figure 36-8
Schematic diagram of a proposed mechanism for the Na^+-K^+ pump. The upper sequence of reactions depicts the extrusion of three Na^+ ions, whereas the lower reactions show the entry of two K^+ ions. The E_1 (yellow) and E_2 (blue) forms are shown here as having very different conformations. The actual conformational difference may be quite small.

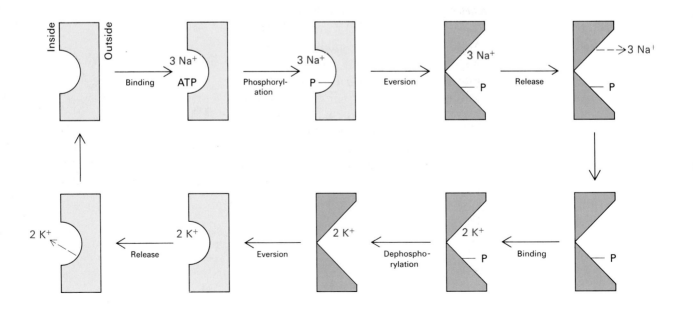

The two conformations are the E_1 and E_2 forms described earlier. It is assumed that (1) the ion-binding cavity in E_1 faces the inside of the cell, whereas in E_2 the cavity faces the outside, and that (2) E_1 has a high affinity for Na^+, whereas E_2 has a high affinity for K^+. The model also makes use of two established facts: (1) Na^+ triggers phosphorylation, whereas K^+ triggers dephosphorylation; and (2) E_2 is stabilized by phosphorylation, whereas E_1 is stabilized by dephosphorylation. In Figure 36-8, E_1 and E_2 have very different shapes. However, it must be stressed that the conformational differences between these forms need not be large. A shift of a few atoms by a distance of 2 Å might suffice to alter the relative affinity of the cavity for Na^+ and K^+ and change its orientation. There is ample precedent for assuming that phosphorylation can readily produce changes of this magnitude. Recall the effect of phosphorylation on the properties of glycogen phosphorylase and synthetase and the change in oxygen affinity elicited by the noncovalent binding of diphosphoglycerate to hemoglobin.

CARDIOTONIC STEROIDS ARE SPECIFIC INHIBITORS OF THE SODIUM-POTASSIUM ATPase AND PUMP

Certain steroids derived from plants are potent inhibitors of the Na^+-K^+ ATPase and pump. The inhibition of both processes is half-maximal at an inhibitor concentration of the order of 10^{-8} M. Digitoxigenin and ouabain are members of this class of inhibitors, which are known as *cardiotonic steroids* because of their profound effects on the heart (Figure 36-9). The activities of the cardiotonic steroids depend on the presence of a five- or six-membered unsaturated lactone ring in the β configuration at C-17. Also, a hydroxyl group at C-14 and a *cis* fusion of rings C and D are essential. Ouabain and some other cardiotonic steroids contain a sugar residue attached to C-3, but this sugar does not contribute to the inhibition of the ATPase by these compounds.

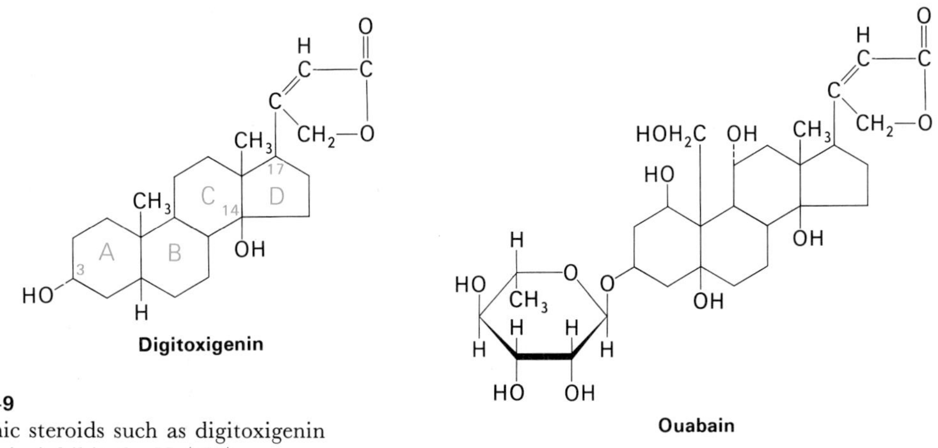

Figure 36-9
Cardiotonic steroids such as digitoxigenin and ouabain inhibit the Na^+-K^+ pump.

$$E—P + H_2O \xrightarrow{\quad K^+ \quad} E + P_i$$

Inhibited by
cardiotonic steroids

Figure 36-10
A foxglove plant. [From A. Krochmal and C. Krochmal, *A Guide to the Medicinal Plants of the United States* (Quadrangle Books, 1973), p. 243. Courtesy of Dr. James Hardin.]

As mentioned earlier, cardiotonic steroids inhibit the dephosphorylation reaction of the Na^+-K^+ ATPase. Inhibition occurs only if the cardiotonic steroid is located on the *outside* face of the membrane. Thus, inhibition of dephosphorylation by these steroids has the same spatial asymmetry as the activation of dephosphorylation by K^+.

Cardiotonic steroids such as *digitalis* are of great clinical significance. Digitalis increases the force of contraction of heart muscle, which makes it a choice drug in the treatment of congestive heart failure. Inhibition of the Na^+-K^+ pump by digitalis leads to a higher level of Na^+ inside the cell. The subsequent increase in the intracellular level of Ca^{2+} enhances the contractility of cardiac muscle. It is interesting to note that digitalis was effectively used long before the discovery of the Na^+-K^+ ATPase. In 1785, William Withering, a physician and botanist, published "An Account of the Foxglove and Some of its Medicinal Uses." He described how he first learned about the use of digitalis to cure congestive heart failure:

> In the year 1775, my opinion was asked concerning a family receipt for the cure of the dropsy. I was told that it had long been kept a secret by an old woman in Shropshire, who had sometimes made cures after the more regular practitioners had failed. . . . This medicine was composed of twenty or more different herbs; but it was not very difficult for one conversant in these subjects to perceive that the active herb could be no other than the foxglove. . . . It has a power over the motion of the heart to a degree yet unobserved in any other medicine, and this power may be converted to salutary ends.

CALCIUM IS TRANSPORTED BY A DIFFERENT ATPase

Calcium ion plays an important role in the regulation of muscle contraction (p. 826) and in many other physiological processes. Skeletal muscle contains an intricate network of membrane-bound tubules and vesicles. This membrane system, called the *sarcoplasmic reticulum*, regulates the Ca^{2+} concentration surrounding the contractile fibers of the muscle. At rest, Ca^{2+} is pumped into the sarcoplasmic reticulum so that the Ca^{2+} concentration around the muscle fibers is very low. Excitation of the sarcoplasmic-reticulum membrane by a nerve impulse leads to a sudden release of large amounts of Ca^{2+}, which triggers muscle contraction. In other words, *Ca^{2+} is the intermediary between the nerve impulse and muscle contraction.*

The transport of Ca^{2+} by the sarcoplasmic reticulum is driven by the hydrolysis of ATP. There is an ATPase in the sarcoplasmic reticulum that is activated by Ca^{2+}. This Ca^{2+} ATPase is an integral part of the Ca^{2+} pump, just as the Na^+-K^+ ATPase is part of the Na^+-K^+ pump. The Ca^{2+} ATPase is also transiently phosphorylated by ATP.

$$E + ATP \xrightleftharpoons{Ca^{2+},\ Mg^{2+}} E-P + ADP$$
$$E-P + H_2O \xrightarrow{Mg^{2+}} E + P_i$$

A cycle of conformational changes driven by phosphorylation and dephosphorylation transports two Ca^{2+} for each ATP hydrolyzed. The very high affinity of this ATPase for Ca^{2+} ($K \sim 10^{-7}$ M) enables it to effectively transport Ca^{2+} from the cytosol (where $[Ca^{2+}] < 10^{-5}$ M) into the sarcoplasmic reticulum (where $[Ca^{2+}] \sim 10^{-2}$ M).

The density of Ca^{2+} pumps in the sarcoplasmic reticulum membrane, about 20,000 per μm^2, is very high. In fact, the Ca^{2+} ATPase constitutes more than 80% of the integral membrane protein and takes up a third of the surface area. The large subunit (100 kdal) of the Ca^{2+} pump traverses the membrane and contains the phosphorylation site, which is a specific aspartate side chain, as in the Na^+-K^+ pump. Another similarity is that a 55-kdal glycoprotein is associated with the large subunit.

A functional Ca^{2+} pump was reconstituted from purified Ca^{2+} ATPase and phospholipids. The Ca^{2+} ATPase was isolated from sarcoplasmic-reticulum membranes following solubilization by cholate, a detergent. This solubilized enzyme was added to phospholipids obtained from soybeans. Membrane vesicles formed when the detergent was removed by dialysis. *These reconstituted vesicles pumped Ca^{2+} at a rapid rate when provided with ATP and Mg^{2+}.*

THE FLOW OF Na^+ DRIVES THE ACTIVE TRANSPORT OF SUGARS AND AMINO ACIDS INTO ANIMAL CELLS

Many transport processes are not directly driven by the hydrolysis of ATP. Instead, these transport processes are coupled to the flow of an ion down its electrochemical gradient. For example, glucose is pumped into many animal cells by the simultaneous entry of Na^+. In fact, sodium ion and glucose bind to a specific transport protein and enter together. A concerted movement of two species is called *cotransport*. A *symport* carries two species in the same direction, whereas an *antiport* carries them in opposite directions. Sodium ions that enter a cell together with glucose molecules by means of a symport are pumped out of the cell by a Na^+-K^+ ATPase (Figure 36-12). The rate and extent of glucose transport depends on the Na^+ concentration gradient across the membrane.

Symports driven by Na^+ are widely used by animal cells to accumulate amino acids. Some cells, such as the brush border of intesti-

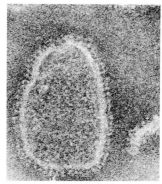

Figure 36-11
Membrane vesicles formed from purified Ca^{2+} ATPase. The globular particles on the surface of the membrane are part of the ATPase molecule, which extends across the membrane. [From P. S. Stewart and D. H. MacLennan. *J. Biol. Chem.* 249(1974):987.]

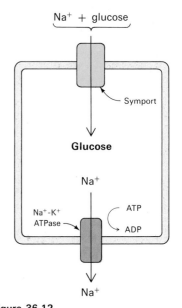

Figure 36-12
A Na^+ gradient drives the active transport of glucose. This symport system is present in the plasma membrane of intestinal cells and kidney cells.

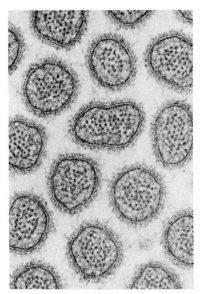

Figure 36-13
Electron micrograph of a cross section of intestinal microvilli, which greatly increase the surface area for transport. [Courtesy of Dr. George Palade.]

nal microvilli (Figure 36-13), are also rich in symports for the active transport of sugars. The small intestine has a specialized Na⁺-driven symport for the uphill transport of Cl^-. Sodium ions are also the driving force in antiports that extrude calcium ions from a variety of cells. *Most symports and antiports in animal cells are driven by Na⁺ gradients that are generated by the Na⁺-K⁺ ATPase.*

THE FLOW OF PROTONS DRIVES MANY BACTERIAL TRANSPORT PROCESSES

Symports and antiports are ancient molecular machines. Many bacterial transport systems are driven by the flow of protons across the plasma membrane. The best-understood symport system in bacteria is the one in *E. coli* for lactose (Figure 36-14). This denizen

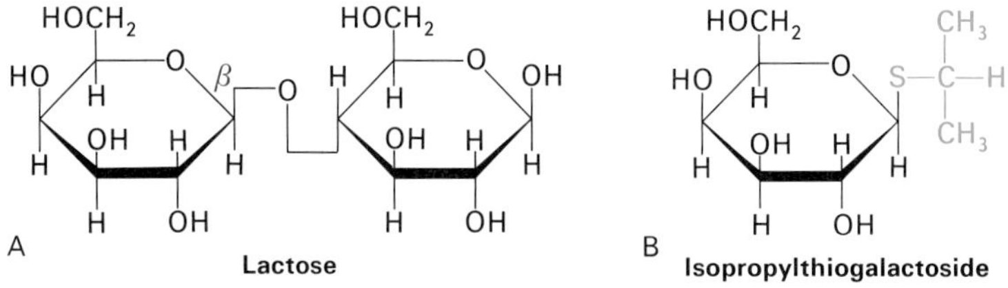

Figure 36-14
Lactose permease transports β-galactosides such as (A) lactose and (B) isopropylthiogalactoside. Thiogalactosides have been useful in studies of this system because they are transported by lactose permease but are not hydrolyzed by β-galactosidase.

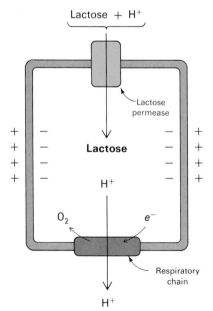

Figure 36-15
A proton gradient drives the active transport of some sugars and amino acids into bacteria. This proton gradient is generated by the flow of electrons through the respiratory chain.

of the mammalian lower intestinal tract has evolved a highly efficient mechanism for concentrating lactose. The pump for this disaccharide is a single 30-kdal polypeptide chain called *lactose permease* (or the M protein), which has been isolated by Eugene Kennedy. This integral membrane protein is encoded by the *y* gene of the *lac* operon (p. 671). It comprises about 4% of the membrane protein of induced cells.

Analyses of *y*-gene mutants have contributed to an understanding of the mechanism of this pump. Insight into its action has also come from studies of vesicles formed from bacterial membranes. These vesicles are attractive for transport studies because they are much simpler than whole bacteria. They contain the machinery for oxidative phosphorylation as well as other membrane-bound proteins but are devoid of the cytoplasmic constituents of the intact cell. These membrane vesicles alone do not accumulate lactose but can be stimulated to do so by adding substrates that can transfer their high-potential electrons to the respiratory chain. Alternatively, an imposed pH gradient, with the outside made acid, leads to the accumulation of lactose. A membrane potential created by a K⁺ gradient can also result in the pumping of lactose. *These experiments revealed that the driving force for the active transport of lactose is the proton-*

motive force across the plasma membrane. The transport of a lactose molecule is coupled to the movement of a proton into the cell. Under physiological conditions, the proton gradient needed for this active transport is generated by the flow of electrons from high-potential donors (such as NADH) through the respiratory chain. This *lactose-proton symport* exemplifies the "transduction of energy by proticity," a unifying concept developed by Peter Mitchell (p. 330).

THE ACTIVE TRANSPORT OF SOME SUGARS IS COUPLED TO THEIR PHOSPHORYLATION

Symports are not the only kinds of pumps for the active transport of sugars. Some bacteria accumulate sugars by coupling their entry to phosphorylation. For example, glucose is converted into glucose 6-phosphate during its transport into many bacteria. The distinctive feature of this kind of transport, called *group translocation,* is that the *solute is changed* during transport. The cell membrane is impermeable to sugar phosphates, and so they accumulate inside the bacterium.

The best-understood group-translocation processes are the ones carried out by the *phosphotransferase system* (*PTS*), which was discovered by Saul Roseman. An unusual feature of this system is that *phosphoenolpyruvate is the phosphoryl donor,* rather than ATP or another nucleoside triphosphate. The overall reaction carried out by the phosphotransferase system is

$$\text{Sugar}_{\text{outside}} + \text{phosphoenolpyruvate} \xrightarrow[\text{Mg}^{2+}]{\text{PTS}}$$

$$\text{sugar-phosphate}_{\text{inside}} + \text{pyruvate}$$

Four proteins participate in this group translocation: HPr, enzyme I, enzyme II, and enzyme III. Enzyme II, an integral membrane protein, forms the transmembrane channel and catalyzes the phosphorylation of the sugar. The phosphoryl group of phosphoenolpyruvate is not transferred directly to the sugar. Rather, it is transferred from phosphoenolpyruvate to enzyme I and then to a specific histidine residue of HPr, a small heat-stable protein (Figure 36-16). This phosphohistidine intermediate has a high group-transfer potential, intermediate between those of phosphoenolpyruvate and ATP. The phosphorylated form of HPr then transfers its phosphoryl group to enzyme III, a peripheral membrane protein that interacts with enzyme II, the actual channel. The final step is the

Phosphohistidine

Table 36-1

Sugars transported by the phosphotransferase system in *E. coli*

Glucose

Fructose

Mannose

N-Acetylglucosamine

Mannitol

Sorbitol

Galactitol

Lactose

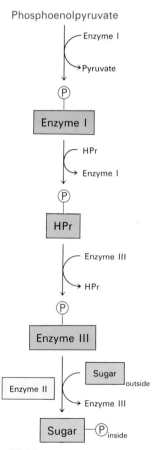

Figure 36-16

Flow of phosphoryl groups from phosphoenolpyruvate to the sugar that is translocated by the phosphotransferase system.

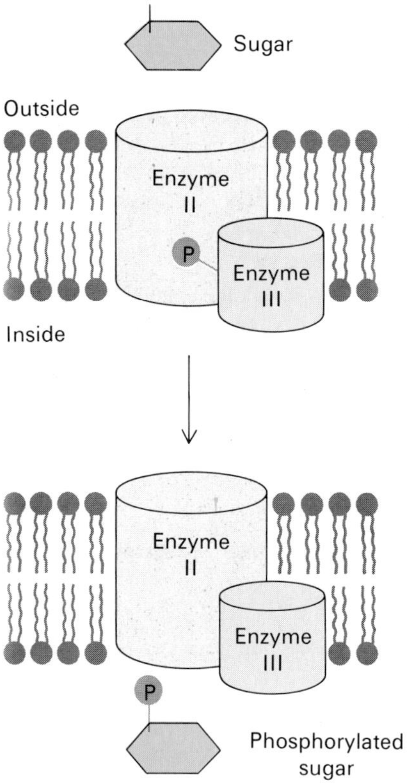

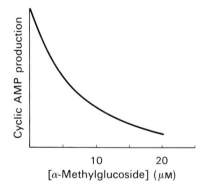

Figure 36-17
Proposed mechanism of group translocation by the phosphotransferase system.

transfer of a phosphoryl group from enzyme III to the sugar that is being translocated (Figure 36-17). A functional enzyme complex has been reconstituted from these four purified proteins.

Some of the proteins in the phosphotransferase system are general, whereas others are specific. HPr and enzyme I, which are soluble proteins in the cytosol, participate in the transport of all sugars that are translocated by this system. In contrast, enzymes II and III are specific for particular sugars. For example, there are different enzymes II and III for the transport of glucose, lactose, and fructose. Genetic studies have confirmed these conclusions. Mutants defective in HPr or enzyme I are unable to transport many different sugars, whereas mutants defective in enzymes II or III are unable to transport a particular sugar. In the translocation of hexitols such as galactitol, enzyme III does not participate; instead, a phosphoryl group is directly transferred from HPr to the sugar.

Why is the phosphotransferase system so complex compared with transporters such as lactose permease? A likely reason is that the phosphotransferase system has regulatory roles in addition to mediating the transport of sugars. An abundance of any of the sugars taken up by the phosphotransferase system strongly inhibits the active transport of sugars brought in by other systems. This effect seems to be mediated by changes in the level of cyclic AMP. Specifically, a high concentration of a sugar translocated by the phosphotransferase system leads to decreased production of cyclic AMP (Figure 36-18). The transcription of several inducible operons then

Figure 36-18
α-Methylglucoside, a sugar transported by the phosphotransferase system, inhibits cyclic AMP production.

stops. Recall that the expression of inducible catabolic operons such as the *lac* and *gal* operons is markedly enhanced by the binding of a complex of cyclic AMP and the CAP protein to their promoter sites (p. 674). Thus, *the phosphotransferase system regulates the acquisition of carbon sources.*

TRANSPORT ANTIBIOTICS INCREASE THE IONIC PERMEABILITY OF MEMBRANES

Some microorganisms synthesize compounds that make membranes permeable to certain ions. These small molecules, called *transport antibiotics,* are valuable experimental tools, in addition to being sources of insight into the mechanism of ion binding. For example, *valinomycin* interferes with oxidative phosphorylation in mitochondria by making them permeable to K^+. The result is that mitochondria use the energy generated by electron transport to accumulate K^+ rather than to make ATP. Valinomycin is a repeating cyclic molecule made of four kinds of residues (A, B, C, and D) taken three times (Figure 36-19). The four kinds of residues are alternately joined by ester and peptide bonds.

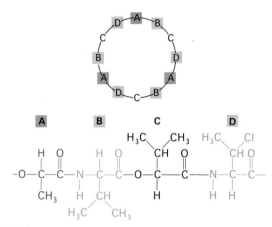

Figure 36-19
Valinomycin is a repeating cyclic molecule made up of L-lactate (A), L-valine (B), D-hydroxyisovalerate (C), and D-valine (D) residues.

Gramicidin A is another transport antibiotic that has been studied extensively (Figure 36-20). Gramicidin is an open-chain polypeptide consisting of fifteen amino acid residues. A noteworthy feature of the structure is that L- and D-amino acids alternate. Also, the amino and carboxyl termini of this polypeptide are modified. As will be discussed shortly, gramicidin A and valinomycin transport ions in quite different ways.

Ion transport by these antibiotics is readily studied in well-defined model systems such as lipid bilayer vesicles (p. 214) and planar bilayer membranes (p. 215). Vesicles containing a radioactive ion such as $^{42}K^+$ are prepared by sonicating membranes in the presence of this ion and then removing untrapped $^{42}K^+$ by gel filtration. The rate of efflux of this radioactive ion is then measured in the presence and absence of a transport antibiotic. Another way to obtain information about the ionic permeability of a bilayer membrane is to measure electrical properties such as membrane resistance and potential. For example, the resistance of a bilayer membrane to K^+ drops by a factor of more than 10,000 in the presence of 10^{-7} M valinomycin or 10^{-9} M gramicidin, when there is a gradient of 0.02 M KCl across the membrane.

O   H
 \\ /
  C
  | L   L   D   L
HN-Val-Gly-Ala-Leu-Ala-5

    D   L   D   L   D
   Val-Val-Val-Trp-Leu-10
                         O
   L   D   L   D   L   ‖
  Trp-Leu-Trp-Leu-Trp-C
                        |
                       N—H
                        |
                       CH₂
                        |
                       CH₂
                        |
                       OH

Figure 36-20
Structure of gramicidin A.

TRANSPORT ANTIBIOTICS ARE CARRIERS OR CHANNEL FORMERS

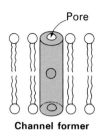

Pore

Channel former

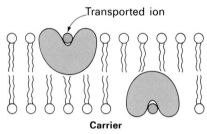

Transported ion

Carrier

Figure 36-21
Schematic diagram comparing a channel-forming with a carrier transport antibiotic. All known protein transporters are of the channel-forming type.

Transport antibiotics make membranes permeable to ions in two contrasting ways (Figure 36-21). Some of these antibiotics (e.g., gramicidin A) form channels that traverse the membrane. Ions enter such a channel at one surface of the membrane and diffuse through it to the other side of the membrane. The channel former itself need not move for ion transport to occur. The other group of antibiotics (e.g., valinomycin) functions by carrying ions through the hydrocarbon region of the membrane. Diffusion of these transport antibiotics is essential for their activities.

Carriers and *channel formers* can be experimentally distinguished in the following way. The ionic conductance of a synthetic lipid bilayer membrane containing the transport antibiotic is measured as a function of temperature through the transition temperature of the lipid. The hydrocarbon interior of the membrane changes from an essentially frozen state to a highly fluid one in this thermal transition. A channel former need not diffuse in order to mediate ion transport across the membrane. Hence, freezing of the hydrocarbon core should have little effect on its ion-transport activity. In contrast, freezing should markedly diminish the efficiency of a carrier because a carrier must diffuse through the hydrocarbon region to be active. The experimental data for valinomycin and gramicidin A distinguish between these mechanisms (Figure 36-22). The con-

Figure 36-22
Comparison of the temperature dependence of the conductance of a lipid bilayer membrane containing a channel-forming and a carrier transport antibiotic.

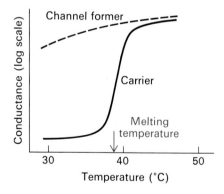

ductance of a bilayer membrane containing valinomycin increases more than a thousandfold as the membrane is melted. In contrast, the transport activity of gramicidin A is relatively unaffected by the melting of the membrane.

It is important to note that *all naturally occurring transport systems studied thus far are channels.* As discussed in an earlier chapter (p. 224), integral membrane proteins flip-flop very slowly or not at all. Hence, they cannot serve as diffusional carriers.

Ion-carrying transport antibiotics have some common structural features, as shown by x-ray crystallographic and spectroscopic studies. The carriers studied thus far have a shape like that of a donut. *A single metal ion is coordinated to several oxygen atoms that surround a central cavity.* The number of oxygen atoms that bind the metal ion is typically six or eight. *The periphery of the carrier consists of hydrocarbon groups* (Figure 36-23).

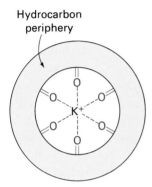

Hydrocarbon periphery

Figure 36-23
Schematic diagram of the chelation of K^+ by oxygen atoms of a carrier transport antibiotic.

The roles of the central oxygen atoms of the hydrocarbon exterior are evident. In an aqueous medium, a metal ion such as K^+ binds several water molecules through their oxygen atoms. A carrier competes with water for binding the ion by chelating it to several appropriately arranged oxygen atoms in its central cavity. The hydrocarbon periphery makes the ion-carrier complex soluble in the lipid interior of the membranes. In essence, *these antibiotics catalyze the transport of ions across membranes by making them soluble in lipid.*

The structures of valinomycin and its complex with K^+ are shown in Figure 36-24. The K^+ ion is coordinated to six oxygen

Figure 36-24
Models of (A) valinomycin and (B) its complex with K^+. The conformation of the antibiotic changes upon binding K^+. [After atomic coordinates kindly provided by Dr. William Duax (for valinomycin) and by Dr. Larry Steinrauf (for the valinomycin-K^+ complex).]

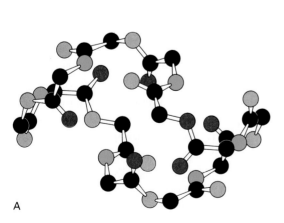

A

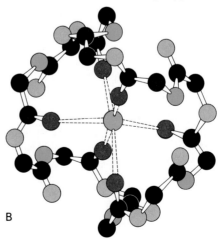

B

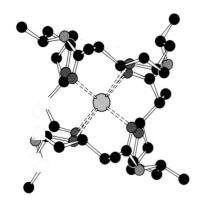

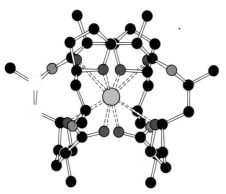

Figure 36-25
Two views of a model of the K⁺ complex of nonactin.

atoms, which are arranged octahedrally around the center of the molecule. These oxygen atoms come from the carbonyl groups of the six valine residues in the antibiotic. The methyl and isopropyl side chains constitute the hydrocarbon periphery of valinomycin.

The structure of the K⁺ complex of nonactin, another ion-carrying antibiotic, also is of interest (Figure 36-25). The ion is bound to eight oxygen atoms in the center of the molecule. Four of these oxygens come from carboxyl groups, whereas the other four are ether oxygens.

VALINOMYCIN BINDS K⁺ A THOUSAND TIMES AS STRONGLY AS Na⁺

How do ion transporters distinguish between such related ions as Na⁺ and K⁺? Valinomycin binds K⁺ about a thousand times as strongly as Na⁺. This degree of discrimination is even higher than that achieved by the Na⁺-K⁺ ATPase, and so it is instructive to examine its basis. Spectroscopic studies indicate that the structure of the Na⁺ complex of valinomycin closely resembles that of the K⁺ complex. In particular, these complexes have similar bond distances between their central cation and coordinating oxygen atoms. Why then does valinomycin bind K⁺ much more tightly than Na⁺? The reason for this selectivity is that water has less attraction for K⁺ than for Na⁺. The free energy of solvation of Na⁺ is 17 kcal/mol more favorable than that of K⁺ (Table 36-2). *Thus, valinomycin binds*

Table 36-2
Properties of alkali cations

Ion	Atomic number	Ionic radius (Å)	Hydration free energy (kcal/mol)
Li⁺	3	0.60	−98
Na⁺	11	0.95	−72
K⁺	19	1.33	−55
Rb⁺	37	1.48	−51
Cs⁺	55	1.69	−47

K⁺ more tightly because it is energetically more costly to pull Na⁺ away from water.

A transporter that markedly prefers Na⁺ has not yet been analyzed in atomic detail, but a key aspect of its structure can be surmised. Such a transporter must take advantage of the smaller ionic radius of Na⁺ ($r = 0.95$ Å) compared with that of K⁺ ($r = 1.33$ Å). Negatively charged oxygen atoms can approach Na⁺ more closely than K⁺, and so the higher cost of removing Na⁺ from water could be overcome. *The prediction then is that Na⁺-specific sites have a high density of negative charge and a geometry that allows Na⁺ to fit snugly in such a cluster. In contrast, K⁺-specific sites are expected to have a lower density of negative charge and a geometry less conducive to the formation of very short bonds between the cation and coordinating groups.*

The flexibility of the valinomycin molecule (see Figure 36-24) is also noteworthy. Chelation of K^+ is a step-by-step process in which water molecules in its hydration shell are successively displaced by oxygen atoms of the antibiotic. New bonds are being made as old ones are being broken, and so the activation barrier for binding is small. Likewise, the energy of activation for release, the reverse process, is small. Hence, valinomycin picks up and unloads K^+ many times a second. The importance of flexibility is evident here, as in enzyme action.

THE FLOW OF IONS THROUGH A SINGLE CHANNEL IN A MEMBRANE CAN BE DETECTED

The conductance of a planar bilayer membrane containing a very small amount of gramicidin A is not constant. Denis Haydon discovered that the conductance of this membrane to Na^+ fluctuates with time in a quantized manner (Figure 36-26). These steps in

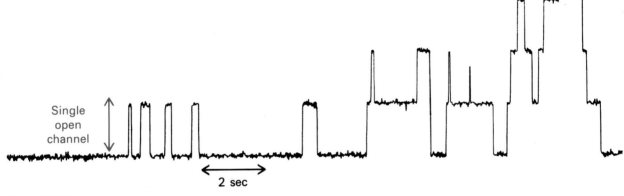

Figure 36-26
The conductance of a lipid bilayer membrane containing a few molecules of gramicidin A fluctuates in a step-by-step manner. The smallest step in conductance arises from a single open channel. [Courtesy of Dr. Olaf Anderson.]

conductance arise from the spontaneous opening and closing of gramicidin channels. A single channel stays open for about a second. This channel is highly permeable to monovalent cations but not to divalent cations or anions. *In fact, more than 10^7 ions can traverse a single channel in a second.* This transport rate is only a factor of ten less than that of diffusion through pure water. In contrast, the maximal transport rate of diffusional carriers is less than 10^3 ions per second because rotation and translation of these transporters in the membrane takes at least a millisecond.

Spectroscopic and x-ray–crystallographic studies have revealed that a transmembrane channel is formed from two molecules of

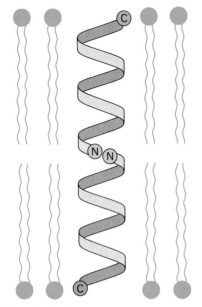

Figure 36-27
Schematic diagram of a gramicidin-A channel formed by the association of two polypeptides at their *N*-formyl ends. Each chain is folded into a β helix, which resembles a rolled-up β pleated sheet. This model was proposed by Dan Urry. [After S. Weinstein, B. A. Wallace, E. R. Blout, J. S. Morrow, and W. Veatch, *Proc. Nat. Acad. Sci.* 76(1979):4230.]

gramicidin A (Figure 36-27). This conducting helical dimer is in equilibrium in the membrane with nonconducting monomers. In fact, the steps in conductance in Figure 36-26 correspond to the formation and dissociation of dimers. The dimer contains a 4-Å–diameter aqueous channel that is surrounded by polar peptide groups. In contrast, the hydrophobic side chains are at the periphery of the channel, in contact with hydrocarbon chains of membrane phospholipids. The carboxyl groups surrounding the aqueous pore transiently coordinate the cation as it passes through the channel. X-ray studies indicate that the channel becomes shorter and wider on binding cations, which again emphasizes the dynamic nature of transport molecules.

GAP JUNCTIONS ALLOW IONS AND SMALL MOLECULES TO FLOW BETWEEN COMMUNICATING CELLS

Large aqueous channels for passive transport are present in many procaryotic and eucaryotic biological membranes. For example, the outer membrane of a gram-negative bacterium contains channels formed by *porin*, a 37-kdal transmembrane protein (p. 785). These 10-Å–diameter channels enable polar molecules with a mass of less than about 600 daltons to readily enter the periplasmic space. These molecules are then transported into the cytosol by symports, group translocators, and other permeases. Likewise, the outer membranes of mitochondria and chloroplasts contain large aqueous channels that are formed by molecules analogous to porin.

The best-understood aqueous channel in eucaryotes is the *gap junction,* which is also known as the *cell-to-cell channel* because it serves as a passageway between the interiors of many contiguous cells. Gap junctions, which were discovered by Jean-Paul Revel and Morris Karnovsky, are clustered in discrete regions of the plasma membranes of apposed cells. Electron micrographs of sheets of gap junctions (Figure 36-28) indicate that they are made up of six subunits surrounding a 15- to 20-Å–diameter hole. A tangential view (Figure

Figure 36-28
Electron micrograph of isolated gap junction sheets. The cylindrical connexons are organized in a hexagonal lattice with a unit-cell length of 85 Å. The densely stained central hole has a diameter of about 20 Å. [Courtesy of Dr. Nigel Unwin and Dr. Guido Zampighi.]

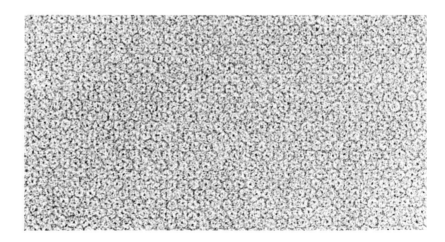

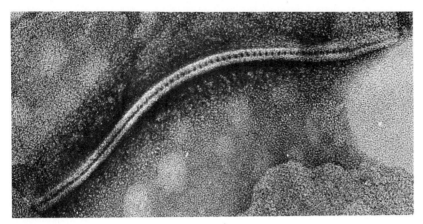

Figure 36-29
Electron micrograph of a tangential view of apposed cell membranes that are joined by gap junctions. [From E. L. Hertzberg and N. B. Gilula, *J. Biol. Chem.* 254(1979):2143.]

36-29) shows that these hexamers span the intervening space, or gap, between apposed cells (hence, the name gap junction). The bore size of gap junctions was determined by microinjecting a series of fluorescent molecules into cells and observing their passage into adjoining cells. Werner Loewenstein found that all polar molecules with a mass of less than about 1 kdal can readily pass through these cell-to-cell channels. In other words, *inorganic ions and most metabolites (e.g., sugars, amino acids, and nucleotides) can flow between the interiors of cells joined by gap junctions.* In contrast, proteins, nucleic acids, and polysaccharides are too large to traverse these channels.

Gap junctions are important for intercellular communication. Cells in some excitable tissues, such as heart muscle, are coupled by the rapid flow of ions through these junctions, which assure a rapid and synchronous response to stimuli. Gap junctions are also essential for the nourishment of cells that are distant from blood vessels, as in the lens and bone. These communicating channels are also likely to be important in regulating development and differentiation.

The permeability of gap junctions is regulated by calcium ion. *Elevation of the intracellular level of Ca^{2+} leads to a graded closure of gap junctions.* These channels are fully open when the Ca^{2+} level is less than 10^{-7} M and are shut when the level of this ion is higher than 5×10^{-5} M. As the concentration of Ca^{2+} increases in this range, the effective diameter of gap junctions decreases so that they become impermeable first to larger molecules. Three-dimensional image-reconstruction analyses of sheets of gap junctions in two quaternary states provide a structural basis for this graded closure. These structural studies show that a gap junction consists of two apposed cylindrical units, called *connexons,* each made up of six subunits. The 75-Å–long connexon traverses the plasma membrane. A 20-Å–long segment of the connexon protrudes into the extracellular space, where it joins the connexon from the contiguous cell. The long axis

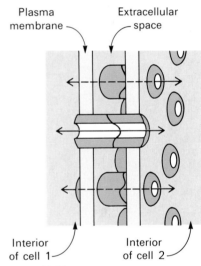

Figure 36-30
Schematic diagram of a gap junction.

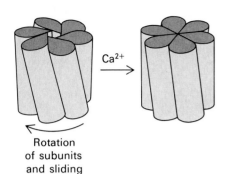

Rotation
of subunits
and sliding

Figure 36-31
Model for the graded closure of gap junctions by Ca^{2+}. [After a drawing kindly provided by Dr. Nigel Unwin and Dr. Guido Zampighi.]

of each of the subunits in a connexon is tilted with respect to the transmembrane axis. The channel closes when these subunits slide past each other to decrease their tilt (Figure 36-31). The conformational change is largest at the center of the channel, where subunits of apposed connexons slide past each other by a distance of about 11 Å, which corresponds to a rotation of 28 degrees. A higher-resolution analysis should reveal how Ca^{2+} induces this sliding and rotation.

SUMMARY

Molecules and ions are transported across biological membranes by oriented transmembrane proteins that form channels. A transport process is passive if ΔG for the transported species is negative, whereas it is active if ΔG is positive. The free-energy charge depends on the concentration ratio of the transported species and on the membrane potential if it is charged. Active transport requires an input of free energy. The most ubiquitous transport system in animal cells is the Na^+-K^+ pump, which hydrolyzes a molecule of ATP to pump out three Na^+ ions and to pump in two K^+ ions. In the presence of Na^+, ATP phosphorylates an aspartate side chain of the α subunit of the $\alpha_2\beta_2$ enzyme complex. This phosphorylated intermediate is hydrolyzed in the presence of K^+. The latter reaction is inhibited by cardiotonic steroids, such as digitalis, which are highly specific inhibitors of this pump. A cycle of conformational changes driven by phosphorylation and dephosphorylation results in the transport of Na^+ and K^+ ions. Calcium ion, which plays an important role in the control of muscle contraction, is transported by a different ATPase system in the sarcoplasmic-reticulum membrane. An aspartate residue is phosphorylated during the transport cycle, as in the Na^+-K^+ pump. Both the Ca^{2+} and the Na^+-K^+ transport systems have been reconstituted from purified ATPases and phospholipids.

Some transport systems are directly driven by ionic gradients rather than by the hydrolysis of ATP. For example, the active transport of glucose and amino acids into some animal cells requires the simultaneous entry of Na^+, a process called cotransport. Na^+ and glucose enter together on a specific symport. In turn, the Na^+ gradient for this coupled entry is maintained by the Na^+-K^+ pump. Bacteria generally use H^+ instead of Na^+ to drive symports and antiports. For example, the active transport of lactose by lactose permease is coupled to the movement of a proton into the bacterium. The driving force is the proton-motive force generated by the flow of electrons through the respiratory chain. A different type of transport process in bacteria is group translocation, in which a solute is changed during its transport. The phosphotransferase system phosphorylates several sugars (e.g., glucose to glucose 6-phosphate) as it brings them into the cell. Phosphoenolpyruvate is the phosphoryl donor in this process, which is mediated by three kinds of enzymes and a small phosphoryl-carrier protein (HPr).

Procaryotic and eucaryotic cells also contain aqueous channels that allow ions and small polar molecules to passively diffuse across membranes. For example, porin channels in the outer membrane of gram-negative bacteria have a diameter of about 10 Å. Many contiguous cells in higher organisms are joined by gap junctions. These 20-Å–diameter channels enable ions and most metabolites (e.g., monosaccharides, amino acids, and nucleotides) to flow from the interior of one cell to the adjoining one. Gap junctions are closed by Ca^{2+}. These conducts between apposed cells are important for intercellular communication.

Transport antibiotics make membranes permeable to certain ions by serving as carriers (e.g., valinomycin) or as channel formers (e.g., gramicidin A). A carrier antibiotic is a donut-shaped molecule that binds a single metal ion in its central cavity. The hydrocarbon periphery enables the complex to traverse the hydrocarbon interior of the membrane. Channel-forming antibiotics create an aqueous pore that traverses the membrane.

SELECTED READINGS

WHERE TO START

Hobbs, A. S., and Albers, R. W., 1980. The structure of proteins involved in active membrane transport. *Ann. Rev. Biophys. Bioeng.* 9:259–291.

Wilson, D. B., 1978. Cellular transport mechanisms. *Ann. Rev. Biochem.* 47:933–965.

Harold, F. M., 1978. Vectorial metabolism. *In* Ornston, L. N., and Sokatch, J. R., (eds.), *The Bacteria,* vol. 6, pp. 463–521. Academic Press.

Saier, M. H., Jr., 1979. The role of the cell surface in regulating the internal environment. *In* Sokatch, J. R., and Ornston, L. N., (eds.), *The Bacteria,* vol. 7, pp. 167–227, Academic Press.

Estes, J. W., and White, P. D., 1965. William Withering and the purple foxglove. *Sci. Amer.* 212(6):110–117. [An interesting historical account of digitalis.]

BOOKS

Andreoli, T. E., Hoffman, J. F., and Fanestil, D. D., (eds.), 1978. *Physiology of Membrane Disorders.* Plenum. [Contains excellent articles on many aspects of membrane transport.]

Rosen, B. P., (ed.), 1978. *Bacterial Transport.* Dekker.

TRANSPORT OF Na⁺ AND K⁺ IONS

Skou, J. C., and Norby, J. G., (eds.), 1979. *Na, K-ATPase: Structure and Kinetics.* Academic Press.

Fishman, M. C., 1979. Endogenous digitalis-like activity in mammalian brain. *Proc. Nat. Acad. Sci.* 76:4661–4663.

Sweadner, K. J., and Goldin, S. M., 1980. Active transport of sodium and potassium ions: mechanism, function, and regulation. *New Engl. J. Med.* 302:777–783.

Cantley, L. C., Jr., Resh, M. D., and Guidotti, G., 1978. Vanadate inhibits the red cell-(Na^+, K^+)-ATPase from the cytoplasmic side. *Nature* 272:552–554.

Akera, T., 1977. Membrane adenosinetriphosphatase: a digitalis receptor? *Science* 198:569–574. [Review of evidence showing that the Na^+-K^+ ATPase is a target for the pharmacologic effect of digitalis.]

Jardetzky, O., 1966. Simple allosteric model for membrane pumps. *Nature* 211:969.

TRANSPORT OF CALCIUM IONS

deMeis, L., and Vianna, A. L., 1979. Energy interconversion by the Ca^{2+}-dependent ATPase of the sarcoplasmic reticulum. *Ann. Rev. Biochem.* 48:275–292.

MacLennan, D. H., and Campbell, K. P., 1979. Structure, function, and biosynthesis of sarcoplasmic reticulum proteins. *Trends Biochem. Sci.* 4:148–151.

Tada, M., Yamamoto, T., and Tonomura, Y., 1978. Molecular mechanism of active calcium transport by sarcoplasmic reticulum. *Physiol. Rev.* 58:1–79.

Racker, E., 1972. Reconstitution of a calcium pump with phospholipids and a purified Ca^{2+}-adenosine triphosphatase from sarcoplasmic reticulum. *J. Biol. Chem.* 247:8198–8200.

Wasserman, R. H., Fullmer, C. S., and Taylor, A. N., 1978. The vitamin D-dependent calcium binding proteins. *In* Lawson, D. E. M., (ed.), *Vitamin D,* Academic Press.

Kretsinger, R. H., 1976. Calcium-binding proteins. *Ann. Rev. Biochem.* 45:239–266.

BACTERIAL TRANSPORT SYSTEMS

Kaback, H. R., Ramos, S., Robertson, D. E., Stroobant, P., and Tokuda, H., 1977. Energetics and molecular biology of active transport in bacterial membrane vesicles. *J. Supramol. Struc.* 7:443–461.

Saier, M. H., Jr., 1977. Bacterial phosphoenolpyruvate-sugar phosphotransferase systems: structural, functional, and evolutionary interrelationships. *Bacteriol. Rev.* 41:856–871.

Simoni, R. D., and Postma, P. W., 1975. The energetics of bacterial active transport. *Ann. Rev. Biochem.* 44:523–554.

CHANNELS BETWEEN CELLS

Loewenstein, W. R., Kanno, Y., and Socolar, S. J., 1978. The cell-to-cell channel. *Fed. Proc.* 37:2645–2650.

Unwin, P. N. T., and Zampighi, G., 1980. Structure of the junction between communicating cells. *Nature* 283:545–549.

Hertzberg, E. L., and Gilula, N. B., 1979. Isolation and characterization of gap junctions from rat liver. *J. Biol. Chem.* 254:2138–2147.

Staehelin, L. A., and Hull, B. E., 1978. Junctions between living cells. *Sci. Amer.* 238(5):140–152. [Available as *Sci. Amer.* Offprint 1388.]

TRANSPORT ANTIBIOTICS

Urban, B. W., Hladky, S. B., and Haydon, D. A., 1978. The kinetics of ion movements in the gramicidin channel. *Fed. Proc.* 37:2628–2632.

Ovchinnikov, Y. A., 1979. Physico-chemical basis of ion transport through biological membranes: ionophores and ion channels. *Eur. J. Biochem.* 94:321–336.

Laüger, P., 1972. Carrier-mediated ion transport. *Science* 178:24–30.

Krasne, S., Eisenman, G., and Szabo, G., 1971. Freezing and melting of lipid bilayers and the mode of action of nonactin, valinomycin, and gramicidin. *Science* 174:412–415.

Koeppe, R. E., Berg, J. M., Hodgson, K. O., and Stryer, L., 1979. Gramicidin A crystals contain two cation binding sites per channel. *Nature* 279:723–725.

CHAPTER 37

EXCITABLE MEMBRANES
AND SENSORY SYSTEMS

Many cell membranes can be excited by specific chemical or physical stimuli. The responses of a nerve-axon membrane to an electrical stimulus, of a synapse to a transmitter substance, of a retinal rod cell to light, and of motile cells to attractant molecules are mediated by excitable assemblies in membranes. The common features of these processes and others carried out by excitable assemblies are:

1. The stimulus is detected by a highly specific protein receptor, which is an *integral component of the excitable membrane.*

2. The specific stimulus elicits a *conformational change* in the receptor. As a result, the permeability of the membrane or the activity of a membrane-bound enzyme changes. Many of the responses are *highly amplified.*

3. The conformational change in the receptor and the resulting alterations in function are *reversible.* There are mechanisms that take the receptor back to its resting state and restore its excitability.

This chapter deals with four types of excitable assemblies. The sodium channel in nerve-axon membranes will be considered first. This voltage-sensitive gate participates in the action potential of nerves. We will then turn to a channel that is chemically regulated. The acetylcholine receptor enables nerve impulses to be transmitted at certain synapses. The focus will then shift to the retinal rod cell, an exquisitely sensitive detector of light. The role of rhodopsin, the photoreceptor protein, in transducing light into a nerve signal

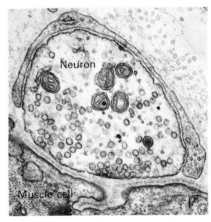

Figure 37-1
Electron micrograph of a synapse.
[Courtesy of Dr. U. Jack McMahan.]

will be discussed. The final topic of this chapter is chemotaxis, the movement of cells toward attractant molecules and away from repellents. Much has been learned in recent years about the coupling of chemoreceptor molecules to the motor apparatus of bacteria.

ACTION POTENTIALS ARE MEDIATED BY TRANSIENT CHANGES IN Na⁺ AND K⁺ PERMEABILITY

Nerve impulses are electrical signals produced by the flow of ions across the plasma membrane of neurons. The interior of a neuron, like those of most other cells, has a high concentration of K^+ and a low one of Na^+. These ionic gradients are generated by the Na^+-K^+ ATPase pump (p. 862). In the resting state, the nerve-axon membrane is much more permeable to K^+ than to Na^+, and so the membrane potential is largely determined by the ratio of the internal to the external concentration of K^+ (Figure 37-2A). The membrane potential of -60 mV in unstimulated axons is near the -75-mV level (the K^+ equilibrium potential) that would be given by a membrane permeable to only K^+. A nerve impulse or *action potential* is generated when the membrane potential is depolarized beyond a critical threshold value (i.e., from -60 to -40 mV). The membrane potential becomes positive within about a millisecond and attains a value of about $+30$ mV before turning negative again. This amplified depolarization is propagated along the nerve terminal. The giant axons of squids played an important role in the discovery of the nature of the action potential. Electrodes can be readily inserted into these unusually large axons, which have a diameter of about a millimeter, and so squid axons have been favorite objects of inquiry.

What is the mechanism of the action potential? Alan Hodgkin and Andrew Huxley carried out ingenious studies that revealed that *the action potential arises from large, transient changes in the permeability of the axon membrane to Na^+ and K^+ ions* (Figure 37-2B). The conductance of the membrane to Na^+ changes first. Depolarization of the membrane beyond the threshold level leads to an opening of Na^+ channels. Sodium ions begin to flow into the cell because of the large electrochemical gradient across the plasma membrane. The entry of Na^+ further depolarizes the membrane, and so more gates for Na^+ are opened. This positive feedback between depolarization and Na^+ entry leads to a very rapid and large change in membrane potential, from about -60 mV to $+30$ mV in a millisecond. The entry of Na^+ stops at about $+30$ mV because this is the Na^+ equilibrium potential. In other words, the thermodynamic driving force for the entry of Na^+ vanishes when this potential is attained. The Na^+ channels spontaneously close and the K^+ gates begin to open at about this time (Figure 37-2B). Consequently, potassium ions flow inward and so the membrane potential returns to a negative value. At about 2 msec, the membrane potential is -75 mV, the

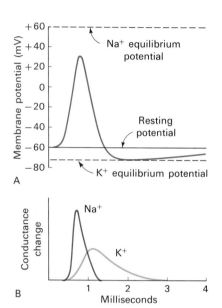

Figure 37-2
Depolarization of a nerve-axon membrane results in an action potential. Time course of (A) the change in membrane potential and (B) the change in Na^+ and K^+ conductances.

K+ equilibrium potential. The resting level of -60 mV is restored in a few milliseconds as the K+ conductance decreases to the value characteristic of the unstimulated state. It is important to stress that a very small proportion of the sodium and potassium ions in a nerve cell, of the order of one in a million, flows across the plasma membrane during the action potential. In other words, only a tiny fraction of the Na+-K+ gradient is dissipated by a single nerve impulse. Clearly, the action potential is a very efficient means of signalling over large distances.

The Na+ channel favors the passage of Na+ over K+ by a factor of 11. How is this selectivity achieved? A definitive answer to this question will have to await a high-resolution analysis of the structure of the channel. However, electrophysiological studies of the relative permeabilities of alkali cations and organic cations provide some clues. The dependence of permeability on ionic size (Table 37-1) indicates that the channel is narrow. Ions having diameters greater than 5 Å are excluded. However, conductance is not governed solely by size. For example, methylamine ($H_3CNH_3^+$) has nearly the same dimensions as hydrazine ($H_2NNH_3^+$) and hydroxylamine ($HONH_3^+$), yet it is much less permeable. A likely reason for this difference is that the methyl group of methylamine, in contrast with the amino group of hydrazine or the hydroxyl group of hydroxylamine, cannot form a hydrogen bond with an oxygen atom in the channel. Hence, methylamine does not fit. Another significant finding is that the conductance of the Na+ channel to all permeant cations decreases markedly when the pH is lowered. In fact, the relative permeability follows a titration curve for an acid with a pK of 5.2, which suggests that the active form of the channel contains a negatively charged carboxylate group. Thus, *the Na+ channel selects for Na+ by providing a negatively charged site with a small radius.* A K+ ion cannot readily pass through this region because it is larger than Na+ (Figure 37-3).

Table 37-1
Relative permeabilities of the sodium and potassium channels in axon membranes

Ion	Na+ channel	K+ channel
Li+	0.93	<0.01
Na+	1.00	<0.01
K+	0.09	1.00
Rb+	<0.01	0.91
Cs+	<0.01	<0.08
NH_4^+	0.16	0.13
$OHNH_3^+$	0.94	<0.03
$H_2NNH_3^+$	0.59	<0.03
$H_3CNH_3^+$	<0.01	<0.02

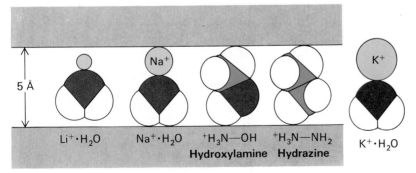

Figure 37-3
The ionic selectivity of the sodium channel partly depends on steric factors. Na+ and Li+, together with a water molecule, fit in the channel, as do hydroxylamine and hydrazine. In contrast, K+ with a water molecule is too large. [After R. D. Keynes. Ion-channels in the nerve-cell membrane. © 1979 by Scientific American, Inc. All rights reserved.]

TETRODOTOXIN AND SAXITOXIN BLOCK SODIUM CHANNELS IN NERVE-AXON MEMBRANES

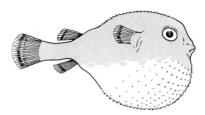

A puffer fish, regarded as a culinary delicacy in Japan.

Tetrodotoxin, a highly potent poison from the puffer (fugu) fish, blocks the conduction of nerve impulses along axons and in excitable membranes of nerve fibers, which leads to respiratory paralysis. The lethal dose for a mouse is about 0.01 μg. Tetrodotoxin is a very useful probe because it is highly specific. It binds very tightly ($K \sim 10^{-9}$ M) to the Na^+ channel and blocks the flow of sodium ions but has no effect on the K^+ channel. *Saxitoxin,* which is produced by a marine dinoflagellate, acts in the same way. Shellfish, particularly clams and mussels, feeding on dinoflagellates are poisonous. Indeed, a small mussel may contain enough saxitoxin to kill fifty humans! A common feature of tetrodotoxin and saxitoxin is the presence of a guanido group (Figure 37-4). This positively

Figure 37-4

Poisons of the Na^+ channel.

Tetrodotoxin   **Saxitoxin**

charged group of the toxin interacts with a negatively charged carboxylate at the mouth of the channel on the extracellular side of the membrane. In effect, these toxins are competitive inhibitors of Na^+.

Tetrodotoxin and saxitoxin are very useful probes because of their specificity and high affinity for the Na^+ channel. For example, the densities of Na^+ channels in a variety of excitable membranes have been determined by measuring the binding of highly radioactive tetrodotoxin. *Unmyelinated nerve fibers, which are devoid of an insulating sheath of myelin, typically have low densities of Na^+ channels, of the order of twenty per square micrometer.* Na^+ channels in these axonal membranes are about 2000 Å apart. *In contrast, myelinated nerve fibers have a very high density of channels, of the order of 10^4 per μm^2, in specialized regions called nodes of Ranvier.* These nodes, spaced at intervals of 2 mm, are the only sites at which the axonal membrane of a myelinated nerve is exposed to the extracellular fluid. The axonal membrane between nodes has a very low density of channels and does not participate in conduction. Rather, the action potential jumps from node to node, and so the impulse is transmitted more rapidly and efficiently than in an unmyelinated fiber. The presence of 10^4 channels per μm^2 at a node of Ranvier means that much of the membrane area in this region is occupied by Na^+ channels.

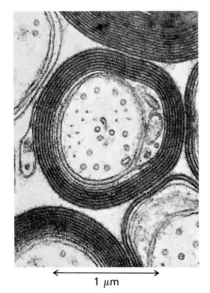

Figure 37-5

Electron micrograph of a myelinated axon from the spinal cord. Myelin, a membranous wrapping around the axon, acts as an insulating layer. Myelinated nerves have a much higher conduction velocity than unmyelinated ones of the same diameter. [Courtesy of Dr. Cedric Raine.]

The specific binding of tetrodotoxin has also been exploited in the purification of the Na^+ channel. Integral membrane proteins of excitable membranes have been solubilized by the addition of a detergent and separated by ion-exchange chromatography. The binding of tetrodotoxin to the solubilized Na^+ channel made it possible to assay for this protein during its purification. The Na^+ channel isolated in this way is a 230-kdal protein consisting of several kinds of subunits. A likely reason for part of the complexity of the channel is that it contains a *voltage-sensing unit* in addition to a highly selective pore. Charged groups in this unit respond to the change in membrane potential during an action potential and convey this information to the pore. In fact, a *gating current* arising from the movement of charged groups in the protein prior to the flow of sodium ions has been detected.

ACETYLCHOLINE IS A NEUROTRANSMITTER

Nerve cells interact with other nerve cells at junctions called *synapses* (Figure 37-6). Nerve impulses are communicated across most synapses by chemical transmitters, which are small, diffusible molecules such as acetylcholine and norepinephrine. Acetylcholine is also the transmitter at motor end plates (neuromuscular junctions), which are the junctions between nerve and striated muscle.

The *presynaptic membrane* of a *cholinergic synapse*—that is, one that uses acetylcholine as the neurotransmitter—is separated from the *postsynaptic membrane* by a gap of about 500 Å, called the *synaptic cleft.* The end of the presynaptic axon is filled with *synaptic vesicles* containing acetylcholine. The arrival of a nerve impulse leads to the release of acetylcholine into the cleft. The acetylcholine molecules then diffuse to the postsynaptic membrane, where they combine with specific receptor molecules. This produces a *depolarization of the postsynaptic membrane,* which is propagated along the electrically excitable membrane of the second nerve cell. Acetylcholine is *hydrolyzed by acetylcholinesterase* and the polarization of the postsynaptic membrane is restored.

Acetylcholine is synthesized near the presynaptic end of axons by the transfer of an acetyl group from acetyl CoA to choline. This reaction is catalyzed by *choline acetyltransferase* (choline acetylase). Some of the acetylcholine is then taken up by synaptic vesicles, whereas the remainder stays in the cytosol. A cholinergic synaptic vesicle, which is typically 400 Å in diameter, contains about 10,000 acetylcholine molecules. The study of synaptic function has been facilitated by the isolation of *synaptosomes* from homogenates of nerve tissue. Synaptosomes are presynaptic endings that have resealed after being pinched off. They consist of mitochondria, cytosol, and synaptic vesicles surrounded by an intact presynaptic membrane.

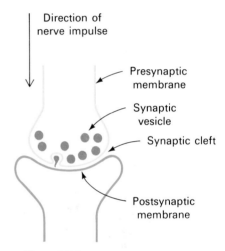

Figure 37-6
Schematic diagram of a cholinergic synapse.

ACETYLCHOLINE OPENS CATION GATES IN THE POSTSYNAPTIC MEMBRANE

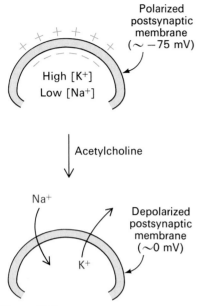

Figure 37-7
Acetylcholine depolarizes the postsynaptic membrane by increasing the conductance of Na^+ and K^+.

The resting potential of a postsynaptic membrane or of a motor end plate is about -75 mV. The interaction of acetylcholine with specific receptors produces a large change in the permeability properties of these membranes (Figure 37-7). *The conductance of both Na^+ and K^+ increases markedly within 0.1 msec, and there is a large inward current of Na^+ and a smaller outward current of K^+.* The inward Na^+ current depolarizes the postsynaptic membrane and triggers an action potential in the adjacent axon or muscle membrane. *Acetylcholine opens a single kind of cation channel, which is almost equally permeable to Na^+ and K^+.* The flow of Na^+ is larger than the flow of K^+ because the electrochemical gradient across the membrane is steeper for Na^+.

Two molecules of acetylcholine bind to a receptor and produce a conformational change that opens a channel. A kinetic scheme that is consistent with experimental data is

$$2\,A + R \rightleftharpoons A_2R \rightleftharpoons A_2R^*$$

in which A is an acetylcholine molecule, R is a closed channel, and R^* is an open one. About 10^4 ions flow through a single open channel during its millisecond lifetime. Prolonged exposure of the receptor to acetylcholine results in its *desensitization*—that is, the channel closes and does not respond to acetylcholine for a long interval.

ACETYLCHOLINE IS RELEASED IN PACKETS

Studies of nerve-impulse transmission at neuromuscular junctions by Bernard Katz have revealed that acetylcholine is released from the presynaptic membrane in the form of packets containing of the order of 10^4 molecules. The evidence for *quantal release of acetylcholine* comes from an analysis of the membrane potential of the motor end plate, which shows spontaneous electrical activity even when the associated nerve is not stimulated. Depolarizing pulses that are 0.5 mV in amplitude and last about 20 msec occur intermittently. These *miniature end-plate potentials* occur randomly with a probability that stays constant for long periods. A miniature end-plate potential is caused by the spontaneous release of a single synaptic vesicle. The full depolarization of the end plate elicited by an action potential results from the synchronous release of about a hundred packets of acetylcholine in less than a millisecond. *The number of packets released depends on the potential of the presynaptic membrane.* In other words, *the release of acetylcholine is an electrically controlled form of secretion.* Release of acetylcholine depends on the presence of Ca^{2+} in the extracellular fluid. The depolarization of the presynaptic membrane leads to the entry of Ca^{2+}, which promotes a transient fusion of the synaptic vesicle membrane and the presynaptic membrane.

Considerable progress has recently been made in the purification of the acetylcholine receptor and in the reconstitution of functional membrane vesicles. A choice starting material is the *electric organ* of an electric fish such as *Torpedo*, which is very rich in cholinergic postsynaptic membranes. Electric organs are made of columns of cells called *electroplaxes*. One side of the cell (called the innervated face) receives a nerve ending and is electrically excitable. The other side, called the noninnervated face, is highly folded and not electrically excitable. The voltage produced by a stimulated electroplax results from the asymmetry of the responses of its two faces. In an electric fish such as *Electrophorus*, the membrane potential of the innervated side changes from -90 to $+60$ mV on excitation, whereas the noninnervated side stays at -90 mV. Consequently, *the potential between the outside of the two faces is 150 mV at the peak of the*

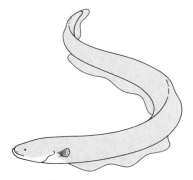

Electrophorus electricus, the electric eel.

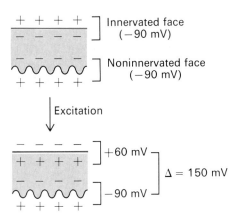

Figure 37-8
Generation of a voltage by an electroplax cell of the electric eel.

action potential (Figure 37-8). Electroplaxes in an electric organ are arranged in series so that their voltages are additive. An organ containing 5,000 rows of electroplaxes can therefore generate a 750-V discharge. It is interesting to note that the electroplaxes of electric eels have evolved from muscle cells. They retain the electrically excitable outer membrane of muscle but have lost the contractile apparatus. The electric organ of *Electrophorus* is an excellent source of sodium channels.

Another exotic biological material has been invaluable in the isolation of acetylcholine receptors. The receptor must be *specifically labeled* so that it can be unequivocally identified in a mixture of macromolecules. This has been accomplished by using neurotoxins from snakes, such as *α-bungarotoxin*, which comes from the venom of a Formosan snake, or *cobratoxin*. These neurotoxins block neuromuscular transmission by binding to acetylcholine receptors on motor end plates or on the innervated face of the electroplax in the electric organ. They are small basic proteins (7 kdal). These neurotoxins can be made highly radioactive by iodinating them with ^{125}I or by forming a Schiff base with pyridoxal phosphate and then reducing

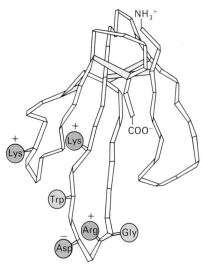

Figure 37-9
Three-dimensional structure of a neurotoxin inhibitor of the acetylcholine receptor. This toxin comes from a sea snake. [Courtesy of Dr. Demetrius Tsernoglou and Dr. Gregory Petsko.]

it with ^{3}H-borohydride. Labeled cobratoxin binds tightly to the acetylcholine receptor with a dissociation constant of 10^{-9} M. Most important, it does not bind appreciably to other macromolecules in the postsynaptic membrane. Thus, the acetylcholine receptor can be specifically marked with a radioactive label.

The acetylcholine receptor of the electric organ has been solubilized by adding a nonionic detergent (such as Tween 80, a derivative of polyoxyethylene) to membrane fragments. The soluble extract was fractionated by gel-filtration chromatography and ion-exchange chromatography. Affinity chromatography on a column containing covalently attached cobratoxin was the final step in achieving a 10,000-fold purification of the receptor. The acetylcholine receptor is a 270-kdal complex of four kinds of subunits. The 40-kdal subunit is affinity labeled by reactive compounds containing a trimethylammonium group, which indicates that it possesses the binding site for acetylcholine. Membrane vesicles containing purified acetylcholine receptor were formed by adding phospholipids to a solution of the receptor and then removing the detergent by dialysis. Radioactive sodium ion (^{22}Na$^+$) trapped in the inner aqueous space of the reconstituted membrane vesicles was released on addition of acetylcholine or analogs such as carbamoylcholine (Figure 37-10). Release was blocked by bungarotoxin and by known antagonists of acetylcholine, which showed that it was mediated by a specific interaction of acetylcholine with its membrane-bound receptor.

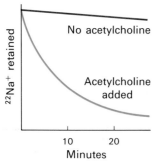

Figure 37-10
Acetylcholine releases Na$^+$ from reconstituted membrane vesicles containing the acetylcholine receptor.

ACETYLCHOLINE IS RAPIDLY HYDROLYZED AND THE END PLATE REPOLARIZES

The depolarizing signal must be switched off to restore the excitability of the postsynaptic membrane. Acetylcholine is hydrolyzed to acetate and choline by *acetylcholinesterase,* an enzyme discovered by David Nachmansohn in 1938. The permeability properties of the postsynaptic membrane are restored, and the membrane becomes repolarized.

$$H_3C-\overset{O}{\underset{\|}{C}}-O-CH_2-CH_2-\overset{+}{N}-(CH_3)_3 + H_2O \rightleftharpoons H_3C-C\overset{O}{\underset{O^-}{\diagdown}} + HO-CH_2-CH_2-\overset{+}{N}-(CH_3)_3 + H^+$$

Acetylcholine **Acetate** **Choline**

Acetylcholinesterase is located in the synaptic cleft, where it is bound to a network of collagen and glycosaminoglycans derived from the postsynaptic cell. This 260-kdal enzyme, which has an $\alpha_2\beta_2$ structure, can be readily separated from the acetylcholine receptor. A striking feature of acetylcholinesterase is its very high turnover number of 25,000 sec^{-1}, which means that it cleaves an acetylcholine molecule in 40 μsec. The high turnover number of the

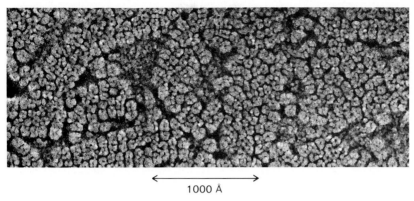

Figure 37-11
Electron micrograph of acetylcholine receptors in a postsynaptic membrane.
[Courtesy of Dr. John Heuser and Dr. Steven Salpeter.]

enzyme is essential for the rapid restoration of the polarized state of the postsynaptic membrane. Synapses can transmit 1,000 impulses per second only if the postsynaptic membrane recovers its polarization within a fraction of a millisecond.

The catalytic mechanism of acetylcholinesterase resembles that of chymotrypsin. Acetylcholine reacts with a specific serine residue at the active site of acetylcholinesterase to form a covalent acetyl-enzyme intermediate, and choline is released. The acetyl-enzyme intermediate then reacts with water to form acetate and regenerate the free enzyme (Figure 37-12).

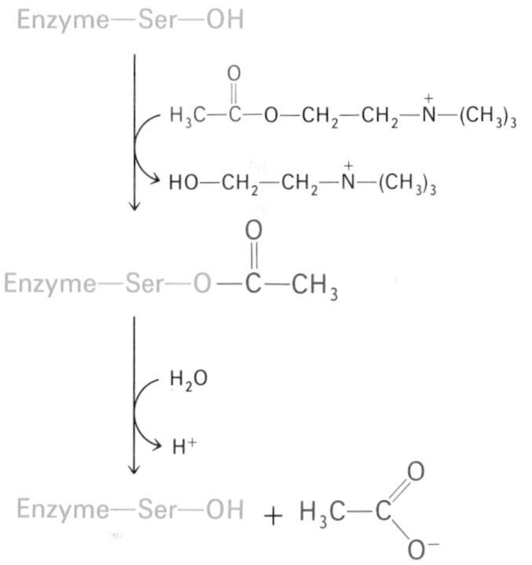

ACETYLCHOLINESTERASE INHIBITORS ARE USED AS DRUGS AND POISONS

The therapeutic and toxic properties of acetylcholinesterase inhibitors are of considerable practical importance. *Physostigmine* (also called eserine) is an alkaloid derived from the Calabar bean, which was once used as an ordeal poison in witchcraft trials. Physostigmine and related inhibitors such as neostigmine are *carbamoyl esters* (Figure 37-13). They inhibit acetylcholinesterase by forming a covalent intermediate that is hydrolyzed very slowly. Neostigmine binds to acetylcholinesterase so that its positively charged trimethylammonium group is at the anionic site of the enzyme, whereas its

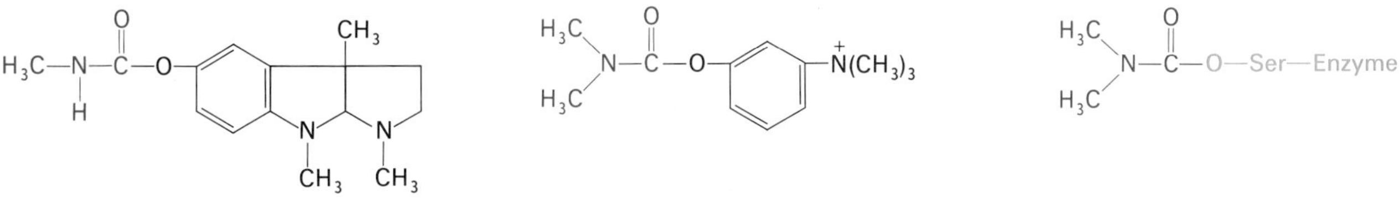

Enzyme—Ser—OH

$H_3C-\overset{O}{\overset{\|}{C}}-O-CH_2-CH_2-\overset{+}{N}-(CH_3)_3$

$HO-CH_2-CH_2-\overset{+}{N}-(CH_3)_3$

Enzyme—Ser—$O-\overset{O}{\overset{\|}{C}}-CH_3$

H_2O

H^+

Enzyme—Ser—OH $+ H_3C-\overset{O}{\overset{\|}{C}}\overset{}{\underset{O^-}{}}$

Figure 37-12
Catalytic mechanism of acetylcholinesterase.

Physostigmine

Neostigmine

Carbamoyl-enzyme complex

Figure 37-13
Physostigmine and neostigmine inhibit acetylcholinesterase by carbamoylating the active-site serine.

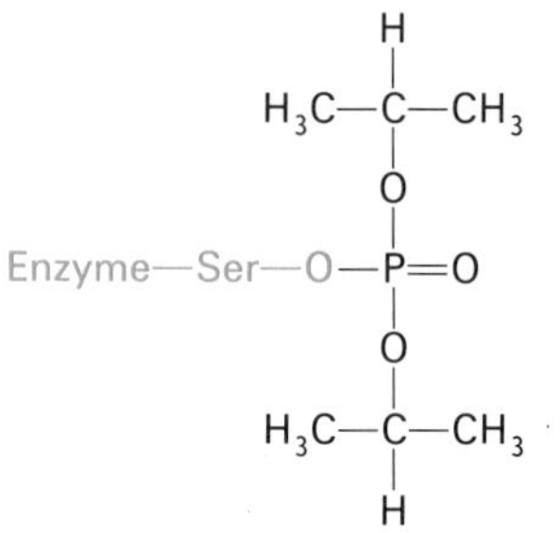

DIPF-inhibited enzyme

Figure 37-14
DIPF-inhibited acetylcholinesterase.

Tabun

Sarin

Parathion

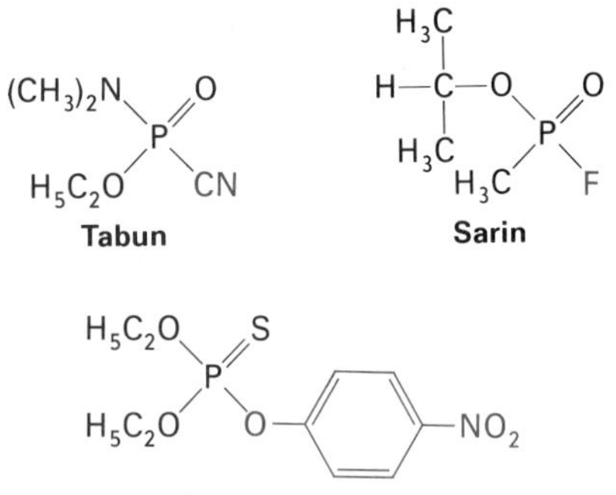

Figure 37-15
Structures of some organic phosphate inhibitors of acetylcholinesterase.

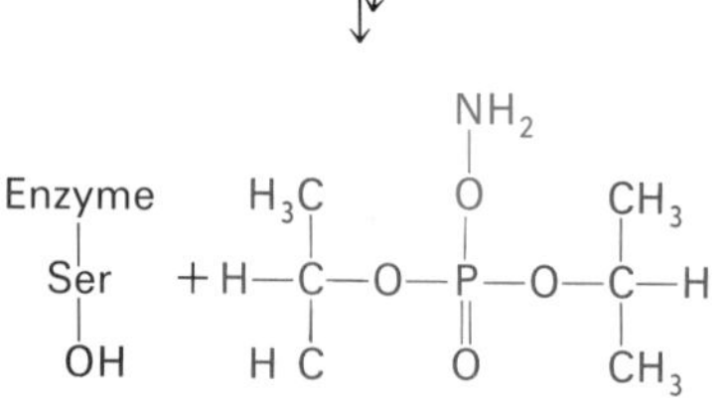

Figure 37-16
Reactivation of DIPF-inhibited acetylcholinesterase by hydroxylamine.

carbamoyl group is adjacent to the reactive serine residue at the esteratic site. The enzyme is then carbamoylated and the alcohol moiety is released. *The carbamoyl-enzyme intermediate is subsequently hydrolyzed at a very slow rate, in contrast with the acetyl-enzyme intermediate.* Thus, the active site of the enzyme is effectively blocked. Neostigmine is used to treat glaucoma, an eye disease characterized by abnormally high intraocular pressure. The therapeutic rationale is that neostigmine inhibits acetylcholinesterase and thereby enhances the effects of acetylcholine.

Even more potent inhibitors are the *organic fluorophosphates,* such as diisopropyl phosphofluoridate (DIPF). These compounds react with acetylcholinesterase to form *very stable covalent phosphoryl-enzyme complexes* (Figure 37-14). The phosphoryl group becomes bonded to the active-site serine, as in serine proteases that have reacted with DIPF.

Many organic phosphate compounds have been synthesized for use as *agricultural insecticides* or as *nerve gases* for chemical warfare (Figure 37-15). These compounds can be lethal by causing respiratory paralysis. Tabun and sarin are among the most toxic of them. Parathion has been widely used as an agricultural insecticide.

The number of acetylcholinesterase molecules in the end plate of mouse-diaphragm muscle has been counted using radioactive DIPF as a label. The density is 12,000 μm^{-2}, which is nearly the same as was found for acetylcholine receptors using radioactive neurotoxins. Thus, *the postsynaptic membrane is very densely packed with both acetylcholinesterase and the acetylcholine receptor.* Only a small fraction of the acetylcholinesterase is required for the transmission of nerve impulses at low frequencies. However, most of the enzyme molecules must be active to sustain transmission at high rates of firing.

A DESIGNED ANTIDOTE FOR ORGANOPHOSPHORUS POISONING

Acetylcholinesterase inhibited by organic phosphates such as DIPF can be reactivated by derivatives of hydroxylamine (NH_2OH). The starting point in the development of an antidote was the finding by Irwin Wilson that hydroxylamine releases the phosphoryl group that is attached to the serine residue in the inhibited enzyme and thereby reactivates it (Figure 37-16). The challenge was to make this reaction therapeutically useful. Hydroxylamine could not be used in vivo because it is toxic at the high concentrations required to reactivate DIPF-inhibited acetylcholinesterase. Wilson's approach was to search for a compound that retains the chemical reactivity of hydroxylamine but has specificity and very high affinity for acetylcholinesterase. The enzyme was known to have an anionic site for the positively charged choline moiety of acetylcholine, and so it seemed worthwhile to synthesize a hydroxylamine derivative containing a quaternary ammonium group. Should the substituent be on the oxygen or nitrogen atom of hydroxylamine? It was found that *O*-methylhydroxylamine was inactive as a reactivator, whereas *N*-methylhydroxylamine was active. The hydroxyl group of the reactivator cannot be substituted because it is the anionic form that attacks the phosphorus atom in the inhibited enzyme.

The next step, a crucial one, was to place the quaternary ammonium group at an appropriate distance from the nucleophilic oxygen atom. Furthermore, the orientation of these groups had to be complementary to that of the anionic site and the phosphorus atom of the inhibited enzyme.

Numerous compounds containing a quaternary ammonium group and a hydroxylamine function were tested as reactivators. The most effective one was *2-pyridine aldoxime methiodide* (PAM). This compound has a well-defined geometry because of the double bond in the oxime, which tends to fix the oxygen atom in the plane of the ring.

PAM reactivates DIPF-inhibited acetylcholinesterase in much the same way as does hydroxylamine. However, 10^{-6} M PAM is as effective as 1 M hydroxylamine. *This millionfold enhancement makes PAM therapeutically useful in the treatment of organophosphorus poisoning.* The remarkable effectiveness of PAM arises from the optimal positioning of the quaternary ammonium group in relation to the nucleophilic oxygen atom. The development of PAM is a landmark in the rational design of drugs.

Pyridine aldoxime methiodide (PAM)

INHIBITORS OF THE ACETYLCHOLINE RECEPTOR

Neuromuscular transmission can also be impaired by *compounds that act directly on the acetylcholine receptor. Curare* has been used for centuries by South American Indians. Soon after Columbus returned, d'Anghera wrote *De Orbe Novo* in which he noted that "the natives poisoned their arrows with the juice of a death-dealing herb. . . ." One of the active components of curare is d-*tubocurarine* (Figure 37-17). *Tubocurarine inhibits the depolarization of the end plate by competing with acetylcholine for binding to the acetylcholine receptor.* α-Bungarotoxin and cobratoxin have a similar action. In contrast, compounds such as *decamethonium* bind to the receptor and cause a *persistent depolarization* of the end plate.

$(CH_3)_3$—$\overset{+}{N}$—$(CH_2)_{10}$—$\overset{+}{N}$—$(CH_3)_3$
Decamethonium

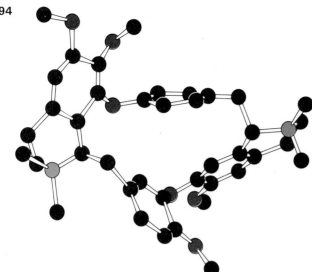

Figure 37-17
Formula and model of *d*-tubocurarine.

Succinylcholine, an analog of acetylcholine, is used to produce muscular relaxation in surgical procedures.

$$(CH_3)_3\overset{+}{-N}-CH_2-CH_2-O-\overset{\overset{O}{\|}}{C}-CH_2-CH_2-\overset{\overset{O}{\|}}{C}-O-CH_2-CH_2-\overset{+}{N}-(CH_3)_3$$
Succinylcholine

Succinylcholine is hydrolyzed very slowly by acetylcholinesterase in the postsynaptic membrane. Hence, it causes a persistent depolarization of the end plate. Succinylcholine is hydrolyzed by less specific cholinesterases in the plasma and in the liver, which are called *plasma cholinesterase* or *pseudocholinesterase* to distinguish them from acetylcholinesterase of the postsynaptic membrane. An attractive feature of succinylcholine is that neuromuscular transmission resumes soon after the infusion of this drug is stopped. In a small proportion of patients, however, muscular relaxation and respiratory paralysis persist for many hours. In such patients, succinylcholine is hydrolyzed at a very slow rate because their plasma cholinesterase has a markedly reduced affinity for succinylcholine. Sensitivity to succinylcholine, like sensitivity to pamaquine (p. 343), is an example of a genetically determined drug idiosyncrasy.

Rabbits exhibit muscle weakness and fatigability several weeks after being immunized with purified acetylcholine receptor. They produce antibodies that react with acetylcholine receptors in their neuromuscular junctions. The resulting decrease in the number of functional acetylcholine receptors leads to defective neuromuscular transmission. The symptoms displayed by these rabbits are very much like those of *myasthenia gravis,* a disease of humans. Indeed, the sera of myasthenic patients contain antibody directed against their own acetylcholine receptors. In other words, myasthenia gravis is an *autoimmune disorder,* in which self becomes the target of the immune system.

CATECHOLAMINES AND γ-AMINOBUTYRATE (GABA) ALSO SERVE AS NEUROTRANSMITTERS

A number of neurotransmitters in addition to acetylcholine have been identified. An established neurotransmitter meets several criteria. First, microinjection of the proposed transmitter into the synaptic cleft must elicit the same response as does excitation of the presynaptic nerve. Second, the presynaptic nerve terminals must be rich in this substance. The isolation of synaptic vesicles containing the putative transmitter is the strongest evidence in this regard. Third, the presynaptic nerve must release the postulated transmitter at the right time and in a quantity sufficient to act on the postsynaptic nerve.

Several catecholamines meet these criteria. For example, norepinephrine is the transmitter at smooth-muscle junctions that are innervated by sympathetic nerve fibers, in contrast with parasympathetic junctions in which acetylcholine is the transmitter. Epinephrine and dopamine are two other catecholamine transmitters. In fact, these catecholamines are synthesized from tyrosine in sympathetic-nerve terminals and in the adrenal gland (Figure 37-18). The first step, which is rate limiting, is the hydroxylation of tyrosine to form 3,4-dihydroxyphenylalanine (*dopa*). This reaction is catalyzed by tyrosine hydroxylase, which is like phenylalanine hydroxylase. Molecular oxygen is activated by tetrahydrobiopterin, the cofactor. The second step is the decarboxylation of dopa by dopa decarboxylase, a pyridoxal phosphate enzyme, to yield 3,4-dihydroxyphenylethylamine (*dopamine*). Dopamine is then hydroxylated to *norepinephrine* by a copper-containing hydroxylase. Finally, *epinephrine* is formed by the methylation of norepinephrine by a transmethylase that utilizes *S*-adenosylmethionine.

Catecholamine neurotransmitters are inactivated by methylation of the 3-hydroxyl group of the catechol ring. This reaction is catalyzed by *catechol-O-methyltransferase,* which uses *S*-adenosylmethionine as the methyl donor. Alternatively, these neurotransmitters can be inactivated by oxidative removal of their amino group by *monoamine oxidase* (Figure 37-19).

Another amino acid derivative has been shown to be a neurotransmitter. γ-*Aminobutyrate* (also called gamma-aminobutyric acid, or GABA) increases the permeability of postsynaptic membranes to

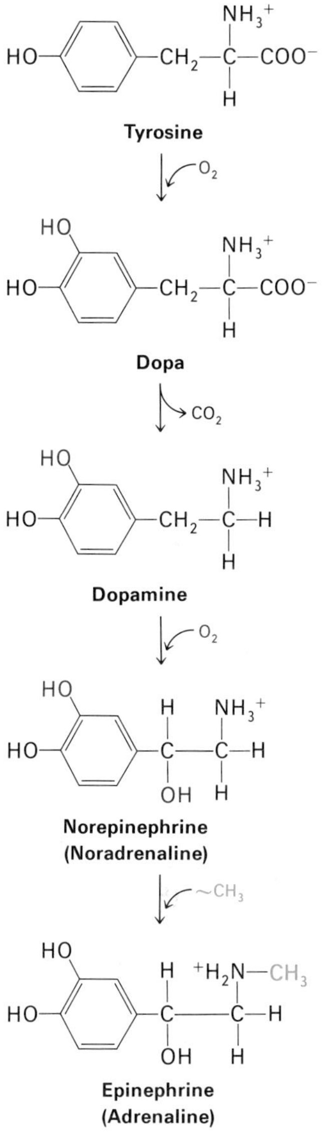

Figure 37-18
Pathway for the synthesis of catecholamine neurotransmitters.

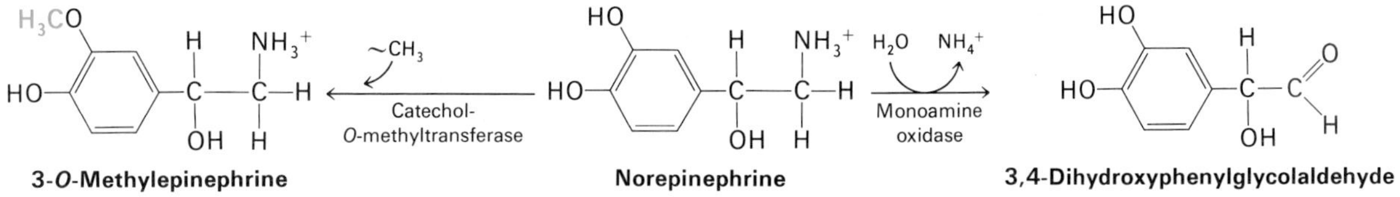

Figure 37-19
Inactivation of norepinephrine.

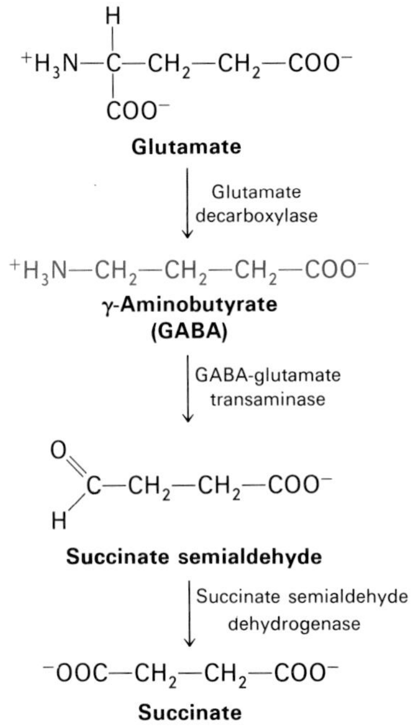

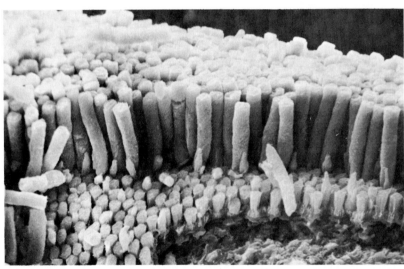

Glutamate

Glutamate decarboxylase

^+H_3N—CH_2—CH_2—CH_2—COO^-

**γ-Aminobutyrate
(GABA)**

GABA-glutamate transaminase

Succinate semialdehyde

Succinate semialdehyde dehydrogenase

^-OOC—CH_2—CH_2—COO^-

Succinate

Figure 37-20
Synthesis and inactivation of γ-aminobutyrate.

K^+. Hence, γ-aminobutyrate drives the membrane potential away from the threshold for triggering the action potential, and so it acts as an inhibitory transmitter. γ-Aminobutyrate is formed by the decarboxylation of glutamate, a reaction catalyzed by glutamate decarboxylase (Figure 37-20). As might be expected, pyridoxal phosphate is the prosthetic group of this decarboxylase. γ-Amino-butyrate is inactivated by transamination to succinate semialde-hyde, which is then oxidized to succinate.

A RETINAL ROD CELL CAN BE EXCITED BY A SINGLE PHOTON

We now turn to excitable receptors that are activated by *light.* There are two kinds of photoreceptor cells in humans, called rods and cones because of their distinctive shapes. *Cones* function in

Figure 37-21
Scanning electron micrograph of retinal rod cells. [Courtesy of Dr. Deric Bownds.]

bright light and are responsible for color vision, whereas *rods* func-tion in dim light but do not perceive color. There are three million cones and a billion rods in a human retina. These photoreceptor cells *convert light into atomic motion and then into a nerve impulse.* Rods and cones form synapses with bipolar cells, which in turn interact with other nerve cells in the retina. The electrical signals generated by the photoreceptors are processed by an intricate array of nerve cells within the retina and then transmitted to the brain by the fibers of the optic nerve. Thus, the retina has a dual function: to transform light into nerve impulses and to integrate visual informa-tion.

In 1938, Selig Hecht discovered that *a human rod cell can be excited by a single photon.* Let us explore the molecular basis of the exquisite

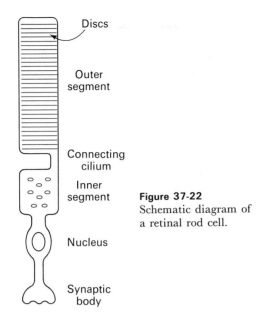

Figure 37-22
Schematic diagram of
a retinal rod cell.

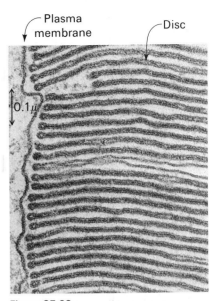

Figure 37-23
Electron micrograph of a rod outer
segment, showing the stack of discs.
[From John E. Dowling. The organiza-
tion of vertebrate visual receptors. In
*Molecular Organization and Biological
Function,* John M. Allen, ed. (Harper &
Row, 1967).]

sensitivity of these cells. Rods are slender, elongated structures, typically 1 μm in diameter and 40 μm in length. The major functions
of a rod cell are compartmentalized in a very distinctive way (Figure 37-22). *The outer segment of a rod is specialized for photoreception.* It
contains a stack of about 1,000 *discs* (Figure 37-23), which are
closed, flattened sacs about 160 Å thick. These membranous structures are densely packed with photoreceptor molecules. The disc
membranes are separate from the plasma membrane of the outer
segment. A slender cilium joins the outer segment and the *inner
segment,* which is rich in mitochondria and ribosomes. The inner
segment generates ATP at a very rapid rate and is highly active in
synthesizing proteins. The discs in the outer segment have a life of
only ten days and are continually renewed. The inner segment is
contiguous with the nucleus, which is next to the *synaptic body.*
Many synaptic vesicles are present in the synaptic body, which
forms a synapse with a bipolar cell.

RHODOPSIN IS THE PHOTORECEPTOR PROTEIN OF RODS

Light must be absorbed to stimulate a photoreceptor cell. In addition, the light-absorbing group (called a *chromophore*) must undergo
a conformational change after it has absorbed a photon. The photosensitive molecule in rods is *rhodopsin,* which consists of *opsin,* a
protein, and *11*-cis-*retinal,* a prosthetic group (Figure 37-24). Rhodopsin is a 38-kdal transmembrane protein. Its amino terminus is
located in the intradiscal aqueous space, whereas its carboxyl terminus is positioned on the other side of the disc membrane, in the
cytosol. The amino-terminal region of rhodopsin contains two oligosaccharide units that are covalently attached to asparagine side
chains. These sugars are probably important in guiding rhodopsin

11-*cis*-Retinal

All-*trans*-retinal

**All-*trans*-retinol
(Vitamin A)**

Figure 37-24
Structure of 11-*cis*-retinal, all-*trans*-retin-
al, and all-*trans*-retinol (Vitamin A).

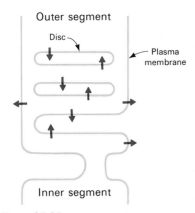

Figure 37-25
Formation of discs by invagination of the plasma membrane. The arrows denote the polarity of rhodopsin molecules.

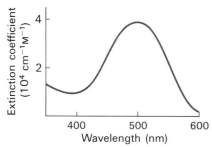

Figure 37-26
Absorption spectrum of rhodopsin.

from the inner segment to the discs. Rhodopsin, like other eucaryotic membrane proteins, is synthesized by ribosomes that are attached to the endoplasmic reticulum. The newly synthesized protein then travels to the Golgi apparatus before reaching the plasma membrane. New discs are formed at the base of the outer segment by invagination of the plasma membrane, which accounts for the fact that the sugar units of rhodopsin are located inside the discs, whereas they face the extracellular space while still on the plasma membrane (Figure 37-25).

Opsin, like other proteins devoid of prosthetic groups, does not absorb visible light. The color of rhodopsin and its responsiveness to light depend on the presence of 11-*cis*-retinal, which is a very effective chromophore. 11-*cis*-Retinal gives rhodopsin a broad absorption band in the visible region of the spectrum with a peak at 500 nm, which nicely matches the solar output. The intensity of the visible absorption band of rhodopsin is also noteworthy. The extinction coefficient of rhodopsin at 500 nm is 40,000 cm^{-1} M^{-1}, a high value (Figure 37-26). The integrated absorption strength of the visible absorption band of rhodopsin approaches the maximum value attainable by organic compounds. 11-*cis*-Retinal has these favorable chromophoric properties because it is a *polyene*. Its six alternating single and double bonds constitute a long, unsaturated electron network.

11-*cis*-Retinal is attached to rhodopsin by a *Schiff-base linkage*. The aldehyde group of 11-*cis*-retinal is linked to the ε-amino group of a specific lysine residue of opsin. The spectral properties of rhodopsin indicate that the Schiff base is protonated.

$$R-\overset{O}{\underset{H}{\overset{\|}{C}}} + H_2N-(CH_2)_4-Opsin \overset{H^+}{\rightleftharpoons} R-\overset{H}{\underset{H}{C}}=\overset{+}{N}-(CH_2)_4-Opsin + H_2O$$

11-*cis*-Retinal   **Lysine side chain**   **Protonated Schiff base**

The precursor of 11-*cis*-retinal is *all*-trans-*retinol* (Vitamin A), which cannot be synthesized de novo by mammals. All-*trans*-retinol (Figure 37-24) is converted into 11-*cis*-retinal in two steps. The alcohol group is oxidized to an aldehyde by *retinol dehydrogenase,* which uses $NADP^+$ as the electron acceptor. Then, the double bond between C-11 and C-12 is isomerized from a *trans* to a *cis* configuration by *retinal isomerase*. A deficiency of Vitamin A leads to *night blindness* and eventually to the deterioration of the outer segments of rods.

LIGHT ISOMERIZES 11-*cis*-RETINAL

The primary event in visual excitation is known. George Wald showed that *light isomerizes the 11-*cis-*retinal group of rhodopsin to all-*trans-*retinal*. This isomerization markedly alters the geometry of

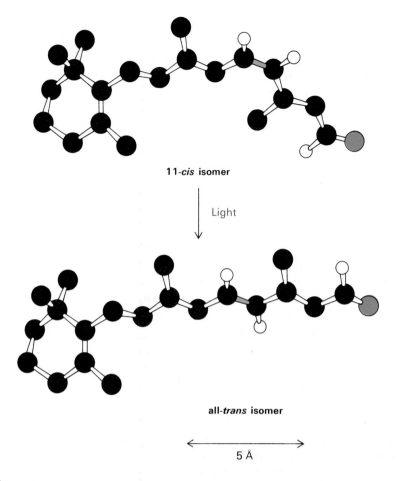

11-*cis* isomer

Light

all-*trans* isomer

5 Å

Figure 37-27
The primary event in visual excitation is the isomerization of the 11-*cis* isomer of the Schiff base of retinal to the all-*trans* form. The C11–C12 double bond is shown in green.

retinal (Figure 37-27). The Schiff-base linkage of retinal moves approximately 5 Å in relation to the ring portion of the chromophore. In essence, *a photon has been converted into atomic motion.*

Much of the isomerization of retinal takes place within a few picoseconds of the absorption of a photon, as shown by the appearance of a new absorption band following an intense laser pulse. This photolytic intermediate, called *bathorhodopsin* (or prelumirhodopsin), contains a strained all-*trans* form of the chromophore. Both retinal and the protein continue to change their conformations, as reflected in the formation of a series of transient intermediates with distinctive spectral properties (Figure 37-28). The Schiff-base linkage becomes deprotonated in the transition from metarhodopsin I to II, which takes about a millisecond. The unprotonated Schiff base in metarhodopsin II is hydrolyzed in about a minute to yield opsin and all-*trans*-retinal, which diffuses away from the protein because it does not fit into the binding site for the 11-*cis* isomer. All-*trans*-retinal is isomerized in the dark to 11-*cis*-retinal, which associates with opsin to regenerate rhodopsin. The hydrolysis of the Schiff base, in contrast with the preceding reactions in the photolysis of rhodopsin, is too slow to be pertinent to the generation of a nerve impluse.

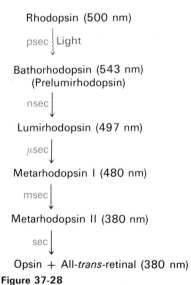

Rhodopsin (500 nm)

psec | Light

Bathorhodopsin (543 nm)
(Prelumirhodopsin)

nsec

Lumirhodopsin (497 nm)

μsec

Metarhodopsin I (480 nm)

msec

Metarhodopsin II (380 nm)

sec

Opsin + All-*trans*-retinal (380 nm)

Figure 37-28
Intermediates in the photolysis of rhodopsin. The wavelength of the absorption maximum of each species and the time constant of each transition are given.

LIGHT HYPERPOLARIZES THE PLASMA MEMBRANE OF THE OUTER SEGMENT

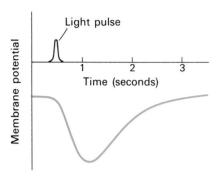

Figure 37-29
Light hyperpolarizes the plasma membrane of a retinal rod cell.

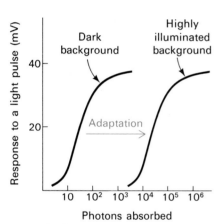

Figure 37-30
The sensitivity of a retinal rod cell to a light pulse depends on the background light level.

The *cis-trans* isomerization of retinal and the consequent conformational changes in rhodopsin are the primary events in visual excitation. An important later step in the generation of a nerve impulse has been elucidated by electrophysiological studies of intact retinas. *The plasma membrane of the outer segment becomes transiently hyperpolarized following a light pulse* (Figure 37-29). The kinetics of hyperpolarization depend on the intensity of the light pulse and on the level of steady background illumination. The response to a single photon takes place in about a second, whereas an intense pulse hyperpolarizes the plasma membrane in a few milliseconds. The rod cell does not have action potentials. Instead, its response to light is graded. The signal sent from the outer segment to the synapse depends on the number of absorbed photons. The hyperpolarization of a fully sensitive dark-adapted rod is half-maximal when only 30 photons are absorbed by an outer segment containing 40×10^6 rhodopsin molecules (Figure 37-30). A single photon absorbed by a dark-adapted rod leads to a hyperpolarization of about 1 mV, which is sensed by the synapse and conveyed to other neurons of the retina. Exquisite sensitivity is not the only remarkable property of the rod cell. Another striking characteristic is that the response of this photodetector to a light pulse depends on the background light level. Many more photons are needed to excite a constantly illuminated rod than one in the dark (Figure 37-30). This property, called *adaptation*, enables the rod cell to perceive contrast over many orders of magnitude of background light intensity.

What is the ionic basis of this light-induced hyperpolarization? The plasma membrane of the rod outer segment is highly permeable to Na^+ in the dark. Sodium ions rapidly flow into the outer segment in the dark because of the existence of a large Na^+ concentration gradient across the plasma membrane. This gradient is maintained by a Na^+-K^+ ATPase pump located in the plasma membrane of the inner segment. Thus, in the dark, sodium ions flow into the outer segment, diffuse to the inner segment, and are then extruded by an ATP-driven pump. *Light somehow blocks the Na^+ channels in the plasma membrane of the outer segment.* Consequently, the inward flow of Na^+ decreases, and so the membrane becomes more negative on the inside. In other words, the membrane potential of an illuminated rod shifts towards the K^+ equilibrium potential. This light-induced hyperpolarization in the vicinity of the illuminated discs is then passively transmitted by the plasma membrane to the synaptic body.

The change in permeability of the plasma membrane to Na^+ and the consequent hyperpolarization are *highly amplified responses* of the outer segment. The flow of more than a *million* sodium ions is blocked by the absorption of a *single* photon by a dark-adapted rod. What is the mechanism of this remarkable amplification? The first point to note is that the disc membranes, which contain most of the rhodopsin molecules, are not continuous with the plasma membrane of the rod cell. Also, these membranes are not coupled electrically. Furthermore, a rhodopsin molecule that absorbs a photon may be several thousand angstroms away from sodium channels in the plasma membrane, and so a direct interaction between these sites is clearly not feasible. Rather, the signal is almost certainly carried from photolyzed rhodopsins in disc membranes to the plasma membrane by diffusible transmitters. Indeed, a large number of transmitters must be formed (or destroyed) by the action of a single photolyzed rhodopsin to account for the high degree of amplification that is observed.

The identity of the transmitter is still a mystery, but the search is closing in on two attractive candidates: calcium ion (Figure 37-31) and cyclic GMP (Figure 37-32). *The experimental findings pointing to*

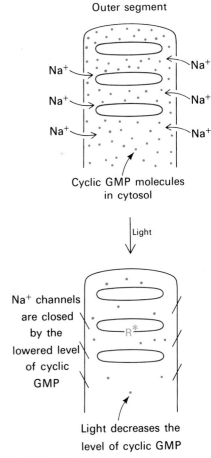

Cyclic GMP molecules
in cytosol

Light

Na^+ channels are closed by the lowered level of cyclic GMP

R^*

Light decreases the level of cyclic GMP

Figure 37-32
Schematic diagram depicting the hypothesis that cyclic GMP is the transmitter in visual excitation.

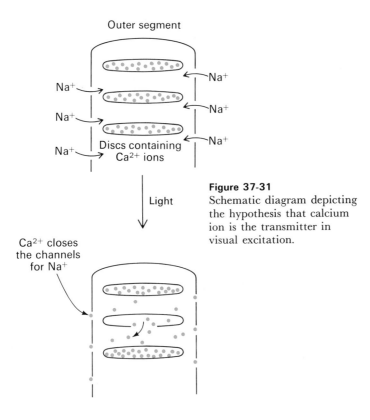

Figure 37-31
Schematic diagram depicting the hypothesis that calcium ion is the transmitter in visual excitation.

Ca^{2+} as a transmitter are:

1. Sodium channels in the plasma membrane close when the level of Ca^{2+} in the cytosol is raised, and they open when the level of Ca^{2+} is lowered.

2. Chelating agents that specifically bind Ca^{2+} lower the sensitivity of the rod cell to light when they are introduced into the cytosol. This desensitization suggests that photolysis of a single rhodopsin leads to the release of several hundred Ca^{2+} into the cytosol.

3. Many Ca^{2+} are extruded from an illuminated rod outer segment following a light pulse.

Other experiments show that cyclic GMP may be a transmitter. The key observations are:

1. Sodium channels in the plasma membrane open when the level of cyclic GMP in the cytosol is raised, and they close when the level of this nucleotide is lowered.

2. The level of cyclic GMP is regulated by light. Specifically, light activates a phosphodiesterase that hydrolyzes cyclic GMP, as will be described shortly.

3. Photolysis of a single rhodopsin molecule leads to the rapid hydrolysis of more than 10^5 molecules of cyclic GMP.

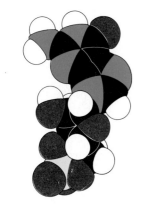

Figure 37-33
Molecular model of cyclic GMP.

LIGHT LOWERS THE CYCLIC GMP LEVEL BY ACTIVATING A PHOSPHODIESTERASE

These experimental findings show that the excitability of rods is profoundly affected by both Ca^{2+} and cyclic GMP. The interplay of these species may be critical in visual excitation. Little is yet known about the molecular mechanism of the light-induced release of Ca^{2+} into the cytosol. However, considerable progress has recently been made in determining how light controls the level of cyclic GMP in rod outer segments. *Guanyl cyclase,* the enzyme that catalyzes the synthesis of cyclic GMP, does not seem to be markedly influenced by light.

$$GTP \xrightarrow{\text{guanyl cyclase}} \text{cyclic GMP} + PP_i$$

In contrast, the phosphodiesterase that hydrolyzes cyclic GMP is controlled by light in a striking way.

$$\text{Cyclic GMP} + H_2O \xrightarrow{\text{phosphodiesterase}} GMP + H^+$$

The activity of this phosphodiesterase increases several hundred times on illumination. The stimulation of this enzyme by photolyzed rhodopsin is mediated by a regulatory protein called *transducin*. In the dark, transducin contains a tightly bound GDP molecule. On illumination, photolyzed rhodopsin forms a complex with GDP-transducin and catalyzes the exchange of GTP for GDP (Figure 37-34). The resulting complex of GTP-transducin activates the phosphodiesterase. *The important point is that a single photolyzed rhodopsin catalyzes the exchange of GTP for GDP on several hundred molecules of transducin, which in turn activate hundreds of molecules of the phosphodiesterase.* Each phosphodiesterase has a turnover number of about 10^3 sec^{-1}, and so more than 10^5 cyclic GMP are hydrolyzed per second per photolyzed rhodopsin. This system is restored to the dark state by a built-in GTPase activity. GTP bound to transducin is slowly hydrolyzed to yield GDP-transducin, which no longer activates the phosphodiesterase. Thus, the free energy that drives this amplification cycle comes from the hydrolysis of GTP. We see here the use of ∼ P to amplify a signal. This cyclic nucleotide cascade in vision (Figure 37-35) is reminiscent of the one that mediates the action of β-adrenergic hormones such as epinephrine (p. 843).

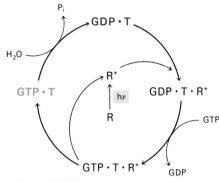

Figure 37-34
Photolyzed rhodopsin catalyzes the formation of the GTP complex of transducin, which in turn activates the cyclic GMP phosphodiesterase. [After B. K.-K. Fung and L. Stryer, *Proc. Nat. Acad. Sci.* 77(1980):2503.]

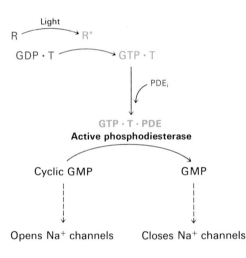

Figure 37-35
Proposed cyclic nucleotide cascade in vision. Abbreviations used are: R, rhodopsin; R*, photolyzed rhodopsin; T, transducin; PDE$_i$, inactive phosphodiesterase; PDE, active phosphodiesterase.

COLOR VISION IS MEDIATED BY THREE KINDS OF PHOTORECEPTORS

In 1802, Thomas Young proposed that color vision is mediated by *three fundamental receptors*. Spectrophotometric studies of intact retinas more than a century and a half later have revealed that there are in fact *three types of cone cells: blue-, green-, and red-absorbing.* The absorption spectra of these three photoreceptor pigments have been obtained by illuminating cones with a beam of light having a diameter of 1 μm (Figure 37-36). Furthermore, microelectrodes have been inserted into many cone cells. The action spectra for the hyperpolarization of their plasma membranes fall into three groups with maxima in the blue, green, and red regions of the visible

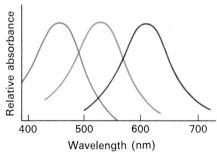

Figure 37-36
Absorption spectra of the three color receptors.

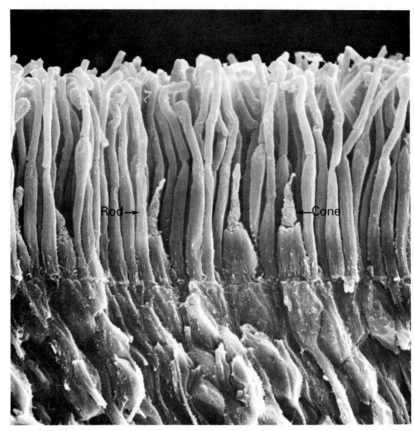

Figure 37-37
Scanning electron micrograph of rod and cone cells in the photoreceptor layer of the retina. [Courtesy of Dr. William Miller.]

spectrum. In goldfish, the absorption maxima of the three color receptors are at 455, 530, and 625 nm, whereas that of rhodopsin is at 500 nm.

The chromophore in all three kinds of cones is 11-cis-retinal. The protonated Schiff base of 11-*cis*-retinal in the absence of a protein has an absorption maximum at 380 nm. Thus, groups on opsin have a large effect on the chromophoric properties of the bound 11-*cis*-retinal. The dependence of the absorption properties of this chromophore on its protein environment exemplifies a general principle: *the properties of a prosthetic group are modulated by its interaction with the protein.* Another example is that the heme group functions as an oxygen carrier in hemoglobin, an electron carrier in cytochrome *c*, and a catalyst in peroxidase.

Most forms of color blindness are caused by a sex-linked recessive mutation. About 1% of men are red-blind, and 2% are green-blind. Spectral measurements of the intact eye have shown that these people lack either the red-absorbing or the green-absorbing photoreceptor molecule or they have an altered pigment with a shifted absorption spectrum. Thus, *color blindness is caused by an absence or defect of one kind of cone opsin.*

11-*cis*-RETINAL IS THE CHROMOPHORE IN ALL KNOWN VISUAL SYSTEMS

Image-resolving eyes are found in only three phyla: molluscs, arthropods, and vertebrates. The three kinds of eyes are anatomically quite different and are thought to have evolved independently. However, all three use 11-*cis*-retinal as the chromophore in their photoreceptor molecules. *This is a striking example of convergent evolution.* What is so special about 11-*cis*-retinal? First, it has an intense absorption band that can be readily shifted into the visible region of the spectrum. Second, light isomerizes 11-*cis*-retinal with high efficiency. Moreover, its rate of isomerization in the dark is very low. Third, the structural change produced by isomerization is large. Light is converted into atomic motion of sufficient magnitude to trigger the generation of a nerve impulse. Finally, the ultimate precursors of 11-*cis*-retinal are the carotenes (p. 480), which have a very broad biological distribution.

CHEMORECEPTORS ON BACTERIA DETECT SPECIFIC MOLECULES AND SEND SIGNALS TO FLAGELLA

In the late nineteenth century, Wilhelm Pfeffer, a German botanist, showed that motile bacteria cluster near the mouth of a thin capillary tube containing an attractant such as a sugar (Figure 37-38). In contrast, they move away from the tube if it contains a repellent, which is usually a harmful substance or a bacterial excretory product. This directed movement of bacteria toward some specific substances and away from others is called *chemotaxis.* In the 1960s, Julius Adler began to investigate the molecular basis of bacterial chemotaxis. Biochemical, genetic, and structural analyses carried out by him and numerous other investigators have revealed much about this process. Chemotaxis begins with the detection of chemicals by specific *chemoreceptors* on the cell surface. Information from these sensors is transmitted to a *processing system* that analyzes and integrates many stimuli. This sensory processing system then sends signals to the *motors* that drive the *flagella.* These signals determine whether a bacterium continues to swim smoothly in a straight line or abruptly alters its course.

About twenty different chemoreceptors have been identified in *E. coli.* Each of these proteins is located either in the plasma membrane or in the periplasmic space. A chemoreceptor has a recognition component and a signaling component. The recognition component for positive chemotaxis (attraction) turns out to be the binding protein for transport of that substance into the cell. For example, the galactose-binding protein, a soluble protein in the periplasmic space, is the recognition protein for positive chemotaxis to galactose in addition to being part of the pump that actively

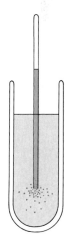

Figure 37-38
Chemotaxis of bacteria to a capillary containing an attractant such as glucose.

transports it into the cell. The chemoreceptor for glucose is a component of the membrane-bound phosphotransferase system for the active uptake of this sugar (p. 871). The surface of *E. coli* also contains chemoreceptors for attractants such as serine, cysteine, alanine, and glycine. Transport and chemotaxis are closely related, but transport is not required for chemotaxis. Some mutants that are defective in the transport of a particular sugar or amino acid retain the capacity to move toward it. A variety of chemoreceptors for negative chemotaxis are also present. Fatty acids, alcohols, hydrophobic amino acids, indole, H^+ (pH < 6.5), OH^- (pH > 7.5), and sulfide repel bacteria by interacting with specific chemosensors.

BACTERIAL FLAGELLA ARE ROTATED BY REVERSIBLE MOTORS AT THEIR BASES

Bacteria swim by rotating flagella that emerge from their surfaces. These thin helical filaments are built from 53-kdal *flagellin* subunits. An *E. coli* bacterium has about six flagella, which are 150 Å in diameter and 10 μm long. Bacterial flagella are much smaller and far less complex than eucaryotic flagella and cilia (p. 834). A bacterial flagellum is unable to actively bend because it lacks a contractile apparatus. Instead, it is rotated by a motor located at the junction of the filament and cell envelope. The isolation of flagella with the basal structure still attached has contributed to an understanding of these assemblies, which consist of a *filament*, a *hook*, and a *rod*. In *E. coli*, four rings are mounted on the rod. The outer ring is attached to the outer membrane, and the inner ring is attached to the plasma membrane of the cell envelope. The basal body (Figure 37-40) anchors the flagellum to the cell envelope and actively drives its rotation. *This motor is powered by the proton-motive force across the*

Figure 37-39
Electron micrograph of *S. typhimurium* showing flagella in a bundle. [Courtesy of Dr. Daniel Koshland, Jr.]

Figure 37-40
Basal body of the flagellum of *E. coli*. [After J. Adler. The sensing of chemicals by bacteria. Copyright © 1976 by Scientific American, Inc. All rights reserved.]

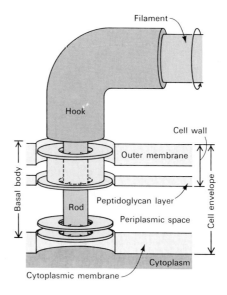

plasma membrane rather than by the hydrolysis of ATP. In fact, the angular velocity of rotation is directly proportional to the proton-motive force. An intriguing feature of this motor is that it can run clockwise or counterclockwise.

An *E. coli* bacterium typically swims smoothly in nearly a straight line for about a second. In this interval, it travels a distance of about 30 μm, which is equal to about fifteen body lengths. The bacterium then abruptly alters its course by tumbling (Figure 37-41). The average change in direction is about 60 degrees. What determines whether a bacterium swims smoothly or tumbles? When the flagella rotate counterclockwise, the helical filaments form a coherent bundle and the cell swims smoothly. In contrast, the bundle flies apart when the flagella abruptly rotate clockwise. Each flagellum then pulls in a different direction, and so the cell tumbles.

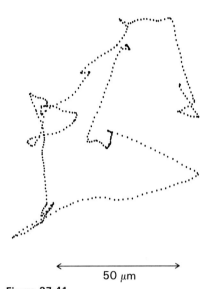

← 50 μm →

Figure 37-41
A projection of the track of an *E. coli* bacterium obtained with a microscope that automatically follows its motion in three dimensions. The points are 80 msec apart. [From H. C. Berg, *Nature* 254(1975):390.]

BACTERIA DETECT TEMPORAL RATHER THAN INSTANTANEOUS SPATIAL GRADIENTS

The regulation of tumbling frequency is central to chemotaxis. When a bacterium moves toward an increasing concentration of an attractant, tumbling becomes less frequent. In contrast, when it moves away from an attractant, tumbling becomes more frequent. Repellents have opposite effects on the tumbling frequency. Thus, if a bacterium is moving toward an attractant or away from a repellent it will swim smoothly for a longer time than if it is moving in the opposite direction. Tumbling serves to reorient a misdirected bacterium.

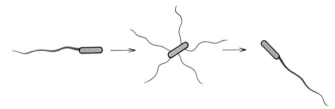

Figure 37-42
Tumbling is caused by an abrupt reversal of the flagellar motor, which disperses the flagellar bundle. [After a drawing kindly provided by Dr. Daniel Koshland, Jr.]

Does a bacterium compare the concentration of an attractant at one end of the cell with that at the other end or does it compare the concentration of attractant now with that of a few moments ago? In other words, is the sensing mechanism spatial or temporal? Daniel Koshland, Jr., and Robert Macnab answered this important question by carrying out an ingeniously simple experiment. They rapidly mixed a suspension of bacteria in a medium devoid of an attractant with a solution containing an attractant and observed the tumbling frequency of the bacteria. The striking finding was that tumbling was suppressed within a second after mixing. The bacteria swam for long distances in a straight line in this rapidly mixed solution, which was devoid of a spatial gradient. Hence, this experiment demonstrated that *bacteria possess a temporal sensing mechanism.* In other words, a bacterium detects a spatial gradient of attractant not by comparing the concentration at its head and tail, but by *traveling through space and comparing its observations through time.* In essence, *bacterial chemotaxis is a biased random walk.* Purposeful behavior in bacteria arises from the *selection* of random movements.

INFORMATION FLOWS THROUGH METHYLATED PROTEINS IN BACTERIAL CHEMOTAXIS

The response of a bacterium to a particular concentration gradient of an attractant is not fixed. The suppression of tumbling elicited by an increase in the concentration of an attractant is transient. After a lag period, the tumbling rate returns to the unstimulated level. In effect, the bacterium has become insensitive to the continued presence of the attractant. This desensitization, called *adaptation,* enables bacteria to sense gradients over a wide range of attractant concentrations. As mentioned earlier, retinal rod cells respond analogously to light (Figure 37-30).

Genetic dissection of bacterial chemotaxis has revealed that information funnels from at least twelve chemoreceptors into three proteins, which are products of the *tsr, tar,* and *trg* genes (Figure 37-43). For example, signals from the ribose and galactose receptors

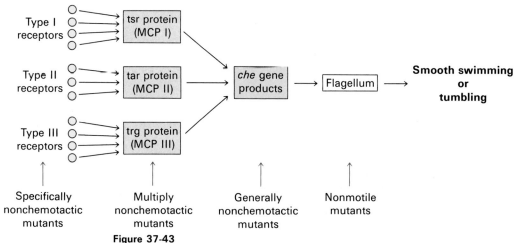

Figure 37-43
Information flow in bacterial chemotaxis. The types of mutants that have been isolated are noted in the lower part of the diagram. MCP refers to methyl-accepting chemotaxis protein. [After R. S. Springer, M. F. Coy, and J. Adler, *Nature* 280(1979):280.]

converge on the trg protein. These three proteins are reversibly methylated, and so they are called *methyl-accepting chemotaxis proteins* (*MCPs*). Several glutamate side chains on each of these proteins are methylated by a methyl transferase that uses *S*-adenosylmethionine as the activated donor. These covalent modifications are reversed by a specific esterase. The level of methylation of the MCPs is

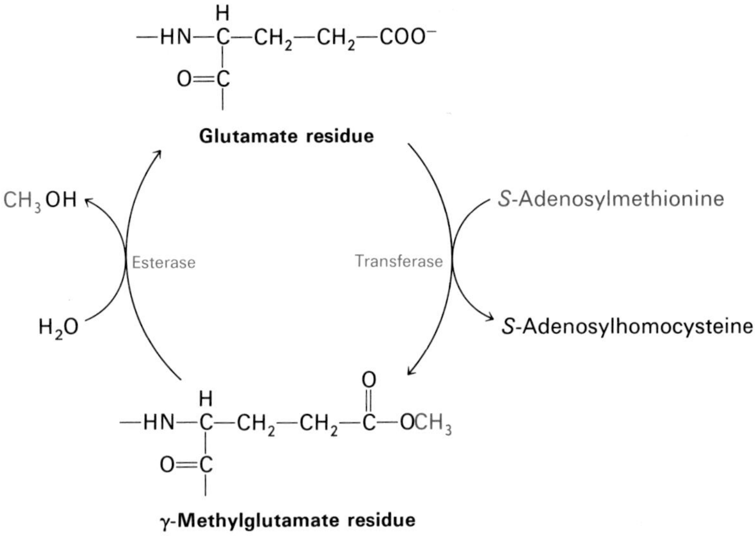

Figure 37-44
Reversible methylation of the methyl-accepting chemotaxis proteins.

increased by attractants and decreased by repellents. *In fact, adaptation is probably mediated by the degree of the methylation of these proteins.*

Information from the three methyl-accepting proteins then converges on products on the *che* genes. It is not yet known how these molecules determine the direction of rotation of flagella. However, it is already evident that a bacterium possesses a primitive sensory system built from the products of some thirty genes that processes and integrates information about nutrients and poisons in its environment. For bacteria, the quintessential question is whether to tumble!

SUMMARY

Action potentials in nerve-axon membranes are mediated by transient changes in Na^+ and K^+ permeability. Both the Na^+ and the K^+ channels are regulated by the transmembrane voltage. Tetrodotoxin and saxitoxin specifically block the Na^+ channel. Nerve impulses are communicated across most synapses by chemical transmitters. Acetylcholine, the neurotransmitter at many synapses and at motor end plates, is synthesized from acetyl CoA and choline. Acetylcholine molecules are packaged in synaptic vesicles, which fuse with the presynaptic membrane upon arrival of a nerve impulse. The released acetylcholines diffuse across the synaptic cleft and combine with specific protein receptors on the postsynaptic membrane. The consequent increase in the permeability of Na^+ and K^+ leads to the depolarization of this membrane. The acetylcholine receptor has high affinity for α-bungarotoxin and other neurotoxins, a characteristic that has facilitated the purification of this integral-membrane protein. The postsynaptic membrane repolarizes following the hydrolysis of acetylcholine by acetylcholinesterase. This enzyme is inhibited by organic phosphates such as DIPF, which can be lethal by causing respiratory paralysis. Catecholamines (epinephrine, norepinephrine, and dopamine) and γ-aminobutyrate (GABA) also serve as neurotransmitters.

A retinal rod cell can be excited by a single photon. The outer segment of a rod contains a stack of about a thousand discs, which are closed bilayer membranes densely packed with molecules of rhodopsin, a transmembrane protein. The chromophore of rhodopsin, the photoreceptor protein, is 11-*cis*-retinal, which is derived from all-*trans*-retinol (Vitamin A). 11-*cis*-Retinal forms a Schiff-base linkage with a specific lysine residue of opsin. The primary event in visual excitation is the isomerization of the 11-*cis*-retinal group to all-*trans*-retinal. A late step in visual excitation is the hyperpolarization of the plasma membrane of the outer segment, which results from a decrease in its permeability to Na^+. The flow of more than a million sodium ions is blocked by the absorption of a single photon by a dark-adapted rod. Transmitters (possibly Ca^{2+} or cyclic GMP) carry the signal from photolyzed rhodopsin to the plasma membrane. Color vision is mediated by three kinds of photoreceptors, each containing 11-*cis*-retinal. In fact, 11-*cis*-retinal is the chromophore in all known visual systems, which is a striking example of convergent evolution.

Motile bacteria can move toward attractants (such as glucose) and away from repellents (such as fatty acids). The sensors for chemotaxis are located in the periplasmic space or plasma membrane. Bacterial flagella are rotated by motors that are powered by the flow of protons across the plasma membrane. Bacteria swim smoothly for about a second and then tumble when the sense of rotation of their flagella abruptly changes from counterclockwise to clockwise. Tumbling is suppressed when a bacterium moves toward an attractant or away from a repellent. Bacteria detect temporal rather than spatial gradients. Information flows from the chemosensors to three methyl-accepting chemotaxis proteins (MCPs) and then to the flagella. Reversible methylation of the MCPs controls the sensitivity of the detection system.

SELECTED READINGS

WHERE TO START

Lester, H. A., 1977. The response to acetylcholine. *Sci. Amer.* 236(2):106–118. [Available as *Sci. Amer.* Offprint 1352.]

Keynes, R. D., 1979. Ion channels in the nerve-cell membrane. *Sci. Amer.* 240(3):126–135. [Offprint 1423.]

Rushton, W. A. H., 1975. Visual pigments and color blindness. *Sci. Amer.* 232(3):64–74. [Offprint 1317.]

Adler, J., 1976. The sensing of chemicals by bacteria. *Sci. Amer.* 234(4):40–47. [Offprint 1337.]

Berg, H., 1975. How bacteria swim. *Sci. Amer.* 233(2):36–44.

BOOKS

Kuffler, S. W., and Nicholls, J. G., 1976. *From Neuron to Brain.* Sinauer. [Part II of this fine book contains an excellent account of mechanisms for neuronal signalling.]

Katz, B., 1966. *Nerve, Muscle, and Synapse.* McGraw-Hill. [A superb concise introduction.]

Cone, R. A., and Dowling, J. E., (eds.), 1979. *Membrane Transduction Mechanisms.* Raven Press. [Contains concise and lucid overviews of a variety of excitable membrane systems.]

Aidley, D. J., 1977. *The Physiology of Excitable Cells* (2nd ed.). Liverpool University Press.

NERVE-AXON CONDUCTION

Hille, B., 1978. Ionic channels in excitable membranes: current problems and biophysical approaches. *Biophys. J.* 22:283–294.

Barchi, R. L., Cohen, S. A., and Murphy, L. E., 1980. Purification from rat sarcolemma of the saxitoxin-binding component of the excitable membrane sodium channel. *Proc. Nat. Acad. Sci.* 77:1306–1310.

Morell, P., and Norton, W. T., 1980. Myelin. *Sci. Amer.* 242(5):88–118. [Offprint 1469.]

Ritchie, J. M., and Rogart, R. B., 1977. Density of sodium channels in mammalian myelinated nerve fibers and nature of the axonal membrane under the myelin sheath. *Proc. Nat. Acad. Sci.* 74:211–215.

SYNAPTIC TRANSMISSION

Axelrod, J., 1974. Neurotransmitters. *Sci. Amer.* 230(6):59–71. [Offprint 1297.]

Heidmann, T., and Changeux, J.-P., 1978. Structural and functional properties of the acetylcholine receptor protein in its purified and membrane-bound states. *Ann. Rev. Biochem.* 47:317–357.

Raftery, M. A., Hunkapiller, M. W., Strader, C. D., and Hood, L. E., 1980. Acetylcholine receptor: complex of homologous subunits. *Science* 208:1454–1457.

Klymkowsky, M. W., and Stroud, R. M., 1979. Immunospecific identification and three-dimensional structure of a membrane-bound acetylcholine receptor from *Torpedo california. J. Mol. Biol.* 128:319–334.

Moore, H. H., Hartig, P. R., and Raftery, M. A., 1979. Correlation of polypeptide composition with functional events in acetylcholine receptor-enriched membranes from *Torpedo california. Proc. Nat. Acad. Sci.* 76:6265–6269.

Fulpius, B. W., Miskin, R., and Reich, E., 1980. Antibodies from myasthenic patients that compete with cholinergic agents for binding to nicotinic receptors. *Proc. Nat. Acad. Sci.* 77:4326–4330.

Hess, G. P., Lipkowitz, S., and Struve, G. E., 1978. Acetylcholine-receptor mediated ion flux in electroplax membrane microsacs (vesicles): change in mechanism produced by asymmetrical distribution of sodium and potassium ions. *Proc. Nat. Acad. Sci.* 75:1703–1707.

Anglister, L., and Silman, I., 1978. Molecular structure of elongated forms of electric eel acetylcholinesterase. *J. Mol. Biol.* 125:293–311.

VISION

Miller, W. H., (ed.), 1981. *Molecular Mechanisms of Photoreceptor Transduction.* Academic Press.

Hagins, W. A., 1979. Excitation in vertebrate photoreceptors. *In* Schmitt, F. O., and Worden, F. G., (eds.), *The Neurosciences: Fourth Study Program,* pp. 183–191. MIT Press.

Hubbell, W. L., and Bownds, M. D., 1979. Visual transduction in vertebrate photoreceptors. *Ann. Rev. Neurosci.* 2:17–34.

Wald, G., 1968. The molecular basis of visual excitation. *Nature* 219:800–807. [This Nobel Lecture contains an interesting account of the discovery of the primary event in vision.]

Young, R. W., 1970. Visual cells. *Sci. Amer.* 223(4):80–91. [Offprint 1201.]

Dowling, J. E., 1966. Night blindness. *Sci. Amer.* 215(4):78–84. [Offprint 1053.]

MacNichol, E. F., Jr., 1964. Three-pigment color vision. *Sci. Amer.* 211(6):48–56. [Offprint 197.]

Yoshikami, S., George, J. S., and Hagins, W. A., 1980. Light-induced calcium fluxes from outer segment layer of vertebrate retinas. *Nature* 286:395–398.

Pober, J. S., and Bitensky, M. W., 1979. Light-regulated enzymes of vertebrate retinal rods. *Advan. Cyclic Nucleotide Res.* 11:265–301.

Liebman, P. A., and Pugh, E. N., Jr., 1979. The control of phosphodiesterase in rod disk membranes: kinetics, possible mechanisms, and significance for vision. *Vision Res.* 19:375–380.

Fung, B. K.-K., and Stryer, L., 1980. Photolyzed rhodopsin catalyzes the exchange of GTP for bound GDP in retinal rod outer segments. *Proc. Nat. Acad. Sci.* 77:2500–2504.

912 CHEMOTAXIS

Koshland, D. E., Jr., 1979. A model regulatory system: bacterial chemotaxis. *Physiol. Rev.* 59:811–862.

Macnab, R., and Koshland, D. E., Jr., 1972. The gradient-sensing mechanism in bacterial chemotaxis. *Proc. Nat. Acad. Sci.* 69:2509–2512. [These experiments revealed that bacteria detect a temporal gradient rather than a spatial one.]

Springer, M. S., Goy, M. F., and Adler, J., 1979. Protein methylation in behavioural control mechanisms and in signal transduction. *Nature.* 280:279–284.

Silverman, M. R., and Simon, M. I., 1974. Flagellar rotation and the mechanism of bacterial motility. *Nature* 249:73–74.

Manson, M. D., Tedesco, P., Berg, C., Harold, F. M., and van der Drift, C., 1977. A protomotive force drives bacterial flagella. *Proc. Nat. Acad. Sci.* 74:3060–3064.

Manson, M. D., Tedesco, P. M., and Berg, H. C., 1980. Energetics of flagellar rotation in bacteria. *J. Mol. Biol.* 138:541–561.

Khan, S., and Macnab, R. M., 1980. Proton chemical potential, proton electrical potential and bacterial motility. *J. Mol. Biol.* 138:599–614.

Stock, J. B., and Koshland, D. E., Jr., 1978. A protein methylesterase involved in bacterial sensing. *Proc. Nat. Acad. Sci.* 75:3659–3663.

Kondoh, H., Ball, C. B., and Adler, J., 1979. Identification of a methyl-accepting chemotaxis protein for the ribose and galactose chemoreceptors of *Escherichia coli.* *Proc. Nat. Acad. Sci.* 76:260–264.

Berg, H. C., and Purcell, E. M., 1977. Physics of chemoreception. *Biophys. J.* 20:193–219.

APPENDIXES
ANSWERS TO PROBLEMS
INDEX

PHYSICAL CONSTANTS AND CONVERSION OF UNITS

Values of physical constants

Physical constant	Symbol	Value
Atomic mass unit (dalton)	amu	1.661×10^{-24} g
Avogadro's number	N	6.022×10^{23} mol^{-1}
Boltzmann's constant	k	1.381×10^{-23} J deg^{-1} 3.298×10^{-24} cal deg^{-1}
Electron volt	eV	1.602×10^{-19} J 3.828×10^{-20} cal
Faraday constant	F	9.649×10^{4} C mol^{-1} 2.306×10^{4} cal volt^{-1} eq^{-1}
Curie	Ci	3.70×10^{10} disintegrations sec^{-1}
Gas constant	R	8.314 J mol^{-1} deg^{-1} 1.987 cal mol^{-1} deg^{-1}
Planck's constant	h	6.626×10^{-34} J sec 1.584×10^{-34} cal sec
Speed of light in a vacuum	c	2.998×10^{10} cm sec^{-1}

Abbreviations: C, coulomb; cal, calorie; cm, centimeter; deg, degree Kelvin; eq, equivalent; g, gram; J, joule; mol, mole; sec, second.

Mathematical constants

$\pi = 3.14159$

$e = 2.71828$

$\log_e x = 2.303 \log_{10} x$

Conversion factors

Physical quantity	Equivalent
Length	1 cm $= 10^{-2}$ m $= 10$ mm $= 10^4$ μm $= 10^7$ nm 1 cm $= 10^8$ Å $= 0.3937$ inch
Mass	1 g $= 10^{-3}$ kg $= 10^3$ mg $= 10^6$ μg 1 g $= 3.527 \times 10^{-2}$ ounce (avoirdupoir)
Volume	1 cm$^3 = 10^{-6}$ m$^3 = 10^3$ mm^3 1 ml $= 1$ cm$^3 = 10^{-3}$ l $= 10^3$ μl 1 cm$^3 = 6.1 \times 10^{-2}$ in$^3 = 3.53 \times 10^{-5}$ ft^3
Temperature	K $= °C + 273.15$ $°C = 5/9$ $(°F - 32)$
Energy	1 J $= 10^7$ erg $= 0.239$ cal $= 1$ watt sec
Pressure	1 torr $= 1$ mm Hg $(0°C)$ $= 1.333 \times 10^2$ newton/m^2 $= 1.333 \times 10^2$ pascal $= 1.316 \times 10^{-3}$ atmospheres

Standard prefixes

Prefix	Symbol	Factor
kilo	k	10^3
hecto	h	10^2
deca	da	10^1
deci	d	10^{-1}
centi	c	10^{-2}
milli	m	10^{-3}
micro	μ	10^{-6}
nano	n	10^{-9}
pico	p	10^{-12}

ATOMIC NUMBERS AND WEIGHTS OF THE ELEMENTS

Element	Symbol	Atomic number	Atomic weight
Actinium	Ac	89	227.03
Aluminum	Al	13	26.98
Americium	Am	95	243.06
Antimony	Sb	51	121.75
Argon	Ar	18	39.95
Arsenic	As	33	74.92
Astatine	At	85	210.99
Barium	Ba	56	137.34
Berkelium	Bk	97	247.07
Beryllium	Be	4	9.01
Bismuth	Bi	83	208.98
Boron	B	5	10.81
Bromine	Br	35	79.90
Cadmium	Cd	48	112.40
Calcium	Ca	20	40.08
Californium	Cf	98	249.07
Carbon	C	6	12.01
Cerium	Ce	58	140.12
Cesium	Cs	55	132.91
Chlorine	Cl	17	35.45
Chromium	Cr	24	52.00
Cobalt	Co	27	58.93
Copper	Cu	29	63.55
Curium	Cm	96	245.07
Dysprosium	Dy	66	162.50
Einsteinium	Es	99	254.09
Erbium	Er	68	167.26
Europium	Eu	63	151.96
Fermium	Fm	100	252.08
Fluorine	F	9	18.99
Francium	Fr	87	223.02
Gadolinium	Gd	64	157.25
Gallium	Ga	31	69.72
Germanium	Ge	32	72.59
Gold	Au	79	196.97
Hafnium	Hf	72	170.49
Helium	He	2	4.00
Holmium	Ho	67	164.93
Hydrogen	H	1	1.01
Indium	In	49	114.82
Iodine	I	53	126.90
Iridium	Ir	77	192.22
Iron	Fe	26	55.85
Khurchatovium	Kh	104	260
Krypton	Kr	36	83.80
Lanthanum	La	57	138.91
Lawrencium	Lr	103	256
Lead	Pb	82	207.20
Lithium	Li	3	6.94
Lutetium	Lu	71	174.97
Magnesium	Mg	12	24.31
Manganese	Mn	25	54.94
Mendelevium	Md	101	255.09
Mercury	Hg	80	200.59
Molybdenum	Mo	42	95.94
Neodymium	Nd	60	144.24
Neon	Ne	10	20.18
Neptunium	Np	93	237.05
Nickel	Ni	28	58.71
Niobium	Nb	41	92.91
Nitrogen	N	7	14.01
Nobelium	No	102	255
Osmium	Os	76	190.20
Oxygen	O	8	16.00
Palladium	Pd	46	106.40
Phosphorus	P	15	30.97
Platinum	Pt	78	195.09
Plutonium	Pu	94	242.06
Polonium	Po	84	208.98
Potassium	K	19	39.10
Praseodymium	Pr	59	140.91
Promethium	Pm	61	145
Protactinium	Pa	91	231.04
Radium	Ra	88	226.03
Radon	Rn	86	222.02
Rhenium	Re	75	186.20
Rhodium	Rh	45	102.91
Rubidium	Rb	37	85.47
Ruthenium	Ru	44	101.07
Samarium	Sm	62	150.40
Scandium	Sc	21	44.96
Selenium	Se	34	78.96
Silicon	Si	14	28.09
Silver	Ag	47	107.87
Sodium	Na	11	22.99
Strontium	Sr	38	87.62
Sulfur	S	16	32.06
Tantalum	Ta	73	180.95
Technetium	Tc	43	98.91
Tellurium	Te	52	127.60
Terbium	Tb	65	158.93
Thallium	Tl	81	204.37
Thorium	Th	90	232.04
Thulium	Tm	69	168.93
Tin	Sn	50	118.69
Titanium	Ti	22	47.90
Tungsten	W	74	183.85
Uranium	U	92	238.03
Vandadium	V	23	50.94
Xenon	Xe	54	131.30
Ytterbium	Yb	70	173.04
Yttrium	Y	39	88.91
Zinc	Zn	30	65.37
Zirconium	Zr	40	91.22

pK' VALUES OF SOME ACIDS

Acid	pK' (at 25°C)
Acetic acid	4.76
Acetoacetic acid	3.58
Ammonium ion	9.25
Ascorbic acid, pK'_1	4.10
pK'_2	11.79
Benzoic acid	4.20
n-Butyric acid	4.81
Cacodylic acid	6.19
Carbonic acid, pK'_1	6.35
pK'_2	10.33
Citric acid, pK'_1	3.14
pK'_2	4.77
pK'_3	6.39
Ethylammonium ion	10.81
Formic acid	3.75
Glycine, pK'_1	2.35
pK'_2	9.78
Imidazolium ion	6.95

Acid	pK' (at 25°C)
Lactic acid	3.86
Maleic acid, pK'_1	1.83
pK'_2	6.07
Malic acid, pK'_1	3.40
pK'_2	5.11
Phenol	9.89
Phosphoric acid, pK'_1	2.12
pK'_2	7.21
pK'_3	12.67
Pyridinium ion	5.25
Pyrophosphoric acid, pK'_1	0.85
pK'_2	1.49
pK'_3	5.77
pK'_4	8.22
Succinic acid, pK'_1	4.21
pK'_2	5.64
Trimethylammonium ion	9.79
Tris (hydroxymethyl) aminomethane	8.08
Water	14.0

STANDARD BOND LENGTHS

Bond	Structure	Length (Å)
C—H	R_2CH_2	1.07
	Aromatic	1.08
	RCH_3	1.10
C—C	Hydrocarbon	1.54
	Aromatic	1.40
C=C	Ethylene	1.33
C≡C	Acetylene	1.20
C—N	RNH_2	1.47
	O=C—N	1.34
C—O	Alcohol	1.43
	Ester	1.36
C=O	Aldehyde	1.22
	Amide	1.24
C—S	R_2S	1.82
N—H	Amide	0.99
O—H	Alcohol	0.97
O—O	O_2	1.21
P—O	Ester	1.56
S—H	Thiol	1.33
S—S	Disulfide	2.05

ANSWERS TO PROBLEMS

CHAPTER 2

1. (a) Phenyl isothiocyanate.
   (b) Dansyl chloride.
   (c) Urea; β-mercaptoethanol to reduce disulfides.
   (d) Chymotrypsin.
   (e) CNBr.
   (f) Trypsin.
2. (a) 3; (b) 12; (c) 4.28; (d) 9.8; and (e) 2.4.
3. 0.01; 0.1; 1; 10; and 100.
4. 477 Å (318 residues per strand, 1.5 Å per residue).
5. Each amino acid residue, except the carboxyl-terminal one, gives rise to a hydrazide on reaction with hydrazine. The carboxyl-terminal residue can be identified because it yields a free amino acid.
6. (a) Approximately +1.
   (b) Two peptides.
7. The S-aminoethylcysteine side chain resembles that of lysine. The only difference is a sulfur atom in place of a methylene group.
8. The native conformation of insulin is not the thermodynamically most stable form. See page 847 for further discussion.

CHAPTER 3

1. (a) 2.96×10^{-11} g.
   (b) 2.71×10^8 molecules.
   (c) No. There would be 3.22×10^8 hemoglobin molecules in a red cell if they were packed in a cubic crystalline array. Hence, the actual packing density is about 84% of the maximum possible.
2. 2.65 g (or 4.75×10^{-2} moles) of Fe.
3. (a) In humans, 1.44×10^{-2} g (4.49×10^{-4} moles) of O_2 per kg of muscle. In sperm whale, 0.144 g (4.49×10^{-3} moles) of O_2 per kg.
   (b) 128.
4. (a) Approximately +36 at pH 2, +4 at pH 7, and +2 at pH 9.
   (b) Approximately 10. However, there are interactions between titratable groups. The actual isoelectric point is 8.2. For an excellent treatment of this problem, see S. J. Shire, G. I. H. Hanania, and F. R. N. Gurd, *Biochemistry* 13(1974):2967.
5. (a) Three peptides, consisting of residues 1 to 55, 56 to 131, and 132 to 153.

(b) Short α helices are marginally stable in aqueous solution. They are stabilized by tertiary interactions in myoglobin.

6.
$$\log \frac{[\text{MbO}_2]}{[\text{Mb}]} = \log p\text{O}_2 - \log P_{50}$$

7. (a) $k_{\text{off}} = 20 \text{ sec}^{-1}$.

$$K = \frac{[\text{Mb}][\text{O}_2]}{[\text{MbO}_2]} = \frac{k_{\text{off}}}{k_{\text{on}}}$$

   (b) Mean duration is 0.05 sec (the reciprocal of k_{off}).
8. A 1 mg/ml solution of myoglobin (17.8 kdal) corresponds to 5.62×10^{-5} M. The absorbance of a 1-cm path length is 0.84, which corresponds to an I_0/I ratio of 6.96. Hence, 14.4% of the incident light is transmitted.

CHAPTER 4

1. (a) Increased; (b) decreased; (c) decreased; and (d) increased oxygen affinity.
2. (a and b) Fewer H^+ bound.
3. Inositol hexaphosphate.
4. (a) $\Delta Y = .977 - .323 = .654$.
   (b) $\Delta Y = .793 - .434 = .359$.
5. The pK is (a) lowered, (b) raised, and (c) raised.
6. The additional methyl group hinders the movement of the iron atom into the plane of the porphyrin on oxygenation. This model compound mimics the T state of hemoglobin. For an interesting discussion, see J. P. Collman, J. I. Brauman, K. M. Doxsee, T. R. Halbut, and K. S. Suslick, *Proc. Nat. Acad. Sci.* 75(1978):564.
7. (a) Yes. $K_{\text{AB}} = K_{\text{BA}} (K_{\text{B}}/K_{\text{A}}) = 2 \times 10^{-5}$ M.
   (b) The presence of A enhances the binding of B, and so the presence of B enhances the binding of A.
8. Carbon monoxide bound to one heme alters the oxygen affinity of the other hemes in the same hemoglobin molecule. Specifically, CO increases the oxygen affinity of hemoglobin and thereby decreases the amount of O_2 released in actively metabolizing tissues. Carbon monoxide stabilizes the quaternary structure characteristic of oxyhemoglobin. In other words, CO mimics O_2 as an allosteric effector.

9. (a) For maximal transport, $K = 10^{-5}$ M. In general, maximal transport is achieved when $K = \sqrt{L_A L_B}$.
   (b) For maximal transport, $P_{50} = 44.7$ torrs, which is considerably higher than the physiological value of 26 torrs. However, it must be stressed that this calculation ignores cooperative binding and the Bohr effect.

CHAPTER 5

1. (a) Lysine or arginine at position 6.
   (b) This hemoglobin migrates least rapidly toward the anode at pH 8 because it has the smallest net negative charge of the three hemoglobins.
2. (a) Hb C; (b) Hb D; (c) Hb J; and (d) Hb N.
3. Mutations in the α gene affect all three hemoglobins because their subunit structures are $\alpha_2\beta_2$, $\alpha_2\delta_2$, and $\alpha_2\gamma_2$, respectively. Mutations in the β, δ, or γ genes affect only one of them.
4. The reaction

$$2(\alpha_2\beta\beta^{\text{S}}) \rightleftharpoons \alpha_2\beta_2 + \alpha_2\beta_2{}^{\text{S}}$$

is rapid compared with the duration of an electrophoresis experiment. The separation of $\alpha_2\beta_2$ and $\alpha_2\beta_2{}^{\text{S}}$ by the electric field shifts the equilibrium to the right.

5. Deoxy Hb A contains a complementary site, and so it can add on to a fiber of deoxy Hb S. The fiber cannot then grow further because the terminal deoxy Hb A molecule lacks a sticky patch.
6. The aspartate–tyrosine hydrogen bond at the $\alpha_1\beta_2$ interface of deoxyhemoglobin A is missing in deoxyhemoglobin Kempsey (p. 99). Hence, the deoxy form of this mutant should dissociate into dimers more readily than does normal hemoglobin. In contrast, the oxygenated forms of these hemoglobins should dissociate to nearly the same extent because neither is stabilized by this hydrogen bond.

CHAPTER 6

1. (a) 31.1×10^{-6} moles.
   (b) 5×10^{-8} moles.
   (c) 622 sec^{-1}.
2. (a) Yes. $K_M = 5.2 \times 10^{-6}$ M.
   (b) $V_{max} = 6.84 \times 10^{-10}$ moles.
   (c) 337 sec^{-1}.
3. (a) In the absence of inhibitor, V_{max} is 47.6 μmole/min and K_M is 1.1×10^{-5} M. In the presence of inhibitor, V_{max} is the same, and the apparent K_M is 3.1×10^{-5} M.
   (b) Competitive.
   (c) 1.1×10^{-3} M.
   (d) f_{ES} is 0.243 and f_{EI} is 0.488.
   (e) f_{ES} is 0.73 in the absence of inhibitor and 0.49 in the presence of 2×10^{-3} M inhibitor. The ratio of these values, 1.49, is the same as the ratio of the reaction velocities under these conditions.

4. (a) V_{max} is 9.5 μmole/min. K_M is 1.1×10^{-5} M, the same as without inhibitor.
   (b) Noncompetitive.
   (c) 2.5×10^{-5} M.
   (d) 0.73, in the presence or absence of this noncompetitive inhibitor.
5. (a) $V = V_{max} - (V/[S])\,K_M$.
   (b) Slope $= -K_M$, y-intercept $= V_{max}$, x-intercept $= V_{max}/K_M$.
   (c)

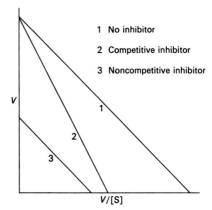

    1 No inhibitor
    2 Competitive inhibitor
    3 Noncompetitive inhibitor

6. The binding of a competitive inhibitor to one active site of an allosteric enzyme may increase the substrate-binding affinity of the other sites on the same enzyme molecule. In the concerted model, such a competitive inhibitor acts in this way by promoting the T $\rightarrow$ R transition.
7. Potential hydrogen-bond donors at pH 7 are the side chains of the following residues: arginine, asparagine, glutamine, histidine, lysine, serine, threonine, tryptophan, and tyrosine.
8. The rates of utilization of A and B are given by

$$V_A = \left(\frac{k_3}{K_M}\right)_A [E][A]$$

and

$$V_B = \left(\frac{k_3}{K_M}\right)_B [E][B]$$

Hence, the ratio of these rates is

$$V_A/V_B = \left(\frac{k_3}{K_m}\right)_A [A] \Big/ \left(\frac{k_3}{K_M}\right)_B [B]$$

Thus, an enzyme discriminates between competing substrates on the basis of their values of k_3/K_M rather than of K_M alone.

CHAPTER 7

1. The fastest is (b) and the slowest is (a).
2. (a) B-C; (b) A-B or E-F; and (c) A-B-C (one sugar residue does not interact with the enzyme, thus avoiding site D, which is energetically unfavorable).
3. The ^{18}O emerges in the C-4 hydroxyl of di-NAG (residues E-F).

4. This analog lacks a bulky substituent at C-5 and so it can probably bind to site D without being strained. Consequently, the binding of residue D of this analog is likely to be energetically favorable, whereas the binding of residue D of tetra-NAG costs free energy. See P. van Eikeren and D. M. Chipman, *J. Am. Chem. Soc.* 94(1972):4788.

5. (a) In oxymyoglobin, Fe is bonded to five nitrogens and one oxygen. In carboxypeptidase A, Zn is bonded to two nitrogens and two oxygens.

   (b) In oxymyoglobin, one of the nitrogens bonded to Fe comes from the proximal histidine residue, whereas the other four come from the heme. The oxygen atom linked to Fe is that of O_2. In carboxypeptidase A, the two nitrogen atoms coordinated to Zn come from histidine residues. One of the oxygen ligands is from a glutamate side chain, the other from a water molecule.

   (c) Aspartate, cysteine, and methionine.

6. H_2O labeled with ^{18}O is formed when carboxypeptidase A catalyzes the *synthesis* of a peptide bond between *N*-benzoylglycine and L-phenylalanine. ^{18}O is not transferred in the presence of L-β-phenyllactic acid because this compound lacks an α-amino group and so a peptide bond cannot be formed. See R. Breslow and D. L. Wernick, *Proc. Nat. Acad. Sci.* 74(1977):1303, for a discussion of these experiments.

CHAPTER 8

1. (a) Carboxypeptidase A.
   (b) Lysozyme and carboxypeptidase A.
   (c) Chymotrypsin.
   (d) Carboxypeptidase A and chymotrypsin.

2. (a) Yes.
   (b) Histidine 57 in chymotrypsin, glutamic 35 in lysozyme, and tyrosine 248 in carboxypeptidase A.

3. Serine 195 in chymotrypsin and glutamate 270 (or a hydroxyl ion) in carboxypeptidase A.

4. Precise positioning of catalytic residues and substrates, geometrical strain (distortion of the substrate), electronic strain, and desolvation of the substrate.

5. (a) Tosyl-L-lysine chloromethyl ketone (TLCK).
   (b) First, determine whether substrates protect trypsin from inactivation by TLCK and, second, ascertain whether the D-isomer of TLCK inactivates trypsin.
   (c) Thrombin.

6. (a) Serine.
   (b) A hemiacetal between the aldehyde of the inhibitor and the hydroxyl group of the active site serine.

7. The boron atom becomes bonded to the oxygen atom of the active-site serine. This tetrahedral intermediate has a geometry similar to that of the transition state.

8. Activated Factor X remains bound to blood platelet membranes, which accelerates its activation of prothrombin.

9. Antithrombin III is probably a transition-state analog, and so its interaction with thrombin depends on the presence of a fully formed active site.

10. Residues *a* and *d* are located in the interior of an α-helical coiled coil, near the axis of the superhelix. Hydrophobic interactions between these side chains contribute to the stability of the coiled coil. For a discussion, see G.E. Schulz and R.H. Schirmer, *Principles of Protein Structure* (Springer-Verlag, 1978), p. 79.

CHAPTER 9

1. (a) Every third residue in each strand of a collagen triple helix must be glycine because there is insufficient space for a larger residue.
   (b) Poly (Gly-Pro-Gly) melts at a lower temperature than poly (Gly-Pro-Pro).
   (c) No. Glycine does not occupy every third position.

2. (a and b)

-Gly-Leu-Pro-Gly-Pro-Pro-Gly-Ala-Pro-Gly

Susceptible peptide bond

3. (a) Disulfides.
   (b) None.
   (c) Peptide bonds between specific glutamine and lysine side chains.
   (d) Aldol cross-link, histidine–aldol cross-link, and lysinonorleucine.
   (e) Aldol cross-link, lysinonorleucine, and desmosine.

4. The decarboxylation of α-ketoglutarate is half of the physiological reaction. It seems likely that α-ketoglutarate is first attacked by oxygen to form a peroxy acid, which then reacts with the prolyl substrate. See D. F. Counts, G. J. Cardinale, and S. Udenfriend, *Proc. Nat. Acad. Sci.* 75(1978):2145, for a discussion of this experiment and its mechanistic implications.

CHAPTER 10

1. 2.86×10^6 molecules.

2. Cyclopropane rings interfere with the orderly packing of hydrocarbon chains, and so they increase membrane fluidity.

3. 2×10^{-7} cm, 6.32×10^{-6} cm, and 2×10^{-4} cm.

4. The radius of this molecule is 3.08×10^{-7} cm and its diffusion coefficient is 7.37×10^{-9} cm^2/sec. The average distances traversed are 1.72×10^{-7} cm in 1 μsec, 5.42×10^{-6} in 1 msec, and 1.72×10^{-4} cm in 1 sec.

5. The initial decrease in the amplitude of the paramagnetic resonance spectrum results from the reduction of spin-labeled phosphatidyl cholines in the outer leaflet of the bilayer. Ascorbate does not traverse the membrane under these experimental conditions, and so it does not reduce the phospholipids in the inner leaflet. The slow decay of the residual spectrum is due to the reduction of phospholipids that have flipped over to the outer leaflet of the bilayer.

CHAPTER 11

1. Reactions (a) and (c), to the left; reactions (b) and (d), to the right.
2. None whatsoever.
3. (a) $\Delta G^{\circ\prime} = +7.5$ kcal/mol and $K'_{eq} = 3.16 \times 10^{-6}$.
   (b) 3.16×10^4.
4. $\Delta G^{\circ\prime} = 1.7$ kcal/mol. The equilibrium ratio is 17.8.
5. (a) $+0.2$ kcal/mol.
   (b) -7.8 kcal/mol. The hydrolysis of PP_i drives the reaction toward the formation of acetyl CoA.
6. (a) $\Delta G^{\circ} = 2.303\, RT\, pK$.
   (b) -6.53 kcal/mol at 25°C.
7. An ADP unit (or a closely related derivative, in the case of CoA).
8. The activated form of sulfate in most organisms is 3′-phosphoadenosine 5′-phosphosulfate. See P. W. Robbins and F. Lipmann, *J. Biol. Chem.* 229(1957):837.

CHAPTER 12

1. (a) Aldose-ketose; (b) epimers; (c) aldose-ketose; (d) anomers; (e) aldose-ketose; and (f) epimers.
2. The methyl carbon of pyruvate is ^{14}C-labeled.
3. (a) $\Delta G^{\circ\prime}$ is -29.5 kcal/mol for the reaction

$$\text{Glucose} + 2\,P_i + 2\,\text{ADP} \longrightarrow 2\,\text{lactate} + 2\,\text{ATP}$$

   (b) $\Delta G' = -27.2$ kcal/mol.
4. 3.06×10^{-5}.
5. The equilibrium concentrations of fructose 1,6-diphosphate, dihydroxyacetone phosphate, and glyceraldehyde 3-phosphate are 7.76×10^{-4} M, 2.24×10^{-4} M, and 2.24×10^{-4} M, respectively.
6. All three carbon atoms of 2,3-DPG are ^{14}C-labeled. The phosphorus atom attached to the C-2 hydroxyl is ^{32}P-labeled.
7. Hexokinase has a low ATPase activity in the absence of a sugar because it is in a catalytically inactive conformation (p. 271). The addition of xylose closes the cleft between the two lobes of the enzyme. However, xylose lacks a hydroxymethyl at C-6, and so it cannot be phosphorylated. Instead, a water molecule at the site normally occupied by the C-6 hydroxymethyl group acts as the phosphoryl acceptor from ATP.
8. X-ray crystallographic studies [S. I. Winn, H. C. Watson, R. N. Harkins, and L. A. Fothergill, *Biochem. Soc. Trans.* 5(1978):657–659] suggest that 3-phosphoglycerate binds to the active site of the enzyme, in which histidine 184 is phosphorylated and histidine 8 is free. Phosphohistidine 184 transfers its phosphoryl group to the substrate, which results in the formation of 2,3-diphosphoglycerate. The phosphoryl group on the 3-position of this intermediate is then transferred to hisitidine 8, which yields 2-phosphoglycerate. The original form of the enzyme is regenerated by a subsequent transfer of the phosphoryl group from histidine 8 to histidine 184. Note the similarity of this catalytic mechanism to that of diphosphoglycerate mutase (p. 277).

CHAPTER 13

1. (a) After one round of the citric acid cycle, the label emerges in C-2 and C-3 of oxaloacetate.
   (b) After one round of the citric acid cycle, the label emerges in C-1 and C-4 of oxaloacetate.
   (c) The label emerges in CO_2 in the formation of acetyl CoA from pyruvate.
   (d and e) Same fate as in (a).
2. No, because two carbon atoms are lost in the two decarboxylation steps of the cycle. Hence, there is no *net* synthesis of oxaloacetate.
3. 0.90, 0.03, and 0.07.
4. -9.8 kcal/mol.
5. The coenzyme stereospecificity of glyceraldehyde 3-phosphate dehydrogenase is the opposite of that of alcohol dehydrogenase (type B versus type A, respectively.)
6. Thiamine thiazolone pyrophosphate is a transition-state analog. The sulfur-containing ring of this analog is uncharged, and so it closely resembles the transition state of the normal coenzyme in thiamine-catalyzed reactions (e.g., the uncharged resonance form of hydroxyethyl-TPP, p. 291). See J. A. Gutowski and G. E. Lienhard, *J. Biol. Chem.* 251(1976):2863, for a discussion of this analog.
7. The ratio of malate to oxaloacetate must be greater than 1.75×10^4 for oxaloacetate to be formed.

CHAPTER 14

1. (a) 15; (b) 2; (c) 38; (d) 16; (e) 36; and (f) 19.
2. (a) $\Delta E_0'$ is $+1.05$ V and $\Delta G^{\circ\prime}$ is -48.4 kcal/mol for the reaction

$$2\,\text{G—SH} + \tfrac{1}{2}\,O_2 \rightleftharpoons \text{G—S—S—G} + H_2O$$

   (b) $\Delta E_0' = +0.09$ V and $\Delta G^{0\prime}$ is -4.15 kcal/mol.
3. (a) Blocks electron transport and proton pumping at site 3.
   (b) Blocks electron transport and ATP synthesis by inhibiting the exchange of ATP and ADP across the inner mitochondrial membrane.
   (c) Blocks electron transport and proton pumping at site 1.
   (d) Blocks ATP synthesis without inhibiting electron transport by dissipating the proton gradient.
   (e) Blocks electron transport and proton pumping at site 3.
   (f) Blocks electron transport and proton pumping at site 2.
4. Oligomycin inhibits ATP formation by interfering with the utilization of the proton gradient. It does not block electron transport.
5. $\Delta G^{\circ\prime}$ is $+16.1$ kcal/mol for oxidation by NAD^+ and $+1.4$ kcal/mol for oxidation by FAD. The reduction of succinate by NAD^+ is not thermodynamically feasible.
6. Cyanide can be lethal because it binds to the ferric form of cytochrome ($a + a_3$) and thereby inhibits oxidative phosphorylation. Nitrite converts ferrohemo-

globin to ferrihemoglobin, which also binds cyanide. Thus, ferrihemoglobin competes with cytochrome $(a + a_3)$ for cyanide. This competition is therapeutically effective because the amount of ferrihemoglobin that can be formed without impairing oxygen transport is much greater than the amount of cytochrome $(a + a_3)$.

7. The available free energy from the translocation of 2, 3, and 4 protons is -9.23, -13.8, and -18.5 kcal, respectively. The free energy consumed in synthesizing a mole of ATP under standard conditions is 7.3 kcal. Hence. the residual free energy of -1.93, -6.5, and -11.2 kcal can drive the synthesis of ATP until the $[ATP]/[ADP][P_i]$ ratio is 26.2, 6.51×10^4, and 1.62×10^8, respectively. Suspensions of isolated mitochondria synthesize ATP until this ratio is greater than 10^4, which shows that the number of protons translocated per ATP synthesized is at least three.

CHAPTER 15

1. (a) 5 Glucose 6-phosphate + ATP $\longrightarrow$
$\qquad$ 6 ribose 5-phosphate + ADP + H$^+$

   (b) Glucose 6-phosphate + 12 NADP$^+$ +
$\qquad$ 7 H$_2$O $\longrightarrow$ 6 CO$_2$ +
$\qquad\qquad$ 12 NADPH + 12 H$^+$ + P$_i$

2. The label emerges at C-5 of ribulose 5-phosphate.

3. Oxidative decarboxylation of isocitrate to α-ketoglutarate. A β-keto acid intermediate is formed in both reactions.

4. C-1 and C-3 of fructose 6-phosphate are labeled, whereas erythrose 4-phosphate is not labeled.

5. Reactions b and e would be blocked.

6. Form a Schiff base between a ketose substrate and transaldolase, reduce it with tritiated NaBH$_4$, and fingerprint the labeled enzyme.

CHAPTER 16

1. Galactose + ATP + UTP + H$_2$O + glycogen$_n$ $\longrightarrow$
$\qquad$ glycogen$_{n+1}$ + ADP + UDP + 2 P$_i$ + H$^+$

2. Fructose + 2 ATP + 2 H$_2$O $\longrightarrow$
$\qquad\qquad\qquad$ glucose + 2 ADP + 2 P$_i$

3. There is a deficiency of the branching enzyme.

4. The concentration of glucose 6-phosphate is elevated in von Gierke's disease. Consequently, the phosphorylated D form of glycogen synthetase is active.

5. Glucose is an allosteric inhibitor of phosphorylase a. Hence, crystals grown in its presence are in the T state. The addition of glucose 1-phosphate, a substrate, shifts the R $\rightleftharpoons$ T equilibrium toward the R state. The conformational differences between these states are sufficiently large that the crystal shatters unless it is stabilized by chemical cross-links. The shattering of a crystal caused by an allosteric transition was first observed by Haurowitz in the oxygenation of crystals of deoxyhemoglobin.

6. H. G. Hers [*Ann. Rev. Biochem.* 45(1976):167] suggested that these kinetics would ensure a lag in the dephosphorylation of subunit B, which would allow glycogen to be degraded before phosphorylase kinase is inactivated by its phosphatase.

CHAPTER 17

1. (a) Glycerol + 2 NAD$^+$ + P$_i$ + ADP $\longrightarrow$
$\qquad$ pyruvate + ATP + H$_2$O + 2 NADH + H$^+$

   (b) Glycerol kinase and glycerol phosphate dehydrogenase.

2. Stearate + ATP + 13$\frac{1}{2}$ H$_2$O + 8 FAD +
$\qquad$ 8 NAD$^+$ $\longrightarrow$ 4$\frac{1}{2}$ acetoacetate + 12$\frac{1}{2}$ H$^+$ +
$\qquad\qquad$ 8 FADH$_2$ + 8 NADH + AMP + 2 P$_i$

3. (a) Oxidation in mitochondria, synthesis in the cytosol.
   (b) Acetyl CoA in oxidation, acyl carrier protein for synthesis.
   (c) FAD and NAD$^+$ in oxidation, NADPH for synthesis.
   (d) L-isomer of 3-hydroxyacyl CoA in oxidation, D-isomer in synthesis.
   (e) Carboxyl to methyl in oxidation, methyl to carboxyl in synthesis.
   (f) The enzymes of fatty acid synthesis, but not those of oxidation, are organized in a multienzyme complex.

4. (a) Palmitoleate; (b) linoleate; (c) linoleate; (d) oleate; (e) oleate; and (f) linolenate.

5. C-1 is more radioactive (see p. 655 for a discussion of this experimental approach in the elucidation of the direction of synthesis of a polypeptide chain).

6. (a) Yes.

$$2 \text{ Acetyl CoA} + 3 \text{ H}_2\text{O} + \text{FAD} + 2 \text{ NAD}^+ \longrightarrow$$
$$\text{oxaloacetate} + 2 \text{ CoA} + \text{FADH}_2 +$$
$$2 \text{ NADH} + 4 \text{ H}^+$$

   (b) Yes, because glucose can be synthesized from oxaloacetate by the gluconeogenic pathway.

CHAPTER 18

1. (a) Pyruvate; (b) oxaloacetate; (c) α-ketoglutarate; (d) α-ketoisocaproate; (e) phenylpyruvate; and (f) hydroxyphenylpyruvate.

2. Aspartate + α-ketoglutarate + GTP + ATP +
$\qquad$ 2 H$_2$O + NADH + H$^+$ $\longrightarrow$
$\qquad$ $\frac{1}{2}$ glucose + glutamate + CO$_2$ + ADP +
$\qquad\qquad\qquad$ GDP + NAD$^+$ + 2 P$_i$

3. Aspartate + CO$_2$ + NH$_4^+$ + 3 ATP + NAD$^+$ +
$\qquad$ 4 H$_2$O $\longrightarrow$ oxaloacetate + urea +
$\qquad$ 2 ADP + 4 P$_i$ + AMP + NADH + H$^+$

4. (a) Label the methyl carbon atom of L-methylmalonyl CoA with ^{14}C. Determine the location of ^{14}C in succinyl CoA. The group transferred is the one bonded to the labeled carbon atom.

(b) Label the methyl carbon atom of L-methylmalo-nyl CoA with ^{14}C and the sulfur atom of its CoA moiety with ^{35}S. The transfer of the —CO—S—CoA group is intramolecular if succinyl CoA contains both ^{14}C and ^{35}S.

(c) The proton that is abstracted from the methyl group of L-methylmalonyl CoA is directly transferred to the adjacent carbon atom.

5. Thiamine pyrophosphate.

CHAPTER 19

1. $\Delta E'_0 = +0.28$ V and $\Delta G^{\circ\prime} = -12.9$ kcal/mol.
2. Aldolase participates in the Calvin cycle, whereas transaldolase participates in the pentose phosphate pathway.
3. The concentration of 3-phosphoglycerate would increase, whereas that of ribulose 1,5-diphosphate would decrease.
4. The concentration of 3-phosphoglycerate would decrease, whereas that of ribulose 1,5-diphosphate would increase.
5. Phycoerythrin and phycocyanin serve as antenna molecules. They absorb light in the spectral region in which chlorophyll a has little absorbance and then transfer their electronic excitation energy to chlorophyll a.
6. (a) It expresses a key aspect of photosynthesis—namely, that water is split by light. The evolved oxygen in photosynthesis comes from water.

   (b) The Van Niel equation for respiration expresses the fact that the combustion of glucose requires the input of six molecules of H_2O. For an interesting discussion, see G. Wald, On the nature of cellular respiration, *Current Aspects of Biochemical Energetics*, N. O. Kaplan and E. P. Kennedy, eds. (Academic Press, 1966), pp. 27–37.
7. The addition of pyridine increases the proton storage capacity of the thylakoid space. More pumped protons can then flow through the ATP-synthesizing complex in the dark. For a discussion of this experiment, see M. Avron, *Ann. Rev. Biochem.* 46(1977):145.
8. DCMU inhibits electron transfer between Q and plastoquinone in the link between photosystems II and I. O_2 evolution can occur in the presence of DCMU if an artificial electron acceptor such as ferricyanide can accept electrons from Q.

CHAPTER 20

1. Glycerol + 4 ATP + 3 fatty acids + 4 H_2O $\longrightarrow$ triacylglycerol + ADP + 3 AMP + 7 P_i + 4 H^+

2. Glycerol + 3 ATP + 2 fatty acids + 2 H_2O + CTP + serine $\longrightarrow$ phosphatidyl serine + CMP + ADP + 2 AMP + 6 P_i + 3 H^+

3. (a) CDP-diacylglycerol; (b) CDP-ethanolamine; (c) acyl CoA; (d) CDP-choline; (e) UDP-glucose or UDP-galactose; (f) UDP-galactose; and (g) geranyl pyrophosphate.

4. (a and b) None, because the label is lost as CO_2.

CHAPTER 21

1. Glucose + 2 ADP + 2 P_i + 2 NAD^+ + 2 glutamate $\longrightarrow$ 2 alanine + 2 α-ketoglutarate + 2 ATP + 2 NADH + H^+

2. $N_2 \longrightarrow NH_4^+ \longrightarrow$ glutamate $\longrightarrow$ serine $\longrightarrow$ glycine $\longrightarrow$ δ-aminolevulinate $\longrightarrow$ porphobilinogen $\longrightarrow$ heme

3. (a) Tetrahydrofolate; (b) tetrahydrofolate; and (c) N^5-methyltetrahydrofolate.

4. γ-Glutamyl phosphate may be a reaction intermediate.

5. The administration of glycine led to the formation of isovalerylglycine. This water-soluble conjugate, in contrast with isovaleric acid, is excreted very rapidly by the kidneys. See R. M. Cohn, M. Yudkoff, R. Rothman, and S. Segal, *New Engl. J. Med.* 299(1978):996.

6. H-D exchange points to the existence of a diimide intermediate. See W. A. Bulen, *Proc. Int. Symp. N_2 Fixation* (Washington State University Press, 1976).

7. They carry out nitrogen fixation. The absence of photosystem II provides an environment in which O_2 is not produced. Recall that the nitrogenase is very rapidly inactivated by O_2. See R. Y. Stanier, E. A. Adelberg, and J. L. Ingraham, *The Microbial World*, 4th ed. (Prentice-Hall, 1976), pp. 542–544, for a discussion of heterocysts.

CHAPTER 22

1. Glucose + 2 ATP + 2 $NADP^+$ + H_2O $\longrightarrow$ PRPP + CO_2 + ADP + AMP + 2 NADPH + H^+

2. Glutamine + aspartate + CO_2 + 2 ATP + NAD^+ $\longrightarrow$ orotate + 2 ADP + 2 P_i + glutamate + NADH + H^+

3. (a, c, d, and e) PRPP; (b) carbamoyl phosphate.

4. PRPP and formylglycinamide ribonucleotide.

5. dUMP + serine + NADPH + H^+ $\longrightarrow$ dTMP + $NADP^+$ + glycine

6. There is a deficiency of N^{10}-formyltetrahydrofolate. Sulfanilamide inhibits the synthesis of folate by acting as an analog of p-aminobenzoate, one of the precursors of folate.

7. PRPP is the activated intermediate in the synthesis of (a) phosphoribosylamine in the de novo pathway of purine formation, (b) purine nucleotides from free bases by the salvage pathway, (c) orotidylate in the

formation of pyrimidines, (d) nicotinate ribonucleo-tide, (e) phosphoribosyl-ATP in the pathway leading to histidine, and (f) phosphoribosyl-anthranilate in the pathway leading to tryptophan.

8. It seems likely that glutamine yields ammonia as a result of the catalytic action of the small subunit. The nascent ammonia would then react with an activated form of CO_2 that is formed by the large subunit. The bicarbonate-dependent ATPase activity implies that this activated species is carbonyl phosphate. Reaction of this carbonic-phosphoric mixed anhydride with NH_3 yields carbamate, which would then react with ATP to give carbamoyl phosphate. For a discussion of this enzymatic mechanism, see C. Walsh, *Enzymatic Reaction Mechanisms* (Freeman, 1979), pp. 150–154.

CHAPTER 24

1. (a) TTGATC; (b) GTTCGA; (c) ACGCGT; and (d) ATGGTA.
2. (a) [T] + [C] = 0.46.
   (b) [T] = 0.30, [C] = 0.24, and [A] + [G] = 0.46.
3. 5.88×10^3 base pairs.
4. After 1.0 generation, one-half of the molecules would be ^{15}N-^{15}N, the other half ^{14}N-^{14}N. After 2.0 generations, one-quarter of the molecules would be ^{15}N-^{15}N, the other three-quarters ^{14}N-^{14}N. Hybrid ^{14}N-^{15}N molecules would not be observed in conservative replication.
5. FAD, CoA, NADP+.
6. DNA ligase relaxes supertwisted DNA by catalyzing the cleavage of a phosphodiester bond in a DNA strand. The attacking group is AMP, which becomes attached to the 5′-phosphoryl group at the site of scission. AMP is required because this reaction is the reverse of the final step in the joining of pieces of DNA (see Figure 24-32 on p. 580).
7. 5′-GGCATAC-3′.

CHAPTER 25

1. (a) DNA polymerase I is a single chain, whereas RNA polymerase has the subunit structure $\alpha_2\beta\beta'\sigma$.
   (b) Deoxyribonucleoside triphosphates versus ribonucleoside triphosphates.
   (c) 5′ → 3′ for both.
   (d) DNA polymerase I has 5′ → 3′ and 3′ → 5′ nuclease activities, whereas RNA plymerase has none.
   (e) Semiconserved for DNA polymerase I, conserved for RNA polymerase.
   (f) DNA polymerase I needs a primer, whereas RNA polymerase does not.
   (g) Both are driven by the hydrolysis of pyrophosphate.
2. 5′-UAACGGUACGAU-3′.
3. The 2′-OH group in RNA acts as an intramolecular catalyst. A 2′-3′ cyclic intermediate is formed in the alkaline hydrolysis of RNA.

4. Cordycepin terminates RNA synthesis. An RNA chain containing cordycepin lacks a 3′-OH group.
5. (a) pGCp, AGUp, ACp, Up, GUp, and C.
   (b) pGp, CAGp, UACUGp, and UC.
   (c) pGp, CAp, Gp, UAp, CUGp, and UC.
   (d) pGp, CAp, Gp, Up, Ap, CUp, C.
6. UAGCCUGAAUp.

CHAPTER 26

1. Leu-Pro-Ser-Asp-Trp-Met-
2. Poly (Leu-Leu-Thr-Tyr).
3. (a) Pro (CCC), Ser (UCC), Leu (CUC), and Phe (UUC). Alternatively, the last base of each of these codons could be U.
   (b) These C → U mutations were produced by nitrous acid.
4. (a) It required two base changes.
   (b) Arg, Asn, Gln, Glu, Ile, Met, or Thr.
5. (a) No, it was produced by the deletion of the first base in the sequence shown below and the insertion of another base at the end of this sequence.
   (b) -AGUCCAUCACUUAAU-
6. The three sequences are: -Lys-stop; -Met-Arg-; and -Asn-Glu-.

CHAPTER 27

1. (a) No; (b) no; and (c) yes.
2. Four bands: light, heavy, a hybrid of light 30S and heavy 50S, and a hybrid of heavy 30S and light 50S.
3. About 799 high-energy phosphate bonds are consumed—400 to activate the 200 amino acids, 1 for initiation, and 398 to form 199 peptide bonds.
4. (b, c, and f) Type 1; (a, d, and e) type 2.
5. The simplest hypothesis is that the CCA anticodon of a tryptophan tRNA has mutated to UCA, which is complementary to UGA. However, analysis of this altered tRNA produces a surprise. Its anticodon is unaltered. Rather, there is a substitution of A for G at position 24. Thus, a residue far from the anticodon in the linear base sequence can influence the fidelity of codon recognition.
6. One approach is to synthesize a tRNA charged with a reactive amino acid analog. For example, bromo-acetyl-phenylalanyl-tRNA is an affinity-labeling reagent for the P site of *E. coli* ribosomes. See H. Oen, M. Pellegrini, D. Eilat, and C. R. Cantor, *Proc. Nat. Acad. Sci.* 70(1973):2799.
7. The sequence GAGGU is complementary to a sequence of five bases at the 3′ end of 16S rRNA and is located several bases on the 5′ side of an AUG codon. Hence this region is a start signal for protein synthesis. The replacement of G by A would be expected to weaken the interaction of this mRNA with the 16S rRNA and thereby diminish its effectiveness as an initiation signal. In fact, this mutation results in a tenfold decrease in the rate of synthesis of the protein specified by this mRNA. See J. J. Dunn, E. Buzash-Pollert, and F. W. Studier, *Proc. Nat. Acad. Sci.*

75(1978):2741, for a discussion of this informative mutant.

8. Both involve hydrolytic reactions: the $3' \rightarrow 5'$ exonuclease action of DNA polymerase I and the hydrolysis of an erroneous amino acid-AMP intermediate by an aminoacyl-tRNA synthetase.

CHAPTER 28

1. (a) The *lac* repressor is missing in an i^- mutant. Hence, this mutant is constitutive for the proteins of the *lac* operon.

   (b) This mutant is constitutive for the proteins of the *trp* operon because the *trp* repressor is missing.

   (c) The arabinose operon is not expressed in this mutant because the P2 form of the *araC* protein is needed to activate transcription.

   (d) This mutant is lytic but not lysogenic because it cannot synthesize the λ repressor.

   (e) This mutant is lysogenic but not lytic because it cannot synthesize the N protein, a positive control factor in transcription.

2. One possibility is that an i^s mutant produces an altered *lac* repressor that has almost no affinity for inducer but normal affinity for the operator. Such a *lac* repressor would bind to the operator and block transcription even in the presence of inducer.

3. This mutant has an altered *lac* operator that fails to bind the repressor. Such a mutator is called O^c (operator constitutive).

4. The cyclic AMP binding protein (CAP) is probably defective or absent in this mutant.

5. An *E. coli* cell bearing a λ prophage contains λ repressor molecules, which also block the transcription of the immediate-early genes of other λ viruses.

6. (a) Translation of the P_{RE} transcript is from five to ten times as rapid as that of the P_{RM} transcript because it contains the full protein synthesis initiation signal (as discussed on p. 657).

   (b) The more effective translation of the P_{RE} transcript provides a burst of λ repressor molecules needed to establish the lysogenic state. See M. Ptashe, K. Backman, Z. Humagun, A. Jeffrey, R. Maurer, B. Meyer, and R. T. Sauer, *Science* 194(1976):156, for a discussion.

INDEX

References to structural formulas are given in **boldface** type.